Equilibrium Statistical Mechanics of Lattice Models

Theoretical and Mathematical Physics

The series founded in 1975 and formerly (until 2005) entitled *Texts and Monographs in Physics* (TMP) publishes high-level monographs in theoretical and mathematical physics. The change of title to *Theoretical and Mathematical Physics* (TMP) signals that the series is a suitable publication platform for both the mathematical and the theoretical physicist. The wider scope of the series is reflected by the composition of the editorial board, comprising both physicists and mathematicians.

The books, written in a didactic style and containing a certain amount of elementary background material, bridge the gap between advanced textbooks and research monographs. They can thus serve as basis for advanced studies, not only for lectures and seminars at graduate level, but also for scientists entering a field of research.

More information about this series at http://www.springer.com/series/720

David A. Lavis

Equilibrium Statistical Mechanics of Lattice Models

Springer

David A. Lavis
Department of Mathematics
King's College London
London
UK

ISSN 1864-5879 ISSN 1864-5887 (electronic)
Theoretical and Mathematical Physics
ISBN 978-94-017-9429-9 ISBN 978-94-017-9430-5 (eBook)
DOI 10.1007/978-94-017-9430-5

Library of Congress Control Number: 2014955152

Springer Dordrecht Heidelberg New York London

Printed on acid-free paper

Springer Science+Business Media B.V. Dordrecht is part of Springer Science+Business Media
(www.springer.com)

To the memory of George MacDonald Bell (1925–1993)

Preface

This book was originally planned as an updated one-volume edition of the two works co-authored with the late George Bell (Lavis and Bell 1999a, b). However, as the project developed this became a substantially different book under my sole authorship and I am very grateful to Vanessa and Debbie Bell for permission to make use of some of the material from my work with their father.

This is not a book on the foundations of statistical mechanics; attention is restricted to equilibrium systems, with equilibrium defined simply in terms of the Gibbs distribution. The material is also restricted to deterministic systems on lattices. Within these constraints, most interesting and difficult problems arise from cooperative effects, that is from interactions between the constituent particles of the system. This leads frequently to the occurrence of phase transitions and much of the work of equilibrium statistical mechanics concentrates on the investigation of their properties, particularly their classification into universality classes. The book is structured into parts. Part I contains the basic elements of equilibrium thermodynamics and statistical mechanics, phenomenological scaling theory and a survey of a range of different models, many of which are studied in greater detail later in the book. A broad division of the rest of the material can be made between approximation methods and exact results. Parts II and IV are concerned with the former, II being devoted to classical, mean-field, methods and IV to series approximations and the renormalization group. Exact results and methods are presented in Part III. The last part of the book contains a collection of mathematical appendices on a range of relevant topics. For convenience, they have been divided into material related to graphs and lattices and (perhaps rather inexactly) algebra and analysis. The intention is to minimise the need to refer to other mathematical sources.

Some parts of the material were originally written for an M.Sc. course at King's College London and I am grateful to the many students who, by their questions and comments, helped to reduce errors and obscurities. The exchange of ideas and information with friends and co-workers has contributed to this book. In particular I owe a great deal to the interaction with my two long-term collaborators, the late George Bell, to whom this work is dedicated, and Byron Southern of the University of Manitoba. I should also like to thank Reimer Kühn, who has always been prepared to spend time in discussion of points of substance and presentation. As was the case with our previous books I acknowledge with gratitude the continued support and encouragement of Prof. Elliott Lieb. During the final stages of production, the book was read by three referees, for whose comments and suggestions I am very grateful. Finally, I wish to express my thanks to Prof. Wolf Beiglböck, Dr. Aldo Rampioni and the staff at Springer for their work in the production of this book.

London, June 2014 David A. Lavis

Contents

Part II Classical Approximation Methods

Part I
Thermodynamics, Statistical Mechanical Models and Phase Transitions

Introduction

Thermodynamics is concerned with the laws governing the macroscopic properties of systems subject to thermal change. The only explicit recognition of the atomic nature of matter which occurs in the theory is the inclusion, for systems with a variable number of particles, of the chemical potential which is the energy a particle brings on entry to the system. Otherwise a thermodynamic system is regarded as a 'black box' with a set of variables related by differential relationships which apply when the system is in *equilibrium*. This is defined operationally by supposing that the system is left so that it "eventually reaches a state (the equilibrium state) in which no further change is perceptible, no matter how long one waits" (Pippard 1957, p. 6). So when the system is in equilibrium the variables will all assume time-independent values. The thermodynamic context of this book is that of a system in equilibrium, where changes of thermodynamic state are considered to be quasi-static ('infinitely slow') and reversible.[1]

The principle aim of equilibrium statistical mechanics is to derive the macroscopic properties of a system from the laws governing the microscopic behaviour of its constituent particles. In particular, statistical mechanics gives rise to the equations of thermodynamics once the thermodynamic variables have been given an interpretation within the theory.[2] Since this is done in terms of the expectation

[1] An account of an extension of thermodynamics to non-equilibrium irreversible processes is given by Prigogine (1967) and Lieb and Yngvason (1998, 1999, 2013) (with a summary of the ideas of Lieb and Yngvason (1998, 1999) in Lieb and Yngvason (2000)) have developed a rigorous approach to thermodynamics which in Lieb and Yngvason (2013) expands the entropy concept to non-equilibrium states.

[2] A classic problem in statistical mechanics (Sklar 1993, Chap. 9, Callender 1999) is to understand in what sense thermodynamics can be said to *reduce*, to statistical mechanics. In particular the relationship between the meanings of equilibrium in thermodynamics and in statistical mechanics, together with the extension to non-equilibrium, are questions with no settled resolution. For a recent attempt to develop a solution to these problems see Lavis (2005, 2008, 2011).

values associated with a probability distribution, fluctuations of dependent variables are predicted for finite systems. Exact correspondence to a thermodynamic system at equilibrium occurs only in the thermodynamic limit (Sects. 1.3 and 8.1) of an infinite system, when, except close to a phase transition, the relative fluctuations of the dependent variables disappear.

Most physical systems can exist in a number of different *phases*, distinguished by their different types of molecular or atomic order. This order may be in the spatial configurations of one or more kinds of microsystems or it may be in the orientations or conformations of the microsystems themselves.[3] In the case of the vapour (gas), liquid and solid phases of, for example, water or hydrogen, the order is spatial with no order in the vapour, a short-range clustering type of order in the liquid and long-range lattice order in the solid.[4] The most well-known example of orientational order occurs in magnetic systems where ferromagnetism corresponds to the alignment of the unpaired spins of the microsystems (Sect. 3.5). Although a simple magnetic system may possess just one type of ferromagnetic phase, more complex ferrimagnetic systems can have a large number of such phases. In the case, for example, of cerium antimonide 14 different phases have been identified by neutron diffraction experiments (Fischer et al. 1978) and specific heat analysis (Rossat-Mignod et al. 1980). These phases exhibit different forms of anisotropic spatial ordering. The occurrence of different types of anisotropic ordering is also a feature which distinguishes between the nematic, smectic and columnar phases in liquid crystals (de Gennes and Prost 1993).

Chapter 3 contains a survey of lattice models with an indication of some of the relationships between them. Although the main emphasis of this book is on classical (as distinct from quantum) systems it is of interest to discuss the quantum Heisenberg model not least because of the relationships between d-dimensional quantum systems and $(d + 1)$-dimensional classical systems (Polyakov 1987, Sect. 1.1). These are exemplified (Sect. 3.2) by the d-dimensional quantum Ising model and the $(d + 1)$-dimensional classical Ising model (Suzuki 1976; Björberg and Grimmett 2009), the XXZ (Heisenberg–Ising) chain and the six-vertex model (Lieb 1967a, b, c, d) and the XYZ (general Heisenberg) chain and the eight-vertex model (Sutherland 1970).

The boundaries, or *transition regions*, between phases are characterized by discontinuous or singular behaviour in one or more of the thermodynamic variables associated with the system. A classification of phase transitions according to whether particular derivatives of the free-energy density have singularities or discontinuities across the transition region was proposed by Ehrenfest (1933) (See e.g. Pippard 1957). However, it is now known that many transitions do not fit into this classification. The attempt to place all phase transitions into classes according these kinds of criteria has been abandoned and the more modest classification of Fisher (1967) has been largely adopted. In both Ehrenfest's and Fisher's terminologies regions of *first-order phase*

[3] See Ziman (1979) for a discussion of the wide variety of possible types of order.

[4] For water this is not the complete picture. There are at least nine different ice phases distinguished by their lattice structures and proton configurations (Eisenberg and Kauzmann 1969).

transitions are those across which densities are discontinuous. These can also be identified as regions of phase coexistence. In place of the more detailed proposals of Ehrenfest, Fisher suggested that all transition regions across which the densities are continuous should be called *continuous transitions*. These may, of course, involve discontinuities or singularities in second or higher derivatives of the free-energy density. In Chap. 4 we shall augment Fisher's scheme by referring to continuous transitions for which one or more of the thermodynamic response functions is singular as *second-order* and other continuous transitions as *higher-order*. The onset of ferromagnetism in iron is accompanied by a singularity in the susceptibility and is thus a second-order transition. The freezing of water is accompanied by a discontinuous reduction in density implying a first-order transition. In some of the discussion in Chap. 4 we shall need to make a distinction between different types of transition regions. The term *coexistence region* will be used for a region of first-order transitions and *critical region* for any region in which the transition is continuous.

Ehrenfest's classification of transitions is one attempt to produce a taxonomy of phase behaviour. With a rather different emphasis, the earlier formulation of the law of corresponding states is another. Scaling theory can be viewed in a similar light. It develops from an assumed form for the free-energy density, or equation of state, close to the transition region and the underlying idea is that the probability distribution of microstates is invariant under a uniform change of length scale. The asymptotic behaviour of thermodynamic functions as a transition region is approached can be characterized by a set of *critical exponents*. Using purely thermodynamic arguments, inequalities between these exponents can be derived (see, for example, Buckingham 1972). One of the salient features of scaling theory is that, subject to the validity of its assumptions, these inequalities become equalities, the so-called *scaling laws*. This means that a smaller set of variables, the *scaling exponents*, one for each independent field, are sufficient to characterize the leading asymptotic behaviour at the transition. Although scaling theory is unable to provide numerical values for scaling exponents, the generalization of the theory given by conformal invariance is able to do so in two dimensions. A conformal transformation can be understood as a generalization of a scaling transformation in which the rescaling factor varies continuously over the physical space of the system.

The idea of scaling was proposed independently by a number of authors (Widom 1965; Domb and Hunter 1965; Kadanoff 1975; Patashinskii and Pokrovskii 1966). The presentation used in Chap. 4 is based on the subsequent developments by Hankey and Stanley (1972), Hankey et al. (1972) and Nightingale and 'T Hooft (1974). The application of conformal invariance to critical phenomena was first made by Polyakov (1970) and has been much developed in recent years (Cardy 1987; Christe and Henkel 1993). This approach is of particular importance in the case of two-dimensional systems where models are associated with specific values of c, the *central charge in the Virasoro algebra* of the generators of the conformal transformation. In the final section of Chap. 4 we give a brief introduction to *Schramm–Loewner evolution* indicating how under certain conditions it becomes a standard Brownian motion, for which the diffusion constant κ is related to the central charge of the conformal transformation.

Chapter 1
Thermodynamics

1.1 Formulae and Variables

Thermodynamic variables fall into two classes, intensive and extensive and intensive variables again are of two types, fields such as the temperature[1] $\widetilde{T}$, the pressure $\widetilde{P}$, the magnetic field $\widetilde{\mathcal{H}}$ and the chemical potential μ, and densities which are defined below. For fixed values of the intensive variables, the extensive variables such as entropy $\widetilde{S}$, volume $\widetilde{V}$, magnetic moment $\widetilde{\mathcal{M}}$ and number of molecules M are proportional to the size of the system. For temperature $\widetilde{T}$, pressure $\widetilde{P}$ and magnetic field $\widetilde{\mathcal{H}}$ the fields are[2]

$$T := k_{\mathrm{B}}\widetilde{T}\,, \qquad P := \widetilde{P}v_0\,, \qquad \mathcal{H} := \widetilde{\mathcal{H}}m_0, \tag{1.1.1}$$

where k_{B} is Boltzmann's constant, v_0 is a standard volume and m_0 is the dipole moment of a single molecule so that each quantity is of the dimensions of energy.[3] The extensive variables conjugate to fields are dimensionless, those corresponding to (1.1.1) being

$$S := \widetilde{S}/k_{\mathrm{B}}\,, \qquad V := \widetilde{V}/v_0\,, \qquad \mathcal{M} := \widetilde{\mathcal{M}}/m_0, \tag{1.1.2}$$

respectively. The *internal energy* U of a system is an extensive variable arising from the kinetic energies of the molecules and the energies of interaction between them.

---

[1] The reason for attaching the tilde symbol to these variables (apart from μ and M) is that it is convenient to define all fields so that they have the dimensions of energy and extensive variables so that they are dimensionless; μ and M are already of these types.

[2] To distinguish the use of the equals sign to signify a definition, as distinct from a derived equality, we use the symbol ':=' to represent the former.

[3] In this context it is interesting to note the work of Underwood, Sutton and de Podesta at the NPL (de Podesta 2013) which aims to set a new standardised temperature scale by measuring, through a determination of the speed of sound in a monotonic gas at low densities, the kinetic energy per particle and thus $k_{\mathrm{B}}\widetilde{T}$ for the gas.

D.A. Lavis, *Equilibrium Statistical Mechanics of Lattice Models*,
Theoretical and Mathematical Physics, DOI 10.1007/978-94-017-9430-5_1

For a system with $n_\mathrm{f}+1$ fields $T,\ \xi_i,\ i=1,\dots,n_\mathrm{f}$, with conjugate extensive variables $S,\ Q_i, i=1,\dots,n_\mathrm{f}$,

$$\mathrm{d}U = T\mathrm{d}S + \sum_{i=1}^{n_\mathrm{f}} \xi_i \mathrm{d}Q_i \tag{1.1.3}$$

applies to a quasi-static change between equilibrium states.[4] It follows that[5]

$$T = \left(\frac{\partial U}{\partial S}\right)_{Q_i}, \qquad \xi_i = \left(\frac{\partial U}{\partial Q_i}\right)_{S,Q_j}, \qquad i=1,\dots,n_\mathrm{f}\ \ (j\neq i). \tag{1.1.4}$$

Suppose that, with its internal constitution fixed, the size of the system is increased by a factor λ. Since all extensive variables will increase by the factor λ,

$$U(\lambda Q_i, \lambda S) = \lambda U(Q_i, S). \tag{1.1.5}$$

Differentiation with respect to λ, and substitution from (1.1.4) gives

$$U = TS + \sum_{i=1}^{n_\mathrm{f}} \xi_i Q_i, \tag{1.1.6}$$

when λ is put equal to 1, and

$$S\mathrm{d}T + \sum_{i=1}^{n_\mathrm{f}} Q_i \mathrm{d}\xi_i = 0. \tag{1.1.7}$$

A subset, $Q_1,\dots,Q_\eta$, of the extensive variables can now be replaced as independent variables by their conjugate fields $\xi_1,\dots,\xi_\eta$ to give, using a Legendre transformation, the *enthalpy*

$$H_\eta := U - \sum_{i=1}^{\eta} \xi_i Q_i = TS + \sum_{i=\eta+1}^{n_\mathrm{f}} \xi_i Q_i, \tag{1.1.8}$$

---

[4] In this formula the contributions to the summation are assumed to be such that each field ξ_i causes an increase in the corresponding extensive variable Q_i. As we shall see in Sect. 1.4 this is the case for a magnetic system, when $\xi_i = \mathcal{H}$ and $Q_i = \mathcal{M}$. However, for a fluid system the pressure $\xi_i = P$ leads to a decrease in volume $Q_i = V$. Thus the contribution to (1.1.3) will in this case be $-P\mathrm{d}V$.

[5] Because of the various choices possible for the independent variables in any thermodynamic system, it is often necessary to specify, using subscripts, the variable or variables kept constant during a partial differentiation. In general this is not necessary for free energies which are defined to be functions of a specific set of variables.

with

$$dH_\eta = T dS - \sum_{i=1}^{\eta} Q_i d\xi_i + \sum_{i=\eta+1}^{n_f} \xi_i dQ_i. \tag{1.1.9}$$

The replacement of $Q_1, \ldots, Q_\eta$ by their conjugate fields $\xi_i, \ldots, \xi_\eta$ means that the system is open to energy interchange by way of these fields. The appropriate *free energy* is now[6]

$$F_\eta := H_\eta - TS = U - TS - \sum_{i=1}^{\eta} \xi_i Q_i = \sum_{i=\eta+1}^{n_f} \xi_i Q_i , \tag{1.1.10}$$

$$dF_\eta = -S dT - \sum_{i=1}^{\eta} Q_i d\xi_i + \sum_{i=\eta+1}^{n_f} \xi_i dQ_i. \tag{1.1.11}$$

From (1.1.10) and (1.1.11),

$$\begin{aligned} Q_i &= -\frac{\partial F_\eta}{\partial \xi_i}, \qquad i = 1, \ldots, \eta, \qquad S = -\frac{\partial F_\eta}{\partial T}, \\ \xi_i &= \frac{\partial F_\eta}{\partial Q_i}, \qquad i = \eta + 1, \ldots, n_f, \quad H_\eta = \frac{\partial (F_\eta / T)}{\partial (1/T)}. \end{aligned} \tag{1.1.12}$$

Finally we observe for later reference that

$$F_\eta(Q_{\eta+1}, \ldots, Q_{n_f}, T, \xi, \ldots, \xi_\eta) = F_{\eta-1}(Q_\eta, \ldots, Q_{n_f}, T, \xi, \ldots, \xi_{\eta-1}) - \xi_\eta Q_\eta, \tag{1.1.13}$$

where Q_η on the right-hand side of this equation is given, from (1.1.12), as a solution of

$$\frac{\partial F_{\eta-1}}{\partial Q_\eta} = \xi_\eta. \tag{1.1.14}$$

If we were to include Q_η explicitly on left-hand side of (1.1.13) by replacing $F_\eta(Q_{\eta+1}, \ldots, Q_{n_f}, T, \xi, \ldots, \xi_\eta)$ by $\widetilde{F}_\eta(Q_\eta, Q_{\eta+1}, \ldots, Q_{n_f}, T, \xi, \ldots, \xi_\eta)$, then (1.1.14) is the condition for the required value of Q_η to be a stationary point of $\widetilde{F}_\eta$ with respect to Q_η.

---

[6] For enthalpy and free energies and later in the book for Hamiltonians and partition functions the subscript η (or its particular value) is used to indicate the number of external fields (excluding the temperature). In later chapters where there is no possibility of ambiguity this subscript is omitted.

1.2 The Field-Density Representation

The formulation of thermodynamics, given by (1.1.3)–(1.1.11), relates pairs of variables one of which is a field and the other an extensive variable; it is thus referred to as the *field-extensive variable representation*. We shall now introduce a transformation in which extensive variables are replaced by densities to give the *field-density representation*. Densities have the advantage that they remain finite in the thermodynamic limit of an infinite system (Sect. 1.3).

From (1.1.6) and (1.1.10) we see that, in order for the free energy $F_\eta(T, \xi_1, \ldots, \xi_\eta, Q_{\eta+1}, \ldots, Q_{n_\mathrm{f}})$ not to be identically zero, we must have $\eta < n_\mathrm{f}$; there must be at least one independent extensive variable Q_{n_f}. We take this as a fixed measure $Q := Q_{n_\mathrm{f}}$ of the *size* of the system, with $\mathrm{d}Q_{n_\mathrm{f}} = 0$ in (1.1.3), (1.1.9) and (1.1.11). The remaining independent extensive variables are replaced as arguments of F_η by independent densities

$$\rho_i := Q_i/Q, \qquad i = \eta + 1, \ldots, n_\mathrm{f} - 1, \tag{1.2.1}$$

with the free-energy density

$$f_\eta(Q, \rho_{\eta+1}, \ldots, \rho_{n_\mathrm{f}-1}, T, \xi_1, \ldots, \xi_\eta) := \frac{F_\eta(Q, \rho_{\eta+1}, \ldots, \rho_{n_\mathrm{f}-1}, T, \xi_1, \ldots, \xi_\eta)}{Q}. \tag{1.2.2}$$

From (1.1.6), (1.1.8) and (1.1.10),

$$u = sT + \xi_{n_\mathrm{f}} + \sum_{i=1}^{n_\mathrm{f}-1} \rho_i \xi_i, \tag{1.2.3}$$

$$h_\eta = u - \sum_{i=1}^{\eta} \rho_i \xi_i = sT + \xi_{n_\mathrm{f}} + \sum_{i=\eta+1}^{n_\mathrm{f}-1} \rho_i \xi_i, \tag{1.2.4}$$

$$f_\eta = h_\eta - sT = \xi_{n_\mathrm{f}} + \sum_{i=\eta+1}^{n_\mathrm{f}-1} \xi_i \rho_i, \tag{1.2.5}$$

and, from (1.1.3), (1.1.7), (1.1.9) and (1.1.11),

$$\mathrm{d}u = T\mathrm{d}s + \sum_{i=1}^{n_\mathrm{f}-1} \xi_i \mathrm{d}\rho_i, \tag{1.2.6}$$

$$0 = s\mathrm{d}T + \mathrm{d}\xi_{n_\mathrm{f}} + \sum_{i=1}^{n_\mathrm{f}-1} \rho_i \mathrm{d}\xi_i, \tag{1.2.7}$$

$$dh_\eta = du - \sum_{i=1}^{\eta} \rho_i d\xi_i - \sum_{i=1}^{\eta} \xi_i d\rho_i, \tag{1.2.8}$$

$$df_\eta = -sdT - \sum_{i=1}^{\eta} \rho_i d\xi_i + \sum_{i=\eta+1}^{n_f-1} \xi_i d\rho_i. \tag{1.2.9}$$

Another transformation, which is of importance for the development of scaling theory and the renormalization group, replaces the fields by dimensionless ratios of the fields with respect to T; thus yielding the *coupling-extensive variable* and *coupling-density representations*, which are described in Sect. 2.3.

1.3 The Thermodynamic Limit

The system described above is *finite* with all the independent extensive variables Q_i, $i = \eta + 1, \ldots, N_f$ having finite values. However, for statistical mechanics it is important to consider *infinite* systems since this is a necessary condition for the occurrence of a phase transition (see p. 253). There are two possible ways of dealing with this problem. The one adopted in this book is that in which an infinite system is obtained by taking the *thermodynamic limit* $Q \to \infty$ with the independent densities $\rho_i, i = \eta+1, \ldots, n_f-1$ fixed. The conditions for the existence of the thermodynamic limit for a one-component lattice fluid are given in Sect. 8.1. An alternative to this is to consider an infinite system from the outset. For the sake of comparison a brief summary of this approach is given in Sect. 8.2.

Of course, the free-energy density, as defined by (1.2.2), and its field derivatives, which are the dependent densities s and $\rho_i, i = 1, \ldots, \eta$, are all, in general, functions of Q. The thermodynamic limit can be achieved only if it leads to a finite value $f_\eta(\rho_{\eta+1}, \ldots, \rho_{n_f-1}, T, \xi_1, \ldots, \xi_\eta)$ for the free-energy density, and hence for the dependent densities.

In this book we shall make extensive use of the field-density representation defined in Sect. 1.2 and the coupling-density representation which will be defined in Sect. 2.3. In each case the free-energy density and the other dependent densities will be functions of the size parameter Q until the thermodynamic limit is applied. In many cases the form of the analysis is the same whether or not the limit has been applied and we shall not display the argument Q. In other cases it will be important to include a specific reference to whether the system is finite or infinite. For a system on a lattice $\mathcal{N}$ of $N := |\mathcal{N}|$ sites we often include $\mathcal{N}$ as an argument with $\mathcal{L}$ denoting the corresponding infinite lattice resulting from the limit $N \to \infty$.

1.4 Particular Response Functions

In Chap. 4 critical exponents are introduced. The most commonly used of these are associated with the limiting properties of densities and response functions, such as susceptibility, heat capacity and compressibility. These are given in the field-extensive variable representation by the second-order partial derivatives of the free energy F_η. They can in consequence be expressed easily in the field-density representation in terms of the second-order partial derivatives of the free-energy density f_η. Their forms in the coupling-density representation are more complicated as we shall see in Sect. 2.3. Here the discussion is restricted to the case $n_{\mathrm{f}} = 2$, setting $\rho := \rho_1$ and $\xi := \xi_1$. From (1.2.5) and (1.2.9),

$$\mathrm{d}f_0 = -s\mathrm{d}T + \xi\mathrm{d}\rho, \tag{1.4.1}$$

$$\mathrm{d}f_1 = -s\mathrm{d}T - \rho\mathrm{d}\xi, \tag{1.4.2}$$

$$f_1 = \xi_2 = f_0 - \rho\xi. \tag{1.4.3}$$

There are four principle response functions

$$c_\xi := k_{\mathrm{B}}T\left(\frac{\partial s}{\partial T}\right)_\xi, \tag{1.4.4}$$

$$c_\rho := k_{\mathrm{B}}T\left(\frac{\partial s}{\partial T}\right)_\rho, \tag{1.4.5}$$

$$\varphi_T := T\left(\frac{\partial \rho}{\partial \xi}\right)_T, \tag{1.4.6}$$

$$\alpha_\xi := T\left(\frac{\partial \rho}{\partial T}\right)_\xi. \tag{1.4.7}$$

c_ξ and c_ρ are the *heat capacities* at constant field and density respectively, φ_T has the limiting properties of the *compressibility of a fluid* or the *susceptibility of a magnet* and α_ξ has the limiting properties of the *coefficient of thermal expansion.* The standard formula

$$c_\xi - c_\rho = k_{\mathrm{B}}\frac{\alpha_\xi^2}{\varphi_T}, \tag{1.4.8}$$

relating the heat capacities at constant field and density can now be derived.

1.4.1 Magnetic Systems

Here the size parameter is $Q := M$, the number of magnetic dipoles, with $f_1 = \xi_2 := \mu$, the chemical potential per dipole. Then $\rho = m = \mathcal{M}/M$, the magnetization density and $\xi = \mathcal{H}$.[7] $c_{\mathcal{H}}$ and c_m, given by (1.4.4) and (1.4.5) respectively, are the heat capacities per dipole at constant field and magnetization respectively. The *isothermal magnetic susceptibility* is

$$\chi_T := \frac{C\varphi_T}{T} = C\left(\frac{\partial m}{\partial \mathcal{H}}\right)_T, \qquad \text{where} \qquad C := \frac{m_0^2}{v_0 \mu_0}, \tag{1.4.9}$$

m_0 and v_0 being, as defined in Sect. 1.1 and μ_0 being the permeability of free space, $4\pi 10^{-7}$ Wb/(A m). The inclusion of the constant C, which is of the dimensions of energy, makes the susceptibility dimensionless.

1.4.2 Fluid Systems

Here one choice of size parameter is $Q := M$ the number of particles in the gas, with $f_1 = \xi_2 := \mu$, the chemical potential per particle. Then $v = V/M$ and $\xi = -P$.[8] This is the *constant-pressure distribution* in statistical mechanics and c_P and c_v, given by (1.4.4) and (1.4.5) respectively, are the heat capacities per particle at constant pressure and volume respectively. The *isothermal compressibility* is

$$\mathcal{K}_T := -\frac{\varphi_T}{Tv} = -\frac{1}{v}\left(\frac{\partial v}{\partial P}\right)_T \tag{1.4.10}$$

and the *isobaric coefficient of thermal expansion* is

$$\mathcal{A}_P := -\frac{\alpha_P}{Tv} = -\frac{1}{v}\left(\frac{\partial v}{\partial T}\right)_P. \tag{1.4.11}$$

An alternative form of the field-density representation for fluids is achieved by scaling with respect to volume V, or, in the case of a lattice fluid (Sect. 3.5.2), the number of lattice sites N, rather than the number of particles. The independent field is now the chemical potential μ with conjugate density $\rho := M/V$ and $f_1 := -P$. The expression (1.4.10) for the isothermal compressibility is replaced by

$$\mathcal{K}_T := \frac{1}{\rho^2}\left(\frac{\partial \rho}{\partial \mu}\right)_T. \tag{1.4.12}$$

This choice gives the *grand-canonical distribution* in statistical mechanics.

---

[7] When we consider lattice systems with a dipole on each lattice site $M = N$ the number of lattice sites.

[8] That the field must correspond to *minus* the pressure has already been noted in Sect. 1.1.

Chapter 2
Statistical Mechanics

2.1 Distributions

The formalism described in Chap. 1 can be regarded as a macroscopic perspective on a system with a microscopic structure consisting of a collection of atoms or molecules of one or more types. The numbers of particles of each type, or component, will be among the set of extensive variables $Q_1, \ldots, Q_{n_\mathrm{f}}$.

A configuration of the particles of the system is called a *microstate* and is denoted by $\boldsymbol{\sigma}$. For each microstate there is a value $E(\boldsymbol{\sigma})$ of the *statistical mechanical internal energy* E and values $\widehat{Q}_i(\boldsymbol{\sigma})$, $i = 1, \ldots, \eta$, of the dependent statistical mechanical extensive variables.[1] The Hamiltonian[2]

$$\widehat{H}_\eta(\boldsymbol{\sigma}) := E(\boldsymbol{\sigma}) - \sum_{i=1}^{\eta} \xi_i \widehat{Q}_i(\boldsymbol{\sigma}), \tag{2.1.1}$$

is a crucial quantity in the statistical mechanics of the system with independent variables $Q_{\eta+1}, \ldots, Q_{n_\mathrm{f}}, T, \xi_1, \ldots, \xi_\eta$. It is the microscopic equivalent of the thermodynamic enthalpy defined by (1.1.8).

As indicated above we shall be concerned with the case where the system is in equilibrium and then the probability function for the *distribution* of $\boldsymbol{\sigma}$ is

$$p_\eta(T; \boldsymbol{\sigma}) := \frac{\exp[-\widehat{H}_\eta(\boldsymbol{\sigma})/T]}{Z_\eta(T)}, \tag{2.1.2}$$

---

[1] In cases where we wish to use the same letter for a thermodynamic variable and its classical statistical mechanical counterpart we make the distinction by adding a 'hat' to the statistical mechanical variable. In particular this relationship applies to the enthalpy H_η and the Hamiltonian $\widehat{H}_\eta$ (Eq. 2.1.7).

[2] For a comment concerning the usage of the subscript η, specifying the number of independent fields (and hence the number of independent extensive variables, or densities) see the footnote on page 7. For the sake of brevity it is convenient in most cases to drop reference to the independent fields and extensive variables as arguments, retaining only T.

D.A. Lavis, *Equilibrium Statistical Mechanics of Lattice Models*,
Theoretical and Mathematical Physics, DOI 10.1007/978-94-017-9430-5_2

where

$$Z_\eta(T) := \sum_{\{\sigma\}} \exp[-\widehat{H}_\eta(\sigma)/T], \tag{2.1.3}$$

is the *partition function.*[3] This is usually referred to as the *Gibbs distribution*. In the case when the system is a perfect gas it is the *Maxwell-Boltzmann distribution*. The classic account which justifies the role of (2.1.2) as the equilibrium distribution in statistical mechanics is that of Khinchin (1949), who established the result for the case $\eta = 0$, called the *canonical distribution*. This situation can be pictured as a system in a heat bath which controls the temperature. Other cases, where, for example, $\eta = 1$ and ξ_1 is the chemical potential (called the *grand-canonical distribution*, where in addition to the heat bath the system has permeable walls with the passage of particles controlled by the chemical potential), or $\eta = 1$ and ξ_1 is the pressure (called the *constant-pressure distribution*, where for a fluid system the volume is controlled by a piston used to vary the pressure), can be established in a similar way.[4] Functions of the microstate of the system, like $\widehat{C}(\boldsymbol{\sigma})$, are random variables for which the expectation value

$$\langle \widehat{C}(\boldsymbol{\sigma}) \rangle := \sum_{\{\sigma\}} p_\eta(T; \boldsymbol{\sigma})\widehat{C}(\boldsymbol{\sigma}), \tag{2.1.4}$$

variance and covariance between pairs of such variables play a fundamental role in connecting the distribution to thermodynamics (Sect. 2.1.2).

2.1.1 Quantum Systems

This is a text on *classical* statistical mechanics. However, it is useful to give a brief indication of the corresponding development for quantum systems where the microstate $\boldsymbol{\sigma}$ will be some set of quantum operators. Then the Hamiltonian $\widehat{H}_\eta(\boldsymbol{\sigma})$ is an operator and $p_\eta(T; \boldsymbol{\sigma})$ is the *density operator*, which in terms of a representation of the system becomes the *density matrix*. Equations (2.1.3) and (2.1.4) are replaced by

$$Z_\eta(T) := \text{Trace}\{\exp[-\widehat{H}_\eta(\boldsymbol{\sigma})/T]\}, \tag{2.1.5}$$

---

[3] In this and all subsequent formulae, the summation is over all microstates $\boldsymbol{\sigma}$ compatible with the given values of the independent extensive variables.

[4] The different distributions, for different values of η are usually called *ensembles*. This term is associated with the (relative-frequency flavoured) picture of a collection (or ensemble) of systems whose phase points have the density (2.1.2) in the space of microstates (Gibbs 1902).

and

$$\langle \widehat{C}(\boldsymbol{\sigma}) \rangle := \text{Trace}\{p_\eta(T;\boldsymbol{\sigma})\widehat{C}(\boldsymbol{\sigma})\}. \tag{2.1.6}$$

A *complete solution* to this problem, in both the quantum and statistical mechanical sense, would be a determination of all the eigenvalues and eigenvectors of the Hamiltonian. The complicating feature of a quantum system, as can be seen from the brief account of the Heisenberg model in Sect. 3.2, is that $\boldsymbol{\sigma}$ may contain non-commuting operators.

2.1.2 The Connection to Thermodynamics

This is achieved by the identifications

$$H_\eta := \langle \widehat{H}_\eta \rangle = -\frac{\partial \ln(Z_\eta)}{\partial(1/T)}, \tag{2.1.7}$$

$$Q_i := \langle \widehat{Q}_i \rangle = T\frac{\partial \ln(Z_\eta)}{\partial \xi_i}, \quad i = 1, \ldots, \eta, \tag{2.1.8}$$

$$\xi_i := \left\langle \frac{\partial \widehat{H}_\eta}{\partial Q_i} \right\rangle = -T\frac{\partial \ln(Z_\eta)}{\partial Q_i}, \quad i = \eta + 1, \ldots, n_{\mathrm{f}}. \tag{2.1.9}$$

The 'averaging' relations (2.1.7)–(2.1.9) become identical to (1.1.12) if we set

$$F_\eta(T) := -T \ln[Z_\eta(T)]. \tag{2.1.10}$$

In the special case of the canonical distribution, $\eta = 0$, $\widehat{H}_0(\boldsymbol{\sigma})$ is the energy $E(\boldsymbol{\sigma})$, H_0 is the internal energy U and F_0 is the Helmholtz free energy, usually denoted by A. The second derivatives of F_η with respect to the independent variables $T, \xi_1, \ldots, \xi_\eta$ are the *response functions*. It is not difficult to show that the response function with respect to the variables ξ_i, ξ_j is related to the covariance of the random variables $\widehat{Q}_i$ and $\widehat{Q}_j$ by

$$\frac{\partial^2 F_\eta}{\partial \xi_i \partial \xi_j} = -\frac{\partial Q_j}{\partial \xi_i} = -\frac{\partial Q_i}{\partial \xi_j} = -\frac{1}{T}\text{Cov}[\widehat{Q}_i, \widehat{Q}_j]. \tag{2.1.11}$$

This is a *fluctuation-response function relation.*

It follows from (2.1.3) and (2.1.10) that

$$F_\eta(T) = -T \ln\left\{\sum_{\{Q_\eta\}} \exp[\{\xi_\eta Q_\eta - F_{\eta-1}(T)\}/T]\right\}. \tag{2.1.12}$$

Differentiating with respect to T and using the expression for S given by (1.1.10) gives

$$F_\eta(T) = \langle F_{\eta-1}(T) \rangle - \xi_\eta \langle Q_\eta \rangle, \tag{2.1.13}$$

where the expectations are taken with respect to the η distribution. This formula is the statistical mechanical analogue of (1.1.13).

2.2 Variations of the Probability Function

For the Hamiltonian (2.1.1), let

$$\mathfrak{F}_\eta(\Pi) := \sum_{\{\sigma\}} \Pi(\sigma) \widehat{H}_\eta(\sigma) + T \sum_{\{\sigma\}} \Pi(\sigma) \ln[\Pi(\sigma)], \tag{2.2.1}$$

where $\Pi(\sigma) \geq 0$ satisfies the normalization condition

$$\sum_{\{\sigma\}} \Pi(\sigma) = 1. \tag{2.2.2}$$

It is easy to verify, from (2.1.2) and (2.1.10), that

$$\mathfrak{F}(p_\eta) = F_\eta \tag{2.2.3}$$

and thus, from (1.1.10) and (2.1.7), that

$$S = -\sum_{\{\sigma\}} p_\eta(\sigma) \ln[p_\eta(\sigma)]. \tag{2.2.4}$$

Now consider variations of $\Pi(\sigma)$ in (2.2.1), subject to the normalization condition (2.2.2), for which we use the undetermined multiplier λ. Then

$$\begin{aligned} \delta \mathfrak{F}_\eta(\Pi) = \sum_{\{\sigma\}} \Bigg\{ & [\widehat{H}_\eta(\sigma) + T \ln \Pi(\sigma) + T + \lambda] \delta \, \Pi \, (\sigma) \\ & + T \frac{(\delta \, \Pi \, (\sigma))^2}{2 \Pi \, (\sigma)} + \mathrm{O}\left((\delta \, \Pi \, (\sigma))^3 \right) \Bigg\}, \end{aligned} \tag{2.2.5}$$

yielding a minimum given by

$$\widehat{H}_\eta(\sigma) + T \ln \Pi(\sigma) + T + \lambda = 0, \tag{2.2.6}$$

which is the unique solution $\Pi(\sigma) = p_\eta(\sigma)$.

2.3 Coupling Representations

In general the statistical mechanical internal energy is a linear function of a set of energy parameters ε_ℓ of the form

$$E(\varepsilon_\ell;\boldsymbol{\sigma}) := \sum_{\ell=1}^{n_e} E_\ell(\boldsymbol{\sigma})\varepsilon_\ell. \tag{2.3.1}$$

These parameters can be regarded as *internal fields* and it will be seen that the formulae of Sect. 2.1 can be simplified if the temperature is absorbed into the ratios

$$K_\ell := \varepsilon_\ell/T, \quad \ell = 1,\ldots,n_e, \qquad L_i := \xi_i/T, \quad i = 1,\ldots,n_f, \tag{2.3.2}$$

called *internal* and *external couplings* respectively. This then gives

$$\widehat{H}_\eta(K_\ell, L_i;\boldsymbol{\sigma}) = \sum_{\ell=1}^{n_e} E_\ell(\boldsymbol{\sigma})K_\ell - \sum_{i=1}^{\eta} \widehat{Q}_i(\boldsymbol{\sigma})L_i. \tag{2.3.3}$$

The discussion of scaling in Chap. 4 shows that the distinction between the internal couplings K_ℓ and the external couplings L_i is physical rather than mathematical.[5] The latter arise from interactions between the system and external fields, like the magnetic field $\mathcal{H}$ or the chemical potential μ, whereas the former arise from the energies of interaction between the microsystems. The external couplings are, in general, open to change by the experimenter, either by individual variations of the fields ξ_i or together by variations of T. The internal couplings can be changed only by means of a change of T, so that

$$\mathrm{d}K_\ell = -K_\ell\,\mathrm{d}T/T, \qquad \ell = 1,\ldots,n_e. \tag{2.3.4}$$

From (2.1.3),[6]

$$Z_\eta(K_\ell, L_i) := \sum_{\{\sigma\}} \exp[-\widehat{H}_\eta(K_\ell, L_i;\boldsymbol{\sigma})]. \tag{2.3.5}$$

Then, from (2.1.8)–(2.1.10),

$$\Phi_\eta(K_\ell, L_i) := F_\eta(T,\varepsilon_\ell,\xi_i)/T = -\ln[Z_\eta(K_\ell, L_i)] \tag{2.3.6}$$

---

[5] Although we do need to be careful about the consequences of the difference of sign in the two terms on the right of (2.3.10).

[6] Here it is clearer if we explicitly display representative members K_ℓ and L_i of the n_e internal couplings (all independent) and the η independent external couplings.

is the *dimensionless free energy*, with

$$Q_i = \langle \widehat{Q}_i(\boldsymbol{\sigma}) \rangle = -\frac{\partial \Phi_\eta}{\partial L_i}, \quad i = 1, \ldots, \eta, \qquad L_i = \frac{\partial \Phi_\eta}{\partial Q_i}, \quad i = \eta + 1, \ldots, n_{\mathrm{f}}, \tag{2.3.7}$$

$$\langle E_\ell(\boldsymbol{\sigma}) \rangle = \frac{\partial \Phi_\eta}{\partial K_\ell}, \quad \ell = 1, \ldots, n_{\mathrm{e}}. \tag{2.3.8}$$

Since $\Phi_\eta(K_\ell, L_i)$ is a function of T only through the couplings this must also be the case for the expectations $\langle \widehat{Q}_i(\boldsymbol{\sigma}) \rangle$ and $\langle E_\ell(\boldsymbol{\sigma}) \rangle$ and thus, from (2.3.1) and (2.3.3),

$$U(K_\ell) := \sum_{\ell=1}^{n_{\mathrm{e}}} \langle E_\ell(\boldsymbol{\sigma}) \rangle K_\ell, \tag{2.3.9}$$

$$H_\eta(K_\ell, L_i) := \sum_{\ell=1}^{n_{\mathrm{e}}} \langle E_\ell(\boldsymbol{\sigma}) \rangle K_\ell - \sum_{i=1}^{\eta} \langle \widehat{Q}_i(\boldsymbol{\sigma}) \rangle L_i. \tag{2.3.10}$$

The *coupling-density representation*, like the field-density representation, is obtained by forming densities by dividing the independent extensive variables by a size parameter which we again take to be $Q := Q_{n_{\mathrm{f}}}$. Displaying a representative independent density ρ_i and extensive variable $Q_i := \rho_i Q$ explicitly, the *dimensionless free-energy density*[7]

$$\phi_\eta(\rho_i, K_\ell, L_i) := \frac{\Phi_\eta(Q, \rho_i Q, K_\ell, L_i)}{Q} \tag{2.3.11}$$

is a function of the independent couplings and densities, without explicit reference to the temperature, making them convenient tools for the formulation of scaling theory and renormalization group methods.

We can also, of course, defined densities $\hat{\rho}_i := \widehat{Q}_i / Q$, for the dependent extensive variables, $i = 1, 2, \ldots, \eta$. These are random variables with $\rho_i := \langle \hat{\rho}_i \rangle$ defining thermodynamic densities and, from (2.1.11),

$$\mathrm{Var}[\hat{\rho}_i] = -\frac{1}{Q} \frac{\partial^2 \phi_\eta}{\partial L_i^2} \qquad i = 1, 2, \ldots, \eta. \tag{2.3.12}$$

In the thermodynamic limit $Q \to \infty$, unless the response function $\partial^2 \phi_\eta / \partial L_i^2$ diverges faster that Q, fluctuations in $\hat{\rho}_i$ will tend to zero and the distinction between dependent and independent densities, and hence between the different choices of

[7] In later work, when the context makes it unnecessary we shall omit the qualification 'dimensionless'.

distribution (canonical, grand-canonical etc.), will disappear. We have a *thermodynamic equivalence of distributions* away from the exceptional points where the response functions diverge, which are associated with phase transitions (Sect. 4.1).

As in Sect. 1.3 we now omit explicit reference to Q unless we need to be specific about the system being finite. With

$$\mathrm{u}_\ell := \langle E_\ell(\boldsymbol{\sigma})\rangle/Q, \qquad \upsilon := u/T = \sum_{\ell=1}^{n_\mathrm{e}} \mathrm{u}_\ell K_\ell, \tag{2.3.13}$$

from (1.2.3)–(1.2.5),

$$\upsilon = s + L_{n_\mathrm{f}} + \sum_{i=1}^{n_\mathrm{f}-1} \rho_i L_i, \tag{2.3.14}$$

$$\phi_\eta = L_{n_\mathrm{f}} + \sum_{i=\eta+1}^{n_\mathrm{f}-1} \rho_i L_i \tag{2.3.15}$$

and, from (1.2.6)–(1.2.9) and (2.3.4),

$$\mathrm{d}s = \sum_{\ell=1}^{n_\mathrm{e}} K_\ell \mathrm{d}\mathrm{u}_\ell - \sum_{i=1}^{n_\mathrm{f}-1} L_i \mathrm{d}\rho_i, \tag{2.3.16}$$

$$0 = \sum_{\ell=1}^{n_\mathrm{e}} \mathrm{u}_\ell \mathrm{d}K_\ell - \mathrm{d}L_{n_\mathrm{f}} - \sum_{i=1}^{n_\mathrm{f}-1} \rho_i \mathrm{d}L_i, \tag{2.3.17}$$

$$\mathrm{d}\phi_\eta = \sum_{\ell=1}^{n_\mathrm{e}} \mathrm{u}_\ell \mathrm{d}K_\ell - \sum_{i=1}^{\eta} \rho_i \mathrm{d}L_i + \sum_{i=\eta+1}^{n_\mathrm{f}-1} L_i \mathrm{d}\rho_i. \tag{2.3.18}$$

2.3.1 The Case $n_\mathrm{f} = 2$

As in Sect. 1.4 we set $\rho := \rho_1$ with the one independent external coupling $L := L_1$. In place of the temperature we now have K_ℓ, $\ell = 1, \dots, n_\mathrm{e}$ internal couplings. To reduce these to one variable which corresponds to thermal variation we set $\varepsilon := \varepsilon_1$, $K := \varepsilon/T$ with the parameters $\kappa_\ell := \varepsilon_\ell/\varepsilon$ kept constant and

$$\upsilon = \mathrm{u}\, K, \qquad \mathrm{u} := \sum_{\ell=1}^{n_\mathrm{e}} \mathrm{u}_\ell \kappa_\ell. \tag{2.3.19}$$

Now there are two free-energy densities: $\phi_0(K, \rho)$ and $\phi_1(K, L)$ with

$$d\phi_0 = \mathrm{u}\,dK + L\,d\rho, \tag{2.3.20}$$

$$d\phi_1 = \mathrm{u}\,dK - \rho\,dL. \tag{2.3.21}$$

They are related by the Legendre transformation

$$\phi_1 = \phi_0 - L\rho, \tag{2.3.22}$$

where $\rho = \rho(K, L)$ is the solution of

$$\frac{\partial \phi_0}{\partial \rho} = L. \tag{2.3.23}$$

Alternatively

$$\phi_0 = \phi_1 + \rho L, \tag{2.3.24}$$

where $L = L(K, \rho)$ is the solution of

$$\frac{\partial \phi_1}{\partial L} = -\rho. \tag{2.3.25}$$

In Sects. 1.4.1 and 1.4.2 we considered two examples of $n_\mathrm{f} = 2$, a magnetic system where the two extensive variables are the magnetization $\mathcal{M}$ and the number of dipoles M and a fluid system where the two extensive variables are the volume V and the number of particles M. In the former case there is an obvious choice for scaling; taking the magnetization per dipole as the density, conjugate to the magnetic field. For the fluid the choice is less obvious: volume per particle, with conjugate field $-P$, or particles per unit volume, with conjugate field μ. It is useful to list the relationship between these two choices. Denoting one using a prime,

$$\begin{aligned} &\rho' = 1/\rho, \quad \mathrm{u}' = \rho'\mathrm{u}, && \phi'(K, \rho') = \rho'\phi_0(K, \rho), \\ &L' = \phi_1(K, L), && L = \phi'(K, L'). \end{aligned} \tag{2.3.26}$$

It is now easy to check that Eqs. (2.3.20)–(2.3.25) take exactly the same form in the primed quantities. An example of the interchange between different choices of scaling variable is given for the van der Waals equation in Sect. 5.2.

We shall, in Chap. 4, need forms for the response functions, defined in Sect. 1.4, in terms of the coupling-density representation. From (2.3.20) and (2.3.21),

$$\mathrm{u} = \frac{\partial \phi_0}{\partial K} = \frac{\partial \phi_1}{\partial K} \tag{2.3.27}$$

and, from (2.3.15),

$$s = -\phi_0 + uK = -\phi_1 + uK - L\rho. \tag{2.3.28}$$

Then from (1.4.5)–(1.4.7), (2.3.20)–(2.3.22) and (2.3.25),

$$c_\rho = -k_B K \left(\frac{\partial s}{\partial K}\right)_\rho = -k_B K^2 \left(\frac{\partial u}{\partial K}\right)_\rho = -k_B K^2 \frac{\partial^2 \phi_0}{\partial K^2}, \tag{2.3.29}$$

$$\begin{aligned} c_\xi &= -k_B \left\{ K \left(\frac{\partial s}{\partial K}\right)_L + L \left(\frac{\partial s}{\partial L}\right)_K \right\} \\ &= -k_B \left\{ K^2 \left(\frac{\partial u}{\partial K}\right)_L - 2KL \left(\frac{\partial \rho}{\partial K}\right)_K - L^2 \left(\frac{\partial \rho}{\partial L}\right)_K \right\} \\ &= -k_B \left(K^2 \frac{\partial^2 \phi_1}{\partial K^2} + 2KL \frac{\partial^2 \phi_1}{\partial K \partial L} + L^2 \frac{\partial^2 \phi_1}{\partial L^2} \right), \end{aligned} \tag{2.3.30}$$

$$\varphi_T = \left(\frac{\partial \rho}{\partial L}\right)_K = -\frac{\partial^2 \phi_1}{\partial L^2}, \tag{2.3.31}$$

$$\alpha_\xi = -K \left(\frac{\partial \rho}{\partial K}\right)_L - L \left(\frac{\partial \rho}{\partial L}\right)_K = L \frac{\partial^2 \phi_1}{\partial L^2} + K \frac{\partial^2 \phi_1}{\partial L \partial K}. \tag{2.3.32}$$

Alternatively c_ρ can be obtained from (2.3.30)–(2.3.32) using (1.4.8).

2.4 Lattice Systems

In Sect. 16.2.1 we define what is meant by a lattice, give diagrams for the regular lattices and consider the question of finite lattices and periodic boundary conditions. Here we need simply to assume that the lattice, denote by $\mathcal{N}$, is one of the regular types with periodic boundary conditions and $N := |\mathcal{N}|$ sites, which is taken to be the one independent extensive variable Q_{n_f}.[8]

Chapter 3 contains a survey of lattice models, some, but not all, of which are treated in this book. They are distinguished by the types of cooperative interaction which contribute to the Hamiltonian. These fall broadly into two classes, those for which the state variables are associated with lattice sites and those for which they are associated with lattice edges (connecting first-neighbour pairs of sites).

[8] Of course, all the extensive quantities in the system, including the partition function and Hamiltonian are functions of N. However, it, or the lattice signifier $\mathcal{N}$, will be displayed explicitly only when it seems useful to do so. In the case of densities like ϕ it is also sometimes useful to indicate the underlying lattice by including the infinite lattice signifier $\mathcal{L}$.

2.4.1 Site-Variable Models

There is a microsystem at each site of the lattice and the state of the microsystem at site $\boldsymbol{r}$ is specified by the variable $\sigma(\boldsymbol{r})$. The symbol $\boldsymbol{\sigma}$ represents the collection or vector specifying a set of values of the states of all the microsystems on the lattice. The simplest contributions to the Hamiltonian are single-site terms with the extensive variables $\widehat{Q}_i$ expressed in the form

$$\widehat{Q}_i(\boldsymbol{\sigma}) = \sum_{\{\boldsymbol{r}\}} \mathfrak{q}_i(\boldsymbol{r};\boldsymbol{\sigma}), \qquad i = 1,\ldots,n_{\mathrm{f}}-1, \tag{2.4.1}$$

where $\mathfrak{q}_i(\boldsymbol{r};\boldsymbol{\sigma})$ is a function of the one component $\sigma(\boldsymbol{r})$ of $\boldsymbol{\sigma}$ at $\boldsymbol{r}$. Unless the states of the microsystems are very complicated there will be only a rather small number of independent extensive variables and different functions $\mathfrak{q}_i$.

The extensive internal energy variables E_ℓ in the Hamiltonian involve interactions between groups of more than one site. They can be expressed in the form

$$E_\ell(\boldsymbol{\sigma}) = \sum_{\{\boldsymbol{r}\}} \mathfrak{e}_\ell(\boldsymbol{r};\boldsymbol{\sigma}), \qquad \ell = 1,\ldots,n_{\mathrm{e}}, \tag{2.4.2}$$

where $\mathfrak{e}_\ell(\boldsymbol{r};\boldsymbol{\sigma})$ is a function of $\sigma(\boldsymbol{r})$ and the state variables of some set of sites around $\boldsymbol{r}$. In most cases of interest the functions $\mathfrak{e}_\ell(\boldsymbol{r};\boldsymbol{\sigma})$ are taken to be *local* in the sense that they involve only sites quite close to the site $\boldsymbol{r}$, although in Sects. 3.2 and 3.3 we consider cases where interactions, involving pairs of sites, are allowed to have an arbitrary range.

The simplest *pair-interaction models* are those where the terms are restricted to first-neighbour sites and this is the case for most site-variable models considered here. However, it is sometimes of interest to include second-neighbour interactions, particularly when there is 'conflict' between the types of ordering encouraged by the two different interactions (Sect. 3.5.1.1).

Another type of site-variable model which has attracted interest is the *face model* where the interaction term involves, in an indecomposable way, the variables $\sigma(\boldsymbol{r})$ for all sites $\boldsymbol{r}$ around a face of the lattice.[9] Such a face model is considered in Sect. 3.7.2.

2.4.2 Edge-Variable Models

In this case a variable is attached to each edge of the lattice. The edge-decorated Ising models discussed in Chap. 10 are obvious examples of this, although in that case there are also state variables at the sites of the lattice and the model can equally

[9] It is obvious that the Hamiltonian of a pair-interaction model can be represented (in a decomposable way) as a sum over lattice faces and this is often useful in cases where the Hamiltonian also includes an indecomposable face interaction.

well be thought of as a site-variable model on a lattice modified by the addition of a site on each lattice edge. The two types of true edge-variable models discussed in this book are the vertex models introduced in Sect. 3.8 and analyzed in Sect. 9.3 and Chap. 12, and the dimer model, introduced in Sect. 3.9 and analyzed in Chap. 13. In each model the edge variable has two states, which for the vertex model corresponds to the two directions of an arrow along the edge and for the dimer model indicates the presence or absence of a dimer on the edge.

Although vertex models can be formed on any regular two-dimensional lattice[10] we shall restrict attention to the two two-dimensional regular lattices with coordination number four; that is the square and kagomé lattices (Sect. 16.2.1). There are then sixteen arrow arrangements around a lattice site. However, the cases which have been solved are the six-vertex model (Sect. 3.8.2) where the arrow configuration is constrained so that there is the same number of arrows pointing towards and pointing away from the lattice site and the eight-vertex model (Sect. 3.8.1) where there is an even number of arrows pointing towards the lattice site.

The dimer configuration is also constrained by a condition at each lattice site. In this case there can be at most (and in the case of a complete covering of the lattice with dimers, exactly) one dimer end at each lattice site (Sect. 3.9).

2.5 Correlation Functions and Symmetry Properties

2.5.1 A General Hamiltonian

The coupling-density representation gives a particularly elegant formulation of thermodynamics in the special case $\eta = n_\mathrm{f} - 1$, when there are no independent densities. The function $\phi_{n_\mathrm{f}-1}$ is dependent only on couplings and, from (2.3.15), is itself equal to the coupling L_{n_f}. We shall concentrate on this case, omitting subscripts indicating the number of independent external couplings.

The values of the two sets of quantities $\mathfrak{q}_i(\boldsymbol{r};\boldsymbol{\sigma})$, $i = 1,\ldots,n_\mathrm{f}-1$ and $\mathfrak{e}_\ell(\boldsymbol{r};\boldsymbol{\sigma})$, $\ell = 1,\ldots,n_\mathrm{e}$ are determined by the microstate $\boldsymbol{\sigma}$ of the system. We shall therefore refer to them respectively as *external* and *internal state operators.* It is clear from the above discussion that the sets $\{Q_i, L_i, \widehat{Q}_i(\boldsymbol{\sigma}), \mathfrak{q}_i(\boldsymbol{r};\boldsymbol{\sigma})\}$ and $\{U_i, K_i, E_i(\boldsymbol{\sigma}), \mathfrak{e}_i(\boldsymbol{r};\boldsymbol{\sigma})\}$ play the same mathematical role in the development of the theory.[11] We shall, therefore, use the generic set $\{C_j, \zeta_j, \widehat{C}_j(\boldsymbol{\sigma}), \mathfrak{c}_j(\boldsymbol{r};\boldsymbol{\sigma})\}$ to stand for members of either the external or internal sets with

$$\widehat{C}_j(\boldsymbol{\sigma}) = \sum_{\{\boldsymbol{r}\}} \mathfrak{c}_j(\boldsymbol{r};\boldsymbol{\sigma}). \qquad (2.5.1)$$

---

[10] Or indeed on an arbitrary planar lattice (Baxter 1978) or a three-dimensional lattice.

[11] Although, of course, an internal coupling K_i is simply replaced by a ζ_j, whereas the coupling corresponding to an external coupling L_i is a $-\zeta_j$.

The state operators $\mathfrak{c}_j(\boldsymbol{r};\boldsymbol{\sigma})$, $j = 1, \ldots, n := n_{\mathrm{f}} + n_{\mathrm{e}} - 1$, with $\boldsymbol{r}$ ranging over all N sites of the lattice form a set of nN random variables distributed according to the probability function (2.1.2). In using (2.5.1) to express a general form for the Hamiltonian it is convenient, to include a configuration-independent term. Thus we set

$$\widehat{H}(N, \zeta_j; \boldsymbol{\sigma}) := N\zeta_0 + \sum_{j=1}^{n} \widehat{C}_j(\boldsymbol{\sigma})\zeta_j. \qquad (2.5.2)$$

This definition could, of course, be expressed in a more compact form by extending the summation to the range $j = 0, 1, \ldots, n$ and setting $\widehat{C}_0(\boldsymbol{\sigma}) = N$. It will also be observed that

$$\phi(\zeta_0, \zeta_1, \ldots, \zeta_n) = \zeta_0 + \phi(0, \zeta_1, \ldots, \zeta_n), \qquad (2.5.3)$$

and that the calculation of an expectation value (2.1.4) using the probability formula (2.1.2) is unaffected by the presence of ζ_0. For this reason it is often called the *trivial coupling*, although, as we shall see in Sects. 4.3.6 and 15.4, it plays a crucial role in scaling and renormalization group theory.

It is convenient to generalize this scheme further by supposing that the coupling is different at each site of the lattice with the corresponding generalized Hamiltonian

$$\widehat{H}^{(\mathrm{G})}(N, \zeta_j(\boldsymbol{r}); \boldsymbol{\sigma}) = \sum_{j=0}^{n} \sum_{\{\boldsymbol{r}\}} \mathfrak{c}_j(\boldsymbol{r};\boldsymbol{\sigma})\zeta_j(\boldsymbol{r}), \qquad (2.5.4)$$

where $\mathfrak{c}_0(\boldsymbol{r};\boldsymbol{\sigma}) = 1$, for all $\boldsymbol{r}$. With a similar notation for other quantities,

$$\langle \mathfrak{c}_j(\boldsymbol{r};\boldsymbol{\sigma}) \rangle = -\frac{1}{Z^{(\mathrm{G})}} \frac{\partial Z^{(\mathrm{G})}}{\partial \zeta_j(\boldsymbol{r})} = N \frac{\partial \phi^{(\mathrm{G})}}{\partial \zeta_j(\boldsymbol{r})}. \qquad (2.5.5)$$

2.5.2 Correlation Functions

Let $\mathfrak{c}_1(\boldsymbol{r}_1;\boldsymbol{\sigma}), \ldots, \mathfrak{c}_\tau(\boldsymbol{r}_\tau;\boldsymbol{\sigma})$ be some choice of a subset of these variables at τ sites of the lattice. In terms of these a number of different τ-point correlation functions can be defined:

(i) The *total correlation function* $\langle \mathfrak{c}_1(\boldsymbol{r}_1;\boldsymbol{\sigma}) \cdots \mathfrak{c}_\tau(\boldsymbol{r}_\tau;\boldsymbol{\sigma}) \rangle$.
(ii) The *fluctuation correlation function* $\langle \delta\mathfrak{c}_1(\boldsymbol{r}_1;\boldsymbol{\sigma}) \cdots \delta\mathfrak{c}_\tau(\boldsymbol{r}_\tau;\boldsymbol{\sigma}) \rangle$ where

$$\delta\mathfrak{c}_j(\boldsymbol{r};\boldsymbol{\sigma}) := \mathfrak{c}_j(\boldsymbol{r};\boldsymbol{\sigma}) - \langle \mathfrak{c}_j(\boldsymbol{r};\boldsymbol{\sigma}) \rangle, \qquad j = 1, 2, \ldots, \tau, \qquad (2.5.6)$$

is a measure of the fluctuation of $\mathfrak{c}_j$ at site $\boldsymbol{r}$ about its mean.

(iii) The *net* or *connected correlation function*, which is defined recursively by

$$\Gamma_\tau(\mathfrak{c}_1(\boldsymbol{r}_1)\cdots\mathfrak{c}_\tau(\boldsymbol{r}_\tau)) := \langle \mathfrak{c}_1(\boldsymbol{r}_1;\boldsymbol{\sigma})\cdots\mathfrak{c}_\tau(\boldsymbol{r}_\tau;\boldsymbol{\sigma})\rangle - \sum \Gamma_{p_1}(\mathfrak{c}_{j_1}(\boldsymbol{r}_{j_1}),\ldots,\mathfrak{c}_{p_1}(\boldsymbol{r}_{p_1})) \cdots \Gamma_{p_k}(\mathfrak{c}_{j_k}(\boldsymbol{r}_{j_k}),\ldots,\mathfrak{c}_{p_k}(\boldsymbol{r}_{p_k})), \tag{2.5.7}$$

$$\Gamma_1(\mathfrak{c}_1(\boldsymbol{r}_1)) := \langle \mathfrak{c}_1(\boldsymbol{r}_1;\boldsymbol{\sigma})\rangle, \tag{2.5.8}$$

where the sum in (2.5.7) is over products of the connected correlation functions for all partitions of the τ lattice sites.

For $\tau = 1$ the total and connected correlation functions are the same and the fluctuation correlation function is zero. For $\tau = 2$,

$$\begin{aligned}\Gamma_2(\mathfrak{c}_1(\boldsymbol{r}_1),\mathfrak{c}_2(\boldsymbol{r}_2)) &= \langle \mathfrak{c}_1(\boldsymbol{r}_1;\boldsymbol{\sigma})\mathfrak{c}_2(\boldsymbol{r}_2;\boldsymbol{\sigma})\rangle - \langle \mathfrak{c}_1(\boldsymbol{r}_1;\boldsymbol{\sigma})\rangle\langle \mathfrak{c}_2(\boldsymbol{r}_2;\boldsymbol{\sigma})\rangle,\\ &= \langle[\mathfrak{c}_1(\boldsymbol{r}_1;\boldsymbol{\sigma}) - \langle \mathfrak{c}_1(\boldsymbol{r}_1;\boldsymbol{\sigma})\rangle][\mathfrak{c}_2(\boldsymbol{r}_2;\boldsymbol{\sigma}) - \langle \mathfrak{c}_2(\boldsymbol{r}_2;\boldsymbol{\sigma})\rangle]\rangle\end{aligned} \tag{2.5.9}$$

and, for $\tau = 3$,

$$\begin{aligned}&\Gamma_3(\mathfrak{c}_1(\boldsymbol{r}_1),\mathfrak{c}_2(\boldsymbol{r}_2),\mathfrak{c}_3(\boldsymbol{r}_3)) = \langle \mathfrak{c}_1(\boldsymbol{r}_1;\boldsymbol{\sigma})\mathfrak{c}_2(\boldsymbol{r}_2;\boldsymbol{\sigma})\mathfrak{c}_3(\boldsymbol{r}_3;\boldsymbol{\sigma})\rangle\\ &- \langle \mathfrak{c}_1(\boldsymbol{r}_1;\boldsymbol{\sigma})\rangle\Gamma_2(\mathfrak{c}_2(\boldsymbol{r}_2;\boldsymbol{\sigma}),\mathfrak{c}_3(\boldsymbol{r}_3;\boldsymbol{\sigma})) - \langle \mathfrak{c}_2(\boldsymbol{r}_2;\boldsymbol{\sigma})\rangle\Gamma_2(\mathfrak{c}_3(\boldsymbol{r}_3;\boldsymbol{\sigma}),\mathfrak{c}_1(\boldsymbol{r}_1;\boldsymbol{\sigma}))\\ &- \langle \mathfrak{c}_3(\boldsymbol{r}_3;\boldsymbol{\sigma})\rangle\Gamma_2(\mathfrak{c}_1(\boldsymbol{r}_1;\boldsymbol{\sigma}),\mathfrak{c}_2(\boldsymbol{r}_2;\boldsymbol{\sigma})) - \langle \mathfrak{c}_1(\boldsymbol{r}_1;\boldsymbol{\sigma})\rangle\langle \mathfrak{c}_2(\boldsymbol{r}_2;\boldsymbol{\sigma})\rangle\langle \mathfrak{c}_3(\boldsymbol{r}_3;\boldsymbol{\sigma})\rangle\\ &= \langle[\mathfrak{c}_1(\boldsymbol{r}_1;\boldsymbol{\sigma}) - \langle \mathfrak{c}_1(\boldsymbol{r}_1;\boldsymbol{\sigma})\rangle][\mathfrak{c}_2(\boldsymbol{r}_2;\boldsymbol{\sigma}) - \langle \mathfrak{c}_2(\boldsymbol{r}_2;\boldsymbol{\sigma})\rangle][\mathfrak{c}_3(\boldsymbol{r}_3;\boldsymbol{\sigma}) - \langle \mathfrak{c}_3(\boldsymbol{r}_3;\boldsymbol{\sigma})\rangle]\rangle.\end{aligned} \tag{2.5.10}$$

The two- and three-point fluctuation and connected correlation functions are the same. That this identity does not persist for large values of τ can be seen if we expand the general τ-point fluctuation correlation function

$$\langle \delta\mathfrak{c}_1(\boldsymbol{r}_1;\boldsymbol{\sigma})\cdots\delta\mathfrak{c}_\tau(\boldsymbol{r}_\tau;\boldsymbol{\sigma})\rangle = \sum_{\{j_1,\ldots,j_s\}} (-1)^{\tau-s}\langle \mathfrak{c}_{j_1}(\boldsymbol{r}_{j_1};\boldsymbol{\sigma})\cdots\mathfrak{c}_{j_s}(\boldsymbol{r}_{j_s};\boldsymbol{\sigma})\rangle \times \langle \mathfrak{c}_{j_{s+1}}(\boldsymbol{r}_{j_{s+1}};\boldsymbol{\sigma})\rangle\cdots\langle \mathfrak{c}_{j_\tau}(\boldsymbol{r}_{j_\tau};\boldsymbol{\sigma})\rangle, \tag{2.5.11}$$

where the sum is over all subsets $\{j_1,\ldots,j_s\}$ of the indices $\{1,2,\ldots,\tau\}$. Each term in the expansion is of the form of an s-point total correlation function multiplied by $\tau - s$ one-point functions. As can be seen from (2.5.7), the expansion for $\Gamma_\tau(\mathfrak{c}_1(\boldsymbol{r}_1)\cdots\mathfrak{c}_\tau(\boldsymbol{r}_\tau))$ in terms of total correlation functions will contain terms which are products of p- and q-point total correlation functions where *both* p and q are greater than unity when $\tau > 3$.

Using the site-specific couplings defined above,

$$\langle \mathfrak{c}_1(\boldsymbol{r}_1;\boldsymbol{\sigma})\cdots\mathfrak{c}_\tau(\boldsymbol{r}_\tau;\boldsymbol{\sigma})\rangle = \frac{(-1)^\tau}{Z^{(\mathrm{G})}}\frac{\partial^\tau Z^{(\mathrm{G})}}{\partial\zeta_1(\boldsymbol{r}_1)\cdots\partial\zeta_\tau(\boldsymbol{r}_\tau)}. \tag{2.5.12}$$

It can also be shown by induction (Binney et al. 1993) that

$$\Gamma_\tau(\mathfrak{c}_1(\boldsymbol{r}_1),\ldots,\mathfrak{c}_\tau(\boldsymbol{r}_\tau)) = N(-1)^{\tau+1}\frac{\partial^\tau\phi^{(\mathrm{G})}}{\partial\zeta_1(\boldsymbol{r}_1)\cdots\partial\zeta_\tau(\boldsymbol{r}_\tau)}, \tag{2.5.13}$$

for connected correlation functions.

Returning now to the situation where the couplings ζ_j are site independent

$$\begin{aligned}\frac{\partial^\tau\phi}{\partial\zeta_1\cdots\partial\zeta_\tau} &= \sum_{\{\boldsymbol{r}_1,\ldots,\boldsymbol{r}_\tau\}}\frac{\partial^\tau\phi^{(\mathrm{G})}}{\partial\zeta_1(\boldsymbol{r}_1)\cdots\partial\zeta_\tau(\boldsymbol{r}_\tau)}\\ &= \frac{(-1)^{\tau+1}}{N}\sum_{\{\boldsymbol{r}_1,\ldots,\boldsymbol{r}_\tau\}}\Gamma_\tau(\mathfrak{c}_1(\boldsymbol{r}_1),\ldots,\mathfrak{c}_\tau(\boldsymbol{r}_\tau)),\end{aligned} \tag{2.5.14}$$

where, in the second summation, the reduction to site-independent couplings is applied after differentiation. In the case of the connected two-point or *pair* correlation function this result can be derived directly without the use of (2.5.13). From (2.1.11), (2.5.1), and (2.5.9)

$$\frac{\partial^2\phi}{\partial\zeta_1\zeta_2} = -\frac{1}{N}\mathrm{Cov}[\widehat{C}_1,\widehat{C}_2] = -\frac{1}{N}\sum_{\{\boldsymbol{r}_1,\boldsymbol{r}_2\}}\Gamma_2(\mathfrak{c}_1(\boldsymbol{r}_1),\mathfrak{c}_2(\boldsymbol{r}_2)). \tag{2.5.15}$$

Like (2.1.11) this is a *fluctuation-response function relation.* Such relations play an important role in statistical mechanics, relating microscopic and macroscopic properties of a system. Of course, when the two variables are the same the covariance becomes the variance. Thus, for example, from (2.3.31), for the set $\{Q, L, \widehat{Q}(\boldsymbol{\sigma}), \mathfrak{q}(\boldsymbol{r};\boldsymbol{\sigma})\}$,

$$\varphi_T = -\frac{\partial^2\phi}{\partial L^2} = \frac{1}{N}\mathrm{Var}[\widehat{Q}] = \frac{1}{N}\sum_{\{\boldsymbol{r}_1,\boldsymbol{r}_2\}}\Gamma_2(\mathfrak{q}(\boldsymbol{r}_1),\mathfrak{q}(\boldsymbol{r}_2)). \tag{2.5.16}$$

2.5.3 Symmetry Properties

Suppose that the lattice is d-dimensional hypercubic with periodic boundary conditions and lattice vectors given by (18.1.1). The τ-point connected correlation function then has (Sect. 18.1.1) the Fourier transform

$$\begin{aligned}\Gamma^*_\tau(\mathfrak{c}_1,\ldots,\mathfrak{c}_\tau;\boldsymbol{k}_1,\ldots,\boldsymbol{k}_\tau) = &\sum_{\{\boldsymbol{r}_1,\ldots,\boldsymbol{r}_\tau\}}\Gamma_2(\mathfrak{c}_1(\boldsymbol{r}_1),\ldots,\mathfrak{c}_\tau(\boldsymbol{r}_\tau))\\ &\times\exp[-\mathrm{i}(\boldsymbol{k}_1\cdot\boldsymbol{r}_1+\cdots+\boldsymbol{k}_\tau\cdot\boldsymbol{r}_\tau)],\end{aligned} \tag{2.5.17}$$

with

$$\begin{aligned}\Gamma_\tau(\mathfrak{c}_1(\boldsymbol{r}_1),\ldots,\mathfrak{c}_\tau(\boldsymbol{r}_\tau)) = \frac{1}{N}&\sum_{\{\boldsymbol{k}_1,\ldots,\boldsymbol{k}_\tau\}}\Gamma^*_\tau(\mathfrak{c}_1,\ldots,\mathfrak{c}_\tau;\boldsymbol{k}^{(1)},\ldots,\boldsymbol{k}^{(\tau)})\\ &\times\exp[\mathrm{i}(\boldsymbol{k}_1\cdot\boldsymbol{r}_1+\cdots+\boldsymbol{k}_\tau\cdot\boldsymbol{r}_\tau)],\end{aligned} \tag{2.5.18}$$

where the wave vectors are given by (18.1.6). It follows from (2.5.14) and (2.5.17) that

$$\frac{\partial^{\tau}\phi}{\partial\zeta_1\cdots\partial\zeta_{\tau}} = \frac{(-1)^{\tau+1}}{N}\Gamma^{*}_{\tau}(\mathfrak{c}_1,\ldots,\mathfrak{c}_{\tau};\mathbf{0},\ldots,\mathbf{0}). \tag{2.5.19}$$

In many cases of interest the system has the property of invariance under *translations* of the form

$$\boldsymbol{r}\to\boldsymbol{r}+a[m^{(1)}\hat{\boldsymbol{r}}^{(1)}+m^{(2)}\hat{\boldsymbol{r}}^{(2)}+\cdots+m^{(d)}\hat{\boldsymbol{r}}^{(d)}], \tag{2.5.20}$$

for any integers $m^{(1)},\ldots,m^{(d)}$. The difference between the translation given by (18.1.2), (18.1.3) and (2.5.20), is that the former can be applied to each of the vectors $\boldsymbol{r}_1,\ldots,\boldsymbol{r}_{\tau}$ of $\Gamma_{\tau}(\mathfrak{c}_1(\boldsymbol{r}_1),\ldots,\mathfrak{c}_{\tau}(\boldsymbol{r}_{\tau}))$ independently, whereas (2.5.20) must be applied to them all simultaneously. One translation of the form (2.5.20) takes $\boldsymbol{r}_1$ to the origin to give

$$\Gamma_{\tau}(\mathfrak{c}_1(\boldsymbol{r}_1),\ldots,\mathfrak{c}_{\tau}(\boldsymbol{r}_{\tau}))=\Gamma_{\tau}(\mathfrak{c}_1(\mathbf{0}),\mathfrak{c}_2(\bar{\boldsymbol{r}}_1),\ldots,\mathfrak{c}_{\tau}(\bar{\boldsymbol{r}}_{\tau-1})), \tag{2.5.21}$$

where $\bar{\boldsymbol{r}}_j=\boldsymbol{r}_{j+1}-\boldsymbol{r}_1$, $j=1,\ldots,\tau-1$ are the relative lattice vectors. The one-point connected correlation function, given from (2.5.8) by $\langle\mathfrak{c}_1(\boldsymbol{r}_1)\rangle$ is independent of the site location $\boldsymbol{r}_1$. We also see, from (2.5.7) and (2.5.11), that the total and fluctuation correlation functions have a similar dependence on relative vectors. With translational invariance one of the summations in the final form of (2.5.14) can be performed to give

$$\frac{\partial^{\tau}\phi}{\partial\zeta_1\cdots\partial\zeta_{\tau}} = (-1)^{\tau+1}\sum_{\{\bar{\boldsymbol{r}}_1,\ldots,\bar{\boldsymbol{r}}_{\tau-1}\}}\Gamma_{\tau}(\mathfrak{c}_1(\mathbf{0}),\mathfrak{c}_2(\bar{\boldsymbol{r}}_1),\ldots,\mathfrak{c}_{\tau}(\bar{\boldsymbol{r}}_{\tau-1})), \tag{2.5.22}$$

and, if $\bar{\Gamma}^{*}_{\tau}(\mathfrak{c}_1,\ldots,\mathfrak{c}_{\tau};\bar{\boldsymbol{k}}_1,\ldots,\bar{\boldsymbol{k}}_{\tau-1})$ is the Fourier transform defined with respect to the relative vectors,

$$\frac{\partial^{\tau}\phi}{\partial\zeta_1\cdots\partial\zeta_{\tau}} = (-1)^{\tau+1}\bar{\Gamma}^{*}_{\tau}(\mathfrak{c}_1,\ldots,\mathfrak{c}_{\tau};\mathbf{0},\ldots,\mathbf{0}). \tag{2.5.23}$$

Thus, for example, from (2.5.16),

$$\varphi_T=\bar{\Gamma}^{*}_{2}(\mathfrak{q};\mathbf{0}). \tag{2.5.24}$$

Finally we note that if, in additional to translational invariance, the system has $N^{(1)}=N^{(2)}=\cdots=N^{(d)}$ and all its properties invariant under *rotation* between lattice directions the two-point correlation function is given by

$$\Gamma_2(\mathfrak{c}_1(\boldsymbol{r}_1),\mathfrak{c}_2(\boldsymbol{r}_2))=\Gamma_2(\mathfrak{c}_1(\mathbf{0}),\mathfrak{c}_2(\boldsymbol{r}_2-\boldsymbol{r}_1))=\Gamma_2(\mathfrak{c}_1,\mathfrak{c}_2;|\boldsymbol{r}_2-\boldsymbol{r}_1|). \tag{2.5.25}$$

Chapter 3
A Survey of Models

3.1 Upper and Lower Critical Dimensions

One of the main investigative interests of statistical mechanics is in the possibility of phase transitions and, as is seen in many places in this text, this depends on the lattice dimension d of the system in two ways:

(i) It is one of the determining factors which specifies the universality class and hence the critical exponents of a phase transition (Sect. 4.2.1). As is seen in Sect. 5.6 these critical exponents assume dimensionally-independent values (their mean-field values) when the lattice is of dimension $d \geq d_{\text{UC}}$, the *upper critical*[1] *dimension.*

(ii) It affects whether an ordered phase can exist. Such a phase, which corresponds to some kind of long-range spatial order, can, by fluctuations of the microstates, be broken up into domains with their own local order (Fig. 4.4). The free-energy cost of this will be 'paid' at the domain boundary with the boundary to bulk ratios increasing with lattice dimension. Thus it may be anticipated that the temperature range $[0, T_c]$ in which an ordered phase can exist will decrease monotonically with decreasing d, with a *lower critical dimension* d_{LC} such that no long-range ordered phase can occur at non-zero temperature when $d \leq d_{\text{LC}}$.[2] The Peierls method of Sect. 8.3 establishes the existence of a phase transition for the two-dimensional Ising model and for the one-dimensionally infinite Ising model there is no phase transition at non-zero temperature (Sect. 11.3.2), giving $d_{\text{LC}} = 1$, a result which applies more generally for models with discrete symmetry. Fluctuations have a greater effect for models with continuous symmetry. For the quantum Heisenberg model $d_{\text{LC}} = 2$ (Sect. 3.2) and this results carries over to

[1] These 'critical' dimensions are sometimes called 'marginal' or 'borderline' (see, for example, Fisher 1983, Sect. 4.6).

[2] Although there may, as in the case of the Kosterlitz–Thouless transition (Sect. 3.3), be a transition to a different type of phase.

D.A. Lavis, *Equilibrium Statistical Mechanics of Lattice Models*,
Theoretical and Mathematical Physics, DOI 10.1007/978-94-017-9430-5_3

classical models with continuous symmetry (Sect. 3.3). Simple arguments based on the free-energy cost of destabilising the ordered phase can be advanced for these conclusions (Cardy 1996, Chap. 6).

3.2 The Quantum Heisenberg Model

Ferromagnetism and other forms of permanent magnetism are due to exchange interaction between unpaired spins occurring either directly or through conduction electrons or non-magnetic component atoms, (Martin 1967; Sinha and Kumar 1980; Mattis 1981). The standard model for this is the *quantum Heisenberg model* for which, in its general anisotropic form, there are three exchange energies $J_x(\boldsymbol{r}-\boldsymbol{r}')$, $J_y(\boldsymbol{r}-\boldsymbol{r}')$ and $J_z(\boldsymbol{r}-\boldsymbol{r}')$ between a pair of sites $\boldsymbol{r},\boldsymbol{r}' \in \mathcal{N}$ with the Hamiltonian taking the form[3]

$$\widehat{H}(J_x,J_y,J_z,\mathcal{H};\boldsymbol{S}) := -\frac{1}{s^2}\sum_{\{\boldsymbol{r},\boldsymbol{r}'\}}\big[J_x(\boldsymbol{r}-\boldsymbol{r}')S_x(\boldsymbol{r})S_x(\boldsymbol{r}')$$

$$+J_y(\boldsymbol{r}-\boldsymbol{r}')S_y(\boldsymbol{r})S_y(\boldsymbol{r}')+J_z(\boldsymbol{r}-\boldsymbol{r}')S_z(\boldsymbol{r})S_z(\boldsymbol{r}')\big]-\frac{\mathcal{H}}{s}\sum_{\{\boldsymbol{r}\}}S_z(\boldsymbol{r}). \tag{3.2.1}$$

$S_x(\boldsymbol{r})$, $S_y(\boldsymbol{r})$ and $S_z(\boldsymbol{r})$ are the component operators associated with the spin on site $\boldsymbol{r}$ where s takes integer or integer + $\frac{1}{2}$ values and commutation rules

$$[S_a(\boldsymbol{r}),S_b(\boldsymbol{r})]=\mathrm{i}S_c(\boldsymbol{r}), \tag{3.2.2}$$

where a,b,c is any of the three cyclic orderings of x,y,z, apply.[4]

From (3.2.2) it follows that each of the three component operators commutes with $\boldsymbol{S}^2(\boldsymbol{r}) = S_x^2(\boldsymbol{r})+S_y^2(\boldsymbol{r})+S_z^2(\boldsymbol{r})$. So a choice can be made of simultaneous eigenstates for $\boldsymbol{S}^2(\boldsymbol{r})$ and one of the three component operators. Conventionally this is taken to be $S_z(\boldsymbol{r})$ with the external field coupled to that operator. In most magnetic materials with localized spins there is a direction of *easy magnetization* (taken to be the z-direction) so that J_x and J_y are very much smaller than J_z. A field in the direction of easy magnetisation, as shown in (3.2.1) is called *longitudinal* and one orthogonal to that direction (replacing $S_z(\boldsymbol{r})$ in the final term in (3.2.1) by $S_x(\boldsymbol{r})$ or $S_y(\boldsymbol{r})$) is said to be *transverse*.

The question of the occurrence of ferromagnetic ordering (magnetism with $\mathcal{H}=0$) or antiferromagnetic ordering (sublattice-order in the case of a bipartite lattice)

---

[3] The factors of s^2 and s in the denominators of the first and second terms are included for expediting the limit $s\to\infty$ and the Landé factor and Bohr magnon have been absorbed into the field $\mathcal{H}$.

[4] In most investigations of the model the exchange interactions are restricted to first-neighbour pairs, although Zener (1951) has suggested that a ferromagnetic exchange between more distant neighbours, mediated by the conduction electrons, can account for ferromagnetic behaviour in cases where the first-neighbour interaction alone would yield antiferromagnetism.

is important. This is symmetry breaking and the *Mermin–Wagner theorem* (Mermin and Wagner 1966) shows that, in the isotropic case ($J_x(\boldsymbol{r}) = J_y(\boldsymbol{r}) = J_z(\boldsymbol{r}) = J(\boldsymbol{r})$) this cannot occur for $d \leq 2$, when the second moment

$$M_2(J) := \sum_{\{\boldsymbol{r}\}} J(\boldsymbol{r})|\boldsymbol{r}|^2 \tag{3.2.3}$$

is finite.[5] In particular, of course, this applies to the model with only first-neighbour interactions for which it was shown by Kennedy (1985) that, with any degree of anisotropy of the form $0 < J_x = J_y := \alpha J, 0 < J_z := \delta J$, with $\alpha \neq \delta$, there is spontaneous order for $d \geq 2$ and sufficiently low temperatures.

A case of extreme isotropy is the Ising model with $\alpha = 0, \delta = 1$. The Hamiltonian is now diagonal in the representation in which the commuting operators $S_z(\boldsymbol{r})$ and $\boldsymbol{S}^2(\boldsymbol{r})$ are diagonal, as long as the field term in (3.2.1) is longitudinal, and they can be regarded as classical scalars with the range of values $-s, -s+1, \ldots, s-1, s$. This is the *general-s Ising model* which is discussed in Sect. 3.5. The quantum nature of the model (and the name 'quantum Ising model') is retained when a transverse field is added to the Hamiltonian (or replaces the longitudinal field). An important result for this model, shown for $s = \frac{1}{2}$ by Suzuki (1976) and for $s = 1$ by Björnberg and Grimmett (2009),[6] is that the expression for the ground-state[7] energy of the quantum Ising model, with both transverse and longitudinal fields, on a d-dimensional lattice, is equivalent to the free energy of a classical Ising model on a $(d+1)$-dimensional lattice. This implies the existence of a phase transition in the transverse-field quantum Ising model for all d-dimensional hypercubic lattices *including* $d = 1$, a result which has been confirmed by high-temperature series expansions.

3.2.1 One-Dimensional Chains

Apart from the Ising models, exact results for cases of the Heisenberg model have been obtained only for the one-dimensional lattice (chain) and then mainly for the

---

[5] More generally the Mermin–Wagner theorem states that, subject to a bound on the interactions like (3.2.3), there cannot be spontaneous breaking of a continuous symmetry in systems with $d \leq 2$. This applies to the isotropic Heisenberg model which has O(3) symmetry; that is symmetry under arbitrary rotations in three dimensions. It should be noted that the Mermin–Wagner conditions are sufficient but not necessary to ensure the absence of a phase transition. This means, in general, for such systems $d_{\mathrm{LC}} \geq 2$, although for the isotropic Heisenberg model $d_{\mathrm{LC}} = 2$. A lower critical dimension of two has also been established for the superfluid and superconducting phases in Bose and Fermi liquids by Hohenberg (1967) and in quantum field theory in the absence of Goldstone modes by Coleman (1973).

[6] But generally believed to be true for all s.

[7] In this chapter the 'ground state' is understood in the informal sense of being the microstate which gives the smallest value for the Hamiltonian. A more formal definition of the concept is given in Sect. 8.3.

zero-field, $s = \frac{1}{2}$, case, when the Hamiltonian, in the representation in which $S_z(\boldsymbol{r})$ and $\boldsymbol{S}^2(\boldsymbol{r})$ are diagonal, has the matrix form

$$\hat{\boldsymbol{H}}(\{J_\zeta\}; \boldsymbol{\Sigma}) := -\sum_{\{\zeta\}} \sum_{\mathrm{i}=1}^{N} J_\zeta \boldsymbol{\Sigma}_\zeta(\mathrm{i}) \boldsymbol{\Sigma}_\zeta(\mathrm{i}+1), \tag{3.2.4}$$

where the outer summation is over $\zeta = x, y, z$,

$$\boldsymbol{\Sigma}_\zeta(\mathrm{i}) := \boldsymbol{I} \otimes \cdots \otimes \boldsymbol{I} \otimes \boldsymbol{\Sigma}_\zeta \otimes \boldsymbol{I} \otimes \cdots \otimes \boldsymbol{I}, \tag{3.2.5}$$

$\boldsymbol{I}$ is the two-dimensional unit vector and $\boldsymbol{\Sigma}_\zeta$, which lies at the ith place in the product, is the appropriate Pauli spin matrix

$$\boldsymbol{\Sigma}_x := \begin{pmatrix} 0 & +1 \\ +1 & 0 \end{pmatrix}, \quad \boldsymbol{\Sigma}_y := \begin{pmatrix} 0 & -\mathrm{i} \\ +\mathrm{i} & 0 \end{pmatrix}, \quad \boldsymbol{\Sigma}_z := \begin{pmatrix} +1 & 0 \\ 0 & -1 \end{pmatrix}. \tag{3.2.6}$$

Bethe (1931) found the eigenvalues of the *isotropic*, $J_x = J_y = J_z := J$, model and determined the elementary excitation spectrum (in the form of m magnons for $m = 1, 2, \ldots$) for the ferromagnetic ($J > 0$) case using a method now called the *Bethe ansatz*.[8] For the antiferromagnetic ($J < 0$) case the ground state was determined by Hulthén (1938) and the elementary excitations by des Cloizeaux and Pearson (1962). In the ferromagnetic case, for general s, the ground-state and one-magnon and two-magnon excitations were obtained by Wortis (1963). A slightly different approach was used by Babujian (1982), who generalized the terms of the Hamiltonian (3.2.1) to polynomials in $\boldsymbol{S}(\boldsymbol{r}) \cdot \boldsymbol{S}(\boldsymbol{r}')$ and showed that, for a particular choice of polynomial coefficients the complete spectrum could be obtained.

It was shown by Yang and Yang (1966a) that, if one is interested only in the eigenvalues of the Hamiltonian then the signs of J_x, J_y and J_z, as long as they are of equal magnitude, are irrelevant. This is, of course, no longer the case for the more general *Heisenberg–Ising XXZ model* where $J_x = J_y := J$, $J_z := \delta J$, with $|\delta| \neq 1$. For this Lieb et al. (1961) obtained the ground-state and all the elementary excitations and Yang and Yang (1966b, c, d) showed rigorously that the Bethe ansatz is correct, deriving the minimum eigenvalue in the limit of an infinite chain.[9] The parameter δ represents the anisotropy of the system and that and the sign of J are the quantities which characterize the system. With J positive/negative ferromagnetic/antiferromagnetic ordering predominate in the x–y plane with a similar effect in the z direction for $J\delta$ positive or negative if $|\delta| > 1$. When Lieb (1967a, b, c, d) solved the six-vertex model, he found that the Heisenberg–Ising Hamiltonian and the six-vertex transfer matrix commuted implying

[8] For an historical account of the many uses of the Bethe ansatz, including in the solution of the six-vertex model, see Batchelor (2007).

[9] For detailed accounts of the solution see Izyumov and Skryabin (1988, Chap. 5) or Jimbo and Miwa (1995, Chap. 1).

that they share the same eigenvectors (Sects. 12.3.4.2 and 12.3.10.2). He was, therefore, able to exploit many of the results of Yang and Yang.

For $|\delta| < 1$ the XXZ model is paramagnetic and the limit $\delta \to 0$ yields the *XX model* solved by Katsura (1962). This is the special case $\gamma = 0$ of the *XY model* for which $J_z := 0$, $J_x := \frac{1}{2}J(1+\gamma)$, $J_y := \frac{1}{2}J(1-\gamma)$. This was solved by Lieb et al. (1961) for a range of antiferromagnetic cases. With a magnetic field Jh, coupled to the $\mathbf{\Sigma}_z(\mathrm{i})$, Kurmann et al. (1982) obtained the zero-temperature phase diagram in the γ–h plane. Within the unit circle the ground-state is of the classic Néel antiferromagnetic two-sublattice type and the line $h = 1$ separates ordered and disordered phases. For $\gamma = 1$ this transition is in the universality class of the quantum Ising model in a transverse field and otherwise it is in the universality class of the XX model.

In a result parallel to that for the XXZ and six-vertex models, Sutherland (1970) showed that the general XYZ Hamiltonian (3.2.4) and the eight-vertex transfer matrix commute (Sect. 12.3.4.2) implying that they share the same eigenvectors, and Baxter (1972a) (see also, Baxter 1982b, Sect. 10.14) showed that the Hamiltonian is effectively a logarithmic derivative of the eight-vertex transfer matrix, meaning that the ground-state energy can be found. A complete investigation of this case was carried out by Takhtadzhyan and Faddeev (1979).

For later reference we note that (3.2.4) can be re-expressed in the form

$$\hat{\boldsymbol{H}}(J_x, J_y, J_z; \mathbf{\Sigma}) = -\tfrac{1}{4}(J_x + J_y)\sum_{\mathrm{i}=1}^{N}\big[\{\mathbf{\Sigma}_+(\mathrm{i})\mathbf{\Sigma}_+(\mathrm{i}+1) + \mathbf{\Sigma}_-(\mathrm{i})\mathbf{\Sigma}_-(\mathrm{i}+1)\}$$
$$+ I_0\{\mathbf{\Sigma}_+(\mathrm{i})\mathbf{\Sigma}_-(\mathrm{i}+1) + \mathbf{\Sigma}_-(\mathrm{i})\mathbf{\Sigma}_+(\mathrm{i}+1)\} + 2I_z\mathbf{\Sigma}_z(\mathrm{i})\mathbf{\Sigma}_z(\mathrm{i}+1)\big], \qquad (3.2.7)$$

where $\mathbf{\Sigma}_\pm(\mathrm{i})$ are of the form (3.2.5) with

$$\mathbf{\Sigma}_\pm := \mathbf{\Sigma}_x \pm \mathbf{\Sigma}_y, \qquad (3.2.8)$$

and

$$I_0 := \frac{J_x - J_y}{J_x + J_y}, \qquad I_z := \frac{2J_z}{J_x + J_y}. \qquad (3.2.9)$$

3.3 Classical Vector Models

If, for the quantum Heisenberg model, $s_a(\boldsymbol{r}) := S_a(\boldsymbol{r})/s$, $a = x, y, z$, and the limit $s \to \infty$ is taken, it follows, from (3.2.2), that the three operators $s_x(\boldsymbol{r})$, $s_y(\boldsymbol{r})$ and $s_z(\boldsymbol{r})$ commute and become equivalent to the three components of the three-dimensional unit vector $\boldsymbol{s}(r) := (s_x(\boldsymbol{r}), s_y(\boldsymbol{r}), s_z(\boldsymbol{r}))$. In the case $J_x(\boldsymbol{r},\boldsymbol{r}') = J_y(\boldsymbol{r},\boldsymbol{r}') = J_z(\boldsymbol{r},\boldsymbol{r}') = J(\boldsymbol{r}-\boldsymbol{r}')$ the Hamiltonian (3.2.1) becomes

$$\widehat{H}(J(\boldsymbol{r}-\boldsymbol{r}'),\mathcal{H};\boldsymbol{s}(\boldsymbol{r})) := -\sum_{\{\boldsymbol{r},\boldsymbol{r}'\}} J(\boldsymbol{r}-\boldsymbol{r}')\boldsymbol{s}(\boldsymbol{r})\cdot\boldsymbol{s}(\boldsymbol{r}') - \sum_{\{\boldsymbol{r}\}} \boldsymbol{s}(\boldsymbol{r})\cdot\mathcal{H}, \tag{3.3.1}$$

where the 'special status' of the z-direction is now removed and $\mathcal{H}$ can be in any direction. This is the *classical Heisenberg model*, for which a generalization is the *n-vector model* proposed by Stanley (1968a)[10] where, for all $\boldsymbol{r} \in \mathcal{N}$, $\boldsymbol{s}(\boldsymbol{r})$ is replaced by an n-dimensional unit vector and $\mathcal{H}$ is also, of course, n-dimensional.

The symmetry group when $\mathcal{H} = \boldsymbol{0}$ is $\mathrm{O}(n)$, the group of $n \times n$ orthogonal matrices (Sect. 17.2.7). Apart from the classical Heisenberg model with symmetry group $\mathrm{O}(3)$, the class of $\mathrm{O}(n)$ models contains for $n = 2$ the classical XY model, for $n = 1$ the spin-$\frac{1}{2}$ Ising model and, in the limit $n \to 0$, the *self-avoiding walk* (de Gennes 1972; see also, Cardy 1996, Sect. 9.3). The Mermin–Wagner theorem holds when $n > 1$ and is independent of s applying therefore to the limit $s \to \infty$.

Of particular interest is the $\mathrm{O}(2)$ model, where the spins rotate in a plane, with

$$\boldsymbol{s}(\boldsymbol{r}) = (\cos[\theta(\boldsymbol{r})], \sin[\theta(\boldsymbol{r})]), \tag{3.3.2}$$

$$\boldsymbol{s}(\boldsymbol{r})\cdot\boldsymbol{s}(\boldsymbol{r}') = \cos(\theta(\boldsymbol{r}) - \theta(\boldsymbol{r}')). \tag{3.3.3}$$

This is the *classical XY model*, also called the *plane rotator model*. The spin-spin term (3.3.3) is of the state-difference form discussed in Sect. 3.6 and we reserve further discussion of this model to that section.

3.4 The Gaussian and Spherical Models

The spin-$\frac{1}{2}$ Ising model is, in the context of the last section, simply the one-vector model which escapes the application of the Mermin–Wagner theorem because the group $\mathrm{O}(1)$ does not represent a continuous symmetry. There is, however, another interesting connection between the Ising and n-vector models.

Following Onsager's solution of the two-dimensional Ising model in 1944, Berlin and Kac (1952) proposed two models, the *Gaussian model* and the *spherical model*, which were intended to capture the essential qualities of the Ising model while avoiding the complexities of its solution. The starting point for each of these models is to denote, as in the case of the spin-$\frac{1}{2}$ Ising model, the $n = 1$ version of the vector $\boldsymbol{s}(\boldsymbol{r})$ by $\sigma(\boldsymbol{r})$. Thus the Hamiltonian (3.3.1) becomes

$$\widehat{H}(J(\boldsymbol{r}-\boldsymbol{r}'),\mathcal{H};\boldsymbol{\sigma}) := -\sum_{\{\boldsymbol{r},\boldsymbol{r}'\}} J(\boldsymbol{r}-\boldsymbol{r}')\sigma(\boldsymbol{r})\sigma(\boldsymbol{r}') - \mathcal{H}\sum_{\{\boldsymbol{r}\}} \sigma(\boldsymbol{r}), \tag{3.4.1}$$

---

[10] In fact he first used ν in place of n and took the vectors to be of length $\sqrt{\nu}$. Subsequently (Stanley 1974) he replaced 'ν' by 'D'.

where $\sigma(\boldsymbol{r}) = \pm 1$ and, of course, $\sigma^2(\boldsymbol{r}) = 1$. Both the Gaussian and spherical models are predicated on a relaxation of this latter condition and an extension of the range of $\sigma(\boldsymbol{r})$ to all real values.

For the *Gaussian model* the weakened constraint on each of the spin variables is

$$\langle \sigma^2(\boldsymbol{r}) \rangle = 1, \qquad \text{with a term} \qquad \tfrac{1}{2} \sum_{\{\boldsymbol{r}\}} \sigma^2(\boldsymbol{r}) \tag{3.4.2}$$

added to the Hamiltonian, resulting in a Gaussian integral form for the partition function. Berlin and Kac (1952) showed that, for the first-neighbour case of this model and for hypercubic lattices of dimensions one, two and three, there is a critical temperature $T_c = 2Jd$. However, below the critical temperature there is no well-defined ordered phase since the free energy becomes complex.

For the *spherical model* the constraint condition is

$$\sum_{\{\boldsymbol{r}\}} \sigma^2(\boldsymbol{r}) = N \tag{3.4.3}$$

and the first-neighbour case was solved in one, two and three dimensions by Berlin and Kac (1952).[11] However, it can be solved for lattices of all dimensions (Joyce 1972; Baxter 1982b, Chap. 5). Although this model was introduced by reference to the Ising model ($n = 1$) it can be seen that the constraint (3.4.3), if applied to vectors of any dimension, would effectively combines the roles of N and n. In fact the properties of the model are more similar to those of the Heisenberg model than the Ising model and, for the first-neighbour case it has been shown (Stanley 1968b; Kac and Thompson 1971; Pearce and Thompson 1977) to be equivalent to the $n \to \infty$ limit of the n-vector model. Thus the Mermin–Wagner theorem implies that the lower critical dimension $d_{\text{LC}} = 2$. In this context, it is worth anticipating the discussion of Sect. 4.4.1 by quoting the critical exponents, which are (see, Baxter 1982b, Sect. 5.6)

$$\alpha = -\frac{4-d}{d-2}, \qquad \beta = \frac{1}{2}, \qquad \gamma = \frac{2}{d-2}, \qquad \delta = \frac{d+2}{d-2}, \tag{3.4.4}$$

for $2 < d \leq 4$. At $d = 2$ the exponents α, γ and δ diverge and the critical temperature is zero. The critical exponents at $d = 4$ take their classical (Landau) values given by (5.3.30) and as in the Gaussian approximation to Ginzburg–Landau theory they retain these values for all $d > 4$ so that the upper critical dimensions $d_{\text{UC}} = 4$.

Joyce (1966) investigated the behaviour of a ferromagnetic spherical model for $d = 1, 2, 3$ with a long-range interaction of the form $J(\boldsymbol{r}) \sim 1/|\boldsymbol{r}|^{d+\varepsilon}$ ($\varepsilon > 0$). For such an interaction it is simple to show that the second moment $M_2(J)$ exists, implying from the Mermin–Wagner theorem the non-existence of a phase transition

[11] For an account of the origins of the spherical model see Kac (1964).

for $d = 1, 2$, when $\varepsilon > 2$. In fact Joyce showed that there is a phase transition for $d = 1, 2$ only when $0 < \varepsilon < d$. There is also, as expected, a transition for $d = 3$, and for $d = 1, 2, 3$ the exponents α, γ and δ are functions of ε. In particular

$$\gamma = \begin{cases} 1, & \text{for } 0 < \varepsilon < \frac{1}{2}d, \\ \dfrac{\varepsilon}{d-\varepsilon}, & \text{for } \frac{1}{2}d < \varepsilon < \min(2, d). \end{cases} \tag{3.4.5}$$

When $\varepsilon = \frac{1}{2}d$ the susceptibility has a logarithmic singularity and when $d = 3$, $\gamma = 2$, for $\varepsilon > 2$, coinciding with the value given in (3.4.4) for the first-neighbour model. The correspondence between the n-vector model and the spherical model, established by Stanley (1968b) in the limit $n \to \infty$ for the first-neighbour model, was extended to the case $n = 3$ by Joyce (1969) when the exchange interaction is $J(\boldsymbol{r})$. He showed that the isotropic classical Heisenberg model with such an exchange cannot be ferromagnetic or antiferromagnetic if the spherical model with exchange $|J(\boldsymbol{r})|$ does not exhibit a phase transition. This means that the isotropic classical Heisenberg model with $J(\boldsymbol{r}) \sim 1/|\boldsymbol{r}|^{d+\varepsilon}$ cannot be ferromagnetic or antiferromagnetic for $d < 3$ if $\varepsilon \geq d$.

3.5 Ising Models

As has been seen above the general spin-s Ising model arises as the special case $J_x(\boldsymbol{r}, \boldsymbol{r}') = J_y(\boldsymbol{r}, \boldsymbol{r}') = 0$ of the quantum Heisenberg model and the spin-$\frac{1}{2}$ Ising model as the special case $n = 1$ of the n-vector model. We shall initially concentrate on the latter case, with only a first-neighbour interaction.

3.5.1 The Spin-$\frac{1}{2}$ Ising Model

For $n = 1$ the sole component of the vector $\boldsymbol{s}(\boldsymbol{r})$ in (3.3.1) is denoted by $\sigma(\boldsymbol{r})$ with values ± 1. Given that the field is in the direction of this component and the interaction energy is purely first neighbour the Hamiltonian becomes[12]

$$\widehat{H}(K, L; \boldsymbol{\sigma}) := \widehat{H}(J, \mathcal{H}; \boldsymbol{\sigma})/T = -K \sum_{\{\boldsymbol{r},\boldsymbol{r}'\}}^{(1)} \sigma(\boldsymbol{r})\sigma(\boldsymbol{r}') - L \sum_{\{\boldsymbol{r}\}} \sigma(\boldsymbol{r}), \tag{3.5.1}$$

where $K := J/T$ and $L := \mathcal{H}/T$, with $J > 0$ and $J < 0$ corresponding to ferro- and antiferromagnetism. When $\mathcal{H} = L = 0$, $\widehat{H}(K, 0; \boldsymbol{\sigma})$ is invariant under the symmetry group $\mathcal{S}_2^{(m)} := \{\mathrm{I}, \mathrm{Q}^{(m)}\}$, where $\mathrm{Q}^{(m)}$ is the spin-inversion operator

---

[12] Throughout this text the superscript '(k)' indicates that a summation is confined to kth neighbour pairs.

$$Q^{(m)} : +\sigma(\boldsymbol{r}) \to -\sigma(\boldsymbol{r}), \qquad \forall\ \boldsymbol{r} \in \mathcal{N}. \tag{3.5.2}$$

When $L \neq 0$ we must also consider the group $\mathcal{S}_2^{(L)} := \{\mathrm{I}, \mathrm{Q}^{(L)}\}$, where

$$Q^{(L)} : +L \to -L. \tag{3.5.3}$$

However, from the group $\mathcal{S}_2^{(m)} \times \mathcal{S}_2^{(L)} = \{\mathrm{I}, \mathrm{Q}^{(m)}, \mathrm{Q}^{(L)}, \mathrm{Q}^{(m)}\mathrm{Q}^{(L)}\}$ the Hamiltonian is invariant only for the subgroup $\mathcal{S}_2^{(mL)} := \{\mathrm{I}, \mathrm{Q}^{(m)}\mathrm{Q}^{(L)}\}$. This system is often referred to simply as 'the Ising model' since this was the case which was solved for $d = 1$ by Ising (1925).[13] The extensive variable conjugate to $\mathcal{H}$ is the statistical mechanical magnetization

$$\widehat{\mathcal{M}} := \sum_{\{\boldsymbol{r}\}} \sigma(\boldsymbol{r}), \tag{3.5.4}$$

with statistical mechanical and thermodynamic magnetization densities

$$\hat{m} := \widehat{\mathcal{M}}/N, \qquad m := \langle \hat{m} \rangle, \tag{3.5.5}$$

respectively and the partition function

$$Z(K, L) = \sum_{\{\boldsymbol{\sigma}\}} \exp\{-\widehat{H}(K, L; \boldsymbol{\sigma})\}. \tag{3.5.6}$$

Then from (2.3.7), (3.5.4) and (3.5.5)

$$m(K, L) = \langle \sigma(\boldsymbol{r}) \rangle = -\frac{\partial \phi(K, L)}{\partial L}. \tag{3.5.7}$$

Given that the lattice is one of the regular types described in Sect. 16.2.1, with coordination number z, there are $\frac{1}{2}Nz$ first-neighbour pairs and, from (2.3.27),

$$\mathrm{u}(K, L) := u/J = -\tfrac{1}{2}z\langle \sigma(\boldsymbol{r})\sigma(\boldsymbol{r}') \rangle^{(1)} = \frac{\partial \phi(K, L)}{\partial K}, \tag{3.5.8}$$

where $[\boldsymbol{r}, \boldsymbol{r}']$ is any first-neighbour pair of sites. Since, in general, we can consider the correlation $\langle \sigma(\boldsymbol{r}_1)\sigma(\boldsymbol{r}_2) \rangle$ between the spins on any pair of sites, the superscript '(1)' is used to emphasis the fact that the correlation in this case is between first neighbours. It should also be noted that the relationship between the internal energy per site and the first-neighbour correlation is valid in the form (3.5.8) only when, as in the case of (3.5.1), the interaction term in the Hamiltonian consists

---

[13] In fact this model was suggested to Ising by his research director Wilhelm Lenz and historians of science like Brush (1967) and Niss (2005) often call it the *Lenz-Ising model*. Interested readers are referred to these authors for accounts of historic developments related to the model.

Table 3.1 Critical parameters for the Ising model ferromagnet of dimension d and coordination number z

Source	Lattice or Approximation	d	z	$T_c/(zJ)$	$-2u_c/z$	$[\ln(2) - s_c]/\ln(2)$
(1)	Square exact	2	4	0.5673	0.7071	0.5579
(2)	Honeycomb exact	2	3	0.5062	0.7698	0.6181
(2)	Triangular exact	2	6	0.6068	0.6667	0.5235
(2)	Diamond series	3	4	0.6761	0.4370	0.2640
(3)	Simple cubic series	3	6	0.7518	0.3308	0.1951
(3)	B. C. cubic series	3	8	0.7942	0.2732	0.1603
(3)	F. C. cubic series	3	12	0.8163	0.2475	0.1485
(4)	Mean-field approx.	–	–	1.0	0.0	0.0
(5)	Bethe-pair approx.	–	3	0.6068	0.5000	0.2831
(5)	Bethe-pair approx.	–	4	0.7213	0.3333	0.1634
(5)	Bethe-pair approx.	–	6	0.8221	0.2000	0.0871
(5)	Bethe-pair approx.	–	8	0.8690	0.1429	0.0591
(5)	Bethe-pair approx.	–	12	0.9141	0.0909	0.0358

The sources of the data are (1) Sect. 9.2.2, (2), (3), (4) Sect. 6.1, (5) Sect. 7.3. The critical temperature and critical internal energy are scaled by the coordination number z in order that the table may include the results of classical approximations where the lattice type enters only in this way

only of first-neighbour terms with equal couplings. The generalization of (3.5.8) to the case of unequal couplings in different lattice directions or to include second- or further-neighbour interactions is obvious. It follows from the symmetry property $\widehat{H}(J,\mathcal{H};\boldsymbol{\sigma}) = \mathrm{Q}^{(m)}\mathrm{Q}^{(L)}\widehat{H}(J,\mathcal{H};\boldsymbol{\sigma}) = \widehat{H}(J,-\mathcal{H};-\boldsymbol{\sigma})$ that

$$m(K,-L) = -m(K,L), \qquad \mathfrak{u}(K,-L) = \mathfrak{u}(K,L). \tag{3.5.9}$$

Thus, if the magnetization is single-valued on the zero-field axis, that value must be zero. However, as we shall see in detailed calculations throughout this book, the reduction of temperature in zero field leads to the onset of a spontaneous magnetism at a temperature T_c, called the *Curie temperature*. Consistency then requires that the magnetism will take one of two values $\pm m_s(K)$, $(m_s(K) > 0)$. Critical data and curves of zero-field magnetization derived from calculations in different chapters of the book are given here in Table 3.1 and Fig. 3.1 respectively.

The simple (spin-$\frac{1}{2}$) Ising model can be generalized in various ways; second-neighbour pair interactions (as indicated above) or multi-spin interactions around each face can, for example, be included. However, if contributions are restricted to first-neighbour pairs and single-site terms, (3.5.1) is the most general Hamiltonian for the Ising model with $\sigma(\boldsymbol{r}) = \pm 1$. This still leaves the possibilities of interpreting the model as representing systems other than ferromagnets. Examples of this will be discussed in the following sections.

3.5.1.1 Antiferromagnetism

In the discussion of the ferromagnetic Ising model all lattice sites have been regarded as 'equivalent'. This means that, as for all regular lattices listed in Sect. 16.2.1, the

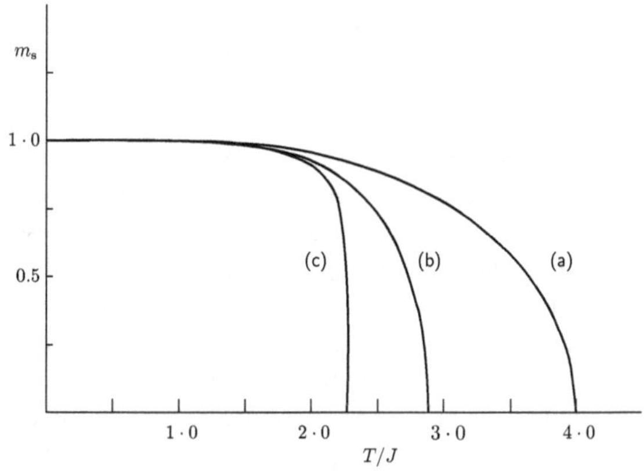

Fig. 3.1 Spontaneous magnetization plotted against temperature for the ferromagnetic Ising model on the square lattice ($z = 4$) in (a) the mean-field approximation (Sect. 6.1), (b) the Bethe-pair approximation (Sect. 7.3), (c) the exact result (Sect. 12.3.9)

unoccupied lattice looks exactly the same when viewed from any site with each site having the same number z_p of pth neighbour sites for any p. The interaction of a microsystem with an external field is the same for all sites and the interaction energy of any two microsystems depends only on the relative position of the lattice sites they occupy.[14] This last condition could easily be extended from two-body (pairwise) interactions to n-body interactions.

The equivalence of the lattice sites means that it may be possible to divide the lattice into two equivalent sublattices, labelled $\mathfrak{a}$ and $\mathfrak{b}$. The sublattices are *interpenetrating* if all first-neighbour sites of any $\mathfrak{a}$-site are $\mathfrak{b}$-sites and vice-versa. A lattice divisible into two equivalent interpenetrating sublattices is called *loose-packed* (or *bipartite*), whereas a lattice for which a division of this kind is not possible is *close-packed*. The $d = 1$ lattice is loose-packed since it can be divided into interpenetrating sublattices simply by labelling alternate sites $\mathfrak{a}$ and $\mathfrak{b}$. The $\mathfrak{a}$ and $\mathfrak{b}$ sublattices (identified respectively as the black and white sites) for the loose-packed square, honeycomb, simple cubic, body-centred cubic and diamond lattices are shown in Sect. 16.2.1.

The Hamiltonian of an antiferromagnet with first-neighbour interactions only on a loose-packed lattice is obtained by adapting (3.5.1) and is given by

$$\widehat{H}(K, L; \boldsymbol{\sigma}) = K \sum_{\{\boldsymbol{r},\boldsymbol{r}'\}}^{(1)} \sigma^{(\mathfrak{a})}(\boldsymbol{r})\sigma^{(\mathfrak{b})}(\boldsymbol{r}') - L\Big[\sum_{\{\boldsymbol{r}\}} \sigma^{(\mathfrak{a})}(\boldsymbol{r}) + \sum_{\{\boldsymbol{r}'\}} \sigma^{(\mathfrak{b})}(\boldsymbol{r}')\Big], \quad (3.5.10)$$

$\sigma^{(\mathfrak{a})}(\boldsymbol{r}')$ and $\sigma^{(\mathfrak{b})}(\boldsymbol{r}')$ respectively denoting the spin variables for a site $\boldsymbol{r}$ of sublattice $\mathfrak{a}$ and a site $\boldsymbol{r}'$ of sublattice $\mathfrak{b}$. The parameter K of (3.5.1) has been changed to $-K$ ($K := J/T > 0$) and the fact that all first-neighbour pairs consist of an $\mathfrak{a}$-site and a $\mathfrak{b}$-site has been used. For $\mathcal{H} = L = 0$ (3.5.10) can be written

[14] 'Relative position' involves the direction as well as the length of the lattice vector connecting the two sites.

$$\widehat{H}(K,0;\boldsymbol{\sigma}) = -K \sum_{\{\boldsymbol{r},\boldsymbol{r}'\}}^{(1)} \sigma^{(\mathfrak{a})}(\boldsymbol{r})[-\sigma^{(\mathfrak{b})}(\boldsymbol{r}')]. \tag{3.5.11}$$

This is the Hamiltonian of a ferromagnet with spin variables $\sigma^{(\mathfrak{a})}(\boldsymbol{r})$ and $-\sigma^{(\mathfrak{b})}(\boldsymbol{r}')$. There is a one-to-one correspondence between microstates of equal energy in the zero-field ferromagnet and antiferromagnet. These microstates are connected by the operation of reversing the spins on one sublattice, leaving those on the other sublattice unchanged, while changing the sign of the interaction parameter. It follows that in zero field the partition function for the antiferromagnet is equal to that for the ferromagnet and the free energy, entropy, internal energy and heat capacity are the same. The critical temperature T_c, where there is a heat capacity singularity at $\mathcal{H} = 0$, is the Curie temperature for the ferromagnet and is termed the *Néel point* for the antiferromagnet. Experimentally, the phenomenon of antiferromagnetism is commoner than that of ferromagnetism (Martin 1967).

The type of long-range order in the antiferromagnet below the critical temperature is quite different from that in the ferromagnet. We define a relative magnetization for each sublattice

$$\begin{aligned} m_{\mathfrak{a}} &:= \langle \hat{m}_{\mathfrak{a}} \rangle, \qquad \hat{m}_{\mathfrak{a}} := \frac{2}{N}\sum_{\{\boldsymbol{r}\}} \sigma^{(\mathfrak{a})}(\boldsymbol{r}), \\ m_{\mathfrak{b}} &:= \langle \hat{m}_{\mathfrak{b}} \rangle, \qquad \hat{m}_{\mathfrak{b}} := \frac{2}{N}\sum_{\{\boldsymbol{r}'\}} \sigma^{(\mathfrak{b})}(\boldsymbol{r}'), \end{aligned} \tag{3.5.12}$$

where the sums are taken over all sites of sublattice $\mathfrak{a}$ and sublattice $\mathfrak{b}$ respectively. The overall relative magnetization m is then given by

$$m = \tfrac{1}{2}(m_{\mathfrak{a}} + m_{\mathfrak{b}}). \tag{3.5.13}$$

For the zero-field ferromagnet above the critical temperature T_c, $m_{\mathfrak{a}} = m_{\mathfrak{b}} = m = 0$ whereas below T_c, $m_{\mathfrak{a}} = m_{\mathfrak{b}} = \pm m_s$, $(m_s > 0)$. Now, from (3.5.11), the zero-field equilibrium state of the antiferromagnet can be derived from that of the ferromagnet by reversing the magnetization of one of the sublattices. Hence, above the Néel point $(T > T_c)$, the magnetizations of both sublattices are zero, but below it either

$$\left.\begin{aligned} m_{\mathfrak{a}} = m_s, &\qquad m_{\mathfrak{b}} = -m_s, \\ &\text{or} \\ m_{\mathfrak{a}} = -m_s, &\qquad m_{\mathfrak{b}} = m_s, \end{aligned}\right\} \qquad m = 0. \tag{3.5.14}$$

The magnetizations of the two sublattices are equal and opposite, giving zero overall magnetization. This is an example of *sublattice ordering*. The degeneracy displayed in (3.5.14) for $\mathcal{H} = 0$, $T < T_c$, parallels the degeneracy $m = \pm m_s$ of the ferromagnet (Sect. 6.1). As $T/J \to 0$, the Ising model antiferromagnet approaches the state $m_{\mathfrak{a}} = \pm 1$, $m_{\mathfrak{b}} = \mp 1$ where all first-neighbour spin pairs are antiparallel.

Since, from (3.5.14), m is a continuous function of $\mathcal{H}$ as the interval $(0, T_c)$ of the $\mathcal{H} = 0$ axis is crossed, this interval is no longer a first-order transition line, as it is in the ferromagnet (Sect. 6.1). The antiferromagnetic ground states are examples of commensurate phases.[15]

For $\mathcal{H} \neq 0$ there is no correspondence between the states of the antiferromagnet and those of the ferromagnet. From the equivalence of sites in the antiferromagnetic model it may be expected that, if either the magnetic field or thermal disordering is strong enough to offset the antiferromagnetic spin interaction, the equilibrium state will be one in which $m_{\mathfrak{a}} = m_{\mathfrak{b}} = m$. Such a state is called *paramagnetic* as opposed to the antiferromagnetic state in which the spins on sublattices $\mathfrak{a}$ and $\mathfrak{b}$ are oppositely oriented. The transition curve in the $\mathcal{H}$–T plane on which the change of symmetry from the paramagnetic state to the antiferromagnetic state takes place for $d > 1$ has to be investigated by approximation or series methods even for $d = 2$, since no exact results are available for either the ferromagnet or antiferromagnet with $\mathcal{H} \neq 0$ and $T > 0$. However, the zero-field results given above indicate that the transition curve meets the $\mathcal{H} = 0$ axis at $T = T_c$ and is symmetric about $\mathcal{H} = 0$. Also, the points where it meets the $T = 0$ axis can be determined quite easily. It follows from (2.1.2) that as $T \to 0$ the probability distribution is dominated by the term or terms that give the ground-state Hamiltonian, which in this limit is equal to the enthalpy. In the paramagnetic ground state, all $\sigma^{(\mathfrak{a})}(\boldsymbol{r}) = \sigma^{(\mathfrak{b})}(\boldsymbol{r}') = \pm 1$, according as $\mathcal{H} \gtrless 0$ and, in the antiferromagnetic ground state, all $\sigma^{(\mathfrak{a})}(\boldsymbol{r}) = 1$ and all $\sigma^{(\mathfrak{b})}(\boldsymbol{r}') = -1$ (or vice-versa). If $\widehat{H}_{\mathrm{P}}$ and $\widehat{H}_{\mathrm{A}}$ denote the values of the Hamiltonian in the paramagnetic and antiferromagnetic states respectively, by (3.5.10),

$$\widehat{H}_{\mathrm{P}} = \tfrac{1}{2} zNJ - N|\mathcal{H}|, \qquad \widehat{H}_{\mathrm{A}} = -\tfrac{1}{2} zNJ, \tag{3.5.15}$$

and $\widehat{H}_{\mathrm{A}} < \widehat{H}_{\mathrm{P}}$, making the antiferromagnetic state stable if

$$-\mathcal{H}_c < \mathcal{H} < \mathcal{H}_c := zJ, \tag{3.5.16}$$

where the last relation defines the *critical field* $\mathcal{H}_c$.

This simple antiferromagnetic model can be generalized by introducing a ferromagnetic interaction (i.e. tending to spin parallelism) between spins on the same sublattice. In a loose-packed lattice (Sect. 16.2.1) both sites of any second-neighbour pair lie on the same sublattice, and we introduce energy terms $-J_2\sigma^{(\mathfrak{a})}(\boldsymbol{r})\sigma^{(\mathfrak{a})}(\boldsymbol{r}')$ and $-J_2\sigma^{(\mathfrak{b})}(\boldsymbol{r})\sigma^{(\mathfrak{b})}(\boldsymbol{r}')$ for each such $\mathfrak{a} - \mathfrak{a}$ and $\mathfrak{b} - \mathfrak{b}$ pair. With $K_1 := J_1/T$ and $K_2 := J_2/T$, the Hamiltonian (3.5.10) is modified to

---

[15] A phase is *commensurate* if its translational periodicity in each lattice direction is a rational (but not necessarily the same) multiple of the periodicity of the lattice. A homogeneous phase like the ferromagnetic ground state will be commensurate on all lattices. Of course, the construction of a commensurate antiferromagnetic phase is not confined to bipartite lattices. Thus, for example, on the triangular lattice the phase with a honeycomb of sites in one state and the centres of the honeycomb in the other state is both a commensurate and a lowest-energy state. However, it is one of a large number of arrangements with the same energy, leading to a macroscopic zero-point entropy, with no non-zero critical temperature to a long-range ordered phase (Sect. 9.4.2.2).

$$\widehat{H}(K_1, K_2, L; \boldsymbol{\sigma}) = K_1 \sum_{\{\boldsymbol{r},\boldsymbol{r}'\}}^{(1)} \sigma^{(\mathfrak{a})}(\boldsymbol{r})\sigma^{(\mathfrak{b})}(\boldsymbol{r}') - K_2 \sum_{\{\boldsymbol{r},\boldsymbol{r}'\}}^{(2)} \sigma^{(\mathfrak{a})}(\boldsymbol{r})\sigma^{(\mathfrak{a})}(\boldsymbol{r}')$$

$$- K_2 \sum_{\{\boldsymbol{r},\boldsymbol{r}'\}}^{(2)} \sigma^{(\mathfrak{b})}(\boldsymbol{r}')\sigma^{(\mathfrak{b})}(\boldsymbol{r}') - L\Big(\sum_{\{\boldsymbol{r}\}} \sigma^{(\mathfrak{a})}(\boldsymbol{r}') + \sum_{\{\boldsymbol{r}'\}} \sigma^{(\mathfrak{b})}(\boldsymbol{r}')\Big), \quad (3.5.17)$$

where J_1 and J_2 are taken to be positive. At $\mathcal{H} = 0$ the correspondence with the ferromagnet is the same as for the case $J_2 = 0$ except that now the equivalent ferromagnetic model has both first- and second-neighbour interactions. The ground-state Hamiltonians for the paramagnetic and antiferromagnetic states are

$$\widehat{H}_{\mathrm{P}} = \tfrac{1}{2} z_1 N J_1 - \tfrac{1}{2} z_2 N J_2 - N|\mathcal{H}|,$$
$$\widehat{H}_{\mathrm{A}} = -\tfrac{1}{2} z_1 N J_1 - \tfrac{1}{2} z_2 N J_2 \quad (3.5.18)$$

and the critical field is

$$\mathcal{H}_{\mathrm{c}} = z_1 J_1. \quad (3.5.19)$$

The simple antiferromagnet is recovered by putting $J_2 = 0$, $J_1 = J$ and $z_1 = z$. It can be seen that the critical field is not altered by the introduction of the second-neighbour interaction. This is because the second-neighbour energy contribution is the same for the paramagnetic and antiferromagnetic ground states.

3.5.1.2 The Simple Lattice Fluid

A lattice fluid is simply a lattice model where a site is either occupied by a fluid particle or unoccupied (a 'hole'). The lattice site variable is $\tau(\boldsymbol{r})$, which take the value one if the site is occupied by a particle and zero otherwise. Each particles has chemical potential μ and particles on first-neighbour lattice sites have an interaction energy $-\varepsilon$, which either encourages or discourages first-neighbour occupation according as $\varepsilon > 0$ or $\varepsilon < 0$. The Hamiltonian is

$$\widehat{H}(\varepsilon, \mu; \boldsymbol{\tau}) = -\varepsilon \sum_{\{\boldsymbol{r},\boldsymbol{r}'\}}^{(1)} \tau(\boldsymbol{r})\tau(\boldsymbol{r}') - \mu \sum_{\{\boldsymbol{r}\}} \tau(\boldsymbol{r}). \quad (3.5.20)$$

By taking $\tau(\boldsymbol{r}) := \frac{1}{2}[1+\sigma(\boldsymbol{r})]$, where $\sigma(\boldsymbol{r})$ is the Ising spin variable it can be shown that (3.5.20) is equivalent to (3.5.1) with an additional contribution

$$C := -\tfrac{1}{4}[\varepsilon + 4(\mu/z)], \quad (3.5.21)$$

per first-neighbour pair and

$$J := \tfrac{1}{4}\varepsilon, \qquad \mathcal{H} := \tfrac{1}{4}(z\varepsilon + 2\mu). \tag{3.5.22}$$

The fluid density is

$$\rho := \langle \tau(\boldsymbol{r}) \rangle = \tfrac{1}{2}[1 + \langle \sigma(\boldsymbol{r}) \rangle]. \tag{3.5.23}$$

It follows from the symmetry properties of the Ising model that, on the line $\mu = -\frac{1}{2}z\varepsilon$ in the μ–T plane, the density is one-half at high temperatures and, if in the corresponding Ising model spontaneous magnetization $\pm m_\mathrm{s}(T)$ occurs for $T < T_\mathrm{c}$, this line becomes a line of coexistence between phases (call them 'liquid' and 'vapour') with densities

$$\rho_\ell := \tfrac{1}{2}[1 + m_\mathrm{s}(T)], \qquad \rho_\mathrm{v} := \tfrac{1}{2}[1 - m_\mathrm{s}(T)]. \tag{3.5.24}$$

In general for any fluid system in a state with coexistent vapour and liquid phases the *diameter of the coexistence region* is

$$\rho_\mathrm{d}(T) := \tfrac{1}{2}[\rho_\ell(T) + \rho_\mathrm{v}(T)] \tag{3.5.25}$$

and the *width of the coexistence region* is

$$\rho_\mathrm{w}(T) = \rho_\ell(T) - \rho_\mathrm{v}(T). \tag{3.5.26}$$

In the simple lattice fluid $\rho_\mathrm{d} = \frac{1}{2}$. This is a special case of the *rectilinear diameter law* which states that $\rho_\mathrm{d}(T)$ is an asymptotically linear function of temperature in a neighbourhood of the critical point. It is now known that this law is not generally true. There are a number of models (Hemmer and Stell 1970; Widom and Rowlinson 1970; Mermin 1971a, b; Wheeler 1977) in which the coexistence diameter is singular at the critical point. One type of such model is treated in Sect. 10.6. In fact, it may be shown (Mermin and Rehr 1971), on the basis of some reasonable assumptions, that, unless the model has some special symmetry, as is the case for the Ising model, $\mathrm{d}\rho_\mathrm{d}/\mathrm{d}T$ is at least as singular as the constant-volume heat capacity c_ρ.

This equivalence between a lattice fluid and the spin-$\frac{1}{2}$ Ising model, which was first shown by Lee and Yang (1952), holds between any one-component lattice fluid of particles of chemical potential μ and a spin-$\frac{1}{2}$ system on the same lattice in a magnetic field $\mathcal{H}$.

3.5.1.3 A Binary Mixture

We consider a mixture of two types of atoms A and B on a lattice. The chemical potentials are μ_A and μ_B and there are first-neighbour interaction energies ε_AA, ε_AB and ε_BB between A–A, A–B and B–B pairs. By identifying $\sigma(\boldsymbol{r}) = +1$ and $\sigma(\boldsymbol{r}) = -1$ with types A and B respectively it is easy to show that the system is equivalent to the spin-$\frac{1}{2}$ Ising model with Hamiltonian (3.5.1) with an additional constant contribution

$$C := \frac{1}{4}(2\varepsilon_{\mathsf{AB}} + \varepsilon_{\mathsf{AA}} + \varepsilon_{\mathsf{BB}}) - \frac{1}{z}(\mu_{\mathsf{A}} + \mu_{\mathsf{B}}) \tag{3.5.27}$$

to each pair of sites and

$$J := \frac{1}{4}(2\varepsilon_{\mathsf{AB}} - \varepsilon_{\mathsf{AA}} - \varepsilon_{\mathsf{BB}}), \quad \mathcal{H} := \frac{1}{4}[2(\mu_{\mathsf{A}} - \mu_{\mathsf{B}}) - z(\varepsilon_{\mathsf{AA}} - \varepsilon_{\mathsf{BB}})]. \tag{3.5.28}$$

Unlike the lattice fluid, which has phase separation and is equivalent to a ferromagnetic Ising model, the binary mixture can be equivalent to either ferromagnetic or antiferromagnetic models. In the former case phase separation will occur to A– and B–rich phases. The latter case yields sublattice ordering with one lattice preferentially occupied by As and the other by Bs. Such a phenomenon occurs in, for example, FeAl on a body-centered cubic lattice (Bragg and Williams 1934) and AuCu on a face-centred cubic lattice (Shockley 1938).

3.5.2 The Spin-1 Ising Model

For the case where $\sigma(\boldsymbol{r})$ at any lattice site $\boldsymbol{r}$ can take the three values ± 1 and 0 the Hamiltonian with only first-neighbour and single-site terms has additional contributions and takes the most general form

$$\widehat{H}(J, \varepsilon, \varepsilon', \mathcal{H}, \mu; \boldsymbol{\sigma}) = -J \sum_{\{\boldsymbol{r},\boldsymbol{r}'\}}^{(1)} \sigma(\boldsymbol{r})\sigma(\boldsymbol{r}') - \varepsilon \sum_{\{\boldsymbol{r},\boldsymbol{r}'\}}^{(1)} \sigma^2(\boldsymbol{r})\sigma^2(\boldsymbol{r}')$$

$$-\varepsilon' \sum_{\{\boldsymbol{r},\boldsymbol{r}'\}}^{(1)} [\sigma^2(\boldsymbol{r})\sigma(\boldsymbol{r}') + \sigma(\boldsymbol{r})\sigma^2(\boldsymbol{r}')] - \mathcal{H} \sum_{\{\boldsymbol{r}\}} \sigma(\boldsymbol{r}) - \mu \sum_{\{\boldsymbol{r}\}} \sigma^2(\boldsymbol{r}). \tag{3.5.29}$$

There are now two densities m, conjugate to $\mathcal{H}$ and

$$\rho := \langle \hat{\rho} \rangle, \qquad \hat{\rho} := \frac{1}{N} \sum_{\{\boldsymbol{r}\}} \sigma^2(\boldsymbol{r}), \tag{3.5.30}$$

conjugate to the chemical potential μ. This is the *spin-1* Ising model, which is also called the *Blume–Emery–Griffiths model* in reference to the work of Blume et al. (1971) (see also, Ellis et al. 2005), who used the model to investigate the λ-transition and phase separation in ^{3}He–^{4}He mixtures.[16]

The spin-1 Ising model Hamiltonian (3.5.29), with $J = \varepsilon' = \mathcal{H} = 0$ and the replacement $\sigma^2(\boldsymbol{r}) \to \tau(\boldsymbol{r})$, also gives (3.5.20) and is thus equivalent to a lattice

[16] They solved the model in the mean-field approximation, with $\mathcal{H} = \varepsilon' = 0$. In that case the Hamiltonian, like (3.5.1), with $\mathcal{H} = 0$, is invariant under spin inversion on all the sites of the lattice.

fluid.[17] In general it is used to model a dilute magnetic system, where the dilution state $\sigma(\boldsymbol{r}) = 0$ is either a vacancy, giving a lattice fluid of magnetic particles, or a second non-magnetic component, in which case the conflicting effects of magnetic and sublattice ordering can be studied (Thompson and Lavis 1967). It can also be used to investigate ternary mixtures (Mukamel and Blume 1974).

3.6 State-Difference Models

A very general class of site-variable models has Hamiltonians of the form

$$\widehat{H}(\boldsymbol{\varepsilon}) = -\sum_{\{\boldsymbol{r},\boldsymbol{r}'\}}^{(1)} \mathfrak{e}(\varepsilon(\boldsymbol{r}) - \varepsilon(\boldsymbol{r}')), \tag{3.6.1}$$

where $\varepsilon(\boldsymbol{r})$ is a state variable at $\boldsymbol{r} \in \mathcal{N}$ and $\mathfrak{e}$ is an energy function. In this section we shall suppose that $\mathfrak{e}(\cdot)$ is an even function so that the ordering of the site vectors $\boldsymbol{r}$ and $\boldsymbol{r}'$ in (3.6.1) is immaterial. In Sect. 3.7 we introduce the idea of *chirality* for which this condition is relaxed.

Here the important distinction is between models where $\varepsilon(\boldsymbol{r})$ takes a continuum of values over some interval $[\varepsilon_0, \varepsilon_1)$ and those where the values of ε are integers $n_0, n_0 + 1, \ldots, n_0 + \nu - 1$. The zero-field spin-$\frac{1}{2}$ Ising model is an example of the latter with

$$\varepsilon(\boldsymbol{r}) := \tfrac{1}{2}[1 - \sigma(\boldsymbol{r})], \qquad \mathfrak{e}(\varepsilon) := K\cos(\pi\varepsilon), \tag{3.6.2}$$

when the Hamiltonian (3.5.1) take the form (3.6.1). We now consider a model with continuously-valued state variables, before turning to discrete-valued cases in Sects. 3.6.2–3.7.1.

3.6.1 *The Classical XY Model*

It can be seen from (3.3.3) and (3.6.2) that the Hamiltonian

$$\widehat{H}(K, \boldsymbol{\theta}) = -\sum_{\{\boldsymbol{r},\boldsymbol{r}'\}}^{(1)} K\cos[\theta(\boldsymbol{r}) - \theta(\boldsymbol{r}')], \tag{3.6.3}$$

could be taken to represent either the Ising model or the classical XY model, the difference being that in the Ising case $\theta(\boldsymbol{r}) = \pm\frac{1}{2}\pi$ and in the XY case $\theta(\boldsymbol{r}) \in [0, 2\pi)$.

---

[17] Except that now the particles have a state degeneracy of two (the states ± 1) which will contribute a trivial term to the entropy.

Here we explore the second of these. For low temperatures we expect $|\theta(\boldsymbol{r}) - \theta(\boldsymbol{r}')|$ to be small with

$$\widehat{H}(K, \boldsymbol{\theta}) = \tfrac{1}{4}K \sum_{\{\boldsymbol{r},\boldsymbol{r}'\}} \Delta(\boldsymbol{r} - \boldsymbol{r}')\{[\theta(\boldsymbol{r}) - \theta(\boldsymbol{r}')]^2 + \mathrm{O}([\theta(\boldsymbol{r}) - \theta(\boldsymbol{r}')]^4\}, \qquad (3.6.4)$$

where the summation is over all $\boldsymbol{r}$ and $\boldsymbol{r}' \in \mathcal{N}$ with $\Delta(\boldsymbol{r} - \boldsymbol{r}') := 1$ or 0, according as $[\boldsymbol{r}, \boldsymbol{r}']$ is or is not a first-neighbour pair, and a constant term has been neglected. For simplicity we suppose that $\mathcal{N}$ is a d-dimensional hypercubic lattice of lattice spacing a, with $z = 2d$ and linear dimension $L = aN^{1/d}$. Let $x_1, x_2, \ldots, x_d$ be cartesian coordinates in the lattice space with unit lattice vectors $\boldsymbol{i}_1, \boldsymbol{i}_2, \ldots, \boldsymbol{i}_d$. Then, in the continuum limit $N \to \infty$, $a \to 0$ and L fixed,

$$\theta(\boldsymbol{r} \pm a\boldsymbol{i}_j) - \theta(\boldsymbol{r}) \simeq \pm a \frac{\partial \theta(\boldsymbol{r})}{\partial x_j}; \qquad (3.6.5)$$

giving

$$\sum_{\{\boldsymbol{r}'\}} \Delta(\boldsymbol{r} - \boldsymbol{r}')\{[\theta(\boldsymbol{r}) - \theta(\boldsymbol{r}')]^2 = 2a^2[\nabla\theta(\boldsymbol{r})]^2. \qquad (3.6.6)$$

Truncating (3.6.4) at the quadratic (harmonic) term gives

$$\widehat{H}(K, \boldsymbol{\theta}) = \frac{K}{2a^{d-2}} \int_{\mathcal{V}} \mathrm{d}\mathcal{V}\, [\nabla\theta(\boldsymbol{r})]^2, \qquad (3.6.7)$$

where $\mathcal{V}$ is the hypercube of edge length L which has replaced the lattice. With periodic boundary conditions applied to $\mathcal{V}$, $\theta(\boldsymbol{r})$ is related to its Fourier transform by formulae like (18.1.14) and (18.1.15). That is

$$\theta(\boldsymbol{r}) = \sum_{\{\boldsymbol{k}\}} \theta^*(\boldsymbol{k}) \exp(\mathrm{i}\boldsymbol{k} \cdot \boldsymbol{r}), \quad \theta^\star(\boldsymbol{k}) := \frac{1}{L^d} \int_{\mathcal{V}} \mathrm{d}\mathcal{V}\, \theta(\boldsymbol{r}) \exp(-\mathrm{i}\boldsymbol{k} \cdot \boldsymbol{r}), \qquad (3.6.8)$$

where $\overline{\theta^*(\boldsymbol{k})} = \theta^*(-\boldsymbol{k})$ and the values of $\boldsymbol{k}$ are given by (18.1.16) with $L^{(1)} = \cdots = L^{(d)} = L$. Substituting into (3.6.7) gives

$$\widehat{H}(K, \boldsymbol{\theta}^*) = \frac{L^d K}{a^{d-2}} \sum_{\{\boldsymbol{k}\}} |\theta^*(\boldsymbol{k})|^2 k^2. \qquad (3.6.9)$$

Then

$$\langle \cos(\theta(\boldsymbol{r})) \rangle = \langle \cos(\theta(\boldsymbol{0})) \rangle = \Re \left\{ \frac{\displaystyle\int_{\Theta^*} \mathrm{d}\Theta^* \exp(-\widehat{H}'(K, \boldsymbol{\theta}^*))}{\displaystyle\int_{\Theta^*} \mathrm{d}\Theta^* \exp(-\widehat{H}(K, \boldsymbol{\theta}^*))} \right\}, \qquad (3.6.10)$$

where Θ^* is the space of the variables $\boldsymbol{\theta}^*$ and

$$\widehat{H}'(K, \boldsymbol{\theta}^*) := \widehat{H}(K, \boldsymbol{\theta}^*) - \mathrm{i}\,\theta(\mathbf{0}). \tag{3.6.11}$$

Let $\theta^*(\boldsymbol{k}) := \theta^*_{\mathrm{R}}(\boldsymbol{k}) + \mathrm{i}\,\theta^*_{\mathrm{I}}(\boldsymbol{k})$, where $\theta^*_{\mathrm{R}}(\boldsymbol{k})$ and $\theta^*_{\mathrm{I}}(\boldsymbol{k})$ are, respectively, even and odd real functions of $\boldsymbol{k}$. Then

$$\widehat{H}'(K, \boldsymbol{\theta}^*) = \frac{L^d K}{a^{d-2}} \sum_{\{\boldsymbol{k}\}} \left\{ \theta^*_{\mathrm{R}}(\boldsymbol{k}) - \frac{\mathrm{i}\, a^{d-2}}{2L^d K k^2} \right\}^2 k^2 + \frac{a^{d-2}}{4K L^d} \sum_{\{\boldsymbol{k}\}} k^{-2} + \frac{L^d K}{a^{d-2}} \sum_{\{\boldsymbol{k}\}} \left\{\theta^*_{\mathrm{I}}(\boldsymbol{k})\right\}^2 k^2. \tag{3.6.12}$$

The integrands are Gaussian which are steep for large L allowing the ranges of integration to be extended to $(-\infty, \infty)$. It follows that

$$\langle \cos(\theta(\boldsymbol{r})) \rangle = \exp[\eta(d; K) I_1(d)], \tag{3.6.13}$$

where

$$\eta(d; K) := (2\pi a^{2-d} K)^{-1} = \frac{T a^{d-2}}{2\pi J}, \tag{3.6.14}$$

$$I_1(d) := \frac{\pi}{2L^d} \sum_{\{\boldsymbol{k}\}} k^{-2}. \tag{3.6.15}$$

At this point it is useful to introduce two scale parameter $L_0 \sim L$ for the size of the system and $a_0 \sim a$, the minimum resolution length. Then

$$\ell_0 := 2\pi/L_0, \qquad k_0 := 2\pi/a_0, \tag{3.6.16}$$

become natural low- and high-frequency boundaries with the summation (3.6.15) approximated by an integral over the hyperspherical shell $\ell_0 < k < k_0$. From (18.1.28) and (18.1.35)

$$\begin{aligned} I_1(d) &= \frac{(2\sqrt{\pi})^{2-d}}{4\,\Gamma\left(\frac{d}{2}\right)} \int_{\ell_0}^{k_0} \mathrm{d}k\, k^{d-3}, \\ &= \begin{cases} \dfrac{(2\sqrt{\pi})^{2-d}}{4(d-2)\,\Gamma\left(\frac{d}{2}\right)} \left[1 - \left(\dfrac{\ell_0}{k_0}\right)^{d-2}\right], & d \neq 2, \\ \frac{1}{4}\ln(k_0/\ell_0), & d = 2; \end{cases} \end{aligned} \tag{3.6.17}$$

which, when $k_0 \gg \ell_0$, gives

$$\langle \boldsymbol{s}(\boldsymbol{r})\rangle = \begin{cases} \exp\left(-\dfrac{\eta(d;K)(k_0/2\sqrt{\pi})^{d-2}}{4\,\Gamma\left(\frac{d}{2}\right)(d-2)}\right), & d>2,\\ \left(\dfrac{\ell_0}{k_0}\right)^{\frac{1}{4}\eta(2;K)}, & d=2,\\ \exp\left(-\dfrac{\eta(1;K)}{2\ell_0}\right), & d=1. \end{cases} \tag{3.6.18}$$

In a similar way

$$\begin{aligned}\langle \boldsymbol{s}(\boldsymbol{r}+\boldsymbol{r}')\cdot\boldsymbol{s}(\boldsymbol{r}')\rangle &= \langle\cos[\theta(\boldsymbol{r})-\theta(\boldsymbol{0})]\rangle\\ &= \Re\left\{\frac{\displaystyle\int_{\Theta^*}\mathrm{d}\Theta^*\exp\{-\widehat{H}''(K,\boldsymbol{\theta}^*)\}}{\displaystyle\int_{\Theta^*}\mathrm{d}\Theta^*\exp(-\widehat{H}(K,\boldsymbol{\theta}^*))}\right\},\end{aligned} \tag{3.6.19}$$

where

$$\begin{aligned}\widehat{H}''(K,\boldsymbol{\theta}^*) &:= \widehat{H}(K,\boldsymbol{\theta}^*) - \mathrm{i}[\theta(\boldsymbol{r})-\theta(\boldsymbol{0})]\\ &= -Kd\left(\frac{L}{a}\right)^d + \frac{L^dK}{a^{d-2}}\sum_{\{\boldsymbol{k}\}}\left\{\theta^*_{\mathrm{R}}(\boldsymbol{k}) - \frac{\mathrm{i}\,a^{d-2}[\cos(\boldsymbol{r}\cdot\boldsymbol{k})-1]}{2L^dKk^2}\right\}^2 k^2\\ &\quad + \frac{L^dK}{a^{d-2}}\sum_{\{\boldsymbol{k}\}}\left\{\theta^*_{\mathrm{I}}(\boldsymbol{k}) + \frac{\mathrm{i}\,a^{d-2}\sin(\boldsymbol{r}\cdot\boldsymbol{k})}{2L^dKk^2}\right\}^2 k^2\\ &\quad + \frac{a^{d-2}}{2KL^d}\sum_{\{\boldsymbol{k}\}}k^{-2}[1-\exp(\mathrm{i}\boldsymbol{r}\cdot\boldsymbol{k})].\end{aligned} \tag{3.6.20}$$

Thus

$$\langle\cos[\theta(\boldsymbol{r})-\theta(\boldsymbol{0})]\rangle = \exp[\eta(d;K)I_2(d;\boldsymbol{r})], \tag{3.6.21}$$

where

$$I_2(d;\boldsymbol{r}) := \frac{\pi}{L^d}\sum_{\{\boldsymbol{k}\}}k^{-2}[\exp(\mathrm{i}\,\boldsymbol{r}\cdot\boldsymbol{k})-1]. \tag{3.6.22}$$

From (18.1.28), (18.1.35) and (18.1.36),

$$\begin{aligned}I_2(k;\boldsymbol{r}) &= \frac{(2\sqrt{\pi})^{2-d}}{2\pi\,\Gamma\left(\frac{d}{2}\right)}\int_{\ell_0}^{k_0}\mathrm{d}k\,k^{d-3}\int_0^{2\pi}\mathrm{d}\psi[\exp(\mathrm{i}\,rk\cos\psi)-1]\\ &= \frac{(2\sqrt{\pi})^{2-d}}{\Gamma\left(\frac{d}{2}\right)}\int_{\ell_0}^{k_0}\mathrm{d}k\,k^{d-3}[\mathrm{J}_0(rk)-1]\end{aligned}$$

$$= -\frac{(2r\sqrt{\pi})^{2-d}}{\Gamma\left(\frac{d}{2}\right)} \mathfrak{F}\,(r\ell_0, rk_0; d-3)\,, \tag{3.6.23}$$

where

$$\mathfrak{F}(\ell_0, k_0; x) := \int_{\ell_0}^{k_0} k^x\{1 - \mathrm{J}_0(k)\}\,\mathrm{d}k. \tag{3.6.24}$$

Interest in this integral can be restricted to $x = -2, -1, 0, 1, \ldots$ and in Sect. 18.1.4 we show, by dividing the range of integration in parts that the limit $\ell_0 \to 0$ can be taken in all cases. Substituting from (18.1.49)–(18.1.54), for $|\boldsymbol{r} - \boldsymbol{r}'| := r \to \infty$, with $k_0 \gg 0$,

$$\langle \boldsymbol{s}(\boldsymbol{r}) \cdot \boldsymbol{s}(\boldsymbol{r}')\rangle \simeq \begin{cases} \exp\left(-\dfrac{\eta(d;\, K)(k_0/2\sqrt{\pi})^{d-2}}{\Gamma\left(\frac{d}{2}\right)(d-2)}\right), & d \geq 3, \\ (rk_0)^{-\eta(2;K)}, & d = 2, \\ \exp[-2\,\eta\,(1;\, K)r], & d = 1. \end{cases} \tag{3.6.25}$$

For $d > 2$ the spin-spin correlation function is non-zero in the limit $r \to \infty$, indicating the presence of a long-range ordered phase. This conclusion is supported in three dimensions where high-temperature series analysis yields a second-order transition in the universality class of the λ-transition in liquid ^{4}He (Butera et al. 1993). For $d = 1$ the correlation function decays exponentially relative to a correlation length $1/2\,\eta\,(1;\, K) = \pi a J/T$ which diverges as $T \to 0$, and $\langle \boldsymbol{s}(\boldsymbol{r})\rangle \to 0$ as $L \to \infty$, for all $T > 0$, indicating that for non-zero temperature there is no long-range order.

However, it is the two-dimensional case which makes this model of particular interest. In the theoretical sense this arises from its archetypal status as a representative of the universality class of $\mathrm{O}(2)$ models which exhibit a *Kosterlitz–Thouless transition*. The experimental importance of this transition, and hence of the two-dimensional XY model, is that it models the superfluid transition in ^{4}He films (Bishop and Reppy 1978; Minnhagen 1987; Chaikin and Lubensky 1995, pp. 551–555) and the transition in weakly-disordered two-dimensional superconductors (Resnick et al. 1981; Erez and Meir 2010).

3.6.1.1 The Two-Dimensional XY Model: Kosterlitz–Thouless Transition

Moore (1969) concluded from high-temperature series calculations that the evidence for a phase transition in the two-dimensional XY model "is as strong in two dimensions as it is in three". However, the Mermin–Wagner theorem excludes the existence of a long-range ordered phase, a conclusion supported by (3.6.18) which yields $\langle \boldsymbol{s}(\boldsymbol{r})\rangle = 0$ for all temperatures in the limit $L_0 \to \infty$ of an infinite system. On the other hand the correlation function is algebraic, like the critical-point correlation in standard scaling theory (Eq. (4.5.55)), with a temperature-dependent exponent $\eta(2; K)$ indicative of a line of fixed points.[18] It is not to be expected that the correlation function can persist in this form as $T \to \infty$; there one expects exponential decay on the scale of the correlation length (Eq. (5.6.37)) and this is borne out by the renormalization group analysis of Sect. 15.6. Thus some kind of phase transition must be anticipated, but from a phase which is not excluded by the conditions of the Mermin–Wagner theorem. This puzzle was resolved by the work of Berezinskii (1971, 1972) and Kosterlitz and Thouless (1972, 1973). They showed that there is indeed a phase transition (now called the *Kosterlitz–Thouless transition*) to a form of *quasi-long range order* very different from that for the three-dimensional model.

To understand the nature of this phase we must re-examine the assumptions underlying the analysis given above. The most obvious of these is that $|\theta(\boldsymbol{r})-\theta(\boldsymbol{r}')|$, for a pair of first-neighbour sites, is small. This assumption, valid at low temperatures, is the basis for the replacement of the Hamiltonian (3.6.3) first by (3.6.4) and then in the continuum limit by (3.6.7), where the set of periodic angle variables $\theta(\boldsymbol{r})$ for all $\boldsymbol{r} \in \mathcal{N}$ becomes a *single* scalar field $\theta(\boldsymbol{r})$ defined at all $\boldsymbol{r} \in \mathcal{V}$.

Whilst we may choose to accept the harmonic assumption, what must be focused on is the inherited property arising from the periodicity, and, therefore, non-single-valued nature, of the angle variables. Suppose $\boldsymbol{r}^\star$ is a point at the centre of one of the squares of $\mathcal{N}$. That is to say, at a site of the dual lattice $\mathcal{N}^\star$ (Sect. 16.2.2.1). And suppose that all the angle variables are aligned in the same (clockwise or anticlockwise) direction along tangents to circles centre $\boldsymbol{r}^\star$. In the continuum limit this becomes a scalar field for which we define the annulus

$$\mathcal{R}(\boldsymbol{r}^\star) := \{\boldsymbol{r} : a \le |\boldsymbol{r} - \boldsymbol{r}*| \le R\} \tag{3.6.26}$$

and

$$\theta^{(\mathrm{v})}(\boldsymbol{r}^\star;\boldsymbol{r}) := \begin{cases} \omega \arccos\left[\dfrac{(\boldsymbol{r}-\boldsymbol{r}^\star)\cdot\boldsymbol{r}^\star}{|\boldsymbol{r}^\star|\,|\boldsymbol{r}-\boldsymbol{r}^\star|}\right], & \boldsymbol{r} \in \mathcal{R}(\boldsymbol{r}^\star), \\ 0, & \boldsymbol{r} \notin \mathcal{R}(\boldsymbol{r}^\star), \end{cases} \tag{3.6.27}$$

where $\omega = \pm 1$ according as the orientation about $\boldsymbol{r}^\star$ is anticlockwise or clockwise. This field satisfies

$$\oint_C \nabla\theta^{(\mathrm{v})}(\boldsymbol{r}^\star;\boldsymbol{r})\cdot \mathbf{d}\boldsymbol{r} = 2\pi\omega, \tag{3.6.28}$$

where C is any simple curve, in the anti-clockwise direction, enclosing $\boldsymbol{r}^\star$ and lying in $\mathcal{R}(\boldsymbol{r}^\star)$. $\theta^{(\mathrm{v})}(\boldsymbol{r}^\star;\boldsymbol{r})$ is the field of a *vortex* v centre $\boldsymbol{r}^\star$ with *winding number* ω, which measures the number of anticlockwise rotations of the field around a circle centre $\boldsymbol{r}^\star$ (so it can, henceforth, be taken to be any positive or negative integer). We have introduced via the annulus $\mathcal{R}(\boldsymbol{r}^\star)$ a core of radius the lattice distance a and

[18] The coupling J/T is marginal like the four-spin coupling in the spin-representation of the eight-vertex model (Sect. 12.3.9).

an outer radius R. Ignoring these constrictions, $\theta^{(v)}(\boldsymbol{r}^\star;\boldsymbol{r})$, given by the first line in (3.6.27), has a branch-point at $\boldsymbol{r}=\boldsymbol{r}^\star$ and satisfies

$$\nabla^2\theta^{(v)}(\boldsymbol{r}^\star;\boldsymbol{r}) = 2\pi\omega\,\delta^{\mathrm{D}}(\boldsymbol{r}-\boldsymbol{r}^\star). \tag{3.6.29}$$

From (3.6.7) the energy of v is

$$\widehat{H}^{(v)}(K) = \tfrac{1}{2}K\int_{\mathcal{R}(\boldsymbol{r}^\star)} \mathrm{d}\mathcal{V}\,[\nabla\theta^{(v)}(\boldsymbol{r}^\star;\boldsymbol{r})]^2. \tag{3.6.30}$$

With polar variables (q,ψ) with respect $\boldsymbol{r}^\star$ and the unit axis vector $\hat{\boldsymbol{i}}$,

$$\theta^{(v)}(\boldsymbol{r}^\star;\boldsymbol{r}) = \omega\,\psi, \qquad \nabla\theta^{(v)}(\boldsymbol{r}^\star;\boldsymbol{r}) = \frac{\omega}{q}\,\hat{\boldsymbol{\psi}}. \tag{3.6.31}$$

Substituting into (3.6.30) gives

$$\widehat{H}^{(v)}(K) = \pi\omega^2 K\int_a^R \frac{\mathrm{d}q}{q} = -\pi\,\omega^2\,K\ln\left(\frac{a}{R}\right). \tag{3.6.32}$$

Consider now the case of a second vortex of winding number $\tilde{\omega}$ located at $\boldsymbol{r}^\star + a\hat{\boldsymbol{i}}$. It is not difficult to show that

$$\tilde{\theta}^{(v)}(\boldsymbol{r}^\star + a\hat{\boldsymbol{i}};\boldsymbol{r}) = \tilde{\omega}\left\{\psi + \frac{|\sin\psi|}{q}a - \frac{|\sin\psi|\cos\psi}{2q^2}a^2 + \mathrm{O}(a^3)\right\}, \tag{3.6.33}$$

giving for the vortex pair $v+\tilde{v}$

$$\widehat{H}^{(v)}(K) = \begin{cases} 2\pi\,K[(\omega+\tilde{\omega})^2 + \mathrm{O}(a)]\,\ln\left(\dfrac{R}{a}\right), & \tilde{\omega}\neq-\omega, \\[2ex] \pi\,K\left[1-\left(\dfrac{a}{R}\right)^2\right] + \mathrm{O}(a^3), & \tilde{\omega}=\omega. \end{cases} \tag{3.6.34}$$

Whilst the energy of a single vortex is $\mathrm{O}(K\ln[R/a])$, that of a bound pair with opposite winding numbers on neighbouring sites is $\mathrm{O}(K)$. However, of course, the thermodynamic quantity of interest is the free energy and one expects bound pairs of vortices to dissociate if there is a free-energy advantage in doing so. A free vortex has entropy of the order of $\ln[(R/a)^2]$. So the free energy released when a single vortex of winding number ω becomes unbound is of the order of $[\pi\omega^2 - 2K^{-1}]\ln(R/a)$. This decreases as T increases becoming negative for $T > \pi\omega^2 J/2$. For

$$T < T_{\mathrm{KT}} := \pi J/2 \simeq 1.5708J, \tag{3.6.35}$$

all free vortices are suppressed and as temperature increases unbinding of vortices of higher winding number becomes advantageous. The *Kosterlitz–Thouless transition* is this unbinding transition. As will be seen in the real-space renormalization treatment of Sect. 15.6 it occurs at a temperature on a critical line controlled by a fixed point at the *Kosterlitz–Thouless temperature* $T_{\rm KT}$.

Consider now the case of a number of vortices with[19]

$$\theta(\boldsymbol{r}) = \sum_{j=1}^{N_{\rm v}} \theta_j^{(\rm v)}(\boldsymbol{r}_j^\star; \boldsymbol{r}), \tag{3.6.36}$$

where $\theta_j^{(\rm v)}(\boldsymbol{r}_j^\star; \boldsymbol{r})$ is the field of a vortex v_j, centre $\boldsymbol{r}_j^\star \in \mathcal{N}^\star$ and winding number ω_j. From (3.6.29),

$$\nabla^2\theta(\boldsymbol{r}) = 2\pi\rho(\boldsymbol{r}), \tag{3.6.37}$$

where

$$\rho(\boldsymbol{r}) := \sum_{j=1}^{N_{\rm v}} \omega_j \delta^{\rm D}(\boldsymbol{r} - \boldsymbol{r}_j^\star) \tag{3.6.38}$$

is the *vortex distribution function.* Imposing the condition that the total vorticity of the system is zero gives

$$\int_{\mathcal{V}} \rho(\boldsymbol{r}) \mathrm{d}\mathcal{V} = \sum_{j=1}^{N_{\rm v}} \omega_j = 0. \tag{3.6.39}$$

Using (18.1.19) to take the Fourier transform of (3.6.37) gives[20]

$$k^2\theta^*(\boldsymbol{k}) = -\frac{2\pi}{L^2} \sum_{j=1}^{N_{\rm v}} \omega_j \exp(-\mathrm{i}\boldsymbol{k} \cdot \boldsymbol{r}_j^\star). \tag{3.6.40}$$

Then using the *lattice Green's function* $G(2; r)$ defined in Sect. 18.1.5, from (3.6.9),

---

[19] In fact the total field of the system will contain a conservative single-valued term $\theta^{(\rm s)}(\boldsymbol{r})$, which thus satisfies (3.6.28) with $\omega = 0$ (Kosterlitz and Thouless 1973). It makes an additively separate contribution to the Hamiltonian and will, therefore, for the sake of simplicity be ignored, although it perturbative effect is the cause of the absence of long-range below the critical point.

[20] The asterisk '$*$' for the Fourier transform should not be confused with the star '$\star$' for the dual lattice.

$$
\begin{aligned}
\widehat{H}(K) &= \frac{4\pi^2 K}{L^2}\sum_{j=1}^{N_v}\sum_{j'=1}^{N_v}\omega_j\omega_{j'}\sum_{\{k\}}\frac{\exp(-\mathrm{i}\boldsymbol{k}\cdot[\boldsymbol{r}^\star_{j'}-\boldsymbol{r}^\star_j])}{k^2} \\
&= 4\pi^2 K\sum_{j=1}^{N_v}\sum_{j'=1}^{N_v}\omega_j\omega_{j'}G(2;|\boldsymbol{r}^\star_j-\boldsymbol{r}^\star_{j'}|) \\
&= -\pi K\sum_{\{j\neq j'\}}\omega_j\omega_{j'}\ln\left|\frac{\boldsymbol{r}^\star_j-\boldsymbol{r}^\star_{j'}}{a}\right| - K\mu\sum_{j=1}^{N_v}\omega_j^2,
\end{aligned}
\tag{3.6.41}
$$

where in the first summation each pair is counted only once and the last term is the sum of the energies of isolated vortices. The last line of (3.6.32) is of the form of the Hamiltonian of a *Coulomb gas* on the dual lattice $\mathcal{N}^\star$ (Minnhagen 1987; Nienhuis 1987) but slightly restricted since the second summation, which corresponds to the fugacity of the gas is related via the vortex winding numbers to the charge energies given by the coefficients in the first summation.

Other models related to the XY model by duality transformations are the *sold-on-solid model* (Knops 1977; see also Cardy 1996, Sect. 6.3) and the *Villain model* (Villian 1975; José et al. 1977; see also Chaikin and Lubensky 1995, Appendix 9B). Further models exhibiting a Kosterlitz–Thouless are discussed by Nelson (1983). A real-space renormalization group treatment of the Kosterlitz–Thouless transition is given in Sect. 15.6.

3.6.2 The Ashkin–Teller Model

Ashkin and Teller (1943) proposed a model with four types of microsystems, A, B, C and D, on the sites of the lattice $\mathcal{N}$ with first-neighbour interaction energies w_0 for AA, BB, CC and DD pairs, w_1 for AB and CD pairs, w_2 for AC and BD pairs and w_3 for AD and BC pairs. In the special case $w_3 = w_1$, if the states A, B, C and D at $\boldsymbol{r} \in \mathcal{N}$ are identified respectively with $\varepsilon(\boldsymbol{r}) := \sigma(\boldsymbol{r}) = 0, 1, 2, 3$, the Ashkin–Teller model has the Hamiltonian (3.6.1) with $\mathfrak{e}(0) = w_0$, $\mathfrak{e}(\pm 1) = \mathfrak{e}(\pm 3) = w_1$, $\mathfrak{e}(\pm 2) = w_2$.

Fan (1972) formulated the model in terms of two spin-$\frac{1}{2}$ Ising variables $\tilde{\sigma}(\boldsymbol{r})$ and $\tilde{\tau}(\boldsymbol{r})$ at each site $\boldsymbol{r} \in \mathcal{N}$ with a site occupation identified by $(\tilde{\sigma},\tilde{\tau})$ with A $:= (+1,+1)$, B $:= (+1,-1)$, C $:= (-1,+1)$ and D $:= (-1,-1)$. The Hamiltonian then takes the form

$$
\begin{aligned}
\widehat{H}(K_0,K_1,K_2,K_3;\tilde{\boldsymbol{\sigma}},\tilde{\boldsymbol{\tau}}) := -K_0 - &\sum_{\{\boldsymbol{r},\boldsymbol{r}'\}}^{(1)}[K_1\tilde{\sigma}(\boldsymbol{r})\tilde{\sigma}(\boldsymbol{r}') + K_2\tilde{\tau}(\boldsymbol{r})\tilde{\tau}(\boldsymbol{r}') \\
&+ K_3\tilde{\sigma}(\boldsymbol{r})\tilde{\tau}(\boldsymbol{r})\tilde{\sigma}(\boldsymbol{r}')\tilde{\tau}(\boldsymbol{r}')],
\end{aligned}
\tag{3.6.42}
$$

where $K_i := J_i/T$, $i = 0, \ldots, 3$, and

$$J_0 := -\tfrac{1}{4}(w_0 + w_1 + w_2 + w_3), \quad J_1 := -\tfrac{1}{4}(w_0 + w_1 - w_2 - w_3),$$
$$J_2 := -\tfrac{1}{4}(w_0 - w_1 + w_2 - w_3), \quad J_3 := -\tfrac{1}{4}(w_0 - w_1 - w_2 + w_3). \tag{3.6.43}$$

This result applies to any lattice. However, if a square lattice $\mathcal{N}$ is chosen it is clear that (3.6.43) represents two first-neighbour Ising models linked by a four-site interaction. Using this result and the spin formulation of the eight-vertex model developed by Kadanoff and Wegner (1971) and Wu (1971) (Sect. 3.8.1.3), Wegner (1972) showed that the Ashkin–Teller and eight-vertex models are related by a duality relation.

3.6.3 Potts Models

The problem suggested to Renfrey Potts by his research supervisor Cyril Domb was to generalize the Ising condition (3.6.2) to one in which the vector states can point in ν directions, expressed by 'recycling' the Ising variable $\sigma(\boldsymbol{r})$ to give

$$\theta(\boldsymbol{r}) = 2\pi\sigma(\boldsymbol{r})/\nu, \qquad \sigma(\boldsymbol{r}) = 0, 1, \ldots, \nu - 1. \tag{3.6.44}$$

With $L = 0$ the Hamiltonian now takes the form

$$\widehat{H}(K; \boldsymbol{\sigma}) = -K \sum_{\{\boldsymbol{r},\boldsymbol{r}'\}}^{(1)} \cos\left(\frac{2\pi}{\nu}[\sigma(\boldsymbol{r}) - \sigma(\boldsymbol{r}')]\right). \tag{3.6.45}$$

Domb (1974a) called this the *planar Potts model* and it is also called the *vector Potts model* or *clock model*. Potts (1952) obtained the critical point on the square lattice for $\nu = 2, 3, 4$.

An obvious generalization to (3.6.45) is to adopt the Hamiltonian (3.6.1) with $\varepsilon(\boldsymbol{r}) := \sigma(\boldsymbol{r}) = 0, 1, \ldots, \nu - 1$ and

$$\mathfrak{e}(\sigma + n\nu) = \mathfrak{e}(\sigma), \qquad n = 0, \pm 1, \pm 2, \ldots \tag{3.6.46}$$

which is invariant under the cyclic group $\mathfrak{C}_\nu$, where $\mathrm{C}_\nu : \sigma \to \sigma + 1 \pmod{\nu}$. An additional property, satisfied by some but not all of the Potts models (Sect. 3.7) is that $\mathfrak{e}(\sigma(\boldsymbol{r}) - \sigma(\boldsymbol{r}'))$ is invariant under the interchange of states on $\boldsymbol{r}$ and $\boldsymbol{r}'$. This implies that

$$\mathfrak{e}(\nu - \sigma) = \mathfrak{e}(\sigma). \tag{3.6.47}$$

The Ashkin–Teller model with $w_1 = w_3$ satisfies both (3.6.46) and (3.6.47) with $\nu = 4$. From (3.6.46) and (18.1.9),

$$\mathfrak{e}(\sigma) = \frac{1}{\nu}\sum_{k=0}^{\nu-1} \kappa_k \exp\left(\mathrm{i}\frac{2\pi k\sigma}{\nu}\right). \tag{3.6.48}$$

with (3.6.47) giving $\kappa_{\nu-k} = \kappa_k$. For the special case of the clock-model Hamiltonian (3.6.45) $\kappa_1 = \kappa_{\nu-1} = \frac{1}{2}K\,\nu$ and $\kappa_k = 0$ otherwise.

3.6.4 *The Standard Potts Model*

One possibility suggested by Potts (1952) is when the interaction between a first-neighbour pair has one value when the states are the same and another value when they are different, irrespective of what that difference may be. This model, called by Domb (1974a) the *standard Potts model*, is the one most usually associated with the name of Potts, to the extent that it is commonly (and henceforth in this text) referred to simply as *the* Potts model (Wu 1982; Martin 1991),

Without loss of generality let the energy of a like pair be $-R$ and of an unlike pair be zero. Then in (3.6.1), with $\varepsilon(\boldsymbol{r}) := \sigma(\boldsymbol{r}) = 0, 1, \ldots, \nu-1$, $\mathfrak{e}(\sigma(\boldsymbol{r}) - \sigma(\boldsymbol{r}')) := K\delta^{\mathrm{Kr}}(\sigma(\boldsymbol{r}) - \sigma(\boldsymbol{r}'))$ and $K := R/T$,

$$\widehat{H}(K;\boldsymbol{\sigma}) = \widehat{H}(R;\boldsymbol{\sigma})/T := -K\sum_{\{\boldsymbol{r},\boldsymbol{r}'\}}^{(1)} \delta^{\mathrm{Kr}}(\sigma(\boldsymbol{r}) - \sigma(\boldsymbol{r}')). \tag{3.6.49}$$

From (18.1.11), (3.6.48) applies with $\kappa_k = 1$, $k = 0, 1, \ldots, \nu-1$. The Hamiltonian (3.6.49) is invariant under the symmetric group $\mathfrak{S}_\nu$, which permutes the ν microstates among themselves. As in the case of the Ising model, it is said to be *ferromagnetic* when $R > 0$, encouraging the members of a first-neighbour pair to be in the same state and *antiferromagnetic* when $R < 0$, encouraging the members of a first-neighbour pair to be in different states. The Ashkin–Teller model with $w_1 = w_2 = w_3$ is equivalent to the (standard) four-state Potts model and Temperley and Lieb (1971) showed the equivalence between this model on a square lattice with different interactions in the two lattice direction and a staggered six-vertex model (Sect. 3.8.2.1).

Another form for the partition function can be obtained using

$$\exp[K\delta^{\mathrm{Kr}}(\sigma(\boldsymbol{r}) - \sigma(\boldsymbol{r}'))] = 1 + v\delta^{\mathrm{Kr}}(\sigma(\boldsymbol{r}) - \sigma(\boldsymbol{r}')), \tag{3.6.50}$$

where

$$v := \exp(K) - 1. \tag{3.6.51}$$

This gives

$$Z(\mathcal{N}, v) = \sum_{\{\sigma\}} \prod_{\{r,r'\}}^{(1)} \left[1 + v\delta^{\mathrm{Kr}}(\sigma(\boldsymbol{r}) - \sigma(\boldsymbol{r}'))\right]$$
$$= \sum_{\{\mathfrak{g}\}}^{(\mathrm{AG})} \nu^{\mathrm{c}(\mathfrak{g})} v^{\mathrm{e}(\mathfrak{g})} = W(\nu, v), \tag{3.6.52}$$

where, as defined Sect. 16.4.1, the summation is over augmented graphs and $W(\nu, v)$ is the Whitney polynomial defined in (16.4.2). This is the *cluster form* for the Potts partition function, the components of $\mathfrak{g}$ being *Fortuin–Kasteleyn clusters* (Kasteleyn and Fortuin 1969).[21]

Although (3.6.49) represents the 'pure' Potts model it is often useful to regard it as a kind of lattice fluid with $\sigma = 0$ and $\sigma \neq 0$ corresponding respectively to a hole and a particle. Then the Hamiltonian is modified to

$$\widehat{H}(K, L; \boldsymbol{\sigma}) := \widehat{H}(R, \mu; \boldsymbol{\sigma})/T := -K \sum_{\{r,r'\}}^{(1)} \delta^{\mathrm{Kr}}(\sigma(\boldsymbol{r}) - \sigma(\boldsymbol{r}'))$$
$$- L \sum_{\{r\}} [1 - \delta^{\mathrm{Kr}}(\sigma(\boldsymbol{r})], \tag{3.6.53}$$

where $L := \mu/T$ and μ is the chemical potential of a particle. Of course, the Hamiltonian (3.6.53) no longer has $\mathcal{S}_\nu$ symmetry. In the next sections we consider equivalences with Ising models.

3.6.4.1 $\nu = 2$: The Spin-$\frac{1}{2}$ Ising Model

The spin-$\frac{1}{2}$ Ising variable which we now denote by $\tilde{\sigma}(\boldsymbol{r})$, is related to the Potts variable by[22]

$$\tilde{\sigma}(r) := 2\sigma(\boldsymbol{r}) - 1, \tag{3.6.54}$$

with

$$\begin{aligned} \delta^{\mathrm{Kr}}(\sigma(\boldsymbol{r}) - \sigma(\boldsymbol{r}')) &= \tfrac{1}{2}[1 + \tilde{\sigma}(\boldsymbol{r})\tilde{\sigma}(\boldsymbol{r}')], \\ 1 - \delta^{\mathrm{Kr}}(\sigma(\boldsymbol{r})) &= \tfrac{1}{2}[1 + \tilde{\sigma}(\boldsymbol{r})]. \end{aligned} \tag{3.6.55}$$

[21] It should be noted that these clusters are not identical to the set of graphs obtained by joining sites with the same microstate. The difference is that they also include graphs derived from the latter set by omitting one or more edges. A small calculation for a lattice consisting of a triangle of sites illustrates this point.

[22] Here and in the rest of the chapter we shall need some 'temporary' notation to express relations between different models.

Substituting from (3.6.55) into (3.6.49) and comparing with (3.5.1) we see that the Hamiltonians and partition functions for the two-state Potts model (PM2) and spin-$\frac{1}{2}$ Ising model (IMH) are related by

$$\widehat{H}^{(\mathrm{PM2})}(R,\mu;\boldsymbol{\sigma}) = -\tfrac{1}{4}N(zR+2\mu) + \widehat{H}^{(\mathrm{IMH})}\left(\tfrac{1}{2}R,\tfrac{1}{2}\mu;\tilde{\boldsymbol{\sigma}}\right), \tag{3.6.56}$$

$$Z^{(\mathrm{PM2})}(N,K,L) = \exp\left(\tfrac{1}{4}N[zK+2L]\right) Z^{(\mathrm{IMH})}\left(N,\tfrac{1}{2}K,\tfrac{1}{2}L\right), \tag{3.6.57}$$

where $K := R/T$ and $L := \mu/T$.

3.6.4.2 $\nu = 3$: The Spin-1 Ising Model

The spin-1 Ising variable, which we again denote by $\tilde{\sigma}(\boldsymbol{r})$, is related to the Potts variable by

$$\tilde{\sigma}(\boldsymbol{r}) := \sigma(\boldsymbol{r}) - 1, \tag{3.6.58}$$

with

$$\begin{aligned}\delta^{\mathrm{Kr}}(\sigma(\boldsymbol{r})-\sigma(\boldsymbol{r}')) &= 1 + \tfrac{1}{2}\tilde{\sigma}(\boldsymbol{r})\tilde{\sigma}(\boldsymbol{r}') + \tfrac{3}{2}\tilde{\sigma}^2(\boldsymbol{r})\tilde{\sigma}^2(\boldsymbol{r}')\\ &\qquad -\tilde{\sigma}^2(\boldsymbol{r}) - \tilde{\sigma}^2(\boldsymbol{r}'),\\ 1-\delta^{\mathrm{Kr}}(\sigma(\boldsymbol{r})) &= \tilde{\sigma}^2(\boldsymbol{r}).\end{aligned} \tag{3.6.59}$$

Substituting from (3.6.59) into (3.6.49) and comparing with (3.5.29) we see that the Hamiltonians and partition functions for the three-state Potts model (PM3) and spin-1 Ising model (IMO) are related by

$$\widehat{H}^{(\mathrm{PM3})}(R,\mu;\boldsymbol{\sigma}) = -\tfrac{1}{2}NzR + \widehat{H}^{(\mathrm{IMO})}\left(\frac{1}{2}R,\frac{3}{2}R,0,0,\mu-zR;\tilde{\boldsymbol{\sigma}}\right), \tag{3.6.60}$$

$$Z^{(\mathrm{PM3})}(N,K,L) = \exp\left(\tfrac{1}{2}zKN\right)Z^{(\mathrm{IMO})}\left(N,\tfrac{1}{2}K,\tfrac{3}{2}K,0,0,L-zK\right). \tag{3.6.61}$$

This result can also be derived by imposing $\mathcal{S}_3$ symmetry on (3.5.29).

3.7 Chirality

The term 'chirality' was introduced by Lord Kelvin (William Thomson) (Thomson 1904) who chose to "call any geometrical figure, or group of points, chiral, and say it has chirality, if its image in a plane mirror, ideally realized, cannot be brought to coincide with itself" (ibid, p. 619). This idea, suitably interpreted (see, for example, Le Guennec 1998), now plays an important role in physics, chemistry and mathematics.

In giving a definition of chirality taylored to our present needs it should first be noted that Kelvin's definition is couched in terms of the absence of something (call it 'achirality'). In statistical mechanics, an achiral system is one which, expressed in rather an imprecise way, has a Hamiltonian which is invariant under orientation reversal. To make this idea more precise we concentrate on two-dimensional lattices with the orientation in question being the ordering (clockwise or anticlockwise) of sites around a lattice face with respect to a chosen direction of the axis orthogonal to the plane. As we saw in Sect. 17.3 the lattice symmetry group for operations in the plane consists of translations and a point group $\mathcal{C}_{n\mathrm{v}}$ for some n. Of these group operations, translations and rotations are direct isometries (preserving orientation) and the reflections are indirect isometries (reversing orientation). Then

Definition 3.7.1 The system is *achiral* if the value of the Hamiltonian is unchanged under the operation of all the indirect isometries in the lattice symmetry group.

Thus a system is chiral if there is at least one indirect isometry in the lattice symmetry group which alters the value of the Hamiltonian.

3.7.1 Chiral Potts Models

From the general Potts interaction $\mathfrak{e}(\sigma(\boldsymbol{r})-\sigma(\boldsymbol{r}'))$ defined in Sect. 3.6.3, the invariant condition under interchange of the states on sites $\boldsymbol{r}$ and $\boldsymbol{r}'$ is removed. This is most simply represented by giving a vector-direction $\boldsymbol{r}-\boldsymbol{r}'$ to the lattice edge and replacing $\mathfrak{e}(\sigma)$ with $\mathfrak{e}(\sigma;\boldsymbol{r}-\boldsymbol{r}')$, so that (3.6.47) is replaced by

$$\mathfrak{e}(\nu-\sigma;\boldsymbol{r})=\mathfrak{e}(\sigma;-\boldsymbol{r}) \tag{3.7.1}$$

and (3.6.48) becomes

$$\mathfrak{e}(\sigma;\boldsymbol{r})=\frac{1}{\nu}\sum_{k=0}^{\nu-1}\kappa_k(\boldsymbol{r})\exp\left(\mathrm{i}\frac{2\pi k\sigma}{\nu}\right), \tag{3.7.2}$$

with $\kappa_{\nu-k}(\boldsymbol{r})=\kappa_k(-\boldsymbol{r})$.

$$\widehat{H}(K;\boldsymbol{\sigma})=-\sum_{\{\boldsymbol{r},\boldsymbol{r}'\}}^{(1)}\mathfrak{e}(\sigma(\boldsymbol{r})-\sigma(\boldsymbol{r}');\boldsymbol{r}-\boldsymbol{r}') \tag{3.7.3}$$

is the most general form for the Hamiltonian of the *chiral Potts model*. A special case is the *chiral clock model*

$$\widehat{H}(K,\boldsymbol{\Delta};\boldsymbol{\sigma}):=-K\sum_{\{\boldsymbol{r},\boldsymbol{r}'\}}^{(1)}\cos\left(\frac{2\pi}{\nu}[\sigma(\boldsymbol{r})-\sigma(\boldsymbol{r}')-\boldsymbol{\Delta}\cdot(\boldsymbol{r}-\boldsymbol{r}')]\right), \tag{3.7.4}$$

for which $\kappa_1(\boldsymbol{r}) = \frac{1}{2}K\nu\exp(-2\pi\mathrm{i}\boldsymbol{\Delta}\cdot\boldsymbol{r}/\nu)$, $\kappa_{\nu-1}(\boldsymbol{r}) = \frac{1}{2}K\nu\exp(2\pi\mathrm{i}\boldsymbol{\Delta}\cdot\boldsymbol{r}/\nu)$ and $\kappa_k(\boldsymbol{r}) = 0$ otherwise. This model was first proposed by Ostlund (1981) and Huse (1981) to investigate a commensurate-incommensurate phase transition. For the square lattice, with $\boldsymbol{r} = a(n_x\hat{\boldsymbol{x}} + n_y\hat{\boldsymbol{y}})$ and $\boldsymbol{\Delta} = \Delta\hat{\boldsymbol{x}}$ the ground states will consist of lines of sites in the y direction in the same state. Corresponding to the $\sigma = 0$ ground state chosen from the ν ferromagnetic ground states of the non-chiral ($\boldsymbol{\Delta} = \boldsymbol{0}$) clock model we now choose the line $\boldsymbol{r}' = an_y\hat{\boldsymbol{y}}$ of sites to be in state $\sigma = 0$ and the line of site $\boldsymbol{r} = a(n_x\hat{\boldsymbol{x}} + n_y\hat{\boldsymbol{y}})$ will be in the state $\sigma = s \pmod{\nu}$ for which $s - \Delta n_x a$ has the least value for all integer values of s. These ground states identify long-range ordered commensurate phases from which there are phase transitions to incommensurate and disordered phases (Ostlund 1981; Huse 1981; Duxbury et al. 1984; Houlrik and Knak Jensen 1986).

The search for a form for the chiral Potts interaction $\mathfrak{e}(\sigma;\boldsymbol{r})$, which yields a solvable model has attracted a great deal of interest. Such a case (now often simply referred to as 'the chiral Potts model') was obtained by Baxter et al. (1988). Over the next two years the free energy of the model was derived (Baxter 1988, 1990; Baxter et al. 1990)[23] and in 2005 Baxter obtained the order parameter (Baxter 2005a, b, 2006).

3.7.2 An Extended 3-State Potts Model on the Triangular Lattice

A different type of chirality involving a three-site term for the 3-state Potts model on a triangular lattice arose as a 'biproduct' of the development of a lattice model for water by Bell and Lavis (1970).

Suppose that the triangular lattice $\mathcal{N}_\mathrm{T}$ is divided into three equivalent sublattices $\mathfrak{a}$, $\mathfrak{b}$ and $\mathfrak{c}$, which can be identified respectively with the black, grey and white sites in Fig. 16.4. It will be noted that the triangles of $\mathcal{N}_\mathrm{T}$ fall into two sets: those for which the sublattice order $\mathfrak{a} \to \mathfrak{b} \to \mathfrak{c} \to \mathfrak{a} \to \cdots$ is clockwise around the triangle and those for which the ordering is anticlockwise; we refer to these two types as *right-handed* and *left-handed* respectively. The symmetry group of the lattice is $\mathfrak{C}_{3\mathrm{v}}$, where the axis of rotation is orthogonal to the lattice plane and through the centre of an arbitrarily chosen triangle and the three planes of reflection $\Pi_\mathfrak{x}$ contain the axis and the vertex of the triangle $\mathfrak{x}$, for $\mathfrak{x} = \mathfrak{a}, \mathfrak{b}, \mathfrak{c}$. It will be seen that the handedness of the triangles are invariant under $\mathfrak{C}_3$ but are interchanged by the three reflections.

The full Bell–Lavis water model, which can be represented as a spin-1 Ising model with additional terms, has attracted considerable attention.[24] However, the interest

---

[23] For a review of this work together with related developments see Au-Yang and Perk (1995).

[24] Its water-like properties have been investigated by Lavis (1973, 1975) and Lavis and Christou (1977) using cluster-variation methods and Lavis (1976) using a transfer matrix method. More recently exact calculations on the Husimi cactus have been used by Barbosa and Henriques (2008)

here is the special case which arises when, as in Sect. 3.6.4.2, the spin-1 formulation reduces to a 3-state Potts model. This takes the form

$$\widehat{H}^{(\mathrm{EP})}(K_1, K_2, K_3; \boldsymbol{\sigma}) := -K_1 \sum_{\{\boldsymbol{r},\boldsymbol{r}'\}}^{(1)} \delta^{\mathrm{Kr}}(\sigma(\boldsymbol{r}) - \sigma(\boldsymbol{r}'))$$

$$- \sum_{\{\boldsymbol{r},\boldsymbol{r}',\boldsymbol{r}''\}}^{\triangle} \{K_2[\mathfrak{C}_{\mathrm{c}}(\sigma(\boldsymbol{r}), \sigma(\boldsymbol{r}'), \sigma(\boldsymbol{r}''))]^2 + K_3 \mathfrak{C}_{\mathrm{c}}(\sigma(\boldsymbol{r}), \sigma(\boldsymbol{r}'), \sigma(\boldsymbol{r}''))\}, \quad (3.7.5)$$

where the ordering of sites $\boldsymbol{r} \to \boldsymbol{r}' \to \boldsymbol{r}''$ in the second summation is clockwise on right-handed triangles and anticlockwise on left-handed triangles, with

$$\mathfrak{C}_{\mathrm{c}}(\tilde{\sigma}, \tilde{\sigma}', \tilde{\sigma}'') := \tfrac{1}{2}[\tilde{\sigma} - \tilde{\sigma}'][\tilde{\sigma}' - \tilde{\sigma}''][\tilde{\sigma}'' - \tilde{\sigma}]. \quad (3.7.6)$$

The presence of this term means that this is a face model in the sense defined in Sect. 2.4.1. It is also chiral in that $\mathfrak{C}_{\mathrm{c}}(\tilde{\sigma}, \tilde{\sigma}', \tilde{\sigma}'')$ is invariant under the rotational elements of the lattice symmetry group but reversed in sign by the reflections. This model has been investigated using a finite-lattice RSRG method by Young and Lavis (1979) and using Monte-Carlo simulations by Wang et al. (2003). The K_2 term in the Hamiltonian, which can be conveniently represented as the square of the chiral term is simply a pure Potts antiferromagnetic three-site interaction. Its inclusion is necessary for the renormalization calculations and means that the results obtained by Young and Lavis (1979) became a generalization of those of Schick and Griffths (1977).

3.8 Vertex Models

A vertex model is one where the microstate of the system is represented by placing an arrow on each edge of the lattice $\mathcal{N}$. If no restrictions are applied, the number of arrow arrangements around a lattice site is 2^{z}, where z is the coordination number. Vertex models on three-dimensional lattices have been investigated (Thibaudier and Villain 1972; Hu and Hou 1996) and also on the honeycomb (Wu 1974) and triangular (Sacco and Wu 1975) lattices. However, most interest has been directed to the square lattice, with the extension, by Baxter (1978), to an arbitrary plane lattice consisting of straight lines intersecting at points with coordination number four. Here the most general case, the *sixteen-vertex model*, which can be shown to be equivalent to an Ising model with two-, three- and four-site interactions and with an external field (Suzuki and Fisher 1971), is unsolved.[25]

(Footnote 24 continued)
to compare the properties of the Bell–Lavis model to those of the antiferromagnetic Blume–Emery–Griffiths model and a Monte-Carlo study has been carried out by Fiore et al. (2009).

[25] As may be expected, since the Ising model with just a two-site interaction and an external field is unsolved.

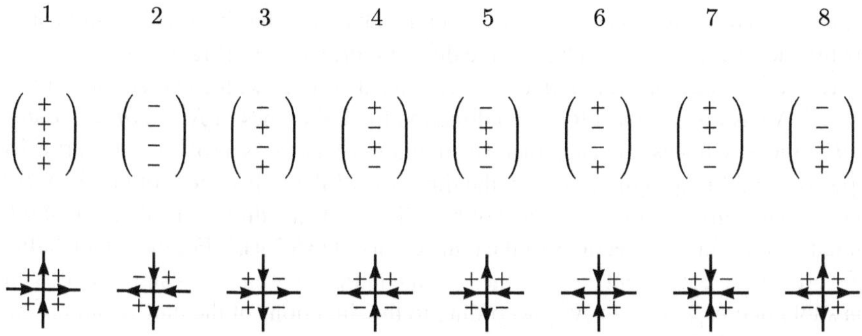

Fig. 3.2 The eight-vertex configurations showing, in row two, the edge variables clockwise in a column starting at the *top* with the southern vertical edge and in row three the *arrow* directions assigned according to the direction rule of Baxter (1972b)

The first square-lattice vertex model to be solved was the *six-vertex model* (Lieb 1967a, b, c, d). Here the vertices are subject to the *ice-rule*[26] which restricts the arrow arrangements around a vertex to those (labelled 1–6 in Fig. 3.2) where there are two inward-pointing and two outward-pointing arrows. This model is explored in more detail in Sect. 3.8.2. If the ice rule is relaxed to one in which there is an even number of arrows pointing in and out, two additional vertex-types are added to give the *eight-vertex model.*

3.8.1 The Eight-Vertex Model

The square-lattice zero-field partition function for this model was derived by Baxter (1972b). In this approach the eight allowed configurations of arrows at a vertex were labelled using a prescribed orientation of the lattice with lattice lines horizontal and vertical (Baxter 1982b, Chap. 10) and toroidal boundary conditions.[27] The other four-coordinated regular lattice is the kagomé lattice and the square-lattice results can be

[26] A water molecule can be regarded as having two negative and two positive 'arms' (Eisenberg and Kauzmann 1969, Chap. 1) and two molecules are hydrogen-bonded by the junction of a positive arm from one and a negative arm from the other. In a fully hydrogen-bonded structure like Ice I (ibid, Chap. 3), the positive arms of each molecule, that is those corresponding to the positions of the two protons, can be labelled with a single arrow on the segment between the centres of the bonded pair. A completely hydrogen-bonded network is thus represented by an oriented graph (Sect. 16.1.4) with the centres of the molecules at the vertices with inward arrows on two of the four edges meeting at each vertex and outward arrows on the other two. As well as water, certain other substances, such as KH_2PO_4 (potassium dihydrogen phosphate), crystallize as four-coordinated hydrogen-bonded networks. KH_2PO_4 is a *ferroelectric* with a spontaneous electrical polarization in zero field below a critical temperature.

[27] Toroidal boundary conditions strain the notion of horizontal/vertical, top/bottom, right/left a little but the problem can be resolved by choosing directions for one site and applying them in a consistent way to the other sites.

extended to this case (Baxter 1982b, Chap. 11), but now the situation is complicated by the fact that a vertex can have three different orientations (Fig. 16.6).

Let $\mathcal{N}^{\circledast}$ consist of straight lines intersecting at vertices with coordination number four.[28] A base line is used to give a direction to all the lines of $\mathcal{N}^{\circledast}$. *For the square lattice* the base line is chosen so that the directions of the two sets of lines are vertically upwards and to the right, that is in the direction of the unit vectors of (16.2.2). *For the kagomé lattice* a suitable choice of base line gives the directions of the three sets of lattice lines in the directions of the unit vectors of (16.2.4).[29] For the square lattice all vertices are equivalent, and for the kagomé lattice they can be designated as lying on sublattice $\mathcal{N}^{\circledast}_{13}$, $\mathcal{N}^{\circledast}_{32}$ or $\mathcal{N}^{\circledast}_{13}$ according to the directions of the unit vectors of the lines for which they are the intersection.

A property of $\mathcal{N}^{\circledast}$ (Sect. 16.3) is that some of its faces can be shaded so the every edge is a boundary between a shaded face and an unshaded face. Suppose that a vertex is oriented so that the directed lines passing through it are pointing north and east as shown in Fig. 16.16 and label the edges clockwise starting with the south-pointing edge. Then the sites are divided into two sublattices: $\mathcal{N}^{\circledast}_{\text{E}}$ consisting of sites for which the first (south-pointing) edge has shading in the clockwise direction and $\mathcal{N}^{\circledast}_{\text{O}}$ consisting of sites for which that edge has shading in the anticlockwise direction.[30] For the square lattice shading is a convenient mechanism for the division into two sublattices and the possible introduction of different Boltzmann weights. For the kagomé lattice one choice is to shade the triangular faces giving $\mathcal{N}^{\circledast}_{\text{E}} = \mathcal{N}^{\circledast}_{13} + \mathcal{N}^{\circledast}_{32}$, $\mathcal{N}^{\circledast}_{\text{E}} = \mathcal{N}^{\circledast}_{12}$, with the alternative, shading of the honeycomb faces, interchanging $\mathcal{N}^{\circledast}_{\text{E}}$ and $\mathcal{N}^{\circledast}_{\text{O}}$. Although, as in the square lattice, shading could be used here to introduce different Boltzmann weights, it relies on a rather arbitrary combination of two of the three sublattice. A proper treatment of this problem Baxter (1982b, Chap. 11) involves different weights on all three sublattices.

Arrow arrangements giving the eight vertex types are shown in the third line of Fig. 3.2 and, differentiating between sublattices, in the third and fourth lines of Fig. 3.3. Edge variables ς_i, $i = 1, 2, 3, 4$ are assigned around a vertex.

Definition 3.8.1 The ***direction rule***, used by Baxter for the square lattice (see, Baxter 1982b, Chap. 10) and the kagomé lattice (see, Baxter 1982b, Chap. 11), is equivalent to setting $\varsigma_i = +1$ or -1 according as the arrow points in the direction the same as, or opposite to, the line direction. These edge-variable assignments are shown as a vertical column (with index increasing downwards) in the second line of Fig. 3.2.

Definition 3.8.2 The ***shading rule***, used by Baxter (1978), sets $\varsigma_i = +1$ or -1 according as the arrow is pointing in the direction with shading on the left or right respectively. With arrows reversed between $\mathcal{N}^{\circledast}_{\text{E}}$ and $\mathcal{N}^{\circledast}_{\text{O}}$, the vertical column of edge-variable assignments shown in the second line of Fig. 3.3 apply to both sublattices.

---

[28] The use of this symbol for the 'arrow-bearing' lattice is for conformity with Sect. 16.3, since it is a *rapidity lattice*.

[29] That is 'generally to the right' as needed in Baxter (1982b, Chap. 11).

[30] This notation, in which E and O stands respectively for 'even' and 'odd', is in keeping with the labelling of sublattices in Chap. 12 (Fig. 12.3).

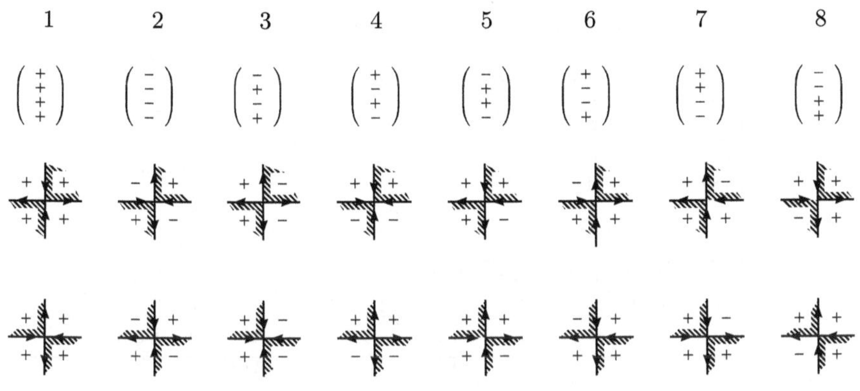

Fig. 3.3 The eight-vertex configurations with the edge variables in row two as in Fig. 3.2. The *arrow* directions which are different for sublattices $\mathcal{N}_{\mathrm{E}}^{\circledast}$ and $\mathcal{N}_{\mathrm{O}}^{\circledast}$ (and different from those given by the direction rule) are assigned using the shading rule of Baxter (1978) and shown in *lines* three and four, respectively, for the two sublattices

In terms of edge variables, assigned by either rule, the eight-vertex constraint condition is

$$\varsigma_1\varsigma_2\varsigma_3\varsigma_4 = 1. \tag{3.8.1}$$

Sites can be placed on the shaded faces of $\mathcal{N}^{\circledast}$ to form a lattice $\mathcal{N}$ and on the unshaded faces to form its dual $\mathcal{N}^{\star}$. Spins are now assigned to the sites of $\mathcal{N}+\mathcal{N}^{\star}$ with the spins around a vertex labelled σ_1, σ_2, σ_3, σ_4 in the clockwise direction starting with the spin on the face immediately following the first-labelled edge. It is a consequence of the eight-vertex rule (3.8.1) that spin-variable values can be related to edge-variable values, assigned by either rule, to satisfy

$$\varsigma_i = \sigma_{i-1}\sigma_i, \quad \Longrightarrow \quad \varsigma_i\varsigma_{i+1} = \sigma_{i-1}\sigma_{i+1} \ \text{ and } \ \varsigma_i\varsigma_{i+2} = \sigma_1\sigma_2\sigma_3\sigma_4,$$
$$i = 1, \ldots, 4 \quad (\mathrm{mod}\ 4). \tag{3.8.2}$$

To see that this is the case, suppose that, for any arrow configuration, the configuration graph $\mathfrak{g}$ is formed on $\mathcal{N}^{\circledast}$ by placing a graph edge on all lattice edges with $\varsigma_i = -1$. From (3.8.1) every vertex of $\mathfrak{g}$ is even and every edge is a boundary between regions of $\mathcal{N}+\mathcal{N}^{\star}$ with different homogeneous spin assignments. It is clear that the spin assignment for a particular assignment of edge-variable values is unique to within an overall reversal of spins on the lattice. The spin assignments shown in line three of Fig. 3.2 and lines three and four of Fig. 3.3 are one possibility.

Both the direction rule and shading rule for vertex assignments will be useful for our discussion of the eight-vertex model and the problem is to retain and compare them with a minimum of duplication. There are two possibilities:

(i) To give the same ordering of configurations with respect to arrow directions. This has the advantage of retaining the labelling of ferroelectric/antiferroelectric ground states, but looses the correspondence for spins between ferromagnetic

Table 3.2 Notation comparison between that adopted here and in Baxter (1978)

	Direction Rule	Shading Rule	Baxter (1978)
(1,2)	a	a	c
(3,4)	b	b	d
(5,6)	c	c	a
(7,8)	d	d	b
	Δ	Δ	$-\Delta$
	Γ	Γ	$-\Gamma$
I	FEN&F/FMU&S	AFEL/FMU&S	FEN&F/AFML
II	FEV&H/FMV&H	AFER/FMV&H	FEV&H/AFMR
III	AFER/AFMR	FEN&F/AFMR	AFER/FMU&S
IV	AFEL/AFML	FEV&H/AFML	AFEL/FMV&H

/antiferromagnetic ground states. It also means that the Boltzmann weights for the two rules have different forms in terms of spins and edge variables.

(ii) To give the same ordering of configurations in terms of edge variables and spins. The advantages and disadvantages in terms of electric and magnetic ground states are reversed, but the eight Boltzmann weights in terms of spins or edge variables now have the same form for each rule.

Baxter (1978)[31] uses (i), but we use (ii). Table 3.2 provides a comparison of notation.

Let the Boltzmann weight of the vertex $\boldsymbol{r} \in \mathcal{N}^{\circledast}$ be $\mathfrak{X}(\boldsymbol{r}; \varsigma_1, \varsigma_2, \varsigma_3, \varsigma_4)$, in terms of edge variables, and $\mathfrak{Y}(\boldsymbol{r}; \sigma_1, \sigma_2, \sigma_3, \sigma_4)$, in terms of spins. Then for the eight vertex configurations

$$\begin{aligned}
\mathfrak{X}_1(\boldsymbol{r}) &:= \mathfrak{X}(\boldsymbol{r}; +1, +1, +1, +1) = \mathfrak{Y}(\boldsymbol{r}; \pm1, \pm1, \pm1, \pm1),\\
\mathfrak{X}_2(\boldsymbol{r}) &:= \mathfrak{X}(\boldsymbol{r}; -1, -1, -1, -1) = \mathfrak{Y}(\boldsymbol{r}; \pm1, \mp1, \pm1, \mp1),\\
\mathfrak{X}_3(\boldsymbol{r}) &:= \mathfrak{X}(\boldsymbol{r}; +1, -1, +1, -1) = \mathfrak{Y}(\boldsymbol{r}; \pm1, \pm1, \mp1, \mp1),\\
\mathfrak{X}_4(\boldsymbol{r}) &:= \mathfrak{X}(\boldsymbol{r}; -1, +1, -1, +1) = \mathfrak{Y}(\boldsymbol{r}; \mp1, \pm1, \pm1, \mp1),\\
\mathfrak{X}_5(\boldsymbol{r}) &:= \mathfrak{X}(\boldsymbol{r}; -1, +1, +1, -1) = \mathfrak{Y}(\boldsymbol{r}; \pm1, \pm1, \pm1, \mp1),\\
\mathfrak{X}_6(\boldsymbol{r}) &:= \mathfrak{X}(\boldsymbol{r}; +1, -1, -1, +1) = \mathfrak{Y}(\boldsymbol{r}; \pm1, \mp1, \pm1, \pm1),\\
\mathfrak{X}_7(\boldsymbol{r}) &:= \mathfrak{X}(\boldsymbol{r}; -1, -1, +1, +1) = \mathfrak{Y}(\boldsymbol{r}; \pm1, \pm1, \mp1, \pm1),\\
\mathfrak{X}_8(\boldsymbol{r}) &:= \mathfrak{X}(\boldsymbol{r}; +1, +1, -1, -1) = \mathfrak{Y}(\boldsymbol{r}; \mp1, \pm1, \pm1, \pm1).
\end{aligned} \tag{3.8.3}$$

[31] Except for an unimportant interchange of (3,4) and (5,6).

3.8.1.1 A Staggered Eight-Vertex Model

Let us step back for a moment from the details so far presented for the analysis of this model and ask more directly the question: "Given that there are eight arrangements of arrows at a vertex, how do we distinguish between them?" In the absence of an orientation to the lattice, directions to the lattice lines or shading, and using the listing of vertex types in Fig. 3.2, the groups $\mathbb{T}_1 := \{1, 2, 3, 4\}$, $\mathbb{T}_2 := \{5, 6\}$ and $\mathbb{T}_3 := \{7, 8\}$ are of indistinguishable vertex types. The initial approach, which was used by Baxter, and which distinguished all eight types, was to give a physical orientation to the lattice (square or kagomé). Our proposed direction rule, giving directions to the lattice lines, is effectively equivalent to that and if *both* directions to the lattice lines *and* shading are included the effect is to distinguish sixteen different vertex types as shown in lines three and four of Fig. 3.3. Here we address the question posed by the abandonment of physical orientation (and line direction) and the reliance simply on shading. All vertex types equivalent under rotation must be counted together and the only distinguishing feature is the orientation of arrows with respect to the two shaded and unshaded faces. This will group together the arrow configurations p_{E} and p_{O}, $p = 1, 2, \ldots, 8$, of lines three and four, respectively, of Fig. 3.3 as follows:

$\mathbb{S}_1 := \{5_{\mathrm{E}}, 6_{\mathrm{E}}, 7_{\mathrm{O}}, 8_{\mathrm{O}}\}$: arrows on the edges of the unshaded regions follow each other round one clockwise and the other anticlockwise.

$\mathbb{S}_2 := \{5_{\mathrm{O}}, 6_{\mathrm{O}}, 7_{\mathrm{E}}, 8_{\mathrm{E}}\}$: arrows on the edges of the shaded regions follow each other round one clockwise and the other anticlockwise.

$\mathbb{S}_3 := \{2_{\mathrm{E}}, 2_{\mathrm{O}}\}$: arrows on the edges of both shaded and unshaded regions follow each other round clockwise on the shaded regions and anticlockwise on the unshaded regions.

$\mathbb{S}_4 := \{1_{\mathrm{E}}, 1_{\mathrm{O}}\}$: arrows on the edges of both shaded and unshaded regions follow each other round clockwise on the unshaded regions and anticlockwise on the shaded regions.

$\mathbb{S}_5 := \{3_{\mathrm{O}}, 4_{\mathrm{E}}\}$: arrows on the edges of both shaded and unshaded regions point in.

$\mathbb{S}_6 := \{3_{\mathrm{E}}, 4_{\mathrm{O}}\}$: arrows on the edges of both shaded and unshaded regions point out.

This is a *staggered eight-vertex model* with the Boltzmann weights on the two sublattices interchanged between 3 and 4 and pairs (5,6) and (7,8). The effect of shading is to divide $\mathbb{T}_1$ into $\mathbb{S}_1$ and $\mathbb{S}_2$, $\mathbb{T}_2$ in $\mathbb{S}_3$ and $\mathbb{S}_4$, and $\mathbb{T}_3$ in $\mathbb{S}_5$ and $\mathbb{S}_6$.

3.8.1.2 The Regular Square-Lattice Model

The Boltzmann weights are now site and sublattice independent and denoted by $\mathfrak{X}_p$. The assumption of sublattice independence is, as can be seen by comparing lines 3 and 4 of Fig. 3.3, the assumption that the vertex weights are unchanged under arrow reversal.[32] From this it then follows that

$$\begin{aligned} \mathfrak{X}_1 &= \mathfrak{X}_2 =: \mathfrak{a}, & \mathfrak{X}_3 &= \mathfrak{X}_4 =: \mathfrak{b}, \\ \mathfrak{X}_5 &= \mathfrak{X}_6 =: \mathfrak{c}, & \mathfrak{X}_7 &= \mathfrak{X}_8 =: \mathfrak{d}. \end{aligned} \tag{3.8.4}$$

The partition function for the model is

$$Z^{(8\text{vM})}(\mathcal{N}^{\circledast}, \mathfrak{a}, \mathfrak{b}, \mathfrak{c}, \mathfrak{d}) := \sum^{[8]} \mathfrak{a}^{n_{12}} \mathfrak{b}^{n_{34}} \mathfrak{c}^{n_{56}} \mathfrak{d}^{n_{78}}, \tag{3.8.5}$$

where n_{pq} is the numbers of p and q vertices and the sum is over all allowed arrow configurations.

The number of arrow configurations is the *entropy of the eight-vertex ice*. This is obtained by setting all the vertex energies in (3.8.5) to zero. However, this is the same as counting the number of spin configurations. Taking into account that one arrow configuration corresponds to two spin configurations, the eight-vertex ice entropy is

$$Z^{(8\text{vM})}(\mathcal{N}^{\circledast}, 1, 1, 1, 1) = 2^{N^{\circledast}-1}. \tag{3.8.6}$$

For large $N^{\circledast}$, in the absence of interactions, the reduced entropy per site s, defined by Eqs. (1.1.2) and (1.2.1) (with $Q := N^{\circledast}$), is thus given by

$$s = \frac{1}{N^{\circledast}} \ln Z^{(8\text{vM})}(N^{\circledast}, 1, 1, 1, 1) - \ln(2) - \frac{1}{N^{\circledast}} \ln[\Omega(N^{\circledast})], \tag{3.8.7}$$

where

$$\Omega(N^{\circledast}) = 2^{N^{\circledast}} = \left(\frac{8}{4}\right)^{N^{\circledast}} \tag{3.8.8}$$

is the number of possible configurations with the eight-vertex types shown in Fig. 3.2. The numerator of this expression $8^{N^{\circledast}}$ would be the value of $\Omega(N^{\circledast})$ if the eight vertex types could be freely distributed over the $N^{\circledast}$ lattice sites. The restriction that any first-neighbour pair of vertices must consist of types with compatible arrows along the connecting edge therefore has the effect of reducing the number of vertex configurations by a factor $4^{N^{\circledast}}$.

---

[32] Vertices 5–8 are polarized with directions of polarization given by the vector sum of the arrows, so an electric field could be introduced which would make different contributions to the energies of vertices. For these vertices the assumption is that the electric field is zero.

3.8.1.3 Ising Model Equivalences

Following Wu (1971) and Kadanoff and Wegner (1971), diagonal second-neighbour Ising interactions with energies $-J^{(\mathrm{f})}(\boldsymbol{r})$ from bottom-left to top-right and $-J^{(\mathrm{b})}(\boldsymbol{r})$ from bottom-right to top-left are introduced between these spins[33] with a four-spin interaction $-J^{(4)}(\boldsymbol{r})$ connecting all four spins around a site of $\mathcal{N}^{\circledast}$. Denoting couplings in the usual way and including a constant coupling $K^{(0)}$, the Boltzmann weights in terms of spins are

$$\mathfrak{Y}(\boldsymbol{r};\sigma_1,\sigma_2,\sigma_3,\sigma_4)=\exp[K^{(0)}+K^{(\mathrm{f})}(\boldsymbol{r})\sigma_1\sigma_3+K^{(\mathrm{b})}(\boldsymbol{r})\sigma_2\sigma_4 + K^{(4)}(\boldsymbol{r})\sigma_1\sigma_2\sigma_3\sigma_4] \tag{3.8.9}$$

and, in terms of the edge variables, are

$$\mathfrak{X}(\boldsymbol{r};\varsigma_1,\varsigma_2,\varsigma_3,\varsigma_4)=\exp[K^{(0)}+K^{(\mathrm{f})}(\boldsymbol{r})\varsigma_2\varsigma_3+K^{(\mathrm{b})}(\boldsymbol{r})\varsigma_3\varsigma_4+K^{(4)}(\boldsymbol{r})\varsigma_1\varsigma_3]. \tag{3.8.10}$$

This yields a site-dependent and lattice-independent version of (3.8.4) with

$$\begin{aligned}
\mathfrak{a}(\boldsymbol{r})&=\mathfrak{X}(\boldsymbol{r};\pm1,\pm1,\pm1,\pm1)=\exp[K^{(0)}+K^{(\mathrm{f})}(\boldsymbol{r})+K^{(\mathrm{b})}(\boldsymbol{r})+K^{(4)}(\boldsymbol{r})],\\
\mathfrak{b}(\boldsymbol{r})&=\mathfrak{X}(\boldsymbol{r};\pm1,\mp1,\pm1,\mp1)=\exp[K^{(0)}-K^{(\mathrm{f})}(\boldsymbol{r})-K^{(\mathrm{b})}(\boldsymbol{r})+K^{(4)}(\boldsymbol{r})],\\
\mathfrak{c}(\boldsymbol{r})&=\mathfrak{X}(\boldsymbol{r};\pm1,\mp1,\mp1,\pm1)=\exp[K^{(0)}+K^{(\mathrm{f})}(\boldsymbol{r})-K^{(\mathrm{b})}(\boldsymbol{r})-K^{(4)}(\boldsymbol{r})],\\
\mathfrak{d}(\boldsymbol{r})&=\mathfrak{X}(\boldsymbol{r};\pm1,\pm1,\mp1,\mp1)=\exp[K^{(0)}-K^{(\mathrm{f})}(\boldsymbol{r})+K^{(\mathrm{b})}(\boldsymbol{r})-K^{(4)}(\boldsymbol{r})].
\end{aligned} \tag{3.8.11}$$

For later reference we define

$$\begin{aligned}
\Delta(\boldsymbol{r}) &:= \sinh[2K^{(\mathrm{f})}(\boldsymbol{r})]\sinh[2K^{(\mathrm{b})}(\boldsymbol{r})] \\
&\qquad + \tanh[2K^{(4)}(\boldsymbol{r})]\cosh[2K^{(\mathrm{f})}(\boldsymbol{r})]\cosh[2K^{(\mathrm{b})}(\boldsymbol{r})] \\
&= \frac{\mathfrak{a}^2(\boldsymbol{r})+\mathfrak{b}^2(\boldsymbol{r})-\mathfrak{c}^2(\boldsymbol{r})-\mathfrak{d}^2(\boldsymbol{r})}{2[\mathfrak{a}(\boldsymbol{r})\mathfrak{b}(\boldsymbol{r})+\mathfrak{c}(\boldsymbol{r})\mathfrak{d}(\boldsymbol{r})]},
\end{aligned} \tag{3.8.12}$$

and

$$\Gamma(\boldsymbol{r}) := \tanh[2K^{(4)}(\boldsymbol{r})] = \frac{\mathfrak{a}(\boldsymbol{r})\mathfrak{b}(\boldsymbol{r})-\mathfrak{c}(\boldsymbol{r})\mathfrak{d}(\boldsymbol{r})}{\mathfrak{a}(\boldsymbol{r})\mathfrak{b}(\boldsymbol{r})+\mathfrak{c}(\boldsymbol{r})\mathfrak{d}(\boldsymbol{r})}. \tag{3.8.13}$$

For the regular model the partition functions of the eight-vertex and Ising models satisfy

$$Z^{(8\mathrm{vM})}(\mathcal{N}^{\circledast},\mathfrak{a},\mathfrak{b},\mathfrak{c},\mathfrak{d})=\tfrac{1}{2}\exp(N^{\circledast}K^{(0)})Z^{(\mathrm{IM})}(\mathcal{N}+\mathcal{N}^{\star},K^{(\mathrm{f})},K^{(\mathrm{b})},K^{(4)}). \tag{3.8.14}$$

---

[33] The notation refers to the directions of the interactions being respectively those of the forward and backward slash punctuation marks.

If we impose the condition $K^{(4)} = 0$ then, from (3.8.11),

$$\mathfrak{ab} = \mathfrak{cd} \tag{3.8.15}$$

and the model decouples into first-neighbour Ising models on $\mathcal{N}$ and $\mathcal{N}^{\star}$, each with unequal couplings $K^{(\mathrm{f})}$ and $K^{(\mathrm{b})}$ and (3.8.14) reduces to

$$Z^{(8\mathrm{vM})}(\mathcal{N}, \mathfrak{a}, \mathfrak{b}, \mathfrak{c}, \mathfrak{d}) = \tfrac{1}{2}\exp(N^{\circledast}K^{(0)}) \times Z^{(\mathrm{IM})}(\mathcal{N}, K^{(\mathrm{f})}, K^{(\mathrm{b})}) Z^{(\mathrm{IM})}(\mathcal{N}^{\star}, K^{(\mathrm{f})}, K^{(\mathrm{b})}). \tag{3.8.16}$$

(3.8.14)–(3.8.16) readily generalize to the case of site-dependent couplings.

If the couplings are site independent except that $K^{(\mathrm{f})}(\boldsymbol{r}) = K^{(\mathrm{b})}(\boldsymbol{r}') = K$, $K^{(\mathrm{b})}(\boldsymbol{r}) = K^{(\mathrm{f})}(\boldsymbol{r}') = K^{\star}$, for $\boldsymbol{r} \in \mathcal{N}^{\circledast}_{\mathrm{E}}$ and $\boldsymbol{r}' \in \mathcal{N}^{\circledast}_{\mathrm{O}}$, then $\mathfrak{a}(\boldsymbol{r}) = \mathfrak{a}(\boldsymbol{r}') =: \mathfrak{a}$, $\mathfrak{b}(\boldsymbol{r}) = \mathfrak{b}(\boldsymbol{r}') =: \mathfrak{b}$, $\mathfrak{c}(\boldsymbol{r}) = \mathfrak{d}(\boldsymbol{r}') =: \mathfrak{c}$, $\mathfrak{d}(\boldsymbol{r}) = \mathfrak{c}(\boldsymbol{r}') =: \mathfrak{d}$. The model is staggered and the partition functions of the eight-vertex and Ising models satisfy

$$Z^{(8\mathrm{vM})}(\mathcal{N}^{\circledast}, \mathfrak{a}, \mathfrak{b}, \mathfrak{c}, \mathfrak{d}) = \tfrac{1}{2}\exp(N^{\circledast}K^{(0)}) Z^{(\mathrm{IM})}(\mathcal{N} + \mathcal{N}^{\star}, K, K^{\star}, K^{(4)}). \tag{3.8.17}$$

If $K^{(4)} = 0$, then the model decomposes into isotropic Ising models on $\mathcal{N}$ and $\mathcal{N}^{\star}$ with first-neighbour couplings K and $K^{\star}$ respectively.

$$Z^{(8\mathrm{vM})}(\mathcal{N}^{\circledast}, \mathfrak{a}, \mathfrak{b}, \mathfrak{c}, \mathfrak{d}) = \tfrac{1}{2}\exp(N^{\circledast}K^{(0)}) Z^{(\mathrm{IM})}(\mathcal{N}, K) Z^{(\mathrm{IM})}(\mathcal{N}^{\star}, K^{\star}). \tag{3.8.18}$$

3.8.1.4 Ground States of the Regular Square-Lattice Model

The essence of our approach is to take the edge variables, or equivalently the spin variables, as the definers of the vertex weights. The arrows are then simply representations, with the arrow configurations, for the direction and shading rules related by *a reversal of arrows on all edges where the line direction gives shading on the right.* Having said that, the ground states of the system can be described either in terms of a perfect ordering of arrow directions, which can be spoken of in ferroelectric language, or a perfect ordering of spins, which can be spoken of in ferromagnetic terms. The matching between these two will of course differ according to the rule used for arrow assignment.

For the discussion of patterns of arrow assignment it is convenient to translate to a configuration graph $\mathfrak{g}$, where a graph edge is drawn for every edge carrying an arrow in the opposite direction to the line direction (that is downwards or to the left). Since this is the procedure used above for the graph constructed for the direction rule it is convenient to use, for reference purposes, the numbering of configurations in Fig. 3.2. The eight-vertex rule implies that the valency of every vertex of $\mathfrak{g}$ is even and the possibilities are:

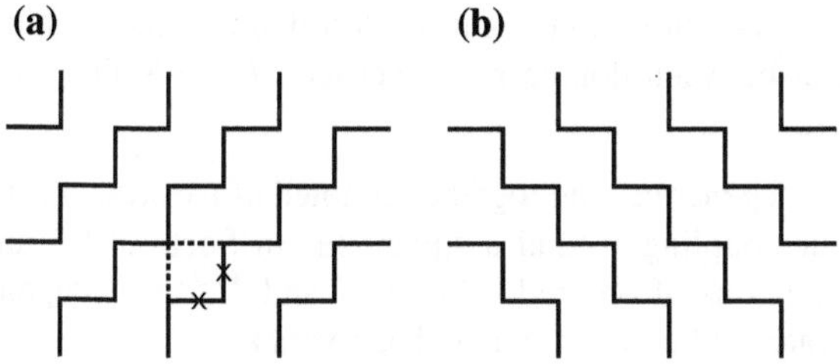

Fig. 3.4 For the square lattice the **a** BL-TR and **b** BR-TL staircase graphs. A perturbation of BL-TR is given by removing the *crossed lines* and replacing them with the *broken lines*

FEN: The null graph; all vertices are of type 1.

FEF: The full lattice graph; all vertices are of type 2.

FEV: The graph consisting of vertical lines; all vertices are of type 3.

FEH: The graph consisting of horizontal lines; all vertices are of type 4.

AFER: A bottom-left-top-right (BL-TR) graph of staircases; vertices 5 and 6 alternate on each staircase.

AFEL: A bottom-right-top-left (BR-TR) graph of staircases; vertices 7 and 8 alternate on each staircase.

FEN, FEF, FEV and FEH can be understood as ferroelectric states. Since vertex energies are unchanged under arrow reversal, they are energetically degenerate in pairs which we denote as FEN&F and FEV&H. AFER and AFEL, which are sublattice-ordered antiferroelectric states, are shown as graphs Fig. 3.4. The effect of arrow reversal is to translate the staircases by one edge.

For spin-ordering the vertex numbering is the same for both rules and the ordered states are:

FMU: Spins aligned uniformly; all vertices are of type 1.

FMS: Spins aligned in different directions on the sublattices $\mathcal{N}$ and $\mathcal{N}^\star$; all vertices are of type 2.

FMV: Spins aligned in vertical rows with antiparallel alignment between neighbouring rows; all vertices are of type 3.

FMH: Spins aligned in horizontal rows with antiparallel alignment between neighbouring rows; all vertices are of type 4.

AFMR: Spins aligned in bottom-left-top-right double rows of spins with antiparallel alignment between double rows; vertices 5 and 6 alternate along BL-TR diagonal lines.

AFML: Spins aligned in bottom-right-top-left double rows of spins with antiparallel alignment between double rows; vertices 7 and 8 alternate along BR-TL diagonal lines.

FMU and FMS are degenerate, and together denoted as FMU&S, since, in the absence of first-neighbour couplings round a square (as in Sect. 3.8.1.5 and as shown in Fig. 12.5) the directions of ordered spin on $\mathcal{N}$ and $\mathcal{N}^{\star}$ are independent. FMV and FMH are degenerate, and together denoted as FMV&H.

Here and in Sect. 9.3.2 it is convenient to expose the temperature-dependence of the Boltzmann weights by using vertex energies $\varepsilon_{\mathfrak{q}} > 0$, with

$$\mathfrak{q} =: \exp(-\varepsilon_{\mathfrak{q}}/T), \qquad \mathfrak{q} = \mathfrak{a}, \mathfrak{b}, \mathfrak{c}, \mathfrak{d}. \tag{3.8.19}$$

The phase space of $(\mathfrak{a}, \mathfrak{b}, \mathfrak{c}, \mathfrak{d})$ is divided into four parts associated with ground states, which can be identified with the different spin-ordered states and, depending on the rule of assignment used, with arrow-ordered states. The regions are:

I: $\{\varepsilon_{\mathfrak{a}} < \varepsilon_{\mathfrak{b}}, \varepsilon_{\mathfrak{c}}, \varepsilon_{\mathfrak{d}}\} \Rightarrow \{\mathfrak{a} > \mathfrak{b}, \mathfrak{c}, \mathfrak{d}\}$,

II: $\{\varepsilon_{\mathfrak{b}} < \varepsilon_{\mathfrak{c}}, \varepsilon_{\mathfrak{d}}, \varepsilon_{\mathfrak{a}}\} \Rightarrow \{\mathfrak{b} > \mathfrak{c}, \mathfrak{d}, \mathfrak{a}\}$,

III: $\{\varepsilon_{\mathfrak{c}} < \varepsilon_{\mathfrak{d}}, \varepsilon_{\mathfrak{a}}, \varepsilon_{\mathfrak{b}}\} \Rightarrow \{\mathfrak{c} > \mathfrak{d}, \mathfrak{a}, \mathfrak{b}\}$,

IV: $\{\varepsilon_{\mathfrak{d}} < \varepsilon_{\mathfrak{a}}, \varepsilon_{\mathfrak{b}}, \varepsilon_{\mathfrak{c}}\} \Rightarrow \{\mathfrak{d} > \mathfrak{a}, \mathfrak{b}, \mathfrak{c}\}$,

and the associations with ordered states are given in Table 3.2.

3.8.1.5 The Free-Fermion Model

In this section we consider a regular model with the same vertex weights at every site, but do not assume that the lattice is necessarily square. An alternative Ising-spin representation to that given above, in which the four-spin coupling is replaced by interactions with a spin σ_0 on the vertex site itself was developed by Jüngling and Obermair (1974) and Jüngling (1975).

Let the forward couplings to σ_0 below and above the site be $K^{(\mathrm{f})}_{\mathrm{B}}$ and $K^{(\mathrm{f})}_{\mathrm{A}}$ with a similar notation for the backward couplings.[34] Also included are horizontal couplings $\frac{1}{2}K^{(\mathrm{h})}_{\mathrm{B}}$ and $\frac{1}{2}K^{(\mathrm{h})}_{\mathrm{A}}$ for the pairs of sites below and above σ_0. The vertical couplings between the pairs to the left and right of σ_0 are denoted by $K^{(\mathrm{v})}_{\mathrm{L}}$ and $K^{(\mathrm{v})}_{\mathrm{R}}$. (This means that in the lattice as a whole the horizontal and vertical couplings between pairs of sites on $\mathcal{N}+\mathcal{N}^{\star}$ are $\frac{1}{2}[K^{(\mathrm{h})}_{\mathrm{B}} + K^{(\mathrm{h})}_{\mathrm{A}}]$ and $\frac{1}{2}[K^{(\mathrm{v})}_{\mathrm{L}} + K^{(\mathrm{v})}_{\mathrm{R}}]$, respectively.) Summing over the states of σ_0 the Boltzmann weight is

[34] The use of the designations 'below', 'above', 'horizontal' and 'vertical' refer to the orientation in which edge number 1 (the south-pointing edge) is vertically downwards and, as for the square lattice the east-west line is horizontal. This is just topological convenience and this description of the free-fermion model is not restricted to that the square lattice. 'B' and 'A' (signifying *below* and *above* the site of $\mathcal{N}^{\circledast}$) is a usage in conformity with the transfer matrix labelling in Sect. 12.2.

$$\mathfrak{Y}(\sigma_1,\sigma_2,\sigma_3,\sigma_4) = 2\exp\{K^{(0)} + K^{(f)}\sigma_1\sigma_3 + K^{(b)}\sigma_2\sigma_4 \\ + \tfrac{1}{2}[K^{(h)}_{B}\sigma_1\sigma_4 + K^{(v)}_{L}\sigma_1\sigma_2 + K^{(h)}_{A}\sigma_2\sigma_3 + K^{(v)}_{R}\sigma_3\sigma_4]\} \\ \times \cosh(K^{(f)}_{B}\sigma_1 + K^{(b)}_{A}\sigma_2 + K^{(f)}_{A}\sigma_3 + K^{(b)}_{B}\sigma_4). \tag{3.8.20}$$

With $\mathfrak{Y}_{\mathrm{p}}$, $\mathrm{p} = 1, 2, \ldots, 8$ given using (3.8.3),[35] the primary interest in this section is in models which satisfy

$$\mathfrak{Y}_1\mathfrak{Y}_2 + \mathfrak{Y}_3\mathfrak{Y}_4 = \mathfrak{Y}_5\mathfrak{Y}_6 + \mathfrak{Y}_7\mathfrak{Y}_8. \tag{3.8.21}$$

This is the *free-fermion condition* (Fan and Wu 1970; Jüngling 1975) The free-fermion model on a square lattice is solvable using Pfaffians (Fan and Wu, ibid) or directly by diagonalization of the transfer matrix (Felderhof 1973a), even in presence of a field (Felderhof 1973b).

In terms of the spin couplings it will be observed that the values of the horizontal and vertical coupling do not affect this, since they cancel in (3.8.21). However, this condition imposes constraints on the remaining couplings. Two cases of note are:

- If $K^{(f)} = K^{(b)} = 0$ the free-fermion condition (3.8.21) is satisfied by the vertex weights given by (3.8.3) and (3.8.20). When $\mathcal{N}^{\circledast}$ is the square lattice, this is a *checkerboard Ising model* on $\mathcal{N} + \mathcal{N}^{\star}$ (Baxter 1986, and Sect. 12.2). We have thus shown that the checkerboard Ising model is equivalent to the square-lattice free-fermion model. Allowing complex couplings, a special case of some interest (Eq. (12.2.60)) is when

$$K^{(f)}_{A} = -K^{(b)}_{B}, \qquad K^{(b)}_{A} = K^{(f)}_{B} + \tfrac{1}{2}\mathrm{i}\pi, \tag{3.8.22}$$

giving

$$\mathfrak{Y}_5 = \mathfrak{Y}_7 = 0, \qquad \mathfrak{Y}_1 = -\mathfrak{Y}_3, \qquad \mathfrak{Y}_2 = \mathfrak{Y}_4. \tag{3.8.23}$$

- When $K^{(f)}_{A} = K^{(b)}_{B} = K^{(f)}_{B} = K^{(b)}_{A} = 0$ the Boltzmann weights are of the form $\mathfrak{Y}_{\mathrm{p}} = \exp(-\varepsilon_{\mathrm{p}}/T)$ in terms of vertex energies ε_{p}, $\mathrm{p} = 1, 2, \ldots, 8$ (not necessarily satisfying (3.8.4)) and the free-fermion condition can be satisfied at all temperatures only if

$$\varepsilon_1 + \varepsilon_2 = \varepsilon_5 + \varepsilon_6 \quad \text{and} \quad \varepsilon_3 + \varepsilon_4 = \varepsilon_7 + \varepsilon_8 \\ \text{or} \\ \varepsilon_1 + \varepsilon_2 = \varepsilon_7 + \varepsilon_8 \quad \text{and} \quad \varepsilon_3 + \varepsilon_4 = \varepsilon_5 + \varepsilon_6. \tag{3.8.24}$$

  With $K^{(h)}_{B} = K^{(h)}_{A}$ and $K^{(v)}_{L} = K^{(v)}_{R}$ these conditions become respectively $K^{(b)} = 0$ and $K^{(f)} = 0$ which in each case, when $\mathcal{N}^{\circledast}$ is a square lattice, is a *triangular Ising model* with different couplings in the three lattice directions.

---

[35] $\mathfrak{Y}_{\mathrm{p}}$ rather that $\mathfrak{X}_{\mathrm{p}}$ is used here since the interest is in the spin formulation.

Although the validity of the free-fermion condition is not affected by the values of the horizontal and vertical couplings, a case of particular interest is when $K^{(\mathrm{h})} := K^{(\mathrm{h})}_{\mathrm{B}} = -K^{(\mathrm{h})}_{\mathrm{A}}$ and $K^{(\mathrm{v})} := K^{(\mathrm{v})}_{\mathrm{L}} = -K^{(\mathrm{v})}_{\mathrm{R}}$ (Sect. 12.2) since these terms cancel in the partition function of the whole lattice. Then it is not difficult to show that:

(i) if $K^{(\mathrm{h})}$ and $K^{(\mathrm{v})}$ are given by[36]

$$\tanh(2K^{(\mathrm{h})}) = \frac{\sinh(2K^{(\mathrm{f})}_{\mathrm{A}})\sinh(2K^{(\mathrm{b})}_{\mathrm{A}}) - \sinh(2K^{(\mathrm{f})}_{\mathrm{B}})\sinh(2K^{(\mathrm{b})}_{\mathrm{B}})}{\cosh(2K^{(\mathrm{f})}_{\mathrm{A}})\cosh(2K^{(\mathrm{b})}_{\mathrm{A}}) + \cosh(2K^{(\mathrm{f})}_{\mathrm{B}})\cosh(2K^{(\mathrm{b})}_{\mathrm{B}})}, \tag{3.8.25}$$

$$\tanh(2K^{(\mathrm{v})}) = \frac{\sinh(2K^{(\mathrm{f})}_{\mathrm{A}})\sinh(2K^{(\mathrm{b})}_{\mathrm{B}}) - \sinh(2K^{(\mathrm{f})}_{\mathrm{B}})\sinh(2K^{(\mathrm{b})}_{\mathrm{A}})}{\cosh(2K^{(\mathrm{f})}_{\mathrm{A}})\cosh(2K^{(\mathrm{b})}_{\mathrm{B}}) + \cosh(2K^{(\mathrm{f})}_{\mathrm{B}})\cosh(2K^{(\mathrm{b})}_{\mathrm{A}})}, \tag{3.8.26}$$

$$\mathfrak{Y}_5 = \mathfrak{Y}_6, \qquad \mathfrak{Y}_7 = \mathfrak{Y}_8. \tag{3.8.27}$$

Relationships (3.8.25) and (3.8.26) satisfy the transformations:

$$\begin{aligned}
&\text{(a)} \quad (K^{(\mathrm{f})}_{\mathrm{B}}, K^{(\mathrm{b})}_{\mathrm{B}}) \longleftrightarrow (K^{(\mathrm{f})}_{\mathrm{A}}, K^{(\mathrm{b})}_{\mathrm{A}}) \\
&\qquad\qquad \Longrightarrow \quad (K^{(\mathrm{h})}, K^{(\mathrm{v})}) \to (-K^{(\mathrm{h})}, -K^{(\mathrm{v})}), \\
&\text{(b)} \quad (K^{(\mathrm{f})}_{\mathrm{B}}, K^{(\mathrm{f})}_{\mathrm{A}}) \longleftrightarrow (K^{(\mathrm{b})}_{\mathrm{B}}, K^{(\mathrm{b})}_{\mathrm{A}}) \\
&\qquad\qquad \Longrightarrow \quad (K^{(\mathrm{h})}, K^{(\mathrm{v})}) \to (K^{(\mathrm{h})}, -K^{(\mathrm{v})}).
\end{aligned} \tag{3.8.28}$$

(ii) if $\{K^{(\mathrm{f})}_{\mathrm{A}} = K^{(\mathrm{f})}_{\mathrm{B}}$ and $K^{(\mathrm{b})}_{\mathrm{A}} = -K^{(\mathrm{b})}_{\mathrm{B}}\}$ or $\{K^{(\mathrm{f})}_{\mathrm{A}} = -K^{(\mathrm{f})}_{\mathrm{B}}$ and $K^{(\mathrm{b})}_{\mathrm{A}} = K^{(\mathrm{b})}_{\mathrm{B}}\}$

$$\mathfrak{Y}_1 = \mathfrak{Y}_2, \qquad \mathfrak{Y}_3 = \mathfrak{Y}_4. \tag{3.8.29}$$

The conditions (3.8.27) and (3.8.29) are independent, but if they are both satisfied the model is equivalent to the regular eight-vertex model with vertex weights $\mathfrak{a}$, $\mathfrak{b}$, $\mathfrak{c}$ and $\mathfrak{d}$ and free-fermion condition

$$\mathfrak{a}^2 + \mathfrak{b}^2 = \mathfrak{c}^2 + \mathfrak{d}^2. \tag{3.8.30}$$

3.8.2 The Six-Vertex Model

Here the arrow assignments are made using the direction rule; that is as shown in Fig. 3.2. The regular six-vertex model on the square lattice corresponds to the

[36] To compare with the formulation of Baxter (1986) it should be noted that his M and P correspond respectively to $K^{(\mathrm{v})}$ and $-K^{(\mathrm{h})}$.

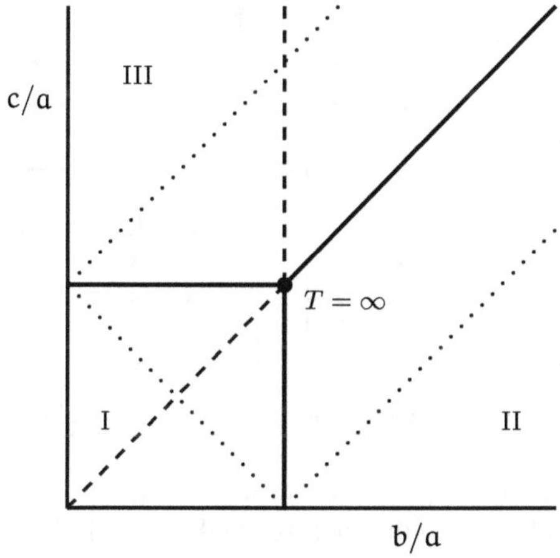

Fig. 3.5 Parameter space for the six-vertex model, showing the ferroelectric regions I and II and the antiferroelectric region III. The *broken lines* $\mathfrak{a} > \mathfrak{b} = \mathfrak{c}$ and $\mathfrak{c} > \mathfrak{a} = \mathfrak{b}$ correspond respectively to the KDP and F models. The *dotted lines* represent the transitions discussed in Sect. 9.3.2.2

limiting case $\varepsilon_{\mathfrak{d}} \to \infty$, ($\mathfrak{d} \to 0$) of the corresponding eight-vertex model. The 'ice-rule' applies and vertices 7 and 8 are eliminated. The partition function is

$$Z^{(6\text{vM})}(\mathcal{N}, \mathfrak{a}, \mathfrak{b}, \mathfrak{c}) := Z^{(8\text{vM})}(\mathcal{N}, \mathfrak{a}, \mathfrak{b}, \mathfrak{c}, 0) = \sum^{[6]} \mathfrak{a}^{n_{12}} \mathfrak{b}^{n_{34}} \mathfrak{c}^{n_{56}}, \tag{3.8.31}$$

where the sum is now over all arrow configurations. If these are pictured as configuration graphs $\mathfrak{g}$, drawn as described in the last section, top-right and bottom-left corners which make up the BR-TL staircase shown in Fig. 3.4b are excluded. As we saw above, the calculation for the entropy of 'eight-vertex ice' (which does not obey the ice-rule) was made quite simple by the mapping from arrows to spins. For the six-vertex model, where six-vertex (square) ice does obey the ice rule, the calculation is much more difficult (Lieb 1967a, b and Sect. 12.3.10.4).

The space of the variables ($\mathfrak{a}, \mathfrak{b}, \mathfrak{c}$) is divided into regions I, II and III of the set for the eight-vertex model (Sect. 3.8.1.4), These are shown in Fig. 3.5 in the plane of the variables $\mathfrak{b}/\mathfrak{a}$ and $\mathfrak{c}/\mathfrak{a}$.

The special cases of the Slater KDP model of a ferroelectric,[37] for which $\varepsilon_{\mathfrak{a}} < \varepsilon_{\mathfrak{b}} = \varepsilon_{\mathfrak{c}}$ ($\mathfrak{a} > \mathfrak{c} = \mathfrak{b}$), and the Rys F model for an antiferroelectric,[38] for which $\varepsilon_{\mathfrak{a}} = \varepsilon_{\mathfrak{b}} > \varepsilon_{\mathfrak{c}}$ ($\mathfrak{a} = \mathfrak{b} < \mathfrak{c}$) were solved by Lieb (1967d) and Lieb (1967c), respectively. The continuation of the line $\varepsilon_{\mathfrak{b}} = \varepsilon_{\mathfrak{c}}$ with $\varepsilon_{\mathfrak{b}} < \varepsilon_{\mathfrak{a}}$ ($\mathfrak{a} < \mathfrak{c}$) along the boundary between regions II and III gives the inverse KDP (IKDP) model and

[37] Slater (1941) introduced a model for a hydrogen-bonded ferroelectric called the *KDP model* since it was intended to represent KH_2PO_4 and similar substances (Nagle 1966b).

[38] The *F model* was originally proposed by Rys (1963) as an interesting statistical mechanical model. However, it is now viewed (Nagle 1966b) as a reasonable model for a hydrogen-bonded antiferroelectric.

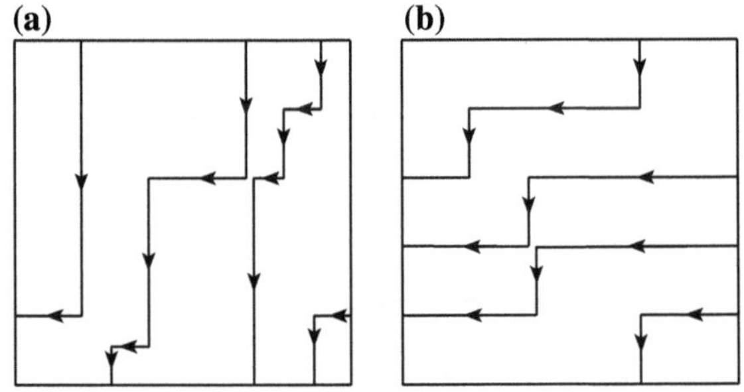

Fig. 3.6 Configuration graphs for the six-vertex model, with the lattice on a vertical cylinder

the continuation of the line $\varepsilon_{\mathfrak{a}} = \varepsilon_{\mathfrak{b}}$ with $\varepsilon_{\mathfrak{a}} < \varepsilon_{\mathfrak{b}}$ ($\mathfrak{a} < \mathfrak{c}$) along the boundary between regions I and II gives the inverse F (IF) model. The more general model with $\varepsilon_{\mathfrak{a}} = 2\varepsilon - \delta$, $\varepsilon_{\mathfrak{b}} = 2\varepsilon + \delta$, $\varepsilon_{\mathfrak{c}} = \varepsilon$ ($\varepsilon \geq 0$), which includes the ice model ($\varepsilon = \delta = 0$), the KDP model ($\varepsilon = \delta$, with the interchange $(1, 2) \leftrightarrow (3, 4)$[39]) and the F model ($\delta = 0$) was solved by Sutherland (1967).

To consider configuration graphs $\mathfrak{g}$ for the six-vertex model it is convenient first just to impose periodic boundary conditions in the horizontal direction so that the lattice forms a vertical cylinder. The only boundary sites are now those on the top and bottom rows. Also we adopt the convention that the 'cross' pattern of vertex 2 is to be regarded as a pair of 'corners' like those at vertices 5 and 6. A typical graph is shown on Fig. 3.6a. It consists of three paths which can be regarded as followed in the sense of the arrows, which is the same as that of the original arrows on the lines making up the path.[40] From the permitted vertex patterns, no sequence of lines can terminate except at a boundary site. Only steps downward or to the left are allowed so that a path leaving the top row of sites must finish at the bottom. So any path passes through a given row of vertical edges exactly once. This has an important consequence. If n_{d} denotes the number of downward arrows on a row of vertical edges in the original arrow configuration, then in the configuration graph the number of paths passing through this row is n_{d}, meaning that n_{d} is the same for all rows of vertical edges. In Fig. 3.6a, for instance, $n_{\mathrm{d}} = 3$ and Fig. 3.6b shows a configuration graph consisting of a single path spiralling round the cylinder so that $n_{\mathrm{d}} = 1$. This preservation property for n_{d} means that, with an electric field $\mathcal{E}$, an electrostatic field term $\mathcal{E}(N - 2n_{\mathrm{d}})$ can be included in the energy (Sutherland 1967; Lieb 1969). Any closed path must encircle the cylinder. Since no upward steps are allowed, no downward steps can occur either, and the path is confined to a single ring of horizontal edges, with $n_{\mathrm{d}} = 0$ and all vertices of type 4. If one ring graph exists all the sites of the lattice must be parts of ring graphs or of type 1.

The ferroelectric ordered states, are the same as those for the eight-vertex model described in Sect. 3.8.1.4. However, there is an important difference between the two

[39] Allowed by the six-vertex version of the symmetry transformation $\mathcal{T}_0$ of (9.3.2).

[40] The apparently isolated piece of path in the bottom right-hand corner is a continuation on the cylinder of the left-hand path.

models. The smallest perturbation of any of the ordered ferroelectric states, FEN, FEF, FEV or FEH, takes the form of a square of sites occupied by vertices 5, 6, 7 and 8. Since vertices 7 and 8 are excluded from the six-vertex model this has the effect of 'freezing in' the ferroelectric order. This contrasts with the Ising ferromagnetic ordered state where the complete spin alignment can be perturbed by overturning one spin. The antiferroelectric ordered state has configuration graph shown in Fig. 3.4a where perturbations like that shown by the broken lines can still occur.

3.8.2.1 The Staggered Six-Vertex Model and the Potts Model

The equivalence, on the square lattice, between a staggered six-vertex model with different vertex energies on the two equivalent sublattices and the ν–state Potts model was first demonstrated by Temperley and Lieb (1971). This relationship was generalized to any (not necessarily regular) four-coordinated planar lattice by Baxter et al. (1976). We shall adopt the graphical approach of this latter work. However, for simplicity, we restrict attention to the cases where the Potts model resides on a square, triangular or honeycomb lattice $\mathcal{N}$,[41] showing the equivalence to a six-vertex model on the medial (rapidity) lattice $\mathcal{N}^{\circledast}$, which (Sect. 16.2.2.2) is for the square lattice a square lattice and for triangular and honeycomb lattices a kagomé lattice.

The formula (3.6.52) can be generalized to include different couplings in different lattice direction. For the square lattice there are two such directions (horizontal and vertical) and for both the triangular and honeycomb lattices there are three. Thus

$$Z^{(\mathrm{PM})}(\mathcal{N},\{\upsilon^{(k)}\}) = \sum_{\{\mathfrak{g}\}}^{(\mathrm{AG})} \nu^{\mathrm{c}(\mathfrak{g})} \prod_{k=1}^{\eta} (\upsilon^{(k)})^{\mathrm{e}^{(k)}(\mathfrak{g})} = W(\nu,\{\upsilon^{(k)}\}), \tag{3.8.32}$$

where $W(\nu,\{\upsilon^{(k)}\})$ is the obvious generalization of the Whitney polynomial (Sect. 16.4.1) and $\eta = 2$ for the square lattice and $\eta = 3$ for the triangular and honeycomb lattices.

Consider one of the graphs $\mathfrak{g}$ on $\mathcal{N}$, augmented so all disconnected points of $\mathcal{N}$ are counted as single-site components. From (16.1.1), the cyclomatic number of $\mathfrak{g}$ is

$$\mathfrak{C}(\mathfrak{g}) = \mathrm{c}(\mathfrak{g}) + \sum_{k=1}^{\eta} \mathrm{e}^{(k)}(\mathfrak{g}) - N. \tag{3.8.33}$$

A site of the medial lattice $\mathcal{N}^{\circledast}$ lies either on an edge of $\mathcal{N}$ which is also an edge of $\mathfrak{g}$ or on edge of $\mathcal{N}$ which is not an edge of $\mathfrak{g}$. $\mathcal{N}^{\circledast}$ is transformed into a set of polygons by cutting the edges at the vertices and reconnecting them to form polygons, so that each connected component of $\mathfrak{g}$ is bounded on the outside by a polygon and each component consisting of a cycle is also bounded on the inside by a polygon. Each

[41] No periodic boundary conditions are assume.

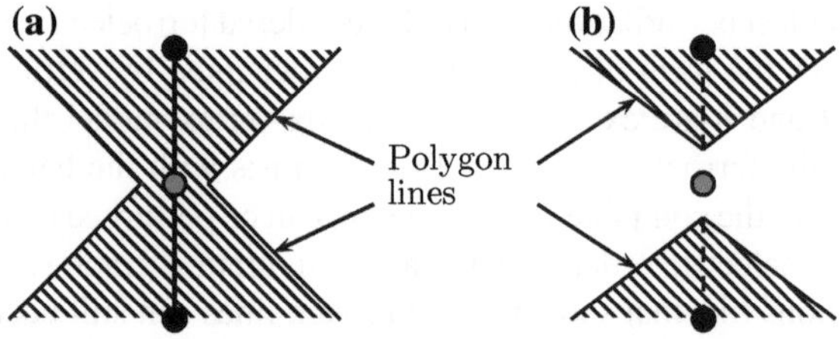

Fig. 3.7 Polygon construction on $\mathcal{N}$ (*black sites*) from the medial lattice $\mathcal{N}^{\circledast}$ (*grey sites*), **a** corresponds to the cases where a (*heavy black*) line of $\mathfrak{g}$ passes through the site of $\mathcal{N}^{\circledast}$, making it an interior site and **b** where it does not, making the $\mathcal{N}^{\circledast}$ site an exterior site

region bounded by a polygon and containing one component of $\mathfrak{g}$ is shaded, as is shown in Fig. 3.7 for the square lattice, and the number of polygons is

$$\mathfrak{N}(\mathfrak{g}) = \mathrm{c}(\mathfrak{g}) + \mathfrak{C}(\mathfrak{g}). \tag{3.8.34}$$

It is clear that any augmented graph $\mathfrak{g} \subset \mathcal{N}$ is in one-to-one correspondence with a polygon decomposition ϖ of $\mathcal{N}^{\circledast}$, for which sites can be classified as *interior* if they lie on an edge of $\mathfrak{g}$ (and hence in the shaded area of the plane) or *exterior* if they do not lie on an edge of $\mathfrak{g}$. Both $\mathfrak{e}^{(k)}(\mathfrak{g})$ and $\mathfrak{N}(\mathfrak{g})$ can be unambiguously relabelled as functions of ϖ and the summation in (3.8.32) taken as that over a polygon decomposition ϖ. Then, from (3.8.33) and (3.8.34),

$$Z^{(\mathrm{PM})}(\mathcal{N}, \{u^{(k)}\}) = \nu^{\frac{1}{2}N} \sum_{\{\varpi\}} \nu^{\frac{1}{2}\mathfrak{N}(\varpi)} \prod_{k=1}^{\eta} (u^{(k)})^{\mathfrak{e}^{(k)}(\varpi)}, \tag{3.8.35}$$

where

$$u^{(k)} := v^{(k)}/\sqrt{\nu} = [\exp(K^{(k)}) - 1]/\sqrt{\nu}. \tag{3.8.36}$$

To re-express this sum in the form of a staggered six-vertex model on $\mathcal{N}^{\circledast}$ with η different sets of vertex energies (depending on the orientations of the corresponding edges of $\mathcal{N}$), the summation must be over terms local to the sites of $\mathcal{N}^{\circledast}$. This is already true for the product term and the problem is to 'localize' $\nu^{\frac{1}{2}\mathfrak{N}(\varpi)}$.

We first choose a direction round each of the polygons. One possibility for the case of the square lattice is shown by the arrows in Fig. 3.8. At each vertex of $\mathcal{N}^{\circledast}$ the directions along the polygons gives a configuration of arrows and these, together with the shading, can be used (when the polygons are deformed back onto the edges) to identify the groups of vertex types $\mathbb{S}_1, \ldots, \mathbb{S}_6$ listed in Sect. 3.8.1.1. However, there are two differences that must be noted. The first is that, because of the polygon construction, arrows at a vertex follow each other round the shaded or unshaded regions (or both). This means that the six-vertex ice-rule is satisfied and $\mathbb{T}_3 = \mathbb{S}_5 \cup \mathbb{S}_6$ is excluded. The second caveat is that it is necessary to distinguish between the cases

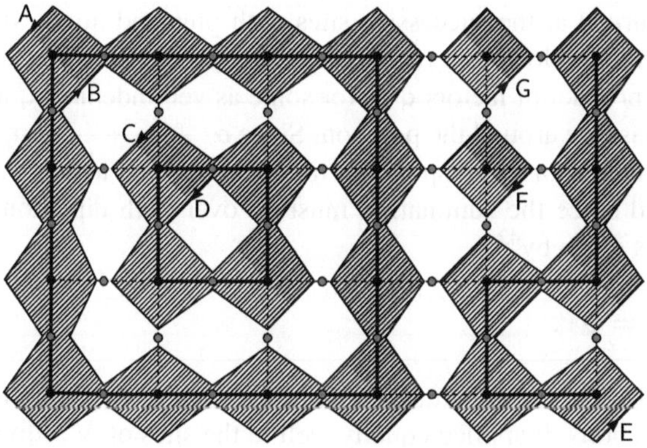

Fig. 3.8 Polygons formed on the square lattice

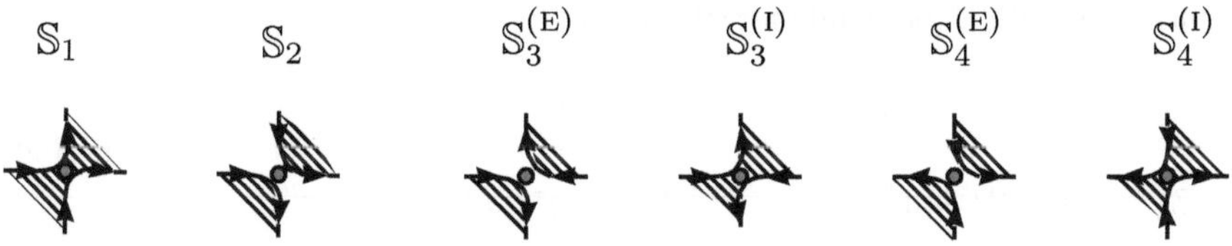

Fig. 3.9 The six vertex types for an interior and an exterior site of $\mathcal{N}^{\circledast}$

of shaded regions close to an interior site and to an exterior site. Since the arrows must follow each other around the shaded region, vertices of type $\mathbb{S}_1$ can occur only at an interior site, and vertices of type $\mathbb{S}_2$ can occur only at an exterior site; $\mathbb{S}_3$ and $\mathbb{S}_4$ type vertices can occur in both ways, giving the six possibilities shown in Fig. 3.9, where exterior and interior cases are distinguish by an appropriate superscript. As an example, the polygons in Fig. 3.9, labelled A–F, are assigned arrow directions. Clearly the interior sites within polygon E must be $\mathbb{S}_3^{(I)}$ or $\mathbb{S}_4^{(I)}$ irrespective of the arrow direction and in a similar way the exterior site between parts of E must be $\mathbb{S}_3^{(E)}$ or $\mathbb{S}_4^{(E)}$. All other sites have the possibility of being one of two types. With the arrow directions shown, the interior C/D sites are $\mathbb{S}_1$ and the interior A/B sites are $\mathbb{S}_3^{(I)}$ or $\mathbb{S}_4^{(I)}$; the exterior B/C, A/E, A/G and E/G sites are $\mathbb{S}_3^{(E)}$ or $\mathbb{S}_4^{(E)}$, and the exterior A/F. E/F and G/F sites are $\mathbb{S}_2$.

Consider a particular single polygon with a particular direction of traverse (arrow direction). As an exterior site of $\mathcal{N}^{\circledast}$ is passed the polygon makes a clockwise or anticlockwise turn through an angle $\pi - \theta$ and as it passes an interior site it makes a clockwise or anticlockwise turn through θ, where θ is just the angle subtended at a site by the shade region; that is $\theta = \frac{1}{2}\pi$ when $\mathcal{N}$ is the square lattice, $\theta = \frac{1}{3}\pi$ when $\mathcal{N}$ is the honeycomb lattice and $\theta = \frac{2}{3}\pi$ when $\mathcal{N}$ is the triangular lattice. Given that, in the passage around the polygon, m sites are passed, let α_j, $j = 1, 2, \ldots, m$

be the angles turned at the successive sites with plus and minus values attributed to clockwise and anticlockwise turns, respectively. Then replace the factor $\sqrt{\nu}$ in (3.8.35) by the product of factors $\mathfrak{q}^{\alpha_j}$, for some as yet undefined $\mathfrak{q}$ and all the sites visited by the passage around the polygon. Since $\alpha_1 + \alpha_2 + \cdots + \alpha_m = \pm 2\pi$, with the plus and minus signs corresponding to clockwise and anticlockwise passages, respectively and since the summation must be over both directions of passage it follows that $\mathfrak{q}$ is given by[42]

$$\mathfrak{q}^{2\pi} + \mathfrak{q}^{-2\pi} = \sqrt{\nu}. \tag{3.8.37}$$

The index $k = 1, 2, \ldots, \eta$ was attached to the edges of $\mathcal{N}$ to distinguish orientations and Potts interactions. It applied equally well to the sites of $\mathcal{N}^{\circledast}$ dividing them into sublattices $\mathcal{N}_k^{\circledast}$. The Boltzmann weights for the vertices of $\mathcal{N}_k^{\circledast}$ are denoted by $\mathfrak{Z}_j^{(k)}$, $j = 1, 2, 3, 4$, $k = 1, 2, \ldots, \eta$, for the four types of vertices shown in Fig. 3.9. These arise from contributions from the two polygons at the site, together with, the edge contribution $\{u^{(k)}\}$, for an interior site. Thus

$$\begin{aligned} \mathfrak{Z}_1^{(k)} &= u^{(k)}, & \mathfrak{Z}_2^{(k)} &= 1, \\ \mathfrak{Z}_3^{(k)} &= \mathfrak{q}^{2\theta} u^{(k)} + \mathfrak{q}^{2\theta - 2\pi}, & \mathfrak{Z}_4^{(k)} &= \mathfrak{q}^{-2\theta} u^{(k)} + \mathfrak{q}^{2\pi - 2\theta}, \end{aligned} \tag{3.8.38}$$

giving

$$Z^{(\mathrm{PM})}(\mathcal{N}, \{u^{(k)}\}) = \nu^{\frac{1}{2}N} Z^{(6\mathrm{vM})}(\mathcal{N}^{\circledast}, \{\mathfrak{Z}_j^{(k)}\}), \tag{3.8.39}$$

where

$$Z^{(6\mathrm{vM})}(\mathcal{N}^{\circledast}, \{\mathfrak{Z}_{\mathrm{p}}^{(k)}\}) := \sum^{[6]} \prod_{j=1}^{4} \prod_{k=1}^{\eta} (\mathfrak{Z}_j^{(k)})^{n_j^{(k)}}, \tag{3.8.40}$$

with the summation over all arrow configurations satisfy the six-vertex ice rule. Since $|\mathcal{N}^{\circledast}| = 2N$

$$\phi^{(\mathrm{PM})}(\mathcal{N}, \{u^{(k)}\}) = -\tfrac{1}{2}\ln(\nu) + 2\phi^{(6\mathrm{vM})}(\mathcal{N}^{\circledast}, \{\mathfrak{Z}_j^{(k)}\}). \tag{3.8.41}$$

We pause to note that this result is, with reference to the first six arrow configurations of Fig. 3.2, for a six-vertex model with the Boltzmann weights of 1 and 2 (equivalent to $\mathbb{S}_1$), and 3 and 4 (equivalent to $\mathbb{S}_2$) set equal but not 5 and 6 (which here are equivalent to $\mathbb{S}_3$ and $\mathbb{S}_4$, respectively) and also with different weights on different sublattices.

[42] As pointed out by Baxter et al. (1976) this formula can be made to apply to the case where the lattice lies on a vertical cylinder, by choosing a vertical line on the cylinder and adding/subtracting a factor of 2π to the sum $\alpha_1 + \alpha_2 + \cdots$ whenever the polygon crosses the line from the left/right.

It is clear that the argument leading to (3.8.39) could equally well be carried out for the dual lattice $\mathcal{N}^\star$. The angle θ is replaced by $\theta^\star = \pi - \theta$ and the parameters $\{u^{(k)}\}$ by $\{u^{\star(k)}\}$. This gives

$$Z^{(\mathrm{PM})}(\mathcal{N}^\star, \{u^{\star(k)}\}) = \nu^{\frac{1}{2}N^\star} Z^{(6\mathrm{vM})}(\mathcal{N}^{\circledast}, \{\mathfrak{z}_j^{\star(k)}\}), \tag{3.8.42}$$

where

$$\begin{aligned} \mathfrak{z}_1^{\star(k)} &= u^{\star(k)}, & \mathfrak{z}_2^{\star(k)} &= 1, \\ \mathfrak{z}_3^{\star(k)} &= \mathfrak{q}^{2\theta^\star} u^{\star(k)} + \mathfrak{q}^{2\theta^\star - 2\pi}, & \mathfrak{z}_4^{\star(k)} &= \mathfrak{q}^{-2\theta^\star} u^{\star(k)} + \mathfrak{q}^{2\pi - 2\theta^\star}. \end{aligned} \tag{3.8.43}$$

The aim, of course, is to have the same six-vertex partition function appearing in (3.8.39) and (3.8.42). This is achieved by taking $u^{\star(k)} = 1/u^{(k)}$, or equivalently from (3.8.36),

$$[\exp(K^{(k)}) - 1][\exp(K^{\star(k)}) - 1] = \nu \tag{3.8.44}$$

and interchanging $\mathbb{S}_1$ with $\mathbb{S}_2$. Extracting a factor of $u^{\star(k)}$ from each vertex weight $\mathfrak{z}_j^{(k)}$ in the six-vertex partition function in (3.8.42) and reverting to the use of couplings in the Potts model yields

$$Z^{(\mathrm{PM})}(\mathcal{N}^\star, \{K^{\star(k)}\}) = \left\{ \prod_{k=1}^{\eta} \frac{[\exp(K^{\star(k)}) - 1]^{N/\eta}}{[\exp(K^{(k)}) - 1]^{N^\star/\eta}} \right\} Z^{(\mathrm{PM})}(\mathcal{N}, \{K^{(k)}\}). \tag{3.8.45}$$

Except in two special cases the staggered square-lattice ($\eta = 2, \theta = \frac{1}{2}\pi$) six-vertex model with weights (3.8.38) is unsolved. These two cases correspond to

$$\xi^{(1)}\xi^{(2)} = \pm 1, \qquad \text{where} \qquad \xi^{(k)} := \frac{u^{(k)}\mathfrak{q}^{2\pi} + 1}{u^{(k)} + \mathfrak{q}^{2\pi}} \tag{3.8.46}$$

and we consider them separately.

- The case $\xi^{(1)}\xi^{(2)} = 1$ was solved by Baxter (1973d). It corresponds to $u^{(1)}u^{(2)} = 1$ which is equivalent to

$$[\exp(K^{(1)}) - 1][\exp(K^{(2)}) - 1] = \nu \tag{3.8.47}$$

  and the vertex weights satisfy the conditions

$$\frac{\mathfrak{z}_1^{(1)}}{\mathfrak{z}_2^{(2)}} = \frac{\mathfrak{z}_2^{(1)}}{\mathfrak{z}_1^{(2)}} = \frac{\mathfrak{z}_3^{(1)}}{\mathfrak{z}_4^{(2)}} = \frac{\mathfrak{z}_4^{(1)}}{\mathfrak{z}_3^{(2)}} = u^{(1)}. \tag{3.8.48}$$

This represents a uniform proportionality between vertex weights which, when $u^{(1)}$ is positive, makes the model effectively equivalent to the regular (non-staggered) six-vertex model. Comparing (3.8.47) and (3.8.44) shows that the Potts model is self-dual. When $K^{(1)} = K^{(2)} = R/T > 0$, (3.8.47) gives the critical temperature

$$T_c = R/\ln(\sqrt{\nu}+1), \tag{3.8.49}$$

of the ferromagnetic Potts model (Sect. 9.2.1) at which, for the corresponding six-vertex model, from (3.8.37) and (3.8.38),

$$\mathrm{a} = \mathrm{b} = 1, \qquad \mathrm{c} = \mathrm{q}^{\pi} + \mathrm{q}^{-\pi} = (2+\sqrt{\nu})^{\frac{1}{2}}. \tag{3.8.50}$$

The staggered property is now absent and it is a case of the F model.

- The case $\xi^{(1)}\xi^{(2)} = -1$ was solved by Baxter (1982a). This constraint reduces to

$$(u^{(1)}\sqrt{\nu}+2)(u^{(2)}\sqrt{\nu}+2) = 4-\nu, \tag{3.8.51}$$

which fixes the temperature of the anisotropic Potts model with the condition

$$4-\nu = [\exp(K^{(1)})+1][\exp(K^{(2)})+1]. \tag{3.8.52}$$

Although it is of interest to consider non-physical situations (Baxter 1982a), (3.8.52) has a physical solution only when $\nu \leq 3$ and then only when at least one of $K^{(1)}$ and $K^{(2)}$ (that is of $R^{(1)}$ and $R^{(2)}$) is negative. When $R^{(1)} = R^{(2)} = R < 0$ (3.8.52) gives the critical temperature

$$T_c = R/\ln(\sqrt{4-\nu}-1), \qquad R < 0 \tag{3.8.53}$$

of the antiferromagnetic square-lattice Potts model. For $\nu = 3$ the critical temperature of the antiferromagnetic 3-state Potts model is $T_c = 0$. Unlike the ferromagnetic case the staggered nature of the underlying six-vertex model does not disappear at the critical temperature and evaluation of the free energy in this case is rather more complicated (Baxter 1982a).

3.9 Dimer Models

Most of the work in this book is on models where each microsystem occupies one site and where two microsystems can occupy any pair of sites. In such systems there can be no provision for size and shape effects although, as in the water models of Bell and Lavis (1970) and Bell (1972) (Sect. 7.6) directional bonding can be included. In the lattice fluid picture (Sect. 3.5.1.2), one possible modification is to introduce an infinite repulsion between microsystems extending as far as a certain range of neighbours;

such models are called *exclusion models*. The hard-square and hard-hexagon models (Sect. 8.4.3) are of this type.

A mathematically similar approach is to suppose that the microsystem itself is a connected subgraph of the lattice $\mathcal{N}$ (Sect. 16.1.1). An *arrangement* of microsystems on the lattice then consists of a subgraph each of whose components is a microsystem. A *covering* is an arrangement which is a *spanning graph* (Sect. 16.4), that is one in which every lattice site is a vertex of a microsystem. In this context a microsystem with a single vertex and no lines is called a *monomer* and a microsystem consisting of a single line and two vertices of valency one is a *dimer*. In this section and in Chap. 13 we are concerned with *pure dimer assemblies*, for which each arrangement or *configuration* is a covering of the lattice consisting entirely of dimers. For a lattice $\mathcal{N}$, it is clear that a dimer covering is possible only when $N := |\mathcal{N}|$ is even, a situation henceforth assumed, and then the number of dimers in the covering is $\frac{1}{2}N$. Figure 3.10 shows a pure dimer configuration on the square lattice with configurations on the honeycomb lattice shown in Figs. 3.11 and 3.12a.

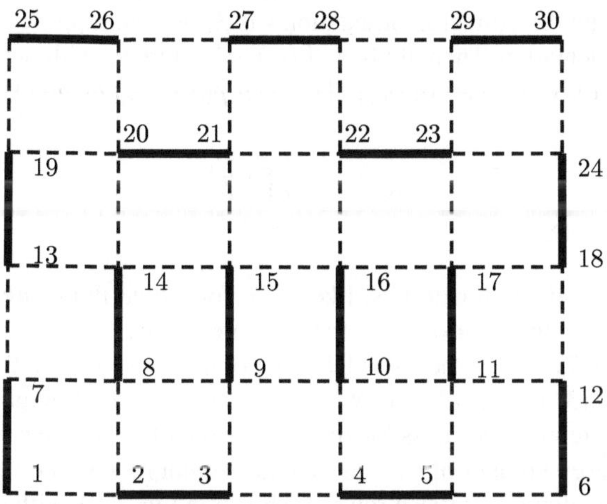

Fig. 3.10 A pure dimer configuration on the square lattice with a numbering of the lattice sites

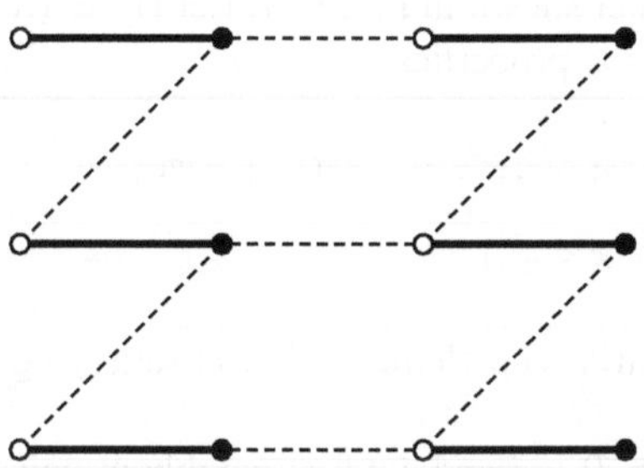

Fig. 3.11 A deformation of the honeycomb lattice with a pure dimer configuration

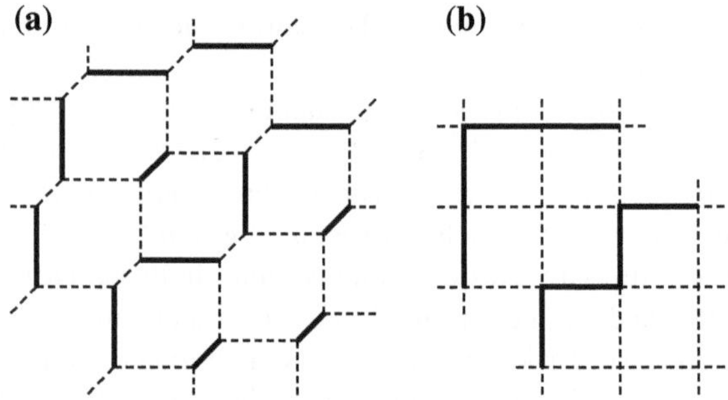

Fig. 3.12 **a** A pure dimer configuration on the honeycomb lattice. **b** The square lattice achieved by contracting the diagonal lines of (**a**) to zero

For the regular plane lattices the edges can be divided into η types according to spatial orientation. For the square lattice $\eta = 2$, with horizontal and vertical edge orientations, for the honeycomb, triangular and kagomé lattices $\eta = 3$. The Boltzmann weight for a dimer in orientation k is $\mathfrak{Z}_k$ and we suppose that there are n_k dimers in this orientation. Then, if $\Omega(\mathcal{N};\{n_k\})$ is the number of dimer configurations on the lattice for given values of $\{n_k\}$, the *dimer partition function* is

$$Z(\mathcal{N},\{\mathfrak{Z}_k\}) = \sum_{\{\sum_{k=1}^{\eta} n_k = \frac{1}{2}N\}} \Omega(\mathcal{N};\{n_k\}) \prod_{k=1}^{\eta} \mathfrak{Z}_k^{n_k}. \tag{3.9.1}$$

The Boltzmann weights for dimers, like those for lattice fluids are usually called *fugacities*; they can have various physical interpretations.

Let $\Pi_0 := (1, 2, \ldots, N)$ be some labelling of the sites of $\mathcal{N}$ so that the partition into pairs $(1, 2), (3, 4), \ldots, (N-1, N)$, with (i, j) being a first-neighbour pair occupied by a dimer, represents a possible dimer covering of $\mathcal{N}$. In the case of the square lattice with an even number of sites in each horizontal row, Π_0 could be a labelling from left to right along successive rows, as shown in Fig. 3.10, with the corresponding covering having all the dimers horizontal. A similar horizontal arrangement of dimers can be effected on the triangular lattice and on the honeycomb lattice if it is deformed in the manner shown in Fig. 3.11. Let $\Pi := (\pi_1, \pi_2, \ldots, \pi_N)$ be some permutation of Π_0 with the properties

$$\begin{gathered} \pi_1 < \pi_2, \qquad \pi_3 < \pi_4, \ldots\ldots, \qquad \pi_{N-1} < \pi_N, \\ \pi_1 < \pi_3 < \pi_5 < \cdots < \pi_{N-1} \end{gathered} \tag{3.9.2}$$

and $\mathbf{\Lambda}$ be an $N \times N$ matrix with elements $\lambda(i, j)$ satisfying the conditions

$$\lambda(i, j) := \begin{cases} \lambda(j, i) > 0, & i \text{ and } j \text{ a first-neighbour pair}, \\ 0, & i \text{ and } j \text{ not a first-neighbour pair}. \end{cases} \tag{3.9.3}$$

Then $(\pi_1, \pi_2), (\pi_3, \pi_4), \ldots, (\pi_{N-1}, \pi_N)$ denotes a dimer covering of $\mathcal{N}$ if and only if

$$\lambda(\pi_1, \pi_2)\lambda(\pi_3, \pi_4) \cdots \lambda(\pi_{N-1}, \pi_N) \neq 0. \tag{3.9.4}$$

As an example, in the case of the square lattice with the dimer configuration shown in Fig. 3.10, it can be seen that the corresponding partition

$$\Pi = \{1, 7, 2, 3, 4, 5, 6, 12, 8, 14, 9, 15, 10, 16, 11, 17, 13, 19, \\ 18, 24, 20, 21, 22, 23, 25, 26, 27, 28, 29, 30\} \tag{3.9.5}$$

satisfies (3.9.2). For the matrix $\mathbf{\Lambda}$, let

$$Z(\mathcal{N}; \mathbf{\Lambda}) := \sum_{\{\Pi\}} \lambda(\pi_1, \pi_2)\lambda(\pi_3, \pi_4) \cdots \lambda(\pi_{N-1}, \pi_N). \tag{3.9.6}$$

In view of (3.9.4) a term in the summation in (3.9.6) not corresponding to a dimer covering is zero and the summation can therefore be restricted to dimer coverings. Then (3.9.6) is a expression for the dimer partition function (3.9.1) when $\lambda(i, j)$ is equal to the same fugacity value for all parallel dimers in any particular principal lattice direction.

3.9.1 The Modified KDP Model Equivalence

Figure 3.12a shows a pure dimer distribution on the honeycomb lattice. Suppose that the lattice is contracted by reducing the lengths of the diagonal edges to zero and letting the two vertices at the ends of each such edge coalesce into one. The resulting square lattice is shown in Fig. 3.12b and it can be seen that the possible vertex line patterns deriving from the horizontal and vertical dimers in Fig. 3.12a are those in Fig. 3.2, labelled 1, 3, 4, 5 and 6. The vertex line patterns, 2, 7 and 8 are absent because they would correspond to overlapping dimers on the original honeycomb lattice. With suitable relations between dimer fugacities and vertex energies the honeycomb lattice dimer model is thus equivalent to a square-lattice *five-vertex model.* This model can be regarded as a special case of the six-vertex model, with vertex 2 absent and different energies for vertices 3 and 4. Of course, this in turn is a special case of the eight-vertex model. So, with $\mathfrak{X}_{\mathrm{p}} := \exp(-\varepsilon_{\mathrm{p}}/T)$, $\mathrm{p} = 1, 2, \ldots, 8$,[43]

[43] The properties of this model are identical to those of the model where ε_2 is finite and $\varepsilon_1 = \infty$. However, for that situation the derivation of the equivalence between the dimer model and the five-vertex model is more complicated and involves an intermediate lattice. This is a consequence of the convention for vertex line configurations used in vertex models.

$$\begin{aligned}&\varepsilon_1 = \varepsilon_{\mathfrak{a}}, \quad \varepsilon_3 = \varepsilon_{\mathfrak{b}}, \quad \varepsilon_4 = \varepsilon_{\mathfrak{b}'}, \quad \varepsilon_5 = \varepsilon_6 = \varepsilon_{\mathfrak{c}}, \quad \varepsilon_2 = \varepsilon_7 = \varepsilon_8 = \infty,\\ &\mathfrak{X}_1 = \mathfrak{a}, \quad \mathfrak{X}_3 = \mathfrak{b}, \quad \mathfrak{X}_4 = \mathfrak{b}', \quad \mathfrak{X}_5 = \mathfrak{X}_6 = \mathfrak{c}, \quad \mathfrak{X}_2 = \mathfrak{X}_7 = \mathfrak{X}_8 = 0.\end{aligned} \tag{3.9.7}$$

In the honeycomb dimer model shown in Fig. 3.12a, let the fugacities $\mathfrak{z}_1$, $\mathfrak{z}_2$ and $\mathfrak{z}_3$ be associated respectively with diagonal, vertical and horizontal edges. Since the contraction of a dimer-occupied diagonal edge results in a type 1 vertex on the square lattice we must set

$$\mathfrak{X}_{\mathfrak{a}} = \mathfrak{z}_1. \tag{3.9.8}$$

For each vertical dimer the fugacity $\mathfrak{z}_2$ can be decomposed into two factors $\sqrt{\mathfrak{z}_2}$ and one of these assigned to each of the two sites occupied by the dimer segments. Treating the fugacity $\mathfrak{z}_3$ in a similar way and comparing Figs. 3.12a, b and 3.2, we have

$$\mathfrak{X}_{\mathfrak{b}} = \mathfrak{z}_2, \qquad \mathfrak{X}_{\mathfrak{b}'} = \mathfrak{z}_3, \qquad \mathfrak{X}_{\mathfrak{c}} = \sqrt{\mathfrak{z}_2\mathfrak{z}_3}, \tag{3.9.9}$$

from which it follows that a condition for equivalence between the dimer and five-vertex models is that

$$\varepsilon_{\mathfrak{b}} + \varepsilon_{\mathfrak{b}'} = 2\varepsilon_{\mathfrak{c}}. \tag{3.9.10}$$

As was seen in Sect. 3.8.2, the ferroelectric six-vertex model is one where the ground-state vertices are of type 1, 2, 3 or 4 and the antiferroelectric model is one where the ground-state vertices are of type 5 or 6. For the latter to be the case it must be that $\varepsilon_{\mathfrak{c}} < \varepsilon_{\mathfrak{b}}$ and $\varepsilon_{\mathfrak{c}} < \varepsilon_{\mathfrak{b}'}$, a situation which is excluded by (3.9.10). When (3.9.10) holds the partition function of the five-vertex model on the square lattice $\mathcal{N}_{\mathrm{S}}$ is related to that of a dimer model on the honeycomb lattice $\mathcal{N}_{\mathrm{H}}$ by

$$Z^{(5\mathrm{vM})}(\mathcal{N}_{\mathrm{S}}, \mathfrak{X}_{\mathfrak{a}}, \mathfrak{X}_{\mathfrak{b}}, \mathfrak{X}_{\mathfrak{b}'}, \mathfrak{X}_{\mathfrak{c}}) = Z^{(\mathrm{DM})}(\mathcal{N}_{\mathrm{H}}, \mathfrak{z}_1, \mathfrak{z}_2, \mathfrak{z}_3), \tag{3.9.11}$$

where $|\mathcal{N}_{\mathrm{S}}| = \frac{1}{2}N$, $|\mathcal{N}_{\mathrm{H}}| = N$.

The case of the six-vertex model for which the only non-zero vertex energies are those of type 1 and 2, which are equal and negative, is the *KDP model* (Sect. 3.8.2). The case of the present model, where

$$\varepsilon_{\mathfrak{a}} < 0, \qquad \varepsilon_{\mathfrak{b}} = \varepsilon_{\mathfrak{b}'} = \varepsilon_{\mathfrak{c}} = 0, \tag{3.9.12}$$

differs from the KDP model in that the energy of a type 2 vertex is infinity rather than $\varepsilon_{\mathfrak{a}}$. It is called the *modified KDP model.*

3.9.2 The Ising Model Equivalence

Using the relationship

$$\exp[K\sigma(\boldsymbol{r})\sigma(\boldsymbol{r}')] = \cosh(K)(1 + v\sigma(\boldsymbol{r})\sigma(\boldsymbol{r}')), \tag{3.9.13}$$

where[44]

$$v := \tanh(K), \tag{3.9.14}$$

the partition function (3.5.6) with zero field becomes

$$\begin{aligned} Z^{(\mathrm{IM})}(\mathcal{N}, v) &= [\cosh(K)]^{\frac{1}{2}zN} \sum_{\{\sigma\}} [1 + \sigma(\boldsymbol{r})\sigma(\boldsymbol{r}')\, v] \\ &= [\cosh(K)]^{\frac{1}{2}zN} \sum_{\{\sigma\}} \sum_{\{\mathfrak{g}\}} v^{\mathrm{e}(\mathfrak{g})} \prod^{\mathfrak{g}} \sigma(\boldsymbol{r})\sigma(\boldsymbol{r}'), \end{aligned} \tag{3.9.15}$$

where the inner sum is over all graphs $\mathfrak{g}$ on $\mathcal{N}$ (Sect. 16.1) and the product is over all spin pairs on the edges of $\mathfrak{g}$. It is clear that on summation over $\{\sigma(\boldsymbol{r})\}$ every term containing any spin variable to an odd power will cancel, giving

$$Z^{(\mathrm{IM})}(\mathcal{N}, v) = 2^N [\cosh(K)]^{\frac{1}{2}zN} \sum_{\{\mathfrak{g}\}}^{(\mathrm{ZF})} v^{\mathrm{e}(\mathfrak{g})}, \tag{3.9.16}$$

where the graphs are restricted to those which have only even vertices. For reasons explained in Sect. 14.2.2 these are called *zero-field graphs*. Given any such graph $\mathfrak{g}$ *placed in a particular way on the lattice*, the set of $\frac{1}{2}zN - \mathrm{e}(\mathfrak{g})$ edges of the lattice which do not belong to $\mathfrak{g}$ is called the *complement* of $\mathfrak{g}$ and denoted by $\mathfrak{g}'$. With a sum over these graphs (3.9.16) can be re-expressed in the form

$$Z^{(\mathrm{IM})}(\mathcal{N}, v) = 2^N [\sinh(K)]^{\frac{1}{2}zN} \sum_{\{\mathfrak{g}'\}}^{(\mathrm{CZF})} (v^{-1})^{\mathrm{e}(\mathfrak{g}')}. \tag{3.9.17}$$

This result can be generalized to an anisotropic Ising model. Suppose the lattice is divided into η classes of edges with s_k edges of class k, where $s_1 + s_2 + \cdots + s_\eta = \frac{1}{2}zN$. Let the first-neighbour coupling on the edges of class k be K_k with $v_k := \tanh(K_k)$. Then in place of (3.9.17) we have

---

[44] This variable features in the duality relationship (9.2.22) ($\mathfrak{X}^\star = 1/v$, with $\nu = 2$) and as the appropriate variable for the high-temperature series expansion in Sect. 14.2.2 (again with $\nu = 2$).

$$Z^{(\mathrm{IM})}(\mathcal{N},\{v_k\}) = 2^N\left\{\prod_{k=1}^{\eta}[\sinh(K_k)]^{\mathrm{s}_k}\right\}\sum_{\{\mathfrak{g}'\}}^{(\mathrm{CZF})}\left[\prod_{k=1}^{\eta}(v_k^{-1})^{\mathrm{e}_k(\mathfrak{g}')}\right], \tag{3.9.18}$$

where $\mathrm{e}_k(\mathfrak{g}')$ is the number of edges of class k in the graph $\mathfrak{g}'$. The equivalence between pure dimer and Ising models is established by finding a relationship between complementary graphs and dimer graphs, formed, as in Figs. 3.10 and 3.12a, by placing a line on every edge occupied by a dimer. However, for dimer graphs, exactly one line is incident at each vertex, while for zero-field graphs, or their complements, several lines can be incident. The way to deal with this is indicated by the procedure of Sect. 3.9.1, where the five-vertex graph of Fig. 3.12b is obtained from the dimer graph of Fig. 3.12a by deleting edges and coalescing sites. The reverse process would be the expansion of the square lattice of Fig. 3.12b by bifurcating each site into two sites connected by a new edge. Fisher (1966) gave a general prescription for splitting sites and inserting edges so as to yield a one-to-one relationship between complementary graphs on the original lattice and dimer graphs on the new lattice. The simplest application of Fisher's method is to the honeycomb lattice Ising model and an outline of this procedure will now be given.

Since all vertices of zero-field graphs have even valency, those on the honeycomb lattice must have either no lines or two lines incident at each vertex. A complementary graph must, therefore, have either one or three lines incident at each vertex. Suppose that each honeycomb site is split into three sites connected by a triangle of new edges, each • site in Fig. 3.11 being replaced by a downward pointing triangle and each ∘ site by an upward pointing triangle. This new lattice has two types of elementary polygons, the new triangles and dodecagons with six sides originating from the honeycomb hexagons and six sides from the new triangles. Hence the term *3–12 lattice* used by Nagle et al. (1989). The six sites A, B, C, D, E and F shown in Fig. 3.13 can be regarded as a *unit cell* in the new lattice, the sites A′, B′, E′ and F′ being first-neighbour sites in adjacent unit cells. Now consider a dimer graph on this lattice. If none of the new edges AB, BC and CA is occupied by a line of the dimer graph then the three edges E′A, F′B and CD originating from the honeycomb lattice will all be so occupied. Otherwise not more than one of the three edges AB, BC and CA can be occupied by a dimer line. If, say, there is a dimer line on AB then there will also be one on CD, but not on E′A or F′B. Similar arguments apply for BC and

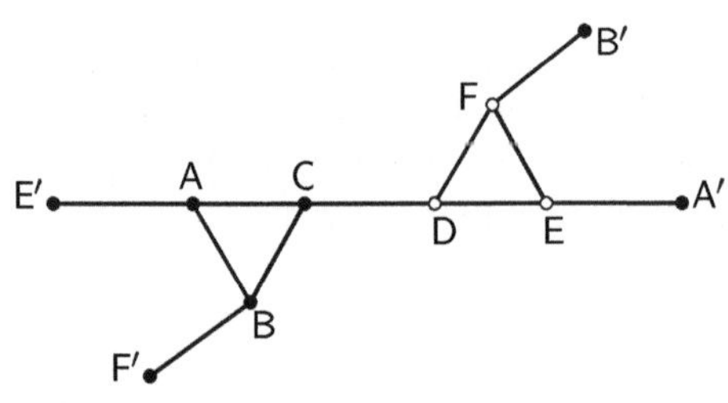

Fig. 3.13 The 3–12 lattice obtained by splitting the sites of a honeycomb lattice

CA. Hence, if the honeycomb lattice is recovered by replacing each triangle by a single site, then either three lines or one line are incident at this site. It follows that there is a one-to-one correspondence between dimer graphs on the new 3–12 lattice denoted by $\mathcal{N}_{\text{TT}}$, where $|\mathcal{N}_{\text{TT}}| = 3N$, and complementary graphs on the honeycomb lattice $\mathcal{N}_{\text{H}}$ of N sites.

For the dimer assembly on $\mathcal{N}_{\text{TT}}$ let the fugacities on the edges of the triangles be unity and those on the honeycomb edges of which CD, EA′ and FB′ are representative be $\mathfrak{z}_1$, $\mathfrak{z}_2$ and $\mathfrak{z}_3$ respectively. Because of the one-to-one correspondence between dimer graphs on the $\mathcal{N}_{\text{TT}}$ and complementary graphs on the $\mathcal{N}_{\text{H}}$, the dimer partition function on $\mathcal{N}_{\text{TT}}$ is given by

$$Z^{(\text{DM})}(\mathcal{N}_{\text{TT}}, \mathfrak{z}_1, \mathfrak{z}_2, \mathfrak{z}_3) = \sum_{\{\mathfrak{g}'\}}^{(\text{CZF})} \mathfrak{z}_1^{\mathrm{e}_1(\mathfrak{g}')} \mathfrak{z}_2^{\mathrm{e}_2(\mathfrak{g}')} \mathfrak{z}_3^{\mathrm{e}_3(\mathfrak{g}')}, \tag{3.9.19}$$

where the summation is over all complementary graphs on $\mathcal{N}_{\text{H}}$. In fact this summation is just the $\eta = 3$ case of the summation in (3.9.18) with $\mathfrak{z}_k := v_k^{-1}$, $k = 1, 2, 3$. For the honeycomb lattice $\mathrm{s}_1 = \mathrm{s}_2 = \mathrm{s}_3 = \frac{1}{2}N$ and, from (3.9.18) and (3.9.19),

$$Z^{(\text{IM})}(\mathcal{N}_{\text{H}}, \{v_k\}) = \left[4 \prod_{k=1}^{3} \sinh(K_k)\right]^{N/2} Z^{(\text{DM})}(\mathcal{N}_{\text{TT}}, v_1^{-1}, v_2^{-1}, v_3^{-1}). \tag{3.9.20}$$

Chapter 4
Phase Transitions and Scaling Theory

4.1 The Geometry of Phase Transitions

We consider a system in the coupling-density representation in which, as in Sect. 2.5, there are no independent densities ($\eta = n_{\mathrm{f}} - 1$).[1] It will also be supposed, until the discussion of finite-size scaling in Sect. 4.10, that the thermodynamic limit has been taken with respect to the one independent extensive variable. From (2.3.15) and (2.3.18),

$$\phi - L_{n_{\mathrm{f}}} = f/T, \tag{4.1.1}$$

$$\phi = -s + \sum_{\ell=1}^{n_{\mathrm{e}}} \mathrm{u}_\ell K_\ell - \sum_{i=1}^{n_{\mathrm{f}}-1} L_i \rho_i, \tag{4.1.2}$$

$$\mathrm{d}\phi = \sum_{\ell=1}^{n_{\mathrm{e}}} \mathrm{u}_\ell \mathrm{d}K_\ell - \sum_{i=1}^{n_{\mathrm{f}}-1} \rho_i \mathrm{d}L_i. \tag{4.1.3}$$

The system has $n = n_{\mathrm{f}} + n_{\mathrm{e}} - 1$ independent couplings and the free-energy density ϕ can be plotted as an n-dimensional hypersurface in the $(n+1)$-dimensional space of the variables $\phi, K_1, \ldots, K_{n_{\mathrm{e}}}, L_1, \ldots, L_{n_{\mathrm{f}}-1}$. This space and hypersurface are unaffected by the choice of $Q_{n_{\mathrm{f}}}$ in Sect. 2.1.[2] From (4.1.3) it can be seen that components of the gradient of the surface at any point correspond to the densities and the elements of the curvature matrix are thermodynamic response functions like the one appearing on the left-hand side of equation (2.5.15).

---

[1] In such general cases the subscript will be omitted from $\phi_{n_{\mathrm{f}}-1}$ (and $f_{n_{\mathrm{f}}-1}$), it being retained only when n_{f} has some small specified value (e.g. $n_{\mathrm{f}} = 2$ for a critical point and $n_{\mathrm{f}} = 3$ for a tricritical point).

[2] Since $\phi = L_{n_{\mathrm{f}}}$ a different choice of an independent extensive variable would correspond simply to a relabelling of the axes.

D.A. Lavis, *Equilibrium Statistical Mechanics of Lattice Models*,
Theoretical and Mathematical Physics, DOI 10.1007/978-94-017-9430-5_4

The free-energy density $\phi(K_1, \ldots, K_{n_e}, L_1, \ldots, L_{n_f-1})$ is, for most values of its arguments, a smooth function.[3] The exceptional points are associated with some kind of critical or phase-coexistence behaviour. Phase boundaries correspond to $(n-1)$-dimensional regions on the hypersurface of ϕ in which some kind of non-smooth behaviour occurs. These are the transition regions of highest dimension. In general a transition region may have any dimension between $n-1$ and zero (a point). The projection of all these transition regions into the n-dimensional space of couplings yields the phase diagram of the system. The simplest form of a phase boundary is when some or all of the densities are discontinuous across the boundary and we have a *first-order transition or region of two-phase coexistence*. In a similar way a region of p-phase coexistence is in general of dimension $n+1-p$. Since this region must have a non-negative dimension we might conclude that largest number of phases which can coexist for n independent couplings is $n+1$. This restriction is the *Gibbs phase rule*.

There are exceptions, which arise when the phase diagram has some special symmetry. One of these is the triangular ferrimagnetic Ising model for which the phase diagram has reflectional symmetry about the zero-field axis. A compensation point[4] occurs on this axis where four first-order transitions meet. However, mean-field (Bell 1974a) and renormalization group (Lavis and Quinn 1983, 1987) calculations indicate that, at the compensation point, the discontinuities in density are zero, with the phases forming two pairs of the same ordering structure. Such a compensation point, at the coincidence of four first-order transitions, also occurs in the decorated lattice model of Sect. 10.3 and a further example of the 'violation' of the Gibbs phase rule is provided by the metamagnet with a staggered field used as an example in Sect. 4.1.2 (see also Sect. 6.2). Here the phase diagram is three-dimensional with reflectional symmetry about the plane on which the staggered field is zero. Mean-field calculations indicate (Kincaid and Cohen 1975) that, for a certain parameter range, this plane contains a line of four-phase coexistence (Fig. 4.3a, b); one more phase than would be allowed by the Gibbs phase rule.

4.1.1 A Two-Dimensional Phase Space

We here consider the case $n_f = 2, n_e = 1, n = 2$, with free-energy densities $\phi_0(K, \rho)$ and $\phi_1(K, L)$ satisfying the differential forms (2.3.20) and (2.3.21), respectively, and related by the Eqs. (2.3.22)–(2.3.25).

For this case phases can be separated by lines of transitions. A *critical point* occurs at a second-order transition terminating a line of first-order transitions (Fig. 4.1a).[5]

[3] See Definition 17.1.1.

[4] For a discussion of ferrimagnetism and compensation points see Martin (1967) pp. 38–44.

[5] In this and subsequent figures first-order transitions are indicated by broken lines and second-order transitions by continuous lines.

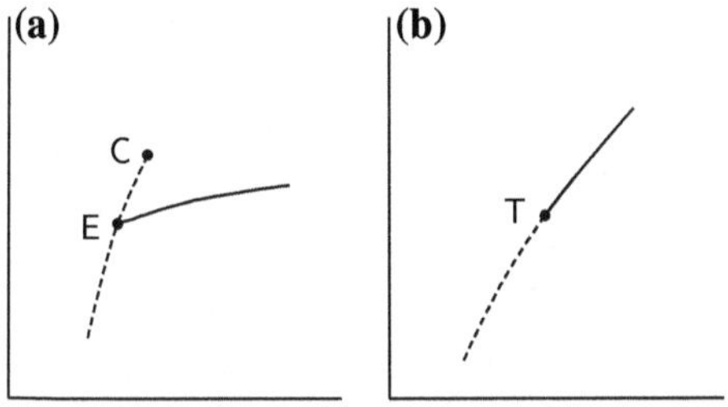

Fig. 4.1 **a** A critical point C terminating a line of first-order transitions and a critical end-point E where a line of second-order transitions terminates on a line of first-order transitions. **b** A tricritical point T where a line of first-order transitions meets a line of second-order transitions

Examples of this are the Curie point in a magnet and the termination of the vapour-liquid transition in a fluid system.

A line of first-order transitions shows as a discontinuity of gradient in the surface of ϕ_1 plotted as a function of K and L (Fig. 4.2a). This line can also be regarded as representing the coexistence of two phases with different densities given, from (2.3.25), by the negatives of the gradients of ϕ_1 with respect to L on each side of the line. In the corresponding picture (Fig. 4.2b) for ϕ_0 plotted as a function of ρ the points A and B represent the two sides of the phase transition. These points are joined by a double tangent with slope L^* and, from (2.3.24),

$$\phi_0^{(+)} - \phi_0^{(-)} = L^*[\rho^{(+)} - \rho^{(-)}]. \tag{4.1.4}$$

As K is varied A and B describe two lines on the ϕ_0 surface bordering a flat region of phase coexistence. Typically the two lines will meet at a critical point (Sect. 4.4). If (2.3.23) is used to plot L as a function of ρ along an isotherm, a first-order transition appears as a flat part of the curve (Fig. 4.2c). A *triple point* occurs at the confluence of three lines of first-order transitions; an example of this is the point where the first-order transitions between ice, liquid water and steam meet.[6] Other types of behaviour, which are grouped under the general heading of *multicritical phenomena* can also occur. When a line of first-order transitions is continued by a line of second-order transitions the two lines are said to meet at a *tricritical point* (Fig. 4.1b). When a line of second-order transitions terminates on a line of first-order transitions at an intermediate point this meeting point is called a *critical end-point* (Fig. 4.1a). Tricritical points and critical end-points occur for different parameter ranges when mean-field theory is applied to a metamagnetic model (Sect. 6.2). In that model spins are arranged in parallel planes with a ferromagnetic interaction within planes and an antiferromagnetic interaction between neighbouring planes. $FeCl_2$ is an experimental example of this kind of system. In this case the

[6] Both the Kelvin and Celsius temperature scales are defined by the triple point of water: 273.16 K and 0.01 °C respectively. (But see the footnote on p. 5 for an alternative way to define the temperature scale.)

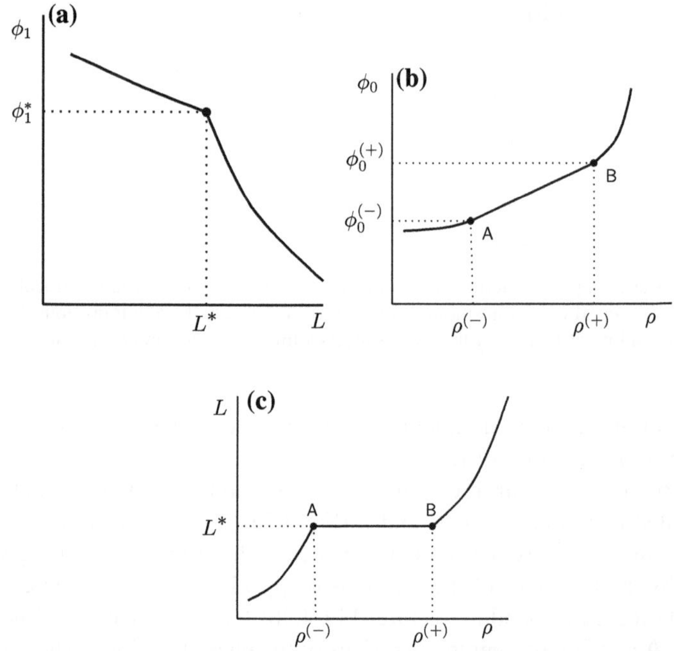

Fig. 4.2 **a** A first-order transition showing as a discontinuity of slope in an isothermal section of the ϕ_1 surface. **b** A first-order transition showing as the double tangent AB in an isothermal section of the ϕ_0 surface. **c** A first-order transition showing as a horizontal part of an isotherm of L plotted against ρ

second-order transition line terminates at the *Néel point* on the zero-field axis and meets a first-order transition line at a point in the plane, with this latter terminating on the zero-temperature axis. The meeting of the two transition curves can be a tricritical point or a critical end-point[7] according to the relative strengths of the ferromagnetic and antiferromagnetic interactions. As can be see from Fig. 4.3a (see also Sect. 6.2.5) the bicritical end-point in the metamagnetic model, occurs *inside* the antiferromagnetic phase. This differs from the behaviour in the *Blume-Emery-Griffiths model* (Blume et al. 1971) where the phase topology is similar to that shown in Fig. 4.1a.

A further type of multicritical behaviour which can occur in a two-dimensional phase space is when the critical lines of two phase transitions meet at a point, which is also the end of a coexistence line of first-order phase transitions separating the two ordered phases. The meeting point is called a *bicritical point*. Bicritical points have been observed most clearly in weakly anisotropic antiferromagnets such as

[7] The name 'tricritical point' in the context seems slightly odd since we are apparently considering a situation with only two phases present. A full understanding of tricritical points needs a phase space of more dimensions (Sect. 4.1.2). The same applies to the designation of the critical end-point as a *bicritical end-point*.

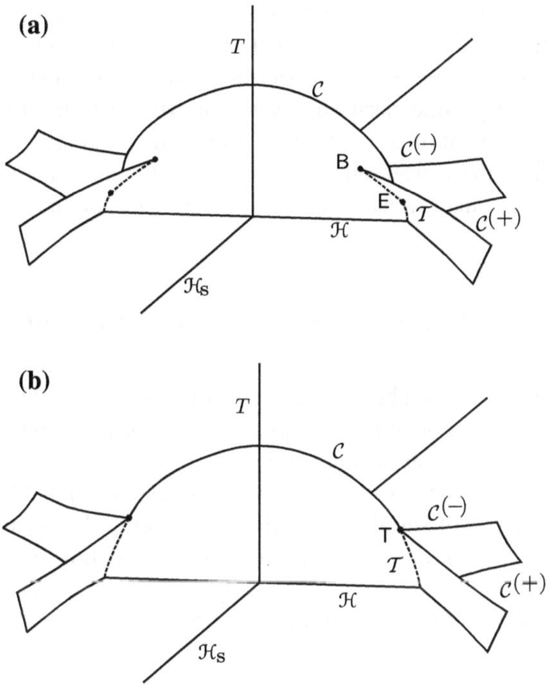

Fig. 4.3 **a** Phase diagram for a metamagnetic model in the space of temperature T, magnetic field $\mathcal{H}$ and staggered field $\mathcal{H}_s$. The critical curve $\mathcal{C}$ meets the triple line $\mathcal{T}$ at the critical end-point E. The critical curves $\mathcal{C}^{(+)}$ and $\mathcal{C}^{(-)}$ meet on the plane $\mathcal{H}_s = 0$ at the bicritical end-point B. (Reprinted from Kincaid and Cohen (1975), p. 114, by permission of the publisher Elsevier Science.) **b** As for (**a**) but now the critical end-point and bicritical end-point have coalesced to form a triple-point T. (Reprinted from Kincaid and Cohen (1975), p. 110, by permission of the publisher Elsevier Science.)

$GdAlO_3$ and MnF_2 in carefully oriented magnetic fields (Rohrer 1975). A theoretical realization of a bicritical point occurs in the temperature–electric field plane of a six-vertex ferroelectric (Lieb and Wu 1972, Fig. 30).

4.1.2 A Three-Dimensional Phase Space

When the phase space acquires a third dimension, so that $n = 3$, the boundaries between pairs of phases become surfaces. A first-order surface may be bounded by a line of critical points or *critical line*. Three first-order surfaces, separating three distinct phases can meet at a *triple line*. With these basic components phase diagrams can exhibit a rich diversity of structure. Many of these appear when a *staggered field*, alternating in sign between layers, is added to the metamagnetic model referred to above. Figure 4.3a reproduces a phase diagram for this model obtained by Kincaid and Cohen (1975) using mean-field methods (Sect. 6.2). This form of phase diagram occurs for a particular range of the ratio between the ferromagnetic interaction within layers and the antiferromagnetic interaction between layers (Sect. 4.6). Three phase boundaries meet along a triple line $\mathcal{T}$, which lies between the zero-temperature plane and the critical end-point E, where it meets the critical curve $\mathcal{C}$ and a line of *four-phase coexistence* BE. The point B is a *bicritical end-point*, where the critical curves

$\mathcal{C}^{(+)}$ and $\mathcal{C}^{(-)}$ meet on the plane $\mathcal{H}_s = 0$. By varying the parameter ratio so that B and E coalesce, a tricritical point can be achieved and the phase diagram takes the form shown in Fig. 4.3b. The tricritical point is the meeting point of the critical curve and the triple line clarifying the reason why Griffiths (1970) chose the name 'tricritical' for a point of coincidence of a first-order and critical line in a plane. It will be clear from these brief remarks that even within the constraints of a phase space of at most three dimensions not all the possible types of phase behaviour have been covered.[8]

4.2 Universality, Fluctuations and Scaling

As indicated above, one aim of the theory of critical phenomena is to formulate a classification of phase transitions. Criteria are sought in terms of which it can be said that certain phase transitions are 'of the same type'. A early attempt of this kind arose from the observation by van der Waals (1873) that his equation could be expressed in terms of variables scaled with respect to their critical values (Sect. 5.2). A *law of corresponding states* was formulated suggesting that a wide class of physical systems would, in terms of appropriately scaled variables, obey this same equation of state. The experimental fact that the *critical ratio*

$$K_\mathrm{c} := \frac{\widetilde{P}_\mathrm{c}\widetilde{V}_\mathrm{c}}{M_\mathrm{A}\widetilde{T}_\mathrm{c}} = k_\mathrm{B}\frac{P_\mathrm{c}V_\mathrm{c}}{M_\mathrm{A}T_\mathrm{c}}, \tag{4.2.1}$$

where M_A is Avogadro's number, is not the same for all gases caused some concern and was the first sign that the law of corresponding states has only limited applicability. However, as has been pointed out by Levelt Sengers (1974):

> Although this law has limited validity experimentally, and is not obeyed by most other analytic equations of state, its role in the development of the physical, chemical and engineering sciences dealing with fluids can hardly be exaggerated.

It led to many attempts to construct more general analytical equations of state and ultimately to the theory of Landau (1937) described in Chap. 5. The first steps towards the development of this approach were taken by van der Waals (1893). In this article on the theory of capillarity he investigated the asymptotic critical behaviour of his equation by expanding about the critical point in terms of the volume and temperature deviations from their critical values. This yielded the first calculation of the critical exponent, now known as β, for which he obtained the value $\frac{1}{2}$. This exponent measures the curvature of the coexistence curve in the vicinity of the critical point and, towards the end of the nineteenth century, experimental results began to appear indicating disagreement with the theoretical value $\beta = \frac{1}{2}$. The initial reaction was to suppose that the experimental measurements were not taken sufficiently close to

---

[8] For a review of theoretical and experimental work on multicritical phenomena see Pynn and Skjeltorp (1984) and for tricritical phenomena, with further examples, the review by Lawrie and Sarbach (1984).

criticality. In 1900 Verschaffelt was, however, able to show by analyzing the results of Young (1892) for isopentane, that $\beta = 0.3434$, with no evidence of the value tending towards $\frac{1}{2}$ as the critical point was approached. This was the beginning of the realization that critical exponents (both β and the other critical exponents associated with the asymptotic properties of thermodynamic functions, see Sect. 4.4) were crucial for the classification of phase transitions. The development of generalized forms of the equation of state incorporating fractional exponents took another 65 years. By this time there was ample evidence that experimental systems yield critical exponents with values different from those predicted by classical (Landau) theory. Theoretical impetus was provided by the Onsager (1944) solution of the two-dimensional Ising model which demonstrated non-analytic behaviour at the critical point. We shall not, with the benefit of hindsight, attempt to trace the details of the early development of generalized equations of state leading to scaling theory. Rather we shall concentrate on trying to distinguish those properties of a system which affect its critical behaviour.

4.2.1 Universality

At points near a transition region in phase space a system will be in different stable phases. This entails the existence of fluctuations, which can be regarded as the short-time and local appearance of fragments of one of the phases in another. An example of this is the critical opalescence present in a fluid near the critical point of the liquid-vapour transition. Thus fluctuations are the appearance of locally-ordered regions. These persist for a certain mean length of time $\mathfrak{t}$ called the *correlation time* and extend over a mean distance $\mathfrak{r}$ called the *correlation length*. Roughly speaking, dynamic and static critical behaviours are governed by $\mathfrak{t}$ and $\mathfrak{r}$ respectively. Since we are concerned exclusively with static critical behaviour it is the correlation length, measuring the mean extension of correlations, which will be of interest to us. Given that, in a neighbourhood of a transition region, correlations extend over distances large with respect to the detailed nature (short-range interactions, lattice structure etc.) of the system, it is intuitively easy to understand that this microstructure will not affect the characteristics of the system. This led to the formulation of a *universality hypothesis* which can be expressed in the following way (Kadanoff 1976b):
All critical problems can be divided into *universality classes* differentiated by:

(a) the (physical) dimension d of the system,
(b) the symmetry group of the *order parameter*,
(c) perhaps other criteria.

Within each class, the critical properties, and in particular the critical exponents, are supposed to be identical or, at worst, to be continuous functions of very few parameters. Although, in theoretical models, systems of any dimension can be easily realized it might be supposed that experiment is relevant only to the case $d = 3$. However, polymers (linear macromolecules) and solids made up of chains of weakly-coupled molecules provide examples of one-dimensional, or quasi-one-dimensional systems and two-dimensional experimental systems can be realized in films, adsorbed phases, weakly-coupled planes and surface phenomena.

In a rather general way the order parameter of a phase can be defined as the expectation value of a quantity which measures the extent to which the microscopic configuration mirrors a state of perfect ordering characteristic of this phase. However, the order parameter is not always easy to define. In a fluid system undergoing a liquid-gas transition the order parameter is a density difference and of dimension one. In a magnetic transition it is the magnetization which may be of dimension one, two or three according to the constraints imposed on the system. The change of the nature of a phase transition according to the symmetry group of the order parameter is illustrated in the context of Landau theory in Chap. 5. The inclusion of item (c) by Kadanoff could be viewed as a means of providing for exceptional cases which will almost inevitably arise in any kind of classification of this type. An example of this is the bond-dilute Ising model where the order parameter is a scalar quantity, so one would, on the basis of (a) and (b) alone, expect the exponents to be those of the pure Ising model, for the corresponding value of d. It can, however, be shown (Sect. 10.2.1) that instead of the Ising exponents α and γ for the heat capacity and susceptibility respectively the actual exponents α_{r} and γ_{r} are given in terms of their Ising values by $\alpha_{\mathrm{r}} = -\alpha/(1-\alpha)$ and $\gamma_{\mathrm{r}} = \gamma/(1-\alpha)$. This change of exponent, whenever $\alpha \neq 0$, ($\alpha = 0$ for $d = 2$) is called *exponent renormalization* (Fisher 1968). It arises whenever the fraction of secondary sites occupied by solute atoms is treated as an independent variable. In the six-vertex model in an electric field the order parameter is the scalar polarization. However, it can be shown (Lieb and Wu 1972, Sect. 5E) that, at the transition to the polarized state, the heat capacity diverges with an exponent $\alpha = \frac{1}{2}$ compared to the zero value of the corresponding Ising model. These two different types of transition are also exhibited by dimer models. They are discussed in Sect. 13.4 and by Nagle et al. (1989) who called them K-type (for Kasteleyn) and O-type (for Onsager). Finally the eight-vertex model should be mentioned. This yields continuously varying critical exponents (Sect. 12.3.8). The way in which such exponents can arise within the context of scaling theory is investigated in Sect. 4.3.7 and the particular case of the eight-vertex model is considered in Sect. 12.3.11. It should be emphasized that the concept of a universality class is associated with a phase transition and not with a physical or theoretical system as a whole. Systems with sufficient complexity may exhibit different phase transitions, in different parts of their phase spaces, lying in different universality classes.

In the presentation of scaling theory in Sect. 4.3 the approach used is that of Hankey and Stanley (1972). This method has the virtues of mathematical elegance and generality. The basic premises do, however, seem to be rather arbitrary without the provision of some intuitive insight. This we shall now try to provide, using the example of the ferromagnetic Ising model, following the ideas of Kadanoff (1966) (see also, Kadanoff et al. 1967).

4.2.2 Scaling for the Ising Model

Consider the spin-$\frac{1}{2}$ Ising model on a d-dimensional hypercubic lattice $\mathcal{N}$ of N sites and lattice spacing a with Hamiltonian given by (3.5.1). The density conjugate to L is the magnetization density $m := \mathcal{M}/N$, where $\mathcal{M} := \langle\widehat{\mathcal{M}}\rangle$ and $\widehat{\mathcal{M}}$ is given by (3.5.4). The ferromagnetic Ising model is believed to have, for all $d > 1$, a first-order phase transition along the zero-field axis ($L = 0$) in the interval $K_c < K \leq \infty$, where the critical point is $K = K_c$, $L = L_c = 0$.[9] The order parameter for this transition region is m and the variance of $\hat{m}$ is given, from (2.5.16), by

$$\mathrm{Var}[\hat{m}] = \frac{1}{N}\sum_{\{\boldsymbol{r}_1,\boldsymbol{r}_2\}} \Gamma_2(\sigma(\boldsymbol{r}_1), \sigma(\boldsymbol{r}_2)), \tag{4.2.2}$$

where

$$\Gamma_2(\sigma(\boldsymbol{r}_1), \sigma(\boldsymbol{r}_2)) = \langle\sigma(\boldsymbol{r}_1)\sigma(\boldsymbol{r}_2)\rangle - \langle\sigma(\boldsymbol{r}_1)\rangle\langle\sigma(\boldsymbol{r}_2)\rangle \tag{4.2.3}$$

is the *spin-spin (connected pair) correlation function*. When the system has translational and rotational symmetry this correlation function satisfies (2.5.25) (with $\mathfrak{c}_1 = \mathfrak{c}_2 = \sigma$) and is a measure of the amount of short-range order (spin alignment) between spins separated by a distance r. It is, therefore, proportional to the probability of a cluster of aligned spins forming with spatial extent r. The correlation length $\mathfrak{r}$, given by

$$\mathfrak{r}^2 := c(\mathfrak{d})\frac{\sum_{\{\boldsymbol{r}_1,\boldsymbol{r}_2\}} |\boldsymbol{r}_2 - \boldsymbol{r}_1|^2 \Gamma_2(\sigma(\boldsymbol{r}_1), \sigma(\boldsymbol{r}_2))}{\sum_{\{\boldsymbol{r}_1,\boldsymbol{r}_2\}} \Gamma_2(\sigma(\boldsymbol{r}_1), \sigma(\boldsymbol{r}_2))}, \tag{4.2.4}$$

is a measure of the largest clusters which will form. The prefactor $c(\mathfrak{d})$ depends on the number of dimensions $\mathfrak{d} \leq d$ in which the system is infinite (Sect. 4.10). Its exact form is unimportant for most purposes and it is sometimes omitted (Ziman 1979, p. 26). However, in Sect. 5.6.3, it is shown that, within the Gaussian approximation, the choice $c(\mathfrak{d}) = 1/(2\mathfrak{d})$ is necessary for consistency.

Consider now a path in the phase diagram with small positive field (L small and positive) beginning at a low temperature (K large). At this point thermal effects will be small and most spins will be aligned in the positive direction. As K is decreased thermal fluctuations will mean that random spin flips will show an increasing tendency to occur and the interaction energy between spins will encourage the formation of clusters based around these spins. As long as the correlation length is quite small these clusters will appear and disappear at random. However, as K continues to decrease towards the critical point, the spin-spin correlation function for

[9] For a mean-field treatment which yields this property see Sect. 6.1.

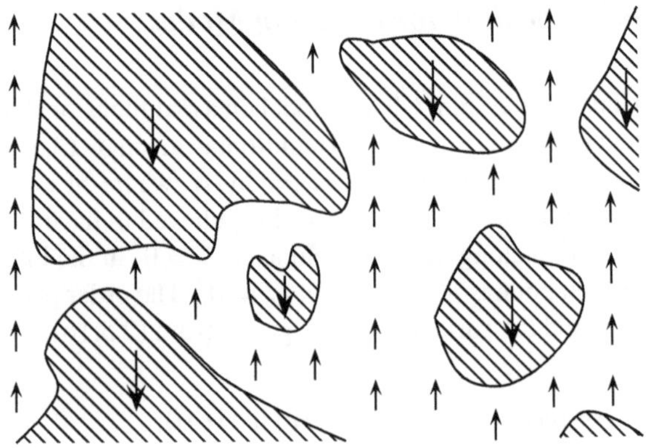

Fig. 4.4 Clusters of down spins forming in a sea of up spins. Clusters form of all sizes up to lengths of the order of $\mathfrak{r}$

a particular fixed r will increase. Correlations will grow stronger over larger distances and the correlation length and largest cluster size of down spins will become very large (Fig. 4.4), finally diverging at the critical point.

Near the critical point the correlation length is much greater than a, so it is possible to find a number λ, such that $\mathfrak{r} >> \lambda a >> a$. Ignoring boundaries or assuming periodic boundary conditions, let the lattice $\mathcal{N}$ be divided into cells, λa on each side and containing λ^d sites (Fig. 4.5), so that the set of cells can be represented by points $\tilde{\boldsymbol{r}} \in \widetilde{\mathcal{N}}$, where $\widetilde{\mathcal{N}}$ is of the same lattice structure as $\mathcal{N}$, but with lattice spacing $\tilde{a}$. This is a transformation from the lattice to the cell lattice given by

$$\begin{aligned} \boldsymbol{r} &\to \tilde{\boldsymbol{r}} := \boldsymbol{r}/\lambda, \\ N &\to \tilde{N} := N/\lambda^d, \qquad (4.2.5) \\ a &\to \tilde{a} := a\lambda. \end{aligned}$$

Suppose that all interactions between different cells are turned off, so that each acts independently. Then there will be no correlations over distances larger than the cell size and no divergences in thermodynamic quantities. Correlations between cells are the necessary ingredient for thermodynamic behaviour near the critical point and this behaviour can be supposed to be independent of the details of the interactions within the cells. The interaction between the cell $\mathcal{B}(\tilde{\boldsymbol{r}})$ centred $\tilde{\boldsymbol{r}}$ and the other cells is modelled in terms of a single spin variable $\tilde{\sigma}(\tilde{\boldsymbol{r}}) = \pm 1$ related to the spins in the cell by

$$\sum_{\boldsymbol{r} \in \mathcal{B}(\tilde{\boldsymbol{r}})} \sigma(\boldsymbol{r}) = w(\lambda)\tilde{\sigma}(\tilde{\boldsymbol{r}}), \qquad (4.2.6)$$

for some $w(\lambda)$. The correlation length is the same for the spins $\tilde{\sigma}(\tilde{\boldsymbol{r}})$ on $\widetilde{\mathcal{N}}$ as for $\sigma(\boldsymbol{r})$ on $\mathcal{N}$, if it is measured in units of a, but, if it is now measured in units of $\tilde{a}$, it has the value $\mathfrak{r}/\lambda$. The thermodynamic state of the original model is given by values of

Fig. 4.5 Lattice of spacing a divided into cells, of linear dimension λa, containing λ^d spins

$\Delta K := K - K_c$ and L. The rescaled model, in which the correlation length is smaller by a factor λ^{-1}, will appear to be further from the critical point. It will, therefore, have effective values of ΔK and L larger than those in the original model. These rescaled values we denote by $\lambda^{y_K}\Delta K$ and $\lambda^{y_L}L$, for some unknown positive scaling exponents y_K and y_L. The important *assumption* made at this stage is that, for points sufficiently close to the critical point in phase space, y_K and y_L are constants. This means that $\mathfrak{r}$ acts as the single measure of the distance from the critical point and, under the sequence of transformations $\mathfrak{r} \to \mathfrak{r}/\lambda \to \mathfrak{r}/\lambda^2 \to \cdots$, ΔK and L transform according to $\Delta K \to \lambda^{y_K}\Delta K \to \lambda^{2y_K}\Delta K \to \cdots$ and $L \to \lambda^{y_L}L \to \lambda^{2y_L}L \to \cdots$ respectively. Thus we have, in a neighbourhood of the critical point,

$$\mathfrak{r}(\lambda^{y_K}\Delta K, \lambda^{y_L}L) = \lambda^{-1}\mathfrak{r}(\Delta K, L). \tag{4.2.7}$$

As we saw in Sect. 4.1, the critical point will correspond to non-smooth behaviour of the free energy of the system. It will, however, itself be finite non-zero and contain a smooth component. The same will be true of ϕ_1 and we suppose that we can make the separation

$$\phi_1 = \phi_{\text{smth}} + \phi_{\text{sing}}, \tag{4.2.8}$$

where

$$\phi_{\text{sing}} = \phi_{\text{sing}}(\Delta K, L) \tag{4.2.9}$$

contains all the non-smooth behaviour of ϕ_1 and $\phi_{\text{sing}}(0,0) = 0$.

That the free-energy density can be divided into smooth and singular parts is an *assumption* of scaling theory (Hankey and Stanley 1972, p. 3519), which is then justified in terms of the form of the Hamiltonian (Sect. 4.3.6). If we are concerned only with the critical behaviour of the system, we can, in most but not all cases (Sect. 4.3.4) ignore the smooth part of ϕ_1.

Given that $\mathfrak{r}$ is the sole measure of the distance from the critical point, ϕ_{sing} can be given as a function of $\mathfrak{r}$. In order for the critical properties of the system to remain unchanged under the scaling transformation the singular part of the free energy must be invariant and, since the number of lattice sites is reduced from N to N/λ^d,

$$\phi_{\text{sing}}(\mathfrak{r}) = \lambda^{-d}\phi_{\text{sing}}(\lambda^{-1}\mathfrak{r}), \tag{4.2.10}$$

or, equivalently,

$$\phi_{\text{sing}}(\lambda^{y_K}\Delta K, \lambda^{y_L}L) = \lambda^{d}\phi_{\text{sing}}(\Delta K, L). \tag{4.2.11}$$

The corresponding expression for the magnetization density is obtain by differentiating (4.2.11) with respect to L and this can also be obtained by summing (4.2.6) over all $\mathcal{B}(\tilde{r})$ and taking the expectation. Thus we conclude that $w(\lambda) = \lambda^{y_L}$.

4.3 General Scaling Formulation

Consider a system of the type described at the beginning of Sect. 4.1, with $n = n_{\text{f}} + n_{\text{e}} - 1$ independent couplings, which, for the sake of brevity, we denote by ζ_k, $k = 1, 2, \ldots, n$, for which, where convenient, we also use $\boldsymbol{\zeta}$ to denote the vector of couplings.

Let there be a transition region $\mathcal{C}$ of dimension $n - s$ in the space of $\boldsymbol{\zeta}$ and suppose that, close to $\mathcal{C}$, there is a set of curvilinear coordinates

$$\theta_j = \theta_j(\Delta\boldsymbol{\zeta}), \qquad j = 1, \ldots, n\,, \qquad \Delta\boldsymbol{\zeta} := \boldsymbol{\zeta} - \boldsymbol{\zeta}^{\star}, \tag{4.3.1}$$

called *scaling fields*, which are smooth functions of the couplings oriented about an origin $\boldsymbol{\zeta} := \boldsymbol{\zeta}^{\star} \in \mathcal{C}$, with $\theta_j^{\star} := \theta_j(0) = 0$, and constructed in such a way that $\mathcal{C}$ is defined by

$$\theta_1 = \theta_2 = \cdots = \theta_s = 0. \tag{4.3.2}$$

The variables $\theta_1, \ldots, \theta_s$ and $\theta_{s+1}, \ldots \theta_n$ are called *relevant* and *irrelevant* scaling fields respectively. The latter give coordinates for points in $\mathcal{C}$.

4.3.1 The Kadanoff Scaling Hypothesis

In this section a version of the scaling hypothesis is introduced, which is essentially that due to Kadanoff (1966), with a treatment which follows the work of Hankey and Stanley (1972). This procedure is sufficient for an analysis of power-law singularities. However, certain systems, in particular the Ising model with $d = 2$ (Sect. 12.2.7.3), exhibit logarithmic singularities. It has been shown by Nightingale and 'T Hooft (1974) that a slight generalization of the Kadanoff scaling hypothesis can be used to analyze these cases (Sect. 4.3.4).

As in the special case described in Sect. 4.2 it is assumed that ϕ can be divided into smooth and singular parts according to equation (4.2.8)[10] with ϕ_{sing} a function of the scaling fields. For these it is supposed that, within a region around $\mathcal{C}$, for all real[11] $\lambda > 1$ and some set of *scaling exponents*[12] y_j, there is a transformation

$$\theta_j \to \tilde{\theta}_j := \lambda^{y_j}\theta_j, \tag{4.3.3}$$

for which

$$\phi_{\text{sing}} \to \tilde{\phi}_{\text{sing}} := \lambda^d \phi_{\text{sing}}. \tag{4.3.4}$$

The origin of the scaling fields is the *fixed point* of the scaling transformation (4.3.3) and (4.3.4) and by extension of $\mathcal{C}$, which is its *basin of attraction.* The scaling transformation is equivalent to the *Kadanoff scaling hypothesis*

$$\phi_{\text{sing}}(\lambda^{y_1}\theta_1, \ldots, \lambda^{y_n}\theta_n) = \lambda^d \phi_{\text{sing}}(\theta_1, \ldots, \theta_n), \tag{4.3.5}$$

which asserts that the singular part of ϕ is a *generalized homogeneous function* of the scaling fields. The scaling exponents fall into two subsets

$$\begin{aligned} &y_j > 0, \quad j = 1, \ldots, s, \\ &y_j < 0, \quad j = s+1, \ldots, n, \end{aligned} \tag{4.3.6}$$

called *relevant* and *irrelevant* respectively, like the corresponding scaling fields. A zero exponent is called *marginal.* We shall at this stage, for the sake of simplicity, assume that none of the exponents is marginal. The significance of marginal exponents in scaling theory is considered in Sect. 4.3.7. It will also be assumed that no exponent is complex. In practice, as will be seen in Sect. 15.5.2, this is not necessarily always the case, but situations arising from complex exponents are not difficult to interpret in particular examples.

Denoting by $\boldsymbol{k}$ a set of non-negative integers $(k_1, \ldots, k_n)$, let $\psi(\boldsymbol{k})$ be the function obtained by differentiating ϕ k_j times with respect to θ_j, for $j = 1, \ldots, n$. Then, from (4.3.5), assuming a division of $\psi(\boldsymbol{k})$ into smooth and singular parts as for ϕ,

$$\psi_{\text{sing}}(\boldsymbol{k}; \lambda^{y_1}\theta_1, \ldots, \lambda^{y_n}\theta_n) = \lambda^{d'(\boldsymbol{k})} \psi_{\text{sing}}(\boldsymbol{k}; \theta_1, \ldots, \theta_n), \tag{4.3.7}$$

---

[10] As we shall see later in this section, this analysis applies both to first-order transitions and to those for which some derivative of the free-energy density diverges in the transition region. The term 'singular' is taken, therefore, to include all types of non-smooth behaviour, including finite discontinuities in derivatives of ϕ.

[11] The parameter λ introduced in Sect. 4.2 is an integer greater than unity. If, however, (4.3.3) and (4.3.4) are defined for all real $\lambda > 1$ it is not difficult to see that (4.3.5) also holds for $1 \geq \lambda > 0$.

[12] The term 'scaling exponent' is used to make the distinction with the critical exponents defined in Sect. 4.4. When there is no danger of confusion they will be referred to simply as 'exponents'.

where

$$d'(\boldsymbol{k}) := d - \sum_{j=1}^{n} k_j y_j. \tag{4.3.8}$$

Setting, for a particular relevant scaling field θ_m,

$$\lambda = |\theta_m|^{-1/y_m} \tag{4.3.9}$$

in (4.3.5) and (4.3.7) gives

$$\phi_{\text{sing}}(\theta_1, \ldots, \theta_n) = |\theta_m|^{\sigma_m} \phi_{\text{sing}}(g_1^{(m)}, \ldots, g_n^{(m)}), \tag{4.3.10}$$

$$\psi_{\text{sing}}(\boldsymbol{k}; \theta_1, \ldots, \theta_n) = |\theta_m|^{\omega_m(\boldsymbol{k})} \psi_{\text{sing}}(\boldsymbol{k}; g_1^{(m)}, \ldots, g_n^{(m)}), \tag{4.3.11}$$

where

$$\omega_m(\boldsymbol{k}) := \frac{d'(\boldsymbol{k})}{y_m}, \qquad \sigma_m := \omega_m(\boldsymbol{0}) = \frac{d}{y_m} \tag{4.3.12}$$

and

$$g_j^{(m)}(\theta_m, \theta_j) := \frac{\theta_j}{|\theta_m|^{\vartheta_j^{(m)}}}, \qquad j = 1, \ldots, n, \tag{4.3.13}$$

with

$$\vartheta_j^{(m)} := \frac{y_j}{y_m} = \frac{1}{\vartheta_m^{(j)}}, \qquad j = 1, \ldots, n. \tag{4.3.14}$$

From (4.3.8), (4.3.12) and (4.3.14)

$$\omega_m(\boldsymbol{k}) = \sigma_m - \sum_{j=1}^{n} k_j \, \vartheta_j^{(m)}. \tag{4.3.15}$$

The exponent $\vartheta_j^{(m)}$ gives the incremental change in the exponent $\omega_m(\boldsymbol{k})$ of the function $\psi_{\text{sing}}(\boldsymbol{k})$ on differentiation with respect to θ_j. These quantities are therefore referred to as *gap exponents*. By extension it is convenient to refer to the quantities $g_j^{(m)}$ as *gap fields*. They satisfy the conditions

$$|g_m^{(\ell)}|^{\vartheta_j^{(m)}} \, |g_j^{(m)}| = |g_j^{(\ell)}|, \qquad j, \ell = 1, \ldots, n, \tag{4.3.16}$$

with $g_m^{(m)} = \pm 1$, according as $\theta_m \gtrless 0$. From (4.3.10), (4.3.11) and (4.3.13),

$$\begin{aligned}\phi_{\text{sing}}(g_1^{(j)}, \ldots, g_n^{(j)}) &= |g_m^{(j)}|^{\sigma_m} \phi_{\text{sing}}(g_1^{(m)}, \ldots, g_n^{(m)}), \\ \psi_{\text{sing}}(\boldsymbol{k}; g_1^{(j)}, \ldots, g_n^{(j)}) &= |g_m^{(j)}|^{\omega_m(\boldsymbol{k})} \psi_{\text{sing}}(\boldsymbol{k}; g_1^{(m)}, \ldots, g_n^{(m)}).\end{aligned} \tag{4.3.17}$$

Now consider the asymptotic forms derived from (4.3.10) and (4.3.11) as $\mathcal{C}$ is approached ($\theta_j \to 0$, $j = 1, \ldots, s$) along some path $\mathcal{P}$. Suppose that

$$\lim_{\mathcal{P}} g_j^{(m)}(\theta_m, \theta_j) = \varpi_j^{(m)}. \tag{4.3.18}$$

It is not difficult to show that, for any path $\mathcal{P}$, it is always possible to find a relevant field θ_m such that $\varpi_j^{(m)}$ is finite for all $j = 1, \ldots, s$.[13] Conversely, as we see below, given a relevant field θ_m, there exists a part of the neighbourhood of $\mathcal{C}$ which contains paths for which $\varpi_j^{(m)}$ are finite for all $j = 1, \ldots, n$. A more difficult problem is to be sure that the amplitudes

$$\begin{aligned}\lim_{\mathcal{P}} \phi_{\text{sing}}(g_1^{(m)}, \ldots, g_n^{(m)}) &= \phi_{\text{sing}}(\varpi_1^{(m)}, \ldots, \varpi_n^{(m)}), \\ \lim_{\mathcal{P}} \psi_{\text{sing}}(\boldsymbol{k}; g_1^{(m)}, \ldots, g_n^{(m)}) &= \psi_{\text{sing}}(\boldsymbol{k}; \varpi_1^{(m)}, \ldots, \varpi_n^{(m)}),\end{aligned} \tag{4.3.19}$$

are finite *and* non-zero. If this is indeed the case then it follows from (4.3.10) and (4.3.11) that

$$\begin{aligned}\phi_{\text{sing}}(\theta_1, \ldots, \theta_n) &\simeq |\theta_m|^{\sigma_m} \phi_{\text{sing}}(\varpi_1^{(m)}, \ldots, \varpi_n^{(m)}), \\ \psi_{\text{sing}}(\boldsymbol{k}; \theta_1, \ldots, \theta_n) &\simeq |\theta_m|^{\omega_m(\boldsymbol{k})} \psi_{\text{sing}}(\boldsymbol{k}; \varpi_1^{(m)}, \ldots, \varpi_n^{(m)}),\end{aligned} \tag{4.3.20}$$

as $\mathcal{C}$ is approached along $\mathcal{P}$. If $\varpi_j^{(m)} \neq 0$ for some j then, with the substitution $|\theta_m| = |\varpi_m^{(j)}||\theta_j|^{\vartheta_m^{(j)}}$ in (4.3.20) and using (4.3.17), the formulae (4.3.20) are recovered with m and j interchanged. For the moment we shall assume that the amplitudes in (4.3.20), given by (4.3.19) are finite and non-zero. The corrections to scaling arising from a breakdown in this assumption are discussed in Sect. 4.8. Since ϕ_{sing} is continuous the amplitude for this asymptotic form is invariant under change of sign of θ_m. For ψ_{sing} this is not in general the case.

---

[13] They are clearly zero for $j = s + 1, \ldots, n$.

The geometry of the path $\mathcal{P}$ close to $\mathcal{C}$ can be understood more clearly if we consider the family of surfaces

$$|\theta_j| = \varpi |\theta_m|^{\vartheta_j^{(m)}}, \qquad \forall\ 0 < \varpi < \infty. \tag{4.3.21}$$

The members of this family are tangential to the surface $\theta_j = 0$ or $\theta_m = 0$ according as $y_j >, < y_m$. If $y_j = y_m$ the members of the family make angles greater than zero and less than $\pi/2$ with both surfaces. If $\varpi_j^{(m)} \neq 0$ the path $\mathcal{P}$ lies asymptotically on the member of the family with $\varpi = \varpi_j^{(m)}$ as $\mathcal{P}$ approaches $\mathcal{C}$. If $\varpi_j^{(m)} = 0$, for any ϖ and points sufficiently close to $\mathcal{C}$, the path $\mathcal{P}$ lies between that member of (4.3.21) and the surface $\theta_j = 0$. The family of surfaces (4.3.21) represent the border between one form of critical or coexistence behaviour and another, since they divide the phase space between paths on which the limiting value $\varpi_j^{(m)}$ of $g_j^{(m)}$ is finite or infinite. They are therefore referred to as *crossover surfaces* and in this context the gap exponent $\vartheta_j^{(m)}$ is called a *crossover exponent*. Looked at from a slightly different point of view the types of scaling behaviour can be classified by considering the different ranges of value of $\vartheta_j^{(m)}$ (Pfeuty and Toulouse 1977, Chap. 3).

4.3.2 First-Order Transitions

Densities, which are the first-order partial derivatives of ϕ with respect to the couplings, can be expressed in terms of its first-order partial derivatives with respect to the scaling fields. These are finite in the transition region and it therefore follows, from (4.3.8), (4.3.12) and (4.3.20), that

$$y_j \leq d, \qquad j = 1, \ldots, n. \tag{4.3.22}$$

In general we see from these same equations that, as $\mathcal{C}$ is approached along the path $\mathcal{P}$, ψ_{sing} is singular, or zero, according to whether

$$d < \sum_{j=1}^{n} k_j y_j, \tag{4.3.23}$$

$$d > \sum_{j=1}^{n} k_j y_j, \tag{4.3.24}$$

respectively. Consider now the possibility of there being a discontinuity in $\psi(\boldsymbol{k})$, as $\mathcal{C}$ is approached along $\mathcal{P}$. This must be in $\psi_{\text{sing}}(\boldsymbol{k}; \varpi_1^{(m)}, \ldots, \varpi_n^{(m)})$, which must be finite and, on at least one side of the coexistence curve, non-zero. This in turn must give a discontinuity in $\psi_{\text{sing}}(\boldsymbol{k}; \theta_1, \ldots, \theta_n)$. It follows from (4.3.20) that a necessary condition for this to be the case is that $\omega_m(\boldsymbol{k})$ and hence $d'(\boldsymbol{k})$ are zero; that is

$$d = \sum_{j=1}^{n} k_j y_j. \tag{4.3.25}$$

It must be emphasized that condition (4.3.25) is a necessary but not sufficient condition for the existence of a discontinuity in $\psi(\boldsymbol{k})$. Indeed it is also the case when the function exhibits a logarithmic singularity (Sect. 4.3.4). The most important case of (4.3.25) is $d = y_j$, for some j. It indicates the possibility of a first-order transition, that is a discontinuity in one or more[14] of the densities (Nienhuis and Nauenberg 1975; Fisher and Berker 1982). In this case the 'evidence' for a discontinuity is rather stronger since the densities do not become infinite as $\mathcal{C}$ is approached, thereby excluding the possibility of a logarithmic singularity. The presence of a first-order transition does not impose any necessary asymptotic properties on the response functions. They may be finite or, as in the case of a *first-order critical point* (Sect. 4.5.6), have singularities.

4.3.3 Effective Exponents

In most cases transition regions do not exist in isolation in phase space. They are normally connected to other transition regions, as in the case of the critical point and coexistence curve discussed in Sects. 4.4 and 4.5. This means that in the neighbourhood of $\mathcal{C}$ governed by the scaling hypothesis there will be singular behaviour apart from that exhibited on $\mathcal{C}$ itself. For simplicity we consider the case where $\mathcal{C}$ (of dimension $n - s$) is connected to a transition region $\mathring{\mathcal{C}}$ of dimension $n - s + 1$ with the scaling fields θ_j for $\mathcal{C}$ and $\mathring{\theta}_j$ for $\mathring{\mathcal{C}}$ related by

$$\mathring{\theta}_j = \theta_j, \quad j \neq s, \qquad \mathring{\theta}_s = 1 - \theta_s. \tag{4.3.26}$$

This means that θ_s is relevant for $\mathcal{C}$ and irrelevant for $\mathring{\mathcal{C}}$. Following Riedel and Wegner (1974) we choose a simple model, with parameter τ and *generalized scaling fields* Θ_j, satisfying

$$\begin{aligned} \frac{\partial \Theta_j(\tau)}{\partial \tau} &= \ln(\lambda)[y_j \Theta_j(\tau) + (\mathring{y}_j - y_j)\Theta_j(\tau)\Theta_s(\tau)], \quad j \neq s, \\ \frac{\partial \Theta_s(\tau)}{\partial \tau} &= \ln(\lambda) y_s \Theta_s(\tau)[1 - \Theta_s(\tau)]. \end{aligned} \tag{4.3.27}$$

---

[14] A discontinuity in one first-order derivative with respect to a scaling field will, in general, lead to discontinuities in more than one 'physical' density (Eq. (4.3.119)).

Linearizing about $\Theta_j = 0$, for $j \neq s$ gives the scaling fields $\theta_j = \mathring{\theta}_j$, and about $\Theta_s = 0$ and $\Theta_s = 1$ the scaling fields θ_s and $\mathring{\theta}_s$, respectively, with $\mathring{y}_s = -y_s$[15] satisfying

$$\frac{\partial \theta_j(\tau)}{\partial \tau} = y_j \ln(\lambda)\theta_j(\tau), \tag{4.3.28}$$

$$\frac{\partial \mathring{\theta}_j(\tau)}{\partial \tau} = \mathring{y}_j \ln(\lambda)\mathring{\theta}_j(\tau). \tag{4.3.29}$$

The solution to (4.3.27) is

$$\begin{aligned} \Theta_j(\tau) &= A_j\lambda^{y_j\tau}[1 + A_s\lambda^{y_s\tau}]^{x_j}, \quad j \neq s, \\ \Theta_s(\tau) &= A_s\lambda^{y_s\tau}[1 + A_s\lambda^{y_s\tau}]^{-1}, \end{aligned} \tag{4.3.30}$$

for some constants A_j, where, with the adaption $\vartheta_j := y_j/y_s$, $\mathring{\vartheta}_j := y_j/y_s$ of the notation of (4.3.14),

$$x_j := \frac{\mathring{y}_j - y_j}{y_s} = \mathring{\vartheta}_j - \vartheta_j. \tag{4.3.31}$$

Since $\theta_j(\tau) = A_j\lambda^{y_j\tau}$ is a solution to (4.3.28),

$$\begin{aligned} \Theta_j(\tau) &= \theta_j(\tau)[1 + \theta_s(\tau)]^{x_j}, \quad j \neq s, \\ \Theta_s(\tau) &= \theta_s(\tau)[1 + \theta_s(\tau)]^{-1}. \end{aligned} \tag{4.3.32}$$

As $\mathcal{C}$ and $\mathring{\mathcal{C}}$ are interchanged by the substitution $\Theta_s \to (1 - \Theta_s)$ in (4.3.27) the solution can also be written

$$\begin{aligned} \Theta_j(\tau) &= \mathring{\theta}_j(\tau)[1 + \mathring{\theta}_s(\tau)]^{x_j}, \quad j \neq s, \\ \Theta_s(\tau) &= [1 + \mathring{\theta}_s(\tau)]^{-1}. \end{aligned} \tag{4.3.33}$$

From (4.3.32) and (4.3.33),

---

[15] Although this is not necessarily so, it is a good approximation in some cases. Thus, for example, as we see in Sect. 5.6.3, the Gaussian fixed point is unstable for $d < 4$. In the context of the ϵ-expansion version of renormalization group theory in $d = 4 - \epsilon$, this leads to the appearance of a neighbouring stable fixed point with the related exponents, for this and the Gaussian fixed point given by $\pm\epsilon + O(\epsilon^2)$ (Ma 1976a, Chap. 7).

$$\theta_j(\tau) = \Theta_j(\tau)[1 - \Theta_s(\tau)]^{x_j}, \qquad \mathring{\theta}_j(\tau) = \Theta_j(\tau)[\Theta_s(\tau)]^{x_j},$$
$$\theta_s(\tau) = [\mathring{\theta}_s(\tau)]^{-1} = \Theta_s(\tau)[1 - \Theta_s(\tau)]^{-1}. \tag{4.3.34}$$

The complete specification of all the scaling and generalized scaling fields is effected by setting values for the constants A_j. Since $\Theta_j(\tau) = 0$, for all j, are constants of the system (4.3.27), we impose the conditions that, for some $\tau = \tau_1$,

$$\Theta_1(\tau_1) = 1, \qquad \Theta_s(\tau_1) = \alpha_1, \qquad \Theta_j(\tau_1) = 0, \quad j \neq 1, s, \tag{4.3.35}$$

assuming that these values are outside the critical region. Then

$$\begin{aligned}
\theta_s(\tau) &= \theta_s(\tau_1)\lambda^{y_s(\tau-\tau_1)}, & \theta_s(\tau_1) &= \alpha_1/(1+\alpha_1),\\
\mathring{\theta}_s(\tau) &= \mathring{\theta}_s(\tau_1)\lambda^{y_s(\tau_1-\tau)}, & \mathring{\theta}_s(\tau_1) &= (1+\alpha_1)/\alpha_1,\\
\theta_1(\tau) &= \theta_1(\tau_1)\lambda^{y_1(\tau-\tau_1)}, & \theta_1(\tau_1) &= (1-\alpha_1)^{x_1},\\
\mathring{\theta}_1(\tau) &= \mathring{\theta}_1(\tau_1)\lambda^{y_1(\tau-\tau_1)}, & \mathring{\theta}_1(\tau_1) &= \alpha_1^{x_1},
\end{aligned} \tag{4.3.36}$$

with the corresponding formulae for the generalized scaling fields obtained by substituting in (4.3.32) or (4.3.33). Neglecting the zero fields, the formula

$$\psi_{\text{sing}}(\boldsymbol{k}; \Theta_1(\tau), \Theta_s(\tau)) = \lambda^{d'(\boldsymbol{k})(\tau-\tau_1)}\psi_{\text{sing}}(\boldsymbol{k}; \Theta_1(\tau_1), \Theta_s(\tau_1)) \tag{4.3.37}$$

extends the Kadanoff scaling hypothesis (4.3.7) for $\psi_{\text{sing}}(\boldsymbol{k})$. From (4.3.30),

$$\lambda^{\mathring{y}_1(\tau-\tau_1)} = \Theta_1(\tau)[\Theta_s(\tau)/\alpha_1]^{x_1} \tag{4.3.38}$$

and, substituting into (4.3.37), using the symmetry relationship between $\mathcal{C}$ and $\mathring{\mathcal{C}}$ and the notation of (4.3.12), gives

$$\psi_{\text{sing}}(\boldsymbol{k}; \Theta_1, \Theta_s) = \begin{cases} [\Theta_1]^{\mathring{\omega}_1(\boldsymbol{k})}[\Theta_s/\alpha_1]^{x_1\mathring{\omega}_1(\boldsymbol{k})}\psi_{\text{sing}}(\boldsymbol{k}; 1, \alpha_1),\\ [\Theta_1]^{\omega_1(\boldsymbol{k})}[(1-\Theta_s)/\alpha_1]^{x_1\omega_1(\boldsymbol{k})}\psi_{\text{sing}}(\boldsymbol{k}; 1, \alpha_1). \end{cases} \tag{4.3.39}$$

We now consider the variation of this quantity along a line of constant Θ_2. A constant of the motion of (4.3.32) is given by

$$(1-\alpha_1)^{\mathring{\vartheta}_1}\alpha_1^{-\vartheta_1} = \Theta_1(1-\Theta_s)^{\mathring{\vartheta}_1}\Theta_s^{-\vartheta_1}. \tag{4.3.40}$$

But, along the line of constant Θ_2, α_1 must vary. To indicate this we replace α_1 with $\alpha(\Theta_1)$ giving

$$[1-\alpha(\Theta_1)]^{\mathring{\vartheta}_1}\alpha(\Theta_1)^{-\vartheta_1} = \Theta_1[1-\alpha(1)]^{\mathring{\vartheta}_1}\alpha(1)^{-\vartheta_1} \tag{4.3.41}$$

and

$$\frac{\mathrm{d}\ln[\alpha(\Theta_1)]}{\mathrm{d}\ln|\Theta_1|} = -\left\{\mathring{\vartheta}_1\left(\frac{\alpha(\Theta_1)}{1-\alpha(\Theta_1)}\right) - \vartheta_1\right\}^{-1}. \tag{4.3.42}$$

To measure the variation of $\psi_{\text{sing}}(\boldsymbol{k})$ we adopt the *effective exponent*

$$\Omega_1(\boldsymbol{k};\Theta_1,\Theta_s) := \frac{\partial\ln\psi_{\text{sing}}(\boldsymbol{k})}{\partial\ln|\Theta_1|}. \tag{4.3.43}$$

introduced by Riedel and Wegner (1974). In the first line of (4.3.39), replacing α_1 with $\alpha(\Theta_1)$ and Θ_s with $\alpha(1)$ gives

$$\Omega_1(\boldsymbol{k};\Theta_1,\alpha(1)) = \frac{\mathring{\omega}_1(\boldsymbol{k})\mathring{\vartheta}_1[2\alpha(\Theta_1)-1]}{\mathring{\vartheta}_1\alpha(\Theta_1)-\vartheta_1[1-\alpha(\Theta_1)]} + Q(\boldsymbol{k};\Theta_1), \tag{4.3.44}$$

where

$$Q(\boldsymbol{k};\Theta_1) := \left\{\frac{\mathring{\vartheta}_1}{1-\alpha(\Theta_1)} - \frac{\vartheta_1}{\alpha(\Theta_1)}\right\}^{-1}\frac{\mathrm{d}\ln[\psi_{\text{sing}}(\boldsymbol{k};1,\alpha(\Theta_1)]}{\mathrm{d}\alpha(\Theta_1)}. \tag{4.3.45}$$

In the two limits $\alpha(\theta_1) \to 0$ and $\alpha(\theta_1) \to 1$ the limits of the first term on the right of (4.3.45) are $\omega_1(\boldsymbol{k})$ and $\mathring{\omega}_1(\boldsymbol{k})$, respectively. In each case $Q(\Theta_1) \to 0$ if, as we assume, the second term on the right of (4.3.45) has a finite, or more weakly singular limit, than the first. From (4.3.41), $\Theta_1 \to 0$ gives $\alpha(\Theta_1) \to 1$, with $\Omega_1(\boldsymbol{k};\Theta_1,\alpha(1)) \to \mathring{\omega}_1(\boldsymbol{k})$, except when $\Theta_s = \alpha(\Theta_1) = \alpha_1 = 0$, an invariant line of (4.3.27), when $\Omega_1(\boldsymbol{k};\Theta_1,\alpha(1)) \to \omega_1(\boldsymbol{k})$. That the whole of the interval $\Theta_s \in (0,1]$ is attractive to $\Theta_s = 1$ is, of course, a consequence of the sign of y_s. However, it is of interest to investigate how close to $\Theta_1 = 0$ one needs to be before the attractive effects of $\Theta_s = 1$ become evident. An indicator of this is the extent of *crossover region*[16] in the space $0 \le \Theta_1 \le 1, 0 \le \Theta_s \le 1$, defined by Riedel and Wegner (1974) as the condition[17]

$$|\,\Omega_1\,(\boldsymbol{k};\Theta_1,\alpha(1)) - \mathring{\omega}_1(\boldsymbol{k})| = \left|\frac{\mathring{\omega}(\boldsymbol{k})[1-\alpha(\Theta_1)][\vartheta_1-\mathring{\vartheta}_1]}{\mathring{\vartheta}\alpha(\Theta_1)-\vartheta_1[1-\alpha(\Theta_1)]}\right| > \delta, \tag{4.3.46}$$

for some δ. An example of a crossover region is shown in Fig. 4.6.

[16] This is different, but related, type of 'crossover' from that discussed in Sect. 4.3.1. There the change is between different types of asymptotic behaviour along different paths to a fixed point. Here it is a question of competition between two fixed points.

[17] To compute this formula the second term in (4.3.44) has been neglected.

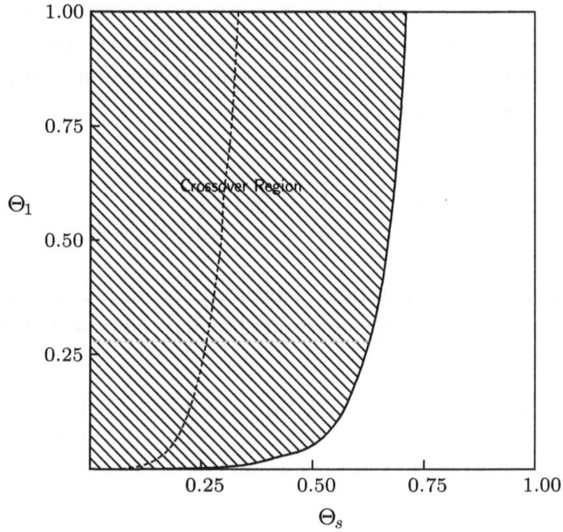

Fig. 4.6 The crossover region for $d = 2$, $\boldsymbol{k} = \boldsymbol{0}$, $y_1 = 0.5$, $\mathring{y}_1 = 1.0$, $y_s = 0.25$ and $\delta = 0.5$. The crossover curve with $y_1 = 0.9$ is shown as a *broken line*

4.3.4 The Nightingale–'T Hooft Scaling Hypothesis

Here we begin with the whole of the free-energy density ϕ and not just its singular part, replacing the Kadanoff scaling hypothesis (4.3.5) with the *Nightingale–'T Hooft scaling hypothesis* (Nightingale and 'T Hooft 1974) which supposes that, in a neighbourhood of the transition region $\mathcal{C}$

$$\phi(\lambda^{y_1}\theta_1, \ldots, \lambda^{y_n}\theta_n) = \lambda^d \phi(\theta_1, \ldots, \theta_n) - \phi_\lambda(\theta_1, \ldots, \theta_n), \tag{4.3.47}$$

for all real $\lambda > 1$. Iterating $\ell - 1$ times

$$\phi(\theta_1, \ldots, \theta_n) = \lambda^{-\ell d} \phi(\lambda^{\ell y_1}\theta_1, \ldots, \lambda^{\ell y_n}\theta_n) + \lambda^{-d} \phi_\lambda^{(\ell)}(\theta_1, \ldots, \theta_n), \tag{4.3.48}$$

where

$$\phi_\lambda^{(\ell)}(\theta_1, \ldots, \theta_n) := \sum_{r=0}^{\ell-1} \lambda^{-rd} \phi_\lambda(\lambda^{r y_1}\theta_1, \ldots, \lambda^{r y_n}\theta_n) \tag{4.3.49}$$

and, of course, $\phi_\lambda^{(1)} = \phi_\lambda$.

We pause to note that (4.3.48) is the result of the repeated application of the transformation (4.3.3), resulting, for an initially chosen set of values, in a trajectory in the space of scaling fields. Of course, a similar iteration could be applied to the Kadanoff relationship (4.3.5). In the case of a relevant scaling field $\theta_j \lambda^{\ell y_j} \to \infty$ and for an irrelevant scaling field $\theta_j \lambda^{\ell y_j} \to 0$. The problem with the former is clear since the validity of the scaling hypotheses have been assumed only within some neighbourhood of $\mathcal{C}$; this is discussed in Sect. 4.3.5. Problems in the latter category arise for so-called *'dangerous' irrelevant exponents* which are discussed in Sect. 4.8.

Since $\phi_\lambda(\theta_1,\ldots,\theta_n)$ is a smooth function of the scaling fields, it has a power series expansion of the form

$$\phi_\lambda(\theta_1,\ldots,\theta_n) = \sum_{\{\boldsymbol{q}\}} B_\lambda(\boldsymbol{q}) \prod_{j=1}^{n} \frac{\theta_j^{q_j}}{q_j!}, \tag{4.3.50}$$

where the summation is over $q_j = 0, 1, 2, \ldots$, for all $j = 1, \ldots, n$, the set $(q_1, \ldots, q_n)$ being represented by the vector $\boldsymbol{q}$. Substituting into (4.3.49) gives

$$\phi_\lambda^{(\ell)}(\theta_1,\ldots,\theta_n) = \sum_{\{\boldsymbol{q}\}} B_\lambda(\boldsymbol{q}) \Lambda_\lambda^{(\ell)}(d'(\boldsymbol{q})) \prod_{j=1}^{n} \frac{\theta_j^{q_j}}{q_j!}, \tag{4.3.51}$$

where

$$\Lambda_\lambda^{(\ell)}(x) := \sum_{r=0}^{\ell-1} \lambda^{-rx} = \begin{cases} \ell, & x = 0, \\ \dfrac{1-\lambda^{-x\ell}}{1-\lambda^{-x}}, & x \neq 0, \end{cases} \tag{4.3.52}$$

and $d'(\boldsymbol{q})$ is as defined in (4.3.8).

For the function $\psi(\boldsymbol{k})$ obtained by differentiation ϕ, k_j times with respect to θ_j, for $j = 1, \ldots, n$,

$$\begin{aligned} \psi(\boldsymbol{k}; \theta_1,\ldots,\theta_n) = \lambda^{-\ell d'(\boldsymbol{k})} \psi(\boldsymbol{k}; \lambda^{\ell y_1}\theta_1, \ldots, \lambda^{\ell y_n}\theta_n) \\ + \lambda^{-d} \psi_\lambda^{(\ell)}(\boldsymbol{k}; \theta_1,\ldots,\theta_n), \end{aligned} \tag{4.3.53}$$

where

$$\psi_\lambda^{(\ell)}(\boldsymbol{k}; \theta_1,\ldots,\theta_n) = \sum_{\{\boldsymbol{q}\}}{}' B_\lambda(\boldsymbol{q}) \Lambda_\lambda^{(\ell)}(d'(\boldsymbol{q})) \prod_{j=1}^{n} \frac{\theta_j^{q_j-k_j}}{(q_j-k_j)!}, \tag{4.3.54}$$

with the summations in each case now restricted to $q_j \geq k_j$, for $j = 1, \ldots, n$.

4.3.5 Constraints on Scaling

Having developed the formalism of scaling it is necessary to consider the limits of its applicability. As indicated above these arise when we consider the repeated application of the transformation (4.3.3). This will result in the arguments in the first term on the right-hand side of (4.3.48) tending to zero and infinity for irrelevant and relevant scaling fields, respectively. The former will cause a problem if $\phi(\theta_1, \ldots, \theta_n)$ is singular when an irrelevant scaling field is zero. For the present discussion we shall assume that this possibility, which is discussed in Sect. 4.8, does not arise and consider only the need to remain within the neighbourhood of validity of the scaling hypothesis. For any λ this imposes an upper bound on the number of iterations of (4.3.3). Suppose that the neighbourhood of validity of the scaling hypothesis (4.3.47) is given by

$$\mu(\theta_1, \ldots, \theta_s) \leq R, \tag{4.3.55}$$

where

$$\mu(\theta_1, \ldots, \theta_s) := [|\theta_1|^{2/y_1} + \cdots + |\theta_s|^{2/y_s}]^{\frac{1}{2}} \tag{4.3.56}$$

is the metric which measures distances from $\mathcal{C}$ and which scales according to the formula

$$\mu(\lambda^{y_1}\theta_1, \ldots, \lambda^{y_s}\theta_s) = \lambda\mu(\theta_1, \ldots, \theta_s). \tag{4.3.57}$$

Then the iterative process remains in the scaling neighbourhood if

$$\lambda^{\ell} \leq \frac{R}{\mu(\theta_1, \ldots, \theta_s)}. \tag{4.3.58}$$

Given a particular set of values for the relevant scaling fields, suppose that (4.3.57) is iterated until

$$\begin{aligned} \ell = L(\lambda; \theta_1, \theta_2, \ldots, \theta_s) &:= \frac{1}{\ln(\lambda)} \ln\left[\frac{R}{\mu(\theta_1, \ldots, \theta_s)}\right] \\ &= \frac{1}{\ln(\lambda)} \ln\left[\frac{\mu(\tilde{\theta}_1, \ldots, \tilde{\theta}_s)}{\mu(\theta_1, \ldots, \theta_s)}\right], \end{aligned} \tag{4.3.59}$$

when a point on the boundary is achieved with (4.3.58) an equality[18] and the relevant scaling fields with values $\tilde{\theta}_j = \lambda^{Ly_j}\theta_j$, $j = 1, 2, \ldots, s$.

In the interests of brevity it is convenient to remember that $\phi(\theta_1, \ldots, \theta_n) = \psi(\mathbf{0}; \theta_1, \ldots, \theta_n)$, so that the discussion can concentrate simply on the function $\psi(\boldsymbol{k}; \theta_1, \ldots, \theta_n)$. Substituting into (4.3.53) for $\ell = L$ gives

[18] The assumption here is that λ is such that equality in (4.3.58) can be achieved for an integer ℓ.

$$\psi(\boldsymbol{k};\theta_1,\dots,\theta_n) = \left[\frac{\mu(\theta_1,\dots,\theta_s)}{\mu(\tilde{\theta}_1,\dots,\tilde{\theta}_s)}\right]^{d'(\boldsymbol{k})}\psi(\boldsymbol{k};\tilde{\theta}_1,\dots,\tilde{\theta}_n) + \lambda^{-d}\psi_\lambda^{(L)}(\boldsymbol{k};\theta_1,\dots,\theta_n). \tag{4.3.60}$$

The first term on the right-hand side of this equation is a generalized homogeneous functions of the relevant scaling fields and $\psi(\boldsymbol{k};\theta_1,\dots,\theta_n)$ must be independent of λ. To achieve this a specific form must be imposed on the coefficients $B_\lambda(\boldsymbol{q})$. To see what this must be we choose some relevant scaling field θ_m when, from (4.3.13), (4.3.56) and (4.3.59),

$$L(\lambda;\theta_1,\theta_2,\dots,\theta_s) = -\frac{\ln|\theta_m|}{y_m\ln(\lambda)} + \frac{1}{\ln(\lambda)}\ln\left[\frac{\mu(\tilde{\theta}_1,\dots,\tilde{\theta}_s)}{\mu(g_1^{(m)},\dots,g_s^{(m)})}\right]. \tag{4.3.61}$$

Now consider the path $\mathcal{P}$ towards $\mathcal{C}$, described in Sect. 4.3.1, where the limiting values of the gap fields are all finite. From (4.3.52) and (4.3.61),

$$\Lambda_\lambda^{(L)}(x) = \begin{cases} -A_m^x|\theta_m|^{x/y_m}/(1-\lambda^{-x}), & \text{if } x<0,\\ -\ln|\theta_m|/[y_m\ln(\lambda)], & \text{if } x=0,\\ 1/(1-\lambda^{-x}), & \text{if } x>0, \end{cases} \tag{4.3.62}$$

where

$$A_m := \frac{\mu(g_1^{(m)},\dots,g_s^{(m)})}{\mu(\tilde{\theta}_1,\dots,\tilde{\theta}_s)}, \tag{4.3.63}$$

which is non-zero since $g_m^{(m)} = \pm 1$, according to the direction of approach to $\mathcal{C}$. Then substituting into (4.3.51) and using (4.3.62) it follows that $\lambda^{-d}\psi_\lambda^{(L)}(\boldsymbol{k};\theta_1,\dots,\theta_n)$ is independent of λ if

$$B_\lambda(\boldsymbol{q}) := \bar{B}(\boldsymbol{q}) \times \begin{cases} \lambda^d(1-\lambda^{-d'(\boldsymbol{q})}), & \text{if } d'(\boldsymbol{q}) \neq 0,\\ \lambda^d\ln(\lambda), & \text{if } d'(\boldsymbol{q}) = 0, \end{cases} \tag{4.3.64}$$

where the coefficients $\bar{B}(\boldsymbol{q})$ are independent of λ. Substituting into (4.3.4) for $\ell = 1$ gives

$$\lambda^{-d}\psi_\lambda(\boldsymbol{k};\theta_1,\dots,\theta_n) = \bar{\psi}(\boldsymbol{k};\theta_1,\dots,\theta_n) - \lambda^{-d'(\boldsymbol{k})}\bar{\psi}(\boldsymbol{k};\lambda^{y_1}\theta_1,\dots,\lambda^{y_n}\theta_n) + \ln(\lambda)\xi(\boldsymbol{k};\theta_1,\dots,\theta_n), \tag{4.3.65}$$

where

$$\bar{\psi}(\boldsymbol{k};\theta_1,\ldots,\theta_n) := \sum_{\{\boldsymbol{q}\}}{}' \bar{B}(\boldsymbol{q}) \prod_{j=1}^{n} \frac{\theta_j^{q_j-k_j}}{(q_j-k_j)!}, \tag{4.3.66}$$

$$\xi(\boldsymbol{k};\theta_1,\ldots,\theta_n) := \sum_{\{\boldsymbol{q}\}}^{[d'(\boldsymbol{q})=0]'} \bar{B}(\boldsymbol{q}) \prod_{j=1}^{n} \frac{\theta_j^{q_j-k_j}}{(q_j-k_j)!}, \tag{4.3.67}$$

are smooth functions of the scaling fields.[19] Substituting into (4.3.3), again with $\ell = 1$, and identifying

$$\psi_{\mathrm{smth}}(\boldsymbol{k};\theta_1,\ldots,\theta_n) := \bar{\psi}(\boldsymbol{k};\theta_1,\ldots,\theta_n), \tag{4.3.68}$$

shows that

$$\psi_{\mathrm{sing}}(\boldsymbol{k};\theta_1,\ldots,\theta_n) := \psi(\boldsymbol{k};\theta_1,\ldots,\theta_n) - \bar{\psi}(\boldsymbol{k};\theta_1,\ldots,\theta_n), \tag{4.3.69}$$

for which

$$\psi_{\mathrm{sing}}(\boldsymbol{k};\lambda^{y_1}\theta_1,\ldots,\lambda^{y_n}\theta_n) = \lambda^{d'(\boldsymbol{k})}\big[\psi_{\mathrm{sing}}(\boldsymbol{k};\theta_1,\ldots,\theta_n) - \ln(\lambda)\xi(\boldsymbol{k};\theta_1,\ldots,\theta_n)\big], \tag{4.3.70}$$

will contain all the singular character of $\psi(\boldsymbol{k})$. The division of $\psi(\boldsymbol{k})$ into smooth and singular parts is represented by (4.3.68) and (4.3.69). The Kadanoff scaling (4.3.7) is replaced by (4.3.70) and is satisfied in its original form only when

$$\xi(\boldsymbol{k};\theta_1,\ldots,\theta_n) \equiv 0. \tag{4.3.71}$$

Setting λ according to (4.3.9), from (4.3.12)–(4.3.14), (4.3.69) and (4.3.70)

$$\begin{aligned}\psi_{\mathrm{sing}}(\boldsymbol{k};\theta_1,\ldots,\theta_n) &= |\theta_m|^{\omega_m(\boldsymbol{k})}\psi_{\mathrm{sing}}(\boldsymbol{k};g_1^{(m)},\ldots,g_n^{(m)})\\ &\quad - |\theta_m|^{\omega_m(\boldsymbol{k})}\ln|\theta_m|y_m^{-1}\sum_{\{\boldsymbol{q}\}}^{[d'(\boldsymbol{q})=0]'} \bar{B}(\boldsymbol{q})\prod_{j=1}^{n}\frac{[g_j^{(m)}]^{q_j-k_j}}{(q_j-k_j)!},\end{aligned} \tag{4.3.72}$$

replaces (4.3.11), with, from (4.3.66) and (4.3.69),

[19] It should be noted that the subset of terms included in (4.3.67) can also be included in (4.3.66) since they cancel out in (4.3.65).

$$\psi(\boldsymbol{k};\theta_1,\ldots,\theta_n) = |\theta_m|^{\omega_m(\boldsymbol{k})}\psi_{\text{sing}}(\boldsymbol{k};g_1^{(m)},\ldots,g_n^{(m)})$$
$$+\,|\theta_m|^{\omega_m(\boldsymbol{k})}\sum_{\{\boldsymbol{q}\}}{}' \bar{B}(\boldsymbol{q})|\theta_m|^{-\omega_m(\boldsymbol{q})}\prod_{j=1}^{n}\frac{[g_j^{(m)}]^{q_j-k_j}}{(q_j-k_j)!}$$
$$-\,|\theta_m|^{\omega_m(\boldsymbol{k})}\ln|\theta_m|y_m^{-1}\sum_{\{\boldsymbol{q}\}}^{[d'(\boldsymbol{q})=0]'}\bar{B}(\boldsymbol{q})\prod_{j=1}^{n}\frac{[g_j^{(m)}]^{q_j-k_j}}{(q_j-k_j)!}. \tag{4.3.73}$$

An explicit asymptotic form for $\psi(\boldsymbol{k};\theta_1,\ldots,\theta_n)$ as $\mathcal{C}$ is approached along the path $\mathcal{P}$ can be derived from (4.3.60). Using (4.3.13)–(4.3.15) and (4.3.18),

$$\left[\frac{\mu(\theta_1,\ldots,\theta_s)}{\mu(\tilde{\theta}_1,\ldots,\tilde{\theta}_s)}\right]^{d'(\boldsymbol{k})}\psi(\boldsymbol{k};\tilde{\theta}_1,\ldots,\tilde{\theta}_n)\simeq\ |\theta_m|^{\omega_m(\boldsymbol{k})}A_m^{d'(\boldsymbol{k})}\psi(\boldsymbol{k};\tilde{\theta}_1,\ldots,\tilde{\theta}_n). \tag{4.3.74}$$

This term reproduces the power-law behaviour obtained in (4.3.20). The other contribution is, from (4.3.54) and (4.3.62)–(4.3.64),

$$\lambda^{-d}\psi_L^{(\lambda)}(\boldsymbol{k};\theta_1,\ldots,\theta_n) = -|\theta_m|^{\omega_m(\boldsymbol{k})}\sum_{\{\boldsymbol{q}\}}^{[d'(\boldsymbol{q})>0]'}\bar{B}(\boldsymbol{q})\prod_{j=1}^{n}\frac{[g_j^{(m)}]^{q_j-k_j}}{(q_j-k_j)!}$$
$$+\,|\theta_m|^{\omega_m(\boldsymbol{k})}\ln|\theta_m|y_m^{-1}\sum_{\{\boldsymbol{q}\}}^{[d'(\boldsymbol{q})=0]'}\bar{B}(\boldsymbol{q})\prod_{j-1}^{n}\frac{[g_j^{(m)}]^{q_j-k_j}}{(q_j-k_j)!}$$
$$-\,|\theta_m|^{\omega_m(\boldsymbol{k})}\sum_{\{\boldsymbol{q}\}}^{[d'(\boldsymbol{q})<0]'}|\theta_m|^{-\omega_m(\boldsymbol{q})}\bar{B}(\boldsymbol{q})\prod_{j=1}^{n}\frac{[g_j^{(m)}]^{q_j-k_j}}{(q_j-k_j)!}. \tag{4.3.75}$$

A term in these sums is zero if $q_j \neq k_j$, for any j for which $g_j^{(m)} = 0$. In particular to have a non-zero term q_j must be equal to k_j for all $j = s+1,\ldots,n$, since $g_j^{(m)} \to \varpi_j^{(m)} = 0$ for all irrelevant fields. The residual terms in the first sum give contributions of the same power-law type as (4.3.74) and can therefore, in general, be ignored, except in the case when the amplitudes in (4.3.74) are not non-zero and finite, when they may provide the only terms with this asymptotic behaviour. The exponent of $|\theta_m|$ for a term in the third sum is positive and less that $\omega_m(\boldsymbol{k})$, so this contribution will be negligible when $\psi(\boldsymbol{k})$ has a power-law singularity as $\mathcal{C}$ is approached along $\mathcal{P}$. The interesting and important terms are, as in (4.3.73), in the second sum. These will occur only when

$$d = \sum_{j=1}^{n} q_j y_j \tag{4.3.76}$$

has a solution, or solutions, for some set of non-negative integers $q_1, \ldots, q_n$. If we choose $k_1, \ldots, k_n$ to be such a solution, then $\psi(\boldsymbol{k})$ is such that (4.3.25) is satisfied and

$$\psi(\boldsymbol{k}; \theta_1, \ldots, \theta_n) \simeq \bar{B}(\boldsymbol{k}) y_m^{-1} \ln |\theta_m|. \qquad (4.3.77)$$

This particular derivative of ϕ exhibits a logarithmic singularity as $\mathcal{C}$ is approached along $\mathcal{P}$, as long as $\bar{B}(\boldsymbol{k}) \neq 0$.[20] The occurrence of logarithmic divergences in thermodynamic response functions are the first indication that the scaling ideas of Sect. 4.3 should be treated with some caution, and need modification in particular circumstances. Further modifications, arising from zero or singular amplitude factors, are considered in Sect. 4.8.

By construction

$$\mathcal{K}_0(\theta_1, \ldots, \theta_n) := \lambda^{-d} \phi_L^{(\lambda)}(\theta_1, \ldots, \theta_n) \qquad (4.3.78)$$

is independent of λ except through the definition (4.3.59) of L. Substituting into (4.3.48)

$$\lambda^{-Ld} \phi(\lambda^{\ell y_1} \theta_1, \ldots, \lambda^{L y_n} \theta_n) = \phi(\theta_1, \ldots, \theta_n) - \mathcal{K}_0(\theta_1, \ldots, \theta_n). \qquad (4.3.79)$$

With

$$\lambda^L = \frac{R}{\mu(\theta_1, \ldots, \theta_s)} > 1 \qquad (4.3.80)$$

replaced by λ this is the linearized version of the scaling formula for the free-energy density derived from the renormalization group (Sect. 15.4 and Niemeijer and van Leeuwen 1976, p. 430).

4.3.6 Scaling Operators and Dimensions

For the sake of brevity we adopt the notation of Sect. 2.5, where ζ_j, $j = 1, 2, \ldots, n$, represents the complete set of internal and external couplings. Suppose that the formulae (4.3.1) are invertible near $\mathcal{C}$ to give

$$\zeta_j = \zeta_j(\theta_1, \theta_2, \ldots, \theta_n), \qquad j = 1, 2, \ldots, n, \qquad (4.3.81)$$

[20] From a renormalization group point of view this divergence arises from the self-energy of the cells, whereas the more usual power-law singularities arise from the interaction between cells.

with

$$\zeta_{jc}(\theta_{s+1}, \dots, \theta_n) := \zeta_j(0, \dots, 0, \theta_{s+1}, \dots, \theta_n), \tag{4.3.82}$$

$$\zeta_j^\star := \zeta_{jc}(0, \dots, 0). \tag{4.3.83}$$

Close to $\mathcal{C}$, to linear terms,

$$\zeta_j(\theta_1, \theta_2, \dots, \theta_n) = \zeta_{jc}(\theta_{s+1}, \dots, \theta_n) + \sum_{i=1}^{s} \theta_i \zeta_{jc}^{(i)}(\theta_{s+1}, \dots, \theta_n), \tag{4.3.84}$$

where

$$\zeta_{jc}^{(i)}(\theta_{s+1}, \dots, \theta_n) := \left(\frac{\partial \zeta_j}{\partial \theta_i}\right)_{\theta_1 = \cdots = \theta_s = 0}. \tag{4.3.85}$$

Substituting from (4.3.84) into the Hamiltonian (2.5.2) gives, to linear terms in the relevant scaling fields,

$$\widehat{H}(N, \zeta_0, \theta_1, \dots, \theta_n; \boldsymbol{\sigma}) = \widehat{H}(N, \zeta_0, \zeta_{jc}; \boldsymbol{\sigma}) + \sum_{i=1}^{s} \Phi_i(\boldsymbol{\sigma})\theta_i \tag{4.3.86}$$

where, from (2.5.1), the coefficients

$$\Phi_i(\boldsymbol{\sigma}) := \sum_{j=1}^{n} \widehat{C}_j(\boldsymbol{\sigma}) \zeta_{jc}^{(i)}(\theta_{s+1}, \dots, \theta_n), \tag{4.3.87}$$

are functions of the irrelevant scaling fields. Close to the fixed point these functions can be replaced by $\zeta_j^{(i)\star} := \zeta_{jc}^{(i)}(0, \dots, 0)$ and the Hamiltonian in $\mathcal{C}$ expanded to give

$$\widehat{H}(N, \zeta_0, \zeta_{jc}; \boldsymbol{\sigma}) = \widehat{H}(N, \zeta_0, 0, \dots, 0, \theta_{s+1}, \dots, \theta_n; \boldsymbol{\sigma}) = N\zeta_0 + \sum_{i=s+1}^{n} \Phi_i(\boldsymbol{\sigma})\theta_i. \tag{4.3.88}$$

Defining

$$\widehat{H}_{\text{sing}}(N, \theta_1, \dots, \theta_n; \boldsymbol{\sigma}) := \sum_{i=1}^{n} \Phi_i(\boldsymbol{\sigma})\theta_i \tag{4.3.89}$$

yields a potential ϕ satisfying condition (2.5.3) with $\phi_{\text{smth}} = \zeta_0$. From (4.3.88) and (4.3.89),

$$\langle \widehat{H} \rangle = N\zeta_0 + \sum_{i=1}^{n} \langle \Phi_i(\boldsymbol{\sigma}) \rangle \theta_i = N\zeta_0 + \sum_{i=1}^{n} \theta_i \frac{\partial \phi_{\text{sing}}}{\partial \theta_i}. \tag{4.3.90}$$

Denoting by $\boldsymbol{i}_i$ the n-dimensional unit vector with one in the ith place, from (4.3.11)–(4.3.14),

$$\sum_{i=1}^{n} \theta_i \frac{\partial \phi_{\text{sing}}}{\partial \theta_i} = |\theta_m|^{d/y_m} \sum_{i=1}^{n} g_i^{(m)} \psi_{\text{sing}}(\boldsymbol{i}_i; g_1^{(m)}, \ldots, g_n^{(m)}). \tag{4.3.91}$$

By approaching $\mathcal{C}$ along the path $\mathcal{P}$ where the limit (4.3.18) applies it can be seen that[21]

$$\phi_{\text{smth}}(\theta_{s+1}, \ldots, \theta_n) := \zeta_0(\theta_{s+1}, \ldots, \zeta_n) = \langle \widehat{H} \rangle_{\text{c}}/N \tag{4.3.92}$$

with

$$\phi_{\text{sing}}(\theta_1, \ldots, \theta_n) := \phi(0, \theta_1, \ldots, \theta_n), \tag{4.3.93}$$

$$\phi(\zeta_0, \theta_1, \ldots, \theta_n) = \phi_{\text{smth}}(\theta_1, \ldots, \theta_n) + \phi_{\text{sing}}(\theta_1, \ldots, \theta_n). \tag{4.3.94}$$

Substituting from (2.5.1) into (4.3.59) and then into (4.3.89) gives

$$\widehat{H}_{\text{sing}}(N, \theta_1, \ldots, \theta_n; \boldsymbol{\sigma}, \{\boldsymbol{r}\}) := \sum_{i=1}^{n} \sum_{\{\boldsymbol{r}\}} \phi_i(\boldsymbol{r}; \boldsymbol{\sigma}) \theta_i \tag{4.3.95}$$

where

$$\phi_i(\boldsymbol{r}; \boldsymbol{\sigma}) := \sum_{j=1}^{n} \mathfrak{c}_j(\boldsymbol{r}; \boldsymbol{\sigma}) \zeta_j^{(i)\star} \tag{4.3.96}$$

are *scaling operators*. In (4.3.95) the explicit dependence of the Hamiltonian on the set of lattice sites has been exposed. The scaling of the lattice vector $\boldsymbol{r}$ is given by (4.2.5). Suppose that the scaling operators transform according to

$$\phi(\boldsymbol{r}; \boldsymbol{\sigma}) \to \tilde{\phi}(\boldsymbol{r}; \boldsymbol{\sigma}) := \phi(\tilde{\boldsymbol{r}}; \boldsymbol{\sigma}) = \lambda^{x_i} \phi(\boldsymbol{r}; \boldsymbol{\sigma}). \tag{4.3.97}$$

To determine the scaling behaviour of $\widehat{H}_{\text{sing}}$, needed to give the scaling formula for ϕ_{sing}, it must be remembered that it is more directly related to $\Phi_{\text{sing}} = N\phi_{\text{sing}}$, and when the sum over lattice sites is transformed by the Jacobian λ^{-d} the Hamiltonian will be unchanged. Thus

---

[21] It should be noted that, if (4.3.11) is replaced by (4.3.72) this result still applies since the scaling exponents satisfy (4.3.22).

$$\widehat{H}_{\text{sing}}(N, \theta_1, \ldots, \theta_n; \boldsymbol{\sigma}, \{\boldsymbol{r}\}) \to \widehat{H}_{\text{sing}}(N, \lambda^{y_1}\theta_1, \ldots, \lambda^{y_n}\theta_n; \boldsymbol{\sigma}, \{\boldsymbol{r}/\lambda\})$$

$$= \sum_{i=1}^{n} \sum_{\{\boldsymbol{r}\}} \lambda^{x_i + y_i - d}\, \phi_i\, (\boldsymbol{r}; \boldsymbol{\sigma})\theta_i = \widehat{H}_{\text{sing}}(N, \theta_1, \ldots, \theta_n; \boldsymbol{\sigma}, \{\boldsymbol{r}\}), \quad (4.3.98)$$

giving the *scaling dimensions* of the scaling operators

$$x_j := d - y_j. \quad (4.3.99)$$

4.3.7 Correlation Functions

The development of Sect. 2.5 can be carried out for scaling fields and scaling operators just as well as for couplings and state operators. Thus for an arbitrary set of τ scaling operators[22]

$$\langle \phi_1(\lambda^{-1}\boldsymbol{r}_1) \cdots \phi_\tau\, (\lambda^{-1}\boldsymbol{r}_\tau)\rangle = \lambda^z \langle \phi_1(\boldsymbol{r}_1) \cdots \phi_\tau\, (\boldsymbol{r}_\tau)\rangle, \quad (4.3.100)$$

$$\Gamma_\tau(\phi_1(\lambda^{-1}\boldsymbol{r}_1), \ldots, \phi_\tau(\lambda^{-1}\boldsymbol{r}_\tau)) = \lambda^z \Gamma_\tau(\phi_1(\boldsymbol{r}_1), \ldots, \phi_\tau(\boldsymbol{r}_\tau)), \quad (4.3.101)$$

$$\Gamma^*_\tau(\phi_1, \ldots, \phi_\tau; \lambda\boldsymbol{k}_1, \ldots, \lambda\boldsymbol{k}_\tau) = \lambda^{z - \tau d} \Gamma^*_\tau(\phi_1, \ldots, \phi_\tau; \boldsymbol{k}_1, \ldots, \boldsymbol{k}_\tau),$$

$$(4.3.102)$$

where

$$z = \sum_{j=1}^{\tau} x_j. \quad (4.3.103)$$

It is straightforward to check that these scaling relationships agree with the scaling for ϕ_{sing} through the forms

$$\sum_{\{\boldsymbol{r}_1, \ldots, \boldsymbol{r}_\tau\}} \Gamma_\tau(\phi_1(\boldsymbol{r}_1), \ldots, \phi_\tau(\boldsymbol{r}_\tau)) = -N \frac{\partial^\tau \phi_{\text{sing}}}{\partial\theta_1 \cdots \partial\theta_\tau}, \quad (4.3.104)$$

$$\Gamma^*_\tau(\phi_1, \ldots, \phi_\tau; \boldsymbol{0}, \ldots, \boldsymbol{0}) = -\frac{1}{N} \frac{\partial^\tau \phi_{\text{sing}}}{\partial\theta_1 \cdots \partial\theta_\tau} \quad (4.3.105)$$

of (2.5.14) and (2.5.22). When the system has translational symmetry the two-point and three-point correlation functions satisfy

---

[22] For easy of presentation, we have omitted the dependence on the scaling fields.

$$\Gamma_2(\phi_1(\mathbf{0}), \phi_2(\bar{\boldsymbol{r}}_1/\lambda)) = \lambda^{x_1+x_2}\Gamma_2(\phi_1(\mathbf{0}), \phi_2(\bar{\boldsymbol{r}}_1)), \tag{4.3.106}$$

$$\Gamma_3(\phi_1(\mathbf{0}), \phi_2(\bar{\boldsymbol{r}}_1/\lambda)), \phi_3(\bar{\boldsymbol{r}}_2/\lambda)) = \lambda^{x_1+x_2+x_3}\Gamma_3(\phi_1(\mathbf{0}), \phi_2(\bar{\boldsymbol{r}}_1), \phi_3(\bar{\boldsymbol{r}}_2)), \tag{4.3.107}$$

in terms of the relative lattice vectors $\bar{\boldsymbol{r}}_1 := \boldsymbol{r}_2 - \boldsymbol{r}_1$ and $\bar{\boldsymbol{r}}_2 := \boldsymbol{r}_3 - \boldsymbol{r}_1$. Making the substitution $\lambda = r := |\bar{\boldsymbol{r}}_1|$ in (4.3.106) and (4.3.107) gives

$$\Gamma_2(\phi_1(\mathbf{0}), \phi_2(\bar{\boldsymbol{r}}_1)) = \frac{1}{r^{x_1+x_2}}\Gamma_2(\phi_1(\mathbf{0}), \phi_2(\bar{\boldsymbol{r}}_1/r)), \tag{4.3.108}$$

$$\Gamma_3(\phi_1(\mathbf{0}), \phi_2(\bar{\boldsymbol{r}}_1), \phi_3(\bar{\boldsymbol{r}}_2)) = \frac{1}{r^{x_1+x_2+x_3}}\Gamma_3(\phi_1(\mathbf{0}), \phi_2(\bar{\boldsymbol{r}}_1/r), \phi_3(\bar{\boldsymbol{r}}_2/r)). \tag{4.3.109}$$

On the transition surface the asymptotic dependence of Γ_2 on r, in the limit of large r, is given by setting $\theta_1 = \cdots = \theta_s = 0$. As long as the amplitude factor on the right-hand side of (4.3.108) is non-zero and finite we have

$$\Gamma_2(\phi_1(\mathbf{0}), \phi_2(\bar{\boldsymbol{r}}_1)) \simeq \frac{G_2(\phi_1, \phi_2)}{r^{x_1+x_2}}, \tag{4.3.110}$$

where $G_2(\phi_1, \phi_2)$ is a constant dependent only on the choice of scaling operators. A similar expression

$$\bar{\Gamma}_2^*(\phi_1, \phi_2; \bar{\boldsymbol{k}}_1) \simeq \frac{G_2^*(\phi_1, \phi_2)}{k^{y_1+y_2-d}} \tag{4.3.111}$$

can be obtained for the Fourier transform taken with respect to the relative lattice vectors and for the total pair correlation function

$$\langle\phi_1(\mathbf{0})\, \phi_2\, (\bar{\boldsymbol{r}}_1)\rangle \simeq \frac{C_2(\phi_1, \phi_2)}{r^{x_1+x_2}}. \tag{4.3.112}$$

In this context it is a common terminology (Kadanoff 1969b) to say that ϕ_j *scales as* $1/r^{x_j}$. This simply means that, in the expectation value each scaling operator, ϕ_j contributes a scaling factor $1/r^{x_j}$.

4.3.8 Variable Scaling Exponents

Throughout the discussion in the previous sections we have assumed that the scaling exponents $y_1, \ldots, y_n$ are constants. There are, however, a number of models, the most well-known being the eight-vertex model, in which the critical exponents are functions of one of the couplings (Sects. 12.3.8 and 12.3.11). The way that this behaviour can be reconciled with the ideas of scaling has been explored by Kadanoff and Wegner (1971), using the *reduction hypothesis* of Kadanoff (1969a, b). We shall now present a simplified version of their argument.

If $|\bar{\boldsymbol{r}}_1| \sim |\bar{\boldsymbol{r}}_2|$, the length-scale dependence of Γ_3 on the right-hand side of (4.3.109) can be neglected, giving

$$\Gamma_3(\phi_1(\mathbf{0}), \phi_2(\bar{\boldsymbol{r}}_1), \phi_3(\bar{\boldsymbol{r}}_2)) \sim \frac{1}{r^{x_3}} \Gamma_2(\phi_1(\mathbf{0}), \phi_2(\bar{\boldsymbol{r}}_1)). \tag{4.3.113}$$

From the formulae equivalent to (2.5.13) and (2.5.14) for scaling fields and operators

$$\frac{\partial \Gamma_2(\phi_1(\mathbf{0}), \phi_2(\bar{\boldsymbol{r}}_1))}{\partial \theta_3} = \sum_{\{\bar{\boldsymbol{r}}_2\}} \Gamma_3(\phi_1(\mathbf{0}), \phi_2(\bar{\boldsymbol{r}}_1), \phi_3(\bar{\boldsymbol{r}}_2)), \tag{4.3.114}$$

and using (18.1.35) to replace the summation by an integral over $[0, r]$ we have

$$\frac{\partial \Gamma_2(\phi_1(\mathbf{0}), \phi_2(\bar{\boldsymbol{r}}_1))}{\partial \theta_3} \sim \begin{cases} \dfrac{1}{r^{x_1+x_2-y_3}}, & \text{if } y_3 \neq 0, \\ \dfrac{\ln(r)}{r^{x_1+x_2}}, & \text{if } y_3 = 0. \end{cases} \tag{4.3.115}$$

The case for $y_3 \neq 0$ is easily understood on differentiating (4.3.108) with respect to θ_3. On the right-hand side θ_3 appears in the combination $r^{y_3}\theta_3$ and the leading term in the derivative of $\Gamma_2(\phi_1(\mathbf{0}), \phi_2(\bar{\boldsymbol{r}}_1/r))$ contributes a factor r^{y_3}. When $y_3 = 0$, ϕ_3 scales as $1/r^d$. It would appear that in this case differentiation of (4.3.108) leaves the scaling dependence on r unchanged. Suppose however that y_1 and y_2 are functions of θ_3. Then

$$\frac{\partial \Gamma_2(\phi_1(\mathbf{0}), \phi_2(\bar{\boldsymbol{r}}_1))}{\partial \theta_3} \sim \frac{\ln(r)}{r^{x_1+x_2}} \left(\frac{\partial y_1}{\partial \theta_3} + \frac{\partial y_2}{\partial \theta_3} \right) \tag{4.3.116}$$

and the required logarithmic dependence is obtained. A scaling field with zero exponent is marginal (Sect. 4.3) and in renormalization group analysis is often associated with a surface or line of fixed points. We have now shown that, if such a field is present, it is not inconsistent with scaling theory that the other scaling exponents should be functions of that field. A marginal scaling field or coupling represents an 'underlying' parameter of the system. Scaling theory can proceed without taking it into account. Once the phase diagram is understood the additional dimension represented by a marginal parameter simply increases the dimension of the phase space, without affecting the type of phenomena observed. The application of this analysis to the eight-vertex model is considered in Sect. 12.3.11.

4.3.9 Densities and Response Functions

Before considering the application of scaling to some particular cases it is useful to relate the general ideas of Sect. 4.3 to the scaling properties of more familiar thermodynamic functions. The differential form (4.1.3) contains two types of densities: ρ_i, $i = 1, \ldots, n_{\mathrm{f}} - 1$ are associated with external fields and are quantities like magnetization density or particle density; $\mathfrak{u}_\ell$, $\ell = 1, \ldots, n_{\mathrm{e}}$ are the energy-densities associated with the types of inter-particle interaction. The linear sum in (2.3.13) gives the total dimensionless internal energy density υ, and from (4.1.3)

$$\mathfrak{u}_\ell = -\frac{\partial \phi}{\partial K_\ell}, \qquad \rho_i = -\frac{\partial \phi}{\partial L_i}. \tag{4.3.117}$$

Let

$$\rho_{i\mathrm{c}} := -\frac{\partial \phi_{\mathrm{smth}}}{\partial L_i}, \qquad \mathfrak{u}_{\ell\mathrm{c}} := -\frac{\partial \phi_{\mathrm{smth}}}{\partial K_\ell}. \tag{4.3.118}$$

Then

$$\begin{aligned} \rho_i - \rho_{i\mathrm{c}} &= -\frac{\partial \phi_{\mathrm{sing}}}{\partial L_i} = -\sum_{j=1}^{n} \frac{\partial \phi_{\mathrm{sing}}}{\partial \theta_j} \frac{\partial \theta_j}{\partial L_i}, \\ \mathfrak{u}_\ell - \mathfrak{u}_{\ell\mathrm{c}} &= -\frac{\partial \phi_{\mathrm{sing}}}{\partial K_\ell} = -\sum_{j=1}^{n} \frac{\partial \phi_{\mathrm{sing}}}{\partial \theta_j} \frac{\partial \theta_j}{\partial K_\ell}. \end{aligned} \tag{4.3.119}$$

The scaling properties of the deviations of the densities from their transition values are determined by the scaling properties of the first-order partial derivatives of ϕ_{sing} with respect to the scaling fields θ_j. If $\mathcal{C}$ is approached along a path $\mathcal{P}$ such that, according to the second of equations (4.3.19),

$$\frac{\partial \phi_{\mathrm{sing}}}{\partial \theta_j}(\varpi_1^{(m)}, \ldots, \varpi_n^{(m)}) \quad \text{is finite and non-zero,} \tag{4.3.120}$$

it follows from (4.3.12), (4.3.15) and (4.3.20) that

$$\frac{\partial \phi_{\mathrm{sing}}}{\partial \theta_j} \sim |\theta_m|^{(d-y_j)/y_m}. \tag{4.3.121}$$

The scaling behaviour in Eq. (4.3.119) is dominated by the term in the summation on the right-hand side which corresponds to the largest y_j for which the derivative of θ_j with respect to the appropriate coupling is non-zero on $\mathcal{C}$. We denote these exponents by y_i^{f} and y_ℓ^{e} respectively for the first and second of equations (4.3.119). Then, supposing that a path is chosen so that the finite non-zero amplitude condition (4.3.120) is satisfied,

$$\rho_i - \rho_{ic} \sim |\theta_m|^{(d-y_i^{\mathrm{f}})/y_m}, \qquad \mathrm{u}_\ell - \mathrm{u}_{\ell c} \sim |\theta_m|^{(d-y_\ell^{\mathrm{e}})/y_m}. \tag{4.3.122}$$

The response functions, given by the second-order partial derivatives of ϕ with respect to the internal and external couplings, can be analyzed in the same way. If it is again assumed that the appropriate amplitudes are finite and non-zero, the scaling forms can be seen to be

$$\frac{\partial^2\phi}{\partial L_i \partial L_j} \sim |\theta_m|^{(d-y_i^{\mathrm{f}}-y_j^{\mathrm{f}})/y_m}, \tag{4.3.123}$$

$$\frac{\partial^2\phi}{\partial K_\ell \partial K_j} \sim |\theta_m|^{(d-y_\ell^{\mathrm{e}}-y_j^{\mathrm{e}})/y_m}, \tag{4.3.124}$$

$$\frac{\partial^2\phi}{\partial L_i \partial K_\ell} \sim |\theta_m|^{(d-y_i^{\mathrm{f}}-y_\ell^{\mathrm{e}})/y_m}. \tag{4.3.125}$$

4.4 Critical Point and Coexistence Curve

In Sect. 4.2 we discussed, in an intuitive way, the ideas of scaling as they relate to the critical point in the ferromagnetic Ising model. The case of a critical point, terminating a line of first-order transitions, is the situation which has received most attention both in discussions of scaling theory and more generally in the field of critical phenomena. The ferromagnetic Ising model has the virtue, from the point of view of simplicity, that the first-order transition and critical point lie on the zero-field axis, about which the phase diagram is symmetric. In this section we return to the case of a critical point, but assume no intrinsic symmetry. This means that the discussion is applicable to both fluid and magnetic models, these special cases being described in detail below.

The phase space is a plane ($n = 2$), the internal energy parameter is denoted by ε and the one independent external field is ξ, with couplings $K := \varepsilon/T$ and $L := \xi/T$. The free-energy density is $\phi_1 = \phi_1(K, L)$ with

$$\mathrm{d}\phi_1 = \mathrm{u}\mathrm{d}K - \rho\mathrm{d}L. \tag{4.4.1}$$

The first-order transition appears as a curve in the L–K plane terminating at the critical point (L_c, K_c) (Fig. 4.7). The phase transition makes its appearance in the ρ–K plane as a coexistence region bounded by two curves (Fig. 4.8)

$$\rho = \rho^{(\pm)}(K), \tag{4.4.2}$$

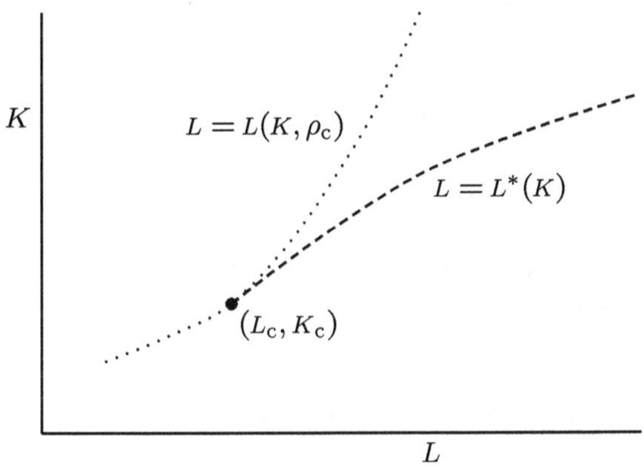

Fig. 4.7 A critical point (L_c, K_c) in the L–K plane. The first-order transition (coexistence curve) $L = L^*(K)$ is represented by a *broken line* and the critical isochore $L = L(K, \rho_c)$ by a *dotted line* (The fact that these two curves have a common tangent at the critical point is established by scaling theory)

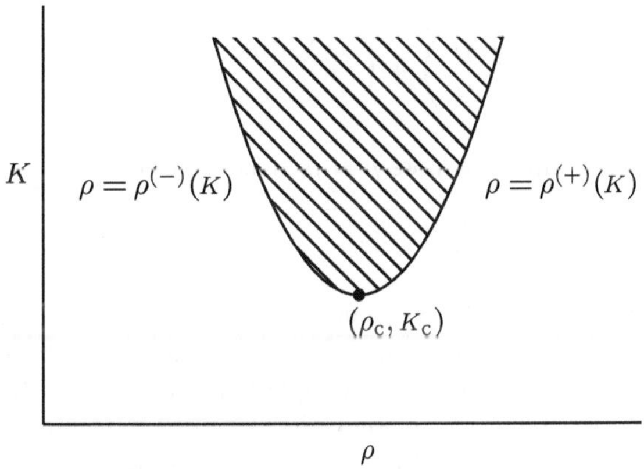

Fig. 4.8 The coexistence region (*shaded*) in the ρ–K plane

where

$$\rho^{(+)}(K) > \rho^{(-)}(K), \qquad K > K_c, \tag{4.4.3}$$

$$\rho^{(+)}(K_c) = \rho^{(-)}(K_c) = \rho_c, \tag{4.4.4}$$

$$\left(\frac{d\rho^{(+)}}{dK}\right)_c = -\left(\frac{d\rho^{(-)}}{dK}\right)_c = \infty, \tag{4.4.5}$$

$$\left(\frac{d^2\rho^{(+)}}{dK^2}\right)_c = -\left(\frac{d^2\rho^{(-)}}{dK^2}\right)_c. \tag{4.4.6}$$

The Eqs. (4.4.3)–(4.4.5) are necessary consequences of the definition of the critical point. Equation (4.4.6) is not a necessary condition but is true in most cases of critical points. The equation of state of the system in terms of the variables K, L and ρ is given, from (4.4.1), by

$$\rho = -\frac{\partial \phi_1}{\partial L}, \tag{4.4.7}$$

yielding an equation of the form

$$L = L(K, \rho), \tag{4.4.8}$$

which is smooth outside the coexistence region. The projection of the coexistence curve into the L–K plane, which is the line of first-order transitions shown in Fig. 4.7, is given, for $K \geq K_c$, by

$$L = L^*(K) := L(K, \rho^{(\pm)}(K)). \tag{4.4.9}$$

From (4.4.9),

$$\frac{dL^*}{dK} = \left(\frac{\partial L}{\partial K}\right)_\rho + \left(\frac{\partial L}{\partial \rho}\right)_K \frac{d\rho^{(\pm)}}{dK}. \tag{4.4.10}$$

The critical isochore $\rho = \rho_c$ makes the curve

$$L = L(K, \rho_c) \tag{4.4.11}$$

in the L–K plane passing through the critical point. The gradients of the critical isochore and coexistence curve differ at the critical point by the critical point value of the second term on the right-hand side of (4.4.10). This term contains an infinite factor at the critical point (Eq. (4.4.5)). The fact that this term as a whole is zero will be established in Sect. 4.5.6. In terms of the scaling theory described in Sect. 4.3 the situation here is one in which we have two transition regions: the critical point of zero dimension, with $s = n = 2$ and two relevant scaling fields, and the coexistence curve, with $s = n - 1 = 1$ and one relevant and one irrelevant scaling field.

4.4.1 Critical Exponents

In the standardized notation of critical phenomena (Buckingham 1972) the critical exponents α, α', β, δ, γ and γ' are defined by

$$c_\rho \sim \begin{cases} (T - T_c)^{-\alpha}, & \text{along the critical isochore}, \ T > T_c, \\ (T_c - T)^{-\alpha'}, & \text{along the coexistence curve}, \ T < T_c, \end{cases} \tag{4.4.12}$$

$$\rho - \rho_c \sim (T_c - T)^\beta, \ \text{along the coexistence curve}, \ T < T_c, \tag{4.4.13}$$

$$\varphi_T \sim \begin{cases} (T - T_c)^{-\gamma}, & \text{along the critical isochore, } T > T_c, \\ (T_c - T)^{-\gamma'}, & \text{along the coexistence curve, } T < T_c, \end{cases} \tag{4.4.14}$$

$$\xi - \xi_c \sim (\rho - \rho_c)|\rho - \rho_c|^{\delta - 1}, \text{ along the critical isotherm.} \tag{4.4.15}$$

It also convenient to define the exponents σ and σ' according to

$$c_\xi \sim \begin{cases} (T - T_c)^{-\sigma}, & \text{along the critical isochore, } T > T_c, \\ (T_c - T)^{-\sigma'}, & \text{along the coexistence curve, } T < T_c. \end{cases} \tag{4.4.16}$$

4.4.2 Exponent Inequalities

The heat capacities c_ρ and c_ξ in (1.4.8) are both positive. From this it follows that, if $\varphi_T > 0$, then c_ξ dominates both c_ρ and α_ξ^2/φ_T as $T \to T_c$. This means that

$$\sigma \geq \alpha, \qquad \sigma' \geq \alpha', \tag{4.4.17}$$

$$\sigma' + 2\beta + \gamma' \geq 2. \tag{4.4.18}$$

The condition $\varphi_T > 0$ is true for a magnetic system and in this case the inequality (4.4.18) was first established by Rushbrooke (1963). The stronger condition

$$\alpha' + 2\beta + \gamma' \geq 2 \tag{4.4.19}$$

was obtained by Griffiths (1965) for both ferromagnetic and fluid systems using the convexity properties of the free energy. In fact, as we shall see in Sect. 4.5, for systems with the special symmetry of the ferromagnet $\alpha' = \sigma'$ and inequalities (4.4.18) and (4.4.19) become identical. Griffiths (1965) also derived a number of other inequalities. In particular

$$\gamma' \geq \beta(\delta - 1). \tag{4.4.20}$$

4.5 Scaling for a Critical Point

From the discussion in Sect. 4.3 it is clear that there is a degree of latitude in the definition of the singular part of ϕ. Given that it should contain all the non-smooth part of ϕ, it is convenient, for the critical point, to take

$$\phi_{\text{sing}}(\Delta K, \Delta L) := \phi_1(K, L) - \phi_1(K_c, L_c) - u_c\Delta K + \rho_c\Delta L, \tag{4.5.1}$$

where $\Delta K := K - K_c$ and $\Delta L := L - L_c$, so that

$$\phi_{\text{sing}}(0,0) = 0, \tag{4.5.2}$$

$$\frac{\partial \phi_{\text{sing}}}{\partial \Delta K} = \Delta u, \tag{4.5.3}$$

$$\frac{\partial \phi_{\text{sing}}}{\partial \Delta L} = -\Delta \rho. \tag{4.5.4}$$

4.5.1 Scaling Fields for the Critical Point

In the light of Sect. 4.3, we see that (4.2.11) is the scaling hypothesis for a magnetic system using scaling fields $\Delta L = L$, ($L_c = 0$), and ΔK. The special feature of the magnetic system which is relevant here is that the first-order phase transition lies along the line $L = 0$. The scaling fields were chosen to pick out this special line. In the general case, before defining scaling fields, we must define new coordinates oriented to select the special direction represented by the tangent to the coexistence curve at the critical point. Assuming that the coexistence curve at the critical point is not parallel to the L-axis, $L^*(K)$ can be expressed in the form

$$L^*(K) = L_c + a\Delta K + \mathfrak{L}(K), \tag{4.5.5}$$

where a is the (finite) slope of the coexistence curve at the critical point and

$$\mathfrak{L}(K_c) = 0, \qquad \mathfrak{L}'(K_c) = 0, \qquad \mathfrak{L}(K) \simeq b|\Delta K|^{\psi}, \tag{4.5.6}$$

b being zero if the coexistence curve is linear, and ψ (> 1) being called the *shift exponent*. We use the orthogonal coordinates

$$\Theta_1 := L - aK, \qquad \Theta_2 := K + aL, \tag{4.5.7}$$

where $\Theta_1 = \Theta_{1c} = L_c - aK_c$ gives the tangent to the coexistence curve. Along the coexistence curve $\Delta K \geq 0$ and, assuming that $0 \leq a < \infty$, with $0 \leq \mathfrak{L}(K) < \infty$ for $\Delta K \geq 0$, this curve will leave the critical point $(\Theta_{1c}, \Theta_{2c})$ and then satisfy the conditions $\Theta_1 \geq \Theta_{1c}$ and $\Theta_2 > \Theta_{2c}$. In terms of the orthogonal scaling fields

$$\theta_1 := \Theta_1 - \Theta_{1c} = \Delta L - a\Delta K, \quad \theta_2 := \Theta_2 - \Theta_{2c} = \Delta K + a\Delta L, \tag{4.5.8}$$

the scaling hypothesis takes the form

$$\phi_{\text{sing}}(\lambda^{y_1}\theta_1, \lambda^{y_2}\theta_2) = \lambda^d \phi_{\text{sing}}(\theta_1, \theta_2). \tag{4.5.9}$$

Following the procedure of Sect. 4.3, set $\lambda = |\theta_1|^{-1/y_1}$, giving

$$\phi_{\text{sing}}(\theta_1, \theta_2) = |\theta_1|^{d/y_1} \phi_{\text{sing}}(\pm 1, g_2), \tag{4.5.10}$$

and $\lambda = |\theta_2|^{-1/y_2}$, giving

$$\phi_{\text{sing}}(\theta_1, \theta_2) = |\theta_2|^{d/y_2} \phi_{\text{sing}}(g_1, \pm 1). \tag{4.5.11}$$

For the sake of brevity, the gap exponent is denoted as $\vartheta := y_1/y_2$ and the gap fields as

$$g_1(\theta_1, \theta_2) := \theta_1/|\theta_2|^{\vartheta}, \qquad g_2(\theta_1, \theta_2) := \theta_2/|\theta_1|^{1/\vartheta}, \tag{4.5.12}$$

where $|g_1| = |g_2|^{-\vartheta}$. Both (4.5.10) and (4.5.11) are valid in a neighbourhood of the critical point and thus

$$\begin{aligned} \phi_{\text{sing}}(g_1, \pm 1) &= |g_1|^{d/y_1} \phi_{\text{sing}}(\pm 1, g_2), \\ \phi_{\text{sing}}(\pm 1, g_2) &= |g_2|^{d/y_2} \phi_{\text{sing}}(g_1, \pm 1), \end{aligned} \tag{4.5.13}$$

(cf. Eq. (4.3.17)). Along the coexistence curve $\phi_{\text{sing}}(\theta_1, \theta_2)$ exhibits non-smooth behaviour; specifically one, or both, of the first-order partial derivatives is discontinuous. Since $\theta_2 \neq 0$ along the coexistence curve, apart from at the critical point, it follows from (4.5.11) that the coexistence curve corresponds to singular behaviour in $\phi_{\text{sing}}(g_1, +1)$. Supposing that, in a neighbourhood of the critical point, the only singular behaviour corresponds to the critical point and coexistence curve, $\phi_{\text{sing}}(g_1, +1)$ will be singular at only one value

$$g_1(\theta_1, \theta_2) = \varpi_0, \tag{4.5.14}$$

where $0 \leq \varpi_0 < \infty$ and (4.5.14) is the equation of the coexistence curve in a neighbourhood of the critical point. The constant will be zero only if $\mathfrak{L}(K) \equiv 0$, when the coexistence curve lies along the $\theta_1 = 0$ axis. Except in this special case the coexistence curve is tangential to $\theta_1 = 0$ and it follows that

$$\vartheta > 1, \qquad y_1 > y_2. \tag{4.5.15}$$

The curves, defined by

$$|\theta_1| = \varpi |\theta_2|^{\vartheta}, \tag{4.5.16}$$

for fixed finite non-zero ϖ, are tangential to the $\theta_1 = 0$ axis at the critical point, picking it out as a special direction when (4.5.15) is satisfied. Experimental evidence indicates that, in all cases, there is one special direction at the critical point which is the direction tangential to the coexistence curve. This supports the conclusion that, even when the coexistence curve lies along $\theta_1 = 0$, (4.5.15) must hold and that the curves (4.5.16), with non-zero, finite ϖ are tangential to $\theta_1 = 0$. The exponents y_1 and y_2 are referred to as the *strong* and *weak* exponents respectively.

Given that ϕ_{sing} is a continuous function we assume that the amplitudes

$$\lim_{g_2 \to 0} \phi_{\text{sing}}(\pm 1, g_2) := A_{\infty}^{(\pm)}, \qquad \lim_{g_1 \to 0} \phi_{\text{sing}}(g_1, \pm 1) := A_0^{(\pm)} \tag{4.5.17}$$

are finite and non-zero.[23] Then, from (4.5.13),

$$\begin{aligned} \lim_{g_2 \to \infty} \phi_{\text{sing}}(\pm 1, g_2) &\simeq |g_2|^{d/y_2} A_0^{(\pm)}, \\ \lim_{g_1 \to \infty} \phi_{\text{sing}}(g_1, \pm 1) &\simeq |g_1|^{d/y_1} A_{\infty}^{(\pm)}. \end{aligned} \tag{4.5.18}$$

4.5.2 Approaches to the Critical Point

These can be divided into three classes:

(i) Paths on which $|g_1| \to \infty$ ($|g_2| \to 0$), including the axis $\theta_2 = 0$, for which, from (4.5.10) and (4.5.17),

$$\phi_{\text{sing}}(\theta_1, \theta_2) \simeq |\theta_1|^{d/y_1} A_{\infty}^{(\pm)}. \tag{4.5.19}$$

(ii) Paths on which $|g_1| \to 0$, including the axis $\theta_1 = 0$, for which, from (4.5.11) and (4.5.17),

$$\phi_{\text{sing}}(\theta_1, \theta_2) \simeq |\theta_2|^{d/y_2} A_0^{(\pm)}. \tag{4.5.20}$$

(iii) Paths on which $|g_1| \to \pi_1$, where π_1 is finite and non-zero, when

$$\phi_{\text{sing}}(\theta_1, \theta_2) \simeq |\theta_2|^{d/y_2} A_{\pi_1}^{(\pm)}, \tag{4.5.21}$$

with

$$\lim_{g_1 \to \pi_1} \phi_{\text{sing}}(g_1, \pm 1) = A_{\pi_1}^{(\pm)} \tag{4.5.22}$$

assumed to be finite and non-zero.

It is, of course, the case that an alternative to (4.5.21) could be given in terms of θ_1, when $\pi_1 \neq 0$. However, it is convenient to class cases (ii) and (iii) together with (4.5.21) and (4.5.22) applying for all finite values of π_1. All these approaches will

[23] See Sect. 4.8 for a discussion of circumstances where this assumption breaks down.

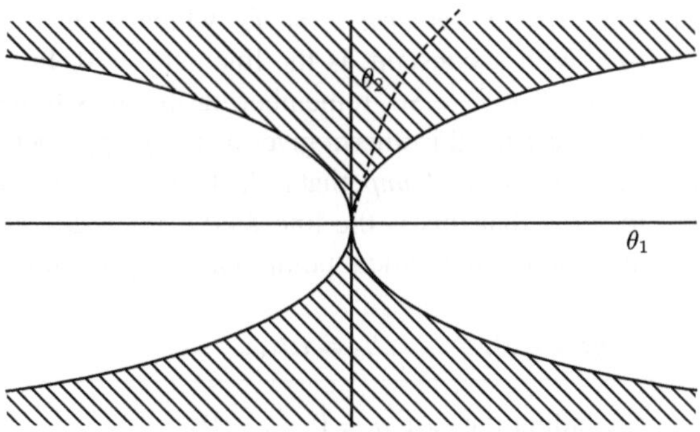

Fig. 4.9 The θ_1–θ_2 plane in a neighbourhood of the critical point. The line $\theta_1 = 0, \theta_2 \geq 0$ is tangential to the coexistence curve and the crossover curves $|\theta_1| = \varpi|\theta_2|^{\vartheta}$ separate weak (*shaded*) and strong (*unshaded*) scaling approaches

be called *weak* as they differ only in the value of the amplitude $A^{(\pm)}_{\pi_1}$. Approaches in class (i) are called *strong*. Equation (4.5.16), for some finite non-zero ϖ, gives two crossover curves in the θ_1–θ_2 plane, which separate strong and weak scaling approaches (Fig. 4.9).[24]

It is clear that an approach to the critical point along the coexistence curve itself is a weak approach with $\pi_1 = \varpi_0$. From (4.5.8),

$$\theta_1\left(1 + a\frac{\theta_2}{\theta_1}\right) = (1 + a^2)\Delta L, \qquad \theta_2\left(1 - a\frac{\theta_1}{\theta_2}\right) = (1 + a^2)\Delta K \tag{4.5.23}$$

and since, by (4.5.16) with $\varpi = \varpi_0$, $\theta_1/\theta_2 \to 0$ as the critical point is approached along the coexistence curve

$$\theta_1 \simeq \varpi_0(1 + a^2)^{\vartheta}|\Delta K|^{\vartheta}. \tag{4.5.24}$$

Then, from (4.5.5) and (4.5.8),

$$\mathcal{L}(K) \simeq \varpi_0(1 + a^2)^{\vartheta}|\Delta K|^{\vartheta}, \tag{4.5.25}$$

showing that the gap exponent $\vartheta = \psi$, the shift exponent defined in (4.5.6).

4.5.3 Experimental Variables

To link this analysis with experiment it is necessary to translate the asymptotic formulae, given above, into formulae expressed in thermodynamic variables. Two approaches to the critical point are of particular interest:

---

[24] It will be seen from (4.5.16) that the actual value of ϖ, as long as it is finite and non-zero, is irrelevant to the asymptotic behaviour of ϕ_{sing}, representing only a multiplicative scaling of θ_1.

(a) The *critical isotherm* $T = T_c$, which in terms of the couplings, where $T = \varepsilon/K$ with ε constant, is the line $\Delta K = 0$. From (4.5.8) and (4.5.12) $|g_1| = 1/|a|^{\vartheta}|\Delta L|^{\vartheta-1}$. So, remembering that a is finite and possibly zero, it is clear that the critical isotherm gives a strong approach to the critical point.
(b) The *critical isochamp*, that is the line $\xi = \xi_c$, where $\xi = TL$. It is straightforward to show that this is the line $K_c \Delta L = L_c \Delta K$ and this also corresponds to an approach to the critical point in a strong scaling direction unless

$$\Theta_{2c} = 0, \qquad L_c = aK_c, \tag{4.5.26}$$

(including the case $a = 0$, $L_c = 0$).

For a weak scaling approach $\theta_1/\theta_2 \to 0$. So, from (4.5.20) and (4.5.23),

$$\phi_{\text{sing}}(\theta_1, \theta_2) \sim |\Theta_2 - \Theta_{2c}|^{d/y_2} \sim |\Delta K|^{d/y_2}. \tag{4.5.27}$$

Strong scaling approaches include all those where $\theta_2/\theta_1 \to 0$, for which (4.5.19) can be replaced by

$$\phi_{\text{sing}}(\theta_1, \theta_2) \sim |\Theta_1 - \Theta_{1c}|^{d/y_1} \sim |\Delta L|^{d/y_1}. \tag{4.5.28}$$

They also include approaches where θ_2/θ_1 remains finite. This affects the amplitude of the asymptotic form, but (4.5.28) still applies. We should, however, bear in mind the case where $\theta_1/\theta_2 \to 0$ with a curvature larger than that of the crossover curve. This is a somewhat pathological strong scaling approach. Equations (4.5.27) and (4.5.28) link the weak and strong exponents y_2 and y_1 with the couplings K and L and, as in Sect. 4.2, they could be denoted by y_K and y_L respectively. One could go further than this and link these exponents to thermodynamic variables. Since K is an internal coupling with ε fixed the only physically variable parameter is the temperature T. The exponent y_2 can, therefore, be referred to as the *thermal exponent* often denoted as y_T. In a similar way, L is an external coupling, which can be changed by altering the field ξ. So y_1, can be referred to as the *field exponent* and denoted by y_ξ. We shall, however, in the remainder of this section continue to use the notation y_1 and y_2 for the exponents.

4.5.4 The Density and Response Functions

It is now necessary to express the density and the response functions in terms of the partial derivatives of ϕ_{sing} with respect to the scaling fields (Eqs. (4.3.119)). From (2.3.30), (2.3.31), (4.5.4), (4.5.7) and (4.5.8)

$$\rho - \rho_c = \Delta\rho = -\left[\frac{\partial \phi_{\text{sing}}}{\partial \theta_1} + a\frac{\partial \phi_{\text{sing}}}{\partial \theta_2}\right], \tag{4.5.29}$$

$$\varphi_T = -\left[\frac{\partial^2\phi_{\text{sing}}}{\partial\theta_1^2} + 2a\frac{\partial^2\phi_{\text{sing}}}{\partial\theta_1\partial\theta_2} + a^2\frac{\partial^2\phi_{\text{sing}}}{\partial\theta_2^2}\right], \tag{4.5.30}$$

$$c_\xi = -k_{\text{B}}\left[\Theta_1{}^2\frac{\partial^2\phi_{\text{sing}}}{\partial\theta_1{}^2} + 2\Theta_1\Theta_2\frac{\partial^2\phi_{\text{sing}}}{\partial\theta_1\partial\theta_2} + \Theta_2{}^2\frac{\partial^2\phi_{\text{sing}}}{\partial\theta_2{}^2}\right]. \tag{4.5.31}$$

For c_ρ the expression (1.4.8) is used with

$$\alpha_\xi = \Theta_1\frac{\partial^2\phi_{\text{sing}}}{\partial\theta_1^2} + (a\Theta_1 + \Theta_2)\frac{\partial^2\phi_{\text{sing}}}{\partial\theta_1\partial\theta_2} + a\Theta_2\frac{\partial^2\phi_{\text{sing}}}{\partial\theta_2^2}. \tag{4.5.32}$$

Substituting from (4.5.30)–(4.5.32) into (1.4.8)

$$c_\rho = -\frac{k_{\text{B}}}{\varphi_T}(a\Theta_1 - \Theta_2)^2\left[\left(\frac{\partial^2\phi_{\text{sing}}}{\partial\theta_1\partial\theta_2}\right)^2 - \frac{\partial^2\phi_{\text{sing}}}{\partial\theta_1^2}\frac{\partial^2\phi_{\text{sing}}}{\partial\theta_2^2}\right]. \tag{4.5.33}$$

4.5.5 Asymptotic Forms

To obtain asymptotic forms it is necessary to determine the dominant contributions in (4.5.29)–(4.5.31) and (4.5.33), using the reasoning of Sect. 4.3.9. Since $y_1 > y_2$, these correspond to the terms with the largest number of differentiations with respect to θ_1. On this basis we see that the terms in (4.5.29)–(4.5.31) are ordered in descending dominance and those in (4.5.33) have equal weight. Repeating the reasoning in (4.5.9)–(4.5.22) for the partial derivatives of ϕ_{sing}, replaces d by the corresponding $d - k_1y_1 - k_2y_2$. For the weak scaling direction the formulae corresponding to (4.5.22), when $\pi_1 = \varpi_0$, that is across the coexistence curve, do not give an unambiguous limit, yielding possibly different values as $g_1 \to \varpi_0$ from the two sides. This means that the amplitudes for approaches along the two branches of the coexistence curve can be different. Except in the special case where one is zero and corrections to scaling need to be considered (Sect. 4.8), this will not affect the exponent in the asymptotic form. Without further comment we shall now assume that all the appropriate amplitude factors given by the second of equations (4.3.19) are non-zero and, using the arguments of Sect. 4.3.9, it follows, from (4.5.29) and (4.5.30), that

$$\rho - \rho_c \sim \begin{cases} |\Delta L|^{(d-y_1)/y_1}, & \text{strong direction,} \\ |\Delta K|^{(d-y_1)/y_2}, & \text{weak direction,} \end{cases} \tag{4.5.34}$$

$$\varphi_T \sim \begin{cases} |\Delta L|^{(d-2y_1)/y_1}, & \text{strong direction,} \\ |\Delta K|^{(d-2y_1)/y_2}, & \text{weak direction.} \end{cases} \tag{4.5.35}$$

To consider the asymptotic properties of c_ξ it is necessary to take account of the possibility that $\Theta_1(L_c, K_c) = 0$, which is the condition given by (4.5.26). $\Theta_2(L_c, K_c) \neq 0$ if $\Theta_1(L_c, K_c) = 0$ and we have, from (4.5.31), two cases

$$c_\xi \sim \begin{cases} |\Delta L|^{(d-2y_1)/y_1}, & \text{strong direction,} \\ |\Delta K|^{(d-2y_1)/y_2}, & \text{weak direction,} \end{cases} \qquad \text{if } L_c \neq aK_c \tag{4.5.36}$$

and

$$c_\xi \sim \begin{cases} |\Delta L|^{(d-2y_2)/y_1}, & \text{strong direction,} \\ |\Delta K|^{(d-2y_2)/y_2}, & \text{weak direction,} \end{cases} \qquad \text{if } L_c = aK_c. \tag{4.5.37}$$

From (4.5.33) it can be seen that the asymptotic expressions for c_ρ will involve a combination of amplitude factors. Unless these cancel

$$c_\rho \sim \begin{cases} |\Delta L|^{(d-2y_2)/y_1}, & \text{strong direction,} \\ |\Delta K|^{(d-2y_2)/y_2}, & \text{weak direction.} \end{cases} \tag{4.5.38}$$

4.5.6 Critical Exponents and Scaling Laws

Given that an approach to the critical point along the coexistence curve is a weak approach and along the critical isotherm is a strong approach it follows from a comparison of (4.4.12)–(4.4.16) with (4.5.34)–(4.5.38) that

$$\alpha' = (2y_2 - d)/y_2, \tag{4.5.39}$$

$$\beta = (d - y_1)/y_2, \tag{4.5.40}$$

$$\gamma' = (2y_1 - d)/y_2, \tag{4.5.41}$$

$$\delta = y_1/(d - y_1), \tag{4.5.42}$$

$$\sigma' = \begin{cases} (2y_1 - d)/y_2 = \gamma', & \text{if } L_c \neq aK_c, \\ (2y_2 - d)/y_2 = \alpha', & \text{if } L_c = aK_c. \end{cases} \tag{4.5.43}$$

From (4.5.43) and (4.4.17) it follows that, when $L_c \neq aK_c$, $\gamma' \geq \alpha'$, a result which can be established without the aid of scaling theory (Buckingham 1972).

The magnetic system, described in Sect. 4.2 has the first-order transition line lying along the zero-field axis. So $L_c = 0, a = 0$ and $\sigma' = \alpha'$. It is for this reason that, when defining critical exponents for fluids and magnets, it is common practice (Stanley 1987; Yeomans 1992) to use α' for the heat capacity at constant density for the fluid and constant magnetic field for the magnet. We can now understand that the reason that the same exponent is appropriate to both cases is due to the special symmetry properties of the magnet, which also makes the analysis rather simpler.

Equations (4.5.39)–(4.5.42) give formulae for the exponents α', β, γ' and δ in terms of y_1 and y_2. They are, therefore, not independent and two relationships exist between them. These can be expressed in the form

$$\alpha' + 2\beta + \gamma' = 2, \tag{4.5.44}$$

called the *Essam–Fisher scaling law* (Essam and Fisher 1963), which is a strengthening of the inequality (4.4.19) and

$$\gamma' = \beta(\delta - 1), \tag{4.5.45}$$

called the *Widom scaling law* (Widom 1964) which is a strengthening of the inequality (4.4.20).

All the information necessary to consider the second term on the right-hand side of (4.4.10) is now available. Since this is the case of a weak direction approach to the critical point,

$$\frac{d\rho^{(\pm)}}{dK} \sim |\Delta K|^{-1+(d-y_1)/y_2}, \qquad \left(\frac{\partial L}{\partial \rho}\right)_K \sim |\Delta K|^{-(d-2y_1)/y_2},$$

giving

$$\left(\frac{\partial L}{\partial \rho}\right)_K \frac{d\rho^{(\pm)}}{dK} \sim |\Delta K|^{\vartheta-1}. \tag{4.5.46}$$

As $\vartheta > 1$, this establishes the result that the critical isochore and coexistence curve share a common tangent (given by $\theta_1 = 0$) at the critical point. The critical isochore is thus a direction of weak approach to the critical point. This means that, in general,

$$\alpha = \alpha', \qquad \gamma = \gamma', \qquad \sigma = \sigma'. \tag{4.5.47}$$

However, of course, the amplitudes associated with the asymptotic forms for approaches to the critical point along the critical isochore and the coexistence curve are, in general different. If one of these were zero then this would lead to a correction to scaling as described in Sect. 4.8 with a corresponding breakdown in the equality given by (4.5.47).

The special case $y_1 = 0$, giving $\beta = 0$, $\delta = \infty$, allows for the possibility of discontinuity in ρ at the critical point. This is called a *first-order critical point* (Fisher and Berker 1982).

As indicated above the critical point and the coexistence curve can be regarded as two connected critical regions $\mathcal{C}$ and $\mathring{\mathcal{C}}$, of respective dimensions zero and one. The exponents y_1 and y_2 of $\mathcal{C}$ are both relevant and the exponent $\mathring{y}_1$ and $\mathring{y}_2$ of $\mathring{\mathcal{C}}$ are respectively relevant and irrelevant. This situation can be treated with the model of Sect. 4.3.3. There the assumption is made that $\mathring{y}_2 = -y_2$ and here the first-order transition across the coexistence curve implies the condition $\mathring{y}_1 = 1$. An effective exponent and the inequality (4.3.46) are used to defined the (shaded) crossover region shown in Fig. 4.6. This contracts to the axis $\Theta_2 = 0$ as $y_1 \to 1$, when the critical point becomes a first-order critical point.

4.5.7 Correlation Scaling at a Critical Point

In this section we consider scaling for the connected two-point correlation function $\Gamma_2(\mathfrak{q}(\boldsymbol{r}_1), \mathfrak{q}(\boldsymbol{r}_2))$ associated with the coupling L and defined by (2.5.9) with $\mathfrak{c}_1 = \mathfrak{c}_2 = \mathfrak{q}$. As before a d-dimensional hypercubic lattice is assumed with periodic boundary conditions and translational invariance so that the correlation function is dependent on $\boldsymbol{r}_1$ and $\boldsymbol{r}_2$ through their difference $\bar{\boldsymbol{r}} := \boldsymbol{r}_2 - \boldsymbol{r}_1$.

In Sect. 4.3.7 the scaling behaviour of correlation functions of scaling operators was developed. To translate these results into the scaling behaviour of state operators like $\mathfrak{q}$ the reasoning of Sect. 4.3.9 is used to determine the leading scaling behaviour associated with the coupling L. With the definitions (4.5.8) for the scaling fields we showed in Sect. 4.5 that L scales like θ_1. It follows that the two-point $\mathfrak{q}$ correlation function scales like the two-point ϕ_1–ϕ_1 correlation function. Thus from (4.3.99), (4.3.101) and (4.3.102)

$$\Gamma_2(\lambda^{y_1}\theta_1, \lambda^{y_2}\theta_2; \mathfrak{q}(\mathbf{0}), \mathfrak{q}(\lambda^{-1}\bar{\boldsymbol{r}})) = \lambda^{2x_1}\Gamma_2(\theta_1, \theta_2; \mathfrak{q}(\mathbf{0}), \mathfrak{q}(\bar{\boldsymbol{r}})) \tag{4.5.48}$$

and

$$\bar{\Gamma}_2^*(\lambda^{y_1}\theta_1, \lambda^{y_2}\theta_2; \mathfrak{q}; \lambda\bar{\boldsymbol{k}}) = \lambda^{d-2y_1}\bar{\Gamma}_2^*(\theta_1, \theta_2; \mathfrak{q}; \bar{\boldsymbol{k}}), \tag{4.5.49}$$

where $x_1 := d - y_1$ is the scaling dimension of $\mathfrak{q}$.

The correlation length is defined by (4.2.4) and it is easily seen that this can be expressed in the form

$$\mathfrak{r}^2(\mathfrak{q}) = -c(\mathfrak{d})\frac{\nabla_{\bar{k}}^2 \bar{\Gamma}_2^*(\mathfrak{q}; \mathbf{0})}{\bar{\Gamma}_2^*(\mathfrak{q}; \mathbf{0})}. \tag{4.5.50}$$

Thus, from (4.5.50), the scaling form for the correlation length is

$$\mathfrak{r}(\lambda^{y_1}\theta_1, \lambda^{y_2}\theta_2; \mathfrak{q}) = \lambda^{-1}\mathfrak{r}(\theta_1, \theta_2; \mathfrak{q}). \tag{4.5.51}$$

According to the standard notation the exponents ν and ν' are defined by

$$\mathfrak{r}(\mathfrak{q}) \sim \begin{cases} (T - T_c)^{-\nu}, & \text{along the critical isochore, } T > T_c, \\ (T_c - T)^{-\nu'}, & \text{along the coexistence curve, } T < T_c, \end{cases} \tag{4.5.52}$$

and thus, taking into account that both the coexistence curve and the critical isochore are directions of weak approach,

$$\nu = \nu' = 1/y_2. \tag{4.5.53}$$

This result and (4.5.39) with (4.5.47) yields the *Josephson hyper-scaling law*

$$d\nu = 2 - \alpha, \tag{4.5.54}$$

which was proved by Josephson (1967) as an inequality with '$\geq$' replacing '$=$'. At the critical point, for large $|\boldsymbol{r}_2 - \boldsymbol{r}_1|$, the standard asymptotic form for the connected two-point correlation function is given in terms of the exponent η by

$$\Gamma_2(\mathfrak{q}(\boldsymbol{r}_1), \mathfrak{q}(\boldsymbol{r}_2)) \sim \frac{1}{|\boldsymbol{r}_2 - \boldsymbol{r}_1|^{d-2+\eta}}. \tag{4.5.55}$$

From (4.3.110) this means that

$$\eta = 2 - 2y_1 + d = 2 + 2x_1 - d, \tag{4.5.56}$$

if it is assumed that the appropriate amplitude factor is non-zero and finite. It now follows from (4.5.49) that, at the critical point for small $|\bar{\boldsymbol{k}}|$,

$$\bar{\Gamma}_2^*(\mathfrak{q}; \bar{\boldsymbol{k}}) \sim |\bar{\boldsymbol{k}}|^{\eta-2}. \tag{4.5.57}$$

The *Fisher scaling law*

$$(2 - \eta)\nu = \gamma, \tag{4.5.58}$$

follows from (4.5.41), (4.5.47), (4.5.53) and (4.5.56). This was proved by Fisher (1969) as an inequality with '$\geq$' replacing '$=$'.

From the definition of the gap exponent ϑ, given below Eqs. (4.5.11), and (4.5.40)

$$\vartheta = \beta\,\delta\,. \tag{4.5.59}$$

From (4.5.16) and Fig. 4.9 it can be seen that the larger the value of ϑ the more of the θ_1–θ_2 plane in a neighbourhood of the critical point corresponds to weak scaling approaches. We now see, from (4.5.59), that, for fixed δ, larger values of ϑ correspond to larger values of β, corresponding to a narrowing of the coexistence region in the ρ–K plane (Fig. 4.8) and a corresponding increase in the part of the plane corresponding to strong scaling approaches. This apparently paradoxical situation is understood if we project the crossover curves into the ρ–K plane where they share a common tangent with the coexistence region. So in the θ_1–θ_2 (or equivalently L–K) plane the region of weak scaling approaches is wedge shaped at the critical point, whereas in the ρ–K plane it is the region of strong scaling approaches which is wedge shaped. This interesting point was discussed by Griffiths and Wheeler (1970), who observe that weak scaling approaches to the critical point will be more easily achieved by adjustments of ρ and K, than by adjustments of L and K.

4.6 Tricritical Point

Section 4.1 contains a brief description of the kinds of critical phenomena which may be encountered in phase spaces of dimensions $n = 2$ and $n = 3$. We have described in Sects. 4.5 and 4.5.7 the application of scaling theory to the most well-known example of critical behaviour for $n = 2$, a critical point terminating a line of first-order transitions. We shall now consider the case of a tricritical point, an example of which occurs in the mean-field treatment of a metamagnet in Sect. 6.2.[25] For ease of discussion we shall in this section and Sect. 4.7 adopt a magnetic model with the curves of second-order transitions $\mathcal{C}$ and first-order transitions $\mathcal{T}$ lying in the T–$\mathcal{H}$ plane where they meet at the tricritical point T (Fig. 4.10). Of course, for this model there is no symmetry associated with $\mathcal{H} = 0$ axis, as there is with simple magnetic models. However, we shall assume the existence of a second *staggered field*[26] $\mathcal{H}_\mathrm{s}$ with a phase diagram with symmetry with respect to the plane $\mathcal{H}_\mathrm{s} = 0$. When the third dimension corresponding to the staggered field is included $\mathcal{C}$ becomes a critical line marking the edge of a surface of first-order transitions in the $\mathcal{H}_\mathrm{s} = 0$ plane, and $\mathcal{T}$ is a triple line. Two further critical lines $\mathcal{C}^{(+)}$ and $\mathcal{C}^{(-)}$, bounding first-order surfaces, symmetrically placed about the plane $\mathcal{H}_\mathrm{s} = 0$, also meet at the tricritical point T. For this system the free-energy density with no independent densities is $\phi_2(K, L, L_\mathrm{s})$ with

---

[25] In fact, as was indicated in Sect. 4.1, a tricritical point is a special case of a more general type of phase behaviour with a critical end-point and possibly a bicritical end-point (Fig. 4.3a). For the metamagnet an adjustable parameter is needed to achieve a tricritical point.

[26] The reason for this designation is made clear in Sect. 6.2.

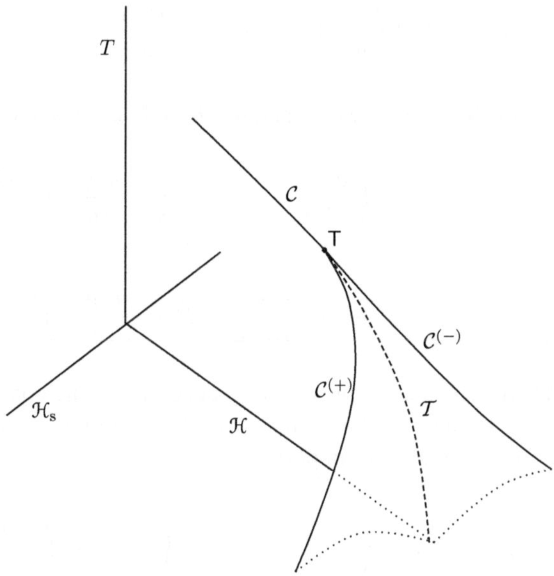

Fig. 4.10 A tricritical point in a magnetic system

$$\mathrm{d}\phi_2 = \mathrm{u}\mathrm{d}K - m\mathrm{d}L - q\mathrm{d}L_\mathrm{s}, \tag{4.6.1}$$

where u, the (dimensionless) internal energy per lattice site, and m, the magnetization per lattice site, are the densities associated with the internal thermal coupling K and the external field coupling L. The magnetic density q measures ordering with respect to the staggered field coupling $L_\mathrm{s} := \mathcal{H}_\mathrm{s}/T$. The critical curve corresponds to the onset of this ordering. So $q = 0$ in the disordered large temperature or field region and also on the critical curve itself. The ordered ($q \neq 0$) region below the critical curve in the plane $\mathcal{H}_\mathrm{s} = 0$ represents, as is the case on the zero-field axis below the Curie temperature for the Ising ferromagnet, coexistence of two phases with equal and opposite values of q.

The response function relations (2.3.29)–(2.3.32) need to be generalized to take into account the presence of two external couplings. The functions which measure the response of the magnetic densities to their conjugate fields are

$$\varphi_T := T\left(\frac{\partial m}{\partial \mathcal{H}}\right)_{T,\mathcal{H}_\mathrm{s}} = \left(\frac{\partial m}{\partial L}\right)_{K,L_\mathrm{s}} = -\frac{\partial^2 \phi_2}{\partial L^2}, \tag{4.6.2}$$

$$\psi_T := T\left(\frac{\partial q}{\partial \mathcal{H}_\mathrm{s}}\right)_{T,\mathcal{H}} = \left(\frac{\partial q}{\partial L_\mathrm{s}}\right)_{K,L} = -\frac{\partial^2 \phi_2}{\partial L_\mathrm{s}^2}. \tag{4.6.3}$$

There are now four heat capacities corresponding to the choices of maintaining constant either the density or coupling in the pairs $(m, \mathcal{H})$ and $(q, \mathcal{H}_\mathrm{s})$. Two of these are the m and q constant heat capacity

$$c_{m,q} := k_{\mathrm{B}} T \left(\frac{\partial s}{\partial T} \right)_{m,q} = -k_{\mathrm{B}} K^2 \frac{\partial^2 \phi_0}{\partial K^2}, \tag{4.6.4}$$

where $\phi_0 = \phi_0(K, m, q)$ and the $\mathcal{H}$ and $\mathcal{H}_{\mathrm{s}}$ constant heat capacity

$$c_{\mathcal{H},\mathcal{H}_{\mathrm{s}}} := k_{\mathrm{B}} T \left(\frac{\partial s}{\partial T} \right)_{\mathcal{H},\mathcal{H}_{\mathrm{s}}} = -k_{\mathrm{B}} \left(K^2 \frac{\partial^2 \phi_2}{\partial K^2} + L^2 \frac{\partial^2 \phi_2}{\partial L^2} + L_{\mathrm{s}}^2 \frac{\partial^2 \phi_2}{\partial L_{\mathrm{s}}^2} \right.$$

$$\left. + 2KL \frac{\partial^2 \phi_2}{\partial K \partial L} + 2K L_{\mathrm{s}} \frac{\partial^2 \phi_2}{\partial K \partial L_{\mathrm{s}}} + 2L L_{\mathrm{s}} \frac{\partial^2 \phi_2}{\partial L \partial L_{\mathrm{s}}} \right). \tag{4.6.5}$$

There are in addition two heat capacities at one constant density and one constant field. Because of the symmetry of the system about the plane $\mathcal{H}_{\mathrm{s}} = 0$ the only one of these which is of importance is

$$c_{m,\mathcal{H}_{\mathrm{s}}} := k_{\mathrm{B}} T \left(\frac{\partial s}{\partial T} \right)_{m,\mathcal{H}_{\mathrm{s}}} = -k_{\mathrm{B}} \left(K^2 \frac{\partial^2 \phi_1}{\partial K^2} + L_{\mathrm{s}}^2 \frac{\partial^2 \phi_1}{\partial L_{\mathrm{s}}^2} + 2K L_{\mathrm{s}} \frac{\partial^2 \phi_1}{\partial K \partial L_{\mathrm{s}}} \right), \tag{4.6.6}$$

where $\phi_1 = \phi_1(K, m, L_{\mathrm{s}})$. The generalizations of the standard formula (1.4.8) relating heat capacities now need

$$\vartheta_T := \left\{ \begin{array}{l} T \left(\dfrac{\partial m}{\partial \mathcal{H}_{\mathrm{s}}} \right)_{T,\mathcal{H}} = \left(\dfrac{\partial m}{\partial L_{\mathrm{s}}} \right)_{K,L} \\ T \left(\dfrac{\partial q}{\partial \mathcal{H}} \right)_{T,\mathcal{H}_{\mathrm{s}}} = \left(\dfrac{\partial q}{\partial L} \right)_{K,L_{\mathrm{s}}} \end{array} \right\} = -\frac{\partial^2 \phi_2}{\partial L \partial L_{\mathrm{s}}} \tag{4.6.7}$$

$$\alpha_{\mathcal{H},\mathcal{H}_{\mathrm{s}}} := T \left(\frac{\partial m}{\partial T} \right)_{\mathcal{H},\mathcal{H}_{\mathrm{s}}} = -K \left(\frac{\partial m}{\partial K} \right)_{L,L_{\mathrm{s}}} - L \left(\frac{\partial m}{\partial L} \right)_{K,L_{\mathrm{s}}} - L_{\mathrm{s}} \left(\frac{\partial m}{\partial L_{\mathrm{s}}} \right)_{K,L}$$

$$= K \frac{\partial^2 \phi_2}{\partial K \partial L} + L \frac{\partial^2 \phi_2}{\partial L^2} + L_{\mathrm{s}} \frac{\partial^2 \phi_2}{\partial L \partial L_{\mathrm{s}}}, \tag{4.6.8}$$

$$\beta_{\mathcal{H},\mathcal{H}_{\mathrm{s}}} := T \left(\frac{\partial q}{\partial T} \right)_{\mathcal{H},\mathcal{H}_{\mathrm{s}}} = -K \left(\frac{\partial q}{\partial K} \right)_{L,L_{\mathrm{s}}} - L \left(\frac{\partial q}{\partial L} \right)_{K,L} - L_{\mathrm{s}} \left(\frac{\partial q}{\partial L_{\mathrm{s}}} \right)_{K,L_{\mathrm{s}}}$$

$$= K \frac{\partial^2 \phi_2}{\partial K \partial L_{\mathrm{s}}} + L \frac{\partial^2 \phi_2}{\partial L_{\mathrm{s}} \partial L} + L_{\mathrm{s}} \frac{\partial^2 \phi_2}{\partial L_{\mathrm{s}}^2}, \tag{4.6.9}$$

and it can be shown that

$$c_{\mathcal{H},\mathcal{H}_s} - c_{m,\mathcal{H}_s} = k_B \frac{\alpha^2_{\mathcal{H},\mathcal{H}_s}}{\varphi_T}, \tag{4.6.10}$$

$$c_{\mathcal{H},\mathcal{H}_s} - c_{m,q} = k_B \frac{\psi_T \alpha^2_{\mathcal{H},\mathcal{H}_s} + \varphi_T \beta^2_{\mathcal{H},\mathcal{H}_s} - 2\vartheta_T \alpha_{\mathcal{H},\mathcal{H}_s} \beta_{\mathcal{H},\mathcal{H}_s}}{\varphi_T \psi_T - \vartheta_T^2}. \tag{4.6.11}$$

We shall assume that the critical line $\mathcal{C}$ and the triple line $\mathcal{T}$ are asymptotically parallel at the tricritical point T.[27] Let the critical curve $\mathcal{C}$ have the equation $\mathcal{H} = \mathcal{H}^*(T)$ with the tricritical point T given by the coordinates $T = T_t$, $\mathcal{H} = \mathcal{H}_t = \mathcal{H}^*(T_t)$ or alternatively, in terms couplings, $L = L^*(K)$ with the tricritical point $K = K_t$, $L = L_t = L^*(K_t)$.

Unlike the situation described in Sect. 4.5, the phase diagram here has reflectional symmetry; so the L_s axis is a special direction at the tricritical point. The only other special direction is the line of the tangent to the curves $\mathcal{C}$ and $\mathcal{T}$. Assuming that the curve $\mathcal{C}$ at the tricritical point is not parallel to the L-axis, $L^*(K)$ is expressed in the form

$$L^*(K) = L_t + a\Delta K + \mathcal{L}(K), \qquad \Delta K = K - K_t, \tag{4.6.12}$$

where a is the (finite) slope of $\mathcal{C}$ at the tricritical point and

$$\mathcal{L}(K_t) = 0, \qquad \mathcal{L}'(K_t) = 0, \qquad \mathcal{L}(K) \simeq b|\Delta K|^{\psi_t}, \tag{4.6.13}$$

ψ_t (>1) being the *tricritical shift exponent* (cf. Eqs. (4.5.5) and (4.5.6)). Then the orthogonal coordinates

$$\Theta_1 := L_s, \qquad \Theta_2 := L - aK, \qquad \Theta_3 := K + aL \tag{4.6.14}$$

are such that $\Theta_2 = \Theta_{2t} = L_t - aK_t$ is the equation of the tangent to $\mathcal{C}$ at T and $\Theta_3 = \Theta_{3t} = L_t + aK_t$ is the equation of an orthogonal line, with $\Theta_2 > \Theta_{2t} = L_t - aK_t$ in the disordered region and $\Theta_2 < \Theta_{2t} = L_t - aK_t$ in the ordered region. We now define critical exponents for the tricritical point. One exponent is defined for the direction orthogonal to the plane $\Theta_1 = L_s = 0$, namely

$$L_s \sim \quad q|q|^{\delta_t - 1}, \text{ along } \Theta_2 = \Theta_{2t}, \Theta_3 = \Theta_{3t} \tag{4.6.15}$$

and the rest are for directions in the plane $\Theta_1 = L_s = 0$[28]:

---

[27] This is supported in Sect. 6.2.6 by mean-field calculations with the Hamiltonian (6.2.1).

[28] Lawrie and Sarbach (1984) define only two 'α-type' tricritical exponents, adopting the commonly-used scaling ansatz for the free energy ((Fisher (1983) and Eq. (4.9.2)), whereby $2-\alpha_t$ is the scaling exponent for the free energy in the direction of the temperature axis, with another exponent, which

$$c_{m,q} \sim \begin{cases} |\Theta_2 - \Theta_{2t}|^{-\alpha_t}, & \text{along } \Theta_3 = \Theta_{3t}, \Theta_2 > \Theta_{2t}, \\ |\Theta_2 - \Theta_{2t}|^{-\alpha'_t}, & \text{along } \Theta_3 = \Theta_{3t}, \Theta_2 < \Theta_{2t}, \\ |\Theta_3 - \Theta_{3t}|^{-\alpha_{tc}}, & \text{along } \Theta_2 = \Theta_{2t}, \end{cases} \tag{4.6.16}$$

$$c_{m,\mathcal{H}_s} \sim \begin{cases} |\Theta_2 - \Theta_{2t}|^{-\alpha_{2t}}, & \text{along } \Theta_3 = \Theta_{3t}, \Theta_2 > \Theta_{2t}, \\ |\Theta_2 - \Theta_{2t}|^{-\alpha'_{2t}}, & \text{along } \Theta_3 = \Theta_{3t}, \Theta_2 < \Theta_{2t}, \\ |\Theta_3 - \Theta_{3t}|^{-\alpha_{2tc}}, & \text{along } \Theta_2 = \Theta_{2t}, \end{cases} \tag{4.6.17}$$

$$c_{\mathcal{H},\mathcal{H}_s} \sim \begin{cases} |\Theta_2 - \Theta_{2t}|^{-\sigma_t}, & \text{along } \Theta_3 = \Theta_{3t}, \Theta_2 > \Theta_{2t}, \\ |\Theta_2 - \Theta_{2t}|^{-\sigma'_t}, & \text{along } \Theta_3 = \Theta_{3t}, \Theta_2 < \Theta_{2t}, \\ |\Theta_3 - \Theta_{3t}|^{-\sigma_{tc}}, & \text{along } \Theta_2 = \Theta_{2t}, \end{cases} \tag{4.6.18}$$

$$q \sim \begin{cases} |\Theta_2 - \Theta_{2t}|^{\beta_t}, & \text{along } \Theta_3 = \Theta_{3t}, \Theta_2 < \Theta_{2t}, \\ |\Theta_3 - \Theta_{3t}|^{\beta_{tc}}, & \text{along } \Theta_2 = \Theta_{2t}, \end{cases} \tag{4.6.19}$$

$$(m - m_t) \sim \begin{cases} |\Theta_2 - \Theta_{2t}|^{\beta_{2t}}, & \text{along } \Theta_3 = \Theta_{3t}, \Theta_2 > \Theta_{2t}, \\ |\Theta_2 - \Theta_{2t}|^{\beta'_{2t}}, & \text{along } \Theta_3 = \Theta_{3t}, \Theta_2 < \Theta_{2t}, \\ |\Theta_3 - \Theta_{3t}|^{\beta_{2tc}}, & \text{along } \Theta_2 = \Theta_{2t}, \end{cases} \tag{4.6.20}$$

$$\psi_T \sim \begin{cases} |\Theta_2 - \Theta_{2t}|^{-\gamma_t}, & \text{along } \Theta_3 = \Theta_{3t}, \Theta_2 > \Theta_{2t}, \\ |\Theta_2 - \Theta_{2t}|^{-\gamma'_t}, & \text{along } \Theta_3 = \Theta_{3t}, \Theta_2 < \Theta_{2t}, \\ |\Theta_3 - \Theta_{3t}|^{-\gamma_{tc}}, & \text{along } \Theta_2 = \Theta_{2t}, \end{cases} \tag{4.6.21}$$

$$\varphi_T \sim \begin{cases} |\Theta_2 - \Theta_{2t}|^{-\gamma_{2t}}, & \text{along } \Theta_3 = \Theta_{3t}, \Theta_2 > \Theta_{2t}, \\ \Theta_2 - \Theta_{2t}|^{-\gamma'_{2t}}, & \text{along } \Theta_3 = \Theta_{3t}, \Theta_2 < \Theta_{2t}, \\ |\Theta_3 - \Theta_{3t}|^{-\gamma_{2tc}}, & \text{along } \Theta_2 = \Theta_{2t}. \end{cases} \tag{4.6.22}$$

(Footnote 28 continued)
they call simply α, giving the scaling exponent $2 - \alpha$ along the tangent to $\mathcal{C}$. Their calculations agree with (4.7.36) for α_t with $\alpha = \alpha_{2tc}$, given by (4.7.35).

4.7 Scaling for a Tricritical Point

Following a procedure similar to that of Sect. 4.5, the singular part of ϕ can be taken to be

$$\phi_{\text{sing}}(\Delta K, \Delta L, L_{\text{s}}) = \phi_2(K, L, L_{\text{s}}) - \phi_2(K_{\text{t}}, L_{\text{t}}, 0) - \text{u}_{\text{t}}\Delta K + m_{\text{t}}\Delta L, \tag{4.7.1}$$

where $\Delta K = K - K_{\text{t}}$, $\Delta L = L - L_{\text{t}}$, and u_{t} and m_{t} are the values of the internal energy and magnetization densities at the tricritical point.[29] Thus we have

$$\phi_{\text{sing}}(0, 0, 0) = 0, \tag{4.7.2}$$

$$\frac{\partial \phi_{\text{sing}}}{\partial \Delta K} = \Delta \text{u}, \tag{4.7.3}$$

$$\frac{\partial \phi_{\text{sing}}}{\partial \Delta L} = -\Delta m, \tag{4.7.4}$$

$$\frac{\partial \phi_{\text{sing}}}{\partial L_{\text{s}}} = -q. \tag{4.7.5}$$

4.7.1 Scaling Fields for the Tricritical Point

We use the orthogonal scaling fields

$$\theta_1 := \Theta_1, \qquad \theta_2 := \Theta_2 - \Theta_{2\text{t}}, \qquad \theta_3 := \Theta_3 - \Theta_{3\text{t}}, \tag{4.7.6}$$

with the triad of orthogonal unit vectors

$$\hat{\boldsymbol{\theta}}_1 := \begin{pmatrix} 1 \\ 0 \\ 0 \end{pmatrix}, \quad \hat{\boldsymbol{\theta}}_2 := \frac{1}{\sqrt{1+a^2}} \begin{pmatrix} 0 \\ 1 \\ -a \end{pmatrix}, \quad \hat{\boldsymbol{\theta}}_3 := \frac{1}{\sqrt{1+a^2}} \begin{pmatrix} 0 \\ a \\ 1 \end{pmatrix}. \tag{4.7.7}$$

[29] At the tricritical point $L_{\text{s}} = 0$ and $q = 0$.

Setting $\theta_1 = 0$ gives the plane of reflectional symmetry, with $\hat{\boldsymbol{\theta}}_3$ giving the direction of the tangent to $\mathcal{C}$ and $\mathcal{T}$ at the tricritical point. The scaling hypothesis takes the form

$$\phi_{\text{sing}}(\lambda^{y_1}\theta_1, \lambda^{y_2}\theta_2, \lambda^{y_3}\theta_3) = \lambda^d \phi_{\text{sing}}(\theta_1, \theta_2, \theta_3), \tag{4.7.8}$$

for $\lambda > 0$, (Riedel 1972; Hankey et al. 1972) and the general shapes of the critical curves in a neighbourhood of the tricritical point can be inferred from the relative magnitudes of y_1, y_2 and y_3. A consequence of (4.7.8) is that, if ϕ has non-smooth behaviour of a particular type at $(\theta_1, \theta_2, \theta_3) = (\kappa_1, \kappa_2, \kappa_3)$, then it will also have the same non-smooth behaviour at

$$(\theta_1, \theta_2, \theta_3) = (\kappa_1\lambda^{y_1}, \kappa_2\lambda^{y_2}, \kappa_3\lambda^{y_3}). \tag{4.7.9}$$

It follows that the critical lines $\mathcal{C}$, $\mathcal{C}^{(+)}$ and $\mathcal{C}^{(-)}$ and the triple line $\mathcal{T}$ must all have parametric equations of the form (4.7.9), for some different sets of parameters $(\kappa_1, \kappa_2, \kappa_3)$ and $\lambda > 0$. We shall take the set of parameters for $\mathcal{C}$ to be $(\kappa_1, \kappa_2, \kappa_3)$, for $\mathcal{C}^{(\pm)}$ to be $(\kappa_1^{(\pm)}, \kappa_2^{(\pm)}, \kappa_3^{(\pm)})$ and for $\mathcal{T}$ to be (τ_1, τ_2, τ_3). It follows from the symmetry of the model and the choice of scaling fields that $\kappa_1 = \tau_1 = 0$, $\kappa_1^{(+)} = -\kappa_1^{(-)} > 0$, $\kappa_2^{(+)} = \kappa_2^{(-)}$ and $\kappa_3^{(+)} = \kappa_3^{(-)}$. It will be assumed that in each of these triplets not more than one member is zero.[30] For $\mathcal{C}$ to approach the tricritical point in the direction of $\hat{\boldsymbol{\theta}}_3$ it must be the case that the *tricritical gap exponent*

$$\vartheta_{\text{t}} := y_2/y_3 > 1. \tag{4.7.10}$$

It follows, by eliminating λ from (4.7.9), that the equation of $\mathcal{C}$ in a neighbourhood of the tricritical point is

$$\theta_2 = \varpi_0 |\theta_3|^{\vartheta_{\text{t}}}, \tag{4.7.11}$$

where $\varpi_0 = \kappa_2/|\kappa_3|^{\vartheta_{\text{t}}}$. From Eq. (4.7.6),

$$\theta_2\left(1 + a\frac{\theta_3}{\theta_2}\right) = (1 + a^2)\Delta L, \qquad \theta_3\left(1 - a\frac{\theta_2}{\theta_3}\right) = (1 + a^2)\Delta K \tag{4.7.12}$$

and, by the same reasoning as that employed in Sect. 4.5, it follows, as in the case of a critical point, that the shift exponent $\psi_{\text{t}} = \vartheta_{\text{t}}$. The way in which the curves $\mathcal{C}^{(\pm)}$ approach the tricritical point is determined by the magnitude of y_1 relative to y_2 and y_3 and whether one pair of $\kappa_2^{(\pm)}$ or $\kappa_3^{(\pm)}$ is zero.

---

30 Otherwise, with the definitions (4.7.6) for the scaling fields, the corresponding curve would be linear and in the direction of one of the unit vectors (4.7.7).

4.7.2 Tricritical Exponents and Scaling Laws

We now consider the asymptotic forms for the densities m and q and the response functions φ_T, ψ_T, $c_{m,q}$, $c_{m,\mathcal{H}_s}$ and $c_{\mathcal{H},\mathcal{H}_s}$ at the tricritical point. Since we are, for all except the exponent δ_t defined by (4.6.15), considering asymptotic forms in the plane $\Theta_1 = 0$ we shall, for simplicity, in the following formulae apply this condition.[31] From (4.6.2), (4.6.3), (4.6.5) and (4.7.4)–(4.7.6),

$$m - m_t = -\left[\frac{\partial \phi_{\text{sing}}}{\partial \theta_2} + a\frac{\partial \phi_{\text{sing}}}{\partial \theta_3}\right], \tag{4.7.13}$$

$$q = -\frac{\partial \phi_{\text{sing}}}{\partial \theta_1}, \tag{4.7.14}$$

$$\varphi_T = -\left[\frac{\partial^2 \phi_{\text{sing}}}{\partial \theta_2^2} + 2a\frac{\partial^2 \phi_{\text{sing}}}{\partial \theta_2 \partial \theta_3} + a^2\frac{\partial^2 \phi_{\text{sing}}}{\partial \theta_3^2}\right], \tag{4.7.15}$$

$$\psi_T = -\frac{\partial^2 \phi_{\text{sing}}}{\partial \theta_1^2}, \tag{4.7.16}$$

$$c_{\mathcal{H},\mathcal{H}_s} = -k_B\left[\Theta_2^2\frac{\partial^2 \phi_{\text{sing}}}{\partial \theta_2^2} + 2\Theta_2\Theta_3\frac{\partial^2 \phi_{\text{sing}}}{\partial \theta_2 \partial \theta_3} + \Theta_3^2\frac{\partial^2 \phi_{\text{sing}}}{\partial \theta_3^2}\right]. \tag{4.7.17}$$

For $c_{m,\mathcal{H}_s}$ and $c_{m,q}$ we use (4.6.10) and (4.6.11). Thus we need the expressions

$$\vartheta_T = a\frac{\partial^2 \phi_{\text{sing}}}{\partial \theta_1 \partial \theta_2} - \frac{\partial^2 \phi_{\text{sing}}}{\partial \theta_1 \partial \theta_3}, \tag{4.7.18}$$

$$\alpha_{\mathcal{H},\mathcal{H}_s} = \Theta_2\frac{\partial^2 \phi_{\text{sing}}}{\partial \theta_2^2} + (a\Theta_2 + \Theta_3)\frac{\partial^2 \phi_{\text{sing}}}{\partial \theta_2 \partial \theta_3} + a\Theta_3\frac{\partial^2 \phi_{\text{sing}}}{\partial \theta_3^2}, \tag{4.7.19}$$

$$\beta_{\mathcal{H},\mathcal{H}_s} = \Theta_2\frac{\partial^2 \phi_{\text{sing}}}{\partial \theta_1 \partial \theta_2} + \Theta_3\frac{\partial^2 \phi_{\text{sing}}}{\partial \theta_1 \partial \theta_3} \tag{4.7.20}$$

giving

$$c_{m,\mathcal{H}_s} = -\frac{k_B}{\varphi_T}(a\Theta_2 - \Theta_3)^2\left[\left(\frac{\partial^2 \phi_{\text{sing}}}{\partial \theta_2 \partial \theta_3}\right)^2 - \frac{\partial^2 \phi_{\text{sing}}}{\partial \theta_2^2}\frac{\partial^2 \phi_{\text{sing}}}{\partial \theta_3^2}\right]. \tag{4.7.21}$$

The exponents associated with m, q and the response functions for different paths of approach are given by Eqs. (4.6.15)–(4.6.22). Following the procedure used for the

[31] The formula (4.7.14), for q is unaffected by this assumption.

general scaling formulation in Sect. 4.3 and for the critical point in Sect. 4.5 we take $\lambda = |\theta_1|^{-1/y_1}$, $\lambda = |\theta_2|^{-1/y_2}$ and $\lambda = |\theta_3|^{-1/y_3}$ in Eq. (4.7.8) to give respectively

$$\phi_{\text{sing}}(\theta_1, \theta_2, \theta_3) = |\theta_1|^{d/y_1} \phi_{\text{sing}}(\pm 1, g_2^{(1)}, g_3^{(1)}), \tag{4.7.22}$$

$$\phi_{\text{sing}}(\theta_1, \theta_2, \theta_3) = |\theta_2|^{d/y_2} \phi_{\text{sing}}(g_1^{(2)}, \pm 1, g_3^{(2)}), \tag{4.7.23}$$

$$\phi_{\text{sing}}(\theta_1, \theta_2, \theta_3) = |\theta_3|^{d/y_3} \phi_{\text{sing}}(g_1^{(3)}, g_2^{(3)}, \pm 1), \tag{4.7.24}$$

where the gap fields $g_j^{(m)}(\theta_m, \theta_j)$ and exponents $\vartheta_j^{(m)}$, $m, j = 1, 2, 3$, are given by (4.3.13) and (4.3.14). It follows from (4.7.11) that, in a neighbourhood of the tricritical point, the equation of $\mathcal{C}$ in the plane $\theta_1 = 0$ is $g_2^{(3)}(\theta_2, \theta_3) = \varpi_0$. From (4.7.22), with $\theta_2 = \theta_3 = 0$, using similar reasoning to that used in Sect. 4.5, it follows that the exponent δ_{t}, defined in (4.6.15), is given by

$$\delta_{\text{t}} = y_1/(d - y_1). \tag{4.7.25}$$

A classification of the paths of approach to the tricritical point in the plane $L_{\text{s}} = \Theta_1 = 0$ can now be made in a similar way to that for the critical point in Sect. 4.5. The two types of approach are:

(i) Paths on which $|g_3^{(2)}| \to 0$, including the direction of the vector $\hat{\boldsymbol{\theta}}_2$, for which, from (4.7.23),

$$\phi_{\text{sing}}(0, \theta_2, \theta_3) \simeq |\theta_2|^{d/y_2} A_\infty^{(\pm)}, \tag{4.7.26}$$

where

$$\lim_{g_3^{(2)} \to 0} \phi_{\text{sing}}(0, \pm 1, g_3^{(2)}) = A_\infty^{(\pm)} \tag{4.7.27}$$

is assumed to be finite and non-zero.

(ii) Paths on which $|g_2^{(3)}| \to \pi_2$, where π_2 is finite (possibly zero), when

$$\phi_{\text{sing}}(0, \theta_2, \theta_3) \simeq |\theta_3|^{d/y_3} A_{\pi_2}^{(\pm)}, \tag{4.7.28}$$

where

$$\lim_{g_2^{(3)} \to \pi_2} \phi_{\text{sing}}(0, g_2^{(3)}, \pm 1) = A_{\pi_2}^{(\pm)} \tag{4.7.29}$$

is assumed to be finite and non-zero.

Since $\vartheta_{\text{t}} > 1$, all curves in class (ii), including $\mathcal{C}$, are tangential to the direction of the unit vector $\hat{\boldsymbol{\theta}}_3$. Some curves in class (i) are also tangential to $\hat{\boldsymbol{\theta}}_3$, but all curves not tangential to this vector are certainly in class (i). The asymptotic forms in equations

(4.6.16)–(4.6.21) with $\Theta_3 = \Theta_{3t}$ fall into class (i) and those with $\Theta_2 = \Theta_{2t}$ into class (ii). Thus from (4.6.19), (4.6.21), (4.7.14) and (4.7.16)

$$\beta_t = (d - y_1)/y_2, \qquad \beta_{tc} = \vartheta_t \beta_t = (d - y_1)/y_3, \tag{4.7.30}$$

$$\gamma'_t = (2y_1 - d)/y_2, \qquad \gamma_{tc} = \vartheta_t \gamma_t = (2y_1 - d)/y_3. \tag{4.7.31}$$

Since the density m and response functions φ_T and $c_{\mathcal{H},\mathcal{H}_s}$ arise as derivatives of ϕ with respect to couplings K and L, which appear as linear combinations in the scaling field θ_2 and θ_3, we must consider, as we did in Sect. 4.3.9, the relative importance of different terms. From (4.6.18), (4.6.20), (4.6.22), (4.7.13), (4.7.15), (4.7.17) and (4.7.21), since $y_2 > y_3$,

$$\beta'_{2t} = (d - y_2)/y_2, \qquad \beta_{2tc} = \vartheta_t \beta_{2t} = (d - y_2)/y_3, \tag{4.7.32}$$

$$\gamma'_{2t} = (2y_2 - d)/y_2, \qquad \gamma_{2tc} = \vartheta_t \gamma_{2t} = (2y_2 - d)/y_3, \tag{4.7.33}$$

$$\sigma'_{2t} = (2y_2 - d)/y_2, \qquad \sigma_{tc} = \vartheta_t \sigma_t = (2y_2 - d)/y_3, \tag{4.7.34}$$

$$\alpha'_{2t} = (2y_3 - d)/y_2, \qquad \alpha_{2tc} = \vartheta_t \alpha_{2t} = (2y_3 - d)/y_3. \tag{4.7.35}$$

That $\sigma'_t = \gamma'_{2t}$ and $\sigma_{tc} = \gamma_{2tc}$ could be anticipated from our calculations for the critical point, in Sect. 4.5.6, since the system has no symmetry with respect to the conjugate variables m and $\mathcal{H}$. The formula we used for deriving (4.7.21) and thus the exponents (4.7.35) was (4.6.10). The difference between the exponent for $c_{\mathcal{H},\mathcal{H}_s}$ and $c_{m,\mathcal{H}_s}$ is a consequence of the cancellation of the most singular term in $c_{\mathcal{H},\mathcal{H}_s}$ with that contained in the right-hand side of (4.6.10). We apply a similar procedure, using (4.6.11), to obtain the exponents for $c_{m,q}$. In this case, some calculation shows that no cancellation occurs. The exponents are the same as those for $c_{\mathcal{H},\mathcal{H}_s}$, namely,

$$\alpha'_t = (2y_2 - d)/y_2, \qquad \alpha_{tc} = \vartheta_t \alpha'_t = (2y_2 - d)/y_3. \tag{4.7.36}$$

From these formulae relating critical exponents to scaling exponents it follows that

$$2 = \alpha'_t + 2\,\beta_t + \gamma'_t, \qquad 2 = \alpha_{2tc} + 2\,\beta_{2tc} + \gamma_{2tc}, \tag{4.7.37}$$

which are versions of the Essam–Fisher scaling law (4.5.44), and

$$\gamma'_t = \beta_t(\delta_t - 1), \qquad \gamma_{tc} = \beta_{tc}(\delta_t - 1), \tag{4.7.38}$$

which are versions of the Widom scaling law (4.5.45). We also have

$$\gamma'_{2t} = \beta_{2t}(\delta_{2t} - 1), \qquad \gamma_{2tc} = \beta_{tc}(\delta_{2t} - 1), \tag{4.7.39}$$

where $\delta_{2t} = 1/\beta_{2t}$ is the exponent which would appear in a formula like (4.6.15) with $L - L_t$ replacing L_s and $m - m_t$ replacing q.

Since, in the plane $\Theta_3 = \Theta_{3t}$, both the critical isochore $q = 0$ and the coexistence curve lie along the line $\Theta_1 = 0$, scaling predicts the equality of the exponents in the ordered and disordered regions. That is[32]

$$\gamma_t = \gamma'_t, \qquad \beta_{2t} = \beta'_{2t}, \qquad \gamma_{2t} = \gamma'_{2t}, \qquad \alpha_{2t} = \alpha'_{2t}. \tag{4.7.40}$$

In terms of the scaling formulation of Sect. 4.3, the tricritical point is a transition region connected to seven other transition regions; the critical curves $\mathcal{C}$, $\mathcal{C}^{(+)}$ and $\mathcal{C}^{(-)}$, the triple curve $\mathcal{T}$ and the three first-order surfaces. An analysis of scaling along these curves can be made using effective exponents as described in Sect. 4.3.3.

4.8 Corrections to Scaling

In Sect. 4.3.4 we considered one case where the simple power-law asymptotic behaviour of equations (4.3.20) broke down. This arose from the exponent ω_m being zero, and led to a logarithmic divergence in the corresponding derivative of ϕ. In this section modifications to scaling which arise from the amplitude factors are considered. For simplicity we concentrate on the case $n = 3$ with a transition region $\mathcal{C}$ with $s = 2$. That is to say, we have a system with two relevant fields θ_1 and θ_2 and one irrelevant field θ_3. From (4.3.10)–(4.3.13),

$$\phi_{\text{sing}}(\theta_1, \theta_2, \theta_3) = |\theta_1|^{\sigma_1} \phi_{\text{sing}}\left(\pm 1, \frac{\theta_2}{|\theta_1|^{\vartheta_2^{(1)}}}, \frac{\theta_3}{|\theta_1|^{\vartheta_3^{(1)}}}\right), \tag{4.8.1}$$

$$\phi_{\text{sing}}(\theta_1, \theta_2, \theta_3) = |\theta_2|^{\sigma_2} \phi_{\text{sing}}\left(\frac{\theta_1}{|\theta_2|^{\vartheta_1^{(2)}}}, \pm 1, \frac{\theta_3}{|\theta_2|^{\vartheta_3^{(2)}}}\right). \tag{4.8.2}$$

In Sect. 4.3 we considered the asymptotic forms which can be derived from (4.8.1) and (4.8.2). In each case it was necessary to choose a path of approach to the transition region so that the arguments of the functions on the right-hand sides tend to finite values. Even then, in order to obtain the simple form of scaling described in Sect. 4.3 and utilized for a critical point and tricritical point in Sects. 4.5 and 4.7 respectively, it is necessary that the amplitude factors are finite and non-zero. We begin by considering the functions

$$f^{(\pm)}(u, v) := \phi_{\text{sing}}(\pm 1, u, v), \tag{4.8.3}$$

[32] However, of course, the same warning applies as that given for the critical point in Sect. 4.5.6. A zero amplitude in an asymptotic form can cause the equality of the exponents to break down. It is also the case that if the model were generalized to eliminate the symmetry about $\Theta_1 = 0$ the equality of the exponents would need to be established analytically as it was for the critical point.

which appear on the right-hand side of (4.8.1). The limit $u \to 0$ with v small and non-zero can be achieved by taking $\theta_2 \to 0$ with θ_1 and θ_3 non-zero. Excluding the possibility that $\theta_2 = 0$ is some transition region connected to $\mathcal{C}$, $f^{(\pm)}(u, v)$ have power-series expansions of the form

$$f^{(\pm)}(u, v) = w_0^{(\pm)}(v) + u w_1^{(\pm)}(v) + \cdots . \tag{4.8.4}$$

If $w_0^{(\pm)}(0)$ is finite and non-zero then the power-law behaviour is that predicted by simple scaling theory, although the magnitude of $w_1^{(\pm)}(0)$ will affect the range over which the leading scaling behaviour can be observed. If $w_0^{(\pm)}(0) = 0$ there is a breakdown of simple scaling and the power-law predictions rest on the first non-zero term in the expansion. If this is given by $w_1^{(\pm)}(0)$ being finite and non-zero, then the singular behaviour is not only weaker but will depend on the next layer of detail of the path of approach $\mathcal{P}$. A possible $\mathcal{P}$ consistent with the condition $u \to 0$ is $\theta_2 \sim |\theta_1|^{\tau}$, where $\tau > \vartheta_2^{(1)}$. Then from (4.8.1)

$$\phi_{\text{sing}}(\theta_1, \theta_2, \theta_3) \sim |\theta_1|^{\sigma_1 + \tau - \vartheta_2^{(1)}}, \tag{4.8.5}$$

which is a weaker power-law behaviour. The same analysis can be applied to (4.8.2).

Consider now the possibility of $f^{(\pm)}(0, 0)$ being infinite. It is clear that, for an isolated transition region, $f^{(\pm)}(0, v)$ cannot be infinite and thus, any singular behaviour must be arise from $w_0^{(\pm)}(v)$, therefore involving the irrelevant variable θ_3. That this is a possibility was pointed out by Fisher (1983), who classified irrelevant variables as *dangerous* or *harmless* according as they do or do not yield such divergent behaviour. Suppose that

$$w_0^{(\pm)}(v) \sim |v|^{-\tau}, \tag{4.8.6}$$

where $\tau > 0$. Then, from (4.8.1)–(4.8.4),

$$\phi_{\text{sing}}(\theta_1, \theta_2, \theta_3) \sim |\theta_1|^{\bar{\sigma}_1}, \tag{4.8.7}$$

where

$$\bar{\sigma}_1 := \frac{\bar{d}}{y_1}, \tag{4.8.8}$$

and

$$\bar{d} := d - \tau |y_3|. \tag{4.8.9}$$

The effect of the dangerous irrelevant variable θ_3 can be viewed as an effective reduction in the dimension of the system from d to $\bar{d}$, provided that all paths to the transition region produce the same change. This means that we must be satisfied that a parallel analysis of (4.8.2) yields the same formula (4.8.9). In Sect. 5.6.3 the significance of the presence of a dangerous irrelevant variable for the interpretation of Gaussian critical exponents in relation to scaling theory is discussed.

4.9 Scaling and Universality

The universality hypothesis, which is described in Sect. 4.2.1, and the scaling hypotheses of Sects. 4.3.1 and 4.3.4 both apply to a particular transition region rather than to a physical system or theoretical model as a whole. Different transition regions of the same system may be in different universality classes and may have different scaling functions and exponents. Discussion of the relationship between scaling and universality is made easier if we take a particular type of transition region. The simplest example to choose is a critical point with power-law scaling. In the weak scaling direction, from (4.5.16), $\theta_1/\theta_2 \to 0$ as the critical point is approached and thus, from (4.5.8) and (4.5.23),

$$\theta_1 = \Delta L - a\Delta K, \qquad \theta_2 \simeq (1+a^2)\Delta K. \tag{4.9.1}$$

From Eqs. (4.5.11), (4.5.12), (4.5.40), (4.5.39) and (4.5.59), the forms for the singular part of ϕ, the density difference from its critical value and the response function φ_T can be expressed as

$$\phi_{\text{sing}} = A_0|\Delta K|^{2-\alpha} W_0^{(\pm)}\left(C\,\frac{[\Delta L - a\Delta K]}{|\Delta K|^{\beta\delta}}\right), \tag{4.9.2}$$

$$\Delta\rho = -\frac{\partial\phi_{\text{sing}}}{\partial\Delta L} = A_1^{(\pm)}|\Delta K|^{\beta} W_1^{(\pm)}\left(C\,\frac{[\Delta L - a\Delta K]}{|\Delta K|^{\beta\delta}}\right), \tag{4.9.3}$$

$$\varphi_T = -\frac{\partial^2\phi_{\text{sing}}}{\partial\Delta L^2} = A_2^{(\pm)}|\Delta K|^{-\gamma} W_2^{(\pm)}\left(C\,\frac{[\Delta L - a\Delta K]}{|\Delta K|^{\beta\delta}}\right), \tag{4.9.4}$$

where C, A_0, $A_1^{(\pm)}$ and $A_2^{(\pm)}$ are constants and

$$A_s^{(\pm)} W_s^{(\pm)}(x) = -C^s A_0 \frac{\mathrm{d}^s W_0^{(\pm)}(x)}{\mathrm{d}x^s}, \qquad s = 1, 2. \tag{4.9.5}$$

By imposing the conditions

$$W_0^{(\pm)}(0) = W_1^{(\pm)}(0) = W_2^{(\pm)}(0) = 1, \tag{4.9.6}$$

we ensure that A_0, $A_1^{(\pm)}$ and $A_2^{(\pm)}$ are the critical amplitudes with

$$A_s^{(\pm)} = C^s A_0 Q_s^{(\pm)},$$
$$Q_s^{(\pm)} = -\left(\frac{\mathrm{d}^s W_0^{(\pm)}(x)}{\mathrm{d}x^s}\right)_{x=0} \qquad s = 1, 2. \tag{4.9.7}$$

Each of the Eqs. (4.9.2)–(4.9.4) represents a family of functions in two variables, $(\phi_{\text{sing}}, \Delta L)$, $(\Delta\rho, \Delta L)$, and $(\varphi_T, \Delta L)$ respectively, parameterized by ΔK. The scaling-hypothesis implication of the form of these equations is that each family of isotherms (ΔK constant) 'collapses' onto one curve when plotted in the space of $(\Delta L - a\Delta K)/|\Delta K|^{\beta\delta}$ against $\phi_{\text{sing}}/|\Delta K|^{2-\alpha}$, $\Delta\rho/|\Delta K|^{\beta}$ and $\varphi_T|\Delta K|^{\gamma}$ respectively. Equation (4.9.3) is the equation of state. For a ferromagnetic critical point $a = 0$, $\Delta L := \mathcal{H}/T \simeq \mathcal{H}/T_c$, $\Delta K := K - K_c \simeq J(T_c - T)/T_c^2$ and $\Delta\rho = m$, the magnetization per lattice site. The validity of an equation of the form (4.9.3) has been investigated for nickel by Arrott and Noakes (1967), Kouvel and Rodbell (1967) and Kouvel and Comly (1968), for CrO_2 by Kouvel and Rodbell (1967) and for $CrBr_3$ by Ho and Litster (1969). Similar tests for a number of fluid systems have been made by Green et al. (1967). In all cases the experimental data gives strong support for the scaling hypothesis.

Although the critical exponent equalities (4.5.47) and the Essam–Fisher and Widom scaling laws (4.5.44) and (4.5.45) are analytic consequences of (4.9.2)–(4.9.4), they can be used as a direct test of scaling for theoretical models and experimental data. The well-known classical values for the purely thermodynamic exponents given in (5.3.30) satisfy these scaling laws. They arise for systems of any dimension when the free energy is assumed to be smooth in a neighbourhood of the critical point. In particular they follow from Landau theory, as we shall see in Sect. 5.3. Exact calculations for the two-dimensional Ising model give $\alpha = \alpha' = 0$, $\beta = \frac{1}{8}$ and $\gamma = \gamma' = \frac{7}{4}$, where the value of α corresponds to a logarithmic singularity (Sects. 12.2.7.3 and 12.3.9). These values satisfy the Essam–Fisher scaling law (4.5.44). If the Widom scaling law (4.5.45) is assumed then $\delta = 15$, which is the generally accepted value (Baxter 1982b). Ho and Litster (1969) obtained the values $\gamma = \gamma' = 1.215 \pm 0.015$, $\beta = 0.368 \pm 0.005$ and $\delta = 4.28 \pm 0.1$ for $CrBr_3$, which yields $\gamma' - \beta(\delta - 1) = 0.00746 \pm 0.0682$.

The universality hypothesis was presented in Sect. 4.2.1 as a list of criteria for dividing critical problems into universality classes. This hypothesis can be tested when it is clear what is understood by the critical properties of two problems being 'the same'. This is usually taken to mean that the problems have the same values for critical exponents and the same scaling functions to within metric factors (Fisher 1983). For Eq. (4.9.2)–(4.9.4), this means that the functions $W_s^{(\pm)}(x)$, $s = 0, 1, 2$ and the exponents α, β, γ and δ are the same for all problems in the same universality class but the constants C, $A_s^{(\pm)}$, $s = 0, 1, 2$ may be different. As indicated above, the work of Green et al. (1967) represents a test of scaling. It can also be taken as a test of universality, which is clearly supported by their results which show that the branches of the coexistence curves of a number of fluids lie on a single universal curve.

So far in this section we have concentrated the discussion on thermodynamic functions and variables. Now we consider the correlation function and correlation length for the statistical mechanical variable $\mathfrak{q}(\boldsymbol{r})$ associated with the coupling L. Let $\Delta\mathfrak{q}(\boldsymbol{r}) := \mathfrak{q}(\boldsymbol{r}) - \rho_c$. Then

$$\langle \Delta\mathfrak{q}(\boldsymbol{r})\rangle = \langle \mathfrak{q}(\boldsymbol{r})\rangle - \rho_c = \rho - \rho_c = \Delta\rho, \tag{4.9.8}$$

$$\langle \Delta\mathfrak{q}(\boldsymbol{r}_1)\Delta\mathfrak{q}(\boldsymbol{r}_2)\rangle = \langle \mathfrak{q}(\boldsymbol{r}_1)\mathfrak{q}(\boldsymbol{r}_2)\rangle - 2\rho\rho_c + \rho_c^2 \tag{4.9.9}$$

and, from (2.5.9) with $\mathfrak{c}_1 = \mathfrak{c}_2 = \mathfrak{q}$,

$$\Gamma_2(\mathfrak{q}(\boldsymbol{r}_1), \mathfrak{q}(\boldsymbol{r}_2)) = \langle \Delta\mathfrak{q}(\boldsymbol{r}_1)\Delta\mathfrak{q}(\boldsymbol{r}_2)\rangle - \langle \Delta\mathfrak{q}(\boldsymbol{r}_1)\rangle\langle \Delta\mathfrak{q}(\boldsymbol{r}_2)\rangle. \tag{4.9.10}$$

As in Sect. 4.5.7 we assume that the system is translational invariant so that the connected two-point correlation function is dependent on the single spatial variable $\bar{\boldsymbol{r}} = \boldsymbol{r}_2 - \boldsymbol{r}_1$. Then, from (4.5.48), (4.5.49) and (4.9.1),

$$\Gamma_2(\mathfrak{q}(\boldsymbol{0}), \mathfrak{q}(\bar{\boldsymbol{r}})) = B^{(\pm)}|\Delta K|^{d\nu-\gamma} X^{(\pm)}\left(C\,\frac{[\Delta L - a\Delta K]}{|\Delta K|^{\beta\delta}}, \frac{\bar{\boldsymbol{r}}|\Delta K|^{\nu}}{R_0}\right), \tag{4.9.11}$$

$$\bar{\Gamma}_2^*(\mathfrak{q}; \bar{\boldsymbol{k}}) = B^{(\pm)}(R_0^d/v_0)|\Delta K|^{-\gamma} X^{*(\pm)}\left(C\,\frac{[\Delta L - a\Delta K]}{|\Delta K|^{\beta\delta}}, \frac{R_0\bar{\boldsymbol{k}}}{|\Delta K|^{\nu}}\right), \tag{4.9.12}$$

where v_0 is the volume per lattice site and

$$X^{*(\pm)}(x; \bar{\boldsymbol{k}}) = \sum_{\{\bar{\boldsymbol{r}}\}} X^{(\pm)}(x; \bar{\boldsymbol{r}}) \exp(-\mathrm{i}\bar{\boldsymbol{k}} \boldsymbol{.} \bar{\boldsymbol{r}}). \tag{4.9.13}$$

That the metric factor C in (4.9.11) and (4.9.12) is the same as that in (4.9.2)–(4.9.4) follows from (2.5.16) and (2.5.24), from which it can also be seen that

$$B^{(\pm)} = A_2^{(\pm)} v_0/R_0^d, \tag{4.9.14}$$

$$X^{*(\pm)}(x; \boldsymbol{0}) = W_2^{(\pm)}(x). \tag{4.9.15}$$

From (4.5.50),

$$\mathfrak{r}(\mathfrak{q}) = R_0 Q_3^{(\pm)}|\Delta K|^{-\nu} Y^{(\pm)}\left(C\,\frac{[\Delta L - a\Delta K]}{|\Delta K|^{\beta\delta}}\right), \tag{4.9.16}$$

where

$$Q_3^{(\pm)}\, Y^{(\pm)}(x) = \sqrt{-c(d)\frac{\nabla_{\bar{k}}^2 X^{*(\pm)}(x, \bar{\boldsymbol{0}})}{W_2^{(\pm)}(x)}}. \tag{4.9.17}$$

By imposing the condition

$$Y^{(\pm)}(0) = 1, \tag{4.9.18}$$

we have

$$Q_3^{(\pm)} = \sqrt{-c(d)\nabla_k^2 X^{*(\pm)}(0, \bar{\mathbf{0}})}. \tag{4.9.19}$$

The functions $X^{(\pm)}(x; \bar{\boldsymbol{r}})$ are supposed to be universal making $X^{*(\pm)}(x; \bar{\boldsymbol{k}})$ and $Y^{(\pm)}(x)$ also universal. It thus follows, from (4.9.19), that $Q_3^{(\pm)}$, like $Q_1^{(\pm)}$ and $Q_2^{(\pm)}$ defined in (4.9.7), are universal constants.

It would seem that in each of the regions $\Delta K >, < 0$ there are three independent non-universal metric parameters C, A_0 and R_0. It has, however, been argued by Stauffer et al. (1972) that, at the critical point, the singular part of the dimensionless free energy, in a region of the dimension of the correlation length, is a universal constant. In our notation this quantity is the critical point value of $\mathfrak{r}^d(\mathrm{q})\phi_{\mathrm{sing}}/v_0$. From (4.9.2), (4.9.6), (4.9.16), (4.9.18) and (4.9.19)

$$\frac{\mathfrak{r}^d(\mathrm{q})\phi_{\mathrm{sing}}}{v_0} \sim A_0 R_0^d v_0^{-1} |\Delta K|^{2-\alpha-d\nu} [Q_3^{(\pm)}]^d \tag{4.9.20}$$

at the critical point. When the Josephson hyper-scaling law (4.5.54) is satisfied the right-hand side of (4.9.20) is a constant and the proposition, referred to as the *two-scale factor* or *hyper-universality hypothesis*, is that the quantity $A_0 R_0^d / v_0$ is universal. In Sect. 4.8 we discussed the idea of dangerous irrelevant variables. In Sect. 5.6.3 these ideas will be amplified in relation to the Gaussian critical exponents to show that the hyper-scaling law is satisfied with the 'real' physical dimension d when $d \leq d_{\mathrm{u}}$, the upper borderline dimension, which is four for a critical point. For $d > d_{\mathrm{u}}$ the hyper-scaling law is satisfied (at least for the Gaussian exponents) when d is replaced by $\bar{d}$, the dimension reduced by the presence of a dangerous irrelevant variable (Eq. (4.8.9)). The hyper-universality hypothesis is thus expected to be true for $d \leq 4$. This is supported by the results for theoretical models and experimental systems given by Stauffer et al. (1972). Later work (Ferer et al. 1973) shows good agreement for systems with different lattice structures, which on other grounds are taken to be in the same universality class, and Aharony (1974) used a renormalization group ϵ-expansion to establish its validity to order ϵ^2.

The argument of Privman and Fisher (1984) for the truth of the hyper-universality hypothesis is also based on renormalization group ideas. They conclude that $\langle \Delta \mathrm{q}(\boldsymbol{r}_1) \Delta \mathrm{q}(\boldsymbol{r}_2) \rangle$ and $\Gamma_2(\mathrm{q}; \bar{\boldsymbol{r}})$ have scaling forms which may differ in their universal scaling functions, but will have the same non-universal metric factors. Thus

$$\langle \Delta \mathrm{q}(\boldsymbol{r}_1) \Delta \mathrm{q}(\boldsymbol{r}_2) \rangle = B^{(\pm)} |\Delta K|^{d\nu-\gamma} Z^{(\pm)} \left(C \frac{[\Delta L - a\Delta K]}{|\Delta K|^{\beta\delta}}, \frac{\bar{\boldsymbol{r}} |\Delta K|^{\nu}}{R_0} \right), \tag{4.9.21}$$

and, taking the limit $|\bar{r}| \to \infty$, it follows that

$$(\Delta\rho)^2 = B^{(\pm)}|\Delta K|^{d\nu-\gamma} Z^{(\pm)}\left(C\,\frac{[\Delta L - a\Delta K]}{|\Delta K|^{\beta\delta}}, \infty\right). \tag{4.9.22}$$

Comparing Eqs. (4.9.3) and (4.9.22) gives

$$B^{(\pm)} = [A_1^{(\pm)}]^2 \tag{4.9.23}$$

and, from (4.9.7) and (4.9.14),

$$A_0 R_0^d / v_0 = Q_2^{(\pm)}/[Q_1^{(\pm)}]^2. \tag{4.9.24}$$

In Sect. 5.6.3 we compute $A_0 R_0^d / v_0$ in the Gaussian approximation.

The universality hypothesis for the equation of state (4.9.3) is rather similar to the law of corresponding states (Sect. 4.2). One difference is that now the universal scaling function is not analytic in the thermodynamic variables but in non-analytic combinations of them. It also contains non-universal metric parameters which are not simply the critical point parameters. Another important difference is that (4.9.3) applies to the locality of the critical point rather than to the system as a whole. In line with the quote from Levelt Sengers (1974) on p. 94, one may nevertheless regard the universality hypothesis as a result of the same 'unifying' impulse which led to the earlier formulation of the law of corresponding states.

4.10 Finite-Size Scaling

All the discussion in this chapter until now has assumed that the system of interest is one for which the thermodynamic limit has been applied, as described in Sect. 1.3 and defined with greater mathematical precision for a one-component lattice fluid in Sect. 8.1. Here we consider a system on the finite d-dimensional hypercubic lattice $\mathcal{N}$ defined in Sect. 18.1.1. The thermodynamic limit when $\mathcal{N} \to \mathcal{L}$, the corresponding d-dimensional infinite system, is achieved by taking $N_\ell \to \infty$ simultaneously for every $\ell = 1, 2, \ldots, d$, in such a way that the van Hove criterion (8.1.11) is satisfied. Consider now the case where the thermodynamic limit is taken only for the variables $N_\ell, \ell \le \eth < d$. This is a system which is infinite in $\eth$ dimensions and finite in $d - \eth$ dimensions. The parameter

$$\aleph := \left[\prod_{\ell=\eth+1}^{d} N_\ell\right]^{1/(d-\eth)} \tag{4.10.1}$$

can be understood as the *thickness* of the system. We denote this system by $\mathcal{L}_\eth(\aleph)$. The thermodynamic limit $\mathcal{L}_\eth(\aleph) \to \mathcal{L}$ now means that $\aleph \to \infty$ with the *shape ratios* $N_\ell/\aleph$, $\ell = \eth + 1, \ldots, d$ kept constant.

Two distinguishable phenomena are of interest for $\mathcal{L}_{\mathfrak{d}}(\aleph)$, surface effects and finite-size effects. The former, which are present only in the absence of periodic boundary conditions, concern the effect on thermodynamic and statistical mechanical properties of the presence of a surface, including the occurrence of surface phase transitions.[33] This is not, however, the subject of this section which will be solely concerned with the changes to bulk scaling which arise from the finite value of $\aleph$. This subject was initiated by Fisher (1971) and Fisher and Barber (1972).[34] We have seen in Sect. 4.5.7 that bulk scaling ideas can, with suitable modification, be applied to first-order transitions. Similar developments can also be made to finite-size scaling methods (Fisher and Berker 1982; Privman and Fisher 1983; Binder and Landau 1984; Fisher and Privman 1985). We shall, however, in this section restrict our attention to continuous transitions.

If $\mathfrak{d} = 0$ the system $\mathcal{L}_0(\aleph) = \mathcal{N}$ is completely finite. In terms of some suitable variable like the fugacity the partition function will be a polynomial with positive coefficients and no phase transitions can occur (see p. 253). In spite of this such systems play an important role as approximations to infinite systems in computer simulation methods and in real-space renormalization group finite-lattice methods (Sect. 15.8). When $\mathfrak{d} > 0$, $\mathcal{L}_{\mathfrak{d}}(\aleph)$ can be regarded as a perhaps somewhat complicated infinite lattice system of $\mathfrak{d}$ dimensions and the type of phase transition behaviour will be characterized by $\mathfrak{d}$ rather than d. For $\mathfrak{d} = 1$ and short-range interactions transfer matrix methods of the type described in Chap. 11 can be applied yielding no phase transitions at non-zero temperatures. The correlation length may, however, (for reasonably larger values of $\aleph$) attain a steep maximum at some values of the couplings. These are termed *pseudocritical values*, (Fisher 1971; Fisher and Ferdinand 1967) associated with an *incipient phase transition* (Sect. 11.2). Many authors (Ree and Chesnut 1966; Runnels and Combs 1971; Lavis 1976) have used such maxima and similar maxima in response functions as indicating the approximate location of the phase transition in the corresponding infinite system. The pseudocritical values of the couplings will be functions of the thickness $\aleph$ of the system. When $\mathfrak{d} > 1$ the maxima associated with pseudocritical coupling values will be replaced by genuine critical values. These will be dependent on both $\aleph$ and $\mathfrak{d}$, although, according to the discussion of universality in Sects. 4.2 and 4.9, the corresponding exponents will be dependent solely on $\mathfrak{d}$.

4.10.1 The Finite-Size Scaling Field

Consider the situation described in Sect. 4.3 of a system with n couplings and a transition region $\mathcal{C}$ of dimension $n-s$, defined in terms of the scaling fields $\theta_1, \theta_2, \ldots, \theta_n$ by (4.3.2). The basic assumption of finite-size scaling theory is that, for any specified boundary conditions, the variable $\theta_{n+1} = 1/\aleph$ can be treated as another scaling field.

---

33 For a review, see Binder (1983).

34 For a review see Barber (1983) and, for a collection of papers on finite-size scaling, Cardy (1988).

The defining condition (4.3.2) for $\mathcal{C}$ must then be augmented by

$$\theta_{n+1} = 0. \tag{4.10.2}$$

It follows from the block-spin idea of Kadanoff, described in Sect. 4.2, that the scaling exponent y_{n+1} associated with the finite-size scaling field θ_{n+1} is unity. It is therefore a relevant scaling field. In the case $\mathfrak{d} > 1$ there is a transition region $\mathcal{C}_{\mathfrak{d}}(\aleph)$ for $\mathcal{L}_{\mathfrak{d}}(\aleph)$, which can be regarded as the projection of $\mathcal{C}$ into the extended phase space. Since $\mathfrak{d} \neq d$, the critical exponents for this projection will have different critical exponents from those of $\mathcal{C}$, which is consistent with θ_{n+1} being a relevant scaling field.

For a d-dimensionally infinite system, with $d > 1$, the singular part of ϕ can be identified unambiguously. For a d-dimensional system with $\mathfrak{d} < 2$ this is not the case. To cover both situations Privman and Fisher (1984) proposed that $\phi_{\text{sing}}(K_1, \ldots, L_1, \ldots, \aleph)$, the singular part of the free-energy density of a system of thickness $\aleph$, is *defined* by

$$\phi_{\text{sing}}(K_1, \ldots, L_1, \ldots, \aleph) = \phi(K_1, \ldots, L_1, \ldots, \aleph) - \phi_{\text{smth}}(K_1, \ldots, L_1, \ldots), \tag{4.10.3}$$

where the second term on the right is the smooth component of ϕ for the $\aleph$-infinite system. For this singular part of ϕ, and $\psi(\boldsymbol{k})$ obtained by differentiating it k_j times with respect to θ_j, for $j = 1, 2, \ldots, n$, (4.3.5) and (4.3.7) must be modified to give

$$\phi_{\text{sing}}(\lambda^{y_1}\theta_1, \ldots, \lambda^{y_n}\theta_n, \lambda\theta_{n+1}) = \lambda^d \phi_{\text{sing}}(\theta_1, \ldots, \theta_n, \theta_{n+1}), \tag{4.10.4}$$

$$\psi_{\text{sing}}(\boldsymbol{k}; \lambda^{y_1}\theta_1, \ldots, \lambda^{y_n}\theta_n, \lambda\theta_{n+1}) = \lambda^{d'(\boldsymbol{k})} \psi_{\text{sing}}(\boldsymbol{k}; \theta_1, \ldots, \theta_n, \theta_{n+1}), \tag{4.10.5}$$

where d' is given by (4.3.8). In a similar way the formula (4.5.51) for the correlation length generalizes to

$$\mathfrak{r}(\mathfrak{q}; \lambda^{y_1}\theta_1, \ldots, \lambda^{y_n}\theta_n, \lambda\theta_{n+1}) = \lambda^{-1}\mathfrak{r}(\mathfrak{q}; \theta_1, \ldots, \theta_n, \theta_{n+1}). \tag{4.10.6}$$

Substituting $\lambda = 1/\theta_{n+1} = \aleph$ in Eqs. (4.10.4)–(4.10.6) gives

$$\phi_{\text{sing}}(\theta_1, \ldots, \theta_{n+1}) = \aleph^{-d} \phi_{\text{sing}}(\theta_1 \aleph^{y_1}, \ldots, \theta_n \aleph^{y_n}, 1), \tag{4.10.7}$$

$$\psi_{\text{sing}}(\boldsymbol{k}; \theta_1, \ldots, \theta_{n+1}) = \aleph^{-d'(\boldsymbol{k})} \psi_{\text{sing}}(\boldsymbol{k}; \theta_1 \aleph^{y_1}, \ldots, \theta_n \aleph^{y_n}, 1), \tag{4.10.8}$$

$$\mathfrak{r}(\mathfrak{q}; \theta_1, \ldots, \theta_{n+1}) = \aleph\mathfrak{r}(\mathfrak{q}; \theta_1 \aleph^{y_1}, \ldots, \theta_n \aleph^{y_n}, 1) \tag{4.10.9}$$

and it is also useful to note that

$$\frac{\partial \mathfrak{r}}{\partial \theta_m}(\mathfrak{q}; \theta_1, \ldots, \theta_{n+1}) = \aleph^{1-y_m} \frac{\partial \mathfrak{r}}{\partial \theta_m}(\mathfrak{q}; \theta_1 \aleph^{y_1}, \ldots, \theta_n \aleph^{y_n}, 1),$$
$$m = 1, 2, \ldots, n. \tag{4.10.10}$$

Equations (4.10.8) and (4.10.10) encapsulate the important practical value of finite-size scaling. If the transition region $\mathcal{C}$ is approached along any path for which the function on the right-hand side has a finite non-zero limit then, from (4.3.8),

$$\lim_{\aleph \to \infty} \frac{\ln[\psi_{\text{sing}}(\boldsymbol{k})]}{\ln(\aleph)} = -d'(\boldsymbol{k}) = \sum_{j=1}^{n} k_j y_j - d, \tag{4.10.11}$$

$$\lim_{\aleph \to \infty} \frac{\ln\left(\dfrac{\partial \mathfrak{r}}{\partial \theta_m}\right)}{\ln(\aleph)} = 1 - y_m, \qquad m = 1, 2, \ldots, n. \tag{4.10.12}$$

From the second of equations (4.3.12) it can be seen that the variable $d'(\boldsymbol{k})$ is the numerator of the critical exponent $\omega_m(\boldsymbol{k})$ associated with the response function (or density) $\psi(\boldsymbol{k})$ as $\mathcal{C}$ is approached along a path $\mathcal{P}$ which satisfies (4.3.18) with finite values of $\varpi_j^{(m)}$, $j = 1, 2, \ldots, s$. The corresponding denominator y_m appears in formula (4.10.12). This means that the critical exponents for the infinite system $\mathcal{L}$ can be obtained approximately from investigating systems of finite thickness. Just as in the case of scaling for infinite systems, modifications to this analysis are needed for the case of logarithmic singularities (Sect. 4.3.4) and when zero amplitude factors imply the need for considering corrections to scaling (Sect. 4.8).

4.10.2 The Shift and Rounding Exponents

We now investigate in more detail the relationship between the transition region $\mathcal{C}$ for the d-dimensionally infinite system and the transition region $\mathcal{C}_{\mathfrak{d}}(\aleph)$ for the system of thickness $\aleph$. For simplicity we confine our attention to the case of a critical point, with two couplings L and K, where the approach to the critical point is in the weak scaling direction. For the infinite system $\mathcal{L}$ the scaling fields θ_1 and θ_2 are given by Eqs. (4.9.1) and from (4.9.4) the response function φ_T is given by

$$\varphi_T = -\frac{\partial^2 \phi_{\text{sing}}}{\partial \theta_1^2}. \tag{4.10.13}$$

Thus, from (4.3.8) and (4.10.8),

$$\varphi_T(\theta_1, \theta_2, \theta_3) = \aleph^{2y_1 - d} \varphi_T(\theta_1 \aleph^{y_1}, \theta_2 \aleph^{y_2}, 1). \tag{4.10.14}$$

When $\mathfrak{d} > 1$, φ_T will exhibit some kind of singular behaviour in the transition region $\mathcal{C}_{\mathfrak{d}}(\aleph)$. Since $\aleph$ is finite and non-zero on $\mathcal{C}_{\mathfrak{d}}(\aleph)$ it follows that this transition region for systems of finite width must be given by some finite values ϖ_j for the arguments $\theta_j \aleph^{y_j}$, $j = 1, 2$ of the function on the right-hand side of (4.10.14). The transition region is, therefore, given by

$$[L_c(\mathfrak{d};\aleph) - L_c] - a[K_c(\mathfrak{d};\aleph) - K_c] \simeq \varpi_1 \aleph^{-y_1}, \tag{4.10.15}$$

$$K_c(\mathfrak{d};\aleph) - K_c \simeq (1 + a^2)^{-1} \varpi_2 \aleph^{-y_2}, \tag{4.10.16}$$

where $L_c(\mathfrak{d};\aleph)$ and $K_c(\mathfrak{d};\aleph)$ are the critical values of L and K for the system of dimension $\mathfrak{d}$ and thickness $\aleph$. In this context the exponent y_2 which appears on the right-hand side of (4.10.16) is called the *shift exponent* (Fisher and Ferdinand 1967; Fisher 1971) and is usually denoted by λ. According to (4.5.53)

$$\lambda = y_2 = 1/\nu. \tag{4.10.17}$$

In the case $\mathfrak{d} = 1$, $L_c(\mathfrak{d};\aleph)$ and $K_c(\mathfrak{d};\aleph)$ are replaced by $L_{max}(\mathfrak{d};\aleph)$ and $K_{max}(\mathfrak{d};\aleph)$, the values where φ_T attains a maximum. A second value of K which is of interest (Fisher and Ferdinand 1967) is the *rounding value* $K^*(\mathfrak{d};\aleph)$ at which φ_T for $\mathcal{L}_{\mathfrak{d}}(\aleph)$ differs significantly from its value for $\mathcal{L}$. A *rounding exponent* w can then be defined by[35]

$$|K^*(\mathfrak{d};\aleph) - K_c| \sim \aleph^{-w}. \tag{4.10.18}$$

If, following Fisher and Ferdinand (1967), we suppose that this will occur when $\mathfrak{r}(\mathfrak{q}) \sim \aleph$ then, from (4.5.52) and (4.10.18),

$$w = 1/\nu. \tag{4.10.19}$$

The relationship (4.10.17) between the shift exponent and ν is a consequence of the finite-size scaling assumption (4.10.4) for the singular part of the free-energy density as long as the coefficient ϖ_2 in (4.10.16) is non-zero. If we include the possibility that $\varpi_2 = 0$, then corrections to scaling lead to

$$\lambda \geq 1/\nu. \tag{4.10.20}$$

For the two-dimensional Ising model (Ferdinand and Fisher 1969; Au-Yang and Fisher 1975) $\lambda = w = 1/\nu = 1$. Fisher and Ferdinand (1967) have, however, argued for the possibility that $\lambda > 1/\nu$ in three dimensions and the results of Barber and Fisher (1973a, b) for the spherical model and ideal Bose gas respectively in d dimensions with layer geometries and open boundary conditions gave $\lambda = 1 > 1/\nu$.

[35] The rounding exponent is usually denoted by θ. We have used w to avoid confusion with the use of θ for scaling fields.

This was later understood to be a pathological result arising from constraints which give an effective long-range interaction (Brézin 1983) and (4.10.17) is now thought to be generally true (Barber 1983).

As indicated above, when $\mathfrak{d} > 1$, the transition regions $\mathcal{C}$ and $\mathcal{C}_{\mathfrak{d}}(\aleph)$ are distinct but connected in phase space. In this sense the situation is somewhat similar to the relationship between the first-order line and the critical point in the ferromagnetic Ising model. A situation of this kind is analyzed using effective exponents in Sect. 4.3.3.

4.10.3 Universality and Finite-Size Scaling

As in Sect. 4.9 we concentrate on the case of a critical point. From (4.10.4) and (4.10.6), the scaling forms (4.9.2) and (4.9.16) are replaced by

$$\phi_{\text{sing}} = A_0|\Delta K|^{2-\alpha}\mathring{W}_0^{(\pm)}\left(C\,\frac{[\Delta L - a\Delta K]}{|\Delta K|^{\beta\delta}}, \frac{D}{|\Delta K|^{\nu}\aleph}\right), \tag{4.10.21}$$

$$\mathfrak{r}(\mathfrak{q}) = R_0 Q_3^{(\pm)}|\Delta K|^{-\nu}\mathring{Y}^{(\pm)}\left(C\,\frac{[\Delta L - a\Delta K]}{|\Delta K|^{\beta\delta}}, \frac{D}{|\Delta K|^{\nu}\aleph}\right), \tag{4.10.22}$$

with the bulk forms being recovered, in the limit $\aleph \to \infty$, with

$$\mathring{W}_0^{(\pm)}(x,0) := W_0^{(\pm)}(x), \qquad \mathring{Y}^{(\pm)}(x,0) := Y^{(\pm)}(x). \tag{4.10.23}$$

When the Josephson hyper-scaling law (4.5.54) is valid, (4.10.21) can be rescaled.

$$\begin{aligned}\phi_{\text{sing}} &= \frac{A_0 D^d}{\aleph^d}\left[\frac{|\Delta K|\aleph^{\frac{1}{\nu}}}{D^{\frac{1}{\nu}}}\right]^{2-\alpha} \\ &\times \mathring{W}_0^{(\pm)}\left(\left[\frac{C}{D^{\frac{\beta\delta}{\nu}}}\right][(\Delta L - a\Delta K)\aleph^{\frac{\beta\delta}{\nu}}]\left[\frac{|\Delta K|\aleph^{\frac{1}{\nu}}}{D^{\frac{1}{\nu}}}\right]^{-\beta\delta}, \left[\frac{|\Delta K|\aleph^{\frac{1}{\nu}}}{D^{\frac{1}{\nu}}}\right]^{-\nu}\right),\end{aligned} \tag{4.10.24}$$

gives

$$\phi_{\text{sing}} = A_0 D^d \aleph^{-d} U(C'[\Delta L - a\Delta K]\aleph^{\frac{\beta\delta}{\nu}}, D'\Delta K\aleph^{\frac{1}{\nu}}), \tag{4.10.25}$$

where

$$U(x, \pm y) := y^{2-\alpha}\mathring{W}_0^{(\pm)}(xy^{-\beta\delta}, y^{-\nu}), \tag{4.10.26}$$

$$C' := C/D^{\frac{\beta\delta}{\nu}}, \qquad D' := 1/D^{\frac{1}{\nu}}. \tag{4.10.27}$$

In a similar way (4.10.22) can be re-expressed in the form

$$\mathfrak{r}(\mathfrak{q}) = \frac{R_0 Q_3^{(\pm)}}{D} \aleph V(C'[\Delta L - a\Delta K]\aleph^{\frac{\beta\delta}{\nu}}, D'\Delta K \aleph^{\frac{1}{\nu}}), \tag{4.10.28}$$

where

$$V(x, \pm y) := y^{-\nu} \mathring{Y}^{(\pm)}(xy^{-\beta\delta}, y^{-\nu}). \tag{4.10.29}$$

Equations (4.10.25) and (4.10.28) could be derived directly from (4.10.7) and (4.10.9) except that then no relationship would be established with the metric factors in (4.10.21) and (4.10.22).

As in the case of the d-dimensionally infinite system discussed in Sect. 4.9 the scaling functions $\mathring{W}_0^{(\pm)}(x, y)$, $\mathring{Y}^{(\pm)}(x, y)$, $U(x, y)$ and $V(x, y)$ are supposed to be universal with non-universal metric factors A_0, C, D and R_0. However, we can in this case go further if we adopt a standard assumption of finite-size scaling theory (Barber 1983) that all lengths which diverge at bulk criticality should scale like the bulk correlation length. This means that the second argument in (4.10.21) and (4.10.22) $D/|\Delta K|^{\nu}\aleph \simeq \mathfrak{r}(\mathfrak{q})/v_0^{\frac{1}{d}}\aleph$. From (4.9.16) $D \simeq R_0 Q_3^{(\pm)}/v_0^{\frac{1}{d}}$. Substituting into (4.10.25) and (4.10.28) and using the hyper-universality formula (4.9.24) gives

$$\phi_{\text{sing}} = E_0 \aleph^{-d} U(C'[\Delta L - a\Delta K]\aleph^{\frac{\beta\delta}{\nu}}, D'\Delta K \aleph^{\frac{1}{\nu}}), \tag{4.10.30}$$

$$\mathfrak{r}(\mathfrak{q}) = v_0^{\frac{1}{d}} \aleph V(C'[\Delta L - a\Delta K]\aleph^{\frac{\beta\delta}{\nu}}, D'\Delta K \aleph^{\frac{1}{\nu}}), \tag{4.10.31}$$

where

$$E_0 := A_0 D^d = Q_2^{(\pm)}[Q_3^{(\pm)}]^d/[Q_1^{(\pm)}]^2. \tag{4.10.32}$$

That the scaling forms (4.10.30) and (4.10.31) have leading universal amplitude factors is a result derived by Privman and Fisher (1984). At the bulk critical point $\Delta L = \Delta K = 0$ (4.10.30) and (4.10.31) becomes

$$\phi_{\text{sing}} = \aleph^{-d} E_0 U(0, 0), \tag{4.10.33}$$

$$\mathfrak{r}(\mathfrak{q})/v_0^{\frac{1}{d}} = \aleph V(0, 0). \tag{4.10.34}$$

As we shall see in Sect. 4.11 the value of the universal constants $V(0, 0)$ and $E_0 U(0, 0)$ can in the case $d = 2$, $\mathfrak{d} = 1$ be derived from conformal invariance.

4.11 Conformal Invariance

As indicated in the introduction this part of the book the application of conformal invariance to critical phenomena began with the work of Polyakov (1970). Since then it has developed into a powerful tool which provides results beyond those given by scaling theory. In this section we indicate briefly some of the main features of the theory and prove a famous result due to Cardy (1984), which establishes the value of the universal constant $V(0,0)$ of Eq. (4.10.34) in two dimensions.[36]

4.11.1 From Scaling to the Conformal Group

As we have seen in Sects. 4.5 and 4.7, scaling theory yields scaling laws for critical and tricritical points. Stronger assumptions are needed to give the possibility of extracting further information. Suppose that the elements of the scaling transformation given by (4.2.5), (4.3.3) and (4.3.97) are 'localized' to

$$\begin{aligned} \boldsymbol{r} &\to \tilde{\boldsymbol{r}} &&= [\lambda(\boldsymbol{r})]^{-1}\boldsymbol{r}, \\ \theta_j &\to \tilde{\theta}_j &&= [\lambda(\boldsymbol{r})]^{y_j}\theta_j, \\ \phi_j(\boldsymbol{r}) &\to \phi_j(\tilde{\boldsymbol{r}}) &&= [\lambda(\boldsymbol{r})]^{x_j}\,\phi_j(\boldsymbol{r}). \end{aligned} \tag{4.11.1}$$

This modification to scaling, which can be justified with the same kind of argument that was used in Sect. 4.2, will leave (4.3.98) intact with $\lambda(\boldsymbol{r})$ replacing λ as long as the spatial transformation has the Jacobian $[\lambda(\boldsymbol{r})]^{-d}$. It is not difficult to confirm that this will be the case if either $\lambda(\boldsymbol{r})$ is a constant, which is the usual dilatation transformation of scaling, or $\lambda(\boldsymbol{r}) = 1/|\boldsymbol{r}|^2$, which is the inversion transformation. As we describe in Sect. 17.4 the group generated by translations, rotations, dilatations and inversions is the *conformal group*.

4.11.2 Correlation Functions for *d* ≥ 2

Given that the transformation (4.11.1) belongs to the conformal group, the analysis of Sect. 4.3.7 leading to the scaling form (4.3.100) for the total correlation function is not affected by the Jacobian having a spatial dependence and thus we now have the generalization

$$\langle\phi_1(\tilde{\boldsymbol{r}}_1)\cdots\phi_\tau(\tilde{\boldsymbol{r}}_\tau)\rangle = \left\{\prod_{j=1}^{\tau}[\lambda(\boldsymbol{r}_j)]^{x_j}\right\}\langle\phi_1(\boldsymbol{r}_1)\cdots\phi_\tau(\boldsymbol{r}_\tau)\rangle \tag{4.11.2}$$

---

[36] For more detailed discussions of conformal invariance the reader is referred to the review of Cardy (1987) and the books of Itzykson (1988), Christe and Henkel (1993) and Henkel (1999).

of (4.3.100). In the case of the two-point correlation function, rotation and translation can be used to interchange $\boldsymbol{r}_1$ and $\boldsymbol{r}_2$, giving

$$\begin{aligned}\langle \phi_1(\tilde{\boldsymbol{r}}_1)\, \phi_2\, (\tilde{\boldsymbol{r}}_2)\rangle &= [\lambda(\boldsymbol{r}_1)]^{x_1}[\lambda(\boldsymbol{r}_2)]^{x_2}\langle \phi_1(\boldsymbol{r}_1)\, \phi_2\, (\boldsymbol{r}_2)\rangle,\\ \langle \phi_1(\tilde{\boldsymbol{r}}_2)\, \phi_2\, (\tilde{\boldsymbol{r}}_1)\rangle &= [\lambda(\boldsymbol{r}_2)]^{x_1}[\lambda(\boldsymbol{r}_1)]^{x_2}\langle \phi_1(\boldsymbol{r}_2)\, \phi_2\, (\boldsymbol{r}_1)\rangle,\end{aligned} \tag{4.11.3}$$

with

$$\begin{aligned}\langle \phi_1(\tilde{\boldsymbol{r}}_1)\, \phi_2\, (\tilde{\boldsymbol{r}}_2)\rangle &= \langle \phi_1(\tilde{\boldsymbol{r}}_2)\, \phi_2\, (\tilde{\boldsymbol{r}}_1)\rangle,\\ \langle \phi_1(\boldsymbol{r}_1)\, \phi_2\, (\boldsymbol{r}_2)\rangle &= \langle \phi_1(\boldsymbol{r}_2)\, \phi_2\, (\boldsymbol{r}_1)\rangle.\end{aligned} \tag{4.11.4}$$

For a non-zero two-point correlation function it would appear that these results must be inconsistent unless $x_1 = x_2$, leading to the conclusion that the two-point total correlation function of two operators with different scaling dimensions is zero. In fact it is clear that this conclusion must be modified to take into account two-point correlation functions like

$$\langle \phi(\boldsymbol{r}_1)\nabla_{\boldsymbol{r}_2}\, \phi\, (\boldsymbol{r}_2)\rangle = \nabla_{\boldsymbol{r}_2}\langle \phi(\boldsymbol{r}_1)\, \phi\, (\boldsymbol{r}_2)\rangle, \tag{4.11.5}$$

which will not, in general, be zero in spite of the fact that the scaling dimensions differ by two. This argument has been analyzed by Schäfer (1976) (see also Cardy 1987) who concluded that

$$\langle \phi_1(\boldsymbol{r}_1)\, \phi_2\, (\boldsymbol{r}_2)\rangle = 0, \qquad \text{if } x_1 - x_2 \neq \text{integer}. \tag{4.11.6}$$

When x_1 and x_2 differ by an integer, the two-point correlation function is sometimes but not always non-zero. Analysing the case $x_1 = x_2 = x$, Schäfer (1976) derived conditions for this to be the case. Then, within the context of conformal invariance, (4.3.112) in the form

$$\langle \phi(\boldsymbol{r}_1)\, \phi\, (\boldsymbol{r}_2)\rangle = \frac{C_2(\phi, \phi)}{|\boldsymbol{r}_2 - \boldsymbol{r}_1|^{2x}} \tag{4.11.7}$$

is exact. In the case of the three-point correlation function, the operations of the conformal group can be used to map three arbitrary points $\boldsymbol{r}_1$, $\boldsymbol{r}_2$ and $\boldsymbol{r}_3$ into any three chosen points $\boldsymbol{\alpha}, \boldsymbol{\beta}, \boldsymbol{\gamma}$ It has been shown by Polyakov (1970) that

$$\begin{aligned}&\langle \phi_1(\boldsymbol{r}_1)\, \phi_2\, (\boldsymbol{r}_2)\, \phi_3\, (\boldsymbol{r}_3)\rangle\\ &= \frac{C_3(\phi_1, \phi_2, \phi_3)}{|\boldsymbol{r}_1 - \boldsymbol{r}_2|^{x_1+x_2-x_3}|\boldsymbol{r}_2 - \boldsymbol{r}_3|^{x_2+x_3-x_1}|\boldsymbol{r}_3 - \boldsymbol{r}_1|^{x_3+x_1-x_2}},\end{aligned} \tag{4.11.8}$$

where $C_3(\phi_1, \phi_2, \phi_3)$ is a constant dependent only on the choice of scaling operators. An expression can also be obtained for the four-point correlation function, although since it is not possible to map four arbitrary points into four prescribed points with the operations of the conformal group an arbitrary function is involved (Polyakov 1970).

4.11.3 Universal Amplitudes for d = 2

We now consider the square lattice with lattice spacing a. In two dimensions the conformal group is infinite dimensional and consists, for $\boldsymbol{r} = a(\mathrm{x}, \mathrm{y})$, of transformations given by analytic functions[37] of the complex variable $\mathrm{z} = \mathrm{x} + i\mathrm{y}$ (Sect. 17.4). For the transformation $\mathrm{z} \to \mathrm{Z}(\mathrm{z})$ the Jacobian is $|\mathrm{Z}'(\mathrm{z})|^2$. So, for the two-point correlation function of the scaling operator ϕ_1, (4.11.2) gives

$$\langle \phi(\mathrm{z}_1)\, \phi\, (\mathrm{z}_2)\rangle = |\mathrm{Z}'(\mathrm{z}_1)|^x |\mathrm{Z}'(\mathrm{z}_2)|^x \langle \phi(\mathrm{Z}(\mathrm{z}_1))\, \phi\, (\mathrm{Z}(\mathrm{z}_2))\rangle . \tag{4.11.9}$$

The transformation given by

$$\mathrm{Z}(\mathrm{z}) = \frac{\aleph}{2\pi} \ln(\mathrm{z}), \tag{4.11.10}$$

is analytic for all $\mathrm{z} \neq 0$ and maps the z-plane onto the strip $0 \leq \tilde{\mathrm{y}} < \aleph$. Thus, from (4.11.7), (4.11.9) and (4.11.10), the two-point correlation function on the strip of width $\aleph$ lattice sites with periodic boundary conditions is

$$\langle \phi(\tilde{\boldsymbol{r}})\, \phi\, (\tilde{\boldsymbol{r}}_2)\rangle = \frac{C_2(\phi, \phi)(2\pi/\aleph a)^{2x}}{\left\{2\cosh\left[\frac{2\pi(\tilde{\mathrm{x}}_2 - \tilde{\mathrm{x}}_1)}{\aleph}\right] - 2\cos\left[\frac{2\pi(\tilde{\mathrm{y}}_2 - \tilde{\mathrm{y}}_1)}{\aleph}\right]\right\}^x}, \tag{4.11.11}$$

where $\tilde{\boldsymbol{r}} = a(\tilde{\mathrm{x}}, \tilde{\mathrm{y}})$. When $|\tilde{\mathrm{z}}_2 - \tilde{\mathrm{z}}_1| \ll \aleph$,

$$\langle \phi(\tilde{\boldsymbol{r}})\, \phi\, (\tilde{\boldsymbol{r}}_2)\rangle \simeq \frac{C_2(\phi, \phi)}{|\tilde{\boldsymbol{r}}_1 - \tilde{\boldsymbol{r}}_2|^{2x}}, \tag{4.11.12}$$

but, when $|\tilde{\mathrm{x}}_2 - \tilde{\mathrm{x}}_1| \gg \aleph$,

$$\langle \phi(\tilde{\boldsymbol{r}})\, \phi\, (\tilde{\boldsymbol{r}}_2)\rangle \simeq C_2(\phi, \phi)\left(\frac{2\pi}{\aleph a}\right)^{2x} \exp\left[-\frac{2\pi x(\tilde{\mathrm{x}}_2 - \tilde{\mathrm{x}}_1)}{\aleph}\right]. \tag{4.11.13}$$

---

[37] In complex variable theory a *holomorphic function* is one which is complex differentiable in a neighbourhood of every point in its domain and this should more strictly be our usage here. The term 'analytic' which has already been used in the context of real-valued functions denotes a function (real or complex) which is equal to its Taylor series in a neighbourhood of each point in its domain. The fact that the class of complex analytic functions coincides with the class of holomorphic functions is a theorem of complex analysis. A function which is holomorphic/analytic throughout the complex plane is an *entire function*.

We now consider the case of finite-size scaling at a critical point discussed in Sect. 4.10. If the leading behaviour of the state operator $\mathfrak{q}$ is given by the scaling operator ϕ with $\langle\phi(\boldsymbol{r})\rangle = 0$ at the critical point, (4.11.13) can be substituted into (4.2.4) (with σ replaced by $\mathfrak{q}$) to give the correlation length $\mathfrak{r}(\mathfrak{q})$. In the limit of small lattice spacing the sums become integrals and

$$\mathfrak{r}^2(\mathfrak{q}) = \frac{1}{2}a^2 \frac{\displaystyle\int_0^\infty r^2 \exp\left(-\frac{2\pi x r}{\aleph}\right) \mathrm{d}r}{\displaystyle\int_0^\infty \exp\left(-\frac{2\pi x r}{\aleph}\right) \mathrm{d}r}, \tag{4.11.14}$$

giving

$$\mathfrak{r}(\mathfrak{q}) = \frac{\aleph a}{2\pi x}. \tag{4.11.15}$$

This result was first obtained by Cardy (1984). Substituting into (4.10.34) with $v_0^{\frac{1}{d}} = a$ gives

$$V(0,0) = \frac{1}{2\pi x}. \tag{4.11.16}$$

The detailed analysis which leads to the determination of the critical amplitude $E_0\, U(0,0)$ in (4.10.33), for $d = 2$, is beyond the scope of this book. We simply state the result, obtained by Blöte et al. (1986) and Affleck (1986), that

$$E_0 U(0,0) = -\tfrac{1}{6}\pi \mathrm{c}, \tag{4.11.17}$$

where c is the *conformal anomaly number* defined by $\frac{1}{2}\mathrm{c}$ being the coefficient of the leading term of the operator product expansion of the stress tensor with itself.[38] From this it can be shown that c is the *central charge in the Virasoro algebra* of the generators of the conformal transformation. For models where the transfer matrix is symmetric (or can be made symmetric at criticality) the Virasoro algebra is unitary and, for $\mathrm{c} < 1$, the only allowed values are given by

$$\mathrm{c}(m) = 1 - \frac{6}{m(m+1)}, \qquad m = 3, 4, 5, \ldots. \tag{4.11.18}$$

If a particular model can be identified with a value of m and c the scaling dimensions of its primary scalar operators, indexed by i and j with $1 \le j \le i \le m-1$, are given by the *Kac formula*[39]

[38] For a full explanation of this and all the following discussion see Cardy (1987).

[39] These formulae are usually stated (Henkel 1999, p. 93) in terms of $\Delta_{ij} := \frac{1}{2}x_{ij}$.

$$x_{ij}(m) = \frac{[i(m+1) - jm]^2 - 1}{2m(m+1)}. \tag{4.11.19}$$

The (critical) Ising model has been identified with the case $m = 3$, $c = \frac{1}{2}$, the tricritical Ising model with $m = 4$, $c = \frac{7}{10}$, and the 3-state Potts model with $m = 5$, $c = \frac{4}{5}$. The known models with continuously varying exponents correspond to the limiting case $m \to \infty$, $c = 1$. These include the two-dimensional XY model (Sect. 3.6.1.1) and the Ashkin–Teller Model (Sect. 3.6.2), which is related to the eight-vertex model by a duality transformation.
For the Ising model the three allowed pairs of values for the parameters are:

- $i = j = 1$ giving $x_{11} = 0$.
  This corresponds to an operator present in all models. It is similar to the trivial coupling in renormalization group theory.
- $i = 2$, $j = 1$ giving $x_{21} = 1$.
  This gives the scaling exponent $y_{21} = d - x_{21} = 1$. By identifying y_{21} with the scaling exponent y_2 of Sect. 4.5 and using (4.5.39) we obtain the value $\alpha = 0$, in agreement with the exact result (12.3.280). Since by the fluctuation-response function relation (2.5.15) the heat capacity is associated with the energy-energy correlation function and we are able to identify the operator as the energy density.
- $i = j = 2$ giving $x_{22} = \frac{1}{8}$.
  This gives the scaling exponent $y_{22} = d - x_{22} = \frac{15}{8}$. By identify y_{22} with the exponent y_1 of Sect. 4.5 and using (4.5.40) we obtain the value $\gamma = \frac{7}{4}$ in agreement with the exact result (12.3.280). In this case the operator is the magnetization density.

In the cases of the tricritical Ising model and 3-state Potts model there are a larger number of allowed parameter values. It is only with the use of additional information that it becomes possible to identify the physical nature of the operators. Thus, for example, for the 3-state Potts model the known values of α and γ given in (12.3.322) allow the identification of the cases $i = 2$, $j = 1$ $x_{21} = \frac{4}{5}$ and $i = j = 3$, $x_{33} = \frac{2}{15}$ with the energy density and magnetization density respectively.

4.11.4 Schramm–Loewner Evolution

Schramm–Loewner evolution (SLE) originated with the application by Schramm (2000) of *Loewner's equation* (Löwner 1923) to the study of the continuum limit of a loop-erased random walk and developed into a general method for obtaining rigorous information about two-dimensional critical behaviour. Here we give a brief outline of the ideas involved. For more detailed accounts the reader is referred to Cardy (2005) or Lawler (2011) and the bibliographies included therein.

The starting point for SLE for a particular model is the identification of a type of non-crossing path passing from one point on the boundary of the lattice to another.

For the Ising model on the honeycomb lattice such a path is a domain boundary between regions of aligned spins on the dual triangular lattice. An example for the Potts model on a square lattice is given by Fig. 3.8, where the paths are drawn on the medial lattice as the boundaries of the shaded regions containing connected clusters of spins. Of course, the method of constructing these curves, after the spins have been fixed, will not just produce curves passing between points on the boundary. It will, as can be seen in Fig. 3.8, also produce closed loops. Part of the insight of SLE is that the universal properties of the system can be extracted by concentrating solely on the open paths, which for each of the examples given must terminate on the boundary. The other initiating aspect of the method is to regard the curves as being constructed step-by-step using a random process. Consider for simplicity the finite honeycomb lattice $\mathcal{N}_{\mathrm{HC}}$ and suppose that a path γ is constructed beginning at a boundary site $\boldsymbol{r}_{\mathrm{s}}$ and ending at a boundary site $\boldsymbol{r}_{\mathrm{f}}$. As each site is reached there is a choice between two for the next step. With an initial degeneracy of two, and as the path is generated, spin states are assigned to sites of the dual triangular lattice $\mathcal{N}_{\mathrm{T}}$ making γ a domain boundary, and the probability of turning left or right at a lattice site is calculated from the expected values for the spin on the site of $\mathcal{N}_{\mathrm{T}}$ immediately ahead, based on the states already assigned to sites of that lattice. Thus the relative probabilities of turning left or right at step n are determined by the lattice and γ_{n-1}, the path up to that point. Let $p(\gamma; \boldsymbol{r}_{\mathrm{s}}, \boldsymbol{r}_{\mathrm{f}}, \mathcal{N}_{\mathrm{HC}})$ be the probability of the path γ being constructed from $\boldsymbol{r}_{\mathrm{s}}$ to $\boldsymbol{r}_{\mathrm{f}}$ by this procedure and suppose that γ_n, the path in place after n steps, has reached the point $\boldsymbol{r}_n$. $\mathcal{N}_{\mathrm{HC}}\backslash\gamma_n$ is the part of $\mathcal{N}_{\mathrm{HC}}$ from which the edges of γ_n (and all dangling edges and exposed spins of $\mathcal{N}_{\mathrm{T}}$) have been removed. Then the method of path-generation implies that

$$p(\gamma - \gamma_n; \boldsymbol{r}_n, \boldsymbol{r}_{\mathrm{f}}, \mathcal{N}_{\mathrm{HC}}|\gamma_n) = p(\gamma; \boldsymbol{r}_n, \boldsymbol{r}_{\mathrm{f}}, \mathcal{N}_{\mathrm{HC}}\backslash\gamma_n). \tag{4.11.20}$$

SLE is concerned with the continuum limit of this situation when the lattice is replaced by a domain $\mathcal{D} \subset \mathbb{C}$ with $\boldsymbol{r}_{\mathrm{s}}$ and $\boldsymbol{r}_{\mathrm{f}}$ points on the boundary[40] and $\gamma_n \in \mathcal{D}$ a non-crossing curve from $\boldsymbol{r}_{\mathrm{s}}$ to $\boldsymbol{r}_n$. The domain boundary γ on the lattice neither crosses *nor touches* itelf. In the continuum case the latter situation can occur. Then the curve creates a *hull* $\bar{\gamma}$, which is γ together with any regions of $\mathcal{D}$ which are surrounded by the curve by virtue of it touching itself. That any such curve γ can be described by a process of continuous growth (*called Loewner evolution*) was proved by Löwner (1923)

The relationship (4.11.20), with the probability $p(\cdot)$ replaced by a measure $\mu(\cdot)$ and $\mathcal{N}_{\mathrm{HC}}$ replaced by $\mathcal{D}$, is one of the defining properties of an SLE. The other concerns a conformal invariance property which must hold for the system. Let $\mathfrak{M}$ be a conformal mapping from the interior of $\mathcal{D}$ to the interior of another domain $\mathcal{D}'$ and suppose that the pair $[\boldsymbol{r}_{\mathrm{s}}, \boldsymbol{r}_{\mathrm{f}}]$ maps to $[\boldsymbol{r}'_{\mathrm{s}}, \boldsymbol{r}'_{\mathrm{f}}]$ on the boundary of $\mathcal{D}'$. The measure μ

[40] This is *chordal SLE*. There is another version, called *radial SLE*, which will not be discussed, where $\boldsymbol{r}_{\mathrm{f}}$ is an interior point of $\mathcal{D}$.

induces a measure $\mathfrak{M} \star \mu$ on the image curves $\mathfrak{M}(\gamma)$ in $\mathcal{D}'$. This must satisfy

$$(\mathfrak{M} \star \mu)(\gamma; \boldsymbol{r}_{\mathrm{s}}, \boldsymbol{r}_{\mathrm{f}}, \mathcal{D}) = \mu(\mathfrak{M}(\gamma); \boldsymbol{r}'_{\mathrm{s}}, \boldsymbol{r}'_{\mathrm{f}}, \mathcal{D}'). \tag{4.11.21}$$

The importance of this conformal property is that $\mathcal{D}$ can be chosen to be $\mathbb{C}^{(+)}$, the upper half of $\mathbb{C}$, with $\boldsymbol{r}_{\mathrm{s}} = 0$ and $\boldsymbol{r}_{\mathrm{f}} = \infty$, when the distribution of paths has the reflectional symmetry $x \to -x$. The curve $\gamma(t)$ is now a function of a continuous parameter t with final point $z(t)$. Since $\mathbb{C}^{(+)} \backslash \bar{\gamma}(t)$ is simply connected it can, by the Riemann mapping theorem, be mapped by a real analytic function $g(t; z)$ into $\mathbb{C}^{(+)}$, while preserving the real axis outside $\gamma(t)$. When $\gamma(t)$ is a non-touching curve ($\bar{\gamma}(t) = \gamma(t)$) it is a cut in $\mathbb{C}^{(+)}$ which is 'opened up' into a section of the real axis bounded at its ends by the two images of the origin and containing a point $a(t)$ which is the image of $z(t)$ The development of $\gamma(t)$ is determined by $a(t)$, which is itself a random function.

Subject to the condition that $g(t; z) \sim z + \mathrm{O}(1/z)$ as $z \to \infty$, the function $g(t; z)$ which describes this situation satisfies *Loewner's equation*

$$\frac{\mathrm{d}g(t; z)}{\mathrm{d}t} = \frac{2}{g(t; z) - a(t)}, \tag{4.11.22}$$

(Löwner 1923). Given an evolving path, $\bar{\gamma}(t)$ can be determined and, in principle, $g(t; z)$ and $a(t)$ be obtained. Conversely, the equation can be integrated, with more or less difficulty (Kager et al. 2004) for given $a(t)$. The point given by $a(t)$ moves randomly on the real axis. A crucial ingredient of SLE is the result due to Schramm (2000) that, if (4.11.20) in its continuum version, (4.11.21) and reflectional symmetry hold then the movement of $a(t)$ is proportional to a *standard Brownian motion* with diffusion constant κ. The nature of $\gamma(t)$ varies with κ, which is now the only parameter of the system. For $\kappa \leq 4$ it is a simple curve, for $4 < \kappa < 8$ it intersects itself so that every point is contained in a loop and for $\kappa \geq 8$ it is space filling. These results can be illuminated by a formula, valid for $\kappa \leq 8$ that γ has the fractal dimension $d_f := 1 + \frac{1}{8}\kappa$, which increases with κ reaching the (space-filling) value two when $\kappa = 8$. The central charge in the Virasoro algebra of the generators of the conformal transformation can be obtain in terms of κ as

$$\mathrm{c}(\kappa) = \frac{(3\kappa - 8)(6 - \kappa)}{2\kappa}. \tag{4.11.23}$$

Solving this formula with (4.11.18) gives

$$\kappa = \begin{cases} \dfrac{4m}{m+1}, & \kappa \leq 4, \\[2ex] \dfrac{4(m+1)}{m}, & \kappa \geq 4 \end{cases} \tag{4.11.24}$$

and, from (4.11.19),

$$x(\kappa) := \frac{6-\kappa}{\kappa} = \begin{cases} x_{21}(\kappa), & \kappa \leq 4, \\ x_{12}(\kappa), & \kappa \geq 4. \end{cases} \tag{4.11.25}$$

Although there are few direct results, these formulae allow identification of values of κ with the models with specified values of c on p. 162. Given the nature of the method it is not surprising that the proven and strongly suggested cases of connections to particular models are those related to random walks and percolation. In particular rigorous results have been obtained for the continuum limit of critical site percolation on a triangular lattice, which is identified with $\kappa = 6$ (Smirnov 2001), and for loop-erased random walks and the boundaries of uniform spanning trees, identified respectively with $\kappa = 2$ and $\kappa = 8$ (Lawler et al. 2004).[41]

[41] For $\kappa = 2$ the SLE is radial rather than chordal.

Part II
Classical Approximation Methods

Introduction

As is clear from the discussion in Sect. 2.1 the ultimate aim, given any statistical mechanical model, is to obtain an exact expression for the partition function. In the absence of this, the oldest tool of statistical mechanics is the use of classical approximations.[1] As we see in Chap. 5 the first step in such methods is the replacement of the Hamiltonian with an approximation which leads to a form for the free-energy density $\phi_{\rm U}$ which is a smooth function of the density or densities. This results in the elimination of all reference to spatially fluctuating variables.[2] The equation of state, like that of van der Waals (Sect. 5.2), becomes a smooth function of the thermodynamic variables. Landau theory (Sects. 5.3–5.5), where the thermodynamic potential has a power-series expansion in some order parameter is a natural extension of this. The absence of spatially fluctuating variables prevents an investigation of the critical properties of any quantities, like the correlation length and pair correlation function, which depend on such fluctuations. In Sect. 5.6 a generalization of Landau theory, due to Ginzburg and Landau (1950), is introduced for a critical point in a magnetic system. In this model spatial fluctuations are incorporated and the partition function is expressed as a functional integral. At the level of the *Gaussian approximation,* considered here, only harmonic fluctuations about Landau values are retained. This is the simplest way to include some fluctuations and, in the context of quantum field theory, it is equivalent to the 'tree approximation' (Amit 1978).[3] In the mean-field approximations of Chap. 6 the summation in the partition function is replaced by a single term with the microstates replaced by their mean values. The further summation then needed to obtain the

---

[1] The methods described in this part are often generically referred to as 'mean-field methods'. For an historic perspective on their place in the theory of phase transitions see Kadanoff (2009).
[2] Although, of course, overall density fluctuations are still present (Sect. 6.1.1).
[3] For a more detailed treatment of generalizations to Landau theory the reader is referred to Luban (1976).

free-energy density per lattice site with no independent densities is replaced by selecting the largest term in the summation. Chapter 6 illustrates this approximation procedure with a number of examples. In these, phases are characterized by densities, associated in the case of lattice systems with the proportion of the lattice sites having a microsystem in a particular microstate. The higher order cluster approximations of Chap. 7 refine this programme by distinguishing the different microstates of small clusters of sites. Increasing the size of the clusters yields a hierarchy of approximation methods which, with increasing difficulty and the need for ingenuity, yields increasingly accurate predictions for some characteristics of the system like critical curves. However, these remain classical approximations, which continue to yield classical (in general, incorrect) values for critical exponents and in some cases give incorrect predictions for the orders of the phase transitions.

Chapter 5
Phenomenological Theory and Landau Expansions

5.1 Classical Methods

In Sect. 4.1, as a preliminary to the discussion of scaling theory, we gave a general heuristic treatment of the geometry of phase transitions. This was predicated on the assumption that, for a system with independent internal couplings $K_1, \ldots, K_{n_e}$ and external couplings $L_1, \ldots, L_{n_f - 1}$, the free-energy density $\phi(K_1, \ldots, K_{n_e}, L_1, \ldots, L_{n_f-1})$ was known and thus could be represented as a continuous surface in the $(n_e + n_f)$–dimensional space of the variables $\phi, K_1, \ldots, K_{n_e}, L_1, \ldots, L_{n_f-1}$. As we have indicated above this ideal situation rarely occurs and some approximate form for ϕ must be derived. Beginning with the work of Bragg and Williams (1934), for the order-disorder transition in alloys, a class of approximation methods was developed, which has been very successful in making qualitative predictions for phase transitions. These *classical methods* include not only mean-field approximations, of the type employed by Bragg and Williams, which are discussed in Chap. 6, but also the more sophisticated cluster-variation approximations, described in Chap. 7, of which the improved approximation used by Bethe (1931, 1935) was the first instance.

For simplicity we here concentrate on the case of a system with $n_f = 2$ and fields $\xi = \xi_1$ and ξ_2 with corresponding extensive variables Q_1 and Q_2. Given that there is a set of internal energy parameters ε_ℓ, $\ell = 1, 2, \ldots, n_e$ we shall adopt the device, described in Sect. 2.3.1, of identifying $\varepsilon := \varepsilon_1$ and setting $\kappa_\ell := \varepsilon_\ell / \varepsilon_1$. Then the system has effectively two couplings $K := \varepsilon / T$ and $L := \xi / T$. At this stage it is convenient to frame the discussion in the coupling–extensive variable representation, with dimensionless free energies $\Phi_0(Q_1, Q_2, K)$ and $\Phi_1(Q_2, K, L)$, where, from (2.3.5) and (2.3.6),

$$\Phi_0(Q_1, Q_2, K) = -\ln Z_0(Q_1, Q_2, K), \tag{5.1.1}$$

and

$$Z_0(Q_1, Q_2, K) = \sum_{\{\sigma\}} \exp[-\widehat{H}_0(Q_1, Q_2, K; \boldsymbol{\sigma})], \tag{5.1.2}$$

D.A. Lavis, *Equilibrium Statistical Mechanics of Lattice Models*,
Theoretical and Mathematical Physics, DOI 10.1007/978-94-017-9430-5_5

the summation being over all microstates compatible with a particular value for Q_1. From (2.1.12),

$$\Phi_1(Q_2, K, L) = -\ln\left\{\sum_{\{\widehat{Q}_1\}} \exp[L\,\widehat{Q}_1 - \Phi_0(\widehat{Q}_1, Q_2, K)]\right\}. \tag{5.1.3}$$

The common feature of all classical methods is that some approximation is made to the Hamiltonian in (5.1.2) leading to an approximation for $\Phi_0(Q_1, Q_2, K)$, denoted by $\Phi_0^{(\mathrm{CL})}(Q_1, Q_2, K)$, which is, for all K, a smooth function of Q_1.[1] Given that the thermodynamic limit can be applied to this approximation with size parameter $Q := Q_2$

$$\Phi_0^{(\mathrm{CL})}(Q_1, Q, K) = Q\phi_0^{(\mathrm{CL})}(\rho, K) + \mathrm{O}(Q^0), \tag{5.1.4}$$

where $\rho := Q_1/Q$. Then, from (5.1.3),

$$\phi_1(K, L) = -\frac{1}{Q}\ln\left\{\sum_{\{\hat{\rho}\}} \exp[Q\{L\hat{\rho} - \phi_0^{(\mathrm{CL})}(\hat{\rho}, K)\} + \mathrm{O}(Q^0)]\right\}. \tag{5.1.5}$$

For large Q the summation is dominated by its maximum term and thus, in the thermodynamic limit, $Q \to \infty$,

$$\phi_1^{(\mathrm{CL})}(K, L; \rho) = \phi_0^{(\mathrm{CL})}(\rho, K) - \rho L, \tag{5.1.6}$$

where ρ is given by minimizing $\phi_1^{(\mathrm{CL})}(K, L; \rho)$.[2] Thus we have the conditions

$$\frac{\partial\phi_1^{(\mathrm{CL})}}{\partial\rho} = \frac{\partial\phi_0^{(\mathrm{CL})}}{\partial\rho} - L = 0, \tag{5.1.7}$$

$$\frac{\partial^2\phi_1^{(\mathrm{CL})}}{\partial\rho^2} = \frac{\partial^2\phi_0^{(\mathrm{CL})}}{\partial\rho^2} = \frac{\partial L}{\partial\rho} > 0. \tag{5.1.8}$$

The last condition in (5.1.8) is understood in the sense that L is given as a function of ρ from (5.1.7). This condition, expressed in the form $L = L(\rho, K)$, is the *equation of state* of the system and, from (5.1.7),

$$\phi_0^{(\mathrm{CL})}(\rho, K) = C(K) + \int L(\rho, K)\mathrm{d}\rho, \tag{5.1.9}$$

where $C(K)$ is some undetermined function of K.

---

[1] See Definition 17.1.1.

[2] Cf. (1.1.13), (1.1.14) and the comment following the latter.

5.1.1 A First-Order Transition

As we have indicated in Sect. 4.1.1, the curve derived from the equation of state $L = L(\rho, K)$, for a system undergoing a first-order phase transition, will have a horizontal section connecting points representing phases in coexistence.[3] However, of course, the classical equation of state derived from (5.1.7) has been assumed to give L as a smooth function of ρ for all K, excluding the possibility of a horizontal section (apart from in the form of a single point).

To understand the resolution of this problem it is pertinent to outline briefly the historic development of the theory of phase transitions. Early work on the subject was mainly concerned with the liquid-vapour transition in fluids and Andrews (1869) was the first to measure pressure and temperature along isotherms, both above and below the critical temperature.[4] He concluded that "the gaseous and liquid states are only distant stages of the same condition of matter, and are capable of passing into one another by a process of continuous change" (Andrews 1869, p. 589). Commenting on this work Thomson (1871, p. 2) suggested "that, although there be a practical breach of continuity in crossing the line of boiling-points from liquid to gas or from gas to liquid, there may exist, in the nature of things, a theoretical continuity across the breach having some real and true significance". He supported this view by observing "that the ordinary liquid state does not necessarily cease abruptly at the line of boiling-points, ... liquids may, with due precautions, be heated considerably beyond the boiling temperature" (op. cit., p. 3.). A smooth continuation between the branches below A and above B in Fig. 4.2c, resulting from a classical approximation of the equation of state would, of course, be dependent on the nature of the approximation. The simplest would be the case where the isotherm crossed the line AB at one intermediate point. Such is the case for van der Waals' equation, which is explored in some detail in Sect. 5.2.[5] There the isotherm has a single loop with a range of pressures for which lines of constant pressure cut the isotherm a three points. The corresponding situation for a classical isotherm in the coupling–density plane ρ–L is shown in Fig. 5.1a.[6] However, now the question arises: if we begin, as van der Waals did, with an equation of state (rather than the free energy), at what pressure (or value of L) does the transition between phases occur? For the van der Waals equation this question was answered by Maxwell (1875), who derived, in terms of the transformation of heat into work, the condition now known as *Maxwell's equal area rule*. In terms of Fig. 5.1a this simply means that the points A and B, between

---

[3] See Fig. 4.2c, where the points representing coexistent phases are denoted by A and B.

[4] For an account of this work see Rowlinson (1969).

[5] The reader is recommended to refer to this example as an illustration of the development of Landau theory with one order parameter.

[6] Ignore for the moment the shaded regions. Remember also that isotherms for the van der Waals equation will be mirror images about a vertical line of those given in Fig. 5.1a, because increase in pressure yields a *decrease* in v: see the comment in the footnote on p. 6.

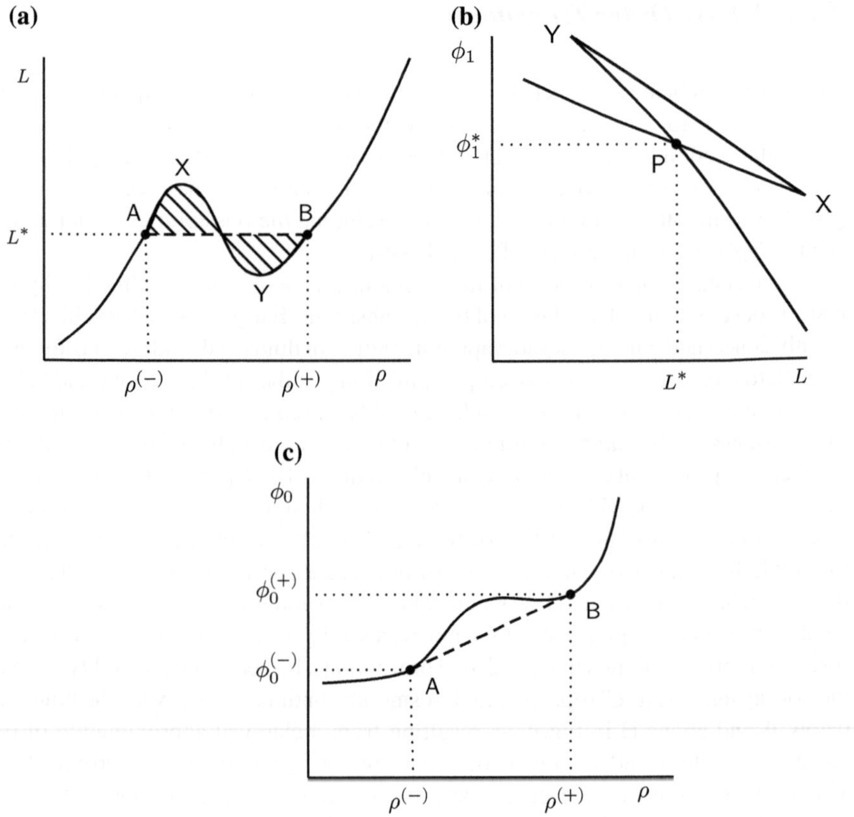

Fig. 5.1 **a** A first-order transition showing as a *horizontal* part of an isotherm of L plotted against ρ. **b** A first-order transition showing as a discontinuity of slope in an isothermal section of the ϕ_1 surface. **c** A first-order transition showing as the *double tangent* AB in an isothermal section of the ϕ_0 surface

which a transition occurs are determined by the equality of the shaded areas. That is

$$\int_{\rho^{(-)}}^{\rho^{(+)}} L(\rho, K)\mathrm{d}\rho = L^*[\rho^{(+)} - \rho^{(-)}]. \tag{5.1.10}$$

From (5.1.9),

$$\int_{\rho^{(-)}}^{\rho^{(+)}} L(\rho, K)\mathrm{d}\rho = \phi_0^{(\mathrm{CL})}(\rho^{(+)}, K) - \phi_0^{(\mathrm{CL})}(\rho^{(-)}, K), \tag{5.1.11}$$

which then gives, from (5.1.6) and (5.1.10),

$$\phi_1^{(\mathrm{CL})}(K, L^*; \rho^{(+)}) = \phi_1^{(\mathrm{CL})}(K, L^*; \rho^{(-)}), \tag{5.1.12}$$

where now the densities are not independent variables but are given by the equation of state (5.1.7). Equality of ϕ_1, which from (2.3.15) is equivalent to equality of the second external coupling L_2, appears in the discussion of the appearance of a first-order transition in Sect. 4.1.1 in an slightly different but equivalent way. There ϕ_1 was assumed to be a continuous function of K and L with a first-order transition signalled by a discontinuity of slope (density). Now the discontinuity of slope in Fig. 4.2a is replaced by a loop in Fig. 5.1b.

A locally stable state, according to (5.1.7)–(5.1.8), corresponds to a minimum of $\phi_1^{(\text{CL})}$ with respect to ρ and this applies to the isotherm in Fig. 5.1a to the left of X and to the right of Y. It can then be seen from Fig. 5.1b that, in the range of L where there are two minima, the lower of the two is given by the isotherm to the left of A and the right of B. A single-phase system in a locally stable but globally unstable state represented by a point on AX or YB is said to be *metastable*. The segment XY consists of *intrinsically unstable* states. The double tangent in Fig. 4.2b appears in Fig. 5.1c as the completion of the convex envelope of $\phi_0^{(\text{CL})}(K, \rho)$; the part of the curve not belonging to the envelope comprises both metastable and intrinsically unstable states. The demarkation between then is given, from (5.1.8), by points where $(\partial^2\phi_0^{(\text{CL})}/\partial\rho^2)_K = 0$.

If we assume that the equation of state (5.1.7) has the same general properties as van der Waals' equation then an isotherm of the form shown in Fig. 5.1a will occur for K greater than that a critical K_c ($T < T_\text{c}$) and on the critical isotherm $K = K_\text{c}$ there will be a critical point (ρ_c, L_c) where

$$\left(\frac{\partial L}{\partial \rho}\right)_K = 0, \qquad \left(\frac{\partial^2 L}{\partial \rho^2}\right)_K = 0. \tag{5.1.13}$$

Plots of the points A and B as K approaches the critical value from above in the $\rho - L$ plane meet at a critical point C and define the boundaries of the coexistence region. In principle we can obtain (at least numerically, if not analytically) a plot for the first-order transition curve $L = L^*(K)$ in the $L - K$ plane similar to that shown in Fig. 4.7. The two branches of the coexistence curve can then be plotted either against K as shown in Fig. 4.8 or against L. We choose the latter, so that we can also include part of the isotherm shown in Fig. 5.1a. The coexistence curve will nevertheless still have the general appearance of that shown in Fig. 4.8.[7] But now we have a feature which did not arise in our analysis in Sect. 4.4 of the exact free-energy density, namely the possibility of a one-phase system being metastable. The region in the ρ–L plane in which this occurs (shaded in Fig. 5.2) lies between the coexistence curve and the *spinodal* curve, which is the line traced out by X and Y as K is varied.

---

[7] Although K increases along the first-order transition curve it is not necessarily the case that L does. However, this will be assumed in line with Fig. 4.7.

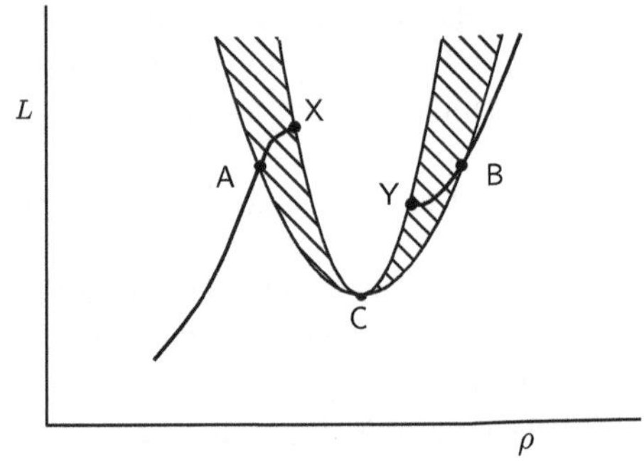

Fig. 5.2 Coexistence *curve* and *spinodal lines* derived from Fig. 5.1a

5.1.2 Metastability

As indicated in the quote from Thomson (1871) given above, metastable states are not simply artifacts of classical theory, but are frequently encountered experimentally. Let us consider, for example, the case of a fluid system where, in Fig. 5.2, the low-density states to the left of X and the high-density states to the right of Y correspond respectively to a vapour and a liquid. The density of the vapour is increased (by increasing L) along the isotherm from the left towards A. If the system remains in thermodynamic equilibrium, any further increase in L will cause a transition to the liquid state. However, when Fig. 5.2 represents the behaviour of a real fluid, a dust-free vapour may progress along the continuous segment AX into the metastable (shaded) region. It is then termed *supersaturated*. Metastable states can also be reached by cooling a liquid below the equilibrium temperature of transition to the crystalline form. Such liquids are called *supercooled*.

We now discuss briefly the physical reasons for the existence of metastable states in real substances, using the supersaturated vapour as an example. For the latter to go over to the liquid state, localized (low-wavelength) density fluctuations of large amplitude must occur, resulting in the formation of nuclei, each containing a large number of molecules. Such a nucleus has a lower free energy than an equal mass of vapour and will grow by accretion. However, a nucleus created by a spontaneous local density fluctuation is likely to be small and because of its relatively large surface area will have a higher free energy than an equal mass of vapour. This is another aspect of the local stability of metastable states. Nevertheless thermodynamically metastable states are different in character from static mechanical metastable states, like that of a ball on a surface at the lowest point of a depression separated by a ridge from a deeper depression.

The degree of permanence of a thermodynamically metastable state depends on the *relaxation time*, the mean waiting time for a sufficiently large fluctuation to initiate the transformation to the stable state. For the mechanical analogy to be appropriate, in fact, the ball must be regarded as subject to impulses of random strength at random intervals. However, relaxation times can be very large. For instance, glasses are supercooled liquids with higher free energies than the crystalline forms whereas diamond is a metastable crystalline form of carbon.[8] In a supersaturated vapour dust particles drastically reduce the relaxation time by facilitating the formulation of large nuclei. Charged particles (ions) have the same effect and this is used in the cloud chamber apparatus. Here the tracks of charged particles through supersaturated water vapour become visible because of the formation of liquid water droplets. Work on first-order transition mechanisms is reviewed by Gunton et al. (1983) and Binder (1987).

5.2 The Van der Waals Equation

In real fluids there are strong repulsive forces between molecules when they are very close together and attractive forces at larger distances. Van der Waals in his thesis of 1873[9] proposed a modification of the perfect gas law $PV = MT$ to take these effects into account. He included a term proportional to[10] M, giving an effective reduction of the volume, and a term proportional to $(M/V)^2$ to increase the pressure, yielding

$$\left(P + \frac{\varepsilon M^2}{V^2}\right)(V - M) = TM. \tag{5.2.1}$$

For $T < T_c := \frac{8}{27}\varepsilon$ the isotherm in the P–V plane has a single loop with a range of pressures for which lines of constant pressure intersect the isotherm at three points and the critical isotherm with $T = T_c$ has a point of inflexion at $P = P_c := \frac{1}{27}\varepsilon$, $V = V_c := 3M$. We note, in reference to the law of corresponding states, discussed in Sect. 4.2, that (5.2.1) can be expressed in the form

$$\left(\frac{P}{P_c} + \frac{3V_c^2}{V^2}\right)\left(\frac{V}{V_c} - \frac{1}{3}\right) = \frac{8T}{3T_c}. \tag{5.2.2}$$

Using $K := \varepsilon/T$, $L' := -P/T$ and $v := V/M$, for which $K_c = \frac{27}{8}$, $L'_c = -\frac{1}{8}$ and $v_c = 3$, (5.2.1) can also be expressed in the coupling-density form

$$L'(K; v) = \frac{K}{v^2} - \frac{1}{v - 1}. \tag{5.2.3}$$

---

[8] Thus, contrary to the common belief, diamonds are not ‘for ever’, just for a very long time.

[9] For a translation see Rowlinson (1988) and for an assessment of the legacy of van der Waals see Rowlinson (1973).

[10] In fact, in terms of the variables introduced in Sect. 1.1, exactly equal to M.

From (5.1.9)[11]

$$\phi_0^{(\mathrm{CL})\prime}(v, K) = C(K) - \ln(v-1) - \frac{K}{v}. \tag{5.2.4}$$

The use of primes in (5.2.3) and (5.2.4) follows the usage in (2.3.26) and is motivated by the intention of using $L := \mu/T$ as independent coupling with density $\rho := 1/v$. From (2.3.22) and (2.3.26),

$$L(K;\rho) = C(K) - 2K\rho - \ln\left(\frac{1-\rho}{\rho}\right) + \frac{1}{1-\rho}, \tag{5.2.5}$$

$$\phi_1^{(\mathrm{CL})}(K, L;\rho) = \rho\, C(K) - K\rho^2 - \rho\ln\left(\frac{1-\rho}{\rho}\right) - L\rho. \tag{5.2.6}$$

Expanding in powers of $\Delta\rho := \rho - \rho_c$ gives

$$\phi_1^{(\mathrm{CL})}(K, L;\rho) = B_0(K, L) + B_1(K, L)\Delta\rho + \sum_{i=2}^{\infty} B_i(K)(\Delta\rho)^i, \tag{5.2.7}$$

where the first few coefficients in the expansion are

$$\begin{aligned} B_0(K, L) &:= \tfrac{1}{3}C(K) - \tfrac{1}{3}\ln(2) - \tfrac{1}{9}K - \tfrac{1}{3}L,\\ B_1(K, L) &:= C(K) + \tfrac{3}{2} - \ln(2) - \tfrac{2}{3}K - L = L(K;\rho_c) - L,\\ B_2(K) &:= \tfrac{27}{8} - K = K_c - K, \qquad B_3(K) := 0, \qquad B_4(K) := \tfrac{243}{64}. \end{aligned} \tag{5.2.8}$$

5.3 Landau Expansions with One Order Parameter

In Chap. 4 we considered the form of ϕ in a neighbourhood of a critical region. The idea underlying the Landau (1937) expansion method for a classical approximation is very similar. Because ϕ is now a smooth function of the density (or densities) it can be expanded as a Taylor series in the differences of the densities from their critical values. These expansion variables are the *order parameters* of the system and the nature of the series (the presence or absence of particular powers in the expansion)

---

[11] By taking the perfect gas $v \to \infty$ limit, when the entropy is given by the Sackur-Tetrode equation, it can be shown (Lavis and Bell 1977) that

$$C(K) = -1 - \ln\left[v_0\left(\frac{2m\,\varepsilon\,\pi}{Kh^2}\right)^{\frac{3}{2}}\right],$$

where m is the mass of a particle and h is Planck's constant.

can be deduced from the symmetry group of the order parameters. In this section we continue the case, discussed in the previous section, of a system with a critical point and a single density, and thus a single order parameter. In Sect. 5.4.2 we consider a case with two order parameters.[12]

The Taylor expansion of $\phi_1^{(\mathrm{CL})}(K, L; \rho)$ about some $\rho = \rho_0$ takes the form[13]

$$\phi_1^{(\mathrm{CL})}(K, L; \rho) := \phi_1^{(\mathrm{CL})}(K, L; \rho_0) + b_1(K, L; \rho_0)(\rho - \rho_0) + \sum_{i=2}^{\infty} b_i(K; \rho_0)(\rho - \rho_0)^i, \tag{5.3.1}$$

where $b_1(K, L; \rho_0)$ is linear in L. The condition $b_1(K, L; \rho) = 0$ is equivalent to (5.1.7), yielding an equilibrium solution $\rho = \rho_e(K, L)$. If $b_2(K; \rho_e) > 0$ then $\rho = \rho_e$ is a minimum (stable equilibrium point) and if $b_2(K; \rho_e) < 0$ then it is a maximum (unstable equilibrium point). If $b_2(K; \rho_e) = 0$ and $b_3(K; \rho_e) \neq 0$, $\rho = \rho_e$ is a point of inflexion. If $b_2(K; \rho_e) = b_3(K; \rho_e) = 0$ the criteria are repeated with the index increased by two. However the values of K and L must, from Eqs. (5.1.6)–(5.1.8) and (5.1.13), now be the critical values. Thus the *critical parameters* K_c, L_c and ρ_c for a critical point terminating a line of first-order transitions in the classical approximation described in Sect. 5.1.1 satisfy the conditions

$$b_1(K, L; \rho) = 0, \tag{5.3.2}$$

$$b_2(K; \rho) = 0, \tag{5.3.3}$$

$$b_3(K; \rho) = 0. \tag{5.3.4}$$

Whether a known ϕ like (5.2.6) is expanded as a Taylor series, or the expansion is of a general form, one important question is immediately evident. Given that we truncate after a certain number of terms, how many are needed to reproduce the local properties of the full free-energy density near to the critical point?[14]

This is a problem which can be discussed most fruitfully in the context of *catastrophe theory* (Sect. 17.1). Let $\rho_0 := \rho_c$ with $\Delta\rho := \rho - \rho_c$ in (5.3.1). Then

$$\phi_1^{(\mathrm{CL})}(K, L; \rho) = B_0(K, L) + B_1(K, L)\Delta\rho + B_2(K)(\Delta\rho)^2 + B_3(K)(\Delta\rho)^3 + \sum_{i=4}^{\infty} B_i(K)(\Delta\rho)^i, \tag{5.3.5}$$

---

[12] Order parameters are often given a vector form, the case of two order parameters being referred to as that of a two-dimensional order parameter.

[13] That the coefficients $b_i(K; \rho_0)$ are not functions of L for $i > 1$ follows from the fact that L appears in $\phi_1^{(\mathrm{CL})}(K, L; \rho)$ only in a term linear in ρ and L; see (5.1.6) and, for the van der Waals equation, (5.2.6).

[14] And, of course, the same question can be posed in relation the neighbourhood of other critical regions.

where, for brevity, we have adopted the notation

$$B_0(K,L) := \phi_1^{(\mathrm{CL})}(K,L;\rho_\mathrm{c}), \qquad B_1(K,L) := b_1(K,L;\rho_\mathrm{c}),$$
$$B_i(K) := b_i(K;\rho_\mathrm{c}), \qquad i = 2,3,\ldots. \tag{5.3.6}$$

With ρ fixed at the critical value ρ_c we have the two equations (5.3.3) and (5.3.4) which must agree on the critical value $K = K_\mathrm{c}$. In some examples like the van der Waals equation $B_3(K)$ is identically zero, but this is not necessarily the case and, in order to formulate the problem in the context of catastrophe theory, the order parameter in (5.3.5) is changed to

$$x := \Delta\rho - \delta, \tag{5.3.7}$$

where δ is chosen so that the cubic term is zero and

$$\phi_1^{(\mathrm{CL})}(K,L;x) = C_0(K,L) + C_1(K,L)x + C_2(K)x^2 + G(K;x), \tag{5.3.8}$$

with

$$G(K;x) := \sum_{i=4}^{\infty} C_i(K)x^i. \tag{5.3.9}$$

Both δ and the coefficients $C_i(K)$ can be derived as series expansions in the parameter

$$y := \frac{B_3(K)}{4B_4(K)} \tag{5.3.10}$$

and it is not difficult to show that the required expression for δ gives

$$x = \Delta\rho + y + \frac{5B_5(K)}{2B_4(K)}y^2 + \mathrm{O}(y^3), \tag{5.3.11}$$

with

$$C_0(K,L) = B_0(K,L) - B_1(K,L)y$$
$$- \frac{5B_1(K,L)B_5(K) - 2B_2(K)B_4(K)}{2B_4(K)}y^2 + \mathrm{O}(y^3), \tag{5.3.12}$$

$$C_1(K,L) = B_1(K,L) - 2B_2(K)y - \frac{5B_2(K)B_5(K)}{B_4(K)}y^2 + \mathrm{O}(y^3), \tag{5.3.13}$$

$$C_2(K) = B_2(K) - 6B_4(K)y^2 + \mathrm{O}(y^3), \tag{5.3.14}$$

$$C_4(K) = B_4(K) - 5B_5(K)y$$
$$+\frac{5(6B_4(K)B_6(K) - 5[B_5(K)]^2)}{2B_4(K)}y^2 + \mathrm{O}(y^3), \tag{5.3.15}$$

with the remaining coefficients of the form $C_i(K) = B_i(K) + \mathrm{O}(y)$. Suppose that $C_4(K) > 0$ in a neighbourhood of $K = K_c$. Then $G(K; x)$ has a degenerate minimum and a critical point[15] at $x = 0$ for all K in a neighbourhood of K_c. It was shown in Example 17.1.3 that a function of this form has strong determinancy of 4 and codimension 2. $G(K; x)$ is strongly equivalent to $C_4(K)x^4$ and the universal unfolding (5.3.8) of $G(K; x)$ is strongly equivalent to

$$\tilde{\phi}_1^{(\mathrm{CL})}(K, L; x) := C_0(K, L) + C_1(K, L)x + C_2(K)x^2 + C_4(K)x^4. \tag{5.3.16}$$

All the local properties of $\phi_1^{(\mathrm{CL})}$ are present in $\tilde{\phi}_1^{(\mathrm{CL})}$. In particular, from Theorem 17.1.3, $\tilde{\phi}_1^{(\mathrm{CL})}$ has at most three equilibrium points near $x = 0$ for small $\Delta K := K - K_c$ and $\Delta L := L - L_c$. These are given by

$$C_1(K, L) = -2C_2(K)x - 4C_4(K)x^3, \tag{5.3.17}$$

with a solution being stable or unstable according as

$$2C_2(K) + 12C_4(K)x^2 >, < 0. \tag{5.3.18}$$

The equal-areas rule is satisfied for the cubic (5.3.17) by the coexistent non-zero solutions on the line

$$C_1(K, L) = 0, \tag{5.3.19}$$

which gives the coexistence curve in the K–L plane. From (5.3.11), (5.3.14) and (5.3.15) the solutions $\rho^{(0)}$ and $\rho^{(\pm)}$ to (5.3.17) and (5.3.19) are given by

$$\rho^{(0)} = \rho_c - y - \frac{5B_5(K)}{2B_4(K)}y^2 + \mathrm{O}(y^3), \tag{5.3.20}$$

$$\rho^{(\pm)} = \rho^{(0)} \pm \sqrt{-\frac{C_2(K)}{2C_4(K)}}. \tag{5.3.21}$$

From (5.3.18), the solution $\rho^{(0)}$ is stable and the solutions $\rho^{(\pm)}$ unstable if $C_2(K) > 0$ with the converse being the case if $C_2(K) < 0$. We also see that the states $\rho^{(\pm)}$ can exist only if $C_2(K)$ and $C_4(K)$ have opposite signs. Thus, for a line of first-order

[15] In both the statistical mechanical and catastrophe sense (Definition 17.1.11 together with the footnote).

transitions to develop from the critical point with a phase coexistence diagram of the form of Fig. 4.8, we must have $C_2(K) <, > 0$ for $K >, < K_c$ and $C_4(K) > 0$ for $K > K_c$ and $< K_c$. The simplest, although not unique way, to achieve this is to suppose that

$$\begin{aligned} B_0(K, L) &\simeq -L\rho_c + B_{00} + B_{01}\Delta K + B_{02}(\Delta K)^2, \\ B_1(K, L) &\simeq -\Delta L + B_{11}\Delta K + B_{12}(\Delta K)^2, \\ B_2(K) &\simeq -B_{21}\Delta K + B_{22}(\Delta K)^2, \\ B_3(K) &\simeq B_{31}\Delta K, \qquad B_4(K) \simeq B_{40} + B_{41}\Delta K, \end{aligned} \tag{5.3.22}$$

where B_{21} and B_{40} (but not necessarily the other constants) are positive.[16] Then, from (5.3.10), (5.3.14), (5.3.15) and (5.3.20)–(5.3.22),

$$\rho^{(0)} \simeq \rho_c - \frac{B_{31}}{B_{40}}\Delta K, \tag{5.3.23}$$

$$\rho^{(\pm)} \simeq \rho^{(0)} \pm \sqrt{\frac{B_{21}}{2B_{40}}}(\Delta K)^{\frac{1}{2}}, \tag{5.3.24}$$

along the coexistence curve ($K > K_c$) and, from (5.3.11), (5.3.13)–(5.3.15), (5.3.17) and (5.3.22),

$$\Delta L \simeq 4B_{40}(\Delta\rho)^3, \tag{5.3.25}$$

along the critical isotherm ($K = K_c$). From (2.3.29)–(2.3.32), (5.3.11), (5.3.13), (5.3.17) and (5.3.22)–(5.3.24)

$$\varphi_T \simeq \begin{cases} -\dfrac{1}{2B_{21}}\Delta K, & \text{along the critical isochore, } K < K_c, \\ \dfrac{1}{4B_{21}}\Delta K, & \text{along the coexistence curve, } K > K_c, \end{cases} \tag{5.3.26}$$

$$c_\rho \simeq \begin{cases} -k_B K_c^2[2B_{02} + O(\Delta K)], & \\ \qquad\qquad \text{along the critical isochore, } K < K_c, & \\ -k_B K_c^2\left[2B_{02} + O([\Delta K]^{\frac{1}{2}})\right], & \\ \qquad\qquad \text{along the coexistence curve, } K > K_c, & \end{cases} \tag{5.3.27}$$

---

16 We have explicitly displayed only those coefficients needed in the calculations of the asymptotic forms for the response functions.

$$c_\xi \simeq \begin{cases} -k_{\mathrm{B}}K_{\mathrm{c}}^2\left[\dfrac{Q^2}{2B_{21}}(\Delta K)^{-1} + 2B_{02} + \dfrac{f_1(Q)}{2B_{21}^2} + \mathrm{O}(\Delta K)\right], \\ \qquad\qquad \text{along the critical isochore, } K < K_{\mathrm{c}}, \\ -k_{\mathrm{B}}K_{\mathrm{c}}^2\left[-\dfrac{Q^2}{4B_{21}}(\Delta K)^{-1} + \dfrac{f_2(Q)}{16B_{21}^2}(\Delta K)^{-\frac{1}{2}}\right. \\ \qquad \left. + 2B_{02} + \dfrac{f_3(Q)}{32B_{40}B_{21}^2} + \mathrm{O}([\Delta K]^{\frac{1}{2}})\right], \\ \qquad\qquad \text{along the coexistence curve, } K > K_{\mathrm{c}}, \end{cases} \tag{5.3.28}$$

where

$$Q := \frac{B_{11}K_{\mathrm{c}} - L_{\mathrm{c}}}{K_{\mathrm{c}}}, \qquad f_1(Q) := Q(4B_{12}B_{21} + QB_{22}),$$

$$f_2(Q) := Q\sqrt{\frac{2B_{21}}{B_{40}}}(8B_{21}^2 + 3QB_{31}), \tag{5.3.29}$$

$$f_3(Q) := -16B_{21}^4 - 16Q(3B_{21}^2B_{31} + 2B_{21}B_{12}B_{40}) + Q^2(4B_{22}B_{40} + 12B_{21}B_{41} - 9B_{31}^2).$$

Comparing (5.3.24)–(5.3.27) with (4.4.12)–(4.4.15) we obtain the *classical critical exponent values*[17]

$$\alpha = \alpha' = 0, \qquad \beta = \tfrac{1}{2}, \qquad \gamma = \gamma' = 1, \qquad \delta = 3, \tag{5.3.30}$$

with c_ρ finite and continuous with a discontinuous gradient at the critical point. Since $f_1(0) = f_2(0) = 0$, $f_3(0) = -16B_{21}^4$, it follows from (4..4.16) and (5.3.28) that

$$\sigma = \sigma' = \begin{cases} \gamma, & \text{if } Q \neq 0, \\ \alpha, & \text{if } Q = 0. \end{cases} \tag{5.3.31}$$

From (5.3.19) and (5.3.22), a which appears in (4.5.5) is equal to B_{11} and (5.3.31) is the result given by (4.5.43).

[17] Sometimes referred to as the mean-field or Landau values.

5.3.1 The Spin–$\frac{1}{2}$ Ising Model

The spin–$\frac{1}{2}$ Ising model, with Hamiltonian (3.5.1) has magnetization density m conjugate to $L := \mathcal{H}/T$. As has been discussed in Sect. 3.5.1 the symmetry group of the Hamiltonian is $\mathcal{S}_2^{(mL)} := \{\mathrm{I}, \mathrm{Q}^{(m)}\mathrm{Q}^{(L)}\}$, where $\mathrm{Q}^{(m)}$ and $\mathrm{Q}^{(L)}$ denote spin and field inversion respectively. The free-energy density $\phi_1^{(\mathrm{CL})}(K, L; m)$ inherits the symmetry properties of the Hamiltonian, with $\mathrm{Q}^{(m)} : m \to -m$. Its most general form would therefore be a linear combination of the invariants $m^i L^j$ of $\mathcal{S}_2^{(mL)}$, for all $i + j$ even. However, from (3.5.1), L can appear only linearly. So apart from a term containing mL, $\phi_1^{(\mathrm{CL})}(K, L; m)$ is an even function of m (see Eqs. (6.1.9) and (6.1.10)). As we showed in Example 17.1.5 the catastrophe theory analysis remains valid, but now $B_{2i+1}(K) = 0$ for $i = 1, 2, \ldots$ with $B_1(K) = -L$. The calculation to remove the cubic term in the expansion is not required, $C_i(K) = B_i(K), i = 0, 1, 2, \ldots$ and, from (5.3.16),

$$\phi_1^{(\mathrm{CL})}(K, L; m) := B_0(K, L) - Lm + B_2(K)m^2 + B_4(K)m^4. \tag{5.3.32}$$

The solutions (5.3.20) and (5.3.21) reduce to $m = 0$ and

$$m^2 = -\frac{B_2(K)}{2B_4(K)} \simeq \frac{B_{21}\Delta K}{B_{40}}. \tag{5.3.33}$$

From (5.3.19), the critical curve in the K–L plane lies along the line $L = 0$.

Landau theory and scaling theory can both be seen as methods which extract information about a whole class of physical systems, based on a minimum of assumptions about their physical structure. The difference between the two is, of course, that, while scaling theory is capable of determining only scaling law relationships between critical exponents, Landau theory allows actual values for the exponents to be obtained. It is an illuminating exercise to apply the procedures of scaling theory to the free-energy density (5.3.32), in order to obtain the critical exponents. The scaling fields in this case are, as for the example of Kadanoff (1966) described in Sect. 4.2 and for the mean-field calculation of Sect. 6.1, $\theta_1 = L$ and $\theta_2 = \Delta K$. Let the smooth part of $\phi_1^{(\mathrm{CL})}(K, L; m)$ be identified as $B_0(K)$. Then

$$\phi_{\mathrm{sing}}(\theta_1, \theta_2; m) = -\theta_1 m - B_{21}\theta_2 m^2 + B_{40}m^4. \tag{5.3.34}$$

Suppose the homogeneity condition, with exponents associated with θ_1 and θ_2 *but also with* m. Then

$$\phi_{\mathrm{sing}}(\lambda^{y_1}\theta_1, \lambda^{y_2}\theta_2; \lambda^z m) = \lambda^d \phi_{\mathrm{sing}}(\theta_1, \theta_2; m). \tag{5.3.35}$$

It follows from (5.3.34) and (5.3.35) that

$$y_1 = \tfrac{3}{4}d, \qquad y_2 = \tfrac{1}{2}d, \qquad z = \tfrac{1}{4}d\,. \tag{5.3.36}$$

The exponent z for the magnetization satisfies, as it must (Eq. (4.3.8)), the condition $z = d - y_1$. Substituting into (4.5.40) and (4.5.39) the classical critical exponent values given in (5.3.30) are recovered. The exponent γ applies to the divergence of the susceptibility as the critical point is approached from above and below along the zero-field axis. However, it can be easily seen that these directions yield different amplitudes. Since $\alpha = 0$ there is no divergence, at the critical point, in the heat capacity $c_{\mathcal{H}}$ at constant field along the zero-field axis. However, there is a discontinuity, given, from (5.3.28) (with $B_{11} = 0$, $L_c = 0$, giving $Q = 0$) by

$$\Delta c_{\mathcal{H}} = \frac{k_B K_c^2 B_{21}^2}{2B_{40}}. \tag{5.3.37}$$

5.4 Landau Expansions with Two Order Parameters

5.4.1 The Spin-1 Ising Model

The Hamiltonian of the spin-1 Ising model is given by (3.5.29) and there are now two densities $m := \langle\sigma(\boldsymbol{r})\rangle$, conjugate to $\mathcal{H}$ and $\rho := \langle\sigma^2(\boldsymbol{r})\rangle$ conjugate to μ. The free-energy density is a function of these densities and four couplings, $K_1 := J/T$, $K_2 := \varepsilon/T$, $L_1 := \mathcal{H}/T$ and $L_2 := \mu/T$. The symmetry group is in general the same as the spin-$\frac{1}{2}$ case. The parameter ρ, is unaffected by spin inversion and is thus related to the one-dimensional symmetric representation of $\mathcal{S}_2$ (Sect. 17.3.4) giving

$$\phi_1^{(\text{CL})}(K_1, K_2, L_1, L_2; m, \rho) = -L_1 m - L_2\rho + B_0(K_1, K_2; \rho) + \sum_{i=1}^{\infty} B_{2i}(K_1, K_2; \rho) m^{2i}. \tag{5.4.1}$$

Since, in general, no new symmetry is associated with the parameter ρ there is no way, within the framework of Landau theory, of obtaining information about a possible transition with ρ as order parameter. There are two exceptions to this. One is the case of the 3-state Potts model, discussed in Sect. 5.4.2, and the other is when $J = \varepsilon' = \mathcal{H} = 0$. Then, as we have indicated in Sect. 3.5.2, the Hamiltonian is equivalent to the lattice gas Hamiltonian (3.5.20). The order parameter $n := \langle\sigma^2(\boldsymbol{r})\rangle - \frac{1}{2}$ is invariant under the operator $\mathrm{Q}^{(n)}$ which changes all site states $|\sigma| = 1$ to zero and sites states zero to $+1$ or -1 with equal probability. The first-order transition line and critical point lie, from (3.5.22), on $\mu = \mu_c = -\frac{1}{2}z\varepsilon$ and the Landau expansion has the same form as that for the spin-$\frac{1}{2}$ model with n replacing m, $(\mu - \mu_c)/T$ replacing L and ε/T replacing K.

5.4.2 The 3-State Potts Model

The ν-*state Potts model* is described in Sect. 3.6.4. Here we use Landau theory to consider the case $\nu = 3$ and $\mu = 0$, which was shown (Sect. 3.6.4.2) to be a special case of the spin-1 Ising model, with Hamiltonian (3.5.29), in zero field. This means that, if the 3-state Potts model has a transition, this will occur in a subspace of the phase space of the spin-1 Ising model.

We first need to reformulate the problem in terms of two order parameters which are zero at high temperatures. This is true for $m := \langle\sigma(\boldsymbol{r})\rangle$, since the states $+1$ and -1 will be equally probable. But now we have three equally probable states with $\langle\sigma^2(\boldsymbol{r})\rangle = \frac{2}{3}$ at high temperatures. So we choose the two order parameters[18]

$$\begin{aligned} p &:= \tfrac{2}{\sqrt{3}}\left(1 - \tfrac{3}{2}\langle\sigma^2(\boldsymbol{r})\rangle\right), \\ m &:= \langle\sigma(\boldsymbol{r})\rangle. \end{aligned} \tag{5.4.2}$$

The group $\mathcal{S}_3$ has six elements and three generators I, $\mathrm{Q}^{(m)}$ and

$$\mathrm{Q}^{(*)} : -1 \to 0 \to +1 \to -1. \tag{5.4.3}$$

The group $\mathcal{S}_2^{(m)}$, which is a subgroup of $\mathcal{S}_3$, has the two-dimensional representation

$$\boldsymbol{R}(\mathrm{I}) = \boldsymbol{I} = \begin{pmatrix} 1 & 0 \\ 0 & 1 \end{pmatrix}, \qquad \boldsymbol{R}(\mathrm{Q}^{(m)}) = \boldsymbol{Q}^{(m)} = \begin{pmatrix} 1 & 0 \\ 0 & -1 \end{pmatrix}, \tag{5.4.4}$$

for operations on the vector

$$\boldsymbol{x} := \begin{pmatrix} p \\ m \end{pmatrix}. \tag{5.4.5}$$

This representation is, of course, reducible, corresponding to the symmetric and antisymmetric representations A and B with operations on the order parameters p and m respectively. From (5.4.2) and (5.4.3),

$$\begin{aligned} \mathrm{Q}^{(*)} &: p \to -\tfrac{1}{2}p + \tfrac{\sqrt{3}}{2}m \to -\tfrac{1}{2}p - \tfrac{\sqrt{3}}{2}m \to p, \\ \mathrm{Q}^{(*)} &: m \to -\tfrac{\sqrt{3}}{2}p - \tfrac{1}{2}m \to \tfrac{\sqrt{3}}{2}p - \tfrac{1}{2}m \to m, \end{aligned} \tag{5.4.6}$$

18 The leading factor of $2/\sqrt{3}$ in p is included in order to make the matrix representations of $\mathrm{Q}^{(m)}$ and $\mathrm{Q}^{(*)}$ unitary. It is omitted in series calculations (see Enting and Wu 1982, and Sect. 14.2.1).

Table 5.1 The character table for $\mathcal{S}_3$

$\mathcal{S}_3$	$\mathcal{Q}_1$	$\mathcal{Q}_2$	$\mathcal{Q}_3$
A	1	1	1
B	1	1	−1
E	2	−1	0

giving the unitary matrix

$$\boldsymbol{R}(\mathrm{Q}^{(*)}) = \boldsymbol{Q}^{(*)} = \begin{pmatrix} -\frac{1}{2} & \frac{\sqrt{3}}{2} \\ -\frac{\sqrt{3}}{2} & -\frac{1}{2} \end{pmatrix}. \tag{5.4.7}$$

$\mathcal{S}_3$ is generated by $\mathrm{Q}^{(m)}$ and $\mathrm{Q}^{(*)}$ and since $\mathrm{Q}^{(m)}\mathrm{Q}^{(*)} = \mathrm{Q}^{(*)2}\mathrm{Q}^{(m)}$ and $\mathrm{Q}^{(m)}\mathrm{Q}^{(*)2} = \mathrm{Q}^{(*)}\mathrm{Q}^{(m)}$ the conjugacy classes of $\mathcal{S}_3$ are $\mathcal{Q}_1 := \mathrm{I}$, $\mathcal{Q}_2 := \{\mathrm{Q}^{(*)}, \mathrm{Q}^{(*)2}\}$ and $\mathcal{Q}_3 := \{\mathrm{Q}^{(m)}, \mathrm{Q}^{(m)}\mathrm{Q}^{(*)}, \mathrm{Q}^{(m)}\mathrm{Q}^{(*)2}\}$ with character Table 5.1 (see e.g. Leech and Newman 1969, p. 109).[19] It follows from (17.3.2) that the matrices (5.4.4) and (5.4.7) are generators of the two-dimensional irreducible representation E.

To obtain the Landau expansion for the 3-state Potts model we need to determine the functions of p and m which are invariant under the operations of $\mathcal{S}_3$. Since

$$\boldsymbol{x}^{\mathrm{T}}\boldsymbol{Q}^{(*)\mathrm{T}}\boldsymbol{Q}^{(*)}\boldsymbol{x} = \boldsymbol{x}^{\mathrm{T}}\boldsymbol{Q}^{(m)\mathrm{T}}\boldsymbol{Q}^{(m)}\boldsymbol{x} = \boldsymbol{x}^{\mathrm{T}}\boldsymbol{x} = p^2 + m^2, \tag{5.4.8}$$

a set of even-order invariants are given by powers of

$$\upsilon_{\mathrm{E}}(p, m) := p^2 + m^2. \tag{5.4.9}$$

Odd-order invariants under the operation $\mathrm{Q}^{(*)}$ are given, from (5.4.6), by observing that the product of the elements of each of these three-cycles will give a third-order invariant under $\mathrm{Q}^{(*)}$. The second of these is odd with respect to m and thus not an invariant under $\mathrm{Q}^{(m)}$. However, the first gives the function

$$\upsilon_{\mathrm{O}}(p, m) := p(p^2 - 3m^2), \tag{5.4.10}$$

which is an odd-order invariant. All other invariants are products of powers of $\upsilon_{\mathrm{E}}(p, m)$ and $\upsilon_{\mathrm{O}}(p, m)$ (Gaite 1992).

[19] $\mathcal{S}_3$ is isomorphic to the dihedral group $\mathcal{D}_3$ and to $\mathcal{C}_{3\mathrm{v}}$ (see the character table in Example 17.3.2).

Applying the analysis of catastrophe theory (Example 17.1.6), the Landau expansion for the 3-state Potts model is equivalent to

$$\begin{aligned}\phi_1^{(\mathrm{CL})}(K;p,m) := {} & B_0(K) + B_2(K)(p^2+m^2) + B_3(K)p(p^2-3m^2) \\ & + B_4(K)(p^2+m^2)^2 ,\end{aligned} \tag{5.4.11}$$

as long as $B_4(K) \neq 0$. Stable equilibrium states are given by minimizing this free-energy density with respect to p and m. The equilibrium conditions are

$$\begin{aligned} 0 &= 2B_2(K)p + 3B_3(K)(p^2-m^2) + 4B_4(K)p(p^2+m^2) , \\ 0 &= 2B_2(K)m - 6B_3(K)pm + 4B_4(K)m(p^2+m^2) , \end{aligned} \tag{5.4.12}$$

and the stability conditions are obtained by considering the determinant

$$\Phi(K;p,m) := \begin{vmatrix} \phi_{pp}(K;p,m) & \phi_{pm}(K;p,m) \\ \phi_{mp}(K;p,m) & \phi_{mm}(K;p,m) \end{vmatrix} , \tag{5.4.13}$$

where

$$\begin{aligned} \phi_{pp}(K;p,m) &:= 2B_2(K) + 6B_3(K)p + 4B_4(K)(3p^2+m^2) , \\ \phi_{pm}(K;p,m) &:= \phi_{mp}(K;p,m) = -6B_3(K)m + 8B_4(K)pm , \\ \phi_{mm}(K;p,m) &:= 2B_2(K) - 6B_3(K)p + 4B_4(K)(p^2+3m^2) . \end{aligned} \tag{5.4.14}$$

A solution to (5.4.12) is a maximum or a minimum of $\tilde{\phi}^{(\mathrm{CL})}(K;p,m)$ if $\Phi(K;p,m) > 0$, otherwise it is a saddle point. In the former case the solution is a maximum or minimum according as $\phi_{pp}(K;p,m)$ is negative or positive. The high-temperature solution $p = m = 0$ is a maximum if $B_2(K) < 0$ and a minimum if $B_2(K) > 0$. Suppose that, to leading order in ΔK,

$$B_2(K) \simeq -B_{21}\Delta K , \qquad B_3(K) \simeq B_{30} , \qquad B_4(K) \simeq B_{40} , \tag{5.4.15}$$

where B_{21} and B_{40} (but not necessarily B_{30}) are positive. Then the high-temperature solution is stable if $K < K_\mathrm{c}, (T > T_\mathrm{c})$. In general, the derivation of non-zero solutions to (5.4.12) is quite complicated. We shall, however, be concerned only with solutions at the critical temperature, which are not in a neighbourhood of the high-temperature solution. In this case it is not difficult to show that, when $K = K_\mathrm{c}$ ($B_2(K_\mathrm{c}) = 0$), there are three such solutions

$$\boldsymbol{x}_1^{(\pm)} := \frac{3B_{30}}{4B_{40}} \begin{pmatrix} \frac{1}{2} \\ \pm\frac{\sqrt{3}}{2} \end{pmatrix} \qquad \boldsymbol{x}_2 := -\frac{3B_{30}}{4B_{40}} \begin{pmatrix} 1 \\ 0 \end{pmatrix} . \tag{5.4.16}$$

Given any solution to (5.4.12), other solutions can be obtained by the operations of the group $\mathcal{S}_3$. In general solutions will occur in sets of six, corresponding to the six elements of $\mathcal{S}_3$. However, in this case

$$\begin{aligned} \boldsymbol{Q}^{(m)}\boldsymbol{x}_1^{(+)} &= \boldsymbol{x}_1^{(-)}, & \boldsymbol{Q}^{(*)}\boldsymbol{x}_1^{(-)} &= \boldsymbol{x}_1^{(+)}, \\ \boldsymbol{Q}^{(m)}\boldsymbol{x}_2 &= \boldsymbol{x}_2, & \boldsymbol{Q}^{(*)}\boldsymbol{x}_1^{(+)} &= x_2, \\ \boldsymbol{Q}^{(m)}\boldsymbol{x}_1^{(-)} &= \boldsymbol{x}_1^{(+)}, & \boldsymbol{Q}^{(*)}\boldsymbol{x}_2 &= \boldsymbol{x}_1^{(-)}, \end{aligned} \tag{5.4.17}$$

and, since the group is generated by $\mathrm{Q}^{(m)}$ and $\mathrm{Q}^{(*)}$, the solutions (5.4.16) form a closed set under the operations of the group. Since

$$\begin{aligned} &\Phi(K_c;\, p_1^{(\pm)}, m_1^{(\pm)}) = \frac{243 B_{30}^4}{16 B_{40}^2}, \qquad \phi_{pp}(K_c;\, p_1^{(\pm)}, m_1^{(\pm)}) = \frac{45 B_{30}^2}{8 B_{40}}, \\ &\tilde{\phi}_1^{(\mathrm{CL})}(K_c;\, p_1^{(\pm)}, m_1^{(\pm)}) = B_0(K_c) - \frac{27 B_{30}^4}{256 B_{40}^3}, \end{aligned} \tag{5.4.18}$$

$$\begin{aligned} &\Phi(K_c;\, p_2, m_2) = \frac{243 B_{30}^4}{16 B_{40}^2}, \qquad \phi_{pp}(K_c;\, p_2, m_2) = \frac{9 B_{30}^2}{4 B_{40}}, \\ &\phi_1^{(\mathrm{CL})}(K_c;\, p_2, m_2) = B_0(K_c) - \frac{27 B_{30}^4}{256 B_{40}^3}, \end{aligned} \tag{5.4.19}$$

it follows that, when $B_{40} > 0$, the solutions (5.4.16) are lower minima of $\phi_1^{(\mathrm{CL})}(K;\, p, m)$ than $p = m = 0$ at, and in a neighbourhood of, the critical temperature. This means that the system will have already passed, with equal probability, to one of these states at some temperature above the critical value. The phase transition will be first-order since it involves a discontinuous change in p and m.

The prediction by Landau theory of a first-order phase transition in the 3-state Potts model is an interesting example of the limitations of this method. The results of Landau theory for the order of phase transitions are the same as those obtained by any mean-field calculation and, for the general ν-state Potts model, these have been shown (Mittag and Stephen 1974) to give a first-order transition for $\nu > 2$ in any dimension. In contradiction to this it has been shown rigorously by Baxter (1973d), for the square lattice and by Baxter et al. (1978), for the triangular lattice, that the ferromagnetic ν-state Potts model has a first-order transition for $\nu > 4$, but a higher-order transition for $\nu \leq 4$. For the antiferromagnetic model the situation is in general more complicated (Wang et al. 1989) because of the presence of infinite ground-state degeneracy. There are two exceptions to this. On the square lattice the antiferromagnetic 3-state Potts model is, like the ferromagnetic case, equivalent to a staggered six-vertex model (Sect. 3.8.2.1) with a higher-order transition although in this case it occurs at zero temperature. On the triangular lattice the ground state is six-fold degenerate (Schick and Griffths 1977) like the triangular spin-$\frac{1}{2}$ Ising antiferromagnet in a field. Monte Carlo simulations (Grest 1981; Saito 1981; Adler et al. 1995), series methods (Enting and Wu 1982) and analytical investigations (Park 1994) indicate that the transition is first-order in agreement with the prediction of Landau theory.

5.5 Landau Theory for a Tricritical Point

In Sect. 4.6 we described a magnetic model which exhibits a critical line meeting a triple line at a tricritical point in the symmetry plane with the staggered field $\mathcal{H}_s = 0$ (Fig. 4.10). In Sect. 6.2 we show, using mean-field theory, that this model can be realized by a metamagnetic generalization of the spin-$\frac{1}{2}$ Ising model for which we have discussed a Landau treatment in Sect. 5.3.1. Points in the symmetry plane $L_s := \mathcal{H}_s/T = 0$ are given in terms of the coupling $K := J/T$ and $L := \mathcal{H}/T$ and ϕ_2 is an even function of q, the order parameter associated with the staggered field. Thus we have

$$\begin{aligned}\phi_2^{(\mathrm{CL})}(K, L, L_s; q) := & -L_s q + B_0(K, L) + B_2(K, L)q^2 \\ & + B_4(K, L)q^4 + G(L, L_s; q),\end{aligned} \tag{5.5.1}$$

where

$$G(L, L_s; q) := \sum_{i=3}^{\infty} B_{2i}(K, L)q^{2i}. \tag{5.5.2}$$

We consider only the case $B_6(K, L) > 0$, for which, as it was shown in Example 17.1.5, (5.5.1) is strongly equivalent to

$$\begin{aligned}\tilde{\phi}_2^{(\mathrm{CL})}(K, L, L_s; q) := & -L_s q + B_0(K, L) + B_2(K, L)q^2 \\ & + B_4(K, L)q^4 + B_6(K, L)q^6.\end{aligned} \tag{5.5.3}$$

The conditions for minimizing $\phi_2^{(\mathrm{CL})}$ as a function of q are

$$2B_2(K, L)q + 4B_4(K, L)q^3 + 6B_6(K, L)q^5 - L_s = 0, \tag{5.5.4}$$

$$2B_2(K, L) + 12B_4(K, L)q^2 + 30B_6(K, L)q^4 > 0\,. \tag{5.5.5}$$

In the symmetry plane when $L_s = 0$ the solutions of (5.5.4) are $q = 0$, which is stable when $B_2(L, K) > 0$, and

$$q^{(\pm)2} = -\frac{B_4(K, L)}{3B_6(K, L)}\left[1 \pm \sqrt{1 - w(K, L)}\right], \tag{5.5.6}$$

where

$$w(K, L) := \frac{3B_2(K, L)B_6(K, L)}{B_4^2(K, L)}\,. \tag{5.5.7}$$

From (5.5.4) and (5.5.5) the stability condition for the solutions (5.5.6) becomes

$$\mp B_4(K,L)\sqrt{1-w(K,L)} > 0. \tag{5.5.8}$$

It follows that the possible stable solutions from (5.5.6) are $\pm q^{(-)}$ or $\pm q^{(+)}$ according as $B_4(K,L) >, < 0$ respectively.

- In the case $B_4(K,L) > 0$, $q^{(-)}$ is real only when $w(K,L) < 0$. The curve

$$L = L^*(K), \tag{5.5.9}$$

given as a solution of

$$B_2(K,L) = 0, \tag{5.5.10}$$

is, as in Sect. 4.7.1, the critical line $\mathcal{C}$ of continuous transitions in the plane $L_s = 0$, with the ordered and disordered phases stable in the regions $B_2(K,L) < 0$ and $B_2(K,L) > 0$ respectively. On physical grounds we expect the ordered phase to occur on the side of $\mathcal{C}$ for which $L < L^*(K)$.

- In the case $B_4(K,L) < 0$, (5.5.6) gives a real value of $q^{(+)}$ as long as $w(K,L) < 1$ and, from (5.5.3),

$$\tilde{\phi}_2^{(\mathrm{CL})}(K,L,0;q^{(+)}) = B_0(K,L) + \tfrac{2}{3}B_4(K,L){q^{(+)}}^4\left[\sqrt{1-w(K,L)}-\tfrac{1}{2}\right]. \tag{5.5.11}$$

The two minima $\pm q^{(+)}$ are therefore metastable, with the stable solution given by $q = 0$, when $w(K,L) < \frac{3}{4}$. A first-order phase transition occurs at $w(K,L) = \frac{3}{4}$, or

$$B_2(K,L) = \frac{B_4^2(K,L)}{4B_6(K,L)}, \tag{5.5.12}$$

from the disordered state to the ordered state, with a discontinuous change of the order parameter to

$$q^{(+)} = \pm\sqrt{\frac{-B_4(K,L)}{2B_6(K,L)}}, \tag{5.5.13}$$

the two values occurring with equal probability. The solution curve $\mathcal{T}$ to the equation (5.5.12) is a line of first-order transitions, which becomes a triple line in the full phase diagram in which the staggered coupling L_s is included (Fig. 4.10). Suppose that $\mathcal{T}$ takes the form

$$L = L^\dagger(K). \tag{5.5.14}$$

Then the critical curve $\mathcal{C}$ given by (5.5.9) and the first-order curve $\mathcal{T}$ given by (5.5.14) meet at a tricritical point T with coordinates $(K_\mathrm{t}, L_\mathrm{t}, L_\mathrm{s} = 0)$ given as the solution of[20]

$$B_2(K, L) = 0\,, \qquad B_4(K, L) = 0, \tag{5.5.15}$$

where

$$L_\mathrm{t} = L^*(K_\mathrm{t}) = L^\dagger(K_\mathrm{t}). \tag{5.5.16}$$

It is straightforward to show that $\mathcal{C}$ and $\mathcal{T}$ together form a curve in the K–L plane with a continuous gradient at T.

5.5.1 Tricritical Exponents

In Sect. 4.6 exponents for the tricritical point T were introduced and orthogonal coordinates defined to represent the directions along and perpendicular to the tangent to $\mathcal{C}$ at T. Here we repeat this procedure but alter the point of interest to an arbitrarily chosen point P on $\mathcal{C}$, whilst also considering the limit as $\mathsf{P} \to \mathsf{T}$. The set of exponents at P are defined as in (4.6.15)–(4.6.22) with the 't' omitted. Let

$$a_\mathrm{p} := \left(\frac{\mathrm{d}L^*}{\mathrm{d}K}\right)_\mathrm{p}, \tag{5.5.17}$$

with orthogonal coordinates

$$\Theta_2^{(\mathrm{p})} := L - a_\mathrm{p} K, \qquad \Theta_3^{(\mathrm{p})} := K + a_\mathrm{p} L. \tag{5.5.18}$$

At the tricritical point the label 'p' is replaced by 't'.[21] Now, following (4.6.12) and (4.6.13), let

$$L^*(K) := L_\mathrm{p} + a_\mathrm{p}\Delta_\mathrm{p}K + \mathcal{L}_\mathrm{p}(K), \qquad \Delta_\mathrm{p}K := K - K_\mathrm{p}, \tag{5.5.19}$$

where

$$\mathcal{L}_\mathrm{p}(K_\mathrm{p}) = 0, \qquad \mathcal{L}'_\mathrm{p}(K_\mathrm{p}) = 0. \tag{5.5.20}$$

---

[20] The free-energy density (5.5.3) is, therefore, the universal unfolding of $B_6(KL)q^6$ at the tricritical point.

[21] It is clear that $(\Theta_2^{(\mathrm{t})}\Theta_3^{(\mathrm{t})})$ are the coordinates (Θ_2, Θ_3) defined in (4.6.14) with $a = a_\mathrm{t}$. It is for this reason that we have retained the labels 2,3 rather that the more obvious 1,2.

In order to satisfy the critical curve condition (5.5.10), the tricritical condition (5.5.15), and with $B_4(K, L)$ and $B_6(K, L)$ having the appropriate signs, we suppose that, to leading order,

$$\begin{aligned} B_2(K, L) &\simeq B_{21}[L - L^*(K)] \simeq B_{21}[\Theta_2^{(\mathrm{p})} - \Theta_{2\mathrm{p}}^{(\mathrm{p})}], \\ B_4(K, L) &\simeq -B_{41}\Delta_{\mathrm{t}}K, \\ B_6(K, L) &\simeq B_{60}, \end{aligned} \tag{5.5.21}$$

where B_{21}, B_{41} and B_{60} are positive constants.[22] From (5.5.7),

$$w(K, L) \simeq \frac{3B_{21}[\Theta_2^{(\mathrm{p})} - \Theta_{2\mathrm{p}}^{(\mathrm{p})}]}{B_{41}^2(\Delta_{\mathrm{t}}K)^2}. \tag{5.5.22}$$

At the point P consider an approach to $\mathcal{C}$ in the orthogonal direction $\Theta_3^{(\mathrm{p})} = \Theta_{3\mathrm{p}}^{(\mathrm{p})}$ within the ordered region ($\Theta_2^{(\mathrm{p})} < \Theta_{2\mathrm{p}}^{(\mathrm{p})}$). From (5.5.6), (5.5.7) and (5.5.21),

$$q^2 = q^{(-)2} \simeq -\frac{B_{21}[\Theta_2^{(\mathrm{p})} - \Theta_{2\mathrm{p}}^{(\mathrm{p})}]}{2B_{41}\Delta_{\mathrm{t}}K} \tag{5.5.23}$$

and it follows from (5.5.23) that $\beta = \frac{1}{2}$. At the tricritical point $\Theta_i^{(\mathrm{p})} = \Theta_i^{(\mathrm{t})}$, $i = 2, 3$ and with $\Theta_3^{(\mathrm{t})} = \Theta_{3\mathrm{t}}^{(\mathrm{t})}$

$$\Delta_{\mathrm{t}}K = -\frac{a_{\mathrm{t}}[\Theta_2^{(\mathrm{t})} - \Theta_{2\mathrm{t}}^{(\mathrm{t})}]}{1 + a_{\mathrm{t}}^2}, \tag{5.5.24}$$

$w(K, L)$ diverges at the tricritical point and

$$q^2 = q^{(-)2} \simeq \sqrt{\frac{B_{21}[\Theta_2^{(\mathrm{t})} - \Theta_{2\mathrm{t}}^{(\mathrm{t})}]}{3B_{60}}}. \tag{5.5.25}$$

Thus, from (4.6.19), $\beta_{\mathrm{t}} = \frac{1}{4}$. Consider now the asymptotic form for q^2 as the tricritical point is approached along the critical curve. From (5.5.5), (5.5.6) and (5.5.10), $w(K, L)$ is identically zero and, from (5.5.18),

$$\Delta_{\mathrm{t}}K = \frac{[\Theta_3^{(\mathrm{t})} - \Theta_{3\mathrm{t}}^{(\mathrm{t})}]}{1 + a_{\mathrm{t}}^2}, \tag{5.5.26}$$

---

[22] On $\mathcal{C}$, $K < K_{\mathrm{t}}$, and, on $\mathcal{T}$, $K > K_{\mathrm{t}}$ (Fig. 4.10).

giving

$$q^2 = q^{(-)2} \simeq -\frac{B_{41}[\Theta_3^{(t)} - \Theta_{3t}^{(t)}]}{3(1+a_t^2)B_{60}}. \tag{5.5.27}$$

It follows from (5.5.26) that $\Theta_3^{(t)} < \Theta_{3t}^{(t)}$ as the tricritical point is approached along the critical curve and, from (4.6.19), $\beta_{tc} = \frac{1}{2}$. From (4.6.3) and (5.5.4)

$$\psi_T := \left(\frac{\partial q}{\partial L_s}\right)_{K,L} \simeq \frac{1}{2B_2(K,L) + B_4(K,L)q^2 + 30B_6q^4}. \tag{5.5.28}$$

In the disordered region where $q = 0$, from (4.6.21) and (5.5.21), $\gamma = \gamma_t = 1$ and in the ordered region, using asymptotic forms for q^2 given by (5.5.23), (5.5.25) and (5.5.27) $\gamma' = \gamma_t' = 1$, $\gamma_{tc} = 2$. From (5.5.4) and (5.5.21), when $K = K_p$

$$L_s \simeq 4B_{41}(K_t - K_p)q^3 + 6B_{60}q^5, \tag{5.5.29}$$

giving, from (4.6.15) $\delta = 3$, $\delta_t = 5$. The exponents β, γ and δ on $\mathcal{C}$ away from the tricritical point are the same as those given by (5.3.30). This means that, if the coincidence of values (but not, of course, the values themselves) applies to the exact and not just classical calculations, the second-order transition on the critical curve is in the universality class of a ordinary critical point with $\alpha' = 0$ (Sect. 4.2.1).

The classical values of the remaining tricritical exponents, defined in (4.6.16)–(4.6.18), and their equivalents on the critical curve could be derived in a similar way.[23] However, if scaling is assumed, it follows from the discussion is Sect. 4.7.2 that, once we have expressions for y_1, y_2 and y_3, all the remaining exponents along $\mathcal{C}$ and in the ordered region can be derived. Using the values we have just derived for δ_t, β_t and β_{tc} and (4.7.25) and (4.7.30)

$$y_1 = \frac{d\delta_t}{1+\delta_t} = \tfrac{5}{6}d, \qquad y_2 = \frac{d - y_1}{\beta_t} = \tfrac{2}{3}d, \qquad y_3 = \frac{d - y_1}{\beta_{tc}} = \tfrac{1}{3}d. \tag{5.5.30}$$

These expressions then, from (4.7.31), return our values for γ_t' and γ_{tc} and, using (4.7.32)–(4.7.36), we can derive the set of classical values for the tricritical exponents defined in (4.6.16)–(4.6.18).

---

[23] For the heat capacity exponents we should need also to assume an asymptotic form for $B_0(K, L)$ as we did for the critical point in (5.3.22).

$$\begin{aligned} \alpha'_{\rm t} &= \tfrac{1}{2}, & \alpha_{\rm tc} &= -1, \\ \beta_{\rm t} &= \tfrac{1}{4}, & \beta_{\rm tc} &= \tfrac{1}{2}, \\ \beta'_{2\rm t} &= \tfrac{1}{2}, & \beta_{2\rm tc} &= 1, \\ \gamma'_{\rm t} &= 1, & \gamma_{\rm tc} &= 2, \\ \gamma'_{2\rm t} &= \tfrac{1}{2}, & \gamma_{2\rm tc} &= 1, \\ \delta_{\rm t} &= 5. \end{aligned} \tag{5.5.31}$$

5.6 Ginzburg–Landau Theory

5.6.1 A Critical Point

Let $\mathcal{V}$ be a d-dimensional hypercube of volume R^d. In terms of some reference volume v_0, $V = R^d/v_0$ is the dimensionless volume of the system. At every point $\boldsymbol{r}$ of $\mathcal{V}$ an n-dimensional vector field $\boldsymbol{\sigma}(\boldsymbol{r})$ is defined, in terms of which the statistical mechanical magnetization density is

$$\hat{\boldsymbol{m}} := \frac{1}{R^d}\int_{\mathcal{V}} \boldsymbol{\sigma}(\boldsymbol{r})\mathrm{d}\mathcal{V}, \tag{5.6.1}$$

(cf. (3.5.4), (3.5.5)). The vector field $\boldsymbol{\sigma}(\boldsymbol{r})$ is the *local* magnetization density with conjugate vector coupling $\boldsymbol{L} := \boldsymbol{\mathcal{H}}/T$, where $\boldsymbol{\mathcal{H}}$ is an n-dimensional magnetic field. The local free-energy density is given by generalizing (5.3.32) to[24]

$$\hat{\phi}_n^{(\mathrm{CL})}(K,\boldsymbol{L};\boldsymbol{\sigma}(\boldsymbol{r})) := -\boldsymbol{L}.\boldsymbol{\sigma}(\boldsymbol{r}) + B_0(K) + B_2(K)|\boldsymbol{\sigma}(\boldsymbol{r})|^2 + B_4(K)|\boldsymbol{\sigma}(\boldsymbol{r})|^4 + C(K)\sum_{j=1}^{n}|\nabla\sigma_j(\boldsymbol{r})|^2 . \tag{5.6.2}$$

Then

$$\phi_n(K,\boldsymbol{L}) := -\frac{1}{V}\ln Z_n(K,\boldsymbol{L}), \tag{5.6.3}$$

[24] Suppose, in the case $n = 1$, we select the form of σ which minimizes $\hat{\phi}_1^{(\mathrm{CL})}$. For $C(K) > 0$ this will occur when σ is uniform and equal to m over $\mathcal{V}$ and, assuming that this distribution dominates the functional integral in (5.6.4) we recover (5.3.32).

where the partition function is given by the functional integral

$$Z_n(K, \boldsymbol{L}) := \int_{\Sigma} \exp\left[-\frac{1}{v_0}\int_{\mathcal{V}} \hat{\phi}_n^{(\mathrm{CL})}(K, \boldsymbol{L}; \boldsymbol{\sigma}(\boldsymbol{r}))\mathrm{d}\mathcal{V}\right]\mathrm{d}\Sigma \tag{5.6.4}$$

and Σ is the space of values of $\boldsymbol{\sigma}$. Suppose that the magnetic field $\mathcal{H}$ is in the direction of the first component σ_1 of $\boldsymbol{\sigma}$. The first term in (5.6.2) is now replaced by $-L\sigma_1$. The first component of $\boldsymbol{\sigma}$ is referred to as *longitudinal*, the other components as *transverse*. Let $\mathring{\boldsymbol{\sigma}}$ be defined according to the high-temperature minimization conditions for $\hat{\phi}_n^{(\mathrm{CL})}$, when the distribution is uniform. Thus

$$\begin{aligned} 0 &= 2B_2(K)\mathring{\sigma}_1 + 4B_4(K)\mathring{\sigma}_1^3 - L, \\ 0 &= \mathring{\sigma}_j, \qquad j \neq 1. \end{aligned} \tag{5.6.5}$$

5.6.2 The Gaussian Approximation

This involves retaining only harmonic fluctuations around the uniform values given by (5.6.5). With

$$\sigma_j'(\boldsymbol{r}) := \sigma_j(\boldsymbol{r}) - \mathring{\sigma}_j\,, \qquad j = 1, \ldots, n, \tag{5.6.6}$$

we have

$$\begin{aligned} \hat{\phi}_n^{(\mathrm{CL})}(K, \boldsymbol{L}; \boldsymbol{\sigma}(\boldsymbol{r})) = \hat{\phi}_n^{(\mathrm{CL})}(K, \boldsymbol{L}; \mathring{\boldsymbol{\sigma}}) + 4B_4(K)\mathring{\sigma}_1^2\sigma_1'^2(\boldsymbol{r}) + D(K)|\boldsymbol{\sigma}'(\boldsymbol{r})|^2 \\ + C(K)\sum_{j=1}^{n}|\nabla\sigma_j'(\boldsymbol{r})|^2, \end{aligned} \tag{5.6.7}$$

where

$$D(K) := B_2(K) + 2B_4(K)\mathring{\sigma}_1^2. \tag{5.6.8}$$

With periodic boundary conditions on the hypercube $\mathcal{V}$, Fourier transforms are defined (Eqs. (18.1.14) and (18.1.15)) by[25]

$$\psi_j(\boldsymbol{k}) := \frac{1}{R^d}\int_{\mathcal{V}} \sigma_j'(\boldsymbol{r})\exp(-\mathrm{i}\boldsymbol{k}\cdot\boldsymbol{r})\mathrm{d}\mathcal{V}, \tag{5.6.9}$$

---

[25] Not to be confused with the derivatives of the free-energy density appearing in other chapters.

with

$$\sigma'_j(\boldsymbol{r}) = \sum_{\{\boldsymbol{k}\}} \psi_j(\boldsymbol{k}) \exp(\mathrm{i}\boldsymbol{k} \cdot \boldsymbol{r}), \tag{5.6.10}$$

where the wave vectors

$$\boldsymbol{k} := \frac{2\pi}{R} \begin{pmatrix} m_1 \\ \vdots \\ m_d \end{pmatrix} \qquad m_i = 0, \pm 1, \pm 2, \ldots . \tag{5.6.11}$$

It can be seen from (5.6.10) that $\psi_j(\boldsymbol{k})$ corresponds to the amplitude fluctuation of $\sigma_j(\boldsymbol{r})$ with wave vector $\boldsymbol{k}$. The larger the magnitude of $\boldsymbol{k}$ the shorter the wave length of the fluctuation. The coefficients $\psi_1(\boldsymbol{k})$ represent fluctuations in the direction of the external field, called *longitudinal fluctuations*; $\psi_j(\boldsymbol{k})$, for $j \neq 1$, correspond to *transverse fluctuations*. Since all the variables $\sigma'_j(\boldsymbol{k})$ are real, $\psi_j(-\boldsymbol{k})$ is the conjugate complex of $\psi_j(\boldsymbol{k})$, with $\psi_j(\boldsymbol{0})$ real. From (5.6.1), (5.6.6) and (5.6.9)

$$\hat{\boldsymbol{m}} = \mathring{\boldsymbol{\sigma}} + \boldsymbol{\psi}(\boldsymbol{0}), \tag{5.6.12}$$

where $\boldsymbol{\psi}(\boldsymbol{0})$ is the n-dimensional vector with components $\psi_j(\boldsymbol{0})$. This term in (5.6.12) represents the fluctuations in magnetization per unit volume. For $\boldsymbol{k} \neq \boldsymbol{0}$ let

$$\psi_j(\pm\boldsymbol{k}) := \psi_j^{(\Re)}(\boldsymbol{k}) \pm \mathrm{i}\psi_j^{(\Im)}(\boldsymbol{k}). \tag{5.6.13}$$

From (5.6.7) and (5.6.10),

$$\begin{aligned}\frac{1}{R^d}\int_{\mathcal{V}} \hat{\phi}_n^{(\mathrm{CL})}(K, \boldsymbol{L}; \boldsymbol{\sigma}(\boldsymbol{r}))\mathrm{d}\mathcal{V} &= \hat{\phi}_n^{(\mathrm{CL})}(K, \boldsymbol{L}; \mathring{\boldsymbol{\sigma}}) + 4B_4(K)\mathring{\sigma}_1^2\psi_1^2(\boldsymbol{0}) + D(K)|\boldsymbol{\psi}(\boldsymbol{0})|^2 \\ &+ 2\sum_{\{\boldsymbol{k}>\boldsymbol{0}\}} [4B_4(K)\mathring{\sigma}_1^2 + D(K) + C(K)k^2][\psi_1^{(\Re)2}(\boldsymbol{k}) + \psi_1^{(\Im)2}(\boldsymbol{k})] \\ &+ 2\sum_{j=2}^{n}\sum_{\{\boldsymbol{k}>\boldsymbol{0}\}} [D(K) + C(K)k^2][\psi_j^{(\Re)2}(\boldsymbol{k}) + \psi_j^{(\Im)2}(\boldsymbol{k})],\end{aligned} \tag{5.6.14}$$

where '$\boldsymbol{k} > \boldsymbol{0}$' means all the wave vectors in half the $\boldsymbol{k}$-vector space excluding $\boldsymbol{k} = \boldsymbol{0}$. The partition function is now given by substituting from (5.6.14) into (5.6.4). The functional integral over the space Σ is replaced by integrals over the variables $\psi_j(\boldsymbol{0})$, $\psi_j^{(\Re)}(\boldsymbol{k})$ and $\psi_j^{(\Im)}(\boldsymbol{k})$, for $j = 1, \ldots, n$, which we take to extend over the range $(-\infty, \infty)$. This gives

$$Z_n(K, \boldsymbol{L}) = \exp[-V\hat{\phi}_n^{(\mathrm{CL})}(K, \boldsymbol{L}; \mathring{\boldsymbol{\sigma}})] \prod_{j=1}^{n} \left[\Psi_j(\boldsymbol{0}) \prod_{\{\boldsymbol{k}>\boldsymbol{0}\}} \Psi_j(\boldsymbol{k})\right], \tag{5.6.15}$$

where

$$\Psi_j(\mathbf{0}) := \begin{cases} \displaystyle\int_{-\infty}^{\infty} \mathrm{d}\psi_1(\mathbf{0}) \exp\{-V[4B_4(K)\mathring{\sigma}_1^2 + D(K)]\psi_1^2(\mathbf{0})\}, & j = 1, \\ \displaystyle\int_{-\infty}^{\infty} \mathrm{d}\psi_j(\mathbf{0}) \exp\{-V D(K)\psi_j^2(\mathbf{0})\}, & j \neq 1, \end{cases} \tag{5.6.16}$$

and, for $\boldsymbol{k} \neq \mathbf{0}$,

$$\Psi_j(\boldsymbol{k}) := \begin{cases} \displaystyle\int_{-\infty}^{\infty} \mathrm{d}\psi_1^{(\Re)}(\boldsymbol{k}) \int_{-\infty}^{\infty} \mathrm{d}\psi_1^{(\Im)}(\boldsymbol{k}) \exp\{-V[4B_4(K)\sigma_1^{\star 2} + D(K) \\ \qquad + C(K)k^2][\psi_1^{(\Re)2}(\boldsymbol{k}) + \psi_1^{(\Im)2}(\boldsymbol{k})]\}, & j = 1, \\ \displaystyle\int_{-\infty}^{\infty} \mathrm{d}\psi_j^{(\Re)}(\boldsymbol{k}) \int_{-\infty}^{\infty} \mathrm{d}\psi_j^{(\Im)}(\boldsymbol{k}) \exp\{-V[D(K) \\ \qquad + C(K)k^2][\psi_j^{(\Re)2}(\boldsymbol{k}) + \psi_j^{(\Im)2}(\boldsymbol{k})]\}, & j \neq 1. \end{cases} \tag{5.6.17}$$

The variables $\psi_j(\mathbf{0})$, $\psi_j^{(\Re)}(\boldsymbol{k})$ and $\psi_j^{(\Im)}(\boldsymbol{k})$, for all j and $\boldsymbol{k}$ are normally distributed and uncorrelated with

$$\langle \psi_j(\boldsymbol{k}) \rangle = 0, \tag{5.6.18}$$

$$\langle |\psi_j(\boldsymbol{k})|^2 \rangle = \begin{cases} g_\parallel(K, \boldsymbol{L}; k), & j = 1, \\ g_\perp(K, \boldsymbol{L}; k), & j \neq 1, \end{cases} \tag{5.6.19}$$

where

$$\begin{aligned} g_\parallel(K, \boldsymbol{L}; k) &:= \{2V[4B_4(K)\mathring{\sigma}_1^2 + D(K) + C(K)k^2]\}^{-1}, \\ g_\perp(K, \boldsymbol{L}; k) &:= \{2V[D(K) + C(K)k^2]\}^{-1}. \end{aligned} \tag{5.6.20}$$

From Eqs. (5.6.5), (5.6.10), (5.6.12) and (5.6.18),

$$\begin{aligned} \langle \boldsymbol{\sigma}'(\boldsymbol{r}) \rangle &= 0, \\ \langle \boldsymbol{\sigma}(\boldsymbol{r}) \rangle &= \langle \mathring{\boldsymbol{\sigma}} \rangle = \mathring{\boldsymbol{\sigma}}, \end{aligned} \tag{5.6.21}$$

$$\boldsymbol{m} = \langle \hat{\boldsymbol{m}} \rangle = \mathring{\boldsymbol{\sigma}} = \mathring{\sigma}_1 \hat{\boldsymbol{h}}, \tag{5.6.22}$$

where $\hat{\boldsymbol{h}}$ is the unit vector in the direction of the external magnetic field. It follows that the thermodynamic magnetization per unit volume is the same as that per lattice site in Landau theory.[26]

Formula (4.2.3) for the pair correlation function applies to the continuum, as well as to a lattice system, and from (5.6.10) using the notation introduced in (2.5.25) for a system with translational and rotational invariance,

$$\Gamma_2(\sigma; |\boldsymbol{r}-\boldsymbol{r}'|) := \langle \boldsymbol{\sigma}'(\boldsymbol{r}).\boldsymbol{\sigma}'(\boldsymbol{r}')\rangle$$
$$= \sum_{j=1}^{n}\sum_{\{\boldsymbol{k}\}} \langle |\psi_j(\boldsymbol{k})|^2\rangle \exp[\mathrm{i}\boldsymbol{k}\,.\,(\boldsymbol{r}-\boldsymbol{r}')]. \qquad (5.6.23)$$

With $r := |\boldsymbol{r}-\boldsymbol{r}'|$, the Fourier transform $\Gamma_2^*(\sigma;k)$ of $\Gamma_2(\sigma;r)$ is defined by the continuum form of (2.5.17) (Eq. (5.6.9)) and thus, from (5.6.20) and (5.6.23),

$$\Gamma_2^*(\sigma;k) = R^d \sum_{j=1}^{n}\langle |\psi_j(\boldsymbol{k})|^2\rangle = R^d[g_{\parallel}(K,\boldsymbol{L};k) + (n-1)g_{\perp}(K,\boldsymbol{L};k)]. \quad (5.6.24)$$

From (4.5.50),[27] (5.6.20) and (5.6.24) the correlation length $\mathfrak{r}$ is given by

$$\mathfrak{r}^2 = 4c(d)dVC(K)\frac{g_{\parallel}^2(K,\boldsymbol{L};0)+(n-1)g_{\perp}^2(K,\boldsymbol{L};0)}{g_{\parallel}(K,L;0)+(n-1)g_{\perp}(K,\boldsymbol{L};0)}. \qquad (5.6.25)$$

To evaluate ϕ, given by (5.6.3) and (5.6.15), it is necessary to insert explicit expressions for the Gaussian integrals (5.6.16) and (5.6.17). From (5.6.19) and (5.6.20) this gives

$$\phi(K,\boldsymbol{L}) = \hat{\phi}_{\mathrm{L}}(K,\boldsymbol{L};\mathring{\boldsymbol{\sigma}})$$
$$-\frac{1}{2V}\sum_{\{\boldsymbol{k}\}}\{\ln[2\pi g_{\parallel}(K,\boldsymbol{L};k)] + (n-1)\ln[2\pi g_{\perp}(K,\boldsymbol{L};k)]\}. \quad (5.6.26)$$

5.6.3 Gaussian Critical Exponents

Adopting the forms for $B_2(K)$ and $B_4(K)$ given by the leading terms in (5.3.22), it can be shown from (5.6.5) and (5.6.22) that the exponents β, γ and δ take their classical values (5.3.30). We now consider the situation along the zero-field axis, $\boldsymbol{L}=\boldsymbol{0}$, and suppose that, to leading order in $\Delta K := K - K_{\mathrm{c}}$,

$$C(K) \simeq C_0 > 0. \qquad (5.6.27)$$

---

[26] Cf. (5.6.5) and (5.3.17) with $\Delta L = L$, $\Delta\rho = m$, $B_1(K) = B_3(K) = 0$.

[27] In this case $\mathfrak{d} = d$, the thermodynamic limit is taken in each of the d dimensions.

From (5.3.22) and (5.6.20),

$$g_{\|}(K,\mathbf{0};k) = g_{\perp}(K,\mathbf{0};k) = [2V(-B_{21}\Delta K + C_0 k^2)]^{-1}, \tag{5.6.28}$$

for $K < K_c$, when $\mathring{\sigma}_1 = 0$, and

$$\begin{aligned} g_{\|}(K,\mathbf{0};k) &= [2V(2B_{21}\Delta K + C_0 k^2)]^{-1}, \\ g_{\perp}(K,\mathbf{0};k) &= (2VC_0 k^2)^{-1}, \end{aligned} \tag{5.6.29}$$

for $K > K_c$, when, from (5.3.33) and (5.6.22), $\mathring{\sigma}_1^2 = B_{21}\Delta K/(2B_{40})$. Thus, from (5.6.25), it can be seen that throughout the magnetized, low-temperature, phase the correlation length is infinite when $n \neq 1$. This behaviour arises from the transverse singular amplitude of the long-wavelength contribution to the pair correlation function. The magnetized phase is not well-defined in the Gaussian approximation. We shall, for the rest of this section, concentrate on the high-temperature phase. From (5.6.25) and (5.6.28)

$$\mathfrak{r} = \left[\frac{2c(d)dC_0}{B_{21}(K_c - K)}\right]^{\frac{1}{2}} = \left[\frac{C_0}{B_{21}(K_c - K)}\right]^{\frac{1}{2}}, \tag{5.6.30}$$

with $c(d) = 1/(2d)$, (a choice which is justified below). From (4.5.52) this gives the value for the exponent $\nu = \frac{1}{2}$. From (5.6.24), (5.6.28) and (5.6.30),

$$\bar{\Gamma}_2^*(\sigma;k) = \frac{nv_0}{2C_0(k^2 + \mathfrak{r}^{\;2})} \tag{5.6.31}$$

and thus

$$\bar{\Gamma}_2^*(\sigma;k) = \frac{nv_0}{2C_0 k^2} \tag{5.6.32}$$

at the critical point. Comparing this formula with the asymptotic form (4.5.57) yields the result $\eta = 0$. As in Sect. 3.6.1 we need to introduce the high-frequency wavelength cut-off k_0 given in terms of the minimum resolution length a_0 by (3.6.16). Then, from (18.1.38),

$$\begin{aligned} \Gamma_2(\sigma;r) &= \lim_{k_0\to\infty} r \int_0^{k_0} \frac{nv_0}{2C_0(k^2+\mathfrak{r}^{-2})}\left(\frac{k}{2\pi r}\right)^{\frac{d}{2}} \mathrm{J}_{\frac{d}{2}-1}(kr)\mathrm{d}k \\ &= \frac{nv_0 r^{2-d}}{2C_0(2\pi)^{\frac{d}{2}}} \lim_{k_0\to\infty} \mathfrak{G}\left(k_0; \tfrac{d}{2}, \tfrac{d}{2}-1; r/\mathfrak{r}\right), \end{aligned} \tag{5.6.33}$$

where

$$\mathfrak{G}(k_0; x, y; z) := \int_0^{k_0} \frac{k^x \mathrm{J}_y(k)}{k^2+z^2}\mathrm{d}k. \tag{5.6.34}$$

From (18.1.41),

$$\Gamma_2(\sigma;r) = \frac{nv_0}{2C_0(2\pi)^{\frac{d}{2}}}(r\mathfrak{r})^{1-\frac{d}{2}}\mathrm{K}_{\frac{d}{2}-1}(r/\mathfrak{r}), \qquad \text{for } d < 5. \tag{5.6.35}$$

Substituting into the continuum form of (4.2.4) and using (18.1.35) for the surface area of a d-dimensional hypersphere, gives

$$\mathfrak{r}^2 = c(d)\frac{\displaystyle\int_0^\infty r^{\frac{d}{2}+2}\mathrm{K}_{\frac{d}{2}-1}(r/\mathfrak{r})\mathrm{d}r}{\displaystyle\int_0^\infty r^{\frac{d}{2}}\mathrm{K}_{\frac{d}{2}-1}(r/\mathfrak{r})\mathrm{d}r}. \tag{5.6.36}$$

The integrals in (5.6.36) are of a standard form, yielding the result $c(d) = 1/(2d)$. In the limit $r \to \infty$, from (18.1.42),

$$\Gamma_2(\sigma;r) \simeq \frac{nv_0}{2^{\frac{d}{2}+\frac{3}{2}}C_0(\pi r\mathfrak{r})^{\frac{d}{2}-\frac{3}{2}}}\frac{\exp(-r/\mathfrak{r})}{r}. \tag{5.6.37}$$

Since (18.1.42) is exact for $x = \pm\frac{1}{2}$, (5.6.37) is exact for $d = 1$ and $d = 3$, when it includes the limiting behaviour as $\mathfrak{r} \to \infty$. From (5.6.35) and (18.1.43),

$$\begin{aligned} \Gamma_2(\sigma;r) &\simeq -\frac{nv_0}{2\pi C_0}\ln\left(\frac{r}{2\mathfrak{r}}\right), && \text{if } d = 2, \\ \Gamma_2(\sigma;r) &\simeq \frac{nv_0}{8\pi^2 C_0}\frac{1}{r^2}, && \text{if } d = 4, \end{aligned} \tag{5.6.38}$$

in the limit $\mathfrak{r} \to \infty$. From (5.6.33) and (18.1.48)

$$\Gamma_2(\sigma;r) \simeq \frac{nv_0}{2C_0 r^{d-2}(2\pi)^{\frac{d}{2}}}\left\{\Omega\left(\tfrac{d}{2},\tfrac{d}{2}-1\right)\right.$$
$$\left.+\sqrt{\tfrac{2}{\pi}}k_0^{\frac{d}{2}-\frac{5}{2}}\cos\left(k_0 - \tfrac{\pi}{4}[d+1]\right)\right\}, \qquad \text{for } d > 4. \tag{5.6.39}$$

So for $d \geq 5$, $\Gamma_2(\sigma;r)$ diverges in the $k_0 \to \infty$, high-frequency limit. From the point of view of critical phenomena it is the long-wavelength, low-frequency, fluctuations which are important. This short-wavelength, high-frequency, divergence is dealt with by supposing that the system has a minimum resolution length a_0, so that fluctuations of frequency greater than $k_0 := 2\pi/a_0$ are dampened by the system. From (5.6.37)–(5.6.39),

$$\Gamma_2(\sigma;r) \sim \frac{1}{r^{d-2}}, \tag{5.6.40}$$

for $d \geq 3$. Comparison with (4.5.55) again shows that $\eta = 0$, but now we have additional information about the cases $d = 1$ and $d = 2$. In the limit $\mathfrak{r} \to \infty$, from (5.6.37) and (5.6.38) respectively,

$$\Gamma_2(\sigma; r) \sim \begin{cases} \mathfrak{r} \exp(-r/\mathfrak{r}), & d = 1, \\ \ln(r/\mathfrak{r}), & d = 2. \end{cases} \tag{5.6.41}$$

The classical values for the exponents β, γ and δ satisfy the Widom scaling law (4.5.45) and these together with the values, derived above for η and ν satisfy the Fisher scaling law (4.5.58).

It is interesting to see from (5.6.35) that, according to this analysis for $d < 5$, the correlation function can be expressed in the form

$$\Gamma_2(\sigma; r) = \frac{f(\mathfrak{r}/r)}{r^{d-2-\eta}}, \tag{5.6.42}$$

given by (4.5.55), where in this case $\eta = 0$ and

$$f(x) := \frac{n v_0 x^{\frac{d}{2}-1}}{2C_0 (2\pi)^{\frac{d}{2}}} \mathrm{K}_{\frac{d}{2}-1}(x). \tag{5.6.43}$$

It is now necessary to determine the exponent associated with $c_{\mathcal{H}}$, which because of the magnetic symmetry is α. Again we consider the thermodynamic limit. The sum in (5.6.26) becomes an integral with the wave vectors forming an effective continuum with $\mathcal{S}_d(k)V(2\pi)^{-d}\mathrm{d}k$ states in the hypershell $k < |\boldsymbol{k}| < k + \mathrm{d}k$, where $\mathcal{S}_d(k)$ is the surface area of a d-dimensional hypersphere of radius k, given by (18.1.35). Introducing the high-frequency cut off k_0, from (5.6.26) and (5.6.28),

$$\begin{aligned} \phi_n(K, \mathbf{0}) = {} & \hat{\phi}_n^{(\mathrm{CL})}(K, \mathbf{0}; \mathbf{0}) \\ & + \frac{n v_0}{\Gamma(d/2) 2^d \pi^{d/2}} \int_0^{k_0} k^{d-1} \ln \left\{ \frac{V[A_2(K_\mathrm{c} - K) + C_0 k^2]}{\pi} \right\} \mathrm{d}k \, , \end{aligned} \tag{5.6.44}$$

for $K < K_\mathrm{c}$. The heat capacity $c_{\mathcal{H}}$ at constant field along the zero-field axis is given, from (2.3.30), by

$$c_{\mathcal{H}} = -k_\mathrm{B} K^2 \frac{\partial^2 \phi_n}{\partial K^2} \tag{5.6.45}$$

The first term on the right-hand side of (5.6.44) is the contribution arising from Landau theory and gives the discontinuity $\Delta c_{\mathcal{H}}$ of (5.3.37). In differentiating the integral in (5.6.44) only the most potentially singular contribution is retained to give

$$c_{\mathcal{H}} \simeq \frac{k_\mathrm{B} n K_\mathrm{c}^2 v_0 B_{21}^2}{\Gamma(d/2) 2^d \pi^{d/2}} \int_0^{k_0} \frac{k^{d-1} \mathrm{d}k}{[B_{21}(K_\mathrm{c} - K) + C_0 k^2]^2}. \tag{5.6.46}$$

Transforming to the variable

$$\kappa(k) := k\mathfrak{r} = k\left[\frac{C_0}{B_{21}(K_c - K)}\right]^{\frac{1}{2}} \tag{5.6.47}$$

with $\kappa_0 = \kappa(k_0)$ gives

$$c_{\mathcal{H}} \simeq \frac{k_B n K_c^2 v_0}{\Gamma(d/2)2^d\pi^{d/2}}\left(\frac{B_{21}}{C_0}\right)^2\left[\frac{C_0}{B_{21}(K_c - K)}\right]^{\frac{4-d}{2}}\int_0^{\kappa_0}\frac{\kappa^{d-1}\mathrm{d}\kappa}{(1+\kappa^2)^2}. \tag{5.6.48}$$

A singularity in $c_{\mathcal{H}}$, at the critical point can arise, either from the integral, or, when $d < 4$, from the prefactor. The integral is bounded above by its limiting value, given by taking $\kappa_0 \to \infty$. This is

$$\int_0^{\infty}\frac{\kappa^{d-1}\mathrm{d}\kappa}{(1+\kappa^2)^2} = \Gamma\left(\tfrac{1}{2}d\right)\Gamma\left(2-\tfrac{1}{2}d\right). \tag{5.6.49}$$

The gamma function is singular at $0, -1, -2, \ldots$ and the integral works as a finite upper bound only when $d < 4$. In this case the prefactor diverges giving $\alpha = (4-d)/2$. When $d = 4$ the prefactor is absent and any singular behaviour must come from the integral. Since the integrand is zero at the lower limit, for K sufficiently close to K_c,

$$\int_0^{\kappa_0}\frac{\kappa^3\mathrm{d}\kappa}{(1+\kappa^2)^2} \sim \int_{\epsilon}^{\kappa_0}\frac{\mathrm{d}\kappa}{\kappa} \sim -\tfrac{1}{2}\ln(K_c - K). \tag{5.6.50}$$

The heat capacity has a logarithmic divergence at the critical point. For $d > 4$ the prefactor does not diverge and, as in the case of $d = 4$, any singularity must arise from the integral in a neighbourhood of its upper limit. But

$$\int_0^{\kappa_0}\frac{\kappa^{d-1}\mathrm{d}\kappa}{(1+\kappa^2)^2} \sim \int_{\epsilon}^{\kappa_0}\kappa^{d-5}\mathrm{d}\kappa = \frac{\kappa_0^{d-4} - \epsilon^{d-4}}{d-4}. \tag{5.6.51}$$

The heat capacity does not diverge, but there is a discontinuity from the Landau term. The exponent $\alpha = 0$.

The results of the combination of Landau theory and the Gaussian approximation applied to Ginzburg–Landau theory can now be summarized as follows:

- When $d \geq 4$, the values for the critical exponents are

$$\begin{aligned} &\alpha = 0, \quad \beta = \tfrac{1}{2}, \quad \delta = 3, \\ &\gamma = 1, \quad \eta = 0, \quad \nu = \tfrac{1}{2}. \end{aligned} \tag{5.6.52}$$

These values satisfy the three scaling laws (4.5.45) (Widom), (4.5.44) (Essam–Fisher) and (4.5.58) (Fisher), but not the Josephson hyper-scaling law (4.5.54), except at $d = 4$.

- When $d < 4$, η and ν are still given by (5.6.52) and to satisfy the Fisher scaling law (4.5.58) we need also to retain the value of γ given in (5.6.52). Since we have shown that now

$$\alpha = \frac{4-d}{2}, \tag{5.6.53}$$

the Josephson hyper-scaling law (4.5.54) is satisfied. However, the Widom and Essam–Fisher scaling laws are not satisfied by α from (5.6.53) and β and δ from (5.6.52). To satisfy these two laws we need the new expressions

$$\beta = \frac{d-2}{4}, \qquad \delta = \frac{d+2}{d-2}. \tag{5.6.54}$$

The classical values for β and δ derived from Landau theory in Sect. 5.3 are intended to be true for all values of d. The analysis, carries over to the Ginzburg–Landau theory and this seems, therefore, to contradict the validity of the values (5.6.54) (or alternatively of scaling theory) unless a reconciliation can be achieved. The key to this can be found in the treatment of Ginzburg–Landau using the renormalization group (see, for example Ma 1976a). In this the parameters B_{21} and B_{40} of Eq. (5.3.34) are also treated as renormalizable (rescalable) parameters. If we modify the scaling approach use in (5.3.34) to include exponents u_2 and u_4 for B_{21} and B_{40} respectively then it follows that

$$z = d - y_1\,, \qquad u_2 = 2y_1 - y_2 - d, \qquad u_4 = 4y_1 - 3d\,. \tag{5.6.55}$$

It is not possible to obtain unique exponent values from these expressions. They are, however, satisfied by

$$y_2 = 2, \qquad y_1 = \tfrac{1}{2}d + 1, \qquad u_2 = 0, \qquad u_4 = 4 - d \tag{5.6.56}$$

and these are the values associated with the *Gaussian fixed point* of renormalization theory (Ma 1976a). From (4.5.40) and (4.5.39) the values of $\gamma = 1$ and α, β and δ given by (5.6.53) and (5.6.54) are then obtained. These values are called the *Gaussian critical exponents*. In this new context we can view the derivation of the exponent values (5.3.36) as based on the tacit assumption that both B_{21} and B_{40} are marginal parameters ($u_2 = u_4 = 0$). The modified derivation leading to (5.6.56) still leaves B_{21} marginal. Indeed, if this were not the case, the derivation of $\nu = \frac{1}{2}$ from (5.6.30) would not be valid. However, for $d < 4$, B_{40} is relevant, for $d = 4$ it is marginal and for $d > 4$ it is irrelevant. A discussion of the case $d < 4$, for which the Gaussian fixed point is unstable requires the full renormalization group analysis. The case of interest here is $d > 4$. Taking the singular part of the free-energy density (5.3.34) on the zero-field axis ($\theta_1 = 0$) and substituting from (5.3.33) and (5.3.22) we have

$$\phi_{\text{sing}}(0, \theta_2, B_{40}) = \frac{3B_{21}^2\theta_2^2}{4B_{40}}. \tag{5.6.57}$$

Using the homogeneity condition

$$\phi_{\text{sing}}(0, \lambda^{y_2}\theta_2, \lambda^{u_4}B_{40}) = \lambda^d\phi_{\text{sing}}(0, \theta_2, B_{40}) \tag{5.6.58}$$

and substituting $\lambda = |\theta_2|^{-1/y_2}$ we have

$$\phi_{\text{sing}}(0, \theta_2, B_{40}) = |\theta_2|^{d/y_2} w_0^{(\pm)}(v), \tag{5.6.59}$$

where $v := B_{40}/|\theta_2|^{u_4/y_2}$ and

$$w_0^{(\pm)}(v) := \frac{3B_{21}^2}{4v} \tag{5.6.60}$$

is a particular realization of the function appearing in Eq. (4.8.4). It follows that, in terms of the discussion in Sect. 4.8, B_{40} is a dangerous irrelevant variable, with $\tau = 1$ and, from (4.8.9), an effective change of dimension in the system to

$$\bar{d} := d - |d - 4| = 4. \tag{5.6.61}$$

Using $\bar{d}$ in place of d in Eqs. (5.6.53) and (5.6.54) gives the mean-field exponents (5.6.52). We have, therefore, obtained a consistent scaling analysis using the Gaussian exponents, with the proviso that, for $d > 4$, account is taken of the dangerous irrelevant variable B_{40}, which leads to the effective dimension of the system being fixed at the value four.[28] Thus, for a simple critical point, the upper critical dimension (Sect. 3.1) $d_{\text{UC}} = 4$.

Although the existence of a singularity in the heat capacity at the critical point, superimposed upon a discontinuity, is dependent only on the dimension d, the ability to detect the singularity experimentally will depend on the range of temperature over which the singularity dominates the discontinuity. This in turn will depend on the relative size of the parameters B_{21}, B_{40} and C_0 and on the dimension n of the order parameter. The *width of the critical region* $\mathcal{W}_c$ can be defined by the condition that

$$\text{msp}\{c_{\mathcal{H}}\} \geq \Delta c_{\mathcal{H}} \qquad \text{if and only if} \qquad \left|1 - \frac{T}{T_c}\right| \leq \mathcal{W}_c, \tag{5.6.62}$$

[28] The discussion presented here is not quite complete. As pointed out in Sect. 4.8 it is necessary to consider other approaches to the critical point to determine whether the change of effective dimension can be consistently applied. This involves consideration of the case $\theta_1 \neq 0$, when the magnetization is given, as in Sect. 6.1, by a root of a cubic equation.

where 'msp' stands for 'most singular part'. For $T > T_c$ the most singular part of $c_{\mathcal{H}}$ is given by substituting from (5.6.49) into (5.6.48) and $\Delta c_{\mathcal{H}}$ is given by (5.3.37). This yields the result

$$\mathcal{W}_c = \frac{1}{K_c B_{21}} \left[\frac{n v_0 \Gamma\left(2 - \frac{1}{2}d\right) B_{40}}{2^{d-1}(C_0 \pi)^{\frac{d}{2}}} \right]^{2/(4-d)} . \tag{5.6.63}$$

Equation (5.6.63) is a form of the *Ginzburg criterion*, (Ginzburg 1960). The singularity in $c_{\mathcal{H}}$ is caused by fluctuations and the critical region is the range of temperature around the critical point within which fluctuations are important. The motivation for Ginzburg's work was to understand why a finite discontinuity was seen at some critical points, like those in superconductors, whereas in others, like the λ-point in liquid helium, a singularity was observed in addition to the discontinuity. On the basis of (5.6.63) it can now be seen that, when $\mathcal{W}_c$ is too small, the critical region cannot be resolved experimentally and the singularity will not be seen. Estimates of $\mathcal{W}_c$ for various systems are given by Ginzburg (1960). He concludes that, for superconductors $\mathcal{W}_c \sim 3 \times 10^{-16}$, whereas in helium $\mathcal{W}_c \sim 0.3$.

An analysis of tricritical behaviour using the Gaussian approximation to Ginzburg–Landau theory can be made on the lines of Sects. 5.6.1 and 5.6.3. This yields $\nu = \frac{1}{2}$ and $\eta = 0$, the same values as for a critical point. The Fisher scaling law (4.5.58) is therefore satisfied. The Josephson hyper-scaling law (4.5.54) is satisfied only when $d = d_{UC} = 3$, which is the upper critical dimension for a tricritical point. It may, however, be shown (Sarbach and Fisher 1978) that the Gaussian critical exponents (5.6.53) and (5.6.54) apply to the tricritical point with the crossover from the classical values governed by a parameter with exponent $3 - d$. This parameter is thus relevant for $d < 3$, leading to the crossover, and marginal for $d = 3$. For $d > 3$ it is dangerously irrelevant, leading to an effective change of dimension in the system to $\bar{d} = 3$.

Chapter 6
Mean-Field Theory

6.1 The Ising Ferromagnet

The zero-field Ising Hamiltonian $\widehat{H}_0(N, K, \mathcal{M}; \boldsymbol{\sigma})$ is just $\widehat{H}(K, 0; \boldsymbol{\sigma})$, given by (3.5.1), with the sum of the microstates constrained by the condition

$$\mathcal{M} = \sum_{\{\boldsymbol{r}\}} \sigma(\boldsymbol{r}). \tag{6.1.1}$$

This can be re-expressed in the form

$$\widehat{H}_0(N, K, \mathcal{M}; \boldsymbol{\sigma}) = -\tfrac{1}{2}K \sum_{\{\boldsymbol{r}\}} \sigma(\boldsymbol{r}) \sum_{\{\boldsymbol{r}'\}}^{(\boldsymbol{r})} \sigma(\boldsymbol{r}'), \tag{6.1.2}$$

where the inner summation over $\boldsymbol{r}'$ is restricted to the z first-neighbours of $\boldsymbol{r}$. In the canonical distribution ($\eta = 0$) considered here the magnetization density $m := \mathcal{M}/N$ is an independent variable with $\langle \sigma(\boldsymbol{r}) \rangle = m$ for every lattice site. If in (6.1.2) $\sigma(\boldsymbol{r}')$ is replaced by its expectation value m it follows from (6.1.1) that (6.1.2) becomes

$$\widehat{H}_0(N, K, \mathcal{M}) = -\frac{1}{2N} zK\, \mathcal{M}^2. \tag{6.1.3}$$

Equivalently this result can be achieved in two steps. First, the first-neighbour interaction terms $\sigma(\boldsymbol{r})\sigma(\boldsymbol{r}')$ in (6.1.2) are replaced by weighted expectation values $\langle \sigma(\boldsymbol{r})\sigma(\boldsymbol{r}') \rangle$ over all pairs of sites. Except at the critical point, spin states on widely separated sites are more-or-less independent. So taking

$$\langle \sigma(\boldsymbol{r})\sigma(\boldsymbol{r}') \rangle = \langle \sigma(\boldsymbol{r}) \rangle \langle \sigma(\boldsymbol{r}') \rangle = m^2 \tag{6.1.4}$$

D.A. Lavis, *Equilibrium Statistical Mechanics of Lattice Models*,
Theoretical and Mathematical Physics, DOI 10.1007/978-94-017-9430-5_6

is a reasonable approximation. Using this assumption we obtain (6.1.3). However, with $\boldsymbol{r}$ and $\boldsymbol{r}'$ close together (in this case first-neighbour sites), the situation is different. For the Ising ferromagnet the spins tend to align, with

$$\langle \sigma(\boldsymbol{r})\sigma(\boldsymbol{r}')\rangle > m^2 . \tag{6.1.5}$$

This represents *short-range order*. As we saw in Sect. 2.5.2,

$$\langle \sigma(\boldsymbol{r})\sigma(\boldsymbol{r}')\rangle - m^2 = \langle [\langle \sigma(\boldsymbol{r}) - \langle \sigma(\boldsymbol{r})\rangle][\langle \sigma(\boldsymbol{r}') - \langle \sigma(\boldsymbol{r}')\rangle]\rangle \tag{6.1.6}$$

is the (fluctuation) pair correlation function. So the assumption (6.1.4), which is applied to achieve the mean-field approximation, is equivalent to assuming that *all* spins are uncorrelated; not just those which are widely separated.

The number of terms in the the partition function $Z_0(N, K, \mathcal{M})$ is

$$W(N, \mathcal{M}) := \frac{N!}{\{\frac{1}{2}(N + \mathcal{M})\}!\,\{\frac{1}{2}(N - \mathcal{M})\}!}, \tag{6.1.7}$$

the number of spin arrangement consistent with (6.1.1). From (5.1.1)–(5.1.6), by applying Stirling's formula[1] to $W(N, \mathcal{M})$,

$$\begin{aligned}\Phi_1^{(\mathrm{CL})}(N, K, L; Nm) = {} & \tfrac{1}{4}[2N(1+m)+1]\ln(1+m)\\ & +\tfrac{1}{4}[2N(1-m)+1]\ln(1-m)\\ & -(N+1)\ln(2) - \tfrac{1}{2}\ln(2\pi N)\\ & -\tfrac{1}{2}NzKm^2 - NLm + \mathrm{O}(1/N)\end{aligned} \tag{6.1.8}$$

and taking the thermodynamic limit, $N \to \infty$, gives

$$\phi_1^{(\mathrm{CL})}(K, L; m) = \mathcal{G}(zK; m) - \ln(2) - Lm, \tag{6.1.9}$$

where

$$\mathcal{G}(X; m) := \tfrac{1}{2}[(1+m)\ln(1+m) + (1-m)\ln(1-m)] - \tfrac{1}{2}Xm^2. \tag{6.1.10}$$

This function and its derivatives, for which we introduce the notation

$$\mathcal{G}^{(i)}(X; m) := \frac{\partial^i \mathcal{G}}{\partial m^i}, \tag{6.1.11}$$

---

[1] For large n

$$\ln(n!) = \left(n + \tfrac{1}{2}\right)\ln(n) - n + \tfrac{1}{2}\ln(2\pi) + \mathrm{O}(1/n).$$

giving

$$\mathcal{G}^{(1)}(X; m) = \tfrac{1}{2} \ln \left[\frac{1+m}{1-m} \right] - Xm, \tag{6.1.12}$$

$$\mathcal{G}^{(2)}(X; m) = \frac{1}{1-m^2} - X, \tag{6.1.13}$$

appears in most mean-field calculations. From (5.1.7) and (5.1.8) the equilibrium and stability conditions are

$$\frac{\partial \phi_1^{(\mathrm{CL})}}{\partial m} = \mathcal{G}^{(1)}(zK; m) - L = 0, \tag{6.1.14}$$

$$\frac{\partial^2 \phi_1^{(\mathrm{CL})}}{\partial m^2} = \mathcal{G}^{(2)}(zK; m) > 0. \tag{6.1.15}$$

Substituting from (6.1.14) back into (6.1.9) shows that, at equilibrium,

$$\phi_1^{(\mathrm{CL})}(K, L; m) = \tfrac{1}{2} \ln(1 - m^2) + \tfrac{1}{2} Kzm^2 - \ln(2). \tag{6.1.16}$$

Curves derived from (6.1.14) are shown in Fig. 6.1. For $K < z^{-1}$ Eq. (6.1.14) has a single real root $m = m(K, L)$ (the broken line) where, $m(K, 0) = 0$. The coupling $K_c := z^{-1}$ is equivalent to $T_c := Jz$, which is the mean-field expression for the critical or *Curie temperature* in the ferromagnetic Ising model. For $K > K_c$ the equation exhibits the characteristic looping effect found in all classical approximations (Fig. 5.1a). Because of the magnetic symmetry it follows from Maxwell's equal areas rule, described in Sect. 5.1.1, that the first-order transition is along the line $L = 0$, where the minima in $\phi_1^{(\mathrm{CL})}(K, 0; m) = \phi_0^{(\mathrm{CL})}(K, m)$ have equal depth

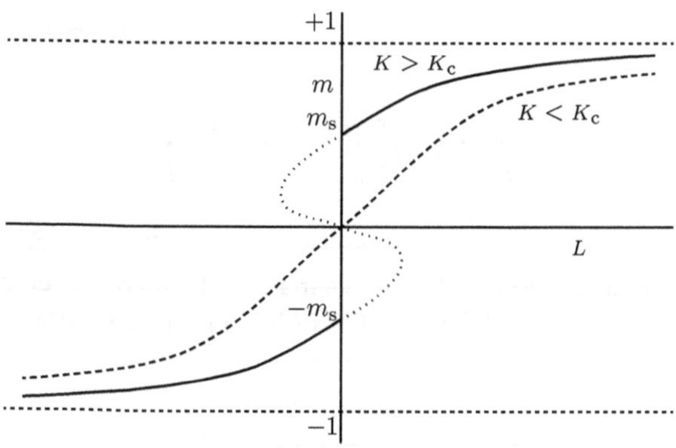

Fig. 6.1 Isotherms in the L–m plane derived from (6.1.14)

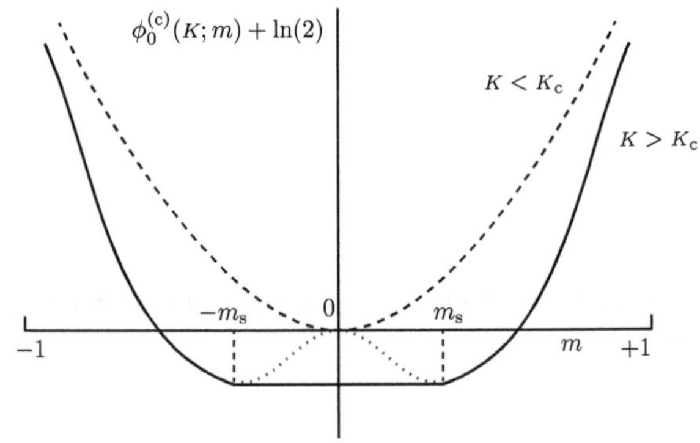

Fig. 6.2 Free energy–magnetization isotherms at zero field

(Fig. 6.2). The unbroken part of the isotherm $K > K_c$ in Fig. 6.1 corresponds to the solution to (6.1.14) which is the lower minimum of the free energy. It follows from (6.1.14) and (6.1.15) that the portion of the broken part of the curve $K > K_c$ with $\partial L/\partial m < 0$ represents unstable solutions and that with $\partial L/\partial m > 0$ metastable solutions (Sect. 5.1.2).

Along the critical isotherm $K = K_c$, from (6.1.14),

$$L \simeq \tfrac{2}{3}m^3 + O(m^5). \tag{6.1.17}$$

The implicit form for the magnetization, derived from (6.1.14), with

$$\lim_{L\to\pm 0} m(K, L) := \pm m_s(K)\,, \tag{6.1.18}$$

has a discontinuity of $2m_s$ along the zero-field axis for $K > K_c$ with

$$m_s^2(K) = 3\left(\frac{K-K_c}{K_c}\right) - \frac{3}{5}\left(\frac{K-K_c}{K_c}\right)^2 + O\left[\left(\frac{K-K_c}{K_c}\right)^3\right], \quad \text{as } K \to K_c+ \tag{6.1.19}$$

and

$$m_s(K) = 1 - 2\exp\left(-\frac{2K}{K_c}\right) + O\left[\frac{2K}{K_c}\exp\left(-\frac{4K}{K_c}\right)\right], \quad \text{as } K \to \infty. \tag{6.1.20}$$

The curve of m_s as a function of K is shown in Fig. 3.1, where it is compared with the corresponding results derived from higher-order cluster expansions. From (2.3.31) and (6.1.14),

$$\varphi_T = \frac{\partial m}{\partial L} = \frac{1}{\mathcal{G}^{(2)}(zK; m)} = \frac{K_c}{K_c(1-m^2)^{-1} - K}, \tag{6.1.21}$$

giving, from (6.1.19),

$$\varphi_T = \begin{cases} \dfrac{1}{\mathcal{G}^{(2)}(zK;0)} = \dfrac{K_c}{K_c - K}, \\ \qquad\qquad \text{on the zero-field axis, for } K < K_c, \\ \dfrac{1}{\mathcal{G}^{(2)}(zK;m_s)} = \dfrac{K_c}{2(K-K_c)}\left\{1 - \dfrac{21}{5}\left(\dfrac{K-K_c}{K_c}\right) + \mathrm{O}\left[\left(\dfrac{K-K_c}{K_c}\right)^2\right]\right\}, \\ \qquad\qquad \text{on the coexistence curve, for } K > K_c. \end{cases} \tag{6.1.22}$$

From (2.3.30), (6.1.9), (6.1.14) and (6.1.21)

$$c_{\mathcal{H}} = \frac{k_B(Km + LK_c)^2}{K_c\mathcal{G}^{(2)}(zK;m)}, \tag{6.1.23}$$

giving, from (6.1.19) and (6.1.22),

$$c_{\mathcal{H}} = \begin{cases} 0, \qquad \text{on the zero-field axis, for } K < K_c, \\ \dfrac{3}{2}k_B K_c\left\{1 - \dfrac{21}{5}\left(\dfrac{K-K_c}{K_c}\right) + \mathrm{O}\left[\left(\dfrac{K-K_c}{K_c}\right)^2\right]\right\}, \\ \qquad \text{on the coexistence curve, for } K > K_c. \end{cases} \tag{6.1.24}$$

Comparing (6.1.17), (6.1.19), (6.1.22) and (6.1.24) with (4.4.12)–(4.4.16) (and remembering that in this case because of the magnetic symmetry $\sigma = \alpha$) we recover, as we might expect, the classical values of the critical exponents given by (5.3.30).

We also note that (6.1.14) can be re-expressed in the form

$$m = \tanh(L + Kzm). \tag{6.1.25}$$

With $J = 0$ (giving $K = 0$) this formula gives the magnetization of a paramagnet in a field $\mathcal{H} = LT$. It can be obtained exactly by summing the partition function with the Hamiltonian (3.5.1) with $J = 0$. For the ferromagnet considered here (6.1.25) is the *Weiss field equation*. It can be understood as the consistency equation resulting from an approximation in which the interaction between a spin and its neighbours is replaced by a molecular field Jzm.

6.1.1 Mean-Field Fluctuations

As was indicated in the discussion in Sect. 4.2.2, the onset of a phase transition is related to the growth in the size of regions of local order. These long-range correlations lead to divergences in some or all of the response functions. In the case of a ferromagnet the susceptibility divergence is thus related to the divergence in the size of the regions of parallel spins. Although the overall numbers of up and down spins remain equal there are large regions of local order or 'patches' of parallel spins. As $K \to K_{\mathrm{c}}$ the patches become infinitely large which means that the correlation length, defined in Sect. 4.2.2, is approaching infinity. In mean-field theory, only overall fluctuations, which change the magnetization density m from its equilibrium value, are recognized, although (6.1.22) shows that these are still sufficient to produce a susceptibility divergence as $K \to K_{\mathrm{c}} \pm 0$.[2]

From (5.1.5) the mean-field probability distribution for $\hat{m}$, along the zero-field axis, is given by

$$p(K;\hat{m}) = \frac{\exp[-N\phi_0^{(\mathrm{CL})}(K,\hat{m})]}{\sum\limits_{\{\hat{m}\}}\exp[-N\phi_0^{(\mathrm{CL})}(K,\hat{m})]}. \tag{6.1.26}$$

Expanding (6.1.9) about $\hat{m} = 0$ gives

$$\begin{aligned}\phi_0^{(\mathrm{CL})}(K,\hat{m}) &= \phi_1^{(\mathrm{CL})}(K,0;\hat{m}) \\ &= -\ln(2) + \left(\frac{K_{\mathrm{c}}-K}{2K_{\mathrm{c}}}\right)\hat{m}^2 + \frac{1}{12}\hat{m}^4 + \mathrm{O}(\hat{m}^6).\end{aligned} \tag{6.1.27}$$

If terms of degree greater than four are neglected, the turning point at $\hat{m} = 0$ and the leading term of (6.1.19) are still correctly reproduced by (6.1.27) when $(K - K_{\mathrm{c}})$ is small and positive. Using (6.1.26) and (6.1.27) the mean-field probability distribution is shown in Fig. 6.3. For $K < K_{\mathrm{c}}$ the distribution shows one peak at $\hat{m} = 0$ and, for $K > K_{\mathrm{c}}$, two peaks at

$$\hat{m} = m^{(\pm)} = \pm\sqrt{\frac{3(K-K_{\mathrm{c}})}{K_{\mathrm{c}}}}, \tag{6.1.28}$$

which corresponds to the leading terms in $\pm m_{\mathrm{s}}(K)$.

For $K < K_{\mathrm{c}}$ the distribution is concentrated around $\hat{m} = 0$ and the quartic term in (6.1.27) can be neglected. Since the distribution is symmetric $\langle\hat{m}\rangle = 0$. To determine the variance we make the substitution

$$x = \hat{m}\sqrt{\frac{N(K_{\mathrm{c}}-K)}{2K_{\mathrm{c}}}} \tag{6.1.29}$$

and, for large N, the sum in the denominator of (6.1.26) can be replaced by an integral over the range $(-\infty, \infty)$. Thus, from (6.1.22),

[2] In an accurate treatment, which must include the effect of local fluctuations, the susceptibility divergence is sharper.

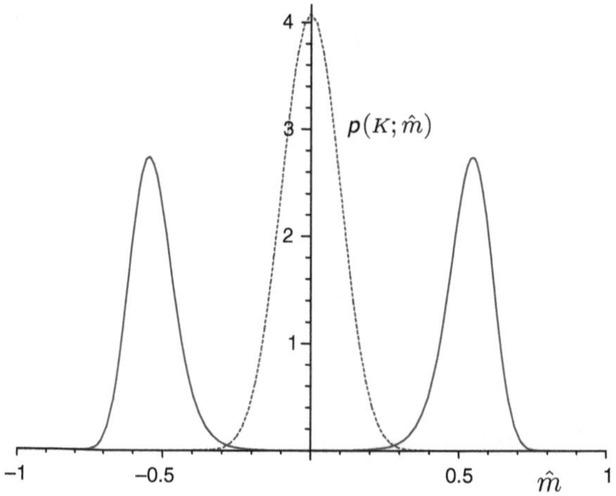

Fig. 6.3 Plots of $p(K;\hat{m})$ with $N = 1000$, $K = 1.1K_c$ (*continuous curve*) and $K = 0.9K_c$ (*broken curve*)

$$\mathrm{Var}[\hat{m}] = \frac{2K_c}{N(K_c - K)} \frac{\displaystyle\int_{-\infty}^{\infty} x^2 \exp(-x^2)\mathrm{d}x}{\displaystyle\int_{-\infty}^{\infty} \exp(-x^2)\mathrm{d}x} = \frac{K_c}{N(K_c - K)} = \frac{1}{N}\varphi_T. \quad (6.1.30)$$

For $K > K_c$ the quartic term must be retained and, from (6.1.27) and (6.1.28),

$$\phi_0^{(\mathrm{CL})}(K,\hat{m}) \simeq -\ln(2) + \frac{(K - K_c)}{K_c}[\hat{m} - m^{(\pm)}]^2, \qquad \hat{m} \gtrless 0. \quad (6.1.31)$$

The method of evaluation of the mean and variance is now similar to that for $K < K_c$ except that the range of the integral is divided into two parts

$$\begin{aligned} &(-\infty, 0) \quad \text{for which the variable is} \quad x_1 = [\hat{m} - m^{(-)}]\sqrt{\frac{N(K-K_c)}{K_c}}, \\ &(0, \infty) \quad \text{for which the variable is} \quad x_2 = [\hat{m} - m^{(+)}]\sqrt{\frac{N(K-K_c)}{K_c}}. \end{aligned} \quad (6.1.32)$$

Each integral is then extended over the whole range $(-\infty, \infty)$ and it is straightforward to show that

$$\langle\hat{m}\rangle = m^{(+)} + m^{(-)} = 0. \quad (6.1.33)$$

Just as in the case $K < K_c$ the mean magnetization is zero, but here the reason is because the two peaks of the probability distribution make equal and opposite contributions. As we have seen, the zero-field line represents a coexistence region between phases with equal and opposite magnetizations. In a similar way

$$\mathrm{Var}[\hat{m}] \simeq \frac{K_c}{N(K-K_c)} \frac{\displaystyle\int_{-\infty}^{\infty} y_1^2 \exp(-x_1^2)\mathrm{d}x_1 + \int_{-\infty}^{\infty} x_2^2 \exp(-x_2^2)\mathrm{d}x_2}{\displaystyle\int_{-\infty}^{\infty} \exp(-x_1^2)\mathrm{d}x_1 + \int_{-\infty}^{\infty} \exp(-x_2^2)\mathrm{d}x_2}$$

$$\simeq \frac{K_c}{2N(K-K_c)} \simeq \frac{1}{N}\varphi_T. \tag{6.1.34}$$

Equations (6.1.30) and (6.1.34) are examples of fluctuation-response function relations (Sect. 2.5.2).

6.2 A Model for Metamagnetism

Metamagnetism arises in situations where a lattice of magnetic dipoles is divides into a layered sublattice structure with ferromagnetic interactions within layers and antiferromagnetic interactions between layers (Martin 1967, p. 58). Such a model can be developed as a generalization of the Ising ferromagnet on a d-dimensional hypercubic lattice. Suppose that one particular axis direction is selected and the $(d-1)$-dimensional hyperplanes orthogonal to this axis are labelled in sequence $\mathfrak{a}, \mathfrak{b}, \mathfrak{a}, \mathfrak{b}, \ldots$ to form a layered sublattice structure with two equivalent sublattices $\mathfrak{a}$ and $\mathfrak{b}$. The Hamiltonian for such a model is given by

$$\widehat{H}(N, J_1, J_2, \mathcal{H}, \mathcal{H}_s; \boldsymbol{\sigma}) = J_1 \sum_{\{\boldsymbol{r},\boldsymbol{r}'\}}^{(\mathfrak{a},\mathfrak{b})} \sigma(\boldsymbol{r})\sigma(\boldsymbol{r}') - J_2 \sum_{\{\boldsymbol{r},\boldsymbol{r}'\}}^{(\mathfrak{a},\mathfrak{a})} \sigma(\boldsymbol{r})\sigma(\boldsymbol{r}')$$

$$-J_2 \sum_{\{\boldsymbol{r},\boldsymbol{r}'\}}^{(\mathfrak{b},\mathfrak{b})} \sigma(\boldsymbol{r})\sigma(\boldsymbol{r}') - (\mathcal{H}+\mathcal{H}_s) \sum_{\{\boldsymbol{r}\}}^{(\mathfrak{a})} \sigma(\boldsymbol{r}) - (\mathcal{H}-\mathcal{H}_s) \sum_{\{\boldsymbol{r}\}}^{(\mathfrak{b})} \sigma(\boldsymbol{r}), \tag{6.2.1}$$

where the first-neighbour pairs and site summations are restricted as indicated by the superscripts, with $J_1, J_2 > 0$. The field $\mathcal{H}$ is uniform over the lattice, but $\mathcal{H}_s$ is a staggered field with a different sign on the two sublattices. The system can be thought of as having four independent fields $T, \mathcal{H}, \mathcal{H}_s$ and one of J_1 or J_2. Of these only T and $\mathcal{H}$ are the theoretical counterparts of experimentally controllable quantities. Here we concentrate mainly on the case where the staggered field is zero,[3] although it will be included in the initial stages of the discussion. The number of first-neighbours to a site which are situated in the other sublattice are denoted by z_1 and in the same sublattice by z_2.[4]

The mean-field free-energy density is now derived using a generalization of the procedure of Sect. 6.1. The Hamiltonian $\widehat{H}_0(N, \kappa_1, \kappa_2, \mathcal{M}_\mathfrak{a}, \mathcal{M}_\mathfrak{b}; \boldsymbol{\sigma})$ for the

---

[3] We are interested is in deriving the part of the phase diagram in the plane $\mathcal{H}_s = 0$ shown in Fig. 4.3a, b.

[4] Of course, with this layered model on a d-dimensional hypercubic lattice $z_1 = 2$, $z_2 = 2(d-1)$, but z_1 and z_2 are used in the interests of generality.

metamagnet is just $\widehat{H}(N, K_1, K_2, 0, 0; \boldsymbol{\sigma})$, given by (6.2.1), with the sum of the microstates on the two sublattices constrained by the conditions

$$\mathcal{M}_{\mathfrak{a}} = \sum_{\{r\}}^{(\mathfrak{a})} \sigma(\boldsymbol{r}), \qquad \mathcal{M}_{\mathfrak{b}} = \sum_{\{r\}}^{(\mathfrak{b})} \sigma(\boldsymbol{r}), \tag{6.2.2}$$

and $K_1 := J_1/T$, $K_2 := J_2/T$. Approximating the interactions in the Hamiltonian in the manner exemplified by (6.1.2) and (6.1.3) gives

$$\widehat{H}_0(N, K_1, K_2, \mathcal{M}_{\mathfrak{a}}, \mathcal{M}_{\mathfrak{b}}) = \frac{1}{N}\{2K_1 z_1 \mathcal{M}_{\mathfrak{a}}\mathcal{M}_{\mathfrak{b}} - K_2 z_2(\mathcal{M}_{\mathfrak{a}}^2 + \mathcal{M}_{\mathfrak{b}}^2)\}. \tag{6.2.3}$$

The number of terms in the partition function $Z_0(N, K_1, K_2, \mathcal{M}_{\mathfrak{a}}, \mathcal{M}_{\mathfrak{b}})$ is the number

$$\begin{aligned} W(N, \mathcal{M}_{\mathfrak{a}}, \mathcal{M}_{\mathfrak{b}}) := & \frac{(\frac{1}{2}N)!}{\{\frac{1}{2}(\frac{1}{2}N + \mathcal{M}_{\mathfrak{a}})\}!\,\{\frac{1}{2}(\frac{1}{2}N - \mathcal{M}_{\mathfrak{a}})\}!} \\ & \times \frac{(\frac{1}{2}N)!}{\{\frac{1}{2}(\frac{1}{2}N + \mathcal{M}_{\mathfrak{b}})\}!\,\{\frac{1}{2}(\frac{1}{2}N - \mathcal{M}_{\mathfrak{b}})\}!} \end{aligned} \tag{6.2.4}$$

of spin arrangements consistent with (6.2.2). Applying Stirling's formula and taking the thermodynamic limit, as we did in Sect. 6.1, gives

$$\begin{aligned} \phi_2^{(\mathrm{CL})}(K, L, L_{\mathrm{s}}; m_{\mathfrak{a}}, m_{\mathfrak{b}}) = & \tfrac{1}{2}\left[\mathcal{G}(K(1+\lambda); m_{\mathfrak{a}}) + \mathcal{G}(K(1+\lambda); m_{\mathfrak{b}})\right] \\ & + \tfrac{1}{4}K(m_{\mathfrak{a}} + m_{\mathfrak{b}})^2 - \tfrac{1}{2}L(m_{\mathfrak{a}} + m_{\mathfrak{b}}) - \tfrac{1}{2}L_{\mathrm{s}}(m_{\mathfrak{a}} - m_{\mathfrak{b}}), \end{aligned} \tag{6.2.5}$$

where $m_{\mathfrak{a}} := 2\mathcal{M}_{\mathfrak{a}}/N$, $m_{\mathfrak{b}} := 2\mathcal{M}_{\mathfrak{b}}/N$. The function $\mathcal{G}$ and its first two derivatives are given by (6.1.10)–(6.1.13) and, for convenience, we now take $K := z_1 K_1 = z_1 J_1/T$ and $\lambda := z_2 K_2/z_1 K_1 = z_2 J_2/z_1 J_1 > 0$, which measures the relative strengths of the ferromagnetic and antiferromagnetic interaction energies.

The Hamiltonian (3.5.17), for the Ising model with first-neighbour antiferromagnetic interactions and second-neighbour ferromagnetic interactions, and the Hamiltonian (6.2.1), for our present metamagnetic model, will be seen to be of identical form when $\mathcal{H}_{\mathrm{s}} = 0$. It follows that the mean-field free-energy densities for these two models will be identical with, in the first case, the parameters J_1, J_2 and the corresponding z_1 and z_2 being reinterpreted as applying to first-neighbour and second-neighbour pairs.[5] As indicated above, for the metamagnetic model λ is positive. However, as we shall see below, the analysis is applicable to the case $\lambda = 0$, which is the first-neighbour antiferromagnetic model. Thus we shall henceforth assume that $\lambda \geq 0$.

It is convenient to reformulate the free-energy density in terms of the magnetization m and sublattice order parameter q, given respectively by

---

[5] For the simple cubic lattice $z_1 = 6$, $z_2 = 12$.

$$m := \tfrac{1}{2}(m_{\mathfrak{a}} + m_{\mathfrak{b}}), \tag{6.2.6}$$

$$q := \tfrac{1}{2}(m_{\mathfrak{a}} - m_{\mathfrak{b}}), \tag{6.2.7}$$

when

$$\phi_2^{(\mathrm{CL})}(K, L, L_{\mathrm{s}}; m, q) = \tfrac{1}{2}\left[\mathcal{G}(K(1+\lambda); m+q) + \mathcal{G}(K(1+\lambda); m-q)\right] + Km^2 - Lm - L_{\mathrm{s}}q\,. \tag{6.2.8}$$

The generalizations of the equilibrium and stability conditions (5.1.7) and (5.1.8) are now

$$2\frac{\partial \phi_2^{(\mathrm{CL})}}{\partial m} = \mathcal{G}^{(1)}(K(1+\lambda); m+q) + \mathcal{G}^{(1)}(K(1+\lambda); m-q) + 4Km - 2L = 0, \tag{6.2.9}$$

$$2\frac{\partial \phi_2^{(\mathrm{CL})}}{\partial q} = \mathcal{G}^{(1)}(K(1+\lambda); m+q) - \mathcal{G}^{(1)}(K(1+\lambda); m-q) - 2L_{\mathrm{s}} = 0, \tag{6.2.10}$$

and

$$2\frac{\partial^2 \phi_2^{(\mathrm{CL})}}{\partial m^2} = \mathcal{G}^{(2)}(K(1+\lambda); m+q) + \mathcal{G}^{(2)}(K(1+\lambda); m-q) + 4K > 0, \tag{6.2.11}$$

$$2\frac{\partial^2 \phi_2^{(\mathrm{CL})}}{\partial q^2} = \mathcal{G}^{(2)}(K(1+\lambda); m+q) + \mathcal{G}^{(2)}(K(1+\lambda); m-q) > 0, \tag{6.2.12}$$

$$\begin{aligned}\Delta &:= \frac{\partial^2 \phi_2^{(\mathrm{CL})}}{\partial m^2}\frac{\partial^2 \phi_2^{(\mathrm{CL})}}{\partial q^2} - \left(\frac{\partial^2 \phi_2^{(\mathrm{CL})}}{\partial m \partial q}\right)^2 \\ &= \mathcal{G}^{(2)}(K(1+\lambda); m+q)\mathcal{G}^{(2)}(K(1+\lambda); m-q) \\ &\quad + K[\mathcal{G}^{(2)}(K(1+\lambda); m+q) + \mathcal{G}^{(2)}(K(1+\lambda); m-q)] > 0\,,\end{aligned} \tag{6.2.13}$$

where the first and second derivatives of the function $\mathcal{G}$ are given by (6.1.12) and (6.1.13). Substituting conditions (6.2.9) into (6.2.8), it follows that, at equilibrium,

$$\phi_2^{(\mathrm{CL})}(K, L, L_{\mathrm{s}}; m, q) = \tfrac{1}{4}\ln[(1-m^2-q^2)^2 - 4m^2q^2] - \ln(2) - \tfrac{1}{4}K[(1-\lambda)m^2 - (1+\lambda)q^2], \tag{6.2.14}$$

and from (4.6.2), (4.6.3), (6.2.9) and (6.2.13),

$$\varphi_T = \left(\frac{\partial m}{\partial L}\right)_{K,L_s} = \frac{\mathcal{G}^{(2)}(K(1+\lambda); m+q) + \mathcal{G}^{(2)}(K(1+\lambda); m-q)}{2\Delta}, \tag{6.2.15}$$

$$\psi_T = \left(\frac{\partial q}{\partial L_s}\right)_{K,L} = \frac{\mathcal{G}^{(2)}(K(1+\lambda); m+q) + \mathcal{G}^{(2)}(K(1+\lambda); m-q) + 4K}{2\Delta}. \tag{6.2.16}$$

We shall now investigate the phase diagram in the plane $\mathcal{H}_s = L_s = 0$.

6.2.1 The Paramagnetic State

The paramagnetic state in the plane $L_s = 0$, where $q = 0$, $m_\mathfrak{a} = m_\mathfrak{b} = m$, is stable for large temperatures (K small) or fields (L large). Condition (6.2.10) is identically satisfied and, from (6.2.9), the magnetization, as a function of K and L, is given by

$$\mathcal{G}^{(1)}(K(1+\lambda); m) + 2Km = L. \tag{6.2.17}$$

From (6.2.11)–(6.2.13), this state is locally stable when

$$\mathcal{G}^{(2)}(K(1+\lambda); m) > 0. \tag{6.2.18}$$

As K increases (T decreases), with $m(K)$ given by (6.2.17), $\mathcal{G}^{(2)}(K(1+\lambda); m)$ decreases monotonically (Fig. 6.4) and the paramagnetic state will become unstable with respect to sublattice ordering when

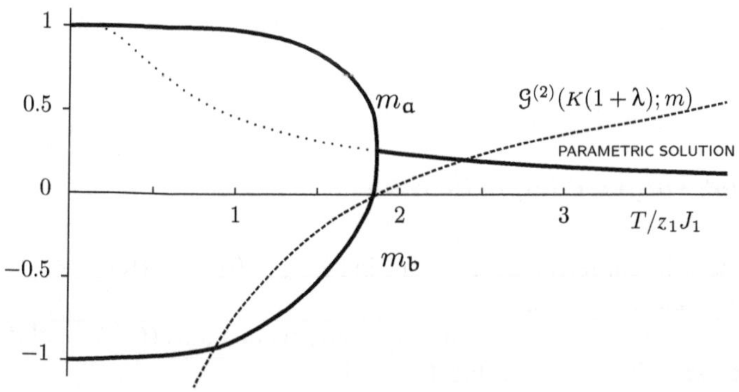

Fig. 6.4 Magnetization plotted against temperature for the metamagnetic for $\lambda = 1, \mathcal{H} = 0.5z_1J_1$. The point on the critical curve separating the paramagnetic and antiferromagnetic solutions is at $T = 1.8625z_1J_1$ where $m = 0.2622$. Below the critical curve the unstable paramagetic solution is denoted by a *dotted line*

$$\mathcal{G}^{(2)}(K(1+\lambda); m) = 0\,. \tag{6.2.19}$$

The paramagnetic solution as a function of temperature along an isochamp (line of constant field) is shown in Fig. 6.4[6] with the unstable part of the solution shown by a dotted line. By eliminating m between (6.2.19) and (6.2.17) we can obtain the equation $\mathcal{H} = \mathcal{H}^*(T)$ of the *critical curve* $\mathcal{C}$ of points, where the paramagnetic state becomes unstable, in the $\mathcal{H}$–T plane or equivalently its form $L = L^*(K)$ in the L–K plane. It follows from (6.1.13) and (6.2.19) that the magnetizations on the two branches of the critical curve are given by

$$m = \pm m^*(\lambda, K), \qquad m^*(\lambda, K) := \left(\frac{K(1+\lambda)-1}{K(1+\lambda)}\right)^{\frac{1}{2}} \tag{6.2.20}$$

and, from (6.2.17),

$$\mathcal{G}^{(1)}(K(1+\lambda); m^*(\lambda, K)) + 2Km^*(\lambda, K) = L^*(K) \tag{6.2.21}$$

gives the critical curve. From (6.2.17), $m = 0$ on the zero-field axis in the paramagnetic state and thus, from (6.2.19)–(6.2.21), the critical curve cuts the zero-field axis orthogonally at the critical (Néel) temperature given by

$$K_c = \frac{1}{1+\lambda}\,, \qquad T_c = z_1 J_1(1+\lambda) = z_1 J_1 + z_2 J_2\,. \tag{6.2.22}$$

This is the point where the paramagnetic state becomes unstable when temperature is reduced at zero field.

From (6.1.12), (6.2.20) and (6.2.21) it can be shown that on the critical curve, as $T \to 0$, $\mathcal{H}^*(T) \to \pm z_1 J_1(1-\lambda)$, which are not equal to the critical field values $\mathcal{H}_c = \pm z_1 J_1$, given by (3.5.19), unless $\lambda = 0$. The implications of this are discussed below.

6.2.2 The Antiferromagnetic State

Consider now the antiferromagnetic state in the $L_s = 0$ plane. Here there is sublattice order, $q \neq 0$ and $m_{\mathfrak{a}} \neq m_{\mathfrak{b}}$.

On the zero-field axis ($L = 0$) the equilibrium equations (6.2.9) and (6.2.10) are satisfied by $m = 0$, $q = \pm m_s$, (Fig. 6.7a) with

[6] Here we encounter the usual problem in statistical mechanics that, whereas the equations take a simpler form in terms of couplings K and L, physical insight is helped by presenting figures in terms of the fields T and $\mathcal{H}$. We shall, therefore, in the discussion make free use of both sets of variables.

$$\mathcal{G}^{(1)}(K(1+\lambda); m_\mathrm{s}) = \tfrac{1}{2}\ln\left(\frac{1+m_\mathrm{s}}{1-m_\mathrm{s}}\right) - K(1+\lambda)m_\mathrm{s} = 0. \tag{6.2.23}$$

With the replacement of Jz by $z_1 J_1 + z_2 J_2$, this equation is identical to (6.1.14)[7] for the spontaneous magnetization in zero field of the simple ferromagnetic Ising model. It is not difficult to show that this replacement is exactly that which occurs when second-neighbour interactions are added to the ferromagnetic model of Sect. 6.1. This together with the correspondence, discussed in Sect. 3.5.1.1, between that model and the one with antiferromagnetic first-neighbour interactions and ferromagnetic second-neighbour interactions and the further correspondence, noted above, between the latter and the metamagnetic model, allows us to understand why the Curie temperature for the ferromagnet is equal to the Néel temperature given by (6.2.22).

To investigate the stability of the antiferromagnetic state it is convenient to use the response function φ_T given by (6.2.15). Since the numerator of φ_T is positive it can be seen, by comparing (6.2.15) with (6.2.11)–(6.2.13), that the homogeneous antiferromagnetic phase is stable if and only if

$$\varphi_T = \left(\frac{\partial m}{\partial L}\right)_K > 0. \tag{6.2.24}$$

6.2.3 A Neighbourhood of the Critical Curve

Here we calculate some but, for the sake of brevity, not all of the critical exponents at a point P with coordinates $(K_\mathrm{p}, L_\mathrm{p})$ and $m_\mathrm{p} = m^*(\lambda, K_\mathrm{p})$ on the critical curve $\mathcal{C}$, defined by (6.2.20) and (6.2.21), and at the tricritical point T. Let

$$\Delta_\mathrm{p}K := K - K_\mathrm{p}, \qquad \Delta_\mathrm{p}L := L - L_\mathrm{p}, \qquad \Delta_\mathrm{p}m := m - m_\mathrm{p}. \tag{6.2.25}$$

Any second-order transitions, that is ones for which m and q are continuous, from the paramagnetic to the antiferromagnetic state, occur on this curve as it is approached along any line in the plane $L_\mathrm{s} = 0$ which is not tangential to the curve. In terms of the orthogonal coordinates defined by (5.5.18)

$$a_\mathrm{p} = \left(\frac{\mathrm{d}L^*}{\mathrm{d}K}\right)_\mathrm{p} = (1-\lambda)m^*(\lambda, K_\mathrm{p}). \tag{6.2.26}$$

So, in the asymptotic forms (4.6.16)–(4.6.22), with 't' replaced by 'p',[8] approaches along $\Theta_3 = \Theta_{3\mathrm{t}}$ can be replaced by either an approach along the isotherm $K = K_\mathrm{p}$ with $\Theta_2 - \Theta_{2\mathrm{p}}$ replaced by $\Delta_\mathrm{p}L$ or along the line $L = L_\mathrm{p}$ with $\Theta_2 - \Theta_{2\mathrm{p}}$ replaced by $\Delta_\mathrm{p}K$.

---

[7] With $\mathcal{H} = 0$, and remembering that in Sect. 6.1 $K := J/T$ not zJ/T.

[8] As indicated in Sect. 5.5, for the exponents we simply omit the 'p'.

Taking the plane $K = K_\mathrm{p}$ and expanding (6.2.9) and (6.2.10) about P gives

$$\triangle_\mathrm{p}L = W_1(K_\mathrm{p})(\triangle_\mathrm{p}m) + W_2(\lambda, K_\mathrm{p})[(\triangle_\mathrm{p}m)^2 + q^2] + W_3(\lambda, K_\mathrm{p})(\triangle_\mathrm{p}m)\,q^2 + W_4(\lambda, K_\mathrm{p})\,q^4 + \mathrm{O}[(\triangle_\mathrm{p}m)^3, q^6], \tag{6.2.27}$$

$$L_\mathrm{s} = 2W_2(\lambda, K_\mathrm{p})(\triangle_\mathrm{p}m)\,q + \tfrac{1}{3}W_3(\lambda, K_\mathrm{p})[3(\triangle_\mathrm{p}m)^2\,q + q^3] + 4W_4(\lambda, K_\mathrm{p})(\triangle_\mathrm{p}m)\,q^3 + W_5(\lambda, K_\mathrm{p})q^5 + \mathrm{O}[(\triangle_\mathrm{p}m)^3, q^6], \tag{6.2.28}$$

where

$$\begin{aligned}
W_1(K) &:= 2K,\\
W_2(\lambda, K) &:= \tfrac{1}{2}\mathcal{G}^{(3)}(K(1+\lambda); m^*(\lambda, K)) = m^*(\lambda, K)K^2(1+\lambda)^2,\\
W_3(\lambda, K) &:= \tfrac{1}{2}\mathcal{G}^{(4)}(K(1+\lambda); m^*(\lambda, K)) = K^2(1+\lambda)^2[4K(1+\lambda) - 3],\\
W_4(\lambda, K) &:= \tfrac{1}{24}\mathcal{G}^{(5)}(K(1+\lambda); m^*(\lambda, K))\\
&= m^*(\lambda, K)K^3(1+\lambda)^3[2K(1+\lambda) - 1],\\
W_5(\lambda, K) &:= \tfrac{1}{120}\mathcal{G}^{(6)}(K(1+\lambda); m^*(\lambda, K))\\
&= \tfrac{1}{5}K^3(1+\lambda)^3[16K^2(1+\lambda)^2 - 20K(1+\lambda) + 5].
\end{aligned} \tag{6.2.29}$$

In the paramagnetic phase ($q = 0$) along the line $L_\mathrm{s} = 0$, (6.2.10) and (6.2.28) are identically satisfied and, from (6.2.27) with $\triangle_\mathrm{p}m < 0$,

$$\triangle_\mathrm{p}L = W_1(K_\mathrm{p})\triangle_\mathrm{p}m + W_2(\lambda, K_\mathrm{p})(\triangle_\mathrm{p}m)^2 + \mathrm{O}[(\triangle_\mathrm{p}m)^3]\,. \tag{6.2.30}$$

In the antiferromagnetic phase ($q \neq 0$) along the line $L_\mathrm{s} = 0$, (6.2.28) gives

$$q^2 = X_1(\lambda, K_\mathrm{p})\triangle_\mathrm{p}m + X_2(\lambda, K_\mathrm{p})(\triangle_\mathrm{p}m)^2 + \mathrm{O}[(\triangle_\mathrm{p}m)^3]\,, \tag{6.2.31}$$

where

$$\begin{aligned}
X_1(\lambda, K) &:= -6\frac{W_2(\lambda, K)}{W_3(\lambda, K)},\\
X_2(\lambda, K) &:= -3 + 72\frac{W_2(\lambda, K)W_4(K)}{[W_3(\lambda, K)]^2} - 108\frac{[W_2(\lambda, K)]^2 W_5(\lambda, K)}{[W_3(\lambda, K)]^3}.
\end{aligned} \tag{6.2.32}$$

Then, from (6.2.27) with $\triangle_\mathrm{p}m > 0$,

$$\triangle_\mathrm{p}L = W_1'(\lambda, K_\mathrm{p})\triangle_\mathrm{p}m + W_2'(\lambda, K_\mathrm{p})(\triangle_\mathrm{p}m)^2 + \mathrm{O}[(\triangle_\mathrm{p}m)^3]\,, \tag{6.2.33}$$

where

$$\begin{aligned} W_1'(\lambda, K) &:= W_1(K) + X_1(\lambda, K) W_2(\lambda, K) \\ &= \frac{6\lambda K K_c (K_t - K)}{K_t (4K - 3K_c)}, \\ W_2'(\lambda, K) &:= [1 + X_2(\lambda, K)] W_2(\lambda, K) \\ &\qquad + X_1(\lambda, K) W_3(\lambda, K) + [X_1(\lambda, K)]^2 W_4(\lambda, K), \end{aligned} \tag{6.2.34}$$

K_c is given by (6.2.22),

$$K_t = \frac{3\lambda}{(1+\lambda)(3\lambda - 1)}, \qquad T_t = \frac{(3\lambda - 1)}{3\lambda} T_c, \tag{6.2.35}$$

and

$$W_2'(\lambda, K_t) = \frac{24\lambda^3 (3\lambda)^{\frac{1}{2}} (3 - 5\lambda)}{5(1+\lambda)^3 (3\lambda - 1)^2}. \tag{6.2.36}$$

Comparing (4.6.20) with (6.2.30) and (6.2.33) gives

$$\beta_2 = 1, \qquad \beta_2' = 1, \qquad \beta_{2t} = 1, \qquad \beta_{2t}' = \tfrac{1}{2}. \tag{6.2.37}$$

Along the line $L = L_p$, (6.2.27) gives

$$\Delta_p m = Y_1(\lambda, K_p) q^2 + Y_2(\lambda, K_p) q^4 + O[q^6], \tag{6.2.38}$$

where

$$\begin{aligned} Y_1(\lambda, K) &:= -\frac{W_2(\lambda, K)}{W_1(K)}, \\ Y_2(\lambda, K) &:= -\frac{W_2(\lambda, K)}{[W_1(K)]^3} \{[W_2(\lambda, K)]^2 - W_1(K) W_3(\lambda, K) \\ &\qquad + W_2(\lambda, K) W_4(\lambda, K)\}. \end{aligned} \tag{6.2.39}$$

Then, from (6.2.28),

$$L_s = W_1''(\lambda, K_p) q^3 + W_2''(\lambda, K_p) q^5 + O[q^7], \tag{6.2.40}$$

where

$$\begin{aligned} W_1''(\lambda, K) &:= 2Y_1(\lambda, K)W_2(\lambda, K) + \tfrac{1}{3}W_3(\lambda, K) \\ &= \frac{\lambda K^2(K_t - K)}{K_t K_c^2}, \end{aligned} \tag{6.2.41}$$

$$\begin{aligned} W_2''(\lambda, K) &:= 2Y_2(\lambda, K)W_2(\lambda, K) + [Y_1(\lambda, K)]^2 W_3(\lambda, K) \\ &\quad + 4Y_1(\lambda, K)W_4(\lambda, K) + W_5(\lambda, K), \end{aligned}$$

and

$$W_2''(\lambda, K_t) = \frac{54\lambda^5(5\lambda - 3)}{5(3\lambda - 1)^5}. \tag{6.2.42}$$

It follows from (4.6.15) and (6.2.40)–(6.2.42) that

$$\delta = 3, \qquad \delta_t = 5. \tag{6.2.43}$$

Now consider the response function φ_T close to the critical curve. From (6.2.30) and (6.2.33),

$$\varphi_T \simeq \begin{cases} [W_1(K_p) + 2W_2(\lambda, K_p)\Delta_p m]^{-1}, \\ \qquad \text{in the paramagnetic phase,} \\ [W_1'(\lambda, K_p) + 2W_2'(\lambda, K_p)\Delta_p m]^{-1}, \\ \qquad \text{in the antiferromagnetic phase.} \end{cases} \tag{6.2.44}$$

When $\frac{3}{4}K_c < K < K_t$, $(T_t < T < \frac{4}{3}T_c)$, from (6.2.30) and (6.2.34), $W_1'(\lambda, K) > 0$ and

$$\varphi_T \simeq \begin{cases} \dfrac{1}{2K_p}, & \text{in the paramagnetic region,} \\ \dfrac{K_t(4K - 3K_c)}{6\lambda K K_c(K_t - K)}, & \text{in the antiferromagnetic region.} \end{cases} \tag{6.2.45}$$

The interesting case, from the point of view of the metamagnet, is when $\lambda > \frac{1}{3}$ and $K_t > 0$. Then, as K is increased through K_t, $W_1'(\lambda, K)$ changes sign and, from (6.2.24) and (6.2.44), the antiferromagnetic state becomes unstable near the critical curve. The situation for $K > K_t$ will be explored in Sect. 6.2.5 where we show that a first-order transition occurs between the antiferromagnetic and paramagnetic states meaning that $K = K_t$, $L = L^*(K_t)$ is a tricritical point. At this point, from (6.2.36), $W_2'(\lambda, K_t) > 0$, if $\lambda > \frac{3}{5}$. The significance of this value of λ will be discussed in

Sect. 6.2.5. For the moment we confine attention to $\frac{3}{5} < \lambda < 1$ where, from (6.2.44) and (6.2.33), in the antiferromagnetic region

$$\varphi_T \simeq \frac{1}{2|W_2'(\lambda, K_t)||\Delta_t m|} \simeq \frac{1}{2|W_2'(\lambda, K_t)|^{\frac{1}{2}}|\Delta_t L|^{\frac{1}{2}}}. \tag{6.2.46}$$

Comparing (4.6.22) with (6.2.45) and (6.2.46) gives

$$\gamma_2 = 0, \qquad \gamma_2' = 0, \qquad \gamma_{2t} = 0, \qquad \gamma_{2t}' = \tfrac{1}{2}. \tag{6.2.47}$$

In the antiferromagnetic phase, from (6.2.31) and (6.2.33),

$$\Delta_p L = Y_1'(\lambda, K_p)q^2 + Y_2'(\lambda, K_p)q^4 + O(q^6), \tag{6.2.48}$$

where

$$\begin{aligned} Y_1'(\lambda, K) &:= \frac{W_1'(\lambda, K)}{X_1(\lambda, K)} = -\frac{m^*(\lambda, K)\lambda K^2(K_t - K)}{K_t(K - K_c)} \\ Y_2'(\lambda, K_t) &:= \frac{W_2'(\lambda, K_t)X_1(\lambda, K_t) - W_1'(\lambda, K_t)X_2(\lambda, K_t)}{[X_1(\lambda, K_t)]^3} \\ &= -\frac{9\lambda^2(3\lambda)^{\frac{1}{2}}(2\lambda^3 + 15\lambda^2 + 54\lambda + 9)}{(3\lambda - 1)^4(1 + \lambda)} \end{aligned} \tag{6.2.49}$$

and comparing (6.2.48) with (4.6.19) gives

$$\beta = \tfrac{1}{2}, \qquad \beta_t = \tfrac{1}{4}. \tag{6.2.50}$$

The values of the exponents β_{2t}', γ_{2t}', δ, δ_t, β and β_t obtained here agree with those derived by a mixture of Landau theory and scaling in Sect. 5.5. However, it is interesting to observe that this mean-field calculation gives $\beta_{2t} \neq \beta_{2t}'$ and $\gamma_{2t} \neq \gamma_{2t}'$. In each case this is a consequence of a vanishing amplitude factor ($W_1(K_t) \neq 0$ but $W_1'(\lambda, K_t) = 0$).

6.2.4 The First-Neighbour Antiferromagnet: $\lambda = 0$

Here we consider briefly the case $\lambda = 0$, which is a first-neighbour antiferromagnetic system. As noted in Sect. 6.2.1, this case is the only one in which the critical curve meets the zero-temperature axis at the critical field. Since the tricritical temperature is now negative the antiferromagnetic state is stable close to the critical temperature for the entire range $0 < T < T_c$ and the transition line is identical to the critical curve which is the $\lambda = 0$ line in Fig. 6.5. Although the finite discontinuity in φ_T at the transition may be an artifact of mean-field theory, the general picture of the phase

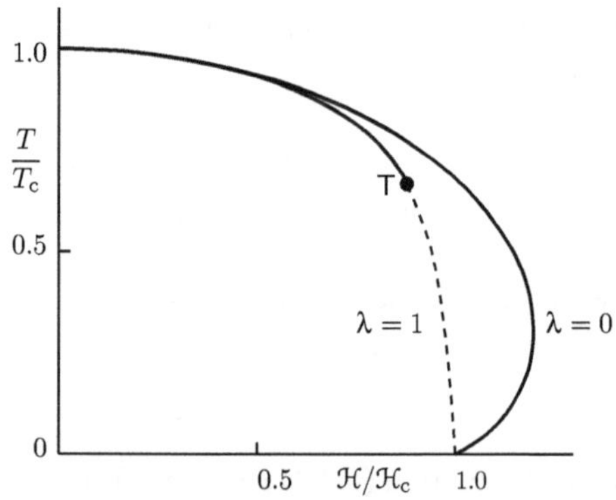

Fig. 6.5 Second-order (*full lines*) and first-order (*broken line*) transition curves in field–temperature plane for the antiferromagnet ($\lambda = 0$) and metamagnet ($\lambda = 1$), where the first and second-order transitions meet at a tricritical point T

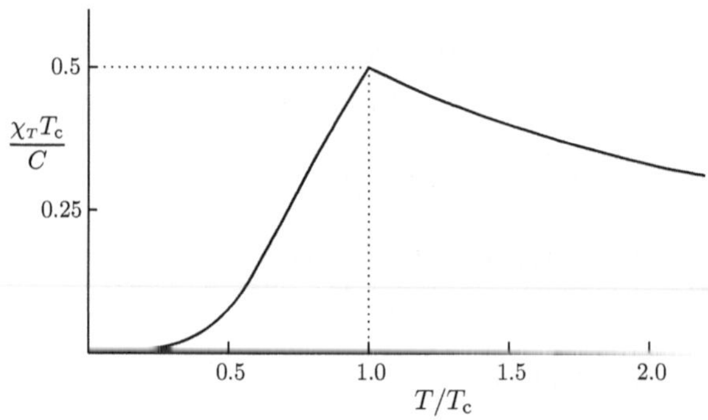

Fig. 6.6 The zero-field susceptibility for the first-neighbour antiferromagnet, ($\lambda = 0$)

diagram for the simple antiferromagnet, with the antiferromagnetic and paramagnetic regions separated by a higher-order transition line, is confirmed by series analysis (Domb 1974c).

Consider now the situation as K (or T) is varied along $L = 0$, where (6.2.9) is satisfied by $m = 0$ for any value of q. From (1.4.9), (6.1.13), (6.2.13) and (6.2.15) the zero-field susceptibility is

$$\chi_T := \frac{C\psi_T}{T} = \frac{C}{T(1-q^2)^{-1} + T_c}. \tag{6.2.51}$$

In the paramagnetic region ($T > T_c$), $q = 0$ and in the antiferromagnetic region ($T < T_c$) q is given by (6.2.10). It follows that χ_T and $\mathrm{d}\chi_T/\mathrm{d}T \to 0$, as $T \to 0$. Whereas $\mathrm{d}\chi_T/\mathrm{d}T > 0$, for $T \le T_c$, from (6.2.51), $\mathrm{d}\chi_T/\mathrm{d}T < 0$, for $T > T_c$. While χ_T is continuous at $T = T_c$, $\mathrm{d}\chi_T/\mathrm{d}T$ has a finite discontinuity, giving a cusp in the χ_T–T curve (Fig. 6.6). This contrasts with series results (Fisher and Sykes 1962;

Domb 1974c) which indicate that $\mathrm{d}\chi_T/\mathrm{d}T$ is continuous at all T and that χ_T has a maximum at a temperature which is about $1.5\,T_\mathrm{c}$ for $d = 2$ and $1.08\,T_\mathrm{c}$ for $d = 3$.

When an experimental plot of χ_T^{-1} against T is asymptotic at high temperatures to a line $A(T - T_\mathrm{pc})$, for some constant A, a *Curie–Weiss law* is said to be obeyed, and T_pc is called a *paramagnetic Curie temperature*. From (6.1.22), $T_\mathrm{pc} = T_\mathrm{c}$ in the mean-field approximation for the Ising model ferromagnet. In a real ferromagnet, the paramagnetic Curie temperature is 10 or 20 K above the ferromagnetic Curie temperature (Martin 1967). From (6.2.51), $T_\mathrm{pc} = -T_\mathrm{c}$ for the mean-field antiferromagnet. A negative paramagnetic Curie temperature is a well-known characteristic of real antiferromagnetic substances for which the susceptibility below the Néel point depends on the direction in which the external magnetic field is applied. The curve shown in Fig. 6.6 corresponds qualitatively to the response to a field along the zero-field magnetization axis of the sublattices. For responses to fields perpendicular to the axis of zero-field magnetization see Fisher (1963), Martin (1967), Mattis (1985).

6.2.5 The First-Order Transition

As indicated above, a positive λ is essential for the metamagnetic model and in that case, on the zero-temperature axis, the end points $\mathcal{H}^*(0) = \pm z_1 J_1(1-\lambda)$ of the critical curves are not coincident with the critical field $\mathcal{H}_\mathrm{c} = \pm z_1 J_1$; at low temperatures there cannot be a second-order transition from the antiferromagnetic state to the paramagnetic state. It has already been shown in Sect. 6.2.3 that, when $\lambda > \frac{1}{3}$, the antiferromgnetic state is unstable close to the critical curve when $T < T_\mathrm{t}$. However, there remains the range of λ values $(0, \frac{1}{3})$ when $T_\mathrm{t} < 0$. Now on the zero-field axis $m_\mathfrak{a} = -m_\mathfrak{b} = q \geq 0$, $m = 0$, and on the critical curve $m_\mathfrak{a} = m_\mathfrak{b} = m > 0$, $q = 0$. Since $m_\mathfrak{b}$ is a monotonically increasing function of L, or $\mathcal{H}$, as the field is increased from zero to its critical value, at constant temperature, there is an intermediate field value where $m_\mathfrak{b} = 0$ $(m = q)$, and at that value it can be shown that, for sufficiently small temperature, $\varphi_T < 0$ for all $0 < \lambda \leq 1$. There is thus an instability loop in the plot of L (or $\mathcal{H}$) against m at constant K similar to that shown in Fig. 5.1a and resulting in phase separation, with the point where $m_\mathfrak{b} = 0$ lying in the region of intrinsic instability (Fig. 6.7a).[9] Hence a line of first-order transitions exists in the low-temperature part of the $\mathcal{H}$–T plane, terminating at $T = 0$, where we would expect it to meet the $\mathcal{H}$-axis at $\mathcal{H} = \mathcal{H}_\mathrm{c}$. This is confirmed by the fact that, if the values of $\mathcal{H}$ and T where $m_\mathfrak{b} = 0$ are plotted in the $\mathcal{H}$–T plane, this curve meets the $\mathcal{H}$–axis at $\mathcal{H} = \mathcal{H}_\mathrm{c}$. The remainder of this section will he concerned with how this first-order transition line terminates at its upper end.[10]

---

[9] The case shown here is for $\lambda = 1$, $T_\mathrm{t} > 0$; the differences that occur when $\lambda < \frac{1}{3}$, $T_\mathrm{t} < 0$ are discussed below.

[10] There is also a mirror-image transition line in the $\mathcal{H} < 0$ half-plane, but we need consider only the $\mathcal{H} > 0$ line explicitly.

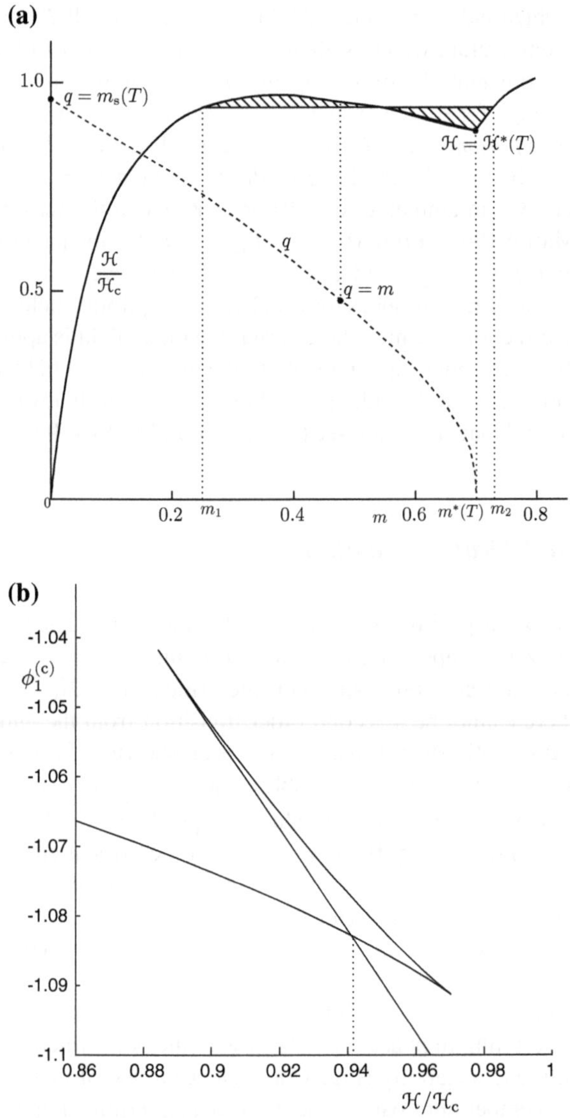

Fig. 6.7 **a** Field–magnetization and q–magnetization plots for $\lambda = 1$, $T = \frac{1}{2}T_c$. The field–magnetization plot shows an instability loop and the equal-areas tie line. **b** A plot, for $\lambda = 1$, $T = \frac{1}{2}T_c$ ($K = 1$) of the equilibrium free energy density given by (6.2.14) against $\mathcal{H}/\mathcal{H}_c = L$. The crossing point of the curve is used to locate the equal-areas tie line in Fig. 6.7a

Since, from (6.2.44), $(\partial L/\partial m)_K = 1/\varphi_T = W_1'(\lambda, K)$ at $m = m^*-0$, the stability condition (6.2.24) is not satisfied on the antiferromagnetic side of the critical curve if $W_1'(\lambda, K) < 0$. From (6.2.34) and (6.2.35), $W_1'(\lambda, K) = 2K_c$ at $T = T_c$ and decreases steadily with T, passing through the value zero at $T = T_t$. For $T_c > T > T_t$

there is a second-order transition on the critical curve, as discussed at the end of Sect. 6.2.2. For $T_t > T > 0$ and λ sufficiently large (in fact, $\lambda > \frac{3}{5}$, as shown below), the plot of $\mathcal{H}$ against m is similar to that in Fig. 6.7a where the loop implies a first-order transition.

The magnetizations m_1 and m_2 in the conjugate phases, correspond to the ends of the equal-areas tie line shown. The value of $\mathcal{H}$ for the tie line can be most easily calculated by using (6.2.14), with $m_\mathfrak{a}$ and $m_\mathfrak{b}$ given by (6.2.9), to plot $\phi_1^{(\mathrm{CL})}$ against $\mathcal{H}/\mathcal{H}_c$ (or L) as shown in Fig. 6.7b. The crossing point of the curve locates the value of $\mathcal{H}$ for the tie line (see Fig. 5.1a, b). The slope discontinuity at $m = m^*$ falls in the instability interval (m_1, m_2) and, for the range $T_t > T > 0$, there is a line of first-order transitions, as shown for the $\lambda = 1$ case in Fig. 6.5, This line ends at the *tricritical point* $T = T_t$, $\mathcal{H} = \mathcal{H}^*(T_t)$, where it is continued by a second-order transition line (see Sects. 4.1.1, 4.6 and 5.5). The full phase diagram for this case (including the staggered field) has been obtained by Kincaid and Cohen (1975) and is shown in Fig. 4.3b, where it can be seen that the line of first-order phase transitions can now be understood as a line of coexistence of three phases, two antiferromagnetic and one paramagnetic.

For small values of λ the first-order transition line starting at $\mathcal{H} = \mathcal{H}_c$ on the zero-temperature axis cannot end at a tricritical point. In fact from (6.2.35), $T_t < 0$ for $\lambda < 1/3$ and there is an critical point given by

$$\left(\frac{\partial L}{\partial m}\right)_K = 0, \qquad \left(\frac{\partial^2 L}{\partial m^2}\right)_K = 0. \tag{6.2.52}$$

Since the condition (6.2.24) is satisfied everywhere in the paramagnetic region this critical point must lie in the antiferromagnetic region, that is inside the critical curve in the $\mathcal{H}$–T plane and it is referred to as a *bicritical end-point*. Figure 6.8 shows part of the $\mathcal{H}$–T plane for $\lambda = 0.3$. For any T between the values corresponding to points B and E, the plot of $\mathcal{H}$ against m will show a loop, but the point $m = m^*$, where a slope discontinuity occurs, lies outside the instability interval associated with the loop. Hence, when $\mathcal{H}$ is increased at constant temperature, there is a first-order transition on the segment BE, and then an antiferromagnetic–paramagnetic second-order transition at $\mathcal{H} = \mathcal{H}^*$ on the critical line. At the temperature corresponding to E the point $m = m^*$ enters the instability interval of the first-order transition, and the line of second-order transitions on the critical curve terminates at the *critical end-point* E (see Sect. 4.1.1). The full phase diagram for this case is shown in Fig. 4.3a, where it can be seen that the line of first-order phase transitions BE is a line of coexistence of four antiferromagnetic phases.

It is now easy to find the value $\lambda = \lambda_1$ where the low-λ critical end-point phase pattern goes over to the high-λ tricritical point phase pattern. Consider Fig. 6.8 and suppose that λ increases steadily. The segment BE will shrink until eventually the critical end-point E coalesces with the bicritical end-point B at $\lambda = \lambda_1$. When coalescence occurs, the conditions (6.2.52) defining E are satisfied on the antiferromagnetic

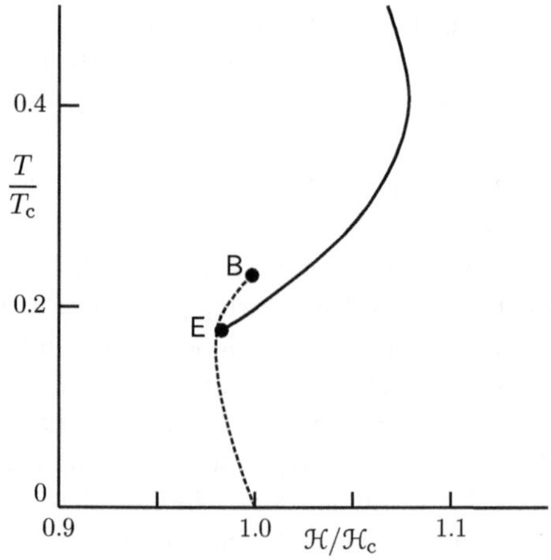

Fig. 6.8 As in Fig. 6.5 for $\lambda = 0.3$, with critical end-point E and bicritical end-point B

side of the critical curve so that $W_2'(\lambda, K_t) = 0$, giving $\lambda_1 = \frac{3}{5}$. (This result was derived by Bidaux et al. (1967), who gave a comprehensive set of diagrams.)

6.2.6 A Neighbourhood of the Tricritical Point

We now consider the tricritical regime again and in particular the situation just below the tricritical temperature where

$$T = T_t - \theta\, z_1 J_1, \qquad \text{or equivalently,} \qquad K = \frac{K_t}{1 - K_t\theta}. \tag{6.2.53}$$

From (6.2.35) the tricritical coupling $K_t = 3\lambda/(3\lambda - 1)(1 + \lambda)$ and $\theta \ll 1$ and the conjugate antiferromagnetic and paramagnetic phases existing for $T < T_t$ are labelled by the indices 1 and 2. From (5.1.6), (5.1.7) and (5.1.12) the phase equilibrium conditions become

$$\begin{aligned} &L(K, m_1) = L(K, m_2)\,, \\ &L(K, m_1)m_1 - L(K, m_2)m_2 = -\int_{m_1}^{m_2} L(K, m)\mathrm{d}m. \end{aligned} \tag{6.2.54}$$

Using the variables

$$\delta := m - m^*(\lambda, K), \qquad \varepsilon(K, \delta) := L(K, m) - L(K, m^*(\lambda, K)), \tag{6.2.55}$$

these conditions give

$$\begin{aligned}\varepsilon(K,\delta_1) &= \varepsilon(K,\delta_2)\,,\\ \varepsilon(K,\delta_1)\,\delta_1 - \int_0^{\delta_1}\varepsilon(K,\delta)\mathrm{d}\delta &= \varepsilon(K,\delta_2)\,\delta_2 - \int_0^{\delta_2}\varepsilon(K,\delta)\mathrm{d}\delta\,,\end{aligned} \tag{6.2.56}$$

with $\delta_1 < 0$ and $\delta_2 > 0$. On the antiferromagnetic side of the critical curve where $\delta < 0$, transcription of (6.2.33) gives

$$\varepsilon(K,\delta) = W_1'(\lambda,K)\delta + W_2'(\lambda,K)\delta^2 + \mathrm{O}(\delta^3). \tag{6.2.57}$$

Since $\theta \ll 1$ we can, from (6.2.34) and (6.2.36) use the leading-term approximations

$$W_1'(\lambda,K) \simeq= -K_{\mathrm{t}}W_{11}\,\theta \qquad W_2'(\lambda,K) \simeq K_{\mathrm{t}}W_{20}\,, \tag{6.2.58}$$

where

$$W_{11} := \frac{6\lambda^2}{(1+\lambda)^2}\,, \qquad W_{20} := \frac{8\lambda^2(3\lambda)^{\frac{1}{2}}(3-5\lambda)}{5(1+\lambda)^2(3\lambda-1)}\,. \tag{6.2.59}$$

For $\delta > 0$ and close to the critical curve, from (6.2.45),

$$\varepsilon(K,\delta) \simeq \frac{\delta}{\varphi_T} = 2K_{\mathrm{t}}\delta \tag{6.2.60}$$

and since, for the tricritical range $\lambda > \frac{3}{5}$, $W_{20} < 0$, (6.2.57) yields a plot of the same general form as Fig. 6.7a. Substitution in the two relations of (6.2.56) yields

$$\begin{aligned}&-W_{11}\theta\delta_1 + W_{20}\delta_1^2 = 2\delta_2\,,\\ &-\tfrac{1}{2}W_{11}\theta\delta_1^2 + \tfrac{2}{3}W_{20}\delta_1^3 = \delta_2^2\,.\end{aligned} \tag{6.2.61}$$

The conjugate magnetizations and coupling

$$\begin{aligned}m_1(\lambda,K) &= m^*(\lambda,K) + \delta_1(\theta) = m^*(\lambda,K) - \frac{3W_{11}}{4|W_{20}|}\theta + \mathrm{O}(\theta^2)\,,\\ m_2(\lambda,K) &= m^*(\lambda,K) + \delta_2(\theta) = m^*(\lambda,K) - \frac{3{W_{11}}^2}{32|W_{20}|}\theta^2 + \mathrm{O}(\theta^3)\,,\end{aligned} \tag{6.2.62}$$

$$L(K) = L^*(K) + \frac{3K_{\mathrm{t}}{W_{11}}^2}{16|W_{20}|}\theta^2 + \mathrm{O}(\theta^3)\,, \tag{6.2.63}$$

can then be derived. From (6.2.63) the first-order transition line and the critical curve have a common tangent at the tricritical point ($K = K_{\mathrm{t}}, \theta = 0$), as shown for $\lambda = 1$ in Fig. 6.5, and from (6.2.62) the paramagnetic branch $m = m_2(K)$ of the coexistence curve in the m–K plane has a common tangent with the critical magnetization curve

$m = m^*(\lambda, K)$ at $K = K_t$, but the antiferromagnetic branch $m = m_1(K)$ meets it at a non-zero angle. Hence the coexistence curve has a cusp at the tricritical point, in contrast to the continuous tangent at a critical point. From (6.2.57) and (6.2.58), at the tricritical point

$$L - L^* \simeq K_t W_{20}(m^* - m)^2, \tag{6.2.64}$$

for $m < m^*$ (that is, on the antiferromagnetic side of the critical curve). Now, from (6.2.59), $W_{20} < 0$ for $\lambda < \frac{3}{5}$, giving $L > L^*$. This confirms that, for $\lambda < \frac{3}{5}$, the tricritical point falls in the instability range of the first-order transition. As we have seen, T_t becomes negative for $\lambda < \frac{1}{3}$.

Chapter 7
Cluster-Variation Methods

7.1 Improving Mean-Field Theory

As we saw in the application of mean-field theory to the Ising model in Sect. 6.1, the assumption that the spins on the lattice sites are uncorrelated is equivalent to replacing the interactions on a single spin with a *mean or molecular field* proportional to the magnetization. Minimising the free energy within the approximation then gives Eq. (6.1.25). Of course, if the partition function with an independent magnetic field were derived exactly, the magnetization would arise as the first derivative of the free energy with respect to the field. However, using the mean-field approximation, the free energy *is a function of the magnetization*. So (6.1.25) can also be regarded as a consistency condition.

It is not difficult to see how to improve on this method. Instead of breaking the correlations between single sites we could divide the lattice into larger uncorrelated clusters of sites. Methods based on a pair of sites (*first-order methods*) were invented independently by various authors, notably Guggenheim (1935), (the *quasi-chemical method*), Bethe (1931, 1935), (the *Bethe-pair method*, see Sect. 7.3)[1] and Kasteleyn and van Kranendonk (1956a, b, c), (the *constant-coupling method*) and many schemes for proceeding to larger groups have been proposed. The generalization of the quasi-chemical method, due to Guggenheim and McGlashan (1951), which allows any suitable group of sites to be used, enables quite complicated forms of ordering and intermolecular attraction to be incorporated in a natural way. However it remains a first-order method in the sense that it does not deal in a systematic way with all the sub-clusters within the chosen group of sites. Kikuchi (1951) developed a method for carrying the process of approximation to a higher level by minimizing the free energy with respect to the occupation probabilities of subgroups of sites, including first-neighbour pairs, as well as those of the main group and this procedure was placed

---

[1] The 'original' Bethe method was based on a cluster consisting of a site and all its z first neighbours. However, it can be show that this yields no greater accuracy that a cluster consisting of a pair of first-neighbour sites.

D.A. Lavis, *Equilibrium Statistical Mechanics of Lattice Models*,
Theoretical and Mathematical Physics, DOI 10.1007/978-94-017-9430-5_7

on a systematic basis by Hijmans and de Boer (1955a, b, c). We shall, therefore, refer to these methods as the Kikuchi-Hijmans-De Boer (KHDeB) hierarchy of approximations. The cluster counting procedures used by Hijmans and de Boer have now been used in the finite-lattice series expansion method of de Neef and Enting (1977) described in Sect. 14.5.

Mean-field approximations, based on clusters consisting of a single site, take their place in the KHDeB hierarchy as *zeroth-order methods*. However, the way that contributions to the Hamiltonian (or enthalpy) are incorporated, both at the mean-field and higher levels of approximation is worthy of some comment. In Sect. 2.3 we derived the coupling-density formulation of thermodynamics, which involved the introduction of a set $L_i := \xi_i/T$, $i = 1, \ldots, n_{\mathrm{f}}$ of external couplings and $K_\ell := \varepsilon_\ell/T$, $\ell = 1, \ldots, n_{\mathrm{e}}$, of internal couplings. The fields ξ_i give contributions to the Hamiltonian (and enthalpy) from single microsystems, either because ξ_i is a single-site energy like the chemical potential, or because ξ_i is a magnetic or electric field representing an interaction with the environment. The internal energies ε_ℓ represent interactions between clusters of sites and one of the principles guiding the use of an appropriate approximation for a lattice model is to try to ensure that each group of sites having an irreducible interaction energy ε_ℓ is small enough to be a subset of a cluster within the chosen cluster approximation. This criterion is satisfied for the simple Ising model by the Bethe-pair approximation (Sect. 7.3). However, it is not satisfied by mean-field methods for any multi-site interactions. In these terms the only 'satisfactory' mean-field results are for models like the paramagnet, when it is exact[2] and, as we shall see in Sect. 7.4, the 'reduction' of a cluster-variation approximation to the corresponding mean-field analysis for a system with more that single-site energies needs an additional assumption. Of course applying the full KHDeB method to a model with irreducible interactions in large groups of sites will be very complicated and one simplification, which is still an advance on the zeroth-order method, is the Guggenheim and McGlashan approximation (see Sects. 7.6 and 16.4.8).

Because they incorporate short-range ordering, cluster-variation methods improve on the mean-field methods as regards the accuracy of phase diagrams and critical parameters. Sometimes they can treat effects which the mean-field method cannot and which are not yet accessible to exact analysis. However, the behaviour they predict near critical points is still 'classical', giving the same critical exponents as in mean-field theory. In fact, this is bound to be true for any method in which the densities depend analytically on thermodynamic variables and order parameters.

---

[2] An interesting twist to the story is that, though first-order theory is approximate for $d = 2$ or $d = 3$ regular lattices, like those shown in Sect. 16.2.1, it is exact for the interior sites of more loosely connected arrays, interacting only through first-neighbour pair interactions, (Baxter 1982b, Chap. 4).

7.2 The KHDeB Hierarchy of Approximations

Consider a lattice $\mathcal{N}$ of N sites and assume it to be so large that boundary effects can be neglected. The KHDeB cluster method begins by taking a particular *connected section graph*[3] $\mathfrak{g}$ on $\mathcal{N}$ and generating the set of section graphs obtained as distinct overlapping subgraphs of two different copies of $\mathfrak{g}$. If any one of these subgraphs is not connected then all of its components are included in the sequence; making each member of the sequence connected. This is an example of the overlap procedure described in Sect. 16.4.6. Suppose that the number of *non-isomorphic* (distinct) graphs generated in this way (including $\mathfrak{g}$) is n and label them as a partially-ordered sequence $\mathcal{S}$ so that if $\mathfrak{g}_i \subset \mathfrak{g}_j$ then $i < j$, with $\mathfrak{g}_n$ being the generating graph $\mathfrak{g}$ and $\mathfrak{g}_1$ being a single lattice point. As shown in Theorem 16.4.4, $\mathcal{S}$ is *complete* (Definition 16.4.1).

7.2.1 Distribution Numbers

A *configuration* $\boldsymbol{\sigma}$ on $\mathcal{N}$ is a particular assignment of microsystem states to the sites of $\mathcal{N}$ and a configuration on $\mathfrak{g}_j$ is defined in the same way. The configurations on two copies of $\mathfrak{g}_j$ are said to be *isomorphic* if, when they are brought into coincidence by some combination of translations, rotations and reflections, the state on each coinciding site is the same. The non-isomorphic configurations on $\mathfrak{g}_j$ are labelled with the index[4] $\boldsymbol{\ell}$ and we denote the graph $\mathfrak{g}_j$ in configuration $\boldsymbol{\ell}$ by $\mathfrak{g}_j(\boldsymbol{\ell})$ with $\lambda_j(\boldsymbol{\ell})$ denoting the number of isomorphic configurations on $\mathfrak{g}_j$ with the same $\boldsymbol{\ell}$. Let

$$\pi_j(\boldsymbol{\ell}\,|\,\boldsymbol{\sigma}) := \mathsf{Prob}[\text{configuration of } \mathfrak{g}_j \text{ is } \boldsymbol{\ell} \mid \text{configuration of } \mathcal{N} \text{ is } \boldsymbol{\sigma}]. \qquad (7.2.1)$$

Then, for large N, the total number of $\mathfrak{g}_j(\boldsymbol{\ell})$ in each of the $\lambda_j(\boldsymbol{\ell})$ isomorphic configurations is $N\,\omega_j\,\pi_j(\boldsymbol{\ell} \mid \boldsymbol{\sigma})$. The set of probabilities $\{\pi_j(\boldsymbol{\ell} \mid \boldsymbol{\sigma})\}$, also known as *distribution numbers*, must satisfy, for all $\boldsymbol{\sigma}$ (as will be implied in all relevant places in the following equations), the normalization conditions

$$1 = \sum_{\{\boldsymbol{\ell}\}} \lambda_j(\boldsymbol{\ell})\, \pi_j\,(\boldsymbol{\ell}\,|\,\boldsymbol{\sigma}), \qquad j = 1, 2, \ldots, n. \qquad (7.2.2)$$

As defined in Sect. 16.4.5, the lattice constant Ω_{ij} is the number of ways of arranging $\mathfrak{g}_i$ on $\mathfrak{g}_j$ and the lattice constant per site ω_i is such that that $N\omega_i$ is the number of

---

[3] For definitions of emphasized lattice graph theoretic terms see Sect. 16.4.

[4] We use a vector variable to label the configurations since they may, in general, contain more than one designating integer. There is further discussion of the meaning of isomorphic, when applied to configurations on graphs is Sect. 7.2.6.

ways of arranging $\mathfrak{g}_i$ on $\mathcal{N}$.[5] We define Λ_{ij} to be the number of $\mathfrak{g}_j$ of which a particular $\mathfrak{g}_i$ is a subgraph. By counting the number of $\mathfrak{g}_i$ on the lattice it follows that

$$\omega_i \Lambda_{ij} = \Omega_{ij}\, \omega_j\,. \tag{7.2.3}$$

Consistency relationships between the distribution numbers can now be obtained by a counting procedure similar to that used to derive (7.2.3). Let $\vartheta_{ij}(\boldsymbol{\ell}' \mid \boldsymbol{\ell})$ be the number of $\mathfrak{g}_i(\boldsymbol{\ell}')$ in $\mathfrak{g}_j(\boldsymbol{\ell})$. Then

$$N\,\omega_i\,\lambda_i(\boldsymbol{\ell}')\,\pi_i\,(\boldsymbol{\ell}'|\boldsymbol{\sigma})\Lambda_{ij} = N\,\omega_j \sum_{\{\boldsymbol{\ell}\}} \vartheta_{ij}(\boldsymbol{\ell}' \mid \boldsymbol{\ell})\,\lambda_j\,(\boldsymbol{\ell})\,\pi_j\,(\boldsymbol{\ell}|\boldsymbol{\sigma}) \tag{7.2.4}$$

is obtained by counting the number of $\mathfrak{g}_i(\boldsymbol{\ell}')$ on the lattice directly (on the right) and as a subgraph of each of the $\mathfrak{g}_j(\boldsymbol{\ell})$ (on the left), when each $\mathfrak{g}_i(\boldsymbol{\ell}')$ is counted Λ_{ij} times. Thus, from (7.2.3), we have the *$i - j$ consistency conditions*

$$\lambda_i(\boldsymbol{\ell}')\,\pi_i\,(\boldsymbol{\ell}'|\boldsymbol{\sigma})\Omega_{ij} = \sum_{\{\boldsymbol{\ell}\}} \vartheta_{ij}(\boldsymbol{\ell}' \mid \boldsymbol{\ell})\,\lambda_j\,(\boldsymbol{\ell})\,\pi_j\,(\boldsymbol{\ell}|\boldsymbol{\sigma}), \qquad \forall\ \boldsymbol{\ell}'. \tag{7.2.5}$$

Summing (7.2.5) over all $\boldsymbol{\ell}'$ and using (7.2.2) gives

$$\Omega_{ij} = \sum_{\{\boldsymbol{\ell}'\}} \vartheta_{ij}(\boldsymbol{\ell}' \mid \boldsymbol{\ell}), \qquad \forall\ \boldsymbol{\ell}. \tag{7.2.6}$$

For later use note that the important equations among those given here are the normalization condition (7.2.2), with $j = n$,

$$1 = \sum_{\{\boldsymbol{\ell}\}} \lambda_n(\boldsymbol{\ell})\,\pi_n\,(\boldsymbol{\ell}|\boldsymbol{\sigma}) \tag{7.2.7}$$

and, from (7.2.5), the j–n consistency conditions

$$\lambda_j(\boldsymbol{\ell}')\,\pi_j\,(\boldsymbol{\ell}'|\boldsymbol{\sigma})\Omega_{jn} = \sum_{\{\boldsymbol{\ell}\}} \vartheta_{jn}(\boldsymbol{\ell}' \mid \boldsymbol{\ell})\,\lambda_n\,(\boldsymbol{\ell})\,\pi_n\,(\boldsymbol{\ell}|\boldsymbol{\sigma}), \qquad \forall\ \boldsymbol{\ell}'. \tag{7.2.8}$$

Using these equations all the distribution numbers can be expressed in terms of the set $\{\pi_n(\boldsymbol{\ell}|\boldsymbol{\sigma})\}$.

---

[5] As $\mathfrak{g}_i$ is connected ω_i is independent of N.

7.2.2 Extensive Quantities

Let $\varepsilon(\mathfrak{g}_j(\boldsymbol{\ell}))$ be a property (like, for example, the energy) of $\mathfrak{g}_j(\boldsymbol{\ell})$ which is additive on non-overlapping copies of $\mathfrak{g}_j$ and which includes all contributions from proper subgraphs of $\mathfrak{g}_j$ in configurations compatible with $\boldsymbol{\ell}$. We now wish to calculate $\mathcal{E}(\mathcal{N}(\boldsymbol{\sigma}))$, which is the sum of ε–contributions from all the graphs of $\mathcal{N}$ in the configurations implied by the configuration $\boldsymbol{\sigma}$ of $\mathcal{N}$. The KHDeB method provides a hierarchy of methods for achieving this to a corresponding level of approximation. In Sect. 7.2.1 we introduced the distribution numbers $\{\pi_i(\boldsymbol{\ell} \mid \boldsymbol{\sigma})\}$, $i = 1, 2, \ldots, n$ and showed that it is necessary to specify only the set $\{\pi_n(\boldsymbol{\ell} \mid \boldsymbol{\sigma})\}$, since the distribution numbers for graphs further down the sequence are given by the consistency conditions (7.2.8). Suppose that n is increased with the sequence $\mathcal{S}$ including larger and larger graphs. Ultimately it will embrace the whole lattice; then $\boldsymbol{\ell} \equiv \boldsymbol{\sigma}$ and $\pi_n(\boldsymbol{\ell} \mid \boldsymbol{\sigma}) = 1$, and in a trivial sense for this limiting sequence the configurations $\{\boldsymbol{\sigma}\}$ can be replaced by the distribution numbers $\{\pi_n(\boldsymbol{\sigma})\}$.

The steps of the KHDeB method, for some more restricted sequence $\mathcal{S}$ and some (usually quite small) n, are:

(1) To suppose that the configuration $\boldsymbol{\sigma}$ can be approximately represented by the distribution numbers $\{\pi_i(\boldsymbol{\ell})\}$, $i = 1, 2, \ldots, n$, from which dependence on $\boldsymbol{\sigma}$ is now removed so that $\mathcal{E}(\mathcal{N}(\boldsymbol{\sigma}))$ becomes a function of the distribution numbers and is written $\mathcal{E}(\{\pi_n(\boldsymbol{\ell})\})$.

(2) To suppose that n is sufficiently large so that every graph $\mathfrak{g}_j$, which makes a contribution $\varepsilon(\mathfrak{g}_j(\boldsymbol{\ell}))$ when its configuration is $\boldsymbol{\ell}$ and which is not already contributed by a proper subgraph of $\mathfrak{g}_j$, is included in $\mathcal{S}$.

(3) The total number of $\mathfrak{g}_j(\boldsymbol{\ell})$ on $\mathcal{N}$ is $N\omega_j\,\lambda_j(\boldsymbol{\ell})\,\pi_j\,(\boldsymbol{\ell})$. So if we were simply to count contributions only from members of $\mathcal{S}$ in their various configurations and ignore any overlaps we would have

$$\mathcal{E}(\{\pi_n(\boldsymbol{\ell})\}) = N\sum_{j=1}^{n}\omega_j\sum_{\{\boldsymbol{\ell}\}}\varepsilon(\mathfrak{g}_j(\boldsymbol{\ell}))\,\lambda_j\,(\boldsymbol{\ell})\,\pi_j\,(\boldsymbol{\ell}). \tag{7.2.9}$$

Formula (7.2.9) is clearly incorrect, since some contributions from $\mathfrak{g}_j(\boldsymbol{\ell})$ will have already been counted in a contribution from $\mathfrak{g}_i(\boldsymbol{\ell})$, where $i > j$. The correct expression must be of the form

$$\mathcal{E}(\{\pi_n(\boldsymbol{\ell})\}) = N\sum_{j=1}^{n}\xi_j\sum_{\{\boldsymbol{\ell}\}}\varepsilon(\mathfrak{g}_j(\boldsymbol{\ell}))\,\lambda_j\,(\boldsymbol{\ell})\,\pi_j\,(\boldsymbol{\ell}), \tag{7.2.10}$$

for some set of numbers ξ_j, $j = 1, 2, \ldots, n$, with $\xi_j \le \omega_j$, $j < n$. These numbers now need to be determined. A reasonable assumption is that they are independent of the property ε. In particular we can choose, for some fixed $\mathfrak{g}_i$,

$\varepsilon(\mathfrak{g}_j(\boldsymbol{\ell}))$ to be equal, for all $\boldsymbol{\ell}$, to the number Ω_{ij} of arrangements of $\mathfrak{g}_i$ on $\mathfrak{g}_j$. Then $\mathcal{E}(\{\pi_n(\boldsymbol{\ell})\}) = N\omega_i$ and (7.2.10) gives

$$\omega_i = \sum_{j=1}^{n} \Omega_{ij}\xi_j, \qquad i = 1, 2, \ldots, n. \tag{7.2.11}$$

Thus, from (16.4.26), $\xi_j = \chi_j^{(n)}$, $j = 1, 2, \ldots, n$, and

$$\mathcal{E}(\{\pi_n(\boldsymbol{\ell})\}) = N\sum_{j=1}^{n} \chi_j^{(n)} \sum_{\{\boldsymbol{\ell}\}} \varepsilon(\mathfrak{g}_j(\boldsymbol{\ell}))\, \lambda_j\,(\boldsymbol{\ell})\, \pi_j\,(\boldsymbol{\ell}). \tag{7.2.12}$$

The numbers $\chi_j^{(n)}$, $j = 1, 2, \ldots, n$, are the weights for the sequence $\mathcal{S}$ (Sect. 16.4.5).

This method of deriving (7.2.12) is not that given by Hijmans and de Boer (1955a) and it is instructive to give a brief account of their method, which consists of disconnecting the lattice into its separate graphs $\mathfrak{g}_j$ at each of the levels $j = 1, 2, \ldots, n$. The set of separate graphs $\mathfrak{g}_j$ is called a j-graph assembly. If disconnection occurs at level j of the lattice $\mathcal{N}$ then there will be $N\omega_j$ members of the j-graph assembly. If disconnection at level j occurs for a lattice which has already been disconnected at a higher level then the number must be made smaller to take into account the graphs already counted at a higher level. We denote by $\mathcal{N}_j(\xi_j)$ a j–graph assembly of $N\xi_j$ members. The approximation of Hijmans and de Boer, as given in (1) above, is to take the largest member of $\mathcal{S}$ as the highest level of disconnection. Thus we begin with an n-assembly $\mathcal{N}_n(\xi_n)$ with, from (16.4.26), $\xi_n = \omega_n = \chi_n^{(n)}$. Now consider $j = n-1$. The number of members $N\xi_{n-1}$ in the $(n-1)$–graph assembly must take account of the fact that $N\,\Omega_{n-1\,n}\,\chi_n^{(n)}$ have already been counted in the n–graph assembly. Thus, again using (16.4.26), $\xi_{n-1} = \omega_{n-1} - \Omega_{n-1\,n}\chi_n^{(n)} = \chi_{n-1}^{(n)}$. Repeating this process gives

$$\xi_j = \omega_j - \sum_{i=j+1}^{n} \Omega_{ji}\chi_i^{(n)} = \chi_j^{(n)}, \qquad j = 1, 2, \ldots, n. \tag{7.2.13}$$

As indicated above, the quantity $\varepsilon_j(\boldsymbol{\ell})$ contains, in general, not only a contribution involving all the sites of $\mathfrak{g}_j(\boldsymbol{\ell})$, but also contributions from proper subgraphs of $\mathfrak{g}_j(\boldsymbol{\ell})$. Let $\tilde{\varepsilon}_j(\boldsymbol{\ell})$ be the contribution to $\varepsilon_j(\boldsymbol{\ell})$ from $\mathfrak{g}_j(\boldsymbol{\ell})$ which would not appear in any subgraph. Then

$$\varepsilon_j(\boldsymbol{\ell}) = \sum_{i=1}^{j} \sum_{\{\boldsymbol{\ell}'\}} \vartheta_{ij}(\boldsymbol{\ell}'\,|\,\boldsymbol{\ell})\tilde{\varepsilon}_i(\boldsymbol{\ell}'). \tag{7.2.14}$$

Substituting into (7.2.12) and using (7.2.5) gives

$$\mathcal{E}(\{\pi_n(\boldsymbol{\ell})\}) = N\sum_{j=1}^{n}\omega_j\sum_{\{\boldsymbol{\ell}\}}\tilde{\varepsilon}_j(\boldsymbol{\ell})\,\lambda_j\,(\boldsymbol{\ell})\,\pi_j\,(\boldsymbol{\ell}). \qquad (7.2.15)$$

Although (7.2.15) has the virtue of simplicity (7.2.12) is more useful for the development of the KHDeB cluster method.

7.2.3 The Hamiltonian and Free Energy

Let the contribution to the Hamiltonian from a graph $\mathfrak{g}_j$ in occupation $\boldsymbol{\ell}$, including all contributions from proper subgraphs of $\mathfrak{g}_j$ in configurations compatible with $\boldsymbol{\ell}$, be $\hat{h}_j(\boldsymbol{\ell})$. Since the Hamiltonian is additive over non-overlapping graphs it follows from (7.2.12), that it takes the form

$$\widehat{H}(\{\pi_n(\boldsymbol{\ell})\}) := N\sum_{j=1}^{n}\chi_j^{(n)}\sum_{\{\boldsymbol{\ell}\}}\hat{h}_j(\boldsymbol{\ell})\,\lambda_j\,(\boldsymbol{\ell})\,\pi_j\,(\boldsymbol{\ell}), \qquad (7.2.16)$$

where an implicit linear dependence on *couplings* (not energy parameters) is assumed for the members of the set $\{\hat{h}_j(\boldsymbol{\ell})\}$, and hence for the whole Hamiltonian.

Let $W(\{\pi_n(\boldsymbol{\ell})\})$ be the number of microstate configurations $\boldsymbol{\sigma}$ consistent with the distribution numbers $\{\pi_n(\boldsymbol{\ell})\}$. Then the dimensionless free energy is

$$\Phi(\{\pi_n(\boldsymbol{\ell})\}) = -\ln\left[\sum_{\{\pi_n(\boldsymbol{\ell})\}} W(\{\pi_n(\boldsymbol{\ell})\})\exp(-\widehat{H}(\{\pi_n(\boldsymbol{\ell})\}))\right]. \qquad (7.2.17)$$

Again, as in Sect. 6.1, the summation in (7.2.17) is approximated by the largest term, giving the cluster-variation free energy

$$\Phi(\{\pi_n(\boldsymbol{\ell})\}) = \widehat{H}(\{\pi_n(\boldsymbol{\ell})\}) - \ln[W(\{\pi_n(\boldsymbol{\ell})\})]. \qquad (7.2.18)$$

7.2.4 The Entropy

In Eq. (7.2.18)

$$S(\{\pi_n(\boldsymbol{\ell})\}) := \ln[W(\{\pi_n(\boldsymbol{\ell})\})], \qquad (7.2.19)$$

is the cluster-variation entropy (cf. Eq. (1.1.10)). For brevity we denote $W(\{\pi_n(\boldsymbol{\ell})\})$ as W_n and, in a similar way, denote by W_j the number of microstate configurations consistent with the distribution numbers $\{\pi_j(\boldsymbol{\ell})\}$. Using the procedure of Hijmans and de Boer (1955a), described in Sect. 7.2.2, let $\mathsf{G}_k(j, \xi_j)$ be the number of microstate configurations on the j–assembly $\mathcal{N}_j(\xi_j)$ consistent with the distribution numbers $\{\pi_k(\boldsymbol{\ell})\}$. It is clear that $k \le j$ and

$$\mathsf{G}_j(j, \xi_j) = \Psi_j(N\xi_j), \quad \text{where} \quad \Psi_j\,(X) := \frac{X!}{\prod_{\{\boldsymbol{\ell}\}} [\{X\,\pi_j\,(\boldsymbol{\ell})\}!]^{\lambda_j(\boldsymbol{\ell})}}. \tag{7.2.20}$$

Using Stirling's formula and neglecting all terms which are negligible in the thermodynamic limit, $N \to \infty$,

$$\Psi_j(N\xi_j) = [\psi_j(N)]^{\xi_j}. \tag{7.2.21}$$

For brevity $\psi_j(N)$ is denoted by ψ_j. Consider now the quantity

$$C_k(j) := \frac{\mathsf{W}_k}{\mathsf{G}_k(j, \omega_j)}, \qquad k \le j. \tag{7.2.22}$$

The numerator is the number of microstate configurations on $\mathcal{N}$ consistent with the distribution numbers $\{\pi_k(\boldsymbol{\ell})\}$ and the denominator is the number of microstate configurations on $\mathcal{N}_j(\omega_j)$ consistent with the distribution numbers $\{\pi_k(\boldsymbol{\ell})\}$. It is clear that $C_k(j) < 1$. The assumption, originating in the work of Kikuchi (1951) and adopted in the derivation of the entropy in Hijmans and de Boer (1955a), is that $C_j(j) = C_{j-1}(j)$, for $< j \le n$. Thus, from (7.2.20)–(7.2.22),

$$\frac{\mathsf{W}_j}{(\psi_j)^{\omega_j}} = \frac{\mathsf{W}_{j-1}}{\mathsf{G}_{j-1}(j, \omega_j)} \qquad j = 2, 3, \ldots, n. \tag{7.2.23}$$

Theorem 7.2.1 *With* $\chi_i^{(j)}$, $i = 1, 2, \ldots, k$ *the components of the vector solution* $\boldsymbol{\chi}^{(j)}$ *of* (16.4.61) *and* ψ_i, $i = 1, 2, \ldots, j$ *given by* (7.2.21), *it follows from* (7.2.23) *that*

$$\mathsf{W}_j = \prod_{i=1}^{j} (\psi_i)^{\chi_i^{(j)}}. \tag{7.2.24}$$

Proof As noted in Sect. 16.4.9, $\chi_j^{(j)} = \omega_j$ and (7.2.24) gives $\mathsf{W}_1 = (\psi_1)^{\omega_1}$, which is true since this is the correct combinatorial factor for the single sites of the lattice. We now prove (7.2.24) by induction, assuming

$$\mathsf{W}_{j-1} = \prod_{i=1}^{j-1} (\psi_i)^{\chi_i^{(j-1)}}. \tag{7.2.25}$$

This formula must also be valid for a j–figure assembly. This is equivalent to replacing, for all $i = 1, 2, \ldots, j-1$ the lattice constant ω_i by $\Omega_{i\,j}\omega_j$. To see the effect on the numbers $\chi_i^{(j-1)}$ refer to (16.4.67), which yields, from (7.2.25),

$$\mathrm{G}_{j-1}(j, \omega_j) = \prod_{i=1}^{j-1} (\psi_i)^{\chi_i^{(j-1)} - \chi_i^{(j)}}. \tag{7.2.26}$$

Substituting from (7.2.25) and (7.2.26) into (7.2.23) and, using $\chi_j^{(j)} = \omega_j$, gives (7.2.24). □

Using Stirling's formula, and neglecting all terms which are negligible in the thermodynamic limit, it follows from (7.2.19) and (7.2.24) that

$$S(\{\pi_n(\boldsymbol{\ell})\}) = -N \sum_{j=1}^{n} \chi_j^{(n)} \sum_{\{\boldsymbol{\ell}\}} \lambda_j(\boldsymbol{\ell})\, \pi_j(\boldsymbol{\ell}) \ln[\pi_j(\boldsymbol{\ell})]. \tag{7.2.27}$$

7.2.5 Minimization

From (7.2.16), (7.2.18) and (7.2.27), the cluster-variation free energy per lattice site

$$\phi(\{\pi_n(\boldsymbol{\ell})\}) = \sum_{j=1}^{n} \chi_j^{(n)} \sum_{\{\boldsymbol{\ell}\}} \lambda_j(\boldsymbol{\ell})\, \pi_j(\boldsymbol{\ell})\{\hat{h}_j(\boldsymbol{\ell}) + \ln[\pi_j(\boldsymbol{\ell})]\}, \tag{7.2.28}$$

in the limit $N \to \infty$, is a function of all the distribution numbers $\{\pi_j(\boldsymbol{\ell})\}$, for $j = 1, 2, \ldots, n$ and it is with respect to variations of all of them that we wish to find an extremum. The distribution numbers $\{\pi_j(\boldsymbol{\ell})\}$ for $j < n$ are given in terms of $\{\pi_n(\boldsymbol{\ell})\}$ by the j–n consistency conditions (7.2.8), which yield

$$\frac{\partial\, \pi_j(\boldsymbol{\ell})}{\partial\, \pi_n(\boldsymbol{\ell}')} = \frac{\vartheta_{jn}(\boldsymbol{\ell}\,|\,\boldsymbol{\ell}')\, \lambda_n(\boldsymbol{\ell}')}{\Omega_{jn}\, \lambda_j(\boldsymbol{\ell})} \tag{7.2.29}$$

and using for the normalization condition (7.2.7) an undetermined multiplier ζ,[6] the extremum conditions are

$$\zeta = \sum_{j=1}^{n} \frac{\chi_j^{(n)}}{\Omega_{jn}} \sum_{\{\boldsymbol{\ell}\}} \{\hat{h}_j(\boldsymbol{\ell}) + \ln[\pi_j(\boldsymbol{\ell})]\}\vartheta_{jn}(\boldsymbol{\ell}\,|\,\boldsymbol{\ell}') \qquad \forall\ \boldsymbol{\ell}'. \tag{7.2.30}$$

---

[6] Into which, from (7.2.6), a constant factor $\sum_{j=1}^{n} \chi_j^{(n)}$ has been absorbed.

Multiplying this equation by $\pi_n(\boldsymbol{\ell}')\,\lambda_n\,(\boldsymbol{\ell}')$ and summing over $\boldsymbol{\ell}'$ gives, from (7.2.5),

$$\zeta = \sum_{j=1}^{n} \chi_j^{(n)} \sum_{\{\boldsymbol{\ell}\}} \pi_j(\boldsymbol{\ell})\,\lambda_j\,(\boldsymbol{\ell})\{\hat{h}_j(\boldsymbol{\ell}) + \ln[\pi_j(\boldsymbol{\ell})]\} \tag{7.2.31}$$

and, from (7.2.28),

$$\phi(\{\pi_n(\boldsymbol{\ell})\}) = \zeta. \tag{7.2.32}$$

7.2.6 Labelling Configurations

From the point of view of the application of this analysis to particular examples the important equations are the normalization condition (7.2.7), the consistency conditions (7.2.8), the equilibrium conditions (7.2.31) and the equilibrium free-energy formula (7.2.32). In each of these the index $\boldsymbol{\ell}$ is used to distinguish between non-isomorphic configurations. Suppose that $\boldsymbol{\ell}$ is replaced by the pair of indices $(\boldsymbol{\ell}; \boldsymbol{k})$, where $\boldsymbol{\ell}$ now distinguishes configurations which make different contributions to the Hamiltonian and $\boldsymbol{k}$ (dependent on $\boldsymbol{\ell}$) distinguishes between non-isomorphic configurations within the set $\boldsymbol{\ell}$. Then $\hat{h}_j(\boldsymbol{\ell}; \boldsymbol{k}) = \hat{h}_j(\boldsymbol{\ell})$ for all $j, \boldsymbol{\ell}$ and $\boldsymbol{k}$. In Eqs. (7.2.7), (7.2.8) and (7.2.31) $\boldsymbol{\ell}$ is replaced by $(\boldsymbol{\ell}; \boldsymbol{k})$ and, it is assumed that:

(i)

$$\pi_j(\boldsymbol{\ell}; \boldsymbol{k}) = \pi_j(\boldsymbol{\ell}), \qquad \forall \;\; j, \boldsymbol{\ell} \text{ and } \boldsymbol{k}; \tag{7.2.33}$$

the probabilities of the configurations of $\mathfrak{g}_j$ are dependent only on $\hat{h}_j(\boldsymbol{\ell})$.

(ii)

$$\vartheta_{ij}(\boldsymbol{\ell}' \,|\, \boldsymbol{\ell}) = \sum_{\{\boldsymbol{k}'\}} \vartheta_{ij}(\boldsymbol{\ell}'; \boldsymbol{k}' \,|\, \boldsymbol{\ell}; \boldsymbol{k}); \tag{7.2.34}$$

the total number of $g_i(\ell')$ (adding over all $\boldsymbol{k}'$) on $\mathfrak{g}_j(\boldsymbol{\ell}; \boldsymbol{k})$ is independent of $\boldsymbol{k}$.

Then, with

$$\lambda_j(\boldsymbol{\ell}) := \sum_{\{\boldsymbol{k}\}} \lambda_j(\boldsymbol{\ell}; \boldsymbol{k}), \tag{7.2.35}$$

the new versions of (7.2.7), (7.2.8) and (7.2.31) containing $\boldsymbol{k}$ can be summed over that index (and, in the case of (7.2.8), also over $\boldsymbol{k}'$) to restore them to their original form with the index $\boldsymbol{\ell}$ for $\mathfrak{g}_j(\ell)$ now distinguishing energy levels with degeneracies $\lambda_j(\boldsymbol{\ell})$. Henceforth, it will be assumed that the index $\boldsymbol{\ell}$ is used in this way.

7.3 The Bethe-Pair Approximation for the Ising Model

This method corresponds to the case where the partially-ordered sequence $\mathcal{S}$ consists of the section graphs $\mathfrak{g}_1$, a single site and $\mathfrak{g}_2$, a single edge (bond). Then with the coordination number of the lattice being z

$$\boldsymbol{\Omega} = \begin{pmatrix} 1 & 2 \\ 0 & 1 \end{pmatrix}, \qquad \boldsymbol{\omega} = \begin{pmatrix} 1 \\ \frac{1}{2}\mathrm{z} \end{pmatrix} \tag{7.3.1}$$

(Example 14.1) and, from (16.4.26), the weights are $\chi_1^{(2)} = 1 - z$ and $\chi_2^{(2)} = \frac{1}{2}\mathrm{z}$. We now label the configurations on both $\mathfrak{g}_1$ and $\mathfrak{g}_2$ according to the aggregate spin and

$$\begin{array}{lll} \lambda_1(-1) = 1, & \lambda_1(+1) = 1, & \\ \lambda_2(-2) = 1, & \lambda_2(0) = 2, & \lambda_2(+2) = 1, \end{array} \tag{7.3.2}$$

$$\begin{array}{lll} \vartheta_{12}(-1\,|-2) = 2, & \vartheta_{12}(-1\,|\,0) = 1, & \vartheta_{12}(-1\,|+2) = 0, \\ \vartheta_{12}(+1\,|-2) = 0, & \vartheta_{12}(+1\,|\,0) = 1, & \vartheta_{12}(+1\,|+2) = 2, \end{array} \tag{7.3.3}$$

$$\begin{array}{lll} \hat{h}_1(-1) = L, & \hat{h}_1(+1) = -L, & \\ \hat{h}_2(-2) = -K + 2L, & \hat{h}_2(0) = K, & \hat{h}_2(+2) = -K - 2L, \end{array} \tag{7.3.4}$$

where $K := J/T$ and $L := \mathcal{H}/T$. The magnetization per lattice site is

$$m = \pi_1(+1) - \pi_1(-1) \tag{7.3.5}$$

and, from (7.2.8), the 1–2 consistency conditions are

$$\tfrac{1}{2}(1 \pm m) = \pi_2(0) + \pi_2(\pm 2). \tag{7.3.6}$$

From (7.2.28), the free energy per lattice site is

$$\begin{aligned} \phi(\{\pi_2(\boldsymbol{\ell})\}) &= mL - \tfrac{1}{2}K\mathrm{z}[\pi_2(-2) - 2\,\pi_2\,(0) + \pi_2(+2)] \\ &+ \tfrac{1}{2}(1 - \mathrm{z})[(1 + m)\ln(1 + m) + (1 - m)\ln(1 - m) - 2\ln(2)] \\ &+ \tfrac{1}{2}\mathrm{z}[\pi_2(-2)\ln\pi_2(-2) + 2\,\pi_2\,(0)\ln\pi_2(0) + \pi_2(+2)\ln\pi_2(+2)] \end{aligned} \tag{7.3.7}$$

and from (7.2.30), the equilibrium conditions are

$$\left[\frac{\pi_2(+2)}{\pi_2(-2)}\right]^{\frac{\mathrm{z}}{2}} \mathfrak{M} = \left(\frac{1 + m}{1 - m}\right)^{(\mathrm{z}-1)}, \tag{7.3.8}$$

$$\frac{[\pi_2(0)]^2}{\pi_2(-2)\,\pi_2(+2)} = \mathfrak{X}^2, \tag{7.3.9}$$

where

$$\mathfrak{X} := \exp(-2K), \qquad \mathfrak{M} := \exp(-2L). \tag{7.3.10}$$

Let

$$\tau := \sqrt{\frac{\pi_2(+2)}{\pi_2(-2)}}. \tag{7.3.11}$$

Then, from (7.3.6)–(7.3.9),

$$\mathfrak{M} = g(\tau), \qquad \text{where} \qquad g(\tau) = \tau^{-1}\left(\frac{\tau+\mathfrak{X}}{1+\tau\mathfrak{X}}\right)^{(z-1)}, \tag{7.3.12}$$

and

$$m = \frac{\tau - \tau^{-1}}{\tau + 2\mathfrak{X} + \tau^{-1}}. \tag{7.3.13}$$

From (7.2.32), (7.3.8) and (7.3.9) the free-energy density is

$$\begin{aligned}\phi_1(K, L; \tau) &= \tfrac{1}{2}zK - \tfrac{1}{2}(z-1)\ln\left(\frac{1-m^2}{4}\right) + \tfrac{1}{2}z\ln[\pi_2(0)] \\ &= \tfrac{1}{2}(z-2)\ln(\tau + 2\mathfrak{X} + \tau^{-1}) \\ &\quad -\tfrac{1}{2}(z-1)\ln[1 + \mathfrak{X}(\tau+\tau^{-1}) + \mathfrak{X}^2] - \tfrac{1}{2}zK. \end{aligned} \tag{7.3.14}$$

In the mean-field analysis of the ferromagnetic Ising model in Sect. 6.1 the order parameter was the magnetization density m. Here it is more appropriate to regard τ as the order parameter. Once a solution $\tau = \tau(\mathfrak{M}, \mathfrak{X})$ has been obtained then the magnetization and free-energy density are given by (7.3.13) and (7.3.14). It is clear from (7.3.12) that $[g(\tau^{-1})]^{-1} = g(\tau)$. So, for fixed $\mathfrak{X}$, solutions to (7.3.12) occur in pairs $(\tau, \mathfrak{M})$ and $(\tau^{-1}, \mathfrak{M}^{-1})$.

When $\mathcal{H} = L = 0$ ($\mathfrak{M} = 1$) the equilibrium values of τ are solutions of $g(\tau) = 1$. One solution is $\tau = 1$, giving $m = 0$. To determine whether other pairs of solutions $\tau^{(\pm)} \neq 1$ exist consider

$$\frac{dg}{d\tau} = -g(\tau)\frac{\mathfrak{X}(\tau+\tau^{-1}-2) + (\mathfrak{X}+1)(2-z+\mathfrak{X}z)}{(\tau+\mathfrak{X})(1+\tau\mathfrak{X})}, \tag{7.3.15}$$

which is a maximum when $\tau = 1$. Ferromagnetic solution $\tau_s^{(\pm)}$ will therefore occur when $\mathfrak{X} < (z-2)/z$ with corresponding magnetizations $\pm m_s$. This indicates that the critical temperature is given by

$$\mathfrak{X}_c = \frac{z-2}{z}, \qquad K_c = \frac{1}{2}\ln[z/(z-2)], \qquad T_c = \frac{2J}{\ln[z/(z-2)]}. \tag{7.3.16}$$

Critical temperatures derived from this formula are given in Table 3.1. However, to confirm that a transition occurs, it must be shown that

$$\phi(K;\tau_s) - \phi(K;1) < 0, \qquad K > K_c, \tag{7.3.17}$$

where it follows from (7.3.14) that $\phi(K;\tau_s^{-1}) = \phi(K;\tau_s)$.

For the square lattice, apart from the solutions $\tau = \pm 1$, $g(\tau) = 1$ yields

$$\tau^{(\pm)} = \frac{1 - 3\mathfrak{X}^2 \pm (1-\mathfrak{X}^2)\sqrt{1-4\mathfrak{X}^2}}{2\mathfrak{X}^3}. \tag{7.3.18}$$

From (7.3.14), with $\tau_s = \tau^{(\pm)}$,

$$\phi(K;\tau_s) - \phi(K;1) = \ln\left[\frac{1-2\mathfrak{X}^2}{2(1-\mathfrak{X})^2}\right]. \tag{7.3.19}$$

For $\mathfrak{X} < \mathfrak{X}_c = \frac{1}{2}$ the argument of the logarithm in this equation is monontonically increasing to unity at the critical value, which establishes the inequality (7.3.17). The spontaneous magnetization derived from (7.3.13) and (7.3.18) is compared with that derived from (6.1.25) for the mean-field approximation in Fig. 3.1.

For the square lattice the next approximation in the hierarchy is to take $\mathcal{S}$ to consist of $\mathfrak{g}_1 := [1,1]$, a single site, $\mathfrak{g}_2 := [2,2]$, a single edge and $\mathfrak{g}_3 := [4,8]$, a single square (Table 16.1). The calculations for $\chi_j^{(3)}$, $j = 1,2,3$ are given in Example 16.4.4 and the critical temperature is $T_c = 2.426J$ (Kikuchi 1951). For the triangular lattice the next approximation in the hierarchy is used in Sect. 7.5 for the 3-state Potts model. For the spin-$\frac{1}{2}$ Ising model the corresponding calculations yield a critical temperature $T_c = 3.916J$ (Hijmans and de Boer 1955c).

7.4 Reduction to the Mean-Field Approximation

Consider, for simplicity, the case where $\mathfrak{g}_1$ is the only single-site member of the sequence $\mathcal{S}$,[7] which is chosen such that $\mathfrak{g}_n$ is sufficiently large to contain any group of sites with an irreducible multi-site interaction. To approximate the Hamiltonian (7.2.15) and entropy (7.2.27) to forms expressible in terms of the single-site

[7] The generalization to the case of a lattice divided into sublattices (Sect. 16.4.7) is straightforward.

probabilities $\{\pi_1(\boldsymbol{\ell})\}$ take

$$\pi_j(\boldsymbol{\ell}) := \prod_{\{\boldsymbol{\ell}'\}} [\pi_1(\boldsymbol{\ell}')]^{\vartheta_{1j}(\boldsymbol{\ell}'|\boldsymbol{\ell})} \qquad j = 2, 3, \ldots, n. \tag{7.4.1}$$

This is a generalization to larger groups of sites of the assumption (6.1.4) that the states on a pair of sites are uncorrelated.

Let $\hat{u}_j(\boldsymbol{\ell})$ be the contribution to $\hat{h}_j(\boldsymbol{\ell})$ from $\mathfrak{g}_j(\boldsymbol{\ell})$ which would not appear in any subgraph. Then applying (7.2.15) to the Hamiltonian and substituting from (7.4.1) gives

$$\widehat{H}(\{\pi_n(\boldsymbol{\ell})\}) = N \sum_{j=1}^{n} \omega_j \sum_{\{\boldsymbol{\ell}\}} \lambda_j(\boldsymbol{\ell}) \hat{u}_j(\boldsymbol{\ell}) \prod_{\{\boldsymbol{\ell}'\}} [\pi_1(\boldsymbol{\ell}')]^{\vartheta_{1j}(\boldsymbol{\ell}'|\boldsymbol{\ell})}. \tag{7.4.2}$$

This expression will contain a term corresponding to any $\hat{u}_j(\boldsymbol{\ell}) \neq 0$. For the entropy, substituting into (7.2.27) and using (7.2.5) gives

$$\begin{aligned}
S(\{\pi_n(\boldsymbol{\ell})\}) &= N \sum_{j=1}^{n} \chi_j^{(n)} \sum_{\{\boldsymbol{\ell}\}} \lambda_j(\boldsymbol{\ell})\, \pi_j\, (\boldsymbol{\ell}) \sum_{\{\boldsymbol{\ell}'\}} \vartheta_{1j}(\boldsymbol{\ell}'\,|\,\boldsymbol{\ell}) \ln[\pi_1(\boldsymbol{\ell}')] \\
&= N \sum_{j=1}^{n} \Omega_{1j}\, \chi_j^{(n)} \sum_{\{\boldsymbol{\ell}\}} \lambda_1(\boldsymbol{\ell})\, \pi_1\, (\boldsymbol{\ell}) \ln[\pi_1(\boldsymbol{\ell})] \\
&= N\, \omega_1 \sum_{\{\boldsymbol{\ell}\}} \lambda_1(\boldsymbol{\ell})\, \pi_1\, (\boldsymbol{\ell}) \ln[\pi_1(\boldsymbol{\ell})] = S_1(\{\pi_1(\boldsymbol{\ell})\}).
\end{aligned} \tag{7.4.3}$$

So the approximation (7.4.1) reduces the entropy of the cluster method with sequence $\mathcal{S}$ to the mean-field (single-site) entropy.

As an example of this analysis consider the case of the Bethe-pair approximation for the Ising model discussed in Sect. 7.3. It is straightforward to show that substituting

$$\begin{aligned}
\pi_2(+2) &= [\pi_1(+1)]^2 = \tfrac{1}{4}(1+m)^2, \\
\pi_2(0) &= \pi_1(+1)\, \pi_1\, (-1) = \tfrac{1}{4}(1-m^2), \\
\pi_2(-2) &= [\pi_1(-1)]^2 = \tfrac{1}{4}(1-m)^2,
\end{aligned} \tag{7.4.4}$$

into (7.3.7) yields the mean-field free energy (6.1.9). However some questions remain. Whilst we do not expect the approximation (7.4.1) to give the equilibrium solution for the cluster approximation based on the sequence $\mathcal{S}$, we do expect it to satisfy the normalization and consistency conditions. Suppose, as we have indicated in Sect. 7.2, that each microsystem on a lattice site can be in any one of ν states. Then the single-site graph, with occupation k, is denoted by $\mathfrak{g}_1(k)$ with occupation number $\pi_1(k)$ and $\lambda_1(k) = 1$, for $k = 1, 2, \ldots, \nu$. Let the graph $\mathfrak{g}_j$ have n_j sites. Its possible configurations are given by a set of numbers $\ell_j(k)$, $k = 1, 2, \ldots, \nu$, where $\ell_j(k)$ denotes the number of sites with microsystem in state k, with

$$\sum_{k=1}^{\nu} \ell_j(k) = n_j. \tag{7.4.5}$$

Then

$$\lambda_j(\{\ell_j(k)\}) = \frac{n_j!}{\prod_{k=1}^{\nu}[\ell_j(k)]!}, \qquad \vartheta_{1j}(k|\{\ell_j(k)\}) = \ell_j(k). \tag{7.4.6}$$

The mean-field assumption (7.4.1) now takes the form

$$\pi_j(\{\ell_j(k)\}) = \prod_{k=1}^{\nu}[\pi_1(k)]^{\ell_j(k)} \tag{7.4.7}$$

and it is straightforward to show that the normalization condition (7.2.2) and the consistency condition (7.2.5) (with $\iota = 1$ and $\Omega_{1j} = n_j$) are satisfied.

7.5 3-State Potts Model on a Triangular Lattice

Here it is convenient to use the Ising spin-1 variable $\sigma = -1, 0, 1$ to denote the three states on a lattice site. The Hamiltonian (3.6.53) is generalized so that the chemical potentials for states ± 1 are $\mu_\pm$ and there is a three-site energy $-R_3$ around every triangle, present only when the states of the sites are the same. We shall concentrate on the 'ferromagnetic' case $R \geq 0$, $R_3 \geq 0$.

Let the partially ordered sequence $\mathcal{S}$ be the section graphs $\mathfrak{g}_1$: a single site, $\mathfrak{g}_2$: a single edge and $\mathfrak{g}_3$: a single triangle. Then $\omega_1 = 1$, $\omega_2 = 3$ and $\omega_3 = 2$ and, as shown in Example 16.4.3, the weights are $\chi_1^{(3)} = 1$, $\chi_2^{(2)} = -3$ and $\chi_3^{(3)} = 2$. Labelling site, pair and triangle states in the obvious way[8]

$$\hat{h}_1(0) = 0, \qquad \hat{h}_1(\pm 1) = -L_\pm,$$

$$\hat{h}_2(0,0) = -K, \qquad \hat{h}_2(\pm 1, \pm 1) = -2L_\pm - K,$$

---

[8] The ordering of the arguments in all the probabilities and variables defined here is arbitrary. In the interests of brevity a list of all the λ's and ϑ's has been omitted; it is easily calculated.

$$\begin{aligned}
&\hat{h}_2(0,\pm1) = -L_{\pm1}, && \hat{h}_2(\pm1,\mp1) = -L_+ - L_-,\\
&\hat{h}_3(0,0,0) = -3K - K_3, && \hat{h}_3(\pm1,\pm1,\pm1) = -3L_\pm - 3K - K_3,\\
&\hat{h}_3(0,\pm1,\pm1) = -2L_{\pm1} - K, && \hat{h}_3(\pm1,\mp1,\mp1) = -L_{\pm1} - 2L_\mp - K,\\
&\hat{h}_3(\pm1,0,0) = -L_\pm - K, && \hat{h}_3(0,+1,-1) = -L_+ - L_-,
\end{aligned} \tag{7.5.1}$$

where $K := R/T$, $K_3 := R_3/T$ and $L := \mu/T$. In terms of the one- and two-site probabilities we now define the five variables

$$\begin{aligned}
\tau(\pm1,\pm1) &= \left[\frac{\pi_2(\pm1,\pm1)}{\pi_2(0,0)}\right]^{\frac{1}{2}}\left[\frac{\pi_1(\pm1)}{\pi_1(0)}\right]^{-\frac{1}{6}} \mathfrak{z}_\pm^{\frac{1}{6}},\\
\tau(\pm1,\mp1) &= \left[\frac{\pi_2(\pm1,\mp1)}{\pi_2(0,0)}\right]^{\frac{1}{2}}\\
&\quad\times\left[\frac{\pi_1(\pm1)}{\pi_1(0)}\right]^{-\frac{1}{12}}\left[\frac{\pi_1(\mp1)}{\pi_1(0)}\right]^{-\frac{1}{12}} (\mathfrak{z}_\pm\mathfrak{z}_\mp)^{\frac{1}{12}},\\
\tau(0,\pm1) &= \left[\frac{\pi_2(0,\pm1)}{\pi_2(0,0)}\right]^{\frac{1}{2}}\left[\frac{\pi_1(\pm1)}{\pi_1(0)}\right]^{-\frac{1}{12}} \mathfrak{z}_\pm^{\frac{1}{12}},
\end{aligned} \tag{7.5.2}$$

where

$$\mathfrak{z}_\pm := \exp(-L_\pm). \tag{7.5.3}$$

Then from the equilibrium conditions (7.2.30),

$$\begin{aligned}
\frac{\pi_3(\pm1,\pm1,\pm1)}{\pi_3(0,0,0)} &= [\tau(\pm1,\pm1)]^3,\\
\frac{\pi_3(0,\pm1,\pm1)}{\pi_3(0,0,0)} &= [\tau(0,\pm1)]^2\tau(\pm1,\pm1)\mathfrak{X}\mathfrak{M},\\
\frac{\pi_3(0,0,\pm1)}{\pi_3(0,0,0)} &= [\tau(0,\pm1)]^2\mathfrak{X}\mathfrak{M},\\
\frac{\pi_3(\pm1,\mp1,\mp1)}{\pi_3(0,0,0)} &= [\tau(\pm1,\mp1)]^2\tau(\mp1,\mp1)\mathfrak{X}\mathfrak{M},\\
\frac{\pi_3(-1,0,+1)}{\pi_3(0,0,0)} &= \tau(-1,0)\tau(0,+1)\tau(+1,-1)\mathfrak{X}^{\frac{3}{2}}\mathfrak{M},
\end{aligned} \tag{7.5.4}$$

where

$$\mathfrak{X} := \exp(-K), \qquad \mathfrak{M} := \exp(-K_3). \tag{7.5.5}$$

Using the consistency conditions, five equations can be obtain relating the five τ's. However, for simplicity we shall consider only the case where $\mathfrak{Z}_+ = \mathfrak{Z}_- = \mathfrak{Z}$ and the solution is in the subspace invariant under spin inversion. Now we have three variables

$$\begin{aligned} &x = \tau(0,+1) = \tau(0,-1), \qquad y = \tau(+1,-1), \\ &z = \tau(+1,+1) = \tau(-1,-1) \end{aligned} \tag{7.5.6}$$

and the 1–3 and 2–3 consistency conditions yield

$$\begin{aligned} z^3 \frac{f_2(x,y,z)}{f_1(x,y,z)} &= \mathfrak{Z}\left[\frac{g_2(x,y,z)}{g_1(x,y,z)}\right]^3, \\ \frac{f_2(x,y,z)}{f_1(x,y,z)} &= \mathfrak{Z}(\mathfrak{X}\mathfrak{M})^6\left[\frac{g_3(x,y,z)}{g_1(x,y,z)}\right]^6, \\ y^3 \frac{f_2(x,y,z)}{f_1(x,y,z)} &= \mathfrak{Z}(\mathfrak{X}\mathfrak{M})^3\left[\frac{g_2(x,y,z)}{g_1(x,y,z)}\right]^3, \end{aligned} \tag{7.5.7}$$

where

$$\begin{aligned} f_1(x,y,z) &:= 1 + 2x^2\mathfrak{X}\mathfrak{M}(2 + z + y\mathfrak{X}^{\frac{1}{2}}), \\ f_2(x,y,z) &:= z^3 + \mathfrak{X}\mathfrak{M}(x^2 + 2zx^2 + 3zy^2 + 2x^2y\mathfrak{X}^{\frac{1}{2}}), \\ g_1(x,y,z) &:= 1 + 2x^2\mathfrak{X}\mathfrak{M}, \\ g_2(x,y,z) &:= z^2 + \mathfrak{X}\mathfrak{M}(x^2 + zy^2), \\ g_3(x,y,z) &:= 1 + z + y\mathfrak{X}^{\frac{1}{2}}, \\ g_4(x,y,z) &:= 2yz + x^2\mathfrak{X}^{\frac{1}{2}}. \end{aligned} \tag{7.5.8}$$

It is not difficult to show that the Eq. (7.5.7) have the disordered solutions

$$x_{\mathrm{D}}^{(\pm)} = y_{\mathrm{D}}^{(\pm)} = \frac{\mathfrak{X}^{\frac{3}{2}}\mathfrak{M} \pm \sqrt{\mathfrak{X}^3\mathfrak{M}^2 - 8\mathfrak{X}\mathfrak{M}(1 - 2\mathfrak{X}\mathfrak{M})}}{4\mathfrak{X}\mathfrak{M}}, \qquad z_{\mathrm{D}} = 1. \tag{7.5.9}$$

At high temperatures these solutions are real with $x_{\mathrm{D}}^{(-)} < 0$ when $T > (R + R_3)/\ln(2)$. Below a temperature given by $\mathfrak{X}^2\mathfrak{M} = 8(1 - 2\mathfrak{X}\mathfrak{M})$ the disordered

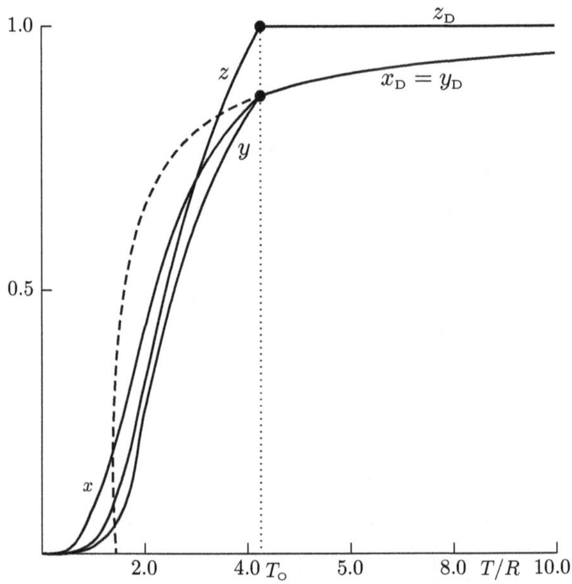

Fig. 7.1 The second-order transition in the 3-state Potts model for the case $R_3 = 0$. The continuation of the disordered solution $x_D = y_D$ below the transition is shown by a *broken line*

solutions no longer exist. A second-order transition to an ordered solution occurs at a point on the disordered solution curve, yielding a temperature T_o, which is a repeated root of the Eq. (7.5.7). Numerical calculations yield the results $T_o = 4.2520\,R$, when $R_3 = 0$ and $T_o = 3.3761\,R_3$, when $R = 0$ and graphs of the variables x, y and z for the case $R_3 = 0$ are shown in Fig. 7.1.

7.6 A Lattice Model for Fluid Water

The feature of the water molecule whose cooperative effect leads to the anomalous properties of liquid and supercritical water is the hydrogen bond.[9] A number of statistical mechanical models, attempting to incorporate this feature have been developed. The simplest of these is the two-dimensional triangular lattice model of Bell and Lavis (1970) (see Sect. 3.7.2). In this section we consider a somewhat more realistic three-dimensional model, based on the body-centred cubic lattice (Fig. 16.8) and due to Bell (1972). This is a lattice fluid model of M molecules on $N(\geq M)$ sites with each molecule having four bonding arms pointing to the vertices of a tetrahedron. These arms are directed towards four of the eight first-neighbour sites and two arms have positive polarity and two negative, implying twelve distinguishable orientations

[9] For an account of the role of hydrogen bonding in water see the footnote on p. 61.

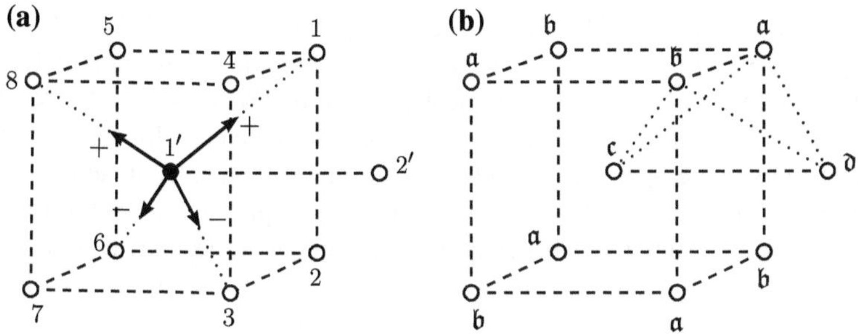

Fig. 7.2 **a** Sites on the body-centred cubic lattice with a water molecule placed on site 1′. (Reprinted from Bell (1972), by permission of the publisher IOP Publishing Ltd.). **b** The division of the body-centred cubic lattice into four sublattices

(Fig. 7.2a). Two first-neighbour molecules are connected by a hydrogen bond when a positive arm of one and a negative arm of the other lie along the segment connecting their sites. This occurs in 18 of the 144 possible orientational states of the two molecules.

The body-centred cubic lattice can be divided into two equivalent simple-cubic lattices. Let one be divided into two sublattices 𝔞 and 𝔟 as shown in Fig. 16.7 and the other into sublattices 𝔠 and 𝔡, resulting in a division of the lattice into four equivalent sublattices (Fig. 7.2b). It can be seen that each of these sublattices forms a face-centred cubic lattice (Fig. 16.9) and the pairs 𝔞–𝔠, 𝔞–𝔡, 𝔟–𝔠, 𝔟–𝔡 combine to form diamond lattices (Fig. 16.10). An open ice Ic structure, with one diamond sublattice of the body-centred cubic vacant and the other occupied by a bonded network, is thus possible. In the present fluid-state theory this arrangement occurs only as short-range order, but it is of crucial importance since it is the breakdown of the local open structure by increasing temperature which produces anomalous effects.[10] Locally-ordered regions with the close-packed ice VII structure, with each diamond sublattice occupied by a fully-bonded structure, will exist at high pressures together with a range of other types of short-range order including the 'cage' structure in which unbonded molecules lie in the interstices of an open bonded network. However, the types of local order in real water are considerably more complex. The reduction of the ten or so observed ice structures to two indicates the approximations made in neglecting hydrogen bond distortion from the tetrahedral angles and the possibility of ordered hydrogen positions.

An interaction energy $-\varepsilon$ ($\varepsilon > 0$) is assumed for each unbonded first-neighbour pair of molecules which changes to $-\varepsilon - w$ ($w > 0$) when a hydrogen bond is formed. To ensure that the ice VII structure is stable only at high pressures, an energy penalty against close packing is needed. This is imposed by associating an energy $\frac{1}{3}u$ ($u > 0$) with the occupation by three molecules of a triad of sites where two second neighbours share a first neighbour (for example, the triads $121'$ or $1'2'1$ in Fig. 7.2a).

[10] For the nomenclature for the various phases of ice (I, Ic, II, etc.) see Eisenberg and Kauzmann (1969, Chap. 3) The difference between the locally similar ice Ic (diamond) and ice I (quartzite) structures is not likely to be significant in the fluid state.

The field conjugate to M is μ the chemical potential and to N is $-P$, where P is the pressure. In our calculations we choose μ to be the independent field, which is equivalent to using the grand-canonical distribution (Sect. 1.4.2) with free energy $-PN$. At zero temperature the free energy is equal to the enthalpy H and we compare its values in the open and closed-packed structures and also in an intermediate structure, which is similar to the open structure except that one of the remaining face-centred cubic sublattices is occupied by unbonded molecules. It is not difficult to show that

$$\begin{aligned} H_1(\text{open structure}) &= -N\left(\varepsilon + w + \tfrac{1}{2}\mu\right), \\ H_1(\text{intermediate structure}) &= -N\left(2\varepsilon + w - u + \tfrac{3}{4}\mu\right), \\ H_1(\text{close-packed structure}) &= -N\left(4\varepsilon + 2w - 4u + \mu\right). \end{aligned} \tag{7.6.1}$$

At high and low values of chemical potential the close-packed and open structures are respectively stable. By determining the boundaries of stability between the three phases it is easy to see that, if $2u > \varepsilon + w$, there will be a range of chemical potentials (and pressures) for which the intermediate phase is stable at $T = 0$. For simplicity we assume that this is not the case. Then there is a chemical potential μ_0 and pressure P_0 given by

$$\mu_0 = 2(4u - w - 3\varepsilon), \qquad P_0 = 2(2u - \varepsilon), \tag{7.6.2}$$

such the open structure is stable at $T = 0$ for $\mu < \mu_0$ and the close packed structure for $\mu > \mu_0$, with equivalent conditions applying to the pressure. If, on physical grounds, the pressure is restricted to positive values then, in order to ensure that both close-packed and open phases can be stable at T,

$$P_0 > 0, \qquad \text{which is equivalent to} \qquad \mu_0 > -2(\varepsilon + w) \tag{7.6.3}$$

and thus

$$\varepsilon < 2u < \varepsilon + w. \tag{7.6.4}$$

To incorporate sublattice ordering we use a cluster approximation based on a tetrahedron of four sites, like $141'2'$ in Fig. 7.2a, one belonging to each of the four sublattices, as shown in Fig. 7.2b. The partially-ordered sequence of subgraphs $\mathfrak{g}_1, \mathfrak{g}_2, \ldots, \mathfrak{g}_{15}$, where $\mathfrak{g}_{15}$ is the tetrahedron, are given in Example 16.4.7 and the parameters for the full KHDeB treatment are given in (16.4.51) and (16.4.52). However, this treatment is rather complicated and we shall present a version of the calculations of Bell (1972) using the Guggenheim and McGlashan (1951) method. For this the parameters $\widetilde{\omega}_j^{(15)}, \widetilde{\chi}_j^{(15)}$, $j = 1, 2, \ldots, 15$ are given in (16.4.59). The states of a single site $(\mathfrak{g}_1, \ldots, \mathfrak{g}_4)$ are denoted by an index $\ell = 0, 1$ for an unoccupied and an occupied site, respectively. Then[11]

$$\begin{aligned} &\lambda_j(0) = 1, \qquad \lambda_j(1) = 12, \\ &\hat{h}_j(0) = 0, \qquad \hat{h}_j(1) = -\mu, \qquad j = 1, \ldots, 4. \end{aligned} \tag{7.6.5}$$

For the energy levels and parameters for $\mathfrak{g}_{15}$ given in Table 7.1 we need to adjust the triad energy according to the prescription given by (16.4.51), (16.4.58) and (16.4.60). That is to say, $\frac{1}{3}u$ is replaced by u.

A number density ρ, equal to 1 for close packing, is

$$\rho := M/N, \tag{7.6.6}$$

in terms of which the single-site probabilities are given by

$$\pi_j(0) = 1 - \rho, \qquad \pi_j(1) = \frac{1}{12}\rho, \qquad j = 1, \ldots, 4. \tag{7.6.7}$$

Using the parameters given here together with those contained in (16.4.59) we have, from (7.2.31), ten equilibrium conditions for $\pi_{15}(\ell)$, $\ell = 1, 2, \ldots, 10$. With μ taken as the independent field, the free energy per lattice site is $-P$ and using the equilibrium condition for $\pi_{15}(4)$, from (7.2.31) and (7.2.32),

$$P = \tfrac{3}{2}T\left[\ln\left(\frac{\rho}{12}\right) + \ln(1-\rho)\right] - T\ln[\pi_{15}(4)] - \varepsilon - w + \tfrac{1}{2}\mu. \tag{7.6.8}$$

A parameter τ is now defined by

$$\tau^2 := \frac{\pi_{15}(8)}{\pi_{15}(4)} = \exp\left(\frac{2\varepsilon - 4u + \mu}{2T}\right)\left[\frac{\rho}{12(1-\rho)}\right]^{3/2}, \tag{7.6.9}$$

where the last expression follows from the equilibrium conditions. Reference to Table 7.1 shows that τ^2 expresses the ratio of ice VII-type to ice Ic-type configurations on the basic group of four sites. The equilibrium conditions can now be used to obtain all the ratios $\pi_{15}(\ell)/\pi_{15}(4)$ and substituting into the two 1–15 consistency conditions gives

$$\begin{aligned} \rho &= \mathfrak{X}^{-4}\mathfrak{M}^{-1}\,\tau^{-2}X(\tau)\,\pi_{15}(4), \\ 1 - \rho &= \mathfrak{X}^{-4}\mathfrak{M}^{-1}\,\tau^{-2}Y(\tau)\,\pi_{15}(4), \end{aligned} \tag{7.6.10}$$

[11] In this section it is convenient to use energies rather that couplings.

Table 7.1 Configurations and energy parameters for the tetrahedron $\mathfrak{g}_{15}$

Configuration index ℓ	Configuration	$\lambda_{15}(\ell)$	$\hat{h}_{15}(\ell)$
1	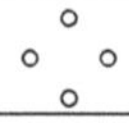	1	0
2	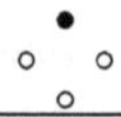	48	$-\mu$
3	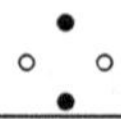	288	-2μ
4	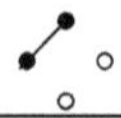	72	$-w-\varepsilon-2\mu$
5	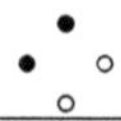	504	$-\varepsilon-2\mu$
6		1728	$-w-2\varepsilon+u-3\mu$
7	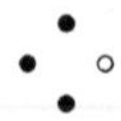	5184	$-2\varepsilon+u-3\mu$
8		648	$-2w-4\varepsilon+4u-4\mu$
9		9072	$-w-4\varepsilon+4u-4\mu$
10		11016	$-4\varepsilon+4u-4\mu$

where

$$\begin{aligned} X(\tau) := {} & 12\mathfrak{X}^3\,\tau[\mathfrak{Y}\mathfrak{M}^3 + 3\mathfrak{X}\mathfrak{M}(1+7\mathfrak{Y}^2+4\mathfrak{X}^2\mathfrak{Y}^2)\tau \\ & + 108\mathfrak{X}^2\mathfrak{Y}(1+3\mathfrak{Y}^2)\tau^2 + 54\mathfrak{X}\mathfrak{M}(1+14\mathfrak{Y}^2+17\mathfrak{Y}^4)\tau^3\}, \end{aligned} \quad (7.6.11)$$

$$\begin{aligned} Y(\tau) := {} & \mathfrak{M}^5 + 36\mathfrak{X}^3\mathfrak{Y}\mathfrak{M}^3\,\tau + 36\mathfrak{X}^4\mathfrak{M}(1+7\mathfrak{Y}^2+4\mathfrak{X}^2\mathfrak{Y}^2)\tau^2 \\ & + 432\mathfrak{X}^5\mathfrak{Y}(1+3\mathfrak{Y}^2)\tau^3 \end{aligned} \quad (7.6.12)$$

and

$$\mathfrak{X} := \exp\left(-\frac{\varepsilon}{2T}\right), \qquad \mathfrak{Y} := \exp\left(-\frac{w}{2T}\right), \qquad \mathfrak{M} := \exp\left(-\frac{u}{T}\right). \tag{7.6.13}$$

Dividing the first relation of (7.6.10) by the second yields

$$\frac{\rho}{1-\rho} = \frac{X(\tau)}{Y(\tau)} \tag{7.6.14}$$

and, from (7.6.8) and (7.6.10),

$$\exp\left(\frac{P}{T}\right) = \frac{[X(\tau)]^3}{\mathfrak{M}^5\,[X(\tau)+Y(\tau)]^2}. \tag{7.6.15}$$

Equations (7.6.14) and (7.6.15) constitute an implicit equation of state giving ρ in terms of P and T. Best results are found when u lies in the middle of the range given by (7.6.4). That is, when $2u = \frac{1}{2}w + \varepsilon$, which gives, from (7.6.2), $P_0 = w$. There is liquid–vapour phase separation, and density maxima occur on (ρ, T) isobars in the liquid state and in the supercritical state for $P < P_0$, as shown in Fig. 7.3. The temperature of maximum density varies very little with pressure. A compressibility minimum also occurs on isobars in the liquid state, and the mean number of first neighbours of a given molecule is between 4 and 5 at low pressures (Bell 1972). It can be seen that reasonable qualitative agreement with experiment is obtained in spite of the simplifications inherent in the model. The same model was used by Lavis and Christou (1977) to study dielectric properties and Wilson and Bell (1978) for an investigation of aqueous solutions. Bell and Salt (1976) used the mean-field approximation on a similar model to discuss transitions to ordered (ice-like) states.

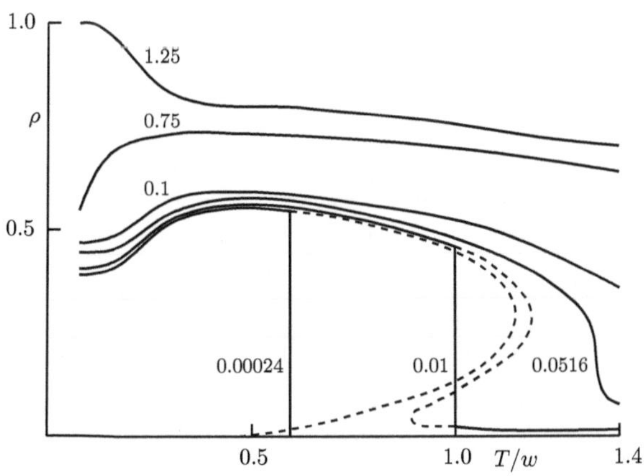

Fig. 7.3 Density–temperature isobars for $\varepsilon/w = 2$ and $u/w = 5/4$. Each isobar is labelled with the value of P/w. (Reprinted from Bell (1972), by permission of the publisher IOP Publishing Ltd.)

Weres and Rice (1972) used a model based on the body-centred cubic lattice, making rather different assumptions and were mainly concerned with local properties. These authors had reservations about Bell's triad term u which they considered made excessively large contributions to the energies of less dense structures. Other authors who used the body-centred cubic lattice with a different approach were O'Reilly (1973) and Fleming and Gibbs (1974) (see also the review by Stillinger 1975). A square lattice model with polarized bonds was discussed by Lavis and Christou (1979). Whitehouse et al. (1984) examined the Bell model using Monte Carlo methods and concluding that 'the cooperativity of the model is seriously underestimated by the first-order approximation, although the general behaviour of the model is correctly represented'. These authors consider that it would be more realistic to have $u = 0$, $\varepsilon < 0$, pointing out that the inequality (7.6.4) could still be satisfied and a positive P_0 obtained. Shinmi and Huckaby (1986) have used a similar model with non-polar bonding arms and repulsive bond energy to represent carbon tetrachloride.

Both the two-dimensional bonded lattice model of Bell and Lavis (1970) and the three-dimensional non-polar version of the Bell (1972) model, introduced by Shinmi and Huckaby (1986), can be expressed in Hamiltonian form as modified versions of the spin-1 Ising model. The modifications consist in the introduction of certain terms involving more that two sites. Based on this approach Southern and Lavis (1980) and Lavis and Southern (1984) used finite-size cluster real-space renormalization group methods to investigate the two models.

Part III
Exact Results

Introduction

For a one-component lattice fluid[1] on a finite lattice of N sites the grand partition function is an N-th degree polynomial in the fugacity $\mathfrak{z} := \exp(\mu/T)$. The only singularities of the free energy are the zeros of the partition function. However, none of these is real and positive since the coefficients of the partition-function polynomial are non-negative. Thus the lattice fluid on a finite lattice $\mathcal{N}$ can have no phase transitions. Since phase transitions do actually occur they must either be an artefact of the passage to the thermodynamic limit or be because the system is intrinsically infinite. The first of these possibilities is the one adopted in this book with the passage to the thermodynamic limit considered in detail in Sect. 8.1. However, it is of interest to consider briefly in Sect. 8.2 the formalism of statistical mechanics for infinite systems.

The *Peierls method* (Peierls 1936), which is described in Sect. 8.3, provides one of the few procedures for proving the existence of a phase transition without explicit calculation of the transition point. The original work of Peierls contains a proof that the Ising model on the square lattice has a spontaneous magnetic moment at sufficiently low temperatures. The argument was given a rigorous formulation by Griffiths (1964) and Dobrushin (1965). It was extended to lattice fluids with first-neighbour repulsion or exclusion by Dobrushin (1968a, b), for hypercubic lattices, and by Heilmann (1972), Abraham and Heilmann (1980) and Heilmann (1980) for triangular and face-centred cubic lattices. The Peierls method has been used to

---

[1] The equivalence between the spin-$\frac{1}{2}$ Ising model with first-neighbour interactions and a simple lattice fluid with first-neighbour pair interactions between particles was shown in Sect. 3.5.1.2 and, as we indicated there, this relationship holds between any one-component lattice fluid of particles of chemical potential μ and a spin-$\frac{1}{2}$ system on the same lattice in a magnetic field $\mathcal{H}$. It follows that the methods and results for the lattice fluid can be 'translated' into a spin-$\frac{1}{2}$ formulation and on that basis it is valid to described the ideas developed in Chap. 8 in terms of a lattice fluid model.

established the existence of phase transitions at sufficiently low temperatures for the bonded lattice fluid models of Bell and Lavis (1970) and Bell (1972) by Heilmann and Huckaby (1979) and Lavis and Southern (1984) respectively.

The location and evolution of partition function zeros is clearly important for the understanding of the occurrence of phase transitions. This approach, which is the subject of Sect. 8.4, was initiated by the important insights contained in the papers of Yang and Lee, (Yang and Lee 1952; Lee and Yang 1952). Their approach is based on the idea of taking the fugacity $\mathfrak{Z}$ to be a complex variable. They proved the well-known *circle theorem* that, for a spin-$\frac{1}{2}$ Ising ferromagnet with pairwise interactions of arbitrary range, the zeros of the grand partition function lie on the unit circle $|\mathfrak{Z}| = 1$ in the complex plane of the variable $\mathfrak{Z} := \exp(-2\mathcal{H}/T)$.[2] This result can be extended to a larger class of models (Asano 1968, 1970; Suzuki and Fisher 1971; Newman 1975; Lieb and Sokal 1981) but this does not include the ν-state Potts model for $\nu > 2$ (Martin and Maillard 1986; O'Rourke et al. 1995). Lieb and Ruelle (1972) considered the distribution of zeros for the antiferromagnetic case and showed that, if the temperature is above the critical temperature, with an arc of the unit circle around $\mathfrak{Z} = 1$ free of zeros, there will also be no zeros in the circle containing this arc and orthogonal to the unit circle. Heilmann and Lieb (1972) were able to derive the circle theorem for both Ising and Heisenberg models from an investigation of the partition function zeros of a monomer–dimer system in the monomer fugacity plane. Nishimori and Griffiths (1983) have investigated the motion of the zeros in the complex plane as the temperature is changed. Ruelle (1971) used the contraction method of Asano (1970) to prove a result which provides, in principle, a method for obtaining regions containing the grand partition function zeros of one-component lattice fluids with finite-range interactions whether or not they lie on $|\mathfrak{Z}| = 1$.

The Yang–Lee circle theorem and Ruelle's theorem are both about the complex plane of the variable $\mathfrak{Z}$, whether it is understood as the Boltzmann factor of the magnetic field $\mathcal{H}$ or of the chemical potential μ. The important point about $\mathfrak{Z}$ is that it is the Boltzmann factor for a field associated with single-site occupations. Any partition function of a lattice system can, however, be expressed as a sum of powers of a number of Boltzmann factors, some of which will arise from particle interactions. These quantities can also, in the spirit of Yang and Lee, be treated as complex variables. Thus, for example, for the spin-$\frac{1}{2}$ ferromagnetic Ising model with Hamiltonian (Sect. 3.5.1) on the finite lattice $\mathcal{N}$ the grand partition function is a polynomial in $\mathfrak{X} := \exp(-2J/T)$ and $\mathfrak{Z} := \exp(-2\mathcal{H}/T)$. As indicated above the distribution of zeros in the complex plane of $\mathfrak{Z}$ satisfies the circle theorem. The distribution of zeros in the complex plane of $\mathfrak{X}$ has also been investigated (Fisher 1965; Jones 1966; Brascamp and Kunz 1974) and it has been shown that in the limit

[2] There is, of course, an equivalent lattice fluid formulation of this result (Sect. 8.4.2), but since it relies on the Ising model hole–particle symmetry condition $E(\mathcal{M}) = E(\mathcal{N} - \mathcal{M})$, where $\mathcal{N} - \mathcal{M}$ is the complement of $\mathcal{M}$ on the lattice $\mathcal{N}$, it has the rather artificial property that two empty sites interact with the same energy as two occupied sites.

of an infinite system, when $\mathcal{H} = 0$, the asymptotic locus of zeros is given by the *Fisher circles*

$$\mathfrak{X} = \sqrt{2}\exp(i\theta) \pm 1\,.$$

For the square-lattice Ising model on a semi-infinite lattice Wood (1985) showed that the Fisher circles could be computed as curves on which transfer matrix eigenvalues are degenerate in modulus and thus that these latter yield the asymptotic locus of partition function zeros.

As we saw in Sect. 3.6.4, the 2-state Potts model is isomorphic to the zero-field spin-$\frac{1}{2}$ Ising model with $R = 2J$. The Fisher circle with the plus sign has been generalized to the two-dimensional ν-state Potts model (Kenna 1998) to show that, in the complex plane of $\mathfrak{X} := \exp(-R/T)$, the asymptotic locus of zeros lies on the Fisher circle

$$\mathfrak{X} = G(\nu;\theta) := \frac{\sqrt{\nu}\exp(i\theta)-1}{\nu-1}\,.$$

The treatment of duality for the ν-state Potts model in Sect. 9.2.1 yields the self-dual circle

$$\mathfrak{X} = F(\nu;\theta) := 1 + \sqrt{\nu}\exp(i\theta)\,.$$

Taking into account the fact that the variable $\mathfrak{X}$ used here is the reciprocal of that used in Sect. 9.2.1, it is not difficult to show that $\mathfrak{X} = G(\nu;\theta)$ is simply the self-dual circle $\mathfrak{X} = 1/F(\nu;\varphi)$, where $\exp(i\varphi) = [\sqrt{\nu} - \exp(i\theta)]/[\sqrt{\nu}\exp(i\theta) - 1]$. The 'extra' Fisher circle with the minus sign is due to the ferromagnetic–antiferromagnetic symmetry present in the spin-$\frac{1}{2}$ Ising model but not in the Potts model when $\nu \neq 2$.

One of the most useful techniques in statistical mechanics is the formulation of a transformation leading to an equivalence between two different models. We have already seen in Chap. 3 that such equivalences exist between different interpretations of the Ising model (ferromagnetic, antiferromagnetic, lattice gas and binary mixture) and between Ising, Potts and vertex models. Underlying the transformation methods described in Chaps. 9 and 10 is a *lattice transformation* of the type described in Sect. 16.2.2. Briefly summarized, for a given lattice the first step is to add a lattice site to some or all of either its faces or edges. The second step is then to either remove the original lattice, leaving only the new sites, or to consider the 'decorated' lattice of new and old sites. The Wegner transformation, incorporating the duality and weak-graph transformations and treated in Sect. 9.2, is a case where sites are added to the faces of the lattice and the original lattice is removed. The star-triangle transformation of Sect. 9.4 is an example of *face decoration*, where in this case sites are added to half the faces of a triangular lattice with the originally lattice retained to give a honeycomb lattice. Chapter 10 considers various models obtained by edge-decorating an Ising model. The aim for all these transformation methods is to obtain relationships between the couplings for the systems on the original and transformed lattices which yield a relationship between their partition functions. This means that if one model is solved then so is the other.

Given a finite lattice $\mathcal{N}$ of dimension d the thermodynamic limit in which $\mathcal{N}$ tends to the infinite lattice $\mathcal{L}$ is normally understood to be a growth in lattice size in all d lattice dimensions (Sect. 8.1) and the critical exponents of a phase transition in the system will be functions of d (Sects. 4.5.6 and 4.7.2). In Sect. 4.10 we considered the modified situation where the thermodynamic limit was taken in only $\mathfrak{d} < d$ dimensions. The resulting lattice $\mathcal{L}_{\mathfrak{d}}$ then has the critical properties of a $\mathfrak{d}$-dimensional system, with no non-analytic behaviour associated with phase transitions unless $\mathfrak{d} > 1$.

Transfer matrices in statistical mechanics are used to grow a finite lattice step-by-step to the infinite limit in one lattice dimension. In principle, this method can be applied to a lattice of any dimension and it has been successfully applied to some three-dimensional systems (Huang et al. 1997; Popkov and Nienhuis 1997). However, most applications have been to one- and two-dimensional lattices and in Chaps. 11 and 12 attention is restricted to these cases.

The first use of transfer matrices is due to Kramers and Wannier (1941a, b). As a preparation for a study of the two-dimensional square-lattice Ising model they first re-derived in transfer matrix form the solution of the one-dimensional Ising model, which was first obtain by Ising (1925) by combinatorial means. They then proceeded to investigate lattices with increasing values of N_2, the number of lines of sites. Although, of course, these models effectively represent increasingly complicated one-dimensional models without true phase transitions, they were able to show that an increasingly steep maximum in the heat capacity gave progressively more accurate estimates of the critical temperature of the two-dimensional system, which they also derived using a duality transformation. The use of systems with finite values of N_2 and *incipient phase transitions* to estimate the critical parameters of truly two-dimensionally infinite systems has proved a valuable tool in statistical mechanics (see, for example, Lavis 1976) and, after an introduction to transfer matrices in Sect. 11.1 this method is developed and used in Sect. 11.2.

An interesting approach due to the late David Wood and his coworkers (Wood 1987; Wood et al. 1987; Wood and Turnbull 1988; Wood et al. 1989; Wood and Ball 1990) combines the investigation of partition function zeros contained in the work of Yang and Lee and Ruelle with an exploitation of transfer matrices. This method, described in Sect. 8.4, is able, by tracing the evolution of partition function zeros with increasing N_2 to obtain estimates for critical points and exponents.

In the context of transfer matrices the 'solution' of a two-dimensional model means the determination of the partition function in the limiting case where $N_2 \to \infty$. Thus the problem is to find some way of handling the limit where the dimension of the transfer matrix becomes infinite. This must involve a calculation of some, or all, of the transfer matrix eigenvalues and ideally also the eigenvectors, as functions of N_2.

Although there are now many different published solutions of the two-dimensional Ising model,[3] the first solution of the square-lattice case by Onsager (1944) was an important landmark in the history of statistical mechanics. By applying

[3] To emphasize this Baxter and Enting (1978) called their method (of which an account is also given in Lavis and Bell 1999a, Chap. 8) the '339th' presentation.

group representation theory he was able to determine the eigenvalues of the transfer matrix as functions of N_2. Subsequent simplification was achieved by Kaufman (1949) who showed that the transfer matrix belongs to the group of spinor operators, with Kaufman and Onsagar (1949) deriving the pair correlation function. Although the work of Onsager and Kaufman is restricted to the square lattice, their work was soon generalized to the triangular and honeycomb lattices (Houtappel 1950; Husimi and Syôzi 1950; Newell 1950; Syôzi 1950; Temperley 1950; Wannier 1950).

A general introduction to vertex models has been given in Sect. 3.8. As was noted there a major breakthrough for vertex models was the derivation of the entropy of the ice-rule (six-vertex) model by Lieb (1967a, b), followed by solutions for the F model (Lieb 1967c) and the KDP model (Lieb 1967d).[4] Unlike Onsager's solution of the zero-field Ising model, with which they are comparable in importance, Lieb's solutions of the six-vertex model include an external field. They also proved fruitful in opening up the search for exact solutions of other models particularly the eight-vertex model which was solved by Baxter (1972b, 1973a, b, c). In Chap. 12 we find it convenient to model our presentation on the account by Baxter (1978) of the eight-vertex model on a rapidity lattice (Sect. 16.3). This yields the *checkerboard Ising model* of Baxter (1986) and the six-vertex model as special cases, although for the latter this approach has the disadvantage that it cannot include an external field.[5]

Section 3.9 contains a brief account of dimer models. In Chap. 13 a method is described for evaluating exactly the dimer partition function. This was first presented for the square-lattice case by Temperley and Fisher (1961), Fisher (1961) and Kasteleyn (1961), but it can be applied to any regular two-dimensional lattice. Vertex and Ising model equivalences to the dimer model are derived is Sect. 13.3. It is of particular interest to note that, although the work of Chap. 12 is restricted to the square-lattice eight-vertex model, the dimer equivalence of Sect. 13.3.2 allows the eight-vertex free-energy densities for the honeycomb and triangular lattice models to be derived.

---

[4] Elliott Lieb (private communication) was alerted to the problem of calculating the entropy of the ice-rule model by reading the paper of Nagle (1966a) in which he used a series method to obtain an approximate value of $\ln(1.543)$ for the entropy per site; this being compared to the exact result of $\frac{3}{2}\ln\left(\frac{4}{3}\right)\simeq\ln(1.5396)$, subsequently obtained by Lieb (Sect. 12.3.10.4).

[5] For this the reader is referred to the original papers or to Lieb and Wu (1972).

[illegible] representation theory he was able to determine the eigenvalues of the transfer matrix as functions of [illegible]. Subsequent simplification was achieved by Kaufman (1949) who showed that the transfer matrix belonged to the group of spinor operators, and Kaufman and Onsager (1949) derived the pair correlation function. Although the work of Onsager and Kaufman is important to the [illegible] their work [illegible] to the [illegible] (see, for example, [illegible] 1964; Domb and Sykes 1956; Montroll 1964; Sykes 1961; Temperley 1959; Wannier 1950).

A general introduction to vertex models has been given in Sect. 5.12. As with the Ising model, there a major breakthrough for vertex models was the derivation of the entropy of the ice-rule (six-vertex) model by Lieb (1967a, b), followed by solutions to the F model (Lieb 1967c) and the KDP model (Lieb 1967d). Since the general solution [illegible] Ising model, with which they are comparable in importance, [illegible]. Lieb's solution of the six-vertex model includes an external field. The [illegible] proved by a [illegible] up the general [illegible] solution of such model [illegible] namely the eight-vertex model which was solved by Baxter (1972b, 1973a, b, c). In Chap. 12 we include a comprehensive modern presentation of the account by Baxter (1982) for the eight-vertex model on a square lattice (Sect. 10.5). This yields the zero-field Ising model of Baxter (1986) and discuss [illegible] special case, although for the latter [illegible] approach has the disadvantage that it cannot include an external field.

Section 12.? contains a brief account of other models. In Chap. 13 a method is described for calculating exactly the dimer partition function. This was first presented for the square lattice case by Temperley and Fisher (1961), Fisher (1961) and Kasteleyn (1961). It can also be applied to any planar [illegible] showing the equivalence of the dimer model [illegible] discussed in Sect. [illegible]. It [illegible] the work of Sect. 12 [illegible] the [illegible]

[illegible]

[1] [illegible] general communication was alerted to the problem of calculating the entropy of the [illegible] rule model by reading the paper of Nagle (1966a) in which he used a series method to obtain an [illegible] to the value [illegible] for the entropy per site, this being compared to the experimental [illegible] and [illegible] subsequently obtained by Lieb (Sect. 10.?).

[2] [illegible] and [illegible] to [illegible] original papers for each [illegible]

Chapter 8
Algebraic Methods

8.1 The Thermodynamic Limit

For a one-component lattice fluid on a finite lattice $\mathcal{N}$ the state of the system can be described by specifying the subset $\mathcal{M} \subseteq \mathcal{N}$ of sites occupied by particles, and the Hamiltonian in the *grand-canonical distribution* (Sects. 1.4.2 and 2.1) is

$$\widehat{H}(\mathcal{N}, \mu; \mathcal{M}) := E(\mathcal{M}) - \mu|\mathcal{M}|, \tag{8.1.1}$$

$E(\mathcal{M})$ being the interaction energy between the particles of $\mathcal{M}$. The equilibrium (Gibbs) distribution is

$$p(\mathcal{N}, T, \mu; \mathcal{M}) = \frac{\exp[-\widehat{H}(\mathcal{N}, \mu; \mathcal{M})/T]}{Z_1(\mathcal{N}, T, \mu)}, \tag{8.1.2}$$

and

$$Z_1(\mathcal{N}, T, \mu) := \sum_{\{\mathcal{M}\}} \exp[-\widehat{H}(\mathcal{N}, \mu; \mathcal{M})/T] \tag{8.1.3}$$

is the grand partition function, with the pressure and density given by

$$P(\mathcal{N}, T, \mu) = \frac{T}{N} \ln Z_1(\mathcal{N}, T, \mu) \tag{8.1.4}$$

and

$$\rho = \frac{1}{N} \sum_{\{\mathcal{M}\}} |\mathcal{M}|\, p(\mathcal{N}, T, \mu; \mathcal{M}) = \frac{\partial P}{\partial \mu}, \tag{8.1.5}$$

D.A. Lavis, *Equilibrium Statistical Mechanics of Lattice Models*,
Theoretical and Mathematical Physics, DOI 10.1007/978-94-017-9430-5_8

respectively. If the number M of particles on $\mathcal{N}$ were fixed, the distribution would be the *canonical distribution* with independent extensive variables $N := |\mathcal{N}|$ and M, and partition function

$$Z_0(\mathcal{N}, M, T) := \sum_{\{|\mathcal{M}|=M\}} \exp[-E(\mathcal{M})/T], \tag{8.1.6}$$

where the sum is over all subsets of $\mathcal{N}$ with M sites. The grand partition function is now given by

$$Z_1(\mathcal{N}, T, \mathfrak{z}) = \sum_{\widehat{M}=0}^{N} Z_0(\mathcal{N}, \widehat{M}, T)\mathfrak{z}^{\widehat{M}}, \tag{8.1.7}$$

where $\mathfrak{z} := \exp(\mu/T)$ is the *fugacity* or *activity*.

We consider the thermodynamic limit of this fluid for an infinite sequence of lattices $\mathcal{N}$ growing so that $|\mathcal{N}| \to \infty$ and $\mathcal{N}$ tends to an infinite lattice $\mathcal{L}$. To achieve a sensible outcome in this procedure it is necessary that

$$\lim_{\mathcal{N}\to\mathcal{L}} P(\mathcal{N}, T, \mathfrak{z}) = P(\mathcal{L}, T, \mathfrak{z}), \tag{8.1.8}$$

where $P(\mathcal{L}, T, \mathfrak{z})$ is the finite pressure in the infinite lattice $\mathcal{L}$. Yang and Lee (1952) established the existence of the thermodynamic limit for the grand-canonical distribution of a one-component fluid of particles with hard cores and finite-range interactions in a volume V. This work was extended by Witten (1954), who relaxed the condition of hard cores. The corresponding theorem for lattice systems is given by Ruelle (1969) (see also Griffiths 1972). Three ingredients are necessary for the existence of the thermodynamic limit:

(i) translational invariance of the interparticle interactions,
(ii) some restriction on the range of the interactions,
(iii) a limitation on the proportional number of boundary sites as the lattice grows.

These conditions can be stated more precisely in the following way:

Given any subset $\mathcal{X}$ of the lattice sites, let $\mathcal{E}(\mathcal{X})$ be the contribution to the energy which would arise if every site of $\mathcal{X}$ were occupied, but which would be absent if one or more of its sites were empty. This means that $\mathcal{E}(\mathcal{X})$ cannot be decomposed in a sum of contributions from proper subsets of $\mathcal{X}$. If $|\mathcal{X}| = X > 1$ and $\mathcal{E}(\mathcal{X}) \neq 0$, then $\mathcal{X}$ is called an *interacting set*. The system is *translationally invariant* if, for all $\mathcal{X}$,

$$\begin{array}{l}\mathcal{E}(\mathcal{X}) \text{ is dependent only on the geometric form of} \\ \mathcal{X} \text{ and not on its location on the lattice.}\end{array} \tag{8.1.9}$$

Now choose a site $\boldsymbol{r}_0$ belonging to every lattice $\mathcal{N}$ in the sequence and let $\{\mathcal{X}\}_0$ be all the subsets, of all possible $\mathcal{N}$, which contain $\boldsymbol{r}_0$. The necessary condition on the range of interactions is that

$$\sum_{\{\mathcal{X}\}_0} \frac{|\mathcal{E}(\mathcal{X})|}{|\mathcal{X}|} < \infty. \tag{8.1.10}$$

Finally suppose that $\mathcal{N}_\ell$ is the set of sites of $\mathcal{N}$ which are within some fixed distance ℓ of the boundary. A limiting procedure $\mathcal{N} \to \mathcal{L}$ is said to be taken *in the sense of van Hove* (Ruelle 1969, p. 14) if

$$\lim_{\mathcal{N}\to\mathcal{L}} \frac{|\mathcal{N}_\ell|}{|\mathcal{N}|} = 0, \qquad \forall \ \ell. \tag{8.1.11}$$

The following theorem can now be proved (ibid p. 22):

Theorem 8.1.1 *The limit* (8.1.8) *exists for* $\mathfrak{Z} > 0$ *with* $P(\mathcal{L}, T, \mathfrak{Z})$ *a continuous non-decreasing function of* $\mathfrak{Z}$ *if* (8.1.9)–(8.1.11) *are satisfied.*

Because of the equivalence between the one-component lattice fluid and spin-$\frac{1}{2}$ lattice models this result can be translated into an equivalent result for the latter, establishing the continuity and monotonicity of the free-energy density as a function of the appropriate variable.

8.2 The Infinite-System Approach

Here, instead of considering a sequence of finite lattices, a system on the infinite lattice $\mathcal{L}$ is adopted at the start, eliminating the need to consider boundary effects.[1] In such a system equilibrium can no longer be described using the Gibbs distribution (2.1.2). There are a number of alternatives available in the literature (Penrose 1979, Sect. 2.2). We shall give a brief summary of the first steps of one of these, based on the work of Dobrushin (1968a) and Lanford and Ruelle (1969), in which the Gibbs distribution is replaced by the more general *Gibbs measure*.

The validity of the distribution (2.1.2), defined for $\boldsymbol{\sigma} \in \Sigma$, is dependent on its numerator being summable (or for a continuous system integrable) over Σ to give the partition function. This breaks down for an infinite system, although as we saw in Sect. 8.1 we can under certain specific conditions obtain convergence for the partition function for a sequence of increasing large systems. The aim here is to replace $p(\boldsymbol{\sigma})$ by a probability (that is to say, normalized) measure $m(\boldsymbol{\sigma})$ on Σ which coincides with $p(\boldsymbol{\sigma})$ for a finite system but retains its meaning for an infinite system.[2]

To be specific consider the lattice fluid of Sect. 8.1 with the set $\mathcal{M}$ of occupied sites replaced by the equivalent designation of $\boldsymbol{\sigma}$ standing for the set of states $\sigma(\boldsymbol{r}) = 0, 1,$

[1] Our discussion will be concerned solely with classical systems. However, it should be noted that another advantage of using the infinite-system approach for quantum systems is that the quasi-periodicity of finite quantum systems arising from the discrete spectrum of the Hamiltonian is absent in the infinite system.

[2] For convenience we drop reference to η and the independent intensive variables.

for all $\boldsymbol{r} \in \mathcal{N}$, where 0 and 1 denote respectively the presence or absence of a fluid particle. Then $|\mathcal{N}| = N$, $|\Sigma| = 2^N$ and we choose a pair-interaction of arbitrary range depending on the distance between the pair of sites. Thus

$$E(\varepsilon;\boldsymbol{\sigma}) = -\sum_{\{\boldsymbol{r},\boldsymbol{r}'\}} \varepsilon(|\boldsymbol{r}-\boldsymbol{r}'|)\sigma(\boldsymbol{r})\sigma(\boldsymbol{r}'). \tag{8.2.1}$$

This is the simple lattice fluid of Sect. 3.5.1.2, except that the interaction is not restricted to first neighbours. Let $\mathcal{N}_{\mathrm{A}}(\neq \emptyset) \subset \mathcal{N}$ with $\mathcal{N}_{\mathrm{B}} := \mathcal{N}\backslash\mathcal{N}_{\mathrm{A}} \neq \emptyset$. The configurations $\boldsymbol{\sigma}_{\mathrm{A}}$ and $\boldsymbol{\sigma}_{\mathrm{B}}$ of spins on $\mathcal{N}_{\mathrm{A}}$ and $\mathcal{N}_{\mathrm{B}}$ can be regarded as events occurring on the respective sublattices with $\boldsymbol{\sigma}_{\mathrm{A}} \wedge \boldsymbol{\sigma}_{\mathrm{B}}$ an event on $\mathcal{N}$. The probability that the event $\boldsymbol{\sigma}_{\mathrm{A}}$ occurs, given that $\boldsymbol{\sigma}_{\mathrm{B}}$ has occurred is

$$\begin{aligned} m(\boldsymbol{\sigma}_{\mathrm{A}}|\boldsymbol{\sigma}_{\mathrm{B}}) &= \frac{m(\boldsymbol{\sigma}_{\mathrm{A}} \wedge \boldsymbol{\sigma}_{\mathrm{B}})}{m(\boldsymbol{\sigma}_{\mathrm{B}})} \\ &= \frac{\exp[-\widehat{H}(\boldsymbol{\sigma}_{\mathrm{A}}|\boldsymbol{\sigma}_{\mathrm{B}})/T]}{\sum_{\{\sigma_{\mathrm{A}}\}} \exp[-\widehat{H}(\boldsymbol{\sigma}_{\mathrm{A}}|\boldsymbol{\sigma}_{\mathrm{B}})/T]} = \frac{\exp[-\widehat{H}_{\mathrm{A}}(\boldsymbol{\sigma}_{\mathrm{A}}|\boldsymbol{\sigma}_{\mathrm{B}})/T]}{Z_{\mathrm{A}}(\boldsymbol{\sigma}_{\mathrm{B}})}, \end{aligned} \tag{8.2.2}$$

where

$$\begin{aligned} \widehat{H}_{\mathrm{A}}(\boldsymbol{\sigma}_{\mathrm{A}}|\boldsymbol{\sigma}_{\mathrm{B}}) := &- \sum_{\{\boldsymbol{r},\boldsymbol{r}'\in\mathcal{N}_{\mathrm{A}}\}} \varepsilon(|\boldsymbol{r}-\boldsymbol{r}'|)\,\sigma(\boldsymbol{r})\sigma(\boldsymbol{r}') \\ &- \sum_{\{\boldsymbol{r}\in\mathcal{N}_{\mathrm{A}}\}} \sigma(\boldsymbol{r})\big[\mu + \sum_{\{\boldsymbol{r}'\in\mathcal{N}_{\mathrm{B}}\}} \varepsilon(|\boldsymbol{r}-\boldsymbol{r}'|)\,\sigma(\boldsymbol{r}')\big], \end{aligned} \tag{8.2.3}$$

$$Z_{\mathrm{A}}(\boldsymbol{\sigma}_{\mathrm{B}}) := \sum_{\{\sigma_{\mathrm{A}}\}} \exp[-\widehat{H}_{\mathrm{A}}(\boldsymbol{\sigma}_{\mathrm{A}}|\boldsymbol{\sigma}_{\mathrm{B}})/T]. \tag{8.2.4}$$

The Hamiltonian $\widehat{H}_{\mathrm{A}}(\boldsymbol{\sigma}_{\mathrm{A}}|\boldsymbol{\sigma}_{\mathrm{B}})$ is that of the system on $\mathcal{N}_{\mathrm{A}}$ with 'boundary conditions' given by the event $\boldsymbol{\sigma}_{\mathrm{B}}$. In fact with any finite-range condition $\varepsilon(|\boldsymbol{r}-\boldsymbol{r}'|) = 0$, if $|\boldsymbol{r}-\boldsymbol{r}'| > R$, imposed on the interaction, the boundary conditions will be specified by only part of $\boldsymbol{\sigma}_{\mathrm{B}}$; so it will remain valid when $\mathcal{N}_{\mathrm{B}}$ and hence $\mathcal{N}$ becomes infinite.

This example encapsulates the essential feature of this approach, which is that, given a system on a infinite lattice $\mathcal{L}$, equilibrium is defined locally in terms of a measure on finite subsets of $\mathcal{L}$, which is conditional on boundary conditions provided by the state on the lattice exterior to the sets in question. To generalize:

Definition 8.2.1 Let $\Sigma := \mathfrak{E}^{\mathcal{S}}$ be a product space where $\mathcal{S}$ is countably infinite and $\mathfrak{E}$ is any measurable space. Then m is a *Gibbs measure* if, for every finite subset $\mathcal{A}$ of $\mathcal{S}$ and m–almost every configuration σ_{B} outside $\mathcal{A}$, the conditional distribution of the configuration in $\mathcal{A}$ given σ_{B} is a Gibbs distribution (2.1.2).

The 'm–almost' caveat in the definition makes provision for exceptional boundary conditions, but apart from that it will be seen that, for any $\mathcal{A}$, the Gibbs measure corresponds to a family $m(\cdot|\sigma_{\rm B})$ of conditional Gibbs distributions.

With the interpretation that $\mathfrak{E} := \{0, 1\}$, the two-point space of states at a single site, $\mathcal{S} := \mathcal{L}$ the infinite lattice and $\mathcal{A} := \mathcal{N}_{\rm A}$ we have, by construction, obtained the Gibbs measure for the lattice fluid. The distribution (8.2.2) is called the *Dobrushin–Lanford–Ruelle (DLR) equilibrium condition* and a distribution on $\mathcal{L}$ which satisfies it is said to be in a *Gibbs state*. For the consequences of this development particularly in relation to the treatment of phase transitions and long-range correlation functions the reader is referred to Georgii (1988), Sewell (1989) and Simon (1993), which are concerned wholly or partly with this approach.

8.3 Lower Bounds for Phase Transitions: The Peierls Method

Phase transitions in statistical mechanical models correspond to the onset of some kind of order. Except in exceptional circumstances,[3] immediately below the phase transition this ordering is imperfect, with the perfectly-ordered state occurring only at zero temperature. An ordered phase can thus be understood by reference to its perfect realization at zero temperature and the types of ordered phases can in turn be understood by defining the concept of a *ground state*[4] and, as indicated in Sect. 4.2.1, the *order parameter* of the phase is the expectation value of a quantity which measures the extent to which the site occupations mirror those of the ground state. For the Ising ferromagnet on the square lattice, with Hamiltonian (3.5.1), the magnetization density is

$$m(\mathcal{N}, K, L) = 2\left[\frac{\langle\widehat{N}(\uparrow)\rangle}{N} - \frac{1}{2}\right], \tag{8.3.1}$$

where $K := J/T$, $L := \mathcal{H}/T$ and $\widehat{N}(\uparrow)$ is the number from the N sites of $\mathcal{N}$ which are in the spin-up state. The order parameter for the all spin-up ferromagnetic phase in zero field on the infinite lattice $\mathcal{L}$ is

$$m(\mathcal{L}, K, 0) = \lim_{L\to 0+}\lim_{\mathcal{N}\to\mathcal{L}} m(\mathcal{N}, K, L). \tag{8.3.2}$$

---

[3] Of which the transition to the frozen ferroelectric state in the six-vertex model is an example (Sect. 3.8.2).

[4] We have already made informal use of the idea of a ground state, beginning with the antiferromagnet in Sect. 3.5.1.1.

Let

$$m_{\uparrow}(\mathcal{N}, K, L) = 2\left[\frac{\langle \widehat{N}(\uparrow)\rangle_{\uparrow}}{N} - \frac{1}{2}\right], \tag{8.3.3}$$

where $\langle\cdot\rangle_{\uparrow}$ denotes the constrained expectation value with all the boundary sites containing up spins. There are two steps in the Peierls argument for the square-lattice Ising ferromagnet (Peierls 1936):

(a) To show that, if there exists an $\alpha > 0$, independent of N, such that

$$m_{\uparrow}(\mathcal{N}, K, 0) \geq \alpha, \qquad \forall \;\; K > K' \text{ and } \mathcal{N}, \tag{8.3.4}$$

then

$$m(\mathcal{L}, K, 0) \geq \alpha, \qquad \forall \;\; K > K'. \tag{8.3.5}$$

(b) To establish the existence of α and K'.

There are many presentations of details of this argument (Thompson 1988). Rather than giving another we shall adopt the more general approach of Heilmann and Huckaby (1979) and concentrate on the derivation of a lower temperature bound for the existence of a phase transition. Although Heilmann and Huckaby (1979) present their development for some bonded lattice fluid models it is not in essence specific to those models and can be applied to any one-component lattice fluid satisfying conditions (8.1.9)–(8.1.11).

Suppose that $E(\mathcal{M})$, defined in Sect. 8.1 is, for all $\mathcal{M}$, a linear function of parameters $\varepsilon_1, \varepsilon_2, \ldots, \varepsilon_{n_e}$. Then the Hamiltonian is a linear function of this set and the chemical potential μ. A *ground state* G is defined to be a configuration of the system for which the set $\mathcal{G} \subseteq \mathcal{N}$ of sites is occupied by particles and for which:

(i) $\mathcal{G}$ is a periodic arrangement of particles on the lattice.
(ii) There is a set of values of $\varepsilon_1, \varepsilon_2, \ldots, \varepsilon_{n_e}, \mu$ for which

$$\widehat{H}(\mathcal{G}) \leq \widehat{H}(\mathcal{M}), \qquad \forall \;\; \mathcal{M} \subseteq \mathcal{N}, \tag{8.3.6}$$

with equality holding, if at all, only when $\mathcal{M}$ is also periodic and thus itself a ground state.

G is said to be a *stable* ground state in the region of parameter space in which it satisfies (8.3.6). The number $\omega(\mathrm{G})$ of ground states which are stable in the region in which G is stable is called the *degeneracy* of G.

Consider now a ground state G_0. For any occupation $\mathcal{M}$, a site of the lattice is said to *belong to* G_0 if it is occupied in the way it would be if the system were in the ground state G_0. For any occupation of the lattice let $N_0(\mathcal{M})$ be the number of sites which belong to G_0 and

$$\rho_0(\mathcal{N}, K_1, \ldots, K_{n_e}, L) := \frac{\langle N_0(\mathcal{M})\rangle_0}{N} - \frac{1}{\omega(\mathrm{G}_0)}, \tag{8.3.7}$$

where now $K_\ell := \varepsilon_\ell/T$, $L := \mu/T$ and $\langle\cdot\rangle_0$ denotes the constrained expectation value taken over all $\mathcal{M}$ for which the boundary sites belong to G_0. When the parameters $\varepsilon_1, \varepsilon_2, \ldots, \varepsilon_{n_e}, \mu$ take values for which G_0 is a stable ground state the limit $\mathcal{N} \to \mathcal{L}$ of $\rho_0(\mathcal{N}, K_1, \ldots, K_{n_e}, L)$ is the order parameter for G_0. The replacement for (8.3.4) is then

$$\rho_0(\mathcal{N}, K_1, \ldots, K_{n_e}, L) \geq \alpha > 0, \tag{8.3.8}$$

(Heilmann and Huckaby 1979; Dobrushin 1968a, b). It is clear that (8.3.8) could never be satisfied if $\omega(\mathrm{G}_0) = 1$; non-degenerate ground states cannot lead to ordered phases. This would allow us, if we chose, to normalize $\rho_0(\mathcal{N}, K_1, \ldots, K_{n_e}, L)$ at $T = 0$ by dividing by $1 - 1/\omega(\mathrm{G}_0)$. It is in fact this normalized quantity which is equivalent to $m_\uparrow(\mathcal{N}, K, 0)$ in the zero-field Ising ferromagnet. To obtain a useful lower bound on phase transitions we must now use (8.3.8) to derive an inequality involving the temperature.

Let the lattice $\mathcal{N}$ be divided into *units* $\mathcal{C}_\iota$, $\iota = 1, 2, \ldots, \eta$ which consist of the smallest subsets of the sites with the following properties:

(a) All the units have the same number of sites $C = |\mathcal{C}_\iota|$ and the same geometrical structure. That is, they can be transformed into each other by rotations and translations.
(b) For any ground state and for some labelling of the sites of the units each unit is identically occupied both with respect to the number of particles and the labelling.
(c) Every interacting set is contained within at least one unit.

It is not difficult to see that it is always possible to divide the lattice into units in the way described here as long as the boundaries of the lattice are constructed in a suitable way. For each unit $\mathcal{C}_\iota$,

$$\mathfrak{C}_\iota := \left\{ \boldsymbol{r} : \sum_{j=1}^{C} \lambda_j \boldsymbol{r}_j^{(\iota)} = \boldsymbol{r},\ \forall\, \lambda_j \geq 0 \text{ such that } \sum_{j=1}^{C} \lambda_j = 1 \right\}, \tag{8.3.9}$$

is a *cell*, where $\boldsymbol{r}_1^{(\iota)}, \boldsymbol{r}_2^{(\iota)}, \ldots, \boldsymbol{r}_C^{(\iota)}$ are the sites of the unit $\mathcal{C}_\iota$, which are called the *vertices* of $\mathfrak{C}_\iota$. The cells of the lattice are closed convex sets of points in the space occupied by the lattice and spanned by their vertices.[5] In a d-dimensional lattice the dimension $d_{\mathfrak{C}}$ of the cells is less than or equal to d. The elements of the surfaces of the cells spanned by sets of $d_{\mathfrak{C}} - 1$ vertices are called *faces*.[6] We shall confine

---

[5] If the points of the unit are linearly independent, as is the case for the models treated by Heilmann and Huckaby (1979), then (8.3.9) is the usual definition of a *simplex* (Munkres 1984).

[6] For a simplex, all such linear combinations will give points on the surface, but this is not generally true for the definition of a cell given here.

our attention to the case where the only overlaps between cells are the points on the faces and where each cell face lies in two cells which will be called *neighbours*. A simply-connected set of cells will be called a *cluster*. The *surface* of a cluster consists of those faces of the cells of the cluster which belong to only one cell of the cluster. The *boundary* of the cluster consists of those cells which have faces on the surface. An *interior site* of a cluster is one which does not lie on its surface. An *interior cell* of the cluster is one which is not part of the boundary. Suppose that all sites of the lattice not lying on the boundary belong to $q_0 \geq 2$ cells. Then a cluster with an interior site must contain at least q_0 cells.

Each cell of the lattice can be occupied in 2^C ways denote by $\mathcal{O}_k, k = 1, 2, \ldots, 2^C$. Among the particle arrangements when each cell is identically occupied will be, by definition, all the ground states of the system. However, it is not necessarily the case that all such configurations are ground states, since there may be some corresponding to particle occupations which do not satisfy the minimization condition (8.3.6) for any set of values of the energy parameters. Any particle occupation $\mathcal{M}$ of the lattice implies the occupation $\mathcal{C}_\iota \cap \mathcal{M}$ for the cell $\mathfrak{C}_\iota$. Each $\mathcal{C}_\iota \cap \mathcal{M}$ is one of the occupations $\mathcal{O}_k$ and if, for occupation $\mathcal{M}$, $\nu_k(\mathcal{M})$ of the cells have occupation $\mathcal{O}_k$ the Hamiltonian can be expressed in the form

$$\widehat{H}(\mathcal{M}) = \sum_{\iota=1}^{\eta} \widehat{H}_{\mathfrak{C}}(\mathcal{C}_\iota \cap \mathcal{M}) = \sum_{k=1}^{2^C} \nu_k(\mathcal{M})\widehat{H}_{\mathfrak{C}}(\mathcal{O}_k), \tag{8.3.10}$$

where $\widehat{H}_{\mathfrak{C}}(\cdot)$ is the contribution to the Hamiltonian from a single cell $\mathfrak{C}$. This contribution will contain all factors of the form $\mathcal{E}(\mathcal{X})$, for every fully occupied interacting set $\mathcal{X} \subseteq \mathcal{C}$. If $\mathcal{X}$ lies in more than one unit the contribution is divided equally between that subset of units. For any occupation $\mathcal{M}$ in which each of the cells is identically occupied each of the terms in the first sum in (8.3.10) will be equal and the second sum will have only one term. For the ground state G_0 in which each cell has occupation $\mathcal{C} \cap \mathcal{G}_0$

$$\widehat{H}(\mathcal{G}_0) = \eta \widehat{H}_{\mathfrak{C}}(\mathcal{C} \cap \mathcal{G}_0). \tag{8.3.11}$$

A face of a cell is said to be G_0-*ordered* (or more briefly *ordered*, since we are concentrating on a fixed ground state) if all its sites belong to G_0. Otherwise it is said to be *disordered*. Given any occupation $\mathcal{M} \neq \mathcal{G}_0$, two cells of the lattice are said to be *linked* if they share a disordered face. Given the specific occupation $\mathcal{M}$ of the lattice a cell $\mathfrak{C}_\iota$ with some occupation $\mathcal{O}_k \neq \mathcal{C}_\iota \cap \mathcal{G}_0$ is called a *contour segment*. The boundary of a cluster is called a *contour* if:

(i) Each of its cells is a contour segment. (Some, but not necessarily all, of the interior cells of the cluster will also be contour segments.)
(ii) The cluster contains at least one interior site not belonging to G_0. (A contour must, therefore, consist of at least q_0 cells.)

(iii) The cells of the contour are not linked to any cell not in the cluster. (The surface of the cluster must consist of ordered faces.)
(iv) The cluster cannot be decomposed into two disjoint clusters, each bounded by a contour.

We denote a particular contour by $[q;\, j]$, where q is the number of segments in the contour and $j = 1, 2, \ldots, m(q)$ will be the variable which differentiates between different contours which can be formed on the lattice with q segments. Let $n_j(q)$ be the number of interior sites of the cluster bounded by the contour $[q;\, j]$.

Suppose that the occupation $\mathcal{M}$ has all the boundary sites belonging to G_0. We are interested in obtaining an upper bound on the number of sites of $\mathcal{M}$ which do not belong to G_0. This will clearly be given by the total number of interior sites contained within the contours present in $\mathcal{M}$. Thus

$$N - N_0(\mathcal{M}) \leq \sum_{\{q\}} \sum_{j=1}^{m(q)} n_j(q)\delta[q;\, j], \tag{8.3.12}$$

where the outer summation is over all possible contour sizes and

$$\delta[q;\, j] := \begin{cases} 1, & \text{if } [q;\, j] \text{ is present in } \mathcal{M}, \\ 0, & \text{otherwise.} \end{cases} \tag{8.3.13}$$

Let $\{\mathcal{M}\}_0$ be the set of lattice occupations for which the boundary sites of $\mathcal{N}$ belong to G_0 and let $\{\mathcal{M}\}_0^{[q;j]}$ be the set which satisfy the additional requirement that the contour $[q;\, j]$ is present. Then

$$\langle \delta[q;\, j] \rangle_0 = \frac{\displaystyle\sum_{\{\mathcal{M}\}_0^{[q;j]}} \exp[-\hat{H}(\mathcal{M})/T]}{\displaystyle\sum_{\{\mathcal{M}\}_0} \exp[-\hat{H}(\mathcal{M})/T]}. \tag{8.3.14}$$

For every $\mathcal{M} \in \{\mathcal{M}\}_0^{[q;j]}$ there is an $\mathcal{M}^* \in \{\mathcal{M}\}_0$, but $\notin \{\mathcal{M}\}_0^{[q;j]}$, which differs from $\mathcal{M}$ only in that all the interior points of $[q;\, j]$ not belonging to G_0 have had their occupations changed to put them into G_0. We denote this new set of occupations by $\{\mathcal{M}^*\}_0^{[q;j]}$. Thus

$$\langle \delta[q;\, j] \rangle_0 < \frac{\displaystyle\sum_{\{\mathcal{M}\}_0^{[q;j]}} \exp[-\widehat{H}(\mathcal{M})/T]}{\displaystyle\sum_{\{\mathcal{M}^*\}_0^{[q;j]}} \exp[-\widehat{H}(\mathcal{M}^*)/T]}$$

$$= \exp\left\{-\sum_{k=1}^{2^C} \theta_k[q;\, j][\widehat{H}_{\mathfrak{C}}(\mathcal{O}_k) - \widehat{H}_{\mathfrak{C}}(\mathcal{C} \cap \mathcal{G}_0)]/T\right\}, \tag{8.3.15}$$

where $\theta_k[q;\, j]$ is the number of cells with occupation $\mathcal{O}_k$ in contour $[q;\, j]$. To find an upper bound to the final term in (8.3.15) it is necessary to replace $\widehat{H}_{\mathfrak{C}}(\mathcal{O}_k)$ by its least possible value for each cell of the cluster bounded by $[q;\, j]$. For the interior cells of the cluster it is possible for $\widehat{H}_{\mathfrak{C}}(\mathcal{O}_k) = \widehat{H}_{\mathfrak{C}}(\mathcal{C} \cap \mathcal{G}_0)$, either because $\mathcal{O}_k = \mathcal{C} \cap \mathcal{G}_0$ or because $\mathcal{O}_k$ is the occupation appropriate to some other ground-state degenerate with G_0. The cells of the contour itself, however, are contour segments with occupations different from $\mathcal{C} \cap \mathcal{G}_0$. They must also be different from the cell occupation of some other ground state degenerate with G_0.[7] It follows that

$$\langle \delta[q;\, j] \rangle_0 < \exp[-\varepsilon_0 q/T], \tag{8.3.16}$$

where

$$\varepsilon_0 := \min_{\widehat{H}_{\mathfrak{C}}(\mathcal{O}_k) \neq \widehat{H}_{\mathfrak{C}}(\mathcal{C} \cap \mathcal{G}_0)} \{\widehat{H}_{\mathfrak{C}}(\mathcal{O}_k)\} - \widehat{H}_{\mathfrak{C}}(\mathcal{C} \cap \mathcal{G}_0). \tag{8.3.17}$$

From (8.3.7), (8.3.12) and (8.3.16),

$$1 - \sum_{\{q\}} Q(q) \exp(-\varepsilon_0 q/T) - \frac{1}{\omega(\mathrm{G}_0)} < \rho_0(\mathcal{N}, K_1, \ldots, K_{n_e}, L), \tag{8.3.18}$$

where $Q(q)$ is some function of q such that

$$\frac{1}{N} \sum_{j=1}^{m(q)} n_j(q) \leq Q(q). \tag{8.3.19}$$

If such a $Q(q)$ can be found such that the series in (8.3.18) is convergent for $T \in [0, \varepsilon_0 t(\mathrm{G}_0)]$, where $t(\mathrm{G}_0)$ is the least positive solution of the equation

$$\sum_{\{q\}} Q(q) \exp\left[-\frac{q}{t(\mathrm{G}_0)}\right] = 1 - \frac{1}{\omega(\mathrm{G}_0)}, \tag{8.3.20}$$

then (8.3.8) is satisfied if

$$T < \varepsilon_0 t(\mathrm{G}_0), \tag{8.3.21}$$

and, in the thermodynamic limit, the system, with parameter values appropriate to the ground state G_0, will be in the corresponding ordered state.

---

[7] If this were not the case it would be possible to 'mix' these two occupations to achieve a ground-state degeneracy of $\mathrm{O}(N)$.

Table 8.1 Cell occupations for first-neighbour pairs for the simple lattice fluid

ι	Occupation	$\widehat{H}_{\mathfrak{C}}(\mathcal{C} \cap \mathcal{G}_\iota)$
1	•—•	$-\varepsilon - 2\mu/z$
2	•—○	$-\mu/z$
3	○—•	$-\mu/z$
4	○—○	0

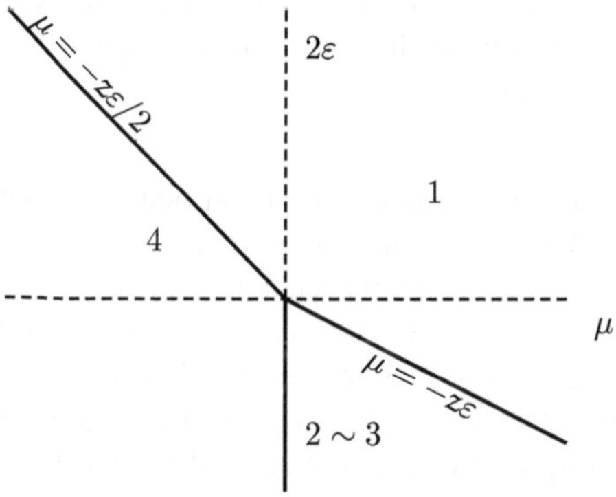

Fig. 8.1 The ground-state diagram for the simple lattice fluid. Regions of stability are labelled by ι, $\iota = 1, \ldots, 4$. The region $2 \sim 3$ is that in which the degenerate ground states 2 and 3 are both stable

8.3.1 The Simple Lattice Fluid

As shown in Sect. 3.5.1.2, the Hamiltonian (3.5.1) for the spin-$\frac{1}{2}$ Ising model can be transformed[8] into the form (3.5.20), for the simple lattice fluid where the lattice-fluid variables ε and μ are related to the Ising variables J and $\mathcal{H}$ by (3.5.22). The cells of the lattice can be taken to be the first-neighbour pairs, of which there are the four possible occupations listed in Table 8.1. They are denoted by $\mathcal{C} \cap \mathcal{G}_\iota$, $\iota = 1, \ldots, 4$, since for each there is a region of parameter space for which it corresponds to a stable ground state. These regions are shown in Fig. 8.1. In the region $2 \sim 3$ the two ground states 2 and 3 are degenerate so it is possible for an ordered phase to occur. Otherwise ordered states can occur only on the phase boundaries. In particular on the boundary $\mu = -\frac{1}{2}z\varepsilon$, $\varepsilon > 0$ the ground states 1 and 4 are degenerate.[9] The method described above is now applied to the model along this phase boundary to investigate the existence of the ordered phase corresponding to G_1. From (8.3.17)

[8] Apart from a constant C, given by (3.5.21), which we can for the present purposes neglect.

[9] This line corresponds to the zero-field axis $\mathcal{H} = 0$ in the spin representation and the model is equivalent to the zero-field Ising ferromagnet.

$\varepsilon_0 = -\mu/z = \varepsilon/2$ and, of course, $\omega(G_1) = 2$, $q_0 = z$. It is now necessary to determine a quantity $Q(q)$ satisfying (8.3.19). This will depend on the lattice structure under consideration.

Example 8.3.1 (The Linear Lattice) Although it is well-known that there is no phase transition in this case (Sect. 11.2) it is of interest to see how the theory can be applied. Consider a linear lattice of N sites with the two boundary sites occupied by particles. The only possible contours are ones for which $q = 2$. Such contours can bound clusters with n interior sites for $n = 1, 2, \ldots, N-2$ and a cluster with n interior sites can be placed on the lattice in $N - n - 1$ ways. So (8.3.19) is satisfied by $Q(2) = \frac{1}{6}N^2$ and, substituting into (8.3.20), gives $t(G_1) = 1/\ln(N/\sqrt{3})$. Since $t(G_1) \to 0$ as $N \to \infty$ the method, as is to be expected, fails to establish the existence of an ordered phase.

Example 8.3.2 (The Square Lattice) For a particular occupation, with all the boundary sites of the lattice occupied by particles, a contour can be represented by drawing a line through the centre of each boundary segment (each first-neighbour pair with occupation 2 or 3), between the centres of the two lattice squares of which the segment is a side. It is clear that an even number of lines will meet at the centre of each square and that the lines form closed polygons. Whenever four lines meet at the centre of a square the lines are cut to disconnected the polygon into two simply-connected parts. The effect of this procedure is to enclose all the interior sites of the clusters bounded by the contours within polygons (Fig. 8.2). Now a contour of q segments is equivalent to a polygon consisting of q line elements drawn between the centres of lattice squares and the possible values of q are $4, 6, 8, \ldots$ A polygon encloses the maximum number of sites when it is a square. It follows that $n_j(q) \le (q/4)^2$. We now need an upper bound on the total number $m(q)$ of closed polygons of q line elements. Ignoring the boundaries of the lattice we chose the centre of one of the N squares and begin to draw a polygon. The first line element can be chosen in four ways and each subsequent line element in three ways.

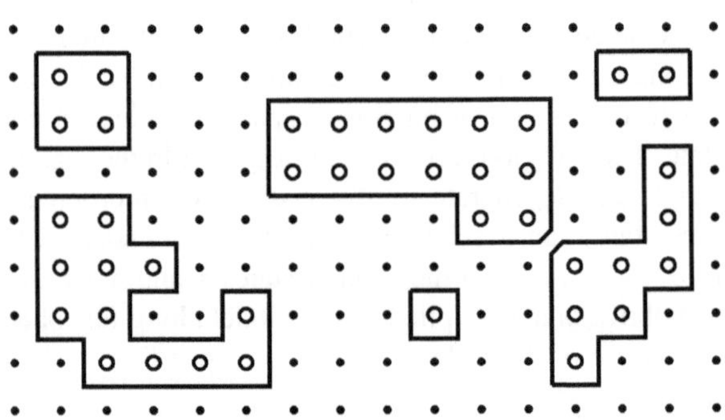

Fig. 8.2 The equivalence between contours and polygons on the square lattice. The maximum number of vacant sites corresponds to the case, as shown here, where all the sites enclosed by the polygons are empty

There are, therefore, $4N3^{q-1}$ ways of completing this construction. Among these will be all closed polygons, including those which cross themselves, and all open arcs. Each simply-connected closed polygon will have been counted q times and thus $m(q) < 4N3^{q-1}/q$. It follows that

$$\frac{1}{N}\sum_{j=1}^{m(q)} n_j(q) < \tfrac{1}{4}q3^{q-1} = Q(q). \tag{8.3.22}$$

Substituting into (8.3.20) gives

$$x^3 + x^2 - 6x + 3 = 0, \tag{8.3.23}$$

as long as $x = 9\exp[-2/t(\mathrm{G}_1)] < 1$. This equation has the single root $x = 0.5936$ in (0, 1) and, from (8.3.21), this establishes the existence of an ordered phase when $T < 0.3678\varepsilon$. This can be compared with the exact result for the critical temperature given, from (9.2.36), by $T_\mathrm{c} = \varepsilon/[2\ln(1+\sqrt{2})] = 0.5673\varepsilon$. The difference between these two values illustrates the point that the importance of the Peierls method consists not in its ability to yield good values for critical points, for which series methods are much better (Chap. 14), but rather in its ability to establish in a rigorous way the existence of an ordered phase.

8.4 Grand Partition Function Zeros and Phase Transitions

Suppose that $\mathbb{O}_{\mathcal{N}}(T) := \{\mathfrak{z}_1, \mathfrak{z}_2, \ldots, \mathfrak{z}_N\}$ is the set of zeros of the grand partition function (8.1.7) in the complex plane $\mathbb{C}_{\mathfrak{z}}$. The connection between the roots of the grand partition function and phase transitions in the system is established by the following theorem, due to Yang and Lee (1952), for which we shall provide a proof:

Theorem 8.4.1 *Let $\mathbb{W}(T)$ be a region in $\mathbb{C}_{\mathfrak{z}}$ which contains a segment of the positive real axis $\mathbb{R}^{(+)}$, but none of the members of $\mathbb{O}_{\mathcal{N}}(T)$, for any of the sequence of lattices involved in the limiting process described in Sect.* 8.1. *Then for all $\mathfrak{z} \in \mathbb{W}(T)$ the limit* (8.1.8) *is uniform and $P(\mathcal{L}, T, \mathfrak{z})$ is an analytic function*[10] *of $\mathfrak{z}$ in $\mathbb{W}(T)$.*

Proof Let $\mathbb{V}(T) \subseteq \mathbb{W}(T)$ be a closed disc of finite radius with its centre on $\mathbb{R}^{(+)}$. We first prove the theorem for $\mathbb{V}(T)$ using the *Vitali convergence theorem.*[11] Consider the sequence of functions $P(\mathcal{N}, T, \mathfrak{z})$ of $\mathfrak{z}$ given by the set of lattices leading to the

[10] See the footnote on p. 161 for the usage of the term 'analytic' for complex-valued functions.

[11] The statement of the theorem is as follows (for a proof see Titchmarch 1939): Let $f_n(\mathfrak{z})$ be a sequence of functions which are analytic in a region $\mathbb{D} \subset \mathbb{C}_{\mathfrak{z}}$ and have the following properties: (i) $|f_n(\mathfrak{z})| \leq B$ for all n and all $\mathfrak{z} \in \mathbb{D}$. (ii) $f_n(\mathfrak{z})$ tends to a limit as $n \to \infty$ at a set of points having a limit point inside $\mathbb{D}$. Then $f_n(\mathfrak{z})$ tends uniformly to a limit in any region bounded by a contour interior to $\mathbb{D}$, the limit being an analytic function of $\mathfrak{z}$.

limit in Theorem 8.1.1. These are analytic functions in $\mathbb{V}(T)$ for the reasons given above. According to Theorem 8.1.1 they tend to a limit on $\mathbb{V}(T) \cap \mathbb{R}^{(+)}$. Now let $\mathfrak{z}^*$ be the right-hand extremity of the diameter of $\mathbb{V}(T)$. The sequence of finite real numbers $P(\mathcal{N}, T, \mathfrak{z}^*)$ tends to the limit $P(\mathcal{L}, T, \mathfrak{z}^*)$. It is therefore bounded above by some number B and since $|\mathfrak{z}| \leq \mathfrak{z}^*$ on $\mathbb{V}(T)$ it follows that $P(\mathcal{N}, T, \mathfrak{z}) \leq B$ for all lattices $\mathcal{N}$ in the sequence and all $\mathfrak{z} \in \mathbb{V}(T)$. This establishes the theorem for $\mathbb{V}(T)$. Now take any disc $\mathbb{V}'(T) \subseteq \mathbb{W}(T)$ with centre in $\mathbb{V}(T)$. The conditions of the Vitali convergence theorem now hold in $\mathbb{V}'(T)$ with $\mathfrak{z}^*$ replaced by the real number equal to the maximum modulus of $\mathfrak{z}$ in $\mathbb{V}'(T)$. By repeating the process with a sequence of discs the whole of $\mathbb{W}(T)$ can be covered. □

Suppose that $\mathbb{Q}(T)$ is a closed region in $\mathbb{C}_\mathfrak{z}$ which contains the set $\mathbb{O}_\mathcal{N}(T)$ of zeros of $\Xi(\mathcal{N}, T, \mathfrak{z})$ for all of the sequence of finite lattices. It follows from Theorem 8.4.1 that $P(\mathcal{L}, T, \mathfrak{z})$ is an analytic function of $\mathfrak{z}$ in any $\mathbb{W}(T) \subseteq \mathbb{C}_\mathfrak{z}$ for which $\mathbb{Q}(T) \cap \mathbb{W}(T) = \emptyset$. Any phase transition of the system at temperature T must therefore occur at a chemical potential μ for which $\exp(\mu/T) \in \mathbb{Q}(T) \cap \mathbb{R}^{(+)}$.

We have now defined the main problem in this approach to the theory of phase transitions, namely, for any particular model leading to a set of canonical partition functions $Z_0(\mathcal{N}, T, M)$, to determine the set $\mathbb{Q}(T)$. This set will not, of course, be unique and different choices may have different advantages according to circumstances. Under rather general conditions a lattice fluid will not have any phase transitions at sufficiently high temperatures. If a set $\mathbb{Q}(T)$ can be found which evolves on to $\mathbb{R}^{(+)}$ as late as possible as the temperature is decreased then a good upper bound on T at any fixed μ will be obtained. On the other hand, if a $\mathbb{Q}(T)$ can be found, for which the range of μ lying in $\mathbb{Q}(T) \cap \mathbb{R}^{(+)}$ is as narrow as possible, then good bounds of μ at constant T will be obtained. These differing aims in the construction of $\mathbb{Q}(T)$ will not, as we shall see below, necessarily lead to the same set, although we can always form the intersection of all known sets of type $\mathbb{Q}(T)$ in order to improve the bounds.

There is one class of one-component lattice fluids for which it is not difficult to obtain some information about $\mathbb{Q}(T) \cap \mathbb{R}^{(+)}$ without any calculation. If

$$Z_0(\mathcal{N}, N - M, T) = Z_0(\mathcal{N}, M, T), \qquad M = 1, 2, \ldots, N - 1, \tag{8.4.1}$$

then, if $\mathfrak{z}_0 \in \mathbb{O}_\mathcal{N}(T)$, so does its conjugate complex $\bar{\mathfrak{z}}_0$ and $1/\mathfrak{z}_0$ and $1/\bar{\mathfrak{z}}_0$. This set of roots in invariant under inversion and conjugation and it follows that $\mathbb{O}_\mathcal{N}(T)$ must also have this property. Thus, if $\mathbb{O}_\mathcal{N}(T)$ is exterior to the disc $|\mathfrak{z}| < R < 1$ it is contained in the annulus $R \leq |\mathfrak{z}| \leq 1/R$. If condition (8.4.1) is replaced by

$$Z_0(\mathcal{N}, N - M, T) = Z_0(\mathcal{N}, M, T)[Z_0(\mathcal{N}, N, T)]^{1-\frac{2M}{N}}, \qquad M = 1, 2, \ldots, N - 1, \tag{8.4.2}$$

it can be shown that the grand partition function can be symmetrized by a change of variable. If the lattice fluid is expressed in terms of Ising spin variables then condition (8.4.2) is equivalent to the restriction that the only odd degree terms in the Hamiltonian are linear. This condition is satisfied for a lattice fluid with only pairwise interactions satisfying conditions (8.1.9) and (8.1.10). If condition (8.4.1) holds then, for there to be a phase transition at temperature T, the condition $1 \in \mathbb{Q}(T) \cap \mathbb{R}^{(+)}$ must be satisfied. In this case the region of phase transitions can be reduced to the largest closed interval $\mathbb{I}(T) \subset \mathbb{Q}(T) \cap \mathbb{R}^{(+)}$ which maps into itself under inversion. If (8.4.1) holds and the interactions are pairwise attractive the circle theorem of Lee and Yang (1952) proves that the set $\mathbb{O}_{\mathcal{N}}(T)$ lies on the unit circle $|\mathfrak{Z}| = 1$. This result is a special case of the theorem of Ruelle (1971), which will now be discussed.

8.4.1 Ruelle's Theorem

Given the set of variables $\{\mathfrak{Z}(\boldsymbol{r})\}$ defined on the sites of $\mathcal{N}$, let

$$\Xi(\mathcal{N}, T, \{\mathfrak{Z}(\boldsymbol{r})\}) := \sum_{\{\mathcal{M}\}} \left\{\exp[-E(\mathcal{M})/T] \prod_{\{\boldsymbol{r}\in\mathcal{M}\}} \mathfrak{Z}(\boldsymbol{r})\right\}. \tag{8.4.3}$$

It is clear, from (8.1.6) and (8.1.7), that

$$\Xi(\mathcal{N}, T, \{\mathfrak{Z}(\boldsymbol{r}) = \mathfrak{Z}\}) = Z_1(\mathcal{N}, T, \mathfrak{Z}). \tag{8.4.4}$$

Suppose now that a set of closed subsets $\mathbb{S}_{\mathcal{N}}(T; \boldsymbol{r})$ of $\mathbb{C}_{\mathfrak{Z}}$ can be found such that

$$\Xi(\mathcal{N}, T, \{\mathfrak{Z}(\boldsymbol{r})\}) \neq 0, \quad \text{if } \mathfrak{Z}(\boldsymbol{r}) \notin \mathbb{S}_{\mathcal{N}}(T; \boldsymbol{r}), \qquad \forall\ \boldsymbol{r} \in \mathcal{N}. \tag{8.4.5}$$

The zeros of $Z_1(\mathcal{N}, T, \mathfrak{Z})$ must lie in the closed set

$$\mathbb{U}_{\mathcal{N}}(T) := \bigcup_{\{\boldsymbol{r}\in\mathcal{N}\}} \mathbb{S}_{\mathcal{N}}(T; \boldsymbol{r}) \tag{8.4.6}$$

and a possible set $\mathbb{Q}(T)$ would be one which satisfied the condition that

$$\mathbb{U}_{\mathcal{N}}(T) \subseteq \mathbb{Q}(T), \qquad \forall \mathcal{N} \text{ in the sequence } \mathcal{N} \to \mathcal{L}. \tag{8.4.7}$$

In the cases discussed in Sects. 8.4.2 and 8.4.3, below, the determination of $\mathbb{Q}(T)$ from $\mathbb{U}_{\mathcal{N}}(T)$ is trivial, since the application of Ruelle's theorem leads to a set $\mathbb{U}_{\mathcal{N}}(T)$, which is independent of $\mathcal{N}$.

We now define a *covering* $\mathfrak{Y}_{\mathcal{N}}$ of the lattice $\mathcal{N}$ to be the set $\mathcal{Y}_\kappa$, $\kappa = 1, 2, \ldots, \nu$ of subsets of $\mathcal{N}$ such that:

(i) Every site of $\mathcal{N}$ belongs to at least one member of $\mathfrak{Y}_{\mathcal{N}}$.
(ii) Every interacting set is contained within at least one member of $\mathfrak{Y}_{\mathcal{N}}$.

It is clear that the set of units, defined in Sect. 8.3, is one possible covering of the lattice, but for present purposes it is convenient to be less specific. For any occupation $\mathcal{M}$ of the lattice we wish to be able to write

$$E(\mathcal{M}) = \sum_{\kappa=1}^{\nu} E_\kappa(\mathcal{Y}_\kappa \cap \mathcal{M}), \tag{8.4.8}$$

where $E_\kappa(\mathcal{Y}_\kappa \cap \mathcal{M})$ is the contribution to the energy from $\mathcal{Y}_\kappa$. To do this we must avoid multiple counting of energies from interacting sets of particles which occur in more than one $\mathcal{Y}_\kappa$. One way to do this is to divide the energy of such a set equally between all $\mathcal{Y}_\kappa$ in which it is contained. Another way would be to define the covering so that every interacting set is a member of $\mathfrak{Y}_\mathcal{N}$ and, in the energy $E_\kappa(\mathcal{Y}_\kappa \cap \mathcal{M})$, to exclude contributions from proper subsets of $\mathcal{Y}_\kappa$. For each $\mathcal{Y}_\kappa \in \mathfrak{Y}_\mathcal{N}$, let

$$\xi(\mathcal{Y}_\kappa, T, \{\mathfrak{z}_\kappa(\boldsymbol{r})\}) := \sum_{\{\mathcal{X}\subseteq\mathcal{Y}_\kappa\}} \left\{\exp[-E_\kappa(\mathcal{X})/T] \prod_{\{\boldsymbol{r}\in\mathcal{X}\}} \mathfrak{z}_\kappa(\boldsymbol{r})\right\}, \tag{8.4.9}$$

where $\{\mathfrak{z}_\kappa(\boldsymbol{r})\}$ is a set of variables defined for $\boldsymbol{r} \in \mathcal{Y}_\kappa$. For later reference we note, from (8.4.8), that, when the members of the covering are disjoint sets,

$$\Xi(\mathcal{N}, T, \{\mathfrak{z}(\boldsymbol{r})\}) = \prod_{\kappa=1}^{\nu} \xi(\mathcal{Y}_\kappa, T, \{\mathfrak{z}_\kappa(\boldsymbol{r}) = \mathfrak{z}(\boldsymbol{r})\}). \tag{8.4.10}$$

Ruelle's theorem can now be expressed in the following way:

Theorem 8.4.2 *If, for all $\mathcal{Y}_\kappa \in \mathfrak{Y}_\mathcal{N}$ and all $\boldsymbol{r} \in \mathcal{Y}_\kappa$, there exists a closed set $\mathbb{X}_\kappa(T;\boldsymbol{r}) \subset \mathbb{C}_3$ such that $0 \notin \mathbb{X}_\kappa(T;\boldsymbol{r})$ and $\xi(\mathcal{Y}_\kappa, T, \{\mathfrak{z}_\kappa(\boldsymbol{r})\}) \neq 0$ when $\mathfrak{z}_\kappa(\boldsymbol{r}) \notin \mathbb{X}_\kappa(T;\boldsymbol{r})$, for all $\boldsymbol{r} \in \mathcal{Y}_\kappa$, then $\Xi(\mathcal{N}, T, \{\mathfrak{z}(\boldsymbol{r})\}) \neq 0$ when*[12]

$$\mathfrak{z}(\boldsymbol{r}) \notin \mathbb{S}_\mathcal{N}(T;\boldsymbol{r}) = -\prod_{\{\kappa\}}[-\mathbb{X}_\kappa(T;\boldsymbol{r})], \qquad \forall\ \boldsymbol{r} \in \mathcal{N}, \tag{8.4.11}$$

where the product in (8.4.11) *is over those values of κ for which $\boldsymbol{r} \in \mathcal{Y}_\kappa$.*

Proof An important step in the proof of this theorem is the derivation of a formula for $\Xi(\mathcal{N}, T, \{\mathfrak{z}(\boldsymbol{r})\})$ in terms of the set $\xi(\mathcal{Y}_\kappa, T, \{\mathfrak{z}_\kappa(\boldsymbol{r})\})$, $\kappa = 1, 2, \ldots, \nu$. When the covering $\mathfrak{Y}_\mathcal{N}$ consists of disjoint sets, this formula is (8.4.10) and the theorem is trivial. This is because each lattice site belongs to exactly one member of the covering, the product in (8.4.11) has one term with $\mathbb{S}_\mathcal{N}(T;\boldsymbol{r}) = \mathbb{X}_\kappa(T;\boldsymbol{r})$ and $\mathfrak{z}(\boldsymbol{r}) = \mathfrak{z}_\kappa(\boldsymbol{r})$. The result follows from the assumption $\xi(\mathcal{Y}_\kappa, T, \{\mathfrak{z}(\boldsymbol{r})\}) \neq 0$ and (8.4.10).

When the covering contains overlapping sets (8.4.10) is no longer valid. Its replaced is given by the *contraction procedure* of Asano (1970):

Lemma 8.4.1 *Let*

$$\phi_\alpha = \phi_{\alpha-1} \circ \xi(\mathcal{Y}_\alpha, T, \{\mathfrak{z}_\alpha(\boldsymbol{r})\}), \qquad \alpha = 2, 3, \ldots, \nu, \tag{8.4.12}$$

---

12 Notation: For subsets $\mathbb{A}$ and $\mathbb{B}$ of $\mathbb{C}_3$, $-\mathbb{A}$ is the set of all points $-\mathfrak{z}_A$, for $\mathfrak{z}_A \in \mathbb{A}$, and $\mathbb{A}\mathbb{B}$ is the set of all points $\mathfrak{z}_A\mathfrak{z}_B$, for $\mathfrak{z}_A \in \mathbb{A}$ and $\mathfrak{z}_B \in \mathbb{B}$.

where

$$\phi_1 = \xi(\mathcal{Y}_1, T, \{\mathfrak{z}_1(\boldsymbol{r})\}), \tag{8.4.13}$$

and the contraction symbol $\circ$ *means that, when the product is taken, terms containing factors of the form* $\prod_{\{\kappa\}} \mathfrak{z}_\kappa(\boldsymbol{r})$ *are deleted unless the product contains all* $\mathfrak{z}_\kappa(\boldsymbol{r})$ *for which* $\boldsymbol{r} \in \mathcal{Y}_\kappa$ *for* $\kappa = 1, 2, \ldots, \alpha$. *Then with the substitution*

$$\mathfrak{z}(\boldsymbol{r}) = \prod_{\{\kappa\}} \mathfrak{z}_\kappa(\boldsymbol{r}), \tag{8.4.14}$$

$$\Xi(\mathcal{N}, T, \{\mathfrak{z}(\boldsymbol{r})\}) = \phi_\nu. \tag{8.4.15}$$

Proof (*of Lemma*) From (8.4.9), each of the terms in ϕ_ν is of the form

$$\prod_{\kappa=1}^{\nu} \left\{ \exp[-E_\kappa(\mathcal{X}_\kappa)/T] \prod_{\{\boldsymbol{r}\in\mathcal{X}_\kappa\}} \mathfrak{z}_\kappa(\boldsymbol{r}) \right\}, \tag{8.4.16}$$

where $\mathcal{X}_\kappa \subseteq \mathcal{Y}_\kappa$. The sum is over certain subsets $\mathcal{X}_\kappa$ of the covering sets $\mathcal{Y}_\kappa$ with a constraint imposed by the contraction procedure of (8.4.12). This can be expressed by the condition that if, for some subsets $\mathcal{Y}_i, \mathcal{Y}_j, \mathcal{Y}_k, \ldots$ of the covering $\mathfrak{Y}_\mathcal{N}$,

$$\mathcal{Y}_i \cap \mathcal{Y}_j \cap \mathcal{Y}_k \cap \cdots \neq \emptyset, \tag{8.4.17}$$

then

$$\mathcal{Y}_i \cap \mathcal{X}_i = \mathcal{Y}_j \cap \mathcal{X}_j = \mathcal{Y}_k \cap \mathcal{X}_k = \cdots. \tag{8.4.18}$$

With this condition we can define

$$\mathcal{M} = \bigcup_{\kappa=1}^{\nu} \mathcal{X}_\kappa, \tag{8.4.19}$$

with

$$\mathcal{X}_\kappa = \mathcal{M} \cap \mathcal{Y}_\kappa, \qquad \kappa = 1, 2, \ldots, \nu, \tag{8.4.20}$$

and the lemma follows from (8.4.3) and (8.4.8). □

The formula (8.4.11) for $\mathbb{S}_\mathcal{N}(T; \boldsymbol{r})$ can now be obtained by induction, using the contraction procedure of Lemma 8.4.1 and the following result, proved by Ruelle (1971):

Lemma 8.4.2 *Let* $\mathbb{A}$ *and* $\mathbb{B}$ *be closed subsets of* $\mathbb{C}_3$ *which do not contain* 0. *Suppose that the complex polynomial*

$$c + a\mathfrak{z}_A + b\mathfrak{z}_B + g\mathfrak{z}_A\mathfrak{z}_B$$

can vanish only when $\mathfrak{z}_A \in \mathbb{A}$ *and* $\mathfrak{z}_B \in \mathbb{B}$. *Then*

$$c + g\mathfrak{z}$$

can vanish only when $\mathfrak{z} \in -\mathbb{AB}$.

To complete the theorem using induction on the contraction procedure we first define $\mathbb{S}_{\mathcal{N}}^{(\alpha)}(T;\boldsymbol{r})$ by a formula like (8.4.11), except that the product is restricted to values of κ in the range $[1,\alpha]$ for which $\boldsymbol{r} \in \mathcal{Y}_\kappa$. The statement of the theorem, applied to the lattice $\mathcal{N}_\alpha = \cup_{\kappa=1}^{\alpha}\mathcal{Y}_\kappa$, is that $\Xi(\mathcal{N}_\alpha, T, \{\mathfrak{z}(\boldsymbol{r})\}) \neq 0$ when $\mathfrak{z}(\boldsymbol{r}) \notin \mathbb{S}_{\mathcal{N}}^{(\alpha)}(T;\boldsymbol{r})$, for all $\boldsymbol{r} \in \mathcal{N}_\alpha$. This is true for $\alpha = 1$, when $\Xi(\mathcal{N}_1, T, \{\mathfrak{z}(\boldsymbol{r})\}) = \xi(\mathcal{Y}_1, T, \{\mathfrak{z}(\boldsymbol{r})\})$, $\mathbb{S}_{\mathcal{N}}^{(1)}(T;\boldsymbol{r}) = \mathbb{X}_1(T;\boldsymbol{r})$. Let it be assumed for $\alpha - 1$. It is now important to note that the substitution (8.4.14) can be made step by step in the contraction procedure with $\phi_{\alpha-1}$ having a single variable $\mathfrak{z}(\boldsymbol{r})$ defined for each site $\boldsymbol{r} \in \mathcal{N}_{\alpha-1}$. Consider now the formation of ϕ_α according to the formula (8.4.12). For a site $\boldsymbol{r}$ in $\mathcal{Y}_\alpha$ but not in $\mathcal{N}_{\alpha-1}$, $\mathbb{S}_{\mathcal{N}}^{(\alpha)}(T;\boldsymbol{r}) = \mathbb{X}_\alpha(T;\boldsymbol{r})$ and for a site $\boldsymbol{r}$ in $\mathcal{N}_{\alpha-1}$ but not in $\mathcal{Y}_\alpha$, $\mathbb{S}_{\mathcal{N}}^{(\alpha)}(T;\boldsymbol{r}) = \mathbb{S}_{\mathcal{N}}^{(\alpha-1)}(T;\boldsymbol{r})$. In both these cases no contraction is necessary for the terms in $\mathfrak{z}(\boldsymbol{r})$ and $\mathfrak{z}_\alpha(\boldsymbol{r})$. For a site which belongs to both $\mathcal{Y}_\alpha$ and $\mathcal{N}_{\alpha-1}$ the contraction procedure is applied. Linear terms in $\mathfrak{z}(\boldsymbol{r})$ and $\mathfrak{z}_\alpha(\boldsymbol{r})$ are deleted and $\mathfrak{z}(\boldsymbol{r})\mathfrak{z}_\alpha(\boldsymbol{r})$ is replaced by $\mathfrak{z}(\boldsymbol{r})$. According to Lemma 8.4.2, $\Xi(\mathcal{N}_\alpha, T, \{\mathfrak{z}(\boldsymbol{r})\})$ vanishes as a function of $\mathfrak{z}(\boldsymbol{r})$ only when $\mathfrak{z}(\boldsymbol{r}) \in -\mathbb{S}_{\mathcal{N}}^{(\alpha-1)}(T;\boldsymbol{r})\mathbb{X}_\alpha(T;\boldsymbol{r}) = \mathbb{S}_{\mathcal{N}}^{(\alpha)}(T;\boldsymbol{r})$. This completes the proof of Ruelle's theorem. □

We now summarize the 'practical' procedure for obtaining the set $\mathbb{U}_{\mathcal{N}}(T)$:

(i) For the lattice $\mathcal{N}$ choose the covering $\mathfrak{Y}_{\mathcal{N}}$ of subsets $\mathcal{Y}_\kappa$, $\kappa = 1, 2, \ldots, \nu$.
(ii) For each $\mathcal{Y}_\kappa$ form the polynomial $\xi(\mathcal{Y}_\kappa, T, \{\mathfrak{z}_\kappa(\boldsymbol{r})\})$.
(iii) For each $\boldsymbol{r} \in \mathcal{Y}_\kappa$ find a subset $\mathbb{X}_\kappa(T;\boldsymbol{r}) \subset \mathbb{C}_3$ and not containing 0 such that $\xi(\mathcal{Y}_\kappa, T, \{\mathfrak{z}_\kappa(\boldsymbol{r})\}) \neq 0$ when $\mathfrak{z}_\kappa(\boldsymbol{r}) \notin \mathbb{X}_\kappa(T;\boldsymbol{r})$ for every $\boldsymbol{r} \in \mathcal{Y}_\kappa$.
(iv) For each $\boldsymbol{r} \in \mathcal{N}$ construct $\mathbb{S}_{\mathcal{N}}(T;\boldsymbol{r})$ according to the formula (8.4.11).
(v) Obtain $\mathbb{U}_{\mathcal{N}}(T)$ by taking the union, as in (8.4.6).

It is clear that the tractability of this procedure depends very much on the covering chosen. The number of geometrically different covering sets should be very small. In practice it is often convenient to use the lattice units, defined in Sect. 8.1, as the elements of the covering.

8.4.2 The Yang–Lee Circle Theorem

Consider the generalization

$$\widehat{H}(\{J(\boldsymbol{r},\boldsymbol{r}')\}, \mathcal{H}; \boldsymbol{\sigma}) := -\sum_{\{\boldsymbol{r},\boldsymbol{r}'\}} J(\boldsymbol{r},\boldsymbol{r}')\sigma(\boldsymbol{r})\sigma(\boldsymbol{r}') - \mathcal{H}\sum_{\{\boldsymbol{r}\}}\sigma(\boldsymbol{r}) \qquad (8.4.21)$$

of the spin-$\frac{1}{2}$ Ising model Hamiltonian (3.5.1), where the first summation is over all pairs of sites, with $J(\boldsymbol{r},\boldsymbol{r}') = J(\boldsymbol{r}',\boldsymbol{r})$ and $J(\boldsymbol{r},\boldsymbol{r}) = 0$. Using the lattice fluid variables $\tau(\boldsymbol{r})$ of Sect. 3.5.1.2 and omitting the constant term, (8.4.21) can be re-expressed in the form

$$\widehat{H}(\{\tilde{\varepsilon}(\boldsymbol{r},\boldsymbol{r}')\},\tilde{\mu};\boldsymbol{\tau}) = -\sum_{\{\boldsymbol{r},\boldsymbol{r}'\}}\tilde{\varepsilon}(\boldsymbol{r},\boldsymbol{r}')[2\tau(\boldsymbol{r})\tau(\boldsymbol{r}') - \tau(\boldsymbol{r}) - \tau(\boldsymbol{r}')] - \tilde{\mu}\sum_{\{\boldsymbol{r}\}}\tau(\boldsymbol{r}), \tag{8.4.22}$$

where $\tilde{\varepsilon}(\boldsymbol{r},\boldsymbol{r}') := 2J(\boldsymbol{r},\boldsymbol{r}')$ and $\tilde{\mu} := 2\mathcal{H}$. The terms in the first summation in (8.4.22) are the energies of a pair of sites. The linear terms are retained in this summation, rather than being absorbed into the chemical potential, in order to maintain the hole-particle symmetry. This implies the rather artificial situation where the occupations ∘—∘ and •—• both have zero interaction energy, with non-zero energies occurring for ∘—• and •—∘. This can be regarded as a mathematical device for simplifying the application of Ruelle's theorem, with a transformation to more natural variables being applied at the end. The interacting sets for this system consist of all pairs of sites for which $\tilde{\varepsilon}(\boldsymbol{r},\boldsymbol{r}') \neq 0$ and thus, from (8.1.10), we must have

$$\frac{1}{2}\sum_{\{\boldsymbol{r}'\}}|\tilde{\varepsilon}(\boldsymbol{r},\boldsymbol{r}')| < \infty \qquad \forall\ \boldsymbol{r} \in \mathcal{N}. \tag{8.4.23}$$

This is achieved by imposing the condition

$$\tilde{\varepsilon}(\boldsymbol{r},\boldsymbol{r}') = 0 \qquad \text{when}\ \ |\boldsymbol{r}-\boldsymbol{r}'| > R, \tag{8.4.24}$$

where R is the *range of interaction*. A particular pair of lattice sites is denoted by $[\boldsymbol{r},\boldsymbol{r}']$. The covering $\mathcal{Y}_{\mathcal{N}}$ can now be taken to be all the pairs $[\boldsymbol{r},\boldsymbol{r}']$ for which $\tilde{\varepsilon}(\boldsymbol{r},\boldsymbol{r}') \neq 0$ and the polynomials (8.4.9) take the form

$$\begin{aligned}\mathfrak{z}([\boldsymbol{r},\boldsymbol{r}'],T,\mathfrak{Z}(\boldsymbol{r}';\boldsymbol{r}),\mathfrak{Z}(\boldsymbol{r};\boldsymbol{r}')) = 1 + \tilde{\mathfrak{X}}(T;\boldsymbol{r},\boldsymbol{r}')[\mathfrak{Z}(\boldsymbol{r}';\boldsymbol{r}) + \mathfrak{Z}(\boldsymbol{r};\boldsymbol{r}')]\\ + \mathfrak{Z}(\boldsymbol{r}';\boldsymbol{r})\mathfrak{Z}(\boldsymbol{r};\boldsymbol{r}'),\end{aligned} \tag{8.4.25}$$

where $\tilde{\mathfrak{X}}(T;\boldsymbol{r},\boldsymbol{r}') := \exp[-\tilde{\varepsilon}(\boldsymbol{r},\boldsymbol{r}')/T]$ and $\mathfrak{Z}(\boldsymbol{r}';\boldsymbol{r})$ is the variable for the site $\boldsymbol{r}$ and the covering set $[\boldsymbol{r},\boldsymbol{r}']$. The analysis of this problem now depends of whether the terms $\tilde{\mathfrak{X}}(T;\boldsymbol{r},\boldsymbol{r}')$ are greater or less than unity. The case to which the Yang–Lee circle theorem applies is when

$$\tilde{\mathfrak{X}}(T;\boldsymbol{r},\boldsymbol{r}') < 1, \qquad \forall\ [\boldsymbol{r},\boldsymbol{r}'], \tag{8.4.26}$$

which in spin formulation corresponds to all interactions being ferromagnetic. $\mathfrak{z}([\boldsymbol{r},\boldsymbol{r}'],T,\mathfrak{Z}(\boldsymbol{r}';\boldsymbol{r}),\mathfrak{Z}(\boldsymbol{r};\boldsymbol{r}'))$ is of the form given by Eq. (17.5.2). The regions $\mathbb{X}_\kappa(T;\boldsymbol{r})$ are then given by (17.5.9) for all covering sets and all lattice sites for any choice of $x \neq 1$. As will be seen in the discussions in the next section, the difficult step in the application of Ruelle's theorem is (iv), the construction of $\mathbb{S}_{\mathcal{N}}(T)$ according to the formula (8.4.11). There is, however, one choice of x for which this process is straightforward, namely $x = 0$, when

$$\mathbb{X}_{\kappa}(T;\boldsymbol{r}) = \mathbb{D}^{(+)} := \{\mathfrak{z} : |\mathfrak{z}| \geq 1\}. \tag{8.4.27}$$

To obtain $\mathbb{S}_{\mathcal{N}}(T;\boldsymbol{r})$ (8.4.11) must be used. In general this can be difficult, particularly since boundary sites are members of fewer covering sets than interior sites and must be considered separately. However, in this case the form of $\mathbb{D}^{(+)}$ means that the product procedure has no effect and $\mathbb{S}_{\mathcal{N}}(T;\boldsymbol{r}) = \mathbb{D}^{(+)}$ for every site of the lattice. It then follows from (8.4.6) that $\mathbb{U}_{\mathcal{N}}(T) = \mathbb{D}^{(+)}$, which is independent of $\mathcal{N}$. Thus $\mathbb{D}^{(+)}$ is a possible choice of $\mathbb{Q}(T)$ for all T. However, in this case the symmetry condition (8.4.1) is satisfied. If some $\mathfrak{z}_0 \in \mathbb{O}_{\mathcal{N}}(T)$, with $|\mathfrak{z}_0| > 1$, then $1/\mathfrak{z}_0 \in \mathbb{O}_{\mathcal{N}}(T)$. This contradicts the result of Ruelle's theorem and it follows that

$$\mathbb{Q}(T) = \{\mathfrak{z} : |\mathfrak{z}| = 1\}, \qquad \forall\ T. \tag{8.4.28}$$

This is the Yang–Lee circle theorem.

For this formulation in lattice fluid terms, for which condition (8.4.26) is equivalent to an attractive potential between pair of particles (and pairs of holes), we have shown that, for an arbitrary but finite range of interaction on any lattice, phase transitions can occur only for $\tilde{\mu} = 0$. In the magnetic spin formulation this is equivalent to the result that the Ising ferromagnet with arbitrary range of interaction can have a phase transition only on the zero field axis. Equation (8.4.22) can be re-expressed in the more usual lattice fluid form (Eq. (3.5.20))

$$\widehat{H}(\{\varepsilon(\boldsymbol{r},\boldsymbol{r}')\},\mu;\boldsymbol{\tau}) = -\sum_{\{\boldsymbol{r},\boldsymbol{r}'\}} \varepsilon(\boldsymbol{r},\boldsymbol{r}')\tau(\boldsymbol{r})\tau(\boldsymbol{r}') - \mu\sum_{\{\boldsymbol{r}\}} \tau(\boldsymbol{r}), \tag{8.4.29}$$

where

$$\begin{aligned} \varepsilon(\boldsymbol{r},\boldsymbol{r}') &:= 2\tilde{\varepsilon}(\boldsymbol{r},\boldsymbol{r}') = 4J(\boldsymbol{r},\boldsymbol{r}'), \\ \mu &:= \tilde{\mu} - \sum_{\{\boldsymbol{r}\}} \varepsilon(\boldsymbol{r},\boldsymbol{r}') = 2\mathcal{H} - 2\sum_{\{\boldsymbol{r}\}} J(\boldsymbol{r},\boldsymbol{r}'). \end{aligned} \tag{8.4.30}$$

In this case it follows that, as long as the summations in (8.4.30) have the same value for each lattice site, a phase transition can occur only on the line

$$\mu = -\tfrac{1}{2}\sum_{\{\boldsymbol{r}\}} \varepsilon(\boldsymbol{r},\boldsymbol{r}'). \tag{8.4.31}$$

8.4.3 Systems with Pair Interactions

For the application of Ruelle's theorem, the choice of the sets $\mathbb{X}_\kappa(T;\boldsymbol{r})$ is, as we have seen above, not unique. For covering sets consisting of pairs of sites this is exemplified by the different allowed values of the parameters x and y of Sect. 17.5. In particular cases one choice may have particular advantages, in either narrowing the intersection of $\mathbb{Q}(T)$ with $\mathbb{R}^{(+)}$ or obtaining a good estimate of the critical temperature by delaying as much as possible, as T is decreased, the value of T at which $\mathbb{Q}(T)$ intersects $\mathbb{R}^{(+)}$. The circle theorem, obtained by the application of Ruelle's theorem and symmetry, narrows $\mathbb{Q}(T)\cap\mathbb{R}^{(+)}$ to the single point $\mathfrak{z}=1$. This is an extreme case, where the value $\tilde{\mu}=0$ for the phase transition (if it occurs) is predicted with complete accuracy. This result is, however, achieved at the expense of a total inability to obtain an upper bound for the critical temperature. In this section we consider other possible choices for $\mathbb{X}_\kappa(T;\boldsymbol{r})$ for systems with pair interactions. We first examine the situation where

$$\tilde{\varepsilon}(\boldsymbol{r},\boldsymbol{r}') := \begin{cases} \tilde{\varepsilon}>0, & \text{for every first-neighbour pair } [\boldsymbol{r},\boldsymbol{r}'],\\ 0, & \text{otherwise,}\end{cases} \tag{8.4.32}$$

which gives

$$\tilde{\mathfrak{X}}(T;\boldsymbol{r},\boldsymbol{r}') = \begin{cases} \tilde{\mathfrak{X}}(T)=\exp(-\tilde{\varepsilon}/T)<1, & \text{for every first-neighbour pair } [\boldsymbol{r},\boldsymbol{r}'],\\ 1, & \text{otherwise.}\end{cases} \tag{8.4.33}$$

This is a special case of the situation to which the circle theorem applies. We can however, from (17.5.9), choose

$$\mathbb{X}_\kappa(T;\boldsymbol{r}) = \mathbb{X}_x(T) := \left\{\mathfrak{z} : |2\tilde{\mathfrak{X}}(T)\mathfrak{z}+x| \le \sqrt{4\tilde{\mathfrak{X}}^2(T)(1-x)+x^2}\right\}, \tag{8.4.34}$$

for every lattice site, every covering set and any $x>1$. The region $\mathbb{X}_x(T)$ is a disc not including $\mathfrak{z}=0$ and with centre on $\mathbb{R}^{(-)}$. From (8.4.11) and (8.4.34),

$$\mathbb{S}_{\mathcal{N}}(T;\boldsymbol{r}) = -[-\mathbb{X}_x(T)]^{z}, \tag{8.4.35}$$

for every interior lattice site, where z is the coordination number of the lattice. For boundary sites the formula is the same as (8.4.35) except that z is replaced by some $z'<z$, this being the number of first-neighbour sites of the particular boundary site. Since, however, $-1\in\mathbb{X}_x(T)$, for all $x>1$, it follows that

$$-[-\mathbb{X}_x(T)]^{z'} \subset -[-\mathbb{X}_x(T)]^{z} \tag{8.4.36}$$

and thus, from (8.4.6),

$$\mathbb{U}_{\mathcal{N}}(T) = -[-\mathbb{X}_x(T)]^{\mathsf{z}}. \tag{8.4.37}$$

Since this $\mathbb{U}_{\mathcal{N}}(T)$ does not involve $\mathcal{N}$ it can be taken to be $\mathbb{Q}(T)$. From (17.5.7) the points $\mathfrak{z}^{(\pm)} = -\tilde{\mathfrak{X}}(T) \pm \mathrm{i}\sqrt{1-\tilde{\mathfrak{X}}^2(T)}$ lie on the boundary of $\mathbb{X}_x(T)$ for all $x > 1$. These points subtend an angle $\arccos[\tilde{\mathfrak{X}}(T)]$ with $\mathbb{R}^{(-)}$. It follows that the arc $\mathrm{Arg}(\mathfrak{z}) \in [\mathsf{z}\{\pi - \arccos[\tilde{\mathfrak{X}}(T)]\}, \mathsf{z}\{\pi + \arccos[\tilde{\mathfrak{X}}(T)]\}]$ of the unit circle $|\mathfrak{z}| = 1$ is contained in $\mathbb{Q}(T)$. Since the zeros of the partition function lie on $|\mathfrak{z}| = 1$ an upper bound $T = T_{\mathrm{u}}$ for the occurrence of a phase transition is given by

$$\tilde{\mathfrak{X}}(T_{\mathrm{u}}) = \cos\left(\frac{\pi}{\mathsf{z}}\right), \qquad T_{\mathrm{u}} := \frac{\tilde{\varepsilon}}{-\ln(\pi/\mathsf{z})} = \frac{\varepsilon}{-2\ln(\pi/\mathsf{z})}, \tag{8.4.38}$$

where $\varepsilon := 2\tilde{\varepsilon}$ is the usually lattice-fluid interaction energy given by the first-neighbour case of (8.4.30). For the square lattice $T_{\mathrm{u}} = 1.4427\varepsilon$, which is consistent with the known critical temperature $T_{\mathrm{c}} = 0.5673\varepsilon$ (formula (9.2.36) with $\varepsilon = 4J$).

Consider now the homogeneous first-neighbour lattice fluid with Hamiltonian (3.5.20). In this case

$$\xi([\boldsymbol{r},\boldsymbol{r}'],T,\mathfrak{z},\mathfrak{z}') = 1 + \mathfrak{z} + \mathfrak{z}' + \mathfrak{X}(T)\mathfrak{z}\mathfrak{z}', \tag{8.4.39}$$

where $\mathfrak{X}(T) := \exp(\varepsilon/T)$, for every covering set, consisting of a pair of first-neighbour sites. The expression (8.4.39) is of the form given by (17.5.12) and the sets $\mathbb{X}_\kappa(T;\boldsymbol{r})$ can be taken to be all the same and given by (17.5.14) or (17.5.15) according as $\mathfrak{X}(T) <, > 1$. These cases correspond respectively to repulsive and attractive first-neighbour interactions between particles.

When the interparticle interaction is attractive the possible covering sets $\mathbb{X}_\kappa(T,\boldsymbol{r})$ are given by (17.5.15). The useful choices are given by $x > 1$ and consist of discs not containing $\mathfrak{z} = 0$. The best upper bound for the critical temperature is given by (8.4.35) with $\mathbb{X}$ replaced by $\mathbb{X}'$ when x is chosen so that $\mathbb{S}_{\mathcal{N}}(T;\boldsymbol{r})$ reaches $\mathbb{R}^{(+)}$ as late as possible as the temperature is decreased. This involves the choice of the disc for which the tangent from the origin to the boundary subtends the smallest angle with $\mathbb{R}^{(-)}$. As indicated in Sect. 17.5 this is given by choosing $x = 2$ with

$$\mathbb{X}_\kappa(T;\boldsymbol{r}) = \mathbb{X}'_2(T) := \left\{\mathfrak{z} : |\mathfrak{z}+1| \le \sqrt{1-\mathfrak{X}^{-1}(T)}\right\}, \tag{8.4.40}$$

for every lattice site and every covering set. The tangent from the origin to the boundary of this disc subtends angles $\pm\arccos[1/\sqrt{\mathfrak{X}(T)}]$ with $\mathbb{R}^{(-)}$. Again, since $-1 \in \mathbb{X}'_2(T)$, we can make the identification $\mathbb{Q}(T) = \mathbb{U}_{\mathcal{N}}(T) = \mathbb{S}_{\mathcal{N}}(T;\boldsymbol{r})$, where $\boldsymbol{r}$ is any interior site of the lattice. Using (8.4.35) the upper bound temperature T_{u} is then given by

$$T_{\mathrm{u}} = -\frac{|\varepsilon|}{2\ln(\pi/\mathsf{z})}, \tag{8.4.41}$$

which, as is to be expected, is the same as the form given by (8.4.38).

In a similar way, when the interparticle interaction is repulsive, the possible sets $\mathbb{X}_\kappa(T;\boldsymbol{r})$ are given by (17.5.14) with the best choice being given by $y = 0$. Thus

$$\mathbb{X}_\kappa(T;\boldsymbol{r}) = \mathbb{Y}'_0(T) := \left\{\mathfrak{z} : |\mathfrak{X}(T)\mathfrak{z} + 1| \le \sqrt{1-\mathfrak{X}(T)}\right\}, \tag{8.4.42}$$

for every lattice site and every covering set. The tangent from the origin to the boundary of the disc subtends angles $\pm\arccos[\sqrt{\mathfrak{X}(T)}]$ with $\mathbb{R}^{(-)}$ and the upper bound temperature is again given by (8.4.42), except that ε is now positive.

A particular case of this analysis, which was investigated by Runnels and Hubbard (1972), is that of the lattice fluid with first-neighbour exclusion. This corresponds to the limit $\mathfrak{X}(T) \to 0$, for all T, when Eq. (8.4.42) takes the form

$$\mathbb{X}_\kappa(T;\boldsymbol{r}) = \{\mathfrak{z} : \Re(\mathfrak{z}) \le -\tfrac{1}{2}\}. \tag{8.4.43}$$

The points on the boundary of this region are given by

$$\mathfrak{z} = -\frac{\exp(\mathrm{i}\theta)}{2\cos(\theta)}, \qquad -\pi \le \theta \le \pi. \tag{8.4.44}$$

The set $\mathbb{Q}$, containing the zeros of the grand partition function, will consist of the whole of the plane $\mathbb{C}_\mathfrak{z}$ except a region containing the origin. The boundary of this region will be obtained from products of z points given by (8.4.44). However, not all such products lie on the boundary. Suppose that $R(\psi)\exp(\mathrm{i}\psi)$ is a point on the boundary. It is clear, from (8.4.35), that

$$R(\psi) = -\frac{1}{2^z}\min\prod_{j=1}^{z}\sec(\theta_j), \tag{8.4.45}$$

where the minimization is subject to the constraint

$$\sum_{j=1}^{z}\theta_j = \psi. \tag{8.4.46}$$

Using a Lagrange undetermined multiplier, it is not difficult to show that the required minimum is achieved when $\theta_j = \psi/z$, for all j, and the boundary curve is given by

$$\mathfrak{z} = -\frac{\exp(\mathrm{i}\psi)}{[2\cos(\psi/z)]^z}, \qquad -\pi \le \psi \le \pi. \tag{8.4.47}$$

The point where the curve (8.4.47) meets $\mathbb{R}^{(+)}$ gives a lower bound $\mathfrak{z}_\mathrm{L}$ for the critical fugacity $\mathfrak{z}_\mathrm{c}$. The square lattice fluid model with first-neighbour exclusion is known as the *hard-square model*. From (8.4.47), setting $\psi = \pi$ and $z = 4$, $\mathfrak{z}_\mathrm{L} = \frac{1}{4}$. The

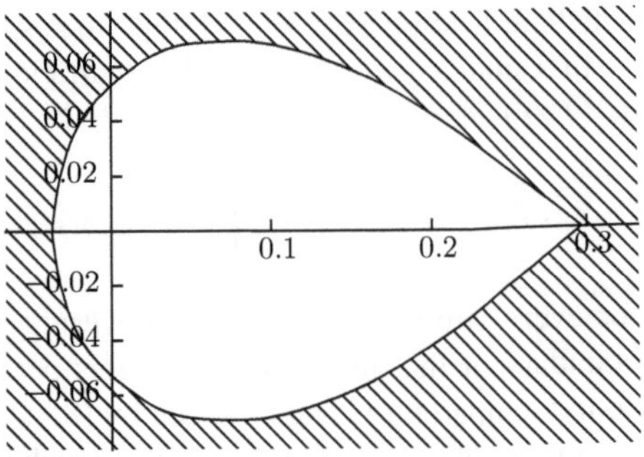

Fig. 8.3 The result given by Eq. (8.4.48) for the hard-hexagon model. The zeros of the grand partition function lie in the shaded region

corresponding model on the triangular lattice is the *hard-hexagon model*. In this case $z = 6$ and $\mathfrak{z}_{\mathrm{L}} = \frac{1}{27}$. The hard-square model has been shown to be equivalent to a sixteen-vertex model on the square lattice (Wood and Goldfinch 1980). Estimates of the critical fugacity were been obtained by transfer matrix (Runnels and Combs 1971) and series (Gaunt and Fisher 1965) methods. Baxter et al. (1980) used corner transfer matrices to obtain the high-density series of the order parameter to twenty-four terms giving an estimate 3.7962 ± 0.0001 for the critical fugacity. The hard-hexagon model is one of the few examples of a model with a known exact solution (Baxter 1980, 1981, 1982b; Joyce 1988a, b), the critical fugacity being $\frac{1}{2}(5\sqrt{5}+11) = 11.09$. As Runnels and Hubbard (1972) observed for the hard-square model, the lower bounds derived by the procedure described here, although of course valid, differ substantially from the critical values. These authors obtained the improvement $\mathfrak{z}_{\mathrm{L}} = \frac{1}{2}$ for the hard-square model by choosing covering sets consisting of half the elementary squares of the lattice with each site belonging to two squares. The same procedure can be applied to the hard-hexagon model. In that case Eq. (8.4.47) is replaced by

$$\mathfrak{z} = -\frac{\exp(\mathrm{i}\psi)}{[3\cos(\psi/3)]^3}, \qquad -\pi \le \psi \le \pi, \tag{8.4.48}$$

which gives $\mathfrak{z}_{\mathrm{L}} = \frac{8}{27}$. The region free from zeros obtained from (8.4.48) is shown in Fig. 8.3.

The analysis of this section has given no unexpected information, showing simply that for first-neighbour lattice fluids with either repulsive or attractive interactions there is no ordered phase at high temperatures and giving consistency with known critical fugacities for models with first-neighbour exclusion. In all but the last example we have chosen covering sets consisting only of pairs of sites. The last example gives some indication that improvements can be gained by increasing the number of sites within a covering.

Chapter 9
Transformation Methods

9.1 Related Systems

We consider a transformation from a model on lattice $\mathcal{N}$ with couplings $\boldsymbol{K}$ and partition function Z, denoted by the triplet $\{\mathcal{N}, \boldsymbol{K}, Z\}$, to a model denoted by $\{\mathring{\mathcal{N}}, \mathring{\boldsymbol{K}}, \mathring{Z}\}$ with all or some of its lattice, couplings and partition function different from those of $\{\mathcal{N}, \boldsymbol{K}, Z\}$. It is supposed that the numbers of lattice sites $N := |\mathcal{N}|$ and $\mathring{N} := |\mathring{\mathcal{N}}|$ are related by $\mathring{N} = n_0 N$, for some rational number n_0 and the relationships between couplings and partition functions are of the form

$$\mathring{\boldsymbol{K}} = \mathfrak{T}(\boldsymbol{K}), \tag{9.1.1}$$

$$\mathring{Z}(\mathring{\mathcal{N}}, \mathring{\boldsymbol{K}}) = \mathcal{Z}(\mathring{\mathcal{N}}, \mathcal{N}, \mathring{\boldsymbol{K}}, \boldsymbol{K}) Z(\mathcal{N}, \boldsymbol{K}). \tag{9.1.2}$$

It is useful to distinguish two different situations in which these equations can arise and be used[1]:

(a) If no singularities are contributed by $\mathcal{Z}(\mathring{\mathcal{N}}, \mathcal{N}, \mathring{\boldsymbol{K}}, \boldsymbol{K})$ to (9.1.2) in the limit $N \to \infty$, then (9.1.1) represents a mapping between critical couplings of $\{\mathcal{N}, \boldsymbol{K}, Z\}$ and $\{\mathring{\mathcal{N}}, \mathring{\boldsymbol{K}}, \mathring{Z}\}$. In particular if the two systems differ only in their coupling values, the fixed points of (9.1.1) will give the critical couplings of the system. Duality for the square lattice ν-state Potts model (Sect. 9.2.2.1) and a combination of duality and the star-triangle transformation for the triangular and honeycomb 2-state Potts model (Sects. 9.4.2.2 and 9.4.2.3) are examples of this, although the partition function of the general ν-state Potts model is unknown. For the square-lattice Ising model Kramers and Wannier (1941a) derived the critical temperature by duality in advance of Onsager's derivation of the partition function in 1944.

[1] They also feature in the real-space renormalization group method of Chap. 15. There the aim is to obtain an exact, or approximate, transformation for which $\mathcal{N}$ and $\mathring{\mathcal{N}}$ are lattices of the same type with $n_0 < 1$. Repeated application of this transformation preserves the critical properties associated with the couplings and leads to trajectories in the space of couplings for which the fixed points can be used to determine critical exponents.

D.A. Lavis, *Equilibrium Statistical Mechanics of Lattice Models*,
Theoretical and Mathematical Physics, DOI 10.1007/978-94-017-9430-5_9

(b) If the model $\{\mathcal{N}, \boldsymbol{K}, Z\}$ is solved (in the sense that its partition function is known), then (9.1.1) and (9.1.2) yield the solution for $\{\mathring{\mathcal{N}}, \mathring{\boldsymbol{K}}, \mathring{Z}\}$. Examples of this are provided by the transformation between the eight-vertex and Ising models (Sect. 3.8.1.3)[2] and the staggered six-vertex model and the ν-state Potts model (Sect. 3.8.2.1). In the latter case the staggered six-vertex model is solved only for particular parameter values which under the transformation entail the Potts model being self-dual, or equivalently at its critical point.

Except where otherwise stated the lattices $\mathcal{N}$ and $\mathring{\mathcal{N}}$ are two-dimensional and of one of the regular types described in Sect. 16.2.1.[3] This dimensionality restriction is not necessary for the edge-decoration transformations discussed in Chap. 10.

9.2 The Wegner Transformation

The are many ways of deriving duality transformations. Here we adopt the general formulation of Wegner (1973) which has the advantage that it incorporates both duality and the weak-graph transformation (Nagle and Temperley 1968), which is needed for the eight-vertex model in Sect. 9.3.1.

Consider a system on the lattice $\mathcal{N}$ with the microstate at $\boldsymbol{r} \in \mathcal{N}$ specified by $\sigma(\boldsymbol{r})$ and the complete microstate of $\mathcal{N}$ denoted by $\boldsymbol{\sigma}$. Suppose that the lattice is divided into n identical (possibly overlapping) subsets of sites,[4] denoted by $\mathcal{Y}_k$, $k = 1, 2, \ldots, n$ and that the Hamiltonian for the system can be expressed in the form

$$\widehat{H}(\mathcal{N}, \boldsymbol{K}; \boldsymbol{\sigma}) := \sum_{k=1}^{n} \hat{h}(\mathcal{Y}_k, \boldsymbol{K}; \boldsymbol{\sigma}_k), \tag{9.2.1}$$

where $\boldsymbol{\sigma}_k$ denotes the subset of the variables $\boldsymbol{\sigma}$ which determine the state of $\mathcal{Y}_k$ and $\boldsymbol{K}$ denotes whatever set of couplings is involved.

With $X(\mathcal{Y}_k, \boldsymbol{K}; \boldsymbol{\sigma}_k) := \exp(-\hat{h}(\mathcal{Y}_k, \boldsymbol{K}; \boldsymbol{\sigma}_k))$ the partition function takes the form

$$Z(\mathcal{N}, \boldsymbol{K}) = \sum_{\{\boldsymbol{\sigma}\}} \prod_{k=1}^{n} X(\mathcal{Y}_k, \boldsymbol{K}; \boldsymbol{\sigma}_k). \tag{9.2.2}$$

---

[2] Although these models were solved independently, it is an interesting and by no means trivial problem (Lavis 1996) to derived the Ising model partition function from that of the eight-vertex model.

[3] We shall also use the graph theory terminology of Sect. 16.1.

[4] This condition can be easily relaxed for specific lattices to include a small number of different kinds of subsets of sites.

Although the components of $\boldsymbol{\sigma}$ can be chosen independently the components of $\boldsymbol{\sigma}_k$, for all k cannot, since in general the subsets $\mathcal{Y}_k$, and in consequence their state variables, overlap. The steps of the Wegner transformation, if it can be applied,[5] are:

(i) To find an alternative state vector $\boldsymbol{\tau}_k$ for $\mathcal{Y}_k$ so that all the distinct values of $\hat{h}(\mathcal{Y}_k, \boldsymbol{K}; \boldsymbol{\sigma}_k)$ for different choices of $\boldsymbol{\sigma}$ are reproduced by a transformed Hamiltonian $\hat{h}'(\mathcal{Y}_k, \boldsymbol{K}; \boldsymbol{\tau}_k)$ for different choices of $\boldsymbol{\tau}_k$. *But*, whereas different $\boldsymbol{\sigma}_k$ may share components, all the components of all the $\boldsymbol{\tau}_k$ are distinct and the state of the whole system, denoted by $\boldsymbol{\tau}$ is just the union of $\boldsymbol{\tau}_k$, for all k. This means that each τ variable is associated with a unique $\mathcal{Y}_k$; a relationship denoted as $\tau \in \mathcal{Y}_k$. Of course, this change of variables comes with a cost. The components of $\boldsymbol{\tau}$ will now be constrained. That is to say, certain $\boldsymbol{\tau}$ will be impossible and ones which are possible may correspond to more than one term in the partition function (9.2.2) which can be re-expressed in the form

$$Z(\mathcal{N}, \boldsymbol{K}) = \sum_{\{\boldsymbol{\tau}\}} \eta(\boldsymbol{\tau}) \prod_{k=1}^{n} X'(\mathcal{Y}_k, \boldsymbol{K}; \boldsymbol{\tau}_k), \tag{9.2.3}$$

where $X'(\mathcal{Y}_k, \boldsymbol{K}; \boldsymbol{\tau}_k) := \exp(-\hat{h}'(\mathcal{Y}_k, \boldsymbol{K}; \boldsymbol{\tau}_k))$ and $\eta(\boldsymbol{\tau})$ is equal to the number of $\boldsymbol{\sigma}$ configurations corresponding to the configuration $\boldsymbol{\tau}$, including, of course, zero if the $\boldsymbol{\tau}$ configuration is impossible.

(ii) For a lattice $\mathring{\mathcal{N}}$, to associate each constraint conditions with a site $\mathring{\boldsymbol{r}} \in \mathring{\mathcal{N}}$ and denote the fact that a particular variable τ belongs to the constraint condition associated with $\mathring{\boldsymbol{r}}$ by writing $\tau \in \mathring{\boldsymbol{r}}$. It should be emphasized that a particular variable τ can in general be associated with more that one constraint condition and thus with more than one site $\mathring{\boldsymbol{r}}$.

(iii) Suppose that on the sites $\mathring{\boldsymbol{r}} \in \mathring{\mathcal{N}}$ variables $\sigma(\mathring{\boldsymbol{r}})$ exist with the same range of values as those on $\mathcal{N}$.

(iv) To find a function υ so that the constraints can be expressed in the form

$$\Upsilon(\mathring{\boldsymbol{r}}; \{\tau \in \mathring{\boldsymbol{r}}\}) := \sum_{\{\sigma(\mathring{\boldsymbol{r}})\}} \prod_{\tau \in \mathring{\boldsymbol{r}}} \upsilon(\sigma(\mathring{\boldsymbol{r}}); \tau) \begin{cases} \neq 0, & \text{if the constraint is satisfied,} \\ = 0, & \text{otherwise.} \end{cases} \tag{9.2.4}$$

(v) To check that Ω exists, independent of $\boldsymbol{\tau}$, so that

$$\eta(\boldsymbol{\tau}) = \Omega \prod_{\{\mathring{\boldsymbol{r}}\}} \Upsilon(\mathring{\boldsymbol{r}}; \{\tau \in \mathring{\boldsymbol{r}}\}). \tag{9.2.5}$$

If each of the steps (i)–(v) can be carried out then, substituting from (9.2.4) and (9.2.5) into (9.2.3), gives

[5] As Wegner says (op.cit. p. 571), "the total class of systems to which the transformation can be applied with meaningful results is not known".

$$Z(\mathcal{N},\boldsymbol{K}) = \Omega\sum_{\{\tau\}}\prod_{\{\mathring{\boldsymbol{r}}\}}\sum_{\{\sigma(\mathring{\boldsymbol{r}})\}}\left[\prod_{\{\tau\in\mathring{\boldsymbol{r}}\}} \upsilon(\sigma(\mathring{\boldsymbol{r}});\tau)\right]\left[\prod_{k=1}^{n} X'(\mathcal{Y}_k,\boldsymbol{K};\boldsymbol{\tau}_k)\right]. \tag{9.2.6}$$

The aim, of course, is to sum over the τ variables in (9.2.6) leaving an expression like (9.2.2), but with a sum over the state $\mathring{\boldsymbol{\sigma}}$, representing the set of state variables on all the sites of $\mathring{\mathcal{N}}$. This means that we need to determine the equivalent groups $\mathring{\mathcal{Y}}_k$ of sites of $\mathring{\mathcal{N}}$. Through the relationship $\tau \in \mathring{\boldsymbol{r}}$ a certain number of the τ variables will be associated with a particular $\mathring{\mathcal{Y}}_k$; a relationship denoted by $\tau \in \mathring{\mathcal{Y}}_k$. But, of course, we already have, from (i) above, the relationship $\tau \in \mathcal{Y}_k$, associating the τ variables with the subsets of $\mathcal{N}$. Thus to complete the procedure we need the final condition that:

(vi) Given a subset $\mathcal{Y}_k$, there exists a unique subset $\mathring{\mathcal{Y}}_j$ such that, if $\tau \in \mathcal{Y}_k$, then $\tau \in \mathring{\mathcal{Y}}_j$. Of course, this does not necessarily establish a one-to-one relationship between the subsets of $\mathcal{N}$ and $\mathring{\mathcal{N}}$. The τ variables in $\mathring{\mathcal{Y}}_j$ may divide into subsets belonging to a number of different $\mathcal{Y}_k$. However, in the cases we consider in Sects. 9.2.1–9.2.3 the relationship is one-to-one and for simplicity this relationship will be assumed. Thus we have $\mathcal{Y}_k \leftrightarrow \mathring{\mathcal{Y}}_k$ with a particular subset of the τ variables associated with this pair of subsets.

Given this final condition, let

$$X(\mathring{\mathcal{Y}}_k,\mathring{\boldsymbol{K}};\mathring{\boldsymbol{\sigma}}_k)\zeta(\boldsymbol{K},\mathring{\boldsymbol{K}}) := \sum_{\{\boldsymbol{\tau}_k\}}\left[\prod_{\{\mathring{\boldsymbol{r}}\in\mathring{\mathcal{Y}}_k\}}\prod_{\{\tau\in\mathring{\boldsymbol{r}}\}} \upsilon(\sigma(\mathring{\boldsymbol{r}});\tau)\right] X'(\mathcal{Y}_k,\boldsymbol{K};\boldsymbol{\tau}_k), \tag{9.2.7}$$

for some function $\zeta(\boldsymbol{K},\mathring{\boldsymbol{K}})$. Then, from (9.2.6),

$$Z(\mathcal{N},\boldsymbol{K}) = \Omega[\zeta(\boldsymbol{K},\mathring{\boldsymbol{K}})]^n Z(\mathring{\mathcal{N}},\mathring{\boldsymbol{K}}), \tag{9.2.8}$$

where $Z(\mathring{\boldsymbol{K}},\mathring{\mathcal{N}})$ is given by (9.2.2) with all variables overscripted with '∘'.

9.2.1 Duality for the ν-State Potts Model

Consider the case where there is just a first-neighbour interaction, no chemical potential term and the Hamiltonian is given by (3.6.49). The subsets of sites $\mathcal{Y}_k$ are first-neighbour pairs which are referenced by their two constituent sites (viz. $[\boldsymbol{r},\boldsymbol{r}']$). Then, in terms of the general notation given, above

$$X([\boldsymbol{r},\boldsymbol{r}'],K;\{\sigma(\boldsymbol{r}),\sigma(\boldsymbol{r}')\}) = \mathfrak{X}^{\delta^{\mathrm{Kr}}(\sigma(\boldsymbol{r})-\sigma(\boldsymbol{r}'))}, \tag{9.2.9}$$

where $K := R/T$, $\mathfrak{X} := \exp(K)$ and $\sigma(\boldsymbol{r}) = 0, 1, \ldots, \nu - 1$. For each $[\boldsymbol{r}, \boldsymbol{r}']$, the *two* pair variables

$$\tau(\boldsymbol{r}, \boldsymbol{r}') := \sigma(\boldsymbol{r}) - \sigma(\boldsymbol{r}') \mod \nu, \tag{9.2.10}$$

are defined and distinguished by the ordering of the pair of sites, with, of course, the consistency conditions

$$\tau(\boldsymbol{r}, \boldsymbol{r}') + \tau(\boldsymbol{r}', \boldsymbol{r}) = 0 \mod \nu. \tag{9.2.11}$$

We now consider the possible boundary conditions for the lattice when, as is our aim here, the Wegner transformation is applied from a lattice to its dual with the groups of subsets being first-neighbour pairs. For this it is necessary to establish a one-to-one relationship between a first-neighbour pair in the lattice and a first-neighbour pair in the dual lattice and obviously there needs to be an equal number of each. Of the four boundary conditions (I)–(IV) listed in Sect. 16.2.2 the obvious choice is the periodic case (I); and indeed duality is often presented in this way. However, for our purposes and for its application in Sect. 14.2.2.1 it is more convenient to use (IV) with duality applied to the lattice $\mathcal{N}^{(+)}$ which includes the linking edges and the terminating sites connecting $\mathcal{N}$ to the boundary, for each of which we set $\sigma = 0$. So, from (16.2.13), $n = N_{\mathrm{E}}^{(+)} = N_{\mathrm{E}}^{\star}$ and the one-to-one mapping between pairs in $\mathcal{N}^{(+)}$ and $\mathcal{N}^{\star}$ is that of orthogonal bisection.

Suppose that $\boldsymbol{r}_0$ is an arbitrarily chosen site of $\mathcal{N}$ and $\boldsymbol{r}_0, \boldsymbol{r}_1, \ldots, \boldsymbol{r}_m$ gives a simply connected graph of lattice edges between $\boldsymbol{r}_0$ and $\boldsymbol{r}_m$. Then,

$$\sigma(\boldsymbol{r}_m) = \tau(\boldsymbol{r}_m, \boldsymbol{r}_{m-1}) + \cdots + \tau(\boldsymbol{r}_1, \boldsymbol{r}_0) + \sigma(\boldsymbol{r}_0), \mod \nu. \tag{9.2.12}$$

If $\boldsymbol{r}_0$ is a boundary site, $\sigma(\boldsymbol{r}_0) = 0$ and all the site states can be derived from the pair variables. There are N state variables and $2N_{\mathrm{E}}^{(+)}$ pair variables, although this number is effectively reduced to $N_{\mathrm{E}}^{(+)}$ by the consistency conditions (9.2.11). Thus we need $N_{\mathrm{E}}^{(+)} - N$ independent constraint conditions. These are provided by the special case of (9.2.12), where $\boldsymbol{r}_0 = \boldsymbol{r}_m$ and the sequence of sites lies on the edges of one of the set of elementary cycles around the faces of $\mathcal{N}$.[6] From (16.2.13) there are $N^{\star} = N_{\mathrm{E}}^{(+)} - N + 1$ elementary cycles; we seem to have one too many constraints, however, as we shall see they are not all independent. Let

$$\alpha(\boldsymbol{r}^{\star}) := \tau(\boldsymbol{r}_{z^{\star}}, \boldsymbol{r}_{z^{\star}-1}) + \cdots + \tau(\boldsymbol{r}_2, \boldsymbol{r}_1) \quad \mod \nu, \tag{9.2.13}$$

where $\boldsymbol{r}_1, \boldsymbol{r}_2, \ldots, \boldsymbol{r}_{z^{\star}}$ are the sites around the face containing $\boldsymbol{r}^{\star}$. Then the constraint condition

$$\alpha(\boldsymbol{r}^{\star}) = 0, \tag{9.2.14}$$

[6] We showed in Sect. 16.1.2 that the number of independent cycles for a section graph is the cyclomatic number, which for a planar graph is the number of faces; in this case it is the number of dual lattice sites.

is associated with the dual-lattice site $\boldsymbol{r}^\star$. As indicated above, there are $N^\star$ conditions, but their independence is reduced by

$$\sum_{\{\boldsymbol{r}^\star\}} \alpha(\boldsymbol{r}^\star) = 0. \tag{9.2.15}$$

The lattice $\mathring{\mathcal{N}}$, used in the general description of the Wegner transformation in Sect. 9.2, is now the dual lattice $\mathcal{N}^\star$ (with $\mathcal{N}^{(+)}$ replacing $\mathcal{N}$) and taking

$$\upsilon(\sigma(\boldsymbol{r}^\star); \tau(\boldsymbol{r}, \boldsymbol{r}')) := \exp\left(\mathrm{i}\frac{2\pi}{\nu}\sigma(\boldsymbol{r}^\star)\tau(\boldsymbol{r}, \boldsymbol{r}')\right), \tag{9.2.16}$$

gives, from (9.2.4), (9.2.13), (9.2.14), (18.1.10) and (18.1.11),

$$\Upsilon(\boldsymbol{r}^\star; \{\tau(\boldsymbol{r}, \boldsymbol{r}')\}) = \sum_{\sigma=0}^{\nu} \exp\left(\mathrm{i}\frac{2\pi}{\nu}\sigma\, \alpha(\boldsymbol{r}^\star)\right) = \nu\delta^{\mathrm{Kr}}(\alpha(\boldsymbol{r}^\star)). \tag{9.2.17}$$

We now use Ω as it appears in (9.2.5) to compensate for the ν, appearing in (9.2.17) for each constraint condition around the faces of $\mathcal{N}^\star$; thus

$$\Omega = 1/\nu^{N^\star}. \tag{9.2.18}$$

Every $\tau(\boldsymbol{r}, \boldsymbol{r}')$, for an edge of $\mathcal{N}^{(+)}$ will belong to two constraint conditions, in one of which it will appear as $\tau(\boldsymbol{r}', \boldsymbol{r})$, and these will be related to two sites $\boldsymbol{r}^\star$ and $\boldsymbol{r}^{\star\prime}$. Thus a pair of sites $(\boldsymbol{r}, \boldsymbol{r}') \in \mathcal{N}^{(+)}$ is in a one-to-one relationship to the pair $(\boldsymbol{r}^\star, \boldsymbol{r}^{\star\prime}) \in \mathcal{N}^\star$; being, of course, pairs of intersecting edges of $\mathcal{N}^{(+)}$ and $\mathcal{N}^\star$. From (9.2.7), (9.2.11), (9.2.16) and (18.1.11)

$$\begin{aligned}(\mathfrak{X}^\star)^{\delta^{\mathrm{Kr}}(\sigma(\boldsymbol{r}^\star)-\sigma(\boldsymbol{r}^{\star\prime}))}\zeta(\mathfrak{X}, \mathfrak{X}^\star) &= \sum_{\tau=0}^{\nu-1} \exp\left(\mathrm{i}\frac{2\pi}{\nu}[\sigma(\boldsymbol{r}^\star) - \sigma(\boldsymbol{r}^{\star\prime})]\tau\right) \mathfrak{X}^{\delta^{\mathrm{Kr}}(\tau)} \\ &= \left\{\mathfrak{X} - 1 + \nu\,\delta^{\mathrm{Kr}}(\sigma(\boldsymbol{r}^\star) - \sigma(\boldsymbol{r}^{\star\prime}))\right\}.\end{aligned} \tag{9.2.19}$$

Making the two choices $\sigma(\boldsymbol{r}^\star) = \sigma(\boldsymbol{r}^{\star\prime})$ and $\sigma(\boldsymbol{r}^\star) \neq \sigma(\boldsymbol{r}^{\star\prime})$ gives

$$\zeta(\mathfrak{X}, \mathfrak{X}^\star) = (\mathfrak{X} - 1), \tag{9.2.20}$$

$$(\mathfrak{X} - 1)(\mathfrak{X}^\star - 1) - \nu. \tag{9.2.21}$$

This last relationship is symmetric under $\mathfrak{X} \leftrightarrow \mathfrak{X}^\star$ and yields both

$$\mathfrak{X}^{\delta^{\mathrm{Kr}}(x)} = 1 + \frac{\nu\,\delta^{\mathrm{Kr}}(x)}{\mathfrak{X}^\star - 1} \qquad \text{and} \qquad (\mathfrak{X}^\star)^{\delta^{\mathrm{Kr}}(x)} = 1 + \frac{\nu\,\delta^{\mathrm{Kr}}(x)}{\mathfrak{X} - 1}. \tag{9.2.22}$$

Substituting from (9.2.18) and (9.2.20) into (9.2.8) gives[7]

$$Z(\mathcal{N}^{(+)}, K, 0) = \nu^{-N^\star}(\mathfrak{X}-1)^{N^\star_{\mathrm{E}}} Z^\star(\mathcal{N}^\star, K^\star, 0), \tag{9.2.23}$$

which, by applying (16.2.13) and (9.2.21), becomes

$$\frac{Z(\mathcal{N}^{(+)}, K, 0)}{(\mathfrak{X}-1)^{N}} = \frac{Z^\star(\mathcal{N}^\star, K^\star, 0)}{(\mathfrak{X}^\star-1)^{N^\star}}. \tag{9.2.24}$$

This relationship has an appearance of symmetric, but it must be remembered that this is not really the case since we have supposed a fixed boundary state on $\mathcal{N}^{(+)}$ and $\mathcal{N}^\star$ with free boundary conditions. However, for large lattices, the boundary conditions become unimportant; N and $N^\star$ are large and also large with respect to N_{L}, the number of boundary sites.[8] From (9.2.24), the free energies per lattice site are related by the symmetric equation

$$z\left\{\ln(\mathfrak{X}^\star - 1) + \phi_1(\mathcal{N}^\star, K^\star)\right\} = z^\star\left\{\ln(\mathfrak{X}-1) + \phi_1(\mathcal{N}, K)\right\}. \tag{9.2.25}$$

Equations (9.2.21) and (9.2.25) are the relationships between couplings and free energies for a pair of dual lattices. In the complex-$\mathfrak{X}$ plane (9.2.21) is satisfied by the points $\mathfrak{X} = F(\nu;\theta)$, $\mathfrak{X} = F(\nu;-\theta)$, which lie on the unit circle

$$\mathfrak{X} = F(\nu;\theta) := 1 + \sqrt{\nu}\exp(\mathrm{i}\theta), \tag{9.2.26}$$

which is invariant under the duality transformation.

For the self-dual square lattice if $\mathfrak{X} = \mathfrak{X}^\star = \mathfrak{X}_{\mathrm{c}}$ then, from (9.2.21),

$$\mathfrak{X}_{\mathrm{c}} = F(\nu, 0) = 1 + \sqrt{\nu}. \tag{9.2.27}$$

In terms of the coupling $K := \ln(\mathfrak{X})$, the transformation (9.2.21) connects any value of K in the high-temperature interval $[0, K_{\mathrm{c}}]$ with one in the low-temperature interval $[K_{\mathrm{c}}, \infty]$, and vice versa. A critical point arises from a singularity in $Z(\mathcal{N}, K)$, and thus from (9.2.21) a critical point in $[0, K_{\mathrm{c}}]$ implies a critical point in $[K_{\mathrm{c}}, \infty]$. If there is a unique critical point, it must given by

$$K = K_{\mathrm{c}}, \qquad T_{\mathrm{c}} = R/\ln(1+\sqrt{\nu}). \tag{9.2.28}$$

---

[7] It should be noted that $Z(\mathcal{N}^{(+)}, K, 0)$ is the partition function for the case where the boundary states have a particular specified value (without loss of generality $\sigma = 0$). If we had simply chosen the condition that all the boundary states were the same the factor $\nu^{-N^\star}$ in (9.2.23) would be replaced by $\nu^{1-N^\star}$.

[8] The limit of a large lattice is in the sense of van Hove; see Sect. 8.1.

9.2.2 Duality for the Spin-$\frac{1}{2}$ Ising Model

As described in Sect. 3.6.4, the $\mu = 0$ 2-state Potts model is equivalent to the zero-field spin-$\frac{1}{2}$ Ising model, with the Ising energy parameter $J = \frac{1}{2}R$. So with $\nu = 2$, $K := J/T$, $\mathfrak{X} := \exp(2K/T)$ and $\mathfrak{X}^\star := \exp(2K^\star/T)$, (9.2.21) gives

$$\sinh(2K)\sinh(2K^\star) = 1 \tag{9.2.29}$$

and applying (3.6.57) to (9.2.25) gives

$$z^\star\left\{\tfrac{1}{2}\ln[2\sinh(2K)] + \phi_1(\mathcal{N}, K, 0)\right\} = z\left\{\tfrac{1}{2}\ln[2\sinh(2K^\star)] + \phi_1^\star(\mathcal{N}^\star, K^\star, 0)\right\}. \tag{9.2.30}$$

Then, from (2.3.27),

$$\frac{\mathrm{u}(\mathcal{N}, K, 0)}{z\coth(2K)} + \frac{\mathrm{u}^\star(\mathcal{N}^\star, K^\star, 0)}{z^\star\coth(2K^\star)} = -\frac{1}{2}. \tag{9.2.31}$$

This duality relationship between internal energy densities is also, from (3.5.8), a relation between first-neighbour correlations.

9.2.2.1 The Square Lattice

This lattice is self-dual and (9.2.29) is a transformation $K_{\mathrm{s}} \leftrightarrow K_{\mathrm{s}}^\star$ between two values of the square-lattice coupling with, from (9.2.29)–(9.2.31),

$$\sinh(2K_{\mathrm{s}}^\star)\sinh(2K_{\mathrm{s}}) = 1, \tag{9.2.32}$$

$$\frac{Z(\mathcal{N}_{\mathrm{s}}, K_{\mathrm{s}}, 0)}{[\sinh(2K_{\mathrm{s}})]^{\frac{1}{2}N}} = \frac{Z(\mathcal{N}_{\mathrm{s}}, K_{\mathrm{s}}^\star, 0)}{[\sinh(2K_{\mathrm{s}}^\star)]^{\frac{1}{2}N}}, \tag{9.2.33}$$

$$\phi_1(\mathcal{N}_{\mathrm{s}}, K_{\mathrm{s}}^\star, 0) + \tfrac{1}{2}\ln[\sinh(2K_{\mathrm{s}}^\star)] = \tfrac{1}{2}\ln[\sinh(2K_{\mathrm{s}})] + \phi_1(\mathcal{N}_{\mathrm{s}}, K_{\mathrm{s}}, 0), \tag{9.2.34}$$

$$\frac{\mathrm{u}(\mathcal{N}_{\mathrm{s}}, K_{\mathrm{s}}, 0)}{\coth(2K_{\mathrm{s}})} + \frac{\mathrm{u}(\mathcal{N}_{\mathrm{s}}, K_{\mathrm{s}}^\star, 0)}{\coth(2K_{\mathrm{s}}^\star)} = -2. \tag{9.2.35}$$

From (9.2.32), when $K_{\mathrm{s}} \to \infty$, $K_{\mathrm{s}}^\star \to 0$ and vice-versa. The transformation is, therefore, between high and low temperatures and it will map between critical points with singular thermodynamic characteristics. Intuitively a single critical temperature seems likely for the Ising ferromagnet since there is only one simple type of long-range order, corresponding to the zero temperature ground state.[9] This must be given by the fixed point $K_{\mathrm{s}} = K_{\mathrm{s}}^\star = K_{\mathrm{sc}}$ of the transformation. Thus we have

$$\sinh^2(2K_{\mathrm{sc}}) = 1, \qquad T_{\mathrm{c}} = J/K_{\mathrm{sc}} = 2J/\ln(1+\sqrt{2}). \tag{9.2.36}$$

[9] That this is the case is evident from Onsager's solution of the model (see Sect. 12.2.7.3).

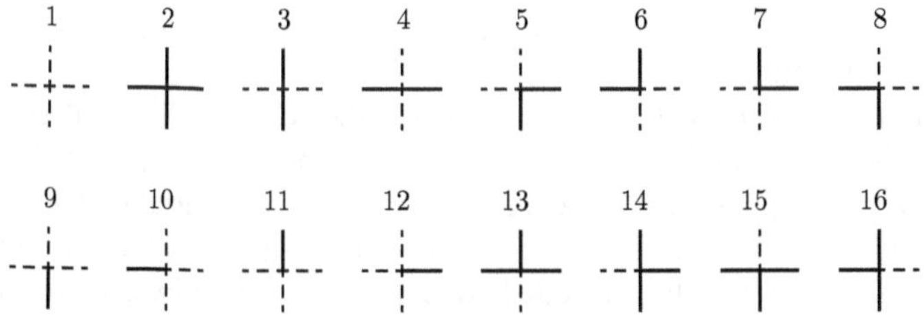

Fig. 9.1 The sixteen vertex types on the square lattice

Substituting $K = K^\star = K_{sc}$ in (9.2.31) or (9.2.35), the critical point values of the first-neighbour correlation (see (3.5.8)) and the internal energy per site for the square lattice are given by

$$\mathfrak{u}_c = -2\langle\sigma(\boldsymbol{r})\sigma(\boldsymbol{r}')\rangle_{1c} = -\coth(2K_{sc}) = -\sqrt{2}. \tag{9.2.37}$$

Numerical values for T_c and $\mathfrak{u}_c$ are given in Table 3.1. It is important to note that the duality form (9.2.34) is not sufficient to determine the critical value $\phi_1(\mathcal{N}_s, K_c, 0)$ for the free-energy density, nor since, from (2.3.28),

$$s = \mathfrak{u}K - \phi_1(\mathcal{N}_s, K, 0), \tag{9.2.38}$$

is it possible from duality to determine the critical entropy density. For this we need the solution of the two-dimensional Ising model of which a version is given in Sect. 12.2.

9.2.3 The Weak-Graph Transformation

Here attention is restricted to a square lattice $\mathcal{N}_s$ with toroidal boundary conditions. Consider all the 2^{2N} graphs obtain by drawing, or not drawing, a graph edge along an edge of $\mathcal{N}_s$. These graphs $\mathfrak{g}$ are weakly embedded in $\mathcal{N}_s$ (Sect. 16.4)[10] and there are 16 vertex types according to whether a particular edge at the vertex contains, or does not contain, an edge of $\mathfrak{g}$. They are shown in Fig. 9.1 with the first eight being those of the eight-vertex model shown in Fig. 3.2. The Boltzmann weights for the different vertices are denoted by $\mathfrak{X}(k)$, $k = 1, 2, \ldots, 16$ and the partition function is

$$Z(\mathcal{N}_s, \boldsymbol{\mathfrak{X}}) = \sum_{\{\mathfrak{g}\}} \prod_{\{\boldsymbol{r}\}} X(\boldsymbol{r}; \mathfrak{g}), \tag{9.2.39}$$

---

[10] Hence the name of this transformation.

where the sum is over all graphs $\mathfrak{g}$ and $X(\boldsymbol{r};\mathfrak{g})$ is the particular $\mathfrak{X}(k)$ at site $\boldsymbol{r}$ when the lattice occupation is $\mathfrak{g}$.

In order to apply the Wegner transformation the edges of $\mathcal{N}_\text{s}$ are decorated with sites with the location of the site on the first-neighbour pair $[\boldsymbol{r},\boldsymbol{r}'] \in \mathcal{N}_\text{s}$ given by $\mathring{\boldsymbol{r}} := (\boldsymbol{r}+\boldsymbol{r}')/2$. Suppose that at each edge site $\mathring{\boldsymbol{r}}$, there is a variable $\sigma(\mathring{\boldsymbol{r}})$, which can take the value 1 or 0, according as the graph $\mathfrak{g}$ does or does not have a edge through $\mathring{\boldsymbol{r}}$. Then the set of all weakly embedded graphs on $\mathcal{N}_\text{s}$ is given by choosing all values of the vector $\boldsymbol{\sigma}$ whose components are the states $\sigma(\mathring{\boldsymbol{r}})$ of all the edge sites.

Now it is necessary to re-express the Boltzman weights in terms of these edge variables. Let $\mathfrak{X}(\sigma_1,\sigma_2,\sigma_3,\sigma_4)$, be the Boltzmann weight at a site, where the neighbouring edge variables are labelled in clockwise order starting at the bottom with σ_1,[11] (e.g. $\mathfrak{X}(1) \to \mathfrak{X}(0,0,0,0)$, $\mathfrak{X}(13) \to \mathfrak{X}(0,1,1,1)$), with the partition function re-expressed in the form

$$Z(\mathcal{N}_\text{s},\mathfrak{X}) = \sum_{\{\boldsymbol{\sigma}\}}\prod_{\{\boldsymbol{r}\}}\mathfrak{X}(\sigma_1(\boldsymbol{r}),\sigma_2(\boldsymbol{r}),\sigma_3(\boldsymbol{r}),\sigma_4(\boldsymbol{r})). \tag{9.2.40}$$

Of course these variables are not all independent: with a lattice spacing of unity and unit horizontal and vertical vectors $\hat{\boldsymbol{x}}$ and $\hat{\boldsymbol{y}}$, $\sigma_1(\boldsymbol{r}) = \sigma_3(\boldsymbol{r}+\hat{\boldsymbol{y}})$, $\sigma_2(\boldsymbol{r}) = \sigma_4(\boldsymbol{r}+\hat{\boldsymbol{x}})$, $\sigma_3(\boldsymbol{r}) = \sigma_1(\boldsymbol{r}-\hat{\boldsymbol{y}})$ and $\sigma_4(\boldsymbol{r}) = \sigma_2(\boldsymbol{r}-\hat{\boldsymbol{x}})$. These constraints are made explicit by replacing, for all $\boldsymbol{r}$, the variables $\sigma_\ell(\boldsymbol{r})$ with the *independent* variables $\tau_\ell(\boldsymbol{r})$, for $\ell = 1,\ldots,4$, with (9.2.40) becoming

$$\begin{aligned} Z(\mathcal{N}_\text{s},\mathfrak{X}) = &\sum_{\{\boldsymbol{\tau}\}}\prod_{\{\boldsymbol{r}\}}\mathfrak{X}(\tau_1(\boldsymbol{r}),\tau_2(\boldsymbol{r}),\tau_3(\boldsymbol{r}),\tau_4(\boldsymbol{r})) \\ &\times \delta^{\text{Kr}}(\tau_1(\boldsymbol{r})-\tau_3(\boldsymbol{r}+\hat{\boldsymbol{y}}))\delta^{\text{Kr}}(\tau_2(\boldsymbol{r})-\tau_4(\boldsymbol{r}+\hat{\boldsymbol{x}})). \end{aligned} \tag{9.2.41}$$

The next step is to factorize the delta functions into a product of contributions from each of the sites in the argument. For an arbitrary real variable q, let

$$\boldsymbol{V}(q) := \frac{1}{\sqrt{1+q^2}}\begin{pmatrix} +1 & q \\ q & -1 \end{pmatrix}, \tag{9.2.42}$$

where, for convenience, we shall index the rows and columns by 0 and 1, rather than 1 and 2. Now suppose that τ and τ' are two variables on one edge and introduce a new edge variable $\mathring{\sigma}$ between them at the corresponding decoration site. Then, since $\boldsymbol{V}(q)$ is a symmetric unitary matrix,

$$\delta^{\text{Kr}}(\tau-\tau') = \sum_{\mathring{\sigma}=0}^{1} V(\mathring{\sigma},\tau;q)V(\mathring{\sigma},\tau';q). \tag{9.2.43}$$

---

[11] See the footnote concerning the relationship between toroidal boundary conditions and lattice orientation on p. 60. This function is essentially the same as that defined in Sect. 3.8.1 except that here it is more convenient to take the argument values 0, 1 rather than −1, +1.

With

$$\mathring{\mathfrak{X}}(\mathring{\sigma}_{\mathrm{T}},\mathring{\sigma}_{\mathrm{R}},\mathring{\sigma}_{\mathrm{B}},\mathring{\sigma}_{\mathrm{L}}) := \sum_{\tau_1=0}^{1}\sum_{\tau_2=0}^{1}\sum_{\tau_3=0}^{1}\sum_{\tau_4=0}^{1}\mathfrak{X}(\tau_1,\tau_2,\tau_3,\tau_4)\prod_{\ell=1}^{4}V(\mathring{\sigma}_\ell,\tau_\ell;q), \tag{9.2.44}$$

$$Z(\mathcal{N}_{\mathrm{S}},\mathfrak{X}) = Z(\mathcal{N}_{\mathrm{S}},\mathring{\mathfrak{X}}), \tag{9.2.45}$$

where

$$Z(\mathcal{N}_{\mathrm{S}},\mathring{\mathfrak{X}}) = \sum_{\{\mathring{\sigma}\}}\prod_{\{\boldsymbol{r}\}}\mathring{\mathfrak{X}}(\mathring{\sigma}_1(\boldsymbol{r}),\mathring{\sigma}_2(\boldsymbol{r}),\mathring{\sigma}_3(\boldsymbol{r}),\mathring{\sigma}_4(\boldsymbol{r})). \tag{9.2.46}$$

Using (9.2.42), (9.2.44) can be re-expressed in the simpler form

$$\mathring{\mathfrak{X}}(k) = \sum_{j=1}^{16}\mathfrak{C}(k,j;q)\mathfrak{X}(j), \tag{9.2.47}$$

where

$$\mathfrak{C}(k,j;q) := \frac{(-1)^{\nu_{kj}}q^{\gamma_k+\gamma_j-2\nu_{kj}}}{(1+q^2)^2}, \tag{9.2.48}$$

γ_k is the valency of a vertex k and ν_{kj} is the number of graph edges which coincide for two superimposed vertices k and j. It is not difficult to show that the matrix with elements $\mathfrak{C}(k,j;q)$ is symmetric and unitary and thus, as in the case of duality, that the weak-graph relationship between two sets of weights is symmetric. It can also be shown that the transformation is satisfied by vertex weights which are functions only of the valencies.

For the eight-vertex model (Sect. 3.8.1) $\mathfrak{X}(k) = \mathfrak{X}(k+1)$, $k = 1,3,5,7$ and $\mathfrak{X}(k) = 0$, $k = 9,10,\ldots,16$. Since, from Fig. 9.1 and (9.2.48), $\mathfrak{C}(k,j;1) = -\mathfrak{C}(k,j+1;1)$, $j = 1,3,5,7$, this condition holds for the weights $\mathring{\mathfrak{X}}(k)$, when $q = 1$ and, as we shall now see, the weak-graph transformation gives a symmetry transformation $\mathcal{T}_{\mathrm{WG}}$ for the eight-vertex model.

9.3 The Regular Square-Lattice Eight-Vertex Model

As in Sect. 3.8.1 the lattice carrying the vertices is denoted by $\mathcal{N}^{\circledast}$ with the spins on the sites of $\mathcal{N}+\mathcal{N}^{\star}$. The vertex weights are expressible in spin and edge-variable forms by (3.8.9) and (3.8.10), where now the couplings are site-independent, as are the eight-vertex Boltzmann weights $\mathfrak{a}$, $\mathfrak{b}$, $\mathfrak{c}$ and $\mathfrak{d}$, given in (3.8.11). In terms of these

$$\Delta := \frac{a^2 + b^2 - c^2 - d^2}{2(ab + cd)}, \qquad \Gamma := \frac{ab - cd}{ab + cd}, \tag{9.3.1}$$

were introduced in (3.8.12) and (3.8.13).

It will be shown, in an investigation of the symmetry properties of the eight-vertex partition function $Z^{(8\text{vM})}(\mathcal{N}^{\circledast}, a, b, c, d)$, that the weak-graph transformation plays a fundamental role and that, as in the case of duality for the Ising model, this leads to a determination of the critical region.

9.3.1 Symmetry Properties and Transformations

Symmetry properties of the partition function are obtained by considering one-to-one transformations between arrow configurations. This can be done using the direction rule and the list of vertex configurations in Fig. 3.2 (see, for example Lavis and Bell 1999b, Sect. 5.4). Here, however, we shall use the shading rule and Fig. 3.3. Then

- Rotation of the lattice through any multiple of 90° represents a translation of the edge variables, while leaving their values unchanged. A clockwise rotation is given by $(\varsigma_1, \varsigma_2, \varsigma_3, \varsigma_4) \to (\varsigma_4, \varsigma_1, \varsigma_2, \varsigma_3)$.[12] Thus we have the transformation

$$\begin{aligned} \mathcal{T}_0 : \quad &(a, b, c, d) \leftrightarrow (a, b, d, c), \\ &\Delta \leftrightarrow \Delta, \qquad \Gamma \leftrightarrow \Gamma, \\ Z^{(8\text{vM})}(\mathcal{N}^{\circledast}, a, b, c, d) &= Z^{(8\text{vM})}(\mathcal{N}^{\circledast}, a, b, d, c). \end{aligned} \tag{9.3.2}$$

  It is evident from Figs. 3.2 and 3.3, that an alternative way to produce the effect of $\mathcal{T}_0$ on Boltzmann weights is to interchange the pair of edge variable on either of the lines through the vertex. That is, for example, $(\varsigma_1, \varsigma_2, \varsigma_3, \varsigma_4) \to (\varsigma_3, \varsigma_2, \varsigma_1, \varsigma_4)$. It is clear that this procedure cannot be carried out over the whole lattice simultaneously. However, it is important for the transfer matrix, which is mapped into its transpose, an element of which is obtained by interchanging c and d.
- Reversing all the arrows on horizontal lines corresponds to $(\varsigma_1, \varsigma_2, \varsigma_3, \varsigma_4) \to (\varsigma_1, -\varsigma_2, \varsigma_3, -\varsigma_4)$, giving

$$\begin{aligned} \mathcal{T}_1 : \quad &(a, b, c, d) \leftrightarrow (b, a, d, c), \\ &\Delta \leftrightarrow \Delta, \qquad \Gamma \leftrightarrow \Gamma, \\ Z^{(8\text{vM})}(\mathcal{N}^{\circledast}, a, b, c, d) &= Z^{(8\text{vM})}(\mathcal{N}^{\circledast}, b, a, d, c). \end{aligned} \tag{9.3.3}$$

  This transformation can, of course, also be derived by considering reversal of vertical arrows.

[12] If the direction rule were used the transformation would correspond to $(\varsigma_1, \varsigma_2, \varsigma_3, \varsigma_4) \to (-\varsigma_4, \varsigma_1, -\varsigma_2, \varsigma_3)$ giving $(a, b, c, d) \to (b, a, c, d)$.

- Suppose, on an arbitrary arrow configuration, a graph in the form of one of the two complete sets of bottom-left-to-top-right (BL-TR) staircases, with step length equal to that of one edge (Fig. 3.4a), is superposed. One way to do this is to choose, for the graph edges, 2 and 3 at sites of $\mathcal{N}^{\circledast}_{\mathrm{E}}$ and 4 and 1 at sites of $\mathcal{N}^{\circledast}_{\mathrm{O}}$. A transformation which reverses the arrows on the edges of the staircases has the effect $(\varsigma_1, \varsigma_2, \varsigma_3, \varsigma_4) \to (\varsigma_1, -\varsigma_2, -\varsigma_3, \varsigma_4)$, on $\mathcal{N}^{\circledast}_{\mathrm{E}}$ and $(\varsigma_1, \varsigma_2, \varsigma_3, \varsigma_4) \to (-\varsigma_1, \varsigma_2, \varsigma_3, -\varsigma_4)$, on $\mathcal{N}^{\circledast}_{\mathrm{O}}$, giving the BL-TR *staircase transformation*

$$\begin{aligned} \mathfrak{T}_2 : \quad & (\mathrm{a}, \mathrm{b}, \mathrm{c}, \mathrm{d}) \leftrightarrow (\mathrm{c}, \mathrm{d}, \mathrm{a}, \mathrm{b}), \\ & \Delta \leftrightarrow -\Delta, \qquad \Gamma \leftrightarrow -\Gamma, \\ Z^{(8\mathrm{vM})}(\mathcal{N}^{\circledast}, \mathrm{a}, \mathrm{b}, \mathrm{c}, \mathrm{d}) &= Z^{(8\mathrm{vM})}(\mathcal{N}^{\circledast}, \mathrm{c}, \mathrm{d}, \mathrm{a}, \mathrm{b}). \end{aligned} \tag{9.3.4}$$

- Referring to Fig. 3.4b, the corresponding bottom-right-to-top-left (BR-TL) *staircase transformation* is $(\varsigma_1, \varsigma_2, \varsigma_3, \varsigma_4) \to (\varsigma_1, \varsigma_2, -\varsigma_3, -\varsigma_4)$, on $\mathcal{N}^{\circledast}_{\mathrm{E}}$ and $(\varsigma_1, \varsigma_2, \varsigma_3, \varsigma_4) \to (-\varsigma_1, -\varsigma_2, \varsigma_3, \varsigma_4)$, on $\mathcal{N}^{\circledast}_{\mathrm{O}}$, giving

$$\begin{aligned} \mathfrak{T}_3 : \quad & (\mathrm{a}, \mathrm{b}, \mathrm{c}, \mathrm{d}) \leftrightarrow (\mathrm{d}, \mathrm{c}, \mathrm{b}, \mathrm{a}), \\ & \Delta \leftrightarrow -\Delta, \qquad \Gamma \leftrightarrow -\Gamma, \\ Z^{(8\mathrm{vM})}(\mathcal{N}^{\circledast}, \mathrm{a}, \mathrm{b}, \mathrm{c}, \mathrm{d}) &= Z^{(8\mathrm{vM})}(\mathcal{N}^{\circledast}, \mathrm{d}, \mathrm{c}, \mathrm{b}, \mathrm{a}). \end{aligned} \tag{9.3.5}$$

It is clear that the transformations $\mathfrak{T}_1$, $\mathfrak{T}_2$ and $\mathfrak{T}_3$ are self inverse with $\mathfrak{T}_3 = \mathfrak{T}_1\mathfrak{T}_2$. In fact, together with the identity, they form a commutative subgroup of the symmetric group $\mathcal{S}_4$.

- Comparing Figs. 3.2 and 3.3 it will be seen that the changes of signs of the edge variables in $\mathfrak{T}_2$ is equivalent to the mapping needed to change between the arrow assignments made using the direction and shading rules; meaning that both sub-lattice arrow configurations, for vertices $p = 1, 2, \ldots, 8$ in Fig. 3.3, map into the corresponding arrow configurations for these vertices in Fig. 3.2. This can thus be used to count the numbers of vertices of each type on the lattice. Let these numbers be $n_p, p = 1, 2, \ldots, 8$. Assuming periodic boundary conditions in both directions, the number of changes of arrow directions left-to-right along a horizontal line must be the same as the number of right-to-left changes. The same must be the case for up-to-down changes on vertical lines. This gives

$$n_5 + n_7 = n_6 + n_8, \qquad n_6 + n_7 = n_5 + n_8, \tag{9.3.6}$$

from which it follows that $n_5 = n_6$, $n_7 = n_8$. The weights c and d appear to an even power in $Z^{(8\mathrm{vM})}(\mathcal{N}^{\circledast}, \mathrm{a}, \mathrm{b}, \mathrm{c}, \mathrm{d})$, and, because of the transformation $\mathfrak{T}_2$ so must a and b. Thus we have the (unphysical) transformations which independently negate the Boltzmann weights

$$\mathcal{T}_{\mathrm{a}} : \mathrm{a} \leftrightarrow -\mathrm{a}, \quad \mathcal{T}_{\mathrm{b}} : \mathrm{b} \leftrightarrow -\mathrm{b}, \quad \mathcal{T}_{\mathrm{c}} : \mathrm{c} \leftrightarrow -\mathrm{c}, \quad \mathcal{T}_{\mathrm{d}} : \mathrm{d} \leftrightarrow -\mathrm{d},$$
$$\Delta \leftrightarrow \Delta, \qquad \Gamma \leftrightarrow 1/\Gamma,$$
$$Z^{(8\mathrm{vM})}(\mathcal{N}^{\circledast}, \mathrm{a}, \mathrm{b}, \mathrm{c}, \mathrm{d}) = Z^{(8\mathrm{vM})}(\mathcal{N}^{\circledast}, \pm\mathrm{a}, \pm\mathrm{b}, \pm\mathrm{c}, \pm\mathrm{d}), \tag{9.3.7}$$

for all choices of signs.

- The weak-graph transformation of Sect. 9.2.3 is applicable to the eight-vertex model. From (9.2.47) and (9.2.48) with $q = 1$

$$\begin{aligned} \mathcal{T}_{\mathrm{WG}}: \quad & (\mathrm{a}, \mathrm{b}, \mathrm{c}, \mathrm{d}) \leftrightarrow (\mathring{\mathrm{a}}, \mathring{\mathrm{b}}, \mathring{\mathrm{c}}, \mathring{\mathrm{d}}), \\ & \Delta \leftrightarrow 1/\Delta, \qquad \Gamma \leftrightarrow \Gamma/\Delta, \\ Z^{(8\mathrm{vM})}(\mathcal{N}^{\circledast}, \mathrm{a}, \mathrm{b}, \mathrm{c}, \mathrm{d}) &= Z^{(8\mathrm{vM})}(\mathcal{N}^{\circledast}, \mathring{\mathrm{a}}, \mathring{\mathrm{b}}, \mathring{\mathrm{c}}, \mathring{\mathrm{d}}), \end{aligned} \tag{9.3.8}$$

where

$$\begin{aligned} \mathring{\mathrm{a}} &:= \tfrac{1}{2}(\mathrm{a} + \mathrm{b} + \mathrm{c} + \mathrm{d}), & \mathring{\mathrm{b}} &:= \tfrac{1}{2}(\mathrm{a} + \mathrm{b} - \mathrm{c} - \mathrm{d}), \\ \mathring{\mathrm{c}} &:= \tfrac{1}{2}(\mathrm{a} - \mathrm{b} + \mathrm{c} - \mathrm{d}), & \mathring{\mathrm{d}} &:= \tfrac{1}{2}(\mathrm{a} - \mathrm{b} - \mathrm{c} + \mathrm{d}). \end{aligned} \tag{9.3.9}$$

Referring to the ground-state regions defined in Sect. 3.8.1.4, the application of $\mathcal{T}_{\mathrm{WG}}$ in region I, is replaced in regions II, III and IV by

$$\begin{aligned} \mathcal{T}_1\mathcal{T}_{\mathrm{WG}}\mathcal{T}_1: \quad & (\mathrm{a}, \mathrm{b}, \mathrm{c}, \mathrm{d}) \to (\mathring{\mathrm{b}}, \mathring{\mathrm{a}}, -\mathring{\mathrm{d}}, -\mathring{\mathrm{c}}), \\ & \Delta \leftrightarrow 1/\Delta, \qquad \Gamma \leftrightarrow \Gamma/\Delta, \\ \mathcal{T}_2\mathcal{T}_{\mathrm{WG}}\mathcal{T}_2: \quad & (\mathrm{a}, \mathrm{b}, \mathrm{c}, \mathrm{d}) \to (\mathring{\mathrm{c}}, -\mathring{\mathrm{d}}, \mathring{\mathrm{a}}, -\mathring{\mathrm{b}}), \\ & \Delta \leftrightarrow 1/\Delta, \qquad -\Gamma \leftrightarrow \Gamma/\Delta, \\ \mathcal{T}_3\mathcal{T}_{\mathrm{WG}}\mathcal{T}_3: \quad & (\mathrm{a}, \mathrm{b}, \mathrm{c}, \mathrm{d}) \to (\mathring{\mathrm{d}}, -\mathring{\mathrm{c}}, -\mathring{\mathrm{b}}, \mathring{\mathrm{a}}), \\ & \Delta \leftrightarrow 1/\Delta, \qquad -\Gamma \leftrightarrow \Gamma/\Delta, \end{aligned} \tag{9.3.10}$$

respectively, where $(\mathring{\mathrm{a}}, \mathring{\mathrm{b}}, \mathring{\mathrm{c}}, \mathring{\mathrm{d}})$ are given by (9.3.9). It follows that the points of the surface:

$$\mathcal{C}(\mathrm{I}) : \quad \mathrm{a} = \mathrm{b} + \mathrm{c} + \mathrm{d} \quad \text{are the fixed points of } \mathcal{T}_{\mathrm{WG}} \text{ in I}, \tag{9.3.11}$$
$$\mathcal{C}(\mathrm{II}) : \quad \mathrm{b} = \mathrm{c} + \mathrm{d} + \mathrm{a} \quad \text{are the fixed points of } \mathcal{T}_1\mathcal{T}_{\mathrm{WG}}\mathcal{T}_1 \text{ in II}, \tag{9.3.12}$$
$$\mathcal{C}(\mathrm{III}) : \quad \mathrm{c} = \mathrm{d} + \mathrm{a} + \mathrm{b} \quad \text{are the fixed points of } \mathcal{T}_2\mathcal{T}_{\mathrm{WG}}\mathcal{T}_2 \text{ in III}, \tag{9.3.13}$$
$$\mathcal{C}(\mathrm{IV}) : \quad \mathrm{d} = \mathrm{a} + \mathrm{b} + \mathrm{c} \quad \text{are the fixed points of } \mathcal{T}_3\mathcal{T}_{\mathrm{WG}}\mathcal{T}_3 \text{ in IV}. \tag{9.3.14}$$

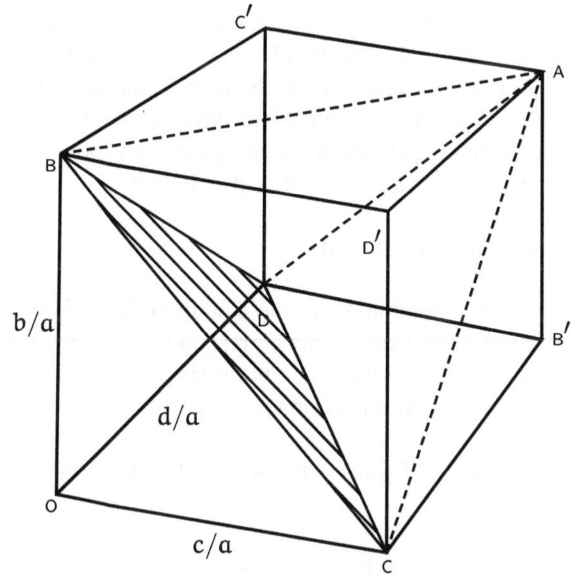

Fig. 9.2 The unit cube corresponding to region I in the space of $(b/a, c/a, d/a)$. The surface $\mathcal{C}(I)$ is hatched

9.3.2 The Case of Region I

This region, in which a is the largest Boltzmann weight, can be represented by the interior of the unit cube OBCDB′C′D′A shown in Fig. 9.2.

9.3.2.1 Transformations and Zones

The cube is divided into two zones[13]

$$\mathcal{R}_L(I): \quad a > b + c + d, \qquad \mathcal{R}_H(I): \quad a < b + c + d, \tag{9.3.15}$$

corresponding respectively to the tetrahedron OBCD and the remaining part of the cube and separated by the shaded planar triangle BCD, which corresponds to the surface $\mathcal{C}(I)$ defined in (9.3.11).

As in (3.8.19) we expose the temperature dependence of the Boltzmann weights by letting $q =: \exp(-\varepsilon_q/T)$ for $q = a, b, c, d$ and consider the trajectory followed, as temperature varies, by the point (a, b, c, d). With the vertex energy satisfying the conditions for region I, the trajectory remains in the unit cube and, as the temperature

[13] For convenience we refer to the different significant subsets of the regions I, II, III and IV as 'zones' and 'subzones'.

is decreased, it passes from A, where $T = \infty$ and $\mathrm{a} = \mathrm{b} = \mathrm{c} = \mathrm{d} = 1$ to O, where $T = 0$ and $\mathrm{b}/\mathrm{a} = \mathrm{c}/\mathrm{a} = \mathrm{d}/\mathrm{a} = 0$. Thus, as the temperature decreases from infinity to zero, the trajectory passes from the high-temperature zone $\mathcal{R}_{\mathrm{H}}(\mathrm{I})$ to the low-temperature zone $\mathcal{R}_{\mathrm{L}}(\mathrm{I})$. It must, therefore, intersect the surface $\mathcal{C}(\mathrm{I})$.

Take a point in $\mathcal{R}_{\mathrm{L}}(\mathrm{I})$ and perform the weak-graph transformation. From (9.3.9) and (9.3.15), $\mathring{\mathrm{a}} > \mathring{\mathrm{b}} > 0$, $\mathring{\mathrm{a}} > \mathring{\mathrm{c}} > 0$ and $\mathring{\mathrm{a}} > \mathring{\mathrm{d}} > 0$ with

$$(\mathring{\mathrm{b}} + \mathring{\mathrm{c}} + \mathring{\mathrm{d}}) - \mathring{\mathrm{a}} = \mathrm{a} - (\mathrm{b} + \mathrm{c} + \mathrm{d}). \tag{9.3.16}$$

It follows that $\mathcal{T}_{\mathrm{WG}}$ maps a point in $\mathcal{R}_{\mathrm{L}}(\mathrm{I})$ into a point in $\mathcal{R}_{\mathrm{H}}(\mathrm{I})$. In general the points are not on the same trajectory (they do not correspond to the same values of the ε_{q}). An exception to this is the case where, in the vertex energies (3.8.9), $K^{(\mathrm{f})} = K^{(\mathrm{b})}$, $K^{(4)} = 0$. Then the weak-graph transformation becomes the duality transformation of Sect. 9.2.2.1.

The image of $\mathcal{R}_{\mathrm{L}}(\mathrm{I})$ under $\mathcal{T}_{\mathrm{WG}}$ is not the whole of $\mathcal{R}_{\mathrm{H}}(\mathrm{I})$. Since a, b, c and d are positive,[14] from (9.3.9),

$$\mathring{\mathrm{a}} + \mathring{\mathrm{b}} > \mathring{\mathrm{c}} + \mathring{\mathrm{d}}, \qquad \mathring{\mathrm{a}} + \mathring{\mathrm{c}} > \mathring{\mathrm{b}} + \mathring{\mathrm{d}}, \qquad \mathring{\mathrm{a}} + \mathring{\mathrm{d}} > \mathring{\mathrm{b}} + \mathring{\mathrm{c}}. \tag{9.3.17}$$

These inequalities define the tetrahedron ABCD of Fig. 9.2, denoted by $\mathcal{R}_{\mathrm{o}}(\mathrm{I})$, and an application of $\mathcal{T}_{\mathrm{WG}}$ to a point $(\mathring{\mathrm{a}}, \mathring{\mathrm{b}}, \mathring{\mathrm{c}}, \mathring{\mathrm{d}})$ in $\mathcal{R}_{\mathrm{o}}(\mathrm{I})$ takes it back to a point (a, b, c, d) in $\mathcal{R}_{\mathrm{L}}(\mathrm{I})$.

However, this is the case only for points in $\mathcal{R}_{\mathrm{o}}(\mathrm{I})$. Since $\mathring{\mathrm{a}} > \mathring{\mathrm{b}}$, $\mathring{\mathrm{a}} > \mathring{\mathrm{c}}$ and $\mathring{\mathrm{a}} > \mathring{\mathrm{d}}$, not more than one of the three inequalities (9.3.17) can be violated at any one point. The zone $\mathcal{R}_{\mathrm{H}}(\mathrm{I})$ can thus be divided into four subzones; referring to Fig. 9.2, they are the tetrahedra:

ABCD, denoted by $\mathcal{R}_{\mathrm{o}}(\mathrm{I})$, where (9.3.1.7) applies,

AB′CD, denoted by $\mathcal{R}_{\mathrm{b}}(\mathrm{I})$, where $\mathring{\mathrm{a}} + \mathring{\mathrm{b}} < \mathring{\mathrm{c}} + \mathring{\mathrm{d}}$,

ABC′D, denoted by $\mathcal{R}_{\mathrm{c}}(\mathrm{I})$, where $\mathring{\mathrm{a}} + \mathring{\mathrm{c}} < \mathring{\mathrm{b}} + \mathring{\mathrm{d}}$,

ABCD′, denoted by $\mathcal{R}_{\mathrm{d}}(\mathrm{I})$, where $\mathring{\mathrm{a}} + \mathring{\mathrm{d}} < \mathring{\mathrm{b}} + \mathring{\mathrm{c}}$.

Although it can easily be verified that the entire high-temperature part of the trajectory of the isotropic Ising model[15] lies in $\mathcal{R}_{\mathrm{o}}(\mathrm{I})$, this is not the case for every trajectory.

Example 9.3.1 Suppose that $\varepsilon_{\mathrm{a}} + \varepsilon_{\mathrm{d}} > \varepsilon_{\mathrm{b}} + \varepsilon_{\mathrm{c}}$, $(\mathrm{ad} < \mathrm{bc})$. By expanding a, b, c and d in inverse powers of T, it can be shown that, after leaving the $T = \infty$ point A, it follows a trajectory which traverses $\mathcal{R}_{\mathrm{d}}(\mathrm{I})$ before entering $\mathcal{R}_{\mathrm{o}}(\mathrm{I})$ and crossing the surface $\mathcal{C}(\mathrm{I})$.

---

[14] The special case of the six-vertex model, where $\mathrm{d} = 0$, is discussed below.

[15] For which $K^{(\mathrm{f})} = K^{(\mathrm{b})} = K_{\mathrm{S}}$, $K^{(4)} = 0$, yielding, from (3.8.11), $\frac{1}{2}(\varepsilon_{\mathrm{a}} + \varepsilon_{\mathrm{b}}) = \varepsilon_{\mathrm{c}} = \varepsilon_{\mathrm{d}}$.

A point in $\mathcal{R}_{\mathrm{d}}(\mathrm{I})$ is mapped by $\mathfrak{T}_{\mathrm{WG}}$ into a 'non-physical' point $(\mathrm{a}, \mathrm{b}, \mathrm{c}, \mathrm{d}')$, with a, b and c positive, but d' negative. The transformation $\mathfrak{T}_{\mathrm{d}}$ defined in (9.3.7) maps this point into $(\mathrm{a}, \mathrm{b}, \mathrm{c}, \mathrm{d})$, with $\mathrm{d} = -\mathrm{d}'$, so that all the coordinates are positive. The transformation $\mathfrak{T}_{\mathrm{d}}\mathfrak{T}_{\mathrm{WG}}$ takes a point in $\mathcal{R}_{\mathrm{d}}(\mathrm{I})$ into a point in $\mathcal{R}_{\mathrm{L}}(\mathrm{I})$ where, from (9.3.9) and $\mathring{\mathrm{a}} > \mathring{\mathrm{b}}$, $\mathring{\mathrm{a}} > \mathring{\mathrm{c}}$, $\mathring{\mathrm{a}} > \mathring{\mathrm{d}}$, it follows that $\mathrm{d} < \mathrm{b}$ and $\mathrm{d} < \mathrm{c}$. The inverse process is effected by the transformation $\mathfrak{T}_{\mathrm{WG}}\mathfrak{T}_{\mathrm{d}}$. More specifically it can be shown that:

(i) points in the (low-temperature) triangle OBC of Fig. 9.2 are mapped by $\mathfrak{T}_{\mathrm{WG}}$ ($\mathfrak{T}_{\mathrm{d}}$ is, of course, unnecessary in this case) into points for which

$$\mathring{\mathrm{a}} + \mathring{\mathrm{d}} = \mathring{\mathrm{b}} + \mathring{\mathrm{c}}, \qquad \mathring{\mathrm{d}} > 0. \tag{9.3.18}$$

This is the triangle ABC, which is the boundary between the (high-temperature) zones $\mathcal{R}_{\mathrm{o}}(\mathrm{I})$ and $\mathcal{R}_{\mathrm{d}}(\mathrm{I})$.

(ii) points in the (high-temperature) triangle D′BC are mapped by $\mathfrak{T}_{\mathrm{d}}\mathfrak{T}_{\mathrm{WG}}$ into DBC (the surface $\mathcal{C}(\mathrm{I})$ given by (9.3.11)).

The invariant points for the transformation $\mathfrak{T}_{\mathrm{WG}}\mathfrak{T}_{\mathrm{d}}$ are

$$\mathring{\mathrm{a}} = \mathring{\mathrm{b}} + \mathring{\mathrm{c}}, \qquad \mathring{\mathrm{d}} = 0. \tag{9.3.19}$$

This is a mapping between the six-vertex model and special cases of the eight-vertex model. The former corresponds, in Fig. 9.2 for region I, to the front face of the cube in the plane $\mathrm{d}/\mathrm{a} = 0$ and to the square region in Fig. 3.5.

The high-temperature subzones $\mathcal{R}_{\mathrm{b}}(\mathrm{I})$ and $\mathcal{R}_{\mathrm{c}}(\mathrm{I})$ are similarly connected to $\mathcal{R}_{\mathrm{L}}(\mathrm{I})$ by the transformations $\mathfrak{T}_{\mathrm{b}}\mathfrak{T}_{\mathrm{WG}}$ and $\mathfrak{T}_{\mathrm{c}}\mathfrak{T}_{\mathrm{WG}}$ respectively. It follows that the partition function $Z^{(8\mathrm{vM})}(\mathcal{N}^{\circledast}, \mathrm{a}, \mathrm{b}, \mathrm{c}, \mathrm{d})$ can be deduced for the entire region I if it is known for the low-temperature zone.

9.3.2.2 Phase Transitions

Suppose that, on some trajectory, a transition occurs at a point in $\mathcal{R}_{\mathrm{L}}(\mathrm{I})$. Making reasonable continuity assumptions, variation of the trajectory (produced by changing the ε_{q}) generates a transition surface in $\mathcal{R}_{\mathrm{L}}(\mathrm{I})$. Since this is a surface of singularities in the free energy, it follows by applying $\mathfrak{T}_{\mathrm{WG}}$, that there is a corresponding surface of singularities in $\mathcal{R}_{\mathrm{o}}(\mathrm{I})$. However, the boundary surface $\mathcal{C}(\mathrm{I})$ between $\mathcal{R}_{\mathrm{L}}(\mathrm{I})$ and $\mathcal{R}_{\mathrm{H}}(\mathrm{I})$ is the invariant surface of $\mathfrak{T}_{\mathrm{WG}}$. So, if it is assumed that there is a unique transition surface in region I (Sutherland 1970) then this surface *may* coincide with $\mathcal{C}(\mathrm{I})$.

However, we know that the transformations $\mathfrak{T}_{\mathrm{WG}}\mathfrak{T}_{\mathrm{b}}$, $\mathfrak{T}_{\mathrm{WG}}\mathfrak{T}_{\mathrm{c}}$ and $\mathfrak{T}_{\mathrm{WG}}\mathfrak{T}_{\mathrm{d}}$ also take points from $\mathcal{R}_{\mathrm{L}}(\mathrm{I})$ to $\mathcal{R}_{\mathrm{H}}(\mathrm{I})$ and it is necessary to examine whether the corresponding invariant surfaces are possible surfaces of transitions. As we have see the invariant points for $\mathfrak{T}_{\mathrm{WG}}\mathfrak{T}_{\mathrm{d}}$ are given by (9.3.19) which specifies the line BC in Fig. 9.2, which is the intersection of $\mathcal{C}(\mathrm{I})$ and the invariant surface $\mathrm{d}/\mathrm{a} = 0$ of $\mathfrak{T}_{\mathrm{d}}$. Now a trajectory cannot cross the plane $\mathrm{d} = 0$, since this would involve entering a region with $\mathrm{d} < 0$. In fact, for $T > 0$, no trajectory can touch this plane except in the limiting case of the six-vertex model, $\varepsilon_{\mathrm{d}} = \infty$, when it lies entirely in the plane. It can similarly be shown that the transformations $\mathcal{T}_{\mathrm{WG}}\mathcal{T}_{\mathrm{b}}$ and $\mathcal{T}_{\mathrm{WG}}\mathcal{T}_{\mathrm{c}}$ do not generate any possible transition surfaces. Hence, if there is only one transition surface it must be the plane surface $\mathcal{C}(\mathrm{I})$, which implies that there is exactly one transition point on any trajectory, occurring at its intersection with $\mathcal{C}(\mathrm{I})$. It also implies the existence of ferroelectric type I order throughout $\mathcal{R}_{\mathrm{L}}(\mathrm{I})$, while $\mathcal{R}_{\mathrm{H}}(\mathrm{I})$ is a zone of disorder. The assumption that there is only one transition surface is confirmed by the analysis of the exact expression for the zero-field free energy of the eight-vertex model in Sect. 12.3.8.

In the high-temperature (disordered) zone $\mathcal{R}_{\mathrm{H}}(\mathrm{I})$ it is of interest to look at the properties of the model on the boundaries between $\mathcal{R}_{\mathrm{o}}(\mathrm{I})$, where all the conditions of (9.3.17) apply, and the other subzones. For the subzone $\mathcal{R}_{\mathrm{d}}(\mathrm{I})$ the boundary in question is the triangle **ABC**, given by (9.3.18) As we have seen $\mathcal{T}_{\mathrm{WG}}$ maps a point on **ABC** into a point of the triangle **OBC**, for which $\mathrm{d} = 0$ and $\mathrm{a} > \mathrm{b} + \mathrm{c}$. This represents a subcritical temperature where the state is one of perfect order in terms of spins or arrows with

$$Z^{(8\mathrm{vM})}(\mathcal{N}^{\circledast}, \mathrm{a}, \mathrm{b}, \mathrm{c}, 0) = Z^{(6\mathrm{vM})}(\mathcal{N}^{\circledast}, \mathrm{a}, \mathrm{b}, \mathrm{c}) = \mathrm{a}^{N^{\circledast}}. \tag{9.3.20}$$

Applying $\mathcal{T}_{\mathrm{WG}}$ gives, for a point on **ABC**,

$$Z^{(8\mathrm{vM})}(\mathcal{N}^{\circledast}, \mathring{\mathrm{a}}, \mathring{\mathrm{b}}, \mathring{\mathrm{c}}, \mathring{\mathrm{d}}) = \frac{1}{4^{N^{\circledast}}}(2\mathring{\mathrm{a}} + 2\mathring{\mathrm{b}} + 2\mathring{\mathrm{c}} + 2\mathring{\mathrm{d}})^{N^{\circledast}}. \tag{9.3.21}$$

Application of the transformation $\mathcal{T}_0\mathcal{T}_1$ shows that this formula also applies on the triangle **ABD** where

$$\mathring{\mathrm{a}} + \mathring{\mathrm{c}} = \mathring{\mathrm{b}} + \mathring{\mathrm{d}}. \tag{9.3.22}$$

Comparison with the eight-vertex ice result (3.8.6) shows that (9.3.21) corresponds to an equilibrium state in which the vertex types are distributed as randomly as possible without breaking the compatibility rules for nearest-neighbour vertices. Hence (9.3.18) and (9.3.22) appear to be surfaces of *disorder points*.[16] That this is indeed the case in the circumstances described here was confirmed by correlation function calculations of Baxter (1982b). Thus trajectories starting in $\mathcal{R}_{\mathrm{c}}(\mathrm{I})$ or $\mathcal{R}_{\mathrm{d}}(\mathrm{I})$ pass through a point of disorder in the range between $T = \infty$ and the critical temperature.

From (3.8.11), the Ising model (with $K^{(4)} = 0$) is compatible with (9.3.18) only when $K^{(\mathrm{b})} = 0$ and the entire trajectory then lies in that surface. Since $K^{(\mathrm{f})} \neq 0$ there are correlations between spins on any line of sites in the forward direction, but $K^{(\mathrm{b})} = 0$ means that the spins on different forward lines are uncorrelated. The system is thus essentially one-dimensional and, from (9.3.18) and (9.3.21) with $K^{(0)} = K^{(\mathrm{b})} = K^{(4)} = 0$,

---

[16] A disorder point in a system is defined to be one, for finite temperature, at which all correlation functions between the states on pairs of sites are zero, as they would be at infinite temperature.

$$\phi(K^{(\mathrm{f})}) = -\ln[2\cosh(K^{(\mathrm{f})})], \tag{9.3.23}$$

which is equivalent to the formula for the free energy of the one-dimensional zero-field Ising model (Domb 1960, Sect. 3.2.1). There is no transition for $T > 0$, which is to be expected since (9.3.18) has no points in common with (9.3.11). The situation on ABD, given by (9.3.22), is similar, with $K^{(\mathrm{f})} = 0$ in the Ising model case. Points on ACD, where $\mathring{a} + \mathring{b} = \mathring{c} + \mathring{d}$, have different properties from those on ABC and ABD since $\mathcal{T}_{\mathrm{WG}}$ maps them into points (a, b, c, d) where $b = 0$ and $a > c + d$. From (9.3.4) the free energy is equal to that of a six-vertex antiferroelectric below its critical temperature. However the six-vertex antiferroelectric state is not one of perfect order and no equation like (9.3.21) can be deduced.

9.3.3 Regions and Variables

Region I is mapped into regions II, III and IV by $\mathcal{T}_1$, $\mathcal{T}_2$ and $\mathcal{T}_3$, respectively so the analysis of Sect. 9.3.2 carries over to these cases with transition surfaces, $\mathcal{C}(\text{II})$, $\mathcal{C}(\text{III})$ and $\mathcal{C}(\text{IV})$, which become, in the six-vertex $d \to 0$ limit, the dotted lines in Fig. 3.5.

When the least two values in the set $\{\varepsilon_a, \varepsilon_b, \varepsilon_c, \varepsilon_d\}$ are degenerate the trajectory of the system phase point for varying temperature lies on the boundary between two regions. For example, suppose that $\varepsilon_a = \varepsilon_b$. Then $a = b$, $c < a$ and $d < a$ at all points on the trajectory, which lies in the boundary between regions I and II. In such cases none of the transition conditions can be satisfied and hence no transition occurs for $T > 0$. This behaviour is similar to that of the IKDP and IF cases of the six-vertex model (Sect. 3.8.2).

Linear pair combinations of the Boltzmann weights, denoted by $w_j, j = 1, 2, 3, 4$, and satisfying

$$w_1 \geq w_2 \geq w_3 \geq |w_4|, \tag{9.3.24}$$

play an important role in the solution of the eight-vertex model. The definitions for these variables in the different regions and their subzones must necessarily be different. However, we now have all the information needed to generate them. One way to do this is to start with the definitions for one zone and then generate the definitions for the rest of the phase space by applying transformations.

Since, in $\mathcal{R}_{\mathrm{L}}(\text{I})$, $a > b+c+d$, the definitions of the w_j in the first line of Table 9.1 satisfy (9.3.24). The definitions for the remaining zones of region I and for the zones of regions II, III and IV are then derived from the first line by the transformations listed in column 1. In each of the regions

$$\mathcal{T}_{\mathrm{WG}}: \quad (w_1, w_2, w_3, w_4) \leftrightarrow (w_1, w_3, w_2, w_4). \tag{9.3.25}$$

Table 9.1 The definitions of the variables $(\mathsf{w}_1, \mathsf{w}_2, \mathsf{w}_3, \mathsf{w}_4)$ in the different regions of the phase space of $(\mathsf{a}, \mathsf{b}, \mathsf{c}, \mathsf{d})$

Zone	Transformation	w_1	w_2	w_3	w_4	W_Δ	W_Γ	Δ^*	Γ^*
$\mathcal{R}_{\mathrm{L}}(\mathrm{I})$		$\frac{1}{2}(\mathsf{a}+\mathsf{b})$	$\frac{1}{2}(\mathsf{a}-\mathsf{b})$	$\frac{1}{2}(\mathsf{c}+\mathsf{d})$	$\frac{1}{2}(\mathsf{c}-\mathsf{d})$	$W(2,3\|1,4)$	$W(3,4\|2,1)$	Δ	Γ
$\mathcal{R}_{\mathsf{o}}(\mathrm{I})$	$\mathcal{T}_{\mathrm{WG}}$	$\frac{1}{2}(\mathsf{a}+\mathsf{b})$	$\frac{1}{2}(\mathsf{c}+\mathsf{d})$	$\frac{1}{2}(\mathsf{a}-\mathsf{b})$	$\frac{1}{2}(\mathsf{c}-\mathsf{d})$	$W(2,3\|4,1)$	$W(2,4\|3,1)$	$1/\Delta$	Γ/Δ
$\mathcal{R}_{\mathsf{b}}(\mathrm{I})$	$\mathcal{T}_{\mathrm{WG}}\mathcal{T}_{\mathsf{b}}$	$\frac{1}{2}(\mathsf{c}+\mathsf{d})$	$\frac{1}{2}(\mathsf{a}+\mathsf{b})$	$\frac{1}{2}(\mathsf{a}-\mathsf{b})$	$\frac{1}{2}(\mathsf{c}-\mathsf{d})$	$W(1,3\|4,2)$	$W(1,4\|3,2)$	$-1/\Gamma$	Δ
$\mathcal{R}_{\mathsf{d}}(\mathrm{I})$	$\mathcal{T}_{\mathrm{WG}}\mathcal{T}_{\mathsf{d}}$	$\frac{1}{2}(\mathsf{a}+\mathsf{b})$	$\frac{1}{2}(\mathsf{c}+\mathsf{d})$	$\frac{1}{2}(\mathsf{c}-\mathsf{d})$	$\frac{1}{2}(\mathsf{a}-\mathsf{b})$	$W(2,4\|3,1)$	$W(2,3\|4,1)$	$1/\Gamma$	Δ/Γ
$\mathcal{R}_{\mathsf{c}}(\mathrm{I})$	$\mathcal{T}_{\mathrm{WG}}\mathcal{T}_{\mathsf{c}}$	$\frac{1}{2}(\mathsf{a}+\mathsf{b})$	$\frac{1}{2}(\mathsf{c}+\mathsf{d})$	$\frac{1}{2}(\mathsf{d}-\mathsf{c})$	$\frac{1}{2}(\mathsf{b}-\mathsf{a})$	$W(2,4\|3,1)$	$W(2,3\|4,1)$	$1/\Gamma$	Δ/Γ
$\mathcal{R}_{\mathrm{L}}(\mathrm{II})$	$\mathcal{T}_1$	$\frac{1}{2}(\mathsf{a}+\mathsf{b})$	$\frac{1}{2}(\mathsf{b}-\mathsf{a})$	$\frac{1}{2}(\mathsf{c}+\mathsf{d})$	$\frac{1}{2}(\mathsf{d}-\mathsf{c})$	$W(2,3\|1,4)$	$W(3,4\|2,1)$	Δ	Γ
$\mathcal{R}_{\mathsf{o}}(\mathrm{II})$	$\mathcal{T}_1\mathcal{T}_{\mathrm{WG}}$	$\frac{1}{2}(\mathsf{a}+\mathsf{b})$	$\frac{1}{2}(\mathsf{c}+\mathsf{d})$	$\frac{1}{2}(\mathsf{b}-\mathsf{a})$	$\frac{1}{2}(\mathsf{d}-\mathsf{c})$	$W(2,3\|4,1)$	$W(2,4\|3,1)$	$1/\Delta$	Γ/Δ
$\mathcal{R}_{\mathsf{a}}(\mathrm{II})$	$\mathcal{T}_1\mathcal{T}_{\mathrm{WG}}\mathcal{T}_1\mathcal{T}_{\mathsf{a}}\mathcal{T}_1$	$\frac{1}{2}(\mathsf{c}+\mathsf{d})$	$\frac{1}{2}(\mathsf{a}+\mathsf{b})$	$\frac{1}{2}(\mathsf{b}-\mathsf{a})$	$\frac{1}{2}(\mathsf{d}-\mathsf{c})$	$W(1,3\|4,2)$	$W(1,4\|3,2)$	$-1/\Gamma$	Δ
$\mathcal{R}_{\mathsf{c}}(\mathrm{II})$	$\mathcal{T}_1\mathcal{T}_{\mathrm{WG}}\mathcal{T}_1\mathcal{T}_{\mathsf{c}}\mathcal{T}_1$	$\frac{1}{2}(\mathsf{a}+\mathsf{b})$	$\frac{1}{2}(\mathsf{c}+\mathsf{d})$	$\frac{1}{2}(\mathsf{d}-\mathsf{c})$	$\frac{1}{2}(\mathsf{b}-\mathsf{a})$	$W(2,4\|3,1)$	$W(2,3\|4,1)$	$1/\Gamma$	Δ/Γ
$\mathcal{R}_{\mathsf{d}}(\mathrm{II})$	$\mathcal{T}_1\mathcal{T}_{\mathrm{WG}}\mathcal{T}_1\mathcal{T}_{\mathsf{d}}\mathcal{T}_1$	$\frac{1}{2}(\mathsf{a}+\mathsf{b})$	$\frac{1}{2}(\mathsf{c}+\mathsf{d})$	$\frac{1}{2}(\mathsf{c}-\mathsf{d})$	$\frac{1}{2}(\mathsf{a}-\mathsf{b})$	$W(2,4\|3,1)$	$W(2,3\|4,1)$	$1/\Gamma$	Δ/Γ
$\mathcal{R}_{\mathrm{L}}(\mathrm{III})$	$\mathcal{T}_2$	$\frac{1}{2}(\mathsf{c}+\mathsf{d})$	$\frac{1}{2}(\mathsf{c}-\mathsf{d})$	$\frac{1}{2}(\mathsf{a}+\mathsf{b})$	$\frac{1}{2}(\mathsf{a}-\mathsf{b})$	$W(1,4\|2,3)$	$W(1,2\|4,3)$	$-\Delta$	$-\Gamma$
$\mathcal{R}_{\mathsf{o}}(\mathrm{III})$	$\mathcal{T}_2\mathcal{T}_{\mathrm{WG}}$	$\frac{1}{2}(\mathsf{c}+\mathsf{d})$	$\frac{1}{2}(\mathsf{a}+\mathsf{b})$	$\frac{1}{2}(\mathsf{c}-\mathsf{d})$	$\frac{1}{2}(\mathsf{a}-\mathsf{b})$	$W(1,4\|3,2)$	$W(1,3\|4,2)$	$-1/\Delta$	Γ/Δ
$\mathcal{R}_{\mathsf{d}}(\mathrm{III})$	$\mathcal{T}_2\mathcal{T}_{\mathrm{WG}}\mathcal{T}_2\mathcal{T}_{\mathsf{d}}\mathcal{T}_2$	$\frac{1}{2}(\mathsf{a}+\mathsf{b})$	$\frac{1}{2}(\mathsf{c}+\mathsf{d})$	$\frac{1}{2}(\mathsf{c}-\mathsf{d})$	$\frac{1}{2}(\mathsf{a}-\mathsf{b})$	$W(2,4\|3,1)$	$W(2,3\|4,1)$	$1/\Gamma$	Δ/Γ
$\mathcal{R}_{\mathsf{b}}(\mathrm{III})$	$\mathcal{T}_2\mathcal{T}_{\mathrm{WG}}\mathcal{T}_2\mathcal{T}_{\mathsf{b}}\mathcal{T}_2$	$\frac{1}{2}(\mathsf{c}+\mathsf{d})$	$\frac{1}{2}(\mathsf{a}+\mathsf{b})$	$\frac{1}{2}(\mathsf{a}-\mathsf{b})$	$\frac{1}{2}(\mathsf{c}-\mathsf{d})$	$W(1,3\|4,2)$	$W(1,4\|3,2)$	$-1/\Gamma$	Δ/Γ
$\mathcal{R}_{\mathsf{a}}(\mathrm{III})$	$\mathcal{T}_2\mathcal{T}_{\mathrm{WG}}\mathcal{T}_2\mathcal{T}_{\mathsf{a}}\mathcal{T}_2$	$\frac{1}{2}(\mathsf{c}+\mathsf{d})$	$\frac{1}{2}(\mathsf{a}+\mathsf{b})$	$\frac{1}{2}(\mathsf{b}-\mathsf{a})$	$\frac{1}{2}(\mathsf{d}-\mathsf{c})$	$W(1,3\|4,2)$	$W(1,4\|3,2)$	$-1/\Gamma$	Δ/Γ
$\mathcal{R}_{\mathrm{L}}(\mathrm{IV})$	$\mathcal{T}_3$	$\frac{1}{2}(\mathsf{c}+\mathsf{d})$	$\frac{1}{2}(\mathsf{d}-\mathsf{c})$	$\frac{1}{2}(\mathsf{a}+\mathsf{b})$	$\frac{1}{2}(\mathsf{b}-\mathsf{a})$	$W(1,4\|2,3)$	$W(1,2\|4,3)$	$-\Delta$	$-\Gamma$
$\mathcal{R}_{\mathsf{o}}(\mathrm{IV})$	$\mathcal{T}_3\mathcal{T}_{\mathrm{WG}}$	$\frac{1}{2}(\mathsf{c}+\mathsf{d})$	$\frac{1}{2}(\mathsf{a}+\mathsf{b})$	$\frac{1}{2}(\mathsf{d}-\mathsf{c})$	$\frac{1}{2}(\mathsf{b}-\mathsf{a})$	$W(1,4\|3,2)$	$W(1,3\|4,2)$	$-1/\Delta$	Γ/Δ
$\mathcal{R}_{\mathsf{c}}(\mathrm{IV})$	$\mathcal{T}_3\mathcal{T}_{\mathrm{WG}}\mathcal{T}_3\mathcal{T}_{\mathsf{c}}\mathcal{T}_3$	$\frac{1}{2}(\mathsf{a}+\mathsf{b})$	$\frac{1}{2}(\mathsf{c}+\mathsf{d})$	$\frac{1}{2}(\mathsf{d}-\mathsf{c})$	$\frac{1}{2}(\mathsf{b}-\mathsf{a})$	$W(2,4\|3,1)$	$W(2,3\|4,1)$	$1/\Gamma$	Δ/Γ
$\mathcal{R}_{\mathsf{a}}(\mathrm{IV})$	$\mathcal{T}_3\mathcal{T}_{\mathrm{WG}}\mathcal{T}_3\mathcal{T}_{\mathsf{a}}\mathcal{T}_3$	$\frac{1}{2}(\mathsf{c}+\mathsf{d})$	$\frac{1}{2}(\mathsf{a}+\mathsf{b})$	$\frac{1}{2}(\mathsf{b}-\mathsf{a})$	$\frac{1}{2}(\mathsf{d}-\mathsf{c})$	$W(1,3\|4,2)$	$W(1,4\|3,2)$	$-1/\Gamma$	Δ/Γ
$\mathcal{R}_{\mathsf{b}}(\mathrm{IV})$	$\mathcal{T}_3\mathcal{T}_{\mathrm{WG}}\mathcal{T}_3\mathcal{T}_{\mathsf{b}}\mathcal{T}_3$	$\frac{1}{2}(\mathsf{c}+\mathsf{d})$	$\frac{1}{2}(\mathsf{a}+\mathsf{b})$	$\frac{1}{2}(\mathsf{a}-\mathsf{b})$	$\frac{1}{2}(\mathsf{c}-\mathsf{d})$	$W(1,3\|4,2)$	$W(1,4\|3,2)$	$-1/\Gamma$	Δ/Γ

So the transition surfaces are given by

$$\mathsf{w}_2 = \mathsf{w}_3. \tag{9.3.26}$$

Any transformation

$$(\mathsf{w}_1, \mathsf{w}_2, \mathsf{w}_3, \mathsf{w}_4) \to (\pm\mathsf{w}_i, \pm\mathsf{w}_j, \pm\mathsf{w}_k, \pm\mathsf{w}_\ell), \tag{9.3.27}$$

where $(\mathsf{w}_i, \mathsf{w}_j, \mathsf{w}_k, \mathsf{w}_\ell)$ is a permutation of $(\mathsf{w}_1, \mathsf{w}_2, \mathsf{w}_3, \mathsf{w}_4)$, and the choices of sign can be made independently, corresponds to some product of the transformations defined above. Let

$$W(i, j|k, \ell) := \frac{\mathsf{w}_i^2 - \mathsf{w}_j^2}{\mathsf{w}_k^2 - \mathsf{w}_\ell^2}. \tag{9.3.28}$$

It is clear that, for these ratio functions, it can be assumed that $i < j$. Then $W(i, j|k, \ell) \geq, \leq 0$ according as $k <, > \ell$. Further if $j - i = 1$, $|k - \ell| = 3$, $|W(i, j|k, \ell)| \leq 1$ or, conversely, if $j - i = 3$, $|k - \ell| = 1$, $|W(i, j|k, \ell)| \geq 1$. There are only two independent ratio functions.[17] which can be chosen by selecting two sets of indices $(i, j|k, \ell)$ and $(i', j'|k', \ell')$ so that

[17] Every ratio function is either the negative or reciprocal of one of $W(1, 2|3, 4)$, $W(1, 4|2, 3)$ and $W(1, 3|2, 4)$, with

$$W(1, 3|2, 4) = [1 + W(1, 2|3, 4)W(1, 4|2, 3)]/[W(1, 2|3, 4) + W(1, 4|2, 3)].$$

$$W(i,j|k,\ell) = W_\Delta := \frac{\Delta - 1}{\Delta + 1} = \frac{(a-b)^2 - (c+d)^2}{(a+b)^2 - (c-d)^2},$$
$$W(i',j'|k',\ell') = W_\Gamma := \frac{\Gamma - 1}{\Gamma + 1} = \frac{(c+d)^2 - (c-d)^2}{(a-b)^2 - (a+b)^2}. \tag{9.3.29}$$

The choices of indices, which vary of course between zones, are shown in Table 9.1. In $\mathcal{R}_L(I)$, $(i,j|k,\ell) = (2,3|1,4)$ and $(i',j'|k',\ell') = (3,4|2,1)$ and using these ratio functions we define

$$\Delta^\star := \frac{1 + W(2,3|1,4)}{1 - W(2,3|1,4)}, \qquad \Gamma^\star := \frac{1 + W(3,4|2,1)}{1 - W(3,4|2,1)}, \tag{9.3.30}$$

so that in that zone $\Delta^\star = \Delta$ and $\Gamma^\star = \Gamma$. Since $0 < W(2,3|1,4) < 1$ and $W(3,4|2,1) < 0$,

$$\Delta^\star \geq 1, \qquad -1 \leq \Gamma^\star \leq 1. \tag{9.3.31}$$

From this and the last two columns of Table 9.1 conditions on the values of Δ and Γ in all the zones can be derived. In particular,

$$\left.\begin{array}{lll} \Delta \geq 1, & -1 \leq \Gamma \leq 1, & \text{in } \mathcal{R}_L(J), \\ 0 \leq \Delta \leq 1, & -\Delta \leq \Gamma \leq \Delta, & \text{in } \mathcal{R}_O(J), \end{array}\right\} \quad \text{for J = I and II,} \tag{9.3.32}$$

and

$$\left.\begin{array}{lll} \Delta \leq -1, & -1 \leq \Gamma \leq 1, & \text{in } \mathcal{R}_L(J), \\ -1 \leq \Delta \leq 0, & \Delta \leq \Gamma \leq -\Delta, & \text{in } \mathcal{R}_O(J), \end{array}\right\} \quad \text{for J = III and IV.} \tag{9.3.33}$$

The transition surfaces $\mathcal{C}(\text{I})$ and $\mathcal{C}(\text{II})$ corresponds to $\Delta = \Delta_c := 1$, with Δ decreasing as temperature is increased across the surfaces. The transition surfaces $\mathcal{C}(\text{III})$ and $\mathcal{C}(\text{IV})$ corresponds to $\Delta = \Delta_c := -1$, with Δ increasing as temperature is increased across the surfaces. From (3.8.12), the coupling forms for the critical surfaces are

$$\sinh(2K^{(f)})\sinh(2K^{(b)}) + \tanh(2K^{(4)})\cosh(2K^{(f)})\cosh(2K^{(b)}) = \Delta_c. \tag{9.3.34}$$

When $K^{(4)} = 0$, this gives

$$\sinh(2K^{(f)})\sinh(2K^{(b)}) = \Delta_c. \tag{9.3.35}$$

In the spin representation this is the formula for the critical temperature of two identical, anisotropic, non-interacting Ising models, with, by reference to the ground-states in Sect. 3.8.1.4, both $K^{(f)}$ and $K^{(b)}$ positive in region I and negative in region II, and with opposite signs in regions III and IV.

9.4 The Star-Triangle Transformation

As we saw in Sects. 9.2.1 and 9.2.2, duality alone can be used to determine the critical temperature for the self-dual square lattice, but is not sufficient for the triangle/honeycomb dual pair.[18] We need the face-decoration star-triangle transformation.

Consider a honeycomb lattice $\mathcal{N}_{\text{H}}$ of $2N$ sites as shown in Fig. 16.14.[19] The lattice is loose-packed and divides into two triangular sublattices, each of N sites, represented by white and black circles. Suppose that there are three unequal couplings $K_{\text{H}}^{(1)}$, $K_{\text{H}}^{(2)}$ and $K_{\text{H}}^{(3)}$ as shown in Fig. 9.3. With the state at $\boldsymbol{r} \in \mathcal{N}_{\text{H}}$ given by $\sigma(\boldsymbol{r})$, the Hamiltonian can be expressed in the form

$$\widehat{H}(\mathcal{N}_{\text{H}}, K_{\text{H}}^{(1)}, K_{\text{H}}^{(2)}, K_{\text{H}}^{(3)}; \boldsymbol{\sigma}) = \sum_{\{\boldsymbol{r}_{\circ}\}} \{K_{\text{H}}^{(1)}\mathfrak{e}(\sigma(\boldsymbol{r}_1), \sigma(\boldsymbol{r}_{\circ}))$$
$$+ K_{\text{H}}^{(2)}\mathfrak{e}(\sigma(\boldsymbol{r}_2), \sigma(\boldsymbol{r}_{\circ})) + K_{\text{H}}^{(3)}\mathfrak{e}(\sigma(\boldsymbol{r}_3), \sigma(\boldsymbol{r}_{\circ}))\}, \tag{9.4.1}$$

where the sum is over all downward pointing triangles identified by the locations $\{\boldsymbol{r}_{\circ}\}$ of their white sites[20] and the partition function is

$$Z(\mathcal{N}_{\text{H}}, K_{\text{H}}^{(1)}, K_{\text{H}}^{(2)}, K_{\text{H}}^{(3)}) = \sum_{\{\boldsymbol{\sigma}\}} \prod_{\{\boldsymbol{r}_{\circ}\}} \left\{ \prod_{\mathrm{x}=1}^{3} X(K_{\text{H}}^{(\mathrm{x})}; \sigma(\boldsymbol{r}_{\mathrm{x}}), \sigma(\boldsymbol{r}_{\circ})) \right\}, \tag{9.4.2}$$

where

$$X(K; \sigma, \sigma') := \exp(-K\mathfrak{e}(\sigma, \sigma')). \tag{9.4.3}$$

The state $\boldsymbol{\sigma}$ can be divided into two parts, the state on the white sites and the state on the black sites and the summation can be performed in two parts, first over the states $\{\sigma(\boldsymbol{r}_{\circ})\}$ on the white sites and then over the states on the black sites. After the first of these has been performed the partition function will be that for the triangular lattice $\mathcal{N}_{\text{T}}$ of N black sites. However, in general it may be of a complicated form. The star-triangle can be carried out if new coupling $K_{\text{T}}^{(1)}$, $K_{\text{T}}^{(2)}$ and $K_{\text{T}}^{(3)}$ (as shown in Fig. 9.3) can be found so that the partition function can, apart from a multiplicative function of the couplings $K_{\text{H}}^{(\mathrm{x})}$, $\mathrm{x} = 1, 2, 3$, be expressed as a sum over first-neighbour pairs on $\mathcal{N}_{\text{T}}$ with the same state variable $\sigma(\boldsymbol{r})$, the same interaction function $\mathfrak{e}(\sigma(\boldsymbol{r}), \sigma(\boldsymbol{r}'))$ and the couplings $K_{\text{T}}^{(1)}$, $K_{\text{T}}^{(2)}$ and $K_{\text{T}}^{(3)}$. This means, as in the case of duality, that three equations can be found relating each triangle coupling to the three honeycomb

[18] Although Wegner (1973) has shown that a version of his transformation can be used to transform the triangular lattice directly into itself.

[19] Toroidal boundary conditions are assumed, although the analysis applies equally well to a bounded lattice with an equal number of white and black sites.

[20] The labels $(\boldsymbol{r}_1, \boldsymbol{r}_2, \boldsymbol{r}_3)$ in Fig. 9.3 for the neighbours of site $\boldsymbol{r}_{\circ}$ are relative to $\boldsymbol{r}_{\circ}$ and are not sublattice labels which can be attached with the same orientation to the black sites in Fig. 16.14. However the coupling designations are applicable to all the downward pointing triangles in Fig. 16.14.

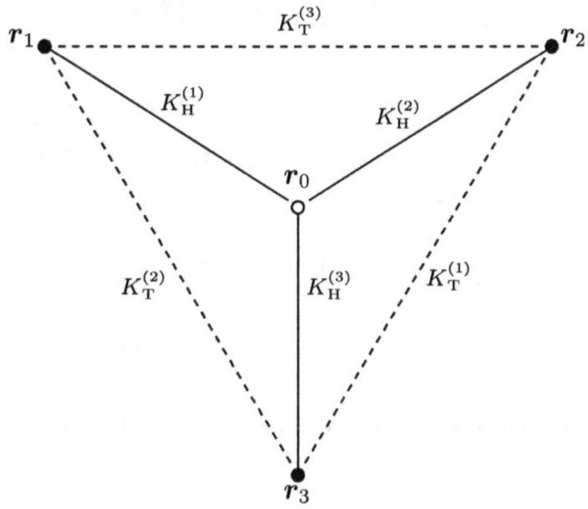

Fig. 9.3 Unequal couplings for the star-triangle transformation

couplings. However, this turns out to be rather more restrictive than was the case for duality.

In order to effect the star-triangle transformation we need to extract the contribution to the partition function from one site $\boldsymbol{r}_\circ$ and its neighbours $(\boldsymbol{r}_1, \boldsymbol{r}_2, \boldsymbol{r}_3)$ and set it equal (apart from a multiplicative factor) to interactions around the triangle $(\boldsymbol{r}_1, \boldsymbol{r}_2, \boldsymbol{r}_3)$. Using the abbreviated notation $\sigma_\circ := \sigma(\boldsymbol{r}_\circ)\ \sigma_\mathrm{x} := \sigma(\boldsymbol{r}_\mathrm{x})$ for $\mathrm{x} = 1, 2, 3$, this is given by

$$\sum_{\{\sigma_\circ\}} X(K_{\mathrm{H}}^{(1)}; \sigma_1, \sigma_\circ) X(K_{\mathrm{H}}^{(2)}; \sigma_2, \sigma_\circ) X(K_{\mathrm{H}}^{(3)}; \sigma_3, \sigma_\circ)$$
$$= \zeta(K_{\mathrm{H}}^{(1)}, K_{\mathrm{H}}^{(2)}, K_{\mathrm{H}}^{(3)}) X(K_{\mathrm{T}}^{(1)}; \sigma_2, \sigma_3) X(K_{\mathrm{T}}^{(2)}; \sigma_3, \sigma_1) X(K_{\mathrm{T}}^{(3)}; \sigma_1, \sigma_2), \qquad (9.4.4)$$

and then

$$Z(\mathcal{N}_{\mathrm{H}}, K_{\mathrm{H}}^{(1)}, K_{\mathrm{H}}^{(2)}, K_{\mathrm{H}}^{(3)}) = [\zeta(K_{\mathrm{H}}^{(1)}, K_{\mathrm{H}}^{(2)}, K_{\mathrm{H}}^{(3)})]^N Z(\mathcal{N}_{\mathrm{T}}, K_{\mathrm{T}}^{(1)}, K_{\mathrm{T}}^{(2)}, K_{\mathrm{T}}^{(3)}). \qquad (9.4.5)$$

As we shall see in Chap. 12, equations of the form (9.4.4) have a remarkable ubiquity in the analysis of statistical mechanical lattice systems. Now often called *Yang-Baxter equations*, they are a key to the solution of many problems quite unrelated to the relationship between the honeycomb and triangular lattice.

A set of transformation equations can be obtained from (9.4.4) by making different choices for the states $(\sigma_1, \sigma_2, \sigma_3)$. Given the need to obtain a form for $\zeta(K_{\mathrm{H}}^{(1)}, K_{\mathrm{H}}^{(2)}, K_{\mathrm{H}}^{(3)})$, the set must contain four independent equations when the couplings are different and two when the couplings are equal. With this in mind, it is of interest to consider some special cases.

9.4.1 The ν-State Potts Model

Now $X(K_{\mathrm{W}}^{(\mathrm{x})};\sigma,\sigma') = [\mathfrak{X}_{\mathrm{W}}^{(\mathrm{x})}]^{\delta^{\mathrm{Kr}}(\sigma-\sigma')}$ and $\mathfrak{X}_{\mathrm{W}}^{(\mathrm{x})} := \exp(-K_{\mathrm{W}}^{(\mathrm{x})})$, $K_{\mathrm{W}}^{(\mathrm{x})} := R_{\mathrm{W}}^{(\mathrm{x})}/T$, $\mathrm{W} = \mathrm{T},\ \mathrm{H}, \mathrm{x} = 1, 2, 3$. There are three distinguishable cases of (9.4.4) for the states $(\sigma_1, \sigma_2, \sigma_3)$:

(a) All states the same giving

$$\mathfrak{X}_{\mathrm{H}}^{(1)}\mathfrak{X}_{\mathrm{H}}^{(2)}\mathfrak{X}_{\mathrm{H}}^{(3)} + \nu - 1 = \zeta(K_{\mathrm{H}}^{(1)}, K_{\mathrm{H}}^{(2)}, K_{\mathrm{H}}^{(3)})\, \mathfrak{X}_{\mathrm{T}}^{(1)}\mathfrak{X}_{\mathrm{T}}^{(2)}\mathfrak{X}_{\mathrm{T}}^{(3)}. \tag{9.4.6}$$

(b) Two the same and one different, giving (when the couplings are not equal) three equations

$$\mathfrak{X}_{\mathrm{H}}^{(\mathrm{x})} + \mathfrak{X}_{\mathrm{H}}^{(\mathrm{y})}\mathfrak{X}_{\mathrm{H}}^{(\mathrm{z})} + \nu - 2 = \zeta(K_{\mathrm{H}}^{(1)}, K_{\mathrm{H}}^{(2)}, K_{\mathrm{H}}^{(3)})\, \mathfrak{X}_{\mathrm{T}}^{(\mathrm{x})}, \qquad \mathrm{x} = 1, 2, 3, \quad \mathrm{y} \neq \mathrm{z} \neq \mathrm{x}. \tag{9.4.7}$$

(c) *Only when* $\nu > 2$; all different giving

$$\mathfrak{X}_{\mathrm{H}}^{(1)} + \mathfrak{X}_{\mathrm{H}}^{(2)} + \mathfrak{X}_{\mathrm{H}}^{(3)} + \nu - 3 = \zeta(K_{\mathrm{H}}^{(1)}, K_{\mathrm{H}}^{(2)}, K_{\mathrm{H}}^{(3)}). \tag{9.4.8}$$

There are, therefore, five equations when the couplings are different and three when the couplings are equal; one too many unless $\nu = 2$ when the numbers reduce to four and two respectively. Thus we conclude that the transformation works, in the sense of yielding the triangular couplings as functions of the honeycomb couplings, only for the 2-state Potts model. For $\nu > 2$ Eqs. (9.4.6)–(9.4.8) can be solved to give

$$\begin{aligned}
&\mathfrak{X}_{\mathrm{T}}^{(1)}\mathfrak{X}_{\mathrm{T}}^{(2)}\mathfrak{X}_{\mathrm{T}}^{(3)} - \mathfrak{X}_{\mathrm{T}}^{(1)}\mathfrak{X}_{\mathrm{T}}^{(2)} - \mathfrak{X}_{\mathrm{T}}^{(2)}\mathfrak{X}_{\mathrm{T}}^{(3)} - \mathfrak{X}_{\mathrm{T}}^{(3)}\mathfrak{X}_{\mathrm{T}}^{(1)}\\
&\qquad - (\nu - 1)[\mathfrak{X}_{\mathrm{T}}^{(1)} + \mathfrak{X}_{\mathrm{T}}^{(2)} + \mathfrak{X}_{\mathrm{T}}^{(3)}] - \nu^2 + 3\nu - 1 = 0,\\
&\mathfrak{X}_{\mathrm{H}}^{(1)}\mathfrak{X}_{\mathrm{H}}^{(2)}\mathfrak{X}_{\mathrm{H}}^{(3)} - \mathfrak{X}_{\mathrm{H}}^{(1)} - \mathfrak{X}_{\mathrm{H}}^{(2)} - \mathfrak{X}_{\mathrm{H}}^{(3)} - \nu + 2 = 0.
\end{aligned} \tag{9.4.9}$$

with their joint solutions satisfying the duality relationship (9.2.21) (with $\mathfrak{X} = \mathfrak{X}_{\mathrm{H}}$ and $\mathfrak{X}^\star = \mathfrak{X}_{\mathrm{T}}$, or vice-versa). Equations (9.4.9) are the critical conditions derived by Hindermann et al. (1978).[21] It must be emphasised that the calculation given here is *not a derivation* of the critical conditions. It simply shows that the star-triangle transformation is valid only at criticality. This is consonant with the fact that the ν–state Potts model is solved only at criticality; this solution arising from that of a special case of the staggered six-vertex model (see Sect. 3.8.2.1). In view of Eq. (3.8.47) this is also referred to as the solution of the 'self dual Potts model'.

[21] The derivation given by Hindermann et al. is for $\nu \geq 4$, but, as remarked by Wu (1982), since it agrees with the $\nu = 2$ (Ising) result (Onsager 1944, see also the comment at the end of Sect. 9.4.2.1), "it is expected the [it] is also exact for $\nu = 3$".

9.4.2 The Spin-$\frac{1}{2}$ Ising Model

Since the two-state Potts model is equivalent to the zero-field spin-$\frac{1}{2}$ Ising model we consider that model, for which $X(K_{\mathrm{W}}^{(\mathrm{x})};\sigma,\sigma') = \exp(-K_{\mathrm{W}}^{(\mathrm{x})}\sigma\sigma')$, $K_{\mathrm{W}}^{(\mathrm{x})} := J_{\mathrm{W}}^{(\mathrm{x})}/T$, $\mathrm{W} = \mathrm{T},\ \mathrm{H}$, $\mathrm{x} = 1, 2, 3$. Then, from (9.4.4), for the two cases (a) and (b),

$$\begin{aligned} 2\cosh(K_{\mathrm{H}}^{(1)} + K_{\mathrm{H}}^{(2)} + K_{\mathrm{H}}^{(3)}) &= \zeta(K_{\mathrm{H}}^{(1)}, K_{\mathrm{H}}^{(2)}, K_{\mathrm{H}}^{(3)}) \\ &\quad \times \exp(K_{\mathrm{T}}^{(1)} + K_{\mathrm{T}}^{(2)} + K_{\mathrm{T}}^{(3)}), \\ 2\cosh(2K_{\mathrm{H}}^{(\mathrm{x})} - K_{\mathrm{H}}^{(1)} - K_{\mathrm{H}}^{(2)} - K_{\mathrm{H}}^{(3)}) &= \zeta(K_{\mathrm{H}}^{(1)}, K_{\mathrm{H}}^{(2)}, K_{\mathrm{H}}^{(3)}) \\ &\quad \times \exp(2K_{\mathrm{T}}^{(\mathrm{x})} - K_{\mathrm{T}}^{(1)} - K_{\mathrm{T}}^{(2)} - K_{\mathrm{T}}^{(3)}) \qquad \mathrm{x} = 1, 2, 3. \end{aligned} \tag{9.4.10}$$

It is convenient at this point to introduce a compact notation for the hyperbolic functions of the honeycomb and triangle couplings[22]

$$\begin{aligned} &\mathrm{ch}_{\mathrm{W}}^{(\mathrm{x})} := \cosh(2K_{\mathrm{W}}^{(\mathrm{x})}), \qquad \mathrm{sh}_{\mathrm{W}}^{(\mathrm{x})} := \sinh(2K_{\mathrm{W}}^{(\mathrm{x})}), \\ &\mathrm{th}_{\mathrm{W}}^{(\mathrm{x})} := \tanh(2K_{\mathrm{W}}^{(\mathrm{x})}), \qquad \mathrm{cth}_{\mathrm{W}}^{(\mathrm{x})} := \coth(2K_{\mathrm{W}}^{(\mathrm{x})}), \\ &\mathrm{sch}_{\mathrm{W}}^{(\mathrm{x})} := \mathrm{sech}(2K_{\mathrm{W}}^{(\mathrm{x})}), \qquad \mathrm{x} = 1, 2, 3, \qquad \mathrm{W} = \mathrm{T},\ \mathrm{H}. \end{aligned} \tag{9.4.11}$$

Then it can be shown from (9.4.10) that

$$\mathrm{sh}_{\mathrm{T}}^{(\mathrm{x})} = \frac{\mathrm{sh}_{\mathrm{H}}^{(\mathrm{y})}\mathrm{sh}_{\mathrm{H}}^{(\mathrm{z})}}{q_{\mathrm{H}}}, \qquad \mathrm{ch}_{\mathrm{T}}^{(\mathrm{x})} = \frac{\mathrm{ch}_{\mathrm{H}}^{(\mathrm{x})} + \mathrm{ch}_{\mathrm{H}}^{(\mathrm{y})}\mathrm{ch}_{\mathrm{H}}^{(\mathrm{z})}}{q_{\mathrm{H}}}, \tag{9.4.12}$$

where in (9.4.12) and subsequent equations $(\mathrm{x}, \mathrm{y}, \mathrm{z})$ is any permutation of $(1, 2, 3)$ and

$$q_{\mathrm{H}} := \tfrac{1}{2}[\zeta(K_{\mathrm{H}}^{(1)}, K_{\mathrm{H}}^{(2)}, K_{\mathrm{H}}^{(3)})]^2. \tag{9.4.13}$$

From (9.4.12),

$$q_{\mathrm{H}}^2 = 2\mathrm{ch}_{\mathrm{H}}^{(\mathrm{a})}\mathrm{ch}_{\mathrm{H}}^{(\mathrm{b})}\mathrm{ch}_{\mathrm{H}}^{(\mathrm{c})} + [\mathrm{ch}_{\mathrm{H}}^{(\mathrm{a})}]^2 + [\mathrm{ch}_{\mathrm{H}}^{(\mathrm{b})}]^2 + [\mathrm{ch}_{\mathrm{H}}^{(\mathrm{c})}]^2 - 1. \tag{9.4.14}$$

It can now be shown that

$$\mathrm{ch}_{\mathrm{H}}^{(\mathrm{x})} = \frac{\mathrm{sh}_{\mathrm{T}}^{(\mathrm{x})}\mathrm{ch}_{\mathrm{T}}^{(\mathrm{y})}\mathrm{ch}_{\mathrm{T}}^{(\mathrm{z})} + \mathrm{ch}_{\mathrm{T}}^{(\mathrm{x})}\mathrm{sh}_{\mathrm{T}}^{(\mathrm{y})}\mathrm{sh}_{\mathrm{T}}^{(\mathrm{z})}}{\mathrm{sh}_{\mathrm{T}}^{(\mathrm{x})}}. \tag{9.4.15}$$

---

[22] This notation, not all of which is needed immediately, will be used throughout this and the following section.

This is the inverse relation giving honeycomb couplings in terms of triangle couplings. From the first relation of (9.4.12),

$$\mathrm{sh}_{\mathrm{T}}^{(\varkappa)}\mathrm{sh}_{\mathrm{H}}^{(\varkappa)} = \mathrm{k}^{-1} \qquad \varkappa = 1, 2, 3, \tag{9.4.16}$$

where

$$\mathrm{k} := \frac{q_{\mathrm{H}}}{\mathrm{sh}_{\mathrm{H}}^{(1)}\mathrm{sh}_{\mathrm{H}}^{(2)}\mathrm{sh}_{\mathrm{H}}^{(3)}}. \tag{9.4.17}$$

This is an expression for k in terms of the honeycomb couplings. From (9.4.15),

$$\mathrm{k} = \frac{[\mathrm{sch}_{\mathrm{T}}^{(1)}\mathrm{sch}_{\mathrm{T}}^{(2)}\mathrm{sch}_{\mathrm{T}}^{(3)}]^2}{4q_{\mathrm{T}}}, \tag{9.4.18}$$

is an expression in terms of triangle couplings, where

$$\begin{aligned} q_{\mathrm{T}}^2 := {} & [\mathrm{th}_{\mathrm{T}}^{(1)} + \mathrm{th}_{\mathrm{T}}^{(2)}\mathrm{th}_{\mathrm{T}}^{(3)}][\mathrm{th}_{\mathrm{T}}^{(2)} + \mathrm{th}_{\mathrm{T}}^{(3)}\mathrm{th}_{\mathrm{T}}^{(1)}][\mathrm{th}_{\mathrm{T}}^{(3)} + \mathrm{th}_{\mathrm{T}}^{(1)}\mathrm{th}_{\mathrm{T}}^{(2)}] \\ & \times [1 + \mathrm{th}_{\mathrm{T}}^{(1)}\mathrm{th}_{\mathrm{T}}^{(2)}\mathrm{th}_{\mathrm{T}}^{(3)}]. \end{aligned} \tag{9.4.19}$$

9.4.2.1 Equal Couplings

We now take the case $K_{\mathrm{W}}^{(1)} = K_{\mathrm{W}}^{(2)} = K_{\mathrm{W}}^{(3)} = K_{\mathrm{W}}$, $\mathrm{W} = \mathrm{H}, \mathrm{T}$ to complete the transformation between the honeycomb and triangular lattices. From (9.4.5) and (9.4.15),

$$\exp(4K_{\mathrm{T}}) = 2\cosh(2K_{\mathrm{H}}) - 1, \tag{9.4.20}$$

$$\phi_1(\mathcal{N}_{\mathrm{T}}, K_{\mathrm{T}}, 0) = 2\phi_1(\mathcal{N}_{\mathrm{H}}, K_{\mathrm{H}}, 0) + \ln[A(K_{\mathrm{T}}, K_{\mathrm{H}})], \tag{9.4.21}$$

$$\frac{\mathrm{u}(\mathcal{N}_{\mathrm{T}}, K_{\mathrm{T}})}{\coth(2K_{\mathrm{T}})} + 1 = B(K_{\mathrm{T}}, K_{\mathrm{H}})\left[\frac{\mathrm{u}(\mathcal{N}_{\mathrm{H}}, K_{\mathrm{H}})}{\coth(2K_{\mathrm{H}})} + 1\right], \tag{9.4.22}$$

where, from (9.4.13), (9.4.14) and (9.4.20),

$$\begin{aligned} A(K_{\mathrm{T}}, K_{\mathrm{H}}) &:= \exp(K_{\mathrm{T}})[\exp(4K_{\mathrm{T}}) + 3]^{\frac{1}{2}} = 2[\cosh(3K_{\mathrm{H}})\cosh^3(K_{\mathrm{H}})]^{\frac{1}{4}}, \\ B(K_{\mathrm{T}}, K_{\mathrm{H}}) &:= \frac{3\exp(4K_{\mathrm{T}})}{\exp(4K_{\mathrm{T}}) + 3} = \frac{2[2\cosh(2K_{\mathrm{H}}) + 1}{\cosh(2K_{\mathrm{H}}) + 1)}. \end{aligned} \tag{9.4.23}$$

Suppose that the duality transformation is performed on a triangular lattice Ising model with N sites and coupling K_{T}', mapping it into a honeycomb lattice Ising model with coupling K_{H}'. As seen in Sect. 16.2.2.1 this gives $N^{\star} = 2N$ and, from (9.2.29)–(9.2.31),

$$\sinh(2K'_{\mathrm{T}})\sinh(2K'_{\mathrm{H}}) = 1, \tag{9.4.24}$$

$$\tfrac{1}{2}\ln[2\sinh(2K'_{\mathrm{T}})] + \phi_1(\mathcal{N}_{\mathrm{T}}, K'_{\mathrm{T}}, 0) = \ln[2\sinh(2K'_{\mathrm{H}})] + 2\phi_1(\mathcal{N}_{\mathrm{H}}, K'_{\mathrm{H}}, 0), \tag{9.4.25}$$

$$\frac{\mathrm{u}(\mathcal{N}_{\mathrm{T}}, K'_{\mathrm{T}})}{\coth(2K'_{\mathrm{T}})} + \frac{2\mathrm{u}(\mathcal{N}_{\mathrm{H}}, K'_{\mathrm{H}})}{\coth(2K'_{\mathrm{H}})} = -3. \tag{9.4.26}$$

We now have two strategies.

9.4.2.2 The Triangular Lattice

A duality transformation is applied to the triangular lattice, then the equal-couplings star-triangle transformation, given by (9.4.20), is performed on the resulting honeycomb lattice. Setting $K'_{\mathrm{H}} = K_{\mathrm{H}}$, and eliminating this quantity between (9.4.20) and (9.4.24) gives

$$[\exp(4K_{\mathrm{T}}) - 1][\exp(4K'_{\mathrm{T}}) - 1] = 4, \tag{9.4.27}$$

which relates the triangular lattice to itself at two different coupling values. In a similar way, from (9.4.21)–(9.4.26),

$$\phi_1(\mathcal{N}_{\mathrm{T}}, K'_{\mathrm{T}}, 0) + \tfrac{1}{2}\ln[\sinh(2K'_{\mathrm{T}})] = \tfrac{1}{2}\ln[\sinh(2K_{\mathrm{T}})] + \phi_1(\mathcal{N}_{\mathrm{T}}, K_{\mathrm{T}}, 0), \tag{9.4.28}$$

$$\frac{\mathrm{u}(\mathcal{N}_{\mathrm{T}}, K_{\mathrm{T}})}{\coth(2K_{\mathrm{T}})} + 1 = -\tfrac{1}{2}Y(K_{\mathrm{T}})\left[\frac{\mathrm{u}(\mathcal{N}_{\mathrm{T}}, K'_{\mathrm{T}})}{\coth(2K'_{\mathrm{T}})} + 1\right]. \tag{9.4.29}$$

The first of these equations is identical in form to the square-lattice duality transformation relation (9.2.34), but the two K_{T} values are now connected by (9.4.27) instead of (9.2.29). Like (9.2.29), Eq. (9.4.27) relates values of K_{T} in a high-temperature interval $[0, K_{\mathrm{Tc}}]$ with those in a low-temperature interval $[K_{\mathrm{Tc}}, \infty]$ where K_{Tc} is now given by

$$[\exp(4K_{\mathrm{Tc}}) - 1]^2 = 4. \tag{9.4.30}$$

By similar reasoning to that used in the square case, if there is a unique critical point it must be at $K_{\mathrm{T}} = K_{\mathrm{Tc}}$ where the two intervals meet, and the critical temperature is then

$$T_{\mathrm{c}} = J/K_{\mathrm{Tc}} = 4J/\ln(3). \tag{9.4.31}$$

It follows from (9.4.29) that u_{c} is given by (9.2.37).

9.4.2.3 The Honeycomb Lattice

The star-triangle transformation is followed by a duality transformation. Now we set $K'_{\mathrm{T}} = K_{\mathrm{T}}$ in (9.4.24) and (9.4.25) and after a similar manipulation to that described for the triangular lattice we have

$$[\cosh(2K_{\mathrm{H}}) - 1][\cosh(2K'_{\mathrm{H}}) - 1] = 1, \tag{9.4.32}$$

$$\phi_1(\mathcal{N}_{\mathrm{H}}, K'_{\mathrm{H}}, 0) + \tfrac{1}{2}\ln[\sinh(2K'_{\mathrm{H}})] = \tfrac{1}{2}\ln[\sinh(2K_{\mathrm{H}})] + \phi_1(\mathcal{N}_{\mathrm{H}}, K_{\mathrm{H}}, 0), \tag{9.4.33}$$

$$\frac{\mathrm{u}(\mathcal{N}_{\mathrm{H}}, K'_{\mathrm{H}})}{\coth(2K'_{\mathrm{H}})} + 1 = -\frac{1}{2}Y(K_{\mathrm{H}})\left[\frac{\mathrm{u}(\mathcal{N}_{\mathrm{H}}, K_{\mathrm{H}})}{\coth(2K_{\mathrm{H}})} + 1\right], \tag{9.4.34}$$

the second equation again being identical in form to the square-lattice duality transformation relation (9.2.34) with the two K_{H} values now being connected by (9.4.32). The meeting point of the high- and low-temperature intervals connected by (9.4.32) is at $K_{\mathrm{H}} = K_{\mathrm{HC}}$, where

$$[\cosh(2K_{\mathrm{HC}}) - 1]^2 = 1 \tag{9.4.35}$$

and, if there is a unique critical point, it occurs at temperature

$$T_{\mathrm{c}} = J/K_{\mathrm{HC}} = 2J/\ln(2 + \sqrt{3}). \tag{9.4.36}$$

Again it follows from (9.4.34) that u_{c} is given by (9.2.37). Numerical values for T_{c} and u_{c} are given in Table 3.1.

Finally we note that, translating the critical point Eqs. (9.4.31) and (9.4.36) into the notation for the 2-state Potts model gives $\mathfrak{X}_{\mathrm{TC}} = \sqrt{3}$, $\mathfrak{X}_{\mathrm{HC}} = 2 + \sqrt{3}$, which satisfy (9.4.9).

Chapter 10
Edge-Decorated Ising Models

10.1 Primary and Secondary Sites

Consider a lattice $\mathcal{N}$ of dimension d, coordination number z, N sites, periodic boundary conditions and an Ising variable $\sigma(\boldsymbol{r}) = \pm 1$ at every $\boldsymbol{r} \in \mathcal{N}$. Let the lattice be edge-decorated by taking $\mathring{\mathcal{N}} = \mathcal{N} + \mathcal{N}^{\circledast}$, where $\mathcal{N}^{\circledast}$ is the *medial lattice* (Sect. 16.2.2) with $|\mathring{\mathcal{N}}| = (1 + \frac{1}{2}z)$. The sites of $\mathcal{N}$ are referred to as *primary sites*, labelling them p, and of $\mathcal{N}^{\circledast}$ as *secondary sites*, labelling them s. Suppose that at the secondary site on the edge connecting the first-neighbour pair of sites $[\boldsymbol{r}, \boldsymbol{r}']$ there is a microsystem whose state is denoted by $\tau(\boldsymbol{r}, \boldsymbol{r}')$. Given that this microsystem interacts only with the microsystems on its neighbouring primary sites and that the primary lattice microsystems form a first-neighbour spin-$\frac{1}{2}$ Ising model the Hamiltonian (3.5.1) is modified to

$$\widehat{H}(\mathring{\mathcal{N}}, K, \mathring{K}, L; \boldsymbol{\sigma}, \boldsymbol{\tau}) = \sum_{\{\boldsymbol{r},\boldsymbol{r}'\}}^{(1)} \left\{\mathring{K}\mathfrak{e}(\sigma(\boldsymbol{r}), \sigma(\boldsymbol{r}'); \tau(\boldsymbol{r}, \boldsymbol{r}')) - K\sigma(\boldsymbol{r})\sigma(\boldsymbol{r}')\right\} - L\sum_{\{\boldsymbol{r}\}} \sigma(\boldsymbol{r}), \tag{10.1.1}$$

where $K := J/T$, $L := \mathcal{H}/T$ and $\mathring{K}$ is the primary-secondary coupling, for which the interaction has the general form $\mathfrak{e}(\sigma(\boldsymbol{r}), \sigma(\boldsymbol{r}'); \tau(\boldsymbol{r}, \boldsymbol{r}'))$. The partition function is now

$$Z(\mathring{\mathcal{N}}, K, \mathring{K}, L) = \sum_{\{\boldsymbol{\sigma},\boldsymbol{\tau}\}} \prod_{\{\boldsymbol{r},\boldsymbol{r}'\}}^{(1)} \left\{\mathfrak{X}^{\sigma(\boldsymbol{r})\sigma(\boldsymbol{r}')} \mathring{\mathfrak{X}}^{-\mathfrak{e}(\sigma(\boldsymbol{r}),\sigma(\boldsymbol{r}');\tau(\boldsymbol{r},\boldsymbol{r}'))} \mathfrak{Z}^{\sigma(\boldsymbol{r})/z} \mathfrak{Z}^{\sigma(\boldsymbol{r}')/z}\right\}, \tag{10.1.2}$$

where $\mathfrak{X} := \exp(K)$, $\mathring{\mathfrak{X}} := \exp(\mathring{K})$, $\mathfrak{Z} := \exp(L)$ and where the single-site magnetic field terms have been divided between each of their z first-neighbour pairs. The next stage of the procedure is similar to that used for the star-triangle transformation in Sect. 9.4; we sum over the states of the secondary sites and demand that the

D.A. Lavis, *Equilibrium Statistical Mechanics of Lattice Models*,
Theoretical and Mathematical Physics, DOI 10.1007/978-94-017-9430-5_10

resulting partition function be equal, within a multiplicative constant, to that of an equivalent (undecorated) Ising model. The effective Ising couplings are $K_e := J_e/T$ and $L_e := \mathcal{H}_e/T$, with $\mathfrak{X}_e := \exp(K_e)$, $\mathfrak{Z}_e := \exp(L_e)$ and, concentrating on the single first-neighbour pair $[\boldsymbol{r}, \boldsymbol{r}']$,

$$\mathfrak{X}^{\sigma(\boldsymbol{r})\sigma(\boldsymbol{r}')}\mathfrak{Z}^{\sigma(\boldsymbol{r})/z}\mathfrak{Z}^{\sigma(\boldsymbol{r}')/z}\,\phi\,(\mathring{K}; \sigma(\boldsymbol{r}), \sigma(\boldsymbol{r}')) = \zeta(\mathring{K})\mathfrak{X}_e^{\sigma(\boldsymbol{r})\sigma(\boldsymbol{r}')}\mathfrak{Z}_e^{\sigma(\boldsymbol{r})/z}\mathfrak{Z}_e^{\sigma(\boldsymbol{r}')/z}, \tag{10.1.3}$$

where

$$\phi(\mathring{K}; \sigma(\boldsymbol{r}), \sigma(\boldsymbol{r}')) := \sum_{\{\tau\}} \mathring{\mathfrak{X}}^{-\mathfrak{e}(\sigma(\boldsymbol{r}),\sigma(\boldsymbol{r}');\tau)}. \tag{10.1.4}$$

By making the three choices $\sigma(\boldsymbol{r}) = \sigma(\boldsymbol{r}') = \pm 1$, $\sigma(\boldsymbol{r}) = -\sigma(\boldsymbol{r}')$, this gives

$$\zeta(\mathring{K}) = \left\{\phi(\mathring{K}; +1, +1)\,\phi\,(\mathring{K}; -1, -1)[\phi(\mathring{K}; +1, -1)]^2\right\}^{\frac{1}{4}}, \tag{10.1.5}$$

$$\mathfrak{X}_e^4 = \mathfrak{X}^4 \frac{\phi(\mathring{K}; +1, +1)\,\phi\,(\mathring{K}; -1, -1)}{[\phi(\mathring{K}; +1, -1)]^2}, \tag{10.1.6}$$

$$\mathfrak{Z}_e^4 = \mathfrak{Z}^4 \left[\frac{\phi(\mathring{K}; +1, +1)}{\phi(\mathring{K}; -1, -1)}\right]^z, \tag{10.1.7}$$

and, from (10.1.2) and (10.1.3),

$$Z(\mathring{\mathcal{N}}, K, \mathring{K}, L) = [\zeta(\mathring{K})]^{\frac{1}{2}Nz} Z^{(\mathrm{IM})}(\mathcal{N}, K_e, L_e), \tag{10.1.8}$$

where the latter partition function is that of the Ising model on $\mathcal{N}$. The general form of these relations is due to Fisher (1959). The equivalent primary lattice Ising model with couplings K_e and L_e is a ferromagnet or antiferromagnet according as K_e is positive or negative. For $K_e > 0$, any critical point or curve in the space of the decorated model thermodynamic variables maps into the point $K_e = K_c$, $L_e = 0$ in the K_e–L_e plane of the Ising model while any first-order transition line or surface maps into the segment $L_e = 0$ $K_e > K_c$.

10.2 Super-Exchange or Bond-Dilution

Consider now the special case of the edge-decorated Ising model where $K = 0$ and the one coupling $\mathring{K}$ is replaced by two couplings $K_p := J_p/T$ and $L_s := \mu/T$ with

$$\mathring{K}\mathfrak{e}(\sigma(\boldsymbol{r}), \sigma(\boldsymbol{r}'); \tau(\boldsymbol{r}, \boldsymbol{r}')) = -\tau(\boldsymbol{r}, \boldsymbol{r}')[L_s + K_p\sigma(\boldsymbol{r})\sigma(\boldsymbol{r}')] \tag{10.2.1}$$

and $\tau(\boldsymbol{r},\boldsymbol{r}')$ able to take two values: zero, meaning that there is no interaction between the Ising spins on $\boldsymbol{r}$ and $\boldsymbol{r}'$ and one, meaning that there is a solute atom, with a chemical potential μ, on the secondary s site which mediates an Ising interaction J_{p} between the two spins on primary p sites. This model can be regarded as one of *super-exchange* mediated by secondary atoms or of bond-dilution (Rapaport 1972). Then, from (10.1.5)–(10.1.8),

$$\exp(2K_{\mathrm{e}}) = \frac{\exp(L_{\mathrm{s}}+K_{\mathrm{p}})+1}{\exp(L_{\mathrm{s}}-K_{\mathrm{p}})+1}, \qquad L_{\mathrm{e}} = L, \tag{10.2.2}$$

$$Z(\mathring{\mathcal{N}},K_{\mathrm{p}},L,L_{\mathrm{s}}) = [(\exp(L_{\mathrm{s}}+K_{\mathrm{p}})+1)(\exp(L_{\mathrm{s}}-K_{\mathrm{p}})+1)]^{\frac{1}{4}Nz} \times Z^{(\mathrm{IM})}(\mathcal{N},K_{\mathrm{e}},L). \tag{10.2.3}$$

From (3.5.7), (3.5.8) and (10.2.1)–(10.2.3), the magnetization per primary lattice site is

$$m := \langle\sigma(\boldsymbol{r})\rangle - -\left(\frac{\partial\phi(\mathring{\mathcal{N}},K_{\mathrm{p}},L,L_{\mathrm{s}})}{\partial L}\right)_{K_{\mathrm{p}},L_{\mathrm{s}}} = m^{(\mathrm{IM}))}(K_{\mathrm{e}},L), \tag{10.2.4}$$

the density of solute atoms is

$$\rho := \langle\tau(\boldsymbol{r},\boldsymbol{r}')\rangle = \frac{2}{z}\left(\frac{\partial\phi(\mathring{\mathcal{N}},K_{\mathrm{p}},L,L_{\mathrm{s}})}{\partial L_{\mathrm{s}}}\right)_{K_{\mathrm{p}},L} = f(K_{\mathrm{p}},K_{\mathrm{e}},L), \tag{10.2.5}$$

and the internal energy per primary site (made dimensionless by dividing by J_{p}) is[1]

$$\begin{aligned} u &:= -\tfrac{1}{2}z\langle\tau(\boldsymbol{r},\boldsymbol{r}')\sigma(\boldsymbol{r})\sigma(\boldsymbol{r}')\rangle_1 = \left(\frac{\partial\phi(\mathring{\mathcal{N}},K_{\mathrm{p}},L,L_{\mathrm{s}})}{\partial K_{\mathrm{p}}}\right)_{L,L_{\mathrm{s}}} \\ &= \tfrac{1}{2}zg(K_{\mathrm{p}},K_{\mathrm{e}},L), \end{aligned} \tag{10.2.6}$$

where

$$f(K_{\mathrm{p}},K_{\mathrm{e}},L) := \mathrm{C}(K_{\mathrm{p}},K_{\mathrm{e}}) - 2\,\mathrm{u}^{(\mathrm{IM})}(K_{\mathrm{e}},L)\mathrm{S}(K_{\mathrm{p}},K_{\mathrm{e}})/z, \tag{10.2.7}$$

$$g(K_{\mathrm{p}},K_{\mathrm{e}},L) := \mathrm{S}(K_{\mathrm{p}},K_{\mathrm{e}}) - 2\,\mathrm{u}^{(\mathrm{IM})}(K_{\mathrm{e}},L)\mathrm{C}(K_{\mathrm{p}},K_{\mathrm{e}})/z, \tag{10.2.8}$$

$$\begin{aligned} \mathrm{C}(K_{\mathrm{p}},K_{\mathrm{e}}) &:= \frac{\sinh(K_{\mathrm{e}})\cosh(K_{\mathrm{p}}-K_{\mathrm{e}})}{\sinh(K_{\mathrm{p}})}, \\ \mathrm{S}(K_{\mathrm{p}},K_{\mathrm{e}}) &:= \frac{\sinh(K_{\mathrm{e}})\sinh(K_{\mathrm{p}}-K_{\mathrm{e}})}{\sinh(K_{\mathrm{p}})} \end{aligned} \tag{10.2.9}$$

---

[1] Where the subscript '1' is now a reminder that $\boldsymbol{r}$ and $\boldsymbol{r}'$ are first-neighbour *primary* sites.

and $m^{(\mathrm{IM})}(K_\mathrm{e}, L)$ and $\mathrm{u}^{(\mathrm{IM})}(K_\mathrm{e}, L)$ are respectively the magnetization and internal energy per primary lattice site on the equivalent Ising model (the latter being made dimensionless by dividing by J_e). From (2.3.29), (10.2.5) and (10.2.6), the heat capacity at constant density per primary site is

$$\begin{aligned} c(K_\mathrm{p}, K_\mathrm{e}, L, \rho^*) &:= -k_\mathrm{B} K_\mathrm{p}^2 \left(\frac{\partial u}{\partial K_\mathrm{p}} \right)_\rho \\ &= -\frac{k_\mathrm{B} z K_\mathrm{p}^2 \{ f_\mathrm{e}(K_\mathrm{p}, K_\mathrm{e}, L) g_\mathrm{p}(K_\mathrm{p}, K_\mathrm{e}, L) - f_\mathrm{p}(K_\mathrm{p}, K_\mathrm{e}, L) g_\mathrm{e}(K_\mathrm{p}, K_\mathrm{e}, L) \}}{2 f_\mathrm{e}(K_\mathrm{p}, K_\mathrm{e}, L)}, \end{aligned} \qquad (10.2.10)$$

where the subscripts 'p' and 'e' attached to f and g denote partial differentiation with respect to K_p and K_e respectively.

10.2.1 Critical Properties and Exponent Renormalization

Equation (10.2.5) is an implicit relation giving K_e in terms of ρ, K_p and L. In the equivalent Ising model there is a second-order transition when K_e passes through the critical value K_c at $L_\mathrm{e} = 0$. A curve of ρ as a function of T/J_p, with $\mathcal{H} = L = 0$ and $K_\mathrm{e} = K_\mathrm{c}$, can be computed from (10.2.5) and is shown in Fig. 10.1. The bond-dilute model is equivalent to the standard Ising model when $\rho = 1$, and it can be verified from (10.2.5) that the critical curve meets the line $\rho = 1$ at $T = T_\mathrm{c}$, the Ising critical temperature. As T decreases from T_c along the critical curve, ρ decreases steadily and $\mathrm{d}\rho/\mathrm{d}T = 0$ when the curve meets the zero-temperature axis at $\rho = \rho_\mathrm{c}$, given by

$$\rho_\mathrm{c} = \frac{1 + \mathrm{u}^{(\mathrm{IM})}(K_\mathrm{c}, 0)}{1 + \coth(K_\mathrm{c})}. \qquad (10.2.11)$$

For $\rho < \rho_\mathrm{c}$ there is no transition at any value of T.

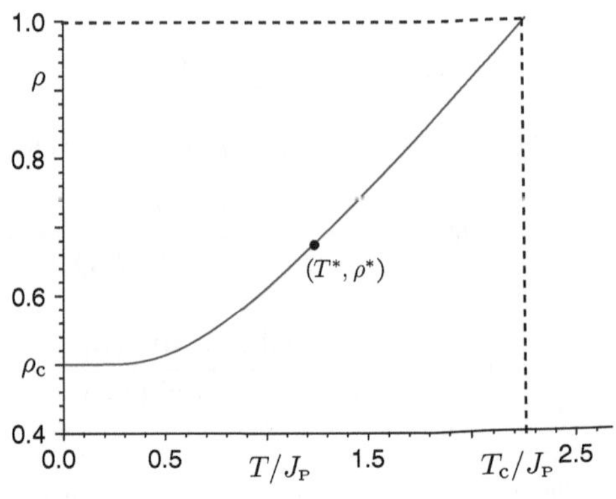

Fig. 10.1 The critical curve for the bond-dilute model. The curve was calculated for the square lattice, using (9.2.36) and (9.2.37), to give $\rho = \frac{1}{2} + \frac{\sqrt{2}}{4}[\coth(K_\mathrm{p}) - 1]$, and for which $\rho_\mathrm{c} = \frac{1}{2}$

We now investigate the critical properties of the system in a neighbourhood of the critical curve. Consider the point (T^*, ρ^*) on this curve approached along the line $\rho = \rho^*$. The critical exponents for the equivalent Ising model are denoted in the usual way (Sect. 4.4.1) as α, β, γ and δ, with primes used to distinguish between approaches from below or above the critical temperature.[2] The corresponding exponents for the approach to $T = T^*$ along the line $\rho = \rho^*$ are denoted by α_r, β_r, γ_r and δ_r. Since $L_e = L$ and, from (10.2.4), the magnetization is the same as that in the equivalent Ising model, $\delta_r = \delta$. It is tempting to suppose that the same reasoning would apply to γ_r and β_r, but, of course, the asymptotic property of the equivalent Ising model is in terms of $\triangle K_e = K_c - K_e$ and that of the full model is in terms of $\triangle K_p = K_p^* - K_p$. So the first task is to relate these two small quantities to each other. This will involve α the exponent for the heat capacity and we must distinguish two cases: $\alpha = 0$, which is the case for $d = 2$, where the heat capacity has a logarithmic singularity (Sect. 12.2.7.3) and $\alpha \neq 0$. Setting aside the case $\alpha = 0$, which we shall consider at the end of the section, from (4.4.12),

$$u^{(\mathrm{IM})}(K_e, 0) \simeq \begin{cases} u_0 + u_1|\triangle K_e|^{1-\alpha}, & K_e < K_c, \\ u_0 + u_1'|\triangle K_e|^{1-\alpha'}, & K_e > K_c, \end{cases} \tag{10.2.12}$$

for some constants u_0, u_1 and u_1'. In order that the internal energy is finite at the critical point and the heat capacity is singular it is necessary that α lies in the interval (0, 1]. In fact we shall impose the more restrictive condition that $0 < \alpha < \frac{1}{2}$, which is certainly true for $d = 3$ (Domb and Hunter 1965) and which has been assumed in the derivation of the asymptotic form (10.2.16). From (10.2.5) and (10.2.7), for small $\triangle K_e > 0$ $(T > T_c)$,

$$\triangle K_p \simeq p_1(K_p^*, K_c)|\triangle K_e|^{1-\alpha} + p_2(K_p^*, K_c)\triangle K_e, \tag{10.2.13}$$

where

$$\begin{aligned} p_1(K_p^*, K_c) &:= \frac{-2u_1 S(K_p^*, K_c)}{zC_p(K_p^*, K_c) - 2u_0 S_p(K_p^*, K_c)}, \\ p_2(K_p^*, K_c) &:= \frac{2u_0 S_e(K_p^*, K_c) - zC_e(K_p^*, K_c)}{zC_p(K_p^*, K_c) - 2u_0 S_p(K_p^*, K_c)}, \end{aligned} \tag{10.2.14}$$

with again the subscripts 'p' and 'e' denoting partial differentiation with respect to K_p and K_e respectively. Since $S(K_c, K_c) = 0$,

$$\triangle K_p \simeq \begin{cases} p_1(K_p^*, K_c)|\triangle K_e|^{1-\alpha}, & \text{when } K_p^* \neq K_c, \\ p_2(K_c, K_c)\triangle K_e, & \text{when } K_p^* = K_c. \end{cases} \tag{10.2.15}$$

---

[2] In the Ising model $\sigma = \alpha$ (Sect. 4.5.6).

Thus we have, when $K_\mathrm{p}^* \neq K_\mathrm{c}$, $\gamma_\mathrm{r} = \gamma/(1-\alpha)$. Similar calculations for the case $\Delta K_\mathrm{e} < 0$ ($T < T_\mathrm{c}$), yield $\gamma_\mathrm{r}' = \gamma'/(1-\alpha')$ and $\beta_\mathrm{r} = \beta/(1-\alpha')$.

To obtain the exponent α_r the asymptotic form for the heat capacity must be determined from (10.2.10). Again consider the two cases where K_p^* is or is not equal to K_c and, retaining only those terms which play a role in the asymptotic form for one of these cases

$$c(K_\mathrm{p}, K_\mathrm{e}, 0, \rho^*) \simeq -k_\mathrm{B} K_\mathrm{p}^2 \frac{(1-\alpha)\mathrm{B}_1(K_\mathrm{p}^*, K_\mathrm{c})|\Delta K_\mathrm{e}|^{-\alpha} + \mathrm{B}_2(K_\mathrm{p}^*, K_\mathrm{c})}{(1-\alpha)\mathrm{B}_3(K_\mathrm{p}^*, K_\mathrm{c})|\Delta K_\mathrm{e}|^{-\alpha} + \mathrm{B}_4(K_\mathrm{p}^*, K_\mathrm{c})}, \tag{10.2.16}$$

where

$$\begin{aligned}
\mathrm{B}_1(K_\mathrm{p}^*, K_\mathrm{c}) &:= \mathrm{u}_1 \left\{\tfrac{1}{2}\mathrm{z}[\mathrm{C}(K_\mathrm{p}^*, K_\mathrm{c})\mathrm{C}_\mathrm{p}(K_\mathrm{p}^*, K_\mathrm{c}) - \mathrm{S}(K_\mathrm{p}^*, K_\mathrm{c})\mathrm{S}_\mathrm{p}(K_\mathrm{p}^*, K_\mathrm{c})]\right.\\
&\qquad \left. + \mathrm{u}_0[\mathrm{S}(K_\mathrm{p}^*, K_\mathrm{c})\mathrm{C}_\mathrm{p}(K_\mathrm{p}^*, K_\mathrm{c}) - \mathrm{C}(K_\mathrm{p}^*, K_\mathrm{c})\mathrm{S}_\mathrm{p}(K_\mathrm{p}^*, K_\mathrm{c})]\right\},\\
\mathrm{B}_2(K_\mathrm{p}^*, K_\mathrm{c}) &:= \left(\tfrac{1}{4}\mathrm{z}^2 - \mathrm{u}_0^2\right)\left\{\mathrm{S}_\mathrm{p}(K_\mathrm{p}^*, K_\mathrm{c})\mathrm{C}_\mathrm{e}(K_\mathrm{p}^*, K_\mathrm{c})\right.\\
&\qquad \left. - \mathrm{S}_\mathrm{e}(K_\mathrm{p}^*, K_\mathrm{c})\mathrm{C}_\mathrm{p}(K_\mathrm{p}^*, K_\mathrm{c})\right\},\\
\mathrm{B}_3(K_\mathrm{p}^*, K_\mathrm{c}) &:= -\mathrm{u}_1\mathrm{S}(K_\mathrm{p}^*, K_\mathrm{c}),\\
\mathrm{B}_4(K_\mathrm{p}^*, K_\mathrm{c}) &:= \tfrac{1}{2}\mathrm{z}\mathrm{C}_\mathrm{e}(K_\mathrm{p}^*, K_\mathrm{c}) - \mathrm{u}_0\mathrm{S}_\mathrm{e}(K_\mathrm{p}^*, K_\mathrm{c}).
\end{aligned} \tag{10.2.17}$$

For $K_\mathrm{p}^* \neq K_\mathrm{c}$ (and $\Delta K_\mathrm{e} > 0$) it follows from (10.2.15) that $\mathrm{p}_1(K_\mathrm{p}^*, K_\mathrm{c}) > 0$ and, with

$$\theta := \frac{\Delta K_\mathrm{p}}{\mathrm{p}_1(K_\mathrm{p}^*, K_\mathrm{c})} > 0, \tag{10.2.18}$$

$$\begin{aligned}
\Delta c(\theta) &:= c(K_\mathrm{p}, K_\mathrm{e}, 0, \rho^*) - c(K_\mathrm{p}^*, K_\mathrm{e}, 0, \rho^*)\\
&\simeq -\frac{k_\mathrm{B} K_\mathrm{p}^2 \mathrm{D}(K_\mathrm{p}^*, K_\mathrm{c})}{\theta^{\alpha_\mathrm{r}}(1-\alpha)},
\end{aligned} \tag{10.2.19}$$

$$\frac{\mathrm{d}\Delta c(\theta)}{\mathrm{d}\theta} \simeq \frac{\alpha_\mathrm{r} k_\mathrm{B} K_\mathrm{p}^2 \mathrm{D}(K_\mathrm{p}^*, K_\mathrm{c})}{\theta^{\alpha_\mathrm{r}+1}(1-\alpha)},$$

where $-1 < \alpha_\mathrm{r} := -\alpha/(1-\alpha) < 0$ and

$$\mathrm{D}(K_\mathrm{p}^*, K_\mathrm{c}) := \frac{[\mathrm{B}_2(K_\mathrm{p}^*, K_\mathrm{c})\mathrm{B}_3(K_\mathrm{p}^*, K_\mathrm{c}) - \mathrm{B}_1(K_\mathrm{p}^*, K_\mathrm{c})\mathrm{B}_4(K_\mathrm{p}^*, K_\mathrm{c})]}{\mathrm{B}_3^2(K_\mathrm{p}^*, K_\mathrm{c})}. \tag{10.2.20}$$

In the case $\Delta K_e < 0$, α' and $p'_1(K_p^*, K_c)$, obtained by replacing u_1 by u'_1 in (10.2.15), replace α and $p_1(K_p^*, K_c)$ respectively. Now it must be the case that $p'_1(K_p^*, K_c) < 0$ and we obtain the expressions equivalent to (10.2.19) but with primed exponents. Collecting together all the exponents

$$\begin{aligned} \alpha_r &= -\frac{\alpha}{1-\alpha}, \qquad \gamma_r = \frac{\gamma}{1-\alpha}, \qquad \delta_r = \delta, \\ \alpha'_r &= -\frac{\alpha'}{1-\alpha'}, \qquad \gamma'_r = \frac{\gamma'}{1-\alpha'}, \qquad \beta_r = \frac{\beta}{1-\alpha'}. \end{aligned} \tag{10.2.21}$$

These transformations are called *exponent renormalization*. From (10.2.21) it is easily verified that, if the Ising exponents satisfy the Essam–Fisher scaling law (4.5.44) and the Widom scaling law (4.5.45), then

$$\begin{aligned} &\alpha'_r + 2\,\beta_r + \gamma'_r = 2 \\ &\gamma'_r = \beta_r(\delta_r - 1). \end{aligned} \tag{10.2.22}$$

These exponents also satisfy the scaling laws with $\alpha'_r = \alpha_r$ and $\gamma'_r = \gamma_r$ if (4.5.47) holds.

For $\rho = 1$, $K_p^* = K_c$ and it follows from (10.2.15) and (10.2.16) that the exponents revert to their Ising values. There are thus finite discontinuities in the exponents, since their limits as $\rho \to 1-0$ are not equal to their values at $\rho = 1$. It is remarkable that for any non-zero value of $1-\rho$, however small, there is a qualitative change in the properties of the model, the infinite discontinuity in the heat capacity being replaced by a cusp with an infinite slope at the critical point. However, the smaller the value of $1-\rho$, the narrower becomes the range around the critical point where the exponent changes have a significant numerical effect (Fisher and Scesney 1970).[3]

The exponent renormalization results from taking ρ as an independent variable which can change at constant T or remain constant when T changes. It is not simply a consequence of the existence of broken bonds. If the latter are regarded as defects with an activation energy $\Delta\varepsilon$, then $L_s = \Delta\varepsilon/T$ and K_e, as given by (10.2.2), depends only on T. Thus $\Delta K_e \sim \Delta K_p$ and no renormalization occurs. The concentration ρ in zero field becomes a function of T. In fact, it is more realistic physically to take ρ as an independent variable in the interpretation of this model as a mixture, where secondary sites are occupied/unoccupied by solute atoms than it is when the two states of a secondary site correspond to a bond which is unbroken/broken. In the mixture case, the mole fraction of solute is $\rho\, z/(2+z)$ which is a legitimate thermodynamic density.

[3] These authors used a different model, but the same general principles must apply here.

As we shall see in Sect. 12.2.7.3, in two-dimensional lattices the Ising model has $\alpha = 0$ with (10.2.12) replaced by[4]

$$u^{(\mathrm{IM})}(K_e, 0) = u_0 + u_1 \Delta K_e \ln|\Delta K_e|. \tag{10.2.23}$$

Then, when $K_p^* \neq K_c$, (10.2.15), (10.2.18) and (10.2.19) are replaced by

$$\theta \simeq -\Delta K_e \ln|\Delta K_e|, \tag{10.2.24}$$

and

$$\Delta c(\theta) \simeq -k_B K_p^2 \frac{D(K_p^*, K_c)}{\ln|\Delta K_e|}. \tag{10.2.25}$$

The appropriate solution to (10.2.24) is

$$\Delta K_e = -\frac{\theta}{\mathcal{W}(-\theta)}, \qquad \ln|\Delta K_e| = \mathcal{W}(-\theta), \tag{10.2.26}$$

where $\mathcal{W}$ is the branch of the Lambert W function chosen so that $\lim_{\theta\to 0} \mathcal{W}(-\theta) = -\infty$ and for which it is the case that $\lim_{\theta\to 0} \sqrt{\theta}\, \mathcal{W}(-\theta) = 0$.[5] Then

$$\left.\begin{aligned} \Delta c(\theta) &\simeq -k_B K_p^2 \frac{D(K_p^*, K_c)}{\mathcal{W}(-\theta)} \to 0, \\ \frac{d\Delta c(\theta)}{d\theta} &\simeq k_B K_p^2 \frac{D(K_p^*, K_c)}{\theta[\mathcal{W}(-\theta)]^2} \to \infty, \end{aligned}\right\} \quad \text{as } \theta \to 0. \tag{10.2.27}$$

There is thus a cusp with vertical tangents at the transition point, as in the $\alpha > 0$ case.

Renormalization relations like (10.2.21) were derived by Essam and Garelick (1967) for the *Syozi dilution model* (Syozi 1972; Essam 1972), which is slightly more elaborate than the one treated here. Widom (1967) derived the expression $\gamma/(1-\alpha)$ for the compressibility exponent in a decorated mixture model with the density of one component fixed. Fisher (1968) used thermodynamic arguments, in conjunction with examples from statistical mechanics, to establish general conditions for exponent renormalization. These were the existence of a 'hidden variable' subject to a constraint, for instance the presence of an equilibrium distribution of a fixed

---

[4] The result in Sect. 12.2.7.3 is just for the square lattice, but it was shown by Houtappel (1950) that (10.2.23) also applied to the triangular and honeycomb lattices.

[5] The Lambert W function (see e.g. Corless et al. 1996) is the multi-valued inverse of the function $z\exp(z)$.

number of impurity atoms. 'Ideal' exponents are obtained when the largest possible number of fields are taken as independent variables, that is $\eta = n - 1$, and are renormalized when a field is replaced as an independent variable by a density.

10.3 A Ferrimagnet

In this type of model, which has been discussed by Syozi and Nakono (1955), Syozi (1972) and Bell (1974b), both the primary p and secondary s sites are occupied by Ising spins but they have different magnet moments νm_0 and m_0 respectively. This is equivalent to taking the magnetic field as $\nu\mathcal{H}$ on primary sites and $\mathcal{H}$ on secondary sites. The other particular feature of this model is that the primary–primary interaction between first-neighbour pairs is ferromagnetic, but the interaction between a secondary spin and its two neighbouring primary spins is antiferromagnetic. The Hamiltonian is (10.1.1) with[6]

$$\mathring{\mathcal{K}e}(\sigma(\boldsymbol{r}), \sigma(\boldsymbol{r}'); \sigma(\boldsymbol{r}, \boldsymbol{r}')) = \mathcal{K}_\mathrm{s}\sigma(\boldsymbol{r}, \boldsymbol{r}')[\sigma(\boldsymbol{r}) + \sigma(\boldsymbol{r}')] - \mathcal{L}_\mathrm{s}\sigma(\boldsymbol{r}, \boldsymbol{r}'), \qquad (10.3.1)$$

and $\mathcal{K}_\mathrm{s} := J_\mathrm{s}/T$, $\mathcal{L}_\mathrm{s} := \mathcal{H}/T$, where $J_s > 0$ is the p–s interaction energy. We shall also, in the interests of conformity, denote the direct p–p Ising interaction, as it appears in (10.1.1), by J_p rather than J with $\mathcal{K}_\mathrm{p} := J_\mathrm{p}/T$, replacing $\mathcal{L}$ by $\mathcal{L}_\mathrm{p} := \nu\mathcal{H}/T$.

We can now carry out a ground-state analysis similar to that for the antiferromagnet in Sect. 3.5.1.1. The number of p sites is N and the number of s sites is $\frac{1}{2}\mathrm{z}N$. In the paramagnetic ground state, which will give the lowest Hamiltonian value for large $\mathcal{H}$, all spins are aligned with the field. The other ground state will be one in which the p spins and s spins have an antiparallel alignment. Since the number of p and s spins differ there will be, unlike in the case of antiferromagnetism, a residual magnetic moment and this phenomenon is call *ferrimagnetism*. The ferrimagnetic ground state has a degeneracy of two for zero field and for small $\mathcal{H}$ the spins on s sites will be aligned with the field if $\frac{1}{2}\mathrm{z} > \nu$, which we shall assume unless otherwise stated. Thus we have

$$\begin{aligned} \widehat{H}_\mathrm{P} &= \tfrac{1}{2}\mathrm{z}NJ_\mathrm{p} - \mathrm{z}NJ_\mathrm{s} - N\left(\tfrac{1}{2}\mathrm{z} + \nu\right)|\mathcal{H}|, \\ \widehat{H}_\mathrm{FI} &= \tfrac{1}{2}\mathrm{z}NJ_\mathrm{p} + \mathrm{z}NJ_s - N\left(\tfrac{1}{2}\mathrm{z} - \nu\right)|\mathcal{H}| \end{aligned} \qquad (10.3.2)$$

and $\widehat{H}_\mathrm{FI} < \widehat{H}_\mathrm{P}$, making the ferrimagnetic state stable if

$$-\mathcal{H}_\mathrm{c} < \mathcal{H} < \mathcal{H}_\mathrm{c} := \mathrm{z}J_\mathrm{s}/\nu. \qquad (10.3.3)$$

[6] We use $\sigma(\boldsymbol{r}, \boldsymbol{r}')$ rather that $\tau(\boldsymbol{r}, \boldsymbol{r}')$ for the secondary site variable since it is an Ising spin variable.

Equations (10.1.5)–(10.1.8) now apply with $\mathring{K}$ replaced by the pair $(K_\mathrm{s}, L_\mathrm{s})$ and

$$\exp(4K_\mathrm{e}) = \frac{\exp(4K_\mathrm{p})\cosh(2K_\mathrm{s}+L_\mathrm{s})\cosh(2K_\mathrm{s}-L_\mathrm{s})}{\cosh^2(L_\mathrm{s})}, \tag{10.3.4}$$

$$\exp(4L_\mathrm{e}) = \exp(4L_\mathrm{p})\left[\frac{\cosh(2K_\mathrm{s}-L_\mathrm{s})}{\cosh(2K_\mathrm{s}+L_\mathrm{s})}\right]^{\mathrm{z}}, \tag{10.3.5}$$

$$Z(\mathring{\mathcal{N}}, K_\mathrm{p}, K_\mathrm{s}, L_\mathrm{p}, L_\mathrm{s}) = [16\cosh(2K_\mathrm{s}+L_\mathrm{s})\cosh(2K_\mathrm{s}-L_\mathrm{s})\cosh^2(L_\mathrm{s})]^{\frac{1}{8}N\mathrm{z}} \times Z^{(\mathrm{IM})}(\mathcal{N}, K_\mathrm{e}, L_\mathrm{e}). \tag{10.3.6}$$

Then the magnetization densities m_p and m_s on the p and s sites are

$$\begin{aligned} m_\mathrm{p} &= -\left(\frac{\partial\phi(\mathring{\mathcal{N}}, K_\mathrm{p}, K_\mathrm{s}, L_\mathrm{p}, L_\mathrm{s})}{\partial L}\right)_{K_\mathrm{p},K_\mathrm{s},L_\mathrm{s}} \\ &= -\left(\frac{\partial\phi(\mathring{\mathcal{N}}, K_\mathrm{p}, K_\mathrm{s}, L, L_\mathrm{s})}{\partial L_\mathrm{e}}\right)_{K_\mathrm{p},K_\mathrm{s},L_\mathrm{s}} = m^{(\mathrm{IM})}(K_\mathrm{e}, L), \end{aligned} \tag{10.3.7}$$

$$m_\mathrm{s} = -\frac{2}{\mathrm{z}}\left(\frac{\partial\phi(\mathring{\mathcal{N}}, K_\mathrm{p}, K_\mathrm{s}, L, L_\mathrm{s})}{\partial L_\mathrm{s}}\right)_{K_\mathrm{p},K_\mathrm{s},L} = f(K_\mathrm{e}, K_\mathrm{s}, L_\mathrm{e}, L_\mathrm{s}), \tag{10.3.8}$$

where

$$\begin{aligned} f(K_\mathrm{e}, K_\mathrm{s}, L_\mathrm{e}, L_\mathrm{s}) &= \tanh(L_\mathrm{s}) \\ &+ \tfrac{1}{4}\left[1-\tfrac{2}{\mathrm{z}}\mathrm{u}^{(\mathrm{IM})}(K_\mathrm{e}, L_\mathrm{e})\right][\tanh(2K_\mathrm{s}+L_\mathrm{s})-\tanh(2K_\mathrm{s}-L_\mathrm{s})-2\tanh(L_\mathrm{s})] \\ &- \tfrac{1}{2}m^{(\mathrm{IM})}(K_\mathrm{e}, L_\mathrm{e})[\tanh(2K_\mathrm{s}+L_\mathrm{s})+\tanh(2K_\mathrm{s}-L_\mathrm{s})]. \end{aligned} \tag{10.3.9}$$

The overall magnetization density is given by

$$m = \frac{2\nu m_\mathrm{p} + \mathrm{z}m_\mathrm{s}}{2\nu + \mathrm{z}}. \tag{10.3.10}$$

We need dimensionless variables to measure temperature and magnetic field. These we take to be $T/J_\mathrm{s} = 1/K_\mathrm{s}$ and $\mathcal{H}/J_\mathrm{s} = L_\mathrm{s}/K_\mathrm{s}$. We already have the relationship $L_\mathrm{p} = \nu L_\mathrm{p}$ and now we take $\xi := J_\mathrm{p}/J_\mathrm{s}$, so that $K_\mathrm{p} = \xi K_\mathrm{s}$. It is clear that this model retains the $\mathcal{H} \leftrightarrow -\mathcal{H}$ symmetry of the Ising model, so attention is restricted to the region $\mathcal{H} \geq 0$. As $T \to 0$, $m \to (\mathrm{z}-2\nu)/(\mathrm{z}+2\nu)$ when $\mathcal{H} < \mathcal{H}_\mathrm{c}$ and $m \to 1$ when $\mathcal{H} > \mathcal{H}_\mathrm{c}$.

10.3.1 The Zero-Field Axis

From (10.3.5), $L_\mathrm{p} = L_\mathrm{s} = 0$ implies $L_\mathrm{e} = 0$ and, from (10.3.4),

$$\exp(2K_\mathrm{e}) = \exp(2\xi K_\mathrm{s})\cosh(2K_\mathrm{s}), \tag{10.3.11}$$

with the spontaneous magnetization given by

$$m_{\mathrm{s}} = \pm m_{\mathrm{s}}^{(\mathrm{IM})}(K_{\mathrm{e}}) \left(\frac{2\nu - z\tanh(2K_{\mathrm{S}})}{2\nu + z} \right), \tag{10.3.12}$$

where $m_{\mathrm{s}}^{(\mathrm{IM})}(K_{\mathrm{e}})$ (>0) is the spontaneous magnetization of the Ising ferromagnet. Since $\mathcal{H} \to +0$ the $\pm$ sign must be chosen to make $m_{\mathrm{s}} > 0$. The critical temperature, above which the spontaneous magnetization is zero, is given by solving (10.3.11) with $K_{\mathrm{e}} = K_{\mathrm{c}}$, the critical parameter value for the Ising model. In particular, for the triangular lattice, from (9.4.31), $K_{\mathrm{c}} = \ln(3)/4 = 0.2747$. Then, when $\xi = 1$ ($J_{\mathrm{P}} = J_{\mathrm{S}}$), $T_{\mathrm{c}} = 4.4353\, J_{\mathrm{S}}$ and, when $\xi = 0$ ($J_{\mathrm{P}} = 0$), $T_{\mathrm{c}} = 1.7449$. When $T > T_{\mathrm{c}}$, $K_{\mathrm{e}} < K_{\mathrm{c}}$ so that $m_{\mathrm{s}}^{(\mathrm{IM})}(K_{\mathrm{e}}) = 0$ while, as T decreases from T_{c} to 0, $\tanh(2K_{\mathrm{S}})$ increases from $\tanh(2J_{\mathrm{S}}/T_{\mathrm{c}})$ to 1. The following cases can be distinguished:

(i) $\nu < \nu_0$, (ii) $\nu = \nu_0$, (iii) $\frac{1}{2}z > \nu > \nu_0$,

(iv) $\nu = \frac{1}{2}z$, (v) $\nu > \frac{1}{2}z$,

where

$$\nu_0 := \tfrac{1}{2}z\tanh(2J_{\mathrm{S}}/T_{\mathrm{c}}). \tag{10.3.13}$$

For (i)–(iii), $\nu < \frac{1}{2}z$ so that, in accordance with our discussion of the ground state,

$$m_{\mathrm{P}} = -m_{\mathrm{s}}^{(\mathrm{IM})} = -1, \qquad m_{\mathrm{s}} = \frac{z - 2\nu}{z + 2\nu} > 0. \tag{10.3.14}$$

For (iv), $m_{\mathrm{s}} = 0$ at $T = 0$ and, for (v),

$$m_{\mathrm{P}} = +m_{\mathrm{s}}^{(\mathrm{IM})} = +1, \qquad m_{\mathrm{s}} = \frac{2\nu - z}{2\nu + z} > 0. \tag{10.3.15}$$

The most interesting case is (iii), where there is a *compensation temperature* $T_{\mathrm{comp}} < T_{\mathrm{c}}$, given by

$$\nu = \tfrac{1}{2}z\tanh(2J_{\mathrm{S}}/T_{\mathrm{comp}}), \tag{10.3.16}$$

where $m_{\mathrm{s}} = 0$, because the first factor on the right-hand side of (10.3.12) is zero. For $T_{\mathrm{comp}} < T < T_{\mathrm{c}}$,

$$m_{\mathrm{P}} = +m_{\mathrm{s}}^{(tI\mathrm{M})}(K_{\mathrm{e}}) > 0, \qquad m_{\mathrm{S}} < 0, \tag{10.3.17}$$

but, for $0 < T < T_{\mathrm{comp}}$,

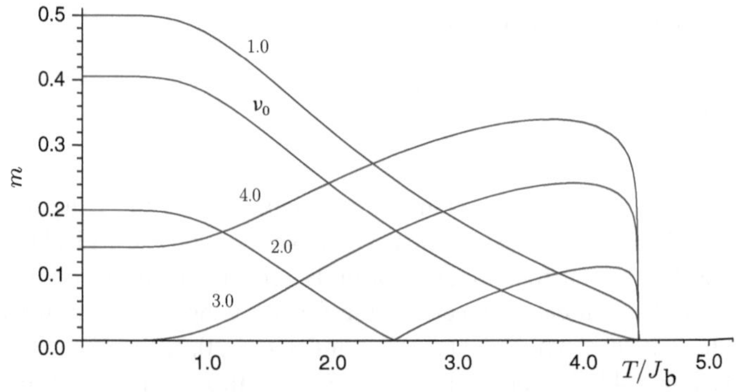

Fig. 10.2 Curves of magnetization density as a function of temperature along the zero-field axis for the triangular lattice with $\xi = 1$, parameterized by the value of ν. From (10.3.13) $\nu_0 = 1.26710$ and for the *curve* $\nu = 2$, from (10.3.16), $T_{\text{comp}} = 2.4853\, J_s$

$$m_p = -m_s^{(\text{IM})}(K_e) < 0, \qquad m_s > 0. \tag{10.3.18}$$

Both m_p and m_s change sign discontinuously at $T = T_{\text{comp}}$ although $|m_p|$ and $|m_s|$ are continuous. The magnetization density (10.3.12) can be calculated for the square, honeycomb and triangular lattices using the formula for the Ising model spontaneous magnetization given in Sect. 12.3.9. Such curves for the triangular lattice with $\xi = 1$ are given in Fig. 10.2.

Since $m_s^{(\text{IM})}(K_e) \sim |\Delta K_e|^\beta$, as $\Delta K_e = K_e - K_c \to 0$, β being the Ising exponent, (10.3.12) gives $m_s \sim (T_c - T)^\beta$, as $T \to T_c - 0$, except in case (ii) when $r - \frac{1}{2}z\tanh(2K_s) \sim T_c - T$, as $T \to T_c - 0$, yielding $m_s \sim (T_c - T)^{\beta+1}$. Since $0 < \beta < 1$ for both $d = 2$ and $d = 3$ lattices it follows that the plot of m_s against T at $\mathcal{H} = 0$ has a vertical tangent at $T = T_c$ except for case (ii) where the tangent is horizontal (see Fig. 10.2). The zero-field susceptibility can be deduced from (10.3.9).

10.3.2 Non-Zero Field

In addition to the segment $[0, T_c]$ of the zero-field axis, where m undergoes a discontinuous change of sign, there are first-order transition lines on which $|\mathcal{H}| \neq 0$ and for which exact relations can be derived (Bell 1974b). This arises from the fact that the mapping $(T, \mathcal{H}) \to (K_e^{-1}, L_e)$ from the plane of this ferrimagnet into the plane of the Ising model is not one-to-one. Equation (10.3.5), with $L_e = 0$, is satisfied by $L_s = 0$, but after some rearrangement it also yields

$$\exp(4K_s) = \frac{\sinh[(1 + 2\nu/z)L_s]}{\sinh[(1 - 2\nu/z)L_s]}, \tag{10.3.19}$$

which has a solution which gives a curve $\mathcal{L}(\nu)$ in the T–$\mathcal{H}$ plane when $\nu < \frac{1}{2}z$. This curve is clearly symmetrical about $\mathcal{H} = 0$ and meets the zero-temperature axis at $\pm\mathcal{H}_c$, where $\mathcal{H}_c$ is the critical field given by (10.3.3). It cuts the zero-field axis orthogonally at a temperature T_I, which, by taking the limit of the right-hand side of (10.3.19) as $L_s \to 0$, is given as the solution of

$$\nu = \frac{1}{2}z\tanh(2K_s). \qquad (10.3.20)$$

For case (i), $T_I > T_c$ and, as can be seen from (10.3.16), for case (ii), $T_I = T_{comp} = T_c$ and, for case (iii), $T_I = T_{comp} < T_c$. From (10.3.4) the curve $\mathcal{C}(\xi)$ of points in the plane T–$\mathcal{H}$ which map into the Ising critical point $(K_c^{-1}, 0)$ is given by

$$\exp[4(K_c - \xi K_s)] - 1 = \frac{\sinh^2(2K_s)}{\cosh^2(L_s)}. \qquad (10.3.21)$$

This curve meets the zero-field axis at $T = T_c$ and, when $\xi \neq 0$, has the infinite-field asymptote $T = J_s/K_c$. When $\xi = 0$ the curve meets the zero-temperature axis $\mathcal{H} = J_s$. The curves $\mathcal{L}(\nu)$ are independent of ξ and the curves $\mathcal{C}(\xi)$ are independent of ν. The intersection of an $\mathcal{L}(\nu)$ and a $\mathcal{C}(\xi)$, if it exists for particular values of ν and ξ, is the mapping of the Ising critical point $(K_c^{-1}, 0)$ and it corresponds to the end-point of a line of first-order transitions along $\mathcal{L}(\nu)$. The case of the triangular lattice with $\xi = 1$, shown in Fig. 10.3a, is typical of the situation where $\xi > 0$ and Fig. 10.3b is the special case $\xi = 0$, where there is no Ising interaction between primary p sites. The first-order transition lines in the T–$\mathcal{H}$ plane are the images of $(K_e^{-1}, 0)$ for $K_e^{-1} < K_c^{-1}$, so lie below the curve $\mathcal{C}(\xi)$.

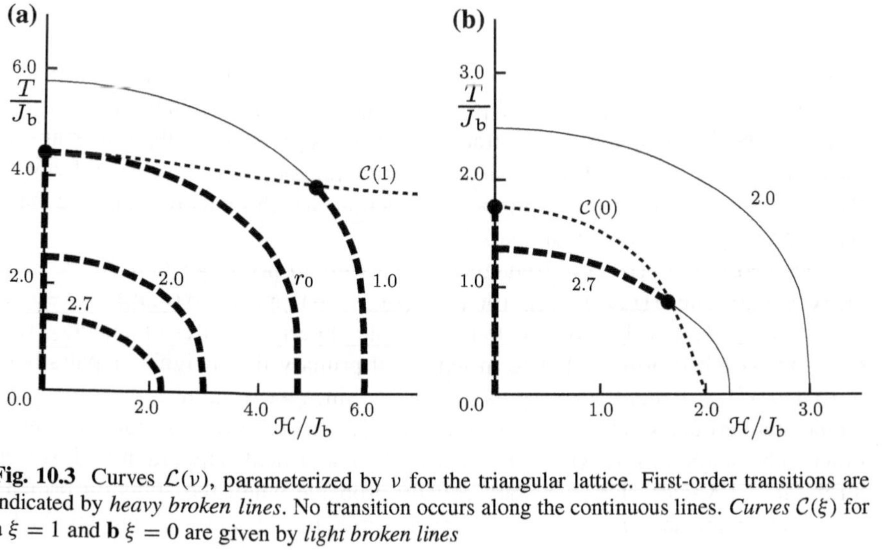

Fig. 10.3 Curves $\mathcal{L}(\nu)$, parameterized by ν for the triangular lattice. First-order transitions are indicated by *heavy broken lines*. No transition occurs along the continuous lines. *Curves* $\mathcal{C}(\xi)$ for **a** $\xi = 1$ and **b** $\xi = 0$ are given by *light broken lines*

For $\xi \neq 0$ (Fig. 10.3a) the curve $\nu = 1$ is an example of case (i), where there are two first-order transition lines which are mirror images of each other in the zero-field axis. The curves $\nu = 2$ and $\nu = 2.7$ are examples of case (iii). They lie entirely in the $K_e > K_c$ region and yield first-order transition lines crossing the $[0, T_c]$ segment on the zero-field at $T = T_{comp}$. The compensation point is thus the intersection of two lines of first-order transitions. The configurational energy is continuous over the segment $[0, T_c]$ of the zero-field axis but, together with the magnetization m, it is discontinuous at first-order transition points for which $\mathcal{H} \neq 0$.

For $\xi = 0$ case (i) curves like $\nu = 1$ contain no transitions except on the zero-field axis. Only case (iii) curves like $\nu = 2$ and $\nu = 2.7$ (Fig. 10.3b) intersect $\mathcal{C}(0)$. The first-order transition segment $[0, T_c]$ on the zero-field axis is crossed at the compensation point by another first-order transition line on $\mathcal{L}(\nu)$. This transition line terminates at two critical points in the regions $\mathcal{H} \gtrless 0$, which are reached by *reducing* the temperature from T_{comp}. They are examples of *lower critical points*, which will be encountered and described in more detail in the next section.

10.4 A Competing-Interaction Magnetic Model

Here we consider only the case $\mathcal{H} = 0$. The primary sites form a spin-$\frac{1}{2}$ *antiferromagnetic* Ising model and there is a ferromagnetic primary-secondary interaction with an Ising spin of magnitude $\frac{1}{2}q$ on the secondary sites, where q can take any positive integer value. The secondary-site variable $\sigma(\boldsymbol{r}, \boldsymbol{r}') = q, q-2, \ldots, -q$ and the secondary-site interaction is $(J/q) > 0$, where the scaling factor q is introduced so that the energies of maximum alignment are $\pm J$. The Hamiltonian is (10.1.1) with

$$\mathring{K}\mathfrak{e}(\sigma(\boldsymbol{r}), \sigma(\boldsymbol{r}'); \sigma(\boldsymbol{r}, \boldsymbol{r}')) = -(K/q)\sigma(\boldsymbol{r}, \boldsymbol{r}')[\sigma(\boldsymbol{r}) + \sigma(\boldsymbol{r}')], \tag{10.4.1}$$

and $K := J/T$ $(J > 0)$, with L, as it appears in (10.1.1), set to zero and K replaced by $-\nu K (1 > \nu > 0)$. Although the primary–secondary interaction term in (10.4.1) is taken to be ferromagnetic, an interaction of either sign tends to align the spins on the primary sites $\boldsymbol{r}$ and $\boldsymbol{r}'$. The primary–secondary interaction is thus in conflict with the antiferromagnetic primary–primary interaction, which tends to make the spins on the primary sites $\boldsymbol{r}$ and $\boldsymbol{r}'$ antiparallel.

The energy of a perfect ferromagnetic state with all primary site $\sigma(\boldsymbol{r}) = 1$ and all secondary site $\sigma(\boldsymbol{r}, \boldsymbol{r}') = q$ (or all $\sigma(\boldsymbol{r}) = -1$ and all $\sigma(\boldsymbol{r}, \boldsymbol{r}') = -q$) is $-\frac{1}{2}zN(2-\nu)J$. It is also possible to have a state of perfect antiferromagnetic order on a loose-packed primary lattice in which all primary first-neighbour pairs have $\sigma(\boldsymbol{r}) + \sigma(\boldsymbol{r}') = 0$ and the energy is $-\frac{1}{2}zN\nu J$. Since we assume $\nu < 1$, the zero-temperature ground state is ferromagnetic. We thus expect ferromagnetic long-range order to be stable in a range $(0, T_c)$, where T_c is a critical temperature. However, all the $q+1$ values of $\sigma(\boldsymbol{r}, \boldsymbol{r}')$ are equally probable when the spins on primary sites $\boldsymbol{r}$ and $\boldsymbol{r}'$ are antiparallel, contributing a term $k_B \ln(q+1)$ to the entropy. States with primary spins antiferromagnetically ordered have a high entropy. Ordered states with this unusual property can be destabilized by either an increase or a decrease in temperature but may be stable in a range $(T_{\rm lc}, T_{\rm uc})$, where $T_{\rm uc}$ and $T_{\rm lc}$ are respectively termed the *upper* and *lower critical temperatures*. For appropriate parameter values, this situation exists for the present model with the antiferromagnetic lower critical temperature $T_{\rm lc} > T_{\rm uc}$. From (10.1.4) and (10.4.1),

$$\begin{aligned}\phi(K; \pm 1, \pm 1) &= \exp(-\nu K)\sum_{\ell=0}^{q} \exp[(q-2\ell)2K/q]\\ &= \exp(-\nu K)\frac{\sinh[2(q+1)K/q]}{\sinh(2G/q)},\end{aligned} \tag{10.4.2}$$

$$\phi(K; +1, -1) = (q+1)\exp(\nu K)$$

and, substituting into (10.1.6) and (10.1.7),

$$\begin{aligned}\exp(2K_{\rm e}) &= \exp(-2\nu K)\frac{\sinh[2(q+1)K/q]}{(q+1)\sinh(2K/q)},\\ L_{\rm e} &= 0.\end{aligned} \tag{10.4.3}$$

From (10.4.3),

$$\exp(2K_{\rm e}) \simeq \frac{\exp[2(1-\nu)K]}{q+1}, \qquad \text{as } K \to \infty. \tag{10.4.4}$$

Hence $K_{\rm e} \to \infty$, as $T \to 0$, in accordance with the previous deduction of a ferromagnetic ground state. However, $K_{\rm e}$ does not increase monotonically with K since it follows, from (10.4.3), that

$$K_{\rm e} \simeq -\nu K + \frac{q+2}{3q}K^2, \qquad \text{as } K \to 0. \tag{10.4.5}$$

For small K, $K_{\rm e} < 0$ but, for large K, $K_{\rm e} > 0$. It is not difficult to show, from (10.4.3), that $\mathrm{d}^2K_{\rm e}/\mathrm{d}K^2 > 0$ for all $K > 0$ and it follows that $K_{\rm e}$ decreases from 0 as K increases from 0 to reach a unique minimum $K_{\rm e} = K_{\rm min} < 0$ at $K = J/T_{\rm min}$. Then $K_{\rm e}$ increases monotonically, passing through 0 at $K = J/T_0$ and through $K_{\rm c}$ (the critical value for the primary lattice Ising ferromagnet) at $K = J/T_{\rm c}$ $(T_{\rm c} < T_0)$ to approach infinity with K as $K \to \infty$.

From the above it follows that $K_{\rm e} > K_{\rm c}$ for the temperature interval $(0, T_{\rm c})$, giving long-range ferromagnetic order when $T < T_{\rm c}$. If $K_{\rm min} < -K_{\rm c}$ then $K_{\rm e} = -K_{\rm c}$ at temperatures $T_{\rm uc}$ and $T_{\rm lc}$, with $T_0 < T_{\rm lc} < T_{\rm min} < T_{\rm uc}$. Thus $K_{\rm e} < -K_{\rm c}$ in the interval $(T_{\rm lc}, T_{\rm uc})$. Since we have specified a loose-packed primary lattice, there is long-range antiferromagnetic order when $T_{lc} < T < T_{uc}$. It follows that T_{uc} and T_{lc} are respectively upper and lower critical points. The value of K_{min}, which depends on q and ν and which is given by

$$\frac{dK_e}{dK} = -\nu + \frac{q+1}{q}\coth\left[\frac{2(q+1)K}{q}\right] - \frac{1}{q}\coth\left(\frac{2K}{q}\right) = 0, \tag{10.4.6}$$

is crucial. It can be shown from (10.4.6) that, with a fixed value of q, K_{min} has a greatest lower bound equal to $-\frac{1}{2}\ln(q+1)$ for the range $0 < \nu < 1$. This bound is approached as $\nu \to 1-0$. Hence antiferromagnetic critical points occur over some range of ν if $\frac{1}{2}\ln(q+1) > K_c$, where K_c, of course, depends only on the primary lattice. With $q = 1$ this condition is satisfied for the $d = 3$ simple cubic and body-centred cubic lattices. With $q = 2$ it is also satisfied for the square and diamond lattices, and with $q = 3$ for the honeycomb lattice as well (Table 3.1). As an example, with a square primary lattice, $q = 3$ and $\nu = 0.8$ we have $T_{uc} = 1.0150\,J$, $T_{lc} = 0.4387\,J$, $T_0 = 0.1443\,J$ and $T_c = 0.0882\,J$. For the general case, since $dK_e/dK \neq 0$ when $K_e = \pm K_c$, $\Delta K \sim \Delta K_e$ and the heat capacity singularity at each critical point is of similar type to that at the critical point of the primary lattice Ising model. The $q = 1$ form of the present model was investigated by Nakono (1968) and Syozi (1972) discusses related models.

The temperature $T = T_0$, where K_e changes sign, is of some interest. Since $K_e = 0$ at $T = T_0$ the primary lattice spins are completely disordered, as they are at $T = \infty$. Suppose that $\langle\sigma(\mathbf{0})\sigma(\boldsymbol{r})\rangle$ is the correlation between primary lattice spins whose relative position is represented by $\boldsymbol{r}$. As the temperature increases through T_0, K_e changes from positive to negative and $\langle\sigma(\mathbf{0})\sigma(\boldsymbol{r})\rangle$ changes from a positive quantity which decreases with $|\boldsymbol{r}|$ to one which alternates in sign, though its magnitude still decreases with $|\boldsymbol{r}|$. At $T = T_0$ itself $\langle\sigma(\mathbf{0})\sigma(\boldsymbol{r})\rangle = 0$ for all $\boldsymbol{r}$. This is an example of a *disorder point*. Early work on these is discussed by Stephenson (1970), who cites the $d = 1$, $q = 1$ case of the present model as one example (see also Georges et al. 1986) As in the present model, disorder points in other cases are due to competing interactions which promote different kinds of short-range order. Disorder points in the eight-vertex model are discussed Sect. 9.3.2.2.

10.5 Decoration with Orientable Molecules

This section is concerned with a binary mixture which has closed-loop coexistence curves in the composition–temperature plane and thus both upper and lower critical solution temperatures. Phenomena of this type occur in real mixtures, for example glycerol-ethylbenzylamine and water-nicotine.

A primary or secondary site can be occupied either by a molecule of component A with chemical potential μ_A or one of component B with chemical potential μ_B. The Ising variable $\sigma(\boldsymbol{r})$ is used with $\sigma(\boldsymbol{r}) = +1$ indicating an A occupation and $\sigma(\boldsymbol{r}) = -1$ indicating a B occupation. We use the binary mixture variables given in

Sect. 3.5.1.3 and the Ising/binary mixture equivalence given by (3.5.27) and (3.5.28). However, we wish to impose symmetry between the two components and this can be achieved, without loss of generality, by taking $\varepsilon_{AA} = \varepsilon_{BB} = 0$ and $\varepsilon_{AB} = \varepsilon$. Thus, from (3.5.28), for the couplings of (10.1.1) we have $K := \varepsilon/(2T)$ and $L := (\mu_A - \mu_B)/(2T)$. The constant coupling $C := [\varepsilon - (2/z)(\mu_A + \mu_B)]/(2T)$, given by (3.5.27), features only in the calculation of the partition function. The existence of a lower critical point is likely to be an entropic phenomenon, as we saw in Sect. 10.4. To achieve the necessary high entropy it is postulated that each A or B molecule on a secondary site has ω distinguishable orientation states, labelled $1, 2, \ldots, \omega$. In state 1 an active *arm* is directed towards the molecule on one of the adjacent primary sites and ε'_{AA}, ε'_{AB} or ε'_{BB} is added to ε_{AA}, ε_{AB} or ε_{BB} according to the A or B occupations of the secondary and the respective primary site. In state 2 the same thing is the case with respect to the other primary site. In the other $\omega - 2$ states the secondary site molecule is said to be *disoriented* and there is no additional energy contribution. Let $\varepsilon'_{AA} = \varepsilon'_{BB} = -\xi\varepsilon$ and $\varepsilon'_{AB} = -\nu\varepsilon$. We again use the Ising equivalence of Sect. 3.5.1.3, but now must include the constant contribution since the disoriented states set the energy zero. Thus, from (3.5.27) and (3.5.28), an addition factor $(\xi - \nu)K$ must be added to the Ising term together with a constant $(\xi + \nu)K$ (unaffected by the spin orientations) for a primary–secondary pair with an active arm. We attach to the secondary site $(\boldsymbol{r}, \boldsymbol{r}')$ two variables $\sigma(\boldsymbol{r}, \boldsymbol{r}') = \pm 1$ signifying, as for the primary sites, A or B occupation and $\tau(\boldsymbol{r}, \boldsymbol{r}') = 1, 2, \ldots, \omega$, which specifies the orientational state. The orientations of the molecules on the primary sites are taken as fixed so that these sites have only two distinguishable states, occupation by A or B respectively. The Hamiltonian is (10.1.1) with

$$\begin{aligned}\mathring{\mathcal{K}}(\sigma(\boldsymbol{r}), \sigma(\boldsymbol{r}'); \sigma(\boldsymbol{r}, \boldsymbol{r}'), \tau(\boldsymbol{r}, \boldsymbol{r}')) &= -L\sigma(\boldsymbol{r}, \boldsymbol{r}') \\ &-K\{1 + (\xi - \nu)\delta^{\mathrm{Kr}}(\tau(\boldsymbol{r}, \boldsymbol{r}') - 1)\}\sigma(\boldsymbol{r}, \boldsymbol{r}')\sigma(\boldsymbol{r}) \\ &-K\{1 + (\xi - \nu)\delta^{\mathrm{Kr}}(\tau(\boldsymbol{r}, \boldsymbol{r}') - 2)\}\sigma(\boldsymbol{r}, \boldsymbol{r}')\sigma(\boldsymbol{r}') \\ &-K(\xi + \nu)\left\{\delta^{\mathrm{Kr}}(\tau(\boldsymbol{r}, \boldsymbol{r}') - 1) + \delta^{\mathrm{Kr}}(\tau(\boldsymbol{r}, \boldsymbol{r}') - 2)\right\}. \end{aligned} \tag{10.5.1}$$

From (10.1.4), summing over the values of both $\sigma(\boldsymbol{r}, \boldsymbol{r}')$ and $\tau(\boldsymbol{r}, \boldsymbol{r}')$ gives

$$\begin{aligned}\phi(K, L; \pm 1, \pm 1) &= \exp(2K + L)[\omega - 2 + 2\exp(2K\xi)] \\ &\quad + \exp(-2K - L)[\omega - 2 + 2\exp(2K\nu)], \end{aligned} \tag{10.5.2}$$

$$\phi(K, L; +1, -1) = 2\cosh(L)[\omega - 2 + \exp(2K\xi) + \exp(2K\nu)].$$

We are interested in phase equilibrium which corresponds in the equivalent Ising model to $L_e = 0$. From (10.5.2) and (10.1.7) this is the case if $L = 0$, that is $\mu_A = \mu_B$. However, we also need the mole fractions x_A and x_B. To obtain these we must compute the partition function for general μ_A and μ_B, using (10.1.5) and (10.1.8), where in the later $g(K, L)$ is, as indicated above, replaced by $\exp(C)g(K, L)$. Now differentiating the resulting free energy with respect to μ_A and μ_B and setting $\mu_A = \mu_B$ yields

$$x_{\mathsf{A}} = \tfrac{1}{2} \pm m_{\mathrm{s}}^{(\mathrm{IM})}(K_{\mathrm{e}})\, X(K), \qquad x_{\mathsf{B}} = \tfrac{1}{2} \mp m_{\mathrm{s}}^{(\mathrm{IM})}(K_{\mathrm{e}})\, X(K), \tag{10.5.3}$$

where

$$X(K) = \frac{1}{z+2}\left\{1 + \frac{z}{2}\,\frac{(\omega-2)\sinh(2K) + \exp[2(\xi+1)K] - \exp[2(\nu-1)K]}{(\omega-2)\cosh(2K) + \exp[2(\xi+1)K] + \exp[2(\nu-1)K]}\right\} \tag{10.5.4}$$

and, by substituting from (10.5.2) into (10.1.6),

$$\exp(2K_{\mathrm{e}}) = \frac{(\omega-2)\cosh(2K) + \exp[2(\xi+1)K] + \exp[2(\nu-1)K]}{\omega - 2 + \exp(2\xi K) + \exp(2\nu K)}. \tag{10.5.5}$$

We now assume that $\varepsilon > 0$, $\xi > 0$ and $\nu > 0$. Then the possibly stable ground states are coexistence between phases all with primary–secondary–primary triplets with one of the occupations:

I AA~B or A~BB with energy $-(\nu-1)\varepsilon$,
II AA~A or B~BB with energy $-\xi\varepsilon$,

where '~' represents the presence of an active arm. We now explore the properties of the curves, derived from (10.5.5), of K_{e} as a function of T. Their limiting properties are given by

$$K_{\mathrm{e}} \simeq \begin{cases} \dfrac{1}{2}\ln\left[\dfrac{\exp[\varepsilon(\xi+1)/T] + \exp[\varepsilon(\nu-1)/T]}{\exp[\varepsilon\xi/T] + \exp[\varepsilon\nu/T]}\right], & \text{as } T \to 0, \\[2ex] \dfrac{\varepsilon^2[\omega - 2(\nu-\xi)]}{4T^2\omega}, & \text{as } T \to \infty, \end{cases} \tag{10.5.6}$$

from which it can be seen that the behaviour in the low-temperature limit is one of three types:

***When* $\xi < \nu - 1$, $K_{\mathrm{e}} \to -\infty$ *as* $T \to 0$:**

The stable ground state is I and on the primary lattice AB first-neighbour pairs are favoured at low temperatures, corresponding to antiferromagnetic local order in the equivalent Ising model. For a loose-packed primary lattice the ground state is one of long-range sublattice order with As occupying one sublattice, Bs the other and ground-state energy $-\frac{1}{2}zN(\nu-1)\varepsilon$. However for a primary–secondary–primary triad of sites there are also occupation states AAA or BBB, with the molecule on the secondary site disoriented, which have zero energy but an entropy $k_{\mathrm{B}}\ln(\omega-2)$. Thus for high enough ω and temperature, AA or BB first-neighbour pairs on the primary lattices will be favoured and separation into A-rich and B-rich phases will occur, corresponding to ferromagnetic order in the equivalent Ising model. This is borne out by the behaviour of K_{e}. From (10.5.6), $K_{\mathrm{e}} < 0$ at low temperature, but at high temperatures, if $\omega > 2(\nu-\xi)$, $K_{\mathrm{e}} > 0$. Plots of K_{e} against temperature are

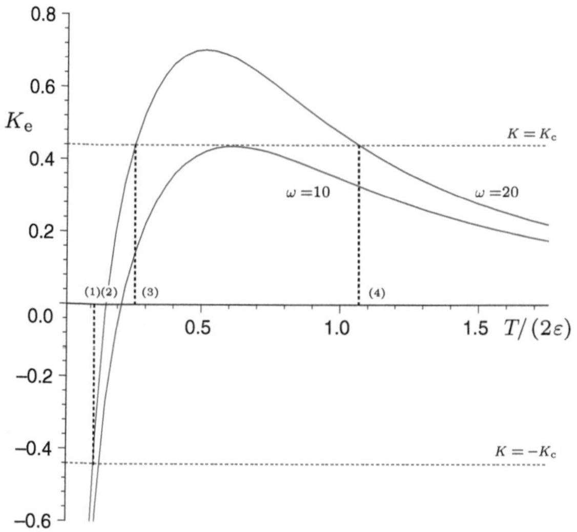

Fig. 10.4 Curves of K_e against temperature derived from (10.5.6) for $\xi = 0.1$, $\nu = 1.2$. The curves are labelled with their value of ω and the lines $K_e = \pm K_c$ are for the square lattice where, from (9.2.36), $K_c = \frac{1}{2}\ln(1 + \sqrt{2})$. The labelled points on the axis are *(1)* $T = T_{sc} = 0.1289(2\varepsilon)$, (2) $T = T_{do} = 0.2029(2\varepsilon)$, *(3)* $T = T_{lc} = 0.2618(2\varepsilon)$, *(4)* $T = T_{uc} = 1.0684(2\varepsilon)$

shown in Fig. 10.4 for a case where $\xi < \nu - 1$ and the high-temperature limit of K_e is positive if $\omega \geq 3$. For sufficiently large ω, the maximum value of $K_e > K_c$, the critical Ising parameter value for the primary lattice, as shown in Fig. 10.4, where for the square lattice ω must be greater than 10. The values of T at which $K_e = K_c$ are lower and upper critical temperatures T_{lc} and T_{uc}. For temperatures in the range (T_{lc}, T_{uc}), $m_s^{(IM)}(K_e) > 0$ and (10.5.3) gives the compositions of A-rich and B-rich conjugate phases. Outside this range of temperature $m_s^{(IM)}(K_e) = 0$ and (10.5.3) gives $x_A = \frac{1}{2}$. Thus in the temperature–mole fraction plane there is a *closed-loop coexistence curve* (similar to the upper part of curve (c) in Fig. 10.6) symmetrical about the line $x_A = \frac{1}{2}$ with lower and upper critical temperatures for phase separation at $T = T_{lc}$ and $T = T_{uc}$ respectively. There is a *disorder point*, with $K_e = 0$, at some temperature $T_{do} < T_{lc}$ (Fig. 10.4) and, for a loose-packed primary lattice, there is a transition at a temperature T_{sc} to a sublattice ordered state when K_e passes through the value $-K_c$.

When $\xi = \nu - 1$, $K_e \to 0$ as $T \to 0$:

The stable ground state is a coexistence of I and II. The (T, K_e) plot assumes the form of curve (b) in Fig. 10.5.

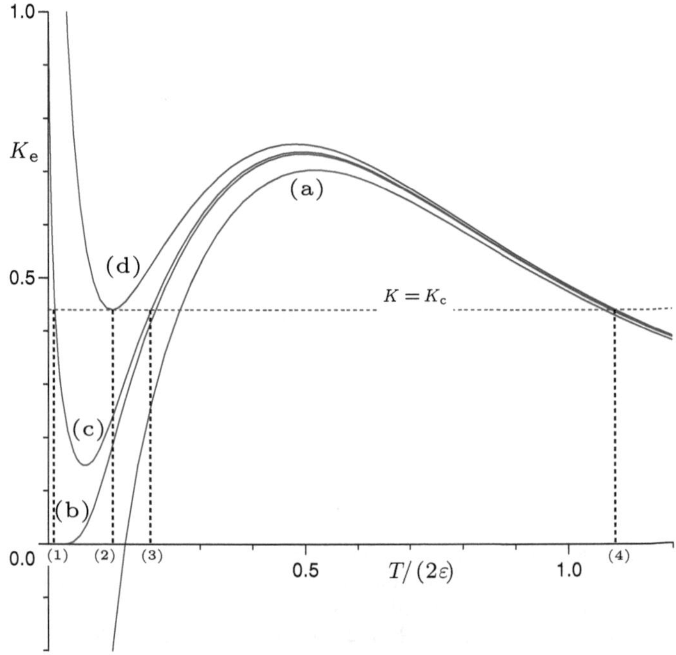

Fig. 10.5 Curves as in Fig. 10.4 for $\nu = 1.2$, with *(a)* $\xi = 0.1 < \nu - 1$, *(b)* $\xi = 0.2 = \nu - 1$, *(c)* $\xi = 0.21 > \nu - 1$, *(d)* $\xi = 0.2451$. The labelled points on the axis are for curve *(c)*: *(1)* $T = T_{auc} = 0.0227(2\varepsilon)$, *(3)* $T = T_{lc} = 0.2079(2\varepsilon)$ and *(4)* $T = T_{uc} = 1.0845(2\varepsilon)$ and for curve *(d)*: (2) $T = T_{dc} = 0.1329(2\varepsilon)$

When $\xi > \nu - 1$, $K_e \to \infty$ as $T \to 0$:

A curve where ξ is slightly greater than $\nu - 1$ is shown as (c) in Fig. 10.5. It remains close to (b) except for low temperatures where $K_e \to \infty$. There is thus a low temperature interval $(0, T_{auc})$ where $K_e > K_c$, so that phase separation occurs. The coexistence diagram, calculated from (10.5.3) and (10.5.4), is shown with a broken line in Fig. 10.6.[7] There is a closed-loop coexistence curve with upper critical temperature T_{uc} and lower critical temperature T_{lc}. Separated from the closed-loop curve by the temperature interval (T_{auc}, T_{lc}), with $x_A = \frac{1}{2}$, is a low-temperature coexistence curve with (another) upper critical temperature T_{auc}. For both intervals (T_{lc}, T_{uc}) and $(0, T_{auc})$ there is separation into A-rich and B-rich phases, but for the upper temperature interval the phase separated state is entropically stabilized, whereas for the lower temperature interval it is energetically stabilized.

As ξ is increased again relative to $\nu - 1$, the minimum of the curve rises until it occurs on the line $K_e = K_c$, as shown in curve (d) of Fig. 10.5. The lower critical point of the upper branch of the coexistence curve and the upper critical point of

[7] For the sake of clarity we have used $\xi = 0.235$, in Fig. 10.6, rather than $\xi = 0.21$ for the (c) curve.

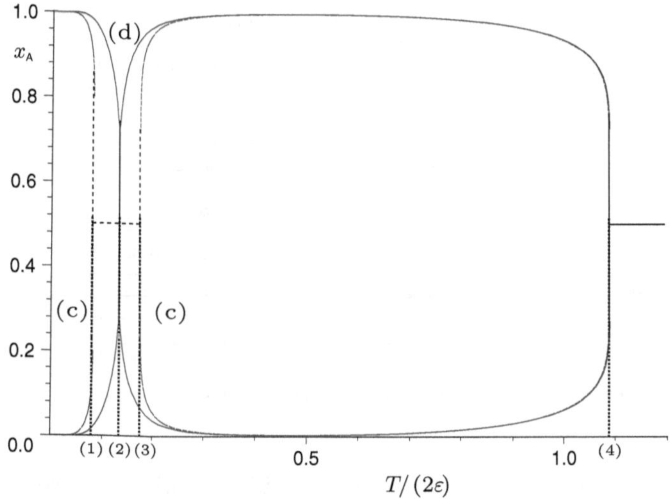

Fig. 10.6 Coexistence curves in the temperature–mole fraction plane for the curves of type (*c*) (*broken line*) and (*d*) (*continuous line*) of Fig. 10.5. Labels on the temperature axis are as in Fig. 10.5

the lower branch have coalesced into a single point at $T = T_{dc}$. This is a *critical double-point*. In Fig. 10.6 the two branches of the coexistence curve are shown (with a continuous line) as meeting with a common tangent at the critical double-point. That this is the case can be shown in the following way. Let Δx_A denote the difference between the values of x_A in the conjugate phases. From (10.5.3), $\Delta x_A \sim m_s^{(IM)}(K_e)$, as $\Delta K_e = K_e - K_c \to +0$, and $m_s^{(IM)}(K_e) \sim (\Delta K_e)^\beta$ where β is the Ising model critical exponent. Now K_e has a minimum at $K_e = K_c$, $T = T_d$ so that $\Delta K_e \sim (T - T_d)^2$. This applies for both positive and negative $T - T_d$, and hence on both sides of the critical double-point we have $\Delta x_A \sim |T - T_d|^{\beta_d}$ where $\beta_d = 2\beta$.[8] Since $\beta < \frac{1}{2}$ for both $d = 2$ and $d = 3$, $\beta_d < 1$ and the two coexistence branches meet with a common tangent parallel to the x_A-axis.[9] If ξ is again increased relative to $\nu - 1$ the minimum in the curve of K_e moves up to a value greater than K_c. The two branches of the coexistence curve are joined to form a rather bizarre 'neck' shape and there is now only one critical temperature.

The $\xi = 0$, $\nu > 1$ case of the above theory was developed by Wheeler (1975, 1977) and Anderson and Wheeler (1978a) fitted theoretical results to the experimental closed-loop curves for some organic mixtures, for example glycerol-ethylbenzylamine. They required the very large value $\omega = 5000$ for the number of secondary site orientation states. Although the model is highly schematic, this value of ω may reflect the large number of configurations available to fairly

[8] This result was derived for a somewhat different decorated solution model by Bartis and Hall (1974).

[9] If it were true that $\beta \geq \frac{1}{2}$, the coexistence curves would form a double cusp at their meeting point.

Table 10.1 The contribution to the Hamiltonian from a primary–secondary–primary triplet of sites

Phase Name	Triplet Occupation	Hamiltonian Contribution
Vapour	∘–∘–∘	0
Intermediate-Primary	•–∘–∘	$-\mu/z$
Intermediate-Secondary	∘–•–∘	$-\mu$
Open	•–∘–•	$-\varepsilon - 2\mu/z$
Partially-Packed	•–•–∘	$-\nu\varepsilon - (1+z)\mu/z$
Close-Packed	•–•–•	$(\xi - 2\nu - 1)\varepsilon - (2+z)\mu/z$

The different configurations are labelled with their phase names

complex organic molecules. Barker and Fock (1953) had earlier obtained a closed-loop coexistence curve from a regular lattice model with orientation dependent interaction energies. This is not equivalent to an Ising model, and a first-order approximation was used. Closed-loop coexistence curves are observed experimentally for some aqueous mixtures (in, for example, water-nicotine) but are highly unsymmetrical with respect to the two components. Anderson and Wheeler (1978b) fitted theoretical results obtained by taking $\varepsilon_{AA} \neq \varepsilon_{BB}$ and $\varepsilon'_{AA} \neq \varepsilon'_{BB}$.

10.6 A Decorated Lattice Fluid

In this section we consider liquid/vapour phase phenomena by using a decorated lattice fluid model. Both primary and secondary lattice sites can either be unoccupied (∘) or occupied (•) by a particle with chemical potential μ. The occupation states of a primary–secondary–primary triad of sites, together with their contributions to the Hamiltonian are shown in Table 10.1. We assume throughout that $\varepsilon > 0$.[10] The special case of this model with $\xi = 0$ was introduced by Mermin (1971a) for the square lattice and, under the name of *Mermin's decorated lattice gas*, has been investgated and generalized to other lattices by Zollweg and Mulholland (1972) and Mulholland and Rehr (1974) (see also Wheeler 1977). For additivity of pair energies $\xi = 0$, but it is useful to include the non-additive case, particularly in order to incorporate as a special case the water-like model of Bell and Sallouta (1975) described in Sect. 10.6.2.

Each of the triplet configuration states represent a possible ground state of the system, when every triplet of the lattice is identically occupied, with the caveat that the intermediate-primary and partially-packed ground states can be perfectly realized only for loose-packed lattices. At zero temperature the Hamiltonian is the enthalpy,

[10] Meaning that the equivalent Ising model on the primary sites is ferromagnetic.

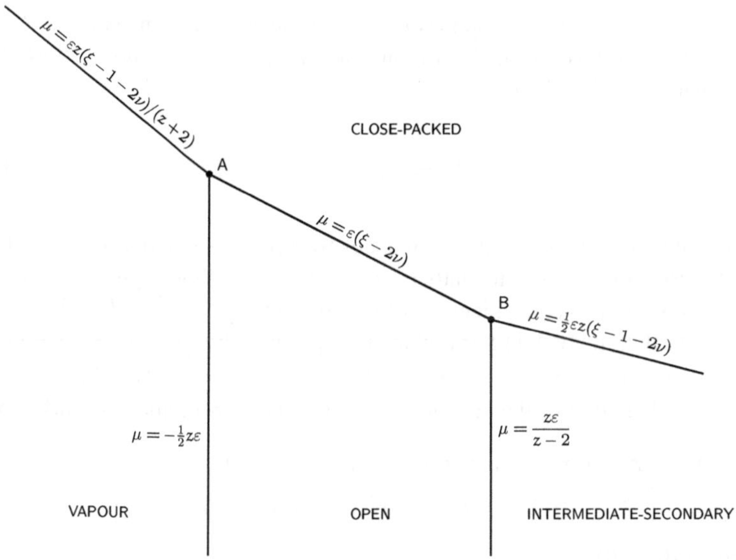

Fig. 10.7 Ground-state diagram in the μ–ν plane for the decorated lattice fluid. The coordinates of the point A of vapour/open/close-packed coexistence and of the point B of open/intermediate-secondary/close-packed coexistence are respectively (μ_A, ν_A) and (μ_B, ν_B)

which from Sect. 1.4.2 is equal to $-PN$. Thus the contributions to the Hamiltonian listed in Table 10.1 are equal to the values of $-2P/z$ in the different phases.[11]

The stable ground state is that with least enthalpy. It is clear that the intermediate-primary state can never be stable, since if $\mu < 0$ the vapour state is more stable and if $\mu > 0$ the intermediate-secondary state is more stable. It is also not difficult to show that if $0 \leq \xi \leq 1$ (a condition which we shall henceforth assume) then the partially-packed state can also never be stable. Figure 10.7 is the ground-state diagram in the μ–ν plane where the coordinates of A and B are respectively

$$\mu_A = -\tfrac{1}{2}z\varepsilon, \qquad \nu_A = \tfrac{1}{4}(z + 2\xi),$$
$$\mu_B = -\frac{z\varepsilon}{z-2}, \qquad \nu_B = \frac{\xi(z-2) - z}{2(z-2)}. \tag{10.6.1}$$

It can be seen that there are three ranges of ν giving three different types of zero-temperature behaviour:

(I) When $\nu > \nu_A$ there is a vapour/close-packed transition at $P = 0$.
(II) When $\nu_A > \nu > \nu_B$ there is a vapour/open transition at $P = 0$ and an open/close-packed transition at

$$P = P_0 := \tfrac{1}{2}\varepsilon(z + 2\xi - 4\nu). \tag{10.6.2}$$

---

[11] We are using the grand-canonical distribution and scaling with respect to the number of primary–secondary–primary triplets rather than the number of primary sites.

(III) When $\nu_{\mathrm{B}} > \nu$ there is a vapour/open transition at $P = 0$, an open/intermediate-secondary transition at $P{=}P_1$ and an intermediate-secondary/close-packed transition at P_2, where

$$P_1 := \frac{\varepsilon z^2}{2(z-2)}, \qquad P_2 = \tfrac{1}{4}\varepsilon z^2(\xi - 1 - 2\nu). \tag{10.6.3}$$

Representing the occupied and vacant states by $\sigma(\boldsymbol{r}) = +1$ and $\sigma(\boldsymbol{r}) = -1$ respectively we use, for the primary lattice, the Ising/lattice fluid equivalence given by (3.5.22). However it is convenient to define the couplings $K := \varepsilon/T$ and $L := \mu/T$, which means that in (10.1.1) K is replaced by $\frac{1}{4}K$ and L by $\frac{1}{4}(2L + zK)$. The constant coupling, given by (3.5.21), which now becomes $C := -\frac{1}{4}(4L + zK)/z$, features only in the calculation of the partition function. The Hamiltonian is (10.1.1) with

$$\begin{aligned}\mathring{K}_{\mathrm{e}}(\sigma(\boldsymbol{r}), \sigma(\boldsymbol{r}'); \sigma(\boldsymbol{r}, \boldsymbol{r}'), \tau(\boldsymbol{r}, \boldsymbol{r}') &= \tfrac{1}{8}[1 + \sigma(\boldsymbol{r}, \boldsymbol{r}')] \\ &\times \left\{4L + K(2\nu - \xi)[2 + \sigma(\boldsymbol{r}) + \sigma(\boldsymbol{r}')] + K\xi[1 - \sigma(\boldsymbol{r})\sigma(\boldsymbol{r}')]\right\}\end{aligned} \tag{10.6.4}$$

and, from (10.2.10),

$$\begin{aligned}&\phi(K, L; +1, +1) = 1 + \mathfrak{z}\mathfrak{X}^{(2\nu-\xi)}, \qquad \phi(K, L; -1, -1) = 1 + \mathfrak{z}, \\ &\phi(K, L; +1, -1) = 1 + \mathfrak{z}\mathfrak{X}^{\nu},\end{aligned} \tag{10.6.5}$$

where $\mathfrak{z} := \exp(L)$ and $\mathfrak{X} := \exp(K)$. Then, from (10.1.6) and (10.1.7), (remembering that the prefactor $\mathfrak{X}^4$ in (10.1.6) must be replaced by $\mathfrak{X}$ and $\mathfrak{z}^4$ in (10.1.7) by $\mathfrak{z}^2\mathfrak{X}^z$)

$$\exp(4K_{\mathrm{e}}) = \frac{\mathfrak{X}(1 + \mathfrak{z})(1 + \mathfrak{z}\mathfrak{X}^{(2\nu-\xi)})}{(1 + \mathfrak{z}\mathfrak{X}^{\nu})^2}, \tag{10.6.6}$$

$$\exp(4L_{\mathrm{e}}) = \mathfrak{z}^2\left\{\frac{\mathfrak{X}(1 + \mathfrak{z}\mathfrak{X}^{(2\nu-\xi)})}{1 + \mathfrak{z}}\right\}^z. \tag{10.6.7}$$

The density ρ (equal to 1 when all primary and secondary sites are occupied) is given by

$$\rho = -\frac{2\mathfrak{z}}{(2 + z)}\left(\frac{\partial\phi(K, L, \tilde{\mathcal{N}})}{\partial\mathfrak{z}}\right)_{\mathfrak{X}}, \tag{10.6.8}$$

where ϕ is the dimensionless free energy per primary lattice site of the decorated model, given in terms of the partition function (10.1.8) by (2.3.6), where, from (10.1.5) (and incorporating the contribution from C)

$$g(\mathfrak{X},\mathfrak{Z}) = \mathfrak{Z}^{1/z}\{\mathfrak{X}(1+\mathfrak{Z}\mathfrak{X}^{(2\nu-\xi)})(1+\mathfrak{Z})(1+\mathfrak{Z}\mathfrak{X}^{\nu})^2\}^{1/4}. \tag{10.6.9}$$

Thus, from (10.6.6)–(10.6.9)

$$\rho = \mathrm{p}(\mathfrak{X},\mathfrak{Z}) + \mathrm{B}(\mathfrak{X},\mathfrak{Z})m^{(\mathrm{IM})}(K_\mathrm{e},L_\mathrm{e}) + \mathrm{C}(\mathfrak{X},\mathfrak{Z})\mathrm{u}^{(\mathrm{IM})}(K_\mathrm{e},L_\mathrm{e}), \tag{10.6.10}$$

where

$$\mathrm{p}(\mathfrak{X},\mathfrak{Z}) = \frac{1}{4(2+\mathrm{z})}\left\{4+\frac{\mathrm{z}\mathfrak{Z}}{1+\mathfrak{Z}}+\frac{\mathrm{z}\mathfrak{Z}\mathfrak{X}^{(2\nu-\xi)}}{1+\mathfrak{Z}\mathfrak{X}^{(2\nu-\xi)}}+\frac{2\mathrm{z}\mathfrak{Z}\mathfrak{X}^{\nu}}{1+\mathfrak{Z}\mathfrak{X}^{\nu}}\right\},$$

$$\mathrm{B}(\mathfrak{X},\mathfrak{Z}) = \frac{1}{2(2+\mathrm{z})}\left\{2+\frac{\mathrm{z}\mathfrak{Z}\mathfrak{X}^{(2\nu-\xi)}}{1+\mathfrak{Z}\mathfrak{X}^{(2\nu-\xi)}}-\frac{\mathrm{z}\mathfrak{Z}}{1+\mathfrak{Z}}\right\}, \tag{10.6.11}$$

$$\mathrm{C}(\mathfrak{X},\mathfrak{Z}) = -\frac{\mathfrak{Z}}{2(2+\mathrm{z})}\left\{\frac{1}{1+\mathfrak{Z}}+\frac{\mathfrak{X}^{(2\nu-\xi)}}{1+\mathfrak{Z}\mathfrak{X}^{(2\nu-\xi)}}-\frac{2\mathfrak{X}^{\nu}}{1+\mathfrak{Z}\mathfrak{X}^{\nu}}\right\}$$

and $m^{(\mathrm{IM})}(K_\mathrm{e},L_\mathrm{e})$ and $\mathrm{u}^{(\mathrm{IM})}(K_\mathrm{e},L_\mathrm{e})$ are respectively the magnetization and internal energy per primary lattice site on the equivalent Ising model (the latter being made dimensionless by dividing by J_e). We denote the curve in the μ–T plane derived from (10.6.6) by setting $K_\mathrm{e} = K_\mathrm{c}$, with specific values of z, ν and ξ, as $\mathcal{C}(\mathrm{z},\nu,\xi)$ and the corresponding curve, derived from (10.6.7) by setting $L_\mathrm{e} = 0$, as $\mathcal{L}(\mathrm{z},\nu,\xi)$. The first-order transition lines in the μ–T plane are the parts of $\mathcal{L}(\mathrm{z},\nu,\xi)$ lying in the region where $K_\mathrm{e} > K_\mathrm{c}$, that is below $\mathcal{C}(\mathrm{z},\nu,\xi)$. This is always a simply-connected curve with asymptotes $T \to 1/(4K_\mathrm{c})$ as $\mu \to \infty$ and $T \to (1-\xi)/(4K_\mathrm{c})$ as $\mu \to -\infty$. Critical temperatures occur at the intersection of $\mathcal{L}(\mathrm{z},\nu,\xi)$ with $\mathcal{C}(\mathrm{z},\nu,\xi)$ We now consider the three cases I–III.

10.6.1 Case I: A Single Vapour/Liquid Transition

Here there is a first-order transition line from the vapour phase to a close-packed liquid phase (Fig. 10.8). The conjugate vapour and liquid densities, ρ_v and ρ_ℓ respectively, are given by taking L_e in (10.6.8) and setting $m^{(\mathrm{IM})}(K_\mathrm{e},0) = -m_\mathrm{s}^{(\mathrm{IM})}(K_\mathrm{e})$ and $m^{(\mathrm{IM})}(K_\mathrm{e},0) = +m_\mathrm{s}^{(\mathrm{IM})}(K_\mathrm{e})$. Since $\mathrm{u}^{(\mathrm{IM})}(K_\mathrm{e},K_\mathrm{e})$ is continuous across $\mathcal{L}(\mathrm{z},\nu,\xi)$, the coexistence width

$$\rho_\mathrm{w} := \rho_\ell - \rho_\mathrm{v} = 2\mathrm{B}(\mathfrak{X},\mathfrak{Z})m_\mathrm{s}^{(\mathrm{IM})}(K_\mathrm{e}). \tag{10.6.12}$$

Since $m_\mathrm{s}^{(\mathrm{IM})}(K_\mathrm{e}) \sim (\Delta K_\mathrm{e})^{\beta}$, as $K_\mathrm{e} \to K_\mathrm{c}+0$, and, since K_e and, L_e are regular functions of μ and T,

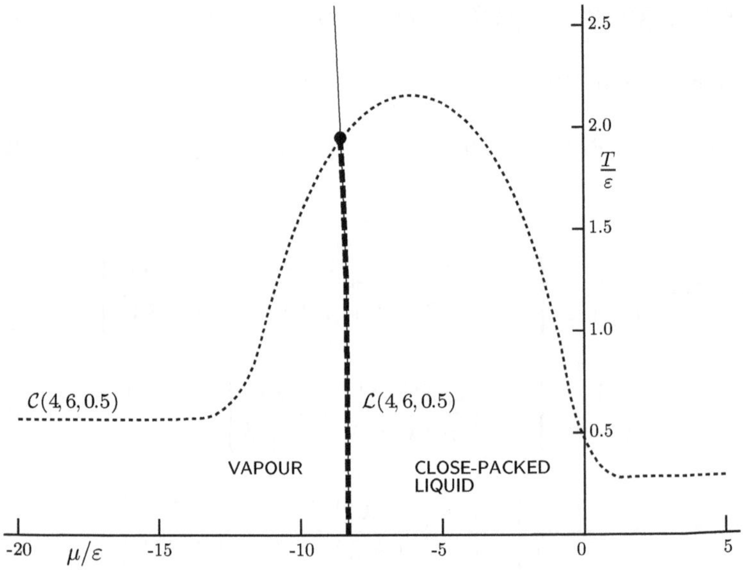

Fig. 10.8 Curves $K_e = K_c$ and $L_e = 0$ for the square lattice with $\nu = 6$ and $\xi = 0.5$

$$\rho_w \sim (\Delta T)^{\beta}, \qquad \text{as } T \to T_c - 0. \tag{10.6.13}$$

This is similar to the behaviour of the simple lattice fluid, but there are important differences, both globally and in the neighbourhood of the critical points. The hole–particle symmetry of the simple lattice fluid, described in Sect. 3.5.1.2, is lost (except for a special case to be discussed in Sect. 10.6.3) and the coexistence curves in the ρ–T plane are no longer symmetrical about $\rho = \frac{1}{2}$. The coexistence diameter

$$\rho_d := \frac{1}{2}(\rho_\ell + \rho_v) \tag{10.6.14}$$

is, from (3.5.24) and (3.5.25), equal to $\frac{1}{2}$ for the simple lattice fluid, but for the decorated model, from (10.6.10) and (10.6.11),

$$\rho_d = \mathrm{p}(\mathfrak{X}, \mathfrak{Z}) + \mathrm{C}(\mathfrak{X}, \mathfrak{Z})\mathrm{u}^{(\mathrm{IM})}(K_e, 0). \tag{10.6.15}$$

The empirical *law of the rectilinear diameter* (Sect. 3.5.1.2) implies that $\mathrm{d}\rho_d/\mathrm{d}T$ remains finite as $T \to T_c - 0$. This law is satisfied by the simple lattice fluid, since $\mathrm{d}\rho_d/\mathrm{d}T = 0$ for all $T < T_c$. However, for the decorated model, the right-hand side of (10.6.15) is not a linear function of T near the critical point. Unless $\mathrm{C}(\mathfrak{X}, \mathfrak{Z}) = 0$, the derivative $\mathrm{d}\rho_d/\mathrm{d}T$ contains a term proportional to $\mathrm{d}\mathrm{u}^{(\mathrm{IM})}(K_e, 0)/\mathrm{d}K_e$ and, from (4.4.12) and (10.2.23),

$$\frac{\mathrm{d}u^{(\mathrm{IM})}(K_\mathrm{e},0)}{\mathrm{d}K_\mathrm{e}} \sim \begin{cases} (\Delta K_\mathrm{c})^{-\alpha'}, & \text{for } d \neq 2, \\ -\ln|\Delta K_\mathrm{e}|, & \text{for } d = 2. \end{cases} \tag{10.6.16}$$

Hence

$$\frac{\mathrm{d}\rho_\mathrm{d}}{\mathrm{d}T} \sim \begin{cases} (\Delta T)^{-\alpha'}, & \text{for } d \neq 2, \\ -\ln|\Delta T|, & \text{for } d = 2, \end{cases} \tag{10.6.17}$$

where, for $d = 3, 0 < \alpha < \frac{1}{2}$. So for $d = 2$ and 3, $\mathrm{d}\rho_\mathrm{d}/\mathrm{d}T \to \infty$ as $T \to T_\mathrm{c}-0$. The special case of this result for the square lattice, with no primary–primary interaction and $\xi = 0$, was obtained by Mermin (1971a) who also showed (Mermin 1971b) that a similar result occurs for a model in which the secondary sites are occupied by bars. This indicates that a non-singular diameter is an artifact of the simple lattice fluid. As pointed out in Sect. 3.5.1.2, more general arguments for the existence of the singularity have been developed. However, experimentally the situation is still not completely resolved (Zollweg and Mulholland 1972; Rowlinson and Swinton 1982).

10.6.2 Case II: A Water-Like Model

This case was investigated by Bell and Sallouta (1975) as a model for water, by taking the open state, with all primary sites occupied and all secondary sites empty, to represent a completely hydrogen-bonded ice-like structure. The interaction energy $-\varepsilon$ of a primary–primary pair of molecules with the intervening secondary site empty is attributed to the formation of a hydrogen bond, whereas the primary–secondary pair energy $-\nu\varepsilon$ is due to the non-bonding interaction. It is assumed that no hydrogen bond can form between a first-neighbour primary pair of molecules with the intervening secondary site occupied so that $\xi = 1$ and, from (10.6.1), case II would correspond to $(z+2)/4 > \nu > -1/(z-2)$. However, we shall follow Bell and Sallouta and take $\nu > 0$, meaning that the secondary–secondary interaction is attractive. The close-packed state is taken to represent the liquid phase.

This model is less realistic structurally then the water model treated in Sect. 7.6 and less satisfactory for fluid phases in that the open structure is imposed *ab initio* rather than arising as short-range order due to bonding between appropriately placed and oriented molecules. However, it does have the advantage that, because of the equivalence to the Ising model, some accurate results can be obtained for comparison with those derived from standard approximations.[12]

---

[12] This decorated model can be looked on as one of the class of *interstitial models* (see, for instance, Perram 1971) in which the open structure is disordered by displacement of molecules from 'framework' onto interstitial sites.

Bearing in mind, as indicated above, that the Hamiltonian contributions in Table 10.1, at $T = 0$, are equal to $-2P/z$, the chemical potentials in the open and close-packed ground states can be expressed in the form

$$\mu^{(\mathrm{o})} = P - \tfrac{1}{2}z\varepsilon \qquad \mu^{(\mathrm{CP})} = \frac{2}{z+2}(P - 2z\nu\varepsilon). \tag{10.6.18}$$

From (10.6.2) and Table 10.7 an open/close-packed transition occurs at $T = 0$ when

$$P = P_0 := \tfrac{1}{2}(z + 2 - 4\nu) > 0, \qquad \mu = \mu_0 := \varepsilon(1 - 2\nu) \tag{10.6.19}$$

and a vapour/open transition at $T = 0$ when

$$P = 0, \qquad \mu = -\tfrac{1}{2}z\varepsilon. \tag{10.6.20}$$

It can be shown from (10.6.7) that the $L_\mathrm{e} = 0$ locus $\mathcal{L}(z, \nu, 1)$ in the T–μ plane meets the zero-temperature axis at $\mu = -\frac{1}{2}z\varepsilon$. However the open/close-packed transition point $\mu = \mu_0$ on the zero-temperature axis is an isolated singularity. It cannot be the terminus of any branch of the $\mathcal{L}(z, \nu, 1)$ because, at low temperatures, the primary lattice is nearly full in both the open and close-packed states and so both correspond to $L_\mathrm{e} > 0$.

Bell and Sallouta (1975) used the method described for case I to determine the vapour/open first-order transition curve and critical point and then used (10.6.10) to compute the densities of the coexistent phases. They obtained results using the exact values of the zero-field magnetization and internal energy for the square and honeycomb lattices and values derived from the series work of Essam and Sykes (1963) and Baker and Gaunt (1967) for the diamond lattice. Their results for this last case are shown as curve A in Fig. 10.9.[13] The terminal points $\rho = 0$ and $\rho = \frac{1}{3}$ correspond to the vapour and open structure phases respectively. The most striking feature, apart from the marked lack of symmetry about $\rho = \frac{1}{2}$, is the pronounced maximum on the liquid density branch. Unfortunately the calculation of (ρ, T) isobaric curves requires Ising model results away from the $\mathcal{H} = 0$ axis. However, near the maximum on the liquid branch of the coexistence curve, the first-order plot is quite close to the accurate one. First-order calculations show maxima on the adjacent isobaric (ρ, T) curves. By considering perturbations from the perfect open structure, Bell and Sallouta (1975) showed that near $T = 0$ the density ρ increases with T on isobars for $0 < P < P_0$, provided that $\nu > \frac{1}{2}$. This implies a density maximum on any continuous isobar in this pressure range. Since P_0 is considerably greater than the critical pressure P_c, density anomalies occur well into the supercritical region, in accordance with experimental results. Compressibility minima are also found.

[13] Curves B and C respectively, given for comparison, show the results of using formulae from first-order and mean-field approximations for $m_\mathrm{s}^{(\mathrm{IM})}(K_\mathrm{e})$ and $u^{(\mathrm{IM})}(K_\mathrm{e}, 0)$.

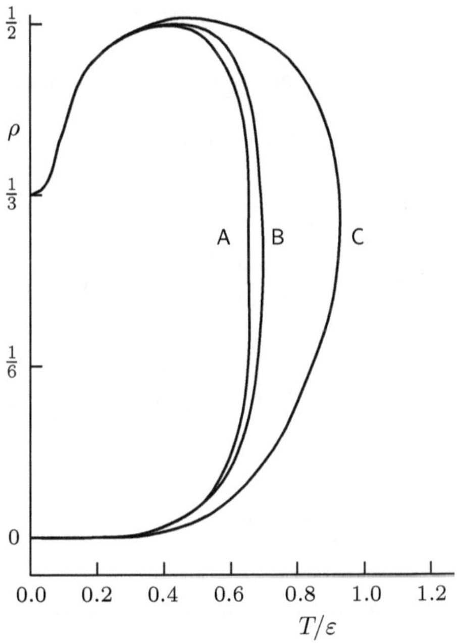

Fig. 10.9 Coexistence curves in the temperature–density plane for the water model for the diamond lattice with $\nu = 1.8$. (Reprinted from Bell and Sallouta 1975, by permission of the publisher Taylor and Francis Ltd.)

There is thus good qualitative agreement between the results for this model and those for the model of Sect. 7.6, where first-order approximation methods had to be used exclusively.

10.6.3 Case III: Maxithermal, Critical Double and Cuspoidal Points

When $\mathcal{C}(z, \nu, \xi)$ and $\mathcal{L}(z, \nu, \xi)$ are both single curves with a unique intersection in the μ–T plane, the phase diagram in this plane is topologically similar to that of a simple lattice fluid, though as we have shown for case I, there are important differences in properties. However, for case III there are three phase transitions at zero temperature and the mapping from the μ–T plane of the decorated model to the K_{e}^{-1}–L_{e} plane of the Ising model is no longer one-to-one and $\mathcal{L}(z, \nu, \xi)$ has several distinct branches. Consider now the case $\xi = 0$, when the pair energies are additive.[14] Defining

$$x := \mathfrak{X}^{\nu} = \exp(\nu\varepsilon/T) \qquad y := \mathfrak{Z}\mathfrak{X}^{\nu} = \exp[(\mu + \nu\varepsilon)/T], \tag{10.6.21}$$

[14] This has been treated extensively by Mulholland and Rehr (1974).

(10.6.6) and (10.6.7) can be re-expressed in the form

$$\exp(4K_e - \varepsilon/T) - 1 = \mathrm{X}(x, y), \tag{10.6.22}$$

$$\exp[4L_e - \varepsilon z(\nu - \nu^*)/(\nu^* T)] = \mathrm{Y}(x, y), \tag{10.6.23}$$

where

$$\mathrm{X}(x, y) := \frac{x + x^{-1} - 2}{y + y^{-1} + 2} = \mathrm{X}(x, y^{-1}), \tag{10.6.24}$$

$$\mathrm{Y}(x, y) := y^2 \left(\frac{1 + xy}{x + y}\right)^z = \frac{1}{\mathrm{Y}(x, y^{-1})} \tag{10.6.25}$$

and

$$\nu = \nu^* := -\frac{z}{z - 2}. \tag{10.6.26}$$

It follows from the symmetry property of $\mathrm{X}(x, y)$ that the curve $\mathcal{C}(z, \nu, 0)$

(i) is symmetrical about the line $\mu = -\nu\varepsilon$ $(y = 1)$,
(ii) has the line $T = \varepsilon/(4K_c)$ as asymptote when $\mu \to \pm\infty$,
(iii) lies entirely above this asymptote,
(iv) has a maximum at $\mu = -\nu\varepsilon$,

as is shown by the example in Fig. 10.10. Although $\mathrm{Y}(x, y)$ also has a symmetry property it will be reflected in the form of the curve $\mathcal{L}(z, \nu, 0)$ derived from (10.6.23) with $L_e = 0$ only when $\nu = \nu^*$ giving

$$L_e(x, y) = -L_e(x, y^{-1}). \tag{10.6.27}$$

The case $\nu = \nu^*$:

The curve $\mathcal{L}(z, \nu^*, 0)$ is given by the intersections of $\mathrm{Y} = \mathrm{Y}(x, y)$ plotted as a function of y for fixed x with the line $\mathrm{Y} = 1$.[15] For high temperatures ($x = \exp(\nu^*\varepsilon/T)$ close to unity) Y is monotonically increasing with only one solution $y = 1$ for $\mathrm{Y}(x, y) = 1$. This solution exists for all x and so one branch of $\mathcal{L}(z, \nu^*, 0)$ is the entire line $\mu = -\nu^*\varepsilon$ in the μ–T plane. As T (and x) decrease towards zero the curve of $\mathrm{Y}(x, y)$ develops a loop with two additional solutions $y(x)$ and $1/y(x)$, giving two other branches of $\mathcal{L}(z, \nu^*, 0)$ symmetrically placed about $\mu = -\nu^*\varepsilon$ in the μ–T plane. This is shown in Fig. 10.10 for the square lattice, although the topology and

[15] There are interesting similarities between this analysis and the discussion of solution of the Bethe-pair approximation for the Ising model using the function $g(\tau)$ in Sect. 7.3.

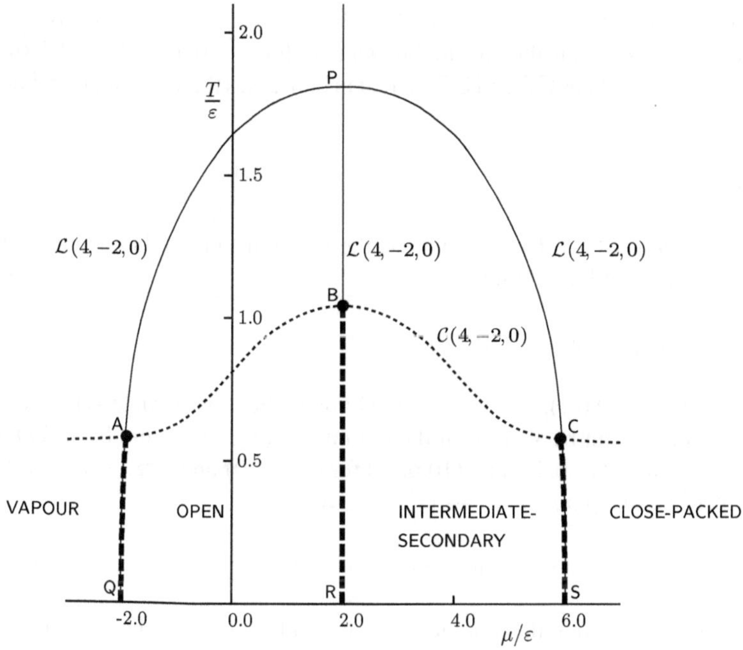

Fig. 10.10 Curves $K_e = K_c$ and $L_e = 0$ for the square lattice with $\nu = \nu^* = -2.0$ and $\xi = 0$

symmetry of the curves are not confined to that lattice. The onset of the symmetric branches of $\mathcal{L}(z, \nu^*, 0)$ at P corresponds to a point inflexion in $\mathrm{Y}(x, y)$ at $y = \mathrm{Y} = 1$ when

$$x = x_0 := \frac{z-2}{z+2}, \qquad \text{giving} \qquad T = T_0 := \nu\varepsilon \left[\ln\left(\frac{z-2}{z+2}\right) \right]^{-1}. \tag{10.6.28}$$

When K_e is calculated at P it proves to be less than K_c for all common lattices. So, as is shown for the square lattice in Fig. 10.10, there are three distinct first-order transition curves, RB on the line $\mu = -\nu^*\varepsilon$ with QA and SC placed symmetrically on either side. Taking the Ising model magnetization and internal energy densities are functions of x and y it follows from (10.6.27) that $m^{(\mathrm{IM})}(K_e, L_e)(x, y^{-1}) = -m^{(\mathrm{IM})}(K_e, L_e)(x, y)$ and $\mathrm{u}^{(\mathrm{IM})}(K_e, L_e)(x, y^{-1}) = \mathrm{u}^{(\mathrm{IM})}(K_e, L_e)(x, y)$. In a similar way, from (10.6.11), $\mathrm{p}(x, y^{-1}) = 1 - \mathrm{p}(x, y)$, $\mathrm{B}(x, y^{-1}) = \mathrm{B}(x, y)$ and $\mathrm{C}(x, y^{-1}) = -\mathrm{C}(x, y)$. It follows from (10.6.10) that

$$\rho(x, y^{-1}) = 1 - \rho(x, y). \tag{10.6.29}$$

Thus there is particle–hole symmetry and the phase pattern in the ρ–T plane is symmetrical about $\rho = \frac{1}{2}$. Since $P = 0$ at Q, QA can be regarded as the vapour/liquid transition line. On RB the conjugate densities are the limits of $\rho(x, y)$ and $\rho(x, y^{-1})$

as $y \to 1$, giving, from (3.5.25) and (10.6.29), $\rho_\mathrm{d} = \frac{1}{2}$. There is no singularity in the diameter which is attributable to the fact that, for $\xi = 0$, $\mathsf{C}(x, 1) = 0$. However, $\mathsf{C}(x, y) \neq 0$ on the lines QA and SC, so that in both cases there is a singular diameter.

The case $\boldsymbol{\nu \neq \nu^*}$**:**

To find the curve $\mathcal{L}(z, \nu, 0)$ we again use the function $\mathsf{Y}(x, y)$, but, from (10.6.23), we are now looking for solutions to

$$\mathsf{Y}(x, y) = \exp[\varepsilon z(\nu - \nu^*)/(\nu^* T)]. \tag{10.6.30}$$

In case III, from (10.6.1), $\nu < \frac{1}{2}\nu^* < 0$. Although the hole-particle symmetry is no longer present, it can be seen from the definitions of x and y in (10.6.21) and the defining equations (10.6.22) and (10.6.30) for $\mathcal{C}(z, \nu, 0)$ and $\mathcal{L}(z, \nu, 0)$ that there is a 'hole-particle mapping' $(\varepsilon, \nu) \leftrightarrow (\varepsilon', \nu')$ given by

$$\nu\varepsilon = \nu'\varepsilon', \qquad (\nu - \nu^*)/\nu = -(\nu' - \nu^*)/\nu', \tag{10.6.31}$$

which reflects μ about the line $\mu = -\nu\varepsilon$ and makes the hole-particle phase interchanges vapour $\leftrightarrow$ close-packed and open $\leftrightarrow$ intermediate-secondary. With $\nu \in [\nu^*, -\infty)$, $\nu' \in (\frac{1}{2}\nu^*, \nu^*]$. So we can concentrate on the case $\nu < \nu^*$, knowing that our conclusions with respect to the topology of the phase diagram map into a case with $\nu^* < \nu < \frac{1}{2}\nu^*$.

As ν decreases from ν^* the right-hand side of (10.6.30) increases from unity and $\mathcal{L}(z, \nu, 0)$ breaks into a closed loop from Q to R with a second branch beginning at S and asymptotic to $\mu = -\varepsilon\nu$. Initially, as shown in Fig. 10.11a,[16] the only effect on the first-order transition lines is to remove the symmetry and RB will like QA and SC now have a singular diameter. A further reduction of ν yields the situation shown in Fig. 10.11b, where $\mathcal{C}(z, \nu, 0)$ passes through the maximum of the Q–R loop of $\mathcal{L}(z, \nu, 0)$. The point B becomes a special type of critical point called a *cusp point*, where both branches of the coexistence curve in the ρ–T plane approach the critical point from densities greater than the critical value. As ν is further reduced $\mathcal{C}(z, \nu, 0)$ cuts the Q–R loop of $\mathcal{L}(z, \nu, 0)$ at two points A and A′ to the left of the maximum point B (Fig. 10.11c). These are both critical points and B becomes a *maxithermal point*, which is like two coinciding cusp points with coexistent regions in the ρ–T plane approaching from above and below the critical density. The maxithermal point persists as ν is further reduced. Figure 10.11d shows the case where $\mathcal{C}(z, \nu, 0)$ touches the Q–R loop of $\mathcal{L}(z, \nu, 0)$ at a single point A, which is the termination of two

[16] The diagrams in Fig. 10.11 are, unlike Fig. 10.10, qualitative rather than quantitative. Accurate calculations for the simple cubic lattice are described by Mulholland and Rehr (1974) who use them to plot diagrams for all these cases in the ρ–T plane. They also investigate the 'antiferromagnetic' case $\varepsilon < 0$, $\xi = 0$.

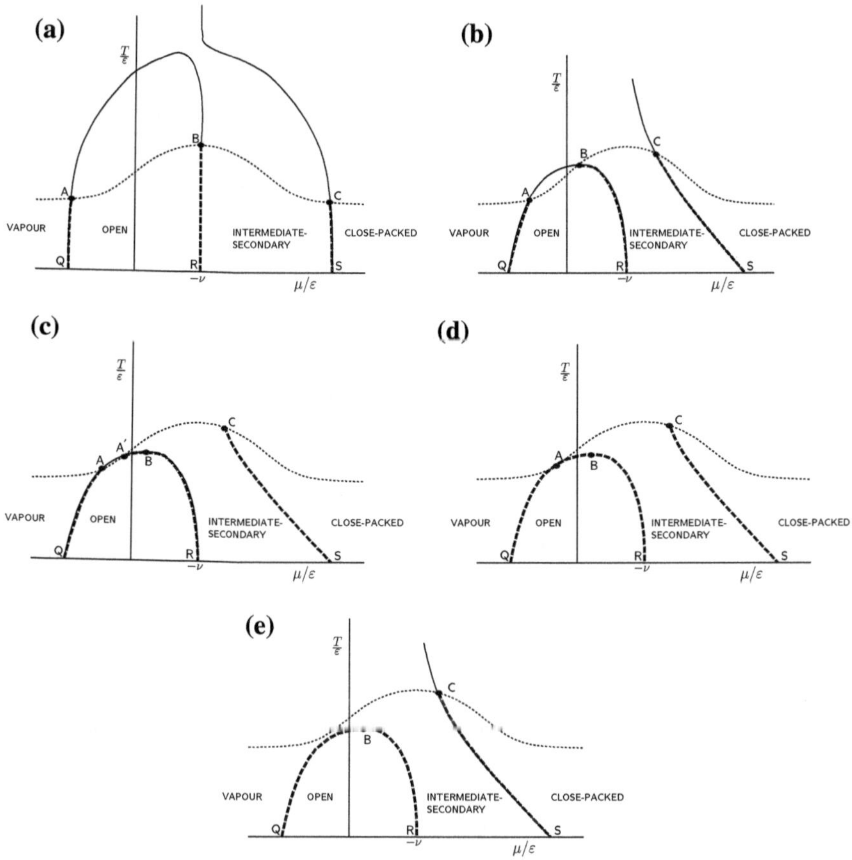

Fig. 10.11 Figures **a–e** show the evolution away from the symmetric phase diagram in Fig. 10.10, as ν decreases from $\nu = \nu^*$

coexistence regions and is called a *critical double-point*. Finally $\mathcal{C}(z, \nu, 0)$ no longer intersects the Q–R loop of $\mathcal{L}(z, \nu, 0)$ and apart from the critical point on SC the only critical point is the maxithermal point at B. However, if $\mathcal{C}(z, \nu, 0)$ passes sufficiently close to $\mathcal{L}(z, \nu, 0)$ it has the effect of producing a 'neck' in the coexistence region for Q–B (Mulholland and Rehr 1974, Fig. 10).

Chapter 11
Transfer Matrices: Incipient Phase Transitions

11.1 The Transfer Matrix Formulation

Let $\mathcal{N}$ be a two-dimensional lattice of N sites with unit vectors $\boldsymbol{i}$ and $\boldsymbol{j}$ in the two lattice directions. It is pictured as consisting of N_1 identical non-overlapping *slices* $\mathcal{N}_{\mathrm{i}}$, $\mathrm{i} = 0, 1, \ldots, N_1 - 1$ parallel to $\boldsymbol{j}$ and consisting of N_2 sites, with $N_1 N_2 = N$. In the simplest case of a square $N_1 \times N_2$ lattice the slices can consist simply of rows of N_2 sites and a one-dimensional lattice is included as the case $N_2 = 1$. The lattice growth achieved by the use of transfer matrices is to the infinite limit for N_1 in the direction of $\boldsymbol{i}$.[1] Let $\boldsymbol{\tau}(\mathrm{i})$ denote the microstate of slice i with interactions restricted to contributions from within slices and between neighbouring slices. Then the Hamiltonian can be expressed in the form

$$\widehat{H}(\boldsymbol{\tau}) := \hat{b}(\boldsymbol{\tau}(N_1 - 1), \boldsymbol{\tau}(0)) + \sum_{\mathrm{i}=0}^{N_1-2} \hat{h}(\boldsymbol{\tau}(\mathrm{i}), \boldsymbol{\tau}(\mathrm{i}+1)), \qquad (11.1.1)$$

where $\hat{h}(\boldsymbol{\tau}(\mathrm{i}), \boldsymbol{\tau}(\mathrm{i}+1))$ contains the contribution in the bulk of the system from interactions between $\mathcal{N}_{\mathrm{i}}$ and $\mathcal{N}_{\mathrm{i}+1}$ and *half* the contributions from within these two slices. The term $\hat{b}(\boldsymbol{\tau}(N_1 - 1), \boldsymbol{\tau}(0))$ contains any boundary contributions at $\mathcal{N}_0$ and $\mathcal{N}_{N_1-1}$, including half the contributions from interactions within these slices. By varying the form of $\hat{b}(\boldsymbol{\tau}(N_1 - 1), \boldsymbol{\tau}(0))$ the model can incorporate the cases where $\boldsymbol{\tau}(0)$ and $\boldsymbol{\tau}(N_1 - 1)$ are fixed or are allowed to vary freely (open boundary condition) or where periodic boundary conditions are applied so that $\hat{b}(\boldsymbol{\tau}(N_1 - 1), \boldsymbol{\tau}(0)) = \hat{h}(\boldsymbol{\tau}(N_1 - 1), \boldsymbol{\tau}(0))$. Let

$$\begin{aligned} V(\boldsymbol{\tau}(\mathrm{i}), \boldsymbol{\tau}(\mathrm{i}')) &:= \exp[-\hat{h}(\boldsymbol{\tau}(\mathrm{i}), \boldsymbol{\tau}(\mathrm{i}'))], \\ B(\boldsymbol{\tau}(\mathrm{i}), \boldsymbol{\tau}(\mathrm{i}')) &:= \exp[-\hat{b}(\boldsymbol{\tau}(\mathrm{i}), \boldsymbol{\tau}(\mathrm{i}'))], \end{aligned} \qquad (11.1.2)$$

[1] Here and in Chap. 12 we picture the lattice oriented so the growth is upwards with the rows of N_2 sites horizontal. In Chap. 14 it is more convenient to picture these rows as vertical columns with growth to the right.

D.A. Lavis, *Equilibrium Statistical Mechanics of Lattice Models*,
Theoretical and Mathematical Physics, DOI 10.1007/978-94-017-9430-5_11

and M be the number of states of a slice, which are denoted by $\boldsymbol{\tau}_m, m = 1, 2, \ldots, M$. Using the notation of Sect. 17.2, the $M \times M$ dimensional *transfer matrix* $\boldsymbol{V}$ and *boundary matrix* $\boldsymbol{B}$ are given by

$$\langle m|\boldsymbol{V}|m'\rangle := V(\boldsymbol{\tau}_m, \boldsymbol{\tau}_{m'}), \quad \langle m|\boldsymbol{B}|m'\rangle := B(\boldsymbol{\tau}_m, \boldsymbol{\tau}_{m'}). \tag{11.1.3}$$

11.1.1 The Eigen Problem

It is not in general the case that $\boldsymbol{V}$ is symmetric. However its elements are real and thus, if it commutes with its transpose, it is normal, and this condition, which we assume (and which needs to be checked in particular applications), is sufficient for it to be diagonalizable (see Sect. 17.2) in the form

$$\boldsymbol{X}^{\dagger}\boldsymbol{V}\boldsymbol{X} = \boldsymbol{\Lambda}, \tag{11.1.4}$$

where $\boldsymbol{X}$ is unitary and has columns which are the right eigenvectors $\{\boldsymbol{x}_m\}$ of $\boldsymbol{V}$ and the diagonal matrix $\boldsymbol{\Lambda}$ has elements $\{\Lambda_m\}$ which are the eigenvalues. The rows of $\boldsymbol{Y} := \boldsymbol{X}^{\dagger}$ are the left eigenvectors $\{\boldsymbol{y}_m^{\mathrm{T}}\}$ of $\boldsymbol{V}$. In general some of the eigenvalues of $\boldsymbol{V}$ will be complex. Since, however, $\boldsymbol{V}$ has only real elements, the eigenvectors corresponding to the conjugate $\overline{\Lambda_m}$ of some complex eigenvalue Λ_m will be $\overline{\boldsymbol{y}_m}$ and $\overline{\boldsymbol{x}_m}$ and the elements of the eigenvectors corresponding to a real eigenvalue will be real. We suppose that the eigenvalues are labelled in order of decreasing magnitude.

If $\boldsymbol{V}$ is a positive matrix it has, according to Perron's theorem (Sect. 17.2.4), a (real) positive eigenvalue larger in magnitude than any other eigenvalue with left and right eigenvectors with positive elements.[2] The transfer matrix of the Ising model satisfies this condition. However, in the case of the eight-vertex model discussed in Chap. 12, the transfer matrix is merely non-negative and also reducible (Sect. 12.3.5.2) so Frobenius generalization of Perron's theorem is inapplicable and we might expect, and in fact find (Sect. 12.3.6.6) cases where the largest eigenvalue is degenerate. In general we shall assume that the transfer matrices of systems of interest are simple and non-negative. For such a matrix a useful result (Sect. 17.2.4) is that it has an eigenvalue equal to its spectral radius. We now introduce a partial ordering of eigenvalues with $|\Lambda_m| \geq |\Lambda_{m'}|$, for $m < m'$, and *principal eigenvalue* $\Lambda_1 = \mathcal{SR}(\boldsymbol{V})$.

---

[2] It is important to emphasize that this non-degeneracy of the largest eigenvalue is for *finite* M. Degeneracy in the limit $N_2 \to \infty$ (and hence $M \to \infty$) is crucial for the occurrence of phase transitions as we shall see in the discussion in Sect. 11.2 and in the treatment of the two-dimensional Ising model in Sect. 12.12.

11.1.2 The Partition Function

The partition function is[3]

$$
\begin{aligned}
Z(\mathcal{N}) &:= \sum_{\{\boldsymbol{\tau}\}} B(\boldsymbol{\tau}(N_1-1), \boldsymbol{\tau}(0)) \prod_{\mathrm{i}=0}^{N_1-2} V(\boldsymbol{\tau}(\mathrm{i}), \boldsymbol{\tau}(\mathrm{i}+1)),\\
&= \mathrm{Trace}\{\boldsymbol{B}\boldsymbol{V}^{N_1-1}\}.
\end{aligned}
\tag{11.1.5}
$$

We shall for the sake of simplicity throughout this chapter assume the application of periodic boundary conditions, so that $\boldsymbol{B} = \boldsymbol{V}$ and

$$
Z(\mathcal{N}) = \mathrm{Trace}\{\boldsymbol{V}^{N_1}\} = \sum_{m=1}^{M} [\Lambda_m]^{N_1}. \tag{11.1.6}
$$

However, it is necessary for development of the finite-lattice series method in Sect. 14.5, to note the case where the lattice has lines of boundary sites at both ends (First and Last) so that

$$
\hat{b}(\boldsymbol{\tau}_m, \boldsymbol{\tau}_{m'}) := \hat{b}^{(\mathrm{F})}(\boldsymbol{\tau}_{m'}) + \hat{b}^{(\mathrm{L})}(\boldsymbol{\tau}_m). \tag{11.1.7}
$$

With the vectors $\boldsymbol{b}^{(\mathrm{F})}$ and $\boldsymbol{b}^{(\mathrm{L})}$ having elements $\exp[-\hat{b}^{(\mathrm{F})}(\boldsymbol{\tau}_m)/T]$ and $\exp[-\hat{b}^{(\mathrm{L})}(\boldsymbol{\tau}_m)/T]$ respectively

$$
\boldsymbol{B} = \boldsymbol{b}^{(\mathrm{L})} \otimes [\boldsymbol{b}^{(\mathrm{F})}]^{(\mathrm{T})} \tag{11.1.8}
$$

and, from (11.1.5),

$$
Z(\mathcal{N}) = [\boldsymbol{b}^{(\mathrm{F})}]^{(\mathrm{T})} \boldsymbol{V}^{N_1-1} \boldsymbol{b}^{(\mathrm{L})}. \tag{11.1.9}
$$

This expression can also be re-expressed in terms of powers of the eigenvalues of $\boldsymbol{V}$, so in both cases the partition function is dependent on N_1 only through the powers of the eigenvalues, although, of course, the number and form of the eigenvalues are dependent on N_2.

We return to the case of periodic boundary conditions where the partition function is given by (11.1.6) and consider three examples:

Case A: V is irreducible and primitive $(p = 1)$:

Λ_1 is the only eigenvalue on the spectral circle $\mathcal{SC}(\boldsymbol{V})$ and, defining scaled eigenvalues

$$
\Omega_m := \Lambda_m / \Lambda_1, \tag{11.1.10}
$$

---

[3] The displayed arguments of functions are treated rather flexibly, with $\mathcal{N}$ sometimes replaced by N_1, N_2 or, in the case where the limit $N_1 \to \infty$ has been taken, $\mathcal{L}_1$, N_2.

in the limit $N_1 \to \infty$, $|\Omega_m|^{N_1} \to 0$, for $m > 1$, giving

$$Z(\mathcal{N}) \simeq [\Lambda_1]^{N_1}. \tag{11.1.11}$$

***Case B: V is irreducible period p > 1*:**

There are p eigenvalues $\Lambda_1 \exp(2\pi \mathrm{i} k/p)$, $k = 0, 1, \ldots, p-1$, on $\mathcal{SC}(\boldsymbol{V})$ and, with M divisible by p, the remaining eigenvalues also fall into sets of p eigenvalues of equal magnitude separated by the same phase factor. $\boldsymbol{V}^{N_1}$ is traceless unless N_1 is divisible by p and restricting N_1 to such values and using (11.1.10), in the limit $N_1 \to \infty$, $|\Omega_m|^{N_1} \to 0$, for $m > p$, giving

$$Z(\mathcal{N}) \simeq p[\Lambda_1]^{N_1}. \tag{11.1.12}$$

***Case C: Λ_1 is of degeneracy ω*:**

A necessary but not sufficient condition for this to be the case is that $\boldsymbol{V}$ is reducible to a block-wise upper triangular matrix with q non-reducible square blocks on the diagonal, where $q \geq \omega$. Since the diagonal blocks are irreducible they each have a non-degenerate largest eigenvalue, of which ω are equal to Λ_1. For simplicity we suppose these blocks are primitive. Using (11.1.10), in the limit $N_1 \to \infty$, $|\Omega_m|^{N_1} \to 0$, for $m > \omega$, giving

$$Z(\mathcal{N}) \simeq \omega[\Lambda_1]^{N_1}. \tag{11.1.13}$$

The factors p in (11.1.12) and ω in (11.1.13) have a negligible effect on the free-energy density, which is given in each case by

$$\phi(\mathcal{L}_1, N_2) = -\lim_{N_1 \to \infty} \frac{\ln[Z(\mathcal{N})]}{N_1 N_2} = -\frac{1}{N_2} \ln[\Lambda_1], \tag{11.1.14}$$

where $\mathcal{L}_1$ indicates that the thermodynamic limit has been taken to a lattice which is infinite in the direction of the interaction between slices.

11.1.3 Correlation Functions and Lengths

The probability that the state of the slice $\mathcal{N}_{\mathrm{i}}$ of the lattice is $\boldsymbol{\tau}_m$ is given by

$$\mathrm{Prob}[\boldsymbol{\tau}(\mathrm{i}) = \boldsymbol{\tau}_m] = \frac{\mathrm{Trace}\{\boldsymbol{V}^{\mathrm{i}-1}|m\rangle\langle m|\boldsymbol{V}^{N_1-\mathrm{i}+1}\}}{\mathrm{Trace}\{\boldsymbol{V}^{N_1}\}} = \frac{\mathrm{Trace}\{\boldsymbol{\Theta}^{(m)}\boldsymbol{\Lambda}^{N_1}\}}{\mathrm{Trace}\{\boldsymbol{\Lambda}^{N_1}\}}, \tag{11.1.15}$$

where

$$\boldsymbol{\Theta}^{(m)} := \boldsymbol{Y}|m\rangle\langle m|\boldsymbol{X}. \tag{11.1.16}$$

As is to be expected this probability is independent of $\mathtt{i}$. Similarly, for $\mathtt{i} \geq \mathtt{i}'$,

$$\mathsf{Prob}\begin{bmatrix} \boldsymbol{\tau}(\mathtt{i}') = \boldsymbol{\tau}_{m'} \\ \boldsymbol{\tau}(\mathtt{i}) = \boldsymbol{\tau}_m \end{bmatrix} = \frac{\mathrm{Trace}\{\boldsymbol{\Theta}^{(m')}\boldsymbol{\Lambda}^{\mathtt{i}-\mathtt{i}'}\boldsymbol{\Theta}^{(m)}\boldsymbol{\Lambda}^{N_1+\mathtt{i}'-\mathtt{i}}\}}{\mathrm{Trace}\{\boldsymbol{\Lambda}^{N_1}\}}. \tag{11.1.17}$$

Formulae can also be obtained for the joint probabilities for the states of other subsets of the slices.[4]

The expectation value of any function $f_{\mathtt{i}} := f(\boldsymbol{\tau}(\mathtt{i}))$ of the state of a single slice $\mathcal{N}_{\mathtt{i}}$ is

$$\langle f_{\mathtt{i}}\rangle = \sum_{m=1}^{M} \mathsf{Prob}[\boldsymbol{\tau}(\mathtt{i}) = \boldsymbol{\tau}_m] f(\boldsymbol{\tau}_m) = \frac{\mathrm{Trace}\{\boldsymbol{\Psi}\boldsymbol{\Lambda}^{N_1}\}}{\mathrm{Trace}\{\boldsymbol{\Lambda}^{N_1}\}} =: \mathcal{F}_1(\mathcal{N}), \tag{11.1.18}$$

given in terms of the Hermitian operator

$$\boldsymbol{\Psi} := \boldsymbol{Y}\boldsymbol{F}\boldsymbol{X}, \quad \text{with} \quad \boldsymbol{F} := \sum_{m=1}^{M} |m\rangle f(\boldsymbol{\tau}_m)\langle m|. \tag{11.1.19}$$

Similarly

$$\begin{aligned}\langle f_{\mathtt{i}'} f_{\mathtt{i}}\rangle &= \sum_{m=1}^{M}\sum_{m'=1}^{M} \mathsf{Prob}\begin{bmatrix} \boldsymbol{\tau}(\mathtt{i}') = \boldsymbol{\tau}_{m'} \\ \boldsymbol{\tau}(\mathtt{i}) = \boldsymbol{\tau}_m \end{bmatrix} f(\boldsymbol{\tau}_{m'}) f(\boldsymbol{\tau}_m) \\ &= \frac{\mathrm{Trace}\{\boldsymbol{\Psi}\boldsymbol{\Lambda}^{\mathtt{i}-\mathtt{i}'}\boldsymbol{\Psi}\boldsymbol{\Lambda}^{N_1+\mathtt{i}'-\mathtt{i}}\}}{\mathrm{Trace}\{\boldsymbol{\Lambda}^{N_1}\}} =: \mathcal{F}_2(\mathcal{N}; \mathtt{i}-\mathtt{i}'). \end{aligned} \tag{11.1.20}$$

The expectation value $\langle f_{\mathtt{i}'} f_{\mathtt{i}}\rangle$ is dependent on $\mathtt{i}$ and $\mathtt{i}'$ only through the difference $\mathtt{i}-\mathtt{i}'$. The pair correlation function between the values of $f(\boldsymbol{\tau})$ at slices $\mathcal{N}_{\mathtt{i}'}$ and $\mathcal{N}_{\mathtt{i}}$ is given (Eq. (2.5.9)) by

$$\Gamma_2(\mathcal{N}; f_{\mathtt{i}'}, f_{\mathtt{i}}) = \langle f_{\mathtt{i}'} f_{\mathtt{i}}\rangle - \langle f_{\mathtt{i}'}\rangle\langle f_{\mathtt{i}}\rangle = \mathcal{F}_2(\mathcal{N}; \mathtt{i}-\mathtt{i}') - [\mathcal{F}_1(\mathcal{N})]^2. \tag{11.1.21}$$

We now obtain expressions for $\Gamma_2(\mathcal{N}; f_{\mathtt{i}'}, f_{\mathtt{i}})$ in the limit $N_1 \to \infty$, for the cases A–C given above. Since the results for case A can be obtained, in the limit $p \to 1$, from B and, when $\omega \to 1$, from C it need not be treated separately. It is straightforward to show that

[4] Given that the energy of the system arises from the interaction between a pair of neighbouring slices, the calculation of the energy-energy correlation function involves probabilities for the states of two pairs of slices. The developments needed to calculate the energy-energy correlation function and a corresponding correlation length are similar to those given in this and the next section.

$$\mathcal{F}_1(\mathcal{L}_1) \simeq \begin{cases} p^{-1} \sum_{m=1}^{p} \langle m|\boldsymbol{\Psi}|m\rangle, & \text{case B}, \\ \omega^{-1} \sum_{m=1}^{\omega} \langle m|\boldsymbol{\Psi}|m\rangle, & \text{case C}, \end{cases} \tag{11.1.22}$$

$$\mathcal{F}_2(\mathcal{L}_1; \mathrm{s}) \simeq \mathcal{F}_2^{(1)}(\mathcal{L}_1; \mathrm{s}) + \mathcal{F}_2^{(2)}(\mathcal{L}_1; \mathrm{s}), \tag{11.1.23}$$

where

$$\mathcal{F}_2^{(1)}(\mathcal{L}_1; \mathrm{s}) = \begin{cases} p^{-1} \sum_{m=1}^{p} \sum_{m'=1}^{p} |\langle m|\boldsymbol{\Psi}|m'\rangle|^2 \exp(2\pi \mathrm{i}\mathrm{s}[m - m']/p), & \text{case B}, \\ \omega^{-1} \sum_{m=1}^{\omega} \sum_{m'=1}^{\omega} |\langle m|\boldsymbol{\Psi}|m'\rangle|^2, & \text{case C}, \end{cases} \tag{11.1.24}$$

$$\mathcal{F}_2^{(2)}(\mathcal{L}_1; \mathrm{s}) = \begin{cases} p^{-1} \sum_{m=p+1}^{M} \sum_{m'=1}^{p} \Omega_m^{\mathrm{s}} \, |\langle m|\boldsymbol{\Psi}|m'\rangle|^2 \exp(-2\pi \mathrm{i}\mathrm{s}[m' - 1]/p), \\ \qquad \text{case B}, \\ \omega^{-1} \sum_{m=\omega+1}^{M} \sum_{m'=1}^{\omega} \Omega_m^{\mathrm{s}} \, |\langle m|\boldsymbol{\Psi}|m'\rangle|^2, \qquad \text{case C}. \end{cases} \tag{11.1.25}$$

Then

$$\Gamma_2(\mathcal{L}_1; f_{\mathrm{i}}, f_{\mathrm{i}'}) = \Gamma_2^{(1)}(\mathcal{L}_1; f_{\mathrm{i}}, f_{\mathrm{i}'}) + \mathcal{F}_2^{(2)}(\mathcal{L}_1; \mathrm{i} - \mathrm{i}'), \tag{11.1.26}$$

where

$$\Gamma_2^{(1)}(\mathcal{L}_1; f_{\mathrm{i}}, f_{\mathrm{i}'}) = \begin{cases} p^{-2} \sum_{m=1}^{p} \sum_{m'=1}^{p} [p|\langle m|\boldsymbol{\Psi}|m'\rangle|^2 \exp(2\pi \mathrm{i}[m - m'][\mathrm{i} - \mathrm{i}']/p) \\ \qquad - \langle m|\boldsymbol{\Psi}|m\rangle\langle m'|\boldsymbol{\Psi}|m'\rangle], \qquad \text{case B}, \\ \omega^{-2} \sum_{m=1}^{\omega} \sum_{m'=1}^{\omega} \{\omega \, |\langle m|\boldsymbol{\Psi}_f|m'\rangle|^2 - \langle m|\boldsymbol{\Psi}_f|m\rangle\langle m'|\boldsymbol{\Psi}_f|m'\rangle\}. \\ \qquad \text{case C}. \end{cases} \tag{11.1.27}$$

In both cases

$$\lim_{|\mathfrak{i}-\mathfrak{i}'|\to\infty} \mathcal{F}_2^{(2)}(\mathcal{L}_1; \mathfrak{i}-\mathfrak{i}') = 0, \tag{11.1.28}$$

but $\Gamma_2^{(1)}(\mathcal{L}_1; f_\mathfrak{i}, f_{\mathfrak{i}'})$ does not decay except in the Case A limit, $p = 1$ or $\omega = 1$, or when it becomes zero because of special symmetry properties.
From (4.2.4), the correlation length with respect to f is

$$\mathfrak{r}^2(\mathcal{L}_1; f) = \frac{a^2}{2} \frac{\sum_{\mathfrak{i}=0}^{\infty} \mathfrak{i}^2 \Gamma_2(\mathcal{L}_1; f_\mathfrak{i}, f_0)}{\sum_{\mathfrak{i}=0}^{\infty} \Gamma_2(\mathcal{L}_1; f_\mathfrak{i}, f_0)}, \tag{11.1.29}$$

where a is a length parameter measuring the distance between $\mathcal{N}_\mathfrak{i}$ and $\mathcal{N}_{\mathfrak{i}+1}$ for any $\mathfrak{i}$, and we have used the result $c(\mathfrak{0}) = c(1) = \frac{1}{2}$ obtained in Sect. 5.6.3. Since the summation is over *all* pairs of slices the implicit assumption, underlying interest in this correlation length, is that the ground state for every slice is the same. That is the ground state is homogeneous, like that of the spin-$\frac{1}{2}$ ferromagnetic Ising model or the ν-state Potts model with $R > 0$.

It is clear from the above discussion that the summations in (11.1.29) converge only when $\Gamma_2^{(1)}(\mathcal{L}_1; f_\mathfrak{i}, f_{\mathfrak{i}'}) = 0$ and, in general this is the case when the transfer matrix is non-reducible and primitive. Then, for $N_1 \to \infty$,

$$\Gamma_2(\mathcal{L}_1; f_\mathfrak{i}, f_0) = \sum_{m=2}^{M} \Omega_m^\mathfrak{i} \, |\langle m|\Psi|1\rangle|^2 \tag{11.1.30}$$

and

$$\mathfrak{r}^2(\mathcal{L}_1; f) = \frac{a^2}{2} \frac{\displaystyle\sum_{m=2}^{M} \frac{\langle m|\Psi|1\rangle|^2 \Omega_m [1+\Omega_m]}{[1-\Omega_m]^3}}{\displaystyle\sum_{m=2}^{M} \frac{|\langle m|\Psi|1\rangle|^2}{1-\Omega_m}}. \tag{11.1.31}$$

If the interest is in a phase transition to a periodic ground state of period η, we need a correlation length based on the correlation between sites at a lattice distance which is an integer multiple of η. The expressions (11.1.29) and (11.1.31) are replaced by

$$\mathfrak{r}^2_\eta(\mathcal{L}_1; f) = \frac{a^2}{2} \frac{\sum_{\mathrm{i}=0}^{\infty} (\mathrm{i}\eta)^2 \Gamma_2(\mathcal{L}_1; f_{\mathrm{i}\eta}, f_0)}{\sum_{\mathrm{i}=0}^{\infty} \Gamma_2(\mathcal{L}_1; f_{\mathrm{i}\eta}, f_0)}, \tag{11.1.32}$$

$$\mathfrak{r}^2_\eta(\mathcal{L}_1; f) = \frac{(\eta a)^2}{2} \frac{\sum_{m=2}^{\infty} \frac{|\langle m|\boldsymbol{\Psi}|1\rangle|^2 (\Omega_k)^\eta \{1 + (\Omega_m)^\eta\}}{\{1 - (\Omega_m)^\eta\}^3}}{\sum_{m=2}^{\infty} \frac{|\langle m|\boldsymbol{\Psi}|1\rangle|^2}{1 - (\Omega_m)^\eta}}. \tag{11.1.33}$$

In the complex plane $\mathbb{C}_\Omega$, the scaled eigenvalues Ω_m are complex numbers lying within the unit-radius spectral circle $\widetilde{SC}(\boldsymbol{V})$. Their number $M = \nu^{N_2}$ increases with increasing N_2 and they will move within the disc as the temperature, or any other thermodynamic variable, is varied. It is clear that the correlation length given by (11.1.31) becomes large when a scaled eigenvalue approaches the point $1 \in \mathbb{C}_\Omega$. This, according to the discussion in Sect. 17.2.4 must mean that $\boldsymbol{V}$ becomes reducible, with two diagonal blocks containing the degenerate eigenvalues Λ_1. In fact, unless this Ω_m is real and the approach is along the real axis the situation will involve a conjugated pair of eigenvalues resulting in a degeneracy of three. The corresponding situation for (11.1.33) occurs when, given that Ω_1 is already equal to one, a scaled eigenvalue approaches $\exp(2\pi \mathrm{i}k/\eta)$, for some $k = 1, \ldots, \eta - 1$. But as we seen in Sect. 17.2.1, this can occur only if eigenvalues tend to all the values $\Lambda_1 \exp(2\pi \mathrm{i}k/\eta)$, $k = 1, \ldots, \eta - 1$. The transfer matrix becomes periodic with period η.

Suppose that a conjugate pair of scaled eigenvalues $\Omega_2 := |\Omega_2| \exp(\mathrm{i}\theta)$ and $\Omega_3 := |\Omega_2| \exp(-\mathrm{i}\theta)$ are such that, for some (small) $\varepsilon > 0$ and $\widetilde{N}$,

$$|\Omega_m - 1| < \varepsilon, \qquad \forall \qquad N_2 > \widetilde{N}, \qquad m = 2, 3. \tag{11.1.34}$$

Then

$$\Omega_m = \exp\{\ln[\Omega_m]\} \simeq 1 + \ln[\Omega_m]. \tag{11.1.35}$$

The summations in (11.1.31) are dominated by these two terms. Since $\langle 2|\boldsymbol{\Psi}|1\rangle$ and $\langle 3|\boldsymbol{\Psi}|1\rangle$ are constructed from the conjugate eigenvectors of $\boldsymbol{V}$, $\langle 3|\boldsymbol{\Psi}|1\rangle = \overline{\langle 2|\boldsymbol{\Psi}|1\rangle}$ giving

$$\mathfrak{r}(f, \mathcal{L}_1) \simeq -a\{\ln|\Omega_2|\}^{-1}, \tag{11.1.36}$$

for any $f(\boldsymbol{\tau})$ which is a function of the state $\boldsymbol{\tau}$ of a single slice.

In the case of the ground-state of periodicity η

$$|\Omega_m - \exp(2\pi \mathrm{i}[m-1]/\eta)| < \varepsilon, \quad \forall \;\; N_2 > \widetilde{N}, \quad m = 2, 3, \ldots, \eta, \tag{11.1.37}$$

giving

$$(\Omega_m)^\eta = \exp\{\ln[(\Omega_m)^\eta]\} \simeq 1 + \ln[(\Omega_m)^\eta]. \tag{11.1.38}$$

The summations in (11.1.33) are dominated by these terms giving

$$\mathfrak{r}_\eta^2(\mathcal{L}_1; f) \simeq (\eta a)^2 \frac{\displaystyle\sum_{m=2}^{\eta} \langle m|\Psi|1\rangle\{\ln[(\Omega_m)^\eta]\}^{-3}}{\displaystyle\sum_{m=2}^{\eta} \langle m|\Psi|1\rangle\{\ln[(\Omega_m)^\eta]\}^{-1}}. \tag{11.1.39}$$

The approach to the spectral circle $\widetilde{\mathcal{SC}}(V)$ will, of course, occur for eigenvalues which are (apart from one approaching the point -1 if η is even) conjugate and approaching conjugate points on the circle. Assuming that they are approaching with equal magnitude $|\Omega_2|$ with

$$(\Omega_m)^\eta = \exp\left[\mathrm{i}\mathrm{Arg}\{(\Omega_m)^\eta\} + \eta \ln|\Omega_2|\right], \tag{11.1.40}$$

and

$$|\mathrm{Arg}\{(\Omega_m)^\eta\}| \ll \eta |\ln|\Omega_2||, \tag{11.1.41}$$

(11.1.39) yields again the formula (11.1.36), which is that used for the correlation length in many investigations which employ transfer matrices (e.g. the phenomenological renormalization method of Sect. 15.10).

11.2 Incipient Phase Transitions

Given that the elements of the transfer matrix can be expressed in terms of some suitable Boltzmann factor $\mathfrak{X}$, it follows that the partition function $Z(\mathcal{N}, \mathfrak{X})$ for finite N_1 and N_2 is a polynomial in $\mathfrak{X}$ with positive coefficients. By the argument used on p. 253 for the simple lattice fluid it follows that the system can have no phase transitions. A modified version of Theorem 8.4.1 can then be used to show that no phase transitions occur in classical one-dimensionally infinite systems with finite-range interaction.[5] Here one must distinguish between the *onset* of long-range order consequent upon the occurrence of a phase transition and the *existence* of a long-range ordered state. As we shall see in Sect. 11.3.2, a state with degenerate eigenvalues, for which the transfer matrix must be reducible, or with other eigenvalues on the

[5] As has been noted in Sect. 3.2.1 a phase transition does occur in the one-dimensional *quantum* Ising model with a transverse field and long-range order is possible in systems with an *infinite-range* interaction (van Enter and Zegarlinski 1997).

spectral circle (particularly an eigenvalue equal to but of opposite sign to the leading eigenvalue) for which the transfer matrix is periodic, is indicative of a state of long-range order with long-range correlations and divergent correlation length. This can occur at zero temperature without it being the result of a finite-temperature phase transition.

Here we are interested in the occurrence of phase transitions at non-zero temperature. Since such do occur, in systems which are more than one-dimensionally infinite, this must involve the double limit when both N_1 and N_2 become large. Ree and Chesnut (1966) have argued that a necessary condition for a phase transition at $\mathfrak{X} = \mathfrak{X}_c$ is that, in the double limit,

$$|\Omega_m(\mathfrak{X})|^{N_2} \to L(\mathfrak{X}), \qquad \text{for some } m \neq 1 \text{ and } L(\mathfrak{X}_c) \neq 0. \tag{11.2.1}$$

To achieve a condition like (11.2.1) it is, of course, necessary to specify the precise way in which the double limit is taken. In any event the condition

$$\lim_{N_2 \to \infty} |\Omega_m(\mathfrak{X}_c)| = 1, \qquad \text{for some } m \neq 1, \tag{11.2.2}$$

is implied by (11.2.1) and is in turn a necessary condition for a phase transition at $\mathfrak{X} = \mathfrak{X}_c$.

We now understand that a phase transition to an ordered phase in a system with interactions only between neighbouring slices $\mathcal{N}_i$ can occur only in the limit of an infinite system, where both N_1 and N_2 become large. It is also clear from (11.1.36) that, for a semi-infinite system ($N_1 \to \infty$), the correlation length becomes large at some $\mathfrak{X} = \mathfrak{X}'$ when one or more of the scaled eigenvalues approaches a point on $\widetilde{SC}(\boldsymbol{V})$. For N_2 finite and the transfer matrix irreducible and primitive all the scaled eigenvalues lie within the circle and the correlation length will be a continuous function of $\mathfrak{X}$ with a maximum at $\mathfrak{X} = \mathfrak{X}'$. A point of closest approach to $\widetilde{SC}(\boldsymbol{V})$ by one or more scaled eigenvalues can be regarded as an *incipient phase transition*. If this eigenvalue $\Omega_m(\mathfrak{X})$ approaches the point 1, then an incipient phase transition is to a homogeneous state. If the approach is to the point $\exp(2\pi\mathrm{i}k/\eta)$, on $\widetilde{SC}(\boldsymbol{V})$, for some k, there will be simultaneous approaches to all the points $\exp(2\pi\mathrm{i}\,k/\eta)$, $k = 1, 2, \ldots, \eta - 1$, indicative of a transition to a ground state of period η.

11.3 Using Symmetry Properties

Let $\mathfrak{Q} := \{\mathrm{I}, \mathrm{Q}^{(1)}, \ldots, \mathrm{Q}^{(n-1)}\}$ be the largest group such that each $\mathrm{Q}^{(s)}$ maps a state $\boldsymbol{\tau}_m$ of a slice of the system into another state $\mathrm{Q}^{(s)}\boldsymbol{\tau}_m$, with $\mathrm{Q}^{(s)}\boldsymbol{\tau}_m \neq \mathrm{Q}^{(s)}\boldsymbol{\tau}_{m'}$ if $\boldsymbol{\tau}_m \neq \boldsymbol{\tau}_{m'}$, and for which $\hat{h}(\boldsymbol{\tau}_m, \boldsymbol{\tau}_{m'})$, and consequently $V(\boldsymbol{\tau}_m, \boldsymbol{\tau}_{m'})$, is left invariant. When periodic boundary conditions are applied in the $\boldsymbol{j}$ direction, so that the lattice forms a cylinder, the symmetry group $\mathfrak{Q}$ will normally include the cyclic group corresponding to rotation about the axis of the cylinder. Other subgroups of $\mathfrak{Q}$ will correspond to any symmetry of the Hamiltonian arising from permutations of the states of the microsystems. For the standard ν-state Potts model $\mathcal{Q}$ will contain the symmetric group $\mathcal{S}_\nu$.

11.3.1 Block-Diagonalization of the Transfer Matrix

The effect of $\mathcal{Q}$ on the states $\{\boldsymbol{\tau}_m\}$ is given, using the representation $\mathcal{R}(\mathcal{Q}) := \{\boldsymbol{I}, \boldsymbol{Q}^{(1)}, \dots, \boldsymbol{Q}^{(\mathrm{n}-1)}\}$ of M-dimensional permutation matrices,[6] by

$$\boldsymbol{Q}^{(\mathrm{s})} \begin{pmatrix} \boldsymbol{\tau}_1 \\ \boldsymbol{\tau}_2 \\ \vdots \\ \boldsymbol{\tau}_M \end{pmatrix} = \begin{pmatrix} \mathrm{Q}^{(\mathrm{s})}\boldsymbol{\tau}_1 \\ \mathrm{Q}^{(\mathrm{s})}\boldsymbol{\tau}_2 \\ \vdots \\ \mathrm{Q}^{(\mathrm{s})}\boldsymbol{\tau}_M \end{pmatrix}, \qquad \mathrm{s} = 0, 1, \dots, \mathrm{n} - 1 \tag{11.3.1}$$

and the symmetry property of the transfer matrix can be represented in the form

$$\boldsymbol{Q}^{(\mathrm{s})} \boldsymbol{V} [\boldsymbol{Q}^{(\mathrm{s})}]^{\dagger} = \boldsymbol{V}, \qquad \text{or equivalently} \qquad [\boldsymbol{Q}^{(\mathrm{s})}, \boldsymbol{V}] = 0,$$
$$\mathrm{s} = 0, 1, \dots, \mathrm{n} - 1. \tag{11.3.2}$$

The irreducible representations of $\mathcal{Q}$ are denoted by $\Theta(\mathrm{k})$, for $\mathrm{k} = 1, 2, \dots, \kappa$, with $\mathrm{e}(\mathrm{k})$ being the dimension of $\Theta(\mathrm{k})$.

Equation (11.3.2) encapsulates the symmetry properties of the transfer matrix with respect to the representation $\mathcal{R}(\mathcal{Q})$ of the group $\mathcal{Q}$, and (17.3.10) asserts that a matrix $\boldsymbol{U}$ exists which reduces all the matrices of $\mathcal{R}(\mathcal{Q})$ to block-diagonal form. It can be shown that, if the matrix $\boldsymbol{U}$ is known, it can be used, together with row and column permutations, to reduce $\boldsymbol{V}$ to a block-diagonal form, each block being one of a set of $\mathrm{m}(\mathrm{k})$ identical matrices identified with the representation $\Theta(\mathrm{k})$. This reduction is to the unique *canonical form* of $\boldsymbol{V}$. However, in general achieving this can be quite difficult and a simpler and often quite sufficient reduction is that given in (17.3.15) by the matrix $\boldsymbol{W}$ since, as can be easily shown:

Theorem 11.3.1 *If* $\boldsymbol{W}$ *reduces each of the matrices of the group* $\mathcal{Q}$ *to block-diagonal form, with one block for each irreducible representation* $\Theta(\mathrm{k})$, *as shown in* (17.3.15), *then the transfer matrix satisfying* (11.3.2) *is reduced in a similar way. That is*

$$\boldsymbol{W}^{\dagger} \boldsymbol{V} \boldsymbol{W} = \widetilde{\boldsymbol{V}} := \widetilde{\boldsymbol{V}}(1) \oplus \widetilde{\boldsymbol{V}}(2) \oplus \cdots \oplus \widetilde{\boldsymbol{V}}(\kappa). \tag{11.3.3}$$

Of course, if $\mathrm{m}(\mathrm{k}) = 1$ for each representation then $\boldsymbol{U} = \boldsymbol{W}$ and reduction to the canonical form is complete. Otherwise a further reduction can be made of the block for $\Theta(\mathrm{k})$ to $\mathrm{m}(\mathrm{k})$ identical sub-blocks.

Any matrix which, like $\boldsymbol{V}$, satisfies (11.3.2) is said to be Q–*invariant*. The problem of reducing such a matrix to block-diagonal form is simplified if the states of the

---

[6] For a brief account of group representation theory and of the notation used here see Sect. 17.3.1.

system are ordered into *equivalence classes*, closed under the operations of the group $\mathcal{Q}$ and not further divisible into closed subsets. Using the notation described in Sect. 17.3.2 with the equivalence classes denoted by $\mathfrak{E}(\alpha)$, $\alpha = 1, 2, \ldots, \gamma$, it can be shown (Theorem 17.3.7) that the equivalence class $\mathfrak{E}(\alpha)$ can be generated by applying all the elements of $\mathcal{Q}$ to any one state of $\mathfrak{E}(\alpha)$ and the number of states $\mathfrak{p}(\alpha)$ in $\mathfrak{E}(\alpha)$ is a divisor of $\mathfrak{n}$. It follows, from (11.3.1), that $\boldsymbol{Q}^{(\mathrm{s})}$ has the block-diagonal form (17.3.23) with $\mathcal{R}(\alpha; \mathcal{Q}) := \{\boldsymbol{I}(\alpha), \boldsymbol{Q}^{(1)}(\alpha), \ldots, \boldsymbol{Q}^{(\mathfrak{n}-1)}(\alpha)\}$, a $\mathfrak{p}(\alpha)$-dimensional representation of $\mathcal{Q}$ operating on the equivalence class $\mathfrak{E}(\alpha)$, for $\alpha = 1, 2, \ldots, \gamma$. The formula (17.3.24) of the same form as (17.3.15) now applies to the matrices $\boldsymbol{Q}^{(\mathrm{s})}(\alpha)$. The matrix $\boldsymbol{W}(\alpha)$ have the effect of reducing $\boldsymbol{Q}^{(\mathrm{s})}(\alpha)$ to block-diagonal form with the blocks corresponding to the irreducible representations in order and in Sect. 17.3.3 we describe a method for deriving it.

Suppose that $\boldsymbol{V}$ is expressed in the form

$$\boldsymbol{V} := \begin{pmatrix} \boldsymbol{V}(1,1) & \boldsymbol{V}(1,2) & \cdots & \boldsymbol{V}(1,\gamma) \\ \boldsymbol{V}(2,1) & \boldsymbol{V}(2,2) & \cdots & \boldsymbol{V}(2,\gamma) \\ \vdots & \vdots & \ddots & \vdots \\ \boldsymbol{V}(\gamma,1) & \boldsymbol{V}(\gamma,2) & \cdots & \boldsymbol{V}(\gamma,\gamma) \end{pmatrix}, \tag{11.3.4}$$

where $\boldsymbol{V}(\alpha, \alpha')$ is the submatrix with row and column numbers in the equivalence classes α and α' respectively. The matrix

$$\check{\boldsymbol{V}} := [\boldsymbol{W}(1) \oplus \boldsymbol{W}(2) \oplus \cdots \oplus \boldsymbol{W}(\gamma)]^{\dagger} \boldsymbol{V} [\boldsymbol{W}(1) \oplus \boldsymbol{W}(2) \oplus \cdots \oplus \boldsymbol{W}(\gamma)] \tag{11.3.5}$$

has submatrices $\check{\boldsymbol{V}}(\alpha, \alpha') = \boldsymbol{W}^{\dagger}(\alpha)\boldsymbol{V}(\alpha, \alpha')\boldsymbol{W}(\alpha')$ which, from (11.3.2), (17.3.23) and (17.3.24), satisfy the equations

$$[\widetilde{\boldsymbol{Q}}^{(\mathrm{s})}(\alpha)]^{\dagger} \check{\boldsymbol{V}}(\alpha, \alpha') \widetilde{\boldsymbol{Q}}^{(\mathrm{s})}(\alpha') = \check{\boldsymbol{V}}(\alpha, \alpha'), \qquad \mathrm{s} = 1, 2, \ldots, \mathfrak{n} - 1. \tag{11.3.6}$$

Using analysis similar to that in the proof of Theorem 11.3.1, it can be shown that the rows and columns of $\check{\boldsymbol{V}}(\alpha, \alpha')$ can be indexed by the pair (k, j) with $j = 1, 2, \ldots, \mathrm{w}(\alpha; \mathrm{k})$ and

$$\langle \mathrm{k}, j | \check{\boldsymbol{V}}(\alpha, \alpha') | \mathrm{k}', j' \rangle = \delta^{\mathrm{Kr}}(\mathrm{k} - \mathrm{k}') \langle \mathrm{k}, j | \check{\boldsymbol{V}}(\alpha, \alpha') | \mathrm{k}, j' \rangle. \tag{11.3.7}$$

For every irreducible representation $\Theta(\mathrm{k})$ which occurs in both $\mathfrak{C}(\alpha)$ and $\mathfrak{C}(\alpha')$ there will be $\mathrm{w}(\alpha; \mathrm{k}) \times \mathrm{w}(\alpha'; \mathrm{k})$ elements of $\check{\boldsymbol{V}}(\alpha, \alpha')$ which can be non-zero. The (possibly) non-zero elements of $\check{\boldsymbol{V}}(\alpha, \alpha')$ are given by

$$\langle \mathrm{k}, j | \check{\boldsymbol{V}}(\alpha, \alpha') | \mathrm{k}, j' \rangle = \sum_{m=1}^{\mathfrak{p}(\alpha)} \sum_{m'=1}^{\mathfrak{p}(\alpha')} \overline{W}_{mj}^{(\mathrm{k})}(\alpha) \langle m | \boldsymbol{V}(\alpha, \alpha') | m' \rangle W_{m'j'}^{(\mathrm{k})}(\alpha'), \tag{11.3.8}$$

where the quantities $W_{mj}^{(k)}(\alpha)$ are defined in Sect. 17.3.3. In particular, from (17.3.43), for the one-dimension symmetric representation $w(\alpha; 1) = 1$; each submatrix contributes exactly one term given by

$$\langle 1, 1|\mathring{V}(\alpha, \alpha')|1, 1\rangle = \frac{1}{\sqrt{\mathfrak{p}(\alpha)\mathfrak{p}(\alpha')}} \sum_{m=1}^{\mathfrak{p}(\alpha)} \sum_{m'=1}^{\mathfrak{p}(\alpha')} \langle m|V(\alpha, \alpha')|m'\rangle. \tag{11.3.9}$$

The permutation matrix $\boldsymbol{\Pi}$ of equation (17.3.25) collects them together so that

$$\boldsymbol{\Pi}\mathring{V}\boldsymbol{\Pi} = \widetilde{V}, \tag{11.3.10}$$

has the block form given in (11.3.3).

11.3.1.1 Eigenvalues and Eigenvectors

The eigenvalues of $\widetilde{V}$ given by (11.3.3), are the same as those of $\boldsymbol{V}$. Each eigenvalue can, therefore, be identified with some irreducible representation $\Theta(\mathrm{k})$ within which an ordering in decreasing magnitude is applied. Thus, for $\boldsymbol{V}$, we replace the notation $(\Lambda_k, \boldsymbol{y}_k, \boldsymbol{x}_k)$, $k = 1, 2, \ldots, M$ for eigenvalues and vectors with $(\Lambda_\ell(\mathrm{k}), \boldsymbol{y}_\ell(\mathrm{k}), \boldsymbol{x}_\ell(\mathrm{k}))$, $\mathrm{k} = 1, 2, \ldots, \kappa$, $\ell = 1, 2, \ldots, w(\mathrm{k})$. From the commutativity condition (11.3.2) and Theorem 17.2.4 it follows that:

Theorem 11.3.2 *If* $(\boldsymbol{y}_\ell(\mathrm{k}), \boldsymbol{x}_\ell(\mathrm{k}))$ *is a non-degenerate eigenvector pair of* $\boldsymbol{V}$ *it is also an eigenvector pair of* $\boldsymbol{Q}^{(\mathrm{s})}$ *with eigenvalue* $\chi^{(\mathrm{k})}(\mathrm{Q}^{(\mathrm{s})})/\mathrm{e}(\mathrm{k})$, *for* $\mathrm{s} = 0, 1, \ldots, \mathfrak{n} - 1$.

The M-dimensional vector $\boldsymbol{v}$ is a *vector-invariant* of Q if

$$\boldsymbol{Q}^{(\mathrm{s})}\boldsymbol{v} = \boldsymbol{v}, \qquad \forall\ \mathrm{s} = 0, 1, \ldots, \mathfrak{n} - 1. \tag{11.3.11}$$

Suppose the elements of $\boldsymbol{v}$ are ordered in equivalence classes with $\{\boldsymbol{Q}^{(\mathrm{s})}\}$ in the block-diagonal form (17.3.23), so that

$$\boldsymbol{v} = \begin{pmatrix} v(1)\boldsymbol{j}(1) \\ v(2)\boldsymbol{j}(2) \\ \vdots \\ v(\gamma)\boldsymbol{j}(\gamma) \end{pmatrix}, \tag{11.3.12}$$

where $\boldsymbol{j}(\alpha)$ is the vector of dimension $\mathfrak{p}(\alpha)$ with every entry unity. Then:

Theorem 11.3.3 *Non-degenerate eigenvectors* $\boldsymbol{y}_\ell(\mathrm{k})$ *and* $\boldsymbol{x}_\ell(\mathrm{k})$ *with* $\mathrm{k} \neq 1$ *are orthogonal to all vector-invariants of the group.*

Proof From Theorem 11.3.2 and (11.3.11), for some vector-invariant $\boldsymbol{v}$,

$$[\boldsymbol{y}_\ell(\mathrm{k})]^{\mathrm{T}}\boldsymbol{v} = [\boldsymbol{y}_\ell(\mathrm{k})]^{\mathrm{T}}\boldsymbol{Q}^{(\mathrm{s})}\boldsymbol{v} = \chi^{(\mathrm{k})}(\mathrm{Q}^{(\mathrm{s})})[\boldsymbol{y}_\ell(\mathrm{k})]^{\mathrm{T}}\boldsymbol{v}/\mathrm{e}(\mathrm{k}), \tag{11.3.13}$$

for $\mathrm{s} = 0, 1, \ldots, \mathrm{n}-1$. Since the only representation for which $\chi^{(\mathrm{k})}(\mathrm{Q}^{(\mathrm{s})}) = \mathrm{e}(\mathrm{k})$, for all $\mathrm{s} = 0, 1, \ldots, \mathrm{n}-1$, is $\Theta(1)$, it follows that $[\boldsymbol{y}_\ell(\mathrm{k})]^{\mathrm{T}}\boldsymbol{v} = 0$, unless $\mathrm{k} = 1$, with the same analysis applying to the right eigenvector $\boldsymbol{x}_\ell(\mathrm{k})$. □

As we have seen, given that $\boldsymbol{V}$ is irreducible and primitive (Case A, Sect. 11.1.2), the principal eigenvalue is positive and of larger magnitude than all other eigenvalues with positive left and right eigenvectors. But such eigenvectors cannot be orthogonal to vector invariants of the form (11.3.12) with $f(\alpha) > 0$ for $\alpha = 1, 2, \ldots, \gamma$. It follows that:

Theorem 11.3.4 *For such a V the principal eigenvalue is in the one-dimensional symmetric representation (with designation* $\Lambda_1(1)$*).*

This is a particular useful result, since, if we need only the largest eigenvalue, it can be obtained by solving the eigenvalue problem for $\widetilde{\boldsymbol{V}}(1)$ and this matrix can be obtained very simply from $\boldsymbol{V}$ using (11.3.9), thus yielding the partition function and the free-energy density in the limit $N_1 \to \infty$.

However, even an approximate calculation of the correlation function, given, from (11.1.30), by

$$\Gamma_2(\mathcal{L}_1; f_{\mathrm{i}}, f_{\mathrm{i}'}) = \sum_{\{\mathrm{k},\ell\}}^{[1,1]} |\langle \mathrm{k}, \ell|\boldsymbol{\Psi}|1\rangle|^2 [\Omega_\ell(\mathrm{k})]^{\mathrm{i}-\mathrm{i}'}, \tag{11.3.14}$$

where '[1, 1]' denotes the fact that the largest eigenvalue is excluded from the summation, and

$$|\langle \mathrm{k}, \ell|\boldsymbol{\Psi}|1\rangle|^2 = \langle 1, 1|\boldsymbol{YFX}|\mathrm{k}, \ell\rangle\langle \mathrm{k}, \ell|\boldsymbol{YFX}|1, 1\rangle, \tag{11.3.15}$$

still requires the computation of the largest sub-maximum eigenvalues. There is, in general, no reason to suppose that these lie in the one-dimensional symmetric matrix block. Even though block-diagonalization still has the advantage of allowing the handling of a set of smaller matrices, it is in principle necessary to compute the eigenvalues for each block. However, there is an exceptional case. If $\boldsymbol{F}$ is Q–invariant, meaning that function $f(\boldsymbol{\tau}_m)$ has the same value for all states in an equivalence class, $\langle 1, 1|\boldsymbol{YFX}|\mathrm{k}, \ell\rangle = 0$ for $\mathrm{k} \neq 1$. The correlation function and the correlation length are given by just the eigenvalues and vectors in $\Theta(1)$. So the problem is reduced to looking for the largest eigenvalues in $\widetilde{\boldsymbol{V}}(1)$, without the need to consider the other matrix blocks.

11.3.2 Applications

As illustrations of the methods described here we consider two examples with a positive transfer matrix, the spin-$\frac{1}{2}$ Ising model and the 3-state Potts model on a square lattice wrapped around a cylinder with N_1 rings of N_2 sites. In each case the symmetry group for lattice rotations and reflections is $\mathcal{C}_{N_2}$ and the symmetry groups arising from the nature of the models are $\mathcal{S}_2$ and $\mathcal{S}_3$ respectively. Given the advances in algebraic computing software in recent years, block-diagonalization of the transfer matrix is not particularly advantageous for small values of N_2. However, as N_2 increases, a factor which determines the virtue of block-diagonal is the difference between the matrix size, ν_2^N, with $\nu = 2$ and 3 for the Ising and Potts models respectively, and the size $\mathfrak{m}(1; N_2)$ of the one-dimension block.

As an indication of the numbers involved, we consider the Ising model, ignoring, for simplicity, reflections, so that the symmetry group is $\mathcal{Q} = \mathcal{C}_{N_2} \times \mathcal{S}_2$.[7] By counting the states left invariant by this group, it is not difficult to see that, for $N_2 \geq 2$,

$$\mathfrak{m}(1; N_2) = \begin{cases} 2^{\frac{1}{2}(N_2-3)} + \dfrac{1}{4N_2} \displaystyle\sum_{s=0}^{N_2-1} 2^{[s,N_2]}, & N_2 \text{ odd}, \\[2ex] 2^{\frac{1}{2}N_2-1} + \dfrac{1}{4N_2} \displaystyle\sum_{s=0}^{N_2-1} 2^{[s,N_2]} \Delta(N_2, s), & N_2 \text{ even}, \end{cases} \tag{11.3.16}$$

where

$$\Delta(N_2, s) = 2 - \text{Parity}\left(\frac{N_2}{[s, N_2]}\right) \tag{11.3.17}$$

and $[s, N_2]$ is the greatest common divisor of s and N_2. As examples: $2^4 = 16$, $\mathfrak{m}(1; 4) = 4$; $2^{10} = 1024$, $\mathfrak{m}(1; 10) = 44$.

Example 11.3.1 For the Ising model the states of pairs, ordered in the equivalence classes, are

$$\left\{\begin{bmatrix} +1 \\ +1 \end{bmatrix}, \begin{bmatrix} -1 \\ -1 \end{bmatrix}\right\}, \left\{\begin{bmatrix} +1 \\ -1 \end{bmatrix}, \begin{bmatrix} -1 \\ +1 \end{bmatrix}\right\} \tag{11.3.18}$$

and, with $K := 2J/T$, $\mathfrak{X} := \exp(K)$, the transfer matrix takes the form

$$\boldsymbol{V} = \begin{pmatrix} \mathfrak{X}^2 & 1 & 1 & 1 \\ 1 & \mathfrak{X}^2 & 1 & 1 \\ 1 & 1 & 1 & \mathfrak{X}^{-2} \\ 1 & 1 & \mathfrak{X}^{-2} & 1 \end{pmatrix}. \tag{11.3.19}$$

[7] Including reflections will, of course, reduce $\mathfrak{m}(1; N_2)$.

For the symmetry group $\mathcal{Q} := \mathcal{S}_2^{(m)} \times \mathcal{C}_2$, the conjugacy classes, irreducible representations and characters are given in Example 17.3.3, with

$$\begin{aligned} \boldsymbol{R}(\mathrm{I}) &= \begin{pmatrix} \boldsymbol{I}_2 & \boldsymbol{0}_2 \\ \boldsymbol{0}_2 & \boldsymbol{I}_2 \end{pmatrix}, & \boldsymbol{R}(\mathrm{Q}^{(m)}) &= \begin{pmatrix} \boldsymbol{J}_2 & \boldsymbol{0}_2 \\ \boldsymbol{0}_2 & \boldsymbol{J}_2 \end{pmatrix}, \\ \boldsymbol{R}(\mathrm{C}_2) &= \begin{pmatrix} \boldsymbol{I}_2 & \boldsymbol{0}_2 \\ \boldsymbol{0}_2 & \boldsymbol{J}_2 \end{pmatrix}, & \boldsymbol{R}(\mathrm{Q}^{(m)}\mathrm{C}_2) &= \begin{pmatrix} \boldsymbol{J}_2 & \boldsymbol{0}_2 \\ \boldsymbol{0}_2 & \boldsymbol{I}_2 \end{pmatrix}, \end{aligned} \tag{11.3.20}$$

where $\boldsymbol{I}_2$ and $\boldsymbol{0}_2$ are the two-dimensional unit and zero matrices, respectively, and

$$\boldsymbol{J}_2 := \begin{pmatrix} 0 & 1 \\ 1 & 0 \end{pmatrix}. \tag{11.3.21}$$

From the character table the multiplicities of the irreducible representations in the four-dimensional representation of states are $\mathfrak{m}(\mathrm{A}) = 2$, $\mathfrak{m}(\mathrm{B}_1) = 1$, $\mathfrak{m}(\mathrm{B}_2) = 0$ and $\mathfrak{m}(\mathrm{B}_3) = 1$. The matrix which reduces the transfer matrix to diagonal form is $\boldsymbol{W} \oplus \boldsymbol{W}$ where

$$\boldsymbol{W} = \begin{pmatrix} \frac{1}{\sqrt{2}} & \frac{1}{\sqrt{2}} \\ \frac{1}{\sqrt{2}} & -\frac{1}{\sqrt{2}} \end{pmatrix}, \tag{11.3.22}$$

yielding

$$\widetilde{\boldsymbol{V}}(\mathrm{A}) = \begin{pmatrix} \mathfrak{X}^2 + 1 & 2 \\ 2 & 1 + \mathfrak{X}^{-2} \end{pmatrix}, \tag{11.3.23}$$

$$\widetilde{\boldsymbol{V}}(\mathrm{B}_1) = \mathfrak{X}^2 - 1, \qquad \widetilde{\boldsymbol{V}}(\mathrm{B}_3) = 1 - \mathfrak{X}^{-2}. \tag{11.3.24}$$

The characteristic equation of the A is

$$\Lambda^2 - (\mathfrak{X} + \mathfrak{X}^{-1})^2 \Lambda + (\mathfrak{X} - \mathfrak{X}^{-1})^2 = 0, \tag{11.3.25}$$

giving

$$\left.\begin{matrix} \Lambda_1(\mathrm{A}; 2, \mathfrak{X}) \\ \Lambda_2(\mathrm{A}; 2, \mathfrak{X}) \end{matrix}\right\} = \tfrac{1}{2}\left[(\mathfrak{X} + \mathfrak{X}^{-1})^2 \pm \sqrt{(\mathfrak{X} + \mathfrak{X}^{-1})^4 - 4(\mathfrak{X} - \mathfrak{X}^{-1})^2}\right], \tag{11.3.26}$$

$$\Lambda_1(\mathrm{B}_1; 2, \mathfrak{X}) = \mathfrak{X}^2 - 1, \tag{11.3.27}$$

$$\Lambda_1(\mathrm{B}_3; 2, \mathfrak{X}) = 1 - \mathfrak{X}^{-2}. \tag{11.3.28}$$

Example 11.3.2 For the 3-state Potts model the Hamiltonian is given by (3.6.49), where $\sigma(\boldsymbol{r})$ takes three consecutive values. In the convention of Sect. 3.6.4 these would be 0, 1, 2. However it is more convenient here to choose, as in Sect. 5.4.2, the spin-1 Ising values $-1, 0, 1$.

With $N_2 = 1$ the symmetry group is $\mathcal{S}_3$:

We choose the same three group generators as in Sect. 5.4.2: I, $Q^{(m)}$, which simply interchanges +1 and −1 and $Q^{(*)}$, given by (5.4.3), which cyclically permutes the states. The three-dimensional representation of $\mathcal{S}_3$ operating on $\boldsymbol{\sigma} := (-1, 0, +1)^{\mathrm{T}}$ has

$$\boldsymbol{R}(\mathrm{I}) = \boldsymbol{I}_3, \qquad \boldsymbol{R}(\mathrm{Q}^{(*)}) = \boldsymbol{S}_3, \qquad \boldsymbol{R}(\mathrm{Q}^{(m)}) = \boldsymbol{J}_3, \tag{11.3.29}$$

where

$$\boldsymbol{S}_3 := \begin{pmatrix} 0 & 1 & 0 \\ 0 & 0 & 1 \\ 1 & 0 & 0 \end{pmatrix}, \qquad \boldsymbol{J}_3 := \begin{pmatrix} 0 & 0 & 1 \\ 0 & 1 & 0 \\ 1 & 0 & 0 \end{pmatrix}. \tag{11.3.30}$$

The characters for the conjugacy classes defined as in Table 5.1 are $\chi(\mathrm{Q}_1) = 3$, $\chi(\mathrm{Q}_2)=0$ and $\chi(\mathrm{Q}_3) = 1$, and it follows from (17.3.13) that the multiplicities of the irreducible representations are $\mathfrak{m}(\mathsf{A})=1$, $\mathfrak{m}(\mathsf{B})=0$, $\mathfrak{m}(\mathsf{E}) = 1$. The reduction matrix

$$\boldsymbol{W}_3 := \frac{1}{\sqrt{3}}\begin{pmatrix} 1 & 1 & 1 \\ 1 & \omega_3 & \omega_3^2 \\ 1 & \omega_3^2 & \omega_3 \end{pmatrix}, \qquad \omega_3 := \exp(2\pi\mathrm{i}/3), \tag{11.3.31}$$

is given by Ree and Chesnut (1966), with

$$\boldsymbol{W}_3^\dagger \boldsymbol{S}_3 \boldsymbol{W}_3 = \begin{pmatrix} 1 & 0 & 0 \\ 0 & \omega_3 & 0 \\ 0 & 0 & \omega_3^2 \end{pmatrix}, \qquad \boldsymbol{W}_3^\dagger \boldsymbol{J}_3 \boldsymbol{W}_3 = \begin{pmatrix} 1 & 0 & 0 \\ 0 & 0 & \omega_3^2 \\ 1 & \omega_3^2 & 0 \end{pmatrix}. \tag{11.3.32}$$

Using the Hamiltonian (3.6.49) the transfer matrix is

$$\boldsymbol{V}_3 = \begin{pmatrix} \mathfrak{X} & 1 & 1 \\ 1 & \mathfrak{X} & 1 \\ 1 & 1 & \mathfrak{X} \end{pmatrix} \tag{11.3.33}$$

where $\mathfrak{X} := \exp(K)$, $K := R/T$ and

$$\boldsymbol{W}_3^\dagger \boldsymbol{V}_3 \boldsymbol{W}_3 = \widetilde{\boldsymbol{V}}_3 = \begin{pmatrix} \mathfrak{X}+2 & 0 & 0 \\ 0 & \mathfrak{X}-1 & 0 \\ 0 & 0 & \mathfrak{X}-1 \end{pmatrix}. \tag{11.3.34}$$

Since there is only one equivalence class the eigenvalue $\Lambda_1(\mathsf{A}; \mathfrak{X}) = \mathfrak{X} + 2$ for the representation A can, according to (11.3.9), also be obtained as a weighted average of the elements of $\boldsymbol{V}_3$. It is also fortunately the case that the two-dimensional block associated with E is diagonal; $\Lambda_1(\mathsf{E}; \mathfrak{X}) = \Lambda_2(\mathsf{E}; \mathfrak{X}) = \mathfrak{X} - 1$ and the right and left eigenvectors of $\boldsymbol{V}_3$ are respectively the columns of $\boldsymbol{W}_3$ and the rows of $\boldsymbol{W}_3^\dagger$. From (11.1.14) the free energy per lattice site, in the limit $N \to \infty$, $\mathcal{N} \to \mathcal{L}_1$, is

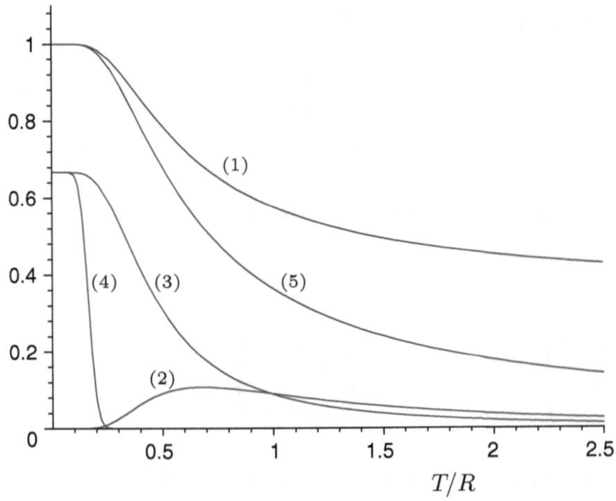

Fig. 11.1 Plots of *(1)* $-\mathrm{u}(\mathcal{L}_1, K)$, *(2)* $c(\mathcal{L}_1, K)/k_\mathrm{B}$, *(3)* $\Gamma_2(\mathcal{L}_1; \sigma(\mathrm{i}), \sigma(\mathrm{i}+2))$ *(4)* $\Gamma_2(\mathcal{L}_1; \sigma(\mathrm{i}), \sigma(\mathrm{i}+100))$ and *(5)* $\Omega_1(\mathrm{E}; K)$, against $T/R = 1/K$

$$\phi_0(\mathcal{L}_1, K) = -\ln[\Lambda_1(\mathrm{A}; \mathfrak{X})] = -\ln[\exp(K) + 2] \tag{11.3.35}$$

and, from (2.3.27) and (2.3.29), the internal energy per lattice site (in units of R) and the heat capacity are respectively

$$\mathrm{u}(\mathcal{L}_1, K) := \frac{\partial \phi_0}{\partial K} = -\frac{\exp(K)}{\exp(K) + 2}, \tag{11.3.36}$$

$$c(\mathcal{L}_1, K) := -k_\mathrm{B} K^2 \left(\frac{\partial \mathrm{u}}{\partial K}\right) = 2k_\mathrm{B} \left[\frac{K}{\exp(K) + 2}\right]^2. \tag{11.3.37}$$

We see that $c(\mathcal{L}_1, K) \to 0$ as $T \to 0$ and as $T \to \infty$. A plot against temperature is shown as plot (2) in Fig. 11.1.

It is straightforward using (11.1.18)–(11.1.20), (11.1.26) and $\boldsymbol{X} := \boldsymbol{W}_3$, $\boldsymbol{Y} := \boldsymbol{W}_3^\dagger$, to show that, in the limit $N \to \infty$,

$$\langle \sigma(\mathrm{i}) \rangle = 0, \qquad \langle \sigma(\mathrm{i})\sigma(\mathrm{i}+\mathrm{s}) \rangle = \tfrac{2}{3}[\Omega_1(\mathrm{E}; K)]^\mathrm{s}, \tag{11.3.38}$$

$$\langle \sigma^2(\mathrm{i}) \rangle = \tfrac{2}{3}, \qquad \langle \sigma^2(\mathrm{i})\sigma^2(\mathrm{i}+\mathrm{s}) \rangle = \tfrac{4}{9} + \tfrac{2}{9}[\Omega_1(\mathrm{E}; K)]^\mathrm{s}, \tag{11.3.39}$$

$$\Gamma_2(\mathcal{L}_1; \sigma(\mathrm{i}), \sigma(\mathrm{i}+\mathrm{s})) = \tfrac{2}{3}[\Omega_1(\mathrm{E}; K)]^\mathrm{s}, \tag{11.3.40}$$

$$\Gamma_2(\mathcal{L}_1; \sigma^2(\mathrm{i}), \sigma^2(\mathrm{i}+\mathrm{s})) = \tfrac{2}{9}[\Omega_1(\mathrm{E}; K)]^\mathrm{s}, \tag{11.3.41}$$

where

$$\Omega_1(\mathsf{E}; K) := \frac{\Lambda_1(\mathsf{E}; \mathfrak{X})}{\Lambda_1(\mathsf{A}; \mathfrak{X})} = \frac{\exp(K) - 1}{\exp(K) + 2}. \tag{11.3.42}$$

From (11.1.29),

$$\mathfrak{r}^2(\mathcal{L}_1; \sigma) = \mathfrak{r}^2(\mathcal{L}_1; \sigma^2) = \tfrac{1}{18} a^2 [\exp(K) - 1][2\exp(K) + 1]. \tag{11.3.43}$$

The first quantities in (11.3.38) and (11.3.39) correspond respectively to the magnetization and density when $\mathcal{H} = \mu = 0$. Thus we see, as expected that this model has no spontaneous magnetization and the expected density of microsystems in each of the three states is $\frac{1}{3}$. Using (3.6.59), (11.3.36), (11.3.38) and (11.3.39) confirms the expected result that $\mathrm{u} = -\langle \delta^{\mathrm{Kr}}(\sigma(\mathrm{i}) - \sigma(\mathrm{i}+1)) \rangle$.

Both correlation functions (11.3.40) and (11.3.41) are, for an sth neighbour pair, independent of position on the lattice and decrease as s increases (cf. plots (3) and (4) in Fig. 11.1). For any fixed value of s, they tend to one as $T \to 0$ and to zero as $T \to \infty$. The picture which emerges is that, although the expectation values of the total numbers of microsystems in each state remain the same, blocks of microsystems in the same state form and steadily increase in length as temperature decreases. In an ordered ground state $\mathrm{u}(\mathcal{L}_1, \infty) = -\frac{1}{2}\mathrm{z} = -1$ and the approach of $\mathrm{u}(\mathcal{L}_1, K)$ to this limiting value corresponds to the number of unlike first-neighbour pairs, which represent block boundaries, becoming very small relative to the number of like pairs at low temperatures. Although there is no *long-range order* in the sense of a majority of microsystems being in the same state, the properties of the system at low temperatures are dominated by the short-range order.

With $N_2 = 2$ ***the symmetry group is*** $\mathfrak{S}_3 \times \mathfrak{C}_2$:

The states of a ring, arranged in equivalence classes, are

$$\left(\left\{\begin{bmatrix}-1\\-1\end{bmatrix}, \begin{bmatrix}0\\0\end{bmatrix}, \begin{bmatrix}+1\\+1\end{bmatrix}\right\},\right.$$
$$\left.\left\{\begin{bmatrix}-1\\0\end{bmatrix}, \begin{bmatrix}0\\+1\end{bmatrix}, \begin{bmatrix}+1\\-1\end{bmatrix}, \begin{bmatrix}-1\\+1\end{bmatrix}, \begin{bmatrix}+1\\0\end{bmatrix}, \begin{bmatrix}0\\-1\end{bmatrix}\right\}\right) \tag{11.3.44}$$

and the irreducible representations and character table for the symmetry group are given in Example 17.3.4. Using a representative matrix for an operator in each of the conjugacy classes, $\boldsymbol{R}(\mathrm{I}) = \boldsymbol{I}_9$ and

$$
\boldsymbol{R}(\mathrm{Q}^{(*)}) = \begin{pmatrix} \boldsymbol{S}_3 & \mathbf{0}_3 & \mathbf{0}_3 \\ \mathbf{0}_3 & \boldsymbol{S}_3 & \mathbf{0}_3 \\ \mathbf{0}_3 & \mathbf{0}_3 & \boldsymbol{S}_3 \end{pmatrix}, \qquad \boldsymbol{R}(\mathrm{Q}^{(m)}) = \begin{pmatrix} \boldsymbol{J}_3 & \mathbf{0}_3 & \mathbf{0}_3 \\ \mathbf{0}_3 & \mathbf{0}_3 & \boldsymbol{S}_3 \\ \mathbf{0}_3 & \boldsymbol{S}_3^2 & \mathbf{0}_3 \end{pmatrix},
$$

$$
\boldsymbol{R}(\mathrm{C}_2) = \begin{pmatrix} \boldsymbol{I}_3 & \mathbf{0}_3 & \mathbf{0}_3 \\ \mathbf{0}_3 & \mathbf{0}_3 & \boldsymbol{J}_3 \\ \mathbf{0}_3 & \boldsymbol{J}_3 & \mathbf{0}_3 \end{pmatrix}, \qquad \boldsymbol{R}(\mathrm{Q}^{(*)}\mathrm{C}_2) = \begin{pmatrix} \boldsymbol{S}_3 & \mathbf{0}_3 & \mathbf{0}_3 \\ \mathbf{0}_3 & \mathbf{0}_3 & \boldsymbol{S}_3\boldsymbol{J}_3 \\ \mathbf{0}_3 & \boldsymbol{S}_3\boldsymbol{J}_3 & \mathbf{0}_3 \end{pmatrix},
$$

$$
\boldsymbol{R}(\mathrm{Q}^{(m)}\mathrm{C}_2) = \begin{pmatrix} \boldsymbol{J}_3 & \mathbf{0}_3 & \mathbf{0}_3 \\ \mathbf{0}_3 & \boldsymbol{S}_3\boldsymbol{J}_3 & \mathbf{0}_3 \\ \mathbf{0}_3 & \mathbf{0}_3 & \boldsymbol{J}_3\boldsymbol{S}_3 \end{pmatrix},
\tag{11.3.45}
$$

where $\boldsymbol{S}_3$ and $\boldsymbol{J}_3$ are given by (11.3.30). Thus the characters for this representation are $\chi(\mathrm{Q}_1) = 9$, $\chi(\mathrm{Q}_2) = 0$, $\chi(\mathrm{Q}_3) = 1$, $\chi(\mathrm{Q}_4) = 3$, $\chi(\mathrm{Q}_5) = 0$ and $\chi(\mathrm{Q}_6) = 3$, and it follows from (17.3.13) that the multiplicities of the irreducible representations are $\mathfrak{m}(\mathrm{A}) = 1$, $\mathfrak{m}(\mathrm{B}_1) = \mathfrak{m}(\mathrm{B}_2) = 0$, $\mathfrak{m}(\mathrm{B}_3) = 1$, $\mathfrak{m}(\mathrm{E}_1) = 3$, $\mathfrak{m}(\mathrm{E}_2) = 0$. The transfer matrix is

$$
\boldsymbol{V}_9 = \left(\begin{array}{ccc|cccccc}
\mathfrak{x}^4 & \mathfrak{x}^2 & \mathfrak{x}^2 & \mathfrak{x}^2 & \mathfrak{x} & \mathfrak{x}^2 & \mathfrak{x}^2 & \mathfrak{x} & \mathfrak{x}^2 \\
\mathfrak{x}^2 & \mathfrak{x}^4 & \mathfrak{x}^2 & \mathfrak{x}^2 & \mathfrak{x}^2 & \mathfrak{x} & \mathfrak{x} & \mathfrak{x}^2 & \mathfrak{x}^2 \\
\mathfrak{x}^2 & \mathfrak{x}^2 & \mathfrak{x}^4 & \mathfrak{x} & \mathfrak{x}^2 & \mathfrak{x}^2 & \mathfrak{x}^2 & \mathfrak{x}^2 & \mathfrak{x} \\
\hline
\mathfrak{x}^2 & \mathfrak{x}^2 & \mathfrak{x} & \mathfrak{x}^2 & 1 & 1 & \mathfrak{x} & \mathfrak{x} & 1 \\
\mathfrak{x} & \mathfrak{x}^2 & \mathfrak{x}^2 & 1 & \mathfrak{x}^2 & 1 & \mathfrak{x} & 1 & \mathfrak{x} \\
\mathfrak{x}^2 & \mathfrak{x} & \mathfrak{x}^2 & 1 & 1 & \mathfrak{x}^2 & 1 & \mathfrak{x} & \mathfrak{x} \\
\mathfrak{x}^2 & \mathfrak{x} & \mathfrak{x}^2 & \mathfrak{x} & \mathfrak{x} & 1 & \mathfrak{x}^2 & 1 & 1 \\
\mathfrak{x} & \mathfrak{x}^2 & \mathfrak{x}^2 & \mathfrak{x} & 1 & \mathfrak{x} & 1 & \mathfrak{x}^2 & 1 \\
\mathfrak{x}^2 & \mathfrak{x}^2 & \mathfrak{x} & 1 & \mathfrak{x} & \mathfrak{x} & 1 & 1 & \mathfrak{x}^2
\end{array}\right).
\tag{11.3.46}
$$

Since the first three and latter six ring-states form equivalence classes under the operations of the group, it follows from (11.3.9) the two-dimensional matrix block for the representation A can be obtained by adding over the elements in the four parts of $\boldsymbol{V}_9$ (indicated by the partitioning lines) and dividing by the appropriated weighting factors. To obtain the one-dimensional block for B_3 and the six-dimensional block for E_1 it is necessary to determine the block-reduction matrix for the group matrices (11.3.45). Here we simply give the eigenvalues which are easily determined, either directly or after block-diagonalization, using an algebraic computing program.

$$\begin{aligned}
&\Lambda_{1,2}(\mathsf{A};\mathfrak{X}) = \tfrac{1}{2}\left[g_1(\mathfrak{X}) \pm \sqrt{g_2(\mathfrak{X})}\right],\\
&\Lambda_1(\mathsf{B}_3;\mathfrak{X}) = (1-\mathfrak{X})^2,\\
&\Lambda_{1,2}(\mathsf{E}_1;\mathfrak{X}) = \Lambda_{3,4}(\mathsf{E}_1;\mathfrak{X}) = \tfrac{1}{2}\mathfrak{X}(\mathfrak{X}-1)\left[g_3(\mathfrak{X}) \pm \sqrt{g_4(\mathfrak{X})}\right],\\
&\Lambda_5(\mathsf{E}_1;\mathfrak{X}) = \Lambda_6(\mathsf{E}_1;\mathfrak{X}) = (\mathfrak{X}+2)(\mathfrak{X}-1),
\end{aligned} \tag{11.3.47}$$

where

$$\begin{aligned}
g_1(\mathfrak{X}) &:= \mathfrak{X}^4 + 3\mathfrak{X}^2 + 2\mathfrak{X} + 3,\\
g_2(\mathfrak{X}) &:= \left(\mathfrak{X}^4 + 2\mathfrak{X}^3 + 5\mathfrak{X}^2 - 2\mathfrak{X} + 3\right)\left(\mathfrak{X}^4 - 2\mathfrak{X}^3 + \mathfrak{X}^2 + 6\mathfrak{X} + 3\right),\\
g_3(\mathfrak{X}) &:= \mathfrak{X}^2 + \mathfrak{X} + 1,\\
g_4(\mathfrak{X}) &:= \mathfrak{X}^4 + 2\mathfrak{X}^3 - \mathfrak{X}^2 - 2\mathfrak{X} + 9.
\end{aligned} \tag{11.3.48}$$

From (11.1.14) the free energy per lattice site in the limit $\mathcal{N} \to \mathcal{L}_1$ is

$$\phi_0(\mathcal{L}_1, K) = -\tfrac{1}{2}\ln[\Lambda_1(\mathsf{A};\mathfrak{X})]. \tag{11.3.49}$$

The formulae for the internal energy and heat capacity are as given in the leading equalities in (11.3.36) and (11.3.37) and these quantities are plotted in Fig. 11.2.

With $N_2 = 3$ the symmetry group is $\mathfrak{S}_3 \times \mathfrak{C}_{3\mathrm{v}}$:

The lattice slices consist of rings of three sites. Appropriately ordered the lattice states now fall into five equivalence classes meaning that there are five eigenvalues

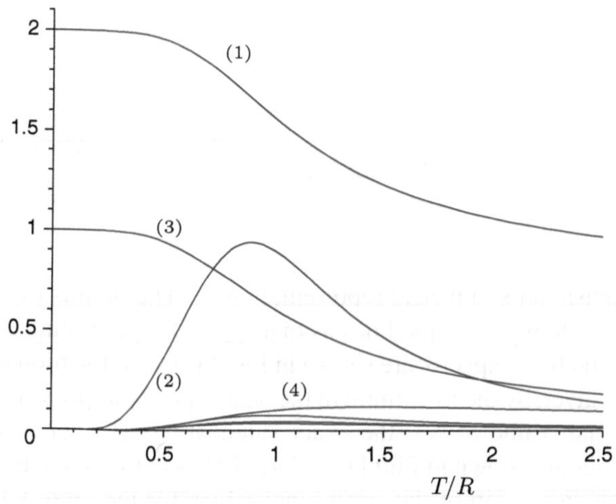

Fig. 11.2 Plots of (*1*) $-\mathrm{u}(\mathcal{L}_1, K)$, (*2*) $c(\mathcal{L}_1, K)/k_{\mathrm{B}}$, (*3*) the largest scaled eigenvalue, (*4*) a group of seven curves representing the remaining scaled eigenvalues, against $T/R = 1/K$

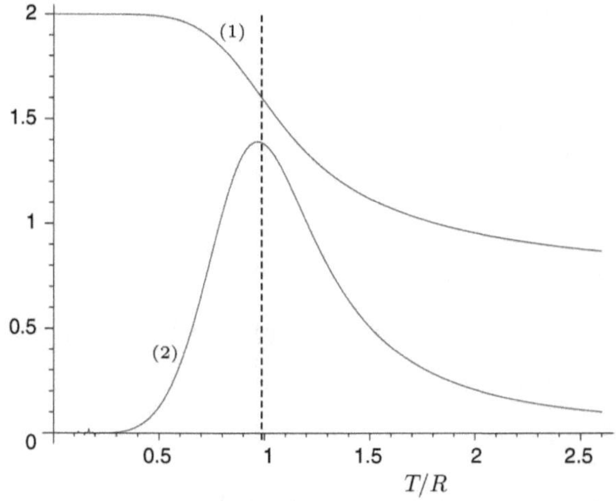

Fig. 11.3 Plots of *(1)* $-\mathrm{u}(\mathcal{L}_1, K)$, *(2)* $c(\mathcal{L}_1, K)/k_{\mathrm{B}}$, against $T/R = 1/K$. The *broken line* indicates the location of the critical temperature $T_c/R = 0.9950$, given by (9.2.28), of the two-dimensional square-lattice model

Fig. 11.4 Plots of *(1)* the two largest scaled eigenvalues, *(2)* the remaining twenty-four scaled eigenvalues, against $T/R = 1/K$

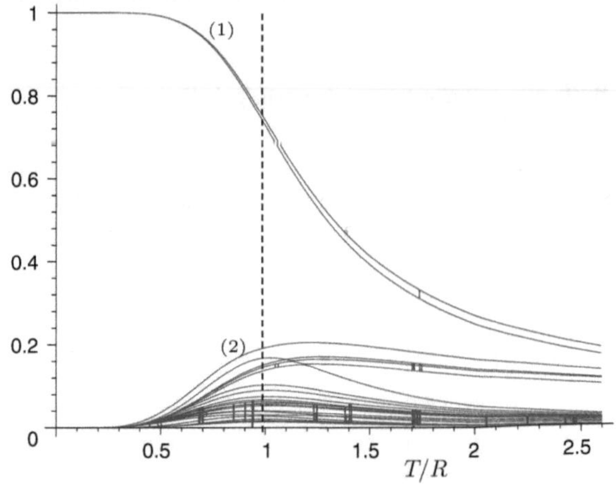

in the one-dimensional symmetric representation A. The details of the analysis of this case, which follows the same lines as that $N_2 = 1, 2$, are omitted. The internal energy density and heat capacity are shown in Fig. 11.3. The location of the incipient phase transition given by the maximum in the heat capacity approximates quite well, even for this small value of N_2, the exact value of the critical temperature. The scaled eigenvalues are shown in Fig. 11.4. Here we begin to see the behaviour which is evident in the discussion of the exact eigenvalues for the spin-$\frac{1}{2}$ Ising model in Sect. 11.4.3. One or more scaled eigenvalues (two in this case) approach the limiting value of unity at a certain temperature, which provides another approximation for the critical temperature, while the remaining scaled eigenvalues peak in the same vicinity.

11.4 Analysis in the Complex Plane: The Wood Method

Here we consider $\mathbb{C}_{\mathfrak{X}}$, the complex plane of a Boltzmann factor $\mathfrak{X}$, associated with an interaction energy (rather than a field) and in terms of which the characteristic equation of the transfer matrix is

$$\mathrm{Det}\{\boldsymbol{I}_M\Lambda - \boldsymbol{V}\} = \mathfrak{F}(N_2, \mathfrak{X}; \Lambda) = 0, \tag{11.4.1}$$

where

$$\mathfrak{F}(N_2, \mathfrak{X}; \Lambda) := \Lambda^M + \mathrm{f}_{M-1}(N_2, \mathfrak{X})\Lambda^{M-1} + \cdots + \mathrm{f}_0(N_2, \mathfrak{X}) \tag{11.4.2}$$

and $\mathfrak{X} := \exp(nK)$, n being chosen for convenience; ($n = 2$ for the case of the spin-$\frac{1}{2}$ Ising model in Sect. 11.4.3).[8] In general the coefficients $\mathrm{f}_k(N_2, \mathfrak{X})$, $k = 0, 1, \ldots, M$, can be sums of monomials in $\mathfrak{X}$ with negative and positive exponents. We can, however, without loss of generality, suppose that a constant has been added to all the interaction energies so that the coefficients are analytic polynomial functions of $\mathfrak{X}$.[9]

As in Sects. 11.1 and 11.2 the eigenvalues of $\boldsymbol{V}$ are, at this point, denoted by $\Lambda_m(N_2, \mathfrak{X})$, $m = 1, 2, \ldots, M$, without distinguishing their association with the irreducible representations of the symmetry group of the system. The partition function is given by (11.1.6) which is exact for finite N_1 and N_2 and can be expanded to give a polynomial in $\mathfrak{X}$ with positive real coefficients. The eigenvalues will be finite in $\mathbb{C}_{\mathfrak{X}}$, except possibly at the point at infinity. They may however, have a complicated branch-point structure. This cancels from the summation in the partition function to produce an analytic function with a finite set of zeros. We denote this set by $\mathbb{O}(N_1, N_2)$.

11.4.1 Evolution of Partition Function Zeros

Any two eigenvalues $\Lambda_m(N_2, \mathfrak{X})$ and $\Lambda_{m'}(N_2, \mathfrak{X})$ will have the same magnitude on a line (if it exists) for which $\mathfrak{X}$ satisfies

$$\Lambda_m(N_2, \mathfrak{X}) = \varpi\Lambda_{m'}(N_2, \mathfrak{X}), \qquad \varpi := \exp(\mathrm{i}\vartheta), \quad 0 \le \vartheta < 2\pi. \tag{11.4.3}$$

---

[8] The system may have a number of couplings, but these can, as we have already seen, be included by using parameter ratios of K.

[9] For an example of this see the discussion of the square-lattice Ising model in Sect. 11.4.3.

In general the order of the relative magnitudes of the eigenvalues will interchange across this line. The set of all curves defined by (11.4.3) for all m and all $m' \neq m$ is denoted by $\mathcal{D}_{N2}$. This forms a complicated network in the plane $\mathbb{C}_{\mathfrak{X}}$. Let $\mathcal{M}_{N2}$ be the subset of $\mathcal{D}_{N2}$ for which the eigenvalues satisfying (11.4.3) are of maximum magnitude. It follows that, for any $\mathfrak{X} \notin \mathcal{M}_{N2}$, there is a unique eigenvalue of maximum magnitude. This we denote by $\Lambda_{\max}(N_2, \mathfrak{X})$.[10] We shall now show that, in the limit $N_1 \to \infty$, the zeros $\mathbb{O}(N_1, N_2)$ converge onto the curves $\mathcal{M}_{N2}$. To make this idea precise the following theorem is needed:

Theorem 11.4.1 *For every $\mathfrak{X} \notin \mathcal{M}_{N2}$ and any ϵ satisfying the condition $|\Lambda_{\max}(N_2, \mathfrak{X})| > \epsilon > 0$, there exists an $N_0 > 0$ such that, for all $N_1 > N_0$, $|Z(N_1, N_2, \mathfrak{X})|^{1/N1} > \epsilon$.*

Proof From (11.1.6),

$$|Z(N_1, N_2, \mathfrak{X})|^{1/N1} = |\Lambda_{\max}(N_2, \mathfrak{X})| \left| \sum_{m=1}^{M} [\Omega_m(N_2, \mathfrak{X})]^{N1} \right|^{1/N1}, \tag{11.4.4}$$

where $\{\Omega_m(N_2, \mathfrak{X})\}$ are the scaled eigenvalues defined in (11.1.10). Since the transfer matrix $\boldsymbol{V}$ is non-singular for any $\mathfrak{X}$, not all of its eigenvalues are zero. In particular it is possible, for the given ϵ, to find a $B > 0$ such that $|\Lambda_{\max}(N_2, \mathfrak{X})| > B > \epsilon$. Since $\mathfrak{X} \notin \mathcal{M}_{N2}$ only one scaled eigenvalue has modulus of unity. Choosing

$$\eta := \frac{B - \epsilon}{B(M-1)} > 0, \tag{11.4.5}$$

it is, therefore, possible to find an N_0 such that

$$|\Omega_k(N_2, \mathfrak{X})|^{N1} < \eta, \qquad \forall\ N_1 > N_0 \tag{11.4.6}$$

and every scaled eigenvalue with modulus less than unity. It follows, from (11.4.4)–(11.4.6), that

$$|Z(N_1, N_2, \mathfrak{X})|^{1/N1} > B[1-(M-1)\eta]^{1/N1} > B[1-(M-1)\eta] = \epsilon. \tag{11.4.7}$$

□

This theorem establishes the result that no point $\mathfrak{X} \notin \mathcal{M}_{N2}$ can be a member of $\mathbb{O}(N_1, N_2)$ for arbitrarily large N_1. Thus that the set

$$\lim_{N1 \to \infty} \mathbb{O}(N_1, N_2) = \mathbb{O}(\mathcal{L}_1, N_2) \in \mathcal{M}_{N2}. \tag{11.4.8}$$

[10] Bearing in mind that the definition is ambiguous on the curves of $\mathcal{M}_{N2}$ across which there is a change of analytic form.

Further insight into this situation can be gained if we consider a curve defined by (11.4.3), for particular values $m = m_0$ and $m' = m'_0$. If this curve belongs to $\mathcal{M}_{N2}$, then, in the limit $N_1 \to \infty$,

$$Z(N_1, N_2, \mathfrak{X}) \simeq [\Lambda_{m_0}(N_2, \mathfrak{X})]^{N_1}[1 + \exp(\mathrm{i}\vartheta N_1)]. \tag{11.4.9}$$

This asymptotic expression is zero at the values $\vartheta = (2r + 1)\pi/N_1$, $r = 0, 1, \ldots, N_1 - 1$, which correspond to a set of points on the curve, with density which increases with N_1. For $M = 2$, as would be the case for the one-dimensional Ising model, this argument establishes the fact that the zeros of the partition function lie on the single curve given by (11.4.3) *for all* N_1. For $M > 2$, it adds understanding rather than rigour to the result established by Theorem 11.4.1.

On $\mathbb{R}^{(+)}$, the positive real axis of $\mathbb{C}_{\mathfrak{X}}$, Perron's theorem applies (Sect. 17.2) and $\Lambda_{\max}(N_2, \mathfrak{X})$ denotes (unambiguously) a real positive non-degenerate eigenvalue. Let the eigenvalues be ordered so that $\Lambda_1(N_2, \mathfrak{X})$ is identified by the condition that $\Lambda_1(N_2, \mathfrak{X}) = \Lambda_{\max}(N_2, \mathfrak{X})$, for $\mathfrak{X} \in \mathbb{R}^{(+)}$. Although curves belonging to $\mathcal{D}_{N_2}$ can cross or touch $\mathbb{R}^{(+)}$, none that do can belong to $\mathcal{M}_{N_2}$. We now define $\mathcal{E}_{N_2}$ as the subset of $\mathcal{M}_{N_2}$ consisting of points which can be reached by a continuous path from some point of $\mathbb{R}^{(+)}$ without first crossing some other member of $\mathcal{M}_{N_2}$. It must be the case that, along the curves of $\mathcal{E}_{N_2}$, $\Lambda_1(N_2, \mathfrak{X})$ is one of the eigenvalues which satisfies (11.4.3).

Suppose that $\mathbb{W}$ is a region in $\mathbb{C}_{\mathfrak{X}}$ containing a segment of $\mathbb{R}^{(+)}$, but none of the members of $\mathbb{O}(N_1, N_2)$ for all N_2 and all N_1 greater than some N_0. Then, by a slight modification of Theorem 8.4.1, it follows that $\phi(\mathcal{L}, \mathfrak{X})$, the free energy per lattice site for the infinite system $\mathcal{L}$ (with both $N_1 \to \infty$ and $N_2 \to \infty$), is a regular function of $\mathfrak{X}$ in $\mathbb{W}$. No singularity of $\phi(\mathcal{L}, \mathfrak{X})$ can occur at a real value of $\mathfrak{X} \in \mathbb{W} \cap \mathbb{R}^{(+)}$. For any fixed value of N_2 it follows from Theorem 11.4.1 that N_0 can be chosen so that, if $\mathbb{W} \cap \mathcal{E}_{N_2} = \emptyset$, then $\mathbb{O}(N_1, N_2) \cap \mathbb{W} = \emptyset$ for $N_1 > N_0$. The final conclusion of this argument is that singularities of the infinite system $\mathcal{L}$ are possible only at real values of $\mathfrak{X}$ to which some points of $\mathcal{E}_{N_2}$ converge as $N_2 \to \infty$.

11.4.2 Connection Curves and Cross-Block Curves

Suppose, as in Sect. 11.3, that the irreducible representations of the symmetry group $\mathcal{Q}$ are $\Theta(\mathrm{k})$, $\mathrm{k} = 1, 2, \ldots, \kappa$. The transfer matrix can be block-diagonalized into blocks corresponding to these representations with, in the case of a representation $\Theta(\mathrm{k})$ of dimension $\mathrm{e}(\mathrm{k}) > 1$, a further reduction into $\mathrm{e}(\mathrm{k})$ identical blocks. The characteristic equation take the form

$$\prod_{\mathrm{k}=1}^{\kappa}[\mathrm{Det}\{\boldsymbol{I}_{\mathrm{m}(\mathrm{k})}\Lambda - \boldsymbol{V}'(\mathrm{k})\}]^{\mathrm{e}(\mathrm{k})} =: \mathcal{F}(N_2, \mathfrak{X}; \Lambda) = 0, \tag{11.4.10}$$

where $\boldsymbol{V}'(\mathrm{k})$ is one of the $\mathrm{e}(\mathrm{k})$ identical blocks obtained by block-diagonalizing the matrix $\widetilde{\boldsymbol{V}}(\mathrm{k})$ which appears in (11.3.3). The factors in the product in (11.4.10) are polynomials of degree $\mathfrak{m}(\mathrm{k})$ with

$$\mathrm{Det}\{\boldsymbol{I}_{\mathfrak{m}(\mathrm{k})}\Lambda - \boldsymbol{V}'(\mathrm{k})\} = \mathcal{G}(\mathrm{k}; N_2, \mathfrak{X}; \Lambda), \tag{11.4.11}$$

where

$$\mathcal{G}(\mathrm{k}; N_2, \mathfrak{X}; \Lambda) := \Lambda^{\mathfrak{m}(\mathrm{k})} + \mathfrak{g}_{\mathfrak{m}(\mathrm{k})-1}(\mathrm{k}; N_2, \mathfrak{X})\Lambda^{\mathfrak{m}(\mathrm{k})-1} + \cdots + \mathfrak{g}_0(\mathrm{k}; N_2, \mathfrak{X}). \tag{11.4.12}$$

Thus we have

$$\mathcal{F}(N_2, \mathfrak{X}; \Lambda) = \prod_{\mathrm{k}=1}^{\kappa}[\mathcal{G}(\mathrm{k}; N_2, \mathfrak{X}; \Lambda)]^{\mathrm{e}(\mathrm{k})}. \tag{11.4.13}$$

The coefficients in (11.4.12) are polynomials in $\mathfrak{X}$ and, if the factorization (11.4.13) were complete in the sense that no further reduction into factors with polynomial coefficients were possible, then the eigenvalue solutions $\Lambda_\ell(\mathrm{k}; N_2, \mathfrak{X})$, $\ell = 1, 2, \ldots, \mathfrak{m}(\mathrm{k})$, of

$$\mathcal{G}(\mathrm{k}; N_2, \mathfrak{X}; \Lambda) = 0 \tag{11.4.14}$$

would be the branches of an (*irreducible*) *algebraic function* (Sect. 17.6). There are two reasons why this might not be the case. Firstly, it is sometimes quite difficult to determine the complete symmetry group of the system and any further 'hidden' symmetry would lead to a factorization of $\mathcal{G}(\mathrm{k}; N_2, \mathfrak{X}; \Lambda)$. Secondly, we have no guarantee that the full symmetry group will lead to a complete factorization of the characteristic equation. An example of this is given, in Sect. 11.4.3, by the square-lattice Ising model. For the moment we shall suppose that $\mathcal{G}(\mathrm{k}; N_2, \mathfrak{X}; \Lambda)$ *is* irreducible. The only modification needed if this were not the case, would be an additional index to distinguish the separate factors. The partition function is now expressible in the form

$$Z(N_1, N_2, \mathfrak{X}) = \sum_{\mathrm{k}=1}^{\kappa}\sum_{\ell=1}^{\mathfrak{m}(\mathrm{k})}[\Lambda_\ell(\mathrm{k}; N_2, \mathfrak{X})]^{N_1} \tag{11.4.15}$$

and the curves in $\mathcal{D}_{N_2}$ are distinguished by their representations. The set for which $\mathfrak{X}$ satisfies the equation

$$\Lambda_\ell(\mathrm{k}; N_2, \mathfrak{X}) = \varpi\Lambda_{\ell'}(\mathrm{k}'; N_2, \mathfrak{X}), \quad \varpi = \exp(\mathrm{i}\vartheta), \quad 0 \le \vartheta < 2\pi, \tag{11.4.16}$$

for all $\ell = 1, 2, \ldots, \mathfrak{m}(\mathrm{k})$, $\ell' = 1, 2, \ldots, \mathfrak{m}(\mathrm{k}')$ is denoted by $\mathcal{C}_{N_2}(\mathrm{k}, \mathrm{k}')$. For reasons made clear in Sect. 17.6, $\mathcal{C}_{N_2}(\mathrm{k}, \mathrm{k})$ are called *connection curves* and $\mathcal{C}_{N_2}(\mathrm{k}, \mathrm{k}')$, with $\mathrm{k} \neq \mathrm{k}'$ are called *cross-block curves*. The total set

$$\bigcup_{\{\mathrm{k},\mathrm{k}'\}} \mathcal{C}_{N2}(\mathrm{k},\mathrm{k}') = \mathcal{D}_{N2}. \tag{11.4.17}$$

On $\mathbb{R}^{(+)}$, $\Lambda_{\max}(N_2, \mathfrak{X}) = \Lambda_1(1; N_2, \mathfrak{X})$. So $\mathcal{E}_{N2}$ consists of curves in the set $\mathcal{C}_{N2}(\mathrm{k}, 1)$ for some k.

In principle an investigation of the evolution of the set $\mathcal{E}_{N2}$ for increasing N_2 should provide some indication of the critical point of the system in the limit $N_2 \to \infty$. However, the dimension M of the transfer matrix will grow very rapidly with N_2, with all the resulting computational difficulties. These can be lessened by using block-diagonalization of the transfer matrix, particularly if $\mathcal{E}_{N2}$ consists of connection curves in $\mathcal{C}_{N2}(1, 1)$ rather than cross-block curves in $\mathcal{C}_{N2}(\mathrm{k}, 1)$ with $\mathrm{k} \neq 1$. Then it will be necessary to calculate only the one-dimensional symmetric block of the transfer matrix, which is straightforward using formula (11.3.9). Unfortunately this is not necessarily the case, as we shall see in Sect. 11.4.3 for the square-lattice first-neighbour spin-$\frac{1}{2}$ Ising model and as was also shown for the square-lattice Ising model with equal first- and second-neighbour interactions by Wood (1987). The interesting and important feature of this method is the argument that, as $N_2 \to \infty$, $\mathcal{E}_{N2}$ converges on to that part $\mathcal{C}^*_{N2}(1, 1)$ of $\mathcal{C}_{N2}(1, 1)$ which involves $\Lambda_1(1; N_2, \mathfrak{X})$. This means that, for finite N_2, the curves of $\mathcal{C}^*_{N2}(1, 1)$ *rather than* $\mathcal{E}_{N2}$ can be used to obtain an approximation to the critical properties of the infinite system. Two arguments can be advanced for the truth of this contention. The first of a general nature and the second based on calculations for particular systems. Our discussion in Sect. 11.4.3 for the square-lattice spin-$\frac{1}{2}$ Ising model falls in the latter category.

In a general context the argument[11] is based on a comparison between two different finite systems, one with the toroidal boundary conditions used here and the other with boundaries and a contribution to the Hamiltonian $\hat{b}(\boldsymbol{\tau}(N_1 - 1), \boldsymbol{\tau}(0))$ as is given in (11.1.1). This results in a modification to the partition function formula (11.4.15) in which N_1 is replaced by $N_1 - 1$ and the terms are weighted by coefficients $\beta(\mathrm{k}; N_2, \mathfrak{X})$ dependent on the boundary conditions. Finally it can be shown that a particular symmetry choice of boundary conditions will yield $\beta(\mathrm{k}; N_2, \mathfrak{X}) = \delta^{\mathrm{Kr}}(\mathrm{k} - 1)$ with

$$Z(N_1, N_2, \mathfrak{X}) = \sum_{\ell=1}^{\mathfrak{m}(1)} [\Lambda_\ell(1; N_2, \mathfrak{X})]^{N_1 - 1}. \tag{11.4.18}$$

Both this partition function and (11.4.15) give the same limit for the free energy per lattice site. This means that the critical properties of the two systems in the limit $N_2 \to \infty$ are identical and are also the same as that of a *reduced* system with toroidal boundary conditions, with the transfer matrix consisting of just the one-dimensional symmetric block of that for the full system, and $N_1 - 1$ rather than N_1 slices. A sequence of finite cases of each of these three systems can be understood to give a succession of approximations to the same infinite system. This is illustrated in Fig. 12.7 for the square-lattice Ising model, with $N_2 = 1000$. It can be see that

[11] Described in detail in Lavis and Bell (1999b, Sects. 4.9 and 4.10)

both the ratio of the two largest eigenvalues and the two largest eigenvalues *in the one-dimensional symmetric representation* approach unity at the critical point.

11.4.3 The Spin-$\frac{1}{2}$ Square-Lattice Ising Model

In this case the symmetry group is $\mathcal{G} := \mathcal{C}_{N_2\mathrm{v}} \times \mathcal{S}_2^{(m)}$, where (Sect. 17.3.4) $\mathcal{C}_{N_2\mathrm{v}}$ is the group of rotations of the lattice and reflections through planes containing the lattice axis and $\mathcal{S}_2^{(m)}$ corresponds to spin inversion. To illustrate the methods described in this section we look ahead to the exact formula for the transfer matrix eigenvalues given in Sect. 12.2.7, restricting attention to the isotropic case $K^{(\mathrm{h})} = K^{(\mathrm{ch})} = K$, $\mathfrak{X}_{\mathrm{h}} = \mathfrak{X}_{\mathrm{ch}} = \mathfrak{X} = \exp(2K)$. To ensure that the coefficients in the characteristic equation (11.4.1) are polynomials in positive powers of $\mathfrak{X}$ we add the constant $-J$ to each first-neighbour pair interaction, giving $-2J$ and zero for a pair of parallel and anti-parallel spins, respectively. This means that each term of the transfer matrix and hence each eigenvalue is multiplied by a factor $\mathfrak{X}^{N_2}$ giving, from (12.2.149),[12]

$$\Lambda^{(s)}(N_2, \mathfrak{X}; \boldsymbol{\gamma}) = [\mathfrak{X}(\mathfrak{X}^2 - 1)]^{\frac{1}{2}N_2} \prod_{\{\psi\}^{(s)}} \exp\left(\tfrac{1}{2}\,\gamma_\psi\,\alpha\,\omega(\mathfrak{X}; \psi)\right), \tag{11.4.19}$$

where, from (18.3.111), (12.2.136) and (12.2.137), $\{\psi\}^{(s)}$ is the set of values

$$\psi_j^{(s)} := \frac{\pi(4j - s - 1)}{4N_2}, \qquad j = 1, 2, \ldots, N_2, \tag{11.4.20}$$

$$\omega(\mathfrak{X}; \psi) := \ln\left[W(\mathfrak{X}; \psi) + \sqrt{W^2(\mathfrak{X}; \psi) - 1}\right], \tag{11.4.21}$$

$$W(\mathfrak{X}; \psi) := \frac{(\mathfrak{X}^2 + 1)^2 - 2\mathfrak{X}(\mathfrak{X}^2 - 1)\cos(2\,\psi)}{2\mathfrak{X}(\mathfrak{X}^2 - 1)} \tag{11.4.22}$$

and $\alpha = \pm 1$ is chosen so that $\alpha\,\omega(\mathfrak{X}; \psi) > 0$ at low temperatures. As is shown in Sect. 12.2.7 this means that $\alpha = 1$ if $\psi \neq 0$ and, for $\psi = 0$, $\alpha = 1$ for $T < T_\mathrm{c}$ and $\alpha = -1$ for $T > T_\mathrm{c}$ where $\mathfrak{X}_\mathrm{c} = \exp(2J/T_\mathrm{c})$ is given by $\mathfrak{X}_\mathrm{c}(\mathfrak{X}_\mathrm{c} - 2) = 1$.

From (11.4.21) and (11.4.22) $\exp[\omega(\mathfrak{X}; \psi)]$ will have branch-points corresponding to zeros of the functions $W(\mathfrak{X}; \psi) \mp 1$ unless they are both perfect squares, which, from (12.2.151), is the case when $\psi = 0$, which is a member of $\{\psi\}^{(-)}$, or $\psi = \frac{1}{2}\pi$, which is a member of $\{\psi\}^{(-)}$ if N_2 is even and $\{\psi\}^{(+)}$ if N_2 is odd.[13] In all other cases the zeros of these quantities give eight branch-points for $\exp[\omega(\mathfrak{X}; \psi)]$. These branch-points will appear in the eigenvalues given by (11.4.19) unless γ_ψ and $\gamma_{\pi-\psi}$

[12] This formula, which is that derived by Onsager (1944) is obtained by neglecting the factors $\ell^{(s)}(\boldsymbol{\gamma})$ and $(\mathrm{k}/2)^{n_2}$ in (12.2.149), which are a consequence of the method of derivation.

[13] For deriving the eigenvalues in Sect. 12.2 we have assumed that N_2 is even but this is not needed for the standard Onsager derivation.

have opposite signs. In that case the branch-points are eliminated. With φ defined in terms of $\mathfrak{X}$ by

$$\mathfrak{X} := \exp(\mathrm{i}\varphi) \pm \sqrt{\exp(2\mathrm{i}\varphi)+1}, \tag{11.4.23}$$

the zeros of $W(\mathfrak{X};\psi) \mp 1$ are given, respectively by[14]

$$\cos\varphi = \begin{cases} \cos^2(\psi) \\ -\sin^2(\psi). \end{cases} \qquad \psi \in \psi_j^{(s)}. \tag{11.4.24}$$

If the characteristic polynomial of any model is expressed, in the form (11.4.13), as a product of irreducible polynomials, then we know that every branch of an algebraic function must be an eigenvalue in the same irreducible representation. However, as we shall now see the converse is not necessarily the case. Not all eigenvalues in the same irreducible representation are branches of the same algebraic function. For the present model the eigenvalues which are the branches of a single algebraic function can be identified in the sets $s = 1$ and $s = -1$ of (11.4.19) from their branch-points. Given any eigenvalue, another branch of the same algebraic function can be obtained by changing the signs of γ_ψ and $\gamma_{\pi-\psi}$. On $\mathbb{R}^{(+)}$ the largest eigenvalue is $\Lambda_{\max}(N_2,\mathfrak{X}) = \Lambda^{(+)}(N_2,\mathfrak{X};\boldsymbol{j})$, where $\boldsymbol{j} = (1,1,\ldots,1)$. This is, therefore, a branch of an algebraic function of degree

$$\mathrm{d}(N_2) = \begin{cases} 2^{\frac{1}{2}N_2}, & N_2 \text{ even}, \\ 2^{\frac{1}{2}(N_2-1)}, & N_2 \text{ odd}. \end{cases} \tag{11.4.25}$$

Using (11.3.16), we see that $\mathfrak{m}(1;N_2) = \mathrm{d}(N_2)$ for $N_2 = 2,3,\ldots,6$, but $\mathfrak{m}(1;7) = 9 > \mathrm{d}(7) = 8$ and $\mathfrak{m}(1;8) = 18 > \mathrm{d}(8) = 16$. For $N_2 > 6$ the one-dimensional symmetric representation contains eigenvalues, with a different branch-point structure from $\Lambda_{\max}(N_2,\mathfrak{X})$. However, its is known that the eigenvalues of any representation are either entirely in the set $s = 1$ or $s = -1$, according respectively to whether the representation in question is formed from the one-dimensional symmetric or anti-symmetric representation of $\mathcal{S}_2^{(m)}$. It follows that all the eigenvalues of the one-dimensional symmetric representation are members of the set $s = 1$.

We now return to (11.4.23) and (11.4.24) which define the branch-points of the eigenvalues. If φ_0 is a solution of the first of the equations (11.4.24) in the interval $[0, \frac{1}{2}\pi]$, then $\varphi_1 \in [\frac{1}{2}\pi, \pi]$ satisfies the second equation when $\cos\varphi_1 = \cos\varphi_0 - 1$. The connection curve in $\mathbb{C}_{\mathfrak{X}}$ described by (11.4.23) with the positive sign and $\varphi \in [\varphi_0, \varphi_1]$ has a cusp at $\varphi = \frac{1}{2}\pi$, $\mathfrak{X} = \mathrm{i}$. Another connection curve is given by taking $\varphi \in [-\varphi_1, -\varphi_0]$. The two connection curves corresponding to the negative sign in (11.4.23) can be obtained by applying the mapping $\mathfrak{X} \to -1/\mathfrak{X}$. It is not difficult to

[14] Meaning that the four roots are given by taking each sign in (11.4.23) and the two roots obtained by solving the quadratic for $\exp(\mathrm{i}\varphi)$ given by the appropriate case of (11.4.24).

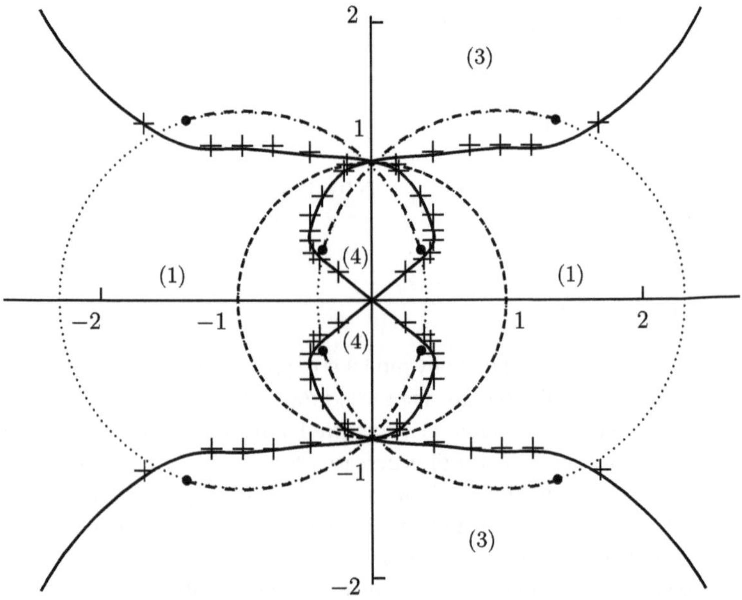

Fig. 11.5 The connection and cross-block curves, in the plane $\mathbb{C}_{\mathfrak{X}}$, for the Ising model with $N_2 = 2$. The zeros of $Z(16, 2, \mathfrak{X})$ are marked by + and the region in which $\Lambda_1(\Bbbk; 2, \mathfrak{X})$ is the eigenvalue of largest magnitude is labelled ($\Bbbk$)

show that these four curves form pairs of crossing curves lying on the *Fisher circles*

$$\mathfrak{X} = \sqrt{2}\exp(\mathrm{i}\theta) \pm 1. \tag{11.4.26}$$

These connection curves for $\psi = \pi/4$, $N_2 = 2$, when $\varphi_0 = \frac{1}{3}\pi$, $\varphi_1 = \frac{2}{3}\pi$ are shown in Fig. 11.5 as chain lines terminating at the eight branch-points which are denoted by •. The Fisher circles (11.4.26) are completed with dotted lines.

The second largest eigenvalue is $\Lambda^{(-)}_{\text{max}}(N_2, \mathfrak{X})$, given by (12.2.167). $\mathcal{E}_{N2}$ therefore consists of cross-block curves. Although it is possible that the zeros of $Z(N_1, N_2, \mathfrak{X})$ all lie on the part of $\mathcal{M}_{N_2}$ which is not $\mathcal{E}_{N_2}$ it is more plausible to expect, (as in the case $N_2 = 2$ investigated below), that some of the zeros $\mathbb{O}(N_1, N_2)$ for any N_1 will lie on $\mathcal{E}_{N_2}$. It can, however, be shown (Fisher 1965; Wood 1985) that the limiting locus of partition function zeros for this model is the pair of Fisher circles (11.4.26). It must, therefore, be concluded that, in the double limit $N_1 \to \infty$, $N_2 \to \infty$, the curve $\mathcal{E}_{N_2}$ closes in on the circles (11.4.26), which according to Theorem 11.4.1 contain the zeros of the partition function $Z^{(1)}(N_1, N_2, \mathfrak{X})$, defined by (11.4.18). These zeros lie on the connection curve $\mathcal{C}^*_{N2}(1, 1)$

$$\Lambda^{(-)}_{\text{max}}(N_2, \mathfrak{X}) = \varpi\,\Lambda^{(-)}_{\text{submax}}(N_2, \mathfrak{X}), \tag{11.4.27}$$

where $\Lambda^{(-)}_{\text{submax}}(N_2, \mathfrak{X})$, given by (12.2.165), is the second largest member of the set (11.4.19) with $s = -1$. This curve has branch-points on the circle $\mathfrak{X} = \sqrt{2}\exp(\mathrm{i}\theta)+1$ at the pair of points

$$\mathfrak{X} = \exp\left\{\pm\mathrm{i}\arccos\left[\cos^2\left(\frac{\pi}{2N_2}\right)\right]\right\} + \sqrt{\exp\left\{\pm 2\mathrm{i}\ \arccos\left[\cos^2\left(\frac{\pi}{2N_2}\right)\right]\right\} + 1}. \qquad (11.4.28)$$

As $N_2 \to \infty$, $\mathfrak{X} \to 1 + \sqrt{2}$, which is equivalent to (9.2.36), the formula for the critical temperature of the square-lattice Ising ferromagnet. All the connection curves form a set of overlapping arcs on the circles (11.4.26) and in this limit branch-points of all the connection curves approach the critical point which becomes for them a point of accumulation. The critical point can also be seen as a result of analytically continuing $\mathcal{C}^*_{N_2}(1, 1)$ through the branch-points to $\mathbb{R}^{(+)}$.

The remarkable feature of this result is that, although the critical point is for the system in the thermodynamic limit of infinite N_1 and N_2, it applies to the eigenvalues for any finite N_2. This can be seen by a more detailed examination of a special case.

Example 11.4.1 The eigenvalues for $N_2 = 2$ are given in Example 11.3.1. Those results can be utilized, but it must be noted that here, in order to ensure that all the elements of the transfer matrix are polynomials in $\mathfrak{X}$, we must multiply the eigenvalue (11.3.28) by $\mathfrak{X}^2$, and the quadratic (11.3.25) must be modified by multiplying the linear term by $\mathfrak{X}^2$ and the constant term by $\mathfrak{X}^4$. This having been done the connection curves $\mathcal{C}_2(1, 1)$ are given, either using (11.4.3) and (11.3.26) or the resultant formula (17.6.19) with the quadratic (11.3.25). They have the structure described above for the general N_2 case and are shown in Fig. 11.5 by chain lines. From (11.3.28) the cross-block curve $\mathcal{C}_2(3, 4)$ is the circle $|\mathfrak{X}| = 1$ (represented by a broken line in Fig. 11.5). From (11.3.26) and (11.3.28) the cross-block curves $\mathcal{C}_2(1, 3)$ and $\mathcal{C}_2(1, 4)$ are both given by the solutions of

$$\mathfrak{X}^2(\mathfrak{X}^2 - 1)\exp(2\mathrm{i}\vartheta) - (\mathfrak{X}^2 + 1)^2\exp(\mathrm{i}\vartheta) + \mathfrak{X}^2 - 1 = 0. \qquad (11.4.29)$$

These curves are represented by continuous lines in Fig. 11.5. On that part of the curves for which $|\mathfrak{X}| > 1$, $|\Lambda_1(\mathrm{A}; 2, \mathfrak{X})| = |\Lambda_1(\mathrm{B}_1; 2, \mathfrak{X})|$ and when $|\mathfrak{X}| < 1$, $|\Lambda_1(\mathrm{A}; 2, \mathfrak{X})| = |\Lambda_1(\mathrm{B}_3; 2, \mathfrak{X})|$. The exterior and interior of the circle $|\mathfrak{X}| = 1$ correspond respectively to the ferromagnetic ($J > 0$) and antiferromagnetic ($J < 0$) models. The inversion symmetry of Fig. 11.5 with respect to this circle corresponds to the invariance of the free energy of the square-lattice Ising model with respect to interchange of the sign of J. It follows from the topological structure of Fig. 11.5 that, for $J > 0$, the curve $|\Lambda_1(\mathrm{A}; 2, \mathfrak{X})| = |\Lambda_1(\mathrm{B}_3; 2, \mathfrak{X})|$ forms the subset $\mathcal{M}_2$ along which there are two eigenvalues of maximum magnitude. In a similar way $|\Lambda_1(\mathrm{A}; 2, \mathfrak{X})| = |\Lambda_1(\mathrm{B}_3; 2, \mathfrak{X})|$ gives the subset $\mathcal{M}_2$ when $J < 0$. The eigenvalues of maximum magnitude in each of the regions formed by the curves $\mathcal{M}_2$ are shown in Fig. 11.5. The curves of $\mathcal{E}_2$ consist of that part of $\mathcal{M}_2$ to the right of the imaginary axis. The conclusion for general N_2 that $\mathcal{E}_{N_2}$ consists of cross-block curves is, therefore, explicitly confirmed in this case. According to Theorem 11.4.1 the zeros of $Z(N_1, 2, \mathfrak{X})$ converge on to $\mathcal{M}_2$ as $N_1 \to \infty$. The zeros of $Z(16, 2, \mathfrak{X})$ are indicated by crosses in Fig. 11.5. Even with this comparatively small value of N_1 the conclusion of the theorem is supported.

We need at this point to distinguish those features of the behaviour of the square-lattice Ising model which can be expected to appear in a wide class of lattice models and those which arise because of its special properties. The characteristic of general validity is the convergence with increasing N_2 of $\mathcal{E}_{N2}$ on to $\mathcal{C}^*_{N2}(1, 1)$, with the analytic continuation of $\mathcal{C}^*_{N_2}(1, 1)$ to $\mathbb{R}^{(+)}$ yielding an approximation to the critical point. A special feature of the Ising model is that $\mathcal{C}^*_{N_2}(1, 1)$ lies on the *same* curves (11.4.26) for all values on N_2. This means that the *exact* critical temperature can be obtain using any finite value of N_2. The reason for this can be found in the self-duality of this model. With

$$\mathfrak{S} := \sinh(2K) = \tfrac{1}{2}(\mathfrak{X} - \mathfrak{X}^{-1}), \tag{11.4.30}$$

the duality transformation (9.2.32) is given by

$$\mathfrak{S} \longleftrightarrow \mathfrak{S}^* = 1/\mathfrak{S} \tag{11.4.31}$$

and the curves given by

$$\mathfrak{S} = \exp(\mathrm{i}\varphi), \qquad -\pi \le \varphi < \pi, \tag{11.4.32}$$

are invariant under the duality transformation, which takes $\varphi \to -\varphi$. Equations (11.4.23) give the solutions of (11.4.32) and it follows that all the branch-points and all the connection curves lie on the self-dual curves (11.4.32) for all N_2.

Duality also leads to the appearance of a curve which contains $\mathcal{C}^*_{N_2}(1, 1)$ for all N_2 in the case of the triangular Ising model and the square-lattice $\mu = 0$ ν-state Potts model (Wood et al. 1987). The interesting question concerning the use of the method to uncover approximate, or hitherto unsuspected symmetries, leading to critical points given in terms of algebraic numbers, has been discussed by Wood and Ball (1990).

11.4.4 Critical Points and Exponents

As has been indicated above, the procedure, in the absence of any special symmetry properties, is to analytically continue the connection curve $\mathcal{C}^*_{N_2}(1, 1)$ to $\mathbb{R}^{(+)}$ to give an approximation to the critical point. Two different approaches can be taken to this task.

In the first it is supposed that the appropriate branch of the connection curve has been identified by

$$\Lambda_1(1; N_2, \mathfrak{X}) = \varpi \Lambda_2(1; N_2, \mathfrak{X}), \tag{11.4.33}$$

with known expressions in terms of $\mathfrak{X}$ for the eigenvalues. With

$$f(\mathfrak{X}) = \frac{\Lambda_1(1; N_2, \mathfrak{X})}{\Lambda_2(1; N_2, \mathfrak{X})}, \tag{11.4.34}$$

the connection curve is given by $|f(\mathfrak{X})| = 1$, with $f(\mathfrak{X}) = 1$ at the branch-points. The analytic continuation of the connection curve through the branch-points, called the *extension curve*, is defined by $\mathrm{Arg}\{f(\mathfrak{X})\} = 0$. Now let

$$\begin{aligned} \mathfrak{X} &:= \mathrm{r} \exp(\mathrm{i}\theta), \\ f(\mathfrak{X}) &:= \mathfrak{F}(\mathrm{r}, \theta) \exp[\mathrm{i}\Theta(\mathrm{r}, \theta)]. \end{aligned} \tag{11.4.35}$$

The extension curve is given by

$$\Theta(\mathrm{r}, \theta) = 0, \tag{11.4.36}$$

with normal in the direction

$$\hat{\boldsymbol{n}}(\mathrm{r}, \theta) = \left(\frac{\partial \Theta}{\partial \mathrm{r}}, \frac{1}{\mathrm{r}} \frac{\partial \Theta}{\partial \theta} \right). \tag{11.4.37}$$

Except at the branch-points the Cauchy–Riemann conditions

$$\frac{\partial \mathfrak{F}}{\partial \mathrm{r}} = \frac{\mathfrak{F}}{\mathrm{r}} \frac{\partial \Theta}{\partial \theta}, \qquad \frac{1}{\mathrm{r}} \frac{\partial \mathfrak{F}}{\partial \theta} = -\mathfrak{F} \frac{\partial \Theta}{\partial \mathrm{r}}, \tag{11.4.38}$$

are satisfied. The approximation to the critical point is given by the intersection of the extension curve and $\mathbb{R}^{(+)}$. If this intersection is orthogonal then, from (11.4.37) and (11.4.38)

$$\frac{\partial \Theta}{\partial \theta} = 0, \qquad \frac{\partial \mathfrak{F}}{\partial \mathrm{r}} = 0. \tag{11.4.39}$$

The second of these conditions, identifies $\mathfrak{X}_c$ as a turning point of $|f(\mathfrak{X})|$ along $\mathbb{R}^{(+)}$.

The second approach, which maps out the whole of $\mathcal{C}_{N_2}(1, 1)$ and its extension curves, is to begin with the defining polynomial $\mathcal{G}(1; N_2, \mathfrak{X}; \Lambda)$, given by (11.4.12), for the eigenvalues in the one-dimensional symmetric representation. According to the method described in Sect. 17.6 we now form the polynomial

$$\mathcal{H}(1; N_2, \mathfrak{X}; \Lambda) := \frac{\mathcal{G}(1; N_2, \varpi\mathfrak{X}; \Lambda) - \mathcal{G}(1; N_2, \mathfrak{X}; \Lambda)}{\Lambda(\varpi - 1)}. \tag{11.4.40}$$

The connection curves and their extensions are given by solving the equation

$$\mathrm{Res}\{\mathcal{X}, \mathcal{G}, \mathcal{H}\} = 0. \tag{11.4.41}$$

Example 11.4.2 Using the second procedure, we consider the hard-square lattice fluid of particles of fugacity $\mathfrak{z} := \exp(\mu/T)$ on the square lattice (Sect. 8.4.3). With $N_2 = 4$ the symmetry group is $\mathcal{C}_{4v}$ and it is not difficult to show that the eigenvalues of the one-dimensional symmetric representation satisfy the cubic equation

$$\Lambda^3 - \Lambda^2(\mathfrak{z}^2 + 3\mathfrak{z} + 1) + \Lambda\mathfrak{z}(\mathfrak{z}^2 - \mathfrak{z} - 1) + \mathfrak{z}^3 = 0. \tag{11.4.42}$$

In this case the resultant equation (11.4.41) gives

$$\begin{aligned} 0 = {} & \mathfrak{z}^8 + 4\mathfrak{z}^7 + 2(9 - 2\varsigma)(1+\varsigma)\mathfrak{z}^6 + 8(5 + 5\varsigma - \varsigma^2)\mathfrak{z}^5 \\ & + 2(41 + 52\varsigma - 4\varsigma^3)\mathfrak{z}^4 + 4(24 + 25\varsigma + 4\varsigma^2)\mathfrak{z}^3 \\ & + 4(13 + 8\varsigma + \varsigma^2)\mathfrak{z}^2 + 4(3+\varsigma)\mathfrak{z} + 1, \end{aligned} \tag{11.4.43}$$

where

$$\varsigma = \tfrac{1}{2}(\varpi + \varpi^{-1}) = \cos\vartheta. \tag{11.4.44}$$

The connection curves $C_4(1,1)$ are the roots of (11.4.43) with $|\varsigma| \le 1$. The extension curves are given by continuing the curves into the region $|\varsigma| > 1$, where ϑ becomes imaginary. These curves are shown in Fig. 11.6. The extensions of the connection curves through the branch-points at $-0.881 \pm \mathrm{i}3.473$ meet $\mathbb{R}^{(+)}$ at $\mathfrak{z}_c = 3.016$, with $\varsigma_c = 5.186$. These values can be computed from the condition that $\mathfrak{z}_c$ is the meeting point of two extension curves approaching from above and below. It, therefore, corresponds to a value of ς for which (11.4.43) has a repeated root. This can be found by the using the resultant with respect to $\mathfrak{z}$ of (11.4.43) and its derivative. The estimate $\mathfrak{z}_c = 3.016$ for the critical fugacity can be compared (rather unfavourably) with the high-density series result $\mathfrak{z}_c = 3.7962 \pm 0.0001$ quoted in Sect. 8.4.3.

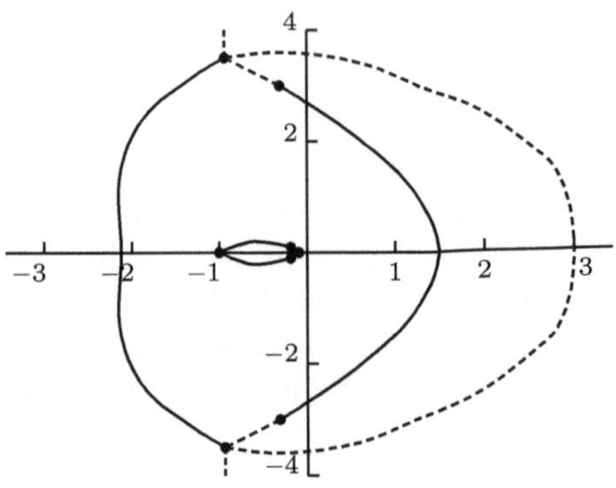

Fig. 11.6 The connection curves $C_4(1,1)$ and their extensions for the hard-square lattice fluid. The connection curves are shown by *continuous lines* and their extensions by *broken lines*

However, it has been shown by Wood et al. (1987) that for $N_2 = 6$ the comparison is much improved with $\mathfrak{z}_c = 3.730$.

In the picture of the development of critical behaviour originating in the work of Yang and Lee the critical point of a one-parameter lattice system corresponds, in the thermodynamic limit, to a point of accumulation of the zeros of the partition function. We have now shown that the critical point can also be identified as the limit of some of the branch-points of $C^*_{N2}(1, 1)$ as $N_2 \to \infty$. Suppose that there are s of these branch-points each with cycle number two. They will all be the termination of connection curves involving $\Lambda_1(1; N_2, \mathfrak{X})$ which coalesce at $\mathfrak{X}_c$ as $N_2 \to \infty$. Wood (1987) has argued that the result is to form a branch-point of $\Lambda_1(1; \infty, \mathfrak{X})$ with cycle number $p \leq s$. It follows, from the discussion of algebraic functions in Sect. 17.6 and of scaling at a critical point in Sect. 4.5, that the scaling exponent y_2 in a two-dimensional system is given by

$$\frac{2}{y_2} = \frac{q}{p}. \tag{11.4.45}$$

Wood (1987) has suggested that p is likely to take its maximum value s and that q should be some multiple t of $2(s-1)$, giving the result

$$y_2 = \frac{s}{t(s-1)}. \tag{11.4.46}$$

This then gives, from (4.5.39) and (4.5.53),

$$\alpha = \frac{2[s - t(s-1)]}{s}, \qquad \nu = \frac{t(s-1)}{s}. \tag{11.4.47}$$

The first of these formulae can give rise to a singularity in heat capacity only when $t = 1$. For this case the hard-hexagon and 3-state Potts models correspond to $s = 6$, $\alpha = \frac{1}{3}$ and the K-type ($\alpha = \frac{1}{2}$) transition exhibited by the dimer model described in Sect. 13.2.2 corresponds to $s = 4$. In the case of the three-spin Ising model on the triangular lattice the polynomial of which $\Lambda_1(1; \mathfrak{X}, \infty)$ is a branch is known (Baxter and Wu 1973, 1974; Baxter 1974; Joyce 1975a, b). This leads directly to the result (Wood 1987) $2/y_2 = \frac{4}{3}$, $\alpha = \frac{2}{3}$ corresponding to $t = 1$, $s = 3$ in formulae (11.4.46) and (11.4.47). As we have seen in Sect. 11.4.3, for the spin-$\frac{1}{2}$ Ising model on a square lattice, the critical point corresponds, in the limit $N_2 \to \infty$, to a point of accumulation of an infinite number of branch-points. Thus we have $s = \infty$, which together with $t = 1$ yields $\alpha = 0$, which is consistent with the logarithmic singularity in the heat capacity.

A theory of finite-size scaling of partition function zeros was developed by Itzykson et al. (1983). For a two-dimensional $N_2 \times N_2$ lattice system they proposed the scaling form $|\mathfrak{X}_0 - \mathfrak{X}_c| \sim N_2^{-z}$, for the zero $\mathfrak{X}_0$ closest to the critical value $\mathfrak{X}_c$ in the complex plane $\mathbb{C}_{\mathfrak{X}}$ of some Boltzmann variable $\mathfrak{X}$ for which the scaling exponent is z. Given that boundary conditions ensure that $\mathcal{E}_{N2}$ involves only eigenvalues in the one-dimensional symmetric representation, the points of this set closest to $\mathbb{R}^{(+)}$ will be branch-points of $\Lambda_1(1; N_2, \mathfrak{X})$, occurring in conjugate pairs. They correspond to bounds on the asymptotic distribution of partition function zeros. Using this insight Williams and Lavis (1996) proposed a modified version of the formula of Itzykson et al. (1983) for a $N_1 \times N_2$ lattice in the limit $N_1 \to \infty$. With the conjugate pair of branch-points closest to $\mathfrak{X}_c$ in $\mathbb{C}_{\mathfrak{X}}$ denoted by $\mathfrak{X}_*$ and $\overline{\mathfrak{X}}_*$, the proposed form was $|\mathfrak{X}_* - \mathfrak{X}_c| \sim N_2^{-z}$, for the calculation of z. Support for this method, when $\mathfrak{X}$ is the thermal Boltzmann factor, is given by the behaviour of the rounding exponent w defined in (4.10.18). The quantity N_2^{-w} is of the order of the size of the neighbourhood in which the correlation length exceeds the lattice width (Fisher and Barber 1972). This should be of the same order as the radius of convergence of a power-series expansion of the free energy about the critical point, which will be given by the location of the nearest branch-point of $\Lambda_1(1; N_2, \mathfrak{X})$. The equality (4.10.19), which in the present notation is equivalent to $w = z$, is thought to be satisfied in a wide class of systems, particularly in the case of a cylinder with free boundary conditions at the ends (Fisher and Ferdinand 1967).

Chapter 12
Transfer Matrices: Exactly Solved Models

12.1 A General Eight-Vertex Model

As described in Sect. 3.8.1 the arrows of this model are on a rapidity lattice $\mathcal{N}^{\circledast}$ consisting of a finite number of straight lines, which here we suppose is bounded by a simple closed curve $\mathfrak{C}$, as shown in Fig. 16.15. Each lattice edge is the boundary between a shaded and an unshaded face these being decorated with one site of the dual lattices $\mathcal{N}$ and $\mathcal{N}^{\star}$, respectively. Boltzmann weights are given in terms of the spins on the sites of $\mathcal{N}$ and $\mathcal{N}^{\star}$ by (3.8.9) and in terms of edge variables on the edges of $\mathcal{N}^{\circledast}$ by (3.8.10).[1]

It is clear that the partition function is invariant under any smooth deformation which, while keeping $\mathfrak{C}$ fixed, distorts the lattice without:

(i) changing the order of the points of intersection of the lattice lines with $\mathfrak{C}$,
(ii) a vertex passing through a lattice line.

Condition (i) is equivalent to preventing vertices passing through $\mathfrak{C}$. The interest here is in systems for which the partition function is invariant under the more 'intrusive' transformation where condition (ii) is no longer demanded.

Definition 12.1.1 The eight-vertex model on $\mathcal{N}^{\circledast}$ is *Z-invariant* if the partition function is unchanged under any orthogonal translations of lattice lines which conserve the order of the intersections on $\mathfrak{C}$.

As we noted in Sect. 3.8.1.3, this eight-vertex model factors into two Ising models when the four-spin interaction is zero and the definition of Z-invariance carries over to the Ising models on $\mathcal{N}$ and $\mathcal{N}^{\star}$. An example of the transformation under which the partition function of a Z-invariant model must remain unchanged is shown in

---

[1] As indicated in Sect. 3.8.1, the edge variables can be used to assign, using the direction and shading rules, two different arrangements of arrows on the edges of $\mathcal{N}^{\circledast}$. Although these arrows give the popular picture of vertex models they play no direct role in the mathematical analysis. This is achieved in terms of edge variables or spins. We shall, therefore, make little reference to the arrows, and to the alternative direction and shading rules, in this chapter.

D.A. Lavis, *Equilibrium Statistical Mechanics of Lattice Models*,
Theoretical and Mathematical Physics, DOI 10.1007/978-94-017-9430-5_12

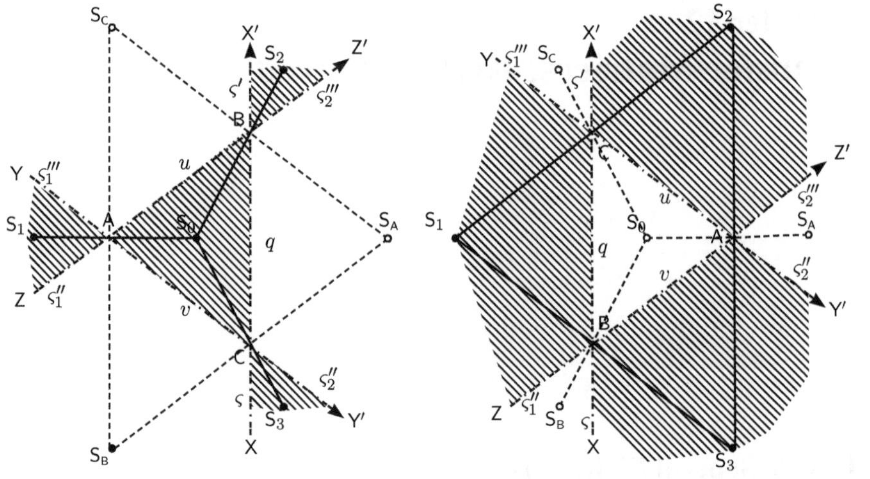

(L) (R)

Fig. 12.1 A translation from (L) to (R) of the line XX′. The lines XX′, YY′ and ZZ′ are given directions and rapidities $\bar{\varrho}$, ϱ and ϱ', respectively

Fig. 12.1 where A, B and C lie within $\mathfrak{C}$ and, in going from (L) to (R), the line XX′ passes through A.

12.1.1 A Generalized Star-Triangle Transformation

In Fig. 12.1, XX′, YY′ and ZZ′ are lines and A, B and C are sites of $\mathcal{N}^{\circledast}$; S_1, S_2 and S_3 are sites of $\mathcal{N}$ and S_A, S_B and S_C are sites of $\mathcal{N}^\star$, with σ_Q the spin on site S_Q, for $\mathsf{Q} = \mathsf{A}, \mathsf{B}, \mathsf{C}$ and σ_j the spin on site S_j, for $j = 1, 2, 3$. The spin on site S_0 is σ_0. The couplings between spins are labelled using the angle bisected by the line between the spins. Thus, for example, K_{ZBX} is the coupling between spins σ_0 and σ_2 in Fig. 12.1 (L) and σ_0 and σ_B in Fig. 12.1 (R). The four-spin couplings are $K^{(4)}_\mathsf{A}$, $K^{(4)}_\mathsf{B}$ and $K^{(4)}_\mathsf{C}$. Neglecting the constant term which does not play a role in the analysis, the Boltzmann weights for the three sites in (L) are

$$
\begin{aligned}
\mathfrak{Y}_\mathsf{A}(\sigma_1, \sigma_\mathsf{C}, \sigma_0, \sigma_\mathsf{B}) &:= \exp(K_{\mathsf{YAZ}}\sigma_1\sigma_0 + K_{\mathsf{YAZ'}}\sigma_\mathsf{B}\sigma_\mathsf{C} + K^{(4)}_\mathsf{A}\sigma_1\sigma_\mathsf{C}\sigma_0\sigma_\mathsf{B}),\\
\mathfrak{Y}_\mathsf{B}(\sigma_0, \sigma_\mathsf{C}, \sigma_2, \sigma_\mathsf{A}) &:= \exp(K_{\mathsf{ZBX}}\sigma_0\sigma_2 + K_{\mathsf{ZBX'}}\sigma_\mathsf{A}\sigma_\mathsf{C} + K^{(4)}_\mathsf{B}\sigma_0\sigma_\mathsf{C}\sigma_2\sigma_\mathsf{A}),\\
\mathfrak{Y}_\mathsf{C}(\sigma_\mathsf{B}, \sigma_0, \sigma_\mathsf{A}, \sigma_3) &:= \exp(K_{\mathsf{XCY}}\sigma_\mathsf{B}\sigma_\mathsf{A} + K_{\mathsf{XCY'}}\sigma_3\sigma_0 + K^{(4)}_\mathsf{C}\sigma_\mathsf{B}\sigma_0\sigma_\mathsf{A}\sigma_3)
\end{aligned}
\tag{12.1.1}
$$

and, distinguishing them with a prime, for (R)

$$\mathfrak{Y}'_{\mathsf{A}}(\sigma_0,\sigma_2,\sigma_{\mathsf{A}},\sigma_3) := \exp(K_{\mathsf{YAZ}}\sigma_0\sigma_{\mathsf{A}} + K_{\mathsf{YAZ}'}\sigma_3\sigma_2 + K^{(4)}_{\mathsf{A}}\sigma_0\sigma_2\sigma_{\mathsf{A}}\sigma_3),$$
$$\mathfrak{Y}'_{\mathsf{B}}(\sigma_{\mathsf{B}},\sigma_1,\sigma_0,\sigma_3) := \exp(K_{\mathsf{ZBX}}\sigma_{\mathsf{B}}\sigma_0 + K_{\mathsf{ZBX}'}\sigma_3\sigma_1 + K^{(4)}_{\mathsf{B}}\sigma_{\mathsf{B}}\sigma_1\sigma_0\sigma_3), \qquad (12.1.2)$$
$$\mathfrak{Y}'_{\mathsf{C}}(\sigma_1,\sigma_{\mathsf{C}},\sigma_2,\sigma_0) := \exp(K_{\mathsf{XCY}}\sigma_1\sigma_2 + K_{\mathsf{XCY}'}\sigma_0\sigma_{\mathsf{C}} + K^{(4)}_{\mathsf{C}}\sigma_1\sigma_{\mathsf{C}}\sigma_2\sigma_0).$$

Apart from representing a translation of the lattice line XX' through the site A, the transformation in Fig. 12.1 from the left-hand figure (L) to right-hand figure (R) can be regarded as a star-triangle transformation on sites of $\mathcal{N}$ in which summation over σ_0 removes the site S_0 together with a triangle-star transformation on sites of $\mathcal{N}^\star$ which introduces the site S_0. Given that the couplings are unchanged by this procedure the system is said to locally satisfy *generalized star-triangle (GST) conditions* if the partition function is unaffected. More specifically, since the rest of the lattice is left unchanged, the contributions to the partition function for the configurations in Fig. 12.1 (L) and (R) with the spins apart from σ_0 fixed must be the same. For (L) this is

$$\begin{aligned}\mathcal{Z}(\sigma_{\mathsf{A}},\sigma_{\mathsf{B}},\sigma_{\mathsf{C}},\sigma_1,\sigma_2,\sigma_3)\\ &:= \sum_{\sigma_0=\pm 1} \tilde{\mathfrak{Y}}_{\mathsf{A}}(\sigma_1,\sigma_{\mathsf{C}},\sigma_0,\sigma_{\mathsf{B}})\tilde{\mathfrak{Y}}_{\mathsf{B}}(\sigma_0,\sigma_{\mathsf{C}},\sigma_2,\sigma_{\mathsf{A}})\tilde{\mathfrak{Y}}_{\mathsf{C}}(\sigma_{\mathsf{B}},\sigma_0,\sigma_{\mathsf{A}},\sigma_3)\\ &= 2\exp(K_{\mathsf{YAZ}'}\sigma_{\mathsf{B}}\sigma_{\mathsf{C}} + K_{\mathsf{ZBX}'}\sigma_{\mathsf{C}}\sigma_{\mathsf{A}} + K_{\mathsf{XCY}}\sigma_{\mathsf{A}}\sigma_{\mathsf{B}})\\ &\qquad \times \cosh(K_{\mathsf{YAZ}}\sigma_1 + K_{\mathsf{ZBX}}\sigma_2 + K_{\mathsf{XCY}'}\sigma_3\\ &\qquad + K^{(4)}_{\mathsf{A}}\sigma_1\sigma_{\mathsf{B}}\sigma_{\mathsf{C}} + K^{(4)}_{\mathsf{B}}\sigma_2\sigma_{\mathsf{C}}\sigma_{\mathsf{A}} + K^{(4)}_{\mathsf{C}}\sigma_3\sigma_{\mathsf{A}}\sigma_{\mathsf{B}}).\end{aligned} \qquad (12.1.3)$$

It is clear from Fig. 12.1 that the transformation between (L) and (R) is equivalent to the interchange $(\sigma_{\mathsf{A}},\sigma_{\mathsf{B}},\sigma_{\mathsf{C}}) \longleftrightarrow (\sigma_1,\sigma_2,\sigma_3)$. So the contribution to the partition function for (R) is

$$\mathcal{Z}'(\sigma_{\mathsf{A}},\sigma_{\mathsf{B}},\sigma_{\mathsf{C}},\sigma_1,\sigma_2,\sigma_3) = \mathcal{Z}(\sigma_1,\sigma_2,\sigma_3,\sigma_{\mathsf{A}},\sigma_{\mathsf{B}},\sigma_{\mathsf{C}}) \qquad (12.1.4)$$

and the GST condition is

$$\mathcal{Z}(\sigma_{\mathsf{A}},\sigma_{\mathsf{B}},\sigma_{\mathsf{C}},\sigma_1,\sigma_2,\sigma_3) = \mathcal{Z}(\sigma_1,\sigma_2,\sigma_3,\sigma_{\mathsf{A}},\sigma_{\mathsf{B}},\sigma_{\mathsf{C}}). \qquad (12.1.5)$$

12.1.2 The Solution to the GST Transformation and the Elliptic Variable Formulation

The formula (12.1.5) gives sixty-four equations, with the number reduced to thirty-two because the equations are unchanged by negating all the spin values. However, there are further symmetries because the sets $\{\sigma_{\mathsf{A}},\sigma_{\mathsf{B}},\sigma_{\mathsf{C}}\}$ and $\{\sigma_1,\sigma_2,\sigma_3\}$ can be negated independently and the members of the pairs $\{\sigma_{\mathsf{A}},\sigma_1\}$, $\{\sigma_{\mathsf{B}},\sigma_2\}$ and $\{\sigma_{\mathsf{C}},\sigma_3\}$ can be interchanged simultaneously. This leaves eight equations of which

Table 12.1 Spin values for the six independent equations derived from (12.1.5)

σ_{A}	σ_{B}	σ_{C}	σ_1	σ_2	σ_3
$+1$	$+1$	$+1$	$+1$	$+1$	-1
$+1$	$+1$	$+1$	$+1$	-1	$+1$
$+1$	$+1$	$+1$	$+1$	-1	-1
$+1$	$+1$	-1	$+1$	-1	$+1$
$+1$	$+1$	-1	$+1$	-1	-1
$+1$	-1	$+1$	$+1$	-1	-1

two are identities. The spin values for the remaining six equations are given in Table 12.1. With the notation $(K_1, K_2, K_3) := (K_{\mathsf{YAZ}}, K_{\mathsf{ZBX}}, K_{\mathsf{XCY'}})$, $(K'_1, K'_2, K'_3) := (K_{\mathsf{YAZ'}}, K_{\mathsf{ZBX'}}, K_{\mathsf{XCY}})$ and $(K_1^{(4)}, K_2^{(4)}, K_3^{(4)}) := (K_{\mathsf{A}}^{(4)}, K_{\mathsf{B}}^{(4)}, K_{\mathsf{C}}^{(4)})$, with the index j being modulo 3, and

$$\begin{aligned} \mathrm{C}(x_1, x_2, x_3, x_4, x_5, x_6) := {} & \cosh(x_1 + x_2 + x_3 + x_4 + x_5 + x_6) \\ & \times \cosh(x_1 - x_2 - x_3 + x_4 + x_5 + x_6), \end{aligned} \tag{12.1.6}$$

these can be simplified to the form

$$\begin{aligned} \exp(4K'_j) &= \frac{\mathrm{C}(K_j, K_{j+1}, K_{j+2}, -K_j^{(4)}, -K_{j+1}^{(4)}, K_{j+2}^{(4)})}{\mathrm{C}(K_j, K_{j+1}, -K_{j+2}, K_j^{(4)}, K_{j+1}^{(4)}, -K_{j+2}^{(4)})} \\ &= \frac{\mathrm{C}(K_j, K_{j+1}, K_{j+2}, -K_j^{(4)}, K_{j+1}^{(4)}, -K_{j+2}^{(4)})}{\mathrm{C}(K_j, K_{j+1}, -K_{j+2}, K_j^{(4)}, -K_{j+1}^{(4)}, K_{j+2}^{(4)})}, \quad j = 1, 2, 3. \end{aligned} \tag{12.1.7}$$

The three equalities between the second and third expressions in (12.1.7) can be taken as defining $K_1^{(4)}$, $K_2^{(4)}$ and $K_3^{(4)}$ as functions of K_1, K_2 and K_3. However, these equations are not independent since they are satisfied by the infinity of solutions given by

$$K_1^{(4)} = K_2^{(4)} = K_3^{(4)} = K^{(4)}, \tag{12.1.8}$$

for all $K^{(4)}$. When $K^{(4)} = 0$, the transformation decomposes into two star-triangle transformations of the form (9.4.10).

No solutions to (12.1.7) are known (and possibly do not exist) except when (12.1.8) is satisfied. We, therefore, follow Baxter (1978) in assuming this condition when (12.1.7) gives, for $j = 1, 2, 3$,

$$\tanh(2K'_j) = \frac{\sinh(2K_{j+1})\sinh(2K_{j+2}) - \sinh(2K_j)\sinh(2K^{(4)})}{\cosh(2K_j)\cosh(2K^{(4)}) + \cosh(2K_{j+1})\cosh(2K_{j+2})}, \tag{12.1.9}$$

and[2]

$$\Delta_1 = \Delta_2 = \Delta_3, \tag{12.1.10}$$

where

$$\Delta_j := \sinh(2K_j)\sinh(2K'_j) + \tanh(2K^{(4)})\cosh(2K_j)\cosh(2K'_j). \tag{12.1.11}$$

Referring to Fig. 12.1 and noting the directions of the lattice lines, it can be seen that, for both (L) and (R), K_1 and K'_1 are the forward and backward couplings at site A and K_2 and K'_2 are the forward and backward couplings at site B. But at site C the situation is reversed with K'_3 and K_3 the forward and backward couplings. This means, by reference to (3.8.11), that the Boltzmann weights $\mathfrak{c}_3$ and $\mathfrak{d}_3$ are interchanged. However, this does not affect the conclusion, using (3.8.12), that

$$\Delta_j = \frac{\mathfrak{a}_j^2 + \mathfrak{b}_j^2 - \mathfrak{c}_j^2 - \mathfrak{d}_j^2}{2(\mathfrak{a}_j\mathfrak{b}_j + \mathfrak{c}_j\mathfrak{d}_j)}. \tag{12.1.12}$$

In a similar way, if

$$\Gamma_j := \tanh(2K_j^{(4)}), \tag{12.1.13}$$

then

$$\Gamma_1 = \Gamma_2 = \Gamma_3, \tag{12.1.14}$$

with, from (3.8.13),

$$\Gamma_j = \frac{\mathfrak{a}_j\mathfrak{b}_j - \mathfrak{c}_j\mathfrak{d}_j}{\mathfrak{a}_j\mathfrak{b}_j + \mathfrak{c}_j\mathfrak{d}_j}. \tag{12.1.15}$$

Conditions (12.1.10) and (12.1.14) are sufficient to ensure that (12.1.5) is satisfied, when Δ_j and Γ_j are expressed in terms of Ising couplings by (12.1.11) and (12.1.13) or in terms of Boltzmann weights by (12.1.12) and (12.1.15). Although the common values Δ and Γ for Δ_j and Γ_j at every site fix the values of $K_j^{(4)} = K^{(4)}$ the values of K_j and K'_j vary from site to site.

It is well-known from the work of Onsager for the Ising model and Baxter for the eight-vertex model that considerable illumination and simplification can achieved if the analysis is formulated in terms of Jacobi elliptic functions. In following the same route it is necessary to place some restrictions on the ranges of the parameters used

[2] The analysis in this chapter is essentially that of Baxter (1978). However, as explained in Sect. 3.8.1, notational changes have been made to bring alignment with Baxter (1982b). These are listed in Table 3.2.

in this discussion. To do so we rely on the results of Sects. 3.8.1.4 and 9.3 for the regular square-lattice model. There it was shown that the condition

$$\mathrm{a} > \mathrm{b} + \mathrm{c} + \mathrm{d}, \tag{12.1.16}$$

defines the low-temperature zone[3] $\mathcal{R}_{\mathrm{L}}(\mathrm{I})$, of frozen ferromagnetic ordering, in the phase space of $(\mathrm{a}, \mathrm{b}, \mathrm{c}, \mathrm{d})$, with $\mathrm{a} = \mathrm{b} + \mathrm{c} + \mathrm{d}$ being the transition surface $\mathcal{C}(\mathrm{I})$ to the disordered high-temperature zone $\mathcal{R}_{\mathrm{H}}(\mathrm{I})$. It was further shown that symmetry transformations, including the low-to-high-temperature weak-graph transformation, allow the partition function throughout phase space to be derived from that within $\mathcal{R}_{\mathrm{L}}(\mathrm{I})$. With this motivation we impose the condition (12.1.16), which, from (3.8.11), becomes in terms of the couplings

$$\exp[2K^{(4)}]\sinh[K^{(\mathrm{f})} + K^{(\mathrm{b})}] \geq \cosh[K^{(\mathrm{f})} - K^{(\mathrm{b})}]. \tag{12.1.17}$$

As was shown in Sect. 9.3.3, $-1 < \Gamma < 1$ throughout the phase space and $\Delta > 1$ in $\mathcal{R}_{\mathrm{L}}(\mathrm{I})$ with $\Delta = 1$ on $\mathcal{C}(\mathrm{I})$. It follows that the (positive) elliptic modulus k defined by[4]

$$\mathrm{k}^2 := \frac{1-\Gamma^2}{\Delta^2 - \Gamma^2} = \frac{1}{\Delta^2\cosh^2(2K^{(4)}) - \sinh^2(2K^{(4)})}, \tag{12.1.18}$$

satisfies the condition $0 < \mathrm{k} < 1$ in $\mathcal{R}_{\mathrm{L}}(\mathrm{I})$, with $\mathrm{k} = 1$ on $\mathcal{C}(\mathrm{I})$. Following Baxter (1978), elliptic parameters λ, α_j and α'_j are defined, in terms of the Jacobi amplitude function (Eq. (18.3.36)), by

$$\begin{aligned} &2K^{(4)} =: \mathrm{i}\,\mathrm{am}(\mathrm{i}[\mathcal{K}' - \lambda]|\mathrm{k}), \\ &2K_j =: -\mathrm{i}\,\mathrm{am}(\mathrm{i}[\mathcal{K}' - \alpha_j]|\mathrm{k}), \qquad 2K'_j =: -\mathrm{i}\,\mathrm{am}(\mathrm{i}[\mathcal{K}' - \alpha'_j]|\mathrm{k}). \end{aligned} \tag{12.1.19}$$

Whatever the value of $\mathrm{k} \in [0, 1]$, and whatever are the real values of the couplings, these equations will have, as is seen in Sect. 18.3.5, a solution in the range $[0, 2\mathcal{K}']$, for each of the elliptic parameters. From (18.3.38) and (18.3.45),

$$\begin{aligned} &\sinh(2K^{(4)}) = -\mathrm{i}\,\mathrm{ns}(\mathrm{i}\lambda|\mathrm{k})/\mathrm{k}, \qquad \cosh(2K^{(4)}) = \mathrm{i}\,\mathrm{ds}(\mathrm{i}\lambda|\mathrm{k})/\mathrm{k}, \\ &\sinh(2K_j) = \mathrm{i}\,\mathrm{ns}(\mathrm{i}\alpha_j|\mathrm{k})/\mathrm{k}, \qquad \cosh(2K_j) = \mathrm{i}\,\mathrm{ds}(\mathrm{i}\alpha_j|\mathrm{k})/\mathrm{k}, \\ &\sinh(2K'_j) = \mathrm{i}\,\mathrm{ns}(\mathrm{i}\alpha'_j|\mathrm{k})/\mathrm{k}, \qquad \cosh(2K'_j) = \mathrm{i}\,\mathrm{ds}(\mathrm{i}\alpha'_j|\mathrm{k})/\mathrm{k}. \end{aligned} \tag{12.1.20}$$

---

[3] As indicated in the footnote on p. 297 the term 'region' is used to refer to the parts I, II, III, IV of the phase space defined in Sect. 3.8.1.4. Subsets of the regions are referred to as 'zones'.

[4] As in Sect. 18.3 we use the notation $\mathcal{K} := \mathcal{K}(\mathrm{k})$ and $\mathcal{K}' := \mathcal{K}(\mathrm{k}')$ to denote the complete elliptic integral and its complement for modulus k.

Then, from (12.1.13) and (12.1.18),

$$\Gamma = -\mathrm{nd}(\mathrm{i}\,\lambda|\mathrm{k}), \tag{12.1.21}$$

$$\Delta = \mathrm{cd}(\mathrm{i}\,\lambda|\mathrm{k}). \tag{12.1.22}$$

Inserting from (12.1.20) and (12.1.22) into (12.1.11) and using (18.3.61),

$$\begin{aligned}\mathrm{k}^2\mathrm{sn}(\mathrm{i}\,\alpha_j|\mathrm{k})\mathrm{sn}(\mathrm{i}\,\alpha_j'|\mathrm{k})\{\mathrm{cn}(\mathrm{i}\,[\alpha_j+\alpha_j']|\mathrm{k}) - \mathrm{cn}(\mathrm{i}\,\lambda|\mathrm{k})\}\\ = \mathrm{dn}(\mathrm{i}\,\lambda|\mathrm{k}) - \mathrm{dn}(\mathrm{i}\,[\alpha_j+\alpha_j']|\mathrm{k}),\end{aligned} \tag{12.1.23}$$

which is satisfied by

$$\alpha_j + \alpha_j' = \lambda. \tag{12.1.24}$$

Substituting from (12.1.20) into (12.1.9) gives

$$\begin{aligned}\mathrm{k}^2\mathrm{cn}(\mathrm{i}\,\alpha_j'|\mathrm{k})\mathrm{sn}(\mathrm{i}\,\alpha_{j+1}|\mathrm{k})\mathrm{sn}(\mathrm{i}\,\alpha_{j+2}|\mathrm{k})\\ = \mathrm{dn}(\mathrm{i}\,\alpha_{j+1}|\mathrm{k})\mathrm{dn}(\mathrm{i}\,\alpha_{j+2}|\mathrm{k}) - \mathrm{dn}(\mathrm{i}\,\alpha_j'|\mathrm{k}).\end{aligned} \tag{12.1.25}$$

Comparing with (18.3.61) it can be seen that (12.1.25) is satisfied by

$$\alpha_j' = \alpha_{j+1} + \alpha_{j+2} \tag{12.1.26}$$

and, from (12.1.24),

$$\alpha_1 + \alpha_2 + \alpha_3 = \lambda, \tag{12.1.27}$$

$$\alpha_1' + \alpha_2' + \alpha_3' = 2\lambda. \tag{12.1.28}$$

The GST transformation (12.1.5) is satisfied if the couplings are related to elliptic variables k, λ, α_j, α_j', for $j = 1, 2, 3$, by (12.1.20) and these satisfy (12.1.24) and (12.1.27), or equivalently (12.1.28). In general there are no other restrictions on the elliptic variables, which can be real or complex.

When $K^{(4)} = 0$ the transformation decomposes into (local) star-triangle transformations on $\mathcal{N}$ and $\mathcal{N}^\star$. We note that, for this case, from (18.3.55), (18.3.56), (12.1.11), (12.1.21) and (12.1.22),

$$\lambda = \mathcal{K}', \qquad \Delta = \sinh(2K_j)\sinh(2K_j') = 1/\mathrm{k}, \qquad \Gamma = 0. \tag{12.1.29}$$

12.1.3 Z-Invariance

Let the elliptic variables be re-labelled in terms of the angle notation initially used for the couplings. Then (12.1.24), becomes

$$\alpha_{\mathsf{YAZ}} = \lambda - \alpha_{\mathsf{YAZ'}}, \qquad \alpha_{\mathsf{ZBX}} = \lambda - \alpha_{\mathsf{ZBX'}}, \qquad \alpha_{\mathsf{XCY'}} = \lambda - \alpha_{\mathsf{XCY}}. \tag{12.1.30}$$

α_{YAZ} and $\alpha_{\mathsf{YAZ}'}$ are supplementary angle variables at the site A, as defined in Sect. 16.3, with α_{ZBX} and $\alpha_{\mathsf{ZBX}'}$, and $\alpha_{\mathsf{XCY}'}$ and α_{XCY} being the corresponding pairs for B and C respectively. Then (12.1.27) and (12.1.28), re-expressed in the form

$$\alpha_{\mathsf{YAZ}} + \alpha_{\mathsf{ZBX}} + \alpha_{\mathsf{XCY}'} = \lambda, \tag{12.1.31}$$

$$\alpha_{\mathsf{YAZ}'} + \alpha_{\mathsf{ZBX}'} + \alpha_{\mathsf{XCY}} = 2\lambda, \tag{12.1.32}$$

are the conditions on the interior and exterior sums, respectively, for the triangle ABC. Of course, (12.1.32) follows from (12.1.30) and (12.1.31), and the latter is the condition that the partition function is left unchanged as A passes through the line.

For the system to be Z-invariant GST conditions must hold for every orthogonal translation of a lattice line through a vertex. Let the lines of $\mathcal{N}^{\circledast}$ be L_i, $i = 1, 2, \ldots$. Not all the vertices V_{ij} of intersections of pairs of lines L_i and L_j belong to $\mathcal{N}^{\circledast}$ (are inside $\mathfrak{C}$), but for those that are there are couplings K_{ij}, $\bar{K}_{ij}$ and $K^{(4)}$ which determine the elliptic parameters λ and k and the supplementary angle variables α_{ij} and $\bar{\alpha}_{ij}$.[5] The eight-vertex model on $\mathcal{N}^{\circledast}$ is Z-invariant if the supplementary angle variables at each vertex and the interior angles of each triangle of $\mathcal{N}^{\circledast}$ sum to λ. It is convenient to extend $\mathcal{N}^{\circledast}$ to $\mathcal{E}(\mathcal{N}^{\circledast})$. This extended lattice is created by lengthening the lines of $\mathcal{N}^{\circledast}$ to include all vertices V_{ij} and beyond to a new boundary $\mathfrak{C}'$. At the new vertices angle variables can be *defined* to satisfy the Z-invariance conditions and then couplings defined so that the eight-vertex model is Z-invariant. The eight-vertex model on $\mathcal{E}(\mathcal{N}^{\circledast})$ is Z-invariant if and only if it is invariant on $\mathcal{N}^{\circledast}$.[6]

In Sect. 16.3, IS(c) was used to denote the sum of the interior angle variables of a polygon c of m edges of $\hat{\mathfrak{g}}$ (now replaced by $\mathcal{N}^{\circledast}$ or $\mathcal{E}(\mathcal{N}^{\circledast})$) and it follows from Theorem 16.3.1 that the eight-vertex model on the graph or lattice is Z-invariant if and only the supplementary angle variables at a vertex sum to λ and $\mathrm{IS(c)} = (m-2)\lambda$, for all polygons formed from edges and vertices of $\mathcal{N}^{\circledast}$. In the proof of this theorem a *base line*, denoted by L_0, was used to assign to each line L_i of the extended graph a number ϱ_i, called the *rapidity*, and a direction (orientation). This relied on the base line being such that it intersected every line of the extended graph. In the present situation there may be a choice of a number of lines of $\mathcal{E}(\mathcal{N}^{\circledast})$ for the role of base line. Any line not parallel to any other line of $\mathcal{E}(\mathcal{N}^{\circledast})$ will suffice, or if such does not exist an extra line will need to be added. In the latter case angle variables α_{i0} and $\bar{\alpha}_{i0}$ at its points of intersection with the lines of $\mathcal{E}(\mathcal{N}^{\circledast})$ are defined as they were when $\mathcal{N}^{\circledast}$ was augmented to $\mathcal{E}(\mathcal{N}^{\circledast})$. With an orientation given to the base line and consequently to all the lines of $\mathcal{E}(\mathcal{N}^{\circledast})$, as described Sect. 16.3, the notation α_{ij} and $\bar{\alpha}_{ij}$ is refined so that these angle variables correspond respectively to the angles interior and exterior to the two orientation directions. Then the rapidities are simply $\varrho_i := \alpha_{i0}$ giving

[5] The bar notation is used rather than primes since the supplementary angles are no longer associated with being interior or exterior angles of a particular triangle.

[6] Although, of course, certain displacements of lattice lines for $\mathcal{E}(\mathcal{N}^{\circledast})$ will not be allowed on $\mathcal{N}^{\circledast}$, since they lead to vertices passing through $\mathfrak{C}$.

$$\alpha_{ij} := \varrho_j - \varrho_i, \qquad \bar{\alpha}_{ij} := \lambda - \varrho_j + \varrho_i. \tag{12.1.33}$$

Of course, the rapidities will depend on the choice of base line. But they will differ by an additive quantity determined by setting the rapidity of the base line to zero.

As an example, if ϱ_{X}, ϱ_{Y} and ϱ_{Z} denote the rapidities of XX', YY' and ZZ' respectively,

$$\begin{aligned}
\alpha_{\mathsf{YAZ}} &= \varrho_{\mathsf{Z}} - \varrho_{\mathsf{Y}}, & \alpha_{\mathsf{YAZ}'} &= \lambda - \varrho_{\mathsf{Z}} + \varrho_{\mathsf{Y}},\\
\alpha_{\mathsf{ZBX}} &= \varrho_{\mathsf{X}} - \varrho_{\mathsf{Z}}, & \alpha_{\mathsf{ZBX}'} &= \lambda - \varrho_{\mathsf{X}} + \varrho_{\mathsf{Z}},\\
\alpha_{\mathsf{XCY}} &= \varrho_{\mathsf{X}} - \varrho_{\mathsf{Y}}, & \alpha_{\mathsf{XCY}'} &= \lambda - \varrho_{\mathsf{X}} + \varrho_{\mathsf{Y}},
\end{aligned} \tag{12.1.34}$$

which satisfy (12.1.30)–(12.1.32).

The conclusion of this analysis, derived through Sects. 12.1.1, 12.1.2, and 16.3 is that:

Theorem 12.1.1 *The eight-vertex model on $\mathcal{N}^{\circledast}$ is Z-invariant if and only if orientations, and rapidities ϱ_i can be chosen for the lines L_i, $i = 1, 2, \ldots$ of $\mathcal{N}^{\circledast}$, so that the couplings $K^{(4)}$, K_{ij} and $\bar{K}_{ij}$ are given at every vertex V_{ij} by*

$$\begin{aligned}
K^{(4)} &= \tfrac{1}{2}\mathrm{i}\,\mathrm{am}(\mathrm{i}[\mathcal{K}' - \lambda]|\mathrm{k}),\\
K_{ij} &= -\tfrac{1}{2}\mathrm{i}\,\mathrm{am}(\mathrm{i}[\mathcal{K}' - \varrho_j + \varrho_i]|\mathrm{k}),\\
\bar{K}_{ij} &= -\tfrac{1}{2}\mathrm{i}\,\mathrm{am}(\mathrm{i}[\mathcal{K}' - \lambda + \varrho_j - \varrho_i]|\mathrm{k}),
\end{aligned} \tag{12.1.35}$$

for some parameter λ and elliptic modulus k.

Suppose that the spins around the vertex V_{ij} are $\sigma_{ij}^{(k)}$, $k = a, b, c, d$, where $k = a, c$ correspond to the spins on a bond like that shown in Fig. 16.16a and $k = b, d$ for a bond like that shown in Fig. 16.16b. Neglecting any constant term, the Hamiltonian is

$$\begin{aligned}
\widehat{H}(K_{ij}, \bar{K}_{ij}, K_{ij}^{(4)}; \boldsymbol{\sigma}) = -\sum_{\{ij\}} \big\{ & K_{ij}\sigma_{ij}^{(a)}\sigma_{ij}^{(c)} + \bar{K}_{ij}\sigma_{ij}^{(b)}\sigma_{ij}^{(d)}\\
& + K_{ij}^{(4)}\sigma_{ij}^{(a)}\sigma_{ij}^{(b)}\sigma_{ij}^{(c)}\sigma_{ij}^{(d)}\big\},
\end{aligned} \tag{12.1.36}$$

where we have (temporarily) generalized[7] to the case of different four-spin couplings on the different sites. Then, for V_{ij}, the local correlations which contribute to the enthalpy are

[7] See Sect. 2.5 for a similar case of generalization.

$$\langle \sigma_{ij}^{(a)} \sigma_{ij}^{(c)} \rangle = -N^{\circledast} \frac{\partial \phi^{(\mathrm{G})}}{\partial K_{ij}}, \qquad \langle \sigma_{ij}^{(b)} \sigma_{ij}^{(d)} \rangle = -N^{\circledast} \frac{\partial \phi^{(\mathrm{G})}}{\partial \bar{K}_{ij}},$$

$$\langle \sigma_{ij}^{(a)} \sigma_{ij}^{(b)} \sigma_{ij}^{(c)} \sigma_{ij}^{(d)} \rangle = -N^{\circledast} \frac{\partial \phi^{(\mathrm{G})}}{\partial K_{ij}^{(4)}}, \tag{12.1.37}$$

where the generalized free energy per vertex is

$$\phi^{(\mathrm{G})}(\{K_{ij}\}, \{\bar{K}_{ij}\}, \{K_{ij}^{(4)}\}); \tag{12.1.38}$$

that is a function of the whole sets of couplings.[8] As $K_{ij}^{(4)} \to K^{(4)}$,

$$\phi^{(\mathrm{G})}(\{K_{ij}\}, \{\bar{K}_{ij}\}, \{K_{ij}^{(4)}\}) \to \phi(\{K_{ij}\}, \{\bar{K}_{ij}\}, K^{(4)}). \tag{12.1.39}$$

Differentiation with respect to $K^{(4)}$ now 'picks up' the equal four-spin correlations at every lattice site and the third of Eqs. (12.1.37) is replaced by

$$\langle \sigma_{ij}^{(a)} \sigma_{ij}^{(b)} \sigma_{ij}^{(c)} \sigma_{ij}^{(d)} \rangle = -\frac{\partial \phi}{\partial K^{(4)}}. \tag{12.1.40}$$

Given that V_{ij} is deep in the lattice, it is reasonable to suppose that these correlations will be independent of the couplings on and position of a lattice line L' far away near $\mathfrak{C}'$.[9] However, for a Z-invariant system, L' can[10] be orthogonally translated to approach V_{ij} without altering the partition function and hence the correlations. Indeed any orthogonal translations can be applied to lattice lines apart from L_i and L_j without changing the correlations. It follows that they must be dependent at most on ϱ_i, ϱ_j, λ and k and be independent of the arrangement of the other lines of the lattice. They must, therefore, be the same as that for the regular eight-vertex model with lattice lines forming parallelograms with rapidities ϱ_i on the lines parallel to L_i and rapidities ϱ_j on the lines parallel to L_j. That is with couplings K_{ij}, $\bar{K}_{ij}$ and $K^{(4)}$ at every vertex. This regular eight-vertex model corresponds to the limits $K_{gh} \to K_{ij}$, $\bar{K}_{gh} \to \bar{K}_{ij}$ for all vertices V_{gh} with

$$\phi(\{K_{ij}\}, \{\bar{K}_{ij}\}, K^{(4)}) \to \phi^{(\mathrm{R})}(K_{ij}, \bar{K}_{ij}, K^{(4)}) =: \phi_{ij}^{(\mathrm{R})} \tag{12.1.41}$$

and the regular lattice correlation functions are given by

---

[8] Which, exceptionally in this case, we have indicated by the enclosing brackets $\{\cdot\}$.

[9] Meaning that in order to get from V_{ij} to any part of L' many other lattice lines need to be crossed.

[10] As long as $\mathfrak{C}'$ is drawn so that no vertices of $\mathcal{E}(\mathcal{N}^{\circledast})$ pass through it in the process.

$$\langle \sigma_{ij}^{(a)} \sigma_{ij}^{(c)} \rangle^{(\mathrm{R})} = -\frac{\partial \phi_{ij}^{(\mathrm{R})}}{\partial K_{ij}}, \qquad \langle \sigma_{ij}^{(b)} \sigma_{ij}^{(d)} \rangle^{(\mathrm{R})} = -\frac{\partial \phi_{ij}^{(\mathrm{R})}}{\partial \bar{K}_{ij}},$$
$$\langle \sigma_{ij}^{(a)} \sigma_{ij}^{(b)} \sigma_{ij}^{(c)} \sigma_{ij}^{(d)} \rangle^{(\mathrm{R})} = -\frac{\partial \phi_{ij}^{(\mathrm{R})}}{\partial K^{(4)}}. \tag{12.1.42}$$

Returning now to the generalized model, with small variations δK_{ij}, $\delta \bar{K}_{ij}$ and $\delta K_{ij}^{(4)}$ in all the couplings,

$$\begin{aligned}\delta\phi^{(\mathrm{G})} &= \sum_{\{ij\}} \left\{ \frac{\partial \phi^{(\mathrm{G})}}{\partial K_{ij}} \delta K_{ij} + \frac{\partial \phi^{(\mathrm{G})}}{\partial \bar{K}_{ij}} \delta \bar{K}_{ij} + \frac{\partial \phi^{(\mathrm{G})}}{\partial K_{ij}^{(4)}} \delta K_{ij}^{(4)} \right\} \\ &= \frac{1}{N^{\circledast}} \sum_{\{ij\}} \left\{ \frac{\partial \phi_{ij}^{(\mathrm{R})}}{\partial K_{ij}} \delta K_{ij} + \frac{\partial \phi_{ij}^{(\mathrm{R})}}{\partial \bar{K}_{ij}} \delta \bar{K}_{ij} + \frac{\partial \phi_{ij}^{(\mathrm{R})}}{\partial K_{ij}^{(4)}} \delta K_{ij}^{(4)} \right\} \\ &= \frac{1}{N^{\circledast}} \sum_{\{ij\}} \delta \phi_{ij}^{(\mathrm{R})}. \end{aligned} \tag{12.1.43}$$

So $N^{\circledast}\phi^{(\mathrm{G})} - \sum_{\{ij\}} \delta\phi_{ij}^{(\mathrm{R})}$ is stationary with respect to these small variations if the system is Z-invariant. However, while keeping the system Z-invariant, the couplings can be increased. In the limit the only term in the partition function both for the general and regular models has all the spins aligned. So for all Z-invariant coupling values it must be the case that

$$\phi^{(\mathrm{G})}(\{K_{ij}\}, \{\bar{K}_{ij}\}, \{K_{ij}^{(4)}\}) = \frac{1}{N^{\circledast}} \sum_{\{ij\}} \phi^{(\mathrm{R})}(K_{ij}, \bar{K}_{ij}, K^{(4)}). \tag{12.1.44}$$

This decomposition of the free-energy density of the general Z-invariant eight-vertex model is the "key result" of Baxter (1978) since it means that the solution of the regular model by Baxter (1972b) implies the solution of the general Z-invariant model.

12.1.4 Edge Variables and Matrix Formulation

The expression for the Boltzmann weight at a site in terms of spin variables can be transformed into one in terms of edge variables using (3.8.2). With edge variables, spins and rapidities as shown in Fig. 12.2a, the pair and four-spin couplings are given, from (12.1.35), by

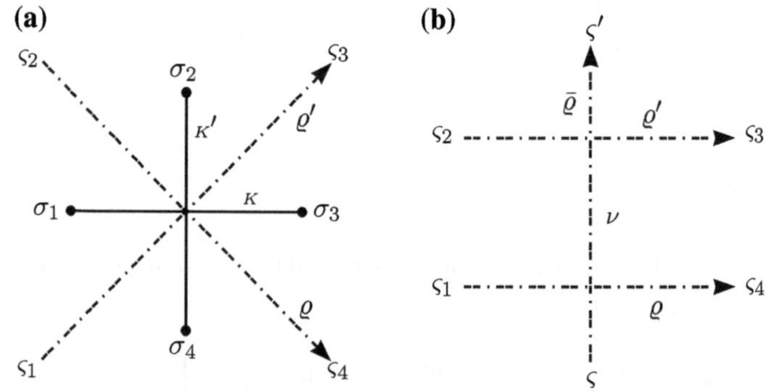

Fig. 12.2 Edges, couplings and spins at a lattice site

$$K = -\tfrac{1}{2}\mathrm{i}\,\mathrm{am}(\mathrm{i}[\mathcal{K}' - \varrho' + \varrho]|\mathrm{k}), \quad K' = -\tfrac{1}{2}\mathrm{i}\,\mathrm{am}(\mathrm{i}[\mathcal{K}' - \lambda + \varrho' - \varrho]|\mathrm{k}),$$
$$K^{(4)} = \tfrac{1}{2}\mathrm{i}\,\mathrm{am}(\mathrm{i}[\mathcal{K}' - \lambda]|\mathrm{k}), \tag{12.1.45}$$

and the four-dimensional matrix $\boldsymbol{U}_0(K, K')$ with elements parameterized by pairs of edge variables is defined by[11]

$$\left\langle {\varsigma_2 \atop \varsigma_1} \middle| \boldsymbol{U}_0(K, K') \middle| {\varsigma_3 \atop \varsigma_4} \right\rangle := \begin{cases} \exp(K^{(0)} + K\varsigma_1\varsigma_4 + K'\varsigma_1\varsigma_2 + K^{(4)}\varsigma_1\varsigma_3), & \\ & \text{if } \varsigma_1\varsigma_2\varsigma_3\varsigma_4 = 1, \\ 0, & \text{if } \varsigma_1\varsigma_2\varsigma_3\varsigma_4 \neq 1. \end{cases} \tag{12.1.46}$$

It is clear that these elements are invariant under a cyclic permutation of the edge variables by two steps and by one step accompanied by an interchange of K and K'. At this point it is convenient to re-represent the matrix $\boldsymbol{U}_0(K, K')$, using (12.1.45) and Fig. 12.2a, in terms of rapidities as $\boldsymbol{U}_0(\varrho' - \varrho)$, when $\boldsymbol{U}_0(K', K)$ becomes $\boldsymbol{U}_0(\lambda - \varrho' + \varrho)$. Then

$$\left\langle {\varsigma_2 \atop \varsigma_1} \middle| \boldsymbol{U}_0(\varrho' - \varrho) \middle| {\varsigma_3 \atop \varsigma_4} \right\rangle = \left\langle {\varsigma_4 \atop \varsigma_3} \middle| \boldsymbol{U}_0(\varrho' - \varrho) \middle| {\varsigma_1 \atop \varsigma_2} \right\rangle = \left\langle {\varsigma_3 \atop \varsigma_2} \middle| \boldsymbol{U}_0(\lambda - \varrho' + \varrho) \middle| {\varsigma_4 \atop \varsigma_1} \right\rangle. \tag{12.1.47}$$

We shall also need the four-dimensional matrix $\boldsymbol{U}_2\left({\varsigma' \atop \varsigma} ; {\bar{\varrho} - \varrho' \atop \bar{\varrho} - \varrho} \right)$ with elements

$$\left\langle {\varsigma_2 \atop \varsigma_1} \middle| \boldsymbol{U}_2\left({\varsigma' \atop \varsigma} ; {\bar{\varrho} - \varrho' \atop \bar{\varrho} - \varrho} \right) \middle| {\varsigma_3 \atop \varsigma_4} \right\rangle := \sum_{\{\nu\}} \left\langle {\varsigma_1 \atop \varsigma} \middle| \boldsymbol{U}_0(\bar{\varrho} - \varrho) \middle| {\nu \atop \varsigma_4} \right\rangle \left\langle {\varsigma_2 \atop \nu} \middle| \boldsymbol{U}_0(\bar{\varrho} - \varrho') \middle| {\varsigma' \atop \varsigma_3} \right\rangle, \tag{12.1.48}$$

[11] For the sake of brevity constant and four-spin couplings are not included as arguments.

as represented by Fig. 12.2b. Then the contribution to the partition functions from the triangle in Fig. 12.1(L) is

$$\begin{aligned}\mathcal{Z}(\varsigma_1'',\varsigma_1''',\varsigma,\varsigma',\varsigma_2'',\varsigma_2''') &= \sum_{\{\nu,\mu\}}\left\langle\begin{matrix}\varsigma_1'''\\ \varsigma_1''\end{matrix}\middle|\boldsymbol{U}_0(\varrho'-\varrho)\middle|\begin{matrix}\mu\\ \nu\end{matrix}\right\rangle\left\langle\begin{matrix}\mu\\ \nu\end{matrix}\middle|\boldsymbol{U}_2\begin{pmatrix}\varsigma' & \bar{\varrho}-\varrho'\\ \varsigma & ;\; \bar{\varrho}-\varrho\end{pmatrix}\middle|\begin{matrix}\varsigma_2'''\\ \varsigma_2''\end{matrix}\right\rangle\\ &= \left\langle\begin{matrix}\varsigma_1''\\ \varsigma_1''\end{matrix}\middle|\boldsymbol{U}_0(\varrho'-\varrho)\boldsymbol{U}_2\begin{pmatrix}\varsigma' & \bar{\varrho}-\varrho'\\ \varsigma & ;\; \bar{\varrho}-\varrho\end{pmatrix}\middle|\begin{matrix}\varsigma_2'''\\ \varsigma_2''\end{matrix}\right\rangle.\end{aligned} \tag{12.1.49}$$

In a similar way the contribution to the partition functions from the triangle in Fig. 12.1(R) is

$$\mathcal{Z}'(\varsigma_1'',\varsigma_1''',\varsigma,\varsigma',\varsigma_2'',\varsigma_2') = \left\langle\begin{matrix}\varsigma_2'''\\ \varsigma_3'\end{matrix}\middle|\boldsymbol{U}_2\begin{pmatrix}\varsigma' & \bar{\varrho}-\varrho\\ \varsigma & ;\; \bar{\varrho}-\varrho'\end{pmatrix}\boldsymbol{U}_0(\varrho'-\varrho)\middle|\begin{matrix}\varsigma_2'''\\ \varsigma_2''\end{matrix}\right\rangle \tag{12.1.50}$$

and the GST transformation is equivalent to

$$\boldsymbol{U}_0(\varrho'-\varrho)\boldsymbol{U}_2\begin{pmatrix}\varsigma' & \bar{\varrho}-\varrho'\\ \varsigma & ;\; \bar{\varrho}-\varrho\end{pmatrix} = \boldsymbol{U}_2\begin{pmatrix}\varsigma' & \bar{\varrho}-\varrho\\ \varsigma & ;\; \bar{\varrho}-\varrho'\end{pmatrix}\boldsymbol{U}_0(\varrho'-\varrho) \quad \forall\varsigma,\varsigma'. \tag{12.1.51}$$

12.1.5 Square-Lattice Models

For simplicity the rest of the chapter is restricted to models on the square lattice. In particular the square-lattice version of the model of Baxter (1978) as shown in Fig. 12.3 and its special cases which are versions of the Ising model. Adopting the notation of Chap. 11, but with $\mathcal{N}$ replaced by $\mathcal{N}^{\circledast}$, the sites of $\mathcal{N}^{\circledast}$ are labelled by the index pairs (i, j), where i $= 1, 2, \ldots, N_1$ and j $= 1, 2, \ldots, N_2$ are respectively the vertical and horizontal indices. Periodic boundary conditions are applied in the horizontal direction, so that the lattice is in the form of a vertical cylinder, and, straining the notions of vertical and horizontal a little,[12] also in the vertical direction, so the indices i and j are, respectively integers modulo N_1 and N_2. Sites of sublattices $\mathcal{N}^{\circledast}_{\mathrm{E}}$ and $\mathcal{N}^{\circledast}_{\mathrm{O}}$, as defined in Sect. 3.8.1, correspond respectively to (i−j) being even or odd and it is clear that in order for the sublattice structure to coexist with the periodic boundary conditions both N_1 and N_2 must be even.[13] The index i labels a slice of the lattice which is either a row (ring) of $\mathcal{N}^{\circledast}$ or equivalently the overlayed zigzag rows of $\mathcal{N}$ and $\mathcal{N}^{\star}$. Given that the system is Z-invariant with the same four-spin coupling

---

[12] See the footnote on p. 60.

[13] For the Ising model in Sect. 12.2 it is necessary to impose the more stringent conditions that N_1 and N_2 are divisibly by four, but for the treatment of the square-lattice eight-vertex model in Sect. 12.3 no restrictions on N_1 and N_2 need be applied.

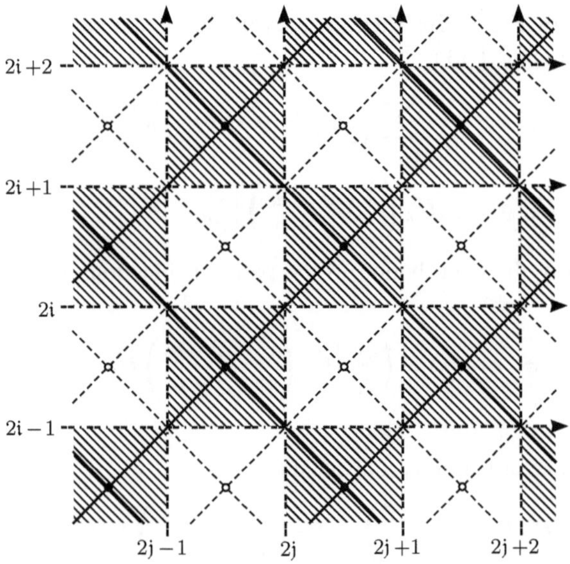

Fig. 12.3 The square lattice version of the model of Baxter (1978). The sites of $\mathcal{N}$ and $\mathcal{N}^{\star}$ are respectively on the *shaded* and *unshaded* faces of $\mathcal{N}^{\circledast}$. The sublattices $\mathcal{N}^{\circledast}_{\mathrm{E}}$ and $\mathcal{N}^{\circledast}_{\mathrm{O}}$ of $\mathcal{N}^{\circledast}$ have (as in Fig. 3.3) *shaded* faces on the *top-right* (and *bottom-left*) and *top-left* (and *bottom-right*) respectively

$$K^{(\mathrm{f})} := \tfrac{1}{2}\mathrm{i}\,\mathrm{am}(\mathrm{i}[\mathcal{K}' - \lambda]|\mathrm{k}), \tag{12.1.52}$$

at every site, directions and rapidities ϱ_{i} and $\bar{\varrho}_{\mathrm{j}}$ can be assigned to the horizontal and vertical lattice lines and, from (12.1.35), the couplings are related to the rapidities by

$$\begin{aligned} K^{(\mathrm{f})}_{\mathrm{ij}} &=: -\tfrac{1}{2}\mathrm{i}\,\mathrm{am}(\mathrm{i}[\mathcal{K}' - \bar{\varrho}_{\mathrm{j}} + \varrho_{\mathrm{i}}]|\mathrm{k}), \\ K^{(\mathrm{b})}_{\mathrm{ij}} &=: -\tfrac{1}{2}\mathrm{i}\,\mathrm{am}(\mathrm{i}[\mathcal{K}' - \lambda + \bar{\varrho}_{\mathrm{j}} - \varrho_{\mathrm{i}}]|\mathrm{k}). \end{aligned} \tag{12.1.53}$$

12.2 Square-Lattice Ising Models

The work on the square-lattice Ising model which evolved from the work of Onsager pictured the lattice with horizontal and vertical edges; this we shall refer to as the *standard square-lattice model*.[14] Of course, the orientation of the lattice does not affect the partition function but it does affect the transfer matrix and its eigenvalues

[14] Baxter (1978, p. 14) calls this the 'regular' model but we shall use this term to refer to the special case which is presented by Baxter (1982b, Chap. 7).

as can be seen from the presentation of Baxter (1982b, Chap. 7) where the lattice edges are at angles of 45° to the horizontal. This can be called the *checkerboard Ising model* because the lattice sites can be viewed as located at the centres of the black or white squares of a checkerboard (draughts board).[15]

The application of transfer matrices to the special case $K^{(4)} = 0$ of the Baxter (1978) model, described in this chapter, presents two possibilities. With the eight-vertex model on a kagomé lattice the spin representation divides into Ising models on triangular and honeycomb lattices and with the eight-vertex model on a square lattice the spin representation divides into two equivalent Ising models on square lattices oriented so that the diagonals of the squares are horizontal and vertical. For simplicity we concentrate solely on the latter case with the method of solution following that of Baxter (1982b, Chap. 7) and Baxter (1986). The lattice of each Ising model consists of the $\frac{1}{2}N_1N_2$ sites on alternate faces of the eight-vertex model, with periodic boundary conditions in both lattice directions and N_1 and N_2 even. So each is a checkerboard Ising model. For reasons which become apparent it will be convenient to assume that both N_1 and N_2 are divisible by four[16] and the integers

$$n_1 := \tfrac{1}{4}N_1, \qquad n_2 := \tfrac{1}{4}N_2 \tag{12.2.1}$$

will be used. As the model is Z-invariant, with the couplings being functions of rapidity differences, it is often more convenient to use the variables[17]

$$u := \bar{\varrho} - \varrho, \qquad u' := \bar{\varrho} - \varrho'. \tag{12.2.2}$$

Now $\lambda = \mathcal{K}'$ and the system is defined in terms of u and the elliptic modulus k,[18] with, from (12.1.53), the forward and backward couplings (at the same site of $\mathcal{N}^{\circledast}$) given by

$$\begin{aligned} \exp[2K^{(\mathrm{f})}(u)] &= \exp\{-\mathrm{i}\,\mathrm{am}(\mathrm{i}[\mathcal{K}' - u]|\mathrm{k})\}, \\ \exp[2K^{(\mathrm{b})}(u)] &= \exp\{-\mathrm{i}\,\mathrm{am}(\mathrm{i}\,u|\mathrm{k})\}. \end{aligned} \tag{12.2.3}$$

From (18.3.89),

$$\begin{aligned} \sinh[2K^{(\mathrm{f})}(u)] &= -\mathrm{i}\,\mathrm{sn}(\mathrm{i}[\mathcal{K}' - u]|\mathrm{k}), \quad \sinh[2K^{(\mathrm{b})}(u)] = -\mathrm{i}\,\mathrm{sn}(\mathrm{i}\,u|\mathrm{k}), \\ \cosh[2K^{(\mathrm{f})}(u)] &= \mathrm{cn}(\mathrm{i}[\mathcal{K}' - u]|\mathrm{k}), \quad \cosh[2K^{(\mathrm{b})}(u)] = \mathrm{cn}(\mathrm{i}\,u|\mathrm{k}), \end{aligned} \tag{12.2.4}$$

---

[15] Although, as we saw in Sect. 3.8.1.5 and as will be evident below, the term is usually used in rather more restrictive circumstances.

[16] See the footnote on p. 393. Since a motivation of this analysis is to develop a method which allow the limit $N_1, N_2 \to \infty$, this is not a serious restriction.

[17] Although, since both the horizontal rapidity ϱ and the vertical rapidity $\bar{\varrho}$ can differ between sites it is sometimes necessary to display them both explicitly.

[18] In the interests of brevity this latter dependence is not always displayed.

where

$$K^{(f)}(u) = K^{(b)}(\mathcal{K}' - u) \tag{12.2.5}$$

and, from (18.3.50),

$$\sinh[2K^{(f)}(u)] \sinh[2K^{(b)}(u)] = 1/\mathrm{k}, \tag{12.2.6}$$

for all u, as has already been shown in the context of the GST transformation in (12.1.29).

Couplings which satisfy (12.2.4)–(12.2.6), with $u = \bar{\varrho} - \varrho$, for interactions between sites on two lattice pairs which pass through the $\mathcal{N}^{\circledast}$ site of intersection of lines with rapidities ϱ and $\bar{\varrho}$, are said to be *conjugate*. In the present context it can be seen that the pairs are one each between sites of $\mathcal{N}$ and $\mathcal{N}^{\star}$,[19] with $K^{(f)}(u)$ and $K^{(b)}(u)$ between site-pairs where, with respect to the directions of the rapidity lines the first divides the interior and the second the exterior angle. From (18.3.52),[20]

$$\begin{aligned}
&K^{(f)}(-u) = K^{(f)}(u) \pm \tfrac{1}{2}\mathrm{i}\pi, && K^{(b)}(-u) = -K^{(b)}(u),\\
&\sinh[2K^{(f)}(-u)] = -\sinh[2K^{(f)}(u)], && \sinh[2K^{(b)}(-u)] = -\sinh[2K^{(b)}(u)],\\
&\cosh[2K^{(f)}(-u)] = -\cosh[2K^{(f)}(u)], && \cosh[2K^{(b)}(-u)] = \cosh[2K^{(b)}(u)],
\end{aligned} \tag{12.2.7}$$

$$\begin{aligned}
&K^{(x)}(u + 2\mathcal{K}') = -K^{(x)}(u) \pm \tfrac{1}{2}\mathrm{i}\pi,\\
&\sinh[2K^{(x)}(u + 2\mathcal{K}')] = \sinh[2K^{(x)}(u)],\\
&\cosh[2K^{(x)}(u + 2\mathcal{K}')] = -\cosh[2K^{(x)}(u)],
\end{aligned} \tag{12.2.8}$$

$$\begin{aligned}
&K^{(x)}(u + 2\mathrm{i}\,\mathcal{K}) = K^{(x)}(u) \pm \tfrac{1}{2}\mathrm{i}\pi,\\
&\sinh[2K^{(x)}(u + 2\mathrm{i}\,\mathcal{K})] = -\sinh[2K^{(x)}(u)],\\
&\cosh[2K^{(x)}(u + 2\mathrm{i}\,\mathcal{K})] = -\cosh[2K^{(x)}(u)],
\end{aligned} \tag{12.2.9}$$

where $\mathrm{x} = \mathrm{f}, \mathrm{b}$. From (18.3.38) and (18.3.44)–(18.3.47),

$$\exp[2K^{(f)}(u)] = W(u|\mathrm{k}), \qquad \exp[2K^{(b)}(u)] = W(\mathcal{K}' - u|\mathrm{k}), \tag{12.2.10}$$

[19] However, when, as below, we take all the horizontal rapidities on the lattice to be equal these relationships imply ones between backward and forward couplings across vertical neighbouring $\mathcal{N}^{\circledast}$ sites for pairs on the same lattice ($\mathcal{N}$ of $\mathcal{N}^{\star}$).

[20] The choice of signs in these formulae arise because couplings occur as the arguments of exponentials.

where

$$W(u|\mathrm{k}) := \frac{\mathrm{i}\,[\mathrm{dn}(\mathrm{i}u|\mathrm{k})+1]}{\mathrm{k}\mathrm{sn}(\mathrm{i}u|\mathrm{k})} = \frac{\mathrm{i}[\sqrt{\mathrm{k}'}\,\Theta\,(\mathrm{i}u+\mathcal{K}|\mathrm{k})+\Theta(\mathrm{i}u|\mathrm{k})]}{\sqrt{\mathrm{k}}\,\mathrm{H}(\mathrm{i}u|\mathrm{k})}, \tag{12.2.11}$$

which is meromorphic with poles at $u = 2\ell\mathcal{K}' + 2\ell'\mathrm{i}\,\mathcal{K}$ for all integer values of ℓ and ℓ'. From (18.3.2), (18.3.6) and (18.3.36),

$$\int_{2K^{(\mathrm{f})}}^{\infty} \frac{\mathrm{d}\omega}{\sqrt{1+\mathrm{k}^2\,\sinh^2(\omega)}} = \int_0^{2K^{(\mathrm{b})}} \frac{\mathrm{d}\omega}{\sqrt{1+\mathrm{k}^2\,\sinh^2(\omega)}} = u, \tag{12.2.12}$$

implying that, if both couplings are real and positive, u is real with $0 < u < \mathcal{K}'$.

We now suppose that the vertical rapidity has one of two values $\bar{\varrho}_{\mathrm{E}}$ and $\bar{\varrho}_{\mathrm{O}}$ according as the index j is even or odd. We shall later make a similar distinction between values of the horizontal rapidities but for the moment it is sufficient simply to use the one variable ϱ.

Referring to Fig. 12.3, the lattice line of spins on sites of $\mathcal{N}$ and $\mathcal{N}^\star$ immediately above the line of $\mathcal{N}^{\circledast}$ labelled i is also labelled with the index i. The spins on the sites of this line are denoted by $\tilde{\boldsymbol{\sigma}}^{(\mathrm{i})} := (\tilde{\sigma}_1^{(\mathrm{i})}, \tilde{\sigma}_2^{(\mathrm{i})}, \ldots, \tilde{\sigma}_{4n_2}^{(\mathrm{i})})$, where $\tilde{\sigma}_{\mathrm{j}}^{(\mathrm{i})}$ is on the site immediately to the right of the vertical lattice line of $\mathcal{N}^{\circledast}$ labelled j. It is now useful to give different variables for the spins on $\mathcal{N}$ and $\mathcal{N}^\star$. This is done by setting

$$\begin{aligned} \tau_j^{(\mathrm{i})} &:= \tilde{\sigma}_{2\mathrm{j}-1}^{(\mathrm{i})}, & \sigma_j^{(\mathrm{i})} &:= \tilde{\sigma}_{2\mathrm{j}}^{(\mathrm{i})}, & &\mathrm{i\,even}, \\ \sigma_j^{(\mathrm{i})} &:= \tilde{\sigma}_{2\mathrm{j}-1}^{(\mathrm{i})}, & \tau_j^{(\mathrm{i})} &:= \tilde{\sigma}_{2\mathrm{j}}^{(\mathrm{i})}, & &\mathrm{i\,odd}. \end{aligned} \tag{12.2.13}$$

The effect of this redesignation (Fig. 12.4) is to have spins labelled with σ's on $\mathcal{N}$ and spins labelled with τ's on $\mathcal{N}^\star$. Around the vertical lattice line 2j of $\mathcal{N}^{\circledast}$ there are zigzag lines of spins on $\mathcal{N}$ and $\mathcal{N}^\star$ all with the index j.

In spin-representation two forms of the eight-vertex transfer matrix $\boldsymbol{V}(\bar{\varrho}_{\mathrm{E}}, \bar{\varrho}_{\mathrm{O}}; \varrho)$ must be distinguished according to whether the row index i is even or odd. These will be denoted respectively by $\boldsymbol{V}_{\mathrm{E}}(\bar{\varrho}_{\mathrm{E}}, \bar{\varrho}_{\mathrm{O}}; \varrho)$ and $\boldsymbol{V}_{\mathrm{O}}(\bar{\varrho}_{\mathrm{E}}, \bar{\varrho}_{\mathrm{O}}; \varrho)$, with[21]

$$\begin{aligned} \langle \boldsymbol{\sigma}, \boldsymbol{\tau} | \boldsymbol{V}_{\mathrm{E}}(\bar{\varrho}_{\mathrm{E}}, \bar{\varrho}_{\mathrm{O}}; \varrho) | \boldsymbol{\sigma}', \boldsymbol{\tau}' \rangle &:= \langle \boldsymbol{\sigma} | \boldsymbol{B}(\bar{\varrho}_{\mathrm{E}}, \bar{\varrho}_{\mathrm{O}}; \varrho) | \boldsymbol{\sigma}' \rangle \, \langle \boldsymbol{\tau} | \boldsymbol{A}(\bar{\varrho}_{\mathrm{E}}, \bar{\varrho}_{\mathrm{O}}; \varrho) | \boldsymbol{\tau}' \rangle, \\ \langle \boldsymbol{\sigma}, \boldsymbol{\tau} | \boldsymbol{V}_{\mathrm{O}}(\bar{\varrho}_{\mathrm{E}}, \bar{\varrho}_{\mathrm{O}}; \varrho) | \boldsymbol{\sigma}', \boldsymbol{\tau}' \rangle &:= \langle \boldsymbol{\sigma} | \boldsymbol{A}(\bar{\varrho}_{\mathrm{E}}, \bar{\varrho}_{\mathrm{O}}; \varrho) | \boldsymbol{\sigma}' \rangle \, \langle \boldsymbol{\tau} | \boldsymbol{B}(\bar{\varrho}_{\mathrm{E}}, \bar{\varrho}_{\mathrm{O}}; \varrho) | \boldsymbol{\tau}' \rangle \end{aligned} \tag{12.2.14}$$

[21] The index i is omitted and spins below and above the lattice line are denoted by $\boldsymbol{\sigma}$ and $\boldsymbol{\tau}$ and $\boldsymbol{\sigma}'$ and $\boldsymbol{\tau}'$ respectively. The use of the letters 'B' and 'A' for matrices signify their locations below and above the lattice line.

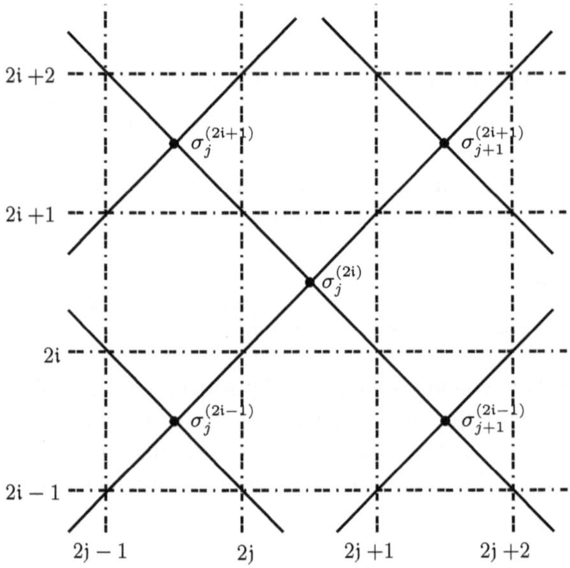

Fig. 12.4 The relabelling of the spins on the sites of $\mathcal{N}$ according to (12.2.13)

and

$$\langle\boldsymbol{\sigma}|\boldsymbol{B}(\bar{\varrho}_{\rm E},\bar{\varrho}_{\rm O};\varrho)|\boldsymbol{\sigma}'\rangle := \prod_{j=1}^{2n_2}\exp\left[K^{(\rm f)}(\bar{\varrho}_{\rm E}-\varrho)\,\sigma_j\sigma_j' + K^{(\rm b)}(\bar{\varrho}_{\rm O}-\varrho)\,\sigma_{j+1}\sigma_j'\right],$$

$$\langle\boldsymbol{\sigma}|\boldsymbol{A}(\bar{\varrho}_{\rm E},\bar{\varrho}_{\rm O};\varrho)|\boldsymbol{\sigma}'\rangle := \prod_{j=1}^{2n_2}\exp\left[K^{(\rm b)}(\bar{\varrho}_{\rm E}-\varrho)\,\sigma_j\sigma_j' + K^{(\rm f)}(\bar{\varrho}_{\rm O}-\varrho)\,\sigma_j\sigma_{j+1}'\right],$$
(12.2.15)

where the index j is an integer modulo $2n_2$. Equation (12.2.14) give the Kronecker-product forms

$$\boldsymbol{V}_{\rm E}(\bar{\varrho}_{\rm E},\bar{\varrho}_{\rm O};\varrho) = \boldsymbol{B}(\bar{\varrho}_{\rm E},\bar{\varrho}_{\rm O};\varrho)\otimes\boldsymbol{A}(\bar{\varrho}_{\rm E},\bar{\varrho}_{\rm O};\varrho),$$
$$\boldsymbol{V}_{\rm O}(\bar{\varrho}_{\rm E},\bar{\varrho}_{\rm O};\varrho) = \boldsymbol{A}(\bar{\varrho}_{\rm E},\bar{\varrho}_{\rm O};\varrho)\otimes\boldsymbol{B}(\bar{\varrho}_{\rm E},\bar{\varrho}_{\rm O};\varrho),$$
(12.2.16)

where the first term in each product is that of $\mathcal{N}$. The partition function of the eight-vertex model is related to that of the disconnected Ising models on $\mathcal{N}$ and $\mathcal{N}^\star$. Given that the value of the horizontal rapidity alternates between rows with $\varrho_{\rm E}$ and $\varrho_{\rm O}$ on even and odd number rows, respectively,

$$\begin{aligned} Z^{(8\text{vM})}(\mathcal{N}^{\circledast}; \bar{\varrho}_{\text{E}}, \bar{\varrho}_{\text{O}}; \varrho_{\text{E}}, \varrho_{\text{O}}) &= \text{Trace}\{[\boldsymbol{V}_{\text{E}}(\bar{\varrho}_{\text{E}}, \bar{\varrho}_{\text{O}}; \varrho_{\text{E}})\boldsymbol{V}_{\text{O}}(\bar{\varrho}_{\text{E}}, \bar{\varrho}_{\text{O}}; \varrho_{\text{O}})]^{2n_1}\} \\ &= Z^{(\text{IM})}(\mathcal{N}; \bar{\varrho}_{\text{E}}, \bar{\varrho}_{\text{O}}; \varrho_{\text{E}}, \varrho_{\text{O}}) Z^{(\text{IM})}(\mathcal{N}^{\star}; \bar{\varrho}_{\text{E}}, \bar{\varrho}_{\text{O}}; \varrho_{\text{E}}, \varrho_{\text{O}}), \end{aligned} \tag{12.2.17}$$

where

$$\begin{aligned} Z^{(\text{IM})}(\mathcal{N}; \bar{\varrho}_{\text{E}}, \bar{\varrho}_{\text{O}}; \varrho_{\text{E}}, \varrho_{\text{O}}) &:= \text{Trace}\{[\boldsymbol{B}(\bar{\varrho}_{\text{E}}, \bar{\varrho}_{\text{O}}; \varrho_{\text{E}})\boldsymbol{A}(\bar{\varrho}_{\text{E}}, \bar{\varrho}_{\text{O}}; \varrho_{\text{O}})]^{2n_1}\}, \\ Z^{(\text{IM})}(\mathcal{N}^{\star}; \bar{\varrho}_{\text{E}}, \bar{\varrho}_{\text{O}}; \varrho_{\text{E}}, \varrho_{\text{O}}) &:= \text{Trace}\{[\boldsymbol{A}(\bar{\varrho}_{\text{E}}, \bar{\varrho}_{\text{O}}; \varrho_{\text{E}})\boldsymbol{B}(\bar{\varrho}_{\text{E}}, \bar{\varrho}_{\text{O}}; \varrho_{\text{O}})]^{2n_1}\}. \end{aligned} \tag{12.2.18}$$

The first of these equations is the partition function of Baxter's (1986) checkerboard Ising model on the lattice $\mathcal{N}$ which we introduced in Sect. 3.8.1.5.

12.2.1 The Modified Checkerboard Ising Model

The star of sites shown in Fig. 12.4, with summation over $\sigma_j^{(2\text{i})}$, is a contribution to $\langle \boldsymbol{\sigma} | \boldsymbol{B}(\bar{\varrho}_{\text{E}}, \bar{\varrho}_{\text{O}}, \varrho_{\text{E}})\boldsymbol{A}(\bar{\varrho}_{\text{E}}, \bar{\varrho}_{\text{O}}, \varrho_{\text{O}})|\boldsymbol{\sigma}'\rangle$. Using

$$\sigma_0 := \sigma_j^{(2\text{i})}, \quad \sigma_1 := \sigma_j^{(2\text{i}-1)}, \quad \sigma_2 := \sigma_j^{(2\text{i}+1)}, \quad \sigma_3 := \sigma_{j+1}^{(2\text{i}+1)}, \quad \sigma_4 := \sigma_{j+1}^{(2\text{i}-1)},$$

this star is reproduced with the respective couplings in Fig. 12.5. Its vertex weight is

$$\begin{aligned} \mathfrak{Y}(\bar{\varrho}_{\text{E}}, \bar{\varrho}_{\text{O}}; \varrho_{\text{E}}, \varrho_{\text{O}}; \sigma_1, \sigma_2, \sigma_3, \sigma_4) = 2\cosh[&K^{(\text{f})}(\bar{\varrho}_{\text{E}} - \varrho_{\text{E}})\sigma_1 + K^{(\text{b})}(\bar{\varrho}_{\text{E}} - \varrho_{\text{O}})\sigma_2 \\ &+ K^{(\text{f})}(\bar{\varrho}_{\text{O}} - \varrho_{\text{O}})\sigma_3 + K^{(\text{b})}(\bar{\varrho}_{\text{O}} - \varrho_{\text{E}})\sigma_4]. \end{aligned} \tag{12.2.19}$$

With the notation

$$\begin{aligned} K_{\text{B}}^{(\text{f})} &:= K^{(\text{f})}(\bar{\varrho}_{\text{E}} - \varrho_{\text{E}}), & K_{\text{B}}^{(\text{b})} &:= K^{(\text{b})}(\bar{\varrho}_{\text{O}} - \varrho_{\text{E}}), \\ K_{\text{A}}^{(\text{f})} &:= K^{(\text{f})}(\bar{\varrho}_{\text{O}} - \varrho_{\text{O}}), & K_{\text{A}}^{(\text{b})} &:= K^{(\text{b})}(\bar{\varrho}_{\text{E}} - \varrho_{\text{O}}), \end{aligned} \tag{12.2.20}$$

and horizontal and vertical couplings and diagonal couplings across the square set to zero, this is the vertex weight (3.8.20) which was shown to satisfy the free-fermion condition (3.8.21). From (12.2.4),

$$\begin{aligned} K_{\text{B}}^{(\text{f})} &= -\tfrac{1}{2}\text{i}\,\text{am}(\text{i}[\mathcal{K}' - \bar{\varrho}_{\text{E}} + \varrho_{\text{E}}]|\text{k}), & K_{\text{B}}^{(\text{b})} &= -\tfrac{1}{2}\text{i}\,\text{am}(\text{i}\,[\bar{\varrho}_{\text{O}} - \varrho_{\text{E}}]|\text{k}), \\ K_{\text{A}}^{(\text{f})} &= -\tfrac{1}{2}\text{i}\,\text{am}(\text{i}[\mathcal{K}' - \bar{\varrho}_{\text{O}} + \varrho_{\text{O}}]|\text{k}), & K_{\text{A}}^{(\text{b})} &= -\tfrac{1}{2}\text{i}\,\text{am}(\text{i}\,[\bar{\varrho}_{\text{E}} - \varrho_{\text{O}}]|\text{k}). \end{aligned} \tag{12.2.21}$$

The four couplings can be chosen independently. Since the rapidities occur as differences they form three variables

$$u_1 := \tfrac{1}{2}[(\bar{\varrho}_{\text{E}}+\bar{\varrho}_{\text{O}})-(\varrho_{\text{E}}+\varrho_{\text{O}})], \quad u_2 := \tfrac{1}{2}(\bar{\varrho}_{\text{E}}-\bar{\varrho}_{\text{O}}), \quad u_3 := \tfrac{1}{2}(\varrho_{\text{E}}-\varrho_{\text{O}}), \tag{12.2.22}$$

(say), which together with the elliptic modulus k are an independent equivalent set.

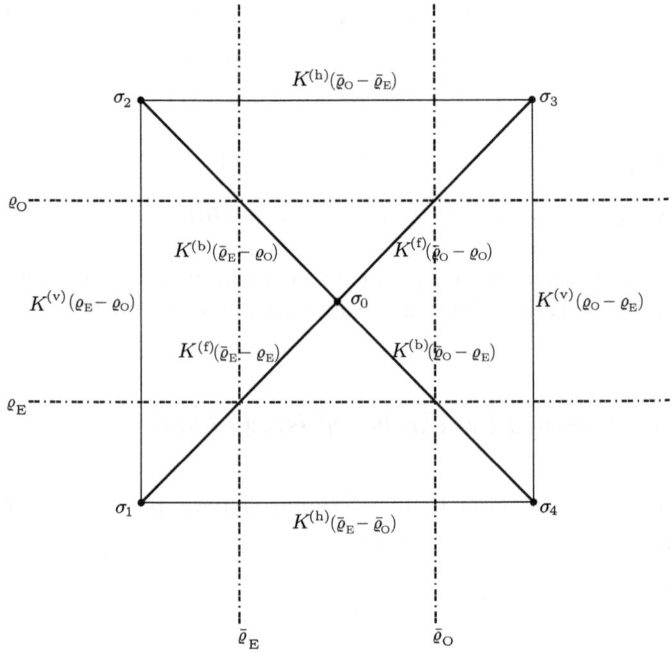

Fig. 12.5 The star of sites with vertex weight $\mathfrak{W}(\bar{\varrho}_O, \bar{\varrho}_E; \varrho_E, \varrho_O; \sigma_1, \sigma_2, \sigma_3, \sigma_4)$ and supplementary vertex weight $\mathfrak{W}'(\bar{\varrho}_O, \bar{\varrho}_E; \varrho_E, \varrho_O; \sigma_1, \sigma_2, \sigma_3, \sigma_4)$

The free-fermion condition continues to hold if couplings between the horizontal and vertical pairs are included and we showed in Sect. 3.8.1.5 that the eight-vertex conditions (3.8.27) can be satisfied if the vertex weight (12.2.19) is modified by including a supplementary multiplicative factor

$$\mathfrak{W}'(\sigma_1, \sigma_2, \sigma_3, \sigma_4) := \exp\{\tfrac{1}{2}[K^{(h)}(\sigma_1\sigma_4 - \sigma_2\sigma_3) + K^{(v)}(\sigma_1\sigma_2 - \sigma_3\sigma_4)]\}, \tag{12.2.23}$$

with $K^{(h)}$ and $K^{(v)}$ satisfying (3.8.25) and (3.8.26). Substituting into the right-hand side of these equations using (12.2.4) it can be shown that

$$\tanh[2K^{(h)}(\bar{\varrho}_E - \bar{\varrho}_O)] = \frac{\begin{bmatrix} \sinh[2K^{(f)}(\bar{\varrho}_O - \varrho_O)]\sinh[2K^{(b)}(\bar{\varrho}_E - \varrho_O)] \\ -\sinh[2K^{(f)}(\bar{\varrho}_E - \varrho_E)]\sinh[2K^{(b)}(\bar{\varrho}_O - \varrho_E)] \end{bmatrix}}{\begin{bmatrix} \cosh[2K^{(f)}(\bar{\varrho}_O - \varrho_O)]\cosh[2K^{(b)}(\bar{\varrho}_E - \varrho_O)] \\ +\cosh[2K^{(f)}(\bar{\varrho}_E - \varrho_E)]\cosh[2K^{(b)}(\bar{\varrho}_O - \varrho_E)] \end{bmatrix}}, \tag{12.2.24}$$

$$\tanh[2K^{(v)}(\varrho_O - \varrho_E)] = \frac{\begin{bmatrix} \sinh[2K^{(f)}(\bar{\varrho}_O - \varrho_O)]\sinh[2K^{(b)}(\bar{\varrho}_O - \varrho_E)] \\ -\sinh[2K^{(f)}(\bar{\varrho}_E - \varrho_E)]\sinh[2K^{(b)}(\bar{\varrho}_E - \varrho_O)] \end{bmatrix}}{\begin{bmatrix} \cosh[2K^{(f)}(\bar{\varrho}_O - \varrho_O)]\cosh[2K^{(b)}(\bar{\varrho}_O - \varrho_E)] \\ +\cosh[2K^{(f)}(\bar{\varrho}_E - \varrho_E)]\cosh[2K^{(b)}(\bar{\varrho}_E - \varrho_O)] \end{bmatrix}}, \tag{12.2.25}$$

where

$$K^{(h)}(u) := -\tfrac{1}{2}\mathrm{i}\,\mathrm{am}(\mathrm{i}\,u|\mathrm{k}), \qquad K^{(v)}(u) := -\tfrac{1}{2}\mathrm{i}\,\mathrm{am}(\mathrm{i}\,u|\mathrm{k}). \tag{12.2.26}$$

Both $K^{(h)}$ and $K^{(v)}$ are of the same functional form as $K^{(b)}$, but it is useful, in the interests of clarity, to retain for them a distinguishing notation. It is also worth repeating (12.2.23) to display the rapidity dependence. Thus

$$\begin{aligned}&\mathfrak{Y}'(\bar{\varrho}_E, \bar{\varrho}_O; \varrho_E, \varrho_O; \sigma_1, \sigma_2, \sigma_3, \sigma_4)\\ &\quad = \exp\{\tfrac{1}{2}[K^{(h)}(\bar{\varrho}_E - \bar{\varrho}_O)(\sigma_1\sigma_4 - \sigma_2\sigma_3) + K^{(v)}(\varrho_O - \varrho_E)(\sigma_1\sigma_2 - \sigma_3\sigma_4)]\}.\end{aligned} \tag{12.2.27}$$

Because $K^{(h)}(u)$ and $K^{(v)}(u)$ are odd functions it can be seen that the changes of sign in the horizontal and vertical pairs of couplings simply correspond to interchanges between the members of their respective pairs of rapidities. It is important to note that the inclusion of the additional Boltzmann factor (12.2.27), whether or not the couplings are given by (12.2.26), does not affect the Hamiltonian on the whole lattice. So the partition function of this modified checkerboard Ising model is exactly the same as that of the unmodified model.

At this point it is worth pausing and considering a curious feature of the couplings $K^{(h)}(u)$ and $K^{(v)}(u)$ as they arise in (12.2.24) and (12.2.25) respectively. This is that, unlike $K^{(f)}(u)$ and $K^{(b)}(u)$, their arguments are the differences of rapidities on *non-intersecting lines of* $\mathcal{N}^{\circledast}$. A conceptual way to resolve this puzzle is to think of extending the vertical and horizontal rapidity lines in Fig. 12.5, respectively, upwards and to the right so that they intersect at infinity with the coupling pairs passing horizontally and vertically through the intersections. Then conjugate couplings

$$K^{(ch)}(u) := \quad -\tfrac{1}{2}\mathrm{i}\,\mathrm{am}(\mathrm{i}[\mathcal{K}' - u]|\mathrm{k}), \qquad K^{(cv)}(u) := -\tfrac{1}{2}\mathrm{i}\,\mathrm{am}(\mathrm{i}[\mathcal{K}' - u]|\mathrm{k}), \tag{12.2.28}$$

can be defined. The coupling $K^{(ch)}(u)$ is between a *vertical pair* and $K^{(cv)}(u)$ between a *horizontal pair*.[22]

[22] Although, of course, they differ from $K^{(v)}(u)$ and $K^{(h)}(u)$ respectively.

We now investigate the existence of a different set of couplings which also satisfy the free-fermion and the eight-vertex conditions. The simplest way to do this is to use the transformations

$$\begin{aligned} \overline{\mathrm{S}}_1 &: \quad (\bar{\varrho}_{\mathrm{E}}, \bar{\varrho}_{\mathrm{O}}) \longrightarrow (\bar{\varrho}_{\mathrm{O}}, \bar{\varrho}_{\mathrm{E}}), & \overline{\mathrm{S}}_2 &: \quad (\bar{\varrho}_{\mathrm{E}}, \bar{\varrho}_{\mathrm{O}}) \longrightarrow (\mathcal{K}' - \bar{\varrho}_{\mathrm{E}}, \mathcal{K}' - \bar{\varrho}_{\mathrm{O}}), \\ \mathrm{S}_1 &: \quad (\varrho_{\mathrm{E}}, \varrho_{\mathrm{O}}) \longrightarrow (\varrho_{\mathrm{O}}, \varrho_{\mathrm{E}}), & \mathrm{S}_2 &: \quad (\varrho_{\mathrm{E}}, \varrho_{\mathrm{O}}) \longrightarrow (-\varrho_{\mathrm{E}}, -\varrho_{\mathrm{O}}). \end{aligned} \tag{12.2.29}$$

$\overline{\mathrm{S}}_1$ and $\overline{\mathrm{S}}_2$ change the sign of $K^{(\mathrm{h})}$ and S_1 and S_2 change the sign of $K^{(\mathrm{v})}$. The coupling transformations (a) and (b) of (3.8.28) are, respectively, equivalent to $\overline{\mathrm{S}}_1\mathrm{S}_1$ and $\overline{\mathrm{S}}_1\overline{\mathrm{S}}_2\mathrm{S}_2$. Consider now three transformations applied to (12.2.24) and (12.2.25):

(i) S_1 which interchanges ϱ_{E} and ϱ_{O}.
(ii) $\overline{\mathrm{S}}_1$ which interchanges $\bar{\varrho}_{\mathrm{E}}$ and $\bar{\varrho}_{\mathrm{O}}$.
(iii) $\overline{\mathrm{S}}_2\mathrm{S}_2$ which has the effect of interchanging $K^{(\mathrm{f})}(u)$ and $K^{(\mathrm{b})}(u)$ for all the arguments u on the right-hand sides of (12.2.24) and (12.2.25) and changing the signs of $K^{(\mathrm{h})}$ and $K^{(\mathrm{v})}$.

These transformations each in a different way (respectively, translating the star of sites a lattice space vertically, translating the star of sites a lattice space horizontally, interchanging the interacting pair at each site of $\mathcal{N}^{\circledast}$) map between $\mathcal{N}$ and $\mathcal{N}^{\star}$. Equations produced by transforming (12.2.24) and (12.2.25) are the same for (i), (ii) and (iii) (after some interchanges in the order of terms). These transformations yield a second solution for a set of couplings which satisfy the eight-vertex condition (3.8.5) and the free-fermion condition (3.8.21), it being precisely the dual solution on $\mathcal{N}^{\star}$.

It is now important to investigate the relationship between the vertex weights for these two sets of couplings. Specifically the problem is to find a function $\mathfrak{C}(\bar{\varrho}_{\mathrm{E}}, \bar{\varrho}_{\mathrm{O}}; \varrho_{\mathrm{E}}, \varrho_{\mathrm{O}})$ so that

$$\frac{\mathfrak{Y}(\bar{\varrho}_{\mathrm{E}}, \bar{\varrho}_{\mathrm{O}}; \varrho_{\mathrm{E}}, \varrho_{\mathrm{O}}; \sigma_1, \sigma_2, \sigma_3, \sigma_4)\mathfrak{Y}'(\bar{\varrho}_{\mathrm{E}}, \bar{\varrho}_{\mathrm{O}}; \varrho_{\mathrm{E}}, \varrho_{\mathrm{O}}; \sigma_1, \sigma_2, \sigma_3, \sigma_4)}{\mathfrak{C}(\bar{\varrho}_{\mathrm{E}}, \bar{\varrho}_{\mathrm{O}}; \varrho_{\mathrm{E}}, \varrho_{\mathrm{O}})} \tag{12.2.30}$$

is invariant, for all σ_1, σ_2, σ_3 and σ_4, under the transformations (i), (ii) and (iii). After some manipulation (Baxter 1986) it can be shown that, to within an arbitrary multiplicative constant,[23]

$$\mathfrak{C}(\bar{\varrho}_{\mathrm{E}}, \bar{\varrho}_{\mathrm{O}}; \varrho_{\mathrm{E}}, \varrho_{\mathrm{O}}) := g^{(\mathrm{f})}(\bar{\varrho}_{\mathrm{E}} - \varrho_{\mathrm{E}})g^{(\mathrm{b})}(\bar{\varrho}_{\mathrm{O}} - \varrho_{\mathrm{E}})g^{(\mathrm{b})}(\bar{\varrho}_{\mathrm{E}} - \varrho_{\mathrm{O}})g^{(\mathrm{f})}(\bar{\varrho}_{\mathrm{O}} - \varrho_{\mathrm{O}}), \tag{12.2.31}$$

where

$$g^{(\mathrm{x})}(u) := \{2\sinh[2K^{(\mathrm{x})}(u)]\}^{\frac{1}{4}}, \qquad \mathrm{x} = \mathrm{f}, \mathrm{b}. \tag{12.2.32}$$

---

[23] For comparison with the account of the uniform case in Baxter (1982b) it should be noted, from (12.2.37), that $\mathfrak{C}(\bar{\varrho}, \bar{\varrho}; \varrho, \varrho) = 2/\sqrt{\mathrm{k}}$.

12.2.2 Properties of the Transfer Matrices

It is convenient at this stage to absorb the quantity $\mathfrak{C}(\bar{\varrho}_{\mathrm{E}}, \bar{\varrho}_{\mathrm{O}}; \varrho_{\mathrm{E}}, \varrho_{\mathrm{O}})$ into $\mathfrak{Y}(\bar{\varrho}_{\mathrm{E}}, \bar{\varrho}_{\mathrm{O}}; \varrho_{\mathrm{E}}, \varrho_{\mathrm{O}}; \sigma_1, \sigma_2, \sigma_3, \sigma_4)$ by dividing the Boltzmann weight with coupling $K^{(\mathrm{x})}(u)$ by $g^{(\mathrm{x})}(u)$, for all u and x = f, b. Denoting the vertex weight with this modification by $\mathring{\mathfrak{Y}}(\bar{\varrho}_{\mathrm{E}}, \bar{\varrho}_{\mathrm{O}}; \varrho_{\mathrm{E}}, \varrho_{\mathrm{O}}; \sigma_1, \sigma_2, \sigma_3, \sigma_4)$ the invariance conditions for (i), (ii) and (iii) take the respective forms

$$\mathring{\mathfrak{Y}}(\bar{\varrho}_{\mathrm{E}}, \bar{\varrho}_{\mathrm{O}}; \varrho_{\mathrm{E}}, \varrho_{\mathrm{O}}; \sigma_1, \sigma_2, \sigma_3, \sigma_4)\mathfrak{Y}'(\bar{\varrho}_{\mathrm{E}}, \bar{\varrho}_{\mathrm{O}}; \varrho_{\mathrm{E}}, \varrho_{\mathrm{O}}; \sigma_1, \sigma_2, \sigma_3, \sigma_4)$$

$$\text{(i)} \quad = \mathring{\mathfrak{Y}}(\bar{\varrho}_{\mathrm{E}}, \bar{\varrho}_{\mathrm{O}}; \varrho_{\mathrm{O}}, \varrho_{\mathrm{E}}; \sigma_1, \sigma_2, \sigma_3, \sigma_4)\mathfrak{Y}'(\bar{\varrho}_{\mathrm{E}}, \bar{\varrho}_{\mathrm{O}}; \varrho_{\mathrm{O}}, \varrho_{\mathrm{E}}; \sigma_1, \sigma_2, \sigma_3, \sigma_4), \tag{12.2.33}$$

$$\text{(ii)} \quad = \mathring{\mathfrak{Y}}(\bar{\varrho}_{\mathrm{O}}, \bar{\varrho}_{\mathrm{E}}; \varrho_{\mathrm{E}}, \varrho_{\mathrm{O}}; \sigma_1, \sigma_2, \sigma_3, \sigma_4)\mathfrak{Y}'(\varrho_{\mathrm{O}}, \varrho_{\mathrm{E}}; \varrho_{\mathrm{E}}, \varrho_{\mathrm{O}}; \sigma_1, \sigma_2, \sigma_3, \sigma_4), \tag{12.2.34}$$

$$\begin{aligned}\text{(iii)} \quad &= \mathring{\mathfrak{Y}}(\mathcal{K}' - \bar{\varrho}_{\mathrm{O}}, \mathcal{K}' - \bar{\varrho}_{\mathrm{E}}; -\varrho_{\mathrm{E}}, -\varrho_{\mathrm{O}}; \sigma_1, \sigma_2, \sigma_3, \sigma_4)\\ &\quad \times \mathfrak{Y}'(\mathcal{K}' - \bar{\varrho}_{\mathrm{O}}, \mathcal{K}' - \bar{\varrho}_{\mathrm{E}}; -\varrho_{\mathrm{E}}, -\varrho_{\mathrm{O}}; \sigma_1, \sigma_2, \sigma_3, \sigma_4).\end{aligned} \tag{12.2.35}$$

These conditions lead to relationships between transfer matrices modified in a similar way, so that

$$\begin{aligned}\langle \boldsymbol{\sigma} | \mathring{\boldsymbol{B}}(\bar{\varrho}_{\mathrm{E}}, \bar{\varrho}_{\mathrm{O}}; \varrho) | \boldsymbol{\sigma}' \rangle &:= \prod_{j=1}^{2n_2} \frac{\exp\left[K^{(\mathrm{f})}(\bar{\varrho}_{\mathrm{E}} - \varrho)\,\sigma_j\sigma_j' + K^{(\mathrm{b})}(\bar{\varrho}_{\mathrm{O}} - \varrho)\,\sigma_{j+1}\sigma_j'\right]}{\{4\sinh[2K^{(\mathrm{f})}(\bar{\varrho}_{\mathrm{E}} - \varrho)]\sinh[2K^{(\mathrm{b})}(\bar{\varrho}_{\mathrm{O}} - \varrho)]\}^{\frac{1}{2}n_2}},\\ \langle \boldsymbol{\sigma} | \mathring{\boldsymbol{A}}(\varrho_{\mathrm{E}}, \varrho_{\mathrm{O}}; \varrho) | \boldsymbol{\sigma}' \rangle &:= \prod_{j=1}^{2n_2} \frac{\exp\left[K^{(\mathrm{b})}(\bar{\varrho}_{\mathrm{E}} - \varrho)\,\sigma_j\sigma_j' + K^{(\mathrm{f})}(\bar{\varrho}_{\mathrm{O}} - \varrho)\,\sigma_j\sigma_{j+1}'\right]}{\{4\sinh[2K^{(\mathrm{f})}(\bar{\varrho}_{\mathrm{O}} - \varrho)]\sinh[2K^{(\mathrm{b})}(\bar{\varrho}_{\mathrm{E}} - \varrho)]\}^{\frac{1}{2}n_2}}.\end{aligned} \tag{12.2.36}$$

Using coupling parameters, which are allowed to have any complex values, it clear from (12.2.36) that

$$\mathring{\boldsymbol{B}}(K^{(\mathrm{b})}, K^{(\mathrm{f})}) = \mathring{\boldsymbol{A}}^{\mathrm{T}}(K^{(\mathrm{f})}, K^{(\mathrm{b})}), \quad \mathring{\boldsymbol{A}}(K^{(\mathrm{b})}, K^{(\mathrm{f})}) = \mathring{\boldsymbol{B}}^{\mathrm{T}}(K^{(\mathrm{f})}, K^{(\mathrm{b})}), \tag{12.2.37}$$

which, using (12.2.5), can be expressed in terms of rapidities as

$$\begin{aligned}\mathring{\boldsymbol{B}}(-\bar{\varrho}_{\mathrm{E}}, -\bar{\varrho}_{\mathrm{O}}; -\mathcal{K}' - \varrho) &= \mathring{\boldsymbol{A}}^{\mathrm{T}}(\bar{\varrho}_{\mathrm{E}}, \bar{\varrho}_{\mathrm{O}}; \varrho),\\ \mathring{\boldsymbol{A}}(-\bar{\varrho}_{\mathrm{E}}, -\bar{\varrho}_{\mathrm{O}}; -\mathcal{K}' - \varrho) &= \mathring{\boldsymbol{B}}^{\mathrm{T}}(\bar{\varrho}_{\mathrm{E}}, \bar{\varrho}_{\mathrm{O}}; \varrho).\end{aligned} \tag{12.2.38}$$

The vertical interactions cancel when the matrices $\mathring{\boldsymbol{B}}(\bar{\varrho}_{\mathrm{E}}, \bar{\varrho}_{\mathrm{O}}; \varrho)$ and $\mathring{\boldsymbol{A}}(\bar{\varrho}_{\mathrm{E}}, \bar{\varrho}_{\mathrm{O}}; \varrho)$ are formed but the horizontal interactions still play a role for (ii) and (iii), so we need the matrix $\boldsymbol{D}(\bar{\varrho}_{\mathrm{E}} - \bar{\varrho}_{\mathrm{O}})$ given by

$$\langle \boldsymbol{\sigma}|\boldsymbol{D}(\bar{\varrho}_{\text{E}}-\bar{\varrho}_{\text{O}})|\boldsymbol{\sigma}'\rangle := \delta^{\text{Kr}}(\boldsymbol{\sigma}-\boldsymbol{\sigma}')\prod_{j=1}^{2n_2}\exp\left[K^{(\text{h})}(\bar{\varrho}_{\text{E}}-\bar{\varrho}_{\text{O}})\sigma_j\sigma_{j+1}\right], \tag{12.2.39}$$

for which

$$\boldsymbol{D}^{-1}(\bar{\varrho}_{\text{E}}-\bar{\varrho}_{\text{O}}) = \boldsymbol{D}(\bar{\varrho}_{\text{O}}-\bar{\varrho}_{\text{E}}). \tag{12.2.40}$$

Using (12.2.37) and (12.2.40) the invariance properties (12.2.33), (12.2.34) and (12.2.35) give

$$\begin{aligned}
&\mathring{\boldsymbol{B}}(\bar{\varrho}_{\text{E}},\bar{\varrho}_{\text{O}};\varrho)\mathring{\boldsymbol{A}}(\bar{\varrho}_{\text{E}},\bar{\varrho}_{\text{O}};\varrho')\\
\text{(i)}\quad &= \mathring{\boldsymbol{B}}(\bar{\varrho}_{\text{E}},\bar{\varrho}_{\text{O}};\varrho')\mathring{\boldsymbol{A}}(\bar{\varrho}_{\text{E}},\bar{\varrho}_{\text{O}};\varrho), &(12.2.41)\\
\text{(ii)}\quad &= \boldsymbol{D}(\bar{\varrho}_{\text{O}}-\bar{\varrho}_{\text{E}})\mathring{\boldsymbol{B}}(\bar{\varrho}_{\text{O}},\bar{\varrho}_{\text{E}};\varrho)\mathring{\boldsymbol{A}}(\bar{\varrho}_{\text{O}},\bar{\varrho}_{\text{E}};\varrho')\boldsymbol{D}(\bar{\varrho}_{\text{E}}-\bar{\varrho}_{\text{O}}), &(12.2.42)\\
\text{(iii)}\quad &= \boldsymbol{D}(\bar{\varrho}_{\text{O}}-\bar{\varrho}_{\text{E}})\mathring{\boldsymbol{B}}(-\bar{\varrho}_{\text{E}},-\bar{\varrho}_{\text{O}};-\mathcal{K}'-\varrho)\mathring{\boldsymbol{A}}(-\bar{\varrho}_{\text{E}},-\bar{\varrho}_{\text{O}};-\mathcal{K}'-\varrho')\boldsymbol{D}(\bar{\varrho}_{\text{E}}-\bar{\varrho}_{\text{O}}). &(12.2.43)
\end{aligned}$$

We now explore some other transformation properties of $\mathring{\boldsymbol{B}}(\bar{\varrho}_{\text{E}},\bar{\varrho}_{\text{O}};\varrho)$ and $\mathring{\boldsymbol{A}}(\bar{\varrho}_{\text{E}},\bar{\varrho}_{\text{O}};\varrho)$:

- Given the periodic boundary conditions in the horizontal direction the system is invariant under the rotation group $\mathcal{C}_{2n_2} := \{\mathrm{I},\mathrm{C},\mathrm{C}^2,\ldots,\mathrm{C}^{2n_2-1}\}$, where C is a rotation about the vertical axis of the lattice cylinder through an angle π/n_2 (equivalent to a translation to the right). For $\mathring{\boldsymbol{B}}$ the lower (unprimed) indices in (12.2.36) remain unchanged and the upper (primed) indices increase by one with the reverse being the case for $\mathring{\boldsymbol{A}}$.[24] The orthogonal matrix $\boldsymbol{C} := \boldsymbol{R}(\mathrm{C})$ given by

$$\langle \boldsymbol{\sigma}|\boldsymbol{C}|\boldsymbol{\sigma}'\rangle := \prod_{j=1}^{2n_2}\delta^{\text{Kr}}(\sigma_{j+1}-\sigma_j'), \tag{12.2.44}$$

is a representational matrix for the 2^{2n_2}-dimensional representation of row states and

$$\boldsymbol{C}\mathring{\boldsymbol{A}}(\bar{\varrho}_{\text{E}},\bar{\varrho}_{\text{O}};\varrho) = \mathring{\boldsymbol{B}}(\bar{\varrho}_{\text{O}},\bar{\varrho}_{\text{E}};\varrho),\qquad \mathring{\boldsymbol{B}}(\bar{\varrho}_{\text{E}},\bar{\varrho}_{\text{O}};\varrho)\boldsymbol{C}^{\text{T}} = \mathring{\boldsymbol{A}}(\bar{\varrho}_{\text{O}},\bar{\varrho}_{\text{E}};\varrho), \tag{12.2.45}$$

giving

$$[\boldsymbol{C},\mathring{\boldsymbol{A}}(\bar{\varrho}_{\text{E}},\bar{\varrho}_{\text{O}};\varrho)] = \boldsymbol{0},\qquad [\boldsymbol{C},\mathring{\boldsymbol{B}}(\bar{\varrho}_{\text{E}},\bar{\varrho}_{\text{O}};\varrho)] = \boldsymbol{0}, \tag{12.2.46}$$

[24] Remember that the matrix ordering in a product from left to right corresponds to bottom to top in terms of the lattice.

and

$$\mathring{\boldsymbol{B}}(\bar{\varrho}_{\rm E}, \bar{\varrho}_{\rm O}; \varrho)\mathring{\boldsymbol{A}}(\bar{\varrho}_{\rm E}, \bar{\varrho}_{\rm O}; \varrho') = \mathring{\boldsymbol{A}}(\bar{\varrho}_{\rm O}, \bar{\varrho}_{\rm E}, \varrho)\mathring{\boldsymbol{B}}(\bar{\varrho}_{\rm O}, \bar{\varrho}_{\rm E}; \varrho'). \tag{12.2.47}$$

From (12.2.41) and (12.2.47)

$$\mathring{\boldsymbol{A}}(\bar{\varrho}_{\rm E}, \bar{\varrho}_{\rm O}; \varrho')\mathring{\boldsymbol{B}}(\bar{\varrho}_{\rm E}, \bar{\varrho}_{\rm O}; \varrho) = \mathring{\boldsymbol{A}}(\bar{\varrho}_{\rm E}, \bar{\varrho}_{\rm O}; \varrho)\mathring{\boldsymbol{B}}(\bar{\varrho}_{\rm E}, \bar{\varrho}_{\rm O}; \varrho'). \tag{12.2.48}$$

- The system is also invariant under the group $\mathcal{S}_2 := \{\mathrm{I}, \mathrm{S}\}$, where S inverts all the spins on a row and the orthogonal and symmetric matrix $\boldsymbol{S} := \boldsymbol{R}(\mathrm{S})$ is given by

$$\langle \boldsymbol{\sigma} | \boldsymbol{S} | \boldsymbol{\sigma}' \rangle := \delta^{\rm Kr}(\boldsymbol{\sigma} + \boldsymbol{\sigma}'). \tag{12.2.49}$$

Its effect is most easily expressed if the rapidities are replaced as arguments by the pair $(K^{(\rm f)}, K^{(\rm b)})$. Then

$$\begin{aligned} \boldsymbol{S}\mathring{\boldsymbol{B}}(K^{(\rm f)}, K^{(\rm b)}) &= \mathring{\boldsymbol{B}}(K^{(\rm f)}, K^{(\rm b)})\boldsymbol{S} = \mathring{\boldsymbol{B}}(-K^{(\rm f)}, -K^{(\rm b)}), \\ \boldsymbol{S}\mathring{\boldsymbol{A}}(K^{(\rm f)}, K^{(\rm b)}) &= \mathring{\boldsymbol{A}}(K^{(\rm f)}, K^{(\rm b)})\boldsymbol{S} = \mathring{\boldsymbol{A}}(-K^{(\rm f)}, -K^{(\rm b)}). \end{aligned} \tag{12.2.50}$$

Equation (12.2.8) can now be used to express this result in terms of rapidities, but to do so we need to know for $\mathring{\boldsymbol{B}}$ that $\sum_{j=1}^{2n_2}(\sigma_j\sigma_j' + \sigma_{j+1}\sigma_j')$ is divisible by four to eliminate the effects of the $\frac{1}{2}\pi\mathrm{i}$ factors. Consider the two sets of spins $\boldsymbol{\sigma}$ and $\boldsymbol{\sigma}'$ ordered in the sequence $\sigma_1, \sigma_1', \sigma_2, \sigma_2', \ldots$ of length $4n_2$. If η and η' are respectively the number of unlike (σ_j, σ_j') and $(\sigma_j', \sigma_{j+1})$ pairs

$$\sum_{j=1}^{2n_2}(\sigma_j\sigma_j' + \sigma_{j+1}\sigma_j') = 4n_2 - 2(\eta + \eta') \tag{12.2.51}$$

and the required result follows from the observation that $\eta + \eta'$ must be even. A similar argument is applicable to $\mathring{\boldsymbol{A}}$ but now the ordered sequence of spins is $\sigma_1', \sigma_1, \sigma_2', \sigma_2, \ldots$ and ζ and ζ' denote respectively the number of unlike (σ_j', σ_j) and $(\sigma_j, \sigma_{j+1}')$ pairs. Then

$$\begin{aligned} \boldsymbol{S}\mathring{\boldsymbol{B}}(\bar{\varrho}_{\rm E}, \bar{\varrho}_{\rm O}; \varrho) &= \mathring{\boldsymbol{B}}(\bar{\varrho}_{\rm E}, \bar{\varrho}_{\rm O}; \varrho - 2\mathcal{K}'), \\ \boldsymbol{S}\mathring{\boldsymbol{A}}(\bar{\varrho}_{\rm E}, \bar{\varrho}_{\rm O}; \varrho) &= \mathring{\boldsymbol{A}}(\bar{\varrho}_{\rm E}, \bar{\varrho}_{\rm O}; \varrho - 2\mathcal{K}'), \end{aligned} \tag{12.2.52}$$

$$[\mathring{\boldsymbol{B}}(\bar{\varrho}_{\rm E}, \bar{\varrho}_{\rm O}; \varrho), \boldsymbol{S}] = [\mathring{\boldsymbol{A}}(\bar{\varrho}_{\rm E}, \bar{\varrho}_{\rm O}; \varrho), \boldsymbol{S}] = \boldsymbol{0}. \tag{12.2.53}$$

- Using η, η', ζ and ζ' as defined above, (12.2.36) yields

$$\langle\boldsymbol{\sigma}|\mathring{\boldsymbol{B}}(K^{(\mathrm{f})},K^{(\mathrm{b})})|\boldsymbol{\sigma}'\rangle=\frac{\exp\left[2(n_2-\eta)K^{(\mathrm{f})}+2(n_2-\eta')K^{(\mathrm{b})}\right]}{\{4\sinh[2K^{(\mathrm{f})}]\sinh[2K^{(\mathrm{b})}]\}^{\frac{1}{2}n_2}},$$

$$\langle\boldsymbol{\sigma}|\mathring{\boldsymbol{A}}(K^{(\mathrm{f})},K^{(\mathrm{b})})|\boldsymbol{\sigma}'\rangle=\frac{\exp\left[2(n_2-\zeta)K^{(\mathrm{b})}+2(n_2-\zeta')K^{(\mathrm{f})}\right]}{\{4\sinh[2K^{(\mathrm{f})}]\sinh[2K^{(\mathrm{b})}]\}^{\frac{1}{2}n_2}},\tag{12.2.54}$$

and thus

$$\mathring{\boldsymbol{B}}\left(K^{(\mathrm{f})}\pm\tfrac{1}{2}\mathrm{i}\pi,K^{(\mathrm{b})}\pm\tfrac{1}{2}\mathrm{i}\pi\right)=\mathring{\boldsymbol{B}}(K^{(\mathrm{f})},K^{(\mathrm{b})}),$$

$$\mathring{\boldsymbol{A}}\left(K^{(\mathrm{f})}\pm\tfrac{1}{2}\mathrm{i}\pi,K^{(\mathrm{b})}\pm\tfrac{1}{2}\mathrm{i}\pi\right)=\mathring{\boldsymbol{A}}(K^{(\mathrm{f})},K^{(\mathrm{b})}),\tag{12.2.55}$$

where the signs in the two arguments can be chosen independently. In terms of rapidities, from (12.2.6), (12.2.10), (12.2.11) and (12.2.20), (12.2.54) becomes

$$\langle\boldsymbol{\sigma}|\mathring{\boldsymbol{B}}(\bar{\varrho}_{\mathrm{E}},\bar{\varrho}_{\mathrm{O}};\varrho)|\boldsymbol{\sigma}'\rangle=\left[\tfrac{1}{2}\Upsilon(\bar{\varrho}_{\mathrm{E}}-\varrho,\bar{\varrho}_{\mathrm{O}}-\varrho|\mathrm{k})\right]^{n_2}$$
$$\times[W(\bar{\varrho}_{\mathrm{E}}-\varrho|\mathrm{k})]^{n_2-\eta}\,[W(\mathcal{K}'-\bar{\varrho}_{\mathrm{O}}+\varrho|\mathrm{k})]^{n_2-\eta'},$$

$$\langle\boldsymbol{\sigma}|\mathring{\boldsymbol{A}}(\bar{\varrho}_{\mathrm{E}},\bar{\varrho}_{\mathrm{O}};\varrho)|\boldsymbol{\sigma}'\rangle=\left[\tfrac{1}{2}\Upsilon(\bar{\varrho}_{\mathrm{O}}-\varrho,\bar{\varrho}_{\mathrm{E}}-\varrho|\mathrm{k})\right]^{n_2}$$
$$\times[W(\bar{\varrho}_{\mathrm{O}}-\varrho|\mathrm{k})]^{n_2-\zeta}\,[W(\mathcal{K}'-\bar{\varrho}_{\mathrm{E}}+\varrho|\mathrm{k})]^{n_2-\zeta'},\tag{12.2.56}$$

where

$$\Upsilon(u',u|\mathrm{k}):=\{-\mathrm{sn}(\mathrm{i}[\mathcal{K}'-u']|\mathrm{k})\,\mathrm{sn}(\mathrm{i}\,u|\mathrm{k})\}^{-\frac{1}{2}}.\tag{12.2.57}$$

From (12.2.9), (12.2.55) is equivalent to

$$\mathring{\boldsymbol{B}}(\bar{\varrho}_{\mathrm{E}},\bar{\varrho}_{\mathrm{O}};\varrho+2\mathrm{i}\,\mathcal{K})=\mathring{\boldsymbol{B}}(\bar{\varrho}_{\mathrm{E}},\bar{\varrho}_{\mathrm{O}};\varrho),\quad\mathring{\boldsymbol{A}}(\bar{\varrho}_{\mathrm{E}},\bar{\varrho}_{\mathrm{O}};\varrho+2\mathrm{i}\,\mathcal{K})=\mathring{\boldsymbol{A}}(\bar{\varrho}_{\mathrm{E}},\bar{\varrho}_{\mathrm{O}};\varrho).\tag{12.2.58}$$

- From (12.2.36), and using the notation of (12.2.20),

$$\langle\boldsymbol{\sigma}|\mathring{\boldsymbol{B}}(K_{\mathrm{A}}^{(\mathrm{f})},K_{\mathrm{A}}^{(\mathrm{b})})\mathring{\boldsymbol{A}}(K_{\mathrm{B}}^{(\mathrm{f})},K_{\mathrm{B}}^{(\mathrm{b})})|\boldsymbol{\sigma}'\rangle$$
$$=\prod_{j=1}^{2n_2}\frac{\cosh[K_{\mathrm{A}}^{(\mathrm{f})}\sigma_j+K_{\mathrm{B}}^{(\mathrm{f})}\sigma'_{j+1}+K_{\mathrm{A}}^{(\mathrm{b})}\sigma_{j+1}+K_{\mathrm{B}}^{(\mathrm{b})}\sigma'_j]}{\{\sinh[2K_{\mathrm{A}}^{(\mathrm{f})}]\sinh[2K_{\mathrm{A}}^{(\mathrm{b})}]\sinh[2K_{\mathrm{B}}^{(\mathrm{f})}]\sinh[2K_{\mathrm{B}}^{(\mathrm{b})}]\}^{\frac{1}{4}}}.\tag{12.2.59}$$

The case where the couplings are related by (3.8.22) is of particular interest since, as can be seen from (3.8.23) and also from (12.2.59), the matrix $\mathring{\boldsymbol{B}}(K_{\mathrm{A}}^{(\mathrm{f})},K_{\mathrm{A}}^{(\mathrm{b})})\mathring{\boldsymbol{A}}(K_{\mathrm{B}}^{(\mathrm{f})},K_{\mathrm{B}}^{(\mathrm{b})})$ is rather sparse. In fact it is not difficult to show that

$$\mathring{\boldsymbol{B}}(K^{(\mathrm{f})}, K^{(\mathrm{b})})\mathring{\boldsymbol{A}}\left(-K^{(\mathrm{b})}, K^{(\mathrm{f})} \pm \tfrac{1}{2}\mathrm{i}\pi\right) = \left[-\frac{\sinh(2K^{(\mathrm{f})})}{\sinh(2K^{(\mathrm{b})})}\right]^{n_2} \boldsymbol{I} + \left[-\frac{\sinh(2K^{(\mathrm{b})})}{\sinh(2K^{(\mathrm{f})})}\right]^{n_2} \boldsymbol{S}. \qquad (12.2.60)$$

Using (12.2.5), (12.2.7) and (12.2.20), (3.8.22) can be translated into relationships between rapidities and then, using (12.2.4) and (18.3.44), (12.2.60) can be re-expressed in the form

$$\mathring{\boldsymbol{B}}(\bar{\varrho}_{\mathrm{E}}, \bar{\varrho}_{\mathrm{O}}; \varrho)\mathring{\boldsymbol{A}}(\bar{\varrho}_{\mathrm{E}}, \bar{\varrho}_{\mathrm{O}}; \varrho - \mathcal{K}') = \left[-\frac{\mathrm{sn}(\mathrm{i}[\mathcal{K}' - \bar{\varrho}_{\mathrm{E}} + \varrho]|\mathrm{k})}{\mathrm{sn}(\mathrm{i}[\bar{\varrho}_{\mathrm{O}} - \varrho]|\mathrm{k})}\right]^{n_2} \boldsymbol{I} + \left[-\frac{\mathrm{sn}(\mathrm{i}[\bar{\varrho}_{\mathrm{O}} - \varrho]|\mathrm{k})}{\mathrm{sn}(\mathrm{i}[\mathcal{K}' - \bar{\varrho}_{\mathrm{E}} + \varrho]|\mathrm{k})}\right]^{n_2} \boldsymbol{S}. \qquad (12.2.61)$$

12.2.3 The Reduction to Regular Ising Models

The regular case of the present model (Baxter 1982b, Chap. 7) corresponds to $K^{(\mathrm{f})}_{\mathrm{A}} = K^{(\mathrm{f})}_{\mathrm{B}}$, $K^{(\mathrm{b})}_{\mathrm{A}} = K^{(\mathrm{b})}_{\mathrm{B}}$, which is achieved by setting $\varrho_{\mathrm{O}} = \varrho_{\mathrm{E}}$, $\bar{\varrho}_{\mathrm{O}} = \bar{\varrho}_{\mathrm{E}}$. In this section we show that interchanges of the rows and columns of $\mathcal{N}^{\circledast}$ can be used to decompose the partition function of the general case to a product of partion functions of four regular models. From (12.2.18),[25]

$$Z^{(\mathrm{IM})}(\mathcal{N}; \bar{\varrho}_{\mathrm{E}}, \bar{\varrho}_{\mathrm{O}}; \varrho_{\mathrm{E}}, \varrho_{\mathrm{O}}) = [\mathfrak{C}(\bar{\varrho}_{\mathrm{E}}, \bar{\varrho}_{\mathrm{O}}; \varrho_{\mathrm{E}}, \varrho_{\mathrm{O}})]^{4n_1n_2} \mathcal{Z}^{(2n_1)}_{2n_2}(\bar{\varrho}_{\mathrm{E}}, \bar{\varrho}_{\mathrm{O}}; \varrho_{\mathrm{E}}, \varrho_{\mathrm{O}}),$$
$$Z^{(\mathrm{IM})}(\mathcal{N}^{\star}; \bar{\varrho}_{\mathrm{O}}, \bar{\varrho}_{\mathrm{E}}; \varrho_{\mathrm{E}}, \varrho_{\mathrm{O}}) = [\mathfrak{C}(\bar{\varrho}_{\mathrm{E}}, \bar{\varrho}_{\mathrm{O}}; \varrho_{\mathrm{O}}, \varrho_{\mathrm{E}})]^{4n_1n_2} \mathcal{Z}^{(2n_1)}_{2n_2}(\bar{\varrho}_{\mathrm{E}}, \bar{\varrho}_{\mathrm{O}}; \varrho_{\mathrm{O}}, \varrho_{\mathrm{E}}), \qquad (12.2.62)$$

where

$$\mathcal{Z}^{(2n_1)}_{2n_2}(\bar{\varrho}_{\mathrm{E}}, \bar{\varrho}_{\mathrm{O}}; \varrho_{\mathrm{E}}, \varrho_{\mathrm{O}}) = \mathcal{Z}^{(2n_1)}_{2n_2}(\bar{\varrho}_{\mathrm{E}}, \bar{\varrho}_{\mathrm{O}}; \varrho_{\mathrm{O}}, \varrho_{\mathrm{E}}) := \mathrm{Trace}\{[\mathring{\boldsymbol{B}}_{2n_2}(\bar{\varrho}_{\mathrm{E}}, \bar{\varrho}_{\mathrm{O}}; \varrho_{\mathrm{E}})\mathring{\boldsymbol{A}}_{2n_2}(\bar{\varrho}_{\mathrm{E}}, \bar{\varrho}_{\mathrm{O}}; \varrho_{\mathrm{O}})]^{2n_1}\}. \qquad (12.2.63)$$

The commutation relations (12.2.41) and (12.2.48) interchange rapidities on rows of the lattice and their repeated application is used to bring together rows with the same rapidity. Thus

[25] As a temporary measure the width $2n_2$ and height $2n_1$ have been explicitly displayed. (The number of lattice sites is, of course, $8n_1n_2$, which can be seen from the fact that the terms in the trace are each equivalent to two rows of sites.)

$$
\begin{aligned}
&[\mathring{\boldsymbol{B}}_{2n_2}(\bar{\varrho}_{\mathrm{E}},\bar{\varrho}_{\mathrm{O}};\varrho_{\mathrm{E}})\mathring{\boldsymbol{A}}_{2n_2}(\bar{\varrho}_{\mathrm{E}},\bar{\varrho}_{\mathrm{O}};\varrho_{\mathrm{O}})]^{2n_1}\\
&\quad=[\mathring{\boldsymbol{B}}_{2n_2}(\bar{\varrho}_{\mathrm{E}},\bar{\varrho}_{\mathrm{O}};\varrho_{\mathrm{E}})\mathring{\boldsymbol{A}}_{2n_2}(\bar{\varrho}_{\mathrm{E}},\bar{\varrho}_{\mathrm{O}};\varrho_{\mathrm{E}})]^{n_1}[\mathring{\boldsymbol{B}}_{2n_2}(\bar{\varrho}_{\mathrm{E}},\bar{\varrho}_{\mathrm{O}};\varrho_{\mathrm{O}})\mathring{\boldsymbol{A}}_{2n_2}(\bar{\varrho}_{\mathrm{E}},\bar{\varrho}_{\mathrm{O}};\varrho_{\mathrm{O}})]^{n_1}\\
&\quad=[\mathring{\boldsymbol{B}}_{2n_2}(\bar{\varrho}_{\mathrm{E}},\bar{\varrho}_{\mathrm{O}};\varrho_{\mathrm{O}})\mathring{\boldsymbol{A}}_{2n_2}(\bar{\varrho}_{\mathrm{E}},\bar{\varrho}_{\mathrm{O}};\varrho_{\mathrm{O}})]^{n_1}[\mathring{\boldsymbol{B}}_{2n_2}(\bar{\varrho}_{\mathrm{E}},\bar{\varrho}_{\mathrm{O}};\varrho_{\mathrm{E}})\mathring{\boldsymbol{A}}_{2n_2}(\bar{\varrho}_{\mathrm{E}},\bar{\varrho}_{\mathrm{O}};\varrho_{\mathrm{E}})]^{n_1}.
\end{aligned}
\tag{12.2.64}
$$

The second and third lines of (12.2.64) are a commutation relation indicating that the factors are simultaneously diagonalizable with eigenvectors independent of ϱ. Thus for large n_1, when the traces are dominated by the largest eigenvalue

$$
\begin{aligned}
&\mathrm{Trace}\{[\mathring{\boldsymbol{B}}_{2n_2}(\bar{\varrho}_{\mathrm{E}},\bar{\varrho}_{\mathrm{O}};\varrho_{\mathrm{E}})\mathring{\boldsymbol{A}}_{2n_2}(\bar{\varrho}_{\mathrm{E}},\bar{\varrho}_{\mathrm{O}};\varrho_{\mathrm{O}})]^{2n_1}\}\\
&\qquad\simeq \mathrm{Trace}\{[\mathring{\boldsymbol{B}}_{2n_2}(\bar{\varrho}_{\mathrm{E}},\bar{\varrho}_{\mathrm{O}};\varrho_{\mathrm{E}})\mathring{\boldsymbol{A}}_{2n_2}(\bar{\varrho}_{\mathrm{E}},\bar{\varrho}_{\mathrm{O}};\varrho_{\mathrm{E}})]^{n_1}\}\\
&\qquad\qquad\times \mathrm{Trace}\{[\mathring{\boldsymbol{B}}_{2n_2}(\bar{\varrho}_{\mathrm{E}},\bar{\varrho}_{\mathrm{O}};\varrho_{\mathrm{O}})\mathring{\boldsymbol{A}}_{2n_2}(\bar{\varrho}_{\mathrm{E}},\bar{\varrho}_{\mathrm{O}};\varrho_{\mathrm{O}})]^{n_1}\}.
\end{aligned}
\tag{12.2.65}
$$

The assumption is simply that the contributions to the free energy from the interfaces between the parts of the system with horizontal rapidities ϱ_{E} and ϱ_{O} is small for large n_1.

The next task is to carry out a similar procedure for columns. Clearly this is possible because of the invariance property of the star contribution (12.2.30) to the partition function under interchange of either row or column rapidities. This means that the two equivalent ways of expressing the partition function as a matrix product, by vertical multiplication of rows (as we have represented it here), or by horizontal multiplication of columns, both allow reordering of the rapidities. A simple way to see the argument is as follows. Think of the system as consisting of connected vertical units. These can either be taken as vertical columns of stars based on a pair of rapidity lines with rapidities $\bar{\varrho}_{\mathrm{E}}$ and $\bar{\varrho}_{\mathrm{O}}$ on the left and right or as vertical columns of diamonds based on a pair of rapidity lines with rapidities $\bar{\varrho}_{\mathrm{O}}$ and $\bar{\varrho}_{\mathrm{E}}$ on the left and right. We know that the rapidities on the stars can be interchanged. That they can also be interchanged on the diamonds follows from the observation that diamonds translated one lattice space upwards or downwards (remembering that now all the horizontal rapidities are the same) become stars. This means that we can freely shuffle the vertical rapidities to bring like rapidities together. The matrix elements of $\mathring{\boldsymbol{B}}_{2n_2}(\bar{\varrho}_{\mathrm{E}},\bar{\varrho}_{\mathrm{O}},\varrho)$ can be reformulated as

$$
\begin{aligned}
\langle\boldsymbol{\sigma}|\mathring{\boldsymbol{B}}_{2n_2}(\bar{\varrho}_{\mathrm{E}},\bar{\varrho}_{\mathrm{O}};\varrho)|\boldsymbol{\sigma}'\rangle &= \prod_{j=1}^{n_2}\mathfrak{X}^{(\mathrm{f})}(\bar{\varrho}_{\mathrm{E}}-\varrho;\sigma_j,\sigma_j')\mathfrak{X}^{(\mathrm{b})}(\bar{\varrho}_{\mathrm{E}}-\varrho;\sigma_j',\sigma_{j+1})\\
&\quad\times\prod_{j=n_2+1}^{2n_2}\mathfrak{X}^{(\mathrm{f})}(\bar{\varrho}_{\mathrm{O}}-\varrho;\sigma_j,\sigma_j')\mathfrak{X}^{(\mathrm{b})}(\bar{\varrho}_{\mathrm{O}}-\varrho;\sigma_j',\sigma_{j+1}).
\end{aligned}
\tag{12.2.66}
$$

where

$$\mathfrak{X}^{(x)}(u;\sigma,\sigma') := \frac{\exp\left[K^{(x)}(u)\sigma\,\sigma'\right]}{g^{(x)}(u)}, \qquad x = f,\ b. \tag{12.2.67}$$

There is one linking term at the end of the first summation tying it to the second and (with j an integer modulo $2n_2$) a term at the end of the second summation linking it to the beginning of the first summation. For large n_2 these linking terms make a negligible contribution to the free energy and we approximate this expression by cutting the links. This simply involved resetting the integer j to modulo n_2 and relabelling the spins for $j = n_2 + 1, n_2 + 2, \ldots, 2n_2$. This gives

$$\mathring{\boldsymbol{B}}_{2n_2}(\bar{\varrho}_E, \bar{\varrho}_O; \varrho) = \mathring{\boldsymbol{B}}_{n_2}(\bar{\varrho}_E, \bar{\varrho}_E; \varrho) \otimes \mathring{\boldsymbol{B}}_{n_2}(\bar{\varrho}_E, \bar{\varrho}_E; \varrho). \tag{12.2.68}$$

In a similar way

$$\mathring{\boldsymbol{A}}_{2n_2}(\bar{\varrho}_E, \bar{\varrho}_O; \varrho) = \mathring{\boldsymbol{A}}_{n_2}(\bar{\varrho}_E, \bar{\varrho}_E; \varrho) \otimes \mathring{\boldsymbol{A}}_{n_2}(\bar{\varrho}_E, \bar{\varrho}_E; \varrho) \tag{12.2.69}$$

and

$$\begin{aligned}
&\mathrm{Trace}\{[\mathring{\boldsymbol{B}}_{2n_2}(\bar{\varrho}_E, \bar{\varrho}_O; \varrho)\mathring{\boldsymbol{A}}_{2n_2}(\bar{\varrho}_E, \bar{\varrho}_O; \varrho)]^{n_1} \\
&= \mathrm{Trace}\{[\mathring{\boldsymbol{B}}_{n_2}(\bar{\varrho}_E, \bar{\varrho}_E; \varrho)\mathring{\boldsymbol{A}}_{n_2}(\bar{\varrho}_E, \bar{\varrho}_E; \varrho)]^{n_1}\} \\
&\qquad \times \mathrm{Trace}\{[\mathring{\boldsymbol{B}}_{n_2}(\bar{\varrho}_E, \bar{\varrho}_E; \varrho)\mathring{\boldsymbol{A}}_{n_2}(\bar{\varrho}_E, \bar{\varrho}_E; \varrho)]^{n_1}\}.
\end{aligned} \tag{12.2.70}$$

Then, from (12.2.63), (12.2.65) and (12.2.70),

$$\begin{aligned}
\mathcal{Z}_{2n_2}^{(2n_1)}(\bar{\varrho}_E, \bar{\varrho}_O; \varrho_O, \varrho_E) &= \mathcal{Z}_{n_2}^{(n_1)}(\bar{\varrho}_E, \bar{\varrho}_E; \varrho_E, \varrho_E)\mathcal{Z}_{n_2}^{(n_1)}(\bar{\varrho}_O, \bar{\varrho}_O; \varrho_E, \varrho_E) \\
&\quad \times \mathcal{Z}_{n_2}^{(n_1)}(\bar{\varrho}_E, \bar{\varrho}_E; \varrho_O, \varrho_O)\mathcal{Z}_{n_2}^{(n_1)}(\bar{\varrho}_O, \bar{\varrho}_O; \varrho_O, \varrho_O).
\end{aligned} \tag{12.2.71}$$

12.2.4 Transfer Matrix Eigenvectors

Assume that $\mathring{\boldsymbol{B}}(\bar{\varrho}_E, \bar{\varrho}_O; \varrho)$, $\mathring{\boldsymbol{A}}(\bar{\varrho}_E, \bar{\varrho}_O; \varrho)$ and the products $\mathring{\boldsymbol{B}}(\bar{\varrho}_E, \bar{\varrho}_O; \varrho)\mathring{\boldsymbol{A}}(\bar{\varrho}_E, \bar{\varrho}_O; \varrho')$ and $\mathring{\boldsymbol{A}}(\bar{\varrho}_E, \bar{\varrho}_O; \varrho)\mathring{\boldsymbol{B}}(\bar{\varrho}_E, \bar{\varrho}_O; \varrho')$ are simple (Sect. 17.2). Let $\boldsymbol{X}_{BA}(\bar{\varrho}_E, \bar{\varrho}_O; \varrho, \varrho')$ and $\boldsymbol{X}_{AB}(\bar{\varrho}_E, \bar{\varrho}_O; \varrho, \varrho')$ be the respective matrices of the right eigenvectors (as columns). It can easily be shown (Example 17.2.1) that the two product matrices have the same eigenvalues, denoted by $\boldsymbol{\Lambda}(\bar{\varrho}_E, \bar{\varrho}_O; \varrho, \varrho')$, and that

$$\boldsymbol{X}_{BA}^{-1}(\bar{\varrho}_E, \bar{\varrho}_O; \varrho, \varrho')\mathring{\boldsymbol{B}}(\bar{\varrho}_E, \bar{\varrho}_O; \varrho)\boldsymbol{X}_{AB}(\bar{\varrho}_E, \bar{\varrho}_O; \varrho', \varrho) \quad \text{is diagonal} \tag{12.2.72}$$

and

$$\boldsymbol{X}_{\mathrm{AB}}^{-1}(\bar{\varrho}_{\mathrm{E}},\bar{\varrho}_{\mathrm{O}};\varrho',\varrho)\mathring{\boldsymbol{A}}(\bar{\varrho}_{\mathrm{E}},\bar{\varrho}_{\mathrm{O}};\varrho')\boldsymbol{X}_{\mathrm{BA}}(\bar{\varrho}_{\mathrm{E}},\bar{\varrho}_{\mathrm{O}};\varrho,\varrho') \quad \text{is diagonal.} \tag{12.2.73}$$

It follows that the eigenvector matrices are respectively independent of ϱ' and ϱ. Taking the argument a step further, by choosing $\varrho = \varrho' = \bar{\varrho}_{\mathrm{O}}$ in $\mathring{\boldsymbol{B}}(\bar{\varrho}_{\mathrm{E}},\bar{\varrho}_{\mathrm{O}};\varrho)$ and $\mathring{\boldsymbol{A}}(\bar{\varrho}_{\mathrm{E}},\bar{\varrho}_{\mathrm{O}};\varrho')$, it follows from (12.2.36) that the eigenvector matrices are functions of the single rapidity variable $\bar{\varrho}_{\mathrm{E}}-\bar{\varrho}_{\mathrm{O}}$.

12.2.5 Notational Changes

From this point we shall specialize to the case with only one horizontal rapidity ϱ so it is convenient to adopt a generalization of the notation of (12.2.2) with

$$u := \tfrac{1}{2}(\bar{\varrho}_{\mathrm{E}}+\bar{\varrho}_{\mathrm{O}})-\varrho, \qquad \bar{u} := \tfrac{1}{2}(\bar{\varrho}_{\mathrm{E}}-\bar{\varrho}_{\mathrm{O}}), \tag{12.2.74}$$

modifying the arguments of the matrices and the eigenvalues and eigenvectors accordingly. In particular we need to express the function $\Upsilon(u,u')$, defined by (12.2.57), in terms of $\bar{u}$ and u (Eq. (18.3.100)).

It is also useful to bring together (12.2.4) and (12.2.20) to give expressions for the couplings in Fig. 12.5 now that $\varrho_{\mathrm{E}}=\varrho_{\mathrm{O}}=\varrho$. From (12.2.20), for the lower row, related to the matrix $\boldsymbol{B}(\bar{u};u)$,

$$K_{\mathrm{B}}^{(\mathrm{f})}=K^{(\mathrm{f})}(u+\bar{u}), \qquad K_{\mathrm{B}}^{(\mathrm{b})}=K^{(\mathrm{b})}(u-\bar{u}), \tag{12.2.75}$$

with

$$\begin{aligned}
\sinh(2K_{\mathrm{B}}^{(\mathrm{f})}) &= \frac{\mathrm{i}}{\mathrm{k}\,\mathrm{sn}(\mathrm{i}[u+\bar{u}]|\mathrm{k})}, \quad \sinh(2K_{\mathrm{B}}^{(\mathrm{b})}) = -\mathrm{i}\,\mathrm{sn}(\mathrm{i}[u-\bar{u}]|\mathrm{k}),\\
\cosh(2K_{\mathrm{B}}^{(\mathrm{f})}) &= \frac{\mathrm{i}\,\mathrm{dn}(\mathrm{i}[u+\bar{u}]|\mathrm{k})}{\mathrm{k}\,\mathrm{sn}(\mathrm{i}[u+\bar{u}]|\mathrm{k})}, \quad \cosh(2K_{\mathrm{B}}^{(\mathrm{b})}) = \mathrm{cn}(\mathrm{i}[u-\bar{u}]|\mathrm{k})
\end{aligned} \tag{12.2.76}$$

and for the upper row, related to the matrix $\boldsymbol{A}(\bar{u};u)$,

$$K_{\mathrm{A}}^{(\mathrm{f})}=K^{(\mathrm{f})}(u-\bar{u}), \qquad K_{\mathrm{A}}^{(\mathrm{b})}=K^{(\mathrm{b})}(u+\bar{u}), \tag{12.2.77}$$

with

$$\begin{aligned}
\sinh(2K_{\mathrm{A}}^{(\mathrm{f})}) &= \frac{\mathrm{i}}{\mathrm{k}\,\mathrm{sn}(\mathrm{i}[u-\bar{u}]|\mathrm{k})}, \quad \sinh(2K_{\mathrm{A}}^{(\mathrm{b})}) = -\mathrm{i}\,\mathrm{sn}(\mathrm{i}[u+\bar{u}]|\mathrm{k}),\\
\cosh(2K_{\mathrm{A}}^{(\mathrm{f})}) &= \frac{\mathrm{i}\,\mathrm{dn}(\mathrm{i}[u-\bar{u}]|\mathrm{k})}{\mathrm{k}\,\mathrm{sn}(\mathrm{i}[u-\bar{u}]|\mathrm{k})}, \quad \cosh(2K_{\mathrm{A}}^{(\mathrm{b})}) = \mathrm{cn}(\mathrm{i}[u+\bar{u}]|\mathrm{k}).
\end{aligned} \tag{12.2.78}$$

As we have observed above, in the general case $\varrho_{\mathrm{E}} \neq \varrho_{\mathrm{O}}$ the four couplings $K^{(\mathrm{f})}_{\mathrm{B}}$, $K^{(\mathrm{f})}_{\mathrm{A}}$, $K^{(\mathrm{b})}_{\mathrm{B}}$, $K^{(\mathrm{b})}_{\mathrm{A}}$ are independent with the corresponding set of combinations of rapidities u_1, u_2, u_3, given by (12.2.22), forming together with the elliptic modulus k an equivalent independent set. Now that the condition $\varrho_{\mathrm{E}} = \varrho_{\mathrm{O}}$, giving $u_1 = u$, $u_2 = \bar{u}$, $u_3 = 0$, has been imposed, the couplings are, from (12.2.76) and (12.2.78), related by the conditions

$$\sinh(2K^{(\mathrm{f})}_{\mathrm{B}})\sinh(2K^{(\mathrm{b})}_{\mathrm{A}}) = \sinh(2K^{(\mathrm{f})}_{\mathrm{A}})\sinh(2K^{(\mathrm{b})}_{\mathrm{B}}) = 1/\mathrm{k}, \tag{12.2.79}$$

$$\frac{\sinh(2K^{(\mathrm{b})}_{\mathrm{B}})}{\sinh(2K^{(\mathrm{f})}_{\mathrm{B}})} = \frac{\sinh(2K^{(\mathrm{b})}_{\mathrm{A}})}{\sinh(2K^{(\mathrm{f})}_{\mathrm{A}})} = -\mathrm{P}(\bar{u}, u|\mathrm{k}), \tag{12.2.80}$$

where

$$\mathrm{P}(\bar{u}, u|\mathrm{k}) := \mathrm{k}\,\mathrm{sn}(\mathrm{i}[u+\bar{u}]|\mathrm{k})\mathrm{sn}(\mathrm{i}[u-\bar{u}]|\mathrm{k}) = \frac{\mathrm{P_H}(\bar{u}, u|\mathrm{k})}{\mathrm{P_\Theta}(\bar{u}, u|\mathrm{k})}, \tag{12.2.81}$$

with $\mathrm{P_H}(\bar{u}, u|\mathrm{k})$ and $\mathrm{P_\Theta}(\bar{u}, u|\mathrm{k})$ defined in Sect. 18.3.7. The relationship (12.2.79) can be compared to (12.2.6), which, applying in the general case, is between coupling on $\mathcal{N}$ and its dual lattice $\mathcal{N}^\star$. Thus the condition $\varrho_{\mathrm{E}} = \varrho_{\mathrm{O}}$ imposes a general form of the self-duality of the square-lattice Ising model (Sect. 9.2.2.1).

With $\varrho_{\mathrm{E}} = \varrho_{\mathrm{O}}$, the couplings on the vertical sides of the square in Fig. 12.5 are zero. The horizontal couplings $K^{(\mathrm{h})}$ and $-K^{(\mathrm{h})}$ on the lower and upper sides of the square, respectively, with

$$K^{(\mathrm{h})} = K^{(\mathrm{h})}(2\,\bar{u}), \qquad K^{(\mathrm{ch})} = K^{(\mathrm{h})}(\mathcal{K}' - 2\bar{u}), \tag{12.2.82}$$

and

$$\begin{aligned} \sinh(2K^{(\mathrm{h})}) &= -\mathrm{i}\,\mathrm{sn}(2\,\mathrm{i}\,\bar{u}|\mathrm{k}),\ \sinh(2K^{(\mathrm{ch})}) = \frac{\mathrm{i}}{\mathrm{k}\,\mathrm{sn}(2\,\mathrm{i}\,\bar{u}|\mathrm{k})}, \\ \cosh(2K^{(\mathrm{h})}) &= \mathrm{cn}(2\,\mathrm{i}\,\bar{u}|\mathrm{k}), \quad \cosh(2K^{(\mathrm{ch})}) = \frac{\mathrm{i}\,\mathrm{dn}(2\,\mathrm{i}\,\bar{u}|\mathrm{k})}{\mathrm{k}\,\mathrm{sn}(2\,\mathrm{i}\,\bar{u}|\mathrm{k})}, \end{aligned} \tag{12.2.83}$$

are present, although, as we have indicated above, their presence does not alter the partition function. Their role is to facilitate the calculation and their presence in the final form for the eigenvalues is as dependent quantities, with the conjugate pair $K^{(\mathrm{h})}$ and $K^{(\mathrm{ch})}$ related by (12.2.82) and, from (18.3.45)–(18.3.50), (18.3.62) and (12.2.83),

$$\cosh(2K^{(\mathrm{ch})}) = \coth(2K^{(\mathrm{h})})\sinh(2K^{(\mathrm{f})}_{\mathrm{B}})\sinh(2K^{(\mathrm{b})}_{\mathrm{B}}) + \cosh(2K^{(\mathrm{f})}_{\mathrm{B}})\cosh(2K^{(\mathrm{b})}_{\mathrm{B}}). \tag{12.2.84}$$

Changing the sign of $\bar{u}$ gives

$$\cosh(2K^{(\mathrm{ch})}) = \coth(2K^{(\mathrm{h})})\sinh(2K^{(\mathrm{f})}_{\mathrm{A}})\sinh(2K^{(\mathrm{b})}_{\mathrm{A}}) - \cosh(2K^{(\mathrm{f})}_{\mathrm{A}})\cosh(2K^{(\mathrm{b})}_{\mathrm{A}}) \tag{12.2.85}$$

and equating the right-hand sides of these two equations gives, in this modified notation, (12.2.24).

12.2.6 *Transfer Matrix Eigenvalues*

12.2.6.1 The Separate Matrices

Using (12.2.41) and (12.2.47),

$$\boldsymbol{X}(\bar{u}) := \boldsymbol{X}_{\mathrm{BA}}(\bar{u}), \qquad \boldsymbol{X}_{\mathrm{AB}}(\bar{u}) = \boldsymbol{X}(-\bar{u}) = \boldsymbol{D}(\bar{u})\boldsymbol{X}(\bar{u}) \tag{12.2.86}$$

and (12.2.72) and (12.2.73) take the form

$$\begin{aligned} \boldsymbol{X}^{-1}(\bar{u})\mathring{\boldsymbol{B}}(\bar{u}, u)\boldsymbol{X}(-\bar{u}) &= \boldsymbol{\Lambda}_{\mathrm{B}}(\bar{u}, u), \\ \boldsymbol{X}^{-1}(-\bar{u})\mathring{\boldsymbol{A}}(\bar{u}, u)\boldsymbol{X}(\bar{u}) &= \boldsymbol{\Lambda}_{\mathrm{A}}(\bar{u}, u), \end{aligned} \tag{12.2.87}$$

where $\boldsymbol{\Lambda}_{\mathrm{B}}(\bar{u}, u)$ and $\boldsymbol{\Lambda}_{\mathrm{A}}(\bar{u}, u)$ are diagonal matrices. From (12.2.45) and (12.2.52),

$$\begin{aligned} \boldsymbol{X}^{-1}(\bar{u})\boldsymbol{C}\boldsymbol{X}(\bar{u})\boldsymbol{\Lambda}_{\mathrm{A}}(\bar{u}, u) &= \boldsymbol{\Lambda}_{\mathrm{B}}(\ \ \bar{u}, u), \\ \boldsymbol{X}^{-1}(\bar{u})\boldsymbol{S}\boldsymbol{X}(\bar{u})\boldsymbol{\Lambda}_{\mathrm{Z}}(\bar{u}, u) &= \boldsymbol{\Lambda}_{\mathrm{Z}}(\bar{u}, u + 2\mathcal{K}'), \end{aligned} \tag{12.2.88}$$

where here and in later equations $\mathrm{Z} = \mathrm{B}, \mathrm{A}$. So $\boldsymbol{X}(\bar{u})$ is the matrix of right eigenvectors of $\boldsymbol{C}$ and $\boldsymbol{S}$ with

$$\boldsymbol{X}^{-1}(\bar{u})\boldsymbol{C}\boldsymbol{X}(\bar{u}) = \boldsymbol{\Xi}, \tag{12.2.89}$$

$$\boldsymbol{X}^{-1}(\bar{u})\boldsymbol{S}\boldsymbol{X}(\bar{u}) = \boldsymbol{\Sigma}, \tag{12.2.90}$$

where $\boldsymbol{\Xi}$ is the diagonal matrix with elements which are complex $2n_2$–th roots of unity and $\boldsymbol{\Sigma}$ is the diagonal matrix with elements equal to $+1$ or -1. Thus we have, from (12.2.58) and (12.2.88)–(12.2.90),

$$\boldsymbol{\Xi}\boldsymbol{\Lambda}_{\mathrm{A}}(\bar{u}, u) = \boldsymbol{\Lambda}_{\mathrm{B}}(-\bar{u}, u), \tag{12.2.91}$$

$$\boldsymbol{\Sigma}\boldsymbol{\Lambda}_{\mathrm{Z}}(\bar{u}, u) = \boldsymbol{\Lambda}_{\mathrm{Z}}(\bar{u}, u + 2\mathcal{K}'), \tag{12.2.92}$$

$$\boldsymbol{\Lambda}_{\mathrm{Z}}(\bar{u}, u) = \boldsymbol{\Lambda}_{\mathrm{Z}}(\bar{u}, u + 2\mathrm{i}\,\mathcal{K}). \tag{12.2.93}$$

As was noted in Sect. 18.3.2 the four theta functions are entire. So, from (12.2.11) and (12.2.56),

$$\langle\boldsymbol{\sigma}|\boldsymbol{B}(\bar{u},u)|\boldsymbol{\sigma}'\rangle=\frac{f_{\mathrm{B}}(\bar{u},u;\boldsymbol{\sigma},\boldsymbol{\sigma}'|\mathrm{k})}{[\mathrm{Q}(\bar{u},u|\mathrm{k})]^{n_2}},\qquad\langle\boldsymbol{\sigma}|\boldsymbol{A}(\bar{u},u)|\boldsymbol{\sigma}'\rangle=\frac{f_{\mathrm{A}}(\bar{u},u;\boldsymbol{\sigma},\boldsymbol{\sigma}'|\mathrm{k})}{[\mathrm{Q}(-\bar{u},u|\mathrm{k})]^{n_2}},\tag{12.2.94}$$

where $f_{\mathrm{B}}(\bar{u},u;\boldsymbol{\sigma},\boldsymbol{\sigma}'|\mathrm{k})$ and $f_{\mathrm{A}}(\bar{u},u;\boldsymbol{\sigma},\boldsymbol{\sigma}'|\mathrm{k})$ are entire functions of u and $\mathrm{Q}(v,u|\mathrm{k})$ is defined in (18.3.94). The 2^{2n_2} eigenvalues of $\boldsymbol{B}(\bar{u},u)$ appear in a certain order in the diagonal of $\boldsymbol{\Lambda}_{\mathrm{B}}(\bar{u},u)$, thus associating them each with an element $s=+1$ or -1 of $\boldsymbol{\Sigma}$; this is the *parity* of the eigenvalue. The eigenvalues will be indexed by the pair $\{s,m\}$, with $m=1,2,\ldots,2^{2n_2-1}$, and written as $\Lambda_{\mathrm{B}}^{(s)}(\bar{u},u;m|\mathrm{k})$ with a corresponding notation for the eigenvalues of $\boldsymbol{A}(\bar{u},u)$. It follows from (12.2.87) that $\Lambda_{\mathrm{B}}^{(s)}(\bar{u},u;m|\mathrm{k})$ and $\Lambda_{\mathrm{A}}^{(s)}(\bar{u},u;m|\mathrm{k})$ are linear combinations of the elements of $\boldsymbol{B}(\bar{u},u)$ and $\boldsymbol{A}(\bar{u},u)$, respectively, with coefficients dependent on $\bar{u}$ and k but not on u. The eigenvalues $\Lambda_{\mathrm{B}}^{(s)}(\bar{u},u;m|\mathrm{k})$ and $\Lambda_{\mathrm{A}}^{(s)}(\bar{u},u;m|\mathrm{k})$ are associated with a complex $2n_2$–th root of unity denoted by $\xi_m^{(s)}$, which is the corresponding element of $\boldsymbol{C}$, and, to satisfy (12.2.91), we let

$$\begin{aligned}\Lambda_{\mathrm{B}}^{(s)}(\bar{u},u;m|\mathrm{k})&:=[\xi_m^{(s)}]^{\frac{1}{2}}[\mathrm{Y}(\bar{u},u|\mathrm{k})]^{n_2}\Upsilon^{(s)}(\bar{u},u;m|\mathrm{k}),\\\Lambda_{\mathrm{A}}^{(s)}(\bar{u},u;m|\mathrm{k})&:=[\xi_m^{(s)}]^{-\frac{1}{2}}[\mathrm{Y}(-\bar{u},u|\mathrm{k})]^{n_2}\Upsilon^{(s)}(-\bar{u},u;m|\mathrm{k}).\end{aligned}\tag{12.2.95}$$

Then from (12.2.92), (12.2.93), (18.3.102) and (18.3.103)

$$\Upsilon^{(s)}(\bar{u},u;m|\mathrm{k})=\Upsilon^{(s)}(\bar{u},u+2\mathrm{i}\,\mathcal{K};m|\mathrm{k}),\tag{12.2.96}$$

$$\Upsilon^{(s)}(\bar{u},u;m|\mathrm{k})=s\Upsilon^{(s)}(\bar{u},u+2\mathcal{K}';m|\mathrm{k}).\tag{12.2.97}$$

So $\Upsilon^{(s)}(\bar{u},u;m)\in\mathrm{DP}(2\mathrm{i}\mathcal{K},2\mathcal{K}';1,s)$.[26] A small shift ϵ in the real direction, so that the period rectangle becomes

$$\mathbb{W}'(2\mathrm{i}\mathcal{K},2\mathcal{K}'):=\{u:\epsilon<\Re(u)<2\mathcal{K}'+\epsilon,0<\Im(u)<2\mathcal{K}\},\tag{12.2.98}$$

is necessary only when it is needed to satisfy (12.2.99) if $\bar{u}$ is an integer. It follows from (12.2.94) that it is meromorphic with poles of order n_2 at $u=(2p_1-1)\mathcal{K}'+2p_2\mathcal{K}\mathrm{i}+\bar{u}$ and $u=2q_1\mathcal{K}'+2q_2\mathcal{K}\mathrm{i}-\bar{u}$, for all integer values of p_1,p_2,q_1 and q_2. $\mathbb{W}'(2\mathrm{i}\mathcal{K},2\mathcal{K}')$ will contain two poles, from these sets with $p_2=q_2=0$ and $p_1=p_1^\star(\bar{u})$ and $q_1=q_1^\star(\bar{u})$, the unique integers satisfying

$$\begin{aligned}&\epsilon-\bar{u}<2\mathcal{K}'p_1^\star(\bar{u})<2\mathcal{K}'-\bar{u}+\epsilon,\\&\epsilon+\bar{u}-\mathcal{K}'<2\mathcal{K}'q_1^\star(\bar{u})<\mathcal{K}'+\bar{u}+\epsilon.\end{aligned}\tag{12.2.99}$$

---

[26] The class of doubly-periodic functions is defined in Sect. 18.2.

Then, from Theorem 18.2.3, $\Upsilon^{(s)}(\bar{u}, u; m|\mathrm{k})$ has exactly $2n_2$ zeros (including each according to its multiplicity) in the interior of $\mathbb{W}'(2\mathrm{i}\mathcal{K}, 2\mathcal{K}')$ with none on the boundary. Let these zeros be $u_1^{(s)}(\bar{u}; m), u_2^{(s)}(\bar{u}; m), \ldots, u_{2n_2}^{(s)}(\bar{u}; m)$. Then, from Theorem 18.3.1 and (18.3.96),

$$\Upsilon^{(s)}(\bar{u}, u; m|\mathrm{k}) = A_1^{(s)}(\bar{u}; m) \exp\left(\frac{\pi \aleph^{(s)}(\bar{u}; m)u}{2\mathcal{K}}\right) \prod_{j=1}^{2n_2} \frac{\mathrm{H}(\mathrm{i}\,[u - u_j^{(s)}(\bar{u}; m)]|\mathrm{k})}{\sqrt{\mathrm{Q}(\bar{u}, u|\mathrm{k})}}. \tag{12.2.100}$$

To satisfy (12.2.96) and (12.2.97) $\aleph^{(s)}(\bar{u}; m) - n_2$ must be an even integer with[27]

$$\frac{1}{\mathcal{K}'} \sum_{j=1}^{2n_2} \Re\{u_j^{(s)}(\bar{u}; m)\} = \aleph^{(s)}(\bar{u}; m), \tag{12.2.101}$$

$$\frac{1}{\mathcal{K}} \sum_{j=1}^{2n_2} \Im\{u_j^{(s)}(\bar{u}; m)\} \equiv -\tfrac{1}{2}(1 - s) \pmod{2}. \tag{12.2.102}$$

From (18.3.23),

$$\begin{aligned} \Upsilon^{(s)}(-\bar{u}, u + \mathcal{K}'; m|\mathrm{k}) &= (-1)^{n_2} A_1^{(s)}(-\bar{u}; m) \exp(-\pi\, n_2 \mathcal{K}'/2\mathcal{K}) \\ &\times \exp\left(\frac{\pi[\aleph^{(s)}(-\bar{u}; m) - 2n_2]u}{2\mathcal{K}}\right) \prod_{j=1}^{2n_2} \frac{\mathrm{H}(\mathrm{i}\,[u + \mathcal{K}' - u_j^{(s)}(-\bar{u}; m)]|\mathrm{k})}{\sqrt{\mathrm{Q}(\bar{u}, u|\mathrm{k})}}. \end{aligned} \tag{12.2.103}$$

12.2.6.2 The Product Formula

From (12.2.61), (12.2.90) and (12.2.91), and using the notation of Sect. 18.3.7 for combinations of theta functions,

$$\boldsymbol{\Lambda}_{\mathrm{B}}(\bar{u}, u)\boldsymbol{\Lambda}_{\mathrm{A}}(\bar{u}, u + \mathcal{K}') = \frac{[\mathrm{P}_{\Theta}(\bar{u}, u|\mathrm{k})]^{2n_2}\boldsymbol{I} + [\mathrm{P}_{\mathrm{H}}(\bar{u}, u|\mathrm{k})]^{2n_2}\boldsymbol{\Sigma}}{[\mathrm{Q}(\bar{u}, u|\mathrm{k})\mathrm{Q}(-\bar{u}, u|\mathrm{k})]^{n_2}}. \tag{12.2.104}$$

From (18.3.100)–(18.3.108),

$$\Lambda_{\mathrm{B}}^{(s)}(\bar{u}, u; m|\mathrm{k})\Lambda_{\mathrm{A}}^{(s)}(\bar{u}, u + \mathcal{K}'; m|\mathrm{k}) = \frac{s[\mathrm{P}_{\Theta}(\bar{u}, u|\mathrm{k})]^{2n_2}}{[\mathrm{Q}(\bar{u}, u|\mathrm{k})\mathrm{Q}(-\bar{u}, u|\mathrm{k})]^{n_2}} \mathrm{G}^{(s)}(\bar{u}, u|\mathrm{k}), \tag{12.2.105}$$

[27] Using (18.3.96) these results also follow directly from Theorem 18.3.1 with $\tau = 1$, $\tau' = s$.

where $\mathrm{G}^{(s)}(\bar{u}, u|\mathrm{k})$ is given by (18.3.109) and, from (12.2.95), (18.3.101)–(18.3.103),

$$\Upsilon^{(s)}(\bar{u}, u; m|\mathrm{k})\Upsilon^{(s)}(-\bar{u}, u + \mathcal{K}'; m|\mathrm{k}) = \frac{s}{\mathrm{k}^{n_2}} \left[\frac{\mathrm{P}_\Theta(\bar{u}, u|\mathrm{k})}{\mathrm{Q}(\bar{u}, u|\mathrm{k})}\right]^{2n_2} \mathrm{G}^{(s)}(\bar{u}, u|\mathrm{k}). \tag{12.2.106}$$

A factorization of $\mathrm{G}^{(s)}(\bar{u}, u|\mathrm{k})$ is derived in Sect. 18.3.7, which, together with (12.2.106), yields

$$\begin{aligned}\Upsilon^{(s)}(\bar{u}, u; m|\mathrm{k})\Upsilon^{(s)}(-\bar{u}, u + \mathcal{K}'; m|\mathrm{k}) &= \frac{1}{\mathrm{k}^{n_2}} C^{(s)}(\bar{u}) \\ \times \prod_{j=1}^{2n_2} \frac{\mathrm{H}(\mathrm{i}\,u - \zeta(\bar{u}; \psi_j^{(s)}) + \frac{1}{2}\mathrm{i}\,\mathcal{K}'|\mathrm{k})\mathrm{H}(\mathrm{i}\,u + \zeta(\bar{u}; \psi_j^{(s)}) - \frac{1}{2}\mathrm{i}\,\mathcal{K}'|\mathrm{k})}{\mathrm{Q}(\bar{u}, u|\mathrm{k})}&,\end{aligned} \tag{12.2.107}$$

where

$$\psi_j^{(s)} := \frac{\pi(4j - s - 1)}{8n_2}, \qquad j = 1, 2, \ldots, 2n_2, \tag{12.2.108}$$

and, from (18.3.119),

$$C^{(s)}(\bar{u}) = s \prod_{j=1}^{2n_2} \frac{\Theta(2\,\mathrm{i}\,\bar{u}|\mathrm{k})\,\Theta\,(0|\mathrm{k})}{\mathrm{H}(\mathrm{i}\,\bar{u} - \zeta(\bar{u}; \psi_j^{(s)}) + \frac{1}{2}\,\mathrm{i}\,\mathcal{K}'|\mathrm{k})\mathrm{H}(\mathrm{i}\,\bar{u} + \zeta(\bar{u}; \psi_j^{(s)}) - \frac{1}{2}\,\mathrm{i}\,\mathcal{K}'|\mathrm{k})}. \tag{12.2.109}$$

From (18.3.104) and (18.3.115), $\zeta(\bar{u}; \psi)$ is given by

$$\mathrm{k}\,\mathrm{sn}\big(\mathrm{i}\,\bar{u} - \zeta(\bar{u}; \psi) + \tfrac{1}{2}\mathrm{i}\,\mathcal{K}'|\mathrm{k}\big)\mathrm{sn}\big(\mathrm{i}\,\bar{u} + \zeta(\bar{u}; \psi) - \tfrac{1}{2}\mathrm{i}\,\mathcal{K}'|\mathrm{k}\big) = -\exp(2\,\mathrm{i}\,\psi). \tag{12.2.110}$$

The solution of this equation is a crucial part of the determination of the eigenvalues.

12.2.6.3 Determining Zeros

The product of (12.2.100) and (12.2.103) must be equivalent to (12.2.107). The task is to match terms in the numerator of the product together with a contribution from the exponential factor with terms in the numerator of (12.2.107). From (12.2.108) and (18.3.116),

$$\psi^{(+1)}_{2n_2-j+1} = \pi - \psi^{(+1)}_j, \quad \zeta(\bar{u}; \psi^{(+1)}_{2n_2-j+1}) = -\zeta(\bar{u}; \psi^{(+1)}_j),$$
$$j = 1, 2, \ldots, n_2, \tag{12.2.111}$$

$$\psi^{(-1)}_{2n_2-j} = \pi - \psi^{(-1)}_j, \quad \zeta(\bar{u}; \psi^{(-1)}_{2n_2-j}) = -\zeta(\bar{u}; \psi^{(-1)}_j),$$
$$j = 1, 2, \ldots, n_2 - 1, \tag{12.2.112}$$

$$\psi^{(-1)}_{n_2} = \tfrac{1}{2}\pi, \quad \zeta(\bar{u}; \psi^{(-1)}_{n_2}) = 0, \qquad \psi^{(-1)}_{2n_2} = \pi, \quad \zeta(\bar{u}; \psi^{(-1)}_{2n_2}) = \mathcal{K}. \tag{12.2.113}$$

Apart from the terms $\psi = \frac{\pi}{2}$ and $\psi = \pi$ which appear in the product for $s = -1$, the terms are in pairs $(\psi, \pi - \psi)$. In the numerator in (12.2.107) for a particular value of ψ we select four roots

$$\begin{aligned} u &= -\mathrm{i}\,\zeta(\bar{u}; \psi) - \tfrac{1}{2}\mathcal{K}', \qquad & u &= +\mathrm{i}\,\zeta(\bar{u}; \psi) + \tfrac{1}{2}\mathcal{K}', \\ u &= +\mathrm{i}\,\zeta(\bar{u}; \psi) - \tfrac{1}{2}\mathcal{K}', \qquad & u &= -\mathrm{i}\,\zeta(\bar{u}; \psi) + \tfrac{1}{2}\mathcal{K}'. \end{aligned} \tag{12.2.114}$$

If we take $u^{(s)}_j(\bar{u}; m) = -\mathrm{i}\,\zeta(\bar{u}; \psi^{(s)}_j) \pm \frac{1}{2}\mathcal{K}'$ in the numerator in (12.2.100), then, since $\zeta(-\bar{u}; \psi^{(s)}_j) = \zeta(\bar{u}; \psi^{(s)}_j)$, the root in $\mathbb{W}'(2\mathrm{i}\mathcal{K}, 2\mathcal{K}')$ in the numerator in (12.2.103) is $u = -\mathrm{i}\,\zeta(\bar{u}; \psi^{(s)}_j) \mp \frac{1}{2}\mathcal{K}'$. Thus we have an independent choice of signs for each $\psi^{(s)}_j$. Let

$$u^{(s)}_j(\bar{u}; m) = -\mathrm{i}\,\zeta(\bar{u}; \psi^{(s)}_j) - \tfrac{1}{2}\gamma_j \mathcal{K}', \qquad j = 1, 2, \ldots, 2n_2, \tag{12.2.115}$$

for some particular choice of γ_j, equal to -1 or $+1$, for $j = 1, 2, \ldots, 2n_2$. Noting that the choice of γ_j for the term with $\pi - \psi^{(s)}_j$ is independent of that for $\psi^{(s)}_j$, an eigenvalue is identified by a set of γ's. The number of choices is subject to the constraint (12.2.101), which now takes the form

$$\frac{1}{2}\sum_{j=1}^{2n_2}(1 - \gamma_j) = n_2 + \aleph^{(s)}(\bar{u}; m). \tag{12.2.116}$$

So $n_2 + \aleph^{(s)}(\bar{u}; m)$ and $n_2 - \aleph^{(s)}(\bar{u}; m)$ are respectively the even number of $\gamma_j = -1$ and $+1$. To satisfy this equation, $2n_2 - 1$ γ_j's can be chosen independently. And (together with s) the eigenvalues can be indexed by a appropriately constrained set $\{\gamma_j\}$, which, for convenience we represent by the $2n_2$-dimensional vector $\boldsymbol{\gamma}$ that *henceforth replaces the parameter m*. Now consider the sum of the imaginary parts of the roots $u^{(s)}_j(\bar{u}; \boldsymbol{\gamma})$. For $s = 1$ they cancel in pairs and for $s = -1$ they cancel in pairs except for the terms $\zeta(\bar{u}; \psi^{(-)}_{n_2}) = 0$ and $\zeta(\bar{u}; \psi^{(-)}_{2n_2}) = \mathcal{K}$. Thus

$$\sum_{j=1}^{2n_2} \zeta(\bar{u}; \psi_j^{(s)}) = \tfrac{1}{2}(1-s)\mathcal{K}. \tag{12.2.117}$$

This is just a stronger version of (12.2.102). The condition

$$\prod_{\{\psi\}^{(s)}} \mathrm{k}\,\mathrm{sn}\big(\mathrm{i}\,\bar{u} \pm \zeta(\bar{u}; \psi) + \tfrac{1}{2}\mathrm{i}\,\mathcal{K}'|\mathrm{k}\big)\mathrm{sn}\big(\mathrm{i}\,\bar{u} \pm \zeta(\bar{u}; \psi) - \tfrac{1}{2}\mathrm{i}\,\mathcal{K}'|\mathrm{k}\big) = 1, \tag{12.2.118}$$

where the product is over the values of ψ given by (12.2.108), follows directly from (18.3.50) when the choice of sign in each term is the same. However, the condition $\zeta(\bar{u}; \pi - \psi) = -\zeta(\bar{u}; \psi)$ allows independent choices of signs in the two terms. From (12.2.100), (12.2.115) and (12.2.116),

$$\Upsilon^{(s)}(\bar{u}, u; \boldsymbol{\gamma}|\mathrm{k}) = A_1^{(s)}(\bar{u}; \boldsymbol{\gamma}) \prod_{\{\psi\}^{(s)}} \exp\left(-\frac{\pi \gamma_\psi u}{4\mathcal{K}}\right) \frac{\mathrm{H}(\mathrm{i}\,u - \zeta(\bar{u}; \psi) + \frac{1}{2}\gamma_\psi\,\mathrm{i}\,\mathcal{K}'|\mathrm{k})}{\sqrt{\mathrm{Q}(\bar{u}, u|\mathrm{k})}} \tag{12.2.119}$$

and, from (18.3.25),

$$\Upsilon^{(s)}(\bar{u}, u; \boldsymbol{\gamma}|\mathrm{k}) = A_2^{(s)}(\bar{u}; \boldsymbol{\gamma}) \prod_{\{\psi\}^{(s)}} \exp\left(\frac{\pi \gamma_\psi u}{4\mathcal{K}}\right) \frac{\Theta(\mathrm{i}\,u - \zeta(\bar{u}; \psi) - \frac{1}{2}\gamma_\psi\,\mathrm{i}\,\mathcal{K}'|\mathrm{k})}{\sqrt{\mathrm{Q}(\bar{u}, u|\mathrm{k})}}, \tag{12.2.120}$$

where

$$A_2^{(s)}(\bar{u}; \boldsymbol{\gamma}) = (-1)^{n_2} A_1^{(s)}(\bar{u}; \boldsymbol{\gamma}) \prod_{\{\psi\}^{(s)}} \exp\left(\frac{\pi\,\mathrm{i}\,\zeta(\bar{u}; \psi)\gamma_\psi}{2\mathcal{K}}\right). \tag{12.2.121}$$

From (12.2.95), (18.3.44) and (18.3.100),

$$\Upsilon^{(s)}(\bar{u}, u; \boldsymbol{\gamma}|\mathrm{k})\Upsilon^{(s)}(-\bar{u}, u; \boldsymbol{\gamma}|\mathrm{k}) = X_1^{(s)}(\bar{u}, u; \boldsymbol{\gamma}|\mathrm{k}) \times \prod_{\{\psi\}^{(s)}} \sqrt{\mathrm{k}}\,\mathrm{sn}(\mathrm{i}\,u - \zeta(\bar{u}; \psi) + \tfrac{1}{2}\gamma_\psi\,\mathrm{i}\,\mathcal{K}'|\mathrm{k}), \tag{12.2.122}$$

where

$$X_1^{(s)}(\bar{u}, u; \boldsymbol{\gamma}|\mathrm{k}) := A_1^{(s)}(\bar{u}; \boldsymbol{\gamma})A_2^{(s)}(-\bar{u}; \boldsymbol{\gamma}) \times \prod_{\{\psi\}^{(s)}} \frac{\Theta(\mathrm{i}\,u - \zeta(\bar{u}; \psi) + \frac{1}{2}\mathrm{i}\,\mathcal{K}'|\mathrm{k})\,\Theta\,(\mathrm{i}\,u - \zeta(\bar{u}; \psi) - \frac{1}{2}\mathrm{i}\,\mathcal{K}'|\mathrm{k})}{\sqrt{\mathrm{Q}(\bar{u}, u|\mathrm{k})\mathrm{Q}(-\bar{u}, u|\mathrm{k})}}. \tag{12.2.123}$$

$X_1^{(s)}(\bar{u},u;\boldsymbol{\gamma}|\mathrm{k}) \in \mathrm{DP}(2\mathrm{i}\,\mathcal{K}, 2\mathcal{K}'; 1, 1)$ and it can be seen from (18.3.108), (18.3.110) and (18.3.118) that it has the same poles and zeros as $\mathrm{F}^{(s)}(\bar{u},u|\mathrm{k})$, defined by (18.3.106). Thus, from Theorem 18.2.4 and (18.3.107), $X_1^{(s)}(\bar{u},u;\boldsymbol{\gamma}|\mathrm{k})$ can be replaced by

$$X_2^{(s)}(\bar{u},u;\boldsymbol{\gamma}|\mathrm{k}) := A_3^{(s)}(\bar{u};\boldsymbol{\gamma})\left\{\left[\frac{\mathrm{sn}(\mathrm{i}[\mathcal{K}'-u-\bar{u}]|\mathrm{k})}{\mathrm{sn}(\mathrm{i}[u-\bar{u}]|\mathrm{k})}\right]^{n_2} + s\left[\frac{\mathrm{sn}(\mathrm{i}[u-\bar{u}]|\mathrm{k})}{\mathrm{sn}(\mathrm{i}[\mathcal{K}'-u-\bar{u}]|\mathrm{k})}\right]^{n_2}\right\}. \tag{12.2.124}$$

From (12.2.107)–(12.2.109), (12.2.122) and (12.2.124)

$$\Upsilon^{(s)}(\bar{u};\bar{u};\boldsymbol{\gamma}|\mathrm{k})\Upsilon^{(s)}(-\bar{u};\bar{u};\boldsymbol{\gamma}|\mathrm{k})\Upsilon^{(s)}(\bar{u};\bar{u}+\mathcal{K}';\boldsymbol{\gamma}|\mathrm{k})\Upsilon^{(s)}(-\bar{u};\bar{u}+\mathcal{K}';\boldsymbol{\gamma}|\mathrm{k})$$

$$\simeq \begin{cases} \dfrac{1}{\mathrm{k}^{2n_2}}\left[\dfrac{\Theta(0|\mathrm{k})\,\Theta\,(2\mathrm{i}\,\bar{u}|\mathrm{k})}{\mathrm{H}(0|\mathrm{k})\mathrm{H}(\mathrm{i}\,\bar{u}|\mathrm{k})}\right]^{2n_2} \\ \qquad \text{and} \\ \dfrac{s}{\mathrm{k}^{2n_2}}[A_3^{(s)}(\bar{u};\boldsymbol{\gamma})]^2\left[\dfrac{\Theta(0|\mathrm{k})\,\Theta\,(2\mathrm{i}\,\bar{u}|\mathrm{k})}{\mathrm{H}(0|\mathrm{k})\mathrm{H}(\mathrm{i}\,\bar{u}|\mathrm{k})}\right]^{2n_2}, \end{cases} \tag{12.2.125}$$

from which it follows that $A_3^{(s)}(\bar{u};\boldsymbol{\gamma}) = 1/\sqrt{s}$. So, from (12.2.95), (12.2.122), (18.3.100) and (12.2.124),

$$\Lambda_{\mathrm{A}}^{(s)}(\bar{u},u;\boldsymbol{\gamma}|\mathrm{k})\Lambda_{\mathrm{B}}^{(s)}(\bar{u},u;\boldsymbol{\gamma}|\mathrm{k}) = \frac{(-1)^{n_2}}{\sqrt{s}}\,\mathrm{F}^{(s)}(\bar{u},u|\mathrm{k})\prod_{\{\psi\}^{(s)}}\mathrm{S}(\bar{u},u;\psi;\alpha,\gamma_\psi|\mathrm{k}), \tag{12.2.126}$$

where

$$\mathrm{S}(\bar{u};v;\psi;\alpha,\gamma) := \sqrt{\mathrm{k}}\,\mathrm{sn}(\alpha\,\mathrm{i}\,v-\zeta(\bar{u};\psi)+\tfrac{1}{2}\gamma\,\mathrm{i}\,\mathcal{K}'|\mathrm{k}), \quad \alpha,\gamma=\pm 1. \tag{12.2.127}$$

The parameter $\alpha = \pm 1$ can be chosen with either sign in (12.2.126) (see (12.2.129)).

12.2.6.4 The General Case

The remaining task is to obtain an expression for $\mathrm{S}(\bar{u};v;\psi;\alpha,\gamma|\mathrm{k})$ using (12.2.110). Let

$$\mathrm{C}(\bar{u};v;\psi;\alpha,\gamma|\mathrm{k}) := \mathrm{cds}(\alpha\,\mathrm{i}\,v-\zeta(\bar{u};\psi)+\tfrac{1}{2}\gamma\,\mathrm{i}\,\mathcal{K}'|\mathrm{k}), \quad \alpha,\gamma=\pm 1. \tag{12.2.128}$$

From (18.3.50), (18.3.51) and (18.3.116),[28]

$$\begin{aligned} S(\bar{u};\, v;\, \psi;\, \alpha, -\gamma|k) &= [S(\bar{u};\, v;\, \psi;\, \alpha, \gamma|k)]^{-1}, \\ S(\bar{u};\, v;\, \pi-\psi;\, \alpha, -\gamma|k) &= -\mathcal{C}\{S(\bar{u};\, v;\, \psi;\, \alpha, -\gamma|k)\} \\ &= -S(\bar{u};\, v;\, \psi;\, -\alpha, \gamma|k), \end{aligned} \tag{12.2.129}$$

$$\begin{aligned} C(\bar{u};\, v;\, \psi;\, \alpha, -\gamma|k) &= -C(\bar{u};\, v;\, \psi;\, \alpha, \gamma|k), \\ C(\bar{u};\, v;\, \pi-\psi;\, \alpha, -\gamma|k) &= -\mathcal{C}\{C(\bar{u};\, v;\, \psi;\, \alpha, -\gamma|k)\} \\ &= -C(\bar{u};\, v;\, \psi;\, -\alpha, \gamma|k). \end{aligned} \tag{12.2.130}$$

Equation (12.2.110), which can now be expressed in the form[29]

$$\overline{S}(\bar{u};\, \psi;\, \alpha, \gamma|k)\overline{S}(\bar{u};\, \psi;\, -\alpha, \gamma|k) = \exp(2\gamma\, i\, \psi), \tag{12.2.131}$$

is expanded using (18.3.57) to give

$$\begin{aligned} \overline{C}(\bar{u};\, \psi;\, \alpha, \gamma|k) &= \alpha\, \mathrm{cds}(2i\,\bar{u}|k) \\ &\quad + \alpha\, \frac{\exp(2\gamma\, i\, \psi)[k\, \mathrm{sn}^2(2i\,\bar{u}|k) - \overline{S}^2(\bar{u};\, \psi;\, \alpha, -\gamma|k)]}{\mathrm{sn}(2i\,\bar{u}|k)}. \end{aligned} \tag{12.2.132}$$

Using the first of Eqs. (12.2.129) and (12.2.130) yields a quadratic for $\overline{S}^2(\bar{u};\, \psi;\, \alpha, \gamma|k)$ and hence a formula for $\overline{C}(\bar{u};\, \psi;\, \alpha, \gamma|k)$. The solution of the former must be chosen to satisfy (12.2.129) and hence

$$\overline{S}(\bar{u};\, \psi;\, \alpha, \gamma|k) = \exp(\gamma\, i\, \psi)\left\{W(\bar{u};\, \psi|k) + \alpha\gamma\sqrt{W^2(\bar{u};\, \psi|k) - 1}\right\}^{\frac{1}{2}}, \tag{12.2.133}$$

$$\overline{C}(\bar{u};\, \psi;\, \alpha, \gamma|k) = \frac{\gamma\{\alpha\, i\, k\, \mathrm{sn}^2(2i\,\bar{u}|k)\sin(2\psi) + \sqrt{W^2(\bar{u};\, \psi|k) - 1}\}}{\mathrm{sn}(2i\,\bar{u}|k)}, \tag{12.2.134}$$

where

$$W(\bar{u};\, \psi|k) := \mathrm{cn}(2i\,\bar{u}|k)\mathrm{dn}(2i\,\bar{u}|k) + k\, \mathrm{sn}^2(2i\,\bar{u}|k)\cos(2\psi). \tag{12.2.135}$$

---

[28] To avoid confusion with the overline notation the complex conjugate is here denoted by $\mathcal{C}\{\cdot\}$.

[29] The notation is simplified for S and C by denoting them by $\overline{S}$ and $\overline{C}$, respectively, when the first two arguments (compressed to one) are $\bar{u}$.

Translating into coupling form using (12.2.75)–(12.2.78),[30]

$$W(K^{(h)}, K^{(ch)}; \psi) = \frac{\cosh(2K^{(h)})\cosh(2K^{(ch)}) - \sinh(2K^{(h)})\cos(2\psi)}{\sinh(2K^{(ch)})}. \tag{12.2.136}$$

With

$$\begin{aligned}\omega(K^{(h)}, K^{(ch)}; \psi) &:= \ln\left[W(K^{(h)}, K^{(ch)}; \psi) + \sqrt{W^2(K^{(h)}, K^{(ch)}; \psi) - 1}\right],\\ &= \operatorname{arccosh}[W(K^{(h)}, K^{(ch)}; \psi)],\end{aligned} \tag{12.2.137}$$

(12.2.133) and (12.2.134) become

$$\overline{S}(K^{(h)}, K^{(ch)}; \psi; \alpha, \gamma) = \exp\left(\gamma\left[i\psi + \tfrac{1}{2}\alpha\,\omega(K^{(h)}, K^{(ch)}; \psi)\right]\right), \tag{12.2.138}$$

$$\overline{C}(K^{(h)}, K^{(ch)}; \psi; \alpha, \gamma) = \gamma\left\{\frac{\sinh(\omega(K^{(h)}, K^{(ch)}; \psi))}{\sinh(2K^{(h)})} - \frac{\alpha\sin(2\psi)}{\sinh(2K^{(ch)})}\right\}. \tag{12.2.139}$$

From (18.3.57) and (12.2.127),

$$S(\bar{u}, u; \psi; 1, \gamma|k) = \overline{S}(\bar{u}; \psi; \alpha, \gamma|k)D(\bar{u}, u; \psi; \alpha, \gamma|k), \tag{12.2.140}$$

for both $\alpha = \pm 1$, where

$$D(\bar{u}, u; \psi; \alpha, \gamma|k) := \frac{\left\{\begin{array}{l}\mathrm{cn}(i[u - \alpha\bar{u}]|k)\mathrm{dn}(i[u - \alpha\bar{u}]|k)\\ +\overline{C}(\bar{u}; \psi; \alpha, \gamma|k)\mathrm{sn}(i[u - \alpha\bar{u}]|k)\end{array}\right\}}{1 - k\,\overline{S}^2(\bar{u}; \psi; \alpha, \gamma|k)\,\mathrm{sn}^2(i[u - \alpha\bar{u}]|k)}. \tag{12.2.141}$$

Then, from (12.2.129) and (12.2.140),

$$D(\bar{u}, u; \psi; \alpha, \gamma|k)D(\bar{u}, u; \psi; \alpha, -\gamma|k) = 1,$$

$$D(\bar{u}, u; \psi; \alpha, \gamma|k)D(\bar{u}, u; \psi; -\alpha, -\gamma|k) = -\frac{1}{\overline{S}(\bar{u}; \psi; \alpha, \gamma|k)\overline{S}(\bar{u}; \pi - \psi; \alpha, \gamma|k)}. \tag{12.2.142}$$

Translating into coupling form using (12.2.75)–(12.2.78),

[30] Henceforth we shall interchange freely between elliptic and coupling arguments.

$$\mathsf{D}(K_{\mathrm{A}}^{(\mathrm{f})}, K_{\mathrm{B}}^{(\mathrm{b})}, K^{(\mathrm{h})}, K^{(\mathrm{ch})}; \psi; 1, 1) = \frac{\left\{\begin{matrix} \cosh(2K_{\mathrm{A}}^{(\mathrm{f})})\cosh(2K_{\mathrm{B}}^{(\mathrm{b})}) \\ -\mathrm{i}\,\overline{\mathsf{C}}(K^{(\mathrm{h})}, K^{(\mathrm{ch})}; \psi; 1, 1)\sinh(2K_{\mathrm{A}}^{(\mathrm{f})})\sinh(2K_{\mathrm{B}}^{(\mathrm{b})}) \end{matrix}\right\}}{\sinh(2K_{\mathrm{A}}^{(\mathrm{f})}) + \overline{\mathsf{S}}^2(K^{(\mathrm{h})}, K^{(\mathrm{ch})}; \psi; 1, 1)\sinh(2K_{\mathrm{B}}^{(\mathrm{b})})}, \tag{12.2.143}$$

$$\mathsf{D}(K_{\mathrm{B}}^{(\mathrm{f})}, K_{\mathrm{A}}^{(\mathrm{b})}, K^{(\mathrm{h})}, K^{(\mathrm{ch})}; \psi; -1, 1) = \frac{\left\{\begin{matrix} \cosh(2K_{\mathrm{B}}^{(\mathrm{f})})\cosh(2K_{\mathrm{A}}^{(\mathrm{b})}) \\ -\mathrm{i}\,\overline{\mathsf{C}}(K^{(\mathrm{h})}, K^{(\mathrm{ch})}; \psi; -1, 1)\sinh(2K_{\mathrm{B}}^{(\mathrm{f})})\sinh(2K_{\mathrm{A}}^{(\mathrm{b})}) \end{matrix}\right\}}{\sinh(2K_{\mathrm{B}}^{(\mathrm{f})}) + \overline{\mathsf{S}}^2(K^{(\mathrm{h})}, K^{(\mathrm{ch})}; \psi; -1, 1)\sinh(2K_{\mathrm{A}}^{(\mathrm{b})})}. \tag{12.2.144}$$

Finally, from (12.2.133) and (12.2.140),

$$\mathsf{S}(\bar{u}, u; \psi; 1, \gamma|\mathrm{k}) := \mathrm{i}\exp(\gamma\,\mathrm{i}\,\psi)\{\mathsf{D}(\bar{u}, u; \psi; 1, 1|\mathrm{k})\mathsf{D}(\bar{u}, u; \psi; -1, 1|\mathrm{k})\}^{\frac{1}{2}\gamma}, \tag{12.2.145}$$

which yields an explicit form for $\Lambda_{\mathrm{B}}^{(s)}(\bar{u}, u; \boldsymbol{\gamma}|\mathrm{k})\Lambda_{\mathrm{A}}^{(s)}(\bar{u}, u; \boldsymbol{\gamma}|\mathrm{k})$ either in terms of elliptic variables or couplings.

Baxter (1986, p. 22) gives separate formulae for $\Lambda_{\mathrm{B}}^{(s)}$ and $\Lambda_{\mathrm{A}}^{(s)}$ in the more general case $\varrho_{\mathrm{E}} \neq \varrho_{\mathrm{U}}$. To compare with the results given here it is necessary to combine the terms for ψ and $\pi - \psi$ in the product in (12.2.126) and expand $\mathsf{F}^{(s)}(\bar{u}, u|\mathrm{k})$ using (18.3.108)–(18.3.110).

The results given by Baxter (1982b, Sect. 7.9) for the regular model $\bar{u} = 0$ can be easily extracted. It can also be shown, as a confirmation of the results of Sect. 12.2.3, that the general model is given, for large n_2, as a combination of regular models with parameter u replaced by $u + \bar{u}$ and $u - \bar{u}$ respectively.

12.2.7 The Standard Model

The eigenvalues for the standard Ising model first solved by Onsager (1944) can be extracted from this analysis if, in one of the two pairs of diagonal couplings $(K_{\mathrm{B}}^{(\mathrm{f})}, K_{\mathrm{A}}^{(\mathrm{b})})$ and $(K_{\mathrm{A}}^{(\mathrm{f})}, K_{\mathrm{B}}^{(\mathrm{b})})$ in Fig. 12.5, one member is taken to the limit of zero and the other to infinity. The latter has the effect of binding the spin on the centre site to that of the corner site, being an effective merging of the two sites. This produces a square-lattice model with horizontal coupling $K^{(\mathrm{h})}$ and vertical coupling $K^{(\mathrm{ch})}$ except that the 'vertical' lattice lines are translated so that they wind around the lattice-column. The lattice now has $2n_1$ rows of $2n_2$ sites and we return to the notation of Chap. 11 by using the variables $N_1 := 2n_1$ and $N_2 := 2n_2$. In order for there to be a periodic boundary condition in the vertical direction it must be the case that N_1 is some multiple of N_2.

The reduction to the standard model can be carried out in four ways. Referring to (12.2.76), (12.2.78) and (12.2.83):

$$\begin{aligned}
&\text{(i) } u \to \bar{u}: && K^{(\mathrm{f})}_{\mathrm{A}} \to \infty,\ K^{(\mathrm{b})}_{\mathrm{B}} \to 0,\ K^{(\mathrm{f})}_{\mathrm{B}} \to K^{(\mathrm{ch})},\ K^{(\mathrm{b})}_{\mathrm{A}} \to K^{(\mathrm{h})};\\
&\text{(ii) } u \to -\bar{u}: && K^{(\mathrm{f})}_{\mathrm{B}} \to \infty,\ K^{(\mathrm{b})}_{\mathrm{A}} \to 0,\ K^{(\mathrm{f})}_{\mathrm{A}} \to K^{(\mathrm{ch})},\ K^{(\mathrm{b})}_{\mathrm{B}} \to K^{(\mathrm{h})};\\
&\text{(iii) } u \to \mathcal{K}' - \bar{u}: && K^{(\mathrm{b})}_{\mathrm{A}} \to \infty,\ K^{(\mathrm{f})}_{\mathrm{B}} \to 0,\ K^{(\mathrm{b})}_{\mathrm{B}} \to K^{(\mathrm{ch})},\ K^{(\mathrm{f})}_{\mathrm{A}} \to K^{(\mathrm{h})};\\
&\text{(iv) } u \to \mathcal{K}' + \bar{u}: && K^{(\mathrm{b})}_{\mathrm{B}} \to \infty,\ K^{(\mathrm{f})}_{\mathrm{A}} \to 0,\ K^{(\mathrm{b})}_{\mathrm{A}} \to K^{(\mathrm{ch})},\ K^{(\mathrm{f})}_{\mathrm{B}} \to K^{(\mathrm{h})}.
\end{aligned}$$

Of course, this limit will lead to an infinite contribution to the eigenvalues of the transfer matrix.

In case (i), from (12.2.10), (12.2.11) and (18.3.16), the infinite contribution is

$$\exp(N_2 K^{(\mathrm{f})}_{\mathrm{A}}) \simeq \left[\frac{-8\mathcal{K}\,\mathrm{k}'}{\pi\,\mathrm{k}\,\mathrm{H}^2(0|\mathrm{k})}\right]^{\frac{1}{4}N_2}. \tag{12.2.146}$$

From (12.2.79), the elliptic modulus is

$$\mathrm{k}(K^{(\mathrm{h})}, K^{(\mathrm{ch})}) = [\sinh(2K^{(\mathrm{h})})\sinh(2K^{(\mathrm{ch})})]^{-1} \tag{12.2.147}$$

and, from (18.3.106),

$$\mathrm{F}^{(s)}(\bar{u}, u|\mathrm{k}) \simeq \exp(N_2 K^{(\mathrm{f})}_{\mathrm{A}}) \left[-\frac{\mathrm{k}\,\sinh(2K^{(\mathrm{ch})})}{2}\right]^{\frac{1}{2}N_2}. \tag{12.2.148}$$

The eigenvalues are the products of the A and B eigenvalues of the general model. From (12.2.126), (12.2.138) and (12.2.146),

$$\begin{aligned}
\Lambda^{(s)}(N_2, K^{(\mathrm{h})}, K^{(\mathrm{ch})}; \boldsymbol{\gamma}) &:= \frac{\Lambda^{(s)}_{\mathrm{B}}(\bar{u}; \bar{u}; \boldsymbol{\gamma}|\mathrm{k})\,\Lambda^{(s)}_{\mathrm{A}}(\bar{u}; \bar{u}; \boldsymbol{\gamma}|\mathrm{k})}{\exp(N_2 K^{(\mathrm{f})}_{\mathrm{A}})}\\
&= \ell^{(s)}(\boldsymbol{\gamma})\left(\frac{\mathrm{k}}{2}\right)^{\frac{1}{2}N_2} [2\,\sinh(2K^{(\mathrm{ch})})]^{\frac{1}{2}N_2} \prod_{\{\psi\}^{(s)}} \exp\left(\tfrac{1}{2}\,\gamma_\psi\,\alpha\,\omega(K^{(\mathrm{h})}, K^{(\mathrm{ch})}; \psi)\right),
\end{aligned} \tag{12.2.149}$$

where

$$\ell^{(s)}(\boldsymbol{\gamma}) := \prod_{\{\psi\}^{(s)}} \exp([\gamma_\psi - 1]\mathrm{i}\,\psi). \tag{12.2.150}$$

This is the result derived by Onsager (1944) apart from:

(1) The constant factors $\ell^{(s)}(\boldsymbol{\gamma})$, for $s = \pm 1$ and all $\boldsymbol{\gamma}$, which are the eigenvalues of the symmetry matrix $\boldsymbol{C}$ defined in Sect. 12.2.2. Their presence is a consequence of the winding of the lattice lines around the vertical cylinder.[31] Since N_1 is a multiple of N_2, $[\ell^{(s)}(\boldsymbol{\gamma})]^{N_1} = 1$, so this factor does not affect the free energy of the system.

(2) The factor $(\mathrm{k}/4)^{\frac{1}{2}N_2}$ is a consequence of the division of the matrix elements by $\mathfrak{C}(\bar{\varrho}_{\mathrm{E}}, \bar{\varrho}_{\mathrm{O}}; \varrho_{\mathrm{E}}, \varrho_{\mathrm{O}})$ (see the footnote on p. 402).

12.2.7.1 Analysis of the Functions $\omega(K^{(\mathrm{h})}, K^{(\mathrm{ch})}; \psi)$

The sign of α in (12.2.149) is chosen so that $\alpha\,\omega(K^{(\mathrm{h})}, K^{(\mathrm{ch})}; \psi)$ is positive at low temperatures and has a continuous derivative for all temperatures.[32] It is now convenient to use the variables $\mathfrak{X}_{\mathrm{h}} := \exp(2K^{(\mathrm{h})})$ and $\mathfrak{X}_{\mathrm{ch}} := \exp(2K^{(\mathrm{ch})})$.[33] From (12.2.136),

$$W(\mathfrak{X}_{\mathrm{h}}, \mathfrak{X}_{\mathrm{ch}}; \psi) \pm 1 = \begin{cases} \dfrac{(\mathfrak{X}_{\mathrm{h}}\mathfrak{X}_{\mathrm{ch}} + \mathfrak{X}_{\mathrm{h}} \pm \mathfrak{X}_{\mathrm{ch}} \mp 1)^2}{2\mathfrak{X}_{\mathrm{h}}(\mathfrak{X}_{\mathrm{ch}}^2 - 1)} - \dfrac{\mathfrak{X}_{\mathrm{ch}}(\mathfrak{X}_{\mathrm{h}}^2 - 1)}{\mathfrak{X}_{\mathrm{h}}(\mathfrak{X}_{\mathrm{ch}}^2 - 1)}[\cos(2\psi) + 1], \\ \text{or} \\ \dfrac{(\mathfrak{X}_{\mathrm{h}}\mathfrak{X}_{\mathrm{ch}} - \mathfrak{X}_{\mathrm{h}} \pm \mathfrak{X}_{\mathrm{ch}} \pm 1)^2}{2\mathfrak{X}_{\mathrm{h}}(\mathfrak{X}_{\mathrm{ch}}^2 - 1)} - \dfrac{\mathfrak{X}_{\mathrm{ch}}(\mathfrak{X}_{\mathrm{h}}^2 - 1)}{\mathfrak{X}_{\mathrm{h}}(\mathfrak{X}_{\mathrm{ch}}^2 - 1)}[\cos(2\psi) - 1]. \end{cases} \tag{12.2.151}$$

It follows that the quantity under the square root sign defining $\omega(\mathfrak{X}_{\mathrm{h}}, \mathfrak{X}_{\mathrm{ch}}; \psi)$ in (12.2.137) is a perfect square if $\psi = 0$ or $\psi = \frac{1}{2}\pi$. We observe that

$$\mathfrak{X}_{\mathrm{h}}\mathfrak{X}_{\mathrm{ch}} - \mathfrak{X}_{\mathrm{h}} \pm \mathfrak{X}_{\mathrm{ch}} \mp 1 > 0, \qquad \mathfrak{X}_{\mathrm{h}}\mathfrak{X}_{\mathrm{ch}} - \mathfrak{X}_{\mathrm{h}} + \mathfrak{X}_{\mathrm{ch}} + 1 > 0 \tag{12.2.152}$$

and

$$\mathfrak{X}_{\mathrm{h}}\mathfrak{X}_{\mathrm{ch}} - \mathfrak{X}_{\mathrm{h}} - \mathfrak{X}_{\mathrm{ch}} - 1 \gtrless 0, \qquad T \lessgtr T_{\mathrm{c}}(\tau), \tag{12.2.153}$$

where, with $\tau = K^{(\mathrm{ch})}/K^{(\mathrm{h})}$, $T_{\mathrm{c}}(\tau)$ is the critical temperature given by

$$\sinh(2K^{(\mathrm{h})})\sinh(2K^{(\mathrm{ch})}) = 1, \tag{12.2.154}$$

---

[31] If $\boldsymbol{V}$ is the transfer matrix of the standard spin-$\frac{1}{2}$ Ising model (Sect. 11.4.3) then the transfer matrix of the present model is (apart from the factors $(\mathrm{k}/2)^{\frac{1}{2}N_2}$) $\boldsymbol{V}\boldsymbol{C}^{\dagger}$.

[32] And (to emphasis an obvious, but important, point) the surd in the first version of (12.2.137) is the *positive* square root.

[33] Henceforth, for the functions displayed, we freely interchange between arguments $(K^{(\mathrm{h})}, K^{(\mathrm{ch})})$ and $(\mathfrak{X}_{\mathrm{h}}, \mathfrak{X}_{\mathrm{ch}})$, whichever seems more appropriate.

for the square-lattice model with unequal couplings.[34] It is also evident from (12.2.147) that $\mathrm{k} = 1$ corresponds to the critical point in the formulation in terms of elliptic variables, with $0 < \mathrm{k} \le 1$ corresponding to $T \le T_{\mathrm{c}}$. From (12.2.137), (12.2.151) and (12.2.153),

$$\omega(\mathfrak{X}_{\mathrm{h}}, \mathfrak{X}_{\mathrm{ch}}; 0) = \pm \ln\left[\frac{\mathfrak{X}_{\mathrm{h}}(\mathfrak{X}_{\mathrm{ch}} - 1)}{\mathfrak{X}_{\mathrm{ch}} + 1}\right], \tag{12.2.155}$$

where the plus and minus signs apply respectively to the regions $T < T_{\mathrm{c}}(\tau)$ and $T > T_{\mathrm{c}}(\tau)$. This means that $\omega(\mathfrak{X}_{\mathrm{h}}, \mathfrak{X}_{\mathrm{ch}}; 0) \ge 0$ for all temperatures with minimum values of zero with a cusp at the critical temperature. However, we need $\alpha\,\omega(\mathfrak{X}_{\mathrm{h}}, \mathfrak{X}_{\mathrm{ch}}; 0)$ to be a smooth function of the couplings (or temperature) which is positive at low temperatures. Thus we need to choose $\alpha = 1$ for $T < T_{\mathrm{c}}(\tau)$ and $\alpha = -1$ for $T > T_{\mathrm{c}}(\tau)$ giving

$$\alpha\,\omega(\mathfrak{X}_{\mathrm{h}}, \mathfrak{X}_{\mathrm{ch}}; 0) = \ln\left[\frac{\mathfrak{X}_{\mathrm{h}}(\mathfrak{X}_{\mathrm{ch}} - 1)}{\mathfrak{X}_{\mathrm{ch}} + 1}\right], \tag{12.2.156}$$

In the case of $\psi = \frac{1}{2}\pi$, the situation is simpler since $\omega(\mathfrak{X}_{\mathrm{h}}, \mathfrak{X}_{\mathrm{ch}}; \frac{1}{2}\pi) > 0$ and, with $\alpha = 1$,

$$\alpha\,\omega\left(\mathfrak{X}_{\mathrm{h}}, \mathfrak{X}_{\mathrm{ch}}; \tfrac{1}{2}\pi\right) = \ln\left[\frac{\mathfrak{X}_{\mathrm{h}}(\mathfrak{X}_{\mathrm{ch}} + 1)}{\mathfrak{X}_{\mathrm{ch}} - 1}\right]. \tag{12.2.157}$$

It is not difficult to show that $\omega(\mathfrak{X}_{\mathrm{h}}, \mathfrak{X}_{\mathrm{ch}}; \psi) > 0$, with $\psi > 0$, is positive with a single minimum with respect to variations of T, tending to infinity, for $T \to 0$ and $T \to \infty$ and, as a function of ψ, is increasing for $0 \le \alpha \le \frac{1}{2}\pi$ with $\omega(\mathfrak{X}_{\mathrm{h}}, \mathfrak{X}_{\mathrm{ch}}; \psi) = \omega(\mathfrak{X}_{\mathrm{h}}, \mathfrak{X}_{\mathrm{ch}}; \pi - \psi)$ (Fig. 12.6).

In terms of the variable $\mathfrak{X}_{\mathrm{h}}$ and $\mathfrak{X}_{\mathrm{ch}}$ the duality transformation (9.2.32)[35] takes the form

$$\mathfrak{X}_{\mathrm{h}} \leftrightarrow \mathfrak{X}^{\star}_{\mathrm{h}} = \frac{\mathfrak{X}_{\mathrm{ch}} + 1}{\mathfrak{X}_{\mathrm{ch}} - 1}, \qquad \mathfrak{X}_{\mathrm{ch}} \leftrightarrow \mathfrak{X}^{\star}_{\mathrm{ch}} = \frac{\mathfrak{X}_{\mathrm{h}} + 1}{\mathfrak{X}_{\mathrm{h}} - 1} \tag{12.2.158}$$

and, from (12.2.136) and (12.2.137),

$$\omega(\mathfrak{X}_{\mathrm{h}}, \mathfrak{X}_{\mathrm{ch}}; \psi) = \omega(\mathfrak{X}^{\star}_{\mathrm{h}}, \mathfrak{X}^{\star}_{\mathrm{ch}}; \psi). \tag{12.2.159}$$

Thus, from this and from (12.2.156),

---

[34] In the case $K^{(\mathrm{ch})} = K^{(\mathrm{cv})} = K_{\mathrm{S}}$, (12.2.154) reduces to the critical coupling equation (9.2.36) with solution $T_{\mathrm{c}}(1) = 2.269185\,J$.

[35] Generalized to the case $K^{(\mathrm{h})} \ne K^{(\mathrm{ch})}$.

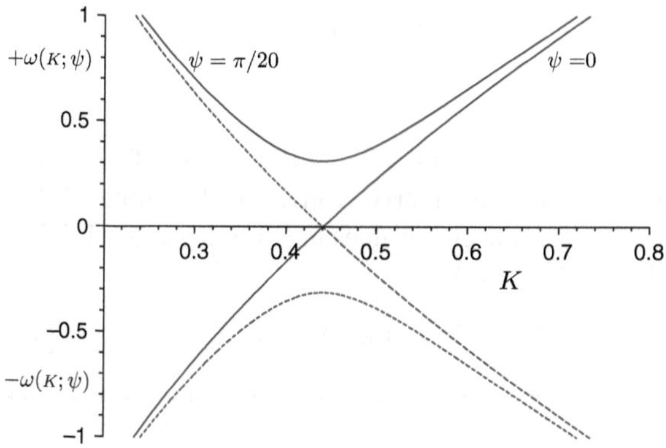

Fig. 12.6 Plots of $+\omega(K^{(\mathrm{h})},K^{(\mathrm{ch})};\psi)$ and $-\omega(K^{(\mathrm{h})},K^{(\mathrm{ch})};\psi)$ for $K^{(\mathrm{h})}=K^{(\mathrm{ch})}=K$. The branches which are $\alpha\,\omega(\mathfrak{X}_{\mathrm{h}},\mathfrak{X}_{\mathrm{ch}};\psi)$ are given by continuous lines

$$\begin{aligned}\alpha\,\omega(\mathfrak{X}_{\mathrm{h}},\mathfrak{X}_{\mathrm{ch}};\psi)&=\alpha\,\omega(\mathfrak{X}^{\star}_{\mathrm{h}},\mathfrak{X}^{\star}_{\mathrm{ch}};\psi), \qquad \psi\neq 0,\\ \alpha\,\omega(\mathfrak{X}_{\mathrm{h}},\mathfrak{X}_{\mathrm{ch}};0)&=-\alpha\,\omega(\mathfrak{X}^{\star}_{\mathrm{h}},\mathfrak{X}^{\star}_{\mathrm{ch}};0).\end{aligned} \tag{12.2.160}$$

It follows that the eigenvalues, given by (12.2.149) for $s=+1$ where the set $\{\psi\}^{(s)}$ does not include $\psi=0$, satisfy the condition

$$\frac{\Lambda^{(+)}(N_2,K^{(\mathrm{h})},K^{(\mathrm{ch})};\boldsymbol{\gamma})}{[\mathrm{k}\,\sinh(2K^{(\mathrm{ch})})]^{\frac{1}{2}N_2}}=\frac{\Lambda^{(+)}(N_2,K^{(\mathrm{h})\star},K^{(\mathrm{ch})\star};\boldsymbol{\gamma})}{[\mathrm{k}^{\star}\,\sinh(2K^{(\mathrm{ch})\star})]^{\frac{1}{2}N_2}}, \tag{12.2.161}$$

where the elliptic modulus k is the function of $K^{(\mathrm{h})}$ and $K^{(\mathrm{ch})}$ given by (12.2.147) and $\mathrm{k}^{\star}$ is the corresponding function of $K^{(\mathrm{h})\star}$ and $K^{(\mathrm{ch})\star}$.[36] The presence of the term $\psi=0$ in the product in (12.2.149) for $s=-1$ prevents a duality relationship like (12.2.161) applying in that case.

The symmetry group used in the analysis of the checkerboard Ising model in Sect. 12.2.2 is $\mathcal{C}_{N_2}\times\mathcal{S}^{(m)}_2$ and this is inherited by the present special case.[37] Block diagonalization of transfer matrices as described in Sect. 11.3 allows the division of eigenvalues into sets corresponding to the irreducible representations of the symmetry group. In this case, although this has not been explicitly evident in our method of analysis, the sets corresponding to the one-dimensional symmetric and antisymmetric representations of $\mathcal{S}^{(m)}_2$ are given respectively by $s=1$ and $s=-1$. It then follows from Theorem 11.3.4, that the largest eigenvalue is in the set $s=1$. Since, for this case, $\alpha\,\omega(K^{(\mathrm{h})},K^{(\mathrm{ch})};\psi)>0$, for all $\{\psi\}^{(+)}$,

[36] The usual Onsager result is obtained by setting $\mathrm{k}=\mathrm{k}^{\star}=2$.

[37] However, the larger symmetry group $\mathcal{C}_{N_2\mathrm{v}}\times\mathcal{S}^{(m)}_2$ of the square-lattice spin-$\frac{1}{2}$ Ising model in Sect. 11.4.3 with rings of N_2 sites and symmetry reflections through planes containing the lattice axis no longer applies because of the winding of the lattice around the cylinder.

$$\Lambda_{\max}(N_2, K^{(h)}, K^{(ch)}) = \Lambda^{(+)}(N_2, K^{(h)}, K^{(ch)}; \boldsymbol{j}), \tag{12.2.162}$$

where $\boldsymbol{j} := (1, 1, \ldots, 1)$.

Example 12.2.1 For $N_2 = 2$ and $K^{(h)} = K^{(ch)} = K$, $\mathfrak{X} = \exp(2K)$ the block-diagonalization of the transfer matrix is given in Example 11.3.1, where, in the notation used in Chap. 11, the eigenvalues in (11.3.26) and (11.3.28) are labelled by representation. From (12.2.149),

$$\Lambda^{(+)}(2, \mathfrak{X}; (+,+)) = \tfrac{1}{2}\mathrm{k}\,\Lambda_1(\mathrm{A}; 2, \mathfrak{X}), \quad \Lambda^{(+)}(2, \mathfrak{X}; (-,-)) = \tfrac{1}{2}\mathrm{k}\,\Lambda_2(\mathrm{A}; 2, \mathfrak{X}),$$
$$\Lambda^{(-)}(2, \mathfrak{X}; (+,+)) = \tfrac{1}{2}\mathrm{k}\,\Lambda_1(\mathrm{B}_1; 2, \mathfrak{X}), \quad \Lambda^{(-)}(2, \mathfrak{X}; (-,-)) = -\tfrac{1}{2}\mathrm{k}\,\Lambda_1(\mathrm{B}_3; 2, \mathfrak{X}). \tag{12.2.163}$$

The change in sign in the last relationship arises from $\ell^{(+)}(\pm, \pm) = \ell^{(-)}(+,+) = 1$, $\ell^{(-)}(-,-) = -1$. It is easy to see that the eigenvalues for $s = +1$ satisfy the duality relationship (12.2.161), but that those for $s = -1$ do not satisfy a corresponding relationship.

12.2.7.2 Eigenvalue Degeneracy

To determine an asymptotic expression for $\Lambda_{\max}(K^{(h)}, K^{(ch)})$ when $N_2 \to \infty$ we apply the Euler-Maclaurin asymptotic formula[38] to give

$$\sum_{j=1}^{N_2} \omega(K^{(h)}, K^{(ch)}; \pi[2j-1]/[2N_2])$$
$$\simeq \frac{N_2}{\pi}\int_0^{\pi} \omega(K^{(h)}, K^{(ch)}; \psi)\mathrm{d}\psi - 2\sum_{k=1}^{\infty} \frac{B_{2k}}{(2k)!}\omega^{(2k-1)}(K^{(h)}, K^{(ch)}; 0). \tag{12.2.164}$$

As we have observed above a consequence of Perron's theorem is that the largest eigenvalue $\Lambda_{\max}(N_2, K^{(h)}, K^{(ch)})$, for finite N_2, is non-degenerate; meaning (Sect. 11.2) that the system, even in the limit $N_1 \to \infty$, cannot exhibit a phase transition. Thus the only possibility for the occurrence of a phase transition is that, at some value

---

[38] If $w(z)$ is a smooth function for $0 \le z \le n$ then

$$\sum_{m=0}^{n} w(m) \simeq \int_0^n w(z)\mathrm{d}z + \frac{1}{2}[w(0) + w(n)] + \sum_{k=1}^{\infty} \frac{B_{2k}}{(2k)!}[w^{(2k-1)}(n) - w^{(2k-1)}(0)],$$

where B_{2k} are the even Bernoulli numbers, (Forder 1949; Goodwin 1949).

of the couplings $(K^{(h)}, K^{(ch)})$, $\Lambda_{max}(N_2, K^{(h)}, K^{(ch)})$ becomes degenerate in the limit $N_2 \to \infty$. To investigate this we need the second largest eigenvalue and we have no guarantee that this belongs to the set corresponding to the one-dimensional symmetric representation.[39] There are two possibilities:

***The second largest eigenvalue is in the one-dimensional symmetric representation*:**

For this we use the subscript 'submax' and the eigenvalue is given by reversing the signs of the first and last γ_ψ's in (12.2.149). Thus

$$\Lambda^{(+)}_{submax}(N_2, K^{(h)}, K^{(ch)}) = \Lambda^{(+)}(N_2, K^{(h)}, K^{(ch)}; (-, +, +, \cdots, +, -)), \tag{12.2.165}$$

giving

$$\Omega^{(+)}_{submax}(N_2, K^{(h)}, K^{(ch)}) := \frac{\Lambda^{(+)}_{submax}(N_2, K^{(h)}, K^{(ch)})}{\Lambda^{(+)}_{max}(N_2, K^{(h)}, K^{(ch)})} \simeq \exp\left\{-2\omega(K^{(h)}, K^{(ch)}; 0)\right\}, \quad \text{as} \quad N_2 \to \infty. \tag{12.2.166}$$

It follows from the form of $\omega(K^{(h)}, K^{(ch)}; 0)$, described above and shown in Fig. 12.6 that $\Upsilon^{(+)}_{submax}(N_2, K^{(h)}, K^{(ch)})$ has a maximum at the coupling values which satisfy (12.2.154) and which give the critical temperature $T = T_c(\tau)$, and that this is a single point of degeneracy between $\Lambda^{(+)}_{max}(N_2, K^{(h)}, K^{(ch)})$ and $\Lambda^{(+)}_{submax}(N_2, K^{(h)}, K^{(ch)})$ in the limit $N_2 \to \infty$.

***The largest eigenvalue is in the one-dimensional anti-symmetric representation*:**

This is

$$\Lambda^{(-)}_{max}(N_2, K^{(h)}, K^{(ch)}) = \Lambda^{(-)}(N_2, K^{(h)}, K^{(ch)}; (+, \cdots, +)), \tag{12.2.167}$$

for which we need

[39] This is one of the complicating features of investigations, even for one-dimensional models with large transfer matrices, where the ratio of the two largest eigenvalues is used to locate incipient phase transitions (Lavis 1976).

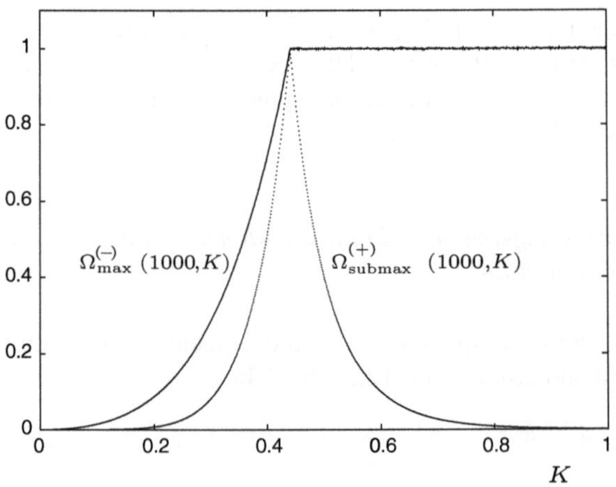

Fig. 12.7 Plots of $\Omega^{(-)}_{\max}(N_2, K^{(h)}, K^{(ch)})$ and $\Omega^{(+)}_{\text{submax}}(N_2, K^{(h)}, K^{(ch)})$ with $K^{(h)} = K^{(ch)} = K$ and $N_2 = 1000$

$$\begin{aligned}\sum_{j=1}^{N_2-1} \omega(K^{(h)}, K^{(ch)}; \pi j/N_2) \pm \omega(K^{(h)}, K^{(ch)}; 0) = \frac{N_2}{\pi}\int_0^{\pi} \omega(K^{(h)}, K^{(ch)}; \psi)\mathrm{d}\psi \\ - \omega(K^{(h)}, K^{(ch)}; 0) \pm \omega(K^{(h)}, K^{(ch)}; 0) \\ - 2\sum_{k=1}^{\infty} \frac{B_{2k}}{(2k)!}\omega^{(2k-1)}(K^{(h)}, K^{(ch)}; 0),\end{aligned} \tag{12.2.168}$$

where the plus and minus signs correspond respectively to $T > T_c(\tau)$ and $T < T_c(\tau)$, giving

$$\begin{aligned}\Omega^{(-)}_{\max}(N_2, K^{(h)}, K^{(ch)}) &:= \frac{\Lambda^{(-)}_{\max}(N_2, K^{(h)}, K^{(ch)})}{\Lambda^{(+)}_{\max}(N_2, K^{(h)}, K^{(ch)})} \\ &\simeq \begin{cases} \exp\{-\omega(K^{(h)}, K^{(ch)}; 0)\}, & T \geq T_c(\tau), \\ 1, & T \leq T_c(\tau), \end{cases} \quad \text{as } N_2 \to \infty.\end{aligned} \tag{12.2.169}$$

Thus, it has been shown that, for large N_2,

$$\Lambda^{(+)}_{\max}(N_2, K^{(h)}, K^{(ch)}) \geq \Lambda^{(-)}_{\max}(N_2, K^{(h)}, K^{(ch)}) > \Lambda^{(+)}_{\text{submax}}(N_2, K^{(h)}, K^{(ch)}), \tag{12.2.170}$$

with the equality applying in the limit $N_2 \to \infty$ when $T \leq T_c(\tau)$. An example of this for $N_2 = 1000$ and $K^{(h)} = K^{(ch)} = K$ is shown in Fig. 12.7, and it can be shown that the inequalities (but not the equality) in (12.2.170) are valid for all $N_2 > 0$.

12.2.7.3 The Free-Energy Density

We are now in the position to obtain the zero-field free-energy density for the two-dimensionally infinite system. However, the lattice differs from the 'normal' square lattice $\mathcal{L}_s$ in that the vertical lattice lines are rotated through an angle of 45°. To signify this we denote the lattice as $\mathcal{L}_{cs}$ retaining the notation $K^{(ch)}$ (rather than $K^{(v)}$) for the coupling in this direction.

Equations (11.1.11) and (11.1.14) give the zero-field free-energy density in the limit $N_1 \to \infty$ when, in the limit $N_2 \to \infty$, the largest eigenvalue of the transfer matrix is non-degenerate. However, it can also be applied to the present case since the degeneracy of the largest eigenvalue below the critical temperature, contributes a vanishingly small contribution $-\ln(2)/N$. Thus we have, from (11.1.14), (12.2.136), (12.2.137), (12.2.147) and (12.2.164),

$$\begin{aligned}\phi_1^{(IM)}(\mathcal{L}_{cs}, K^{(h)}, K^{(ch)}, 0) &= -\lim_{N_2\to\infty}\frac{1}{N_2}\ln[\Lambda^{(+)}(N_2, K^{(h)}, K^{(ch)}; \boldsymbol{j})]\\ &= -\tfrac{1}{2}\ln[\mathrm{k}(K^{(h)}, K^{(ch)})\sinh(2K^{(ch)})]\\ -\frac{1}{2\pi}\int_0^{\pi}\operatorname{arccosh}&\left\{\frac{\cosh(2K^{(h)})\cosh(K^{(ch)}) - \sinh(2K^{(h)})\cos(2\varphi)}{\sinh(2K^{(ch)})}\right\}\mathrm{d}\varphi.\end{aligned} \tag{12.2.171}$$

This formula can be re-expressed in a more symmetric form by using the identity

$$\operatorname{arccosh}(x) = \frac{1}{\pi}\int_0^{\pi}\ln[2x - 2\cos(\varphi)]\mathrm{d}\varphi \tag{12.2.172}$$

to give

$$\begin{aligned}\phi_1^{(IM)}(\mathcal{L}_{cs}, K^{(h)}, K^{(ch)}, 0) &= -\tfrac{1}{2}\ln[2\,\mathrm{k}(K^{(h)}, K^{(ch)})]\\ &\quad -\frac{1}{2\pi^2}\int_0^{\pi}\mathrm{d}\varphi_1\int_0^{\pi}\mathrm{d}\varphi_2\,\ln\{\cosh(2K^{(h)})\cosh(2K^{(ch)})\\ &\quad -\sinh(2K^{(h)})\cos(\varphi_1) + \sinh(2K^{(ch)})\cos(\varphi_2)\}.\end{aligned} \tag{12.2.173}$$

For reasons explained above, the formulae derived from the standard presentation of Onsager's result are given by replacing $\mathrm{k}(K^{(h)}, K^{(ch)})$ by 2.[40] With this identification, in the isotropic case $K^{(h)} = K^{(ch)} = K$, a change of variables $\omega_1 = \frac{1}{2}(\varphi_1 + \varphi_2)$, $\omega_2 = \frac{1}{2}(\varphi_1 - \varphi_2)$ can be used to express (12.2.173) in the form

[40] See, for example, Domb (1960, Sect. 3.5.1) or Lavis and Bell (1999a, Sect. 8.10).

$$\phi_1^{(\mathrm{IM})}(\mathcal{L}_{\mathrm{CS}}, K, 0) = -\ln[2\cosh(2K)] - \frac{1}{\pi^2}\int_0^{\frac{1}{2}\pi} \mathrm{d}\omega_2 \int_0^{\pi} \mathrm{d}\omega_1 \ln[1 - f(K)\cos(\omega_1)\cos(\omega_2)], \tag{12.2.174}$$

where

$$f(K) := \frac{2\sinh(2K)}{\cosh^2(2K)}. \tag{12.2.175}$$

The inner integral in (12.2.174) can be performed by using

$$\int_0^{\pi} \ln[y - x\cos(\omega)]\mathrm{d}\omega = \pi \ln\left[\frac{y + (y^2 - x^2)^{1/2}}{2}\right], \quad x < y, \tag{12.2.176}$$

and a trivial transformation from ω_2 to $\psi = \frac{1}{2}\pi - \omega_2$ gives

$$\phi_1^{(\mathrm{IM})}(\mathcal{L}_{\mathrm{CS}}, K, 0) = -\tfrac{1}{2}\ln[2\cosh^2(2K)] - \frac{1}{\pi}\int_0^{\frac{1}{2}\pi} \ln\left[1 + \{1 - f^2(K)\sin^2(\psi)\}^{1/2}\right]\mathrm{d}\psi. \tag{12.2.177}$$

From (2.3.27), (18.3.3) and (18.3.4), the internal energy per lattice site is

$$\mathrm{u} = -\coth(2K)\left\{1 + \tfrac{2}{\pi}\mathcal{K}(\mathrm{k}_2)[2\tanh^2(2K) - 1]\right\}, \tag{12.2.178}$$

where the elliptic modulus $\mathrm{k}_2(K) = f(K)$,[41] for which, with $K \geq 0$, $0 \leq f(K) \leq 1$, with $f(K_\mathrm{c}) = 1$. It is straightforward to show using (18.3.10) that

$$\tfrac{2}{\pi}\mathcal{K}(\mathrm{k}_2)[2\tanh^2(2K) - 1] \simeq -\frac{1}{\sqrt{2}}\Delta K \ln(\Delta K) \to 0, \quad \text{as } K \to K_\mathrm{c}, \tag{12.2.179}$$

where $\Delta K := K - K_\mathrm{c}$, and hence that $\mathrm{u}_\mathrm{c} = -\sqrt{2}$. From (2.3.28), (2.3.32) and (18.3.8), the zero-field heat capacity is

$$c_{\mathcal{H}} = \frac{k_\mathrm{B}}{\pi}[2K\coth(2K)]^2\{\mathcal{K}(\mathrm{k}_2) - \mathcal{E}(\mathrm{k}_2) - \mathrm{sech}^2(2K)[\tfrac{\pi}{2} + \mathcal{K}(\mathrm{k}_2)[2\tanh^2(2K) - 1]]\}. \tag{12.2.180}$$

---

41 The use of this notation is to give conformity with (12.3.292).

The singular contribution at the critical point is given by the leading term which, using (18.3.10), yields

$$c_{\mathcal{H}} \simeq \frac{32k_{\mathrm{B}}K_{\mathrm{c}}^2}{\pi}\ln|\Delta K|, \qquad \text{as} \quad K \to K_{\mathrm{c}}. \tag{12.2.181}$$

12.3 The Square-Lattice Eight-Vertex Model

In this discussion it will be necessary to make use of a number of different elliptic moduli, k, which is now relabelled k_0, and k_1, k_2 and k_3, related to k_0 by Landen transformations, as described in Sect. 18.3.4.[42]

The lattice is a vertical cylinder with the sites on N_1 rings of N_2 sites and the restriction in Sect. 12.1.5, that these integers must be even, can now be relaxed. In Sect. 12.1.3 it was shown, following Baxter (1978), that the free-energy density of the general Z-invariant eight-vertex model decomposes into a sum of those of regular square-lattice models. For such models all the vertical rapidities are $\bar{\varrho}$. We shall, henceforth, suppose that this is the case, using the variables u and u' given by (12.2.2). Given the couplings $(K^{(\mathrm{f})}, K^{(\mathrm{b})}, K^{(4)})$, it follows from Theorem 12.1.1, that the Z-invariance of the model implies that the equations

$$\begin{aligned} \exp[2K^{(\mathrm{f})}] &= \exp\{-\mathrm{i}\,\mathrm{am}(\mathrm{i}[\mathcal{K}'_0 - u]|\mathrm{k}_0)\}, \\ \exp[2K^{(\mathrm{b})}] &= \exp\{-\mathrm{i}\,\mathrm{am}(\mathrm{i}[\mathcal{K}'_0 - \lambda + u]|\mathrm{k}_0)\}, \\ \exp[2K^{(4)}] &= \exp\{\mathrm{i}\,\mathrm{am}(\mathrm{i}[\mathcal{K}'_0 - \lambda]|\mathrm{k}_0)\}, \end{aligned} \tag{12.3.1}$$

can be solved for u, λ and k_0. Thus the couplings $K^{(\mathrm{f})}$ and $K^{(\mathrm{b})}$ are functions of u, λ and k_0[43] and $K^{(4)}$ of λ and k_0. The Boltzmann weights $(\mathfrak{a}, \mathfrak{b}, \mathfrak{c}, \mathfrak{d})$ given by (3.8.11) have a similar dependence, with, as we have already seen in (12.1.22), Δ and Γ dependent simply on λ and k_0.

12.3.1 The Low-Temperature Zone $\mathcal{R}_{\mathrm{L}}(\mathrm{I})$

It should again be emphasized that the elliptic formulation (12.3.1) is predicated on the condition (12.1.16), which confines its applicability to $\mathcal{R}_{\mathrm{L}}(\mathrm{I})$. In the following we shall show how it can be extended to the rest of phase space. The first step is to use

---

[42] A similar notation is adopted for complete elliptic integrals and nomes.

[43] Or their transformed equivalents. Sometimes, in the interests of brevity, we shall specifically expose dependence on only one variable whose changing values are to be noted.

the variables (w_1, w_2, w_3, w_4) defined in Sect. 9.3.3 and satisfying the inequality (9.3.24). In $\mathcal{R}_L(\mathrm{I})$ these variables are related to the Boltzmann weights by

$$\begin{aligned} w_1 &= \tfrac{1}{2}(\mathrm{a}+\mathrm{b}), \qquad & w_2 &= \tfrac{1}{2}(\mathrm{a}-\mathrm{b}), \\ w_3 &= \tfrac{1}{2}(\mathrm{c}+\mathrm{d}), \qquad & w_4 &= \tfrac{1}{2}(\mathrm{c}-\mathrm{d}), \end{aligned} \tag{12.3.2}$$

(see the first line of Table 9.1) with

$$\Delta^\star = \frac{\mathrm{a}^2+\mathrm{b}^2-\mathrm{c}^2-\mathrm{d}^2}{2[\mathrm{ab}+\mathrm{cd}]} = \Delta, \quad \Gamma^\star = \frac{\mathrm{ab}-\mathrm{cd}}{\mathrm{ab}+\mathrm{cd}} = \Gamma. \tag{12.3.3}$$

Since the Boltzmann weights are functions of the arbitrary constant coupling $K^{(0)}$ their ratios are the quantities of importance and this similarly applies to the w_j's.

In the many accounts of the eight-vertex model, it has proved to be the case that elliptic forms with different moduli are convenient in different circumstances for expressing the ratios $w_1 : w_2 : w_3 : w_4$ and the variables $\Delta^\star$ and $\Gamma^\star$. Here we continue to use the elliptic modulus k_0 and its complement k_0', but from Sect. 12.3.5 onwards a more transparent formulation is achieved by using the Landen-transformed moduli k_1 and k_1'. For the sake of clarity the forms for the w_j's, $\Delta^\star$ and $\Gamma^\star$ in both formulations are given in Table 12.2.

For k_0 and k_0' more succinct expressions are achieved by using

$$\xi := \tfrac{1}{2}\lambda, \qquad z := \tfrac{1}{2}\lambda - u, \tag{12.3.4}$$

rather that λ and u. The forms for the w_j's, $\Delta^\star$ and $\Gamma^\star$ for these moduli are given in the first two columns of Table 12.2. The inequalities (9.3.24) and (9.3.31) are predicated on the quantities involved being real and, as shown in Sect. 18.3.3, this will be the case only if both members of the pair (ξ, z) are real or purely imaginary. Since the analysis of Sect. 12.1 assumes that λ and u are real, then so are ξ and z and the inequalities hold if[44]

$$|z| < \xi < \mathcal{K}_0'. \tag{12.3.5}$$

From (12.3.2), the Boltzmann weights are

[44] Note, however, that the formulation given below in terms of the variable pair $(\hat{v}, \hat{\eta})$ allows for complex values. This implies the extension of (z, ξ) to complex values.

Table 12.2 Low-temperature elliptic expressions for the ratios $w_1 : w_2 : w_3 : w_4$ and the variables $\Delta^\star$ and $\Gamma^\star$ for different moduli

Elliptic Moduli	k_0	k_0'	k_1	k_1'
w_1	1	$\dfrac{\mathrm{cn}(z\vert k_0')}{\mathrm{cn}(\xi\vert k_0')}$	$1-k_1\mathrm{sn}(\eta+v\vert k_1)\mathrm{sn}(\eta-v\vert k_1)$	$1+k_1\mathrm{sc}(i[\eta+v]\vert k_1')\mathrm{sc}(i[\eta-v]\vert k_1')$
w_2	$\dfrac{\mathrm{dn}(i\,z\vert k_0)}{\mathrm{dn}(i\,\xi\vert k_0)}$	$\dfrac{\mathrm{dn}(z\vert k_0')}{\mathrm{dn}(\xi\vert k_0')}$	$1+k_1\mathrm{sn}(\eta+v\vert k_1)\mathrm{sn}(\eta-v\vert k_1)$	$1-k_1\mathrm{sc}(i[\eta+v]\vert k_1')\mathrm{sc}(i[\eta-v]\vert k_1')$
w_3	$\dfrac{\mathrm{cn}(i\,z\vert k_0)}{\mathrm{cn}(i\,\xi\vert k_0)}$	1	$\dfrac{\mathrm{sn}(\eta+v\vert k_1)+\mathrm{sn}(\eta-v\vert k_1)}{\mathrm{sn}(2\eta\vert k_1)}$	$\dfrac{\mathrm{sc}(i[\eta+v]\vert k_1')+\mathrm{sc}(i[\eta-v]\vert k_1')}{\mathrm{sc}(2i\,\eta\vert k_1')}$
w_4	$\dfrac{\mathrm{sn}(i\,z\vert k_0)}{\mathrm{sn}(i\,\xi\vert k_0)}$	$\dfrac{\mathrm{sn}(z\vert k_0')}{\mathrm{sn}(\xi\vert k_0')}$	$\dfrac{\mathrm{sn}(\eta+v\vert k_1)-\mathrm{sn}(\eta-v\vert k_1)}{\mathrm{sn}(2\eta\vert k_1)}$	$\dfrac{\mathrm{sc}(i[\eta+v]\vert k_1')-\mathrm{sc}(i[\eta-v]\vert k_1')}{\mathrm{sc}(2i\,\eta\vert k_1')}$
$\Delta^\star$	$\mathrm{cd}(2i\,\xi\vert k_0)$	$\mathrm{nd}(2\xi\vert k_0')$	$\dfrac{\mathrm{cn}(2\eta\vert k_1)\mathrm{dn}(2\eta\vert k_1)}{1-k_1\,\mathrm{sn}^2(2\eta\vert k_1)}$	$\dfrac{\mathrm{nc}(2i\,\eta\vert k_1')\mathrm{dc}(2i\,\eta\vert k_1')}{1+k_1\,\mathrm{sc}^2(2i\,\eta\vert k_1')}$
$\Gamma^\star$	$-\mathrm{nd}(2i\,\xi\vert k_0)$	$-\mathrm{cd}(2\xi\vert k_0')$	$-\dfrac{1+k_1\,\mathrm{sn}^2(2\eta\vert k_1)}{1-k_1\,\mathrm{sn}^2(2\eta\vert k_1)}$	$-\dfrac{1-k_1\,\mathrm{sc}^2(2i\,\eta\vert k_1')}{1+k_1\,\mathrm{sc}^2(2i\,\eta\vert k_1')}$

$$
\begin{aligned}
a(z,\xi\,|k_0) &= \frac{\mathrm{dn}(i\,\xi\,|k_0)+\mathrm{dn}(i\,z|k_0)}{\mathrm{dn}(i\,\xi\,|k_0)},\\
b(z,\xi\,|k_0) &= \frac{\mathrm{dn}(i\,\xi\,|k_0)-\mathrm{dn}(i\,z|k_0)}{\mathrm{dn}(i\,\xi\,|k_0)},\\
c(z,\xi\,|k_0) &= \frac{\mathrm{sn}(i\,\xi\,|k_0)\mathrm{cn}(i\,z|k_0)+\mathrm{sn}(i\,z|k_0)\mathrm{cn}(i\,\xi\,|k_0)}{\mathrm{sn}(i\,\xi\,|k_0)\mathrm{cn}(i\,\xi\,|k_0)},\\
d(z,\xi\,|k_0) &= \frac{\mathrm{sn}(i\,\xi\,|k_0)\mathrm{cn}(i\,z|k_0)-\mathrm{sn}(i\,z|k_0)\mathrm{cn}(i\,\xi\,|k_0)}{\mathrm{sn}(i\,\xi\,|k_0)\mathrm{cn}(i\,\xi\,|k_0)}.
\end{aligned}
\tag{12.3.6}
$$

12.3.2 The Low-Temperature Zone $\mathcal{R}_L(\mathrm{III})$

This formulation can now be extended to the low-temperature zones $\mathcal{R}_L(\mathrm{II})$, $\mathcal{R}_L(\mathrm{III})$ and $\mathcal{R}_L(\mathrm{IV})$. The forms of the w_j's in these zones will, apart from a multiplicative factor, be the same but there will be a transformation of variables corresponding to $\mathcal{T}_1$, $\mathcal{T}_2$ or $\mathcal{T}_3$, respectively, which maps from the relationships (12.3.2) and (12.3.3) into the corresponding ones, given in Table 9.1, for these low-temperature zones.

In the original work of Baxter and the more recent work of Fabricius and McCoy the 'principal regime' (Baxter 1982b, p. 224) is chosen to be $\mathcal{R}_{\mathrm{L}}(\mathrm{III})$ achieved by the mapping $\mathfrak{T}_2$ from $\mathcal{R}_{\mathrm{L}}(\mathrm{I})$.[45] Here

$$w_1 = \tfrac{1}{2}(\mathrm{c}+\mathrm{d}), \qquad w_2 = \tfrac{1}{2}(\mathrm{c}-\mathrm{d}),$$
$$w_3 = \tfrac{1}{2}(\mathrm{a}+\mathrm{b}), \qquad w_4 = \tfrac{1}{2}(\mathrm{a}-\mathrm{b}), \tag{12.3.7}$$

$$\Delta^\star = \frac{\mathrm{c}^2+\mathrm{d}^2-\mathrm{a}^2-\mathrm{b}^2}{2[\mathrm{cd}+\mathrm{ab}]} = -\Delta, \qquad \Gamma^\star = \frac{\mathrm{cd}-\mathrm{ab}}{\mathrm{cd}+\mathrm{ab}} = -\Gamma. \tag{12.3.8}$$

Let

$$\hat{z} := z + \mathcal{K}_0', \qquad \hat{\xi} := \xi + \mathcal{K}_0'. \tag{12.3.9}$$

Then, from (12.3.53),

$$\mathrm{a}(\hat{z},\hat{\xi}|\mathrm{k}_0) = E_0(z,\xi|\mathrm{k}_0)\mathrm{c}(z,\xi|\mathrm{k}_0), \qquad \mathrm{b}(\hat{z},\hat{\xi}|\mathrm{k}_0) = E_0(z,\xi|\mathrm{k}_0)\mathrm{d}(z,\xi|\mathrm{k}_0),$$
$$\mathrm{c}(\hat{z},\hat{\xi}|\mathrm{k}_0) = E_0(z,\xi|\mathrm{k}_0)\mathrm{a}(z,\xi|\mathrm{k}_0), \qquad \mathrm{d}(\hat{z},\hat{\xi}|\mathrm{k}_0) = E_0(z,\xi|\mathrm{k}_0)\mathrm{b}(z,\xi|\mathrm{k}_0), \tag{12.3.10}$$

where

$$E_0(z,\xi|\mathrm{k}_0) = \frac{\mathrm{sn}(\mathrm{i}\,\xi|\mathrm{k}_0)}{\mathrm{sn}(\mathrm{i}\,z|\mathrm{k}_0)} = \frac{1}{E_0(\hat{z},\hat{\xi}|\mathrm{k}_0)}. \tag{12.3.11}$$

The transformation $\mathfrak{T}_2$ corresponds to the change of variables $(\xi,\eta) \to (\hat{\xi},\hat{\eta})$. Similar changes of variables correspond to the transformations $\mathfrak{T}_1$ and $\mathfrak{T}_3$ into the zones $\mathcal{R}_{\mathrm{L}}(\mathrm{II})$ and $\mathcal{R}_{\mathrm{L}}(\mathrm{IV})$. In the cases of $\mathcal{R}_{\mathrm{L}}(\mathrm{III})$ and $\mathcal{R}_{\mathrm{L}}(\mathrm{IV})$ there is a change of sign of Δ and Γ, but not of $\Delta^\star$ and $\Gamma^\star$, as defined by (9.3.30).

12.3.3 The Transfer Matrix

In this section and in 12.3.4 it is convenient to use the variables ξ and z, given by (12.3.4).[46] To carry out the programme described in Chap. 11 for calculating the partition function in the limit $N_1 \to \infty$ the transfer matrix must be obtained. Suppose

[45] It may be useful to note that, for our variables $(\hat{z}, \hat{\xi}, \mathrm{k}_0', \hat{v}, \hat{\eta}, \mathrm{k}_1)$ in $\mathcal{R}_{\mathrm{L}}(\mathrm{III})$, Baxter (1972b, 1973a, b, c) uses the symbols $(V, \zeta, \ell, v, \eta, k)$, this also being the notation of Fabricius and McCoy (2003, 2005) and McCoy (2010), and Lavis and Bell (1999b, Chap. 5) use $(\mathrm{U}, \zeta, \ell, \mathrm{u}, \xi, \ell_1')$. Baxter (1982b) uses just the variables equivalent to our $(\hat{v}, \hat{\eta}, \mathrm{k}_1)$ for which his notation is $(-\tfrac{1}{2}\mathrm{i}v, \tfrac{1}{2}\mathrm{i}\lambda, k)$.

[46] Beginning with Sect. 12.3.5 the variables associated with k_1, and, on occasion, k_2 and k_3 are more useful choices.

for the moment that the horizontal rapidities are different, so that the transfer matrix on the ith row is a function of $z_i = \frac{1}{2}\lambda - u_i$, where $u_i := \bar{\varrho} - \varrho_i$. From (11.1.6),

$$Z(\mathcal{N}^{\circledast}) = \mathrm{Trace}\left\{\prod_{i=1}^{N_1} \boldsymbol{V}(z_i)\right\}. \tag{12.3.12}$$

The next equality in (11.1.6), which reduces the partition function to a sum of powers of eigenvalues, depends on the reduction of $\boldsymbol{V}(z)$ to diagonal form as shown in (11.1.4). In this case we have a product of matrices parameterized by z_i and the reduction can be effected only if, assuming that $\boldsymbol{V}(z)$ is simple for all z, there exists a non-singular matrix $\boldsymbol{X}$, *independent of* z_i, such that

$$\boldsymbol{X}^{-1}\boldsymbol{V}(z_i)\boldsymbol{X} = \boldsymbol{\Lambda}(z_i), \qquad i = 1, 2, \ldots, N_1. \tag{12.3.13}$$

A necessary and sufficient condition for this (Sect. 17.2) is

$$[\boldsymbol{V}(z), \boldsymbol{V}(z')] = \boldsymbol{0}, \qquad \text{for all } z, z'. \tag{12.3.14}$$

A matrix $\boldsymbol{X}$ satisfying (12.3.13) is called an eigenmatrix of $\boldsymbol{V}(z_i)$ and condition (12.3.14) implies that the set $\mathcal{S}_{\mathrm{E}}(\boldsymbol{V})$ of eigenmatrices is the same for all z. It can be shown (Sect. 12.3.5) that $\boldsymbol{V}(-z) = \boldsymbol{V}^{\mathrm{T}}(z)$. So (12.3.14) also implies (Sect. 17.2) that $\boldsymbol{V}(z)$ is normal and the eigenmatrix $\boldsymbol{X}$ is unitary. That condition (12.3.14) holds depends, as we shall now show, on the system being Z-invariant.[47]

An element of $\boldsymbol{V}(z)$ is denoted by $\langle \boldsymbol{\varsigma}|\boldsymbol{V}(z)|\boldsymbol{\varsigma}'\rangle$, where $\boldsymbol{\varsigma} := (\varsigma_1, \ldots, \varsigma_{N_2})$, with ς_j being the edge variable below the jth site of the row; $\boldsymbol{\varsigma}'$ denotes in a similar way the set of edge variables above the line of sites. Let the horizontal edge variable to the right of the jth site be ς_j''. The matrix element is clearly a sum of products of the Boltzmann weights in a row. With the vertical edge variables fixed the sum will consist of only two non-zero terms since once ς_1'' is fixed the remaining horizontal edge variables are fixed by the eight-vertex constraint condition (3.8.1). Alternatively the eight-vertex condition can be used to extend the summation over all 2^{N_2} terms $\varsigma_j'' = \pm 1$, $j = 1, 2, \ldots, N_2$. From (12.1.46),

$$\left\langle \begin{matrix} \varsigma_{j-1}'' \\ \varsigma_j \end{matrix} \middle| \boldsymbol{U}_0(z) \middle| \begin{matrix} \varsigma_j' \\ \varsigma_j'' \end{matrix} \right\rangle = \begin{cases} \mathfrak{X}(z; \varsigma_j, \varsigma_{j-1}'', \varsigma_j', \varsigma_j''), & \text{if } \varsigma_j\varsigma_{j-1}''\varsigma_j'\varsigma_j'' = 1, \\ 0, & \text{if } \varsigma_j\varsigma_{j-1}''\varsigma_j'\varsigma_j'' \neq 1, \end{cases} \tag{12.3.15}$$

where, from (3.8.10),

$$\mathfrak{X}(z; \varsigma_1, \varsigma_2, \varsigma_3, \varsigma_4) = \exp[K^{(0)} + K^{(\mathrm{f})}(z)\varsigma_2\varsigma_3 + K^{(\mathrm{b})}(z)\varsigma_3\varsigma_4 + K^{(4)}\varsigma_1\varsigma_3]. \tag{12.3.16}$$

---

[47] This argument can equally well be represented in terms of edge variables or spins; we shall use the former. The alternative spin formulation is similar to that for the Ising model given by Baxter (1982b, Sect. 7.3).

We shall also need

$$\langle \nu | \boldsymbol{U}_1 \begin{pmatrix} \varsigma' \\ \varsigma \end{pmatrix}; z \Big) |\nu'\rangle := \left\langle \begin{matrix} \nu \\ \varsigma \end{matrix} \middle| \boldsymbol{U}_0(z) \middle| \begin{matrix} \varsigma' \\ \nu' \end{matrix} \right\rangle. \tag{12.3.17}$$

The row and column indices of $\boldsymbol{U}_0$ are ordered in the pairs

$$\begin{pmatrix} +1 \\ +1 \end{pmatrix}, \quad \begin{pmatrix} +1 \\ -1 \end{pmatrix}, \quad \begin{pmatrix} -1 \\ +1 \end{pmatrix}, \quad \begin{pmatrix} -1 \\ -1 \end{pmatrix} \tag{12.3.18}$$

and for $\boldsymbol{U}_1$ by $+1$ and -1. From (3.8.11) and (12.3.15),

$$\boldsymbol{U}_0(z) = \begin{pmatrix} \mathfrak{a}(z) & 0 & 0 & \mathfrak{d}(z) \\ 0 & \mathfrak{c}(z) & \mathfrak{b}(z) & 0 \\ 0 & \mathfrak{b}(z) & \mathfrak{c}(z) & 0 \\ \mathfrak{d}(z) & 0 & 0 & \mathfrak{a}(z) \end{pmatrix}. \tag{12.3.19}$$

When the four-dimensional matrix $\boldsymbol{U}_0$ was introduced in Sect. 12.1.4, it was not with respect to a particular lattice structure and the edge variables were simply labelled in clockwise order. This is still the case in (12.3.17), but now there is a distinction between the vertical variables ς and ς' used as indices for the transfer matrix elements and the horizontal variables used below to construct those elements in the trace in (12.3.23). It will be convenient if our notation reflects this distinction, particular as, in Sect. 12.3.6, the horizontal variables will be allowed to represent a range of values different from ± 1. To achieve this result the ordering of the variables in the bra on the left of $\boldsymbol{U}_0$ in (12.3.17) are interchanged so that both in this bra and (as is already the case) in the ket on the right the upper and lower variables are for the vertical and horizontal variables respectively. Thus

$$\left\langle \begin{matrix} \varsigma \\ \nu \end{matrix} \middle| \boldsymbol{J}\boldsymbol{U}_0(z) \middle| \begin{matrix} \varsigma' \\ \nu' \end{matrix} \right\rangle = \left\langle \begin{matrix} \nu \\ \varsigma \end{matrix} \middle| \boldsymbol{U}_0(z) \middle| \begin{matrix} \varsigma' \\ \nu' \end{matrix} \right\rangle. \tag{12.3.20}$$

where

$$\boldsymbol{J} := \boldsymbol{I}_1 \oplus \boldsymbol{\Sigma}_x \oplus \boldsymbol{I}_1, \tag{12.3.21}$$

with $\boldsymbol{\Sigma}_x$ the Pauli matrix given by (3.2.6), so that

$$\boldsymbol{J}\boldsymbol{U}_0(z) = \begin{pmatrix} \boldsymbol{U}_1\begin{pmatrix} +1 \\ +1 \end{pmatrix}; z \Big) & \boldsymbol{U}_1\begin{pmatrix} -1 \\ +1 \end{pmatrix}; z \Big) \\ \\ \boldsymbol{U}_1\begin{pmatrix} +1 \\ -1 \end{pmatrix}; z \Big) & \boldsymbol{U}_1\begin{pmatrix} -1 \\ -1 \end{pmatrix}; z \Big) \end{pmatrix} = \begin{pmatrix} \mathfrak{a}(z) & 0 & 0 & \mathfrak{d}(z) \\ 0 & \mathfrak{b}(z) & \mathfrak{c}(z) & 0 \\ 0 & \mathfrak{c}(z) & \mathfrak{b}(z) & 0 \\ \mathfrak{d}(z) & 0 & 0 & \mathfrak{a}(z) \end{pmatrix}. \tag{12.3.22}$$

The elements of the transfer matrix are

$$\langle \boldsymbol{\varsigma}|\boldsymbol{V}(z)|\boldsymbol{\varsigma}'\rangle = \text{Trace}\left\{\boldsymbol{U}_1\begin{pmatrix}\varsigma_1' \\ \varsigma_1\end{pmatrix};z\right)\cdots\boldsymbol{U}_1\left(\begin{matrix}\varsigma_{N_2}' \\ \varsigma_{N_2}\end{matrix};z\right)\right\} \tag{12.3.23}$$

and

$$\langle \boldsymbol{\varsigma}|\boldsymbol{V}(z)\boldsymbol{V}(z')|\boldsymbol{\varsigma}'\rangle = \text{Trace}\left\{\boldsymbol{U}_2\left(\begin{matrix}\varsigma_1' \\ \varsigma_1\end{matrix};\begin{matrix}z' \\ z\end{matrix}\right)\cdots\boldsymbol{U}_2\left(\begin{matrix}\varsigma_{N_2}' \\ \varsigma_{N_2}\end{matrix};\begin{matrix}z' \\ z\end{matrix}\right)\right\}, \tag{12.3.24}$$

where, from (12.1.48) and (17.2.14),

$$\boldsymbol{U}_2\left(\begin{matrix}\varsigma' \\ \varsigma\end{matrix};\begin{matrix}z' \\ z\end{matrix}\right) = \sum_{\{\nu\}}\boldsymbol{U}_1\left(\begin{matrix}\nu \\ \varsigma\end{matrix};z\right)\otimes\boldsymbol{U}_1\left(\begin{matrix}\varsigma' \\ \nu\end{matrix};z\right). \tag{12.3.25}$$

To establish the commutation relation (12.3.14) we need a non-singular matrix $\boldsymbol{Y}$, independent of ς and ς', but, in general, dependent on z and z', such that

$$\boldsymbol{U}_2\left(\begin{matrix}\varsigma' \\ \varsigma\end{matrix};\begin{matrix}z' \\ z\end{matrix}\right) = \boldsymbol{Y}^{-1}(z,z')\boldsymbol{U}_2\left(\begin{matrix}\varsigma' \\ \varsigma\end{matrix};\begin{matrix}z \\ z'\end{matrix}\right)\boldsymbol{Y}(z,z'). \tag{12.3.26}$$

12.3.4 Analysis in Terms of Pauli Matrices

For easy of presentation we relabel the Pauli matrices given by (3.2.6) as

$$\boldsymbol{\Sigma}_1 := \boldsymbol{I}_2, \quad \boldsymbol{\Sigma}_2 := \boldsymbol{\Sigma}_z, \quad \boldsymbol{\Sigma}_3 := \boldsymbol{\Sigma}_x, \quad \boldsymbol{\Sigma}_4 := \boldsymbol{\Sigma}_y. \tag{12.3.27}$$

12.3.4.1 The Commutativity Relationship

In $\mathcal{R}_{\text{L}}(\text{I})$ the Boltzmann weights are related to the w_j's by (12.3.2) and it is straightforward to verify that

$$\left\langle\begin{matrix}q \\ \varsigma\end{matrix}\middle|\boldsymbol{U}_0(z)\middle|\begin{matrix}\varsigma' \\ q'\end{matrix}\right\rangle = \sum_{\ell=1}^{4} w_\ell(z)\langle\varsigma|\boldsymbol{\Sigma}_\ell|\varsigma'\rangle\langle q|\boldsymbol{\Sigma}_\ell|q'\rangle. \tag{12.3.28}$$

Then, from (12.3.25),

$$\begin{aligned}\left\langle\begin{matrix}\varsigma_2 \\ \varsigma_1\end{matrix}\middle|\boldsymbol{U}_2\left(\begin{matrix}\varsigma' \\ \varsigma\end{matrix};\begin{matrix}z' \\ z\end{matrix}\right)\middle|\begin{matrix}\varsigma_3 \\ \varsigma_4\end{matrix}\right\rangle = &\sum_{j=1}^{4}\sum_{k=1}^{4} w_j(z)w_k(z')\langle\varsigma|\boldsymbol{\Sigma}_j\boldsymbol{\Sigma}_k|\varsigma'\rangle \\ &\times\langle\varsigma_1|\boldsymbol{\Sigma}_j|\varsigma_4\rangle\langle\varsigma_2|\boldsymbol{\Sigma}_k|\varsigma_3\rangle.\end{aligned} \tag{12.3.29}$$

Suppose now that

$$\left\langle \begin{matrix} \nu \\ \varsigma \end{matrix} \middle| \boldsymbol{Y}(z,z') \middle| \begin{matrix} \varsigma' \\ \nu' \end{matrix} \right\rangle = \sum_{\ell=1}^{4} y_\ell(z,z') \langle \varsigma | \boldsymbol{\Sigma}_\ell | \varsigma' \rangle \langle \nu | \boldsymbol{\Sigma}_\ell | \nu' \rangle. \tag{12.3.30}$$

Then

$$\left\langle \begin{matrix} \varsigma_2 \\ \varsigma_1 \end{matrix} \middle| \boldsymbol{Y}(z,z') \boldsymbol{U}_2 \begin{pmatrix} \varsigma' & ; & z' \\ \varsigma & & z \end{pmatrix} \middle| \begin{matrix} \varsigma_3 \\ \varsigma_4 \end{matrix} \right\rangle = \sum_{i=1}^{4} \sum_{j=1}^{4} \sum_{k=1}^{4} y_i(z,z') w_j(z) w_k(z')$$

$$\times \langle \varsigma | \boldsymbol{\Sigma}_j \boldsymbol{\Sigma}_k | \varsigma' \rangle \langle \varsigma_1 | \boldsymbol{\Sigma}_i \boldsymbol{\Sigma}_k | \varsigma_3 \rangle \langle \varsigma_2 | \boldsymbol{\Sigma}_i \boldsymbol{\Sigma}_j | \varsigma_4 \rangle, \tag{12.3.31}$$

$$\left\langle \begin{matrix} \varsigma_2 \\ \varsigma_1 \end{matrix} \middle| \boldsymbol{U}_2 \begin{pmatrix} \varsigma' & ; & z \\ \varsigma & & z' \end{pmatrix} \boldsymbol{Y}(z,z') \middle| \begin{matrix} \varsigma_3 \\ \varsigma_4 \end{matrix} \right\rangle = \sum_{i=1}^{4} \sum_{j=1}^{4} \sum_{k=1}^{4} y_i(z,z') w_j(z) w_k(z')$$

$$\times \langle \varsigma | \boldsymbol{\Sigma}_k \boldsymbol{\Sigma}_j | \varsigma' \rangle \langle \varsigma_1 | \boldsymbol{\Sigma}_k \boldsymbol{\Sigma}_i | \varsigma_3 \rangle \langle \varsigma_2 | \boldsymbol{\Sigma}_j \boldsymbol{\Sigma}_i | \varsigma_4 \rangle. \tag{12.3.32}$$

Evaluating the elements of the Pauli matrices in (12.3.31) and (12.3.32), it is straightforward to show that (12.3.26) is satisfied if and only if

$$\begin{aligned} &y_i(z,z') w_j(z) w_k(z') - y_j(z,z') w_i(z) w_\ell(z') \\ &\quad + y_k(z,z') w_\ell(z) w_i(z') - y_m(z,z') w_k(z) w_j(z') = 0 \end{aligned} \tag{12.3.33}$$

holds for all permutations (i, j, k, ℓ) of $(1, 2, 3, 4)$. There are six such distinct relationships which can be regarded as linear homogeneous equations in $y_i(z, z')$, $i = 1, \ldots, 4$. A solution exists if the determinant of the coefficients of any four of them is zero. It can be shown that this is the case if and only if the ratio function, defined by (9.3.28), and for which we now introduce dependence on z, satisfies

$$W(i, j|k, \ell; z) = W(i, j|k, \ell; z'), \tag{12.3.34}$$

for all permutations (i, j, k, ℓ) of $(1, 2, 3, 4)$. Since, as was indicated in Sect. 9.3.3, only two of these ratio functions are independent and can be expressed as functions of the elliptic parameters Δ and Γ, (12.3.34) is satisfied and the matrix $\boldsymbol{Y}(z, z')$, given by (12.3.30) exists. This argument, a form of which is given by Baxter (1972b, Appendix B) is included for its illuminating interest. However, here it is strictly unnecessary since, comparing (12.1.51) and (12.3.26) it can be seen that

$$\boldsymbol{Y}(z, z') = \boldsymbol{U}_0(z' - z). \tag{12.3.35}$$

12.3.4.2 The Relationship to the XYZ Chain

In a derivation mirroring that of (12.3.28)

$$\left\langle {\nu \atop \varsigma} \middle| U_0(z) \middle| {\varsigma' \atop \nu'} \right\rangle = \sum_{\ell=1}^{4} \mathrm{g}_\ell(z)\langle \nu|\boldsymbol{\Sigma}_\ell|\varsigma'\rangle\langle\varsigma|\boldsymbol{\Sigma}_\ell|\nu'\rangle, \tag{12.3.36}$$

where

$$\begin{aligned} \mathrm{g}_1(z) &:= \tfrac{1}{2}[w_1(z)+w_2(z)+w_3(z)+w_4(z)],\\ \mathrm{g}_2(z) &:= \tfrac{1}{2}[w_1(z)+w_2(z)-w_3(z)-w_4(z)],\\ \mathrm{g}_3(z) &:= \tfrac{1}{2}[w_1(z)-w_2(z)+w_3(z)-w_4(z)],\\ \mathrm{g}_4(z) &:= \tfrac{1}{2}[w_1(z)-w_2(z)-w_3(z)+w_4(z)]. \end{aligned} \tag{12.3.37}$$

From Table 12.2,

$$\begin{aligned} \mathrm{g}_1(z) &= g(+1,+1,+1;z), \qquad & \mathrm{g}_2(z) &= g(+1,-1,-1;z),\\ \mathrm{g}_3(z) &= g(-1,+1,-1;z), \qquad & \mathrm{g}_4(z) &= g(-1,-1,+1;z), \end{aligned} \tag{12.3.38}$$

where

$$g(p,q,r;z) := \frac{1}{2}\left\{\frac{\mathrm{cn}(z|\mathrm{k}_0')}{\mathrm{cn}(\xi|\mathrm{k}_0')} + p\frac{\mathrm{dn}(z|\mathrm{k}_0')}{\mathrm{dn}(\xi|\mathrm{k}_0')} + q + r\frac{\mathrm{sn}(z|\mathrm{k}_0')}{\mathrm{sn}(\xi|\mathrm{k}_0')}\right\}. \tag{12.3.39}$$

From (18.3.40) and Table 12.2[48]

$$\begin{aligned} \mathrm{g}_1(\xi) &= 2, \qquad & \dot{\mathrm{g}}_1(\xi) &= \mathrm{ds}(2\xi|\mathrm{k}_0')(1+\Delta^\star+\Gamma^\star),\\ \mathrm{g}_2(\xi) &= 0, \qquad & \dot{\mathrm{g}}_2(\xi) &= \mathrm{ds}(2\xi|\mathrm{k}_0')\,\Gamma^\star,\\ \mathrm{g}_3(\xi) &= 0, \qquad & \dot{\mathrm{g}}_3(\xi) &= -\mathrm{ds}(2\xi|\mathrm{k}_0'),\\ \mathrm{g}_4(\xi) &= 0, \qquad & \dot{\mathrm{g}}_4(\xi) &= \mathrm{ds}(2\xi|\mathrm{k}_0')\,\Delta^\star. \end{aligned} \tag{12.3.40}$$

[48] The dot here denotes differentiation with respect to z. The differences between these quantities and those given by Baxter (1972a, Eq. (3.9)) are due to the different ordering of the Boltzmann factors (compare (12.3.2) with Baxter (1972b, Eq. (2.5)).

Then

$$\langle \nu | \boldsymbol{U}_1 \begin{pmatrix} \varsigma' \\ \varsigma \end{pmatrix} ; \xi \Big) | \nu' \rangle = \left\langle \begin{matrix} \nu \\ \varsigma \end{matrix} \middle| \boldsymbol{U}_0(\xi) \middle| \begin{matrix} \varsigma' \\ \nu' \end{matrix} \right\rangle = \delta^{\mathrm{Kr}}(\nu - \varsigma')\delta^{\mathrm{Kr}}(\varsigma - \nu'), \tag{12.3.41}$$

giving, from (12.3.23),

$$\langle \boldsymbol{\varsigma} | \boldsymbol{V}(\xi) | \boldsymbol{\varsigma}' \rangle = \delta^{\mathrm{Kr}}(\varsigma_1 - \varsigma_2') \cdots \delta^{\mathrm{Kr}}(\varsigma_{N_2-1} - \varsigma_{N_2}')\delta^{\mathrm{Kr}}(\varsigma_{N_2} - \varsigma_1'), \tag{12.3.42}$$

$$\langle \boldsymbol{\varsigma} | \boldsymbol{V}^{-1}(\xi) | \boldsymbol{\varsigma}' \rangle = \delta^{\mathrm{Kr}}(\varsigma_2 - \varsigma_1') \cdots \delta^{\mathrm{Kr}}(\varsigma_{N_2} - \varsigma_{N_2-1}')\delta^{\mathrm{Kr}}(\varsigma_1 - \varsigma_{N_2}'). \tag{12.3.43}$$

In a similar way

$$\begin{aligned} \langle \boldsymbol{\varsigma} | \dot{\boldsymbol{V}}(\xi) | \boldsymbol{\varsigma}' \rangle = \sum_{\mathrm{j}=1}^{N_2} \Big\{ & \delta^{\mathrm{Kr}}(\varsigma_1 - \varsigma_2') \cdots \delta^{\mathrm{Kr}}(\varsigma_{\mathrm{j}-1} - \varsigma_{\mathrm{j}}') \\ & \times \left\langle \begin{matrix} \varsigma_{\mathrm{j}-1} \\ \varsigma_{\mathrm{j}} \end{matrix} \middle| \dot{\boldsymbol{U}}_0(\xi) \middle| \begin{matrix} \varsigma_{\mathrm{j}}' \\ \varsigma_{\mathrm{j}+1}' \end{matrix} \right\rangle \delta^{\mathrm{Kr}}(\varsigma_{\mathrm{j}+1} - \varsigma_{\mathrm{j}+2}') \cdots \delta^{\mathrm{Kr}}(\varsigma_{N_2} - \varsigma_1') \Big\}. \end{aligned} \tag{12.3.44}$$

Now consider the matrix

$$\boldsymbol{L}(z) := \boldsymbol{V}^{-1}(z)\dot{\boldsymbol{V}}(z), \tag{12.3.45}$$

for which it can easily be shown that

$$[\boldsymbol{V}(z), \boldsymbol{L}(z')] = \boldsymbol{0}. \tag{12.3.46}$$

From (12.3.43)–(12.3.45),

$$\begin{aligned} \langle \boldsymbol{\varsigma} | \boldsymbol{L}(\xi) | \boldsymbol{\varsigma}' \rangle &= \sum_{\mathrm{j}=1}^{N_2} \Big\{ \delta^{\mathrm{Kr}}(\varsigma_1 - \varsigma_1') \cdots \delta^{\mathrm{Kr}}(\varsigma_{\mathrm{j}-1} - \varsigma_{\mathrm{j}-1}') \\ &\quad \times \left\langle \begin{matrix} \varsigma_{\mathrm{j}} \\ \varsigma_{\mathrm{j}+1} \end{matrix} \middle| \dot{\boldsymbol{U}}_0(\xi) \middle| \begin{matrix} \varsigma_{\mathrm{j}}' \\ \varsigma_{\mathrm{j}+1}' \end{matrix} \right\rangle \delta^{\mathrm{Kr}}(\varsigma_{\mathrm{j}+2} - \varsigma_{\mathrm{j}+2}') \cdots \delta^{\mathrm{Kr}}(\varsigma_{N_2} - \varsigma_1') \Big\} \\ &= \sum_{\ell=1}^{4} \dot{\mathrm{g}}_\ell(\xi) \sum_{\mathrm{j}=1}^{N_2} \Big\{ \delta^{\mathrm{Kr}}(\varsigma_1 - \varsigma_1') \cdots \delta^{\mathrm{Kr}}(\varsigma_{\mathrm{j}-1} - \varsigma_{\mathrm{j}-1}') \\ &\quad \times \langle \varsigma_{\mathrm{j}} | \boldsymbol{\Sigma}_\ell | \varsigma_{\mathrm{j}}' \rangle \langle \varsigma_{\mathrm{j}+1} | \boldsymbol{\Sigma}_\ell | \varsigma_{\mathrm{j}+1}' \rangle \delta^{\mathrm{Kr}}(\varsigma_{\mathrm{j}+2} - \varsigma_{\mathrm{j}+2}') \cdots \delta^{\mathrm{Kr}}(\varsigma_{N_2} - \varsigma_1') \Big\}. \end{aligned} \tag{12.3.47}$$

Thus

$$\boldsymbol{L}(\xi) = N_2 \dot{\mathrm{g}}_1(\xi)\boldsymbol{I} + \sum_{\mathrm{j}=1}^{N_2} \{\dot{\mathrm{g}}_3(\xi)\boldsymbol{\Sigma}_x(\mathrm{j})\boldsymbol{\Sigma}_x(\mathrm{j}+1) + \dot{\mathrm{g}}_4(\xi)\boldsymbol{\Sigma}_y(\mathrm{j})\boldsymbol{\Sigma}_y(\mathrm{j}+1) + \dot{\mathrm{g}}_2(\xi)\boldsymbol{\Sigma}_z(\mathrm{j})\boldsymbol{\Sigma}_z(\mathrm{j}+1)\}\,. \tag{12.3.48}$$

Apart the leading diagonal term, $\boldsymbol{L}(\xi)$ is the Hamiltonian of the zero-field spin-$\frac{1}{2}$ XYZ model given by (3.2.4), with

$$(J_x, J_y, J_z) = \mathrm{ds}(2\xi\,|\mathrm{k}_0')\,(1, -\Delta^\star, -\Gamma^\star). \tag{12.3.49}$$

The commutation relationship (12.3.46) then implies that the eight-vertex transfer matrix and the XYZ Hamiltonian have a complete set of common eigenvectors; the result due to Sutherland (1970). By keeping $\Delta^\star$ and $\Gamma^\star$ fixed and varying the Boltzmann weights it is easy to see that the minimum (ground-state) eigenvalue of the XYZ Hamiltonian corresponds to the maximum eigenvalue of $\boldsymbol{V}$, this, of course, being that which is needed to obtain the free energy.

Special cases arise when two (or more) of J_x, J_y and J_z are equal or one (or more) is zero. The conditions for this can be seen from the expressions for $\Delta^\star$ and $\Gamma^\star$ in Table 12.2. In order for $\Delta^\star$ and $\Gamma^\star$ to be real ξ can be either real or imaginary and we include both cases:

$$\begin{array}{llll}
\text{(i)} & \xi = \mathrm{i}\mathcal{K}_0 & \text{gives} & (J_x, J_y, J_z) = (1, 1, 1), \\
\text{(ii)} & \xi = \mathcal{K}_0' & \text{gives} & (J_x, J_y, J_z) = (1, -1, -1), \\
\text{(iii)} & \xi = 0 & \text{gives} & (J_x, J_y, J_z) = (1, -1, 1), \\
\text{(iv)} & \xi = \frac{1}{2}\mathrm{i}\,\mathcal{K}_0 & \text{gives} & (J_x, J_y, J_z) = (\mathrm{k}_0', 0, 1), \\
\text{(v)} & \xi = \frac{1}{2}\mathcal{K}_0' & \text{gives} & (J_x, J_y, J_z) = (\mathrm{k}_0, -1, 0), \\
\text{(vi)} & \mathrm{k}_0 = 0 & \text{gives} & (J_x, J_y, J_z) = (1, -\cosh(2\xi), 1), \\
\text{(vii)} & \mathrm{k}_0 = 1 & \text{gives} & (J_x, J_y, J_z) = (1, -1, \cos(2\xi)).
\end{array} \tag{12.3.50}$$

These cases of the Heisenberg spin chain are discussed in Sect. 3.2.1. (i) is the isotropic case and (ii) and (iii) with parameters differing in sign, but not magnitude, from this have the same eigenvalues but different ground states. (iv) and (v) are cases of the XY model except when k_0 takes the value 1 or zero, respectively, when the system reduces to the Ising model. (vi) and (vii) are cases of the XYZ model except when $\xi = 0$ or (for (vii)) when $\xi = \frac{1}{2}\pi$, when they reduces to the XXZ model.

12.3.5 Analysis of the Transfer Matrix

12.3.5.1 A Change of Variables and the High-Temperature Zone

The formulation of the rest of the analysis of the eight-vertex model, both in the original papers of Baxter (1972b, 1973a, b, c) and in the recent work of Fabricius and McCoy (2003, 2005, 2007), Fabricius (2007), Bazhanov and Mangazeev (2007), Roan (2007), Fabricius and McCoy (2009), McCoy (2010) uses the elliptic parametrization with the Landen-transformed modulus k_1, related to k_0 by (18.3.75). We do the same, with (v, η) in $\mathcal{R}_{\mathrm{L}}(\mathrm{I})$, transformed to $(\hat{v}, \hat{\eta})$ in $\mathcal{R}_{\mathrm{L}}(\mathrm{III})$.

For k_1 and k_1' it is convenient to use variables initially set to be purely imaginary. Let

$$v := \frac{\mathrm{i}\,z}{1+\mathrm{k}_1} = \frac{\mathrm{i}\,z\,\mathcal{K}_1}{\mathcal{K}_0} = \frac{\mathrm{i}\,z\,\mathcal{K}_1'}{2\mathcal{K}_0'}, \qquad \eta := \frac{\mathrm{i}\,\xi}{1+\mathrm{k}_1} = \frac{\mathrm{i}\,\xi\,\mathcal{K}_1}{\mathcal{K}_0} = \frac{\mathrm{i}\,\xi\,\mathcal{K}_1'}{2\mathcal{K}_0'}, \tag{12.3.51}$$

with, from (12.3.5),

$$|v| < -\mathrm{i}\,\eta < \tfrac{1}{2}\mathcal{K}_1'. \tag{12.3.52}$$

The third and fourth columns of Table 12.2 are then derived from the first two columns using (18.3.82) and (18.3.83). From (12.3.2), in $\mathcal{R}_{\mathrm{L}}(\mathrm{I})$,

$$\begin{aligned} &\mathrm{a}(v,\eta|\mathrm{k}_1) = 1, \qquad && \mathrm{b}(v,\eta|\mathrm{k}_1) = \mathrm{k}_1 \mathrm{sn}(v+\eta|\mathrm{k}_1)\mathrm{sn}(v-\eta|\mathrm{k}_1), \\ &\mathrm{c}(v,\eta|\mathrm{k}_1) = \frac{\mathrm{sn}(v+\eta|\mathrm{k}_1)}{\mathrm{sn}(2\eta|\mathrm{k}_1)}, && \mathrm{d}(v,\eta|\mathrm{k}_1) = -\frac{\mathrm{sn}(v-\eta|\mathrm{k}_1)}{\mathrm{sn}(2\eta|\mathrm{k}_1)}. \end{aligned} \tag{12.3.53}$$

These expressions for the Boltzmann weights are arbitrary to within a multiplicative factor and differ from those given by (12.3.6) not just in the change of variables but also with respect to such a factor. It is useful, at the outset of the rather technical manipulations needed to derive the eigenvalues of the transfer matrix, to choose a 'canonical form' for the Boltzmann weights. This will be done using *Baxter's modified theta functions*[49] defined in Sect. 18.3.8. As we saw in Sect. 9.3.1 the partition function of the eight-vertex model is invariant under the unphysical transformations $\mathcal{T}_{\mathrm{a}}, \mathcal{T}_{\mathrm{b}}, \mathcal{T}_{\mathrm{c}}$ and $\mathcal{T}_{\mathrm{d}}$, which change the sign of the corresponding Boltzmann factors. The use of such a transformation turns out to be crucial for the solution of the model. In zone $\mathcal{R}_{\mathrm{L}}(\mathrm{I})$ the need is to change the sign of d.[50] With this change implemented, and using (12.3.53) and (18.3.44),[51] the Boltzmann weights are given in terms of the

[49] These functions, which include two new integer parameters m and m', were introduced by Baxter (1973a). The earlier work of Baxter (1972b), using the standard theta functions, corresponds to the case $m = 1$, $m' = 0$.

[50] In zone $\mathcal{R}_{\mathrm{L}}(\mathrm{III})$ it is b which Baxter (1972b) negates.

[51] Which applies equally well to Baxter's modified theta functions.

functions $\Psi_{\mathfrak{B}}(\alpha, \beta; v, \eta|\mathrm{k}_1)$, defined in (18.3.150) and (18.3.151), by

$$\mathfrak{a}(v, \eta|\mathrm{k}_1) := \Psi_{\mathfrak{B}}(+1, +1; v, \eta|\mathrm{k}_1), \quad \mathfrak{b}(v, \eta|\mathrm{k}_1) := \Psi_{\mathfrak{B}}(+1, -1; v, \eta|\mathrm{k}_1),$$
$$\mathfrak{c}(v, \eta|\mathrm{k}_1) := \Psi_{\mathfrak{B}}(-1, +1; v, \eta|\mathrm{k}_1), \quad \mathfrak{d}(v, \eta|\mathrm{k}_1) := \Psi_{\mathfrak{B}}(-1, -1; v, \eta|\mathrm{k}_1).$$
(12.3.54)

We now consider the effect of the *Bazhanov–Mangazeev transformation* described in Sect. 18.3.4. It should first be noted, from (18.3.163), that the Boltzmann weights (12.3.54) satisfy (12.3.2) with the identification

$$\begin{aligned}
w_1 &:= W_{\mathfrak{B}}(+1, +1; v, \eta|\mathrm{k}_1) = \mathfrak{a}(v, \eta|\mathrm{k}_1) + \mathfrak{b}(v, \eta|\mathrm{k}_1), \\
w_2 &:= W_{\mathfrak{B}}(+1, -1; v, \eta|\mathrm{k}_1) = \mathfrak{a}(v, \eta|\mathrm{k}_1) - \mathfrak{b}(v, \eta|\mathrm{k}_1), \\
w_3 &:= W_{\mathfrak{B}}(-1, +1; v, \eta|\mathrm{k}_1) = \mathfrak{c}(v, \eta|\mathrm{k}_1) + \mathfrak{d}(v, \eta|\mathrm{k}_1), \\
w_4 &:= W_{\mathfrak{B}}(-1, -1; v, \eta|\mathrm{k}_1) = \mathfrak{c}(v, \eta|\mathrm{k}_1) - \mathfrak{d}(v, \eta|\mathrm{k}_1).
\end{aligned} \tag{12.3.55}$$

Then, from (18.3.165),

$$\begin{aligned}
W_{\mathfrak{B}}(+1, +1; v_3, \eta_3|\mathrm{k}_3) &= \Pi_{\mathfrak{B}}(v, \eta|\mathrm{k}_1)[\mathfrak{c}(v, \eta|\mathrm{k}_1) + \mathfrak{d}(v, \eta|\mathrm{k}_1)], \\
W_{\mathfrak{B}}(+1, -1; v_3, \eta_3|\mathrm{k}_3) &= \Pi_{\mathfrak{B}}(v, \eta|\mathrm{k}_1)[\mathfrak{a}(v, \eta|\mathrm{k}_1) + \mathfrak{b}(v, \eta|\mathrm{k}_1)], \\
W_{\mathfrak{B}}(-1, +1; v_3, \eta_3|\mathrm{k}_3) &= \Pi_{\mathfrak{B}}(v, \eta|\mathrm{k}_1)[\mathfrak{c}(v, \eta|\mathrm{k}_1) - \mathfrak{d}(v, \eta|\mathrm{k}_1)], \\
W_{\mathfrak{B}}(-1, -1; v_3, \eta_3|\mathrm{k}_3) &= \Pi_{\mathfrak{B}}(v, \eta|\mathrm{k}_1)[\mathfrak{a}(v, \eta|\mathrm{k}_1) - \mathfrak{b}(v, \eta|\mathrm{k}_1)],
\end{aligned} \tag{12.3.56}$$

where k_3 is related to k_1 by (18.3.80) and

$$v_3 := \frac{2\,\mathrm{i}\,\mathcal{K}_3[\mathcal{K}_1 - v]}{\mathcal{K}_1'}, \qquad \eta_3 := \frac{2\,\mathrm{i}\,\mathcal{K}_3[\mathcal{K}_1 - \eta]}{\mathcal{K}_1'}. \tag{12.3.57}$$

From Table 9.1 it can be seen that this transformation is $\mathcal{T}_2\mathcal{T}_{\mathrm{WG}}$, which maps first from $\mathcal{R}_{\mathrm{L}}(\mathrm{I})$ into the high-temperature zone $\mathcal{R}_{\mathrm{O}}(\mathrm{I})$ and then into $\mathcal{R}_{\mathrm{O}}(\mathrm{III})$.

12.3.5.2 Symmetry and Commutativity

The rows and columns of the transfer matrix are indexed by the vectors $\boldsymbol{\varsigma}$, which store the edge variables of the row of N_2 vertical edges and each element $\langle\boldsymbol{\varsigma}|\boldsymbol{V}(v, \eta)|\boldsymbol{\varsigma}'\rangle$ of the transfer matrix is a polynomial of two terms of degree N_2 in the Boltzmann weights. In each term the total number of $\mathfrak{c}$'s and $\mathfrak{d}$'s is even, meaning that the total number of $\mathfrak{a}$'s and $\mathfrak{b}$'s is of the parity of N_2. The two terms differ by a change of sign of the horizontal edge variables (a reversal of horizontal arrows). This has the effect of the transformation $\mathcal{T}_1$ of (9.3.3) and, from (18.3.153),

$$V(v \pm i\mathcal{K}_1', \eta) = \{-\mathfrak{E}(-m[i\mathcal{K}_1' \pm 2v]|k_1)\}^{N_2} V(v, \eta), \tag{12.3.58}$$

where

$$\mathfrak{E}(u|k_1) := \exp\left(\frac{i\pi u}{\mho_{\mathfrak{B}1}}\right), \qquad \mho_{\mathfrak{B}1} := \mho(2m, m'|k_1). \tag{12.3.59}$$

The element $\langle \varsigma | V^{T}(v, \eta) | \varsigma' \rangle = \langle \varsigma' | V(v, \eta) | \varsigma \rangle$ of the transposed matrix obtained by interchanging the vertical edge variables at each site of the row has the same effect as the transformation $\mathcal{T}_0$ of (9.3.2), namely an interchange of the c's and d's. From (18.3.152),

$$V^{T}(v, \eta) = (-1)^{N_2} V(v, -\eta) = V(-v, \eta). \tag{12.3.60}$$

Finally we note, also from (18.3.152), that

$$V(v + 2\mathcal{K}_1, \eta) = \{\mathfrak{E}(2m'[\mathcal{K}_1 + v]|k_1)\}^{N_2} V(v, \eta). \tag{12.3.61}$$

It has so far been unnecessary to impose any particular ordering on the index vectors ς. However, now it becomes useful to do so and the method we use to produce the sequence $\varsigma^{(1)}, \varsigma^{(2)}, \ldots, \varsigma^{(2^{N_2})}$ is recursive[52]:

(i) For $N_2 = 1$, $\varsigma^{(1)} := (+1)$, $\varsigma^{(2)} := (-1)$.

(ii) To generate the index variables for $N_2 + 1$ from those for N_2 add $+1$ on the right to each vector and then repeat the process adding -1 on the right.

Example 12.3.1 For $N_2 = 2$,

$$\begin{aligned} \varsigma^{(1)} &= (+1, +1), & \varsigma^{(2)} &= (-1, +1), \\ \varsigma^{(3)} &= (+1, -1), & \varsigma^{(4)} &= (-1, -1), \end{aligned} \tag{12.3.62}$$

and for $N_2 = 3$,

$$\begin{aligned} \varsigma^{(1)} &= (+1, +1, +1), & \varsigma^{(2)} &= (-1, +1, +1), \\ \varsigma^{(3)} &= (+1, -1, +1), & \varsigma^{(4)} &= (-1, -1, +1), \\ \varsigma^{(5)} &= (+1, +1, -1), & \varsigma^{(6)} &= (-1, +1, -1), \\ \varsigma^{(7)} &= (+1, -1, -1), & \varsigma^{(8)} &= (-1, -1, -1). \end{aligned} \tag{12.3.63}$$

The symmetry group of the model is $\mathcal{S}_2^{(e)} \times \mathcal{S}_{N_2}$, where $\mathcal{S}_2^{(e)}$ corresponds to the change of sign of edge variables in a row (reversal of arrow directions) and $\mathcal{S}_{N_2}$ is the group

---

[52] Contrary to our convention we here denote the index vectors as *rows* containing the edge variables listed from left to right.

of all permutations of a row of edge variables (which leaves invariant the number of up and down arrow directions). We treat these two groups separately and consider their 2^{N_2}-dimensional representations of matrices operating on the column vector $(\varsigma^{(1)}\ \varsigma^{(2)}\ \cdots\ \varsigma^{(2^{N_2})})^{\mathrm{T}}$.

- Since

$$\varsigma^{(k)} = -\varsigma^{(2^{N_2}-k+1)}, \qquad k = 1, 2, \ldots, 2^{N_2}, \tag{12.3.64}$$

the representation matrix for the arrow-reversal operation is $\boldsymbol{R}(\mathrm{S}_2) := \boldsymbol{S}_x$, where

$$\langle \varsigma^{(k)} | \boldsymbol{S}_x | \varsigma^{(k')} \rangle := \delta^{\mathrm{Kr}}(k + k' - 2^{N_2} - 1) \tag{12.3.65}$$

and

$$[\boldsymbol{S}_x, \boldsymbol{V}(v, \eta)] = \boldsymbol{0}. \tag{12.3.66}$$

In terms of the Pauli matrices defined in (3.2.6),

$$\boldsymbol{S}_x = \boldsymbol{\Sigma}_x \otimes \boldsymbol{\Sigma}_x \otimes \cdots \otimes \boldsymbol{\Sigma}_x \otimes \boldsymbol{\Sigma}_x. \tag{12.3.67}$$

Since

$$\boldsymbol{W}_2^{\dagger} \boldsymbol{\Sigma}_x \boldsymbol{W}_2 = \boldsymbol{\Sigma}_z, \tag{12.3.68}$$

where

$$\boldsymbol{W}_2 := \begin{pmatrix} \frac{1}{\sqrt{2}} & \frac{1}{\sqrt{2}} \\ \frac{1}{\sqrt{2}} & -\frac{1}{\sqrt{2}} \end{pmatrix}, \tag{12.3.69}$$

it follows that

$$\boldsymbol{W}^{\dagger} \boldsymbol{S}_x \boldsymbol{W} = \boldsymbol{S}_z, \tag{12.3.70}$$

where

$$\boldsymbol{W} := \boldsymbol{W}_2 \otimes \boldsymbol{W}_2 \otimes \cdots \otimes \boldsymbol{W}_2 \otimes \boldsymbol{W}_2 \tag{12.3.71}$$

and

$$\boldsymbol{S}_z := \boldsymbol{\Sigma}_z \otimes \boldsymbol{\Sigma}_z \otimes \cdots \otimes \boldsymbol{\Sigma}_z \otimes \boldsymbol{\Sigma}_z \tag{12.3.72}$$

is diagonal with an equal number of entries $+1$ and -1. This means that this reducible representation of $\mathcal{S}^{(e)}$ contains an equal number of copies of the symmetric and antisymmetric one-dimensional representations of the group. Let $\boldsymbol{P}$ be the permutation matrix with the effect that

$$\boldsymbol{S}'_z := \boldsymbol{P}^\dagger \boldsymbol{S}_z \boldsymbol{P} = \boldsymbol{P}^\dagger \boldsymbol{W}^\dagger \boldsymbol{S}_x \boldsymbol{W} \boldsymbol{P} \tag{12.3.73}$$

has all the symmetric representation $+1$'s appearing before the antisymmetric -1's. It follows from Theorem 11.3.1 that

$$\boldsymbol{V}_x(v, \eta) := \boldsymbol{P}^\dagger \boldsymbol{W}^\dagger \boldsymbol{V}(v, \eta) \boldsymbol{W} \boldsymbol{P} \tag{12.3.74}$$

is block-diagonal in two blocks to which are attached the variable $\tau_x = +1$ and $\tau_x = -1$ corresponding to the symmetric and antisymmetric representations.

- For $\boldsymbol{\varsigma}^{(k)}$, let

$$\pi(\boldsymbol{\varsigma}^{(k)}) := \prod_{\mathrm{j}=1}^{N_2} \varsigma_{\mathrm{j}}^{(k)}. \tag{12.3.75}$$

This variable assigns parities to the vectors $\boldsymbol{\varsigma}^{(k)}$ and it follows from the method of ordering of these vectors given above that

$$\langle \boldsymbol{\varsigma}^{(k)} | \boldsymbol{S}_z | \boldsymbol{\varsigma}^{(k')} \rangle := \delta^{\mathrm{Kr}}(k - k') \pi(\boldsymbol{\varsigma}^{(k)}), \tag{12.3.76}$$

from which it follows that $\boldsymbol{T} \dagger \boldsymbol{S}_z \boldsymbol{T} = \boldsymbol{S}_z$, where $\boldsymbol{T}$ is any 2^{N_2}-dimensional representative matrix of $\mathcal{S}_{N_2}$. An important property of the transfer matrix is:

Theorem 12.3.1

$$\langle \boldsymbol{\varsigma}^{(k)} | \boldsymbol{V}(v, \eta) | \boldsymbol{\varsigma}^{(k')} \rangle \neq 0 \quad \textit{if and only if} \quad \pi(\boldsymbol{\varsigma}^{(k)})\, \pi(\boldsymbol{\varsigma}^{(k')}) = +1. \tag{12.3.77}$$

Proof The easiest way to picture this is using arrows assigned to edges according to the direction rule as shown in Fig. 3.2. The non-zero terms in the trace in (12.3.23) consist of two terms corresponding to reversal of all the horizontal arrows. Around the horizontal lattice ring the arrow direction will change at a vertex with different horizontal, and hence for a non-zero contribution different vertical, edge variables. There can be only an even number of such vertices and, since

$$\pi(\boldsymbol{\varsigma}^{(k)})\, \pi(\boldsymbol{\varsigma}^{(k')}) = \prod_{\mathrm{j}=1}^{N_2} (\varsigma_{\mathrm{j}}^{(k)} \varsigma_{\mathrm{j}}^{(k')}), \tag{12.3.78}$$

the theorem follows. □

From (12.3.77)

$$[\boldsymbol{S}_z, \boldsymbol{V}(v, \eta)] = \boldsymbol{0}, \tag{12.3.79}$$

and (12.3.66) and this equation imply that $\mathcal{S}_{\rm E}(\boldsymbol{S}_x, \boldsymbol{V})$ and $\mathcal{S}_{\rm E}(\boldsymbol{S}_z, \boldsymbol{V})$ are non-empty.[53] However,

$$\boldsymbol{S}_x\boldsymbol{S}_z = (-1)^{N_2}\boldsymbol{S}_z\boldsymbol{S}_x, \tag{12.3.80}$$

although, from (12.3.66) and (12.3.79),

$$[\boldsymbol{S}_z\boldsymbol{S}_x, \boldsymbol{V}(v, \eta)] = \boldsymbol{0}. \tag{12.3.81}$$

All the eigenvalues of $\boldsymbol{S}_x$ and $\boldsymbol{S}_z$, denoted respectively by θ_x and θ_z, are ± 1.

- If N_2 is odd $\boldsymbol{S}_x$ and $\boldsymbol{S}_z$ have no common eigenvectors and all the eigenvalues of $\boldsymbol{V}(v, \eta)$ are degenerate (Sect. 17.2.2).
- If N_2 is even every non-degenerate eigenvector of $\boldsymbol{V}(v, \eta)$ is an eigenvector of $\boldsymbol{S}_x$ and $\boldsymbol{S}_z$. The subspace defined by a k-degenerate eigenvalue of $\boldsymbol{V}(v, \eta)$ contains k independent common eigenvectors of $\boldsymbol{S}_x$ and $\boldsymbol{S}_z$.

Now we have a second block-diagonalization provided by Theorem 12.3.1. The permutation matrix $\boldsymbol{P}$, which collects the diagonal elements of $\boldsymbol{S}_z$ with the same sign, will shuffled the rows and columns of $\boldsymbol{V}(v, \eta)$ so that

$$\boldsymbol{V}_z(v, \eta) := \boldsymbol{P}^\dagger \boldsymbol{V}(v, \eta)\boldsymbol{P}. \tag{12.3.82}$$

The first and second blocks of $\boldsymbol{V}_z$ are labelled with $\tau_z = +1$ and -1 although, in this case, the variable is not simply related to characters of $\mathcal{S}_{N_2}$.[54] Since the reduction to block diagonal form is in this case effected by a permutation matrix, $\boldsymbol{V}$ is reducible, which is a necessary, but not sufficient, condition for the eigenvalue equal to the spectral radius $\mathcal{SR}(\boldsymbol{V})$ to have a multiplicity of at least two (Sect. 17.2.4); this becomes exactly two if the two blocks of $\boldsymbol{V}_z$ can be shown to be irreducible.

12.3.6 The VQ Equation

To this point our analysis has been for zone $\mathcal{R}_{\rm L}$(I), whereas, as we have already noted, most authors have studied $\mathcal{R}_{\rm L}$(III). Further discussion using our formulation exposes the wisdom of their choice[55] and we bring our variables in line with theirs by using the transformation $\mathcal{T}_2$ to $\mathcal{R}_{\rm L}$(III) effected by the change of variables

$$\hat{v} := v + \tfrac{1}{2}\mathrm{i}\,\mathcal{K}_1', \qquad \hat{\eta} := \eta + \tfrac{1}{2}\mathrm{i}\,\mathcal{K}_1'. \tag{12.3.83}$$

[53] For this notation see Sect. 17.2.

[54] Although $\boldsymbol{S}_z$ commutes with $\boldsymbol{V}(v, \eta)$ it is a representative matrix of $\mathcal{S}_{N_2}$ only when N_2 is even.

[55] In particular the elliptic roots of unity condition (12.3.128) in our variables takes the cumbersome form $\eta + \frac{1}{2}\mathrm{i}\,\mathcal{K}_1' = \mho(2m, m'|\mathrm{k}_1)/L$, where L makes its appearance in Sect. 12.3.6.2 as the dimension of the matrices $\boldsymbol{T}_{\rm R1}$.

For the expressions for the Boltzmann factors given by (12.3.54) this gives

$$\begin{aligned}
\mathfrak{a}(v,\eta|\mathrm{k}_1) &= -\mathfrak{E}(-m[\mathrm{i}\mathcal{K}_1' - \hat{v} - 3\hat{\eta}]|\mathrm{k}_1)\hat{\mathfrak{a}}(\hat{v},\hat{\eta}|\mathrm{k}_1),\\
\mathfrak{b}(v,\eta|\mathrm{k}_1) &= -\mathfrak{E}(-m[\mathrm{i}\mathcal{K}_1' - \hat{v} - 3\hat{\eta}]|\mathrm{k}_1)\hat{\mathfrak{b}}(\hat{v},\hat{\eta}|\mathrm{k}_1),\\
\mathfrak{c}(v,\eta|\mathrm{k}_1) &= -\mathfrak{E}(-m[\mathrm{i}\mathcal{K}_1' - \hat{v} - 3\hat{\eta}]|\mathrm{k}_1)\hat{\mathfrak{c}}(\hat{v},\hat{\eta}|\mathrm{k}_1),\\
\mathfrak{d}(v,\eta|\mathrm{k}_1) &= -\mathfrak{E}(-m[\mathrm{i}\mathcal{K}_1' - \hat{v} - 3\hat{\eta}]|\mathrm{k}_1)\hat{\mathfrak{d}}(\hat{v},\hat{\eta}|\mathrm{k}_1),
\end{aligned} \tag{12.3.84}$$

where

$$\begin{aligned}
\hat{\mathfrak{a}}(\hat{v},\hat{\eta}|\mathrm{k}_1) &:= \Psi_{\mathfrak{B}}(-1,+1;\hat{v},\hat{\eta}|\mathrm{k}_1), \qquad \hat{\mathfrak{b}}(\hat{v},\hat{\eta}|\mathrm{k}_1) := \Psi_{\mathfrak{B}}(-1,-1;\hat{v},\hat{\eta}|\mathrm{k}_1),\\
\hat{\mathfrak{c}}(\hat{v},\hat{\eta}|\mathrm{k}_1) &:= \Psi_{\mathfrak{B}}(+1,+1;\hat{v},\hat{\eta}|\mathrm{k}_1), \qquad \hat{\mathfrak{d}}(\hat{v},\hat{\eta}|\mathrm{k}_1) := \Psi_{\mathfrak{B}}(+1,-1;\hat{v},\hat{\eta}|\mathrm{k}_1).
\end{aligned} \tag{12.3.85}$$

We again consider the effect of the Bazhanov–Mangazeev transformation, where (12.3.55) is now replaced by

$$\begin{aligned}
w_1 &:= W_{\mathfrak{B}}(+1,+1;\hat{v},\hat{\eta}|\mathrm{k}_1) = \hat{\mathfrak{c}}(\hat{v},\hat{\eta}|\mathrm{k}_1) + \hat{\mathfrak{d}}(\hat{v},\hat{\eta}|\mathrm{k}_1),\\
w_2 &:= W_{\mathfrak{B}}(+1,-1;\hat{v},\hat{\eta}|\mathrm{k}_1) = \hat{\mathfrak{c}}(\hat{v},\hat{\eta}|\mathrm{k}_1) - \hat{\mathfrak{d}}(\hat{v},\hat{\eta}|\mathrm{k}_1),\\
w_3 &:= W_{\mathfrak{B}}(-1,+1;\hat{v},\hat{\eta}|\mathrm{k}_1) = \hat{\mathfrak{a}}(\hat{v},\hat{\eta}|\mathrm{k}_1) + \hat{\mathfrak{b}}(\hat{v},\hat{\eta}|\mathrm{k}_1),\\
w_4 &:= W_{\mathfrak{B}}(-1,-1;\hat{v},\hat{\eta}|\mathrm{k}_1) = \hat{\mathfrak{a}}(\hat{v},\hat{\eta}|\mathrm{k}_1) - \hat{\mathfrak{b}}(\hat{v},\hat{\eta}|\mathrm{k}_1)
\end{aligned} \tag{12.3.86}$$

and (12.3.56) by

$$\begin{aligned}
W_{\mathfrak{B}}(+1,+1;\hat{v}_3,\hat{\eta}_3|\mathrm{k}_3) &= \Pi_{\mathfrak{B}}(\hat{v},\hat{\eta}|\mathrm{k}_1)[\hat{\mathfrak{a}}(\hat{v},\hat{\eta}|\mathrm{k}_1) + \hat{\mathfrak{b}}(\hat{v},\hat{\eta}|\mathrm{k}_1)],\\
W_{\mathfrak{B}}(+1,-1;v_3,\eta_3|\mathrm{k}_3) &= \Pi_{\mathfrak{B}}(\hat{v},\hat{\eta}|\mathrm{k}_1)[\hat{\mathfrak{c}}(\hat{v},\hat{\eta}|\mathrm{k}_1) + \hat{\mathfrak{d}}(\hat{v},\hat{\eta}|\mathrm{k}_1)],\\
W_{\mathfrak{B}}(-1,+1;\hat{v}_3,\hat{\eta}_3|\mathrm{k}_3) &= \Pi_{\mathfrak{B}}(\hat{v},\hat{\eta}|\mathrm{k}_1)[\hat{\mathfrak{a}}(\hat{v},\hat{\eta}|\mathrm{k}_1) - \hat{\mathfrak{b}}(\hat{v},\hat{\eta}|\mathrm{k}_1)],\\
W_{\mathfrak{B}}(-1,-1;\hat{v}_3,\hat{\eta}_3|\mathrm{k}_3) &= \Pi_{\mathfrak{B}}(\hat{v},\hat{\eta}|\mathrm{k}_1)[\hat{\mathfrak{c}}(\hat{v},\hat{\eta}|\mathrm{k}_1) - \hat{\mathfrak{d}}(\hat{v},\hat{\eta}|\mathrm{k}_1)],
\end{aligned} \tag{12.3.87}$$

where

$$\hat{v}_3 := \frac{2\,\mathrm{i}\,\mathcal{K}_3[\mathcal{K}_1 - \hat{v}]}{\mathcal{K}_1'}, \qquad \hat{\eta}_3 := \frac{2\,\mathrm{i}\,\mathcal{K}_3[\mathcal{K}_1 - \hat{\eta}]}{\mathcal{K}_1'}. \tag{12.3.88}$$

From Table 9.1 it can be seen that this transformation is again $\mathcal{T}_2\mathcal{T}_{\mathrm{WG}}$, but now it maps $\mathcal{R}_{\mathrm{L}}(\mathrm{III})$ into $\mathcal{R}_{\mathrm{O}}(\mathrm{I})$.

We keep track of the prefactors in (12.3.84) by defining

$$\boldsymbol{V}(v,\eta) := [-\mathfrak{E}(-m[\mathrm{i}\mathcal{K}_1' - \hat{v} - 3\hat{\eta}]|\mathrm{k}_1)]^{N_2}\widehat{\boldsymbol{V}}(\hat{v},\hat{\eta}), \tag{12.3.89}$$

for which

$$\widehat{\boldsymbol{V}}^{\mathrm{T}}(\hat{v},\hat{\eta}) = (-1)^{N_2}\widehat{\boldsymbol{V}}(-\hat{v},\hat{\eta}) = \widehat{\boldsymbol{V}}(-\hat{v},\hat{\eta}), \tag{12.3.90}$$

$$\widehat{\boldsymbol{V}}(\hat{v} \pm 2\mathcal{K}_1,\hat{\eta}) = \{-\mathfrak{E}(2m'[\mathcal{K}_1 \pm \hat{v}]|\mathrm{k}_1)\}^{N_2}\widehat{\boldsymbol{V}}(\hat{v},\hat{\eta}), \tag{12.3.91}$$

$$\widehat{\boldsymbol{V}}(\hat{v} \pm \mathrm{i}\mathcal{K}_1',\hat{\eta}) = \{-\mathfrak{E}(-m[\mathrm{i}\,\mathcal{K}_1' \pm 2\hat{v}]|\mathrm{k}_1)\}^{N_2}\widehat{\boldsymbol{V}}(\hat{v},\hat{\eta}). \tag{12.3.92}$$

From (12.3.22) and (12.3.23),

$$\langle\boldsymbol{\varsigma}|\widehat{\boldsymbol{V}}(\hat{v},\hat{\eta})|\boldsymbol{\varsigma}'\rangle = \mathrm{Trace}\left\{\widehat{\boldsymbol{U}}_1\left(\begin{matrix}\varsigma_1'\\ \varsigma_1\end{matrix};\hat{v},\hat{\eta}\right)\cdots\widehat{\boldsymbol{U}}_1\left(\begin{matrix}\varsigma_{N_2}'\\ \varsigma_{N_2}\end{matrix};\hat{v},\hat{\eta}\right)\right\}, \tag{12.3.93}$$

where

$$\widehat{\boldsymbol{U}}_1\left(\begin{matrix}+1\\ +1\end{matrix};\hat{v},\hat{\eta}\right) = \begin{pmatrix}\hat{\mathfrak{a}}(\hat{v},\hat{\eta}) & 0\\ 0 & \hat{\mathfrak{b}}(\hat{v},\hat{\eta})\end{pmatrix},\quad \widehat{\boldsymbol{U}}_1\left(\begin{matrix}-1\\ +1\end{matrix};\hat{v},\hat{\eta}\right) = \begin{pmatrix}0 & \hat{\mathfrak{d}}(\hat{v},\hat{\eta})\\ \hat{\mathfrak{c}}(\hat{v},\hat{\eta}) & 0\end{pmatrix},$$

$$\widehat{\boldsymbol{U}}_1\left(\begin{matrix}+1\\ -1\end{matrix};\hat{v},\hat{\eta}\right) = \begin{pmatrix}0 & \hat{\mathfrak{c}}(\hat{v},\hat{\eta})\\ \hat{\mathfrak{d}}(\hat{v},\hat{\eta}) & 0\end{pmatrix},\quad \widehat{\boldsymbol{U}}_1\left(\begin{matrix}-1\\ -1\end{matrix};\hat{v},\hat{\eta}\right) = \begin{pmatrix}\hat{\mathfrak{b}}(\hat{v},\hat{\eta}) & 0\\ 0 & \hat{\mathfrak{a}}(\hat{v},\hat{\eta})\end{pmatrix}. \tag{12.3.94}$$

Baxter's strategy for determining the eigenvalues of the transfer matrix involved the determination of two matrices $\boldsymbol{Q}_{\mathrm{R}}(\hat{v},\hat{\eta})$ and $\boldsymbol{Q}_{\mathrm{L}}(\hat{v},\hat{\eta})$ such that:

(i) They satisfy

$$\begin{aligned}\widehat{\boldsymbol{V}}(\hat{v},\hat{\eta})\boldsymbol{Q}_{\mathrm{R}}(\hat{v},\hat{\eta}) = {}& [\mathrm{h}_{\mathfrak{B}}(\hat{v}+\hat{\eta}|\mathrm{k}_1)]^{N_2}\boldsymbol{Q}_{\mathrm{R}}(\hat{v}-2\hat{\eta},\hat{\eta})\\ &+ [\mathrm{h}_{\mathfrak{B}}(\hat{v}-\hat{\eta}|\mathrm{k}_1)]^{N_2}\boldsymbol{Q}_{\mathrm{R}}(\hat{v}+2\hat{\eta},\hat{\eta}),\end{aligned} \tag{12.3.95}$$

$$\begin{aligned}\boldsymbol{Q}_{\mathrm{L}}(\hat{v},\hat{\eta})\widehat{\boldsymbol{V}}(\hat{v},\hat{\eta}) = {}& [\mathrm{h}_{\mathfrak{B}}(\hat{v}+\hat{\eta}|\mathrm{k}_1)]^{N_2}\boldsymbol{Q}_{\mathrm{L}}(\hat{v}-2\hat{\eta},\hat{\eta})\\ &+ [\mathrm{h}_{\mathfrak{B}}(\hat{v}-\hat{\eta}|\mathrm{k}_1)]^{N_2}\boldsymbol{Q}_{\mathrm{L}}(\hat{v}+2\hat{\eta},\hat{\eta}),\end{aligned} \tag{12.3.96}$$

where $\mathrm{h}_{\mathfrak{B}}(u|\mathrm{k})$ is given by (18.3.155).

(ii) For at least one value $\hat{v}_0$ of $\hat{v}$, $\boldsymbol{Q}_{\mathrm{L}}(\hat{v}_0,\hat{\eta})$ and $\boldsymbol{Q}_{\mathrm{R}}(\hat{v}_0,\hat{\eta})$ are non-singular for all $\hat{\eta}$.

(iii) That the interchange relationship

$$\boldsymbol{Q}_{\mathrm{L}}(\hat{v},\hat{\eta})\boldsymbol{E}\boldsymbol{Q}_{\mathrm{R}}(\hat{v}',\hat{\eta}) = \boldsymbol{Q}_{\mathrm{L}}(\hat{v}',\hat{\eta})\boldsymbol{E}\boldsymbol{Q}_{\mathrm{R}}(\hat{v},\hat{\eta}), \tag{12.3.97}$$

is satisfied for all $\hat{v}$, $\hat{v}'$ and $\hat{\eta}$, where $\boldsymbol{E}$ is a constant self-inverse matrix which commutes with $\widehat{\boldsymbol{V}}(\hat{v},\hat{\eta})$.

In Sect. 12.3.6.2 a matrix $\boldsymbol{Q}_{\mathrm{R}}$ satisfying (12.3.95) is derived. Then, from (12.3.90), (12.3.95) and (12.3.96),

$$\boldsymbol{Q}_{\mathrm{L}}(-\hat{v},\hat{\eta}) = \boldsymbol{Q}_{\mathrm{R}}^{\mathrm{T}}(\hat{v},\hat{\eta}). \tag{12.3.98}$$

Baxter (1972b, p. 218) (see also Baxter 1982b, p. 220) *assumed* (for $m' = 0$) that there exists a $\hat{v}_0$ such that $\boldsymbol{Q}_{\mathrm{R}}(\hat{v}_0,\hat{\eta})$ is non-singular for all $\hat{\eta}$. More recent work by Fabricius and McCoy (2003) tested this assertion for small values of N_2, L and m with $m' = 0$. They found that $\boldsymbol{Q}_{\mathrm{R}}(\hat{v},\hat{\eta})$ is generically non-singular (singular for only a small number of values of $\hat{v}$) if either L is even or both L and m are odd. When L is odd and m is even singularity depends on the parity of N_2, with $\boldsymbol{Q}_{\mathrm{R}}(\hat{v},\hat{\eta})$ being singular for all L if N_2 is odd and singular if $N_2 \geq L-1$ if N_2 is even. These investigations were extended to the case $m' \neq 0$ by Fabricius and McCoy (2007). They found the same type of behaviour when m and m' were of different parity and that, when m and m' are both even and N_2 is odd, $\boldsymbol{Q}_{\mathrm{R}}(\hat{v},\hat{\eta})$ is singular for all $\hat{v}$. Beginning with the case m and N_2 even and $m' = 0$, a different $\boldsymbol{Q}_{\mathrm{R}}(\hat{v},\hat{\eta})$, adapted to the situation where both m and m' are even, was developed by Fabricius (2007) and extended to $m' \neq 0$ and even by Fabricius and McCoy (2007). This involves the use of the cyclic condition, described on p. 455.[56]

Henceforth we shall suppose that N_2, m, m' and L have values such that $\boldsymbol{Q}_{\mathrm{R}}(\hat{v}_0,\hat{\eta})$ is non-singular and define

$$\boldsymbol{Q}(\hat{v},\hat{\eta}) := \boldsymbol{Q}_{\mathrm{R}}(\hat{v},\hat{\eta})\boldsymbol{Q}_{\mathrm{R}}^{-1}(\hat{v}_0,\hat{\eta}), \tag{12.3.99}$$

so that, from (12.3.95),

$$\begin{aligned}\widehat{\boldsymbol{V}}(\hat{v},\hat{\eta})\boldsymbol{Q}(\hat{v},\hat{\eta}) = {} & [\mathfrak{h}_{\mathfrak{B}}(\hat{v}+\hat{\eta}|\mathrm{k}_1)]^{N_2}\boldsymbol{Q}(\hat{v}-2\hat{\eta},\hat{\eta}) \\ & + [\mathfrak{h}_{\mathfrak{B}}(\hat{v}-\hat{\eta}|\mathrm{k}_1)]^{N_2}\boldsymbol{Q}(\hat{v}+2\hat{\eta},\hat{\eta}).\end{aligned} \tag{12.3.100}$$

This is the *VQ equation*.[57] The programme (i)–(iii) is completed by a proof of the interchange relationship (12.3.97) in Sect. 12.3.6.3, from which it follows that

$$[\boldsymbol{Q}(\hat{v},\hat{\eta}),\boldsymbol{Q}(\hat{v}',\hat{\eta})] = \boldsymbol{0}, \tag{12.3.101}$$

---

[56] In the work of Fabricius and McCoy the approach used here and the cyclic case are distinguished by a superscripts (1) and (2) respectively.

[57] Baxter (1972b) used the symbol $\boldsymbol{T}$ for the transfer matrix and this convention has been adopted in most subsequent work, with the consequent naming of (12.3.100) as the TQ equation.

$$[\widehat{\boldsymbol{V}}(\hat{v},\hat{\eta}),\ \boldsymbol{Q}(\hat{v},\hat{\eta})] = \mathbf{0}. \tag{12.3.102}$$

These commutation relationships imply that the set $\mathbb{S}_{\mathrm{E}}(\boldsymbol{Q})$ of eigenmatrices of $\boldsymbol{Q}(\hat{v},\hat{\eta})$ is, like that of $\widehat{\boldsymbol{V}}(\hat{v},\hat{\eta})$ independent of $\hat{v}$, but dependent on $\hat{\eta}$, and if $\boldsymbol{X}_Q(\hat{\eta}) \in \mathbb{S}_{\mathrm{E}}(\boldsymbol{Q})$ then $\boldsymbol{X}_Q(\hat{\eta}) \in \mathbb{S}_{\mathrm{E}}(\widehat{\boldsymbol{V}})$, so that $\mathbb{S}_{\mathrm{E}}(\widehat{\boldsymbol{V}},\boldsymbol{Q}) = \mathbb{S}_{\mathrm{E}}(\boldsymbol{Q})$.

12.3.6.1 Commutativity and Degeneracy

It follows from (12.3.102) that $\widehat{\boldsymbol{V}}(\hat{v},\hat{\eta})$ and $\boldsymbol{Q}(\hat{v},\hat{\eta})$ have a complete set of common eigenvectors (Sect. 17.2.2). If for one such the respective eigenvalues are $\hat{\Lambda}_V(\hat{v},\hat{\eta})$ and $\Lambda_Q(\hat{v},\hat{\eta})$, from (12.3.100),

$$\hat{\Lambda}_V(\hat{v},\hat{\eta}|\mathrm{k}_1)\Lambda_Q(\hat{v},\hat{\eta}|\mathrm{k}_1) = [\mathrm{h}_{\mathfrak{B}}(\hat{v}+\hat{\eta}|\mathrm{k}_1)]^{N_2}\Lambda_Q(\hat{v}-2\hat{\eta},\hat{\eta}|\mathrm{k}_1) + [\mathrm{h}_{\mathfrak{B}}(\hat{v}-\hat{\eta}|\mathrm{k}_1)]^{N_2}\Lambda_Q(\hat{v}+2\hat{\eta},\hat{\eta}|\mathrm{k}_1). \tag{12.3.103}$$

It is clear that if this equation is satisfied by more than one eigenvalue of $\boldsymbol{Q}(\hat{v},\hat{\eta})$ for a particular eigenvalue of $\widehat{\boldsymbol{V}}(\hat{v},\hat{\eta})$ then that eigenvalue is degenerate. If $\hat{\Lambda}_V(\hat{v},\hat{\eta})$ is non-degenerate then the eigenvalue of $\boldsymbol{Q}(\hat{v},\hat{\eta})$ satisfying (12.3.103) for this $\hat{\Lambda}_V(\hat{v},\hat{\eta})$ is unique.

In Sect. 12.3.5 we discussed the commutation relationships of $\widehat{\boldsymbol{V}}(\hat{v},\hat{\eta})$ with $\boldsymbol{S}_x$ and $\boldsymbol{S}_z$. Consider now the possible commutation relations

$$[\boldsymbol{S}_x,\ \boldsymbol{Q}(\hat{v},\hat{\eta})] = \mathbf{0}, \tag{12.3.104}$$

$$[\boldsymbol{S}_z,\ \boldsymbol{Q}(\hat{v},\hat{\eta})] = \mathbf{0}, \tag{12.3.105}$$

$$[\boldsymbol{S}_x\boldsymbol{S}_z,\ \boldsymbol{Q}(\hat{v},\hat{\eta})] = \mathbf{0}. \tag{12.3.106}$$

If none of the eigenvalues of $\widehat{\boldsymbol{V}}(\hat{v},\hat{\eta})$ is degenerate (for which it must be the case that N_2 is even) then its unique eigenmatrix is also an eigenmatrix of $\boldsymbol{S}_x$, $\boldsymbol{S}_z$, $\boldsymbol{S}_x\boldsymbol{S}_z$ and $\widehat{\boldsymbol{Q}}(\hat{v},\hat{\eta})$, which thus satisfies all of (12.3.104)–(12.3.106). If $\boldsymbol{Q}(\hat{v},\hat{\eta})$ does not satisfy all these three equations then it will satisfy either none or exactly one of them (Sect. 17.2.2). In those two situations it follows that $\widehat{\boldsymbol{V}}(\hat{v},\hat{\eta})$ will have degenerate eigenvalues and $\boldsymbol{Q}(\hat{v},\hat{\eta})$ will not be unique, since although $\boldsymbol{\Lambda}_Q(\hat{v},\hat{\eta})$ is unique $\boldsymbol{X}^\dagger(\hat{\eta})\boldsymbol{\Lambda}_Q(\hat{v},\hat{\eta})\boldsymbol{X}(\hat{\eta})$ is a possible $\boldsymbol{Q}(\hat{v},\hat{\eta})$ for all $\boldsymbol{X}(\hat{\eta}) \in \mathbb{S}_{\mathrm{E}}(\widehat{\boldsymbol{V}})$.

A relationship between the interchange condition (12.3.97) and the commutation relations (12.3.104)–(12.3.106) is encapsulated in the following result:

Theorem 12.3.2 *If $\boldsymbol{E}_1$ are $\boldsymbol{E}_2$ are two of the four matrices $\boldsymbol{I}$, $\boldsymbol{S}_x$, $\boldsymbol{S}_z$ and $\boldsymbol{S}_x\boldsymbol{S}_z$, then $[\boldsymbol{E}_1\boldsymbol{E}_2,\ \boldsymbol{Q}(\hat{v},\hat{\eta})] = \mathbf{0}$ if $\boldsymbol{E}_1$ and $\boldsymbol{E}_2$ each satisfy* (12.3.97).

Proof If $\boldsymbol{E}_1$ and $\boldsymbol{E}_2$ satisfy (12.3.97), from (12.3.99),

$$\boldsymbol{Q}(\hat{v},\hat{\eta}) = \boldsymbol{Q}_{\mathrm{R}}(\hat{v},\hat{\eta})\boldsymbol{Q}_{\mathrm{R}}^{-1}(\hat{v}_0,\hat{\eta}) = \boldsymbol{E}_i\boldsymbol{Q}_{\mathrm{L}}^{-1}(\hat{v}_0,\hat{\eta})\boldsymbol{Q}_{\mathrm{L}}(\hat{v},\hat{\eta})\boldsymbol{E}_i, \qquad i = 1, 2, \tag{12.3.107}$$

and

$$\begin{aligned}(\boldsymbol{E}_1\boldsymbol{E}_2)\boldsymbol{Q}(\hat{v},\hat{\eta}) &= \boldsymbol{E}_1\boldsymbol{Q}_{\mathrm{L}}^{-1}(\hat{v}_0,\hat{\eta})\boldsymbol{Q}_{\mathrm{L}}(\hat{v},\hat{\eta})\boldsymbol{E}_2\\ &= \boldsymbol{E}_1\boldsymbol{Q}_{\mathrm{L}}^{-1}(\hat{v}_0,\hat{\eta})\boldsymbol{Q}_{\mathrm{L}}(\hat{v},\hat{\eta})\boldsymbol{E}_1(\boldsymbol{E}_1\boldsymbol{E}_2)\\ &= \boldsymbol{Q}(\hat{v},\hat{\eta})(\boldsymbol{E}_1\boldsymbol{E}_2).\end{aligned} \tag{12.3.108}$$

□

The matrix $\boldsymbol{Q}(\hat{v},\hat{\eta})$ is a function of the parameters m and m', which enter through the Baxter's modified theta functions. We show below, in Sect. 12.3.6.3, that the interchange condition holds for $\boldsymbol{E} = \boldsymbol{I}$ unless both m and m' are odd. Fabricius and McCoy (2007) showed that the interchange condition holds for $\boldsymbol{E} = \boldsymbol{S}_x$ except if m is odd and m' even, for $\boldsymbol{E} = \boldsymbol{S}_z$ except if m is even and m' odd and $\boldsymbol{E} = \boldsymbol{S}_x\boldsymbol{S}_z$ only when both m and m' are even. It can, therefore, be concluded from Theorem 12.3.2 that:

(i) If m is odd and m' is even $\boldsymbol{Q}(\hat{v},\hat{\eta})$ commutes with $\boldsymbol{S}_z$.
(ii) If m is even and m' is odd $\boldsymbol{Q}(\hat{v},\hat{\eta})$ commutes with $\boldsymbol{S}_x$.
(iii) If m and m' are odd $\boldsymbol{Q}(\hat{v},\hat{\eta})$ commutes with $\boldsymbol{S}_x\boldsymbol{S}_z$.
(iv) If m and m' are even $\boldsymbol{Q}(\hat{v},\hat{\eta})$ commutes with $\boldsymbol{S}_z$, $\boldsymbol{S}_x$ and $\boldsymbol{S}_x\boldsymbol{S}_z$.

A key element of the derivation of $\boldsymbol{Q}(\hat{v},\hat{\eta})$ is whether or not the value of $\hat{\eta}$ is an elliptic root of unity, as specified by (12.3.128). As a digression from the main aim of the paper Baxter (1973a, Sect. 6) presented the derivation of a $\boldsymbol{Q}(\hat{v},\hat{\eta})$ satisfying (12.3.104)–(12.3.106). This matrix is unique, for specified values of $\hat{v}$, k_0 and $\hat{\eta}$, and all the eigenvalues of $\widehat{\boldsymbol{V}}(\hat{v},\hat{\eta})$ are non-degenerate. This case was presented in greater detail by Baxter (1982b, Chap. 10).

In contrast the first solution of the model by Baxter (1972b) adopted the elliptic roots of unity condition for $\hat{\eta}$. Then the matrix $\boldsymbol{Q}(\hat{v},\hat{\eta})$ is not unique, not all of (12.3.104)–(12.3.106) are satisfied and some of the eigenvalues of $\widehat{\boldsymbol{V}}(\hat{v},\hat{\eta})$ have degeneracies. A comprehensive list of all the known $\boldsymbol{Q}$ matrices is given by Fabricius and McCoy (2009) together with conjectured functional equations which they satisfy.

It should be emphasized that all the methods must give the same eigenvalues for $\widehat{\boldsymbol{V}}(\hat{v},\hat{\eta})$, otherwise they would yield different results for the free energy; the differences are evidenced by different degeneracies.

12.3.6.2 The Derivation of $\boldsymbol{Q}_{\mathrm{R}}$ and $\boldsymbol{Q}_{\mathrm{L}}$

The matrices $\boldsymbol{Q}_{\mathrm{R}}(\hat{\upsilon},\hat{\eta})$ and $\boldsymbol{Q}_{\mathrm{L}}(\hat{\upsilon},\hat{\eta})$ can be derived independently, but the relationship (12.3.98) allows us to consider only $\boldsymbol{Q}_{\mathrm{R}}(\hat{\upsilon},\hat{\eta})$. Suppose that

$$\langle \boldsymbol{\varsigma}|\boldsymbol{Q}_{\mathrm{R}}(\hat{\upsilon},\hat{\eta})|\boldsymbol{\varsigma}'\rangle = \mathrm{Trace}\left\{\boldsymbol{T}_{\mathrm{R}1}\begin{pmatrix}\varsigma_1'\\ \varsigma_1\end{pmatrix};\hat{\upsilon},\hat{\eta}\Big)\cdots\boldsymbol{T}_{\mathrm{R}1}\Big(\begin{matrix}\varsigma_{N_2}'\\ \varsigma_{N_2}\end{matrix};\hat{\upsilon},\hat{\eta}\Big)\right\}, \tag{12.3.109}$$

where the matrices $\boldsymbol{T}_{\mathrm{R}1}$ are of dimension L. As in (12.3.24) and (12.3.25),

$$\langle \boldsymbol{\varsigma}|\widehat{\boldsymbol{V}}(\hat{\upsilon},\hat{\eta})\,\boldsymbol{Q}_{\mathrm{R}}(\hat{\upsilon},\hat{\eta})|\boldsymbol{\varsigma}'\rangle = \mathrm{Trace}\left\{\boldsymbol{Y}_{\mathrm{R}}\Big(\begin{matrix}\varsigma_1'\\ \varsigma_1\end{matrix};\hat{\upsilon},\hat{\eta}\Big)\cdots\boldsymbol{Y}_{\mathrm{R}}\Big(\begin{matrix}\varsigma_{N_2}'\\ \varsigma_{N_2}\end{matrix};\hat{\upsilon},\hat{\eta}\Big)\right\}, \tag{12.3.110}$$

where

$$\boldsymbol{Y}_{\mathrm{R}}\Big(\begin{matrix}\varsigma'\\ \varsigma\end{matrix};\hat{\upsilon},\hat{\eta}\Big) = \sum_{\varsigma''=\pm1}\widehat{\boldsymbol{U}}_1\Big(\begin{matrix}\varsigma''\\ \varsigma\end{matrix};\hat{\upsilon},\hat{\eta}\Big)\otimes\boldsymbol{T}_{\mathrm{R}1}\Big(\begin{matrix}\varsigma'\\ \varsigma''\end{matrix};\hat{\upsilon},\hat{\eta}\Big). \tag{12.3.111}$$

From (12.3.94)

$$\boldsymbol{Y}_{\mathrm{R}}\Big(\begin{matrix}\varsigma\\ +1\end{matrix};\hat{\upsilon},\hat{\eta}\Big) = \begin{pmatrix}\hat{\mathfrak{a}}(\hat{\upsilon},\hat{\eta}|\mathrm{k}_1)\boldsymbol{T}_{\mathrm{R}1}\Big(\begin{matrix}\varsigma\\ +1\end{matrix};\hat{\upsilon},\hat{\eta}\Big) & \hat{\mathfrak{d}}(\hat{\upsilon},\hat{\eta}|\mathrm{k}_1)\boldsymbol{T}_{\mathrm{R}1}\Big(\begin{matrix}\varsigma\\ -1\end{matrix};\hat{\upsilon},\hat{\eta}\Big)\\[2ex] \hat{\mathfrak{c}}(\hat{\upsilon},\hat{\eta}|\mathrm{k}_1)\boldsymbol{T}_{\mathrm{R}1}\Big(\begin{matrix}\varsigma\\ -1\end{matrix};\hat{\upsilon},\hat{\eta}\Big) & \hat{\mathfrak{b}}(\hat{\upsilon},\hat{\eta}|\mathrm{k}_1)\boldsymbol{T}_{\mathrm{R}1}\Big(\begin{matrix}\varsigma\\ +1\end{matrix};\hat{\upsilon},\hat{\eta}\Big)\end{pmatrix}, \tag{12.3.112}$$

$$\boldsymbol{Y}_{\mathrm{R}}\Big(\begin{matrix}\varsigma\\ -1\end{matrix};\hat{\upsilon},\hat{\eta}\Big) = \begin{pmatrix}\hat{\mathfrak{b}}(\hat{\upsilon},\hat{\eta}|\mathrm{k}_1)\boldsymbol{T}_{\mathrm{R}1}\Big(\begin{matrix}\varsigma\\ -1\end{matrix};\hat{\upsilon},\hat{\eta}\Big) & \hat{\mathfrak{c}}(\hat{\upsilon},\hat{\eta}|\mathrm{k}_1)\boldsymbol{T}_{\mathrm{R}1}\Big(\begin{matrix}\varsigma\\ +1\end{matrix};\hat{\upsilon},\hat{\eta}\Big)\\[2ex] \hat{\mathfrak{d}}(\hat{\upsilon},\hat{\eta}|\mathrm{k}_1)\boldsymbol{T}_{\mathrm{R}1}\Big(\begin{matrix}\varsigma\\ +1\end{matrix};\hat{\upsilon},\hat{\eta}\Big) & \hat{\mathfrak{a}}(\hat{\upsilon},\hat{\eta}|\mathrm{k}_1)\boldsymbol{T}_{\mathrm{R}1}\Big(\begin{matrix}\varsigma\\ -1\end{matrix};\hat{\upsilon},\hat{\eta}\Big)\end{pmatrix}. \tag{12.3.113}$$

Suppose that a matrix $\boldsymbol{W}(\hat{\eta})$ can be found so that

$$
\boldsymbol{W}^{-1}(\hat{\eta})\boldsymbol{Y}_{\mathrm{R}}\begin{pmatrix}\varsigma\\ \varsigma'\end{pmatrix};\hat{v},\hat{\eta}\Big)\boldsymbol{W}(\hat{\eta}) = \begin{pmatrix} \boldsymbol{A}\left(\begin{matrix}\varsigma\\ \varsigma'\end{matrix};\hat{v},\hat{\eta}\right) & \boldsymbol{0} \\ \boldsymbol{O}\left(\begin{matrix}\varsigma\\ \varsigma'\end{matrix};\hat{v},\hat{\eta}\right) & \boldsymbol{B}\left(\begin{matrix}\varsigma\\ \varsigma'\end{matrix};\hat{v},\hat{\eta}\right)\end{pmatrix}. \tag{12.3.114}
$$

Then (12.3.110) decomposes into the form

$$
\begin{aligned}
\langle\boldsymbol{\varsigma}|\widehat{\boldsymbol{V}}(\hat{v},\hat{\eta})\boldsymbol{Q}_{\mathrm{R}}(\hat{v},\hat{\eta})|\boldsymbol{\varsigma}'\rangle &= \mathrm{Trace}\left\{\boldsymbol{A}\left(\begin{matrix}\varsigma_1'\\ \varsigma_1\end{matrix};\hat{v},\hat{\eta}\right)\cdots\boldsymbol{A}\left(\begin{matrix}\varsigma_{N_2}'\\ \varsigma_{N_2}\end{matrix};\hat{v},\hat{\eta}\right)\right\}\\
&+\mathrm{Trace}\left\{\boldsymbol{B}\left(\begin{matrix}\varsigma_1'\\ \varsigma_1\end{matrix};\hat{v},\hat{\eta}\right)\cdots\boldsymbol{B}\left(\begin{matrix}\varsigma_{N_2}'\\ \varsigma_{N_2}\end{matrix};\hat{v},\hat{\eta}\right)\right\}.
\end{aligned} \tag{12.3.115}
$$

Suppose that

$$
\boldsymbol{W}(\hat{\eta}) := \begin{pmatrix}\boldsymbol{I}_L & \boldsymbol{W}_L(\hat{\eta})\\ \boldsymbol{0} & \boldsymbol{I}_L\end{pmatrix}, \tag{12.3.116}
$$

where $\boldsymbol{I}_L$ is the L-dimensional unit matrix and

$$
\langle k|\boldsymbol{W}_L(\hat{\eta})|k'\rangle = \left\langle\begin{matrix}+1\\ k\end{matrix}\right|\boldsymbol{W}(\hat{\eta})\left|\begin{matrix}-1\\ k'\end{matrix}\right\rangle = \delta^{\mathrm{Kr}}(k-k')W(\hat{\eta};k|\mathrm{k}_1). \tag{12.3.117}
$$

Then

$$
\boldsymbol{W}^{-1}(\hat{\eta}) := \begin{pmatrix}\boldsymbol{I}_L & -\boldsymbol{W}_L(\hat{\eta})\\ \boldsymbol{0} & \boldsymbol{I}_L\end{pmatrix}, \tag{12.3.118}
$$

giving

$$
\begin{aligned}
\boldsymbol{A}\left(\begin{matrix}\varsigma\\ +1\end{matrix};\hat{v},\hat{\eta}\right) &= \hat{\mathfrak{a}}(\hat{v},\hat{\eta}|\mathrm{k}_1)\boldsymbol{T}_{\mathrm{R}1}\left(\begin{matrix}\varsigma\\ +1\end{matrix};\hat{v},\hat{\eta}\right) - \hat{\mathfrak{c}}(\hat{v},\hat{\eta}|\mathrm{k}_1)\boldsymbol{W}_L(\hat{\eta})\boldsymbol{T}_{\mathrm{R}1}\left(\begin{matrix}\varsigma\\ -1\end{matrix};\hat{v},\hat{\eta}\right),\\
\boldsymbol{A}\left(\begin{matrix}\varsigma\\ -1\end{matrix};\hat{v},\hat{\eta}\right) &= \hat{\mathfrak{b}}(\hat{v},\hat{\eta}|\mathrm{k}_1)\boldsymbol{T}_{\mathrm{R}1}\left(\begin{matrix}\varsigma\\ -1\end{matrix};\hat{v},\hat{\eta}\right) - \hat{\mathfrak{d}}(\hat{v},\hat{\eta}|\mathrm{k}_1)\boldsymbol{W}_L(\hat{\eta})\boldsymbol{T}_{\mathrm{R}1}\left(\begin{matrix}\varsigma\\ +1\end{matrix};\hat{v},\hat{\eta}\right),
\end{aligned} \tag{12.3.119}
$$

$$\boldsymbol{B}\left(\begin{matrix}\varsigma\\+1\end{matrix};\hat{v},\hat{\eta}\right)=\hat{\mathfrak{b}}(\hat{v},\hat{\eta}|\mathrm{k}_1)\boldsymbol{T}_{\mathrm{R}1}\left(\begin{matrix}\varsigma\\+1\end{matrix};\hat{v},\hat{\eta}\right)+\hat{\mathfrak{c}}(\hat{v},\hat{\eta}|\mathrm{k}_1)\boldsymbol{T}_{\mathrm{R}1}\left(\begin{matrix}\varsigma\\-1\end{matrix};\hat{v},\hat{\eta}\right)\boldsymbol{W}_L(\hat{\eta}),$$

$$\boldsymbol{B}\left(\begin{matrix}\varsigma\\-1\end{matrix};\hat{v},\hat{\eta}\right)=\hat{\mathfrak{a}}(\hat{v},\hat{\eta}|\mathrm{k}_1)\boldsymbol{T}_{\mathrm{R}1}\left(\begin{matrix}\varsigma\\-1\end{matrix};\hat{v},\hat{\eta}\right)+\hat{\mathfrak{d}}(\hat{v},\hat{\eta}|\mathrm{k}_1)\boldsymbol{T}_{\mathrm{R}1}\left(\begin{matrix}\varsigma\\+1\end{matrix};\hat{v},\hat{\eta}\right)\boldsymbol{W}_L(\hat{\eta}),$$
(12.3.120)

$$\hat{\mathfrak{a}}(\hat{v},\hat{\eta}|\mathrm{k}_1)\boldsymbol{T}_{\mathrm{R}1}\left(\begin{matrix}\varsigma\\+1\end{matrix};\hat{v},\hat{\eta}\right)\boldsymbol{W}_L(\hat{\eta})-\hat{\mathfrak{b}}(\hat{v},\hat{\eta}|\mathrm{k}_1)\boldsymbol{W}_L(\hat{\eta})\boldsymbol{T}_{\mathrm{R}1}\left(\begin{matrix}\varsigma\\+1\end{matrix};\hat{v},\hat{\eta}\right)$$
$$+\hat{\mathfrak{d}}(\hat{v},\hat{\eta}|\mathrm{k}_1)\boldsymbol{T}_{\mathrm{R}1}\left(\begin{matrix}\varsigma\\-1\end{matrix};\hat{v},\hat{\eta}\right)-\hat{\mathfrak{c}}(\hat{v},\hat{\eta}|\mathrm{k}_1)\boldsymbol{W}_L(\hat{\eta})\boldsymbol{T}_{\mathrm{R}1}\left(\begin{matrix}\varsigma\\-1\end{matrix};\hat{v},\hat{\eta}\right)\boldsymbol{W}_L(\hat{\eta})=\boldsymbol{0},$$

$$\hat{\mathfrak{b}}(\hat{v},\hat{\eta}|\mathrm{k}_1)\boldsymbol{T}_{\mathrm{R}1}\left(\begin{matrix}\varsigma\\-1\end{matrix};\hat{v},\hat{\eta}\right)\boldsymbol{W}_L(\hat{\eta})-\hat{\mathfrak{a}}(\hat{v},\hat{\eta}|\mathrm{k}_1)\boldsymbol{W}_L(\hat{\eta})\boldsymbol{T}_{\mathrm{R}1}\left(\begin{matrix}\varsigma\\-1\end{matrix};\hat{v},\hat{\eta}\right)$$
$$+\hat{\mathfrak{c}}(\hat{v},\hat{\eta}|\mathrm{k}_1)\boldsymbol{T}_{\mathrm{R}1}\left(\begin{matrix}\varsigma\\+1\end{matrix};\hat{v},\hat{\eta}\right)-\hat{\mathfrak{d}}(\hat{v},\hat{\eta}|\mathrm{k}_1)\boldsymbol{W}_L(\eta)\boldsymbol{T}_{\mathrm{R}1}\left(\begin{matrix}\varsigma\\+1\end{matrix};\hat{v},\hat{\eta}\right)\boldsymbol{W}_L(\hat{\eta})=\boldsymbol{0}.$$
(12.3.121)

To satisfy (12.3.121) for each matrix element either

$$\langle k|\boldsymbol{T}_{\mathrm{R}1}\left(\begin{matrix}\varsigma\\\pm 1\end{matrix};\hat{v},\hat{\eta}\right)|k'\rangle=0 \qquad (12.3.122)$$

or

$$8\hat{\Delta}(\hat{\eta})W(\hat{\eta};k|\mathrm{k}_1)W(\hat{\eta};k'|\mathrm{k}_1)=[\hat{\Gamma}(\hat{\eta})+1][W(\hat{\eta};k|\mathrm{k}_1)^2+W(\hat{\eta};k'|\mathrm{k}_1)^2]$$
$$+[\hat{\Gamma}(\hat{\eta})-1][1+W(\hat{\eta};k|\mathrm{k}_1)^2W(\hat{\eta};k'|\mathrm{k}_1)^2], \qquad (12.3.123)$$

where $\hat{\Delta}(\hat{\eta})$ and $\hat{\Gamma}(\hat{\eta})$ are given by (9.3.1) with the Boltzmann weights $\mathfrak{a}$, $\mathfrak{b}$, $\mathfrak{c}$ and $\mathfrak{d}$ replaced by $\hat{\mathfrak{a}}$, $\hat{\mathfrak{b}}$, $\hat{\mathfrak{c}}$ and $\hat{\mathfrak{d}}$. In zone $\mathcal{R}_{\mathrm{L}}(\mathrm{I})$, $\Delta=\Delta^\star$ and $\Gamma=\Gamma^\star$. So, using the substitution (12.3.83) in Table 12.2,

$$2\mathrm{cn}(2\hat{\eta}|\mathrm{k}_1)\mathrm{dn}(2\hat{\eta}|\mathrm{k}_1)W(\hat{\eta};k|\mathrm{k}_1)W(\hat{\eta};k'|\mathrm{k}_1)-[W(\hat{\eta};k|\mathrm{k}_1)^2+W(\hat{\eta};k'|\mathrm{k}_1)^2]$$
$$+\mathrm{k}_1\,\mathrm{sn}^2(2\hat{\eta}|\mathrm{k}_1)[1+W(\hat{\eta};k|\mathrm{k}_1)^2W(\hat{\eta};k'|\mathrm{k}_1)^2]=0. \qquad (12.3.124)$$

This equation is satisfied by

$$W(\hat{\eta};k|\mathrm{k}_1)=\sqrt{\mathrm{k}_1}\,\mathrm{sn}(x\pm\hat{\eta}|\mathrm{k}_1),$$
$$W(\hat{\eta};k'|\mathrm{k}_1)=\sqrt{\mathrm{k}_1}\,\mathrm{sn}(x\mp\hat{\eta}|\mathrm{k}_1), \qquad (12.3.125)$$

for any x, with the two solutions coinciding when $x=\mathcal{K}_1+\mho(2\ell,\ell'|\mathrm{k}_1)$ to give

$$W(\hat{\eta}; k|\mathrm{k}_1) = W(\hat{\eta}; k'|\mathrm{k}_1) = \sqrt{\mathrm{k}_1}\,\mathrm{sn}(\mathcal{K}_1 + \mho(2\ell, \ell'|\mathrm{k}_1) \pm \hat{\eta}|\mathrm{k}_1), \quad (12.3.126)$$

where ℓ and ℓ' are any integers and, as defined in (18.3.21), $\mho(2\ell, \ell'|\mathrm{k}_1) := 2\ell\,\mathcal{K}_1 + \ell'\,\mathrm{i}\,\mathcal{K}_1$. The Eq. (12.3.125) cannot be satisfied for all pairs of integers k and k' in the range $[1, L]$. So (12.3.122) is imposed except for a sparse set of matrix elements. Two choices have been shown to satisfy conditions (i)–(iii) of p. 448. Matrix elements are non-zero only when either:

(1) (a) $k = k' = 1$, (b) $k' = k \pm 1$, (c) $k = k' = L$,

or

(2) $k' = k \pm 1$, with $k', k = 1, 2, \ldots, L \pmod L$.

In this chapter attention is confined to case (1). For the cyclic case (2), which was proposed by Fabricius (2007), the reader is referred to Fabricius and McCoy (2007) and McCoy (2010).

Conditions (a) and (b) are satisfied by

$$W(\hat{\eta}; k|\mathrm{k}_1) := \sqrt{\mathrm{k}_1}\,\mathrm{sn}(\mathcal{K}_1 + (2k-1)\hat{\eta}|\mathrm{k}_1) = W(-\hat{\eta}; k|\mathrm{k}_1). \quad (12.3.127)$$

To satisfy (c), from (18.3.48),

$$\hat{\eta} = \frac{\mho(2m, m'|\mathrm{k}_1)}{2L} = \frac{m}{L}\mathcal{K}_1 + \frac{m'}{2L}\mathrm{i}\,\mathcal{K}_1', \quad (12.3.128)$$

for some integers m and m'; this is called the *elliptic roots of unity condition.*

We pause to note the remarkable nature of this result. A parameter $\hat{\eta}$, to which our initial discussions attributed only imaginary values, must now take one of a set of discrete complex values, called the elliptic roots of unity. Further it is the case that the original solution of Baxter (1972b) corresponds to $m' = 0$, when $\hat{\eta}$ becomes real. The effect of $\hat{\eta}$ changing from a purely imaginary to a complex and more particularly to a real value is to shift the zone of validity from $\mathcal{R}_{\mathrm{L}}(\mathrm{III})$. The zone into which it is transformed will become clear when the calculation yields a form for the free-energy density.

Given that the same integer pair (m, m') is used to define $\hat{\eta}$ and the Baxter modified theta functions, so that $\mho(2m, m'|\mathrm{k}_1) := \mho_{\mathfrak{B}1}$, from (18.3.131),

$$\begin{aligned} W(\hat{\eta}; k|\mathrm{k}_1) &= \frac{\mathrm{H}_{\mathfrak{B}}(\mathcal{K}_1 + (2k-1)\hat{\eta}|\mathrm{k}_1)}{\Theta_{\mathfrak{B}}(\mathcal{K}_1 + (2k-1)\hat{\eta}|\mathrm{k}_1)}, \quad k = 1, 2, \ldots, L-1, \\ W(\hat{\eta}; L|\mathrm{k}_1) &= \frac{\mathrm{H}_{\mathfrak{B}}(\mathcal{K}_1 + \mho_{\mathfrak{B}1} - \hat{\eta}|\mathrm{k}_1)}{\Theta_{\mathfrak{B}}(\mathcal{K}_1 + \mho_{\mathfrak{B}1} - \hat{\eta}|\mathrm{k}_1)}. \end{aligned} \quad (12.3.129)$$

From (12.3.121),

$$\begin{aligned}
&\langle k|\boldsymbol{T}_{\mathrm{R}1}\begin{pmatrix}\varsigma\\+1\end{pmatrix};\hat{\upsilon},\hat{\eta}\Big)|k'\rangle\,F(-1,+1;\hat{\upsilon},\hat{\eta};k,k'|\mathrm{k}_1)\\
&\qquad=\langle k|\boldsymbol{T}_{\mathrm{R}1}\Big(\begin{matrix}\varsigma\\-1\end{matrix};\hat{\upsilon},\hat{\eta}\Big)|k'\rangle\,F(+1,-1;\hat{\upsilon},\hat{\eta};k,k'|\mathrm{k}_1),\\
&\langle k|\boldsymbol{T}_{\mathrm{R}1}\Big(\begin{matrix}\varsigma\\+1\end{matrix};\hat{\upsilon},\hat{\eta}\Big)|k'\rangle\,F(+1,+1;\hat{\upsilon},\hat{\eta};k,k'|\mathrm{k}_1)\\
&\qquad=\langle k|\boldsymbol{T}_{\mathrm{R}1}\Big(\begin{matrix}\varsigma\\-1\end{matrix};\hat{\upsilon},\hat{\eta}\Big)|k'\rangle\,F(-1,-1;\hat{\upsilon},\hat{\eta};k,k'|\mathrm{k}_1),
\end{aligned}\tag{12.3.130}$$

where, from (12.3.127), the functions

$$\begin{aligned}
F(+1,\beta;\hat{\upsilon},\hat{\eta};k,k'|\mathrm{k}_1)&:=\Psi_{\mathfrak{B}}(+1,\beta;\hat{\upsilon},\hat{\eta}|\mathrm{k}_1)\\
&\qquad-\Psi_{\mathfrak{B}}(+1,-\beta;\hat{\upsilon},\hat{\eta}|\mathrm{k}_1)W(\hat{\eta};k|\mathrm{k}_1)W(\eta;k'|\mathrm{k}_1),\\
F(-1,\beta;\hat{\upsilon},\hat{\eta};k,k'|\mathrm{k}_1)&:=\Psi_{\mathfrak{B}}(-1,-\beta;\hat{\upsilon},\hat{\eta}|\mathrm{k}_1)W(\hat{\eta};k|\mathrm{k}_1)\\
&\qquad-\Psi_{\mathfrak{B}}(-1,\beta;\hat{\upsilon},\hat{\eta}|\mathrm{k}_1)W(\hat{\eta};k'|\mathrm{k}_1),
\end{aligned}\tag{12.3.131}$$

have the properties

$$F(\alpha,\beta;\hat{\upsilon},\hat{\eta};k',k|\mathrm{k}_1)=\alpha\,F(\alpha,\beta\alpha;\hat{\upsilon},\hat{\eta};k,k'|\mathrm{k}_1),\tag{12.3.132}$$

$$F(\alpha,\beta;-\hat{\upsilon},\hat{\eta};k,k'|\mathrm{k}_1)=\alpha\,F(\alpha,\beta\alpha;\hat{\upsilon},\hat{\eta};k,k'|\mathrm{k}_1),\tag{12.3.133}$$

$$F(\alpha,\beta;\hat{\upsilon},-\hat{\eta};k,k'|\mathrm{k}_1)=-\alpha\,F(\alpha,\beta\alpha;\hat{\upsilon},\hat{\eta};k,k'|\mathrm{k}_1),\tag{12.3.134}$$

from which it follows that the conditions

$$\langle k'|\boldsymbol{T}_{\mathrm{R}1}\Big(\begin{matrix}\varsigma\\\nu\end{matrix};\hat{\upsilon},\hat{\eta}\Big)|k\rangle=\langle k|\boldsymbol{T}_{\mathrm{R}1}\Big(\begin{matrix}\varsigma\\\nu\end{matrix};-\hat{\upsilon},\hat{\eta}\Big)|k'\rangle=\langle k|\boldsymbol{T}_{\mathrm{R}1}\Big(\begin{matrix}\varsigma\\\nu\end{matrix};\hat{\upsilon},-\hat{\eta}\Big)|k'\rangle\tag{12.3.135}$$

can be consistently imposed. From (12.3.85), (12.3.130), (12.3.131), (12.3.135), (18.3.33) and (18.3.34),

$$\begin{aligned}
\langle k|\boldsymbol{T}_{\mathrm{R}1}\Big(\begin{matrix}\varsigma\\+1\end{matrix};\hat{\upsilon},\hat{\eta}\Big)|k+1\rangle&=\mathrm{H}_{\mathfrak{B}}(\mathcal{K}_1-\hat{\upsilon}+2k\hat{\eta})|\mathrm{k}_1)G(\varsigma;k),\\
\langle k|\boldsymbol{T}_{\mathrm{R}1}\Big(\begin{matrix}\varsigma\\-1\end{matrix};\hat{\upsilon},\hat{\eta}\Big)|k+1\rangle&=\Theta_{\mathfrak{B}}(\mathcal{K}_1-\hat{\upsilon}+2k\hat{\eta})|\mathrm{k}_1)G(\varsigma;k),
\end{aligned}\tag{12.3.136}$$

where the pair $(k, k+1)$ is in the range $[1, L]$, and

$$\begin{aligned} \langle 1|\boldsymbol{T}_{\mathrm{R}1}\begin{pmatrix}\varsigma \\ +1\end{pmatrix}; \hat{v}, \hat{\eta}\Big)|1\rangle &= \mathrm{H}_{\mathfrak{B}}(\mathcal{K}_1 - \hat{v}|\mathrm{k}_1) G_0(\varsigma; 1), \\ \langle 1|\boldsymbol{T}_{\mathrm{R}1}\begin{pmatrix}\varsigma \\ -1\end{pmatrix}; \hat{v}, \hat{\eta}\Big)|1\rangle &= \Theta_{\mathfrak{B}}(\mathcal{K}_1 - \hat{v}|\mathrm{k}_1) G_0(\varsigma; 1), \end{aligned} \tag{12.3.137}$$

$$\begin{aligned} \langle L|\boldsymbol{T}_{\mathrm{R}1}\begin{pmatrix}\varsigma \\ +1\end{pmatrix}; \hat{v}, \hat{\eta}\Big)|L\rangle &= \mathrm{H}_{\mathfrak{B}}(\mathcal{K}_1 + \mho_{\mathfrak{B}1} - \hat{v}|\mathrm{k}_1) G_0(\varsigma; L), \\ \langle L|\boldsymbol{T}_{\mathrm{R}1}\begin{pmatrix}\varsigma \\ -1\end{pmatrix}; \hat{v}, \hat{\eta}\Big)|L\rangle &= \Theta_{\mathfrak{B}}(\mathcal{K}_1 + \mho_{\mathfrak{B}1} - \hat{v}|\mathrm{k}_1) G_0(\varsigma; L), \end{aligned} \tag{12.3.138}$$

where $G(\varsigma; k)$, $G_0(\varsigma; 1)$ and $G_0(\varsigma; L)$ are arbitrary constants.[58]

It follows from (12.3.119) and (12.3.120) the $\boldsymbol{A}$ and $\boldsymbol{B}$ matrices have the same zero elements as $\boldsymbol{T}_{\mathrm{R}1}$ with

$$\begin{aligned} \boldsymbol{A}\left(\begin{matrix}\varsigma' \\ \varsigma\end{matrix}; \hat{v}, \hat{\eta}\right) &= \mathrm{h}_{\mathfrak{B}}(\hat{v} - \hat{\eta}|\mathrm{k})\, \boldsymbol{\Theta}_{\mathfrak{B}}^{-1} \boldsymbol{T}_{\mathrm{R}1}\left(\begin{matrix}\varsigma' \\ \varsigma\end{matrix}; \hat{v} + 2\hat{\eta}, \hat{\eta}\right) \boldsymbol{\Theta}_{\mathfrak{B}}, \\ \boldsymbol{B}\left(\begin{matrix}\varsigma' \\ \varsigma\end{matrix}; \hat{v}, \hat{\eta}\right) &= \mathrm{h}_{\mathfrak{B}}(\hat{v} + \hat{\eta}|\mathrm{k})\, \boldsymbol{\Theta}_{\mathfrak{B}} \boldsymbol{T}_{\mathrm{R}1}\left(\begin{matrix}\varsigma' \\ \varsigma\end{matrix}; \hat{v} - 2\hat{\eta}, \hat{\eta}\right) \boldsymbol{\Theta}_{\mathfrak{B}}^{-1}, \end{aligned} \tag{12.3.139}$$

where $\boldsymbol{\Theta}_{\mathfrak{B}}$ is the diagonal matrix with elements

$$\langle k|\,\boldsymbol{\Theta}_{\mathfrak{B}}|k'\rangle := \delta^{\mathrm{Kr}}(k - k')\, \Theta_{\mathfrak{B}}(\mathcal{K}_1 + (2k-1)\hat{\eta}|\mathrm{k}_1), \tag{12.3.140}$$

confirming, from (12.3.115), that the matrix $\boldsymbol{Q}_{\mathrm{R}}(\hat{v}, \hat{\eta})$, defined by (12.3.109), satisfies (12.3.95).

12.3.6.3 The Interchange Relationship for $\boldsymbol{Q}_{\mathrm{R}}$ and $\boldsymbol{Q}_{\mathrm{L}}$

Here we need the companion formula

$$\langle \boldsymbol{\varsigma}|\,\boldsymbol{Q}_{\mathrm{L}}(\hat{v}, \hat{\eta})|\boldsymbol{\varsigma}'\rangle = \mathrm{Trace}\left\{\boldsymbol{T}_{\mathrm{L}1}\left(\begin{matrix}\varsigma_1' \\ \varsigma_1\end{matrix}; \hat{v}, \hat{\eta}\right) \cdots \boldsymbol{T}_{\mathrm{L}1}\left(\begin{matrix}\varsigma_{N_2}' \\ \varsigma_{N_2}\end{matrix}; \hat{v}, \hat{\eta}\right)\right\} \tag{12.3.141}$$

to (12.3.109), where, from (12.3.98),

[58] Of course, this is another *assumption* built into the derived $\boldsymbol{Q}$ matrix; in general there would be nothing to prevent the presence of arbitrary functions.

$$\boldsymbol{T}_{\mathrm{L}1}\begin{pmatrix} \varsigma' \\ \varsigma \end{pmatrix}; \hat{v}, \hat{\eta}\Bigg) = \boldsymbol{T}_{\mathrm{R}1}\begin{pmatrix} \varsigma \\ \varsigma' \end{pmatrix}; -\hat{v}, \hat{\eta}\Bigg). \tag{12.3.142}$$

We consider only the case where $\boldsymbol{E} = \boldsymbol{I}$ in (12.3.97). As in (12.3.110) and (12.3.111)

$$\langle \boldsymbol{\varsigma} | \boldsymbol{Q}_{\mathrm{L}}(\hat{v}, \hat{\eta}) \boldsymbol{Q}_{\mathrm{R}}(\hat{v}', \hat{\eta}) | \boldsymbol{\varsigma}' \rangle = \mathrm{Trace}\left\{ \boldsymbol{Z}\left(\begin{matrix} \varsigma_1' \\ \varsigma_1 \end{matrix}; \begin{matrix} \hat{v}' \\ \hat{v} \end{matrix}, \hat{\eta} \right) \cdots \boldsymbol{Z}\left(\begin{matrix} \varsigma_{N_2}' \\ \varsigma_{N_2} \end{matrix}; \begin{matrix} \hat{v}' \\ \hat{v} \end{matrix}, \hat{\eta} \right) \right\}, \tag{12.3.143}$$

where

$$\boldsymbol{Z}\left(\begin{matrix} \varsigma' \\ \varsigma \end{matrix}; \begin{matrix} \hat{v}' \\ \hat{v} \end{matrix}, \hat{\eta} \right) = \boldsymbol{Z}^{(+)}\left(\begin{matrix} \varsigma' \\ \varsigma \end{matrix}; \begin{matrix} \hat{v}' \\ \hat{v} \end{matrix}, \hat{\eta} \right) + \boldsymbol{Z}^{(-)}\left(\begin{matrix} \varsigma' \\ \varsigma \end{matrix}; \begin{matrix} \hat{v}' \\ \hat{v} \end{matrix}, \hat{\eta} \right), \tag{12.3.144}$$

$$\boldsymbol{Z}^{(\pm)}\left(\begin{matrix} \varsigma' \\ \varsigma \end{matrix}; \begin{matrix} \hat{v}' \\ \hat{v} \end{matrix}, \hat{\eta} \right) = \boldsymbol{T}_{\mathrm{L}1}\left(\begin{matrix} \pm 1 \\ \varsigma \end{matrix}; \hat{v}, \hat{\eta} \right) \otimes \boldsymbol{T}_{\mathrm{R}1}\left(\begin{matrix} \varsigma' \\ \pm 1 \end{matrix}; \hat{v}', \hat{\eta} \right), \tag{12.3.145}$$

are L^2-dimensional matrices. It is convenient to index their rows and columns by the pair $\begin{pmatrix} k \\ j \end{pmatrix}$, $j, k = 1, 2, \ldots, L$, so that

$$\begin{aligned} \left\langle \begin{matrix} k \\ j \end{matrix} \right| \boldsymbol{Z}^{(\pm)}\left(\begin{matrix} \varsigma' \\ \varsigma \end{matrix}; \begin{matrix} \hat{v}' \\ \hat{v} \end{matrix}, \hat{\eta} \right) \left| \begin{matrix} k' \\ j' \end{matrix} \right\rangle &= \langle j | \boldsymbol{T}_{\mathrm{L}1}\left(\begin{matrix} \pm 1 \\ \varsigma \end{matrix}; \hat{v}, \hat{\eta} \right) | j' \rangle \langle k | \boldsymbol{T}_{\mathrm{R}1}\left(\begin{matrix} \varsigma' \\ \pm 1 \end{matrix}; \hat{v}', \hat{\eta} \right) | k' \rangle \\ &= \langle j | \boldsymbol{T}_{\mathrm{R}1}\left(\begin{matrix} \varsigma \\ \pm 1 \end{matrix}; -\hat{v}, \hat{\eta} \right) | j' \rangle \langle k | \boldsymbol{T}_{\mathrm{R}1}\left(\begin{matrix} \varsigma' \\ \pm 1 \end{matrix}; \hat{v}', \hat{\eta} \right) | k' \rangle. \end{aligned} \tag{12.3.146}$$

A sufficient condition for the interchange relationship to hold is that there exists, for some matrix $\boldsymbol{D}$, the similarity transformation

$$\boldsymbol{Z}\left(\begin{matrix} \varsigma' \\ \varsigma \end{matrix}; \begin{matrix} \hat{v} \\ \hat{v}' \end{matrix}, \hat{\eta} \right) = \boldsymbol{D}^{-1}\left(\begin{matrix} \hat{v}' \\ \hat{v} \end{matrix}, \hat{\eta} \right) \boldsymbol{Z}\left(\begin{matrix} \varsigma' \\ \varsigma \end{matrix}; \begin{matrix} \hat{v}' \\ \hat{v} \end{matrix}, \hat{\eta} \right) \boldsymbol{D}\left(\begin{matrix} \hat{v}' \\ \hat{v} \end{matrix}, \hat{\eta} \right). \tag{12.3.147}$$

In general the task of finding such a matrix is difficult and Baxter (1972b) and later authors (see e.g. McCoy 2010) have confined their attention to a diagonal transformation.[59]

[59] Of course, in cases (specifically for different parities of m and m') where the interchange condition is established using a diagonal transformation this is sufficient. However, when a diagonal transformation does not exist it leave open the question as to whether a more general transformation may perform the task.

We first consider the four diagonal elements of $\boldsymbol{Z}$, where the contributions from the elements of the diagonal matrix $\boldsymbol{D}$ cancel,

$$\left\langle {1 \atop 1} \middle| \mathbf{Z}\left({\varsigma' \atop \varsigma} ; {\hat{\upsilon}' \atop \hat{\upsilon}}, \hat{\eta} \right) \middle| {1 \atop 1} \right\rangle = \mathfrak{L}(\hat{\upsilon}, \hat{\upsilon}'|\mathrm{k}_1) G_0(\varsigma; 1) G_0(\varsigma'; 1), \tag{12.3.148}$$

$$\left\langle {L \atop L} \middle| \mathbf{Z}\left({\varsigma' \atop \varsigma} ; {\hat{\upsilon}' \atop \hat{\upsilon}}, \hat{\eta} \right) \middle| {L \atop L} \right\rangle = \mathfrak{L}(\hat{\upsilon} - \mho_{\mathfrak{B}1}, \hat{\upsilon}' - \mho_{\mathfrak{B}1}|\mathrm{k}_1) G_0(\varsigma; L) G_0(\varsigma'; L), \tag{12.3.149}$$

$$\left\langle {L \atop 1} \middle| \mathbf{Z}\left({\varsigma' \atop \varsigma} ; {\hat{\upsilon}' \atop \hat{\upsilon}}, \hat{\eta} \right) \middle| {L \atop 1} \right\rangle = \mathfrak{L}(\hat{\upsilon}, \hat{\upsilon}' - \mho_{\mathfrak{B}1}|\mathrm{k}_1) G_0(\varsigma; 1) G_0(\varsigma'; L), \tag{12.3.150}$$

$$\left\langle {1 \atop L} \middle| \mathbf{Z}\left({\varsigma' \atop \varsigma} ; {\hat{\upsilon}' \atop \hat{\upsilon}}, \hat{\eta} \right) \middle| {1 \atop L} \right\rangle = \mathfrak{L}(\hat{\upsilon} - \mho_{\mathfrak{B}1}, \hat{\upsilon}'|\mathrm{k}_1) G_0(\varsigma; L) G_0(\varsigma'; 1), \tag{12.3.151}$$

where $\mathfrak{L}(v, u|\mathrm{k})$ is given by (18.3.143). It is clear, from (18.3.147), that the matrix elements given by (12.3.148) and (12.3.149) are invariant under the interchange of $\hat{\upsilon}$ and $\hat{\upsilon}'$, but that, from (18.3.148),

$$\left\langle {L \atop 1} \middle| \mathbf{Z}\left({\varsigma' \atop \varsigma} ; {\hat{\upsilon}' \atop \hat{\upsilon}}, \hat{\eta} \right) \middle| {L \atop 1} \right\rangle = (-1)^{mm'} \left\langle {L \atop 1} \middle| \mathbf{Z}\left({\varsigma' \atop \varsigma} ; {\hat{\upsilon} \atop \hat{\upsilon}'}, \hat{\eta} \right) \middle| {L \atop 1} \right\rangle, \tag{12.3.152}$$

$$\left\langle {1 \atop L} \middle| \mathbf{Z}\left({\varsigma' \atop \varsigma} ; {\hat{\upsilon}' \atop \hat{\upsilon}}, \hat{\eta} \right) \middle| {1 \atop L} \right\rangle = (-1)^{mm'} \left\langle {1 \atop L} \middle| \mathbf{Z}\left({\varsigma' \atop \varsigma} ; {\hat{\upsilon} \atop \hat{\upsilon}'}, \hat{\eta} \right) \middle| {1 \atop L} \right\rangle. \tag{12.3.153}$$

The interchange condition is not satisfied if both m and m' are odd. The non-diagonal elements are

$$\begin{aligned}
\left\langle {k \atop j} \middle| \mathbf{Z}\left({\varsigma' \atop \varsigma} ; {\hat{\upsilon}' \atop \hat{\upsilon}}, \hat{\eta} \right) \middle| {k+1 \atop j+1} \right\rangle &= \mathfrak{L}(\hat{\upsilon} + 2j\hat{\eta}, \hat{\upsilon}' - 2k\hat{\eta}|\mathrm{k}_1) G(\varsigma; j) G(\varsigma'; k), \\
\left\langle {k \atop j+1} \middle| \mathbf{Z}\left({\varsigma' \atop \varsigma} ; {\hat{\upsilon}' \atop \hat{\upsilon}}, \hat{\eta} \right) \middle| {k+1 \atop j} \right\rangle &= \mathfrak{L}(\hat{\upsilon} - 2j\hat{\eta}, \hat{\upsilon}' - 2k\hat{\eta}|\mathrm{k}_1) G(\varsigma; j) G(\varsigma'; k), \\
\left\langle {k+1 \atop j} \middle| \mathbf{Z}\left({\varsigma' \atop \varsigma} ; {\hat{\upsilon}' \atop \hat{\upsilon}}, \hat{\eta} \right) \middle| {k \atop j+1} \right\rangle &= \mathfrak{L}(\hat{\upsilon} + 2j\hat{\eta}, \hat{\upsilon}' + 2k\hat{\eta}|\mathrm{k}_1) G(\varsigma; j) G(\varsigma'; k), \\
\left\langle {k+1 \atop j+1} \middle| \mathbf{Z}\left({\varsigma' \atop \varsigma} ; {\hat{\upsilon}' \atop \hat{\upsilon}}, \hat{\eta} \right) \middle| {k \atop j} \right\rangle &= \mathfrak{L}(\hat{\upsilon} - 2j\hat{\eta}, \hat{\upsilon}' + 2k\hat{\eta}|\mathrm{k}_1) G(\varsigma; j) G(\varsigma'; k),
\end{aligned} \tag{12.3.154}$$

$$\left\langle {k \atop 1} \middle| \mathbf{Z}\begin{pmatrix} \varsigma' & \hat{\upsilon}' \\ \varsigma & ; \hat{\upsilon} & , \hat{\eta} \end{pmatrix} \middle| {k+1 \atop 1} \right\rangle = \mathfrak{L}(\hat{\upsilon}, \hat{\upsilon}' - 2k\hat{\eta}|\mathrm{k}_1)G_0(\varsigma; 1)G(\varsigma'; k),$$

$$\left\langle {k+1 \atop 1} \middle| \mathbf{Z}\begin{pmatrix} \varsigma' & \hat{\upsilon}' \\ \varsigma & ; \hat{\upsilon} & , \hat{\eta} \end{pmatrix} \middle| {k \atop 1} \right\rangle = \mathfrak{L}(\hat{\upsilon}, \hat{\upsilon}' + 2k\hat{\eta}|\mathrm{k}_1)G_0(\varsigma; 1)G(\varsigma'; k),$$

$$\left\langle {1 \atop j} \middle| \mathbf{Z}\begin{pmatrix} \varsigma' & \hat{\upsilon}' \\ \varsigma & ; \hat{\upsilon} & , \hat{\eta} \end{pmatrix} \middle| {1 \atop j+1} \right\rangle = \mathfrak{L}(\hat{\upsilon} + 2j\hat{\eta}, \hat{\upsilon}'|\mathrm{k}_1)G(\varsigma; j)G_0(\varsigma'; 1),$$

$$\left\langle {1 \atop j+1} \middle| \mathbf{Z}\begin{pmatrix} \varsigma' & \hat{\upsilon}' \\ \varsigma & ; \hat{\upsilon} & , \hat{\eta} \end{pmatrix} \middle| {1 \atop j} \right\rangle = \mathfrak{L}(\hat{\upsilon} - 2j\hat{\eta}, \hat{\upsilon}'|\mathrm{k}_1)G(\varsigma; j)G_0(\varsigma'; 1), \tag{12.3.155}$$

$$\left\langle {k \atop L} \middle| \mathbf{Z}\begin{pmatrix} \varsigma' & \hat{\upsilon}' \\ \varsigma & ; \hat{\upsilon} & , \hat{\eta} \end{pmatrix} \middle| {k+1 \atop L} \right\rangle = \mathfrak{L}(\hat{\upsilon} - \mho_{\mathfrak{B}1}, \hat{\upsilon}' - 2k\hat{\eta}|\mathrm{k}_1)G_0(\varsigma; L)G(\varsigma'; k),$$

$$\left\langle {k+1 \atop L} \middle| \mathbf{Z}\begin{pmatrix} \varsigma' & \hat{\upsilon}' \\ \varsigma & ; \hat{\upsilon} & , \hat{\eta} \end{pmatrix} \middle| {k \atop L} \right\rangle = \mathfrak{L}(\hat{\upsilon} - \mho_{\mathfrak{B}1}, \hat{\upsilon}' + 2k\hat{\eta}|\mathrm{k}_1)G_0(\varsigma; L)G(\varsigma'; k),$$

$$\left\langle {L \atop j} \middle| \mathbf{Z}\begin{pmatrix} \varsigma' & \hat{\upsilon}' \\ \varsigma & ; \hat{\upsilon} & , \hat{\eta} \end{pmatrix} \middle| {L \atop j+1} \right\rangle = \mathfrak{L}(\hat{\upsilon} + 2j\hat{\eta}, \hat{\upsilon}' - \mho_{\mathfrak{B}1}|\mathrm{k}_1)G(\varsigma; j)G_0(\varsigma'; L),$$

$$\left\langle {L \atop j+1} \middle| \mathbf{Z}\begin{pmatrix} \varsigma' & \hat{\upsilon}' \\ \varsigma & ; \hat{\upsilon} & , \hat{\eta} \end{pmatrix} \middle| {L \atop j} \right\rangle = \mathfrak{L}(\hat{\upsilon} - 2j\hat{\eta}, \hat{\upsilon}' - \mho_{\mathfrak{B}1}|\mathrm{k}_1)G(\varsigma; j)G_0(\varsigma'; L). \tag{12.3.156}$$

Let the diagonal elements of $\boldsymbol{D}$ be $D(\hat{\upsilon}, \hat{\upsilon}', \hat{\eta}; j, k|\mathrm{k}_1)$ then, using (18.3.149), for the twelve elements in (12.3.154)–(12.3.156) the interchange conditions are satisfied if

$$\frac{D(\hat{\upsilon}, \hat{\upsilon}', \hat{\eta}; j+1, k+1|\mathrm{k}_1)}{D(\hat{\upsilon}, \hat{\upsilon}', \hat{\eta}; j, k|\mathrm{k}_1)} = \frac{\mathfrak{H}(\hat{\upsilon}' - \hat{\upsilon} - 2\hat{\eta}[j+k]|\mathrm{k}_1)}{\mathfrak{H}(\hat{\upsilon}' - \hat{\upsilon} + 2\hat{\eta}[j+k]|\mathrm{k}_1)},$$

$$\frac{D(\hat{\upsilon}, \hat{\upsilon}', \hat{\eta}; j, k+1|\mathrm{k}_1)}{D(\hat{\upsilon}, \hat{\upsilon}', \hat{\eta}; j+1, k|\mathrm{k}_1)} = \frac{\mathfrak{H}(\hat{\upsilon}' - \hat{\upsilon} + 2\hat{\eta}[j-k]|\mathrm{k}_1)}{\mathfrak{H}(\hat{\upsilon}' - \hat{\upsilon} - 2\hat{\eta}[j-k]|\mathrm{k}_1)}, \tag{12.3.157}$$

$$\frac{D(\hat{\upsilon}, \hat{\upsilon}', \hat{\eta}; 1, k+1|\mathrm{k}_1)}{D(\hat{\upsilon}, \hat{\upsilon}', \hat{\eta}; 1, k|\mathrm{k}_1)} = \frac{\mathfrak{H}(\hat{\upsilon}' - \hat{\upsilon} - 2\hat{\eta}k|\mathrm{k}_1)}{\mathfrak{H}(\hat{\upsilon}' - \hat{\upsilon} + 2\hat{\eta}k|\mathrm{k}_1)},$$

$$\frac{D(\hat{\upsilon}, \hat{\upsilon}', \hat{\eta}; j+1, 1|\mathrm{k}_1)}{D(\hat{\upsilon}, \hat{\upsilon}', \hat{\eta}; j, 1|\mathrm{k}_1)} = \frac{\mathfrak{H}(\hat{\upsilon}' - \hat{\upsilon} - 2\hat{\eta}j|\mathrm{k}_1)}{\mathfrak{H}(\hat{\upsilon}' - \hat{\upsilon} + 2\hat{\eta}j|\mathrm{k}_1)}, \tag{12.3.158}$$

$$\begin{aligned}\frac{D(\hat{\upsilon},\hat{\upsilon}',\hat{\eta};L,k+1|\mathrm{k}_1)}{D(\hat{\upsilon},\hat{\upsilon}',\hat{\eta};L,k|\mathrm{k}_1)}&=(-1)^{mm'}\frac{\mathfrak{H}(\hat{\upsilon}'-\hat{\upsilon}-2\hat{\eta}[k+L]|\mathrm{k}_1)}{\mathfrak{H}(\hat{\upsilon}'-\hat{\upsilon}+2\hat{\eta}[k+L]|\mathrm{k}_1)},\\ \frac{D(\hat{\upsilon},\hat{\upsilon}',\hat{\eta};j+1,L|\mathrm{k}_1)}{D(\hat{\upsilon},\hat{\upsilon}',\hat{\eta};j,L|\mathrm{k}_1)}&=(-1)^{mm'}\frac{\mathfrak{H}(\hat{\upsilon}'-\hat{\upsilon}-2\hat{\eta}[j+L]|\mathrm{k}_1)}{\mathfrak{H}(\hat{\upsilon}'+\hat{\upsilon}+2\hat{\eta}[j+L]|\mathrm{k}_1)},\end{aligned} \tag{12.3.159}$$

where $\mathfrak{H}(u|\mathrm{k})$ is given by (18.3.146). Let

$$\frac{\mathfrak{T}(\hat{\upsilon},\hat{\upsilon}',\hat{\eta};k+2|\mathrm{k}_1)}{\mathfrak{T}(\hat{\upsilon},\hat{\upsilon}',\hat{\eta};k|\mathrm{k}_1)}:=\frac{\mathfrak{H}(\hat{\upsilon}'-\hat{\upsilon}-2\hat{\eta}k|\mathrm{k}_1)}{\mathfrak{H}(\hat{\upsilon}'-\hat{\upsilon}+2\hat{\eta}k|\mathrm{k}_1)}=\frac{\mathfrak{T}(\hat{\upsilon},\hat{\upsilon}',\hat{\eta};-k|\mathrm{k}_1)}{\mathfrak{T}(\hat{\upsilon},\hat{\upsilon}',\hat{\eta};2-k|\mathrm{k}_1)}. \tag{12.3.160}$$

Then Eqs. (12.3.157) are satisfied by

$$D(\hat{\upsilon},\hat{\upsilon}',\hat{\eta};j,k|\mathrm{k}_1)=\mathfrak{T}(\hat{\upsilon},\hat{\upsilon}',\hat{\eta};j+k|\mathrm{k}_1)\mathfrak{T}(\hat{\upsilon},\hat{\upsilon}',\hat{\eta};j-k+1|\mathrm{k}_1). \tag{12.3.161}$$

It follows, by induction on (12.3.160), that $\mathfrak{T}(\hat{\upsilon},\hat{\upsilon}',\hat{\eta};1+k|\mathrm{k}_1)=\mathfrak{T}(\hat{\upsilon},\hat{\upsilon}',\hat{\eta};1-k|\mathrm{k}_1)$ and thus that (12.3.161) satisfies (12.3.158). With $\mathfrak{T}(\hat{\upsilon},\hat{\upsilon}',\hat{\eta};1|\mathrm{k}_1)$ and $\mathfrak{T}(\hat{\upsilon},\hat{\upsilon}',\hat{\eta};2|\mathrm{k}_1)$ set to unity the recurrence relationship (12.3.161) generates independent sequences of terms for $\mathfrak{T}(\hat{\upsilon},\hat{\upsilon}',\hat{\eta};k|\mathrm{k}_1)$ with k odd and even and the elements of $\boldsymbol{D}$ are formed as products of one term from each sequence. Since $\mathfrak{T}(\hat{\upsilon},\hat{\upsilon}',\hat{\eta};k+\mho_{\mathfrak{B}1}|\mathrm{k}_1)=\mathfrak{T}(\hat{\upsilon},\hat{\upsilon}',\hat{\eta};k-\mho_{\mathfrak{B}1}|\mathrm{k}_1)$, (12.3.161) satisfies (12.3.159) if not both of m and m' are odd and this is then the condition for the interchange condition (12.3.97) to hold when $\boldsymbol{E}=\boldsymbol{I}$.

12.3.6.4 Periodicity Properties of $\boldsymbol{Q}$

Baxter's modified theta functions satisfy the periodicity conditions (18.3.127), (18.3.128) and (18.3.130)–(18.3.136), which can then be applied to (12.3.136)–(12.3.138) to give[60]

$$\langle k|\boldsymbol{T}_{\mathrm{R1}}\left(\begin{smallmatrix}\varsigma\\ \pm1\end{smallmatrix};\hat{\upsilon}+2\mathcal{O}_{\mathfrak{B}1},\hat{\eta}\right)|k'\rangle=(\mp1)^{\overline{m}}(-1)^{\overline{m}\,\overline{m}'}\langle k|\boldsymbol{T}_{\mathrm{R1}}\left(\begin{smallmatrix}\varsigma\\ \pm1\end{smallmatrix};\hat{\upsilon},\hat{\eta}\right)|k'\rangle, \tag{12.3.162}$$

$$\begin{aligned}\langle k|\boldsymbol{T}_{\mathrm{R1}}\left(\begin{smallmatrix}\varsigma\\ \pm1\end{smallmatrix};\hat{\upsilon}+2\mathcal{O}'_{\mathfrak{B}1},\hat{\eta}\right)|k+1\rangle&=(\mp1)^{n}(-1)^{nn'}\\ &\times\exp\left(-\frac{\pi\mathrm{i}\{\mathcal{O}'_{\mathfrak{B}1}+\hat{\upsilon}-2k\hat{\eta}\}}{\mathcal{O}_{\mathfrak{B}1}}\right)\langle k|\boldsymbol{T}_{\mathrm{R1}}\left(\begin{smallmatrix}\varsigma\\ \pm1\end{smallmatrix};\hat{\upsilon},\hat{\eta}\right)|k+1\rangle,\end{aligned} \tag{12.3.163}$$

[60] With $\mathcal{O}_{\mathfrak{B}1}:=\mho_{\mathfrak{B}1}/m_0=\overline{m}\mathcal{K}_1+\mathrm{i}\overline{m}'\mathcal{K}'_1$.

$$\langle 1|\boldsymbol{T}_{\mathrm{R}1}\left(\begin{matrix}\varsigma\\ \pm 1\end{matrix};\hat{v}+2\mathcal{O}'_{\mathfrak{B}1},\hat{\eta}\right)|1\rangle=(\mp 1)^{n}(-1)^{nn'}$$

$$\times\exp\left(-\frac{\pi\mathrm{i}\{\mathcal{O}'_{\mathfrak{B}1}+\hat{v}\}}{\mathcal{O}_{\mathfrak{B}1}}\right)\langle 1|\boldsymbol{T}_{\mathrm{R}1}\left(\begin{matrix}\varsigma\\ \pm 1\end{matrix};\hat{v},\hat{\eta}\right)|1\rangle, \tag{12.3.164}$$

$$\langle L|\boldsymbol{T}_{\mathrm{R}1}\left(\begin{matrix}\varsigma\\ \pm 1\end{matrix};\hat{v}+2\mathcal{O}'_{\mathfrak{B}1},\hat{\eta}\right)|L\rangle=(\mp 1)^{n}(-1)^{nn'}(-1)^{m_0}$$

$$\times\exp\left(-\frac{\pi\mathrm{i}\{\mathcal{O}'_{\mathfrak{B}1}+\hat{v}\}}{\mathcal{O}_{\mathfrak{B}1}}\right)\langle L|\boldsymbol{T}_{\mathrm{R}1}\left(\begin{matrix}\varsigma\\ \pm 1\end{matrix};\hat{v},\hat{\eta}\right)|L\rangle, \tag{12.3.165}$$

$$\langle k|\boldsymbol{T}_{\mathrm{R}1}\left(\begin{matrix}\varsigma\\ \pm 1\end{matrix};\hat{v}+\mathrm{i}\mathcal{K}'_1,\hat{\eta}\right)|k+1\rangle=\mp\exp\left(\frac{\pi\mathrm{i}\,m[4\mathcal{K}_1-\mathrm{i}\mathcal{K}'_1-2\hat{v}+4k\hat{\eta}]}{2\mathcal{U}_{\mathfrak{B}1}}\right)$$

$$\times\langle k|\boldsymbol{T}_{\mathrm{R}1}\left(\begin{matrix}\varsigma\\ \mp 1\end{matrix};\hat{v},\hat{\eta}\right)|k+1\rangle, \tag{12.3.166}$$

$$\langle 1|\boldsymbol{T}_{\mathrm{R}1}\left(\begin{matrix}\varsigma\\ \pm 1\end{matrix};\hat{v}+\mathrm{i}\mathcal{K}'_1,\hat{\eta}\right)|1\rangle=\mp\exp\left(\frac{\pi\mathrm{i}\,m[4\mathcal{K}_1-\mathrm{i}\mathcal{K}'_1-2\hat{v}]}{2\mathcal{U}_{\mathfrak{B}1}}\right)$$

$$\times\langle 1|\boldsymbol{T}_{\mathrm{R}1}\left(\begin{matrix}\varsigma\\ \mp 1\end{matrix};\hat{v},\hat{\eta}\right)|1\rangle, \tag{12.3.167}$$

$$\langle L|\boldsymbol{T}_{\mathrm{R}1}\left(\begin{matrix}\varsigma\\ \pm 1\end{matrix};\hat{v}+\mathrm{i}\mathcal{K}'_1,\hat{\eta}\right)|L\rangle=\mp(-1)^{m}\exp\left(\frac{\pi\mathrm{i}\,m[4\mathcal{K}_1-\mathrm{i}\mathcal{K}'_1-2\hat{v}]}{2\mathcal{U}_{\mathfrak{B}1}}\right)$$

$$\times\langle L|\boldsymbol{T}_{\mathrm{R}1}\left(\begin{matrix}\varsigma\\ \mp 1\end{matrix};\hat{v},\hat{\eta}\right)|L\rangle. \tag{12.3.168}$$

We also have, in the case that $\overline{m}$ is even and $\overline{m}'$ odd,

$$\langle k|\boldsymbol{T}_{\mathrm{R}1}\left(\begin{matrix}\varsigma\\ \pm 1\end{matrix};\hat{v}+\mathcal{O}_{\mathfrak{B}1},\hat{\eta}\right)|k'\rangle=(\mp 1)^{\frac{1}{2}\overline{m}}\exp\left(\tfrac{1}{4}\mathrm{i}\pi\overline{m}\,\overline{m}'\right)$$

$$\times\langle k|\boldsymbol{T}_{\mathrm{R}1}\left(\begin{matrix}\varsigma\\ \mp 1\end{matrix};\hat{v},\hat{\eta}\right)|k'\rangle. \tag{12.3.169}$$

A term in the trace in (12.3.109) is indexed by a sequence $k_1, k_2, \ldots, k_{N_2-1}, k_{N_2}, k_{N_2+1}$, where k_{j+1} is one of k_j-1, k_j or k_j+1 and $k_{N_2+1}=k_1$. Every pair (k_j, k_{j+1}), where $k_j \neq k_{j+1}$, is matched by another pair (k_i, k_{i+1}) where $k_j = k_{i+1}$ and $k_i = k_{j+1}$. So the pre-factor contributions from the index k for the traces constructed from these matrix elements all cancel. Consider now the different choices for the parities of m, m' and N_2.

***m* odd and *m'* even**:

It follows from (18.3.129) that m_0 is even and $\overline{m}$ is odd, but $\overline{m}'$ can be of either parity. From (12.3.76), (12.3.99) and (12.3.162)

$$\boldsymbol{Q}_{\mathrm{R}}(\hat{v}+2\mathcal{O}_{\mathfrak{B}1},\hat{\eta})=(-1)^{N_2\overline{m}'+1}\boldsymbol{S}_z\boldsymbol{Q}_{\mathrm{R}}(\hat{v},\hat{\eta}), \tag{12.3.170}$$

and, from (12.3.98),

$$\boldsymbol{Q}_{\mathrm{L}}(\hat{v}+2\mathcal{O}_{\mathfrak{B}1},\hat{\eta})=(-1)^{N_2\overline{m}'+1}\boldsymbol{Q}_{\mathrm{L}}(\hat{v},\hat{\eta})\boldsymbol{S}_z. \tag{12.3.171}$$

Applying (12.3.99) to (12.3.170) and (12.3.107), with $\boldsymbol{E}=\boldsymbol{I}$, to (12.3.171)[61] gives

$$\begin{aligned}\boldsymbol{Q}(\hat{v}+2\mathcal{O}_{\mathfrak{B}1},\hat{\eta})&=(-1)^{N_2\overline{m}'+1}\boldsymbol{S}_z\boldsymbol{Q}(\hat{v},\hat{\eta})\\&=(-1)^{N_2\overline{m}'+1}\boldsymbol{Q}(\hat{v},\hat{\eta})\boldsymbol{S}_z,\end{aligned} \tag{12.3.172}$$

establishing the commutation relation (12.3.105). From (12.3.163)–(12.3.165),

$$\boldsymbol{Q}(\hat{v}+2\mathcal{O}'_{\mathfrak{B}1},\hat{\eta})=(-1)^{N_2nn'}\exp\left(-\frac{\pi\mathrm{i}N_2\{\mathcal{O}'_{\mathfrak{B}1}+\hat{v}\}}{\mathcal{O}_{\mathfrak{B}1}}\right)(-\boldsymbol{S}_z)^n\boldsymbol{Q}(\hat{v},\hat{\eta}). \tag{12.3.173}$$

When $m'=0$ the modified theta functions reduce to their standard equivalents and, from (18.3.129), (18.3.130), (18.3.133), (18.3.135), (12.3.89) and (12.3.128)

$$m_0=2m,\quad \overline{m}=1,\quad \overline{m}'=0,\quad n=0,\quad n'=1, \tag{12.3.174}$$

$$\begin{aligned}\mathcal{O}_{\mathfrak{B}1}=\mathcal{K}_1,\quad \mathcal{O}'_{\mathfrak{B}1}=\mathrm{i}\,\mathcal{K}'_1,\quad \mathcal{U}_{\mathfrak{B}1}=2m\mathcal{K}_1,\\ \hat{\eta}=m\mathcal{K}_1/L.\end{aligned} \tag{12.3.175}$$

The Eqs. (12.3.172) and (12.3.173) simplify to

$$\boldsymbol{Q}(\hat{v}+2\mathcal{K}_1,\hat{\eta})=-\boldsymbol{S}_z\boldsymbol{Q}(\hat{v},\hat{\eta})=-\boldsymbol{Q}(\hat{v},\hat{\eta})\boldsymbol{S}_z, \tag{12.3.176}$$

$$\boldsymbol{Q}(\hat{v}+2\mathrm{i}\,\mathcal{K}'_1,\hat{\eta})=\frac{1}{\mathfrak{Q}_1^{N_2}}\exp\left(-\frac{\pi\mathrm{i}N_2\hat{v}}{\mathcal{K}_1}\right)\boldsymbol{Q}(\hat{v},\hat{\eta}) \tag{12.3.177}$$

and if, in addition, N_2 is even, from (12.3.166)–(12.3.168),

$$\boldsymbol{Q}(\hat{v}+\mathrm{i}\,\mathcal{K}'_1,\hat{\eta})=-\frac{1}{\mathfrak{Q}_1^{\frac{1}{4}N_2}}\exp\left(-\frac{\pi\,\mathrm{i}N_2\hat{v}}{2\mathcal{K}_1}\right)\boldsymbol{S}_z\boldsymbol{S}_x\boldsymbol{Q}(\hat{v},\hat{\eta}). \tag{12.3.178}$$

***m* even and *m′* odd**:

It follows from (18.3.129) that m_0 and $\overline{m}'$ are odd and $\overline{m}$ is divisible by four. The analysis is the same as that when m is odd and m' even except that (12.3.169) is used in place of (12.3.162) to give

---

[61] Valid since the interchange relationship holds with $\boldsymbol{E}=\boldsymbol{I}$, when m is odd and m' even.

$$\begin{aligned}\boldsymbol{Q}(\hat{v}+\mathcal{O}_{\mathfrak{B}1},\hat{\eta}) &= (-1)^{\frac{1}{4}N_2\overline{m}\,\overline{m}'}\boldsymbol{S}_x\boldsymbol{Q}(\hat{v},\hat{\eta})\\ &= (-1)^{\frac{1}{4}N_2\overline{m}\,\overline{m}'}\boldsymbol{Q}(\hat{v},\hat{\eta})\boldsymbol{S}_x,\end{aligned} \tag{12.3.179}$$

establishing the commutation relation (12.3.104). Because of the factor $(-1)^{m_0}$ in (12.3.165) the period of quasi-periodicity is now doubled with respect to $\mathcal{O}'_{\mathfrak{B}1}$ giving

$$\boldsymbol{Q}(\hat{v}+4\mathcal{O}'_{\mathfrak{B}1},\hat{\eta}) = \exp\left(\frac{-2\mathrm{i}\pi N_2\{\mathcal{O}'_{\mathfrak{B}1}+\hat{v}\}}{\mathcal{O}_{\mathfrak{B}1}}\right)\boldsymbol{Q}(\hat{v},\hat{\eta}). \tag{12.3.180}$$

m and m′ odd:

It follows from (18.3.129) that m_0 and $\overline{m}'$ are odd and $\overline{m}$ is even but not divisible by four, so (12.3.180) remains valid. In place of (12.3.179) we first have

$$\boldsymbol{Q}_{\mathrm{R}}(\hat{v}+\mathcal{O}_{\mathfrak{B}1},\hat{\eta}) = \exp\left(\tfrac{1}{4}N_2\mathrm{i}\pi\overline{m}\,\overline{m}'\right)\boldsymbol{S}_x\boldsymbol{S}_z\boldsymbol{Q}_{\mathrm{R}}(\hat{v},\hat{\eta}), \tag{12.3.181}$$

$$\boldsymbol{Q}_{\mathrm{L}}(\hat{v}+\mathcal{O}_{\mathfrak{B}1},\hat{\eta}) = \exp\left(\tfrac{1}{4}N_2\mathrm{i}\pi\overline{m}\,\overline{m}'\right)\boldsymbol{Q}_{\mathrm{L}}(\hat{v},\hat{\eta})\boldsymbol{S}_z\boldsymbol{S}_x. \tag{12.3.182}$$

The application of (12.3.99) to (12.3.181) is as in the previous cases, but now, according to the results of Fabricius and McCoy (2007), (12.3.107) becomes

$$\boldsymbol{S}_x\boldsymbol{Q}(\hat{v},\hat{\eta})\boldsymbol{S}_x = \boldsymbol{S}_z\boldsymbol{Q}(\hat{v},\hat{\eta})\boldsymbol{S}_z = \boldsymbol{Q}_{\mathrm{L}}^{-1}(\hat{v}_0,\hat{\eta})\,\boldsymbol{Q}_{\mathrm{L}}(\hat{v},\hat{\eta}). \tag{12.3.183}$$

Applying this to (12.3.182) gives

$$\begin{aligned}\boldsymbol{Q}(\hat{v}+\mathcal{O}_{\mathfrak{B}1},\hat{\eta}) &= \exp\left(\tfrac{1}{4}N_2\mathrm{i}\pi\overline{m}\,\overline{m}'\right)\boldsymbol{S}_x\boldsymbol{S}_z\boldsymbol{Q}(\hat{v},\hat{\eta})\\ &= \exp\left(\tfrac{1}{4}N_2\mathrm{i}\pi\overline{m}\,\overline{m}'\right)\boldsymbol{Q}(\hat{v},\hat{\eta})\boldsymbol{S}_x\boldsymbol{S}_z,\end{aligned} \tag{12.3.184}$$

establishing the commutation relation (12.3.106).

12.3.6.5 The Eigenvalues of $\boldsymbol{Q}$ and $\widehat{\boldsymbol{V}}$

For simplicity we concentrate on the case m odd and $m' = 0$ when (12.3.174) and (12.3.175) apply. For accounts of the other parity choices, including the case $\hat{\eta}$ complex ($m' \neq 0$), the reader is referred to Fabricius and McCoy (2007) or McCoy (2010, Sect. 14.3).

The matrix $\boldsymbol{Q}(\hat{v},\hat{\eta})$ with periodicity properties (12.3.170) and (12.3.173) is not unique, since it commutes with $\boldsymbol{S}_z$ but not with $\boldsymbol{S}_x$ and $\boldsymbol{S}_x\boldsymbol{S}_z$. Of course, since $\boldsymbol{S}_z$ is diagonal, with elements $\pi(\varsigma^{(k)}) = \pm 1$, we know its eigenvalues. However, we need to associate these values with the eigenvalues of $\boldsymbol{Q}(\hat{v},\hat{\eta})$. Suppose that $\Lambda_Q(\hat{v},\hat{\eta}|\mathrm{k}_1)$

and θ_z are associated eigenvalues of $\boldsymbol{Q}(\hat{v},\hat{\eta})$ and $\boldsymbol{S}_z$.[62] Then, from (12.3.172) and (12.3.173),

$$\Lambda_Q(\hat{v}+2\mathcal{K}_1,\hat{\eta}|\mathrm{k}_1) = -\theta_z\,\Lambda_Q(\hat{v},\hat{\eta}|\mathrm{k}_1), \tag{12.3.185}$$

$$\Lambda_Q(\hat{v}+2\mathrm{i}\,\mathcal{K}_1',\hat{\eta}|\mathrm{k}_1) = \frac{1}{\mathrm{Q}_1^{N_2}}\exp\left(\frac{-\mathrm{i}\,\pi N_2\hat{v}}{\mathcal{K}_1}\right)\Lambda_Q(\hat{v},\hat{\eta}|\mathrm{k}_1). \tag{12.3.186}$$

According to the definition in Sect. 18.2, $\Lambda_Q(\hat{v},\hat{\eta}|\mathrm{k}_1)$ is a quasi-periodic function in the class $\mathrm{QP}(2\mathcal{K}_1, 2\mathrm{i}\,\mathcal{K}_1'; -\theta_z, 1; N_2, N_2)$. From Theorem 18.2.5 it has N_2 zeros in the period parallelogram $\mathbb{W}(2\mathcal{K}_1, 2\mathrm{i}\,\mathcal{K}_1')$. Let these be $\hat{v}_j$, $j = 1, 2, \ldots, N_2$. Then, from Theorem 18.3.2, an eigenvalue of $\boldsymbol{Q}(\hat{v},\hat{\eta})$, identified by a constant $\aleph$ and the Q*–zero set* $\{\hat{v}_j\}$, is given, to within a multiplicative constant, by

$$\Lambda_Q(\hat{v},\hat{\eta};\,\aleph,\{\hat{v}_j\}|\mathrm{k}_1) = \exp\left(-\frac{\pi\mathrm{i}\,\aleph\,\hat{v}}{2\mathcal{K}_1}\right)\prod_{j=1}^{N_2}\mathrm{H}(\hat{v}-\hat{v}_j|\mathrm{k}_1), \tag{12.3.187}$$

where

$$\aleph = 2\ell' + \tfrac{1}{2}(1-\theta_z) + 1 + N_2, \tag{12.3.188}$$

$$\sum_{j=1}^{N_2}\hat{v}_j = \aleph\,\mathrm{i}\,\mathcal{K}_1' + (N_2 - 2\ell)\mathcal{K}_1, \tag{12.3.189}$$

for some integers ℓ and ℓ'. There are thus four parameters determining the Q–zero set, N_2, $\aleph$, k_1 and m/L.

The corresponding eigenvalue $\hat{\Lambda}_V(\hat{v},\hat{\eta};\,\aleph,\{\hat{v}_j\}|\mathrm{k}_1)$ is now obtained by substituting from (12.3.187) into (12.3.103) to give

$$\hat{\Lambda}_V(\hat{v},\hat{\eta};\,\aleph,\{\hat{v}_j\}) = \mathsf{F}(\hat{v},\hat{\eta};\,\aleph,\{\hat{v}_j\}|\mathrm{k}_1)\mathsf{G}(\hat{v},\hat{\eta};\,\aleph,\{\hat{v}_j\}|\mathrm{k}_1), \tag{12.3.190}$$

where

$$\mathsf{F}(\hat{v},\hat{\eta};\,\aleph,\{\hat{v}_j\}|\mathrm{k}_1) := \exp\left(-\frac{\pi\mathrm{i}\,\aleph\,\hat{\eta}}{\mathcal{K}_1}\right)\left[\mathrm{h}(\hat{v}+\hat{\eta}|\mathrm{k}_1)\right]^{N_2} \times\prod_{j=1}^{N_2}\frac{\mathrm{H}(\hat{v}+2\hat{\eta}-\hat{v}_j|\mathrm{k}_1)}{\mathrm{H}(\hat{v}-\hat{v}_j|\mathrm{k}_1)}, \tag{12.3.191}$$

[62] Note that θ_z satisfies the same relationship (18.2.5) as τ.

$$\mathrm{G}(\hat{\upsilon}, \hat{\eta};\ \aleph, \{\hat{\upsilon}_j\}|\mathrm{k}_1) := \left[\frac{\mathfrak{h}(\hat{\upsilon}-\hat{\eta}|\mathrm{k}_1)}{\mathfrak{h}(\hat{\upsilon}+\hat{\eta}|\mathrm{k}_1)}\right]^{N_2} + \exp\left(\frac{2\pi \mathrm{i} \aleph \hat{\eta}}{\mathcal{K}_1}\right) \times \prod_{j=1}^{N_2} \frac{\mathrm{H}(\hat{\upsilon}-2\hat{\eta}-\hat{\upsilon}_j|\mathrm{k}_1)}{\mathrm{H}(\hat{\upsilon}+2\hat{\eta}-\hat{\upsilon}_j|\mathrm{k}_1)}. \qquad (12.3.192)$$

$\mathrm{F}(\hat{\upsilon}, \hat{\eta};\ \aleph, \{\hat{\upsilon}_j\}|\mathrm{k}_1)$ is a meromorphic function with poles at all members of the Q–zero set not eliminated by cancellation in the numerator and denominator of the product. Since $\hat{\Lambda}_{\mathrm{V}}(\hat{\upsilon}, \hat{\eta};\ \aleph, \{\hat{\upsilon}_j\}|\mathrm{k}_1)$ is an entire function every pole $\hat{\upsilon}_j$ of $\mathrm{F}(\hat{\upsilon}, \hat{\eta};\ \aleph, \{\hat{\upsilon}_j\}|\mathrm{k}_1)$ must satisfy the equation[63]

$$\mathrm{G}(\hat{\upsilon}_j, \hat{\eta};\ \aleph, \{\hat{\upsilon}_i\}|\mathrm{k}_1) = 0, \qquad (12.3.193)$$

which, following Fabricius and McCoy, we refer to as the *Bethe equation*. Apart from eliminations arising from complete *L-strings* as described below, the solutions, for fixed $\aleph$, of this set of N_2 variables would allow the determination of the complete set of eigenvalues of $\widehat{V}(\hat{\upsilon}, \hat{\eta})$. In fact, since we are concerned here only with determining the free-energy density, which is given by the largest eigenvalue, the task is simplified by an investigation of the properties of the Q–zero set $\{\hat{\upsilon}_j\}$.

12.3.6.6 Properties of the Q–Zero Set: Strings

Suppose that, for some $\hat{\upsilon}_\mu$,

$$\hat{\upsilon}_{\mu+k} = \hat{\upsilon}_\mu - 2k\,\hat{\eta} + \varepsilon_k = \hat{\upsilon}_{\mu+k-1} - 2\hat{\eta} + \varepsilon_k - \varepsilon_{k-1}, \quad k = 1, 2, \ldots, \nu - 1, \qquad \text{for } \nu < L, \qquad (12.3.194)$$

where the ε_k are small. Such a set $\hat{\upsilon}_\mu, \ldots, \hat{\upsilon}_{\mu+\nu-1}$ is called a ν*-string* and

$$\hat{\upsilon}_{\mu+k} + 2\hat{\eta} = \hat{\upsilon}_{\mu+k-2} - 2\hat{\eta} + \varepsilon_k - \varepsilon_{k-2}, \qquad k = 2, 3, \ldots, \nu - 1. \quad (12.3.195)$$

The product in (12.3.191) contains terms

---

[63] If the root $\hat{\upsilon}_j$ is of multiplicity s then

$$\left(\frac{\partial^k \mathrm{G}(\hat{\upsilon}, \hat{\eta};\ \aleph, \{\hat{\upsilon}_i\}|\mathrm{k}_1)}{\partial \hat{\upsilon}^k}\right)_{\upsilon=\upsilon_j} = 0, \qquad k = 0, 1, \ldots, s-1.$$

Such solutions would exist, if at all, only for special values of $\hat{\eta}$. They have not been found for small values of N_2, when $m' = 0$ (McCoy 2010, Sect. 14.3.2) and we shall not explore this possibility.

$$\frac{\mathrm{H}(\hat{v}+2\hat{\eta}-\hat{v}_{\mu+k-1}|\mathrm{k}_1)}{\mathrm{H}(\hat{v}-\hat{v}_{\mu+k}|\mathrm{k}_1)}=\frac{\mathrm{H}(\hat{v}-\hat{v}_{\mu+k}-\varepsilon_k+\varepsilon_{k-1}|\mathrm{k}_1)}{\mathrm{H}(\hat{v}-\hat{v}_{\mu+k}|\mathrm{k}_1)},\qquad k=1,2,\ldots,\nu-1, \tag{12.3.196}$$

and that in (12.3.192) the terms

$$\frac{\mathrm{H}(\hat{v}-2\hat{\eta}-\hat{v}_{\mu+k}|\mathrm{k}_1)}{\mathrm{H}(\hat{v}+2\hat{\eta}-\hat{v}_{\mu+k-2}|\mathrm{k}_1)}=\frac{\mathrm{H}(\hat{v}-2\hat{\eta}-\hat{v}_{\mu+k}|\mathrm{k}_1)}{\mathrm{H}(\hat{v}-2\hat{\eta}-\hat{v}_{\mu+k}-\varepsilon_k+\varepsilon_{k-2}|\mathrm{k}_1)},\qquad k=2,3,\ldots.\nu-1. \tag{12.3.197}$$

In each case these terms 'almost' cancel. Although, of course, this would not, even if all the ε_k were zero, eliminate the string-zeros from the product. In (12.3.191) the denominator retains the term in $\hat{v}_\mu$ and the numerator the term in $\hat{v}_{\mu+\nu-1}$. In (12.3.192) the denominator retains the terms in $\hat{v}_{\mu+\nu-1}$ and $\hat{v}_{\mu+\nu-2}$ and the numerator the terms in $\hat{v}_\mu$ and $\hat{v}_{\mu+1}$.

Consider, however, the case where all the ε_k are zero and $\nu = L$ with

$$\hat{v}_{\mu+k}=\hat{v}_\mu-2k\,\hat{\eta},\qquad k=0,1,\ldots,L-1. \tag{12.3.198}$$

This is called a *complete L-string* and $\hat{v}_\mu$ is its *centre*.[64] Suppose that there are N_{21} complete L-strings with centres, $\hat{v}_{\mu_s}$ $s=1,2,\ldots,N_{21}$ with $N_{22}:=N_2-LN_{21}$. Then, from (12.3.189), (12.3.191) and (12.3.192),

$$\sum_{j=1}^{N_{22}}\hat{v}_j=\aleph\,\mathrm{i}\,\mathcal{K}_1'+[N_2-2\ell-mN_{21}(L-1)]\mathcal{K}_1-L\sum_{s=1}^{N_{21}}\hat{v}_{\mu_s}, \tag{12.3.199}$$

$$\mathrm{F}(\hat{v},\hat{\eta};\aleph,\overline{\{\hat{v}_j\}}|\mathrm{k}_1)=(-1)^{N_{21}}\exp\left(-\frac{\pi\mathrm{i}\aleph\hat{\eta}}{\mathcal{K}_1}\right)\left[\mathrm{h}(\hat{v}+\hat{\eta}|\mathrm{k}_1)\right]^{N_2}\times\prod_{j=1}^{N_{22}}\frac{\mathrm{H}(\hat{v}+2\hat{\eta}-\hat{v}_j|\mathrm{k}_1)}{\mathrm{H}(\hat{v}-\hat{v}_j|\mathrm{k}_1)}, \tag{12.3.200}$$

$$\mathrm{G}(\hat{v},\hat{\eta};\aleph,\overline{\{\hat{v}_j\}}|\mathrm{k}_1)=\left[\frac{\mathrm{h}(\hat{v}-\hat{\eta}|\mathrm{k}_1)}{\mathrm{h}(\hat{v}+\hat{\eta}|\mathrm{k}_1)}\right]^{N_2}+\exp\left(\frac{2\pi\mathrm{i}\aleph\hat{\eta}}{\mathcal{K}_1}\right)\times\prod_{j=1}^{N_{22}}\frac{\mathrm{H}(\hat{v}-2\hat{\eta}-\hat{v}_j|\mathrm{k}_1)}{\mathrm{H}(\hat{v}+2\hat{\eta}-\hat{v}_j|\mathrm{k}_1)}, \tag{12.3.201}$$

where $\overline{\{\hat{v}_j\}}$ is the reduced Q–zero set $\hat{v}_j, j=1,2,\ldots,N_{22}$.

[64] Although, since, from (12.3.174), $2\hat{\eta}L=2m\mathcal{K}_1$, it is clear that any member of the string can equally well be taken as its centre.

It is clear that the centres of the L-strings cannot be determined from (12.3.193), which is now reduced to

$$\mathsf{G}(\hat{\upsilon}_j, \hat{\eta};\ \aleph, \overline{\{\hat{\upsilon}_i\}}|\mathrm{k}_1) = 0. \tag{12.3.202}$$

A method for doing this has been conjectured by Fabricius and McCoy (2003) and proved for $L = 2$ by Fabricius and McCoy (2004). These authors also derived numerical results for $L = 3$ and various values of N_2 (see Fig. 14.10, p. 522, McCoy 2010, for the case $N_2 = 8$). It was shown that:

(a) The number of L-strings N_{21} is even, so that the factors $mN_{21}(L-1)$ and $(-1)^{N_{21}}$ are removed from (12.3.199) and (12.3.200) respectively.
(b) For every configuration of string centres $\{\hat{\upsilon}_{\mu_s}\}$ there is another obtained by adding or not adding $\mathrm{i}\,\mathcal{K}_1$ to a member of this set, giving $2^{N_{21}}$ solutions in all.

The transformation $\aleph \to \aleph + L$ and $\hat{\upsilon}_{\mu_s} \to \hat{\upsilon}_{\mu_s} + \mathrm{i}\,\mathcal{K}'_1$, for one $\hat{\upsilon}_{\mu_s}$, leaves (12.3.199) and (12.3.201) unaffected but changes the sign in (12.3.200) and hence of $\hat{\Lambda}_V(\hat{\upsilon}, \hat{\eta};\ \aleph, \{\hat{\upsilon}_j\})$. The presence of the L-strings implies the existence of a pair of eigenvalues of equal magnitude, opposite sign and degeneracy $2^{N_{21}-1}$. From (12.3.188) it follows that, the two degenerate eigenvalues have the same and different parities with respect to $\boldsymbol{S}_z$ according as L is even or odd. From (12.3.187)

$$\Lambda_Q(\hat{\upsilon}, \hat{\eta};\ \aleph, \{\hat{\upsilon}_{\mu_s}\}, \overline{\{\hat{\upsilon}_j\}}|\mathrm{k}_1) = \exp\left(-\frac{\pi\mathrm{i}\aleph\,\hat{\upsilon}}{2\mathcal{K}_1}\right)\mathsf{L}(\hat{\upsilon}, \hat{\eta};\ \{\hat{\upsilon}_{\mu_s}\}|\mathrm{k}_1) \times \prod_{j=1}^{N_{22}} \mathsf{H}(\hat{\upsilon} - \hat{\upsilon}_j|\mathrm{k}_1), \tag{12.3.203}$$

where

$$\mathsf{L}(\hat{\upsilon}, \hat{\eta};\ \{\hat{\upsilon}_{\mu_s}\}|\mathrm{k}_1) := \prod_{s=1}^{N_{21}}\prod_{k=0}^{L-1} \mathsf{H}(\hat{\upsilon} - \hat{\upsilon}_{\mu_s+k}|\mathrm{k}_1) = \prod_{s=1}^{N_{21}}\prod_{k=1}^{L-1} \mathsf{H}(\hat{\upsilon} - \hat{\upsilon}_{\mu_s} + 2k\,\hat{\eta}|\mathrm{k}_1). \tag{12.3.204}$$

Since

$$\mathsf{L}(\hat{\upsilon}, \hat{\eta};\ \{\hat{\upsilon}_{\mu_s}\}|\mathrm{k}_1) = \mathsf{L}(\hat{\upsilon} \pm 2\hat{\eta}, \hat{\eta};\ \{\hat{\upsilon}_{\mu_s}\}|\mathrm{k}_1) = \mathsf{L}(\hat{\upsilon}, \hat{\eta};\ \{\hat{\upsilon}_{\mu_s} \pm 2\hat{\eta}\}|\mathrm{k}_1), \tag{12.3.205}$$

this term cancels in (12.3.203), where it represents the degeneracy of the eigenvalue of $\widehat{\boldsymbol{V}}(\hat{\upsilon}, \hat{\eta})$.

12.3.6.7 Properties of the Q–Zero Set: Bethe Pairs

The analysis in this section is independent of the presence or absence of complete L-strings. In the latter case $\{\hat{v}_{\mu_s}\} = \emptyset$, $\overline{\{\hat{v}_j\}} = \{\hat{v}_j\}$ and $N_{22} = N_2$. So, supposing the possible presence of complete L-strings, (12.3.190) is replaced by

$$\hat{\Lambda}_V(\hat{v}, \hat{\eta};\ \aleph, \overline{\{\hat{v}_j\}}|\mathrm{k}_1) = \mathsf{F}(\hat{v}, \hat{\eta};\ \aleph, \overline{\{\hat{v}_j\}}|\mathrm{k}_1)\mathsf{G}(\hat{v}, \hat{\eta};\ \aleph, \overline{\{\hat{v}_j\}}|\mathrm{k}_1). \tag{12.3.206}$$

From (18.3.25), (18.3.158), (12.3.199)–(12.3.201),[65]

$$\mathfrak{h}(\hat{v} + \mathrm{i}\mathcal{K}_1'|\mathrm{k}_1) = -\frac{1}{\sqrt{\mathcal{Q}_1}} \exp\left(-\frac{\pi\,\mathrm{i}\,\hat{v}}{\mathcal{K}_1}\right) \mathfrak{h}(\hat{v}|\mathrm{k}_1), \tag{12.3.207}$$

$$\begin{aligned}\mathsf{F}(\hat{v} + \mathrm{i}\mathcal{K}_1', \hat{\eta};\ \aleph, \overline{\{\hat{v}_j\}}|\mathrm{k}_1) &= \left(-\frac{1}{\sqrt{\mathcal{Q}_1}}\right)^{N_2} \exp\left(-\frac{\pi\,\mathrm{i}\,N_2[\hat{v} + 2\hat{\eta}]}{\mathcal{K}_1}\right)\\ &\quad\times \mathsf{F}(\hat{v}, \hat{\eta};\ \aleph + N_2, \overline{\{\hat{v}_j + \mathrm{i}\mathcal{K}_1'\}}|\mathrm{k}_1),\end{aligned} \tag{12.3.208}$$

$$\begin{aligned}\mathsf{G}(\hat{v} + \mathrm{i}\mathcal{K}_1', \hat{\eta};\ \aleph, \overline{\{\hat{v}_j\}}|\mathrm{k}_1) &= \exp\left(\frac{2\pi\,\mathrm{i}\,N_2\hat{\eta}}{\mathcal{K}_1}\right)\\ &\quad\times \mathsf{G}(\hat{v}, \hat{\eta};\ \aleph + N_2, \overline{\{\hat{v}_j + \mathrm{i}\mathcal{K}_1'\}}|\mathrm{k}_1),\end{aligned} \tag{12.3.209}$$

$$\begin{aligned}\hat{\Lambda}_V(\hat{v} + \mathrm{i}\mathcal{K}_1', \hat{\eta};\ \aleph, \overline{\{\hat{v}_j\}}|\mathrm{k}_1) &= \left(-\frac{1}{\sqrt{\mathcal{Q}_1}}\right)^{N_2} \exp\left(-\frac{\pi\,\mathrm{i}\,N_2\hat{v}}{\mathcal{K}_1}\right)\\ &\quad\times \hat{\Lambda}_V(\hat{v}, \hat{\eta};\ \aleph + N_2, \overline{\{\hat{v}_j + \mathrm{i}\mathcal{K}_1'\}}|\mathrm{k}_1).\end{aligned} \tag{12.3.210}$$

But, directly from (12.3.92),

$$\begin{aligned}\hat{\Lambda}_V(\hat{v} + \mathrm{i}\mathcal{K}_1', \hat{\eta};\ \aleph, \overline{\{\hat{v}_j\}}|\mathrm{k}_1) &= \left(-\frac{1}{\sqrt{\mathcal{Q}_1}}\right)^{N_2} \exp\left(-\frac{\pi\,\mathrm{i}\,N_2\hat{v}}{\mathcal{K}_1}\right)\\ &\quad\times \hat{\Lambda}_V(\hat{v}, \hat{\eta};\ \aleph, \overline{\{\hat{v}_j\}}|\mathrm{k}_1).\end{aligned} \tag{12.3.211}$$

So

$$\hat{\Lambda}_V(\hat{v}, \hat{\eta};\ \aleph + N_2, \overline{\{\hat{v}_j + \mathrm{i}\mathcal{K}_1'\}}|\mathrm{k}_1) = \hat{\Lambda}_V(\hat{v}, \hat{\eta};\ \aleph, \overline{\{\hat{v}_j\}}|\mathrm{k}_1). \tag{12.3.212}$$

---

65 Since

$$\exp\left(\frac{2\pi\mathrm{i}N_2\hat{\eta}}{\mathcal{K}_1}\right) = \exp\left(\frac{2\pi\mathrm{i}N_{22}\hat{\eta}}{\mathcal{K}_1}\right)$$

the form of these equations is unaffected by the presence of complete L-strings. In particular N_2 could be replaced by N_{22} in (12.3.214). Incomplete ν-strings, $\nu < L$, are included in the set $\overline{\{\hat{v}_j\}}$.

It follows that

$$\mathring{\Lambda}_Q(\hat{v}, \hat{\eta};\ \aleph, \{\hat{v}_j\}|\mathrm{k}_1) := C\,\Lambda_Q(\hat{v}; \hat{\eta};\ \aleph + N_2, \{\hat{v}_{\mu_s}\}, \overline{\{\hat{v}_j + \mathrm{i}\,\mathcal{K}'_1\}}|\mathrm{k}_1) \tag{12.3.213}$$

satisfies (12.3.103) with the same $\hat{\Lambda}_V(\hat{v}, \hat{\eta};\ \aleph, \overline{\{\hat{v}_j\}})|\mathrm{k}_1$, for any $\hat{v}$-independent quantity C. From (12.3.203)

$$\exp\left(\frac{\pi\,\mathrm{i}\,N_2\hat{v}}{2\mathcal{K}_1}\right)\Lambda_Q(\hat{v} + \mathrm{i}\,\mathcal{K}'_1, \hat{\eta};\ \aleph, \{\hat{v}_j\}|\mathrm{k}_1) = (-1)^{N_2}\,\mathcal{Q}_1^{\frac{1}{2}\aleph}$$
$$\times\ \Lambda_Q(\hat{v}, \hat{\eta};\ \aleph + N_2, \{\hat{v}_{\mu_s}\}, \overline{\{\hat{v}_j + \mathrm{i}\,\mathcal{K}'_1\}}|\mathrm{k}_1) \tag{12.3.214}$$

and, from (12.3.185) and (12.3.186),

$$\Lambda_Q(\hat{v} + 2\mathcal{K}_1, \hat{\eta};\ \aleph, \{\hat{v}_j\}|\mathrm{k}_1) = -\theta_z\,\Lambda_Q(\hat{v}, \hat{\eta};\ \aleph, \{\hat{v}_j\}|\mathrm{k}_1), \tag{12.3.215}$$

$$\Lambda_Q(\hat{v} + 2\mathrm{i}\,\mathcal{K}'_1, \hat{\eta};\ \aleph, \{\hat{v}_j\}|\mathrm{k}_1) = \frac{1}{\mathcal{Q}_1^{N_2}}\exp\left(\frac{-\pi\mathrm{i}\,N_2\hat{v}}{\mathcal{K}_1}\right)\Lambda_Q(\hat{v}, \hat{\eta};\ \aleph, \{\hat{v}_j\}|\mathrm{k}_1), \tag{12.3.216}$$

giving, from (12.3.213) and (12.3.214),

$$\mathring{\Lambda}_Q(\hat{v} + 2\mathcal{K}_1, \hat{\eta};\ \aleph, \{\hat{v}_j\}|\mathrm{k}_1) = -\theta_z\,(-1)^{N_2}\,\mathring{\Lambda}_Q(\hat{v}, \hat{\eta};\ \aleph, \{\hat{v}_j\}|\mathrm{k}_1), \tag{12.3.217}$$

$$\mathring{\Lambda}_Q(\hat{v} + 2\mathrm{i}\,\mathcal{K}'_1, \hat{\eta};\ \aleph, \{\hat{v}_j\}|\mathrm{k}_1) = \frac{1}{\mathcal{Q}_1^{N_2}}\exp\left(\frac{-\pi\mathrm{i}\,N_2\hat{v}}{\mathcal{K}_1}\right)\mathring{\Lambda}_Q(\hat{v}, \hat{\eta};\ \aleph, \{\hat{v}_j\}|\mathrm{k}_1). \tag{12.3.218}$$

Remembering that N_2 and N_{22} have the same parity:

***For N_2 odd*:**

$\Lambda_Q(\hat{v}, \hat{\eta};\ \aleph, \{\hat{v}_j\}|\mathrm{k}_1)$ and $\mathring{\Lambda}_Q(\hat{v}, \hat{\eta};\ \aleph, \{\hat{v}_j\}|\mathrm{k}_1)$ with the same complete L-strings and remaining zeros $\overline{\{\hat{v}_j\}}$ and $\overline{\{\hat{v}_j + \mathrm{i}\,\mathcal{K}'_1\}}$, respectively, are distinct since they are associated with opposite eigenvalues of $\boldsymbol{S}_z$. They are, however, associated with the same eigenvalue of $\widehat{\boldsymbol{V}}(\hat{v}, \hat{\eta})$, which must be (at least) doubly degenerate, in addition to the degeneracy associated with complete L-strings.

For N_2 even:

Either

(a) $\Lambda_Q(\hat{v}, \hat{\eta}; \aleph, \{\hat{v}_j\}|\mathrm{k}_1)$ and $\mathring{\Lambda}_Q(\hat{v}, \hat{\eta}; \aleph, \{\hat{v}_j\}|\mathrm{k}_1)$ are distinct eigenvalues of $\boldsymbol{Q}(\hat{v}, \hat{\eta})$ both associated with the same eigenvalue $\hat{\Lambda}_V(\hat{v}, \hat{\eta}; \aleph, \{\hat{v}_j\}|\mathrm{k}_1)$ of $\widehat{\boldsymbol{V}}(\hat{v}, \hat{\eta})$.

(b) $\Lambda_Q(\hat{v}, \hat{\eta}; \aleph, \{\hat{v}_j\}|\mathrm{k}_1) = \mathring{\Lambda}_Q(\hat{v}, \hat{\eta}; \aleph, \{\hat{v}_j\}|\mathrm{k}_1)$. Then, from (12.3.213) and (12.3.214),

$$\Lambda_Q(\hat{v} + \mathrm{i}\mathcal{K}_1', \hat{\eta}; \aleph, \{\hat{v}_j\}|\mathrm{k}_1) = C \exp\left(-\frac{\pi\, \mathrm{i} N_2 \hat{v}}{2\mathcal{K}_1}\right) \mathfrak{Q}_1^{\frac{1}{2}\aleph} \times \Lambda_Q(\hat{v}, \hat{\eta}; \aleph, \{\hat{v}_j\}|\mathrm{k}_1). \tag{12.3.219}$$

From (12.3.178) we see that an eigenvalue relationship of this form applies only when the corresponding eigenvector is an eigenvector of $\boldsymbol{S}_x$, when it take the form

$$\Lambda_Q(\hat{v} + \mathrm{i}\mathcal{K}_1', \hat{\eta}; \aleph, \{\hat{v}_j\}|\mathrm{k}_1) = -\exp\left(-\frac{\pi\, \mathrm{i} N_2 \hat{v}}{2\mathcal{K}_1}\right) \frac{\theta_z \theta_x}{\mathfrak{Q}_1^{\frac{1}{4}N_2}} \times \Lambda_Q(\hat{v}, \hat{\eta}; \aleph, \{\hat{v}_j\}|\mathrm{k}_1), \tag{12.3.220}$$

allowing the identification of C.

The zeros of $\Lambda_Q(\hat{v}, \hat{\eta}; \aleph, \{\hat{v}_j\}|\mathrm{k}_1)$ are in pairs. These *Bethe pairs* $\overline{\{\hat{v}_j, \hat{v}_j + \mathrm{i}\mathcal{K}_1'\}}$ are solutions of (12.3.202) and numerical studies indicate (see, McCoy 2010, Sect. 14.3.2) that a pair either belongs to two ν-strings ($\nu < L$) or has $\Re\{\hat{v}_j\} = 0$ or $\mathcal{K}_1$. In the case of the largest eigenvalue with N_2 even, the Q–zero set consists entirely of Bethe pairs with $\Re\{\hat{v}_j\} = \mathcal{K}_1$, so we set

$$\hat{v} := \mathcal{K}_1 + \mathrm{i}q. \tag{12.3.221}$$

12.3.6.8 The Largest Eigenvalue of $\widehat{V}(\hat{v}, \hat{\eta})$ for N_2 Even

The analysis of Sect. 12.1.5 needed the condition that N_2 is even. Although this is no longer necessary we confine discussion to this case.[66]

Let $N_{23} := \frac{1}{2}N_2$ and, with a suitable numbering of the roots, $q_j \in [-\frac{1}{2}\mathcal{K}_1', \frac{1}{2}\mathcal{K}_1')$, for $j = 1, 2, \ldots, N_{23}$. Of these values suppose that there are $N_{23}^{(-)}$ in the interval

[66] For treatments of N_2 odd the reader is referred to Fabricius and McCoy (2005) and McCoy (2010, Chap. 14).

$[-\frac{1}{2}\mathcal{K}_1', 0)$ and $N_{23}^{(+)}$ in the interval $[0, \frac{1}{2}\mathcal{K}_1')$. Then there will be $N_{23}^{(+)}$ roots of the form $q_j - \mathrm{i}\,\mathcal{K}_1'$ in the interval $[-\mathcal{K}_1', -\frac{1}{2}\mathcal{K}_1')$ and $N_{23}^{(-)}$ roots of the form $q_j + \mathrm{i}\,\mathcal{K}_1'$ in the interval $[\frac{1}{2}\mathcal{K}_1', \mathcal{K}_1')$. From (12.3.189), with $\ell = 0$,

$$\sum_{j=1}^{N_{23}} q_j = \mathcal{K}_1'\big[\aleph + N_{23}^{(+)} - N_{23}^{(-)}\big]. \tag{12.3.222}$$

Since the largest eigenvalue is real the roots must be in conjugate-complex pairs or be real. Numerical studies indicate that real roots occur only when N_{23} is odd when there is exactly one such. We shall assume that this is generally true, giving for N_{23}

(i) even $N_{23}^{(+)} = N_{23}^{(-)}$ and $\aleph = 0$,

(ii) odd $N_{23}^{(+)} = N_{23}^{(-)} + 1$ and $\aleph = -1$.

In each case, from (18.3.156)–(18.3.158), (12.3.191) and (12.3.192) become

$$\mathrm{F}(q, \hat{\eta}; \{q_i\}|\mathrm{k}_1) = \big[\mathfrak{h}(\hat{\eta} + \mathcal{K}_1 + \mathrm{i}\,q|\mathrm{k}_1)\big]^{2N_{23}} \prod_{i=1}^{N_{23}} \frac{\mathfrak{h}(2\hat{\eta} + \mathrm{i}\,q - \mathrm{i}\,q_i|\mathrm{k}_1)}{\mathfrak{h}(\mathrm{i}\,q - \mathrm{i}\,q_i|\mathrm{k}_1)}, \tag{12.3.223}$$

$$\mathrm{G}(q, \hat{\eta}; \{q_i\}|\mathrm{k}_1) = \left[\frac{\mathfrak{h}(\hat{\eta} + \mathcal{K}_1 - \mathrm{i}\,q|\mathrm{k}_1)}{\mathfrak{h}(\hat{\eta} + \mathcal{K}_1 + \mathrm{i}\,q|\mathrm{k}_1)}\right]^{2N_{23}} + (-1)^{N_{23}} \prod_{i=1}^{N_{23}} \frac{\mathfrak{h}(2\hat{\eta} - \mathrm{i}\,q + \mathrm{i}\,q_i|\mathrm{k}_1)}{\mathfrak{h}(2\hat{\eta} + \mathrm{i}\,q - \mathrm{i}\,q_i|\mathrm{k}_1)}. \tag{12.3.224}$$

The members of the Q–zero set $\{q_j\}$ are solutions of the N_{23} Bethe equations $\mathrm{G}_1(q_j, \hat{\eta}; \{q_i\}) = 0, j = 1, 2, \ldots, N_{23}$. The next step is to take the logarithms of these equations. This will involve

$$\ln[(-1)^{N_{23}+1}] = 2\pi\,\mathrm{i}[N_{23} + 1]\big[p + \tfrac{1}{2}\big], \tag{12.3.225}$$

where the *branch number* p can take any integer value. Let p_j be the branch number associated with q_j, that is p is a function of q with

$$I(q) := [N_{23} + 1]\big[p(q) + \tfrac{1}{2}\big] \tag{12.3.226}$$

an integer if N_{23} is odd and an integer plus a half if N_{23} is even. The Bethe equations take the form

$$2N_{23}\,\mathrm{f}(\hat{\eta}, q_j|\mathrm{k}_1) = 2\pi\,\mathrm{i}\,I(q_j) + \sum_{i=1}^{N_{23}} \mathrm{f}(2\hat{\eta} - \mathcal{K}_1, q_j - q_i|\mathrm{k}_1), \tag{12.3.227}$$

where $\mathrm{f}(v, u|\mathrm{k})$, given by (18.3.160), consists of logarithms which are purely imaginary, lie in the interval $[-\pi\,\mathrm{i}, \pi\,\mathrm{i}]$ and are zero when $v = 0$.

In the limit $N_{23} \to \infty$ the values $q_j, j = 1, 2, \ldots, N_{23}$ are dense on the interval $[-\frac{1}{2}\mathcal{K}'_1, \frac{1}{2}\mathcal{K}'_1]$ with a normalized density function $\rho(q)$. The aim now is to use (12.3.226) and $\mathrm{f}(\hat{\eta}, q_j|\mathrm{k}_1)$ to determine this density function and for this it is necessary to suppose a distribution for the members of the Q–zero set. A form for this has not been obtained in general, but it was shown by Yang and Yang (1966b), for the XXZ Heisenberg chain, and hence for the six-vertex model, that, in the limit of large N_{23} the number of members of the Q–zero set in the interval $[q, q + \mathrm{d}q]$ is $I(q + \mathrm{d}q) - I(q)$. We suppose that this result also applies to the eight-vertex model, and, from (12.3.226),[67]

$$\frac{1}{N_{23}} \int_{-\frac{1}{2}\mathcal{K}'_1}^{\frac{1}{2}\mathcal{K}'_1} \dot{I}(q)\mathrm{d}q = \int_{-\frac{1}{2}\mathcal{K}'_1}^{\frac{1}{2}\mathcal{K}'_1} \dot{p}(q)\mathrm{d}q. \tag{12.3.228}$$

So $\dot{p}(q)$ is the normalized density function. Differentiating (12.3.227) and taking the limit $N_{23} \to \infty$, when the sum becomes an integral, gives

$$2\dot{\mathrm{f}}(\hat{\eta}, q|\mathrm{k}_1) = 2\pi\,\mathrm{i}\,\rho(q) + \int_{-\frac{1}{2}\mathcal{K}'_1}^{\frac{1}{2}\mathcal{K}'_1} \rho(x)\dot{\mathrm{f}}(2\hat{\eta} - \mathcal{K}_1, q - x|\mathrm{k}_1)\mathrm{d}x. \tag{12.3.229}$$

Taking the Fourier transform of (12.3.228) and using the convolution formula (18.1.22)

$$2\dot{\mathrm{f}}^\star(\hat{\eta}, m|\mathrm{k}_1) = 2\pi\,\mathrm{i}\,\rho^\star(m) + \rho^\star(m)\mathcal{K}'_1\,\dot{\mathrm{f}}^\star(2\hat{\eta} - \mathcal{K}_1, m|\mathrm{k}_1) \tag{12.3.230}$$

and, substituting from (18.3.162), gives

$$\rho^\star(m) = \frac{1}{\mathcal{K}'_1 \cosh(2m\pi[\hat{\eta} - \mathcal{K}_1]/\mathcal{K}'_1)}. \tag{12.3.231}$$

Reverting to the variable $\hat{v}$, from (12.3.190), (12.3.223) and (12.3.224),

$$\hat{\Lambda}_V(\hat{v}, \hat{\eta}; \{q_j\}|\mathrm{k}_1) = \mathrm{X}(\hat{v}, \hat{\eta}; \{q_i\}|\mathrm{k}_1) + \mathrm{X}(\hat{v}, -\hat{\eta}; \{q_i\}|\mathrm{k}_1), \tag{12.3.232}$$

where

$$\mathrm{X}(\hat{v}, \hat{\eta}; \{q_i\}|\mathrm{k}_1) = \left[\mathrm{h}(\hat{\eta} + \hat{v}|\mathrm{k}_1)\right]^{2N_{23}} \prod_{i=1}^{N_{23}} \frac{\mathrm{h}(\mathrm{i}\,q_i + 2\hat{\eta} - \hat{v} + \mathcal{K}_1|\mathrm{k}_1)}{\mathrm{h}(\mathrm{i}\,q_i - \hat{v} + \mathcal{K}_1|\mathrm{k}_1)}. \tag{12.3.233}$$

In the limit $N_2 \to \infty$, using (18.1.8),

---

[67] Here a dot is used to denote differentiation with respect to q.

$$\frac{\ln[\mathrm{X}(\hat{v},\hat{\eta}|\mathrm{k}_1)]}{N_2} \simeq \mathrm{Y}(\hat{v},\hat{\eta}|\mathrm{k}_1), \tag{12.3.234}$$

where

$$\begin{aligned}\mathrm{Y}(\hat{v},\hat{\eta}|\mathrm{k}_1) &= \ln[\mathrm{h}(\hat{\eta}+\hat{v}|\mathrm{k}_1)] + \frac{1}{2}\int_{-\frac{1}{2}\mathcal{K}_1'}^{\frac{1}{2}\mathcal{K}_1'} \rho(x)\mathrm{f}(\mathrm{i}x+\hat{\eta}-\hat{v};\mathrm{i}\hat{\eta}|\mathrm{k}_1)\,\mathrm{d}x \\ &= \ln[\mathrm{h}(\hat{\eta}+\hat{v}|\mathrm{k}_1)] \\ &\quad + \frac{1}{2}\sum_{m=-\infty}^{\infty}\rho^\star(m)\int_{-\frac{1}{2}\mathcal{K}_1'}^{\frac{1}{2}\mathcal{K}_1'} \exp\left(\frac{2m\pi\mathrm{i}x}{\mathcal{K}_1'}\right)\mathrm{f}(\mathrm{i}x+\hat{\eta}-\hat{v};\mathrm{i}\hat{\eta}|\mathrm{k}_1)\,\mathrm{d}x.\end{aligned} \tag{12.3.235}$$

Inserting (12.3.231) and the series expansion from (18.3.160) gives

$$\begin{aligned}\mathrm{Y}(\hat{v},\hat{\eta}|\mathrm{k}_1) &= \ln[\mathrm{h}(\hat{\eta}+\hat{v}|\mathrm{k}_1)] + \frac{\pi\hat{\eta}(\hat{v}-\hat{\eta})}{\mathcal{K}_1\mathcal{K}_1'} \\ &\quad + \sum_{n=1}^{\infty}\frac{\sinh(2n\pi\hat{\eta}/\mathcal{K}_1')\sinh(2n\pi[\hat{v}-\hat{\eta}]/\mathcal{K}_1')}{n\cosh(2n\pi[\mathcal{K}_1-\hat{\eta}]/\mathcal{K}_1')\sinh(2n\pi\mathcal{K}_1/\mathcal{K}_1')}.\end{aligned} \tag{12.3.236}$$

So

$$\hat{\Lambda}_V(\hat{v},\hat{\eta}|\mathrm{k}_1) \simeq \exp[N_2\mathrm{Y}(\hat{v},\hat{\eta}|\mathrm{k}_1)] + \exp[N_2\mathrm{Y}(\hat{v},-\hat{\eta}|\mathrm{k}_1|\mathrm{k}_1)]. \tag{12.3.237}$$

12.3.7 The Free Energy and Magnetization

For $0 < \hat{\eta} \le \hat{v} \le 2\mathcal{K}_1$ it can be verified that $\mathrm{Y}(\hat{v},-\hat{\eta}|\mathrm{k}_1) < \mathrm{Y}(\hat{v},\hat{\eta}|\mathrm{k}_1)$ and thus, in the limit $N_2 \to \infty$, the first term in (12.3.237) is dominant giving, from (11.1.14), the free-energy density

$$\phi^{(8\mathrm{vM})}(\mathcal{N}^{\circledast},\hat{v},\hat{\eta}|\mathrm{k}_1) = -\ln[\mathrm{h}(\hat{\eta}+\hat{v}|\mathrm{k}_1)] - \frac{\pi\hat{\eta}(\hat{v}-\hat{\eta})}{\mathcal{K}_1\mathcal{K}_1'} - \sum_{n=1}^{\infty}S_n^{(1)}(\hat{v},\hat{\eta}|\mathrm{k}_1), \tag{12.3.238}$$

where

$$S_n^{(1)}(\hat{v},\hat{\eta}|\mathrm{k}_1) := \frac{\sinh\left(\frac{2n\pi\hat{\eta}}{\mathcal{K}_1'}\right)\sinh\left(\frac{2n\pi[\hat{v}-\hat{\eta}]}{\mathcal{K}_1'}\right)}{n\cosh\left(\frac{2n\pi[\mathcal{K}_1-\hat{\eta}]}{\mathcal{K}_1'}\right)\sinh\left(\frac{2n\pi\mathcal{K}_1}{\mathcal{K}_1'}\right)}. \tag{12.3.239}$$

Translated into our notation the form for the free-energy density derived by Baxter (1972b, p. 202) is

$$\phi^{(8\text{vM})}_{\text{Bx}}(\mathcal{N}^{\circledast},\hat{\upsilon},\hat{\eta}|\mathrm{k}_1) = -\ln[\hat{\mathrm{c}}(\hat{\upsilon},\hat{\eta}|\mathrm{k}_1)] - \sum_{n=1}^{\infty} S^{(2)}_n(\hat{\upsilon},\hat{\eta}|\mathrm{k}_1) \tag{12.3.240}$$

where

$$S^{(2)}_n(\hat{\upsilon},\hat{\eta}|\mathrm{k}_1) := \frac{2\sinh^2\left(\frac{n\pi\,\mathrm{i}[2\hat{\eta}-\mathrm{i}\mathcal{K}'_1]}{2\mathcal{K}_1}\right)\left[\cosh\left(\frac{n\pi\,\mathrm{i}\hat{\eta}}{\mathcal{K}_1}\right)-\cosh\left(\frac{n\pi\,\mathrm{i}\hat{\upsilon}}{\mathcal{K}_1}\right)\right]}{n\sinh\left(\frac{n\pi\mathcal{K}'_1}{\mathcal{K}_1}\right)\cosh\left(\frac{n\pi\,\mathrm{i}\hat{\eta}}{\mathcal{K}_1}\right)}. \tag{12.3.241}$$

In the Bazhanov–Mangazeev transformation, defined in Sect. 18.3.4, the elliptic modulus k_1 is replaced by k_3 and the variables $\hat{\upsilon}$ and $\hat{\eta}$ are replaced by $\hat{\upsilon}_3$ and $\hat{\eta}_3$ given by (12.3.87). For this it may be shown that

$$S^{(1)}_n(\hat{\upsilon},\hat{\eta}|\mathrm{k}_1) - S^{(2)}_n(\hat{\upsilon}_3,\hat{\eta}_3|\mathrm{k}_3) = S^{(3)}_n(\hat{\upsilon},\hat{\eta}|\mathrm{k}_1), \tag{12.3.242}$$

where

$$S^{(3)}_n(\hat{\upsilon},\hat{\eta}|\mathrm{k}_1) := \frac{2\sinh\left(\frac{2n\pi\hat{\eta}}{\mathcal{K}'_1}\right)\left[\sinh\left(\frac{2n\pi\hat{\upsilon}}{\mathcal{K}'_1}\right)-\sinh\left(\frac{2n\pi\hat{\eta}}{\mathcal{K}'_1}\right)\right]}{n\sinh\left(\frac{4\pi n\mathcal{K}_1}{\mathcal{K}'_1}\right)} \tag{12.3.243}$$

and

$$\sum_{n=1}^{\infty} S^{(3)}_n(\hat{\upsilon},\hat{\eta}|\mathrm{k}_1) = \ln[\hat{\mathrm{c}}(\hat{\upsilon}_3,\hat{\eta}_3|\mathrm{k}_3)] - \ln[\mathrm{h}(\hat{\upsilon}+\hat{\eta}|\mathrm{k}_1)] - \frac{\pi\hat{\eta}(\hat{\upsilon}-\hat{\eta})}{\mathcal{K}_1\mathcal{K}'_1} + \ln[\Pi(\hat{\upsilon},\hat{\eta}|\mathrm{k}_1)], \tag{12.3.244}$$

where the last term is given by (18.3.167). Thus we have

$$\phi^{(8\text{vM})}_{\text{Bx}}(\mathcal{N}^{\circledast},\hat{\upsilon}_3,\hat{\eta}_3|\mathrm{k}_3) = \phi^{(8\text{vM})}(\mathcal{N}^{\circledast},\hat{\upsilon},\hat{\eta}|\mathrm{k}_1) + \ln[\Pi(\hat{\upsilon},\hat{\eta}|\mathrm{k}_1)]. \tag{12.3.245}$$

The zone into which the choice of a non-imaginary value for $\hat{\eta}$ projected the solution is now identified as $\mathcal{R}_{\text{O}}(\text{I})$. The method of solution of the eight-vertex model which has been developed in this text, which follows the work of Fabricius and McCoy, has been shown to recover, by means of the Bazhanov-Mangazeev transformation Baxter's 1972 expression for the free energy density in $\mathcal{R}_{\text{L}}(\text{III})$. Of course, as we have already observed, all methods of solution which use the elliptic roots of unity condition, which for $m' = 0$ reduces to (12.3.175), impose a restriction on the values of $\hat{\eta}$. But as Baxter remarked Baxter (1972b, p. 215) "as the positive integer L is arbitrary, we can approach arbitrarily close to any desired value."

In Sect. 12.3.2 it was shown that the transformation $\mathcal{T}_2$ from $\mathcal{R}_{\mathrm{L}}(\mathrm{I})$ to $\mathcal{R}_{\mathrm{L}}(\mathrm{III})$ was achieved by the variable change $(z, \xi) \to (\hat{z}, \hat{\xi})$, given by (12.3.9), and this was mirrored for $(v, \eta) \to (\hat{v}, \hat{\eta})$, with these latter variables given by (12.3.83), in Sect. 12.3.6. The ground state for $\mathcal{R}_{\mathrm{L}}(\mathrm{I})$ has all vertices of type 1 or 2 with Boltzmann weight a and the ground state for $\mathcal{R}_{\mathrm{L}}(\mathrm{III})$ has all vertices of type 5 or 6 with Boltzmann weight c. The full free energy of the system involving all the eigenvalues of the transfer matrix is invariant under $\mathcal{T}_2$ and the change of variables $(v, \eta) \to (\hat{v}, \hat{\eta})$. However, this symmetry is broken by the process of selecting the largest eigenvalue which is involved in the direct derivation of (12.3.240) by Baxter (1972b) and in the more indirect derivation given above. Notwithstanding this, the formula (12.3.240) with the $\hat{\mathrm{c}}(\hat{v}, \hat{\eta}|\mathrm{k}_1)$ replaced, according to Table 9.1, by the 'zone-neutral' expression $\mathrm{w}_1 + \mathrm{w}_2$ and $(\hat{v}, \hat{\eta})$ by (v, η) applies, with an appropriate choice of v and η, to all low-temperature zones $\mathcal{R}_{\mathrm{L}}(\mathrm{J})$, $\mathrm{J} = \mathrm{I}, \mathrm{II}, \mathrm{III}, \mathrm{IV}$ (Baxter 1972b, p. 202). This can be expressed in the form

$$\phi^{(8\mathrm{vM})}(\mathcal{N}^{\circledast}, \mathrm{w}_1, \mathrm{w}_2, \mathrm{w}_3, \mathrm{w}_4) = -\ln[\mathrm{w}_1 + \mathrm{w}_2] - \sum_{n=1}^{\infty} \frac{(\mathcal{X}^{2n} - \mathcal{Q}^n)^2 (\mathcal{X}^n + \mathcal{X}^{-n} - \mathcal{Y}^n - \mathcal{Y}^{-n})}{n\mathcal{X}^n (1 - \mathcal{Q}^{2n})(1 + \mathcal{X}^{2n})}, \tag{12.3.246}$$

where

$$\begin{aligned} \mathcal{X} &:= \exp(-\pi\xi/\mathcal{K}_0) = \exp(\mathrm{i}\pi\eta/\mathcal{K}_1), \\ \mathcal{Y} &:= \exp(-\pi z/\mathcal{K}_0) = \exp(\mathrm{i}\pi v/\mathcal{K}_1), \\ \mathcal{Q} &:= \exp(-2\pi\mathcal{K}_0'/\mathcal{K}_0) = \exp(-\pi\mathcal{K}_1'/\mathcal{K}_1), \end{aligned} \tag{12.3.247}$$

$\mathcal{Q}$ being the nome $\mathcal{Q}_1$ of k_1 and the square $\mathcal{Q}_0^2$ of the nome of k_0.

The spontaneous relative magnetization m_{s} in the spin interpretation of the eight-vertex model can be derived using 'corner' transfer matrices (Baxter 1982b) instead of 'row' transfer matrices like those used here and in Chap. 11. It is found that

$$m_{\mathrm{s}} = \prod_{n=1}^{\infty} \frac{1 - \mathcal{X}^{4n-2}}{1 + \mathcal{X}^{4n-2}}. \tag{12.3.248}$$

If k_m is defined by the relation

$$\frac{\mathcal{K}(\mathrm{k}_m')}{\mathcal{K}(\mathrm{k}_m)} := \frac{2\xi}{\mathcal{K}_0}, \tag{12.3.249}$$

then it follows from (12.3.247) and (18.3.11) that $\mathcal{X}^2$ is the nome corresponding to k_m and, from (18.3.13), that

$$m_{\mathrm{s}} = (\mathrm{k}'_m)^{\frac{1}{4}} = (1 - \mathrm{k}_m^2)^{\frac{1}{8}}. \tag{12.3.250}$$

12.3.8 Critical Behaviour

The phase space of the eight-vertex model is three-dimensional, with axes corresponding, as in Sect. 9.3.2, to ratios of Boltzmann weights or to ratios of (w_1, w_2, w_3, w_4). In the low-temperature zone $\mathcal{R}_{\mathrm{L}}(\mathrm{J})$ of any of the regions $\mathrm{J} = \mathrm{I}, \mathrm{II}, \mathrm{III}, \mathrm{IV}$ of phase space, the free-energy density given by (12.3.246) is an entire function of (w_1, w_2, w_3, w_4) which satisfy (9.3.24). Variations of these quantities can be achieved by varying the members of the set (z, ξ, k_0), using the first column of Table 12.2. Alternatively, through (3.8.19), variations of the Boltzmann weights can be understood as a variation of the temperature with fixed values of the parameters $\varepsilon_{\mathfrak{a}}$, $\varepsilon_{\mathfrak{b}}$, $\varepsilon_{\mathfrak{c}}$ and $\varepsilon_{\mathfrak{d}}$.

Singular behaviour will occur (if at all), as T varies, only at points on a boundary of $\mathcal{R}_{\mathrm{L}}(\mathrm{J})$. As we have seen, the weak-graph transformation $\mathcal{T}_{\mathrm{WG}}$ maps from the low-temperature zone $\mathcal{R}_{\mathrm{L}}(\mathrm{J})$ to the high-temperature zone $\mathcal{R}_{\mathrm{0}}(\mathrm{J})$ with the zones separated by the surface of fixed points $\mathcal{C}(\mathrm{J})$, given by $w_2 = w_3$ and $\mathrm{k}_0 = 1$. In Sect. 9.3.2.2 this was proposed as a surface of phase transitions, a claim that can now be investigated using the expression (12.3.246) for the free energy. Let

$$\varphi := \pi z/\mathcal{K}'_0, \qquad \vartheta := \pi\xi/\mathcal{K}'_0, \qquad \varpi := \mathcal{K}'_0/\mathcal{K}_0. \tag{12.3.251}$$

Then (12.3.246) becomes

$$\phi^{(8\mathrm{vM})}(\mathcal{N}^{\circledast}, T) = -\ln(w_1 + w_2) - \varpi \sum_{n=1}^{\infty} \psi(n\varpi; \varphi, \vartheta), \tag{12.3.252}$$

where

$$\psi(x; \varphi, \vartheta) := \frac{2\sinh^2[(\pi - \vartheta)x]\{\cosh(\vartheta x) - \cosh(\varphi x)\}}{u \sinh(2\pi x)\cosh(\vartheta x)}. \tag{12.3.253}$$

Using the Poisson summation formula (18.1.27),

$$\phi^{(8\mathrm{vM})}(\mathcal{N}^{\circledast}, T) = -\ln(w_1 + w_2) - \tfrac{1}{2}\psi^\star(0; \varphi, \vartheta) - \sum_{n=1}^{\infty} \psi^\star(2\pi n/\varpi; \varphi, \vartheta), \tag{12.3.254}$$

where

$$\psi^\star(s;\varphi,\vartheta) := \int_{-\infty}^{\infty} \exp(\mathrm{i}\,x\,s)\psi(x;\varphi,\vartheta)\mathrm{d}\,x \tag{12.3.255}$$

is an even function of s with $\psi(0;\varphi,\vartheta) = 0$. It has simple poles at $x = \mathrm{i}\,m/2$ and $x = \mathrm{i}\pi(2m-1)/(2\vartheta)$, for $m = 1, 2, \ldots$, in the upper half of the complex x-plane. The integral (12.3.255) can, therefore, be evaluated using the residue method to give

$$\psi^\star(s;\varphi,\vartheta) = 2\sum_{m=1}^{\infty} \frac{\exp(-m\,s)[\cos(m\,\vartheta) - (-1)^m]\left[\cos\left(\frac{m\,\vartheta}{2}\right) - \cos\left(\frac{m\,\varphi}{2}\right)\right]}{m\cos\left(\frac{m\,\vartheta}{2}\right)}$$
$$+2\sum_{m=1}^{\infty} \frac{(-1)^m \exp\left[-\frac{(2m-1)\pi s}{2\vartheta}\right]\left\{\cos\left[\frac{(2m-1)\pi^2}{\vartheta}\right]+1\right\}\cos\left[\frac{(2m-1)\pi\varphi}{2\vartheta}\right]}{\left(m-\frac{1}{2}\right)\sin\left[\frac{(2m-1)\pi^2}{\vartheta}\right]}. \tag{12.3.256}$$

On the critical surface $\mathrm{k}_0 = 1$, $\mathrm{k}_0' = 0$, $\vartheta = 2\xi$ and $\varphi = 2z$, giving, from the second column of Table 12.2 and (18.3.41),

$$w_1 : w_2 : w_3 : w_4 = \frac{\cos\left(\frac{1}{2}\varphi\right)}{\cos\left(\frac{1}{2}\vartheta\right)} : 1 : 1 : \frac{\sin\left(\frac{1}{2}\varphi\right)}{\sin\left(\frac{1}{2}\vartheta\right)}. \tag{12.3.257}$$

From (12.3.251) and (18.3.10), $\varpi \to 0$, as $\mathrm{k}_0' \to 0$, and $\psi^\star(2\pi n/\varpi;\varphi,\vartheta) \to 0$ giving, from (12.3.253)–(12.3.255),

$$\phi^{(8\mathrm{vM})}(\mathcal{N}^{\circledast}, T) = \phi^{(8\mathrm{vM})}(\mathcal{N}^{\circledast}, T_\mathrm{c}) = -\ln(w_1 + w_2) - \tfrac{1}{2}\psi^\star(0;\varphi,\vartheta)$$
$$= -\ln(w_1 + w_2) - \int_{-\infty}^{\infty} \frac{\sinh^2[(\pi-\vartheta)x][\cosh(\vartheta\,x) - \cosh(\varphi\,x)]}{\omega\sinh(2\pi\,x)\cosh(\vartheta\,x)}\mathrm{d}\,x, \tag{12.3.258}$$

with ϑ and φ taking the critical values ϑ_c and φ_c, obtained by solving (12.3.257) for some fixed values of ε_a, ε_b, ε_c and ε_d. For such a set of fixed values, as the temperature is increased, the system describes a trajectory in parameter space which crosses the critical surface as some critical temperature T_c. From (9.3.31)–(9.3.33), $\Delta = \Delta^\star > 1$ in $\mathcal{R}_\mathrm{L}(\mathrm{I})$ and $\mathcal{R}_\mathrm{L}(\mathrm{II})$ and $\Delta = -\Delta^\star < -1$ in $\mathcal{R}_\mathrm{L}(\mathrm{III})$ and $\mathcal{R}_\mathrm{L}(\mathrm{IV})$. From column 2 of Table 12.2, on $\mathcal{C}(\mathrm{J})$,

$$\Delta = \Delta_\mathrm{c} := \pm 1, \qquad \Gamma = \Gamma_\mathrm{c} := \mp\cos(\vartheta_\mathrm{c}), \tag{12.3.259}$$

with the upper and lower signs corresponding, respectively, to the pairs of regions (I, II) and (III, IV). At a temperature $T = T_\mathrm{c} - \delta\,T$,

$$\Delta \simeq \Delta_c + \Delta_0|\delta T|, \qquad \Gamma \simeq \Gamma_c + \Gamma_0 \delta T, \tag{12.3.260}$$

where Δ_0 and Γ_0 are some functions of ε_a, ε_b, ε_c and ε_d, with $\Delta_0 \geq 0$ in (I, II) and $\Delta_0 \leq 0$ in (III, IV). From (12.1.18),

$$k_0'^2 = \frac{\Delta^2 - 1}{\Delta^2 - \Gamma^2} \tag{12.3.261}$$

and, substituting from (12.3.260),

$$k_0'^2 \simeq \frac{2|\Delta_0||\delta T|}{\sin^2(\vartheta_c)}, \tag{12.3.262}$$

giving, from (12.3.251) and (18.3.10),

$$\exp(-\pi/\varpi) \simeq \exp[-2\mathcal{K}_0] \simeq \tfrac{1}{16} k_0'^2. \tag{12.3.263}$$

Combining these results with (12.3.254) and (12.3.256) we see that, at the temperature $T = T_c - \delta T$, near to the critical surface, $\phi^{(8vM)}(\mathcal{N}^{\circledast}, T)$ can, as in (4.2.8), be divided into a sum of smooth and singular parts with

$$\phi^{(8vM)}_{smth}(\mathcal{N}^{\circledast}, T) \simeq \phi^{(8vM)}(\mathcal{N}^{\circledast}, T_c) - 4\left[\frac{|\Delta_0||\delta T|}{8\sin^2(\vartheta_c)}\right]^2 \cos\left(\frac{\vartheta_c}{2}\right)\left[\cos\left(\frac{\vartheta_c}{2}\right) + \cos\left(\frac{\varphi_c}{2}\right)\right], \tag{12.3.264}$$

$$\phi^{(8vM)}_{sing}(\mathcal{N}^{\circledast}, T) \simeq 4\left[\frac{|\Delta_0||\delta T|}{8\sin^2(\vartheta_c)}\right]^{\pi/\vartheta_c} \cot\left(\frac{\pi^2}{2\vartheta_c}\right)\cos\left(\frac{\pi\varphi_c}{2\vartheta_c}\right). \tag{12.3.265}$$

Unless π/ϑ_c is an integer

$$\phi^{(8vM)}_{sing}(\mathcal{N}^{\circledast}, T) \sim |\delta T|^{\pi/\vartheta_c}. \tag{12.3.266}$$

Exceptional cases occur if $\vartheta_c = \pi/p$, where p is an integer. If p is even the factor $\cot(\pi^2/2\vartheta_c)$ in (12.3.265) is infinite. This is due to the coincidence of two poles of $\psi(x; \varphi, \vartheta)$. If the residue of the resulting double pole is calculated, (12.3.266) is replaced by

$$\phi^{(8vM)}_{sing}(\mathcal{N}^{\circledast}, T) \sim |\delta T|^{\pi/\vartheta_c} \ln|\delta T|. \tag{12.3.267}$$

If p is odd $\cot(\pi^2/2\vartheta_c) = 0$ and it is necessary to consider the dependence of ϑ on T.

When π/ϑ_c is non-integer (12.3.266) implies that the heat capacity exponents α and α' defined in (4.4.12) are given by

$$\alpha = \alpha' = 2 - \frac{\pi}{\vartheta_c}. \qquad (12.3.268)$$

The critical value ϑ_c of ϑ is finite and non-zero and, from (12.3.251), $\xi_c = \frac{1}{2}\vartheta_c$. It follows from (12.3.249) that, as $k'_0 \to 0$, $k'_m \to 0$ and, using (18.3.10),

$$k'_m \simeq 4\left(\frac{k'_0}{4}\right)^{\pi/2\vartheta_c}. \qquad (12.3.269)$$

From (12.3.250), (12.3.262) and (12.3.269) the magnetic exponent β, defined in (4.4.13) (with $\rho = m_s$, $\rho_c = 0$) is given by

$$\beta = \frac{\pi}{16\vartheta_c}. \qquad (12.3.270)$$

If the Essam–Fisher and Widom scaling laws, given by (4.5.44) and (4.5.45) respectively, are assumed, then it follows from (12.3.269) and (12.3.270) that

$$\gamma = \frac{7\pi}{8\vartheta_c}, \qquad \delta = 15. \qquad (12.3.271)$$

To complete the picture of the critical properties of the eight-vertex model it should be noted that the correlation length $\mathfrak{r}$ can be shown to take the form

$$\mathfrak{r} = \frac{2}{\ln(1/k_m)}, \qquad (12.3.272)$$

(Baxter 1982b). From (12.3.262) and (12.3.269) it then follows that the critical exponents ν and ν′, defined in (4.5.52), are given by

$$\nu = \nu' = \frac{\pi}{2\vartheta_c}. \qquad (12.3.273)$$

From (12.3.268) and (12.3.273) the Josephson hyper-scaling law (4.5.54), with $d = 2$ is satisfied. If the Fisher scaling law (4.5.58) is assumed then, from (12.3.271) and (12.3.273), the exponent η, defined in (4.5.55), is given by

$$\eta = \tfrac{1}{4}. \qquad (12.3.274)$$

Since ϑ_c is given by (12.3.259) it varies continuously as a function of ε_a, ε_b, ε_c and ε_d, This is a situation where the critical exponents, except δ and η, vary as continuous functions of the energy parameters of the model. The implication of this for the concept of universality is discussed in Sect. 12.3.11.

12.3.9 The Coupling Form and the Ising Model Limit

In terms of the couplings $K^{(f)}$, $K^{(b)}$ and $K^{(4)}$ the critical surfaces are given by (9.3.34) with the critical exponents varying as functions of ϑ_c, which is given, from (12.1.13) (dropping the subscript j) and (12.3.259), by

$$\vartheta_c = \begin{cases} \pi - \arccos[\tanh(2K^{(4)})], & \text{for } \mathcal{C}(\text{I}) \text{ and } \mathcal{C}(\text{II}), \\ \arccos[\tanh(2K^{(4)})], & \text{for } \mathcal{C}(\text{III}) \text{ and } \mathcal{C}(\text{IV}). \end{cases} \tag{12.3.275}$$

It has been shown by Kadanoff and Wegner (1971) that the four-spin interaction in the eight-vertex model scales as $1/r^2$. This means, in terms of our discussion in Sect. 4.3.8, that $K^{(4)}$ is a marginal coupling, and the variation of the critical exponents α, β and γ, given by (12.3.268), (12.3.270) and (12.3.271) respectively, as functions of $K^{(4)}$ does not conflict with scaling theory.

Again using upper and lower signs for the pairs of regions (I, II) and (III, IV), respectively, the critical surfaces given by (9.3.34) cut the planes $K^{(f)} = \pm K^{(b)} = K$ with $K^{(4)} \gtrless 0$ in a curve on which[68]

$$K = \tfrac{1}{2}\operatorname{arcsinh}\left(\sqrt{\frac{1-\tanh(2|K^{(4)}|)}{1+\tanh(2|K^{(4)}|)}}\right). \tag{12.3.276}$$

The plot of this, shown in Fig. 12.8, meets the $K^{(4)} = 0$ axis at $K = K_c$ given by $\sinh(2K_c) = 1$, which is the standard formula for the critical temperature of the square-lattice Ising model given in (9.2.36). We also note that, as $K \to 0$ on the curve, $K^{(4)} \to \infty$, showing that there is no phase transition in a spin-$\frac{1}{2}$ model on a square lattice with a purely four-spin interaction.

The case $K^{(4)} = 0$, where the eight-vertex model, in the coupling formulation, reduces to two disconnected Ising models, has been explored in great detail in Sect. 12.2. Here we simply exploit the information acquired from the derivation of the free energy in this special case, restricting attention to the situation $K^{(f)} = K^{(b)} = K$ (in region I or II). Now, from (3.8.12) and (3.8.13),

$$\Delta = \sinh^2(2K), \qquad \Gamma = 0, \tag{12.3.277}$$

from (12.3.1), (12.3.4) and (12.3.251),

$$\begin{aligned} &\lambda = \mathcal{K}_0', \qquad \xi = \tfrac{1}{2}\mathcal{K}_0', \qquad \vartheta = \tfrac{1}{2}\pi, \\ &u = \tfrac{1}{2}\lambda, \qquad z = \varphi = 0, \end{aligned} \tag{12.3.278}$$

[68] Of course, there is also a choice of sign for K but this applies to all the regions.

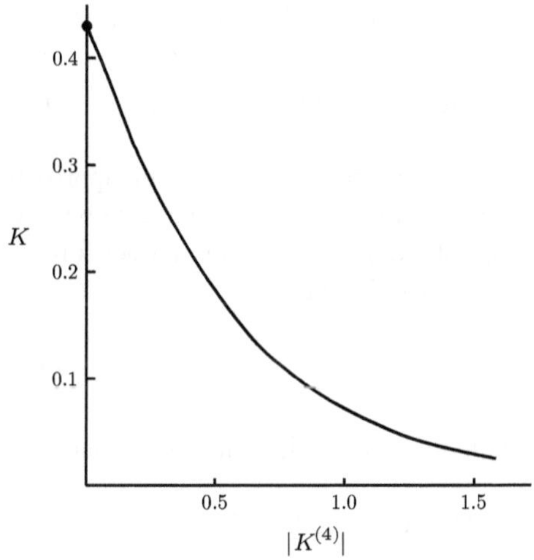

Fig. 12.8 The critical curve for the eight-vertex model, where the critical point of the isotropic Ising model lies on the $K^{(4)} = 0$ axis at $K = 0.4407$

from (12.3.261),

$$\Delta = \begin{cases} 1/\mathrm{k}_0, & K \geq K_\mathrm{c}, \ (T \leq T_\mathrm{c}), \\ \mathrm{k}_0, & K \leq K_\mathrm{c}, \ (T \geq T_\mathrm{c}) \end{cases} \tag{12.3.279}$$

and, from (12.3.268), (12.3.270)–(12.3.274) and (12.3.278),

$$\begin{aligned} \alpha &= 0, & \beta &= \tfrac{1}{8}, & \gamma &= \tfrac{7}{4}, \\ \delta &= 15, & \nu &= 1, & \eta &= \tfrac{1}{4}, \end{aligned} \tag{12.3.280}$$

which are the well-known results for the two-dimensional Ising model. The fact that $\pi/\vartheta_\mathrm{c} = 2$ shows, from the discussion of Sect. 12.3.8, that the heat capacity has a logarithmic singularity at the critical point. From (12.3.249) and (12.3.278),

$$\frac{\mathcal{K}(\mathrm{k}'_m)}{\mathcal{K}(\mathrm{k}_m)} = \frac{\mathcal{K}'_0}{\mathcal{K}_0}, \tag{12.3.281}$$

from which it follows that $\mathrm{k}_m = \mathrm{k}'_0$. Then, from (12.3.250),[69]

$$m_\mathrm{s} = (\mathrm{k}'_m)^{\frac{1}{4}} = (1 - \mathrm{k}'^2_0)^{\frac{1}{8}}, \tag{12.3.282}$$

69 This is a result which applies to the square, triangular and honeycomb lattices (McCoy and Wu 1973).

yielding, from (12.3.277) and (12.3.278).

$$m_s = [1 - \sinh^{-4}(2K)]^{\frac{1}{8}}\,, \qquad T < T_c. \tag{12.3.283}$$

This is show, in comparison with approximate results, in Fig. 3.1.

The series expansion (12.3.246) will now be used to obtain an expression for the internal energy of the Ising model following the procedure of Lavis (1996). From (12.3.278), Eqs. (12.3.247) reduce to

$$\mathcal{Y} = 1\,, \qquad \mathcal{X} = \mathcal{Q}^{\frac{1}{4}} = \exp(-\pi\mathcal{K}_0'/2\mathcal{K}_0). \tag{12.3.284}$$

Using the Landen modulus k_2 defined in Sect. 18.3.4,

$$\mathcal{X} = \exp(-\pi\mathcal{K}_2'/\mathcal{K}_2) \tag{12.3.285}$$

is the nome of k_2. From (3.8.11), Table 9.1, (12.3.246), (12.3.284) and (12.3.285),

$$\phi^{(8\text{vM})}(\mathcal{N}^{\circledast}, K) = \begin{cases} -2|K| - \sum_{n=1}^{\infty} S_n(\mathcal{X})\,, & T < T_c\,, \\ -\ln[\cosh(2K)+1] - \sum_{n=1}^{\infty} S_n(\mathcal{X})\,, & T > T_c\,, \end{cases} \tag{12.3.286}$$

where

$$S_n(\mathcal{X}) = \frac{\mathcal{X}^{2n}(1-\mathcal{X}^{2n})^2(1-\mathcal{X}^n)^2}{n(1-\mathcal{X}^{8n})(1+\mathcal{X}^{2n})}. \tag{12.3.287}$$

The dimensionless internal-energy density u is given by (2.3.27)[70] and, from (12.3.277), (12.3.279), (12.3.286) and (18.3.12),

$$\mathrm{u} = \begin{cases} -2 + \dfrac{B(K)}{\mathcal{K}_2^2}\sum_{n=1}^{\infty}\mathcal{X}\dfrac{\mathrm{d}S_n(\mathcal{X})}{\mathrm{d}\mathcal{X}}\,, & T < T_c\,, \\ -\dfrac{2\sinh(2K)}{1+\cosh(2K)} + \dfrac{B(K)}{\mathcal{K}_2^2}\sum_{n=1}^{\infty}\mathcal{X}\dfrac{\mathrm{d}S_n(\mathcal{X})}{\mathrm{d}\mathcal{X}}\,, & T > T_c\,, \end{cases} \tag{12.3.288}$$

[70] In the notation of Sect. 2.3 ϕ is here $\phi_1(K, 0)$, the free-energy density with one external coupling, $L := \mathcal{H}/T$ set equal to zero.

where

$$B(K) = \frac{\pi^2 \cosh^3(2K)}{\sinh(2K)[\sinh^2(2K) - 1]}. \tag{12.3.289}$$

From (12.3.287)

$$\begin{aligned} \mathcal{X}\frac{\mathrm{d}S_n(\mathcal{X})}{\mathrm{d}\mathcal{X}} &= \frac{\mathcal{X}^n}{1+\mathcal{X}^{2n}} + \frac{2\mathcal{X}^{2n}}{(1+\mathcal{X}^{2n})^2} - \frac{8\mathcal{X}^{3n}}{(1+\mathcal{X}^{2n})^3} \\ &\quad - \frac{\mathcal{X}^n(1+\mathcal{X}^{2n})}{1+\mathcal{X}^{4n}} - \frac{4\mathcal{X}^{4n}}{(1+\mathcal{X}^{4n})^2} + \frac{4\mathcal{X}^{3n}(1+\mathcal{X}^{2n})}{(1+\mathcal{X}^{2n})^3}, \end{aligned} \tag{12.3.290}$$

and hence, from Eqs. (18.3.91)–(18.3.93),

$$\sum_{n=1}^{\infty} \mathcal{X}\frac{\mathrm{d}S_n(\mathcal{X})}{\mathrm{d}\mathcal{X}} = \frac{\mathrm{k}_2'\mathcal{K}_2^2}{\pi^2}\left[\sqrt{2(1+\mathrm{k}_2')} - 1\right] - \frac{2\mathrm{k}_2'^2\mathcal{K}_2^3}{\pi^3}. \tag{12.3.291}$$

Since, from (12.3.277), (12.3.279) and (18.3.78),

$$\mathrm{k}_2(K) = \frac{2\sinh(2K)}{\cosh^2(2K)}, \tag{12.3.292}$$

$$\mathrm{k}_2'(K) = \pm\left[\frac{1-\sinh^2(2K)}{\cosh^2(2K)}\right], \qquad T \gtrless T_{\mathrm{c}}, \tag{12.3.293}$$

by substituting from (12.3.292) and (12.3.293) into (12.3.291) and then into (12.3.288) the formula (12.2.178) for the internal energy per lattice site of the Ising model on the square lattice at all temperatures is obtained.

12.3.10 The Six-Vertex Model

The six-vertex model corresponds to the limiting case $\varepsilon_{\mathrm{d}} \to \infty$ ($\mathrm{d} \to 0$) of the eight-vertex model with, from (9.3.1), $\Gamma = 1$. It can fall in region I, II or III, but not IV.

12.3.10.1 Low-Temperature Zones $\mathcal{R}_{\mathrm{L}}$(I) and $\mathcal{R}_{\mathrm{L}}$(II)

In these zones, from (9.3.31), (9.3.32), Table 12.2 and (18.3.43),

$$\Delta = \Delta^\star = \mathrm{nd}(2\xi\,|\mathrm{k}_0') > 0, \qquad \Gamma = \Gamma^\star = -\mathrm{cd}(2\xi\,|\mathrm{k}_0') = 1. \tag{12.3.294}$$

From (18.3.39) these relationships hold in the limit $k_0 \to 0$ with

$$\Delta^\star = \cosh(2\xi). \tag{12.3.295}$$

This corresponds to a case of the XYX (in standard notation XXZ) linear Heisenberg model (Sects. 3.2.1 and 12.3.4.2, case (vi)). From Table 9.1, for small d, w_4 is positive in $\mathcal{R}_L(I)$ and negative in $\mathcal{R}_L(II)$. So, from Table 12.2, z is positive in $\mathcal{R}_L(I)$ and negative in $\mathcal{R}_L(II)$ with $|z| \le \xi$. From (12.3.247), in the limit $k_0 \to 0$,

$$\begin{aligned}
\mathcal{Q} &= \exp(-2\pi\mathcal{K}_0'/\mathcal{K}_0) \to \mathcal{X}^2 \to 0,\\
\mathcal{X} &= \exp(-\pi\xi/\mathcal{K}_0) \to 0,\\
\mathcal{Y} &= \begin{cases} \exp(-\pi|z|/\mathcal{K}_0) \to \mathcal{X}^2 \to 0 & \text{in } \mathcal{R}_L(I),\\ \exp(\pi|z|/\mathcal{K}_0) \to \mathcal{X}^{-2} \to \infty & \text{in } \mathcal{R}_L(II).\end{cases}
\end{aligned} \tag{12.3.296}$$

From (12.3.246) we have, therefore,

$$\begin{aligned}
\phi^{(6vM)}(\mathcal{N}^{\circledast}, a, b, c) &= -\ln(w_1 + w_2)\\
&= -\ln\left[\tfrac{1}{2}(a+b) + \tfrac{1}{2}|a-b|\right],
\end{aligned} \tag{12.3.297}$$

which is equivalent to (9.3.20), in $\mathcal{R}_L(I)$, where $a > b$. From (12.3.259) $\vartheta_c = \pi$ in this case and, from (12.3.268), $\alpha = \alpha' = 1$. It thus follows that the scaling exponent y_2 of (4.5.39) is equal to $d = 2$, indicative of a first-order transition. This is of course as it should be since the transition is to the frozen completely-ordered ferroelectric state.

12.3.10.2 Low-Temperature Zone $\mathcal{R}_L(III)$

Here, from (9.3.31), (9.3.32), Table 12.2 and (18.3.43),

$$\begin{aligned}
\Delta &= -\Delta^\star = -\mathrm{nd}(2\xi|k_0') < 0,\\
\Gamma &= -\Gamma^\star = \mathrm{cd}(2\xi|k_0') = 1.
\end{aligned} \tag{12.3.298}$$

These relationships again hold in the limit $k_0 \to 0$, with $\Delta^\star$ given by (12.3.295), and the same equivalence to the XXZ model. From Table 12.2,

$$\frac{a}{c} = \frac{\sinh(\xi + z)}{\sinh(2\xi)}, \qquad \frac{b}{c} = \frac{\sinh(\xi - z)}{\sin(2\xi)} \tag{12.3.299}$$

and, from (12.3.247),

$$\mathcal{Q} = 0, \qquad \mathcal{X} = \exp(-2\xi), \qquad \mathcal{Y} = \exp(-2z). \tag{12.3.300}$$

Substituting into (12.3.246) gives

$$\phi^{(6\text{vM})}(\mathcal{N}^{\circledast}, \mathrm{a}, \mathrm{b}, \mathrm{c}) = -\ln(\mathrm{c}) - \sum_{n=1}^{\infty} \frac{\exp(-4n\xi)[\cosh(2n\xi) - \cosh(2nz)]}{n\cosh(2n\xi)} \tag{12.3.301}$$

and, substituting for c from the first of Eqs. (12.3.299) and expanding the resulting logarithmic series, gives the alternative form

$$\phi^{(6\text{vM})}(\mathcal{N}^{\circledast}, \mathrm{a}, \mathrm{b}, \mathrm{c}) = -(\xi - z) - \sum_{n=1}^{\infty} \frac{\exp(-2n\xi)\sinh[2n(\xi - z)]}{n\cosh(2n\xi)}. \tag{12.3.302}$$

12.3.10.3 High-Temperature Zones $\mathcal{R}_{\mathrm{d}}(\mathrm{I})$, $\mathcal{R}_{\mathrm{d}}(\mathrm{II})$ and $\mathcal{R}_{\mathrm{d}}(\mathrm{III})$

In these zones, from Tables 9.1 and 12.2,

$$\begin{aligned} \Delta &= \Gamma^{\star}/\Delta^{\star} = -\mathrm{cn}(2\xi\,|\mathrm{k}_0'), \\ \Gamma &= 1/\Delta^{\star} = \mathrm{dn}(2\xi\,|\mathrm{k}_0') = 1, \end{aligned} \tag{12.3.303}$$

which are satisfied, in the limit $\mathrm{k}_0 \to 1$, when $\vartheta = 2\xi$ and

$$\Delta = -\cos(\vartheta). \tag{12.3.304}$$

Since, in these zones, from Table 9.1 with $\mathrm{d} = 0, w_2 = w_3$, this is the critical surface with

$$\frac{\mathrm{a}}{\mathrm{c}} = \frac{\sin\left[\frac{1}{2}(\vartheta + \varphi)\right]}{\sin(\vartheta)}, \qquad \frac{\mathrm{b}}{\mathrm{c}} = \frac{\sin\left[\frac{1}{2}(\vartheta - \varphi)\right]}{\sin(\vartheta)}, \tag{12.3.305}$$

and the free-energy density is, from (12.3.258),

$$\begin{aligned} \phi^{(6\text{vM})}(\mathcal{N}^{\circledast}, \mathrm{a}, \mathrm{b}, \mathrm{c}) &= -\ln\left[\tfrac{1}{2}(\mathrm{a} + \mathrm{b} + \mathrm{c})\right] \\ &\quad - \int_{-\infty}^{\infty} \frac{\sinh^2[(\pi - \vartheta)x][\cosh(\vartheta x) - \cosh(\varphi x)]}{x\sinh(2\pi x)\cosh(\vartheta x)} \mathrm{d}x. \end{aligned} \tag{12.3.306}$$

This is the expression given by Baxter (1972b). With some manipulation it can be re-expressed in the form

$$\phi^{(6\mathrm{vM})}(\mathcal{N}^{\circledast}, \mathfrak{a}, \mathfrak{b}, \mathfrak{c}) = -\ln(\mathfrak{a}) - \frac{1}{2}\int_{-\infty}^{\infty} \frac{\sinh[(\pi - \vartheta)x]\sinh[(\vartheta - \varphi)x]}{x\sinh(\pi x)\cosh(\vartheta x)}\mathrm{d}x, \tag{12.3.307}$$

given by Lieb and Wu (1972).

12.3.10.4 Square Ice

Here $\mathfrak{a} = \mathfrak{b} = \mathfrak{c} = 1$ imposing, in addition to (12.3.309), the condition that $\sin(\vartheta) = -\sin(2\vartheta)$. This means that $\vartheta = \frac{2}{3}\pi$ in (12.3.310), giving, for the entropy of square ice,

$$\begin{aligned}\ln[\Omega(N^{\circledast})] := -N^{\circledast}\phi^{(6\mathrm{vM})}(\mathcal{N}^{\circledast}, 1, 1, 1) &= N^{\circledast}\ln\left|\frac{4\cot(3\pi/4)}{3\sin(2\pi/3)}\right| \\ &= \tfrac{3}{2}N^{\circledast}\ln\left(\frac{4}{3}\right) = N^{\circledast}\ln(1.5396)\,,\end{aligned} \tag{12.3.308}$$

This is the result obtained by Lieb (1967a, b). It is of interest that it is very close to an estimate obtained by Pauling (1935).[71] He simply took the total unrestricted number of arrow arrangements $2^{2N^{\circledast}}$ and scaled by $(6/16)^{N^{\circledast}}$ to compensate for the fact that only six of the sixteen arrangements around a vertex are permitted; thus giving $\Omega(N^{\circledast}) = (1.5)^{N^{\circledast}}$.

12.3.10.5 The KDP Model

From Sect. 3.8.2, $\mathfrak{a} > \mathfrak{b} = \mathfrak{c}$, giving, from (9.3.1), (12.3.304) and (12.3.305),

$$\sin\left[\tfrac{1}{2}(\vartheta + \varphi)\right] = -\sin(2\vartheta), \qquad \sin\left[\tfrac{1}{2}(\vartheta - \varphi)\right] = \sin(\vartheta), \tag{12.3.309}$$

which have the solutions $(\varphi, \vartheta) = (0, \frac{2}{3}\pi)$ and (π, π). Glasser (1969) showed that when these conditions are satisfied the integral in (12.3.307) can be expressed in terms of elementary functions to give

$$\phi^{(6\mathrm{vM})}(\mathcal{N}^{\circledast}, \mathfrak{a}, \mathfrak{b}, \mathfrak{b}) = -\ln(\mathfrak{a}) - \ln\left|\frac{2\vartheta}{\pi\sin(\vartheta)}\cot\left(\frac{\pi^2}{2\vartheta}\right)\right|. \tag{12.3.310}$$

---

[71] And, as noted above, even closer to $N^{\circledast}\ln(1.543)$ the result of the series approximation by Nagle (1966a).

12.3.10.6 The F Model and Critical Potts Model

From Sect. 3.8.2, $\mathfrak{a} = \mathfrak{b} < \mathfrak{c}$ and, as we saw in Sect. 3.8.2.1, setting the Boltzmann weights to $\mathfrak{a} = \mathfrak{b} = 1$, $\mathfrak{c} = (2+\sqrt{\nu})^{\frac{1}{2}}$ gives equivalence to the square-lattice ferromagnetic ν-state Potts model at its critical temperature. From (9.3.1),

$$\Delta = -\tfrac{1}{2}\sqrt{\nu} \tag{12.3.311}$$

and variations of ν now play the role of temperature variation.

- For $0 < \nu < 4$ $-1 < \Delta < 0$, placing the model in the disordered (high-temperature in the six-vertex sense) zone with, from (12.3.295) and (12.3.305), $\varphi = 0$ and

$$\sqrt{\nu} = 2\cos(\vartheta), \qquad 0 < \vartheta < \tfrac{1}{2}\pi. \tag{12.3.312}$$

  Then, from (3.8.41) and (12.3.307),

$$\phi_{\mathrm{c}}^{(\mathrm{PM})}(\mathcal{N},\nu) = -\tfrac{1}{2}\ln(\nu) - 2\int_0^\infty \frac{\sinh[(\pi-\vartheta)x]\tanh(\vartheta x)}{x\sinh(\pi x)}\mathrm{d}x. \tag{12.3.313}$$

- For $\nu > 4$ $\Delta < -1$, placing the model in $\mathcal{R}_{\mathrm{L}}(\mathrm{III})$. From Table 9.1 $w_4 = 0$ and hence, from Table 12.2, $z = 0$. From (12.3.295),

$$\sqrt{\nu} = 2\cosh(2\xi), \tag{12.3.314}$$

  and, from (12.3.302),

$$\phi_{\mathrm{c}}^{(\mathrm{PM})}(\mathcal{N},\nu) = -\tfrac{1}{2}\ln(\nu) - \xi - \sum_{n=1}^{\infty}\frac{\exp(-2n\xi)\tanh(2n\xi)}{n}. \tag{12.3.315}$$

- For the boundary $\nu = 4$ the free-energy density can be obtained by taking the limit $\vartheta \to 0$ in (12.3.313), or $\xi \to 0$ in (12.3.315), to give

$$\phi_{\mathrm{c}}^{(\mathrm{PM})}(\mathcal{N},\nu) = -\ln(2) - 2\int_0^\infty \exp(-x)\tanh(x)\mathrm{d}x = -8\ln\left[\frac{\Gamma\left(\frac{1}{4}\right)}{2^{\frac{5}{8}}\pi^{\frac{1}{2}}}\right]. \tag{12.3.316}$$

It is clear from these results that the model differs in its behaviour according to whether ν is greater or less that four. One way to understand more clearly the effect on the nature of the phase transition is to calculate the internal energy as the critical point is approached from above and below. For the ferromagnetic case this has been achieved by Baxter (1973d). In our present notation

$$u = -R(1 + \nu^{-\frac{1}{2}})[1 \pm p_0 \tanh(\xi)], \tag{12.3.317}$$

where p_0 is the spontaneous polarization with the $+$ and $-$ signs applying according to whether the critical point is approached from below or above. Baxter (1973e, f) showed that

$$p_0 = \begin{cases} 0, & 0 < \nu \leq 4, \\ \prod_{n=1}^{\infty} \tanh^2(2\,n\,\xi), & \nu > 4. \end{cases} \tag{12.3.318}$$

Twice the second term in (12.3.317), therefore, represents the latent heat at the transition for $\nu > 4$, the transition being first-order. In all cases the first term in (12.3.317) gives the internal energy per site averaged over the values below and above the transition. This was first calculated by Potts (1952).

For $0 < \nu \leq 4$ it has been conjectured by den Nijs (1979) that

$$\alpha = \frac{2\pi - 8\vartheta}{3\pi - 6\vartheta} \tag{12.3.319}$$

and independently by Nienhuis et al. (1980) and Pearson (1980) that

$$\beta = \frac{\pi + 2\vartheta}{12\pi}, \tag{12.3.320}$$

where ϑ is given by (12.3.312). Using the scaling laws (4.5.44), (4.5.45) and (4.5.47) this gives

$$\gamma = \frac{7\pi^2 - 8\vartheta\pi + 4\vartheta^2}{6\pi(\pi - 2\vartheta)}, \qquad \delta = \frac{15\pi^2 - 16\vartheta\pi + 4\vartheta^2}{\pi^2 - 4\vartheta^2}. \tag{12.3.321}$$

The formula (12.3.319) for α has now been establish by Black and Emery (1981) by further exploiting the relationship to the staggered F model. From (12.3.312), with $\nu = 2$, $\vartheta = \pi/4$ yielding the exponent values (12.3.280) for the Ising model, When $\nu = 3$, $\vartheta = \pi/6$,

$$\alpha = \tfrac{1}{3}, \quad \beta = \tfrac{1}{9}, \quad \gamma = \tfrac{13}{9}, \quad \delta = 14. \tag{12.3.322}$$

These values are strongly supported by the finite-size scaling analysis of Blöte and Nightingale (1982) (Sect. 4.10) and by conformal invariance arguments of Friedan et al. (1984) and Cardy (1996) (Sect. 4.11). They are the same as those of the hard-hexagon model (Baxter 1982b) and support the argument of Alexander (1975) that the 3-state Potts model and hard-hexagon model lie in the same universality class. When $\nu = 4$, $\vartheta = 0$,

$$\alpha = \tfrac{2}{3}, \quad \beta = \tfrac{1}{12}, \quad \gamma = \tfrac{7}{6}, \quad \delta = 15, \tag{12.3.323}$$

but this gives only a partial picture of the asymptotic behaviour because of the presence of logarithmic corrections (Cardy et al. 1980; Nauenberg and Scalapino 1980; Salas and Sokal 1997).

12.3.11 The Eight-Vertex Model and Universality

The concept of universality classes was discussed in Sect. 4.2, where we listed Kadanoff's three criteria for the division of phases transitions into classes. It was also pointed out that any particular model may well exhibit phase transitions in different universality classes within different parts of its phase space. We saw in Sect. 5.4.1 that the symmetry group of the zero-field spin-1 Ising model is in general $\mathcal{S}_2$, the same as that of the standard zero-field spin-$\frac{1}{2}$ Ising model and thus, for any particular value of the dimension d, it will exhibit a phase transition in the universality class of the latter. However, when the interaction energy parameters attain certain special values there is a change of symmetry group to $\mathcal{S}_3$, that of the 3-state Potts model (Sect. 5.4.2), for which the transition lies in a different universality class. Examples can also be given when a change in the energy parameters will lead to a change in universality class by means of an effective change in the dimension of the system rather than the symmetry group. We could, for example, consider an anisotropic lattice model on a d-dimensional hypercubic lattice. If the energy parameters connecting microsystems in the direction of one axis tend to zero there will, from the point of view of cooperative phenomena, be an effective change of dimension of the system from d to $d-1$, with a resulting change of universality class for the phase transition.

At this point it is useful to distinguish between the *universality hypothesis*, in whatever form it is represented, and the *scaling hypothesis*, introduced in Sect. 4.3. It is a consequence of the latter and the corresponding hypothesis for correlations of Sect. 4.5.7 that the critical exponents satisfy the four scaling laws (4.5.44), (4.5.45), (4.5.54) and (4.5.58). It is sometimes asserted that the eight-vertex model contradicts the universality hypothesis because its critical exponents vary as a function of a parameter. We have shown in Sect. 4.3.8 that the scaling hypothesis encompasses the possibility of exponents varying as functions of a marginal coupling and in Sect. 12.3.9 it was shown that the four-spin coupling in the spin representation of the eight-vertex model is of this type. The interesting question in relation to universality is whether the variation of the four-spin coupling corresponds to a change of symmetry and thus a change of universal class according to Kadanoff's criterion (b), or whether the symmetry group remains the same. As was pointed out by Kadanoff and Wegner (1971) and Baxter (1982b) there is certainly a change of symmetry when the first two conditions $\mathfrak{X}_1 = \mathfrak{X}_2$, $\mathfrak{X}_3 = \mathfrak{X}_4$ in (3.8.4) are applied. These conditions would be violated if horizontal and vertical spin couplings $K^{(h)}$ and $K^{(v)}$ were added[72] to the Boltzmann weights (3.8.9). With $K^{(h)} = K^{(v)} = 0$ the symmetry group is the direct product $\mathcal{S}_2^{(m)} \times \mathcal{S}_2^{(m)}$, of spin-reversal symmetry applied to the two interpenetrating sublattices independently. With one or both of $K^{(h)}$ and $K^{(v)}$ non-zero the symmetry group is reduced to simply the spin-reversal symmetry group $\mathcal{S}_2^{(m)}$ of the Ising model (on the whole lattice). Although the general eight-vertex model, without the conditions $\mathfrak{X}_1 = \mathfrak{X}_2$, $\mathfrak{X}_3 = \mathfrak{X}_4$ has not been solved, the solution of the special free-fermion case by Fan and Wu (1970) (Sect. 3.8.1.5) gives a strong indication that the critical exponents are those of the Ising model. According to Kadanoff's criteria (a) and (b) for universality classes this is to be expected, since the dimensions and symmetry groups of the models are the same. By the same token we expect the exponents of the eight-vertex model with conditions (3.8.4) to be different. That they vary is understandable in terms of scaling. To set them in the context of Kadanoff's criteria of universality classes two approaches are possible:

(i) The model with varying exponents can be taken as one universality class.
(ii) Criterion (c) can be invoked with the marginal coupling taken as another criterion differentiating between classes.

The latter has some advantages since the concept of constant exponents *within* classes is preserved. Also since, in renormalization group theory (Chap. 15) a marginal exponent is associated with a line of fixed points and a fixed point is usually associated with a universality class, to associate each value of K_4 with a different class has some value.

Finally, the proposal of *weak universality* by Suzuki (1974) should be mentioned. He observed that the exponents in the eight-vertex model which vary are α, β, γ and ν (see Eqs. (12.3.268), (12.3.270), (12.3.271) and (12.3.273) respectively). These are precisely those which represent asymptotic behaviour in terms of $|T - T_c|$ (see Eqs. (4.4.12)–(4.4.14) and (4.5.52)). If, however, asymptotic behaviour is measured in terms of the inverse correlation length $\mathfrak{r}^{-1}$ then this leads to the use of the exponents

$$\hat{\varphi} = (2 - \alpha)/\nu\,, \qquad \hat{\beta} = \beta/\nu\,, \qquad \hat{\gamma} = \gamma/\nu\,. \tag{12.3.324}$$

The four scaling laws (4.5.45), (4.5.44), (4.5.54) and (4.5.58) are replaced by the *weak scaling laws*

$$\begin{aligned} \hat{\gamma} &= \hat{\beta}(\delta - 1)\,, \qquad & 2\hat{\beta} + \hat{\gamma} &= \hat{\varphi}\,, \\ d &= \hat{\varphi}\,, & 2 - \eta &= \hat{\gamma}. \end{aligned} \tag{12.3.325}$$

[72] As in (3.8.20) but now with equal values to left and right and above and below.

For the eight-vertex model the complete set of exponents is now

$$\hat{\varphi} = 2\,, \qquad \hat{\beta} = \tfrac{1}{8}\,, \qquad \hat{\gamma} = \tfrac{7}{4}\,, \qquad \delta = 15\,, \qquad \eta = \tfrac{1}{4}. \tag{12.3.326}$$

These are, of course, the same for the two-dimensional Ising model (Eq. (12.3.280)), which unfortunately obscures the important symmetry difference between the two models.

Chapter 13
Dimer Models

13.1 The Dimer Partition Function

Here we consider a lattice $\mathcal{N}$ with pure dimer coverings, as defined in Sect. 3.9. Let $\Pi_0 := (1, 2, \ldots, N)$ be some labelling of the sites of $\mathcal{N}$ so that the partition into pairs $(1, 2), (3, 4), \ldots, (N-1, N)$, with (i, j) being a first-neighbour pair occupied by a dimer, represents such a covering.[1] Then

$$Z(\mathcal{N}; \boldsymbol{T}) := \sum_{\{\Pi\}} |t(\pi_1, \pi_2)||t(\pi_3, \pi_4)| \cdots |t(\pi_{N-1}, \pi_N)|, \tag{13.1.1}$$

where, as in the dimer partition function (3.9.6), the sum is over all permutations Π of Π_0 satisfying (3.9.2) and $\boldsymbol{T}$ is any matrix with elements $t(i, j) := \langle i|\boldsymbol{T}|j\rangle$ satisfying

$$|t(i, j)| = \begin{cases} |t(j, i)| \neq 0, & i \text{ and } j \text{ a first-neighbour pair}, \\ 0, & i \text{ and } j \text{ not a first-neighbour pair}, \end{cases} \tag{13.1.2}$$

$$t(i, j) = \overline{t(j, i)}, \tag{13.1.3}$$

is a generalization of the formula (3.9.6) for the dimer partition function $Z(\mathcal{N}; \boldsymbol{\Lambda})$. For physical reasons associated with the problem that concerned them, Lieb and Loss (1993) called $\boldsymbol{T}$ a *hopping matrix* and showed that[2]:

Theorem 13.1.1 *If the hopping matrix $\boldsymbol{T}$ satisfies the canonical flux distribution condition* (16.4.8) *then*

---

[1] As observed in Sect. 3.9, for this to be the case N must be even.

[2] Parts (i) and (ii) of this theorem correspond respectively to their Theorems 3.1 and 16.1.

D.A. Lavis, *Equilibrium Statistical Mechanics of Lattice Models*,
Theoretical and Mathematical Physics, DOI 10.1007/978-94-017-9430-5_13

(i) $\mathrm{Det}\{\boldsymbol{T}\} = (-1)^{N/2}[Z(\mathcal{N}; \boldsymbol{T})]^2$, (13.1.4)

(ii) *there exists a gauge transformation* (16.4.5) *with*

$$\boldsymbol{U}^\dagger \boldsymbol{T} \boldsymbol{U} = \mathrm{i}\mathbf{Z}, \tag{13.1.5}$$

where $\mathbf{Z}$ *is a real antisymmetric matrix.*

It follows from Theorem 16.4.2, that a hopping matrix $\boldsymbol{T}$, satisfying the canonical flux distribution, can be constructed from any symmetric matrix $\boldsymbol{\Lambda}$, which satisfies (3.9.3), by taking $t(i, j) := \lambda(i, j)\exp[\mathrm{i}\theta(i, j)]$ with a suitable choice of the antisymmetric arguments $\theta(i, j)$. Then, from (13.1.1)–(13.1.3),

$$Z(\mathcal{N}; \boldsymbol{T}) = Z(\mathcal{N}; \boldsymbol{\Lambda}). \tag{13.1.6}$$

It then follows from Theorem 13.1.1 that there is a matrix $\mathbf{Z}$ related to $\boldsymbol{\Lambda}$ by

$$z(i, j) = \pm\lambda(i, j), \qquad z(j, i) = \mp\lambda(i, j), \tag{13.1.7}$$

for a particular choice of signs, such that

$$\mathrm{Det}\{\mathbf{Z}\} = [Z(\mathcal{N}; \boldsymbol{\Lambda})]^2. \tag{13.1.8}$$

The evaluation of the dimer partition function (3.9.6) for a particular choice of fugacities $\lambda(i, j)$ is thus reduced to:

(i) the determination of the choice of signs for the elements of $\mathbf{Z}$;
(ii) the evaluation of the determinant of $\mathbf{Z}$.

13.2 Superposition Polynomials and Pfaffians

The method used for task (i), the determination of the signs in (13.1.7) for the antisymmetric matrix $\mathbf{Z}$, is that of Kasteleyn (1961), which employs the *Pfaffian*

$$\mathrm{Pf}\{\mathbf{Z}\} := \sum_{\{\Pi\}} \epsilon(\Pi) z(\pi_1, \pi_2) z(\pi_3, \pi_4) \cdots z(\pi_{N-1}, \pi_N) \tag{13.2.1}$$

of $\mathbf{Z}$, where, as in the dimer partition function (3.9.6) and in (13.1.1), the sum is over all permutations Π of Π_0 satisfying (3.9.2), with $\epsilon(\Pi)$ being the *signature*[3] of Π. *Pfaff's theorem* states that:

---

[3] $\epsilon(\Pi) = 1$ or -1 according to whether Π is of even or odd parity, that is whether Π is obtained from $\Pi_0 = (1, 2, \ldots, N)$ by an even or odd number of transpositions. A transposition is an exchange of the position of two indices, with the order of the remaining indices unchanged. Any transposition can be effected by an odd number of transpositions of adjacent indices.

Theorem 13.2.1 *The determinant and Pfaffian of the $N \times N$ antisymmetric matrix* $\mathbf{Z}$ *satisfy the relationship*

$$\mathrm{Det}\{\mathbf{Z}\} = [\mathrm{Pf}\{\mathbf{Z}\}]^2 . \tag{13.2.2}$$

This result was proved in some special cases by Temperley and Fisher (1961), by induction on the order N by Green and Hurst (1964) and directly by Kasteleyn (1967). It follows from (13.1.7), (13.1.8) and (13.2.2) that[4]

$$\mathrm{Pf}\{\mathbf{Z}\} = Z(\mathcal{N}; \mathbf{Z}). \tag{13.2.3}$$

This result is known as *Kastelyn's theorem* (Kasteleyn 1961). Now let

$$\mathcal{P}(\mathbf{Z}; \Pi) := \epsilon(\Pi) z(\pi_1, \pi_2) z(\pi_3, \pi_4) \cdots z(\pi_{N-1}, \pi_N). \tag{13.2.4}$$

This is an arbitrary term of the Pfaffian (13.2.1). The correct choice of signs for the elements of $\mathbf{Z}$ to satisfy (13.1.8) is, from (13.2.3), the one for which $\mathcal{P}(\mathbf{Z}; \Pi) > 0$ for all permutations Π. We have introduced a standard dimer covering of $\mathcal{N}$, labelled so that the dimers are the lines $(1, 2), (3, 4), \ldots, (N-1, N)$ with $\Pi_0 = (1, 2, \ldots, N)$ being the identity permutation. Whatever the eventual choice of signs for the elements of $\mathbf{Z}$ we can choose $\epsilon(\Pi_0)$, so that $\mathcal{P}(\mathbf{Z}; \Pi_0) > 0$ and thus we need to ensure that

$$\mathrm{Sign}\{\mathcal{P}(\mathbf{Z}; \Pi)\} = \mathrm{Sign}\{\mathcal{P}(\mathbf{Z}; \Pi_0)\}, \qquad \forall\ \mathcal{P}(\mathbf{Z}; \Pi) \neq 0. \tag{13.2.5}$$

Now consider the product

$$\epsilon(\mathrm{T}) z(\tau_1, \tau_2) z(\tau_3, \tau_4) \cdots z(\tau_{N-1}, \tau_N), \tag{13.2.6}$$

where $(\tau_1, \tau_2), (\tau_3, \tau_4), \ldots, (\tau_{N-1}, \tau_N)$ is a dimer covering, but the indices in the permutation $\mathrm{T} := (\tau_1, \tau_2, \tau_3, \tau_4, \ldots)$are otherwise unordered. Suppose that for one of the pairs $\tau_{2k} < \tau_{2k-1}$. The transposition of these indices multiplies $\epsilon(\mathrm{T})$ by -1; but, from (13.1.7), $z(\tau_{2k}, \tau_{2k-1}) = -z(\tau_{2k-1}, \tau_{2k})$, so the product (13.2.6) remains unchanged. Again (τ_{2k-1}, τ_{2k}) can be exchanged with (τ_{2j-1}, τ_{2j}) for any k and j without affecting the sign of the product (13.2.6) since it involves two transpositions. It follows that the indices can be rearranged to satisfy (3.9.2) without altering the value of (13.2.6), which is thus equal to the Pfaffian term $\mathcal{P}(\mathbf{Z}; \Pi)$ corresponding to the same dimer configuration on the lattice. This degree of flexibility in the order of the indices is helpful when the assignment of signs to edges is discussed.

The relationship between an arbitrary dimer configuration and the standard configuration can be represented by a *superposition diagram*. In such a diagram full and broken lines respectively are used to denote the dimers of the arbitrary and standard configurations. There will thus be exactly one full and one broken line

[4] In order to achieve the positive sign it may be necessary to change the definition of $\mathbf{Z}$ by multiplying the first row and column by -1.

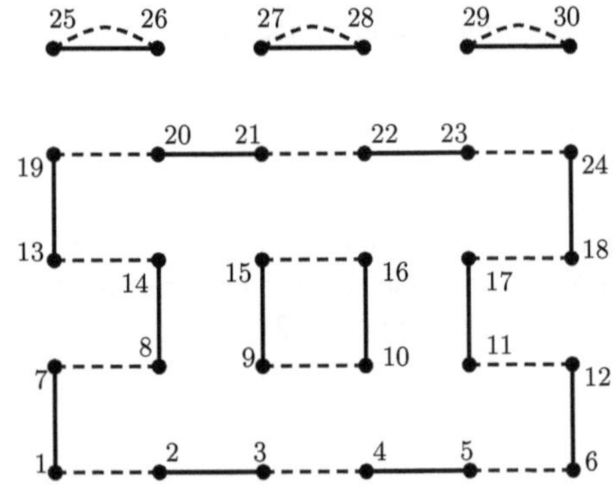

Fig. 13.1 The superposition diagram corresponding to Fig. 3.10

incident at each site. Figure 13.1 shows the superposition diagram corresponding to the dimer configuration of Fig. 3.10. Some first-neighbour site pairs are connected by both a full and a broken line and will be called *doubly connected.* Since the lattice is assumed to have a finite number of sites it is clear that every site of the lattice which is not doubly connected is a vertex of exactly one elementary cycle (polygon) of an even number of lines, which are alternately full and broken. These are called *superposition polygons.* Examples are the polygons through sites (9, 10, 16, 15, 9) and (1, 2, 3, 4, ..., 14, 8, 7, 1) in Fig. 13.1. A superposition diagram consists of doubly-connected first-neighbour pairs and superposition polygons. Any sites enclosed inside a given superposition polygon are arranged in doubly-connected pairs and/or smaller polygons so that the number of such sites is even. The larger polygon in Fig. 13.1 encloses four sites, themselves forming a polygon with two full and two broken edges.

Now set up an alternative permutation M_0 for the standard configuration, starting with the indices of all the doubly-connected pairs of sites and continuing with sequences $(\mu_1, \mu_2, \mu_3, \dots, \mu_{2n})$, representing the pairs connected by *broken* lines in anticlockwise order, for each superposition polygon. For the superposition diagram Fig. 13.1

$$\mathsf{M}_0 = (25, 26|27, 28|29, 30|9, 10, 16, 15|1, 2, 3, 4, \dots, 13, 14, 8, 7). \quad (13.2.7)$$

The cyclic permutation $(\mu_1, \dots, \mu_{2n-1}, \mu_{2n}) \to (\mu_{2n}, \mu_1, \dots, \mu_{2n-1})$ on each polygon, now produces the permutation M with indices of the site pairs connected by *full* lines in anticlockwise order. This represents the arbitrary dimer configuration. For Fig. 13.1

$$\mathsf{M} = (25, 26|27, 28|29, 30|15, 9, 10, 16|7, 1, 2, 3, \dots, 19, 13, 14, 8). \quad (13.2.8)$$

A cycle of $2n$ indices can be achieved by $2n - 1$ adjacent transpositions and thus multiplies the parity factor $\epsilon(\mathsf{M}_0)$ by $(-1)^{2n-1} = -1$. For instance, for the indices of the sites of the small polygon in Fig. 13.1, the transpositions from M_0 to M are

$$(9, 10, 16, 15) \to (9, 10, 15, 16) \to (9, 15, 10, 16) \to (15, 9, 10, 16),$$

yielding a factor $(-1)^3 = -1$. Thus

$$\epsilon(\mathrm{M}) = (-1)^{\mathrm{s}}\epsilon(\mathrm{M}_0), \tag{13.2.9}$$

where s is the number of superposition polygons in the diagram, which can be even or odd. Equation (13.2.5) applies if the product (13.2.6) and the similar product for M_0 have the same sign. For this, a sufficient condition is that, for each superposition polygon,

$$\begin{aligned}&\mathrm{Sign}\{z(\mu_1,\mu_2)z(\mu_3,\mu_4)\cdots z(\mu_{2n-1},\mu_{2n})\}\\&= -\mathrm{Sign}\{z(\mu_{2n},\mu_1)z(u_2,\mu_3)\cdots z(\mu_{2n-2},\mu_{2n-1})\},\end{aligned} \tag{13.2.10}$$

since each factor (-1) in (13.2.9) is then neutralized by another factor (-1). Condition (13.2.10) can be re-expressed in the form

$$\mathrm{Sign}\{z(\mu_1,\mu_2)z(\mu_2,\mu_3)z(\mu_3,\mu_4)\cdots z(\mu_{2n-1},\mu_{2n})z(\mu_{2n},\mu_1)\} = -1, \tag{13.2.11}$$

where $(\mu_1,\mu_2), (\mu_2,\mu_3), \ldots, (\mu_{2n},\mu_1)$ specify the edges, taken in anticlockwise order, of any lattice polygon which can be a superposition polygon.[5] Condition (13.2.11) is sufficient to ensure that all the terms $\mathcal{P}(\boldsymbol{Z};\mathsf{II})$ in (13.2.1) are positive and (13.1.8) is then valid.

We can now develop a topological criterion for ascribing appropriate signs to the non-zero $z(i, j) = \pm\lambda(i, j)$. Put an arrow on the edge connecting sites i and j which indicates the order of the indices for a positive factor. Equation (13.2.11) is equivalent to the condition that any superposition polygon is clockwise odd (Sect. 16.1.4). Suppose that an elementary cycle $\mathfrak{c}$ encloses a subgraph $\mathfrak{g}$ with $\mathrm{f}(\mathfrak{g})$ faces, bounded by face cycles $\mathfrak{c}_k$, and $\mathrm{n}_{\mathrm{i}}(\mathfrak{g})$ interior sites. The number ω_k of clockwise arrows on $\mathfrak{c}_k$ and the number ω of clockwise arrows on $\mathfrak{c}$ are related by (16.1.10). For a superposition polygon $\mathrm{n}_{\mathrm{i}}(\mathfrak{g})$ is an even number since it encloses only other superposition polynomials and doubly-connected lines. It follows that ω is odd if all ω_k are odd. When the lattice $\mathcal{N}$ has arrows indicating the signs of the elements of $\boldsymbol{Z}$ oriented in such a way that all elementary polygons are clockwise odd, then any superposition polygon is clockwise odd and the consequential terms of the Pfaffian are all positive. Thus:

Theorem 13.2.2 *If arrows are assigned to the first-neighbour pairs of $\mathcal{N}$ in such a way that every face cycle is clockwise odd then the assignment of signs to the elements of $z(i, j)$ such that $z(i, j) > 0$ or $z(i, j) < 0$, according as the arrow along (i, j) is in the direction $i \to j$ or $j \to i$, is that which satisfies* (13.1.8).

---

[5] Provided consistency is maintained, either anticlockwise or clockwise order can be used.

Step (ii) is now the evaluation of the determinant of the $N \times N$ matrix $\mathbf{Z}$. Since N is large this is not an obviously simple task. However, in the case where, as in (3.9.1), the fugacities are the same along each principle lattice direction and periodic boundary conditions are applied, the signs of the elements of $\mathbf{Z}$ can be chosen periodically. Then $\mathbf{Z}$ becomes a cyclic matrix and in Sect. 17.7 a procedure is described for evaluating the determinant of a cyclic $N \times N$ matrix in the limit $N \to \infty$. We now consider the cases of the square and honeycomb lattices.

13.2.1 The Square-Lattice Case

On the square lattice $\mathcal{N}_s$, $\lambda(i, j) = \mathfrak{z}_1$ and $\lambda(i, j) = \mathfrak{z}_2$, for all horizontal and vertical first-neighbour pairs respectively, with corresponding elements of $\mathbf{Z}$, $z(i, j) = \pm\mathfrak{z}_1$ and $z(i, j) = \pm\mathfrak{z}_2$. The signs are chosen to match an arrow configuration which makes every face cycle clockwise odd. For Det$\{\mathbf{Z}\}$ to be evaluated the arrow assignment must also be periodic and a suitable one for the square lattice is shown in Fig. 13.2a. With lattice spacing a, and disregarding boundary effects, a translation by a multiple of a in the vertical direction takes the oriented lattice into itself. However, because of the alternating arrow sense on the vertical edges, a translation in the horizontal direction must be by an even multiple of a to achieve this result. The unit cell of the oriented lattice is taken to be a $2a \times a$ rectangle, containing a horizontal first-neighbour pair of sites; such a cell is shown in Fig. 13.2b. The aim, following the general lines of Montroll (1964), is to express $\mathbf{Z}$ as a cyclic block matrix, as defined in Sect.17.7. Suppose that the lattice consists of N_2 rows each of $2N_1$ sites, so that there are N_2 rows each of N_1 unit cells. The cells are now specified using cartesian vectors of the type $\boldsymbol{j} := (j_x, j_y)$, where j_x and j_y are integers and the position of

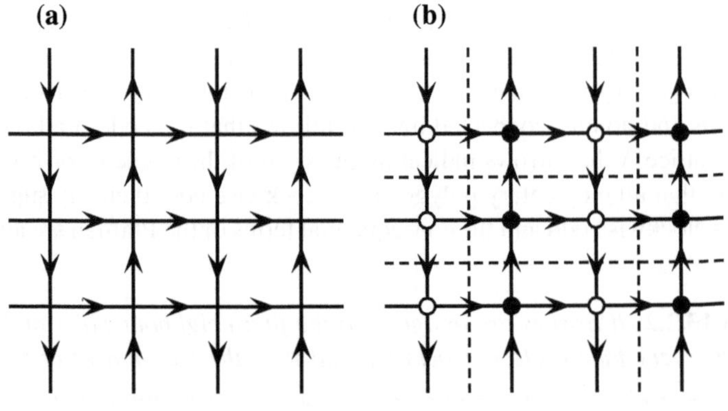

Fig. 13.2 **a** An arrow assignment for the square lattice $\mathcal{N}_s$. **b** A unit cell and sites in the neighbouring unit cells for $\mathcal{N}_s$. The cell boundaries are marked by *broken lines* and L and R sites are represented by • and ∘ respectively

the centre of cell $\boldsymbol{j}$ is $(2aj_x, aj_y)$. The (scalar) labels of the left and right sites of cell $\boldsymbol{j}$ are $\boldsymbol{j}_{\mathrm{L}}$ and $\boldsymbol{j}_{\mathrm{R}}$ respectively. Let

$$z(\boldsymbol{j},\boldsymbol{\ell}) := \begin{pmatrix} z(\boldsymbol{j}_{\mathrm{L}},\boldsymbol{\ell}_{\mathrm{L}}) & z(\boldsymbol{j}_{\mathrm{L}},\boldsymbol{\ell}_{\mathrm{R}}) \\ z(\boldsymbol{j}_{\mathrm{R}},\boldsymbol{\ell}_{\mathrm{L}}) & z(\boldsymbol{j}_{\mathrm{R}},\boldsymbol{\ell}_{\mathrm{R}}) \end{pmatrix}. \tag{13.2.12}$$

Since all cells are identical, $z(\boldsymbol{j},\boldsymbol{\ell})$ depends only on the relative positions of $\boldsymbol{j}$ and $\boldsymbol{\ell}$ and thus

$$z(\boldsymbol{j},\boldsymbol{\ell}) = z(\boldsymbol{\ell}-\boldsymbol{j}). \tag{13.2.13}$$

From (3.9.3) and (13.1.7), $z(\boldsymbol{j},\boldsymbol{\ell}) \neq \boldsymbol{0}$ only if there are first-neighbour links between sites in cells $\boldsymbol{j}$ and $\boldsymbol{\ell}$. From Fig. 13.2b it can be seen that the only non-zero blocks are

$$\begin{gathered} z(\boldsymbol{0}) = \begin{pmatrix} 0 & \mathfrak{z}_1 \\ -\mathfrak{z}_1 & 0 \end{pmatrix}, \\ z(\hat{\boldsymbol{x}}) = \begin{pmatrix} 0 & 0 \\ \mathfrak{z}_1 & 0 \end{pmatrix}, \qquad z(-\hat{\boldsymbol{x}}) = \begin{pmatrix} 0 & -\mathfrak{z}_1 \\ 0 & 0 \end{pmatrix}, \\ z(\hat{\boldsymbol{y}}) = \begin{pmatrix} \mathfrak{z}_2 & 0 \\ 0 & -\mathfrak{z}_2 \end{pmatrix}, \quad z(-\hat{\boldsymbol{y}}) = \begin{pmatrix} -\mathfrak{z}_2 & 0 \\ 0 & \mathfrak{z}_2 \end{pmatrix}, \end{gathered} \tag{13.2.14}$$

where $\hat{\boldsymbol{x}}$ and $\hat{\boldsymbol{y}}$ are respectively the unit horizontal and vertical vectors. Periodic boundary conditions are now imposed on the right and left and top and bottom of the lattice so that

$$z(\boldsymbol{j}+\boldsymbol{n}) = z(\boldsymbol{j}), \qquad \boldsymbol{n} := (n_1N_1, n_2N_2), \tag{13.2.15}$$

where n_1 and n_2 are any integers. Equations (13.2.13) and (13.2.15) imply that conditions (17.7.8) are satisfied and that $\boldsymbol{Z}$ is a cyclic block matrix with blocks of dimension 2×2. The determinant of $\boldsymbol{Z}$ can thus be evaluated using the method of Sect. 17.7. The disadvantage is that extra 'torus-winding' terms have been introduced into $\boldsymbol{Z}$ and, since the discussion in Sect. 13.2 depends on the lattice being planar, this approach cannot be used for finite lattices. However, it may be assumed that the boundary conditions will not affect densities in the thermodynamic limit $N_1 \to \infty$ and $N_2 \to \infty$.[6] Using (17.7.10) with $\theta_1 := 2\pi k_1/N_1$ and $\theta_2 := 2\pi k_2/N_2$,

[6] This is proved by Montroll (1964), Sect. 4.8, using Ledermann's theorem.

$$\zeta(\theta_1, \theta_2) = z(\mathbf{0}) + z(\hat{\boldsymbol{x}})\exp(\mathrm{i}\theta_1) + z(-\hat{\boldsymbol{x}})\exp(-\mathrm{i}\theta_1) + z(\hat{\boldsymbol{y}})\exp(\mathrm{i}\theta_2) + z(-\hat{\boldsymbol{y}})\exp(-\mathrm{i}\theta_2)$$
$$= \begin{pmatrix} \mathfrak{z}_2\exp(\mathrm{i}\theta_2) - \mathfrak{z}_2\exp(-\mathrm{i}\theta_2) & \mathfrak{z}_1 - \mathfrak{z}_1\exp(-\mathrm{i}\theta_1) \\ -\mathfrak{z}_1 + \mathfrak{z}_1\exp(\mathrm{i}\theta_1) & \mathfrak{z}_2\exp(-\mathrm{i}\theta_2) - \mathfrak{z}_2\exp(\mathrm{i}\theta_2) \end{pmatrix} \tag{13.2.16}$$

and thus

$$\mathrm{Det}\{\zeta(\theta_1, \theta_2)\} = 4\left[\mathfrak{z}_1^2\sin^2\left(\tfrac{1}{2}\theta_1\right) + \mathfrak{z}_2^2\sin^2(\theta_2)\right]. \tag{13.2.17}$$

From (13.1.8), (13.2.17) and (17.7.12) the dimensionless free energy per lattice site in the limit $N = 2N_1N_2 \to \infty$, when the finite square lattice $\mathcal{N}_\mathrm{s} \to \mathcal{L}_\mathrm{s}$, is given by

$$\phi(\mathcal{L}_\mathrm{s}, \mathfrak{z}_1, \mathfrak{z}_2) = -\lim_{N_1\to\infty}\lim_{N_2\to\infty}\frac{\ln[Z(\mathcal{N}_\mathrm{s}, \mathfrak{z}_1, \mathfrak{z}_2)]}{2N_1N_2}$$
$$= -\frac{1}{4}\lim_{N_1\to\infty}\lim_{N_2\to\infty}\frac{\ln[\mathrm{Det}\{\mathbf{Z}\}]}{N_1N_2}$$
$$= -\frac{1}{(4\pi)^2}\int_0^{2\pi}\mathrm{d}\theta_1\int_0^{2\pi}\mathrm{d}\theta_2\ln\left\{4\left[\mathfrak{z}_1^2\sin^2\left(\tfrac{1}{2}\theta_1\right) + \mathfrak{z}_2^2\sin^2(\theta_2)\right]\right\}. \tag{13.2.18}$$

With the change of variable $\theta_2 \to \frac{1}{2}\theta_2$, this result can be expressed in the more symmetric form

$$\phi(\mathcal{L}_\mathrm{s}, \mathfrak{z}_1, \mathfrak{z}_2) = -\tfrac{1}{2}\ln(2) + \frac{1}{(4\pi)^2}\int_0^{2\pi}\mathrm{d}\theta_1\int_0^{2\pi}\mathrm{d}\theta_2\ln[g(\mathfrak{z}_1, \mathfrak{z}_2; \cos(\theta_1), \cos(\theta_2))], \tag{13.2.19}$$

where

$$g(\mathfrak{z}_1, \mathfrak{z}_2; x_1, x_2) := [\mathfrak{z}_1^2 + \mathfrak{z}_2^2 - x_1\mathfrak{z}_1^2 - x_2\mathfrak{z}_2^2]^{-1}. \tag{13.2.20}$$

In the thermodynamic limit the numbers ρ_1 and ρ_2 of horizontal and vertical dimers respectively, per lattice site, are given by

$$\rho_i := -\mathfrak{z}_i\frac{\partial\phi}{\partial\mathfrak{z}_i} \tag{13.2.21}$$
$$= \frac{1}{2\pi^2}\int_0^{\pi}\mathrm{d}\theta_1\int_0^{\pi}\mathrm{d}\theta_2\, g(\mathfrak{z}_1, \mathfrak{z}_2; \cos(\theta_1), \cos(\theta_2)), \qquad i = 1, 2. \tag{13.2.22}$$

It can be seen, from these formulae, that $\rho_1 + \rho_2 = \frac{1}{2}$ is necessary for consistency. Using the standard result

$$\int_0^{\pi} \frac{\mathrm{d}\theta}{y - x\cos(\theta)} = \frac{\pi}{\sqrt{y^2 - x^2}},$$

this gives

$$\rho_i = \frac{1}{2\pi} \int_0^{\pi} \frac{\eta_i^2[1-\cos(\theta)]}{\sqrt{[\eta_2^2 + 1 - \eta_i^2\cos(\theta)]^2 - 1}} \mathrm{d}\theta$$

$$= \int_0^{\pi} \frac{\eta_i\sqrt{1-\cos(\theta)}}{\sqrt{\eta_i^2 + 2 - \eta_i^2\cos(\theta)}} \mathrm{d}\theta, \qquad i = 1, 2, \tag{13.2.23}$$

where $\eta_1 := \mathfrak{z}_1/\mathfrak{z}_2$ and $\eta_2 := 1/\eta_1 := \mathfrak{z}_2/\mathfrak{z}_1$. Using the substitution $u = 1-\cos\theta$

$$\rho_i = \frac{1}{2\pi} \int_0^{2} \frac{\mathrm{d}u}{\sqrt{(2-u)(2\eta_i^{-2} + u)}} = \frac{\arctan(\eta_i)}{\pi}. \tag{13.2.24}$$

This remarkably simple result is due to Fisher (1961). When $\mathfrak{z}_1 = 0$ all the dimers are vertical $Z(\mathcal{N}_\mathrm{s}, 0, \mathfrak{z}_2) = \mathfrak{z}_2^{N/2}$, $\phi(\mathcal{L}_\mathrm{s}, 0, \mathfrak{z}_2) = -\frac{1}{2}\ln(\mathfrak{z}_2)$ and similarly, when $\mathfrak{z}_2 = 0$, all the dimers are horizontal and $\phi(\mathcal{L}_\mathrm{s}, \mathfrak{z}_1, 0) = -\frac{1}{2}\ln(\mathfrak{z}_1)$. Using these results with (13.2.21) and (13.2.24), we obtain the alternative forms

$$\phi(\mathcal{L}_\mathrm{s}, \mathfrak{z}_1, \mathfrak{z}_2) = -\tfrac{1}{2}\ln(\mathfrak{z}_2) - \frac{1}{\pi}\int_0^{\mathfrak{z}_1/\mathfrak{z}_2} \frac{\arctan(\eta)\mathrm{d}\eta}{\eta}, \tag{13.2.25}$$

$$\phi(\mathcal{L}_\mathrm{s}, \mathfrak{z}_1, \mathfrak{z}_2) = -\tfrac{1}{2}\ln(\mathfrak{z}_1) - \frac{1}{\pi}\int_0^{\mathfrak{z}_2/\mathfrak{z}_1} \frac{\arctan(\eta)\mathrm{d}\eta}{\eta}. \tag{13.2.26}$$

It follows from these equations that $\phi(\mathcal{L}_\mathrm{s}, \mathfrak{z}_1, \mathfrak{z}_2)$ is a continuous function of $\mathfrak{z}_1$ and $\mathfrak{z}_2$ with continuous derivatives of all orders except at $\mathfrak{z}_1 = \mathfrak{z}_2 = 0$. There can, therefore, be no phase transitions. From (3.9.1) the number of ways $\Omega(N)$ of distributing the $\frac{1}{2}N$ dimers on the lattice $\mathcal{N}_\mathrm{s}$ of N sites is

$$\Omega(N) = \sum_{\{\mathfrak{m}_1+\mathfrak{m}_2=\frac{1}{2}N\}} \Omega(\mathcal{N}_\mathrm{s}; \mathfrak{m}_1, \mathfrak{m}_2) = Z(\mathcal{N}_\mathrm{s}, 1, 1). \tag{13.2.27}$$

For large N, from (13.2.25) or (13.2.26),

$$\frac{\ln\Omega(N)}{N} = -\phi(\mathcal{L}_\mathrm{s}, 1, 1) = \frac{1}{\pi}\int_0^{1} \frac{\arctan(\eta)\mathrm{d}\eta}{\eta}. \tag{13.2.28}$$

Expanding the integrand in a series, valid for $\eta \le 1$, and integrating gives

$$\frac{\ln\Omega(N)}{N} = \frac{1}{\pi}\sum_{k=0}^{\infty} \frac{(-1)^k}{(2k+1)^2} = \frac{\mathrm{G}}{\pi}, \tag{13.2.29}$$

where $\mathrm{G} := 0.915966$ is *Catalan's constant*.

The partition function for the finite $2N_1 \times N_2$ lattice was obtained by Fisher (1961) and Kasteleyn (1961). For even N_2

$$Z(\mathcal{N}_s, \mathfrak{z}_1, \mathfrak{z}_2) = 2^{\frac{1}{2}N} \prod_{k_1=1}^{N_1} \prod_{k_2=1}^{\frac{1}{2}N_2} \left[\mathfrak{z}_1^2 \cos^2\left(\frac{k_1\pi}{2N_1+1}\right) + \mathfrak{z}_2^2 \cos^2\left(\frac{k_2\pi}{N_2+1}\right)\right] \tag{13.2.30}$$

in the present notation. Fisher used an alternative combinatorial method to evaluate the Pfaffian, while Kasteleyn, by some ingenious algebra, was able to dispense with the periodic condition (13.2.15). The formula (13.2.19) for the free energy per site follows from (13.2.30) in the thermodynamic limit, confirming that in this limit intensive quantities are independent of the original boundary conditions. By putting $\mathfrak{z}_1 = \mathfrak{z}_2 = 1$ in (13.2.30), values of $\Omega(N)$ for finite lattices can be derived. The number of ways of arranging 32 symmetric dominos on a chess board corresponds to taking $2N_1 = N_2 = 8$ and gives $\Omega(64) = 12,988,816$ which is surprisingly large. This gives $\ln \Omega(64)/64 = 0.25593$, to be compared with the value 0.291561 obtained from (13.2.29) in the thermodynamic limit.

13.2.2 The Honeycomb-Lattice Case

An arrow assignment for the honeycomb lattice $\mathcal{N}_{\text{H}}$ is shown in Fig. 13.3. There are clockwise arrows on three of the six edges forming an elementary cycle so that the clockwise odd condition is satisfied. The broken lines denote the boundary of a unit cell containing a first-neighbour pair of sites. The notation for the representation of L and R sites is the same as in Fig. 13.2b. To present the cell structure as clearly as possible the usual geometry of the lattice has been distorted, although parallel edges remain parallel. This does not affect the evaluation of the partition function which depends only on the topology of the vertex connections and the fugacities assigned to the lattice directions. These are shown on the bottom right of the figure.

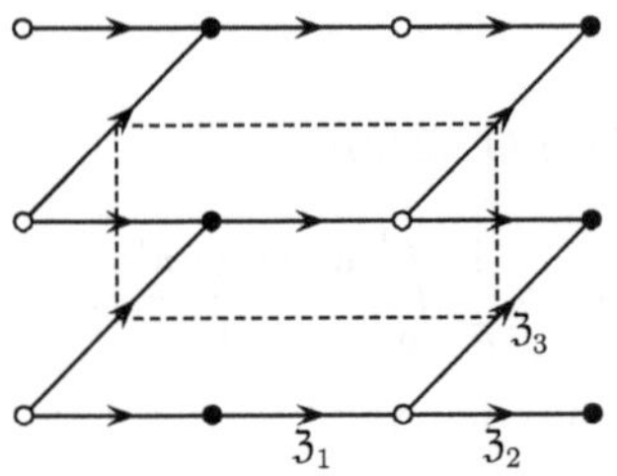

Fig. 13.3 An arrow assignment for the honeycomb lattice $\mathcal{N}_{\text{H}}$. The unit cell boundaries are marked by *broken lines* and L and R sites are represented by • and ∘ respectively

The lattice topology determines an important conservation property of the honeycomb dimer configurations. It can be seen from Fig. 3.11 that the occupation of any member of a row of horizontal edges by a dimer implies that exactly one of the rows of diagonal edges below is unoccupied. So if there are m_1 and m_2 dimers of species 1 and 2 respectively in a given row of horizontal edges then there will be $m_3 = N_1 - m_1 - m_2$ dimers in the row of diagonal edges above and the same number m_3 in the row of diagonal edges below. The number m_3 is thus the same for all rows of diagonal edges. Since the three sets of parallel edges indexed by the numbers 1, 2 and 3 are topologically equivalent a similar conservation rule applies to dimers of species 1 and 2. In the geometry of Fig. 13.3 there is thus the same number of dimers in each column of type 1 edges and in each column of type 2 edges.[7]

The analysis of the honeycomb lattice is similar to that for the square lattice, although $z(\pm\hat{\boldsymbol{y}}) = \boldsymbol{0}$, since, from Fig. 13.3, there are no first-neighbour links between the sites of a given cell and those of the cells immediately above and below it. However, there are diagonal edges, in the direction $\hat{\boldsymbol{d}} := (\hat{\boldsymbol{x}} + \hat{\boldsymbol{y}})/\sqrt{2}$, connecting the R site of a given cell with the L site of the cell above and to the right and the L site to the R site of cell below and to the left. The non-zero blocks are, therefore,

$$
\begin{aligned}
z(\boldsymbol{0}) &= \begin{pmatrix} 0 & \mathfrak{z}_1 \\ -\mathfrak{z}_1 & 0 \end{pmatrix}, \\
z(\hat{\boldsymbol{x}}) &= \begin{pmatrix} 0 & 0 \\ \mathfrak{z}_2 & 0 \end{pmatrix}, \qquad z(-\hat{\boldsymbol{x}}) = \begin{pmatrix} 0 & -\mathfrak{z}_2 \\ 0 & 0 \end{pmatrix}, \\
z(\hat{\boldsymbol{d}}) &= \begin{pmatrix} 0 & 0 \\ \mathfrak{z}_3 & 0 \end{pmatrix}, \qquad z(-\hat{\boldsymbol{d}}) = \begin{pmatrix} 0 & -\mathfrak{z}_3 \\ 0 & 0 \end{pmatrix}.
\end{aligned}
\tag{13.2.31}
$$

From (17.7.10),

$$
\begin{aligned}
\boldsymbol{\zeta}(\theta_1, \theta_2) = {} & z(\boldsymbol{0}) + z(\hat{\boldsymbol{x}}) \exp(\mathrm{i}\theta_1) + z(-\hat{\boldsymbol{x}}) \exp(-\mathrm{i}\theta_1) \\
& + z(\hat{\boldsymbol{d}}) \exp(\mathrm{i}[\theta_1 + \theta_2]) + z(-\hat{\boldsymbol{d}}) \exp(-\mathrm{i}[\theta_1 + \theta_2]),
\end{aligned}
\tag{13.2.32}
$$

and thus

$$
\begin{aligned}
\mathrm{Det}\{\boldsymbol{\zeta}(\theta_1, \theta_2)\} = {} & \mathfrak{z}_1^2 + \mathfrak{z}_2^2 + \mathfrak{z}_3^2 - 2\mathfrak{z}_1\mathfrak{z}_2 \cos(\theta_1) \\
& + 2\mathfrak{z}_2\mathfrak{z}_3 \cos(\theta_2) - 2\mathfrak{z}_1\mathfrak{z}_3 \cos(\theta_1 + \theta_2).
\end{aligned}
\tag{13.2.33}
$$

As in the case of the square lattice this yields the dimensionless free-energy density

$$
\begin{aligned}
\phi(\mathcal{L}_{\mathrm{H}}, \mathfrak{z}_1, \mathfrak{z}_2, \mathfrak{z}_3) = {} & -\frac{1}{(4\pi)^2} \int_0^{2\pi} \mathrm{d}\theta_1 \int_0^{2\pi} \mathrm{d}\theta_2 \ln[\mathfrak{z}_1^2 + \mathfrak{z}_2^2 + \mathfrak{z}_3^2 - 2\mathfrak{z}_1\mathfrak{z}_2 \cos(\theta_1) \\
& + 2\mathfrak{z}_2\mathfrak{z}_3 \cos(\theta_2) - 2\mathfrak{z}_1\mathfrak{z}_3 \cos(\theta_1 + \theta_2)].
\end{aligned}
\tag{13.2.34}
$$

[7] It can be seen from Fig. 3.10 that similar conservation rules do not apply to the square lattice. The reason is that there the number of vacant edges in a row of vertical edges depends not only on the number of dimers in the row of horizontal edges immediately below, but also on the number of dimers in the row of vertical edges below that.

With the change of variable $\theta_1 \to \pi - \theta_1$ this becomes

$$\phi(\mathcal{L}_{\mathrm{H}}, \mathfrak{z}_1, \mathfrak{z}_2, \mathfrak{z}_3) = -\frac{1}{(4\pi)^2}\int_0^{2\pi} \mathrm{d}\theta_1 \int_0^{2\pi} \mathrm{d}\theta_2 \ln[\mathfrak{z}_1^2 + \mathfrak{z}_2^2 + \mathfrak{z}_3^2 + 2\mathfrak{z}_1\mathfrak{z}_2\cos(\theta_1) + 2\mathfrak{z}_2\mathfrak{z}_3\cos(\theta_2) + 2\mathfrak{z}_1\mathfrak{z}_3\cos(\theta_2 - \theta_1)], \qquad (13.2.35)$$

a result given, without derivation, by Wu (1968). As would be expected, this expression is invariant with respect permutations of the fugacities.

So far the analysis has been very similar to that for the square lattice. However, there are important differences in the thermodynamics of the two systems, as appears when (13.2.35) is transformed to a single integral expression. This takes the form

$$\phi(\mathcal{L}_{\mathrm{H}}, \mathfrak{z}_1, \mathfrak{z}_2, \mathfrak{z}_3) = -\frac{1}{(4\pi)}\int_0^{\pi} \ln\left\{Y^{(+)}(\mathfrak{z}_1, \mathfrak{z}_2, \mathfrak{z}_3; \cos(\theta)) + |Y^{(-)}(\mathfrak{z}_1, \mathfrak{z}_2, \mathfrak{z}_3; \cos(\theta))|\right\}\mathrm{d}\theta, \qquad (13.2.36)$$

where

$$Y^{(\pm)}(\mathfrak{z}_1, \mathfrak{z}_2, \mathfrak{z}_3; x) := \tfrac{1}{2}[\mathfrak{z}_1^2 + \mathfrak{z}_2^2 - 2x\,\mathfrak{z}_1\mathfrak{z}_2 \pm \mathfrak{z}_3^2]. \qquad (13.2.37)$$

The fugacities $\mathfrak{z}_1, \mathfrak{z}_2$ and $\mathfrak{z}_3$ are non-negative and, in order to understand the behaviour of this model, we need to determine the analytic nature of $\phi(\mathcal{L}_{\mathrm{H}}, \mathfrak{z}_1, \mathfrak{z}_2, \mathfrak{z}_3)$ in this octant of the fugacity space, which can be divided into four regions:

I: $\mathfrak{z}_1 > \mathfrak{z}_2 + \mathfrak{z}_3,\ \mathfrak{z}_2 < \mathfrak{z}_3 + \mathfrak{z}_1,\ \mathfrak{z}_3 < \mathfrak{z}_1 + \mathfrak{z}_2,$

II: $\mathfrak{z}_2 > \mathfrak{z}_3 + \mathfrak{z}_1,\ \mathfrak{z}_3 < \mathfrak{z}_1 + \mathfrak{z}_2,\ \mathfrak{z}_1 < \mathfrak{z}_2 + \mathfrak{z}_3,$

III: $\mathfrak{z}_3 > \mathfrak{z}_1 + \mathfrak{z}_2,\ \mathfrak{z}_1 < \mathfrak{z}_2 + \mathfrak{z}_3,\ \mathfrak{z}_2 < \mathfrak{z}_3 + \mathfrak{z}_1,$

IV: $\mathfrak{z}_1 < \mathfrak{z}_2 + \mathfrak{z}_3,\ \mathfrak{z}_2 < \mathfrak{z}_3 + \mathfrak{z}_1,\ \mathfrak{z}_3 < \mathfrak{z}_1 + \mathfrak{z}_2.$

The boundaries between regions I, II and III respectively and region IV are the surfaces:

$$\mathcal{C}(\mathrm{I}): \qquad \mathfrak{z}_1 = \mathfrak{z}_2 + \mathfrak{z}_3, \qquad (13.2.38)$$

$$\mathcal{C}(\mathrm{II}): \qquad \mathfrak{z}_2 = \mathfrak{z}_3 + \mathfrak{z}_1, \qquad (13.2.39)$$

$$\mathcal{C}(\mathrm{III}): \qquad \mathfrak{z}_3 = \mathfrak{z}_1 + \mathfrak{z}_2. \qquad (13.2.40)$$

For all $\theta \in [0, \pi]$, $Y^{(-)}(\mathfrak{z}_1, \mathfrak{z}_2, \mathfrak{z}_3; \cos(\theta)) < 0$ in region III and, from (13.2.21) and (13.2.36),

$$\phi(\mathcal{L}_{\mathrm{H}}, \mathfrak{z}_1, \mathfrak{z}_2, \mathfrak{z}_3) = -\tfrac{1}{2}\ln(\mathfrak{z}_3), \qquad \rho_1 = \rho_2 = 0, \qquad \rho_3 = \tfrac{1}{2}. \qquad (13.2.41)$$

So in region III the system is in a 'frozen' state of perfect order, with all dimers parallel in the direction associated with the fugacity $\mathfrak{z}_3$. By symmetry it follows that states of perfect order also exist with the dimers aligned in the other lattice directions. In region I,

$$\phi(\mathcal{L}_{\mathrm{H}}, \mathfrak{z}_1, \mathfrak{z}_2, \mathfrak{z}_3) = -\tfrac{1}{2}\ln(\mathfrak{z}_1), \qquad \rho_1 = \tfrac{1}{2}, \qquad \rho_2 = \rho_3 = 0, \tag{13.2.42}$$

and, in region II,

$$\phi(\mathcal{L}_{\mathrm{H}}, \mathfrak{z}_1, \mathfrak{z}_2, \mathfrak{z}_3) = -\tfrac{1}{2}\ln(\mathfrak{z}_2), \qquad \rho_1 = \rho_3 = 0, \qquad \rho_2 = \tfrac{1}{2}. \tag{13.2.43}$$

These frozen configurations do not admit small (localized) perturbations. Consider, for example, the case of the system in region III, where, in the geometry of Fig. 13.3, all horizontal edges are unoccupied, while all edges in each row of diagonal sites are occupied giving $m_3 = N_1$ for all such rows. Now suppose that one dimer of any row of diagonal dimers is displaced to a horizontal edge. By the conservation rule discussed above, there must be a similar displacement from all other rows of diagonal dimers, so that the disturbance is propagated throughout the lattice. The situation is similar to that in a ferroelectric ground state of the six-vertex model (Sect. 3.8.2). The honeycomb lattice situation contrasts with that on the square lattice, where there can be small disturbances in the basic configuration or in the corresponding configuration with all dimers vertical. Figure 3.10 can, for instance, be regarded as showing a perturbation in a 'corner' of a larger lattice, with all other dimers on this lattice in their basic configuration positions.

Consider the case where the system is in region IV. There is now a real value θ_0 is the range $(0, \pi)$ such that

$$Y^{(-)}(\mathfrak{z}_1, \mathfrak{z}_2, \mathfrak{z}_3; \cos(\theta_0)) = 0. \tag{13.2.44}$$

The range of integration in (13.2.36) can be split into two parts and, using (13.2.37),

$$\phi(\mathcal{L}_{\mathrm{H}}, \mathfrak{z}_1, \mathfrak{z}_2, \mathfrak{z}_3) = -\frac{\theta_0}{2\pi}\ln(\mathfrak{z}_3) - \frac{1}{4\pi}\int_{\theta_0}^{\pi}\ln[\mathfrak{z}_1^2 + \mathfrak{z}_2^2 - 2\mathfrak{z}_1\mathfrak{z}_2\cos(\theta)]d\theta. \tag{13.2.45}$$

From this result it follows that

$$\rho_3 = \frac{\theta_0}{2\pi} = \frac{1}{2\pi}\arccos\left(\frac{\mathfrak{z}_1^2 + \mathfrak{z}_2^2 - \mathfrak{z}_3^2}{2\mathfrak{z}_1\mathfrak{z}_2}\right), \tag{13.2.46}$$

and by symmetry

$$\rho_1 = \frac{1}{2\pi}\arccos\left(\frac{\mathfrak{z}_2^2 + \mathfrak{z}_3^2 - \mathfrak{z}_1^2}{2\mathfrak{z}_2\mathfrak{z}_3}\right), \tag{13.2.47}$$

$$\rho_2 = \frac{1}{2\pi}\arccos\left(\frac{\mathfrak{z}_3^2 + \mathfrak{z}_1^2 - \mathfrak{z}_2^2}{2\mathfrak{z}_3\mathfrak{z}_1}\right). \tag{13.2.48}$$

It can easily be verified that

$$\rho_1 + \rho_2 + \rho_3 = \tfrac{1}{2}. \tag{13.2.49}$$

From (13.2.37) and (13.2.44), as the phase point approaches the surface $\mathcal{C}(\text{III})$, $\theta_0 \to \pi$. It then follows from (13.2.46) that $\rho_3 \to \frac{1}{2}$ and it must be the case that $\rho_2 \to 0$ and $\rho_1 \to 0$. We can also see that $\phi(\mathcal{L}_\text{H}, \mathfrak{z}_1, \mathfrak{z}_2, \mathfrak{z}_3)$, given by (13.2.45), tends to the value $-\frac{1}{2}\ln \mathfrak{z}_3$. The dimer densities and the free-energy density are continuous over the boundary $\mathcal{C}(\text{III})$ between the regions III and IV. They are also continuous over the boundaries $\mathcal{C}(\text{I})$ and $\mathcal{C}(\text{II})$. From (13.2.41)–(13.2.43), in the ordered regions I–III,

$$\frac{\partial \rho_i}{\partial \mathfrak{z}_j} = 0, \qquad i, j = 1, 2, 3, \tag{13.2.50}$$

and, from (13.2.46)–(13.2.48), in the disordered region IV,

$$\frac{\partial \rho_i}{\partial \mathfrak{z}_j} = \begin{cases} \dfrac{\mathfrak{z}_i}{\tau(\mathfrak{z}_1, \mathfrak{z}_2, \mathfrak{z}_3)}, & \text{if } i = j, \\[2ex] \dfrac{\mathfrak{z}_k^2 - \mathfrak{z}_i^2 - \mathfrak{z}_j^2}{2\mathfrak{z}_j \tau(\mathfrak{z}_1, \mathfrak{z}_2, \mathfrak{z}_3)}, & \text{if } i \neq j \text{ and } k \neq i, j, \end{cases} \tag{13.2.51}$$

where

$$\tau(\mathfrak{z}_1, \mathfrak{z}_2, \mathfrak{z}_3) := \pi\sqrt{(\mathfrak{z}_1 + \mathfrak{z}_2 - \mathfrak{z}_3)(\mathfrak{z}_2 + \mathfrak{z}_3 - \mathfrak{z}_1)(\mathfrak{z}_3 + \mathfrak{z}_1 - \mathfrak{z}_2)(\mathfrak{z}_1 + \mathfrak{z}_2 + \mathfrak{z}_3)}. \tag{13.2.52}$$

In view of condition (13.2.49) only six of these nine partial derivatives are independent. From (13.2.38)–(13.2.40) and (13.2.52), it can be seen that, as any one of the boundaries of the disordered region is approached from within that region, the partial derivatives given by (13.2.51) exhibit singular behaviour. Specifically, as surface $\mathcal{C}(\text{I})$ is approached within region IV,

$$\frac{\partial \rho_i}{\partial \mathfrak{z}_j} \sim \frac{1}{\sqrt{\mathfrak{z}_2 + \mathfrak{z}_3 - \mathfrak{z}_1}}, \qquad i, j = 1, 2, 3. \tag{13.2.53}$$

$\mathcal{C}(\text{I})$, $\mathcal{C}(\text{II})$ and $\mathcal{C}(\text{III})$ are surfaces of second-order phase transitions. This result was stated, without derivation, by Kasteleyn (1963).

Consider now the case where the fugacities $\mathfrak{z}_i$ are smooth functions of couplings $K_i := \varepsilon_i/T$, for $i = 1, 2, 3$. The system is controlled by changing the temperature, so we adopt the procedure used in Sect. 2.3.1, setting $K := K_1$ with $\kappa_i := \varepsilon_i/\varepsilon_1$, so that $K_i = \kappa_i K$, $i = 1, 2, 3$. The critical temperature T_c at which the system passes from the disordered state IV to the ordered state I is then given as the solution of

$$\mathfrak{z}_2(T_\text{c}) + \mathfrak{z}_3(T_\text{c}) = \mathfrak{z}_1(T_\text{c}), \tag{13.2.54}$$

and, inside IV and close to the surface $\mathcal{C}(\text{I})$,

$$\mathfrak{z}_2(T) + \mathfrak{z}_3(T) - \mathfrak{z}_1(T) \simeq [\mathfrak{z}_2'(T_c) + \mathfrak{z}_3'(T_c) - \mathfrak{z}_1'(T_c)](T - T_c), \tag{13.2.55}$$

unless

$$\mathfrak{z}_2'(T_c) + \mathfrak{z}_3'(T_c) = \mathfrak{z}_1'(T_c). \tag{13.2.56}$$

From (2.3.18), (2.3.19) and (13.2.21), the internal-energy density

$$\mathfrak{u} = \sum_{i=1}^{3} \kappa_i \frac{\partial \phi}{\partial K_i} = \sum_{i=1}^{3} \kappa_i \frac{\partial \phi}{\partial \mathfrak{z}_i} \frac{\mathrm{d}\mathfrak{z}_i}{\mathrm{d}K_i} = -\sum_{i=1}^{3} \kappa_i \rho_i \frac{\mathrm{d}\ln \mathfrak{z}_i}{\mathrm{d}K_i}. \tag{13.2.57}$$

Since the dimer densities ρ_i are continuous across the critical surfaces this must also be the case for $\mathfrak{u}$ and, from (13.2.41)–(13.2.43),

$$\mathfrak{u} = -\tfrac{1}{2} \kappa_i \frac{\mathrm{d}\ln \mathfrak{z}_i}{\mathrm{d}K_i}, \qquad i = 1, 2, 3, \tag{13.2.58}$$

in the ordered regions I, II and III respectively. The heat capacity c per lattice site, which is given from (2.3.29) by

$$c = k_{\mathrm{B}} \varepsilon_1 \frac{\partial \mathfrak{u}}{\partial T}, \tag{13.2.59}$$

will involve the partial derivatives of the dimer densities with respect to the fugacities. It follows from (13.2.53) and (13.2.55) that

$$c \sim \frac{1}{\sqrt{T - T_c}}, \tag{13.2.60}$$

as $\mathcal{C}$(I) is approached from inside region IV, unless (13.2.56) is satisfied. This gives a critical exponent $\alpha = \frac{1}{2}$.

We now consider a simple interpretation of the fugacities. Suppose that each dimer carries a small electric dipole aligned with its axis and that an electric field $\mathcal{E}$ acts on the system parallel to the direction labelled 3. Given that $\mathcal{E}$ includes the magnitude of the dipole moment, thereby being measured in units of energy, and that dipole-dipole interactions are negligible in comparison with the dipole-field interaction, this model corresponds to the case $\varepsilon_1 = \varepsilon_2 = \varepsilon_3 = \mathcal{E}$ and

$$\mathfrak{z}_3 = \exp(K) + \exp(-K) = 2\cosh(K). \tag{13.2.61}$$

In a similar way

$$\mathfrak{z}_1 = \mathfrak{z}_2 = 2\cosh\left(\tfrac{1}{2}K\right). \tag{13.2.62}$$

Since $\mathfrak{z}_3 > \mathfrak{z}_2 = \mathfrak{z}_1$, a transition at a fixed value of $\mathcal{E}$ will occur on the surface $\mathcal{C}(\text{III})$ at a critical coupling $K_c = \mathcal{E}/T_c$ given by

$$\cosh(K_c) = 2\cosh\left(\tfrac{1}{2}K_c\right), \tag{13.2.63}$$

which can be transformed to

$$\frac{\mathcal{E}}{T_c} = 2\,\text{arccosh}\left(\frac{1+\sqrt{3}}{2}\right). \tag{13.2.64}$$

An equivalent relationship (again without derivation) is given by Kasteleyn (1963), who regards the dimers as polar diatomic molecules adsorbed on to a solid whose surface layer of atoms has the honeycomb form. For this model the internal energy in region III is given, from (13.2.58), by

$$u = -\tfrac{1}{2}\tanh(K). \tag{13.2.65}$$

A second-order transition, where, as in this model, the heat capacity has a square-root singularity for $T > T_c$ and is finite or zero for $T < T_c$ has been given the name *K-type* (K standing for Kasteleyn) by Nagle et al. (1989). These authors also use the name *O-type* (O standing for Onsager) for the kind of transition, which occurs in the two-dimensional Ising model, where the heat capacity has a logarithmic singularity on each side of the critical point (Eq. (12.2.181)). These two types of transition are further discussed in Sect. 13.4.

13.3 Vertex and Ising Model Equivalences

In Sects. 3.9.1 and 3.9.2 equivalences between special cases of dimer models and vertex and Ising models were established. The formulae for the thermodynamic functions of dimer models obtained in this chapter can now be used to investigate this further.

13.3.1 The Five-Vertex Model

Equation (3.9.11) identifies the partition function of a dimer model on the honeycomb lattice with a special case of the six-vertex model on the square lattice which violates the energy equalities (3.8.4) in two respects: (i) ε_1 is finite but $\varepsilon_2 = \infty$, so that the model is effectively a five-vertex model; (ii) $\varepsilon_3 \neq \varepsilon_4$ unless the $\mathfrak{z}_2 = \mathfrak{z}_3$. The relabelling of the vertex energies and Boltzmann weights given in (3.9.7) avoids a clash between vertex and dimer labelling and the relationships between these latter and the dimer fugacities are given by (3.9.8) and (3.9.9). The energy u per lattice site for the five-vertex model on the square lattice of $\frac{1}{2}N$ sites is given, from (13.2.57), by

$$u = 2(\rho_1\varepsilon_{\mathfrak{a}} + \rho_2\varepsilon_{\mathfrak{b}} + \rho_3\varepsilon_{\mathfrak{b}'}). \tag{13.3.1}$$

For the special case of the modified KDP model the vertex energy parameters satisfy (3.9.12) and we define $\varepsilon := -\varepsilon_{\mathfrak{a}} > 0$, $K := \varepsilon/T$, with

$$\mathfrak{z}_1 = \exp(K), \qquad \mathfrak{z}_2 = \mathfrak{z}_3 = 1, \tag{13.3.2}$$

$$u = -2\rho_1\varepsilon. \tag{13.3.3}$$

The only transition surface relevant to this model is $\mathcal{C}(\text{I})$, which, from (13.2.38) and (13.3.2), gives

$$\exp(K_{\text{c}}) = 2, \qquad T_{\text{c}} = \varepsilon/\ln(2), \tag{13.3.4}$$

for the critical temperature.

In region I, *when* $T < T_{\text{c}}$, $\rho_1 = \frac{1}{2}$ and $u = -\varepsilon$, so that for the five-vertex model every vertex is of type 1. Like the original KDP model, this modified model is in a frozen perfectly ordered ferroelectric ground state below the critical temperature. The argument used in Sect. 3.8.2 that the ferroelectric ground state, in which all vertices are of type 1, does not admit small perturbations still applies here. The only difference is that configuration graphs like those shown in Fig. 3.6 cannot any longer 'touch' at a vertex, since such a vertex is of type 2. The equivalent state in the honeycomb dimer model is one in which all the dimers are parallel with fugacity $\mathfrak{z}_1$. It follows that no small perturbations to this state, or the other two fully aligned states, can occur. This confirms the argument of Sect. 13.2.2.

In region IV, *when* $T > T_{\text{c}}$, it follows from (13.2.47) and (13.3.3) that

$$\rho_1 = \frac{1}{2\pi}\arccos\left[1 - \tfrac{1}{2}\exp(2K)\right]. \tag{13.3.5}$$

The zero-field heat capacity c_0 per vertex in the modified KDP model is thus given by

$$c_0 = k_{\text{B}}\frac{\partial u}{\partial T} = 2k_{\text{B}}K^2\frac{\partial \rho_1}{\partial K} = \frac{k_{\text{B}}K^2\exp(K_1)}{\pi\sqrt{1 - \frac{1}{4}\exp(2K_1)}}, \tag{13.3.6}$$

which exhibits a square-root singularity as $T \to T_{\text{c}}$. This same singular behaviour is exhibited by the corresponding six-vertex (KDP) model, except in that case there is a finite discontinuity in u at the critical temperature (see Lieb (1967c) and Lavis and Bell (1999a, Sect. 10.10)).

Now suppose that an electric field with horizontal and vertical components $\mathcal{E}^{(h)}$ and $\mathcal{E}^{(v)}$ in the plane of the lattice is applied to the modified KDP model. Given that these quantities include the magnitude of the horizontal or vertical component of the dipole moment of a polarized vertex, the vertex energies given in (3.9.12) are changed to

$$\varepsilon_{\mathfrak{a}} = -\varepsilon - \mathcal{E}^{(v)} - \mathcal{E}^{(h)}, \qquad \varepsilon_{\mathfrak{b}} = -\mathcal{E}^{(v)} + \mathcal{E}^{(h)},$$
$$\varepsilon_{\mathfrak{b}'} = \mathcal{E}^{(v)} - \mathcal{E}^{(h)}, \qquad \varepsilon_{\mathfrak{c}} = 0. \tag{13.3.7}$$

Since (3.9.10) is still satisfied, the model remains equivalent to the honeycomb dimer model. When

$$-\tfrac{1}{2}\varepsilon < \mathcal{E}^{(v)}, \qquad -\tfrac{1}{2}\varepsilon < \mathcal{E}^{(h)}, \tag{13.3.8}$$

$\varepsilon_{\mathfrak{a}} < \varepsilon_{\mathfrak{b}}$ and $\varepsilon_{\mathfrak{a}} < \varepsilon_{\mathfrak{b}'}$ and again the only relevant transition surface is $\mathcal{C}(\mathrm{I})$. For fixed values of the electric field components the system will go into a perfectly ordered state with all vertices of type 1 at a critical temperature given by

$$\exp\left(\frac{\varepsilon + \mathcal{E}^{(v)} + \mathcal{E}^{(h)}}{T_c}\right) = 2\cosh\left(\frac{\mathcal{E}^{(v)} - \mathcal{E}^{(h)}}{T_c}\right). \tag{13.3.9}$$

When

$$\mathcal{E}^{(h)} < -\tfrac{1}{2}\varepsilon, \qquad \mathcal{E}^{(h)} < \mathcal{E}^{(v)}, \tag{13.3.10}$$

the transition is to a perfectly ordered state with all vertices of type 3 and when

$$\mathcal{E}^{(v)} < -\tfrac{1}{2}\varepsilon, \qquad \mathcal{E}^{(v)} < \mathcal{E}^{(h)}, \tag{13.3.11}$$

the transition is to a perfectly ordered state with all vertices of type 4. In each of these cases the calculation of the transition temperature is similar to that given for the case (13.3.8).

13.3.2 The Honeycomb-Lattice Anisotropic Ising Model

Let the honeycomb lattice $\mathcal{N}_{\mathrm{H}}$ consist of a $N_1 \times N_2$ array of unit cells with $N = 2N_1N_2$. From (3.9.20), it is now necessary to find $Z^{(\mathrm{DM})}(\mathcal{N}_{\mathrm{TT}}, \upsilon_1^{-1}, \upsilon_2^{-1}, \upsilon_3^{-1})$ for the 3-12 lattice $\mathcal{N}_{\mathrm{TT}}$ of $3N$ sites whose unit cell is the six sites A, B, C, D, E and F shown in Fig. 13.4 with A′, B′, E′ and F′ being first-neighbour sites in adjacent unit cells. This lattice was derived from the honeycomb lattice and the arrow assignment for the honeycomb edges remains unchanged, while that for the triangle of new edges makes these triangles clockwise odd. The dodecagons are also clockwise odd.

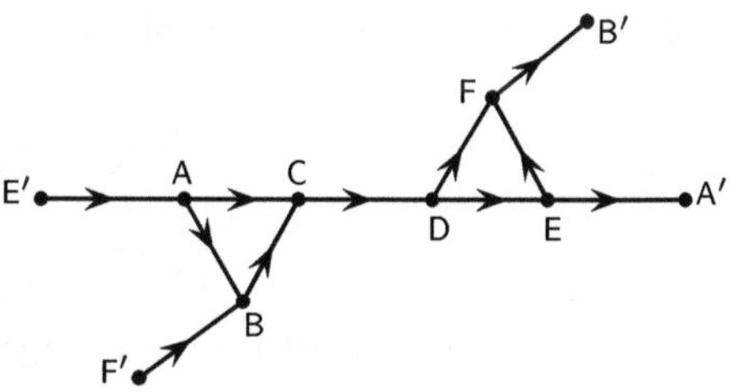

Fig. 13.4 Arrow assignment for the 3-12 lattice $\mathcal{N}_{\mathrm{TT}}$

Equation (13.2.32) can still be used for the diagonal block matrices, but these are now 6×6, since there are six sites in the unit cell. Taking the sites of the unit cell in the literal order A, B, C, D, E and F it follows, using Fig. 13.4, that

$$\boldsymbol{\zeta}(\theta_1, \theta_2) = \begin{pmatrix} 0 & 1 & 1 & 0 & -\overline{x}_2 & 0 \\ -1 & 0 & 1 & 0 & 0 & -\overline{x}_3 \\ -1 & -1 & 0 & x_1 & 0 & 0 \\ 0 & 0 & -x_1 & 0 & 1 & 1 \\ x_2 & 0 & -1 & 0 & 1 & 1 \\ 0 & x_3 & 0 & -1 & -1 & 0 \end{pmatrix}, \tag{13.3.12}$$

where $x_1 := 1/v_1$, $x_2 := \exp(\mathrm{i}\theta_1)/v_2$, $x_3 := \exp[\mathrm{i}(\theta_1 + \theta_2)]/v_3$, giving

$$\begin{aligned} \mathrm{Det}\{\boldsymbol{\zeta}(\theta_1, \theta_2)\} = {} & v_1^{-2} + v_2^{-2} + v_3^{-2} + v_1^{-2} v_2^{-2} v_3^{-2} \\ & - 2 v_1^{-1} v_2^{-1} (v_3^{-2} - 1) \cos(\theta_1) - 2 v_2^{-1} v_3^{-1} (v_1^{-2} - 1) \cos(\theta_2) \\ & - 2 v_3^{-1} v_1^{-1} (v_2^{-2} - 1) \cos(\theta_1 + \theta_2). \end{aligned} \tag{13.3.13}$$

With

$$s_i := \sinh(2K_i) = \frac{2v_i}{1 - v_i^2}, \qquad c_i := \cosh(2K_i) = \frac{1 + v_i^2}{1 - v_i^2}, \qquad i = 1, 2, 3, \tag{13.3.14}$$

$$\begin{aligned} \mathrm{Det}\{\boldsymbol{\zeta}(\theta_1, \theta_2)\} = {} & \tfrac{1}{2}[\sinh(K_1)\sinh(K_2)\sinh(K_3)]^{-2}[c_1 c_2 c_3 + 1 \\ & - s_1 s_2 \cos(\theta_1) - s_2 s_3 \cos(\theta_2) - s_3 s_1 \cos(\theta_1 + \theta_2)]. \end{aligned} \tag{13.3.15}$$

From (13.1.8), (3.9.20), (13.3.15) and (17.7.12), the dimensionless free energy per lattice site of the honeycomb lattice anisotropic Ising model in zero field is

$$\phi^{(\mathrm{IM})}(\mathcal{L}_{\mathrm{H}}, K_1, K_2, K_3) = -\tfrac{1}{2}\ln[4\sinh(K_1)\sinh(K_2)\sinh(K_3)] - \lim_{N_1\to\infty}\lim_{N_2\to\infty}\frac{\ln[Z(3N, v_1^{-1}, v_2^{-1}, v_3^{-1})]}{2N_1N_2}$$
$$= -\tfrac{3}{4}\ln(2) - \frac{1}{16\pi^2}\int_0^{2\pi} d\theta_1 \int_0^{2\pi} d\theta_2 \ln[c_1c_2c_3 + 1 - s_1s_2\cos(\theta_1) - s_2s_3\cos(\theta_2) - s_3s_1\cos(\theta_1+\theta_2)]. \tag{13.3.16}$$

The star-triangle transformation of Sect. 9.4 can be used to show that the corresponding expression for the triangular lattice is

$$\phi^{(\mathrm{IM})}(\mathcal{L}_{\mathrm{T}}, K_1, K_2, K_3) = -\ln(2) - \frac{1}{8\pi^2}\int_0^{2\pi} d\theta_1 \int_0^{2\pi} d\theta_2 \ln[c_1c_2c_3 + s_1s_2s_3 - s_1\cos(\theta_1) - s_2\cos(\theta_2) - s_3\cos(\theta_1+\theta_2)]. \tag{13.3.17}$$

Setting $K_3 = 0, s_3 = 0, c_3 = 1$ gives (with slight changes of notation) the expression (12.2.173) for the anisotropic Ising model on the square lattice.

13.4 K-Type and O-Type Transitions

Considering for simplicity the isotropic case, we have shown in Sects. 3.9.2 and 13.3.2 that the 3-12 dimer model, with fugacity $\mathfrak{z} = v^{-1} = \coth(K) > 1$ on the original honeycomb edges and $\mathfrak{z} = 1$ on the new triangle edges, is equivalent to the honeycomb lattice Ising model ferromagnet. It follows that the transitions in this dimer model are similar to those in a two-dimensional Ising model ferromagnet, and are thus O-type according to the definition at the end of Sect. 13.2.2. With these fugacities the ground state of the 3-12 dimer model has dimers on the honeycomb edges (for example F′B or CD in Fig. 3.12) but no dimers on the triangle edges. This ground state admits localized perturbations. Dimers can, for example, be moved cyclically from the honeycomb to the triangle edges in any of the dodecagons without affecting the dimer configuration on the rest of the lattice. The dimer on F′B in Fig. 3.12 could be moved to BC, that on CD to DE and so on. We have thus encountered three types of critical behaviour in pure dimer assemblies: K-type in the honeycomb lattice model of Sect. 13.2.2, O-type on the 3-12 lattice and no transition in the square lattice model of Sect. 13.2.1. In the last two cases the ground states admit small perturbations, while for the first case the ground state is frozen with small perturbations excluded, so that the heat capacity is zero below the critical temperature.

K-type critical behaviour is not confined to dimer systems. In the six-vertex ferroelectric model with non-zero electric field there is a second-order transition at a field-dependent critical temperature (Baxter 1982b, Sect. 8.12) with a square-root singularity in the heat capacity above the critical temperature. Below the critical temperature the state is frozen, fully polarized and the heat capacity is zero. The square-root singular behaviour above the critical temperature still applies in zero

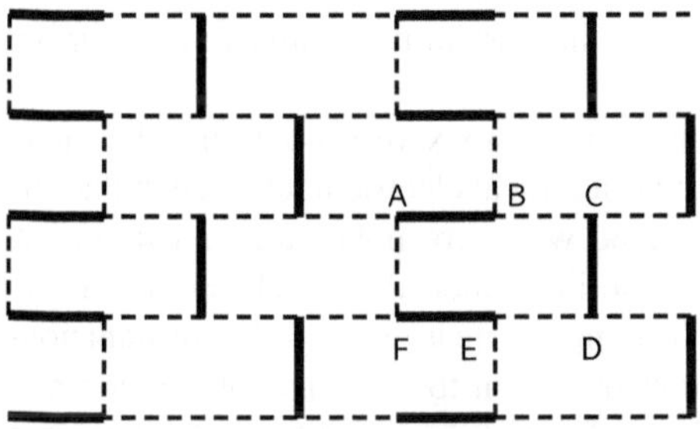

Fig. 13.5 The V_2H_2 dimer model. For the ground state, the positions of the dimers are represented by *full lines* and the empty edges by *broken lines*

field, but the transition to the completely polarized state is first-order with a latent heat (Lieb and Wu 1972). From the evidence presented so far it may seem that a square-root singularity in the heat capacity above a critical temperature is associated with a frozen state below the critical temperature. The situation is, however, not quite so simple, as we shall see.

A sequence of honeycomb lattice dimer models is discussed by Nagle et al. (1989). These models have a less simple assignment of fugacities than that for the models of Sect. 13.2.2, where the fugacities corresponded to the three edge orientations in the usual hexagonal geometry. Using the brick lattice geometry, obtained by rotating the diagonal edges into the vertical position, edge types and fugacity assignments are now made according to the arrangement of dimers in a ground state. One model of this type is called V^2H^2. The ground state for that model is shown in Fig. 13.5, with the full line segments representing the positions of the dimers and the broken lines representing the unoccupied edges. For any arrangement of dimers a fugacity of unity is assigned to a dimer in a ground-state position and a fugacity $\mathfrak{z} = \exp(-\varepsilon/T)$, with $\varepsilon > 0$, to a dimer in any other position. Nagle et al. (1989) also considered dimer models on the 4-8 lattice, formed from a square lattice by replacing each vertex by a quadrangle of edges, and what they called the *SQK model*. In this model extra edges are inserted in the brick lattice to make it into a square lattice, but the dimers on these new vertical edges are given different fugacities from that on the brick lattice edges. Various applications have been made of these types of models, including that to chain conformal systems (Nagle 1975; Bell et al. 1978; Nagle 1986; Nagle et al. 1989).

The V^2H^2 model has some interesting critical properties described by Nagle et al. (1989). First, note that the ground state shown in Fig. 13.5 admits local perturbations. In the hexagon ABCDEF the dimers on AB, CD and EF, can, for example, be shifted to BC, DE and FA, without disturbing the configuration elsewhere on the lattice. Thus, as would be expected, the heat capacity below the critical temperature is not identically zero. It does, however, remain finite as the critical temperature is approached from below and has a square-root singularity as the critical temperature is approached from above. The transition is K-type, where we chose, for the reason just seen, to provide for the possibility of non-zero subcritical values of the heat capacity.

Similar behaviour is found in six-vertex antiferroelectric models (Lieb and Wu 1972), for which there is a critical electric field $\mathcal{E}_c$ such that, for $\mathcal{E} < \mathcal{E}_c$, the ground state is sublattice ordered with zero polarization. This ground state admits small perturbations to the ordering which do not change the overall zero value of the polarization. When the temperature is reduced at a constant non-zero value of $\mathcal{E} < \mathcal{E}_c$ there is a second-order transition to the zero polarization state with a square-root heat capacity singularity. From the possibility of small perturbations we expect the heat capacity to be non-zero below the critical temperature and this is indeed the case. In the zero polarization region the free-energy density $f(\mathcal{E}, T) = f(0, T)$ so that the heat capacity $c(\mathcal{E}, T) = c(0, T)$, which is finite but non-zero. For $\mathcal{E} > \mathcal{E}_c$, antiferroelectric models, apart from the special case of the F model, undergo a transition similar to that in ferroelectric models, while at $\mathcal{E} = 0$ there is yet another kind of critical behaviour. This is a transition of infinite order in which $c(0, T)$ and all its derivatives are continuous.

As we have seen a frozen state below the transition is not a necessary condition for a K-type transition. However, lattice conservation properties may be relevant. The row conservation rule for dimers established at the beginning of Sect. 13.2.2 depends only on the topology of the brick (honeycomb) lattice and not on the assignment of fugacities. So that it applies to any brick lattice dimer system including the V^2H^2 model. Again, six-vertex models have an ice-rule arrow conservation property, which results from the permitted vertex connections and does not depend on the vertex energies. It therefore applies to both ferroelectric and antiferroelectric models.

For the eight-vertex models, the ice-rule does not apply and, as we showed in Chap. 12, there is then a new range of critical properties, including the O-type transition as a special case (Sect. 12.3.9). Both conservation rules and anisotropy play a role in determining the properties of dimer models. The six-vertex models do not have K-type second-order transitions when the field is reduced to zero, although ferroelectric models do show a square-root singularity in the heat capacity as the first-order transition is approached from above. A non-zero field obviously introduces an element of anisotropy. As shown in Sect. 13.3.1, the highly anisotropic five-vertex model has a K-type transition to the completely polarized state, even in zero field. However, the introduction of a degree of spatial anisotropy does not necessarily change the character of transitions in two-dimensional lattice systems. As may be deduced from the expressions obtained in Sect. 13.3.2 the Ising ferromagnets still have O-type transitions even when there are different couplings in different lattice directions. Obviously a similar statement can be made for the equivalent dimer systems.

Part IV
Series and Renormalization Group Methods

Introduction

Although it is possible to identify an early example of the use of a high-temperature series expansion by Opechowski (1937),[1] series expansions became powerful tools in the analysis of phase transitions only in the 1960s. Then, much of the early work was concentrated on the spin-$\frac{1}{2}$ Ising model and undertaken by a group in the Physics Department at King's College London, led by the late Prof. Cyril Domb. At the heart of the enterprise was the development of techniques for counting lattice configurations and the computation of lattice constants for weak-graph embeddings, in the case of high-temperature series, and strong-graph embeddings, in the case of low-temperature series.[2] An early paper by Sykes, Essam and Gaunt (1965)[3] deriving low-temperature expansions on the standard two- and three-dimensional lattices was followed by a systematic presentation by Sykes, Essam, Heap and Hiley (1966) of a procedure (variously called the *partial-sum, shadow* or *code* method) for determining lattice constant for both weakly and strongly embedded graphs. There then followed a long sequence of papers[4] determining low-temperature susceptibility series for the standard two- and three-dimensional lattices. A series for the square lattice was obtained to degree 11 and for the triangular and honeycomb lattices to degree 21. The expansions for the simple cubic, diamond, body-centred cubic and face-centred cubic lattices were to degrees 20, 15, 28 and 40, respectively. The procedure was also applied to the Ising model with spin greater than a half by Fox

[1] Who studied the Heisenberg model by expanding the partition function in inverse powers of temperature.
[2] These elements of graph theory are discussed in Sects. 16.1 and 16.4.
[3] Since much of this work was published by different groups of workers with Dr. Martin Sykes as the lead author, we quote references with the list of authors given *in extenso*.
[4] Sykes, Gaunt, Essam and Hunter (1973); Sykes, Gaunt, Mattingly, Essam and Elliott (1973); Sykes, Gaunt, Martin, Mattingly and Essam (1973); Sykes, Gaunt, Essam, Heap, Elliott and Mattingly (1973); Sykes, Gaunt, Essam and Elliott (1973); Sykes, Watts and Gaunt (1975a, b); Sykes, McKenzie, Watts and Gaunt (1975).

and Guttmann (1973) and to the Ising model on the triangular lattice with a pure triplet interaction by Sykes and Watts (1975).

As will be seen from the presentation of the finite-lattice method in Sect. 14.5 it is, in general, more difficult to obtain a large number of terms of high-temperature series, except when they are provided by duality from low-temperature series. Since duality applies only when the order parameter is zero, this means that, while heat capacities can be derived from duality, this does not apply to field-related response functions (compressibility and susceptibility). Early work by Sykes, Gaunt, Roberts and Wyles (1972a) obtained high-temperature expansions for the susceptibility of the spin-$\frac{1}{2}$ Ising model on the square, triangle and honeycomb lattices to degrees 21, 16 and 32, respectively, and in a second paper Sykes, Gaunt, Roberts and Wyles (1972b) undertook the same calculations for the simple cubic, body-centred cubic and face-centred cubic lattices resulting in series to degrees 17, 15 and 12, respectively. Kim and Joseph (1975) used the finite-cluster method (Sect. 14.4) to derive high-temperature series to degree 8 for the Potts model with $\nu = 3, 4, 5$ on square, triangular and face-centred cubic lattices.

The finite-lattice method was introduced by de Neef (1975) in his doctoral thesis[5] and, according to de Neef and Enting (1977), it is "primarily a technique for obtaining high-temperature expansions for the zero-field free energy of the Ising model and related models on a square lattice". However, de Neef (op. cit.) had already ventured beyond this specification by applying the method to the one-dimensional Heisenberg model and de Neef and Enting (op. cit.) began what has subsequently become the major preoccupation of finite-lattice calculations, which is work on the Potts model for various lattices and values of ν. Another interesting twist to this development is that, whilst de Neef and Enting envisaged the method as mainly a high-temperature procedure, its subsequent development has been largely for the production of low-temperature series. In the list of applications to the Potts model given by Enting (1996), only five are high temperature compared to 20 which are low temperature. As he remarks, in finite-lattice calculations "high-T expansions can be calculated using a low-T style of summation working with high-T variables". The finite-lattice method (with various extensions: Jensen and Guttmann 1996; Arisue and Tabata 1997) is the most widely used method in recent work on series expansions. We shall, therefore, in Chap. 14 mainly concentrate on that method. However, it is useful to set the scene with a preliminary discussion of simple graph-counting in Sect. 14.2 and a treatment of the related finite-cluster method in Sect. 14.4.

The first widely used series-analysis technique was the *ratio method* introduced by Domb and Sykes (1956, 1957a, b) and described in Sect. 14.6.2. Although a number of variants of this method have been developed to improve the approximations to critical parameters and exponents (Guttmann 1989), unless the singularity of interest lies on the circle of convergence ($w_c = r$), the ratios still do not give good estimates for critical parameters and exponents. Smooth monotonic

---

[5] Where he calls it the *finite-lattice cumulant method* (ibid Sect. 5.5).

behaviour of the coefficients is required and, although oscillations arising from another known singularity on the circle of convergence can sometimes be catered for, applications to cases where the critical singularity is not at least as close as any other singularity to the origin do not yield good results.

The use of Padé approximants (Sect. 14.6.3) is due to Baker (1961) (see also, Guttmann 1989). This method has the advantage of being applicable even when $w_c > r$, since Padé approximants effectively provide an analytic continuation of the series to the region $|w| > r$. However, the only singularities it can predict are poles and it is known that many systems have singularities with non-integer exponents. This led to the use of Dlog Padé approximants, which consist of the prior application of the Dlog transformation (Sect. 14.6.4.1) before the calculation of the Padé approximant. This has the advantage of turning singularities with non-integer exponents into poles.

The equation for the Dlog Padé approximant can also be viewed as the lowest level (first-order homogeneous) *differential approximant*. This class of approximants (Sect. 14.6.4.1) is due to Guttmann and Joyce (1972) and is now the most widely used technique of series analysis. The relationship between Dlog Padé and differential approximants is paralleled by a relationship between 'ordinary' Padé approximants and a class of algebraic approximants (Sect. 14.6.4.2).

A Padé approximant can be viewed as a first-degree algebraic equation and Brak and Guttmann (1990) proposed the use of higher-degree algebraic equations. These (for degree greater than one) have the advantage over Padé approximants of being able to predict a singular point with rational non-integer exponents, and the disadvantage compared to differential approximants of being restricted to rational exponents. It is also the case, as is shown in Sect. 17.6, that the solution of an algebraic equation is a solution of a differential equation (but not conversely). However, from a conceptual point of view, the idea that, as for the hard-hexagon model (Example Sect. 18.5.2), a thermodynamic quantity exactly satisfies an algebraic equation is attractive.

In Sect. 14.6 the ratio and various approximant methods are introduced and applied solely to the methods for finding second-order transitions, given by singularities, and formulae for calculating their exponents. Section 14.7 discusses the use of the same approximants for the exploration of situations where the transition may be first order. In the absence of data from other sources, this tends, since it relies on matching of high- and low-temperature series, to be a rather less accurate procedure.

As we have seen, phase transitions can occur only in infinite systems and these are by their nature difficult to handle. Wilson (1971) commented that what "one would like to do is to transform the problem of calculating [the free energy] into a form where it is natural for [it] to have singularities at [the critical point] and where one may hope the nature of the singularity will be more easily seen than from [the partition function]". He then suggested that the appropriate way to solve this problem is in terms of the renormalization group. Although such an idea had already been mooted, in this paper Wilson made the first attempt to lay down the assumptions needed for its implementation.[6]

The approach to the renormalization group in Chap. 15 is usually referred to as the use of the *real-space renormalization group (RSRG)*. This is in contrast to renormalization group methods for continuous spin distributions, using, for example the Ginzburg–Landau local free-energy density of Sect. 5.6.1. These latter methods, in which renormalization is performed in wave-vector space, were initiated by Wilson (1975) and owe much to the influence of quantum field theory. Real-space methods, which are conceptually rather simpler, are based, in many cases, on the block-spin method of Kadanoff (1966) described in Sect. 4.2.[7]

The essential feature of all renormalization group methods is the realization that critical phenomena are associated with long wave length (small wave vector) fluctuations in wave-vector space or long-range correlations in real space. This means that an integration out of large wave vectors or a block averaging of short-range interactions in real space will leave the essential critical features of the system unchanged. The selection of renormalization group transformations contained in Chap. 15 have the common property that they lead to trajectories in the space of coupling which divide the space into regions controlled by fixed points. This parallels the picture of scaling theory presented in Chap. 4, but now we have actual numerical predictions for critical exponents rather than simply scaling-law relations between them. It should, however, be emphasized that, with the exception of the decimation of the one-dimensional Ising model in Sect. 15.5.1, the renormalization group transformation involves some form of approximation or truncation. In consequence, the scaling exponents are also approximations.

---

[6] For a reflective account of the development of the renormalization group see Kadanoff (2006, 2013).

[7] A comprehensive collection of articles on both types of renormalization group methods is contained in PTCP (6) (see the preamble to the References and Author Index for an explanation of this designation) and on real-space methods in the volume edited by Burkhardt and van Leeuwen (1982). For a description of renormalization methods in wave-vector space the reader is referred to Ma (1976a) and Amit (1978). Accounts of both approaches are given by Goldenfeld (1992), Binney et al. (1993) and Cardy (1996).

Chapter 14
Series Expansions

14.1 The Task and the Methods

In this chapter we discuss methods for obtaining coefficients in the expansions of partition functions and free energies, or their derivatives, in a series of powers of temperature-dependent variables. These series are called *low temperature*, or *high temperature*, respectively according to whether the expansion variable $w(T)$ tends to zero as $T \to 0$ or $T \to \infty$. If an expression for the coefficient of the general term were known then a series would constitute an exact solution. However, the phrase 'series methods' refers primarily to the term-by-term calculation of as many coefficients as possible and the deduction of results for the model from this limited information. Low- and high-temperature series expansions provide approximations to the thermodynamic functions of a model in their respective ranges of temperature and this is important for the detection of first-order transitions as we shall show in Sect. 14.7. However, series expansions are more often used to study critical properties for transitions at which the thermodynamic function is singular; in which case the interest is in obtaining critical points and exponents. Suppose that

$$F(w) := \sum_{k=0}^{\infty} f_k \, w^k, \tag{14.1.1}$$

is some thermodynamic function of the model of interest, with $w(T)$ appropriate to a high-temperature or low-temperature series. If

$$\lim_{k \to \infty} |f_k|^{1/k} \tag{14.1.2}$$

exists and is equal to $1/r$, the series converges in the disc $|w| < r$ in the complex w-plane, $\mathbb{C}_w$. There is always a singularity of $F(w)$ on the circle $|w| = r$, which is the *radius of convergence* of the series. If this singularity is on the positive real axis

D.A. Lavis, *Equilibrium Statistical Mechanics of Lattice Models*,
Theoretical and Mathematical Physics, DOI 10.1007/978-94-017-9430-5_14

it can be identified with $w(T_c)$,[1] where T_c is the critical temperature. When this is not the case the analysis is rather more complicated (Sect. 14.6).

In any event, as has been indicated, the methods to be described in this chapter are those for which, for a high- or low-temperature series and some integer ℓ, an ℓ-degree polynomial part

$$F_\ell(w) := \sum_{k=0}^{\ell} f_k\, w^k, \tag{14.1.3}$$

of the expansion (14.1.1) of $F(w)$ is known. In such circumstances, it may seem surprising that any information at all can be obtained about the function $F(w)$, given that the use of (14.1.2) to obtain a good approximation to the radius of convergence needs in general many more coefficients than are available.[2] More sophisticated techniques are needed and these are described in Sect. 14.6. As we shall see, the key to all these methods is the important point made by Guttmann (1989, p. 8) that, in series analysis, it is a question of "fitting series coefficients to an assumed functional form—an assumed form may be implicit or explicit, but is invariably present". Even then, of course, it is possible to obtain only approximate values for the assumed critical parameters. Nevertheless, these methods provide accurate values for critical parameters and exponents for many systems where no exact solution is known. Of course, where exact solutions are known, comparison with deductions from series methods provides a useful test for the latter.

In Sects. 14.2–14.5 we describe methods for obtaining series expansions, and for simplicity we shall concentrate mainly on the case of the ferromagnetic ($R > 0$) ν-state Potts model with Hamiltonian (3.6.53)[3] and partition function $Z(\mathcal{N}, \mathfrak{X}, \mathfrak{Z})$, where $\mathfrak{X} := \exp(K)$ and $\mathfrak{Z} := \exp(L)$, with $K := R/T$ and $L := \mu/T$. In Sect. 14.2 we discuss the determination of suitable variables for low- and high-temperature expansions. Low-temperature expansions are possible only for models (like the Potts model) where there is a well-defined ground state and the successive terms in the expansion represent deviations from that state. The 'obvious' expansion variable for high temperatures is the inverse temperature (equivalently the coupling K) and that is the only choice for n-vector models, with $n > 1$, and for the quantum Heisenberg model. However, as we shall see in Sect. 14.2, there is a more convenient choice for the Potts model.

---

[1] As is discussed in Sect. 14.6, in the case of high-temperature series, a singularity on the negative real axis also corresponds to a critical temperature.

[2] Thus, for example, for $F(w) := (0.5 - w)^{-0.25}$, although the convergence is for a monotonically increasing sequence, the estimate for w_c from $F_{1000}(w)$ still differs from $w_c = 0.5$ by 0.0031577523. Making the same point Guttmann (1989) considers the susceptibility of the spin-$\frac{1}{2}$ Ising model for which Nickel has derived 54 terms of the high-temperature series. In this case the estimate from $F_{54}(w)$ is higher than the exact $w_c = 1 + \sqrt{2}$ by about 5.4 %.

[3] Knowing of course that the spin-$\frac{1}{2}$ Ising model can be recovered as the special case $\nu = 2$ (Sect. 3.6.4).

In Sects. 14.2.1 and 14.2.2 we obtain, by simple graphical reasoning, the first few terms of low- and high-temperature series for $Z(\mathcal{N}, \mathfrak{X}, \mathfrak{Z})$. As we shall see, the main concern of this chapter is to obtain the series for the free energy per lattice site , when $\mathfrak{Z} = 1$, that is for the 'pure' Potts model with Hamiltonian (3.6.49). Contrary to the convention used elsewhere in the text (see the footnote on p. 21), the free energy per lattice site is taken to be the quantity defined *for a finite lattice* $\mathcal{N}$. That is[4]

$$[Z(\mathcal{N}, \mathfrak{X}, \mathfrak{Z})]^{1/N} =: \exp[-\phi(\mathcal{N}, \mathfrak{X}, \mathfrak{Z})]. \tag{14.1.4}$$

We shall also need expansions for the order parameter[5]

$$\widetilde{\rho} := 1 - \frac{\nu}{\nu - 1}\rho = 1 + \frac{\nu}{\nu - 1}\frac{\partial\phi}{\partial L} = 1 + \mathfrak{Z}\frac{\nu}{\nu - 1}\frac{\partial\phi}{\partial\mathfrak{Z}}, \tag{14.1.5}$$

defined in terms of the density ρ so as to be zero at zero temperature, when $\rho = (\nu - 1)/\nu$, and the response functions[6]

$$\frac{c_\rho}{k_{\mathrm{B}}K^2} := -\frac{\partial^2\phi}{\partial K^2} = -\mathfrak{X}\left\{\frac{\partial\phi}{\partial\mathfrak{X}} + \mathfrak{X}\frac{\partial^2\phi}{\partial\mathfrak{X}^2}\right\}, \tag{14.1.6}$$

$$\varphi_T := -\frac{\partial^2\phi}{\partial L^2} = -\mathfrak{Z}\left\{\frac{\partial\phi}{\partial\mathfrak{Z}} + \mathfrak{Z}\frac{\partial^2\phi}{\partial\mathfrak{Z}^2}\right\}, \tag{14.1.7}$$

given respectively by (2.3.29) and (2.3.31). Although our interest is in the case $\mathfrak{Z} = 1$, it is clear, from (14.1.5)–(14.1.7), that we shall need to obtain the free-energy series with general $\mathfrak{Z}$. However, in terms of suitably chosen field variables, it will be sufficient to retain terms only to quadratic degree.

Series for the partition function are called *moment expansions* and, having defined suitable variables, we obtain the first few terms of these, at low and high temperatures for the Potts model in Sect. 14.2. The terms in the expansion of the logarithm of the partition function are called *cumulants*. These can, of course, be obtained by taking the logarithm of the moment expansion and we illustrate this for the low-temperature series in Sect. 14.2.1. However, a more direct approach to the series for the free-energy density (equivalent to the cumulant expansion) is discussed in Sect. 14.3.

For both moment and cumulant series the terms in the expansions can be represented by graphs and, although the mode of their construction at low and high

---

[4] Rather confusingly, this quantity is often simple called 'the partition function' (see, for example Briggs et al. 1994).

[5] The order parameter $\widetilde{\rho}$, which Briggs et al. (1994) call the magnetization, differs from that used (with $\nu = 3$) in the Landau expansion in Sect. 5.4.2 by a factor of $2/\sqrt{3}$.

[6] The response function φ_T, which Briggs et al. call the susceptibility, differs from the compressibility defined by (1.4.12) by a factor of $1/(k_{\mathrm{B}}T\rho^2)$. We shall for the sake of maintaining neutrality between lattice-fluid and magnetic interpretations refer to it as the 'field-response function'. c_ρ is, of course, the heat capacity at constant density. However, since we shall be concerned solely with calculating it for zero μ ($\mathfrak{Z} = 1$) it will be the same as the constant μ heat capacity.

temperatures differ, their edges can be understood to represent sites which are linked by an interaction. However, as we shall see there are two important differences in this respect between the moment and cumulant expansions. These are:

(i) Moments include contributions from both one-component and multi-component graphs, whereas cumulants involve only connected graphs.
(ii) Moments (with the variables we employ) are represented by graphs with at most one edge between a pair of sites whereas cumulants involve *multi-bond graphs.*

A consequence of the occurrence of multi-bond graphs has been the development of *finite-cluster methods* (Sect. 14.4), where the cluster is taken to be the set of all graphs occupying the same set of lattice sites irrespective of the multiplicity of the bonds.[7] A natural development of finite-cluster methods leads to *finite-lattice methods* (Sect. 14.5).

14.2 Moment Expansions

Let the lattice $\mathcal{N}$, with $N := |\mathcal{N}|$ and coordination number $z > 2$, be one of the two- or three-dimensional lattices described in Sect. 16.2.1, with periodic boundary conditions.[8] From (3.6.53) the Potts model partition function takes the form

$$Z(\mathcal{N}, \mathfrak{X}, \mathfrak{Z}) = \sum_{\{\sigma\}} \left\{ \prod_{\{r,r'\}}^{(1)} \mathfrak{X}^{\delta^{\mathrm{Kr}}(\sigma(r)-\sigma(r'))} \right\} \left\{ \prod_{\{r\}} \mathfrak{Z}^{1-\delta^{\mathrm{Kr}}(\sigma(r))} \right\}, \tag{14.2.1}$$

of a polynomial in $\mathfrak{X}$ and $\mathfrak{Z}$. It is, of course, the case that expansions of $Z(\mathcal{N}, \mathfrak{X}, \mathfrak{Z})$ in terms of both high and low-temperature variables (as defined below) will also be polynomials. Since phase transitions occur only in the thermodynamic limit $N \to \infty$ the analysis of finite systems is of use only in so far as it yields a 'sufficient' number of terms which agree with the series which would arise from the corresponding infinite system.

Although this chapter will be concerned exclusively with high- and low-temperature expansions, another possible area of investigations is that of a high(positive)-chemical potential expansion at constant temperature. In the magnetic formulation in which investigations are usually set, this is the *high-field expansion.*[9] For our

[7] Item (ii) needs qualification if high-temperature moment series in terms of the inverse temperature are derived. These give rise to multi-bonded graphs (see Lavis and Bell 1999b, Sect. 7.6) and have led to the proposal for a finite-cluster moment method by de Neef (1975, Sect. 2.3).

[8] We exclude the case of a linear lattice ($z = 2$), for which no phase transition occurs for non-zero temperature.

[9] A case where such an expansion is useful is in the investigation of the zero-field first-order transition in the Ising model (see Sect. 14.7 and Enting and Baxter 1980).

purposes the inclusion of the chemical potential is to provide a means of obtaining density and compressibility expansions (in the magnetic formulation magnetization and susceptibility expansions, respectively).

14.2.1 At Low Temperatures

The variable $\mathfrak{X}$ becomes large at low temperatures and the polynomial in (14.2.1) is dominated by the ground-state term. Assuming, as we shall throughout this section, that $\mu < 0$, the ground state is $\sigma(\boldsymbol{r}) = 0$, for all $\boldsymbol{r}$. Let

$$Z_{\rm LT}(\mathcal{N}, u, p) := Z(\mathcal{N}, \mathfrak{X}, \mathfrak{Z})/(\mathfrak{X}^{N_{\rm E}}) \tag{14.2.2}$$

$$= \sum_{\{\sigma\}} \left\{ \prod_{\{\boldsymbol{r},\boldsymbol{r}'\}}^{(1)} u^{1-\delta^{\rm Kr}(\sigma(\boldsymbol{r})-\sigma(\boldsymbol{r}'))} \right\} \left\{ \prod_{\{\boldsymbol{r}\}} (1-p)^{1-\delta^{\rm Kr}(\sigma(\boldsymbol{r}))} \right\} \tag{14.2.3}$$

$$:= 1 + N \sum_{\mathfrak{m}=1}^{\infty} a_{\mathfrak{m}}(N, p) u^{\mathfrak{m}}, \tag{14.2.4}$$

where

$$u := 1/\mathfrak{X}, \qquad p = 1 - \mathfrak{Z}. \tag{14.2.5}$$

Then, from (16.2.8), (14.2.2) and (14.2.5),

$$\phi(\mathcal{N}, \mathfrak{X}, \mathfrak{Z}) = \phi_{\rm LT}(\mathcal{N}, u, p) + \tfrac{1}{2} z \ln(u). \tag{14.2.6}$$

The variable u has the required property of a low-temperature variable, tending to zero as temperature tends to zero, and the coefficients in (14.2.4) are polynomials in p.[10] However, in order to obtain series, for the free energy, order parameter and response function with $\mu = 0$ we need retain terms only to quadratic degree in p. Thus we require

$$Z_{\rm LT}(\mathcal{N}, u, p) \simeq Z_{\rm LT}^{(0)}(\mathcal{N}, u) + Z_{\rm LT}^{(1)}(\mathcal{N}, u)\, p + Z_{\rm LT}^{(2)}(\mathcal{N}, u)\, p^2, \tag{14.2.7}$$

which will then yield

$$\phi_{\rm LT}(\mathcal{N}, u, p) \simeq \phi_{\rm LT}^{(0)}(\mathcal{N}, u) + \phi_{\rm LT}^{(1)}(\mathcal{N}, u)\, p + \phi_{\rm LT}^{(2)}(\mathcal{N}, u)\, p^2. \tag{14.2.8}$$

[10] In view of the following discussion it is convenient to include explicit linear N-dependence of the coefficients in (14.2.4). However, that does not exclude further non-linear dependence.

The series for the free energy per lattice site with $\mu = 0$ is, of course, given by $\phi_{\mathrm{LT}}^{(0)}(\mathcal{N}, u)$. From (14.1.5)–(14.1.7) the corresponding series for the order parameter and response functions are given as expansions of

$$\widetilde{\rho} = 1 - \frac{\nu}{\nu - 1}\phi_{\mathrm{LT}}^{(1)}, \tag{14.2.9}$$

$$\frac{c_\rho}{k_{\mathrm{B}}K^2} = -u\left\{\frac{\partial\phi_{\mathrm{LT}}^{(0)}}{\partial u} + u\frac{\partial^2\phi_{\mathrm{LT}}^{(0)}}{\partial u^2}\right\}, \tag{14.2.10}$$

$$\varphi_T = \phi_{\mathrm{LT}}^{(1)} - 2\phi_{\mathrm{LT}}^{(2)}. \tag{14.2.11}$$

The series in (14.2.4) results from perturbations of the ground state, so the power of $(1 - p)$ in any term is just the number of perturbed sites and the power of u is the number of pairs in unlike states. In the case $\nu = 2$ (equivalent to the Ising model) such pairs will consist of one ground-state site and one perturbed $\sigma = 1$ site. When $\nu > 2$ there will also be the possibility of pairs of unlike perturbed sites. Thus a term in the series corresponds to a section graph on the lattice formed by drawing edges between all pairs of perturbed sites together with (in the case $\nu > 2$) a particular distribution of perturbed states on the graph. So a particular section graph $[\mathfrak{n}, \mathfrak{k}]$ will, in general, contribute a number of terms in the series each containing a factor $(1-p)^{\mathfrak{n}}$. The number of lattice edges $\mathfrak{l}[\mathfrak{n}, \mathfrak{k}]$, which can be drawn from sites of $[\mathfrak{n}, \mathfrak{k}]$ to ground-state sites, is given by (16.4.9) and we also need $w[\mathfrak{i}\,|\,\mathfrak{n}, \mathfrak{k}]$, the number of arrangements of perturbed states on $[\mathfrak{n}, \mathfrak{k}]$ so that there are exactly $\mathfrak{i}$ edges in different perturbed states.[11] From (16.4.10), a graph $[\mathfrak{n}, \mathfrak{k}]$ containing $\mathfrak{i}$ unlike perturbed pairs of sites makes a contribution to the series in a term $a_{\mathfrak{m}}(N, p)u^{\mathfrak{m}}$ with

$$\mathfrak{m} = \mathfrak{l}[\mathfrak{n}, \mathfrak{k}] + \mathfrak{i} = \mathfrak{n}(\mathfrak{z} - 2) + 2\mathfrak{c}[\mathfrak{n}, \mathfrak{k}] - 2\mathfrak{C}[\mathfrak{n}, \mathfrak{k}] + \mathfrak{i} \tag{14.2.12}$$

and the coefficient is

$$a_{\mathfrak{m}}(N, p) = \sum w[\mathfrak{i}\,|\,\mathfrak{n}, \mathfrak{k}]\varpi[\mathfrak{n}, \mathfrak{k}](1 - p)^{\mathfrak{n}}, \tag{14.2.13}$$

where the summation is over all graphs and configurations which contribute to this term. We now construct the first few terms of the low-temperature series for the ν-state Potts model. This is done by forming a sequence of graphs with increasing values of $\mathfrak{l}[\mathfrak{n}, \mathfrak{k}]$ and then, for each graph taking the possible values of $\mathfrak{i}$. We use the square-lattice graphs in Tables 16.1 and 16.2 as an illustration, but we also consider the triangular and honeycomb lattices; for all of these the cyclomatic number $\mathfrak{C}[\mathfrak{n}, \mathfrak{k}]$ of a graph $[\mathfrak{n}, \mathfrak{k}]$ is the number of its faces. As will be seen, all the section (low-temperature) graphs in Table 16.1 are trees or have components which are trees.[12] So one strategy is to begin with these and then to make modifications to the list which arise when graphs with faces, like those in Table 16.2, make a contribution. It is clear

---

[11] Of course, for all lattices, $w[0\,|\,\mathfrak{n}, \mathfrak{k}] = \nu - 1$.

[12] For simplicity we regard an isolated site as a tree (and also a chain) with no edges.

that, for $\nu > 2$ and a tree with n sites, i can take all values from zero to $n - 1$. The notation $\{n, f, i\}$ is used to denote all the one-component graphs with n sites, f faces and the given value of i. Since we are, at this point, considering only trees, $f = 0$. Thus, for example, $\{4, 0, i\}$ for the square lattice represents all the graphs $[4, k]$, $k = 4, \ldots, 7$ in Table 16.1, with $e[4, k] = 10$. A plus sign is used to indicate a separated graph with particular components.
The first few terms for the sequence for the square lattice are:

- $m = 4$: $\{1, 0, 0\}$.
- $m = 6$: $\{2, 0, 0\}$.
- $m = 7$: $\{2, 0, 1\}$.
- $m = 8$: $\{3, 0, 0\}$; $\{1, 0, 0\} + \{1, 0, 0\}$.
- $m = 9$: $\{3, 0, 1\}$.
- $m = 10$: $\{4, 0, 0\}$; $\{3, 0, 2\}$; $\{1, 0, 0\} + \{2, 0, 0\}$.
- $m = 11$: $\{4, 0, 1\}$; $\{1, 0, 0\} + \{2, 0, 1\}$.

for the triangular lattice are:

- $m = 6$: $\{1, 0, 0\}$.
- $m = 10$: $\{2, 0, 0\}$.
- $m = 11$: $\{2, 0, 1\}$.
- $m = 12$: $\{1, 0, 0\} + \{1, 0, 0\}$.
- $m = 14$: $\{3, 0, 0\}$.
- $m = 15$: $\{3, 0, 1\}$.
- $m = 16$: $\{3, 0, 2\}$; $\{1, 0, 0\} + \{2, 0, 0\}$.

and for the honeycomb lattice are:

- $m = 3$: $\{1, 0, 0\}$.
- $m = 4$: $\{2, 0, 0\}$.
- $m = 5$: $\{3, 0, 0\}$; $\{2, 0, 1\}$.
- $m = 6$: $\{4, 0, 0\}$; $\{3, 0, 1\}$; $\{1, 0, 0\} + \{1, 0, 0\}$.
- $m = 7$: $\{5, 0, 0\}$; $\{4, 0, 1\}$: $\{3, 0, 2\}$; $\{1, 0, 0\} + \{2, 0, 0\}$.
- $m = 8$: $\{6, 0, 0\}$; $\{5, 0, 1\}$; $\{4, 0, 2\}$; $\{1, 0, 0\} + \{2, 0, 1\}$; $\{1, 0, 0\} + \{3, 0, 0\}$; $\{2, 0, 0\} + \{2, 0, 0\}$.
- $m = 9$: $\{7, 0, 0\}$; $\{6, 0, 1\}$; $\{5, 0, 2\}$; $\{4, 3\}$; $\{1, 0, 0\} + \{1, 0, 0\} + \{1, 0, 0\}$; $\{1, 0, 0\} + \{4, 0, 0\}$; $\{1, 0, 0\} + \{3, 0, 1\}$; $\{2, 0, 0\} + \{3, 0, 0\}$; $\{2, 0, 0\} + \{2, 0, 1\}$.

Consider now the role played by the first few graphs with faces ($f \neq 0$). For each lattice the lowest value of m contributed by a graph with a face is a simple cycle around one face (with $i = 0$). For the square lattice the graphs which make contributions with $m \leq 11$ are [4, 8], [5, 1] and [6, 1] in Table 16.2. Thus to the list for the square lattice, given above, we add:

- $m = 8$: $\{4, 1, 0\}$.
- $m = 10$: $\{5, 1, 0\}$; $\{6, 2, 0\}$; $\{4, 1, 2\}$.
- $m = 11$: $\{5, 1, 1\}$; $\{4, 1, 3\}$.

It will be seen from these lists of graphical contributions to the series that graph-specific factors soon play a role in the development. However, it is of interest, particularly in connection with the high-temperature series discussion in Sect. 14.2.2, to develop the lowest-degree terms in a way which applies to all the regular lattices given in Sect. 16.2.1. We have successively the terms with

- $\mathfrak{m} = \mathfrak{z}$: single-site graphs, designated by $\{1, 0, 0\}$, with lattice constant of unity. There are $\nu-1$ ways of choosing the perturbed state so $a_{\mathfrak{z}}(N, p) = (\nu-1)(1-p)$.
- $\mathfrak{m} = 2(\mathfrak{z}-1)$: two-site chains, designated by $\{2, 0, 0\}$, with lattice constant $\frac{1}{2}\mathfrak{z}$ and the sites in the same perturbed state, giving $a_{2(\mathfrak{z}-1)}(N, p) = \frac{1}{2}\mathfrak{z}(\nu - 1)(1 - p)^2$.
- $\mathfrak{m} = 2\mathfrak{z} - 1$: two-site chains, designated by $\{2, 0, 1\}$, with the sites in different perturbed states. There are $(\nu - 1)(\nu - 2)$ ways of choosing the perturbed states so $a_{2\mathfrak{z}-1}(N, p) = \frac{1}{2}\mathfrak{z}(\nu - 1)(\nu - 2)(1 - p)^2$.

Thus we have

$$Z_{\mathrm{LT}}(\mathcal{N}, u, p) \simeq 1 + N(\nu - 1)\Big\{(1 - p)u^{\mathfrak{z}} + \tfrac{1}{2}\mathfrak{z}(1 - p)^2u^{2(\mathfrak{z}-1)} + \tfrac{1}{2}\mathfrak{z}(\nu - 2)(1 - p)^2u^{2\mathfrak{z}-1}\Big\}, \tag{14.2.14}$$

giving[13]

$$\phi^{(0)}_{\mathrm{LT}}(\mathcal{N}, u) = -(\nu - 1)\{u^{\mathfrak{z}} + \tfrac{1}{2}\mathfrak{z}u^{2(\mathfrak{z}-1)}\} + \mathrm{O}(u^{2\mathfrak{z}-1}), \tag{14.2.15}$$

$$\phi^{(1)}_{\mathrm{LT}}(\mathcal{N}, u) = (\nu - 1)\{u^{\mathfrak{z}} + \mathfrak{z}u^{2(\mathfrak{z}-1)}\} + \mathrm{O}(u^{2\mathfrak{z}-1}), \tag{14.2.16}$$

$$\phi^{(2)}_{\mathrm{LT}}(\mathcal{N}, u) = -\tfrac{1}{2}(\nu - 1)\mathfrak{z}u^{2(\mathfrak{z}-1)} + \mathrm{O}(u^{2\mathfrak{z}-1}) \tag{14.2.17}$$

and, from (14.2.9)–(14.2.11), the leading terms of low-temperature expansions of the order parameter and response functions can be derived. However, it is clear that in the term with $\mathfrak{m} = 2\mathfrak{z} - 1$ there is already an omission in the case of the honeycomb lattice, for which the chain of three sites $\{3, 0, 0\}$ also gives this value for $\mathfrak{m}$ with, using (16.4.15), an additional contribution $a'_5(N, p) = 3(\nu - 1)(1 - p)^3$. This illustrates the impossibility of presenting, beyond the first few terms, a general-lattice form for the low-temperature series expansion. We thus restrict our attention to the square lattice and, using (14.2.13) and the lattice constants in Tables 16.1 and 16.2, we have

$$\begin{aligned} Z^{(0)}_{\mathrm{LT}}(\mathcal{N}_{\mathrm{S}}, u) = 1 + N(\nu - 1)\Big\{&u^4 + 2u^6 + 2(\nu - 2)u^7 \\ &+ \tfrac{1}{2}[N(\nu - 1) - 5\nu + 19]\,u^8 \\ &+ 12(\nu - 2)u^9 + 2\Big[N(\nu - 1) + 3\nu^2 - 17\nu + 28\Big]\,u^{10} \\ &+ 2(\nu - 2)\,[N(\nu - 1) - 6\nu + 33]\,u^{11}\Big\} + \mathrm{O}(u^{12}), \end{aligned} \tag{14.2.18}$$

[13] Notation: $g(u) = \mathrm{O}(u^k)$ if $g(u)/u^k$ tends to a finite limit as $u \to 0$. If $g(u) = \mathrm{O}(u^k)$ then $g(u) = \mathrm{O}(u^{k'})$ for $k' \leq k$.

$$Z^{(1)}_{\mathrm{LT}}(\mathcal{N}_{\mathrm{S}}, u) = -N(\nu-1)\left\{u^4 + 4u^6 + 4(\nu-2)u^7 + [N(\nu-1) - 5\nu + 27]u^8 + 36(\nu-2)u^9 + 2[3N(\nu-1) + 9\nu^2 - 48\nu + 98]u^{10} + 2[3N(n-1)(\nu-2) - 16\nu^2 + 160\nu - 256]u^{11}\right\} + \mathrm{O}(u^{12}), \tag{14.2.19}$$

$$Z^{(2)}_{\mathrm{LT}}(\mathcal{N}_{\mathrm{S}}, u) = N(\nu-1)\left\{2u^6 + 2(\nu-2)u^7 + \tfrac{1}{2}[N(\nu-1) - 5\nu + 53]u^8 + 36(\nu-2)u^9 + 2[3N(\nu-1) + 9\nu^2 - 42\nu + 133]u^{10} + 2[3N(\nu-1)(\nu-2) - 12\nu^2 + 214\nu - 380]u^{11}\right\} + \mathrm{O}(u^{12}). \tag{14.2.20}$$

Using the same procedure, series for the triangular and honeycomb lattices can be derived. It will be noted that the two-component graphs [2, 1] and [3, 2] make contributions to $a_8(N, p)$ and to $a_{10}(N, p)$ and $a_{11}(N, p)$ respectively which are dependent on N. This is a case of a result, which is true for all lattices, for both low- and high-temperature series and for both $\mu = 0$ and $\mu \neq 0$; that graphs with c components will make contributions to coefficients of the partition function expansion which are polynomials in N of degree c − 1. However, our interest is not in the partition function itself but in the free energy per lattice site and the order parameter and response functions, for which the series coefficients are N-independence (Sect. 14.3). If it were not the case the extensivity of the free energy would be violated, vitiating the relationship with thermodynamics. In particular, from (14.2.7) and (14.2.18)–(14.2.20), for the square lattice

$$\phi^{(0)}_{\mathrm{LT}}(\mathcal{N}_{\mathrm{S}}, u) = -(\nu-1)\left\{u^4 + 2u^6 + 2(\nu-2)u^7 - \tfrac{1}{2}(5\nu-19)u^8 + 12(\nu-2)u^9 + 2(3\nu^2 - 17\nu + 28)u^{10} - 6(\nu-2)(2\nu-11)u^{11}\right\} + \mathrm{O}(u^{12}), \tag{14.2.21}$$

$$\phi^{(1)}_{\mathrm{LT}}(\mathcal{N}_{\mathrm{S}}, u) = (\nu-1)\left\{u^4 + 4u^6 + 4(\nu-2)u^7 - (5\nu-27)u^8 + 36(\nu-2)u^9 + 2(9\nu^2 - 48\nu + 98)u^{10} - 32(\nu-2)(\nu-8)u^{11}\right\} + \mathrm{O}(u^{12}), \tag{14.2.22}$$

$$\phi^{(2)}_{\mathrm{LT}}(\mathcal{N}_{\mathrm{S}}, u) = -(\nu-1)\left\{2u^6 + 2(\nu-2)u^7 - \tfrac{1}{2}(5\nu-53)u^8 + 36(\nu-2)u^9 + 2(9\nu^2 - 42\nu + 133)u^{10} - 4(\nu-2)(6\nu-95)u^{11}\right\} + \mathrm{O}(u^{12}). \tag{14.2.23}$$

The reasoning used here for the Potts model, whereby the low-temperature series is derived by considering perturbations from a ground state, can be used for the general-spin ferromagnetic Ising model. However, the spatial isotropy of the exchange interaction terms in classical and quantum Heisenberg Hamiltonians lead to important differences in these cases. As we have seen (Sect. 3.6.1) spontaneous magnetization in the two-dimensional classical Heisenberg model is excluded by the Mermin–Wagner theorem. There is, therefore, no Curie temperature.[14] For three-dimensional lattices there is a Curie temperature, but the low-temperature behaviour is radically different from that of the Ising model (see, for example, Ashcroft and Mermin 1976). There are no small localized perturbations from the ground state and, for the quantum Heisenberg model, the excitations are spin waves. It follows that there can be no low-temperature series of the type discussed here for these models. In fact, the changes in the spontaneous magnetization and heat capacity from their zero-temperature values are proportional to $T^{\frac{3}{2}}$. In both the quantum and classical Heisenberg models for $T < T_c$, there is a divergence in the magnetic susceptibility as $\mathcal{H} \to 0$. In the classical Heisenberg model the low-temperature heat capacity has the form $Nk_B + aT$, where a is a constant.

14.2.2 At High Temperatures

The key to the choice of a high-temperature variable for the Potts model is the duality relationship (9.2.21) between $\mathfrak{X}$ and $\mathfrak{X}^\star$ expressible in the form

$$\mathfrak{X} = 1 + \frac{\nu}{\mathfrak{X}^\star - 1} = \frac{\mathfrak{X} + \nu - 1}{\nu}\left\{1 + \frac{\nu - 1}{\mathfrak{X}^\star}\right\}. \tag{14.2.24}$$

As $T \to \infty$, $\mathfrak{X} \to 1$, $\mathfrak{X}^\star \to \infty$, and

$$v := 1/\mathfrak{X}^\star = \frac{\mathfrak{X} - 1}{\mathfrak{X} + \nu - 1} \to 0, \tag{14.2.25}$$

with, from (9.2.22),

$$\mathfrak{X}^{\delta^{\mathrm{Kr}}(\sigma)} = \frac{\mathfrak{X} + \nu - 1}{\nu}[1 + \Delta(\nu;\sigma)v], \tag{14.2.26}$$

where $\Delta(\nu;\sigma)$ is the *Potts delta function* defined and discussed in Sect. 18.4. In a similar way we introduce the high-temperature variable

$$q := \frac{1 - \mathfrak{Z}}{1 + \mathfrak{Z}(\nu - 1)}, \tag{14.2.27}$$

[14] Although there is a Kosterlitz-Thouless transition which can be analysed using a high-temperature series expansion (Oitmaa et al. 2006, Chap. 3).

associated with the chemical potential, with

$$\mathfrak{Z}^{1-\delta^{\mathrm{Kr}}(\sigma)} = \frac{1+\mathfrak{Z}(\nu-1)}{\nu}[1+\Delta(\nu;\sigma)q]. \tag{14.2.28}$$

Substituting into (14.2.1) gives

$$Z_{\mathrm{HT}}(\mathcal{N},v,q) = \nu^{-N} Z(\mathcal{N},\mathfrak{X},\mathfrak{Z})\left\{\frac{\nu}{\mathfrak{X}+\nu-1}\right\}^{N_{\mathrm{E}}}\left\{\frac{\nu}{1+\mathfrak{Z}(\nu-1)}\right\}^{N} \tag{14.2.29}$$

$$= \nu^{-N}\sum_{\{\sigma\}}\left\{\prod_{\{r,r'\}}^{(1)}[1+\Delta(\nu;\sigma(r)-\sigma(r'))v]\right\} \times \left\{\prod_{\{r\}}[1+\Delta(\nu;\sigma(r))q]\right\} \tag{14.2.30}$$

$$:= 1+N_{\mathrm{X}}\sum_{\mathfrak{e}=1}^{\infty}\sum_{\mathfrak{n}=0}^{\infty} b_{\mathfrak{e},\mathfrak{n}}(N_{\mathrm{X}})\, v^{\mathfrak{e}} q^{\mathfrak{n}}, \tag{14.2.31}$$

where N_{X} is an extensive variable to be determined below and the coefficients in the series in (14.2.31) are obtained by expanding out the products in (14.2.30) and summing over the values of the state variables. From (16.2.8), (14.2.25), (14.2.27) and (14.2.29), for the free-energy density,

$$\phi(\mathcal{N},\mathfrak{X},\mathfrak{Z}) = \phi_{\mathrm{HT}}(\mathcal{N},u,q) - \ln(\nu) + \ln[1+(\nu-1)q] + \tfrac{1}{2}z\ln(1-v). \tag{14.2.32}$$

As in the case of the low-temperature series, the aim is to obtain

$$\phi_{\mathrm{HT}}(\mathcal{N},v,q) \simeq \phi_{\mathrm{HT}}^{(0)}(\mathcal{N},v) + \phi_{\mathrm{HT}}^{(1)}(\mathcal{N},v)q + \phi_{\mathrm{HT}}^{(2)}(\mathcal{N},v)q^2, \tag{14.2.33}$$

with approximations to $\phi_{\mathrm{HT}}^{(0)}(\mathcal{N},v)$, $\phi_{\mathrm{HT}}^{(1)}(\mathcal{N},v)$ and $\phi_{\mathrm{HT}}^{(2)}(\mathcal{N},v)$ obtained to as high a power of v as possible. However, as we see below, $\phi_{\mathrm{HT}}^{(1)}(\mathcal{N},v) = 0$ and it follows from (14.1.5)–(14.1.7), (14.2.25) and (14.2.27) that the order parameter $\widetilde{\rho}$ is identically zero, with the response functions given by

$$\frac{c_\rho}{k_{\mathrm{B}}K^2} = \nu^{-2} f(\nu,v)\left\{\frac{1}{2}zg_3(\nu,v) - \frac{\partial\phi_{\mathrm{HT}}^{(0)}}{\partial v}g_2(\nu,v) - \frac{\partial^2\phi_{\mathrm{HT}}^{(0)}}{\partial v^2}f(\nu,v)\right\}, \tag{14.2.34}$$

$$\varphi_T = -\nu^{-2}\{2\phi_{\mathrm{HT}}^{(2)} + 1 - \nu\}, \tag{14.2.35}$$

where

$$f(\nu,v) := (\nu-1)[v(\nu-1)+1], \quad g_n(\nu,v) := nv(\nu-1)+n-\nu. \tag{14.2.36}$$

We have used section graphs in the development of some terms of the low-temperature series. For high-temperature series the criteria for what graphs are relevant are rather different.

A graph is said to be *zero-field* or *magnetic* according to whether it has zero or a non-zero number of odd vertices.[15] This distinction is important for their use in the development of high-temperature series for the $\nu = 2$ (Ising model) case. There contributions to the zero-μ series are made only by zero-field graphs and contributions to the $\mu \neq 0$ terms are made only by magnetic graphs. This distinction no longer holds for general ν. However, an important result proved by Domb (1974a, b) is still valid, namely that the only graphs which contribute to the zero-μ high-temperature series are stars (Sect. 18.4).[16]

14.2.2.1 The Case $\mu = 0$

We first consider the expansion of the first product in (14.2.30) which would yield the series for $q = \mu = 0$, or equivalently the coefficients $b_{\mathrm{e},0}(N_{\mathrm{x}})$, $\mathrm{e} = 1, 2, \ldots$ A term of the expansion which could make a contribution to $b_{\mathrm{e},0}(N_{\mathrm{x}})$ would be a production of e edge Δ-functions

$$\Pi(\mathrm{e},\mathrm{k}) := \nu^{-\mathrm{n}}\Delta(\nu;\sigma(\boldsymbol{r}_1)-\sigma(\boldsymbol{r}'_1))\Delta(\nu;\sigma(\boldsymbol{r}_2)-\sigma(\boldsymbol{r}'_2))$$
$$\times\,\Delta(\nu;\sigma(\boldsymbol{r}_3)-\sigma(\boldsymbol{r}'_3))\cdots\Delta(\nu;\sigma(\boldsymbol{r}_{\mathrm{e}})-\sigma(\boldsymbol{r}'_{\mathrm{e}})), \quad (14.2.37)$$

on the e edges $(\boldsymbol{r}_1, \boldsymbol{r}'_1), (\boldsymbol{r}_2, \boldsymbol{r}'_2), \ldots, (\boldsymbol{r}_{\mathrm{e}}, \boldsymbol{r}'_{\mathrm{e}})$ of a graph (e, k) (Sect. 16.4.3), where $\mathrm{n} := \mathrm{n}(\mathrm{e},\mathrm{k})$ and, as we have shown in Sect. 18.4, the only graphs which need be considered are stars.[17]

At this point we should observe that, although duality between lattices has an unambiguous definition only for two-dimensional lattices,[18] nothing said so far in this section implies two-dimensionality or indeed the existence of a dual lattice. The duality relationship between couplings has been used merely as an ansatz for defining a high-temperature expansion variable. However, in the remaining part of this section our analysis is restricted to a two-dimensional lattice $\mathcal{N}$ with dual lattice $\mathcal{N}^\star$. The number of such sites (and edges) around a face of $\mathcal{N}$ is $\mathrm{z}^\star$, the coordination number of the dual lattice $\mathcal{N}^\star$ (Sect. 16.2.2.1), and the number of faces of $\mathcal{N}$ is $N^\star := |\mathcal{N}^\star|$. So the lattice constants which feature as factors in the coefficients $b_{\mathrm{e},0}(N_{\mathrm{x}})$ are per face of $\mathcal{N}$ (or equivalently per site of $\mathcal{N}^\star$), rather than per site of $\mathcal{N}$, and we take

[15] The distinction is sometimes made between 'magnetic graphs' with two odd vertices and 'hyper-magnetic graphs' with more than two odd vertices; for simplicity we shall refer to all graphs with odd vertices simply as magnetic.

[16] The question of what graphs contribute to the $\mu \neq 0$ series is more complicated, as is illustrated by the examples is Sect. 18.4.

[17] Graphs with type B and C articulation points give a zero contribution and those with a type A articulation point can be considered as cases of disconnected components.

[18] See the remarks at the end of Sect. 16.2.2.1.

$N_{\mathrm{x}} := N^{\star}$. Thus the leading term in the zero-μ part of the expansion (14.2.31) involves one face and configuration number one and the summation over a term like (14.2.37), where (e, k) is the cycle around a single face. From Example 18.4.1,

$$b_{z^{\star},0}(N^{\star}) = (\nu - 1). \tag{14.2.38}$$

The next two terms, each with a lattice constant $\frac{1}{2}z^{\star}$, will correspond to the sites around two faces which share an edge, without and with, respectively a Δ-function for the common edge. In the first instance the two faces of $\mathcal{N}$ give a single face of (e, k) and, in the second, two faces sharing a Δ-function-carrying pair. Thus, from Examples 18.4.1 and 18.4.2,

$$b_{2(z^{\star}-1),0}(N^{\star}) = \tfrac{1}{2}z^{\star}(\nu - 1), \tag{14.2.39}$$

$$b_{(2z^{\star}-1),0}(N^{\star}) = \tfrac{1}{2}z^{\star}(\nu - 1)(\nu - 2) \tag{14.2.40}$$

and, using $N^{\star}/N = z/z^{\star}$,

$$\phi^{(0)}_{\mathrm{HT}}(\mathcal{N}, v) = -(z/z^{\star})(\nu - 1)\left\{v^{z^{\star}} + \tfrac{1}{2}z^{\star}v^{2(z^{\star}-1)}\right\} + \mathrm{O}(v^{2z^{\star}-1}). \tag{14.2.41}$$

14.2.2.2 The Case $\mu \neq 0$

A term of the expansion which could make a contribution to $b_{\mathrm{e},n'}$ would be a product of $\Pi(\mathrm{e}, \mathrm{k})$ given by (14.2.37) and a single-vertex term

$$\Lambda\{\mathfrak{n}', \mathrm{k}'\} := \Delta(\nu; \sigma(\boldsymbol{r}_1))\Delta(\nu; \sigma(\boldsymbol{r}_2)\Delta(\nu; \sigma(\boldsymbol{r}_3)) \cdots \Delta(\nu; \sigma(\boldsymbol{r}_{\mathfrak{n}'})), \tag{14.2.42}$$

which is a product of Δ-functions at $\mathfrak{n}'$ vertices $\boldsymbol{r}_1, \boldsymbol{r}_2, \ldots, \boldsymbol{r}_{\mathfrak{n}'}$ of $\mathcal{N}$ in a configuration denoted by $\{\mathfrak{n}', \mathrm{k}'\}$. In Sect. 18.4 it was shown that, for a non-zero contribution to the series, the sites of $\{\mathfrak{n}', \mathrm{k}'\}$ must be contained within (e, k) with $2 \leq \mathfrak{n}' \leq \mathfrak{n}(\mathrm{e}, \mathrm{k})$.

As we have seen, setting $\mu \neq 0$ for the low-temperature case does not add significantly to the calculation of the series expansion. In contrast $\mu \neq 0$, or equivalently $q \neq 0$, makes the high-temperature case much more complicated. However, as we have seen in the low-temperature case, such a computation is necessary, up to quadratic terms in q, to obtain the response function series. As would be anticipated the order parameter $\widetilde{\rho} = 0$; this follows from (14.1.5), the definition (14.2.27) of q and the fact that none of the series coefficients contains a term linear in q. Thus, as we have already indicated above, $\phi^{(1)}_{\mathrm{HT}}(\mathcal{N}, v) = 0$.

We now consider a number of cases of graphs which contribute to coefficients $b_{\mathrm{e},2}(N^{\star})$. The simplest of these is the graph (1, 1), the single pair of vertices connected by an edge shown in Table 16.1, with a single-vertex term at each end. This will provide the sole contribution to $b_{1,2}(N^{\star})$. If the lattice is not restricted to the square, from (16.2.8), $\Omega(1, 1) = \frac{1}{2}Nz = \frac{1}{2}N^{\star}z^{\star}$. Thus, from Example 18.4.5,

$$b_{1,2}(N^\star) = \tfrac{1}{2}z^\star(\nu - 1). \tag{14.2.43}$$

As was shown in (18.4), intermediate terms of a chain can be effectively removed by summation so that the factor from $\Pi(\mathrm{e},\mathrm{k}) \times \Lambda\{2,\mathrm{k}'\}$ will still be $\nu - 1$. So for $\mathrm{e} = 2$, from (16.4.14),

$$b_{2,2}(N^\star) = \tfrac{1}{2}z^\star(z - 1)(\nu - 1). \tag{14.2.44}$$

After this the lattice constants become more complicated because they have different contributions for different lattices. The lattice with the smallest value of $z^\star$ is the triangular lattice with $z^\star = 3$. Then, according to (16.4.14), we must exclude the chain which closes around a face and calculate the lattice constant around the face separately. In general for any lattice a contribution to $b_{z^\star,2}(N^\star)$ arises from a simple cycle with two sites of the cycle containing single-vertex terms. As was shown in Example 18.4.7 the factor from $\Pi(\mathrm{e},\mathrm{k}) \times \Lambda\{2,\mathrm{k}'\}$ will be $(\nu - 1)(\nu - 2)$ with $\Omega(z^\star, 1) = N^\star$. The number of ways the two single-vertex terms can be distributed on the cycle is $\frac{1}{2}z^\star(z^\star - 1)$. So the contribution to $b_{z^\star,2}(N^\star)$ from the simple cycle round a lattice face is $\frac{1}{2}z^\star(z^\star - 1)(\nu - 1)(\nu - 2)$. Without extending the series into the range in which the lattices need to be distinguished

$$\phi^{(2)}_{\mathrm{HT}}(\mathcal{N}, v) = -\tfrac{1}{2}z(\nu - 1)\{v + (z - 1)v^2\} + \mathrm{O}(v^3). \tag{14.2.45}$$

and, from (14.2.34)–(14.2.36), (14.2.41) and (14.2.45) the expansions of the response functions can be derived.

It is clear from the examples given here and others which could be presented, using more complicated arrangements of chains and cycles (Example 18.4.9), that the calculation of high-temperature series coefficients, especially when $\mu \neq 0$, rapidly become very difficult using the kind of rather ad hoc analysis so far described. However, many systematic procedures exist for facilitating the calculations and, as we have already indicated, we shall mainly concentrate on one, the finite-lattice method described in Sect. 14.5.

14.2.3 Duality for Graphs

Comparison of (14.2.15) and (14.2.41) suggests the relationship

$$z^\star \phi^{(0)}_{\mathrm{LT}}(\mathcal{N}, v) = z\phi^{(0)}_{\mathrm{HT}}(\mathcal{N}^\star, v) \tag{14.2.46}$$

between low- and high-temperature series for the case $\mu = 0$ ($p = 0$ and $q = 0$ respectively), that is that they are related by lattice duality (the low-temperature series for $\mathcal{N}$ is the high-temperature series for $\mathcal{N}^\star$, and vice-versa). This could be inferred from the Euler's relationship (16.2.7), the duality relationship (9.2.24) and the defining relationships (14.2.2) and (14.2.29) for the low- and high-temperature partition functions respectively. However, it is of interest to see how this is realized in terms of the relationship between the graphs for the two series.

The general idea is to begin with a low-temperature (section) graph $\mathfrak{g}$ on $\mathcal{N}$ and to add to it the $\mathfrak{l}[\mathfrak{g}]$ edges and their terminating vertices (lattice sites) which connect $\mathfrak{g}$ to the rest of $\mathcal{N}$; $\mathfrak{g}^{(+)}$ denotes the resulting graph (which is not in general a section graph). The duality formulation of Sect. 9.2.1 can now be applied with the graph $\mathfrak{g}^{(+)}$ replacing the lattice $\mathcal{N}^{(+)}$. The dual high-temperature graph $\mathfrak{g}^{\star}$ is formed by drawing all the edges of $\mathcal{N}^{\star}$ which cut the edges of $\mathfrak{g}^{(+)}$. It is clear that $\Omega(\mathfrak{g}^{(+)}\,|\,\mathcal{N}) = \Omega(\mathfrak{g}^{\star}\,|\,\mathcal{N}^{\star})$.

For the purpose of establishing a duality relationship between the terms of a low- and high-temperature series, this account is not quite sufficient. We need also to be able to incorporate the feature represented by the index $\mathfrak{i}$ in Sect. 14.2.1; that is we need to be able to fix the number of like and unlike pairs of perturbed states on $\mathfrak{g}^{(+)}$. We do this by designating the locations of all the like pairs. (To compute the whole series coefficients it will be necessary to count the number of different locations for these like pairs.) The partition function on $\mathfrak{g}^{(+)}$ is now computed with the restriction that like pairs with zero energy of interaction occur at the prescribed locations. The dual graph $\mathfrak{g}^{\star}$ is now formed by drawing edges only through the edges of $\mathfrak{g}^{(+)}$ not occupied by fixed like pairs. With this modification, (9.2.21) relates the couplings on $\mathfrak{g}^{(+)}$ and $\mathfrak{g}^{\star}$ and (9.2.23) gives

$$Z(\mathfrak{g}^{(+)},\mathfrak{X}) = \nu^{-\mathfrak{n}(\mathfrak{g}^{\star})}(\mathfrak{X}-1)^{\mathfrak{e}(\mathfrak{g}^{\star})}Z(\mathfrak{g}^{\star},\mathfrak{X}^{\star}). \tag{14.2.47}$$

The next step is to relate $Z(\mathfrak{g}^{\star},\mathfrak{X}^{\star})$ to $\Pi(\mathfrak{g}^{\star})$, given by (14.2.37), with $(\mathrm{e},\mathrm{k}) := \mathfrak{g}^{\star}$. The key to this is to note, from (18.4.1), that

$$\Delta(\nu;\sigma) = -(1-\nu)^{\delta^{\mathrm{Kr}}(\sigma)}. \tag{14.2.48}$$

Thus

$$\Pi(\mathfrak{g}^{\star}) = \nu^{-\mathfrak{n}(\mathfrak{g}^{\star})}(-1)^{\mathfrak{e}(\mathfrak{g}^{\star})}Z(\mathfrak{g}^{\star},1-\nu). \tag{14.2.49}$$

Substituting $\mathfrak{X}^{\star} = 1-\nu$ into (9.2.21) gives $\mathfrak{X} = 0$ or equivalently $K = -\infty$. This limiting value for the coupling, when substituted into $Z(\mathfrak{g}^{(+)},\mathfrak{X})$ yields the number of arrangements of states on $\mathfrak{g}^{(+)}$ so that there are no like pairs (apart from those previously designated). This gives precisely that part of the contribution to a low-temperature series coefficient from $\mathfrak{g}^{(+)}$.[19] From (14.2.47) and (14.2.48),

$$Z(\mathfrak{g}^{(+)},0) = \Pi(\mathfrak{g}^{\star}), \tag{14.2.50}$$

[19] Of course, it will then need to be multiplied by the appropriate lattice constant and the number of equivalent locations for the like pairs.

which establishes the relationship between low- and high-temperature series coefficients. It is of interest to check this result by calculating the partition functions for the graph [2, 1] of Table 16.1 in the two cases where the pair is constrained to be in the same state and where this is not the case and the partition functions of the two dual lattices (6, 1) and (7, 1) respectively of Table 16.2.

14.3 Cumulant Expansions

In Sect. 14.2 the first few terms of the low- and high-temperature moment series for the Potts model were derived in terms of an appropriate variable related to the first-neighbour interaction, with a contribution to a moment represented by a graph $\mathfrak{g}$ on the lattice. In the case of low-temperature series $\mathfrak{g}$ is a section graph corresponding to a set of edges drawn between a group of sites in states different from the ground state and terms in the moment arise from unlike pairs between $\mathfrak{g}$ and the ambient ground state and unlike pairs within $\mathfrak{g}$. In the case of high-temperature series the graph $\mathfrak{g}$ is not necessarily a section graph, but each factor in the expansion variable is related directly to an edge of $\mathfrak{g}$. The common feature of both low- and high-temperature graphs is that a contributing graph-edge makes only a linear contribution to the moment; all these graphs are 'single-bonded'.

In Sect. 14.2.1 we used the partition function (moment) series to derive the free energy (cumulant) series and observed that the polynomial N-dependence of the moments arising from graphs of more than one component disappeared. The assertion made there that, in general, for all lattices and both low- and high-temperature expansions, disconnected graphs do not contribute to the cumulants needs now to be examined in more detail. In doing so we shall uncover another feature of the cumulant series, which is the role played by *multi-bonded graphs*. This can already be seen in a limited way when the process by means of which the series (14.2.8) was derived from the series (14.2.7) is considered; certain terms in (14.2.8) are partly produced by multiples of the terms in (14.2.7). This means that a factor of u arising from a particular pair of sites on a graph may be multiplied by itself and thus feature in the series term with a multiplicity greater than one. To follow this process more clearly we shall, in this section adopt a generalized approach in which the interaction variables between different pairs of sites are distinguished.

For the sake of simplicity all reference to the μ-dependence of the system is suppressed and

$$Z_{\square}(\mathcal{N}, w) := 1 + N_{\square}\{f_1 w + f_2 w^2 + \cdots\}, \tag{14.3.1}$$

is taken to represent in generic form either the low-temperature moment expansion (14.2.4) or the high-temperature moment expansion (14.2.31), with the subscript '$\square$' standing for 'LT' with $N_{\square}$ being the number of sites or '$\square$' standing for 'HT' with

$N_\square$ being the number of faces of the lattice.[20] Then

$$\Phi_\square(\mathcal{N}, w) = -\ln\{Z_\square(\mathcal{N}, w)\} := N_\square\{\tilde{f}_1 w + \tilde{f}_2 w^2 + \tilde{f}_3 w^3 + \cdots\} \qquad (14.3.2)$$

is the cumulant expansion for the (dimensionless) free energy.

Suppose that the low- or high-temperature variables u and v are replaced in (14.2.3) and (14.2.30) with different variables, $u(\boldsymbol{r}, \boldsymbol{r}')$ and $v(\boldsymbol{r}, \boldsymbol{r}')$ respectively, for each first-neighbour pair. This generalization does not affect the graphs which contribute to the low- and high-temperature series. It just involves the appearance of different sets of pair variables for each isomorphic graph. Denoting these variables generically by $w(\boldsymbol{r}, \boldsymbol{r}')$, $Z_\square(\mathcal{N}, w)$ in (14.3.1) is replaced by $Z_\square(\mathcal{N}, \{w(\boldsymbol{r}, \boldsymbol{r}')\})$ with the general term $f_{\mathrm{e}} w^{\mathrm{e}}$ replaced by

$$\sum_{\{\mathfrak{g}\}} f(\mathfrak{g}) \prod_{\mathfrak{g}} w(\boldsymbol{r}, \boldsymbol{r}'), \qquad (14.3.3)$$

where the sum is over all graphs which, when the pair variables are all set to w, make a non-zero contribution to $f_{\mathrm{e}} w^{\mathrm{e}}$.

To obtain the cumulant expansion the Maclaurin method is used.[21] This gives

$$\Phi_\square(\mathcal{N}, w) = -\left[\exp\left(w \sum_{\{\boldsymbol{r},\boldsymbol{r}'\}}^{(1)} \mathcal{D}(\boldsymbol{r}, \boldsymbol{r}')\right) \ln[Z_\square(\mathcal{N}, \{w(\boldsymbol{r}, \boldsymbol{r}')\})]\right]_0, \qquad (14.3.4)$$

where

$$\mathcal{D}(\boldsymbol{r}, \boldsymbol{r}') := \frac{\partial}{\partial w(\boldsymbol{r}, \boldsymbol{r}')} \qquad (14.3.5)$$

and the subscript '0' indicates that all $\{w(\boldsymbol{r}, \boldsymbol{r}')\}$ are set equal to zero. Using the expansion

$$\exp\left(w \sum_{\{\boldsymbol{r},\boldsymbol{r}'\}}^{(1)} \mathcal{D}(\boldsymbol{r}, \boldsymbol{r}')\right) = 1 + \sum_{\mathrm{e}=1}^{\infty} \frac{w^{\mathrm{e}}}{\mathrm{e}!}\left[\sum_{\{\boldsymbol{r},\boldsymbol{r}'\}}^{(1)} \mathcal{D}(\boldsymbol{r}, \boldsymbol{r}')\right]^{\mathrm{e}} \qquad (14.3.6)$$

in (14.3.4) gives a power-series expansion in w.[22]

---

[20] The variables $\mathfrak{z}$ and q are reintroduced when the analysis is particularized, respectively, to the low- and high-temperature cases.

[21] The function $f(\boldsymbol{x})$ of the vector $\boldsymbol{x}$ can be expressed in the form

$$f(\boldsymbol{\delta x}) = [\exp(\boldsymbol{\delta x} \cdot \nabla) f(\boldsymbol{x})]_{\boldsymbol{x}=\boldsymbol{0}}$$

at a point $\boldsymbol{\delta x}$ near the origin. The expansion in powers of the components of $\boldsymbol{\delta x}$ is then obtained by using the power-series expansion of the exponential.

[22] Equations (14.3.4) and (14.3.6) provide the starting point for the derivation of *moment-cumulant relations* between the moments $f_1, f_2, \ldots$ and the cumulants $\tilde{f}_1, \tilde{f}_2, \ldots$ (see e.g. Lavis and Bell 1999b, Sect. 7.5.3).

The expression in square brackets in (14.3.6) can be expanded multinomially into a sum of terms each containing a product of $\mathring{e}$ first-neighbour pair factors $\mathcal{D}(\boldsymbol{r}, \boldsymbol{r}')$. A subgraph corresponding to each product is obtained by placing a line between the first-neighbour sites $\boldsymbol{r}$ and $\boldsymbol{r}'$ for each factor $\mathcal{D}(\boldsymbol{r}, \boldsymbol{r}')$ present. However, the class of graphs obtained differs from those for the partition function since a factor $\mathcal{D}(\boldsymbol{r}, \boldsymbol{r}')$ may occur more than once, giving several edges linking a pair of vertices $\boldsymbol{r}$ and $\boldsymbol{r}'$. Thus, as indicated above, we have multi-bonded graphs. Consider the product

$$\mathrm{D}(\mathring{e}, \mathring{k} \,|\, \boldsymbol{R}) := [\mathcal{D}(\boldsymbol{r}_1, \boldsymbol{r}'_1)]^{\alpha_1} [\mathcal{D}(\boldsymbol{r}_2, \boldsymbol{r}'_2)]^{\alpha_2} \cdots [\mathcal{D}(\boldsymbol{r}_e, \boldsymbol{r}'_e)]^{\alpha_e}. \tag{14.3.7}$$

The notation we have used here designates $(\mathring{e}, \mathring{k})$ as a multi-bonded graph with $\mathring{e}$ edges. Underlying this graph is the *silhouette* graph (e, k) obtained by replacing all the multiple edges by single edges. With $\alpha_1, \alpha_2, \ldots \alpha_e$ denoting the multiplicities of the edges in $(\mathring{e}, \mathring{k})$,

$$\mathring{e} = \alpha_1 + \alpha_2 + \cdots + \alpha_e. \tag{14.3.8}$$

Of course (14.3.7) contains information about a specific location for $(\mathring{e}, \mathring{k})$ on the lattice and this is formally represented by $\boldsymbol{R}$. The number of different values for $\boldsymbol{R}$ is the lattice constant $\Omega(e, k)$ of the silhouette graph. However, we also need to include in the lattice constant for $(\mathring{e}, \mathring{k})$ the factor $S(e, k \,|\, \mathring{e}, \mathring{k})$ which is the number of ways of arranging the extra $\mathring{e} - e$ edges on (e, k) to produced the topology $\mathring{k}$. Each graph $(\mathring{e}, \mathring{k})$ with location $\boldsymbol{R}$ will occur $\mathring{e}!\, A(\mathring{e}, \mathring{k})$ times, where

$$A(\mathring{e}, \mathring{k}) := \left(\prod_{i=1}^{e} \alpha_i! \right)^{-1}, \tag{14.3.9}$$

corresponds to all permutations of the differential operators.[23] So, from (14.3.4) and (14.3.6),

$$\Phi_{\square}(\mathcal{N}, w) = -\sum_{\mathring{e}=1}^{\infty} w^{\mathring{e}} \sum_{\{\mathring{k}\}} \Omega(e, k) S(e, k \,|\, \mathring{e}, \mathring{k}) C(\mathring{e}, \mathring{k}) A(\mathring{e}, \mathring{k}), \tag{14.3.10}$$

where

$$C(\mathring{e}, \mathring{k}) := \Big[\mathrm{D}(\mathring{e}, \mathring{k} \,|\, \boldsymbol{R}) \ln\{Z_{\square}(\mathcal{N}, \{w(\boldsymbol{r}, \boldsymbol{r}')\})\} \Big]_0 \tag{14.3.11}$$

is independent of the location $\boldsymbol{R}$ and the order of the operators.

---

23 In quantum systems these operators will not, in general, commute, but here we are concerned only with classical systems.

Before $\mathrm{D}(\mathring{\mathrm{e}}, \mathring{\mathrm{k}} \mid \boldsymbol{R})$ acts on $\ln\{Z_{\square}(\mathcal{N}, \{w(\boldsymbol{r}, \boldsymbol{r}')\})\}$ in (14.3.11) all the variables $w(\boldsymbol{r}, \boldsymbol{r}')$ not corresponding to the edges of (e, k) can be set to zero. Let $Z_{\square}^{(\mathrm{e},\mathrm{k})}(\mathcal{N}, \{w(\boldsymbol{r}, \boldsymbol{r}')\})$ be the partition function for which this has been done. We now consider the implications of this in the low- and high-temperature cases.

14.3.1 The Low-Temperature Case

It follows, from (14.2.3), that all terms in the sum over $\boldsymbol{\sigma}$ are zero except those for which the two sites at the ends of a pair not included in $\mathrm{D}(\mathring{\mathrm{e}}, \mathring{\mathrm{k}} \mid \boldsymbol{R})$ have the same value for σ, which we take to be the ground state $\sigma = 0$. This means that all sites of (e, k) connected by a edge to an exterior site must be in the ground state and the subset of those not in the ground state must form a section graph. It follows that (e, k) must be of the type for which, in Sect. 14.2.3 and at the beginning of this section, we have used the designation $\mathfrak{g}^{(+)}$, that is a section graph $\mathfrak{g}$ to which has been added all the edges connecting it to exterior sites, together with the terminating (ground state) sites on these edges, which we call the *boundary sites of* $\mathfrak{g}^{(+)}$. The partition function $Z_{\mathrm{LT}}(\mathcal{N}, \{u(\boldsymbol{r}, \boldsymbol{r}')\}, p)$ is reduced to

$$Z_{\mathrm{LT}}((\mathrm{e}, \mathrm{k}), \{u(\boldsymbol{r}, \boldsymbol{r}')\}, p) = \sum_{\{\sigma\}} \left\{ \prod_{\{\boldsymbol{r},\boldsymbol{r}'\}}^{(\mathrm{e},\mathrm{k})} u(\boldsymbol{r}, \boldsymbol{r}')^{1-\delta^{\mathrm{Kr}}(\sigma(\boldsymbol{r})-\sigma(\boldsymbol{r}'))} \right\}$$
$$\times \left\{ \prod_{\{\boldsymbol{r}\}}^{(\mathrm{e},\mathrm{k})} (1-p)^{1-\delta^{\mathrm{Kr}}(\sigma(\boldsymbol{r}))} \right\}, \tag{14.3.12}$$

where the summation is now restricted to the sites of the graph $\mathfrak{g} := [\mathfrak{n}, \mathrm{k}]$ which is obtained from $\mathfrak{g}^{(+)} := (\mathrm{e}, \mathrm{k})$ by excluding the boundary sites and a non-zero contribution to $\mathrm{C}(\mathring{\mathrm{e}}, \mathring{\mathrm{k}})$ will occur only from terms of the summation for which each edge is occupied by unlike states.

14.3.2 The High-Temperature Case

Here the partition function summations are performed over the exterior sites and, from (14.2.30), $Z_{\mathrm{HT}}(\mathcal{N}, \{v(\boldsymbol{r}, \boldsymbol{r}')\}, q)$ is reduced to

$$Z_{\mathrm{HT}}((\mathrm{e}, \mathrm{k}), \{v(\boldsymbol{r}, \boldsymbol{r}')\}, q) = \nu^{-\mathfrak{n}(\mathrm{e},\mathrm{k})}$$
$$\times \sum_{\{\sigma\}} \left\{ \prod_{\{\boldsymbol{r},\boldsymbol{r}'\}}^{(\mathrm{e},\mathrm{k})} [1 + \Delta(\nu; \sigma(\boldsymbol{r}) - \sigma(\boldsymbol{r}'))v(\boldsymbol{r}, \boldsymbol{r}')] \prod_{\{\boldsymbol{r}\}}^{(\mathrm{e},\mathrm{k})} [1 + \Delta(\nu; \sigma(\boldsymbol{r}))q] \right\}, \tag{14.3.13}$$

where the summation is over all the sites of (e,k), with a non-zero contribution to $\mathsf{C}(\mathring{\mathsf{e}},\mathring{\mathsf{k}})$, occurring if the corresponding product of Potts Δ-functions is non-zero.

It follows in both the low- and high-temperature cases that, if (e,k) consists of a number of components, the partition function is just a product of those for each component. The logarithm of the partition function is just the corresponding sum of partition functions and operation with $\mathcal{D}(\mathring{\mathsf{e}},\mathring{\mathsf{k}}\,|\,\boldsymbol{R})$ yields the result of zero. Thus $\mathsf{C}(\mathring{\mathsf{e}},\mathring{\mathsf{k}})=0$ if $(\mathring{\mathsf{e}},\mathring{\mathsf{k}})$ is a disconnected graph.

14.4 The Finite-Cluster Method

We now re-express (14.3.10) by denoting a multi-bonded graph by $\mathring{\mathfrak{g}}$ and generalizing $\mathsf{S}(\mathfrak{g}\,|\,\mathring{\mathfrak{g}})$ so that it has the same meaning as before if $\mathfrak{g}$ is the silhouette of $\mathring{\mathfrak{g}}$ and is zero otherwise. Then

$$\Phi_{\square}(\mathcal{N},w)=-\sum_{\{\mathfrak{g}\}}\Omega(\mathfrak{g})\Psi_{\square}(\mathfrak{g},w),\tag{14.4.1}$$

where

$$\Psi_{\square}(\mathfrak{g},w):=\sum_{\{\mathring{\mathfrak{g}}\}}w^{\mathsf{e}(\mathring{\mathfrak{g}})}\mathsf{S}(\mathfrak{g}\,|\,\mathring{\mathfrak{g}})\mathsf{C}(\mathring{\mathfrak{g}})\mathsf{A}(\mathring{\mathfrak{g}}).\tag{14.4.2}$$

Formula (14.4.1) is the *cluster expansion* of the free energy and the cluster function $\Psi_{\square}(\mathfrak{g},w)$ is an infinite series in w, starting with the term $w^{\mathsf{e}(\mathfrak{g})}\mathsf{C}(\mathfrak{g})$, and summing over the cluster of graphs which have the same silhouette $\mathfrak{g}$. If the aim is to obtain a low- or high-temperatures series expansion of $\Phi_{\square}(\mathcal{N},w)$ correct up to and including the terms of degree $\mathsf{e}_{\max}$ in w then graphs can be excluded from (14.4.1) for which $\mathsf{e}(\mathfrak{g})>\mathsf{e}_{\max}$. Thus we can replace (14.4.1) by

$$\overline{\Phi_{\square}(\mathcal{N},w)}^{\,\mathsf{e}_{\max}}:=-\sum_{\{\mathfrak{g}\}}^{\mathsf{e}(\mathfrak{g})\le\mathsf{e}_{\max}}\Omega(\mathfrak{g})\Psi_{\square}(\mathfrak{g},w)\tag{14.4.3}$$

and the task is to obtain the set of cluster functions which appear in this equation. Of course, it is now possible, in principle, to proceed by calculating the terms in (14.4.2) up to degree $\mathsf{e}_{\max}$ in w for each of the cluster functions in (14.4.3).

However, the finite-cluster method of Domb (1960, p. 322; 1974b, pp. 73–76) takes a different approach. The starting point is the observation that (14.4.1) can be generalized by replacing the lattice $\mathcal{N}$ by any graph $\mathfrak{g}$ to give

$$\Phi_{\square}(\mathfrak{g},w)=-\sum_{\{\mathfrak{g}'\}}\Omega(\mathfrak{g}'\,|\,\mathfrak{g})\Psi_{\square}(\mathfrak{g}',w),\tag{14.4.4}$$

which is an explicit realization of Theorem 16.4.3. Consider now the set $\mathcal{S}$ of all graphs $\mathfrak{g}$ such that $\mathsf{e}(\mathfrak{g}) \le \mathsf{e}_{\max}$ and suppose that $\mathcal{S}$ contains n members partially ordered, as described in Sect. 16.4.5, in the sequence $\mathfrak{g}_1, \mathfrak{g}_2, \dots, \mathfrak{g}_n$; that is to say, if $\mathfrak{g}_i \subset \mathfrak{g}_j$ then $i < j$. If $\mathfrak{g}$ in (14.4.4) is a member $\mathfrak{g}_i \in \mathcal{S}$ then the set of graphs in the summation on the right-hand side will consist only of members of $\mathcal{S}$ with $j \le i$. Thus, by varying the graph $\mathfrak{g}$ in (14.4.4) over the members of $\mathcal{S}$, a set of n linear equations in the n cluster functions can be obtained. This set can be expressed in the form

$$[\boldsymbol{\Phi}_{\square}^{(n)}]^{\mathrm{T}} = -[\boldsymbol{\Psi}_{\square}^{(n)}]^{\mathrm{T}}\boldsymbol{\Omega}^{(n)}, \tag{14.4.5}$$

with

$$\boldsymbol{\Phi}_{\square}^{(n)} := \begin{pmatrix} \Phi_{\square}(\mathfrak{g}_1, w) \\ \Phi_{\square}(\mathfrak{g}_2, w) \\ \vdots \\ \Phi_{\square}(\mathfrak{g}_n, w) \end{pmatrix}, \qquad \boldsymbol{\Psi}_{\square}^{(n)} := \begin{pmatrix} \Psi_{\square}(\mathfrak{g}_1, w) \\ \Psi_{\square}(\mathfrak{g}_2, w) \\ \vdots \\ \Psi_{\square}(\mathfrak{g}_n, w) \end{pmatrix}, \tag{14.4.6}$$

where

$$\boldsymbol{\Omega}^{(n)} := \begin{pmatrix} \Omega_{11} & \Omega_{12} & \cdots & \Omega_{1\,n-1} & \Omega_{1\,n} \\ 0 & \Omega_{22} & \cdots & \Omega_{2\,n-1} & \Omega_{2\,n} \\ 0 & 0 & \cdots & \Omega_{3\,n-1} & \Omega_{3\,n} \\ 0 & 0 & \vdots & \vdots & \vdots \\ 0 & 0 & \vdots & 0 & \Omega_{n\,n} \end{pmatrix}, \quad \text{with } \Omega_{ji} := \Omega(\mathfrak{g}_j \,|\, \mathfrak{g}_i), \tag{14.4.7}$$

is the T matrix (Sect. 16.4.5). From (14.4.3),

$$\overline{\phi_{\square}(\mathcal{N}, w)}^{\mathsf{e}_{\max}} = -[\boldsymbol{\Psi}_{\square}^{(n)}]^{\mathrm{T}}\boldsymbol{\omega}^{(n)}, \tag{14.4.8}$$

$\boldsymbol{\omega}^{(n)}$ is the vector of lattice constants for $\mathcal{S}$ and, from (14.4.5),

$$\overline{\phi_{\square}(\mathcal{N}, w)}^{\mathsf{e}_{\max}} = [\boldsymbol{\Phi}_{\square}^{(n)}]^{\mathrm{T}}\boldsymbol{\chi}^{(n)} = \sum_{i=1}^{n} \chi_j^{(n)} \Phi_{\square}(\mathfrak{g}_j, w), \tag{14.4.9}$$

where, as in Sects. 7.2 and 16.4.5, the weights $\chi_j^{(n)}$, $j = 1, 2, \dots, n$, of the sequence $\mathcal{S}$, are defined by

$$\boldsymbol{\chi}^{(n)} := [\boldsymbol{\Omega}^{(n)}]^{-1}\boldsymbol{\omega}^{(n)}. \tag{14.4.10}$$

Example 14.4.1 To calculate the high-temperature series on the square lattice up to v^4 the sequence $\mathcal{S}$ has five members of which the first four $\mathfrak{g}_1, \mathfrak{g}_2, \mathfrak{g}_3, \mathfrak{g}_4$ are simple chains of lengths of one to four edges respectively, shown in Fig. 16.1, and $\mathfrak{g}_5$ is the simple cycle (4, 2) shown in Table 16.2.[24] From Table 16.1 and (16.4.14)

$$\boldsymbol{\Omega}^{(5)} := \begin{pmatrix} 1 & 2 & 3 & 4 & 4 \\ 0 & 1 & 2 & 3 & 4 \\ 0 & 0 & 1 & 2 & 4 \\ 0 & 0 & 0 & 1 & 0 \\ 0 & 0 & 0 & 0 & 1 \end{pmatrix}, \qquad \boldsymbol{\omega}^{(5)} := \begin{pmatrix} 2 \\ 6 \\ 18 \\ 50 \\ 1 \end{pmatrix}. \tag{14.4.11}$$

Thus, from (14.4.9) and (14.4.10),

$$\overline{\phi_{\text{HT}}(\mathcal{N}, v, q)}^{4} = 8\,\Phi_{\text{HT}}(\mathfrak{g}_1, v, q) + 24\,\Phi_{\text{HT}}(\mathfrak{g}_2, v, q) - 86\,\Phi_{\text{HT}}(\mathfrak{g}_3, v, q) + 50\,\Phi_{\text{HT}}(\mathfrak{g}_4, v, q) + \Phi_{\text{HT}}(\mathfrak{g}_5, v, q). \tag{14.4.12}$$

The high-temperature partition functions for the members of $\mathcal{S}$ are given by (14.3.13) with all the parameters $v(\boldsymbol{r}, \boldsymbol{r}')$ replaced by v. The easiest way to obtain them in explicit form is to expand the brackets and use the properties of the Potts Δ-function given in Sect. 18.4 and in particular in (18.4.18) and (18.4.20). Then

$$\Phi_{\text{HT}}(\mathfrak{g}_1, v, q) = -\ln[1 + vq^2(\nu - 1)], \tag{14.4.13}$$

$$\Phi_{\text{HT}}(\mathfrak{g}_2, v, q) = -\ln[1 + v(2+v)q^2(\nu - 1) + v^2q^3(\nu - 1)(\nu - 2)], \tag{14.4.14}$$

$$\Phi_{\text{HT}}(\mathfrak{g}_3, v, q) = -\ln[1 + (3 + 2v + v^2)vq^2(\nu - 1) + 2(1+v)v^2q^3(\nu - 1)(\nu - 2) + v^3q^4(\nu - 1)(\nu - 2)^2 + v^2q^4(\nu - 1)^2], \tag{14.4.15}$$

$$\Phi_{\text{HT}}(\mathfrak{g}_4, v, q) = -\ln[1 + (4 + 3v + 2v^2 + v^3)vq^2(\nu - 1) + (3 + 4v + 3v^2)v^2q^3(\nu - 1)(\nu - 2) + (2 + 3v)v^3q^4(\nu - 1)(\nu - 2)^2 + q^5v^4(\nu - 1)(\nu - 2)^3 + 3v^2q^4(\nu - 1)^2 + 2v^3q^5(\nu - 1)^2(\nu - 2)], \tag{14.4.16}$$

[24] All chains of the same length but different topologies can be grouped together (e.g. graphs (3, 1)–(3, 4) in Table 16.1), since they have the same free energies and cluster functions.

$$\begin{aligned}\Phi_{\mathrm{HT}}(\mathfrak{g}_5, v, q) &= -\ln[1 + 4(1+v+v^2)vq^2(\nu-1)\\ &+ 4(1+2v)v^2q^3(\nu-1)(\nu-2) + 4v^3q^4(\nu-1)(\nu-2)^2\\ &+ q^4v^4(\nu-2)\{(\nu-2)^3+1\} + 4q^3v^4(\nu-2)\{(\nu-2)^2-1\}\\ &+ 6q^2v^4(\nu-2)(\nu-1) + v^4(\nu-1) + 2v^2q^4(\nu-1)^2]. \qquad (14.4.17)\end{aligned}$$

Expanding these free energies to $\mathrm{O}(v^4)$ and substituting in (14.4.12) gives, for the quantities defined in (14.2.33),

$$\phi^{(0)}_{\mathrm{HT}}(\mathcal{N}, v) = -(\nu-1)v^3 + \mathrm{O}(v^5), \qquad (14.4.18)$$

$$\phi^{(2)}_{\mathrm{HT}}(\mathcal{N}, v) = -2(\nu-1)[v + 3v^2 + 9v^3 + (3\nu+19)v^4] + \mathrm{O}(v^5). \qquad (14.4.19)$$

As expected, from duality, (14.4.18) agrees with (14.2.21).[25] There is, of course, no agreement (and none to be expected) between (14.4.19) and (14.2.23).

A more general version of the model in Example 14.4.1, for a range of lattices and with different 'chemical potentials' for each Potts state, was investigated by Kim and Joseph (1975) and coefficients for the square lattice were obtained up to v^8.[26] The finite-cluster method has also be used to obtain high-temperature expansions in terms of the coupling K. A method for deriving partition functions of finite clusters of spins in the classical Heisenberg model was developed by Joyce (1967) and used by Joyce and Bowers (1966) to calculate high-temperature K-expansions on various lattices. Joyce's method was extended by Domb (1976) to give a systematic procedure for calculating the partition functions of star topologies in the n-vector model. This work was used by English et al. (1979) to obtain high-temperature series to K^{12} for general n and to K^{13} for the ($n = 3$) classical Heisenberg model.

14.5 The Finite-Lattice Method

Since the finite-lattice method is very similar to the finite-cluster method it is useful to make a comparison between them:

(i) A *finite cluster* is just a graph. A *finite lattice* is "a cluster that displays certain characteristics of the infinite lattice" (de Neef 1975, p. 25). In particular this means that a finite lattice is a section graph. In the case where $\mathcal{N}$ is square or cubic, the finite lattice is a square or cubic block and for other lattices this result is also achieved by moving the positions of the sites while retaining the first-neighbour connectivity.

(ii) Equation (14.4.9) is modified slightly to

---

[25] The comparative lengths of the series, illustrates the greater difficulty in obtaining high-temperature series, something which we shall see again with the finite-lattice method.

[26] A different expansion variable, equivalent to $\nu\, v/(1-v)$, was used in their calculations.

$$\overset{\mathfrak{e}_{\max}}{\overline{\phi_{\square}(\mathcal{N}, w)}} = \frac{1}{n_c} \sum_{i=1}^{n} \chi_j^{(n)} \Phi_{\square}(\mathfrak{g}_j, w), \tag{14.5.1}$$

where the partially-ordered sequence $\mathcal{S}$ of graphs $\mathfrak{g}_1, \mathfrak{g}_2, \ldots, \mathfrak{g}_n$ is now generated by the overlap procedure described in Sect. 16.4.6 and the factor n_c is included to make provision for the modification of the method to different lattices. In particular, as we shall see, for the honeycomb lattice the basic unit from which the graphs are constructed is a cell of two sites. Thus to obtain the correct free energy *per lattice site* $n_c = 2$. In Theorem 16.4.4 we showed that the sequence is complete, meaning that up to $\mathfrak{g}_n$ there are no missing contributions to the sum in (14.5.1).

(iii) In the *finite-cluster method* the mode of truncation of the series is determined by the cluster functions. In the *finite-lattice method* we need to develop an independent criterion for the selection of the generating set $\mathfrak{S}$ for $\mathcal{S}$ so that the expansion is correct up to a certain specified degree $\mathfrak{e}_{\max}$ in w.

14.5.1 Block-Formation and Accuracy

The crucial element in determining the accuracy of the series expansion arising from (14.5.1) is the choice of the generating set $\mathfrak{S}$ for $\mathcal{S}$. As indicated above the primary application of the finite-lattice method is to models on the square lattice[27] and we shall first consider block-formation and series accuracy in this case before discussing the modifications needed for the triangle and honeycomb lattices.

14.5.1.1 The Square Lattice

The set $\mathfrak{S}$ consists of rectangular blocks of sites ($n_c = 1$) and the members of $\mathcal{S}$ generated by overlap are also rectangular blocks. The sequence is complete. So if every graph on the lattice with $\mathfrak{e} \leq \mathfrak{e}_{\max}$ is a subset of a member of $\mathcal{S}$ then the expansion of the expression on the right-hand side of (14.5.1) will have the required accuracy specified by the left-hand side. The type of blocks and the computation of accuracy depend on the nature of the graphs which feature in the series expansion and since these differ between the low- and high temperature cases they must be considered separately.

The low-temperature case:

Here a relevant graph $\mathfrak{g}$ is a section graph, to which has been attached all the edges joining it to its environment together with their terminating sites, which are in the

[27] For which the argument can easily be extended to the simple cubic lattice.

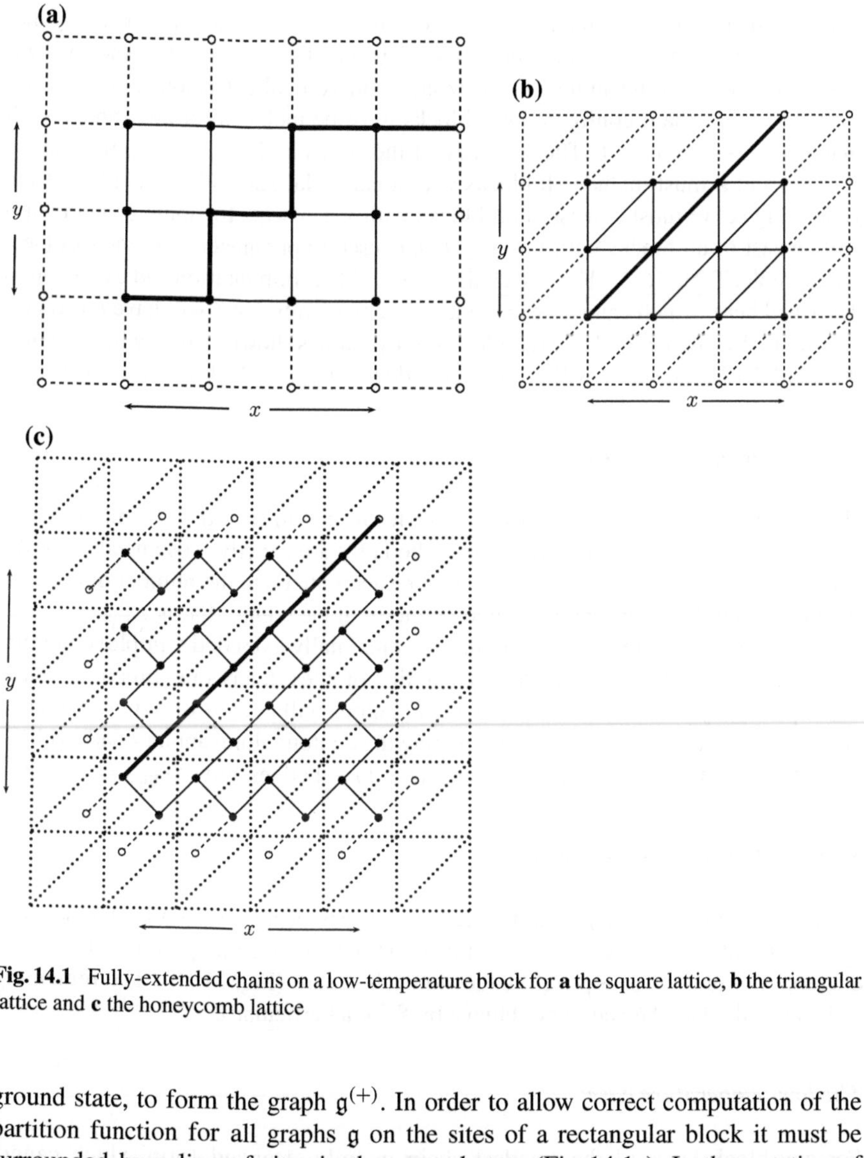

Fig. 14.1 Fully-extended chains on a low-temperature block for **a** the square lattice, **b** the triangular lattice and **c** the honeycomb lattice

ground state, to form the graph $\mathfrak{g}^{(+)}$. In order to allow correct computation of the partition function for all graphs $\mathfrak{g}$ on the sites of a rectangular block it must be surrounded by a line of sites in the ground state (Fig. 14.1a). In the notation of Sect. 16.4.6, a block is denoted by $[x \times y]$, where x and y are respectively the number of horizontal and vertical sites *excluding the surrounding ground-state sites* and the partition function for this block is $Z_{\mathrm{LT}}([x \times y], u, \mathfrak{Z})$. The power of u in $Z_{\mathrm{LT}}([x \times y], u, \mathfrak{Z})$ contributed by the graph $\mathfrak{g}$ is given by (14.2.12) (with $\mathrm{z} = 4$). We need to calculate the lowest power for the most fully-extended graph which cannot be accommodated in the block. This means that the graph must have only one component ($\mathrm{c} = 1$) have all its sites in the same state ($\mathfrak{i} = 0$) and be a chain ($\mathfrak{C} = 0$). The chain-configurations to be considered are then those which start at the bottom left site and are extended on the block in a most extreme form, meaning that they cross any vertical (or horizontal) line of the block exactly once. If such a chain has $\mathfrak{n}$ vertices then the sum of its horizontal and vertical extensions is $\mathfrak{n} - 1$. So, for all possible chain-configurations, blocks must be included for all x and y such that $x + y - 2 = \mathfrak{n} - 1$. The accuracy of the series is thus related to the sum of x and y and $\mathcal{S}$ must include all blocks with a particular sum. We conclude that the generating set $\mathfrak{G}$ must consist of all blocks with $x + y = \ell$, for some ℓ, and then $\mathcal{S}$ will consist of all blocks with $x + y \leq \ell$; ℓ is used to parameterize sequences (see, for example, Table 16.3). The first chain graph, with a configuration and extension as described here, which cannot be embedded in any member of $\mathcal{S}$ will have ℓ vertices. (In Fig. 14.1a where $\ell = 7$ an example of such a chain is shown by heavy lines.) Thus, from (14.2.12), the series will be correct to degree $\mathfrak{e}_{\max} = 2\ell + 2 - 1 = 2\ell + 1$.[28]

***The high-temperature case*:**

Here we must distinguish two cases, the series in v with $\mu = q = 0$ and the series of terms in v which contain a factor q^2. The latter are exactly those needed for the response function series and the series will be accurate up to the number of edges in the most extended chain with a single-site term at each end; that is $e_{\max} = \ell - 2$. For the zero-μ series the limiting graphs are maximally extended (simple) contours. Suppose that the block is extended by one line of sites in any one direction, that is ℓ is replaced by $\ell + 1$. The most extended contour will consist of a most extended chain followed by a second chain completing the contour along another path without any intersections. This contour will have $2(\ell + 1) - 4$ edges and so $e_{\max} = 2\ell - 3$.[29]

14.5.1.2 The Triangular Lattice

Here the argument is very similar, again $\mathcal{S}$ consists of all rectangular blocks with $x + y \leq \ell$ and $n_c = 1$, the topology of the triangular lattice being produced by adding diagonal bottom-left to top-right edges to each square (Fig. 14.1b). The remaining task is to calculate the accuracy achieve by $\mathcal{S}$ for a particular ℓ.

***The low-temperature case*:**

For any block $[x \times y]$ the smallest chain in fully-extended form which cannot be embedded lies along the diagonal edges starting at the bottom left. It cannot be embedded in the block if it has more that the lesser of x and y sites. Applying this to the whole generating set $\mathfrak{G}$, for ℓ even the smallest non-embeddable chain of this type has $\ell/2 + 1$ vertices, and for ℓ odd it has $(\ell + 1)/2$ vertices. (In Fig. 14.1b

[28] For a d-dimensional hypercubic lattice the formula is $\mathfrak{e}_{\max} = 2(d-1)(\ell + 2 - d) + 1$, with the meaning of ℓ generalized in the obvious way.

[29] For a d-dimensional hypercubic lattice the respective formulae are $\mathfrak{e}_{\max} = \ell - d$ and $\mathfrak{e}_{\max} = 2\ell + 1 - 2d$.

where $\ell = 7$ an example of such a chain is shown by heavy lines.) Thus, from (14.2.12) (with $z = 6$), $e_{max} = 2\ell + 5$ if ℓ is even and $2\ell + 3$ if ℓ is odd.

The high-temperature case:

As for the square lattice we must distinguish two cases, the series in v with $\mu = q = 0$ and the series of terms in v which contain a factor q^2. For the latter the series will be accurate up to the most extended chain with a single-site term at each end; that is $e_{max} = (\ell - 2)/2$ if ℓ is even and $(\ell - 3)/2$ if ℓ is odd. For the zero-μ we must consider the maximally extended (simple) contours. Suppose that the block is extended by one line of sites in any one direction, that is ℓ is replaced by $\ell + 1$. If ℓ is odd the most extended chain is along the diagonal which has $(\ell - 1)/2$ edges followed by a return along the subdiagonal which has $(\ell - 3)/2$ edges plus the two connecting edges; thus $e_{max} = \ell - 1$. If ℓ is even the block of size $\ell + 1$ has two diagonals with $\ell/2 - 1$ edges and connecting these with two edges gives again $e_{max} = \ell - 1$.

14.5.1.3 The Honeycomb Lattice

The way that the honeycomb lattice is 'forced' into rectangular form is motivated by the triangle-honeycomb duality. Taking the rectangularized form of the triangular lattice, used in Fig. 14.1b and reproduced as dotted lines in Fig. 14.1c, a honeycomb lattice site is placed in each face. Thus the division of the lattice is into cells each containing two sites and x and y now represent respectively the number of horizontal and vertical cells with $\ell = x + y$, as before, but with $n_c = 2$.

The low-temperature case:

For any block $[x \times y]$ the smallest chain in fully-extended form which cannot be embedded lies along one of the two diagonal edges starting in the bottom-left cell. Either of these diagonals can be chosen if, as in Fig. 14.1c $x = y$. If $x < y$ the lower diagonal must be chosen and the smallest non-embeddable chain has $2x$ vertices; if $y < x$ the upper diagonal must be chosen and the smallest non-embeddable chain has $2y$ vertices. Applying this to the whole set $\mathfrak{S}$, the smallest non-embeddable chain of this type has ℓ vertices for ℓ even and $\ell - 1$ vertices if ℓ is odd. (In Fig. 14.1c where $\ell = 8$ an example of such a chain is shown by heavy lines.) Thus, from (14.2.12) (with $z = 3$), $e_{max} = \ell + 1$ if ℓ is even and ℓ if ℓ is odd.

The high-temperature case:

The series terms in v which contain a factor q^2 will be accurate up to the most extended chain with a single-site term at each end; that is $e_{max} = \ell - 2$ if ℓ is even and $\ell-3$ if ℓ is odd. For zero-μ the maximally extended (simple) contour will contain the two diagonal edges of length $\ell-2$ if ℓ is even and $\ell-3$ if ℓ is odd, starting in the bottom-left cell, and the two edges connecting them. Thus $\mathfrak{e}_{\max} = 2\ell - 2$ if ℓ is even and $2\ell - 4$ if ℓ is odd.

14.5.2 Constructing Block Partition Functions

As we have seen in Sect. 14.5.1 a graph in the sequence $\mathcal{S}$ has sites in an $x \times y$ rectangular block of cells. In the case of the square and triangular lattices each cell contains one site but in the case of the honeycomb lattice it contains two sites, an upper-left site and a lower-right site (Fig. 14.1).

Let $\widetilde{Z}_\square(\ell, w)$ be the product of all partition functions $Z_\square([x \times y], w)$ for graphs in the sequence $\mathcal{S}$ for which $x + y = \ell$. Then, from (14.5.1) and (16.4.47),

$$\overset{\mathfrak{e}_{\max}}{\overline{\phi_\square(\mathcal{N}, w)}} = -\ln[\mathcal{Z}_\square(\ell, w)], \tag{14.5.2}$$

where

$$\mathcal{Z}_\square(\ell, w) := \left[\frac{\widetilde{Z}_\square(\ell-2, w)}{\widetilde{Z}_\square(\ell-1, w)}\right]^3 \frac{\widetilde{Z}_\square(\ell, w)}{\widetilde{Z}_\square(\ell-3, w)} \tag{14.5.3}$$

is an approximation to the Nth root of the partition function.

The great advantage of the finite-lattice method based on rectangular blocks of sites is that, for a fixed y, the partition functions with increasing x can be generated using the transfer matrix methods described in Chap. 11. Here there are two differences from the discussion there and in Chap. 12. Now the slices of the lattice are columns of y sites with the lattice growing by increasing in the horizontal direction the number of columns x. Also there are no periodic boundary conditions in the vertical direction. In the case of the low-temperature series the lattice is of width $x + 2$ with the first and last columns fixed in the ground state $\sigma = 0$. In the same way each column has $y + 2$ sites with the top and bottom cells fixed in the ground state. In the case of the high-temperature series there are x columns of height y with free boundary conditions. The number of states of a column, for both the low- and high-temperature cases, is ν^y for the square and triangular lattices and ν^{2y} for the honeycomb lattice. From (11.1.9), with N_1 and N_2 replaced respectively by x and y,

$$Z_\square([x \times y], w) = \tau[z_{x-1}]^{\mathrm{T}} \boldsymbol{b}^{(\mathrm{L})}, \tag{14.5.4}$$

where,

$$[z_0]^{\mathrm{T}} := [\boldsymbol{b}^{(\mathrm{F})}]^{\mathrm{T}}, \qquad [z_x]^{\mathrm{T}} = [z_{x-1}]^{\mathrm{T}} \boldsymbol{V} \tag{14.5.5}$$

and, in (14.5.4), for the low-temperature case, $\tau = 1$. For the high-temperature case, following (14.2.30), $\tau = 1/\nu^{xy}$ for the square and triangular lattices and $\tau = 1/\nu^{2xy}$ for the honeycomb lattice.

As in Chap. 11, the transfer matrix contains the interaction between two columns of sites and half the interactions within a column. Explicitly,

$$\boldsymbol{V} = \boldsymbol{V}_{\mathrm{w}}^{\frac{1}{2}} \boldsymbol{V}_{\mathrm{b}} \boldsymbol{V}_{\mathrm{w}}^{\frac{1}{2}}, \tag{14.5.6}$$

where $\boldsymbol{V}_{\mathrm{w}}$ is a diagonal matrix. In a similar way the boundary vectors $\boldsymbol{b}^{(\mathrm{F})}$ and $\boldsymbol{b}^{(\mathrm{L})}$ contain half the interactions within the first and last columns respectively and, for the low-temperature case, the extra two columns are included in these vectors. Thus

$$\boldsymbol{b}^{(\mathrm{F})} = \boldsymbol{b}^{(\mathrm{L})} = \boldsymbol{V}_{\mathrm{w}}^{\frac{1}{2}} \boldsymbol{b}_{\square}, \tag{14.5.7}$$

where

$$\boldsymbol{b}_{\mathrm{LT}} = \boldsymbol{V}_{\mathrm{b}} \boldsymbol{V}_{\mathrm{w}} \boldsymbol{i}_0, \qquad \boldsymbol{b}_{\mathrm{HT}} = \boldsymbol{j}, \tag{14.5.8}$$

$\boldsymbol{i}_0$ is the unit vector with one in the first element, fixing the first and last columns in the ground state, and $\boldsymbol{j}$ is the vector with every element of unity, enabling free boundary conditions in first and last columns.

For the square and triangular lattices we denote by $\boldsymbol{\sigma}$ the state of a column, where $[\boldsymbol{\sigma}]^{\mathrm{T}} := (\sigma_1, \sigma_2, \ldots, \sigma_y)$, with the labelling proceeding from the top to the bottom (excluding in the low-temperature case the boundary sites with $\sigma = 0$). For the honeycomb lattices we again denote by $\boldsymbol{\sigma}$ the state of a column, but now $[\boldsymbol{\sigma}]^{\mathrm{T}} := (\sigma_1, \tilde{\sigma}_1, \sigma_2, \tilde{\sigma}_2, \ldots, \sigma_y, \tilde{\sigma}_y)$, where σ_k and $\tilde{\sigma}_k$ are the states on the top-left and bottom-right sites in the kth cell. We now list the formulae for the elements of $\boldsymbol{V}_{\mathrm{w}}$ and $\boldsymbol{V}_{\mathrm{b}}$ for the low- and high-temperature cases and for the square, triangular and honeycomb lattices.

14.5.2.1 The Square Lattice

From (14.2.3) and (14.2.30) the diagonal elements of $\boldsymbol{V}_{\mathrm{w}}$ are

$$(\boldsymbol{V}_{\mathrm{w}})_{\boldsymbol{\sigma}\boldsymbol{\sigma}} = u^{1-\delta^{\mathrm{Kr}}(\sigma_1)} \left\{ \prod_{k=1}^{y-1} u^{1-\delta^{\mathrm{Kr}}(\sigma_k - \sigma_{k+1})} \right\} \left\{ \prod_{k=1}^{y} (1-p)^{1-\delta^{\mathrm{Kr}}(\sigma_k)} \right\} u^{1-\delta^{\mathrm{Kr}}(\sigma_y)} \tag{14.5.9}$$

for the low-temperature case, and

$$(\boldsymbol{V}_{\mathrm{w}})_{\boldsymbol{\sigma}\boldsymbol{\sigma}} = \left\{ \prod_{k=1}^{y-1} [1 + \Delta(\nu; \sigma_k - \sigma_{k+1}) v] \right\} \left\{ \prod_{k=1}^{y} [1 + \Delta(\nu; \sigma_k) q] \right\} \tag{14.5.10}$$

for the high-temperature case. The elements of $\boldsymbol{V}_{\mathrm{b}}$ are

$$(V_{\mathrm{b}})_{\boldsymbol{\sigma}\boldsymbol{\sigma}'} = \prod_{k=1}^{y} u^{1-\delta^{\mathrm{Kr}}(\sigma_k-\sigma'_k)} \tag{14.5.11}$$

for the low-temperature case, and

$$(V_{\mathrm{b}})_{\boldsymbol{\sigma}\boldsymbol{\sigma}'} = \prod_{k=1}^{y} [1 + \Delta(\nu; \sigma_k - \sigma'_k)v] \tag{14.5.12}$$

for the high-temperature case.

14.5.2.2 The Triangular Lattice

Here, as can be seen from Fig. 14.1a and b, the elements of $\boldsymbol{V}_{\mathrm{w}}$ are the same as those for the square lattice, but the elements of $\boldsymbol{V}_{\mathrm{b}}$ contain an additional diagonal interaction giving

$$(V_{\mathrm{b}})_{\boldsymbol{\sigma}\boldsymbol{\sigma}'} = u^{1-\delta^{\mathrm{Kr}}(\sigma_1)} \left\{ \prod_{k=1}^{y} u^{1-\delta^{\mathrm{Kr}}(\sigma_k-\sigma'_k)} \right\} \left\{ \prod_{k=2}^{y} u^{1-\delta^{\mathrm{Kr}}(\sigma_k-\sigma'_{k-1})} \right\} u^{1-\delta^{\mathrm{Kr}}(\sigma'_y)} \tag{14.5.13}$$

for the low-temperature case, and

$$(V_{\mathrm{b}})_{\boldsymbol{\sigma}\boldsymbol{\sigma}'} = \left\{ \prod_{k=1}^{y} [1 + \Delta(\nu; \sigma_k - \sigma'_k)v] \right\} \left\{ \prod_{k=2}^{y} [1 + \Delta(\nu; \sigma_k - \sigma'_{k-1})v] \right\} \tag{14.5.14}$$

for the high-temperature case.

14.5.2.3 The Honeycomb Lattice

In this case the elements of both $\boldsymbol{V}_{\mathrm{w}}$ and $\boldsymbol{V}_{\mathrm{b}}$ are more complicated because of the two sites in each cell. However, by looking a Fig. 14.1c, it is not difficult to see that

$$\begin{aligned}(V_{\mathrm{w}})_{\boldsymbol{\sigma}\boldsymbol{\sigma}} &= u^{1-\delta^{\mathrm{Kr}}(\sigma_1)} \left\{ \prod_{k=1}^{y-1} u^{2-\delta^{\mathrm{Kr}}(\sigma_k-\tilde{\sigma}_k)-\delta^{\mathrm{Kr}}(\tilde{\sigma}_k-\sigma_{k+1})} \right\} \\ &\times \left\{ \prod_{k=1}^{y} (1-p)^{2-\delta^{\mathrm{Kr}}(\sigma_k)-\delta^{\mathrm{Kr}}(\tilde{\sigma}_k)} \right\} u^{2-\delta^{\mathrm{Kr}}(\sigma_y-\tilde{\sigma}_y)-\delta^{\mathrm{Kr}}(\tilde{\sigma}_y)}\end{aligned} \tag{14.5.15}$$

for the low-temperature case,

$$(V_{\mathrm{w}})_{\boldsymbol{\sigma}\boldsymbol{\sigma}} = \prod_{k=1}^{y-1}[1 + \Delta(\nu; \sigma_k - \tilde{\sigma}_k)v][1 + \Delta(\nu; \tilde{\sigma}_k - \sigma_{k+1})v]$$
$$\times\ [1 + \Delta(\nu; \sigma_y - \tilde{\sigma}_y)v]\left\{\prod_{k=1}^{y}[1 + \Delta(\nu; \sigma_k)q][1 + \Delta(\nu; \tilde{\sigma}_k)q]\right\} \tag{14.5.16}$$

for the high-temperature case,

$$(V_{\mathrm{b}})_{\boldsymbol{\sigma}\boldsymbol{\sigma}'} = \prod_{k=1}^{y} u^{1-\delta^{\mathrm{Kr}}(\tilde{\sigma}_k - \sigma'_k)} \tag{14.5.17}$$

for the low-temperature case, and

$$(V_{\mathrm{b}})_{\boldsymbol{\sigma}\boldsymbol{\sigma}'} = \prod_{k=1}^{y}[1 + \Delta(\nu; \tilde{\sigma}_k - \sigma'_k)v] \tag{14.5.18}$$

for the high-temperature case.

14.5.3 Calculating the Series

Using the results displayed above it now possible to obtain series for the terms of the free-energy density to the specified level of accuracy. Techniques for doing this were developed by Enting and coworkers (Enting 1978b, 1980a; Adler et al. 1983) and although there have been some improvements in the programs, most advances shown in the papers cited below are probably due to greatly enhanced computing power. However, for the present heuristic purposes, and for small values of ℓ the reader is recommended to use one of the available computer algebra packages like MAPLE or MATHEMATICA for which no further details are needed here, except to note that, since $Z_{\square}([x \times y], w) = Z_{\square}([y \times x], w)$, it is necessary simply to obtain the more easily calculated of the two partition functions, which is usually that for which $x > y$.

For simplicity we confine our initial discussion to the ν-state Potts model on the (self-dual) square lattice and the triangle-honeycomb dual lattice pair. Then, in the case of the square and triangular lattices, for a particular value of ℓ, the order of the largest transfer matrix is $\nu^{\bar{y}}$, where $\bar{y}$ is the integer part of $\ell/2$. In the case of the honeycomb lattice, since there are two sites per cell, the transfer matrices given in the form shown in Sect. 14.5.2.3 have largest order $\nu^{2\bar{y}}$. From this it is clear that computation time increases with both ν and ℓ, with that for the honeycomb lattice being larger than that for the square and triangular lattices.[30] In the following we restrict our explicit results to the 3-state model with $\ell = 6$ for the square and triangular lattices and $\ell = 5$ for the more time-consuming honeycomb case.

14.5.3.1 The Square Lattice

Using the formulae for the transfer matrices in Sect. 14.5.2.1 it is straightforward to obtain series for the low- and high-temperature free-energy densities and then for the order parameter and response functions.

For the 3-state Potts model and $\ell = 6$ the largest matrices $\boldsymbol{V}_{\rm w}$ and $\boldsymbol{V}_{\rm b}$ needed are of dimension 27, used to compute $Z_{\rm LT}([3 \times 3], u)$ or $Z_{\rm HT}([3 \times 3], v)$. The results for the free-energy densities are

$$\begin{aligned}\phi^{(0)}_{\rm LT}(\mathcal{N}_{\rm s}, u) &= -2u^4 - 4u^6 - 4u^7 - 4u^8 - 24u^9 - 16u^{10} - 60u^{11} \\ &\quad - \tfrac{518}{3}u^{12} - 128u^{13} + 216u^{14} + {\rm O}(u^{15}),\end{aligned} \tag{14.5.19}$$

$$\begin{aligned}\phi^{(1)}_{\rm LT}(\mathcal{N}_{\rm s}, u) &= 2u^4 + 8u^6 + 8u^7 + 24u^8 + 72u^9 + 140u^{10} + 320u^{11} \\ &\quad + 1164 * u^{12} + 1560u^{13} - 892u^{14} + {\rm O}(u^{15}),\end{aligned} \tag{14.5.20}$$

$$\begin{aligned}\phi^{(2)}_{\rm LT}(\mathcal{N}_{\rm s}, u) &= -4u^6 - 4u^7 - 38u^8 - 72u^9 - 352u^{10} - 616u^{11} \\ &\quad - 3340u^{12} - 5276u^{13} + 100u^{14} + {\rm O}(u^{15})\end{aligned} \tag{14.5.21}$$

and

$$\phi^{(0)}_{\rm HT}(\mathcal{N}_{\rm s}, v) = -2v^4 - 4v^6 - 4v^7 - 4v^8 - 24v^9 + 40v^{10} + {\rm O}(u^{11}), \tag{14.5.22}$$

$$\begin{aligned}\phi^{(2)}_{\rm HT}(\mathcal{N}_{\rm s}, v) &= -4v - 12v^2 - 36v^3 - 112v^4 - 192v^5 - 364v^6 - 332v^7 \\ &\quad + 36v^8 + 1736v^9 + 5625v^{10} + {\rm O}(u^{11}).\end{aligned} \tag{14.5.23}$$

Using (14.2.9)–(14.2.11) and (14.5.19)–(14.5.21), the low-temperature series for the order parameter and response functions are

$$\begin{aligned}\widetilde{\rho} &= 1 - 3u^4 - 12u^6 - 12u^7 - 36u^8 - 108u^9 - 210u^{10} - 480u^{11} \\ &\quad - 1746u^{12} - 2340u^{13} + 1338u^{14} + {\rm O}(u^{15}),\end{aligned} \tag{14.5.24}$$

[30] A modification of the honeycomb case can be made by reformulating the transfer matrices into ones between vertical rows of sites. This reduces the order but increases the number of matrix multiplications, still leaving the honeycomb case more expensive in computing time than the square and triangular cases.

$$\frac{c_\rho}{k_B K^2} = 32u^4 + 144u^6 + 196u^7 + 256u^8 + 1944u^9 + 1600u^{10} + 7260u^{11} + 24864u12 + 21632u^{13} - 42336u^{14} + \mathrm{O}(u^{15}), \tag{14.5.25}$$

$$\varphi_T = 2u^4 + 16u^6 + 16u^7 + 100u^8 + 216u^9 + 844u^{10} + 1552u^{11} + 7844u^{12} + 12112u^{13} - 1092u^{14} + \mathrm{O}(u^{15}) \tag{14.5.26}$$

and, using (14.2.34)–(14.2.36) and (14.5.22)–(14.5.23), the corresponding high-temperature series are

$$\frac{c_\rho}{k_B K^2} = \tfrac{8}{3}v + 16v^2 + \tfrac{368}{9}v^3 + \tfrac{896}{9}v^4 + \tfrac{2672}{9}v^5 + 616v^6 + \tfrac{13616}{9}v^7 + \tfrac{16976}{9}v^8 + \mathrm{O}(u^9), \tag{14.5.27}$$

$$\frac{\varphi_T\,\nu^2}{\nu - 1} = 1 + 4v + 12v^2 + 36v^3 + 112v^4 + 192v^5 + 364v^6 + 332v^7 - 36v^8 - 1736v^9 - 5628v^{10} + \mathrm{O}(v^{11}). \tag{14.5.28}$$

De Neef and Enting (1977) obtained the series expansion for the 3-state square-lattice model for the zero-μ free-energy density up to u^{22} and this was extended by Enting (1980a), with the addition of a low-temperature series for the order parameter, to u^{31}. Further extensions, with inclusion of low-temperature series for the field-response function, were made to degree u^{35} by Adler et al. (1983). Briggs et al. (1994) presented results for Potts models with $\nu = 2, 3, \ldots, 10$ which, for $\nu = 3$, extended the series for the order parameter and field response function to u^{37}. These results confirm the degree of accuracy of the low-temperature series given here as $13 = 2\ell + 1$, confirming the prediction in Sect. 14.5.1.1. Arisue and Tabata (1997) used an improvement of the finite-lattice method to extend low-temperature Potts model series on the square lattice to u^{41}, but only for values of ν in the range $[5, 50]$.

Because of self-duality the accuracy of the high-temperature series (14.5.22) can be inferred from (14.5.19) to be up to $9 = 2\ell - 3$, as predicted in Sect. 14.5.1.1. The are no $\mu \neq 0$ high-temperature results for this model using the finite-lattice method.[31] However, the work of Kim and Joseph (1975) using the finite-cluster method shows that the series (14.5.28) is accurate to degree $4 = \ell - 2$, which is again that predicted in Sect. 14.5.1.1.

14.5.3.2 The Triangular and Honeycomb Lattices

For the triangular lattice with $\nu = 3$ and $\ell = 6$

$$\phi_{\mathrm{LT}}^{(0)}(\mathcal{N}_{\mathrm{T}}, u) = -2u^6 - 6u^{10} - 6u^{11} + 10u^{12} - 36u^{14} - 36u^{15} + 54u^{16} + 72u^{17} - \tfrac{794}{3}u^{18} + \mathrm{O}(u^{19}), \tag{14.5.29}$$

[31] See the list of papers in Enting (1996).

$$\phi^{(1)}_{\text{LT}}(\mathcal{N}_{\text{T}}, u) = 2u^6 + 12u^{10} + 12u^{11} - 16u^{12} + 114u^{14} + 108u^{15} - 102u^{16} - 168u^{17} + 1016u^{18} + \mathrm{O}(u^{19}), \tag{14.5.30}$$

$$\phi^{(2)}_{\text{LT}}(\mathcal{N}_{\text{T}}, u) = -6u^{10} - 6u^{11} + 2u^{12} - 126u^{14} - 108u^{15} - 30u^{16} + 72u^{17} - 1602u^{18} + \mathrm{O}(u^{19}) \tag{14.5.31}$$

and

$$\phi^{(0)}_{\text{HT}}(\mathcal{N}_{\text{T}}, u) = -4v^3 - 6u^4 - 18v^5 - 34v^6 - 52v^7 + 38v^8 + \tfrac{1808}{3}v^9 + 2028v^{10} + \mathrm{O}(v^{11}), \tag{14.5.32}$$

$$\phi^{(2)}_{\text{HT}}(\mathcal{N}_{\text{T}}, u) = -6v - 30v^2 - 136v^3 - 524v^4 - 1450v^5 - 2782vu^6 - 44v^7 + 22062v^8 + 92264v^9 + 202888v^{10} + \mathrm{O}(u^{11}). \tag{14.5.33}$$

For the honeycomb lattice with $\nu = 3$ and $\ell = 5$

$$\phi^{(0)}_{\text{LT}}(\mathcal{N}_{\text{HC}}, u) = -2u^3 - 3u^4 - 9u^5 - 19u^6 - 52u^7 + \mathrm{O}(u^8), \tag{14.5.34}$$

$$\phi^{(1)}_{\text{LT}}(\mathcal{N}_{\text{HC}}, u) = 2u^3 + 6u^4 + 24u^5 + 82u^6 + 290u^7 + \mathrm{O}(u^8), \tag{14.5.35}$$

$$\phi^{(2)}_{\text{LT}}(\mathcal{N}_{\text{HC}}, u) = -3u^4 - 21u^5 - 127u^6 - 628u^7 + \mathrm{O}(u^8) \tag{14.5.36}$$

and

$$\phi^{(0)}_{\text{HT}}(\mathcal{N}_{\text{HC}}, v) = -v^6 - 2v^{10} - 2v^{11} + \mathrm{O}(v^{12}), \tag{14.5.37}$$

$$\phi^{(2)}_{\text{HT}}(\mathcal{N}_{\text{HC}}, v) = -3v - 6v^2 - 12v^3 - 22v^4 - 36v^5 - 67v^6 - 64v^7 - 57v^8 - 38v^9 - 46v^{10} + 142v^{11} + \mathrm{O}(v^{12}). \tag{14.5.38}$$

From these results low- and high-temperature series can be obtained for the order parameter and response functions.

Jensen et al. (1997) used the finite-lattice method augmented by the extension procedure of Jensen and Guttmann to obtained low-temperature series for $\nu = 3$ (up to u^{34}) on the honeycomb lattice[32] and Feldmann et al. (1998) used the same extension procedure to obtain the corresponding series for $\nu = 3$ (up to u^{68}) and $\nu = 4$ (up to u^{59}) on the triangular lattice and $\nu = 4$ (up to u^{30}) on the honeycomb lattice. Their results confirm the predicted level of accuracy of (14.5.29)–(14.5.31), (14.5.34)–(14.5.36).

As in the case of the square lattice there is a paucity of high-temperature results. However, the duality relationship (14.2.46) shows that $\phi^{(0)}_{\text{HT}}(\mathcal{N}_{\text{T}}, v)$ is accurate to v^5 as predicted in Sect. 14.5.1.2 ($5 = \ell - 1$) and $\phi^{(0)}_{\text{HT}}(\mathcal{N}_{\text{HC}}, v)$ is accurate to v^6 (for non-zero coefficients) as predicted in Sect. 14.5.1.3 ($6 = 2\ell - 4$).

---

[32] And also, although we have not discussed this, on the kagomé lattice for $\nu = 3$ (up to u^{66}) and $\nu = 4$ (up to u^{56}).

As we have already indicated the finite-lattice method can also be used for three-dimensional lattices. In particular, on the simple cubic lattice, Guttmann and Enting (1993) extended the low-temperature spin-$\frac{1}{2}$ Ising model free-energy series to degree 26 and Guttmann and Enting (1994) obtain corresponding results for the 3-state Potts model to degree 43. Spin-s Ising models for $s > \frac{1}{2}$, do not, even on the square lattice, have duality properties, and it is of particular interest to obtain series approximations for such models. Adler and Enting (1984) obtained 45-term expansions for the magnetization, heat capacity and susceptibility of the spin-1 Ising model on the square lattice. The results were extended to the degree 79 by Enting et al. (1994) and higher-order spin values were investigated by Jensen et al. (1996). The finite-lattice method has also been applied to models with multi-site interactions. In particular Enting and Wu (1982) and Enting (1980b) studied the 3-state Potts model on the triangular lattice with 3-site interactions.

Guttmann and Enting (1993) obtained high-temperature series for the magnetization and susceptibility for the spin-$\frac{1}{2}$ Ising model on the simple cubic lattice up to degree 26 and Guttmann and Enting (1994) obtained corresponding results for the 3-state Potts model up to degree 21.

14.6 The Analysis of Series: Second-Order Transitions

Most analyses of series expansions is directed towards the extraction of critical parameters and exponents associated with singular behaviour of the thermodynamic function under examination at a phase transition[33] and here we give an account of the main methods that have been used for doing this.[34] If the value of either a critical parameter or exponent is known (or assumed to be known) and this value is used to estimate the value of the other quantities, this is known as a *biased* estimate; otherwise estimates are said to be *unbiased.*

Given that $F(w)$ is a function of a variable $w(T)$ which for the low- and high-temperature cases is such that $w(0) = 0$ and $w(\infty) = 0$ respectively, the interest is in the singularities of $F(w)$ in some disc in the complex w plane $\mathbb{C}_w$, centred at the origin. In the case of a low-temperature series, when $w := u$ given by (14.2.5), the physical condition $u > 0$ means that any critical point must be on the positive real axis in $\mathbb{C}_u$. In the case of a high-temperature series, when $w := v$ given by (14.2.25), v can be positive or negative corresponding respectively to $R > 0$ and $R < 0$ and, for the Ising model on loose-packed lattices, the ferromagnetic/antiferromagnetic symmetry $+R \longleftrightarrow -R$ (Sect. 3.5.1.1) guarantees that critical parameter values occur in pairs $\pm v_c$. In the rest of this discussion we shall assume that $F(w)$ has

[33] Rather less attention has been given to the way that series analysis can be used to detect first-order transitions and we shall turn to that topic in Sect. 14.7.

[34] For more detailed accounts the reader is referred to the review articles by Gaunt and Guttmann (1965) and Guttmann (1989).

exactly one singular point $w = w_c$ on the positive real axis of $\mathbb{C}_w$[35] and then, of course, the radius of convergence r of the series (14.1.1) must satisfy the condition $r \le w_c$.

As indicated above the starting point for the series analysis of (14.1.1) is an hypothesized form for $F(w)$ and the truncated expansion (14.1.3) for some ℓ. One way to test the effectiveness of a method is to use a known analytic result, like the magnetization (12.3.283) of the square-lattice Ising model. Another is to use a range of plausible test functions. Such a programme was initiated by Hunter and Baker (1973) in relation to the ratio and Padé methods, developed further by Baker and Hunter (1973) and used to investigate the efficacy of the differential approximant method by Guttmann (1975).

14.6.1 Late-Term Analysis

It can be shown (Hille 1959) that, if $f_k \ge 0$, for all k greater than some $k_0 \ge 0$, then there exists a real positive r with $w = r$ a singular point of $F(w)$ and no singularities with $|w| < r$, giving $w_c = r$.[36] Since one can always choose to analyze $-F(w)$ rather than $F(w)$, the crucial point here is whether after some k_0 all the coefficients have the same sign, when, as indicated in Sect. 14.1, the critical point $w = w_c$ will be the nearest singularity to the origin and $1/w_c$ is given by the limit (14.1.2).[37] These remarks underline the fact that, in inferring the properties of $F(w)$, attention must be paid to what may be called (Dingle 1973, Chap. 7) the 'late terms' of the series (14.1.1). Important properties like sign-periodicity, and the inference just mentioned which can be drawn from all late coefficients being positive, will not, in general, be evident from the first few terms. Hence the need, given a general form for $F(w)$, for an asymptotic expression for the coefficient f_k. This is provided by the work of Darboux (1878, pp. 17–20). In general terms he showed that the asymptotic form for f_k is given by the singularities of $F(w)$ closest to the origin; that is to say on the circle of convergence. Specifically, suppose that, $F(w)$ which is analytic in some disc $|w| < r + \varepsilon$ ($\varepsilon > 0$) in $\mathbb{C}_w$, apart from algebraic singularities at $w_j = r\exp(\mathrm{i}\theta_j)$, $j = 0, 1, \ldots, s-1$, has the form

$$F(w) := \sum_{j=0}^{s-1}\left[A_j(w)\left(1 - \frac{w}{w_j}\right)^{-\lambda_j} + W_j(w)\right], \tag{14.6.1}$$

---

[35] Excluding models like those discussed in Sect. 10.4 which have upper- and lower-critical temperatures.

[36] There may, of course, be other singularities on the circle $|w| = r$.

[37] This provides a 'simple' method for obtaining w_c but, as we have shown by examples in Sect. 14.1, convergence is often very slow, leading to the need for the more sophisticated methods described below.

where $A_j(w)$ and $W_j(w)$ are analytic in a neighbourhood of $w = w_j$. Let

$$f_k = \sum_{j=0}^{s-1} f_{kj}, \tag{14.6.2}$$

where f_{kj} is the contribution to f_k from the singularity at $w = w_j$. Then it can be shown that, in calculating f_{kj}, $W_j(w)$ can be neglected. Replacing $A_j(w)$ by its Taylor expansion truncated at degree n gives

$$A_j(w) = \sum_{m=0}^{n} \left(1 - \frac{w}{w_j}\right)^m \frac{(-w_j)^m A_j^{(m)}(w_j)}{\Gamma(m+1)} \tag{14.6.3}$$

and expanding about $w = 0$ yields

$$f_{kj} = \sum_{m=0}^{n} \frac{(-w_j)^{m-k} A_j^{(m)}(w_j)\, \Gamma(m - \lambda_j + 1)}{\Gamma(k+1)\, \Gamma(m+1)\, \Gamma(m - \lambda_j - k + 1)} + \mathrm{O}\left(\frac{1}{k^{n+2-\lambda_j}}\right). \tag{14.6.4}$$

The approximation in halting the series (14.6.3) and (14.6.4) at n is called the nth *Darboux approximation*. That the error in f_{kj} is $\mathrm{O}(1/k^{n+2-\lambda_j})$ follows from the gamma function's asymptotic property $\Gamma(k+1)\,\Gamma(1+\alpha-k) \sim k^{\alpha+1}$.

This result was generalized to the case where algebraic singularities are confluent by Fields (1967) and Wong and Zhao (2005). Let $w_j = w_c$ with $\lambda_0 = \lambda - \Delta_j$, where $0 = \Delta_0 < \Delta_1 \le \Delta_2 \le \cdots \le \Delta_{s-1} < \lambda$. Then, from (14.6.4),

$$f_{kj} = \sum_{m=0}^{n} \frac{(-w_c)^{m-k} A_j^{(m)}(w_c)\, \Gamma(m + \Delta_j - \lambda + 1)}{\Gamma(k+1)\, \Gamma(m+1)\, \Gamma(m + \Delta_j - \lambda - k + 1)} + \mathrm{O}\left(\frac{1}{k^{n+2+\Delta_j-\lambda}}\right). \tag{14.6.5}$$

The accuracy of these approximations in the case of known functions can be checked. In particular for the case of the Ising model magnetization where the only singularity on the circle of convergence is the critical point.

14.6.2 The Ratio Method

In the special case $s = 1$ of (14.6.1), where

$$F(w) := A(w)\left(1 - \frac{w}{w_c}\right)^{-\lambda} + W(w), \tag{14.6.6}$$

and w_c is real and the singularity of $F(w)$ nearest to the origin, the zeroth Darboux approximation gives

$$f_k = \frac{1}{(-w_c)^k} \frac{A^{(0)}(w_c)\,\Gamma(1-\lambda)}{\Gamma(k+1)\,\Gamma(1-\lambda-k)} \left\{1 + \mathrm{O}\left(\frac{1}{k^{2-\lambda}}\right)\right\} \tag{14.6.7}$$

and

$$\frac{f_k}{f_{k-1}} = \frac{1}{w_c}\left\{1 + \frac{\lambda-1}{k}\right\} + \mathrm{O}\left(\frac{1}{k^{2-\lambda}}\right). \tag{14.6.8}$$

Excluding the correction term, the ratio f_k/f_{k-1} is a linear function of $1/k$ with gradient $(\lambda-1)/w_c$ and intercept $1/w_c$ on the vertical axis. Thus, for sufficiently large k, graphical interpolation of the data can be used to determine approximations to w_c and λ. However, the case of confluent singularities is more problematic. In the zeroth Darboux approximation with one confluent singularity, from (14.6.5),

$$f_k = \frac{1}{(-w_c)^k} \frac{A^{(0)}(w_c)\,\Gamma(1-\lambda)}{\Gamma(k+1)\,\Gamma(1-\lambda-k)} \Big\{1 + B(\lambda, \Delta; k) + \mathrm{O}\left(k^{\lambda-2}\right) + \mathrm{O}\left(k^{\lambda-2-\Delta}\right)\Big\}, \tag{14.6.9}$$

where

$$B(\lambda, \Delta; k) := \frac{A_1^{(0)}(w_c)\,\Gamma(1+\Delta-\lambda)\,\Gamma(1-\lambda-k)}{A_0^{(0)}(w_c)\,\Gamma(1-\lambda)\,\Gamma(1+\Delta-\lambda-k)}. \tag{14.6.10}$$

From the asymptotic properties of the gamma function $B(\lambda, \Delta; k) \sim 1/k^{\Delta}$. So the effect of the confluent singularity is to introduce a term $\mathrm{O}(1/k^{\Delta})$ into (14.6.8). Since typically $0 < \Delta < 1$ the effect is to slow down the converge of the ratio to a range of values for which the linear interpolation is effective.

To obtain improved accuracy in the parameter estimations a number of methods have been employed to accelerate the convergence of the ratios f_k/f_{k-1}. The most widely used of these, when there are no confluent singularities, is the Neville-Aitkin polynomial interpolation scheme (Guttmann 1989, pp. 21–22; see also, Brezinski 2000).

The assumption in (14.6.6), which underpins the ratio method, is that the circle of convergence of $F(w)$ is given by the physical singularity, that is $r = w_c$. The method fails when, as in the case of the low-temperature series for the Ising model (Guttmann 1969; Domb and Guttmann 1970) there are complex singularities w_j with $|w_j| < w_c$, one or more of which fix the circle of convergence. A widely used technique in such cases is the application of a homographic transformation

$$w \longrightarrow \tilde{w} := \frac{bw}{1+aw}, \tag{14.6.11}$$

where a and b are real. This leaves the origin invariant and in the special case of *Euler's transformation*, where $b = a + 1$, it also leaves invariant the point $w = 1$. If there is one complex singularity, a and b in (14.6.11) can be chosen to transform it to infinity. In the case where there is more than one such singularity it is still possible to choose the parameters to project them into the region of $\mathbb{C}_{w'}$ with $|w'| > \tilde{w}_c$.

14.6.3 *Padé Approximants*

Apart from the cases where homographic transformations can be used to extend the applicability of the ratio method, an approach is needed which provides an analytic continuation into the region $|w| > r$. This was provided by the introduction of the use of Padé approximants in statistical mechanics by Baker (1961) which initiated a development in series analysis which ultimately led to the invention of the differential approximant method by Guttmann and Joyce (1972) treated in Sect. 14.6.4.

A Padé approximant $[n/m]$, as given in Definition 18.5.1, is a rational function whose power series agrees with that of $F(w)$ up to and including the term in w^{n+m}. Clearly, rational functions, whose only singularities are poles, can be exactly represented by Padé approximants of sufficiently high order. They can also be expected to give good representations of meromorphic functions which differ from rational functions only at infinity. If the Padé approximant method were restricted to this class of functions it would be of only limited applicability. However, an intrinsic element of their utility is the preparation of the series by a prior transformation so that the expected form of the transformed function is meromorphic.

14.6.3.1 Transformations

The most widely used of these is the *Dlog transformation*[38]

$$F(w) \longrightarrow \mathcal{DL}\{F\}(w) := \frac{\mathrm{d}}{\mathrm{d}w}\ln[F(w)] = \frac{F'(w)}{F(w)}, \tag{14.6.12}$$

which transforms branch-points into poles. Thus

$$F(w) = A\prod_{j=0}^{s-1}\left(1 - \frac{w}{w_j}\right)^{-\lambda_j}, \tag{14.6.13}$$

gives

$$\mathcal{DL}\{F\}(w) = \sum_{j=0}^{s-1}\frac{\lambda_j}{w_j - w}. \tag{14.6.14}$$

[38] The combination of this transformation with a Padé approximant is the *Dlog Padé approximant*.

A modification of this transformation can be used to obtain a biased estimate of λ if w_j is known for some j, since

$$F(w) \longrightarrow (w_j - w)\mathcal{DL}\{F\}(w) \simeq \lambda_j \qquad w \to w_j - 0. \tag{14.6.15}$$

Repeated application of the Dlog transformation yielding the *double Dlog transformation* was used by Guttmann (1978) to treat a Kosterlitz–Thouless singularity (Kosterlitz and Thouless 1973) where

$$F(w) \simeq A \exp\left(B\left[1 - \frac{w}{w_c}\right]^{-\lambda}\right) \qquad w \to w_c - 0, \tag{14.6.16}$$

and

$$\mathcal{DL}\{\mathcal{DL}\{F\}\}(w) \simeq \frac{\lambda + 1}{w_c - w} \qquad w \to w_c - 0. \tag{14.6.17}$$

In the two-dimensional XY model (Sect. 3.6.1), for which the high-temperature susceptibility series has been extensively studied (Betts 1974; Camp and Van Dyke 1975; Guttmann 1978; Ferer and Velgakis 1983; Butera et al. 1989, 1993; Butara and Comi 1994) the data clearly favours methods designed to fit this kind of singularity. Transformations for the special case

$$F(w) \simeq \sum_{j=0}^{s-1} A_j \left(1 - \frac{w}{w_j}\right)^{-\lambda_j} \tag{14.6.18}$$

of (14.6.1) were considered by Baker and Hunter (1973). Here the analysis is rather complicated, but two cases are of interest:

- *All $\lambda_j = \lambda$, which is known.* For this case these authors applied a procedure which they call *exponent renormalization*.[39] They showed that the function

$$G(w) := \sum_{k=0}^{\infty} g_k w^k, \quad \text{where} \quad g_k := f_k \binom{-\lambda}{k}, \tag{14.6.19}$$

has leading power-law behaviour

$$G(w) \simeq \sum_{j=0}^{s-1} A_j \left(1 + \frac{w}{w_j}\right)^{-1}. \tag{14.6.20}$$

---

[39] This terminology is not related to the exponent renormalization described in Sect. 10.2.1.

- *The singularities are confluent (coalesce) at* $w_j = w_c$. This is the case for the heat capacity (Joyce 1975a) and magnetization (Joyce 1975b) of the Ising model with a pure three-spin interaction on a triangular lattice (*the Baxter–Wu Model* Baxter and Wu 1973). The analysis applied here is similar to exponent renormalization.

14.6.3.2 Choosing Approximants

For ease of discussion we shall suppose that our information consists of the coefficients $f_0, \ldots, f_\ell$ of the power series of a function $F(w)$, which have either been obtained by the methods described in the first part of this chapter *or are the result of the prior application of a transformation to such a sequence of coefficients.*

The method for determining Padé approximants is described in Sect. 18.5.1. The task is most straightforward when the power series for $F(w)$ is *normal*, that is $X_0(n/m) \neq 0$ for all n and m. Then it is usually found that the most accurate representations of $F(w)$ are the diagonal approximants $[n/n]$ and near-diagonal approximants $[m-1/m]$, $[n/n-1]$. In general one might suppose that the best choice is to use all the available information, thus taking $n = m = \frac{1}{2}\ell$ if ℓ is even or $n = \frac{1}{2}(\ell \pm 1)$, $m = \frac{1}{2}(\ell \mp 1)$ if ℓ is odd, although there is no guarantee of monotonic improvement in the predictions of critical parameters with increasing n. There has been much discussion of the convergence properties of Padé approximants, particularly of $[n/n]$ for increasing n, in relation to the status of the Baker–Gammel–Wills conjecture for meromorphic functions,[40] (see, for example, Stahl 1997; Baker 2000, 2005, and references therein). However, since in practice it is possible to compute only a finite number of approximants, this is not a question to which we shall give further consideration.

In some cases $f_0, \ldots, f_\ell$, can be obtained as rational numbers and the coefficients of the required approximant obtained without numerical approximation. When the calculations are with real numbers it is, of course, desirable to retain as many significant figures as possible and a rule of thumb proposed by Baker is that the number retained should be at least the number of known series coefficients. A numerically difficult situation is when the required approximant has a zero and a pole very close together and closer to the origin than the physical singularity.[41] This leads to large numerical inaccuracies and corresponds to the situation where the linear equations (18.5.9)–(18.5.11) are ill-conditioned. Such approximants are called *defective* and should be avoided.

As we saw in Sect. 18.5.1, points in the X_0-table for which $X_0(n/m) = 0$ occur in squares, within which the Padé approximants are given by Theorem 18.5.4. There they either do not exist or are equal to those on the boundary. Given a particular

---

[40] This conjectures that, for a function $F(w)$ meromorphic in the unit disc, there is a subsequence of the diagonal approximants which converges to $F(w)$ locally uniformly for all poles in the unit disc.

[41] They cannot, of course, exactly coincide since $P_n^{(m)}(w)$ and $Q_m^{(n)}(w)$ are coprime.

value of ℓ it is an informative task to compute, using (18.5.13), all $X_0(n/m)$, for $n + m \leq \ell$, before completing the calculation of a chosen (non-defective) $[n/m]$ using (18.5.21) and (18.5.23).

14.6.4 Differential and Algebraic Approximants

The common features of Padé, differential and algebraic approximants are:

(i) The known coefficients of the power series of the function of interest $F(w)$ are used to determine the coefficients in a number of polynomials.
(ii) The recurrence relationships used to determine the polynomial coefficients are taken to apply to all the coefficients in the power series of a function $H(w)$ which is taken as an approximant to $F(w)$.

14.6.4.1 Differential Approximants

As we have indicated in Sect. 18.5.3 the procedure for determining the Dlog Padé approximant can be viewed as a bridge between Padé and differential approximants. On the one hand the polynomial quotient $[n/m]_{\mathrm{DL}}$ is useful in its own right since it is an approximant to the function $G(w)$ which is designed to have a simpler structure than $F(w)$ (branch-points replaced by poles). On the other hand the function $H(w)$, which satisfies (18.5.36), is a first-order differential approximant to $F(w)$.[42] In the case of the Dlog Padé approximants the recurrence equations are linear relationships between the known coefficients of $F(w)$.[43] Such relationships were found by M.F. Sykes to exist between the known coefficients of the spontaneous magnetization of the two-dimensional Ising model and used by him as a stable and efficient method of generating further coefficients.[44] Of course, caution is needed in using this method to generate series coefficients and there are well-known examples where it fails. For the F model the coefficients of the zero-field low-temperature polarization series fit the simple recurrence relationship $f_{j+2} = 3f_j + 4$ for $j = 4, 5, \ldots, 12$; however $f_{16} = 3f_{14} - 4$ (Nagle 1969). The recurrence relations may also be non-linear (Joyce and Guttmann 1973; Guttmann and Enting 1988). In such cases the related polynomials are the coefficients of a differential equation, which is of order at least that of the order of the recurrence relations. An illustration of this relationship is given by Example 18.5.1 in Sect. 18.5.3.

---

[42] With a small caveat. Comparing (18.5.36) with (18.5.40), with $k = 1$ and $B(w) = 0$, we see that $\mathring{P}_n^{(m)}(w) = -A_0(w)$ and $\mathring{Q}_n^{(m)}(w) = w\,A_1(w)$, implying, contrary to our assumption for Padé approximants that $\mathring{q}_0 = 0$ and $\mathring{q}_1 = 1$.

[43] Linear in the sense that the index variable appear linearly.

[44] This work by Dr. Sykes is reported by Guttmann (1989) and elsewhere, but does not seem to have been published.

The most general form of a one-variable differential approximant is given by Definition 18.5.3, which includes the possibility of an inhomogeneous term $B(w)$. Of course, as we have already indicated, the type of approximant chosen (the order and the presence or absence of an inhomogeneous term) is influenced by the expected type of singular behaviour. If $m < n$ and the zeros $w_0, w_1, \ldots, w_{m-1}$ of $\mathring{Q}_m^{(n)}(w)$ are all distinct then the general solution of (18.5.36) is of the form (14.6.13) with $s := m$. If it is supposed that $F(w)$ is of this form then polynomials $\mathring{P}_n^{(m)}(w)$ and $\mathring{Q}_m^{(n)}(w)$ are obtained from the known coefficients of the series and the approximant $H(w)$ obtained by solving (18.5.36). This corresponds to the determination of a homogeneous first-order ($k = 1$) differential determinant except that (18.5.40) assumes that the expected form for $H(w)$ has a pole at the origin. Of course, this can be removed, by a little reformulation but the investigations of Rehr et al. (1980) indicate that this does not yield any advantages.

If the supposed form for $F(w)$ differs from (14.6.13) by the addition of an analytic term, then the approximant will not be a solution of (18.5.36) but can be shown to satisfy an inhomogeneous ($B(w) \neq 0$) case of (18.5.40). The use of a first-order inhomogeneous approximant was proposed by Gammel (1973) and has been found to be appropriate for low temperature expansions of the Ising model susceptibility on three-dimensional lattices (Hunter and Baker 1979) for high-temperature expansions of the Ising model susceptibility and heat capacity on three-dimensional lattices (Fisher and Au-Yang 1979) and for low-temperature expansions for the energy, susceptibility and magnetization of the 3-state Potts model on two- and three-dimensional lattices (Bhanot et al. 1993).

In general it is found that an approximant of order greater than one is appropriate when there are confluent singularities (Rehr et al. 1980) as is the case for the Ising model susceptibility on three-dimensional lattices (Guttmann and Enting 1993) However, apart from this consideration there will still be advantages in using differential approximants of some higher order. Exactly what that order should be for any given series is a question discussed by Boukraa et al. (2008). They conclude "*that there is no single unique ODE that describes any given series.* There is a minimum order linear ODE that is unique but this typically contains a very large number of apparent singularities and can only be determined from a corresponding larger number of series coefficients."

14.6.4.2 Algebraic Approximants

The Padé approximant, $H(w)$ is just the solution of the first-order algebraic equation (18.5.6) and just as the generalization of the Dlog Padé approximant led from a first-order ordinary differential equations to one of higher order, it is natural to consider the generalization of the ordinary Padé approximant to algebraic approximants of higher degree. These are described in Sect. 18.5.4 and comparing (18.5.63) and (18.5.65) with the corresponding formulae (18.5.40) and (18.5.43) for the differential approximant it can be seen that they are related, in a formal way, by

$$\{H(w)\}^i \longleftrightarrow \mathcal{D}^{i-1}\{H(w)\}, \quad s \longleftrightarrow k+1, \\ n_i \longleftrightarrow m_{i-1}, \qquad n_0 \longleftrightarrow n. \tag{14.6.21}$$

In fact the relationship is more than formal since it can be shown (Theorem 17.6.1) that the solution $H(w)$ of (18.5.63) satisfies a linear differential equation. Given that this is the case Brak and Guttmann (1990) remark that:

> At first sight ... it might appear that [the method of algebraic approximants] offers no more than the existing differential approximant method. As a mathematical statement this view is correct. As a method of series analysis, however this view is quite misleading ...

This they explain in terms of the need, given a particular number of known coefficients for $F(w)$, for computation of polynomial coefficients. In general fewer are needed for the algebraic approximant. This point can be illustrated by reference to Example 17.6.1. For both the algebraic equation with coefficients (17.6.8), the homogeneous differential equation with coefficients given by (17.6.9) and (17.6.10) and the inhomogeneous differential equation with coefficients given by (17.6.9) and (17.6.12), the total number of coefficients is (taking into account that one coefficient is set to one) the sum of the degrees plus two. For the algebraic equation that number is 11, and for the homogeneous and inhomogeneous differential equations, respectively 50 and 20.

This advantage for algebraic approximants must, however, be balanced against the fact that, for differential approximants of any order, one may, in principle, either derive successive coefficients of the approximant *or* solve the differential equation. In the case of algebraic approximants this is the case only at degrees two, three and four. Beyond this there are, of course, no analytic formulae for solutions of algebraic equations. Successive series coefficients can be obtained using the formulae given in Sect. 18.5.4 and singular points and exponents derived using the method described by Guttmann (1990) and illustrated in Example 18.5.2.

A subject of much interest both for differential and algebraic approximants is the determination of a differential or algebraic equation which gives $F(w)$ exactly. Algebraic approximants yield only rational critical exponents. However, most exactly solved models have exponents of this type[45] and some have been shown to have thermodynamic functions satisfying algebraic equations. An example, which we have used in Sect. 18.5.4, is provided by the work of Joyce (1988b). He has shown (Eq. (10.1)) that the reciprocal activity satisfies a quadratic equation with coefficients which are quartic polynomials in the inverse of the density function. We have used this case as an illustration of the use of algebraic approximants in Example 18.5.2. Beginning with the series expansion to degree 13 the algebraic equation has been obtained, together with its solution, its critical point and exponent. Of course, in this case the complete solution is known so that further series coefficients will simply satisfy the equation. The method advocated by Brak and Guttmann (1990) when

[45] See (12.3.280) for the Ising model, (12.3.322) for the 3-state Potts model and (13.2.2) for the honeycomb dimer model. The case of the eight-vertex model is exceptional since the exponents α, β, γ and ν vary through irrational values (Sect. 12.3.8). However, see the results of the application of the proposal of *weak universality* given in (12.3.326).

the series coefficients represent all the available information, is to successively use an increasing number of these while increasing the degree of the equation and the degrees of the polynomial coefficients. If further series coefficients need no change in the equation one may speculate that it is exact, although there is, of course, always the possibility that further coefficients may show this assumption to be false.

14.7 The Analysis of Series: First-Order Transitions

As described in detail in Sect. 4.1, a second-order phase transition is evidenced by continuous first-order free-energy derivatives (densities) and singular second-order derivatives (response functions), and a first-order transition corresponds to discontinuities in one or all of the densities. In the latter case the response functions may be finite or singular, as in the case of a first-order critical point (Sect. 4.5.6), at the transition point.

When the location and the order of the transition are known the interest is in whether the series for response functions can be analytically continued through the transition point into the metastable region, or whether it is a singularity.[46] In the first case it is to be anticipated that the analytic continuation terminates at a singularity which can be regarded (by analogy with mean-field theory, Sect. 5.1.1) as a point on a *pseudo-spinodal* line. The most well-known system with a known line of first-order transitions is the square-lattice spin-$\frac{1}{2}$ Ising model with the transition line on the zero-field axis terminating at the critical point and there has been considerable interest in the question of whether this transition line is one of singularities. This has been investigated by Enting and Baxter (1980) using a high-field series at fixed temperature. Analysing the series using forms of the ratio method they suggested that the transition line consisted of infinitely differentiable singularities. This conclusion is supported by Isakov (1984), who proved the absence of an analytic continuation for the free energy through the transition line for all $d \geq 2$.

For system for which the transition point is unknown a single series expansion (low or high temperature) is not sufficient for its location. The technique here to use both high- and low-temperature series. A first-order transition approached from above and below will then exhibit a change in slope of the free energy with a corresponding discontinuity in the density.

The ν-state Potts model provides the ideal test-bed for these methods. For the square lattice the transition point, order of the transition and the discontinuity (or otherwise) of the internal-energy density are known (Sects. 9.2.1 and 12.3.10.6) and for the simple cubic lattice they are still a matter of discussion.

[46] Although it is noted by Briggs et al. (1994) that even "with an essential singularity [at the transition point] it may be possible to define the thermodynamic functions in the metastable region by analytic continuation in the complex plane passing *around* the singularity".

Briggs, Enting and Guttmann (1994) used the finite-lattice method to extend the low-temperature series for the square-lattice Potts model for $\nu = 2, 3, \ldots, 10$ (Sect. 14.5.3.1). They then used Dlog Padé approximants to determine analytic continuation through the transition point and to compare the temperature $T_c^\star$, where a singularity occurred, with the transition temperature T_c. For the second-order transition for $\nu = 3$ their calculations gave $T_c^\star = 0.99990\, T_c$ and for the marginal case $\nu = 4$, where the transition is second-order (but with logarithmic corrections to the exponents, Sect. 12.3.10.6) $T_c^\star = 0.99975\, T_c$. For $\nu = 5$, $T_c^\star = 1.00064\, T_c$ and, for $\nu = 10$, $T_c^\star = 1.0104\, T_c$. These latter cases, implying analytic continuation through the critical point, are consonant with the known first-order transition. In the square-lattice Potts model high- and low-temperature series are related by duality. Using these, differential approximant equations were found for the free-energy and internal-energy densities and the order parameter and integrated numerically. The results clearly indicated the order of the phase transition with results for the location of the transition and, for $\nu > 4$, the latent heat and order-parameter discontinuities in good agreement with the known results. With respect to the latter, as might be expected, the agreement is better at $\nu = 10$ than at $\nu = 5$.

The investigations of Briggs et al. (1994) represented a preliminary study for the work of two of the authors (Guttmann and Enting 1994) on the 3-state Potts model on the simple cubic lattice. They again used the finite-lattice method to extend the low- and high-temperature series for the free-energy density. Employing the differential approximant method an estimate for the transition temperature was obtained, in reasonable agreement with Monte-Carlo results, located at a discontinuity of slope, clearly indicating a first-order transition.

Chapter 15
Real-Space Renormalization Group Theory

15.1 The Basic Elements of the Renormalization Group

As in Sect. 16.2.1 we consider a d-dimensional lattice $\mathcal{N}$, with a lattice site $\boldsymbol{r}$ given in terms of unit lattice vectors by (16.2.1). With periodic boundary conditions applied, $n^{(\ell)}$ is an integer modulo $N^{(\ell)}$ and the total number of lattice sites $N := |\mathcal{N}|$ is given by (16.2.5). At each site $\boldsymbol{r}$ there is a microsystem with state specified by the variable $\sigma(\boldsymbol{r})$. For ease of presentation we shall suppose that the state variable at each lattice site is the same and that it can take ν discrete values, although this can be modified in particular applications. As in Sect. 4.3 it is assumed that the system has n independent (non-trivial) couplings. Here we need not distinguish between internal and external couplings and they will be denoted by $\zeta_1, \zeta_2, \ldots, \zeta_n$, with the so-called 'trivial coupling' denoted by ζ_0. The dimensionless Hamiltonian is given by (2.5.2) and the partition function by

$$Z(N, \zeta_i) = \sum_{\{\sigma\}} \exp[-\widehat{H}(N, \zeta_i; \boldsymbol{\sigma})], \tag{15.1.1}$$

where including ζ_0 in the Hamiltonian (2.5.2) effectively sets the zero-point of the Hamiltonian and consequently of the dimensionless free energy Φ, which is defined for the Hamiltonian with ζ_0 set equal to zero; thus

$$\Phi(N, \zeta_i) := -\ln[\exp(N\zeta_0)Z(N, \zeta_i)], \tag{15.1.2}$$

or equivalently,

$$\Phi(N, \zeta_i) := -N\zeta_0 - \ln[Z(N, \zeta_i)]. \tag{15.1.3}$$

An important point about renormalization group theory is that it is, when linearized as in Sect. 15.3, an implementation of the scaling procedure described in Chap. 4. However, while that is able to predict only relations between critical exponents,

D.A. Lavis, *Equilibrium Statistical Mechanics of Lattice Models*,
Theoretical and Mathematical Physics, DOI 10.1007/978-94-017-9430-5_15

renormalization group methods are able to give values for these exponents. This procedure can be carried out in a variety of ways. Even within the context of real-space renormalization group (RSRG) methods it would be very difficult to present a formulation which applies to all possible schemes for all possible systems. We have not attempted this. Instead a block-spin procedure is described, which includes most simple applications of the theory. Although it will need modification in some cases, these changes will not radically affect the development.

The following steps in the development can, in part, be viewed as a formalization and generalization of the ideas introduced in Sect. 4.2.2:

(i) The lattice $\mathcal{N}$ is divided into blocks of the form of d-dimensional parallelepipeds whose edges are given by the vectors $\lambda a^{(\ell)}\hat{\boldsymbol{r}}^{(\ell)}$, for some integer $\lambda > 1$, and such that in each block there are λ^d sites. A lattice $\widetilde{\mathcal{N}}$ is formed of the same structure as $\mathcal{N}$ by associating a lattice site $\tilde{\boldsymbol{r}} \in \widetilde{\mathcal{N}}$ with every block of $\mathcal{N}$. The block of sites in $\mathcal{N}$ associated with $\tilde{\boldsymbol{r}}$ is denoted by $\mathcal{B}(\tilde{\boldsymbol{r}})$ and the number $\widetilde{N}$ of sites in $\widetilde{\mathcal{N}}$ is

$$\widetilde{N} := \lambda^{-d} N, \tag{15.1.4}$$

with lattice spacings

$$\tilde{a}^{(\ell)} := \lambda a^{(\ell)}, \qquad \ell = 1, 2, \ldots, d. \tag{15.1.5}$$

(ii) The size of the lattice $\widetilde{\mathcal{N}}$ is reduced by a length scaling

$$|\tilde{\boldsymbol{r}}| = \lambda^{-1}|\boldsymbol{r}|. \tag{15.1.6}$$

This means that $\widetilde{\mathcal{N}}$ now differs from $\mathcal{N}$ only in the number of lattice sites.

(iii) The conditional probability

$$p(\sigma(\tilde{\boldsymbol{r}})|\sigma(\boldsymbol{r}), \boldsymbol{r} \in \mathcal{B}(\tilde{\boldsymbol{r}})) \geq 0, \tag{15.1.7}$$

that the microstate at $\tilde{\boldsymbol{r}}$ is $\sigma(\tilde{\boldsymbol{r}})$, given a particular configuration of microstates in the block $\mathcal{B}(\tilde{\boldsymbol{r}})$, is called the *weight function* and it satisfies the usual formula

$$\sum_{\{\sigma(\tilde{\boldsymbol{r}})\}} p(\sigma(\tilde{\boldsymbol{r}})|\sigma(\boldsymbol{r}), \boldsymbol{r} \in \mathcal{B}(\tilde{\boldsymbol{r}})) = 1, \tag{15.1.8}$$

where the summation is over all the microstates of $\tilde{\boldsymbol{r}}$. The probability $p(\zeta_i; \sigma(\boldsymbol{r}))$ that there is a particular configuration of microstates on $\mathcal{N}$ is, from (2.1.2),

$$p(\zeta_i; \sigma(\boldsymbol{r})) = \frac{\exp[-\widehat{H}(N, \zeta_i; \sigma(\boldsymbol{r}))]}{Z(N, \zeta_i)}. \tag{15.1.9}$$

Since the mapping is between two lattices which differ only in the number of lattice sites, the Hamiltonian function $\widehat{H}$ and partition function Z will be the same in each case and the probability $p(\tilde{\zeta}_i; \sigma(\tilde{\boldsymbol{r}}))$ that there is a particular configuration of microstates on $\widetilde{\mathcal{N}}$, with coupling constants $\tilde{\zeta}_i$, is

$$p(\tilde{\zeta}_i; \sigma(\tilde{\boldsymbol{r}})) = \frac{\exp[-\widehat{H}(\widetilde{N}, \tilde{\zeta}_i; \sigma(\tilde{\boldsymbol{r}}))]}{Z(\widetilde{N}, \tilde{\zeta}_i,)}. \tag{15.1.10}$$

(iv) The probabilities given by (15.1.9) and (15.1.10) can be related by a conditional probability of the form (15.1.7) using the standard formula of probability theory. In general this means that the conditional probability will be a function of both sets of couplings ζ_i and $\tilde{\zeta}_i$ with no specified relationship between the two. In renormalization group theory, however, the probability formula is used to *impose* the relationship

$$p(\tilde{\zeta}_i; \sigma(\tilde{\boldsymbol{r}})) := \sum_{\{\sigma(\boldsymbol{r})\}} \Big[\prod_{\{\tilde{\boldsymbol{r}}\}} p(\sigma(\tilde{\boldsymbol{r}})|\sigma(\boldsymbol{r}), \boldsymbol{r} \in \mathcal{B}(\tilde{\boldsymbol{r}}))\Big] p(\zeta_i; \sigma(\boldsymbol{r})) \tag{15.1.11}$$

between the two sets of couplings, where the conditional probability is independent of the values of the members of both sets.

(v) Although the probabilities $p(\zeta_i; \sigma(\boldsymbol{r}))$ and $p(\tilde{\zeta}_i; \sigma(\tilde{\boldsymbol{r}}))$, as they are defined in (15.1.9) and (15.1.10), involve the trivial couplings ζ_0 and $\tilde{\zeta}_0$ respectively, this dependence is purely formal. The factor $\exp(-N\zeta_0)$ can be cancelled from the numerator and denominator of (15.1.9) and similarly for $\exp(-\widetilde{N}\tilde{\zeta}_0)$ in (15.1.10). We, therefore, have two arbitrary parameters in (15.1.11) and imposing the condition

$$Z(\widetilde{N}, \tilde{\zeta}_i) = Z(N, \zeta_i) \tag{15.1.12}$$

fixes the difference $\widetilde{N}\tilde{\zeta}_0 - N\zeta_0$.

From (15.1.9)–(15.1.12),

$$\exp[-\widehat{H}(\widetilde{N}, \tilde{\zeta}_i; \sigma(\tilde{\boldsymbol{r}}))] = \mathcal{Z}(N, \zeta_i; \sigma(\tilde{\boldsymbol{r}})), \tag{15.1.13}$$

where

$$\mathcal{Z}(N, \zeta_i; \sigma(\tilde{\boldsymbol{r}})) := \sum_{\{\sigma(\boldsymbol{r})\}} \Big[\prod_{\{\tilde{\boldsymbol{r}}\}} p(\sigma(\tilde{\boldsymbol{r}})|\sigma(\boldsymbol{r}), \boldsymbol{r} \in \mathcal{B}(\tilde{\boldsymbol{r}}))\Big] \exp[-\widehat{H}(N, \zeta_i; \sigma(\boldsymbol{r}))]. \tag{15.1.14}$$

15.2 RG Transformations and Weight Functions

Equation (15.1.13) is the starting point for the development of a renormalization transformation. Ideally, by allowing the microstates of $\widetilde{\mathcal{N}}$ to range over all their possible values and performing the summation on the right-hand side, we hope to obtain *recurrence relationships* of the form

$$\tilde{\zeta}_0 = \lambda^d[\zeta_0 + \mathcal{K}_0(\zeta_j)], \tag{15.2.1}$$

$$\tilde{\zeta}_i = \mathcal{K}_i(\zeta_j), \qquad i = 1, 2, \ldots, n, \tag{15.2.2}$$

where (15.2.1) can be neglected from present consideration as it serves only to obtain thermodynamic properties (Sect. 15.4). If (15.2.2) can be obtained they are used to perform a sequence of iterations indexed by $s = 0, 1, 2, \ldots$ in the space $\mathcal{S}$ of non-trivial couplings. This produces a trajectory $\mathsf{P}_0 \to \mathsf{P}_1 \to \mathsf{P}_2 \to \cdots$ in $\mathcal{S}$. It satisfies the property

$$\zeta_i^{(s+s')} = \mathcal{K}_i^{s'}(\zeta_j^{(s)}) \tag{15.2.3}$$

and is therefore a *semi-group*.[1] As in the case of general dynamic systems (see, for example, Devaney 1989) the behaviour of renormalization group trajectories in $\mathcal{S}$ could be very complicated, exhibiting fixed points, periodic points and even chaos. However, in most simple examples, it might be anticipated that the behaviour is governed by a relatively small set of fixed points. A point $\mathsf{P}^\star \in \mathcal{S}$ with coupling values $\zeta_1^\star, \zeta_2^\star, \ldots,$ is a *fixed point* if

$$\zeta_i^\star = \mathcal{K}_i(\zeta_j^\star), \qquad i = 1, 2, \ldots, n. \tag{15.2.4}$$

If it exists, the set $\mathfrak{B}(\mathsf{P}^\star)$ of all points P_0 which on iteration converge to $\mathsf{P}^\star$ is called the *basin of attraction* of $\mathsf{P}^\star$. If an n-dimensional neighbourhood of $\mathsf{P}^\star$ contains only points of $\mathfrak{B}(\mathsf{P}^\star)$ then $\mathsf{P}^\star$ is called a *sink*. In general fixed points can have neighbourhoods containing points for which they are *attractive*, points for which they are *repulsive* and points for which trajectories iterated neither towards nor away from the fixed point. This last type of fixed point is called *marginal* and this characteristic is often associated with the existence of a region (line or surface) of fixed points.

In Sect. 15.1 we referred to the lattice $\mathcal{N}$ having N sites with the number of sites of $\widetilde{\mathcal{N}}$ being $\widetilde{N} = \lambda^{-d}N < N$. However, it is clear that the development of trajectories in $\mathcal{S}$, leading to fixed points, implicitly assumes that a large (potentially infinite) number of iterations can be applied to the system. This could be the case only if the thermodynamic limit $N \to \infty$ is assumed. The method described in Sect. 15.8, whereby the form of the transformation derived for finite N is applied an indefinite number of times, is in these terms by its very structure an approximation method, even though the calculation of (15.1.13) for finite N may be exact. It is clear that

[1] Not a full group since no inverse transformation is defined.

the form of an transformation will be in part determined by the choice of weight function. Three commonly used forms are:

(a) **The decimation weight function**[2]

Choose a particular site $\mathring{\boldsymbol{r}} \in \mathcal{B}(\tilde{\boldsymbol{r}})$ so that the set $\mathring{\mathcal{N}}$ of all sites $\mathring{\boldsymbol{r}}$ for all the blocks of $\mathcal{N}$ forms a lattice of the same type as $\mathcal{N}$. Then set

$$p(\sigma(\tilde{\boldsymbol{r}})|\sigma(\boldsymbol{r}), \boldsymbol{r} \in \mathcal{B}(\tilde{\boldsymbol{r}})) := \delta^{\mathrm{Kr}}(\sigma(\tilde{\boldsymbol{r}}) - \sigma(\mathring{\boldsymbol{r}})). \tag{15.2.5}$$

It is clear that the effect of choosing this weight function is that the summation on the right-hand side of (15.1.13) is a partial sum over all the sites of the lattice $\mathcal{N}$ except those of $\mathring{\mathcal{N}}$. The effect is to 'thin out' the sites of $\mathcal{N}$ leaving $\mathring{\mathcal{N}}$ which is now identified as $\widetilde{\mathcal{N}}$. For the spin-$\frac{1}{2}$ model ($\nu = 2$, $\sigma(\boldsymbol{r}) = \pm 1$) equation (15.2.5) takes the simple form

$$p(\sigma(\tilde{\boldsymbol{r}})|\sigma(\boldsymbol{r}), \boldsymbol{r} \in \mathcal{B}(\tilde{\boldsymbol{r}})) = \tfrac{1}{2}[1 + \sigma(\tilde{\boldsymbol{r}})\sigma(\mathring{\boldsymbol{r}})]. \tag{15.2.6}$$

(b) **The majority-rule weight function**

This weight function was introduced by Niemeijer and van Leeuwen (1973, 1974). The first step in assigning $\sigma(\tilde{\boldsymbol{r}})$ for the block $\mathcal{B}(\tilde{\boldsymbol{r}})$ can be described in terms of the 'winner takes all' voting procedure used in some democracies. Given that, among the sites of $\mathcal{B}(\tilde{\boldsymbol{r}})$ one of the ν microstates occurs more than any other, $\sigma(\tilde{\boldsymbol{r}})$ is assigned to this value. Unless $\nu = 2$ and the number of sites λ^d in a block is odd it is clear that this is not sufficient to determine $\sigma(\tilde{\boldsymbol{r}})$ for every configuration of the block. A 'tie' can occur in the voting procedure and a strategy must be adopted to deal with such cases. One possibility is to assign to $\sigma(\tilde{\boldsymbol{r}})$ one of these predominating values on the basis of equal probabilities. In the special case of a spin-$\frac{1}{2}$ model this weight function would take the form

$$p(\sigma(\tilde{\boldsymbol{r}})|\sigma(\boldsymbol{r}), \boldsymbol{r} \in \mathcal{B}(\tilde{\boldsymbol{r}})) \simeq \tfrac{1}{2}\Big[1 + \sigma(\tilde{\boldsymbol{r}})S\Big(\sum_{\{\boldsymbol{r}\in\mathcal{B}(\tilde{\boldsymbol{r}})\}} \sigma(\boldsymbol{r})\Big)\Big], \tag{15.2.7}$$

where

$$S(m) = \begin{cases} -1, & \text{if } m < 0, \\ 0, & \text{if } m = 0, \\ 1, & \text{if } m > 0\,. \end{cases} \tag{15.2.8}$$

In some cases this may not, however, be the most appropriate choice. In their work on the Ising model using square first-neighbour blocks Nauenberg and Nienhuis (1974a, b) divided the configurations with equal numbers of up and

---

[2] The term 'decimation' (originating in a form of discipline used in the Roman army) is more properly used when one tenth of the sites is removed; but here it is employed more loosely to denote any proportion.

down spins between block spins up and down with probability one. The rule (one of four) which they chose ensured that the reversal of all the spins in the block reversed the block spin.

(c) **The Kadanoff–Houghton weight function**

For the spin-$\frac{1}{2}$ model the weight function

$$p(\sigma(\tilde{\boldsymbol{r}})|\sigma(\boldsymbol{r}), \boldsymbol{r} \in \mathcal{B}(\tilde{\boldsymbol{r}})) := \frac{\exp\left[\tau\,\sigma(\tilde{\boldsymbol{r}}) \sum_{\{\boldsymbol{r}\in\mathcal{X}(\tilde{\boldsymbol{r}})\}} \sigma(\boldsymbol{r})\right]}{2\cosh\left[\tau \sum_{\{\boldsymbol{r}\in\mathcal{X}(\tilde{\boldsymbol{r}})\}} \sigma(\boldsymbol{r})\right]} \tag{15.2.9}$$

was introduced by Kadanoff and Houghton (1975). In this formula τ is a real number and $\mathcal{X}(\tilde{\boldsymbol{r}})$ is some subset of $\mathcal{B}(\tilde{\boldsymbol{r}})$. It can be shown that, in the limit $\tau \to \infty$, we recover the majority-rule weight function (15.2.7), if $\mathcal{X}(\tilde{\boldsymbol{r}}) = \mathcal{B}(\tilde{\boldsymbol{r}})$, and the decimation weight function (15.2.6), if $\mathcal{X}(\tilde{\boldsymbol{r}})$ is the single site $\mathring{\boldsymbol{r}} \in \mathcal{B}(\tilde{\boldsymbol{r}})$.

In itself the weight function chosen does not lead to any approximation as long as the summation on the left-hand side of (15.1.13) can be evaluated exactly. This means that the calculation of thermodynamic functions (Sect. 15.4) would be exact. What can be lost are fixed points associated with particular parts of phase space. This leads to the inability to map the part of the phase space associated with these fixed points. The usual rule employed is to choose a weight function, which, when applied to the ground state in $\mathcal{N}$ associated with a phase of interest, leads to the same ground state in the lattice $\widetilde{\mathcal{N}}$. This idea will be illustrated by reference to the one-dimensional Ising model in Sect. 15.5.1.

Beginning in the early 1970s the general pattern of development described here was used to construct approximate renormalization group transformations (Niemeijer and van Leeuwen 1976) and to gain a new perspective on the known solutions of some exactly-solvable models (Nelson and Fisher 1975). One problem, recognized at an early stage, was the difficulty of obtaining a consistent set of equations, of the form of (15.2.1) and (15.2.2), from (15.1.13). The number of equations generated by choosing all possible states of the microsystems on $\widetilde{\mathcal{N}}$ is $\nu^{\tilde{N}}$. Because of the symmetries of the system many of these equations will be identical. However, within this set, there must be exactly $n+1$ independent equations which express the new set of couplings as functions of old set. It is not guaranteed that this can be achieved. In fact, as we shall see in Sect. 15.5.2, for the two-dimensional Ising model on a square lattice with first-neighbour couplings alone, using a decimation weight function, this is impossible. Consistent equations can be obtained by adding second-neighbour and four-site interactions in $\widetilde{\mathcal{N}}$. These, however, would in their turn demand the inclusion of further couplings at the next iteration. This *proliferation* of couplings led to the development of a number of approximation methods designed to 'cut-off' the proliferation in a more or less ad hoc manner.

Equations (15.2.1) and (15.2.2) must not only be consistent, but $\mathcal{K}_i$, $i = 0, 1, \ldots, n$, must be smooth functions of the couplings ζ_j, $j = 1, 2, \ldots, n$. For explicit approximation methods for finite systems this is not normally a problem. However, as was first recognized by Griffiths and Pearce (1978, 1979) (see also Griffiths 1981), this condition is not obviously true in the thermodynamic limit. The fundamental point is that, whereas we need, for the existence of phase transitions, non-smooth behaviour in the partition function (15.1.1), we require smooth behaviour from the summation on the left-hand side of (15.1.13), which differs from $Z(N, \zeta_i)$ only by the presence of the weight function. It is by no means assured that the weight function is able to suppress all non-smooth behaviour. As an example Griffiths and Pearce (1979) considered the case, for a spin-$\frac{1}{2}$ model, of the Kadanoff–Houghton weight function, where $\mathcal{X}(\tilde{\boldsymbol{r}}) = \mathring{\boldsymbol{r}} \in \mathcal{B}(\tilde{\boldsymbol{r}})$, yielding

$$p(\sigma(\tilde{\boldsymbol{r}})|\sigma(\boldsymbol{r}), \boldsymbol{r} \in \mathcal{B}(\tilde{\boldsymbol{r}})) = \frac{\exp\left[\tau\, \sigma(\tilde{\boldsymbol{r}})\sigma(\mathring{\boldsymbol{r}})\right]}{2\cosh(\tau)} . \tag{15.2.10}$$

This gives, apart from a multiplicative constant, for the right-hand side of (15.1.13) the partition function of the model with the addition of an external field-type coupling $\tau\, \sigma(\tilde{\boldsymbol{r}})$ on the subset of sites $\mathring{\mathcal{N}}$. Given that the model exhibits phase transitions it is not necessarily the case that the extra external field is sufficient to lead to their suppression. In particular, if the decimation limit $\tau \to \infty$ is chosen, it is possible by a judicious choice of the signs of the states on $\mathring{\mathcal{N}}$ (Griffiths 1981) to cancel out the additional field, leading to zero-field models known to have phase transitions. Transformations which give unwelcome phase transitions in the modified partition function on the right-hand side of (15.1.13) are said by Griffiths and Pearce (1979) to exhibit *peculiarities*. Examples are given by these authors of schemes with and without such behaviour. One simple observation with respect to the decimation weight function can be made. This is that if, as in Sects. 15.5.1 and 15.5.2, the sites of $\mathcal{N} - \mathring{\mathcal{N}}$ form disconnected clusters then no peculiarities will occur. This is clear from the fact that each cluster constitutes an effective finite-lattice system, which cannot have a phase transition.

A number of authors (Wilson 1975; Kadanoff and Houghton 1975; van Leeuwen 1975) have drawn attention to another problem with the decimation weight function. Consider any subset $\{\boldsymbol{r}_1, \boldsymbol{r}_2, \ldots\}$ of $\mathring{\mathcal{N}} \subset \mathcal{N}$. These sites will remain as sites of $\widetilde{\mathcal{N}}$. It is clear that the expectation value $\langle \sigma(\boldsymbol{r}_1)\sigma(\boldsymbol{r}_2)\cdots\rangle$ will have the same value in both $\mathcal{N}$ and $\widetilde{\mathcal{N}}$. In particular the pair correlation function, defined as in (4.2.3), for the states at $\boldsymbol{r}_1$ and $\boldsymbol{r}_2$ satisfies the scaling form

$$\Gamma_2(\theta_1, \ldots, \theta_n; \sigma(\boldsymbol{r}_2 - \boldsymbol{r}_1)) = \Gamma_2(\lambda^{y_1}\theta_1, \ldots, \lambda^{y_n}\theta_n; \sigma(\lambda^{-1}[\boldsymbol{r}_2 - \boldsymbol{r}_1])) . \tag{15.2.11}$$

It follows from (4.5.48) that the scaling exponent associated with the coupling conjugate to the spin variable is equal to d, the physical dimension of the system. This means that everywhere on a critical surface the pair correlation function is equal to its value at the fixed point controlling the surface and, from (4.5.40) and (4.5.56), $\beta = 0$, $\delta = \infty$ and $\eta = 2 - d$. While these results are true for the one-dimensional

Ising model, where the phase transition occurs at zero temperature with a discontinuity in field (Sect. 15.5.1), they are certainly untrue for the two-dimensional Ising model (Eq. (12.3.280)) and conflict with series estimates for the three-dimensional Ising model according to which $\beta \simeq 0.33$, $\delta \simeq 4.8$ and $\eta \simeq 0.04$. Although it may be argued that this failure of an *exact* decimation transformation is hardly relevant to real situations, where in most cases decimation is used in conjunction with further approximation, it is of some interest to understand why it occurs. The basis of scaling and of the use of the renormalization group is, as has already been remarked, the perception that critical behaviour arises from long-range correlations. The application of the renormalization group in most cases averages over the short-range structure of the system leaving the long-range behaviour and thus the critical properties unaffected. As has been observed by Sneddon and Barber (1977), this is not the case with the decimation weight function which sums over some short and some long-range effects. This may be understand as a loss of information about critical behaviour leading to the absence of asymptotic decay of correlations.

15.3 Fixed Points and the Linear Renormalization Group

To analyze the nature of the fixed point given by (15.2.4) we linearize the recurrence relationships (15.2.2) about $\mathsf{P}^\star$. With

$$[\boldsymbol{\Delta\zeta}^{(s)}]^{\mathrm{T}} := (\zeta_1^{(s)} - \zeta_1^\star, \zeta_2^{(s)} - \zeta_2^\star, \ldots, \zeta_n^{(s)} - \zeta_n^\star), \tag{15.3.1}$$

we have

$$\boldsymbol{\Delta\zeta}^{(s+1)} \simeq \boldsymbol{L}^\star \boldsymbol{\Delta\zeta}^{(s)}, \tag{15.3.2}$$

where $\boldsymbol{L}^\star$ is the fixed-point value of

$$\boldsymbol{L} := \begin{pmatrix} \dfrac{\partial \mathcal{K}_1}{\partial \zeta_1} & \dfrac{\partial \mathcal{K}_1}{\partial \zeta_2} & \cdots & \dfrac{\partial \mathcal{K}_1}{\partial \zeta_n} \\ \dfrac{\partial \mathcal{K}_2}{\partial \zeta_1} & \dfrac{\partial \mathcal{K}_2}{\partial \zeta_2} & \cdots & \dfrac{\partial \mathcal{K}_2}{\partial \zeta_n} \\ \vdots & \vdots & \ddots & \vdots \\ \dfrac{\partial \mathcal{K}_n}{\partial \zeta_1} & \dfrac{\partial \mathcal{K}_n}{\partial \zeta_2} & \cdots & \dfrac{\partial \mathcal{K}_n}{\partial \zeta_n} \end{pmatrix}. \tag{15.3.3}$$

In general $\boldsymbol{L}^\star$ is not symmetric and thus we have the left and right eigenvalue equations,

$$\boldsymbol{w}_j^{\mathrm{T}} \boldsymbol{L}^\star = \Lambda_j \boldsymbol{w}_j^{\mathrm{T}}, \tag{15.3.4}$$

$$\boldsymbol{L}^\star \boldsymbol{x}_j = \boldsymbol{x}_j \Lambda_j, \tag{15.3.5}$$

respectively, where $\boldsymbol{w}_j$ and $\boldsymbol{x}_j$ denote the left and right eigenvectors, for the eigenvalue Λ_j. Assuming that $\boldsymbol{L}^\star$ is a simple matrix (Sect. 17.2)

$$\boldsymbol{w}_j \boldsymbol{\cdot} \boldsymbol{x}_{j'} = \delta^{\mathrm{Kr}}(j - j'). \tag{15.3.6}$$

From (15.3.2) and (15.3.4),

$$\boldsymbol{w}_j \boldsymbol{\cdot} \Delta\boldsymbol{\zeta}^{(s+1)} \simeq \Lambda_j \boldsymbol{w}_j \boldsymbol{\cdot} \Delta\boldsymbol{\zeta}^{(s)}. \tag{15.3.7}$$

Suppose, as in Sect. 4.3, that there exist scaling fields

$$\theta_j = \theta_j(\Delta\zeta_i), \tag{15.3.8}$$

which are smooth functions of the couplings (and thus of $\Delta\zeta_j, j = 1, 2, \ldots, n$) and such that

$$\theta_j(0) = 0, \tag{15.3.9}$$

$$\theta_j^{(s+1)} = \Lambda_j \theta_j^{(s)}. \tag{15.3.10}$$

From (15.3.6), (15.3.7) and (15.3.10) it follows that

$$\theta_j \simeq \boldsymbol{w}_j \boldsymbol{\cdot} \Delta\boldsymbol{\zeta}, \tag{15.3.11}$$

$$\Delta\boldsymbol{\zeta} \simeq \sum_{j=1}^{n} \boldsymbol{x}_j \theta_j. \tag{15.3.12}$$

Iterating (15.3.10)

$$\theta_j^{(s)} = \Lambda_j^s \theta_j^{(0)} \tag{15.3.13}$$

and thus, from (15.3.12),

$$\Delta\boldsymbol{\zeta}^{(s)} \simeq \sum_{j=1}^{n} \Lambda_j^s \boldsymbol{x}_j \theta_j^{(0)}. \tag{15.3.14}$$

Given that the scaling fields exist, they are a set of curvilinear coordinates in a region surrounding $\mathsf{P}^\star$ with the following properties:

(i) $\theta_j = 0$ is an $(n-1)$-dimensional subspace in $\mathcal{S}$, which, from (15.3.10), is invariant under the transformation. From (15.3.11) the left eigenvector $\boldsymbol{w}_j$ is normal to this subspace at $\mathsf{P}^\star$.
(ii) $\{\theta_i = 0, \forall\, i \neq j\}$ is an invariant line in $\mathcal{S}$, which, from (15.3.12), is tangential to $\boldsymbol{x}_j$ at $\mathsf{P}^\star$.
(iii) Supposing that all the eigenvalues are real:

(a) If $\Lambda_j > 1$ and $\theta_j^{(0)} \neq 0$ then $\theta_j^{(s)}$ increases as s increases. The trajectory iterates away from $\mathsf{P}^\star$ with a component in the direction of $\boldsymbol{x}_j$ and Λ_j is *relevant*.
(b) If $\Lambda_j < 1$ and $\theta_j^{(0)} \neq 0$ then $\theta_j^{(s)}$ decreases as s increases. The trajectory approaches the subspace $\theta_j = 0$ and Λ_j is *irrelevant*. It will, however, approach $\mathsf{P}^\star$ only if there is no relevant scaling field θ_i with $\theta_i^{(0)} \neq 0$.
(c) If $\Lambda_j = 1$ it is *marginal* and θ_j remains unchanged under iteration.

The exponents y_j for $\mathsf{P}^\star$ are defined by the formulae

$$\Lambda_j := \lambda^{y_j}, \qquad j = 1, 2, \ldots, n \tag{15.3.15}$$

and we recover from (15.3.10) the scaling transformation (4.3.3). Exponents are named in the same way as their respective eigenvalues. Since $\lambda > 1$, an exponent is relevant if it is positive, marginal if it is zero and irrelevant if it is negative (Sect. 4.3). The region around $\mathsf{P}^\star$ with tangent space spanned by the set of right eigenvectors corresponding to irrelevant eigenvalues is the *basin of attraction* $\mathfrak{B}(\mathsf{P}^\star)$. A *sink* has only irrelevant eigenvalues. A fixed point which has a neighbourhood containing no other fixed point is called *isolated*. As indicated above this usually means that it has no marginal eigenvalues.

This analysis is now illustrated by two examples for the case $n = 3$.

(1) $\Lambda_1 > 1, \Lambda_2 < 1, \Lambda_3 < 1$. The region around $\mathsf{P}^\star$ for this case is shown in Fig. 15.1. The local part of $\mathfrak{B}(\mathsf{P}^\star)$ is the surface $\theta_1 = 0$. Trajectories on this surface will converge on $\mathsf{P}^\star$. Any trajectory near to the fixed point, but not on $\theta_1 = 0$ will iterate away from the fixed point, tending towards the curve $\theta_2 = \theta_3 = 0$.

(2) $\Lambda_2 > \Lambda_1 > 1, \Lambda_3 < 1$. Λ_2 and Λ_1 are called respectively the *strong* and *weak* relevant eigenvalues. This is the case shown in Fig. 15.2. The local part of $\mathfrak{B}(\mathsf{P}^\star)$ is the curve $\theta_1 = \theta_2 = 0$. Any trajectory beginning on this curve will converge on $\mathsf{P}^\star$. Any other trajectory will be repelled by $\mathsf{P}^\star$. If it begins at a point where $\theta_2 \neq 0$ then it will tend towards the strong direction $\theta_1 = \theta_3 = 0$. Otherwise it will tend towards the weak direction $\theta_2 = \theta_3 = 0$.

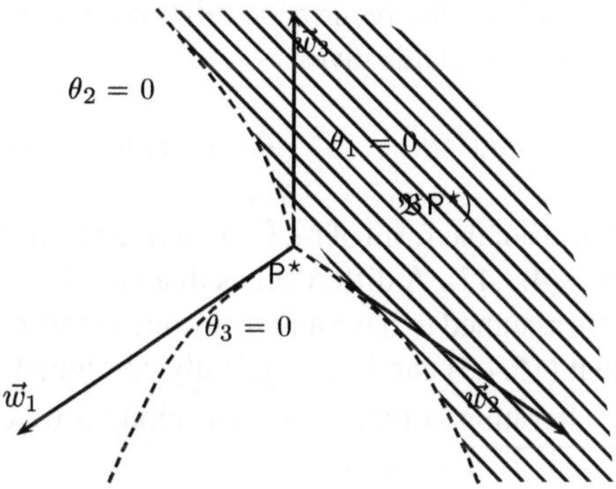

Fig. 15.1 The region around a fixed point $\mathsf{P}^\star$ for $n = 3$, with one relevant and two irrelevant eigenvalues

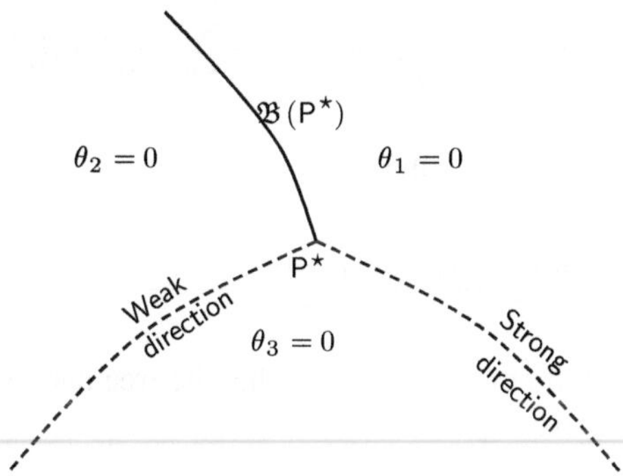

Fig. 15.2 The region around a fixed point $\mathsf{P}^\star$ for $n = 3$, with one irrelevant and two relevant eigenvalues

15.4 Free Energy and Densities

From (15.1.3) and (15.1.12)

$$\Phi(\widetilde{N}, \tilde{\zeta}_i) + \widetilde{N}\tilde{\zeta}_0 = \Phi(N, \zeta_i) + N\zeta_0. \tag{15.4.1}$$

Suppose that free energy per lattice site

$$\phi(\zeta_i) := \Phi(N, \zeta_i)/N, \tag{15.4.2}$$

is, at least for large N, independent of N (Sect. 1.3), then

$$\phi(\tilde{\zeta}_i) + \tilde{\zeta}_0 = \lambda^d \phi(\zeta_i) + \lambda^d \zeta_0, \tag{15.4.3}$$

which, from (15.2.1), becomes

$$\phi(\tilde{\zeta}_i) = \lambda^d[\phi(\zeta_i) - \mathcal{K}_0(\zeta_i)]. \tag{15.4.4}$$

Linearizing the recurrence relations in a neighbourhood of a fixed point in terms of the scaling fields gives

$$\lambda^{-1}\phi(\lambda^{y_1}\theta_1,\ldots,\lambda^{y_n}\theta_n)=\phi(\theta_1,\ldots,\theta_n)-\mathcal{K}_0(\theta_1,\ldots,\theta_n). \tag{15.4.5}$$

This is of the form of (4.3.79) derived from the Nightingale–'T Hooft scaling hypothesis (4.3.47). A difference is that here $\lambda>1$ is an integer, whereas in scaling theory it is assumed to have any real value in this range. This can be resolved if it is assumed that (15.4.5) can be analytically continued into non-integer values of λ.

Iterating equation (15.4.4) along a trajectory from a point with couplings $\zeta_i^{(0)}$, $i=1,2,\ldots,n$ gives

$$\phi(\zeta_i^{(0)})=\frac{1}{\lambda^{kd}}\phi(\zeta_i^{(k)})+\sum_{s=0}^{k-1}\frac{1}{\lambda^{sd}}\mathcal{K}_0(\zeta_i^{(s)}). \tag{15.4.6}$$

If we assume that

$$\lim_{k\to\infty}\frac{1}{\lambda^{kd}}\phi(\zeta_i^{(k)})=0, \tag{15.4.7}$$

it follows from (15.4.6) that the free energy per lattice site, at the initial point $\zeta_i^{(0)}$, is given by

$$\phi(\zeta_i^{(0)})=\sum_{s=0}^{\infty}\frac{1}{\lambda^{sd}}\mathcal{K}_0(\zeta_i^{(s)}). \tag{15.4.8}$$

In practice, it is usually found that this series converges after a very few iterations (see, for example, Southern and Lavis 1980). From (4.3.117) the densities at the initial point on the trajectory are given by

$$\rho_i^{(0)}=-\left(\frac{\partial\phi}{\partial\zeta_i}\right)^{(0)}. \tag{15.4.9}$$

To obtain these functions a chain differentiation must be performed along the trajectory. Let

$$\kappa_{\ell m}:=\frac{\partial\mathcal{K}_m}{\partial\zeta_\ell}, \tag{15.4.10}$$

$$\tau_{im}^{(0)}:=\kappa_{im}(\zeta_j^{(0)}), \tag{15.4.11}$$

$$\tau_{im}^{(k)}:=\sum_{\ell=1}^{n}\tau_{i\ell}^{(k-1)}\kappa_{\ell m}(\zeta_j^{(k)}),\qquad k=1,2,\ldots,s. \tag{15.4.12}$$

Then

$$\rho_i^{(0)} = \sum_{s=0}^{\infty} \frac{1}{\lambda^{sd}} \tau_{i0}^{(s)}. \tag{15.4.13}$$

The response functions, which correspond to the second-order partial derivatives of the free-energy density with respect to the couplings, can be obtained in a similar way.

15.5 Decimation for the Ising Model

15.5.1 In One Dimension

Although the one-dimensional Ising model can be solved quite simply using the transfer matrix methods of Chap. 11 (see, for example, Domb 1960, Sect. 3.2.1), it is illuminating to consider an RSRG treatment of the same problem using a decimation weight function. The only change made to the one-dimensional version of the Hamiltonian (3.5.1) is to add a constant term $-C$ at each lattice site so that the transfer matrix is

$$V := \mathfrak{z}_0 \begin{pmatrix} \mathfrak{z}_1\mathfrak{z}_2 & \mathfrak{z}_2^{-1} \\ \mathfrak{z}_2^{-1} & \mathfrak{z}_1^{-1}\mathfrak{z}_2 \end{pmatrix}, \tag{15.5.1}$$

where $\mathfrak{z}_0 := \exp(C)$ and, for convenience, $\mathfrak{z}_1 := \mathfrak{X}$ and $\mathfrak{z}_2 := \mathfrak{z}$. Then, with periodic boundary conditions applied so $\mathcal{N}$ forms a ring of sites, from (11.1.6),

$$Z(N, C, K, L) = \text{Trace}\{\boldsymbol{V}^N\}. \tag{15.5.2}$$

$\mathcal{N}$ is divided into blocks of λ neighbouring sites and the application of the decimation weight function corresponds to fixing the spin state on one site of each block (the first, say) and performing a partial sum over the remaining spin states. The cases $\lambda = 2$ and $\lambda = 3$ have been considered in detail by Nelson and Fisher (1975). The general case can be most conveniently formulated in terms of the transfer matrix when the partition function for $\widetilde{\mathcal{N}}$ is

$$Z(\widetilde{N}, \widetilde{C}, \widetilde{K}, \widetilde{L}) = \text{Trace}\{\widetilde{\boldsymbol{V}}^{\widetilde{N}}\}, \tag{15.5.3}$$

where

$$\widetilde{\boldsymbol{V}} := \boldsymbol{V}^{\lambda}. \tag{15.5.4}$$

Computing the elements of this formula for any λ yields the recurrence relationships.

We first investigate in detail the case $\lambda = 2$. It is not difficult to see that (15.5.4) yields the three relationships

$$\tilde{\mathfrak{z}}_0\tilde{\mathfrak{z}}_1\tilde{\mathfrak{z}}_2 = \mathfrak{z}_0^2(\mathfrak{z}_1^2\mathfrak{z}_2^2 + \mathfrak{z}_2^{-2}), \tag{15.5.5}$$

$$\tilde{\mathfrak{z}}_0\tilde{\mathfrak{z}}_1^{-1}\tilde{\mathfrak{z}}_2 = \mathfrak{z}_0^2(\mathfrak{z}_1^{-2}\mathfrak{z}_2^2 + \mathfrak{z}_2^{-2}), \tag{15.5.6}$$

$$\tilde{\mathfrak{z}}_0\tilde{\mathfrak{z}}_2^{-1} = \mathfrak{z}_0^2(\mathfrak{z}_1 + \mathfrak{z}_1^{-1}), \tag{15.5.7}$$

which give

$$\tilde{\mathfrak{z}}_0^4 = \mathfrak{z}_0^8(\mathfrak{z}_1^2\mathfrak{z}_2^2 + \mathfrak{z}_2^{-2})(\mathfrak{z}_1^{-2}\mathfrak{z}_2^2 + \mathfrak{z}_2^{-2})(\mathfrak{z}_1 + \mathfrak{z}_1^{-1})^2, \tag{15.5.8}$$

$$\tilde{\mathfrak{z}}_1^4 = \frac{\mathfrak{z}_1^2\mathfrak{z}_2^2 + \mathfrak{z}_2^{-2}}{\mathfrak{z}_1^{-2}\mathfrak{z}_2^2 + \mathfrak{z}_2^{-2}}, \tag{15.5.9}$$

$$\tilde{\mathfrak{z}}_2^4 = \frac{(\mathfrak{z}_1^2\mathfrak{z}_2^2 + \mathfrak{z}_2^{-2})(\mathfrak{z}_1^{-2}\mathfrak{z}_2^2 + \mathfrak{z}_2^{-2})}{(\mathfrak{z}_1 + \mathfrak{z}_1^{-1})^2}. \tag{15.5.10}$$

The formulae (15.5.8)–(15.5.10) are invariant under the transformation $\mathfrak{z}_1 \to 1/\mathfrak{z}_1$, which corresponds to reversing the direction of the magnetic field. In order to obtain recurrence relationships in a form appropriate to the one-dimensional Ising model we must bear in mind that, in this case, the Curie point lies at zero temperature, implying a fixed point on the zero-field axis at $T = 0$. The appropriate variables to choose are ones which are finite at zero temperature. If we concentrate on the region $\mathcal{H} \geq 0$ and $J \geq 0$ ($L \geq 0$, $K \geq 0$) then the variables

$$\mathfrak{u} := \mathfrak{z}_1^{-2} = \exp(-2L), \tag{15.5.11}$$

$$\mathfrak{t} := \frac{\mathfrak{z}_2^2\mathfrak{z}_1 - 1}{\mathfrak{z}_2^2\mathfrak{z}_1 + 1} = \tanh\left(\tfrac{1}{2}L + K\right) \tag{15.5.12}$$

lie in the interval [0, 1] and, from (15.5.9) and (15.5.10),

$$\tilde{\mathfrak{u}} = \frac{\mathfrak{u}^2(1+\mathfrak{t})^2 + (1-\mathfrak{t})^2}{2(1+\mathfrak{t}^2)}, \tag{15.5.13}$$

$$\tilde{\mathfrak{t}} = \frac{4\mathfrak{t}^2 - (1-\mathfrak{u})(\mathfrak{t}^2 - 1)}{4 + (1-\mathfrak{u})(\mathfrak{t}^2 - 1)}. \tag{15.5.14}$$

The first step is to determine the invariant lines. There are three of these:

(a) $\mathfrak{u} = 1$, corresponding to $\mathcal{H} = 0$. On this line there is the one recurrence relationship

$$\tilde{\mathfrak{t}} = \mathfrak{t}^2 \tag{15.5.15}$$

with *fixed points* $\mathfrak{t} = 1, \mathfrak{u} = 1$ and $\mathfrak{t} = 0, \mathfrak{u} = 1$.

(b) $\mathrm{t} = 1$, corresponding to $T = 0$. On this line there is the one recurrence relationship

$$\tilde{\mathrm{u}} = \mathrm{u}^2 \tag{15.5.16}$$

with the *fixed points* $\mathrm{t} = 1, \mathrm{u} = 1$ and $\mathrm{t} = 1, \mathrm{u} = 0$.

(c) Every point on the line

$$\mathrm{u} = \left(\frac{1-\mathrm{t}}{1+\mathrm{t}}\right)^2 \tag{15.5.17}$$

satisfies (15.5.13) and (15.5.14). This is a *line of fixed points* corresponding to $J = 0$.

The invariant lines, fixed points and the fixed line are shown in Fig. 15.3, together with examples of trajectory flows. We now consider the linearized recurrence relationships in the neighbourhood of each fixed point.

- $\mathrm{t} = 1, \mathrm{u} = 1$ is the *ferromagnetic fixed point* at $T = 0$, $\mathcal{H} = 0$. For this

$$\begin{pmatrix} \Delta\tilde{\mathrm{u}} \\ \Delta\tilde{\mathrm{t}} \end{pmatrix} = \begin{pmatrix} 2 & 0 \\ 0 & 2 \end{pmatrix} \begin{pmatrix} \Delta\mathrm{u} \\ \Delta\mathrm{t} \end{pmatrix}. \tag{15.5.18}$$

The eigenvalues, eigenvectors and exponents are given in Table 15.1(a). That the magnetic field exponent $y_1 = y_{\mathcal{H}} = d$ has already been seen in Sect. 15.2 to be a result of the use of the decimation weight function. In this particular case it can also be interpreted as a consequence of the first-order transition at $T = 0$ (Sect. 4.3) when the magnetization reverses direction as the magnetic field passes through zero.

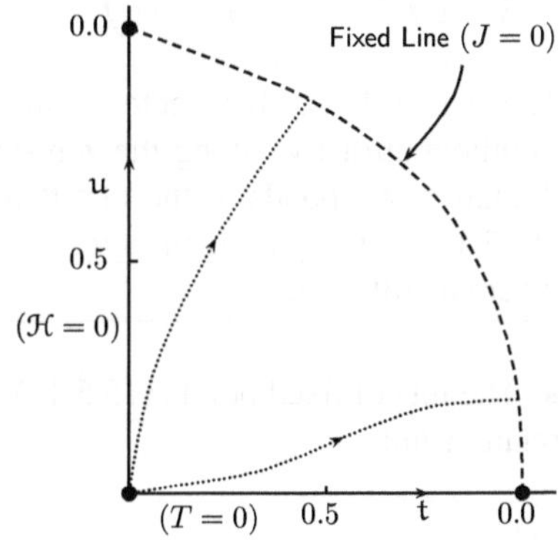

Fig. 15.3 The trajectory flows for the renormalization group transformation of the one-dimensional Ising model. The origin corresponds to $\mathrm{u} = \mathrm{t} = 1$ and *fixed points* are denoted by •

Table 15.1 The eigenvalues Λ_j, exponents y_j, and left and right eigenvectors $\boldsymbol{w}_j$ and $\boldsymbol{x}_j$ for (a) the ferromagnetic fixed point $t = 1, u = 1$, (b) the infinite temperature fixed point $t = 0$, $u = 1$, (c) the infinite field fixed point $t = 1$, $u = 0$, (d) the line of fixed points (15.5.17)

Λ_j	y_j	$\boldsymbol{w}_j$	$\boldsymbol{x}_j$	
2	1	(1,0)	(1,0)	(a)
2	1	(0,1)	(0,1)	
1	0	(1,0)	$(1,-\frac{1}{4})$	(b)
0	$-\infty$	$(\frac{1}{4},1)$	(0,1)	
0	$-\infty$	(1,0)	(1,0)	(c)
1	0	(0,1)	(0,1)	
0	$-\infty$	$h(t)(1, x(t))$	$h(t)(1, y(t))$	(d)
1	0	$h(t)(-y(t), 1)$	$h(t)(-x(t), 1)$	

• $t = 0, u = 1$ is the infinite temperature fixed point on $\mathcal{H} = 0$. For this we have

$$\begin{pmatrix} \Delta\tilde{u} \\ \Delta\tilde{t} \end{pmatrix} = \begin{pmatrix} 1 & 0 \\ -\frac{1}{4} & 2 \end{pmatrix} \begin{pmatrix} \Delta u \\ \Delta t \end{pmatrix}. \tag{15.5.19}$$

The eigenvalues, eigenvectors and exponents are given in Table 15.1(b). The fixed point is infinitely attractive along the zero-field axis. The marginal exponent $y_1 = 0$ corresponds to the fact that this fixed point in on the line of fixed points (15.5.17). The right eigenvector $\boldsymbol{x}_1$ is tangential to the line of fixed points at this fixed point, where $du/dt = -4$.

• $t = 1, u = 0$ is the infinite field fixed point on $T = 0$. For this we have

$$\begin{pmatrix} \Delta\tilde{u} \\ \Delta\tilde{t} \end{pmatrix} = \begin{pmatrix} 0 & 0 \\ 0 & 1 \end{pmatrix} \begin{pmatrix} \Delta u \\ \Delta t \end{pmatrix}. \tag{15.5.20}$$

The eigenvalues, eigenvectors and are given in Table 15.1(c). The fixed point is infinitely attractive along the zero-temperature axis. The marginal exponent $y_2 = 0$ again corresponds to the fact that this fixed point is on the line of fixed points (15.5.17). The right eigenvector $\boldsymbol{x}_2$ is tangential to the fixed line at the fixed point where $du/dt = 0$.

• The line of fixed points (15.5.17) at some $0 \leq t \leq 1$ has linearized recurrence relationships

$$\begin{pmatrix} \Delta\tilde{u} \\ \Delta\tilde{t} \end{pmatrix} = \begin{pmatrix} \dfrac{(1-t)^2}{1+t^2} & \dfrac{-8t(1-t)}{(1+t^2)(1+t)^3} \\ \dfrac{(t^2-1)(1+t)^2}{4(1+t^2)} & \dfrac{2t}{1+t^2} \end{pmatrix} \begin{pmatrix} \Delta u \\ \Delta t \end{pmatrix}. \tag{15.5.21}$$

The eigenvalues, eigenvectors and exponents, as functions of t, are given in Table 15.1(d), where

$$h(t) := \sqrt{\frac{2t}{1+t^2}}, \qquad x(t) := \frac{4(1-t)}{(1+t)^3}, \qquad y(t) := \frac{(1-t)(1+t)^3}{8t}. \tag{15.5.22}$$

The line of fixed points is infinitely attractive to trajectories which approach it. The marginal exponent $y_2 = 0$ has a corresponding right eigenvector $\boldsymbol{x}_2$ which is tangential to the fixed line, where $du/dt = -4(1-t)/(1+t)^3$.

We now calculate the free energy per lattice site. From (15.2.1), (15.5.8), (15.5.11) and (15.5.12)

$$\mathcal{K}_0(t, u) = -\tfrac{1}{8} \ln \left\{ \frac{2(1+u^{-1})^2(1+t^2)[u^2(1+t)^2 + (1-t)^2]}{(1-t^2)^2} \right\}. \tag{15.5.23}$$

The free energy per lattice site at some initial point$(u^{(0)}, t^{(0)})$ is now obtained by inserting this expression into the series (15.4.8), with $\lambda = 2$, $d = 1$. An interesting problem is to show that this series converges to the standard closed-form expression obtained by transfer matrix methods. Translated in the present notation this formula (Domb 1960, Sect. 3.2.1, Eq. 6) becomes

$$\phi_1(t, u) = -\ln \left\{ \frac{1}{2u^{3/4}} \left(\frac{1+t}{1-t} \right)^{1/2} \left[1 + u + \sqrt{(1-u)^2 + 4\left(\frac{1-t}{1+t} \right)^2} \right] \right\}. \tag{15.5.24}$$

It is quite simple to write a computer program, using the recurrence relationships (15.5.13) and (15.5.14) to compute $\phi(u^{(0)}, t^{(0)})$, from (15.5.23) and (15.4.8). This will give convincing evidence that the series converges to the value given by (15.5.24). In general it would appear to be rather difficult to sum the series analytically. In the case $\mathcal{H} = 0$, $(u = 1)$, however, the problem has been solved by Nauenberg (1975). His argument, translated into our notation, is as follows. From (15.5.23),

$$\mathcal{K}_0(t, 1) = -\tfrac{1}{4} \ln \left[4 \left(\frac{1+t^2}{1-t^2} \right) \right] \tag{15.5.25}$$

and hence, from (15.5.14) and (15.4.8),

$$\phi_1(\mathrm{t}, 1) = -\sum_{s=1}^{\infty} \frac{1}{2^s}\left[\ln(2) + \frac{1}{2}\ln\left(\frac{1+\mathrm{t}^{2^s}}{1-\mathrm{t}^{2^s}}\right)\right] = -\ln\left[2\prod_{s=1}^{\infty}\left(\frac{1+\mathrm{t}^{2^s}}{1-\mathrm{t}^{2^s}}\right)^{1/2^{s+1}}\right]. \tag{15.5.26}$$

Using the identity

$$\frac{1}{\sqrt{1-q^2}} = \prod_{s=1}^{\infty}\left(\frac{1+q^{2^s}}{1-q^{2^s}}\right)^{1/2^{s+1}}, \tag{15.5.27}$$

gives

$$\phi_1(\mathrm{t}, 1) = -\ln(2) + \frac{1}{2}\ln(1-\mathrm{t}^2), \tag{15.5.28}$$

which also follows from (15.5.24).

For the ferromagnetic fixed point the exponents given in Table 15.1 can be used together with formulae (4.5.40) and (4.5.39), to give the critical exponents

$$\alpha = 1, \qquad \beta = 0, \qquad \gamma = 1, \qquad \delta = \infty. \tag{15.5.29}$$

The values of β and δ both arise from the first-order phase transition condition $y_1 = d$. Because the matrix in (15.5.18) is diagonal the scaling fields can be identified as

$$\theta_1 = -\Delta\mathrm{u} \simeq 2L, \tag{15.5.30}$$

$$\theta_2 = -\Delta\mathrm{t} \simeq 2\exp(-2K). \tag{15.5.31}$$

From the general analysis of scaling at a critical point described in Sect. 4.5, $\theta_1 \sim \Delta L$ and $\theta_2 \sim \Delta K$. It follows from (15.5.30) that the first of these two formulae still holds in this case since $L_c = 0$. However, the fact that the critical temperature is $T_c = 0$ implies that $K_c = \infty$, leading to the different choice of scaling field θ_2 given by (15.5.31). The formula

$$\frac{\partial m}{\partial \mathcal{H}} = \frac{1}{T}\exp\left(\frac{2J}{T}\right), \tag{15.5.32}$$

(Domb 1960, Sect. 3.2.1, Eqs. 9 and 10) for the response of the magnetization to change of magnetic field, is now consistent with that derived from scaling using the value of γ in (15.5.29) and (15.5.31). A similar analysis is applicable to other one-dimensional models.

For the case $\lambda = 2$ we have considered only the ferromagnetic situation $J > 0$. This is because application of this weight function to the lattice, by removing alternate sites, maps the antiferromagnetic ground state of alternating spins into the ferromagnetic ground state of aligned spins. In general, from (15.5.10), a point in phase space with $\mathfrak{z}_2 < 1$ is mapped in one step into a point with $\mathfrak{z}_2 > 1$. Although this does not affect the evaluation of the free energy which must be correct for either sign of J, (since the method is exact), it does mean that this method does not give fixed points associated with the antiferromagnetic model. In particular it is not possible to obtain the fixed point on the zero-temperature axis associated with the critical field (Sect. 3.5.1.1). On way of achieving this (Nelson and Fisher 1975) is to take $\lambda = 3$. In fact it is obvious that any odd value of λ would suffice. In the special case of zero magnetic field, it is not difficult to show that (15.5.15) generalizes to

$$\tilde{\mathfrak{t}} = \mathfrak{t}^{\lambda}, \tag{15.5.33}$$

which has the ferromagnetic fixed point $\mathfrak{t} = 1$ and the antiferromagnetic fixed point $\mathfrak{t} = -1$.

15.5.2 In Two Dimensions

We now explore the possibility of using the decimation procedure, described in the previous section, for a the first-neighbour Ising model on a square lattice with zero magnetic field. The lattice is divided into black and white sites as shown in Fig. 15.4 and the summation in (15.1.13) is over the states on all the white sites, leaving the spins on the black sites fixed. With the spin states denoted appropriately, summing over the spin state σ_{G} on site G at the centre of the square ABCD gives

$$2\exp(2C)\cosh[K_1(\sigma_{\mathsf{A}} + \sigma_{\mathsf{B}} + \sigma_{\mathsf{C}} + \sigma_{\mathsf{D}})], \tag{15.5.34}$$

where C and $K_1 := J/T$ are the trivial and first-neighbour couplings respectively. The left-hand side of (15.1.13) consists of a product of $N/2$ terms of this form, one

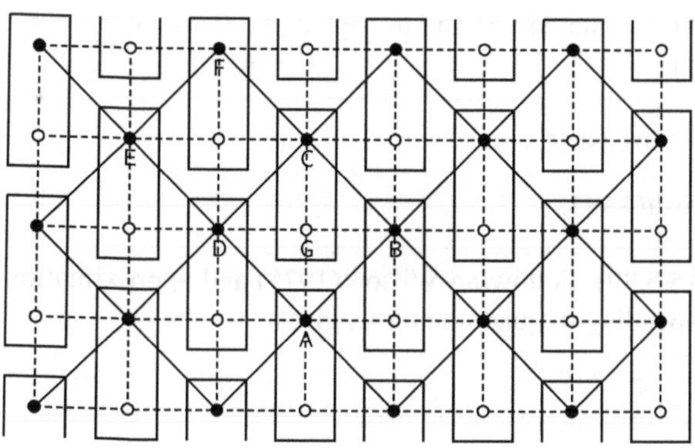

Fig. 15.4 Two site blocks for the first-neighbour Ising model on a square lattice

for each of the squares of type ABCD as shown in Fig. 15.4. It was pointed out by Wilson (1975) that this expression can be rewritten in the form

$$\exp\big[2C_0 + \mathcal{K}_0 + \tfrac{1}{2}\mathcal{K}_1(\sigma_A\sigma_B + \sigma_B\sigma_C + \sigma_C\sigma_D + \sigma_D\sigma_A) \\ + \mathcal{K}_2(\sigma_A\sigma_C + \sigma_B\sigma_D) + \mathcal{K}_3\sigma_A\sigma_B\sigma_C\sigma_D\big], \tag{15.5.35}$$

where

$$\mathcal{K}_0 := \ln(2) + \tfrac{1}{8}\ln\cosh(4K_1) + \tfrac{1}{2}\ln\cosh(2K_1), \tag{15.5.36}$$

$$\mathcal{K}_1 := \tfrac{1}{4}\ln\cosh(4K_1), \tag{15.5.37}$$

$$\mathcal{K}_2 := \tfrac{1}{8}\ln\cosh(4K_1), \tag{15.5.38}$$

$$\mathcal{K}_3 := \tfrac{1}{8}\ln\cosh(4K_1) - \tfrac{1}{2}\ln\cosh(2K_1). \tag{15.5.39}$$

By setting

$$\widetilde{C} := 2C + \mathcal{K}_0(K_1), \tag{15.5.40}$$

$$\widetilde{K}_i := \mathcal{K}_i(K_1), \qquad i = 1, 2, 3, \tag{15.5.41}$$

we have recurrence relationships of the form of (15.2.1) and (15.2.2), with $d = 2$ and $\lambda = \sqrt{2}$, where $\widetilde{K}_1$ is the first-neighbour coupling on the lattice of black sites and $\widetilde{K}_2$ and $\widetilde{K}_3$ are respectively a second-neighbour coupling and a four-site coupling on this new lattice.

The effect of decimation has been to generate new couplings. We can consider the strategy of going back to the original lattice and including second-neighbour and four-site couplings at this stage. These, however, have the effect of coupling white sites together and it is not difficult to see that this will lead to further couplings. This is the beginning of the proliferation of couplings referred to in Sect. 15.2. One possible approximation is to include a second-neighbour interaction only between black sites in the original lattice and to neglect the four-site coupling. This leads to the modification

$$\widetilde{K}_1 = K_2 + \tfrac{1}{4}\ln\cosh(4K_1), \tag{15.5.42}$$

$$\widetilde{K}_2 = \tfrac{1}{8}\ln\cosh(4K_1), \tag{15.5.43}$$

of (15.5.36)–(15.5.39). Following Wilson (1975) and approximating (15.5.42) and (15.5.43) by expanding to quadratic terms gives

$$\widetilde{K}_1 = K_2 + 2K_1^2, \tag{15.5.44}$$

$$\widetilde{K}_2 = K_1^2. \tag{15.5.45}$$

These have the fixed points $K_1 = K_2 = 0$ and $K_1 = K_2 = \infty$, which are both sinks and the fixed point $K_1^\star = \frac{1}{3}$, $K_2^\star = \frac{1}{9}$. Linearizing (15.5.44) and (15.5.45) about this fixed point

$$\begin{pmatrix} \Delta\widetilde{K}_1 \\ \Delta\widetilde{K}_2 \end{pmatrix} = \begin{pmatrix} \frac{4}{3} & 1 \\ \frac{2}{3} & 0 \end{pmatrix} \begin{pmatrix} \Delta K_1 \\ \Delta K_2 \end{pmatrix}. \tag{15.5.46}$$

The eigenvalues of the matrix are $\Lambda_1 = \frac{1}{3}(2 + \sqrt{10}) = 1.721$ and $\Lambda_2 = \frac{1}{3}(2 - \sqrt{10}) = -0.387$. With $\lambda = \sqrt{2}$ it follows from (15.3.15) that Λ_1 gives the relevant (thermal) exponent $y_1 = y_T = 1.566$. The fixed point $(K_1^\star, K_2^\star)$ lies on a critical curve, which cuts the axes in the plane of couplings at $(0, 0.392)$ and $(0.392, 0)$ (Fig. 15.5). We expected these two values to be the same, since an Ising model with only second-neighbour couplings is equivalent to two disconnected Ising models with first-neighbour couplings. Trajectories which begin at points enclosed within the region bounded by the axes and the critical curve iterate ultimately to the sink at the origin. Trajectories not beginning on the critical curve and outside this region iterate to the sink at infinity. The fixed point $(K_1^\star, K_2^\star)$ on the critical curve is attractive to trajectories which begin on the curve. The irrelevant eigenvalue Λ_2 controls the attraction to the fixed point for such trajectories. That it is negative, giving an imaginary exponent y_2, conflicts with the assumption we have made in Sect. 15.3. However, this is not a serious problem. It simply means that the iteration process will jump back forth from one side of the fixed point to the other as shown in Fig. 15.5. The crossing points of the critical curve with the axes give an approximation $T_c/J = 2.550$ to the critical temperature of the two-dimensional Ising model, to be compared with the exact value 2.269 given by (9.2.36). Also, from

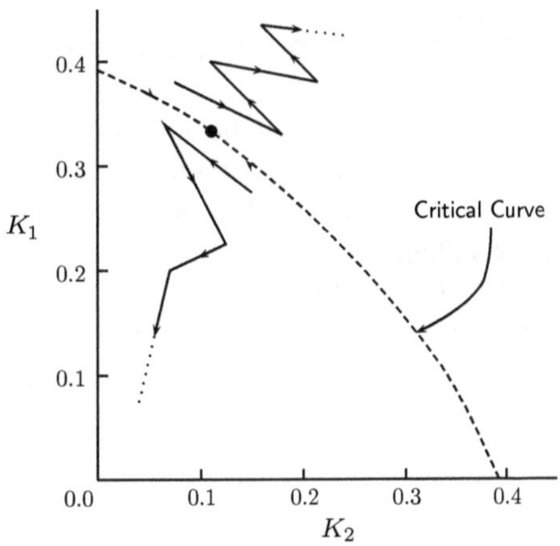

Fig. 15.5 The phase diagram for the approximate treatment of the two-dimensional Ising model using the recurrence relations (15.5.44) and (15.5.45)

(4.5.53) and the value for the exponent given above, we have the approximate value for $\nu = 1/y_T = 0.639$, which should be compared with the exact value of unity (Sect. 12.3.9). Given the crudeness of the approximation the estimate of the critical temperature is surprisingly good although the value of ν is poor. It is interesting to see that if (15.5.42) and (15.5.43) are used rather than (15.5.44) and (15.5.45) the estimates for the critical temperature and ν are worse. The phase diagram again has the same structure but now $T_c/J = 1.582$ and $\nu = 0.556$. This example of ad hoc approximation, applied to a model for which exact results are known, serves to illustrate the rather arbitrary nature of the improvements and deteriorations in results which can occur. It stands as a warning when such methods are applied to models for which exact results are not available.

The problem with proliferating interactions, arises also for the other standard two and three-dimensional lattices, although it can be avoided in hierarchical lattices, where decimation amounts to a reversal of the procedure by which the hierarchical lattice was created (Kaufman and Griffiths 1981). A type of exact decimation procedure for the Ising model on a triangular lattice has been developed by Hilhorst et al. (1978, 1979). This is based on the star-triangle transformation. It avoids the problem of proliferation of interactions, but leads to spatially dependent couplings. Nevertheless the fixed point can be located analytically and the thermal exponent determined exactly.

15.6 The Kosterlitz–Thouless Transition

As we have seen the standard procedure of the renormalization group is to sum or integrate out short-distance details. The scheme of Wilson (1975) works with the Hamiltonian in wave-vector space and integrates out the high-frequency short-wavelength contributions. In the RSRG procedures discussed in Sect. 15.5 short-range detail is eliminated by removing a proportion of the lattice sites leading to a increase in the lattice spacing.

The Kosterlitz–Thouless transition in the two-dimensional XY model was discussed in Sect. 3.6.1.1. Its RSRG analysis by Kosterlitz (1974) (see also, Pierson 1997) has something of the flavour of the Wilson approach, but is performed in real-space and involves integrations which increase the lattice distance from a to $a + \Delta a$. Restoring the partition function to its original form will then involve a rescaling of the couplings.

Attention will be restricted to the case $\omega_j = \pm 1$, with $\boldsymbol{\omega}$ denoting the set of winding numbers of all the vortices. In terms of[3]

$$\zeta_1 := \exp(K\mu), \qquad \zeta_2 := \pi K, \tag{15.6.1}$$

[3] This mixed notation where ζ_2 is a coupling and ζ_1 is a Boltzmann factor turns out the be the most convenient in this case.

and with the inclusion of a trivial coupling ζ_0, the Hamiltonian (3.6.41) takes the form

$$\widehat{H}(\boldsymbol{\zeta}, \boldsymbol{\omega}, a; \{\boldsymbol{r}_j^\star\}) = V\zeta_0/a^2 - \zeta_2 \sum_{\{j \neq i\}} \omega_j \omega_i \ln \left| \frac{\boldsymbol{r}_j^\star - \boldsymbol{r}_i^\star}{a} \right| - N_{\mathrm{v}} \ln(\zeta_1). \tag{15.6.2}$$

The number of lattice sites is $N^\star = N$, which we suppose to be even and, given that the quasi-ordered state consists of bound pairs of vortices, we suppose that there can be $N_{\mathrm{v}} =: 2n$ of these, each of winding number ± 1. At this point we drop the '$\star$' reference to the dual lattice in the Hamiltonian and allow integrations in the partition function to range over all positions $\{\boldsymbol{r}_j\}$ for the vortices in the plane $\mathcal{V}$ subject only to the condition that, for every pair of υ_j and υ_i, $|\boldsymbol{r}_j - \boldsymbol{r}_i| \geq a$. With the order of the variable $\boldsymbol{r}_1, \boldsymbol{r}_2, \ldots, \boldsymbol{r}_{2n}$, fixed, the partition function will involve integration over these variables beginning with $\boldsymbol{r}_1$. At the stage of integration with respect to $\boldsymbol{r}_j$ the only variables remaining are $\boldsymbol{r}_j, \boldsymbol{r}_{j+1}, \ldots, \boldsymbol{r}_{2n}$, so the region of integration is

$$\mathcal{R}_j(a) := \mathcal{V} - \sum_{i=j+1}^{2n} \mathcal{D}(a; \boldsymbol{r}_j - \boldsymbol{r}_i), \tag{15.6.3}$$

where

$$\mathcal{D}(a; \boldsymbol{r}) := \{\boldsymbol{r} : |\boldsymbol{r}| < a\}. \tag{15.6.4}$$

The partition function is

$$\begin{aligned} Z(\boldsymbol{\zeta}, \boldsymbol{\omega}, a) = \exp(-V\zeta_0/a^2) \sum_{n=1}^{N/2} \frac{\zeta_1^{2n}}{(n!)^2} \int_{\mathcal{R}_{2n}(a)} \mathrm{d}\mathcal{V}_{2n} \cdots \int_{\mathcal{R}_1(a)} \mathrm{d}\mathcal{V}_1 \\ \times \exp\left(2\,\zeta_2 \sum_{\{i>j\}} \omega_j \omega_i \ln \left| \frac{\boldsymbol{r}_i - \boldsymbol{r}_j}{a} \right| \right), \end{aligned} \tag{15.6.5}$$

where the factors $n!$ count the rearrangements of the identical vortices and $\mathrm{d}\mathcal{V}_j$ denotes integration with respect to the variable $\boldsymbol{r}_j$. The aim now is to obtain renormalized parameters $\tilde{\zeta}_0$, $\tilde{\zeta}_1$ and $\tilde{\zeta}_2$ so that the invariance condition (15.1.12) holds in the form

$$Z(\tilde{\boldsymbol{\zeta}}, \boldsymbol{\omega}, a + \Delta a) = Z(\boldsymbol{\zeta}, \boldsymbol{\omega}, a) \tag{15.6.6}$$

for this system. In this case the two-dimensional volume V of the system is fixed and the number of lattice sites is reduced by the increase in lattice spacing so that $\tilde{N} = N[a/(a + \Delta a)]^2$. The regions of integration in the two partition functions in (15.6.6) are related by

$$\mathcal{R}_j(a) = \mathcal{R}_j(a + \Delta a) + \sum_{i=j+1}^{2n} \Delta\mathcal{D}(a; \boldsymbol{r}_i - \boldsymbol{r}_j), \tag{15.6.7}$$

where

$$\Delta\mathcal{D}(a; \boldsymbol{r}) := \mathcal{D}(a + \Delta a; \boldsymbol{r}) - \mathcal{D}(a; \boldsymbol{r}), \tag{15.6.8}$$

giving

$$\int_{\mathcal{R}_j(a)} \mathrm{d}\mathcal{V}_j = \int_{\mathcal{R}_j(a+\Delta a)} \mathrm{d}\mathcal{V}_j + \sum_{i=j+1}^{2n} \int_{\Delta\mathcal{D}(a;\boldsymbol{r}_i-\boldsymbol{r}_j)} \mathrm{d}\mathcal{V}_j. \tag{15.6.9}$$

The contribution to the partition function arising from a term in the sum in (15.6.9) is $O(\Delta a)$. So when applying it to (15.6.5) the result to this order will contain only single integrals of this type. Thus there will be a contribution arising from the first term on the right of (15.6.9) plus a set of terms where the integration over $\mathcal{R}_j(a+\Delta a)$ is replaced by integration over $\Delta\mathcal{D}(a; \boldsymbol{r}_i - \boldsymbol{r}_j)$ for every $i > j$. Fixing the pair $[i, j]$ with $i > j$, the integrals over $\boldsymbol{r}_j$ and then $\boldsymbol{r}_i$ can moved rightwards to the beginning of the order of integration when $\mathcal{R}_i(a + \Delta a)$ is replaced by

$$\mathcal{O}_i^{(j)}(a) := \mathcal{V} - \sum_{\{k \neq j\}} \mathcal{D}(a; \boldsymbol{r}_k - \boldsymbol{r}_i). \tag{15.6.10}$$

For the terms in the integrand containing $\boldsymbol{r}_i$ or $\boldsymbol{r}_j$, the factor of interest is

$$\begin{aligned} \mathcal{Z}_{ij}(\zeta_2, \boldsymbol{\omega}, a) = & \int_{\mathcal{O}_i^{(j)}(a)} \mathrm{d}\mathcal{V}_i \int_{\mathcal{D}(a;\boldsymbol{r}_i-\boldsymbol{r}_j)} \mathrm{d}\mathcal{V}_j \\ & \times \exp\Bigg(2\,\zeta_2\Big[\omega_i \sum_{\{k\neq i\}} \omega_k \ln\left|\frac{\boldsymbol{r}_k - \boldsymbol{r}_i}{a}\right| \\ & + \omega_j \sum_{\{k\neq j\}} \omega_k \ln\left|\frac{\boldsymbol{r}_k - \boldsymbol{r}_j}{a}\right| - \omega_i\omega_j \ln\left|\frac{\boldsymbol{r}_j - \boldsymbol{r}_i}{a}\right|\Big]\Bigg). \end{aligned} \tag{15.6.11}$$

The integrand for the whole partition function is dominated by terms with $\omega_j = -\omega_i$ and the inner integration in (15.6.11) imposes the restriction that $\boldsymbol{r}_j = \boldsymbol{r}_i + \boldsymbol{q}$, where $a < |\boldsymbol{q}| < a + \Delta a$, giving

$$\begin{aligned} \mathcal{Z}_{ij}(\zeta_2, \boldsymbol{\omega}, a) = & \int_{\mathcal{O}_i^{(j)}(a)} \mathrm{d}\mathcal{V}_i \int_{\mathcal{D}(a;\boldsymbol{r}_i-\boldsymbol{r}_j)} \mathrm{d}\mathcal{V}_j \\ & \times \exp\left(-\zeta_2\omega_i \sum_{\{k\neq i\}} \omega_k \ln\left|1 + \frac{2q\cos(\alpha_k)}{|\boldsymbol{r}_k - \boldsymbol{r}_i|} + \frac{q^2}{|\boldsymbol{r}_k - \boldsymbol{r}_i|^2}\right|\right), \end{aligned} \tag{15.6.12}$$

where α_k is the angle between $\boldsymbol{r}_k - \boldsymbol{r}_i$ and $\boldsymbol{q}$. Expanding in powers of $q/|\boldsymbol{r}_k - \boldsymbol{r}_i|$ and then expanding the exponential and integrating with respect to q over the range

$(a, a+\Delta a)$ gives

$$\begin{aligned}
\mathcal{Z}_{ij}(\zeta_2, \boldsymbol{\omega}, a) &= \int_{\mathcal{O}_i^{(j)}(a)} \mathrm{d}\mathcal{V}_i \int_{\mathcal{D}(a;\boldsymbol{r}_i-\boldsymbol{r}_j)} \mathrm{d}\mathcal{V}_j \\
&\quad \times \Bigg(1 - \zeta_2\omega_i \sum_{\{k\neq i\}} \omega_k \left[\frac{2q\cos(\alpha_k)}{|\boldsymbol{r}_k - \boldsymbol{r}_i|} + \frac{q^2[1-2\cos^2(\alpha_k)]}{|\boldsymbol{r}_k - \boldsymbol{r}_i|^2}\right] \\
&\quad + 2\zeta_2^2 \sum_{\{k\neq i\}}\sum_{\{\ell\neq i\}} \omega_k\omega_\ell \frac{q^2\cos(\alpha_k)\cos(\alpha_\ell)}{|\boldsymbol{r}_k - \boldsymbol{r}_i||\boldsymbol{r}_\ell - \boldsymbol{r}_i|}\Bigg) \\
&= 2\pi a\Delta a \int_{\mathcal{O}_i^{(j)}(a)} \mathrm{d}\mathcal{V}_i \left\{1 + \frac{\zeta_2^2 a^2}{\pi}\int_0^{2\pi} \mathrm{d}\alpha_k \sum_{\{k\neq i\}}\sum_{\{\ell\neq i\}} \omega_k\omega_\ell \frac{\cos(\alpha_k)\cos(\beta_{k\ell}-\alpha_k)}{|\boldsymbol{r}_k - \boldsymbol{r}_i||\boldsymbol{r}_\ell - \boldsymbol{r}_i|}\right\},
\end{aligned} \tag{15.6.13}$$

where $\beta_{k\ell}$ is the angle between $\boldsymbol{r}_k - \boldsymbol{r}_i$ and $\boldsymbol{r}_\ell - \boldsymbol{r}_i$. Integrating with respect to α_k gives

$$\begin{aligned}
\mathcal{Z}_{ij}(\zeta_2, \boldsymbol{\omega}, a) = 2\pi a\Delta a \int_{\mathcal{O}_i^{(j)}(a)} \mathrm{d}\mathcal{V}_i \Bigg\{1 + \zeta_2^2 a^2 \overbrace{\sum_{\{k\neq i\}}\sum_{\{\ell\neq i\}}}^{k\neq\ell} \omega_k\omega_\ell \frac{(\boldsymbol{r}_k - \boldsymbol{r}_i)\cdot(\boldsymbol{r}_\ell - \boldsymbol{r}_i)}{|\boldsymbol{r}_k - \boldsymbol{r}_i|^2|\boldsymbol{r}_\ell - \boldsymbol{r}_i|^2} \\
+ \zeta_2^2 a^2 \sum_{\{k\neq i\}} \frac{1}{|\boldsymbol{r}_k - \boldsymbol{r}_i|^2}\Bigg\}
\end{aligned} \tag{15.6.14}$$

and a final integration over $\mathcal{O}_i^{(j)}(a)$ gives

$$\mathcal{Z}_{ij}(\zeta_2, \boldsymbol{\omega}, a) = 2\pi a\Delta a\Big[V - 2\pi\zeta_2^2 a^2 \sum_{\{k\neq \ell\}} \omega_k\omega_\ell \ln\Big|\frac{\boldsymbol{r}_k - \boldsymbol{r}_\ell}{a}\Big|\Big], \tag{15.6.15}$$

(Kosterlitz 1974). When substituting into (15.6.9) to obtain the partition function we must note that the contribution from the second term has 'lost' two vortices with winding numbers ± 1. But the factor in (15.6.15) has lost its dependence on i and j so the sum in (15.6.9) and the choice of j compensate for this, the residual effect being just a factor ζ_1^2. The term to insert into the partition function is therefore

$$\begin{aligned}
\zeta_1^2\mathcal{Z}_{ij}(\zeta_2, \boldsymbol{\omega}, a) = \exp\left(2\zeta_1^2\pi a\Delta a\Big[V - 4\pi\zeta_2^2 a^2 \sum_{\{k>\ell\}} \omega_k\omega_\ell \ln\Big|\frac{\boldsymbol{r}_k - \boldsymbol{r}_\ell}{a}\Big|\Big]\right) - 1 \\
+ \mathrm{O}(\Delta a).
\end{aligned} \tag{15.6.16}$$

In applying (15.6.9) to (15.6.5) the a in the denominator of the logarithmic factors must be replaced by $a + \Delta a$ so that

$$\exp\left(2\zeta_2 \sum_{\{i>j\}} \omega_j\omega_i \ln\left|\frac{\boldsymbol{r}_i - \boldsymbol{r}_j}{a}\right|\right) = \exp\left(2\zeta_2 \sum_{\{i>j\}} \omega_j\omega_i \ln\left|\frac{\boldsymbol{r}_i - \boldsymbol{r}_j}{a+\Delta a}\right| - \frac{2n\Delta a\zeta_2}{a}\right). \tag{15.6.17}$$

Applying a similar change to (15.6.16) yields a neglectable factor $\mathrm{O}([\Delta a]^2)$. So from (15.6.5), (15.6.9), (15.6.16) and (15.6.17) the transformation in couplings is given by

$$\frac{V\tilde{\zeta}_0}{(a+\Delta a)^2} = \frac{V\zeta_0}{a^2} - 2V\zeta_1^2 \pi a \Delta a, \tag{15.6.18}$$

$$\tilde{\zeta}_1 (a+\Delta a)^2 = \zeta_1 a^2 \left(1 + \frac{(2-\zeta_2)\Delta a}{a}\right), \tag{15.6.19}$$

$$\tilde{\zeta}_2 = \zeta_2 \left(1 - \frac{(2\pi\zeta_1 a^2)^2 \zeta_2 \Delta a}{a}\right). \tag{15.6.20}$$

In the limit $\Delta a \to 0$ and in terms of the variables

$$\ell := \ln(a), \qquad y := 4\pi\zeta_1 a^2. \qquad x := 2 - \zeta_2, \tag{15.6.21}$$

(15.6.19) and (15.6.20) can be put into the differential form

$$\frac{\mathrm{d}x}{\mathrm{d}\ell} = y^2 \left(1 - \tfrac{1}{2}x\right)^2, \tag{15.6.22}$$

$$\frac{\mathrm{d}y}{\mathrm{d}\ell} = xy, \tag{15.6.23}$$

for which $y = 0$ is a fixed line. In the region close to $x = 0$, where $T \simeq T_{\mathrm{KT}}$, the *Kosterlitz–Thouless temperature* given by (3.6.35), these equations can be linearized in x to give

$$\frac{\mathrm{d}x}{\mathrm{d}\ell} = y^2, \tag{15.6.24}$$

$$\frac{\mathrm{d}y}{\mathrm{d}\ell} = xy, \tag{15.6.25}$$

for which the solutions are two families of rectangular hyperbolae

$$x^2 - y^2 = \pm c^2 \ \text{ together with the lines } y = \pm x. \tag{15.6.26}$$

On physical grounds we are interested only in the region $y \geq 0$ and $x \geq, \leq 0$ correspond respectively to $T \geq, \leq T_{\mathrm{KT}} := \pi J/2$. The scaling parameter ℓ governing the flow increases with a yielding increasing x everywhere and y decreasing or increasing according as $x < 0$ and $x > 0$. The renormalization flow is shown in Fig. 15.6 and the line $x = -y$ is the critical line controlled by the Kosterlitz–Thouless point $x = y = 0$. For the XY model μ is the energy of an isolated vortex leading to a single critical point on the line. For the Coulomb gas the critical line is parameterised by the fugacity μ. For a fixed value of this quantity the physical curve of the system is

$$x(K) = \pi(K_{\mathrm{KT}} - K), \qquad y(K) = 4\pi \exp(\mu K), \tag{15.6.27}$$

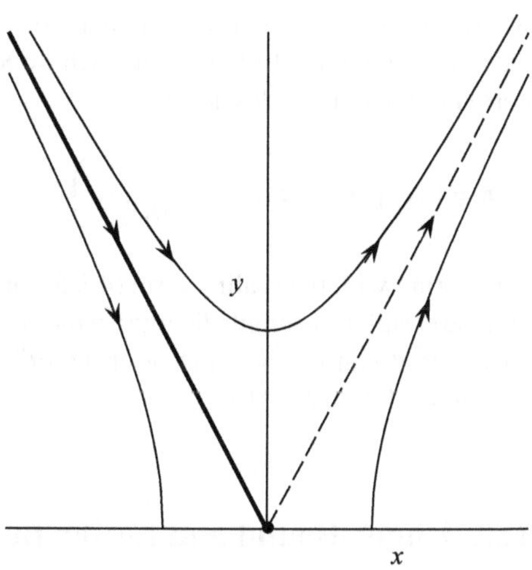

Fig. 15.6 The Kosterlitz–Thouless renormalization group flows. The heavy line $x = -y$ is the critical line terminating at the Kosterlitz–Thouless point at the origin

passing through the critical point $K = K_c$ given by the transcendental equation

$$K_c - K_{KT} = 4\exp(\mu K_c). \tag{15.6.28}$$

In the low-temperature region to the left of the critical line the flow is towards the fixed line $y = 0$. Beginning at a particular (K, μ) the point on the fixed line corresponds to $\mu \to -\infty$ where $K = \bar{K}(K)$. The pair correlation function given by (3.6.25) is algebraic where, from (3.6.14) the exponent is $\eta(2; K, \mu) = K_{KT}/4\bar{K}(K) < 1$, with $\bar{K}(K) > K$. Let $\bar{K}(K) := \rho(K)K/J$ giving $\eta(K) = JK_{KT}/4\rho(K)K$. Since $\bar{K}(K) \geq K$, $\rho(K) \geq J$ with $\rho(K_{KT}) = J$, where $\eta = \frac{1}{4}$. The quantity $\rho(K)$, which vanishes discontinuously at $K = K_{KT}$ is the *spin-wave stiffness* of the system, which is related to the choice of boundary conditions (Cardy 1984, Sect. 6.4; Goldenfeld 1992, Sect. 11.3; Chaikin and Lubensky 1995, Sect. 6.1). Using the continuum form of (4.2.4) it follows from the algebraic form of the pair correlation function that the correlation length $\mathfrak{r} \sim L_0$ the size of the system.

Consider now the high-temperature region to the right of the critical line. Close to the critical point, where $K = K_c - \Delta K$, the physical curve intersects the rectangular hyperbola

$$\begin{aligned} &y^2 - x^2 = \Delta K\, B, \\ &B := 2A(\pi - \mu A), \qquad A := 4\pi\exp(\mu K_c) = \pi(K_c - K_{KT}). \end{aligned} \tag{15.6.29}$$

From (15.6.24) the scaling parameter is given by

$$\ell = \int \frac{dx}{x^2 + B\Delta K} \simeq \frac{\pi}{\sqrt{B\Delta\, K}}, \tag{15.6.30}$$

where, since for small $\triangle K$ the integrand is steeply peaked around $x = 0$, the limits have been extended to infinity in each direction. From (15.6.21) it is clear that the correlation length scales according to

$$\mathfrak{r} = \mathfrak{r}_0 \exp(\ell) = \mathfrak{r}_0 \exp\left(\frac{\pi}{\sqrt{B\triangle K}}\right). \tag{15.6.31}$$

Comparing with the scaling form (4.5.52) it is evident that there is no exponent ν. In fact the correlation length diverges at the critical curve faster than any power of $|\Delta K|$. However, it can be shown (Kosterlitz 1974) that the weak scaling laws (12.3.325) are satisfied with $\delta = 15$, $\eta = \frac{1}{4}$.

15.7 Upper-Bound and Lower-Bound Approximations

The programme for RSRG methods is to use the formula (15.1.13) to obtain the recurrence relationships (15.2.1) and (15.2.2) and then to use the methods described in Sects. 15.3 and 15.4 to obtain the critical exponents and free energy, respectively. As we have seen in Sect. 15.5.1, this scheme can be carried out exactly for some one-dimensional models, but Sect. 15.5.2 describes the problems encountered when a similar approach is attempted in two dimensions. The approximation methods which will be described in the remainder of this chapter are of two broad types. In the methods of Sects. 15.8 and 15.10 the lattice for the system is replaced by a finite subset of sites or a lattice of lower dimensionality. In the method described in Sect. 15.9 an approximation is made to the right-hand side of (15.1.13) so that the summation can be performed. In that case the modified form, when summed over the microstates on $\widetilde{\mathcal{N}}$, will not necessarily reproduce the invariance condition (15.1.12) for the partition function.

There are, however, a number of methods for which this equality can be replaced by an inequality which remains valid throughout the iteration process. Suppose that the approximate recurrence relationships replacing (15.2.1) and (15.2.2) are

$$\tilde{\zeta}_0' = \lambda^d[\zeta_0 + \mathfrak{U}_0(\zeta_j)], \tag{15.7.1}$$

$$\tilde{\zeta}_i' = \mathfrak{U}_i(\zeta_j), \qquad i = 1, 2, \ldots, n, \tag{15.7.2}$$

with (15.1.12) replaced by

$$Z(\widetilde{N}, \tilde{\zeta}_i') \le Z(N, \zeta_i). \tag{15.7.3}$$

Then, in place of (15.4.1) and (15.4.4), we have, respectively,

$$\Phi(\widetilde{N}, \tilde{\zeta}_i') + \widetilde{N}\tilde{\zeta}_0' \ge \Phi(N, \zeta_i) + N\zeta_0, \tag{15.7.4}$$

$$\phi(\tilde{\zeta}_i') \geq \lambda^d[\phi(\zeta_i) - \mathcal{U}_0(\zeta_i)]. \tag{15.7.5}$$

Now suppose that (15.7.1) and (15.7.2) are used to generate a trajectory $\zeta_i^{(0)} \to \zeta_i^{(1)\prime} \to \zeta_i^{(2)\prime} \cdots$ in phase space. By an procedure similar to that of Sect. 15.4 it follows that

$$\phi(\zeta_i^{(0)}) \leq \phi^{(\mathcal{U})}(\zeta_i^{(0)}) = \sum_{s=0}^{\infty} \frac{1}{\lambda^{sd}} \mathcal{U}_0(\zeta_i^{(s)\prime}). \tag{15.7.6}$$

The transformation generates an upper-bound to the free energy. By a similar argument a set of recurrence relationships

$$\tilde{\zeta}_0' = \lambda^d[\zeta_0 + \mathcal{L}_0(\zeta_j)], \tag{15.7.7}$$

$$\tilde{\zeta}_i' = \mathcal{L}_i(\zeta_j), \qquad i = 1, 2, \ldots, n, \tag{15.7.8}$$

with

$$Z(\widetilde{N}, \tilde{\zeta}_i') \geq Z(N, \zeta_i), \tag{15.7.9}$$

will generate a lower bound

$$\phi^{(\mathcal{L})}(\zeta_i^{(0)}) = \sum_{s=0}^{\infty} \frac{1}{\lambda^{sd}} \mathcal{L}_0(\zeta_i^{(s)\prime}) \tag{15.7.10}$$

to the free energy.

15.7.1 An Upper-Bound Method

An approach which leads to an upper bound to the free energy was introduced by Niemeijer and van Leeuwen (1974) (see, also, Niemeijer and van Leeuwen 1976). It is based on a splitting of the Hamiltonian on the right-hand side of (15.1.13) into two parts

$$\widehat{H}(N, \zeta_i, \Delta\zeta; \sigma(\boldsymbol{r})) = \widehat{H}_0(N, \zeta_i; \sigma(\boldsymbol{r})) + \Delta\zeta\, \widehat{C}(N; \sigma(\boldsymbol{r})), \tag{15.7.11}$$

where $\widehat{H}_0$ is referred to as the *zeroth-order part*, in comparison to which the second term is small as a consequence of the small value $\Delta\zeta$ of its single coupling.[4] The zeroth-order average $\langle A \rangle_0$ for any function $A(\sigma(\boldsymbol{r}))$ of the spin states is given by

[4] The argument ζ_i now represents the set of couplings in $\widehat{H}_0$ and any further couplings in the second term in (15.7.11) are absorbed as ratios in $\widehat{C}$.

$$\langle A\rangle_0 := \frac{\displaystyle\sum_{\{\sigma(\boldsymbol{r})\}}\Big[\prod_{\{\tilde{\boldsymbol{r}}\}} p(\sigma(\tilde{\boldsymbol{r}})|\sigma(\boldsymbol{r}), \boldsymbol{r}\in\mathcal{B}(\tilde{\boldsymbol{r}}))\Big]A(\sigma(\boldsymbol{r}))\exp[-\widehat{H}_0(N,\zeta_i;\sigma(\boldsymbol{r}))]}{\mathcal{Z}_0(N,\zeta_i;\sigma(\tilde{\boldsymbol{r}}))}, \tag{15.7.12}$$

where $\mathcal{Z}_0(N,\zeta_i;\sigma(\tilde{\boldsymbol{r}}))$ is the quantity given by (15.1.14), but using only the zeroth-order part of the Hamiltonian. From this it follows that

$$\langle\exp[-\Delta\zeta\,\widehat{C}(N;\sigma(\boldsymbol{r}))]\rangle_0 = \frac{\mathcal{Z}(N,\zeta_i,\Delta\zeta;\sigma(\tilde{\boldsymbol{r}}))}{\mathcal{Z}_0(N,\zeta_i;\sigma(\tilde{\boldsymbol{r}}))} \tag{15.7.13}$$

and (15.1.13) can be expressed in the form

$$\exp[-\widehat{H}(\widetilde{N},\tilde{\zeta}_i,\widetilde{\Delta\zeta};\sigma(\tilde{\boldsymbol{r}}))] = \langle\exp[-\Delta\zeta\,\widehat{C}(N;\sigma(\boldsymbol{r}))]\rangle_0\mathcal{Z}_0(N,\zeta_i;\sigma(\tilde{\boldsymbol{r}}). \tag{15.7.14}$$

Summing over all the spin states $\sigma(\tilde{\boldsymbol{r}})$ gives

$$Z(\widetilde{N},\tilde{\zeta}_i,\widetilde{\Delta\zeta}) = Z(N,\zeta_i,\Delta\zeta) = Z_0(N,\zeta_i)\langle\exp[-\Delta\zeta\,\widehat{C}(N;\sigma(\boldsymbol{r}))]\rangle_0. \tag{15.7.15}$$

These expressions are, of course, exact. The approximation arises when the final factor on the left is replaced by a finite number of terms of its *cumulant expansion*. The simplest way to generate this cluster expansion is to consider

$$\begin{aligned}\ln[\mathcal{Z}(N,\zeta_i,\Delta\zeta;\sigma(\tilde{\boldsymbol{r}}))] &= \left[\exp\left(\Delta\zeta\frac{\partial}{\partial q}\right)\ln[\mathcal{Z}(N,\zeta_i,q;\sigma(\tilde{\boldsymbol{r}}))]\right]_{q=0}\\ &= \ln[\mathcal{Z}_0(N,\zeta_i;\sigma(\tilde{\boldsymbol{r}}))] + \left[\sum_{m=1}^{\infty}(\Delta\zeta)^m\frac{\partial^m}{\partial q^m}\ln[\mathcal{Z}(N,\zeta_i,q;\sigma(\tilde{\boldsymbol{r}}))]\right]_{q=0},\end{aligned} \tag{15.7.16}$$

(cf. (14.3.4)–(14.3.6) and the footnote on p. 537), which gives, from (15.7.13)

$$\begin{aligned}\langle\exp[-\Delta\zeta\widehat{C}(N;\sigma(\boldsymbol{r}))]\rangle_0 = \exp\Big\{&-\Delta\zeta\,\langle\widehat{C}(N;\sigma(\boldsymbol{r}))\rangle_0\\ &+\tfrac{1}{2!}(\Delta\zeta)^2\langle[\widehat{C}(N;\sigma(\boldsymbol{r}))-\langle\widehat{C}(N;\sigma(\boldsymbol{r}))\rangle_0]^2\rangle_0\\ &-\tfrac{1}{3!}(\Delta\zeta)^3\langle[\widehat{C}(N;\sigma(\boldsymbol{r}))-\langle\widehat{C}(N;\sigma(\boldsymbol{r}))\rangle_0]^3\rangle_0+\cdots\Big\}.\end{aligned} \tag{15.7.17}$$

The *first-order cumulant approximation* consists in retaining only the first term in the exponential on the right-hand side of Eq. (15.7.17) for which[5]

---

[5] For any $A(\sigma(\boldsymbol{r}))$

$$\begin{aligned}\langle\exp(A)\rangle_0 &= \exp\langle A\rangle_0\langle\exp(A-\langle A\rangle_0)\rangle_0\\ &= \exp\langle A\rangle_0\langle 1+(A-\langle A\rangle_0)+\tfrac{1}{2}(A-\langle A\rangle_0)^2+\cdots\rangle_0 \geq \exp\langle A\rangle_0.\end{aligned}$$

$$\langle \exp[-\Delta\zeta \widehat{C}(N;\sigma(\boldsymbol{r}))]\rangle_0 \geq \exp[-\Delta\zeta \langle \widehat{C}(N;\sigma(\boldsymbol{r}))\rangle_0]. \tag{15.7.18}$$

Substituting into (15.7.14) and replacing $\tilde{\zeta}_i$ and $\widetilde{\Delta\zeta}$ by $\tilde{\zeta}'_i$ and $\widetilde{\Delta\zeta'}$ gives

$$\exp[-\widehat{H}(\widetilde{N}, \tilde{\zeta}'_i, \widetilde{\Delta\zeta'}; \sigma(\tilde{\boldsymbol{r}}))] = \mathcal{Z}_0(N, \zeta_i; \sigma(\tilde{\boldsymbol{r}})) \exp[\langle -\Delta\zeta \widehat{C}(N;\sigma(\boldsymbol{r}))\rangle_0]. \tag{15.7.19}$$

The corresponding partition function is again obtained by summing over the spin states $\sigma(\tilde{\boldsymbol{r}})$ to give

$$Z(\widetilde{N}, \tilde{\zeta}'_i, \widetilde{\Delta\zeta'}) = Z_0(N, \zeta_i) \exp\langle -\Delta\zeta \widehat{C}(N;\sigma(\boldsymbol{r}))\rangle_0. \tag{15.7.20}$$

The upper-bound condition (15.7.3) then follows from (15.7.15) and (15.7.18).

Example 15.7.1 We return to the problem, discussed in Sect. 15.5.2, of using decimation for the two-dimensional Ising model, adopting a first-order cumulant approximation to construct an upper-bound transformation. The zeroth-order Hamiltonian $\widehat{H}_0$ is taken to include the trivial coupling C and the first-neighbour coupling K_1. Terms corresponding to a second-neighbour coupling K_2 and a four-site coupling K_3 are included in $\Delta\zeta\,\widehat{C}$ with K_2 playing the role of $\Delta\zeta$ and $K_3 = \beta K_2$. Because the four-site term links together pairs of white sites in Fig. 15.4 the method must be based on pairs of neighbouring squares like ABCD and CDEF. If the Hamiltonian is taken to be made up of overlapping pairs of squares, each square is counted four times and thus factors arising from a single square should be reduced by a factor of four. Denoting the contribution to any factor from the pair of squares by the symbol $[..]_{\square\square}$

$$\begin{aligned}&\big[\mathcal{Z}_0(N, C, K_1; \sigma(\tilde{\boldsymbol{r}}))\big]_{\square\square}\\ &\quad = \exp(C)\{4\cosh[K_1(\sigma_{\mathrm{A}}+\sigma_{\mathrm{B}}+\sigma_{\mathrm{C}}+\sigma_{\mathrm{D}})]\cosh[K_1(\sigma_{\mathrm{E}}+\sigma_{\mathrm{F}}+\sigma_{\mathrm{C}}+\sigma_{\mathrm{D}})]\}^{1/4},\end{aligned} \tag{15.7.21}$$

$$\begin{aligned}&[\exp\langle -K_2\widehat{C}(N,\beta;\sigma(\boldsymbol{r}))\rangle_0]_{\square\square} = \exp\Big[K_2\big\{\alpha^{(1)}(\sigma_{\mathrm{A}}\sigma_{\mathrm{B}}+\sigma_{\mathrm{E}}\sigma_{\mathrm{F}})\\ &+\alpha^{(2)}\sigma_{\mathrm{C}}\sigma_{\mathrm{D}}+\alpha^{(3)}(\sigma_{\mathrm{A}}\sigma_{\mathrm{D}}+\sigma_{\mathrm{B}}\sigma_{\mathrm{C}}+\sigma_{\mathrm{E}}\sigma_{\mathrm{D}}+\sigma_{\mathrm{F}}\sigma_{\mathrm{C}})\\ &+(1+\beta\sigma_{\mathrm{C}}\sigma_{\mathrm{D}})\tanh[K_1(\sigma_{\mathrm{A}}+\sigma_{\mathrm{B}}+\sigma_{\mathrm{C}}+\sigma_{\mathrm{D}})]\tanh[K_1(\sigma_{\mathrm{E}}+\sigma_{\mathrm{F}}+\sigma_{\mathrm{C}}+\sigma_{\mathrm{D}})]\big\}\Big],\end{aligned} \tag{15.7.22}$$

where

$$2\alpha^{(1)} + \alpha^{(2)} + 4\alpha^{(3)} = 1. \tag{15.7.23}$$

As in the simpler situation described in Sect. 15.5.2 the product of the right-hand sides of (15.7.21) and (15.7.22) can be re-expressed in the form

$$\begin{aligned}
\exp\{C &+ \tfrac{1}{2}\mathcal{K}_0 + \mathcal{K}_1^{(1)}(\sigma_A\sigma_B + \sigma_E\sigma_F) + \mathcal{K}_1^{(2)}\sigma_C\sigma_D \\
&+ \mathcal{K}_1^{(3)}(\sigma_A\sigma_D + \sigma_B\sigma_C + \sigma_E\sigma_D + \sigma_F\sigma_C) \\
&+ \tfrac{1}{4}\mathcal{K}_2(\sigma_A\sigma_C + \sigma_B\sigma_D + \sigma_C\sigma_E + \sigma_D\sigma_F) \\
&+ \tfrac{1}{4}\mathcal{K}_3\sigma_C\sigma_D(\sigma_A\sigma_B + \sigma_E\sigma_F) \\
&+ \mathcal{K}_4(\sigma_A\sigma_F + \sigma_B\sigma_E) + \mathcal{K}_5(\sigma_A\sigma_E + \sigma_B\sigma_F) + \mathcal{K}_6\sigma_A\sigma_B\sigma_E\sigma_F \\
&+ \mathcal{K}_7\sigma_C\sigma_D(\sigma_A\sigma_E + \sigma_B\sigma_F) + \mathcal{K}_8\sigma_C\sigma_D(\sigma_A\sigma_F + \sigma_B\sigma_E) \\
&+ \mathcal{K}_9[\sigma_A\sigma_B(\sigma_C\sigma_F + \sigma_D\sigma_E) + \sigma_E\sigma_F(\sigma_A\sigma_D + \sigma_B\sigma_C)] \\
&+ \mathcal{K}_{10}[\sigma_A\sigma_B(\sigma_D\sigma_F + \sigma_C\sigma_E) + \sigma_E\sigma_F(\sigma_A\sigma_C + \sigma_B\sigma_D)] \\
&+ \mathcal{K}_{11}\sigma_A\sigma_B\sigma_C\sigma_D\sigma_E\sigma_F\},
\end{aligned} \tag{15.7.24}$$

where, using the compact notation

$$\text{th}_\ell := \tanh(\ell K_1), \qquad \text{lnc}_\ell := \ln\cosh(\ell K_1), \qquad \ell = 2, 4, \tag{15.7.25}$$

and reverting to the use of the coupling $K_3 = \beta K_2$,

$$\mathcal{K}_0 := \tfrac{1}{16}(K_2 + K_3)[\text{th}_4 + 2\,\text{th}_2]^2 + \ln(2) + \tfrac{1}{8}\text{lnc}_4 + \tfrac{1}{2}\text{lnc}_2, \tag{15.7.26}$$

$$\mathcal{K}_1^{(1)} := \tfrac{1}{32}(K_2 + K_3)[\text{th}_4^2 - 4\,\text{th}_2^2] + K_2\alpha^{(1)} + \tfrac{1}{32}\text{lnc}_4, \tag{15.7.27}$$

$$\mathcal{K}_1^{(2)} := \tfrac{1}{32}(K_2 + K_3)[\text{th}_4 + 2\,\text{th}_2]^2 + K_2\alpha^{(2)} + \tfrac{1}{16}\text{lnc}_4, \tag{15.7.28}$$

$$\mathcal{K}_1^{(3)} := \tfrac{1}{32}(K_2 + K_3)\text{th}_4[\text{tanh}_4 + 2\,\text{th}_2] + K_2\alpha^{(3)} + \tfrac{1}{32}\text{lnc}_4, \tag{15.7.29}$$

$$\mathcal{K}_2 := \tfrac{1}{8}(K_2 + K_3)\tanh(4K_1)[\text{th}_4 + 2\,\text{th}_2] + \tfrac{1}{8}\text{lnc}_4, \tag{15.7.30}$$

$$\mathcal{K}_3 := \tfrac{1}{8}[\text{th}_4^2 - 4\,\text{th}_2^2] + \tfrac{1}{8}\text{lnc}_4 - \tfrac{1}{2}\text{lnc}_2, \tag{15.7.31}$$

$$\mathcal{K}_4 = \mathcal{K}_5 := \tfrac{1}{32}[(K_2 + K_3)\text{th}_4 + 4K_2(1-\beta)\text{th}_2], \tag{15.7.32}$$

$$\mathcal{K}_6 := \tfrac{1}{32}(K_2 + K_3)[\text{th}_4 - 2\,\text{th}_2]^2, \tag{15.7.33}$$

$$\mathcal{K}_7 = \mathcal{K}_8 := \tfrac{1}{32}[(K_2 + K_3)\text{th}_4 - 4K_2(1-\beta)\text{th}_2, \tag{15.7.34}$$

$$\mathcal{K}_9 = \mathcal{K}_{10} := \tfrac{1}{32}(K_2 + K_3)\text{th}_4[\text{th}_4 - 2\,\text{th}_2], \tag{15.7.35}$$

$$\mathcal{K}_{11} = \tfrac{1}{32}(K_2 + K_3)[\text{th}_4 - 2\,\text{th}_2]^2. \tag{15.7.36}$$

With

$$\mathcal{K}_1(K_1, K_2, K_3) := 2\mathcal{K}_1^{(1)}(K_1, K_2, K_3) + \mathcal{K}_1^{(2)}(K_1, K_2, K_3) + 4\mathcal{K}_1^{(3)}(K_1, K_2, K_3), \tag{15.7.37}$$

recurrence relationships of the form (15.2.1) and (15.2.2) with $d = 2$, $\lambda = \sqrt{2}$ and $n = 11$ can now be constructed. They reduce to the recurrence relationships given by (15.5.36)–(15.5.41), when $K_2 = K_3 = 0$. $\widetilde{K}_1$ and $\widetilde{K}_2$ are the first- and second-neighbour couplings on a square like ABCD and $\widetilde{K}_3$ is a four-site coupling on such a square. The couplings $\widetilde{K}_i$, $i = 4, 5, \ldots, 11$ have been generated by the transformation and link sites. Sneddon and Barber (1977) have investigated the approximation where these proliferating couplings are neglected and only first-neighbour, second-neighbour and the four-site coupling on the square are retained. They show that the only fixed point not corresponding to either zero or infinite values of the couplings is $K_1^\star = 0.3683$, $K_2^\star = 0.1275$, $K_3^\star = -0.0303$. Linearizing about this fixed point yields the one relevant eigenvalue $\Lambda_1 = 1.4287$, giving the thermal exponent $y_1 = y_T = 1.029$ with $\nu = 1/y_T = 0.9714$. This represent a considerable improvement over the results of the simpler approximation described in Sect. 15.5.2. It is tempting to suppose that cumulant approximations to higher orders would yield further improvement. However this does not seem to be the case. Kadanoff (1976a) reports that unpublished results by Houghton and himself yielded respectively $y_T = 1.002$ and $y_T = 1.7$, for the second- and third-order extension to the method described here. This problem with the cumulant approximation does not appear to be related to the difficulties associated with the decimation weight function. Hsu and Gunton (1977) have used the Kadanoff weight function (15.2.9) to investigate second- and third-order cumulant expansions for the Ising model on the square and simple cubic lattice. They find that the best results are achieved in the decimation limit $\tau \to \infty$ and that, although reasonable results are achieved for the critical temperature and magnetic field exponent, the results for the thermal exponent are poor.

15.7.2 A Lower-Bound Method

A starting point for developing a class of lower-bound transformations can be achieved if the right-hand side of (15.1.13) is modified by replacing $\widehat{H}(N, \zeta_i; \sigma(\boldsymbol{r}))$ by $\widehat{H}(N, \zeta_i; \sigma(\boldsymbol{r})) + \widehat{V}(N, \zeta_i; \sigma(\boldsymbol{r}))$. Then we have

$$\exp[-\widehat{H}(\widetilde{N}, \tilde{\zeta}_i'; \sigma(\tilde{\boldsymbol{r}}))] = \sum_{\{\sigma(\boldsymbol{r})\}} \Bigl[\prod_{\{\tilde{\boldsymbol{r}}\}} p(\sigma(\tilde{\boldsymbol{r}})|\sigma(\boldsymbol{r}), \boldsymbol{r} \in \mathcal{B}(\tilde{\boldsymbol{r}}))\Bigr]$$

$$\times \exp[-\widehat{H}(N, \zeta_i; \sigma(\boldsymbol{r})) - \widehat{V}(N, \zeta_i; \sigma(\boldsymbol{r}))], \qquad (15.7.38)$$

where the functional form of the Hamiltonian on the left-hand side is the same as that in (15.1.13), but the renormalized couplings take different values denoted by $\tilde{\zeta}_i'$. Summing (15.7.38) over the microstates on $\widetilde{\mathcal{N}}$ gives

$$\begin{aligned} Z(\widetilde{N}, \tilde{\zeta}_i') &= Z(N, \zeta_i)\langle \exp[-\widehat{V}(N, \zeta_i; \sigma(\boldsymbol{r}))]\rangle \\ &\geq Z(N, \zeta_i)[1 - \langle \widehat{V}(N, \zeta_i; \sigma(\boldsymbol{r}))\rangle]. \end{aligned} \qquad (15.7.39)$$

If $\widehat{V}(N, \zeta_i; \sigma(\boldsymbol{r}))$ is chosen so that

$$\langle \widehat{V}(N, \zeta_i; \sigma(\boldsymbol{r})) \rangle = 0, \tag{15.7.40}$$

the transformation satisfies the lower-bound condition (15.7.9).

The possibility of achieving a lower-bound transformation with condition (15.7.40) by defining a $\widehat{V}(N, \zeta_i; \sigma(\boldsymbol{r}))$ which moved the interaction between the microsystems of one pair of sites to another pair was first discussed by Kadanoff (1975) (see also, Kadanoff et al. 1976).

Consider the zero-field first-neighbour Ising model on a hypercubic lattice with Hamiltonian

$$\widehat{H}(N, C, K; \sigma(\boldsymbol{r})) := -NC - K \sum_{\{\boldsymbol{r},\boldsymbol{r}'\}}^{(1)} \sigma(\boldsymbol{r})\sigma(\boldsymbol{r}'), \tag{15.7.41}$$

where $K := J/T$. Now choose any two first-neighbour pairs of sites $[\boldsymbol{r}_1, \boldsymbol{r}_1']$ and $[\boldsymbol{r}_2, \boldsymbol{r}_2']$. Remove the interaction (or *bond*) from the first pair and add it to the bond on the second pair. Thus

$$\widehat{V}(N, K; \sigma(\boldsymbol{r})) = K[\sigma(\boldsymbol{r}_1)\sigma(\boldsymbol{r}_1') - \sigma(\boldsymbol{r}_2)\sigma(\boldsymbol{r}_2')] \tag{15.7.42}$$

and

$$\langle \widehat{V}(N, K; \sigma(\boldsymbol{r})) \rangle = K[\langle \sigma(\boldsymbol{r}_1)\sigma(\boldsymbol{r}_1') \rangle - \langle \sigma(\boldsymbol{r}_2)\sigma(\boldsymbol{r}_2') \rangle] = 0. \tag{15.7.43}$$

Of course, this result does not in itself lead to a tractable renormalization scheme. It is necessary to move bonds in a systematic way throughout the lattice so that the summation on the right-hand side of (15.7.38) can be performed, leading to a Hamiltonian $\widehat{H}(\widetilde{N}, \widetilde{C}, \widetilde{K}; \sigma(\tilde{\boldsymbol{r}}))$ of the same form (15.7.41) as $\widehat{H}(N, C, K; \sigma(\boldsymbol{r}))$. This can be achieved using the *Migdal–Kadanoff transformation*, first invented by Migdal (1975) and reinterpreted using bond moving ideas by Kadanoff (1976a). In Kadanoff's formulation of this method bonds are moved so that sites can be eliminated by decimation.

Example 15.7.2 Consider the case of the zero-field first-neighbour Ising model on a square lattice. Taking a block of $\lambda \times \lambda$ squares, as shown in Fig. 15.7 (for the case $\lambda = 3$) and move the horizontal interior bonds to the upper edge and the vertical interior bonds to the left edge. The interior sites do not now interact with their neighbours, so they will each contribute a factor of two to the summation on the left-hand side of (15.7.38). The coupling on each segment of the boundary will be λK. These sites form a one-dimensional chain connecting the corners to which we can apply the decimation formula (15.5.33), which together with (15.5.12) gives

$$\widetilde{K} = \operatorname{arctanh}\{[\tanh(\lambda K)]^{\lambda}\}. \tag{15.7.44}$$

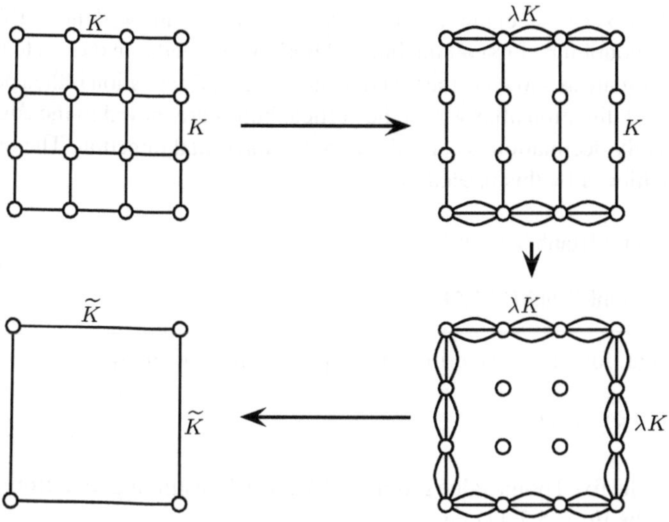

Fig. 15.7 Bond moving and decimation operations on a block of 3 × 3 sites for the Migdal–Kadanoff transformation applied to the zero-field first-neighbour Ising model on a square lattice

The recurrence relation

$$\widetilde{K} = \operatorname{arctanh}\{[\tanh(\lambda^{d-1}K)]^{\lambda}\}, \tag{15.7.45}$$

obtained by Migdal (1975) for the d-dimensional hypercubic lattice is a simple generalization of (15.7.44), which, with $\lambda = 2$, can be re-expressed in the form

$$\widetilde{K} = \tfrac{1}{2}\ln[\cosh(4K)], \tag{15.7.46}$$

giving the critical fixed point $K^\star = K_c = 0.304689$ with $T_c/J = 3.282$ and $\nu = 1/y_T = \ln(2)/\ln[2\tanh(4K_c)] = 1.3383$. Neither the critical coupling nor the exponent compare well with the exact values for the two-dimensional Ising model ($K_c = 0.4407, \nu = 1$, see Sect. 12.3.9) and the comparison deteriorates with increasing λ giving for K_c the values 0.2406 and 0.2018 for $\lambda = 3$ and 4 respectively.

An interesting modification of the method has been proposed by Kadanoff (1976a). He considered the anisotropic Ising model on the d-dimensional hypercubic lattice with different couplings in each axis direction. In the Migdal formulation this would simply lead to the result that each coupling satisfied (15.7.45) and the fixed points for different values of λ would have all the couplings equal. In this modified method, instead of the bond moving for all the axes preceding decimation, bond moving and decimation are performed alternately.

Example 15.7.3 Consider the case of the square lattice K_h and K_v denoting horizontal and vertical couplings. First all the internal horizontal bonds are moved to the edges of the $\lambda \times \lambda$ square as shown in the first line of Fig. 15.7. Decimation is then performed in the vertical direction after which the vertical bonds are moved to the edges of the square. Finally decimation is carried out in the horizontal direction. The recurrence relations achieved by this procedure are

$$\begin{aligned} \widetilde{K}_h &= \operatorname{arctanh}\{[\tanh(\lambda K_h)]^{\lambda}\}, \\ \widetilde{K}_v &= \lambda \operatorname{arctanh}\{[\tanh(K_v)]^{\lambda}\}. \end{aligned} \tag{15.7.47}$$

It is clear that, for any λ, the critical fixed point is now given by

$$K_h^\star = \lambda^{-1} K_v^\star = K^\star, \tag{15.7.48}$$

where $K^\star$ is the fixed point value given by (15.7.44). Substituting for λ from (15.7.48) into one of the Eq. (15.7.47) gives

$$K_v^\star \ln[\tanh(K_v^\star)] = K_h^\star \ln[\tanh(K_h^\star)]. \tag{15.7.49}$$

Equation (15.7.49) has branches[6]

$$K_v^\star = K_h^\star \tag{15.7.50}$$

and

$$\sinh(2K_v^\star)\sinh(2K_h^\star) = 1. \tag{15.7.51}$$

The formula (15.7.50), which is satisfied by the Migdal critical fixed point, disagrees with (15.7.48), except for $\lambda = 1$, when, in the isotropic case, (15.7.51) becomes the formula (9.2.36) which gives the exact critical temperature for the isotropic Ising model. In general (15.7.51) gives the critical curve for the anisotropic Ising model (Baxter 1982b). It is, however, a curious property of the method that the individual points on this curve are achieved by choosing different (including non-integer) values of λ. This is because, for any chosen value of λ, the couplings renormalize independently and thus can yield only a finite number of discrete points, not a continuous curve. It can be shown that both the Migdal method and the Kadanoff modification lead to the same values for the exponent ν, with the 'best' value being achieve in the limit $\lambda \to 1$ when $\nu = 1.326$.

---

[6] Equation (15.7.50) is obvious. Equation (15.7.51), which is equivalent to $K_v^\star = -\frac{1}{2}\ln[\tanh(K_h^\star)]$, follows from the fact that this latter equation is invariant under interchange of $K_v^\star$ and $K_h^\star$.

15.8 Finite-Lattice Approximations

Methods of this type apply the procedure of Sect. 15.1 to a lattice with a small number of sites with the proviso that, in order for the renormalization group transformation to be able to be applied once, it must be the case that $\widetilde{N} \geq \lambda^d$ and thus $N \geq \lambda^{2d}$. Within this framework there is still the choice of weight function and of whether to apply periodic boundary conditions to the lattice. Having made these choices the recurrence relationships can be computed exactly. As has been remarked in Sect. 15.2 the approximation here consists of obtaining the fixed points and phase diagram by iterating the recurrence relationships a large number of times. Early applications of this method to the Ising model were by Nauenberg and Nienhuis (1974a, b) who used four-site blocks on a square lattice and Tjon (1974), who used four three-site blocks on a triangular lattice. The former authors used periodic boundary conditions and the latter did not. Rather than present these calculations, which are described by Niemeijer and van Leeuwen (1976), we shall use as an example the work of Schick et al. (1976, 1977).

Example 15.8.1 Here a nine-site triangular lattice $\mathcal{N}$ with periodic boundary conditions is used to study the antiferromagnetic Ising model in a magnetic field. Given that antiferromagnetism leads to sublattice ordering on the three sublattices A, B and C shown in Fig. 15.8, the blocks are chosen to reflect this phenomenon. As is shown in Fig. 15.8 the three sites A_1, A_2 and A_3 of sublattice A form a block with block-spin site located at $\widetilde{\mathsf{A}}$, with a similar scheme applying to sublattices B and C. The Hamiltonian for this model has the form

$$\widehat{H}(N, C, L, K, M; \sigma) := -NC - \sum_{\triangle} \{\tfrac{1}{6}L(\sigma_{\mathsf{A}} + \sigma_{\mathsf{B}} + \sigma_{\mathsf{C}}) + M\sigma_{\mathsf{A}}\sigma_{\mathsf{B}}\sigma_{\mathsf{C}}$$

$$+ \tfrac{1}{2}K(\sigma_{\mathsf{A}}\sigma_{\mathsf{B}} + \sigma_{\mathsf{B}}\sigma_{\mathsf{C}} + \sigma_{\mathsf{C}}\sigma_{\mathsf{A}})\}, \tag{15.8.1}$$

where C, L and K are, as in Sect. 15.5.1, the trivial, magnetic field and first-neighbour couplings and M is a three-spin coupling. $|\mathcal{N}| =: N = 9$ and the summation is over

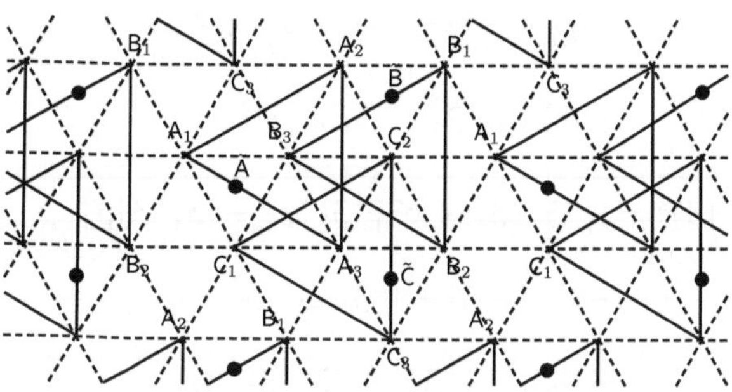

Fig. 15.8 The sublattice and block structure for the triangular lattice showing the form of the periodic boundary conditions

all twenty-four triangles of the lattice. This yields a lattice with $|\widetilde{\mathcal{N}}| =: \widetilde{N} = 3$ and the only triangle in the summation is $\widetilde{\mathsf{A}}\widetilde{\mathsf{B}}\widetilde{\mathsf{C}}$, which must be counted six times because of the periodic boundary conditions. Thus, in this transformation, $\lambda = \sqrt{3}$. Defining

$$\begin{aligned} \mathfrak{z}_0 &:= \exp(C), & \mathfrak{z}_1 &:= \exp(L), \\ \mathfrak{z}_2 &:= \exp(K), & \mathfrak{z}_3 &:= \exp(M), \end{aligned} \tag{15.8.2}$$

(15.1.13) yields the four independent equations

$$[\tilde{\mathfrak{z}}_0]^3 [\tilde{\mathfrak{z}}_1]^3 [\tilde{\mathfrak{z}}_2]^9 [\tilde{\mathfrak{z}}_3]^6 = \mathfrak{z}_0^9 \mathfrak{Z}(+1,+1,+1), \tag{15.8.3}$$

$$[\tilde{\mathfrak{z}}_0]^3 [\tilde{\mathfrak{z}}_1]^{-3} [\tilde{\mathfrak{z}}_2]^9 [\tilde{\mathfrak{z}}_3]^{-6} = \mathfrak{z}_0^9 \mathfrak{Z}(-1,-1,-1), \tag{15.8.4}$$

$$[\tilde{\mathfrak{z}}_0]^3 [\tilde{\mathfrak{z}}_1] [\tilde{\mathfrak{z}}_2]^3 [\tilde{\mathfrak{z}}_3]^{-6} = \mathfrak{z}_0^9 \mathfrak{Z}(+1,+1,-1), \tag{15.8.5}$$

$$[\tilde{\mathfrak{z}}_0]^3 [\tilde{\mathfrak{z}}_1]^{-1} [\tilde{\mathfrak{z}}_2]^{-3} [\tilde{\mathfrak{z}}_3]^6 = \mathfrak{z}_0^9 \mathfrak{Z}(-1,-1,+1), \tag{15.8.6}$$

where $\mathfrak{Z}(\sigma_{\tilde{\mathsf{A}}}, \sigma_{\tilde{\mathsf{B}}}, \sigma_{\tilde{\mathsf{C}}})$ is the expression for the right-hand side of (15.1.13), excluding the trivial coupling, for the designated values of the block spins. These are given (Schick et al. 1977) by

$$\begin{aligned} \mathfrak{Z}(+1,+1,+1) := {} & \mathfrak{z}_1^9 \mathfrak{z}_2^{27} \mathfrak{z}_3^{18} + 9 \mathfrak{z}_1^7 \mathfrak{z}_2^{15} \mathfrak{z}_3^6 + 27 \mathfrak{z}_1^5 \mathfrak{z}_2^7 \mathfrak{z}_3^2 \\ & + 18 \mathfrak{z}_1^3 \mathfrak{z}_2^3 \mathfrak{z}_3^{-2} + 9 \mathfrak{z}_1^3 \mathfrak{z}_2^3 \mathfrak{z}_3^6, \end{aligned} \tag{15.8.7}$$

$$\begin{aligned} \mathfrak{Z}(-1,-1,-1) := {} & \mathfrak{z}_1^{-9} \mathfrak{z}_2^{27} \mathfrak{z}_3^{-18} + 9 \mathfrak{z}_1^{-7} \mathfrak{z}_2^{15} \mathfrak{z}_3^{-6} + 27 \mathfrak{z}_1^{-5} \mathfrak{z}_2^7 \mathfrak{z}_3^{-2} \\ & + 18 \mathfrak{z}_1^{-3} \mathfrak{z}_2^3 \mathfrak{z}_3^2 + 9 \mathfrak{z}_1^{-3} \mathfrak{z}_2^3 \mathfrak{z}_3^{-6}, \end{aligned} \tag{15.8.8}$$

$$\begin{aligned} \mathfrak{Z}(+1,+1,-1) := {} & \mathfrak{z}_1^3 \mathfrak{z}_2^{-9} \mathfrak{z}_3^{-18} + 6 \mathfrak{z}_1 \mathfrak{z}_2^{-9} \mathfrak{z}_3^{-6} + 3 \mathfrak{z}_1^5 \mathfrak{z}_2^3 \mathfrak{z}_3^{-6} \\ & + 9 \mathfrak{z}_1^{-1} \mathfrak{z}_2^{-5} \mathfrak{z}_3^{-2} + 18 \mathfrak{z}_1^3 \mathfrak{z}_2^{-1} \mathfrak{z}_3^{-2} \\ & + 18 \mathfrak{z}_1 \mathfrak{z}_2^{-1} \mathfrak{z}_3^2 + 9 \mathfrak{z}_1 \mathfrak{z}_2^{-1} \mathfrak{z}_3^{-6}, \end{aligned} \tag{15.8.9}$$

$$\begin{aligned} \mathfrak{Z}(-1,-1,+1) := {} & \mathfrak{z}_1^{-3} \mathfrak{z}_2^{-9} \mathfrak{z}_3^{18} + 6 \mathfrak{z}_1^{-1} \mathfrak{z}_2^{-9} \mathfrak{z}_3^6 + 3 \mathfrak{z}_1^{-5} \mathfrak{z}_2^3 \mathfrak{z}_3^6 \\ & + 9 \mathfrak{z}_1^1 \mathfrak{z}_2^{-5} \mathfrak{z}_3^2 + 18 \mathfrak{z}_1^{-3} \mathfrak{z}_2^{-1} \mathfrak{z}_3^2 \\ & + 18 \mathfrak{z}_1^{-1} \mathfrak{z}_2^{-1} \mathfrak{z}_3^{-2} + 9 \mathfrak{z}_1^{-1} \mathfrak{z}_2^{-1} \mathfrak{z}_3^6 . \end{aligned} \tag{15.8.10}$$

From (15.8.3)–(15.8.10),

$$\tilde{\mathfrak{z}}_0 = \mathfrak{z}_0^3\,[\mathcal{Z}(+1,+1,+1)\mathcal{Z}(-1,-1,-1)]^{1/24} \times [\mathcal{Z}(+1,+1,-1)\mathcal{Z}(-1,-1,+1)]^{1/8}\,, \tag{15.8.11}$$

$$\tilde{\mathfrak{z}}_1 = \left[\frac{\mathcal{Z}(+1,+1,+1)\mathcal{Z}(+1,+1,-1)}{\mathcal{Z}(-1,-1,-1)\mathcal{Z}(-1,-1,+1)}\right]^{1/8}, \tag{15.8.12}$$

$$\tilde{\mathfrak{z}}_2 = \left[\frac{\mathcal{Z}(+1,+1,+1)\mathcal{Z}(-1,-1,-1)}{\mathcal{Z}(+1,+1,-1)\mathcal{Z}(-1,-1,+1)}\right]^{1/24}, \tag{15.8.13}$$

$$\tilde{\mathfrak{z}}_3 = \left[\frac{\mathcal{Z}(+1,+1,+1)}{\mathcal{Z}(-1,-1,-1)}\right]^{1/48}\left[\frac{\mathcal{Z}(-1,-1,+1)}{\mathcal{Z}(+1,+1,-1)}\right]^{1/16}. \tag{15.8.14}$$

These are the recurrence relationships for the model. It can be seen that there is not a consistent solution given simply by setting $\mathfrak{z}_3 = 1$, ($M = 0$). However, there is a solution for $\mathfrak{z}_2$ alone when $\mathfrak{z}_1 = \mathfrak{z}_3 = 1$. This is because of the symmetry of the terms of the Hamiltonian. For this type of transformation, once a basic set of sites (a triangle in this case) is chosen, all or none of the possible terms in the Hamiltonian corresponding to a particular symmetry must be included. The term corresponding to K is invariant under spin reversal and it is the only possible function of the spins on a triangle (apart from the trivial coupling C) with this property. The terms corresponding to L and M are the only possible combinations of spins which change sign with spin reversal and either both or neither must be included in the Hamiltonian. When they are present the recurrence relationships are invariant under the simultaneous application of $\mathfrak{z}_1 \leftrightarrow 1/\mathfrak{z}_1$, $\mathfrak{z}_3 \leftrightarrow 1/\mathfrak{z}_3$.

The results derived for this model for the antiferromagnetic case ($K < 0$) in the plane $M = 0$ are presented by Schick et al. (1976) and shown in Fig. 15.9. The transition for $L > 0$ is to the antiferromagnetic phase with ground state $(+1,+1,-1)$ and that for $L < 0$ to the antiferromagnetic phase with ground state $(-1,-1,+1)$. The results show good agreement with the Monte Carlo results of Metcalf (1973). As is well-known there is no non-zero Néel temperature on the zero-field axis and the outer termination points of the curves are at the critical field values $\mathcal{H}_c = \pm 6|J|$. The factor six arises as the coordination number of the lattice. The zero-point entropy per lattice site at $\mathcal{H} = 0$ is obtained by the method described in Sect. 15.4 and yields the value 0.324, which is very close to the exact value 0.32306 (Domb 1960, p. 213). The transitions for $L >, < 0$ are controlled respectively by the fixed points $L^\star = \pm\infty$, $K^\star = -\infty$, $M^\star = \pm 0.113$, $|L^\star| + 6K^\star + 6|M^\star| \sim -0.91$. The two antiferromagnetic phases are each three-fold degenerate corresponding to the three choices for the alignment of spins on the sublattices. This set of six ground states is the same as that of the 3-state antiferromagnetic Potts model (Schick and Griffths 1977). Alexander (1975) has shown that this leads to a cubic invariant in the Landau theory

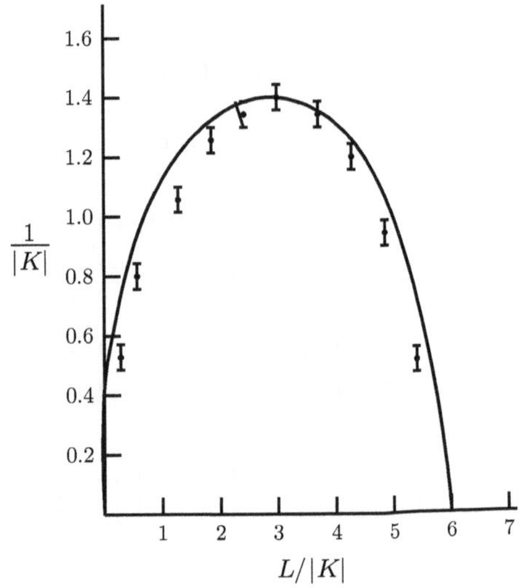

Fig. 15.9 Phase diagram for the continuous phase transition in the plane $M = 0$. The data points are the Monte Carlo results of Metcalf (1973). (Reprinted from Schick et al. (1976), by permission of the publisher Elsevier Science)

formulation of exactly the form exhibited by the 3-state Potts model (Sect. 5.4.2). This implies a first-order transition, a result which for the Potts model is confirmed by other methods (see the discussion in Sect. 5.4.2).

The physical motivation for the work of Schick et al. (1977) is to formulate a model for the order-disorder transition which occurs for gases adsorbed on monolayer films; in particular the case of helium adsorbed on graphite. While the general fit of the curves in Fig. 15.9 with experimental data for this model is good the work by Bretz (1977) for this system yielded a value $\alpha = 0.34$. The thermal exponent derived from these model calculations has the value 0.956, which from Eq. (4.5.39) gives the value $\alpha = -0.093$ for the heat capacity exponent. One might speculate that the exponent -0.26 for this fixed point is associated with a dangerous irrelevant variable (Sect. 4.8). To achieve the value $\alpha = 0.34$ it follows from (4.5.39) that the effective dimension of the system would need to be $\bar{d} = 1.587$. This in turn would imply from Eqs. (4.8.6) and (4.8.9) that the divergence associated with the correction to scaling has exponent 1.5885. This, however, is mere speculation and further calculations would be needed to settle this point. The global phase diagrams for $M \neq 0$ for both signs of K are described in detail by Schick et al. (1977). They show that a very rich phase structure is exhibited with first-order phase transition surfaces, and lines of critical, tricritical and critical end points. This renormalization scheme has been used by Schick and Griffths (1977) for the 3-state Potts model and by Young and Lavis (1979) for a type of bonded lattice fluid which is an extension of the 3-state Potts model. This latter work was extended by Southern and Lavis (1979, 1980) to include the whole class of terms corresponding to the spin-1 model. A spin-1 model renormalized according to this method was used by Lavis et al. (1982) to model phase transitions in monolayers at air/water and oil/water interfaces and Lavis and Quinn (1983) used a spin-$\frac{1}{2}$ model to investigate ferrimagnetic ordering on a triangular lattice.

In general RSRG procedures on finite lattices give qualitatively satisfactory phase diagrams, although the values of critical exponents tend to be rather disappointing particularly for models exhibiting sublattice ordering. Although one might expect the results to improve with the choice of larger lattices, the method then becomes much more difficult to handle. In the work of Lavis and Southern (1984), which treated a spin-1 model with directional bonding on a body-centred cubic lattice of sixteen sites the recurrence relations have thirty-three thousand terms, which placed the computations near the limit of tractability at that time.

15.9 Variational Approximations

As we indicated in Sect. 15.2 a crucial factor in renormalization group calculations is the choice of weight function. Although an exact transformation would not be affected by this choice, the results of approximate transformations can change quite dramatically when the weight function is changed. In Sects. 15.5.1 and 15.5.2 and Examples 15.7.1–15.7.3 cases of the use of the decimation weight function have been explored and, in Sect. 15.8, the majority-rule weight function was employed. The possibility of using a weight function containing a parameter which can be varied to improve the approximation was considered by Kadanoff and Houghton (1975). They used the weight function (15.2.9), with parameter τ, for transformations of the spin-$\frac{1}{2}$ Ising model on the square lattice. The parameter τ was chosen by using the criterion that two different evaluations of the magnetic field exponent agreed with each other.

A form of variational approximation, designed to optimize the lower-bound condition (15.7.9) was developed by Kadanoff (1975) (see also Kadanoff et al. 1976). Known as the *one-hypercube approximation*, it is applicable to a hypercubic lattice of any dimension d. Suppose that the centres of the hypercubes of the lattice are denoted, in terms of the set of independent unit vectors $\hat{\boldsymbol{r}}^{(1)}, \hat{\boldsymbol{r}}^{(2)}, \ldots, \hat{\boldsymbol{r}}^{(d)}$, by vectors of the form

$$\boldsymbol{\rho} := a(n^{(1)}\hat{\boldsymbol{r}}^{(1)} + n^{(2)}\hat{\boldsymbol{r}}^{(2)} + \cdots + n^{(d)}\hat{\boldsymbol{r}}^{(d)}), \tag{15.9.1}$$

where $n^{(\ell)}$ is an integer modulo $N^{(\ell)}$, with the set of *even* integers $N^{(1)}, N^{(2)}, \ldots,$ $N^{(d)}$ satisfying (16.2.5). Let the Hamiltonian be such that all the couplings can be expressed in terms of interactions between microsystems on the sites of a single hypercube. Then

$$\widehat{H}(N, \zeta_i; \sigma(\boldsymbol{r})) := -N\zeta_0 - \sum_{i=1}^{n} \zeta_i \sum_{\{\boldsymbol{\rho}\}} \mathfrak{q}_i(\boldsymbol{\rho}), \tag{15.9.2}$$

where $\mathfrak{q}_i(\boldsymbol{\rho})$ is a function of the $\sigma(\boldsymbol{r})$ at the sites of the cube with centre $\boldsymbol{\rho}$. We assume that the system satisfies the homogeneity condition that $\langle \mathfrak{q}_i(\boldsymbol{\rho}) \rangle$ is independent of $\boldsymbol{\rho}$.

If in (15.7.38) we choose the form

$$\widehat{V}(N, X_i(\boldsymbol{\rho}); \sigma(\boldsymbol{r})) := \sum_{i=1}^{n} \sum_{\{\boldsymbol{\rho}\}} X_i(\boldsymbol{\rho}) \mathfrak{q}_i(\boldsymbol{\rho}), \tag{15.9.3}$$

where

$$\sum_{\{\boldsymbol{\rho}\}} X_i(\boldsymbol{\rho}) = 0, \tag{15.9.4}$$

the condition (15.7.40) is satisfied, leading to the lower-bound condition (15.7.9); (15.7.42) is a particular case of (15.9.3). The quantities $X_i(\boldsymbol{\rho})$ must be chosen to achieve a tractable summation on the right-hand side of (15.7.38) with the renormalized Hamiltonian, now denoted by $\widehat{H}(\widetilde{N}, \widetilde{\zeta}_i; \sigma(\widetilde{\boldsymbol{r}}))$, of the same form as $\widehat{H}(N, \zeta_i; \sigma(\boldsymbol{r}))$. In the one-hypercube approximation this is achieved by dividing the hypercubes of the lattice into subsets. The 'blue' subset, with centres denoted by $\boldsymbol{\rho}_{\mathrm{b}}$ consists of all hypercubes for which all the integers $n^{(\ell)}$ in equation (15.9.1) are odd and the 'red' subset with centres denoted by $\boldsymbol{\rho}_{\mathrm{r}}$, consists of all hypercubes for which all the integers $n^{(\ell)}$ are even. There are $N/2^d$ hypercubes in each of the blue and red subsets and the remaining $N(1-2^{1-d})$ hypercubes, with centres denoted by $\boldsymbol{\rho}_{\mathrm{g}}$ will be called 'green'. It easily be checked that the choice

$$X_i(\boldsymbol{\rho}) := \begin{cases} \zeta_i - Y_i, & \text{if } \boldsymbol{\rho} = \boldsymbol{\rho}_{\mathrm{r}}, \\ \zeta_i, & \text{if } \boldsymbol{\rho} = \boldsymbol{\rho}_{\mathrm{g}}, \\ -(2^d - 1)\zeta_i + Y_i, & \text{if } \boldsymbol{\rho} = \boldsymbol{\rho}_{\mathrm{b}}, \end{cases} \tag{15.9.5}$$

satisfies (15.9.4) for any choice of the new parameters Y_i, which are chosen to ensure that the summation on the right-hand side of (15.7.38) is tractable.

For the one-hypercube approximation applied to a spin-$\frac{1}{2}$ model, the renormalization scheme is constructed with the sites $\widetilde{\boldsymbol{r}} \in \widetilde{\mathcal{N}}$ coinciding with the points $\boldsymbol{\rho}_{\mathrm{r}}$. The scaling parameter $\lambda = 2$, and the Kadanoff–Houghton weight function (15.2.9) is used with $\mathcal{X}(\widetilde{\boldsymbol{r}}) = \mathcal{B}(\widetilde{\boldsymbol{r}})$ taken to be the sites on the red hypercube with centre $\widetilde{\boldsymbol{r}} = \boldsymbol{\rho}_{\mathrm{r}}$. Then

$$\prod_{\{\widetilde{\boldsymbol{r}}\}} \mathrm{p}(\sigma(\widetilde{\boldsymbol{r}})|\sigma(\boldsymbol{r}), \boldsymbol{r} \in \mathcal{B}(\widetilde{\boldsymbol{r}})) = \exp\left\{\sum_{\{\widetilde{\boldsymbol{r}}\}} [\tau\, \sigma(\widetilde{\boldsymbol{r}}) \mathfrak{p}(\widetilde{\boldsymbol{r}}) - \ln(2\cosh(\tau\, \mathfrak{p}(\widetilde{\boldsymbol{r}})))]\right\}, \tag{15.9.6}$$

where $\mathfrak{p}(\widetilde{\boldsymbol{r}})$ is the sum of the variables $\sigma(\boldsymbol{r})$ on all the sites of the red hypercube with centre $\widetilde{\boldsymbol{r}}$. The aim is to reduce the dependence of the argument of the summation on the right-hand side of (15.7.38) to only those terms involving the interactions between the microstates on blue hypercubes. The first term in the exponential on the right-hand side of (15.9.6) is no problem since each site at the corner of one red hypercube is also at the corner of one blue hypercube. The second term can be 'moved' from red to blue hypercubes by imposing the condition

$$\sum_{i=1}^{n} Y_i \mathfrak{q}_i(\tilde{\boldsymbol{r}}) = \ln[2\cosh(\tau\, \mathfrak{p}(\tilde{\boldsymbol{r}}))], \tag{15.9.7}$$

which, when all choices are made for the states on the corners of the cube, is equivalent to 2^d linear equations for the n variables. This means, as was the case of the finite-lattice methods of Sect. 15.8, the number n of non-trivial coupling included in the Hamiltonian must be enough to ensure a consistent solution. A detailed analysis of this method for the square lattice was performed by Burkhardt (1976b) for a Hamiltonian with no magnetic field but with first-neighbour, second-neighbour and four-site couplings. The value of the parameter τ which maximized the free energy for an initial Hamiltonian equal to its critical fixed point value was 0.761. For this he obtained $T_c/J = 2.066$, $\nu = 0.983$ and $\delta = 15.36$. The recurrence relationships for this system have an invariant subspace with equal first- and second-neighbour interactions. Slightly improved results can be achieved by beginning with an initial Hamiltonian with no second-neighbour interactions and performing a preliminary exact decimation to produce a Hamiltonian within this subspace. This was the method employed by Kadanoff (1975) and Kadanoff et al. (1976). A modification of this method applicable to both square and triangular lattices was developed by Southern (1978) and for the body-centred cubic lattice by Burkhardt and Eisenriegler (1978). It has also been generalized to spin-1 models by Burkhardt (1976a) and Burkhardt and Knops (1977) and to the 3-state Potts model by Burkhardt et al. (1976).

15.10 Phenomenological Renormalization

The idea of finite-size scaling was introduced in Sect. 4.10 as an extension of the scaling methods for infinite systems treated in the rest of Chap. 4. A renormalization group justification for this has been given by Suzuki (1977) and Brézin (1982) using momentum-space methods and by Barber (1983) using real-space methods. We now discuss an RSRG method based on finite-size scaling which was developed by Nightingale (1976).

In Sect. 4.10 we described scaling for a d-dimensional system $\mathcal{L}_{\mathfrak{d}}(\aleph)$, which was infinite in $\mathfrak{d}$ dimensions and for which the size of the system for the remaining $d - \mathfrak{d}$ dimensions was represented by a parameter $\aleph$ called the thickness. The essential feature of finite-size scaling is that, for such a system with scaling fields $\theta_1, \theta_2, \ldots, \theta_n$, $1/\aleph$ is treated as a scaling field θ_{n+1}. We now adopt a slight change of notation and express the correlation length $\mathfrak{x}(\theta_1, \ldots, \theta_n, \theta_{n+1})$ for $\mathcal{L}_{\mathfrak{d}}(\aleph)$ as $\mathfrak{x}^{(\aleph)}(\theta_1, \ldots, \theta_n)$. Equation (4.10.6) can now be used to relate the correlation lengths of two similar systems $\mathcal{L}_{\mathfrak{d}}(\aleph)$ and $\mathcal{L}_{\mathfrak{d}}(\widetilde{\aleph})$ with couplings ζ_i and $\tilde{\zeta}_i$ and thicknesses $\aleph$ and $\widetilde{\aleph}$ respectively.

Taking $\aleph > \widetilde{\aleph}$ and making the substitution

$$\lambda = \aleph/\widetilde{\aleph}, \tag{15.10.1}$$

we have

$$\mathfrak{r}^{(\aleph)}(\theta_1, \ldots, \theta_n) = \lambda \mathfrak{r}^{(\widetilde{\aleph})}(\tilde{\theta}_1, \ldots, \tilde{\theta}_n), \tag{15.10.2}$$

where

$$\tilde{\theta}_j = \theta_j(\tilde{\zeta}_i) = \lambda^{y_j}\theta_j(\zeta_i) \tag{15.10.3}$$

relates the scaling fields θ_j and $\tilde{\theta}_j$ for $\mathcal{L}_\partial(\aleph)$ and $\mathcal{L}_\partial(\widetilde{\aleph})$. Equation (15.10.2) forms the basis of Nightingale's phenomenological renormalization method. For the infinite system $\mathcal{L}$, defined by the thermodynamic limit $\mathcal{L}_\partial(\aleph) \to \mathcal{L}$, as $\aleph \to \infty$, the correlation length $\mathfrak{r}$ satisfies an equation similar to (15.10.2) except that the same function appears on both sides of the equation in place of the correlation functions for two systems of different thickness. If, however, we suppose that, in the limit $\aleph, \widetilde{\aleph} \to \infty$ with λ fixed, the correlation lengths for the finite systems approach that of the infinite system, then (15.10.2) can be taken as an approximation to the scaling form for $\mathcal{L}$. This has the advantage, as we shall see below, that the correlation length for $\mathcal{L}_\partial(\aleph)$ can be calculated explicitly as functions of the couplings ζ_i, $i = 1, 2, \ldots, n$, at least for fairly small values of $\aleph$, and we can reasonably suppose that the approximation will increase in accuracy as $\aleph$ is increased. Equation (15.10.2) can be rewritten as an expression relating the couplings for $\mathcal{L}_\partial(\aleph)$ and $\mathcal{L}_\partial(\widetilde{\aleph})$. Thus

$$\mathfrak{r}^{(\aleph)}(\zeta_1, \ldots, \zeta_n) = \lambda \mathfrak{r}^{(\widetilde{\aleph})}(\tilde{\zeta}_1, \ldots, \tilde{\zeta}_n)\,. \tag{15.10.4}$$

If we suppose that there is a renormalization transformation of the form of (15.2.2) relating the coupling of $\mathcal{L}_\partial(\aleph)$ and $\mathcal{L}_\partial(\widetilde{\aleph})$ then a fixed point will satisfy

$$\mathfrak{r}^{(\aleph)}(\zeta_1^\star, \ldots, \zeta_n^\star) = \lambda \mathfrak{r}^{(\widetilde{\aleph})}(\zeta_1^\star, \ldots, \zeta_n^\star)\,. \tag{15.10.5}$$

It is, however, clear that for $n > 1$ this transformation is not completely specified by (15.10.4).

A number of methods for dealing with systems with more than one coupling have been developed and will be described below. We first concentrate on the case $n = 1$, with $\zeta_1 = K := J/T$, where J is some interaction energy between microsystems. The renormalization transformation $K \to \widetilde{K} = \mathcal{K}(K)$ is now defined by

$$\mathfrak{r}^{(\aleph)}(K) = \lambda \mathfrak{r}^{(\widetilde{\aleph})}(\widetilde{K}) \tag{15.10.6}$$

and the fixed points by

$$\mathfrak{r}^{(\aleph)}(K^\star) = \lambda \mathfrak{r}^{(\tilde{\aleph})}(K^\star). \tag{15.10.7}$$

If this transformation is a satisfactory approximation to a system with a non-zero critical temperature T_c, then we expect a finite non-zero critical fixed point solution $K^\star = K_c := J/T_c$ to equation (15.10.7). The scaling field can be chosen as

$$\theta_1 = K - K^\star. \tag{15.10.8}$$

From (15.10.3) the thermal exponent $y_K = 1/\nu$ is given by

$$\lambda^{y_K} = \left(\frac{\partial \widetilde{K}}{\partial K}\right)^\star \tag{15.10.9}$$

and, from (15.10.6),

$$y_K = \frac{\ln\left\{\left(\frac{\partial \mathfrak{r}^{(\aleph)}}{\partial K}\right)^\star \left[\left(\frac{\partial \mathfrak{r}^{(\tilde{\aleph})}}{\partial \widetilde{K}}\right)^\star\right]^{-1}\right\}}{\ln(\lambda)} - 1. \tag{15.10.10}$$

In Chap. 11 we considered the application of the transfer matrix method to a two-dimensional system in which the thermodynamic limit is taken in one dimension. This analysis applies equally well, but with more analytic complexity, to a d-dimensional system infinite in one dimension ($\mathfrak{d} = 1$) and, subject to the assumptions described there, the correlation length for a system, for which the ground state of the ordered phase has a periodicity η, is given in terms of the ratio of the two largest eigenvalues of the transfer matrix by (11.1.36). This formula provides the starting point for most phenomenological renormalization calculations. It is clearly of importance to check the accuracy of the method for models for which the eigenvalues of the transfer method are known exactly for any width of system. This has been done for the isotropic and anisotropic square lattice spin-$\frac{1}{2}$ Ising models by Nightingale (1976) and Sneddon (1978), respectively, for the triangular lattice spin-$\frac{1}{2}$ Ising model by Kinzel and Schick (1981) and for the symmetric eight-vertex model ($K^{(\mathrm{f})} = K^{(\mathrm{b})}$ in Sect. 12.3) by Nightingale (1977). As an example we consider the simplest case of the isotropic square lattice spin-$\frac{1}{2}$ Ising model.

15.10.1 The Square-Lattice Ising Model

Consider a lattice of $N_1 N_2 = N$ sites with the thermodynamic limit being taken for $N_1 \to \infty$ and the thickness of the lattice $\aleph := N_2$. The complete set of eigenvalues for this model is given by Eqs. (12.2.136), (12.2.137) and (12.2.149) with $K^{(\mathrm{h})} = K^{(\mathrm{ch})} = K$. The largest and next largest eigenvalues correspond to taking all plus signs with $s = +1$ and $s = -1$ respectively and in the present notation are given by[7]

$$\Lambda_1(\aleph; K) = [2\sinh(2K)]^{\frac{1}{2}\aleph} \exp\left\{\frac{1}{2}[\omega_1^{(\aleph)}(K) + \cdots + \omega_{2\aleph-1}^{(\aleph)}(K)]\right\} \quad (15.10.11)$$

and

$$\Lambda_2(\aleph; K) := [2\sinh(2K)]^{\frac{1}{2}\aleph} \exp\left\{\frac{1}{2}[\omega_0^{(\aleph)}(K) + \cdots + \omega_{2\aleph-2}^{(\aleph)}(K)]\right\}, \quad (15.10.12)$$

respectively, where

$$\omega_r^{(\aleph)}(K) := \omega(K, K; \frac{r\pi}{2\aleph}), \qquad 0, 1, \ldots, 2\aleph - 1, \quad (15.10.13)$$

and

$$\cosh[\omega_r^{(\aleph)}(K)] := \cosh(2K)\coth(2K) - \cos\left(\frac{r\pi}{\aleph}\right). \quad (15.10.14)$$

From (11.1.36) the correlation length is

$$\mathfrak{r}^{(\aleph)}(K) = -\frac{a}{\sqrt{2}}\left[\ln\left|\frac{\Lambda_2(\aleph; K)}{\Lambda_1(\aleph; K)}\right|\right]^{-1}, \quad (15.10.15)$$

where we have used the result $c(2) = \frac{1}{4}$, obtained in Sect. 5.6.3. Equations (15.10.7) and (15.10.10) can now be used, together with (15.10.15), to obtain K_c and y_K. It follows from these equations that K_c is given by the crossing point of the curves for $g^{(\aleph)}(K)$ and $g^{(\tilde{\aleph})}(K)$ plotted as functions of K, where

$$g^{(\aleph)}(K) :== \left|\frac{\Lambda_2(\aleph; K)}{\Lambda_1(\aleph; K)}\right|^{\aleph}. \quad (15.10.16)$$

Nightingale (1976) considered various values of $\aleph$ and $\tilde{\aleph}$ and showed that the best results were obtained when $\tilde{\aleph} = \aleph - 1$. As may be expected the results improved with increasing $\aleph$, but even with quite small lattice thicknesses the results are remarkably good. Plots for $g^{(10)}(K)$ and $g^{(9)}(K)$ are shown in Fig. 15.10. The crossing point of the curves is at $K_c = 0.440286$ and using (15.10.10) the result $y_K = 1.0067$ is obtained for the critical exponent. The corresponding results for $\tilde{\aleph} = 49$ and $\aleph = 50$ are $K_c = 0.44068412$, $y_K = 1.000233$, the exact results being $K_c = 0.44068679$ and $y_K = 1$. A similar high level of accuracy was also achieved for the other exactly solved models in the work listed above.

---

[7] We omit the factor $\ell^{(s)}(\boldsymbol{\gamma})\left(\frac{k}{2}\right)^{\frac{1}{2}N_2}$, which, as explained in Sect. 12.2.7, is an artifact of our method of derivation.

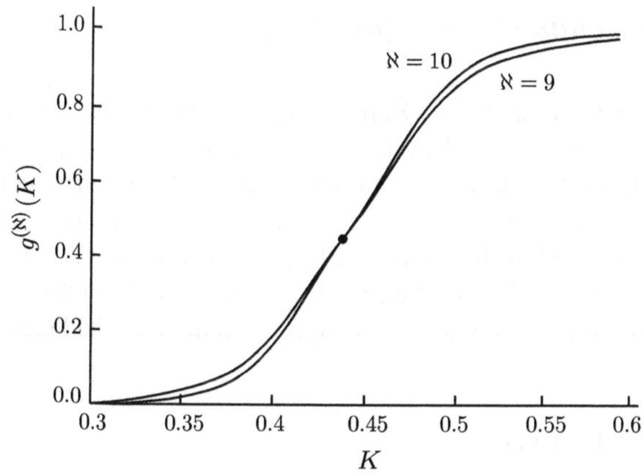

Fig. 15.10 Plots for $g^{(\aleph)}(K)$ with $\aleph = 9, 10$ for the spin-$\frac{1}{2}$ isotropic Ising model on a square lattice

15.10.2 Other Models

The checks on phenomenological renormalization are, of course, more limited for systems which have not been exactly solved. In some cases the critical temperature is known from the transformation methods of Chap. 9 and for the square-lattice Potts model it is given by (9.2.28). This value can also be derived from a mapping to a staggered six-vertex model (Sect. 3.8.2.1), with the value of the exponent α given by exploiting the relationship to the staggered F model. The paper of Sneddon (1978) contains, in addition to the calculations for the spin-$\frac{1}{2}$ anisotropic Ising model, results for spin values $s = 1, \frac{3}{2}, 2, \frac{5}{2}$. For quite small lattice thicknesses ($\aleph = 5, 4, 3$, with $\widetilde{\aleph} = \aleph - 1$) values of the critical temperature correspond quite closely with high-temperature series and lower-bound renormalization group results obtained by Burkhardt and Swendsen (1976). A series of phenomenological renormalization calculations by Nightingale and Blöte (1980), Blöte et al. (1981) and Blöte and Nightingale (1982) for the ν-state Potts model again give encouraging agreement with the exact results available. The method was applied to a triangular Ising ferrimagnetic model by Lavis and Quinn (1987). In that case the sublattice structure gives a periodicity of three rows in the ground state of the ordered phase. By taking $\aleph = 6, \widetilde{\aleph} = 3$ they were able to obtain results for the critical temperature in various cases within 1 % of the exact values obtained by Lavis and Quinn (1983), using transformation methods.

15.10.3 More Than One Coupling

We now consider the case $n > 1$. Here, as was observed above, the transformation cannot be completely specified by the one equation (15.10.4) and the fixed point equation (15.10.5) will yield an $(n-1)$-dimensional surface in the space of couplings. The simplest method is to set $\zeta_i := \omega_i \zeta_1$ for $i = 2, 3, \ldots, n$ and then carry out the renormalization procedure for ζ_1 at fixed values of $\omega_2, \omega_3, \ldots, \omega_n$. This was the approach of Sneddon (1978) for the anisotropic Ising model and Kinzel and Schick (1981) for the triangular Ising antiferromagnet in a field. In the latter case (15.10.4) becomes

$$\mathfrak{r}^{(\aleph)}(K, L) = \lambda \mathfrak{r}^{(\tilde{\aleph})}(\widetilde{K}, \widetilde{L}), \tag{15.10.17}$$

where, as for the zero-field model, $K := J/T$ and $L := \mathcal{H}/T$. If an equation of the form of (15.10.17) is used for a model with a critical temperature on the zero-field axis then the magnetic field exponent y_L can be evaluated using a procedure similar to that contained in Eqs. (15.10.9) and (15.10.10). Since however, in most situations, the correlation length is an even function of $\mathcal{H}$ (or L) it is necessary to use the second derivative, giving

$$y_L = \frac{\ln\left\{\left(\frac{\partial^2 \mathfrak{r}^{(\aleph)}}{\partial L^2}\right)^{\star}\left[\left(\frac{\partial^2 \mathfrak{r}^{(\tilde{\aleph})}}{\partial \widetilde{L}^2}\right)^{\star}\right]^{-1}\right\}}{2\ln(\lambda)} - \frac{1}{2}. \tag{15.10.18}$$

If, as in the case of the antiferromagnetic model of Kinzel and Schick (1981), (15.10.17) is applied by renormalizing K at fixed values of $L/K = \mathcal{H}/J$ then a fixed point corresponds to any point on the critical curve defined by

$$\mathfrak{r}^{(\aleph)}(K, L) = \lambda \mathfrak{r}^{(\tilde{\aleph})}(K, L), \tag{15.10.19}$$

in the plane of couplings. The difference between this method and those based on other procedures, which allow renormalization in more than one coupling, is that here each point of the curve is a fixed point, whereas other methods produce curves containing one fixed point, which controls the whole curve. This has the consequence that the single relevant thermal exponent y_K will be a function of $\mathcal{H}/J$. There is also the problem that, if y_K is evaluated using (15.10.10), it corresponds to the exponent parallel to the L axis, which is not in general the relevant direction with respect to the critical curve. Following Kinzel and Schick (1981), (15.10.10) can be generalized to define the exponent

$$y(\hat{\boldsymbol{\nu}}) := \frac{\ln\left\{(\nabla \mathfrak{r}^{(\aleph)}.\hat{\boldsymbol{\nu}})^{\star}\left[(\nabla \mathfrak{r}^{(\tilde{\aleph})}.\hat{\boldsymbol{\nu}})^{\star}\right]^{-1}\right\}}{\ln(\lambda)} - 1 \tag{15.10.20}$$

for some arbitrary unit vector $\hat{\boldsymbol{v}}$ in the plane, where '$\star$' now simply indicates evaluation at some particularly chosen point $(K^\star, L^\star)$ on the critical curve. If $\hat{\boldsymbol{v}} = (1, 0)$, the unit vector parallel to the L axis, (15.10.20) reduces to (15.10.10). The vector $\boldsymbol{u}^\star :=$ $(\nabla\mathfrak{t}^{(\aleph)} - \lambda\nabla\mathfrak{t}^{(\tilde{\aleph})})^\star$ is orthogonal to the critical curve at $(K^\star, L^\star)$ and, if $\hat{\boldsymbol{v}}$ is tangential to the curve (and thus orthogonal to $\boldsymbol{u}^\star$), $y(\hat{\boldsymbol{v}}) = 0$. This marginal exponent is not surprising given that the whole of the critical curve consists of fixed points. For different $(K^\star, L^\star)$ the exponent $y(\hat{\boldsymbol{v}})$ will vary as the direction of $\hat{\boldsymbol{v}}$ is varied. According to the analysis of the linear renormalization group about a fixed point, described in Sect. 15.3, all scaling directions not along the critical curve should yield the same exponent. For the present method this will be the case only when $\nabla\mathfrak{t}^{(\aleph)}$ and $\nabla\mathfrak{t}^{(\tilde{\aleph})}$ are parallel. For this reason Kinzel and Schick (1981) chose values of $(K^\star, L^\star)$ which satisfied this condition as yielding the 'best' value of y_K, which is now given by

$$y_K = \frac{\ln\left\{|\nabla\mathfrak{t}^{(\aleph)}|^\star \left[|\nabla\mathfrak{t}^{(\tilde{\aleph})}|^\star\right]^{-1}\right\}}{\ln(\lambda)} - 1 . \tag{15.10.21}$$

With $\aleph = 9$, $\tilde{\aleph} = 6$ they obtained the value $y_K = 1.2002$ for the thermal exponent on the antiferromagnetic critical curve for the triangular Ising model. It is supposed that this transition is in the universality class of the 3-state Potts model, (Alexander 1975) for which $\alpha = \frac{1}{3}$. From (4.5.39) $y_K = 2/(2 - \alpha) = \frac{6}{5}$.

Most phenomenological renormalization calculations have concentrated on cases where the phase transitions were known, or expected, to be continuous. Exceptions to this are the ν-state Potts model calculations of Blöte et al. (1981) and Roomany and Wyld (1981). The latter authors reported that the method failed to differentiate between the second-order transition expected for $\nu \leq 4$ and the first-order transition expected for $\nu \geq 5$. Wood and Osbaldstin (1982, 1983) have argued that this inability to distinguish between continuous and first-order transitions is indeed a general characteristic of the method. They developed a general procedure based on phenomenological renormalization for determining an approximated form for phase diagrams including the location of regions of multiphase coexistence. They applied the procedure to the ν-state Potts model and it was applied to a triangular Ising ferrimagnet by Lavis and Quinn (1987).

15.11 Other Renormalization Group Methods

In this chapter we have attempted to outline the basic ideas of the RSRG and to provide a brief account of some of the main methods used. In keeping with the general approach of this book we have omitted applications to dynamic phenomena and quantum systems, for which the reader is referred respectively to the articles by Mazenko and Valls and Pfeuty et al. in Burkhardt and van Leeuwen (1982).

One approach to the statistical mechanics of lattice systems, of which limitations of space have prevented a discussion, is *Monte Carlo simulation* (see, for example, Binder 1986). The possibility of improving the results of a Monte Carlo calculations by combining them with a renormalization group procedure was first suggested by Ma (1976b). In the form developed by Swendsen (1979a, b) this *Monte Carlo renormalization method* uses a standard Monte Carlo simulation to obtain a large number of configurations near to criticality. From these correlation functions can be calculated. A renormalization group transformation can now be applied directly to each configuration as many times as is feasible. The last iteration should leave a lattice still large compared to the range of the fixed point Hamiltonian. Approximations to the matrix $\boldsymbol{L}^{\star}$, defined by Eq. (15.3.3), can now be calculated from correlation functions between the lattices corresponding to each stage of the renormalization procedure and a sequence of estimates of the critical exponents obtained. Swendsen (1979a) applied the method to the two-dimensional spin-$\frac{1}{2}$ Ising model and the 3-state Potts model using an real-space block-spin procedure with the majority-rule weight function. He obtained results which compared well with other renormalization group methods and were better than those obtained by Monte Carlo simulations alone. In a subsequent paper (Swendsen 1979b) he used different weight functions. This demonstrated that the accuracy and reliability of the results can be strongly affected by the choice made. This problem is one experienced by other renormalization group procedures (Sect. 15.9). The method has also been applied to a variety of other models. In particular the application to the three-dimensional Ising model by Pawley et al. (1984) gives results in good agreement with those achieved by series methods.

As indicated on p. 520, RSRG methods represent one of the two ways in which the renormalization group can be applied, the other being the use of renormalization in wave-vector space. This latter method, which involves the elimination of the short-range details of the system by integration over large wave vectors, has led to the *ϵ-expansion method*, (for detailed accounts see, for example, Widom 1965; Ma 1976a), where $\epsilon = 4-d$. For the Ising model the ϵ-expansion confirms the prediction of Ginzburg–Landau theory in Sect. 5.6.3 that the upper critical dimension is at $\epsilon = 0$ ($d_{\mathrm{u}} = 4$). The changeover in critical behaviour in this context can be identified with the change between relevance and irrelevance of one of the parameters at the Gaussian fixed point. Critical exponents are here obtained as expansions in powers of ϵ. For the three-dimensional Ising model the expansion to order ϵ^2 gives good agreement with other methods.

Part V
Mathematical Appendices

Chapter 16
Graphs and Lattices

16.1 Graphs

16.1.1 Introduction

A *graph* $\mathfrak{g}$ is a set of *vertices* (points) some or all of which are connected by *edges* (lines)[1]; the number of vertices of the graph $\mathfrak{g}$ is denoted by $\mathfrak{n}(\mathfrak{g})$ and the number of edges by $\mathfrak{e}(\mathfrak{g})$. The number of edges incident at a vertex of $\mathfrak{g}$ is called its *valency*. A vertex is called *odd* or *even* according to whether its valency is an odd or even number. Since each edge connects two vertices, the sum of all the vertex valencies is equal to $2\mathfrak{e}(\mathfrak{g})$. There is, therefore, either zero or an even number of odd vertices.

A *chain* is a graph where the vertices can be ordered (possibly with repetition) in such a way that each is connected by an edge to the next vertex in the ordering.[2] A chain is *elementary* if each of its vertices is distinct. A *cycle* is a finite chain which begins and ends at the same vertex. A cycle is *elementary* if all its vertices are distinct,[3] and its *length* is the number of its vertices or edges. An elementary cycle is a *polygon*.

A graph is *connected* if, for any pair of vertices, it contains a chain going from one to the other. A *multiply-connected* graph is one for which any two vertices are linked by at least two chains, which have no vertices in common except the end-points. A

---

[1] In this section we present only those elements of graph theory needed for the discussion of statistical mechanical lattice systems. For more general presentations see, for example, Berge (1962) or Wilson and Watkins (1990). In the case of graphs which are lattices (Sect. 16.2) or parts of lattices (Sect. 16.4) the vertices are usually referred to as 'sites' and the edges are 'first-neighbour pairs' or 'second-neighbour pairs' etc., according to whether they are represented by lines joining closest or next-closest sites. The term *bond* is also used in our discussion in Sect. 14.3 of the case where a number of bonds replace a single edge of a graph. For the moment we shall suppose that a pair of vertices is connected by at most one edge.

[2] Of course, if the chain is *finite* this condition does not apply to the last vertex.

[3] The term 'simple' is sometimes used here for both chains and cycles in place of 'elementary'. We follow the usage of Berge (1962) who uses 'simple' for the weaker condition which just disallows repeated edges along the ordering of vertices.

D.A. Lavis, *Equilibrium Statistical Mechanics of Lattice Models*,
Theoretical and Mathematical Physics, DOI 10.1007/978-94-017-9430-5_16

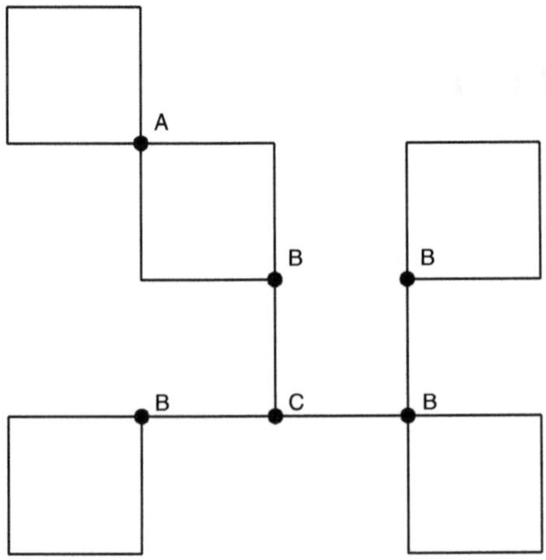

Fig. 16.1 Types of articulation point

graph which is not connected is *separated* and consists of several disjoint connected parts or *components*, where *disjoint* is defined as having no vertices in common. The number of components of the graph $\mathfrak{g}$ is denoted by $c(\mathfrak{g})$.

A connected graph with no cycles is called a *tree*[4] and if all vertices have the same valency, it is a *Bethe lattice*.[5] A Bethe lattice where each vertex is of valency two is an infinite chain.

An *articulation point* is a vertex of a connected graph such that the removal of this vertex and the edges incident there divides the graph into disjoint parts. The alternative name for an articulation point is a *cut vertex* and analogously to this there is a *cut edge* (also called a *bridge*) whose removal from the graph will also divide it into disjoint parts. A connected graph without articulation points is called a *star*. The simplest star is just a single edge (and two vertices) and it can be shown quite easily that a graph with three or more vertices is multiply connected if and only if it is a star. Every vertex of a tree is an articulation point and any graph which is neither a tree nor a star consists of a collection of stars linked by articulation points. Articulation points are of three kinds. Those at which

A: no bridges are incident,
B: bridges and other edges are incident.
C: only bridges are incident.

These three types of articulation point are illustrated in Fig. 16.1.

---

[4] There is unfortunately some variation in the literature in the definition of a tree. Thus Lavis and Bell (1999b, Sect. A.7.2) define a tree to be any connected graph, which is not a star. According to this definition a tree is composed of stars connected at articulation points. Then a tree with no closed cycles of edges is called a *Cayley tree*. To complicate things further the term 'Cayley tree' is sometimes used to signify the graph which we have called a Bethe lattice.

[5] In the case of a finite lattice this definition is modified by excluding from the valency condition sites of valency one (sometimes called *leaves*), which terminate the lattice.

16.1.2 The Cyclomatic Number

For a finite graph $\mathfrak{g}$ which is not a tree it is useful to be able to count the number of cycles it contains; however, this idea needs to be qualified by defining the idea of *independent cycles*. One way to do this (Berge 1962, p. 28) is to *orient* $\mathfrak{g}$ by arbitrarily assigning, by means of an arrow, a direction to each of its edges (Sect. 16.1.4). Then for any cycle of $\mathfrak{e}$ edges, traversed in a chosen direction an $\mathfrak{e}$-dimensional vector can be constructed with its ith entry equal to the number of times the cycle passes through the ith edge in the positive direction minus the number of times it passes through the edge in the negative direction. A basis set of vectors for all cycles on $\mathfrak{g}$ represents a maximal set of independent cycles. The number of cycles in this set is $\mathfrak{C}(\mathfrak{g})$, the *cyclomatic number* of $\mathfrak{g}$ and it can be shown (Berge 1962, p. 29)[6] that

$$\mathfrak{C}(\mathfrak{g}) = \mathfrak{c}(\mathfrak{g}) + \mathfrak{e}(\mathfrak{g}) - \mathfrak{n}(\mathfrak{g}). \tag{16.1.1}$$

Little reference is needed to the space in which these vertices and edges exist, except to say that a *planar graph* is one which can be drawn on a plane without any edges intersecting. In such a case a *face* of the graph is a region of the plane, bounded by an elementary cycle, with no vertices except those on its boundary. The subset of elementary cycles which are the boundaries of faces are called *face cycles* and the number of faces $\mathfrak{f}(\mathfrak{g})$ of a planar graph $\mathfrak{g}$ is equal to its cyclomatic number. For a one-component graph it follows from (16.1.1) that

$$\mathfrak{n}(\mathfrak{g}) + \mathfrak{f}(\mathfrak{g}) - \mathfrak{e}(\mathfrak{g}) = 1. \tag{16.1.2}$$

This is *Euler's relationship* (Wilson and Watkins 1990, p. 219). It can be generalized to surfaces with different topologies. Thus for a graph on the surface of a sphere the exterior of the graph is counted as an additional face giving

$$\mathfrak{n}(\mathfrak{g}) + \mathfrak{f}(\mathfrak{g}) - \mathfrak{e}(\mathfrak{g}) = 2 \tag{16.1.3}$$

and, in general,

$$\mathfrak{n}(\mathfrak{g}) + \mathfrak{f}(\mathfrak{g}) - \mathfrak{e}(\mathfrak{g}) = C, \tag{16.1.4}$$

where, for graphs with simply-connected faces covering a two-dimensional manifold of genus g, $C = 2(1 - g)$.

---

[6] In fact Berge *defines* the cyclomatic number by (16.1.1) and then proves it equal to the number of independent cycles.

16.1.3 Triangulation of Graphs

Consider a one-component planar graph $\mathfrak{g}$ bounded by an elementary cycle denoted by $\mathfrak{c}$. The length of $\mathfrak{c}$ is $\mathfrak{n}(\mathfrak{c}) = \mathfrak{e}(\mathfrak{c})$ and, denoting the number of interior (non-boundary) vertices and edges of $\mathfrak{g}$ by $\mathfrak{n}_{\mathrm{i}}(\mathfrak{g})$ and $\mathfrak{e}_{\mathrm{i}}(\mathfrak{g})$ respectively, it follows from (16.1.2) that

$$\mathfrak{n}_{\mathrm{i}}(\mathfrak{g}) + \mathfrak{f}(\mathfrak{g}) - \mathfrak{e}_{\mathrm{i}}(\mathfrak{g}) = 1. \tag{16.1.5}$$

If $\mathfrak{g}$ is *triangular* (meaning that face cycles are all of length three), every edge of $\mathfrak{c}$ is a side of exactly one triangular face and every interior edge is a side of exactly two faces. So $3\mathfrak{f}(\mathfrak{g}) = \mathfrak{n}(\mathfrak{c}) + 2\mathfrak{e}_{\mathrm{i}}(\mathfrak{g})$ and, substituting into (16.1.5), gives

$$\mathfrak{f}(\mathfrak{g}) = \mathfrak{n}(\mathfrak{c}) + 2\mathfrak{n}_{\mathrm{i}}(\mathfrak{g}) - 2. \tag{16.1.6}$$

In general $\mathfrak{g}$ can be turned into a triangular graph $\mathfrak{g}'$ (*triangulated*) by adding extra edges, but no extra vertices so that each face of $\mathfrak{g}'$ is a triangle. Let $\mathfrak{c}_k$, $k = 1, 2, \ldots, \mathfrak{f}(\mathfrak{g})$ be the face cycles of $\mathfrak{g}$. It is clear that triangulation can be implemented for any plane graph and that it can, in general, be done in a number of different ways. However, from (16.1.6), it follows that the face bounded by $\mathfrak{c}_k$ will be divided into

$$\Upsilon(\mathfrak{c}_k) := \mathfrak{n}(\mathfrak{c}_k) - 2 \tag{16.1.7}$$

faces by any process of triangulation. $\Upsilon(\mathfrak{c}_k)$ is the *triangulation number* of $\mathfrak{c}_k$ and

$$\Upsilon(\mathfrak{g}) = \sum_{k=1}^{\mathfrak{f}(\mathfrak{g})} [\mathfrak{n}(\mathfrak{c}_k) - 2] \tag{16.1.8}$$

is the triangulation number of $\mathfrak{g}$.

16.1.4 Oriented Graphs

As indicated in Sect. 16.1.2 a graph $\mathfrak{g}$ is oriented if each of its edges is given a direction by marking it with an arrow. The face cycles $\mathfrak{c}_k$ have a clockwise and an anticlockwise *orientation* and an arrow on a edge of $\mathfrak{c}_k$ is said to be clockwise or anticlockwise with respect to $\mathfrak{c}_k$ according to whether it is in the direction of the clockwise or anticlockwise orientation of $\mathfrak{c}_k$. Let ω_k be the number of clockwise arrows on $\mathfrak{c}_k$ and ω the number of clockwise arrows on the boundary $\mathfrak{c}$ of $\mathfrak{g}$. Any arrow which is clockwise with respect to $\mathfrak{c}$ is also clockwise with respect to one face cycle. Any interior edge separates two of the faces of $\mathfrak{g}$ and the arrow on this edge is clockwise with respect to one face cycle and anticlockwise with respect to another.

Thus we have, using (16.1.5),

$$\sum_{k=1}^{\mathfrak{f}(\mathfrak{g})} \omega_k = \omega + \mathfrak{e}_{\rm i}(\mathfrak{g}) = \omega + \mathfrak{f}(\mathfrak{g}) + \mathfrak{n}_{\rm i}(\mathfrak{g}) - 1\,, \tag{16.1.9}$$

and it follows that

$$\omega = \sum_{k=1}^{\mathfrak{f}(\mathfrak{g})} [\omega_k - 1] - \mathfrak{n}_{\rm i}(\mathfrak{g}) + 1\,. \tag{16.1.10}$$

16.1.5 The Dual Graph

For any planar graph $\mathfrak{g}$ a *dual graph* $\mathfrak{g}^\star$ can be formed consisting of vertices placed one in each face of $\mathfrak{g}$ with an edge joining the vertices of $\mathfrak{g}^\star$ in adjacent faces. It is clear that the various forms (16.1.2)–(16.1.4) of Euler's relationship, with $\mathfrak{f}(\mathfrak{g})$ replaced by $\mathfrak{n}(\mathfrak{g}^\star)$, apply, with C varying according to the way that the boundary of $\mathfrak{g}$ is handled.

Consider now a finite one-component graph $\mathfrak{g}$ on a plane, bounded by a simple cycle $\mathfrak{c}$ and with no dangling edges (edges terminating at one end in a vertex of valency one). This graph will have $\mathfrak{l}(\mathfrak{g}) := \mathfrak{e}(\mathfrak{c})$ boundary edges and $\mathfrak{e}_{\rm i}(\mathfrak{g}) = \mathfrak{e}(\mathfrak{g}) - \mathfrak{e}(\mathfrak{c})$ interior edges, with a face on either side. Since a site of $\mathfrak{g}^\star$ is placed in each face of $\mathfrak{g}$ and edges are drawn to connect sites in neighbouring faces, $\mathfrak{n}(\mathfrak{g}^\star) = \mathfrak{f}(\mathfrak{g})$, $\mathfrak{e}(\mathfrak{g}^\star) = \mathfrak{e}_{\rm i}(\mathfrak{g})$ and, from (16.1.2),

$$\mathfrak{n}(\mathfrak{g}) + \mathfrak{n}(\mathfrak{g}^\star) - 1 = \mathfrak{e}(\mathfrak{g}) = \mathfrak{e}(\mathfrak{g}^\star) + \mathfrak{l}(\mathfrak{g}). \tag{16.1.11}$$

Now suppose $\mathfrak{g}$ is surrounded by another simple closed curve $\bar{\mathfrak{c}}$ and $\mathfrak{g}^{(+)}$ denotes the graph formed from the union of $\mathfrak{g}$ with edges joining it to $\bar{\mathfrak{c}}$, together with sites on $\bar{\mathfrak{c}}$ terminating these edges. Then

$$\mathfrak{n}(\mathfrak{g}^{(+)}) = \mathfrak{n}(\mathfrak{g}) + \mathfrak{l}(\mathfrak{g}), \qquad \mathfrak{e}(\mathfrak{g}^{(+)}) = \mathfrak{e}(\mathfrak{g}) + \mathfrak{l}(\mathfrak{g}). \tag{16.1.12}$$

The dual graph $\mathfrak{g}^\star$ is now formed by placing a site in every face within $\bar{\mathfrak{c}}$. This increases the number of sites and edges of $\mathfrak{g}^\star$ by $\mathfrak{l}(\mathfrak{g})$ and $2\mathfrak{l}(\mathfrak{g})$, respectively, giving

$$\mathfrak{n}(\mathfrak{g}) + \mathfrak{n}(\mathfrak{g}^\star) - 1 = \mathfrak{e}(\mathfrak{g}) + \mathfrak{l}(\mathfrak{g}) = \mathfrak{e}(\mathfrak{g}^\star). \tag{16.1.13}$$

16.2 Lattices

16.2.1 Types of Regular Lattices

A regular lattice $\mathcal{L}$ in a d-dimensional space is an array of points, called the *sites* of the lattice, which has translational symmetry in all d dimensions. The number z of first-neighbours (or nearest-neighbours) to a site is called the *coordination number*.[7] Second-neighbour and further-neighbour sites are defined in a similar way and if needed we denote the number of nth neighbours by z_n, ($z_1 = z$).

Diagrams of the one-, two- and three-dimensional regular lattices are shown in Figs. 16.2–16.10. Sites are shown by black or white circles and first-neighbour pairs are connected by line segments. The linear, square and simple cubic lattices are the $d = 1, 2, 3$ versions of a *hypercubic lattice*. The body-centred cubic lattice of Fig. 16.8 can be obtained from the simple cubic of Fig. 16.7 by placing a new site at the centre of each cube of eight sites. The face-centred cubic lattice of Fig. 16.9 can be obtained from the simple cubic by placing a new site at the centre of each square of four sites. The diamond lattice of Fig. 16.10 can be obtained from the body-centred cubic by eliminating alternate sites.

A *loose-packed (or bipartite) lattice* is one which can be divided into two equal sublattices in such a way that all the first neighbours of a site of one sublattice belong to the other sublattice; a *close-packed lattice* is one for which this is not possible. The linear, square, honeycomb, simple cubic, body-centred cubic and diamond lattices are loose-packed and the triangular, kagomé and face-centred cubic lattices are close-packed. In the diagrams of the loose-packed lattices the sites of one sublattice are denoted by • and the other by ∘. First-neighbour pairs of sites are linked by full line segments and second-neighbour •–• pairs by broken line segments. The corresponding ∘–∘ links are not shown.

The close-packed triangular and kagomé lattices are divisible into three sublattices, denoted in Figs. 16.4 and 16.6 by black, white and grey disks. Each triangular face contains a site of each sublattice. In the case of the triangular lattice the three sublattices are equivalent and any two of the three sublattices forms a honeycomb lattice. The three different sublattices of the the kagomé lattice are distinguished by the orientations of the edges passing through them and the equivalence is achieved by rotation of the lattice. If, in the case of a d-dimensional lattice, one site is chosen as the origin, all the other sites are given by

$$\boldsymbol{r} = a^{(1)} n^{(1)} \hat{\boldsymbol{r}}^{(1)} + a^{(2)} n^{(2)} \hat{\boldsymbol{r}}^{(2)} + \cdots + a^{(d)} n^{(d)} \hat{\boldsymbol{r}}^{(d)} , \qquad (16.2.1)$$

where the $n^{(\ell)}$ range over all integer values and the $\hat{\boldsymbol{r}}^{(\ell)}$ are the unit lattice vectors with respective lattice spacings $a^{(\ell)}$. For a hypercubic lattice the unit lattice vectors are orthogonal and $a^{(\ell)} = a$, for $\ell = 1, 2, \ldots, d$. For other regular lattices this is not the case.

---

[7] For all the regular lattices, except the dice lattice (Sect. 16.2.2.1), each lattice site has the same coordination number. The dice lattice has sites of coordination numbers three and six.

Fig. 16.2 The linear lattice; $d = 1$, $z = 2$

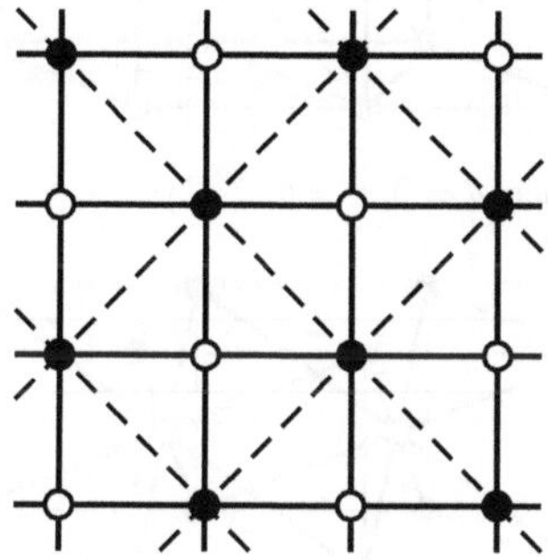

Fig. 16.3 The square lattice; $d = 2$, $z = 4$

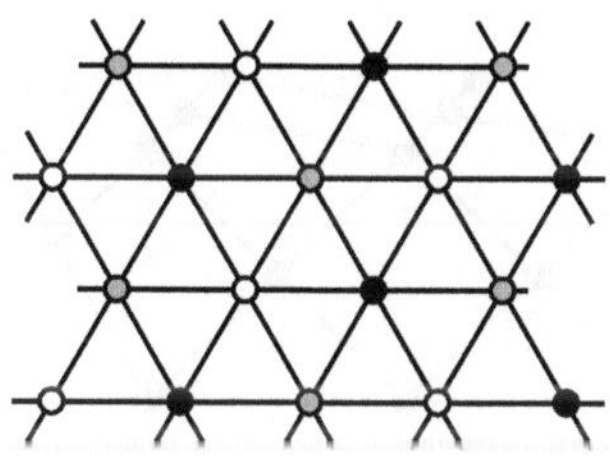

Fig. 16.4 The triangular lattice; $d = 2$, $z = 6$

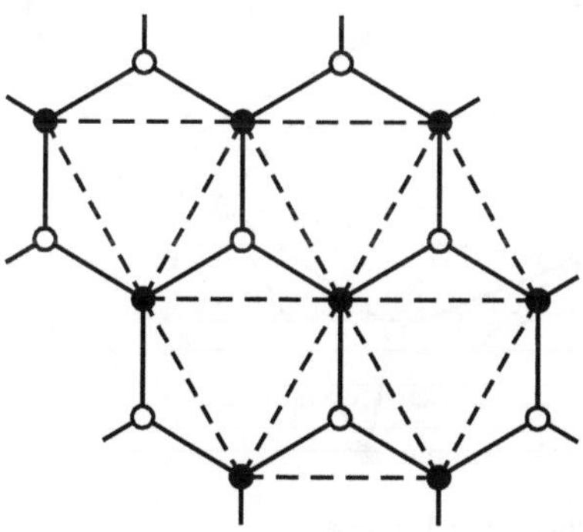

Fig. 16.5 The honeycomb; $d = 2$, $z = 3$

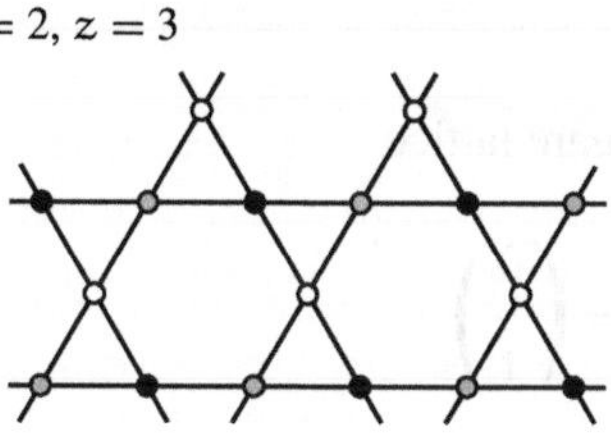

Fig. 16.6 The kagomé lattice; $d = 2$, $z = 4$

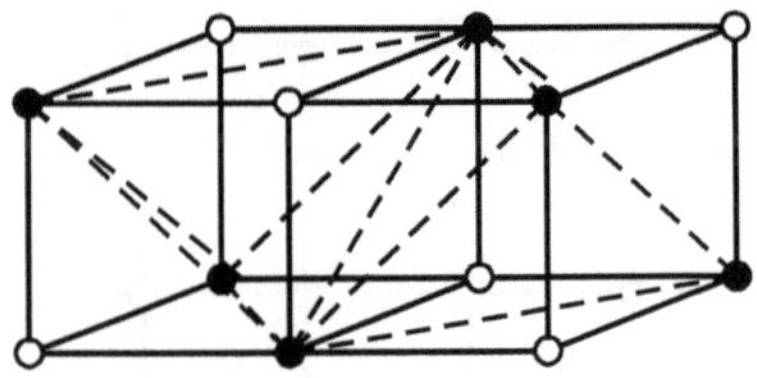

Fig. 16.7 The simple cubic lattice; $d = 3$, $z = 6$

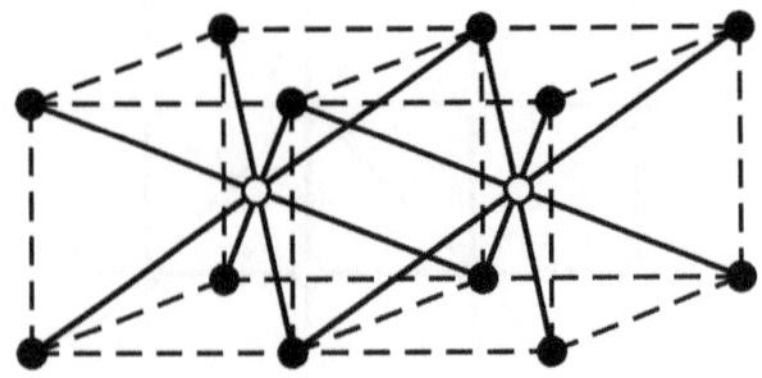

Fig. 16.8 The body-centred cubic lattice; $d = 3$, $z = 8$

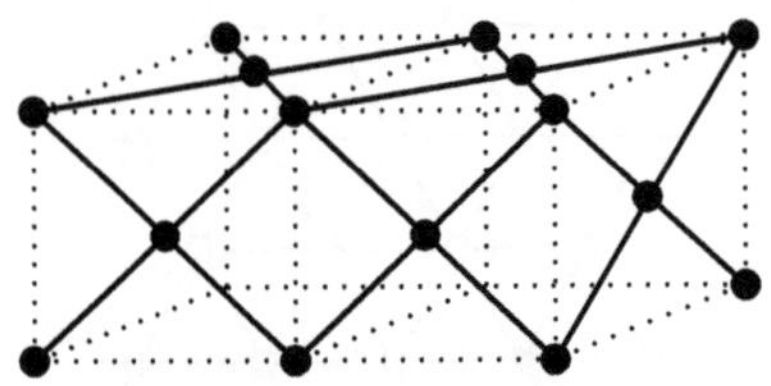

Fig. 16.9 The face-centred cubic lattice; $d = 3$, $z = 12$

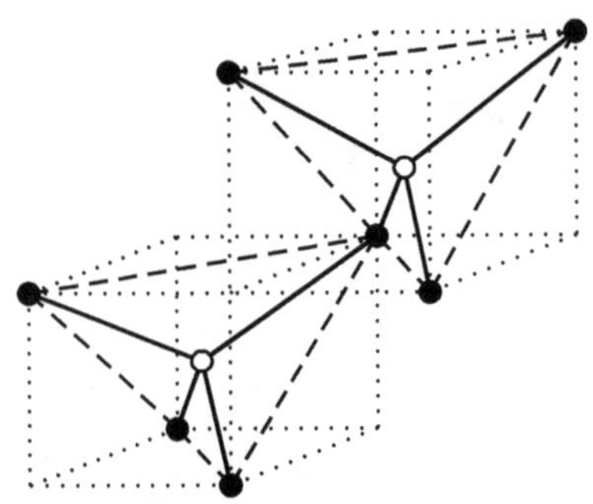

Fig. 16.10 The diamond lattice; $d = 3$, $z = 4$

Example 16.2.1 For the square lattice

$$\hat{\boldsymbol{r}}^{(1)} = \begin{pmatrix} 1 \\ 0 \end{pmatrix}, \qquad \hat{\boldsymbol{r}}^{(2)} = \begin{pmatrix} 0 \\ 1 \end{pmatrix}, \tag{16.2.2}$$

and $a^{(1)} = a^{(2)} = a$.

Example 16.2.2 For the body-centred cubic lattice

$$\hat{\boldsymbol{r}}^{(1)} = \begin{pmatrix} 1 \\ 0 \\ 0 \end{pmatrix}, \qquad \hat{\boldsymbol{r}}^{(2)} = \begin{pmatrix} 0 \\ 1 \\ 0 \end{pmatrix}, \qquad \hat{\boldsymbol{r}}^{(3)} = \frac{1}{\sqrt{3}}\begin{pmatrix} 1 \\ 1 \\ 1 \end{pmatrix} \tag{16.2.3}$$

and, with a the first-neighbour lattice spacing, $a^{(1)} = a^{(2)} = 2a/\sqrt{3}$, $a^{(3)} = a$.

Example 16.2.3 Although the triangular and kagomé lattices are two-dimensional the site locations are most easily expressed in terms of three lattice directions

$$\hat{\boldsymbol{r}}^{(1)} = \frac{1}{\sqrt{3}}\begin{pmatrix} 1 \\ -2 \end{pmatrix}, \quad \hat{\boldsymbol{r}}^{(2)} = \frac{1}{\sqrt{3}}\begin{pmatrix} 1 \\ 2 \end{pmatrix}, \quad \hat{\boldsymbol{r}}^{(3)} = \begin{pmatrix} 1 \\ 0 \end{pmatrix} = \frac{\sqrt{3}}{2}(\hat{\boldsymbol{r}}^{(1)} + \hat{\boldsymbol{r}}^{(2)}), \tag{16.2.4}$$

with $a^{(1)} = a^{(2)} = a^{(3)} = a$. For the triangular lattice with a site chosen as the origin, all the other sites are given by (16.2.1) with $d = 3$. For the kagomé lattice, from a given site other sites, are given in only two of the three lattice directions.

According to the discussion up to this point a lattice will have an infinite number of sites. We shall, however, want to deal with lattices with a finite number of sites. These are denoted by $\mathcal{N}$, with $|\mathcal{N}| := N$, the number of lattice sites, being an extensive variable of the system. There are in general two ways to restrict the size of a lattice. One is simply to truncate it in all dimensions to produce a finite portion of the infinite lattice. This will destroy the translational symmetry of the lattice and produce a subset $\mathcal{N}_{\mathrm{B}}$ of boundary sites which no longer have the same coordination number as the bulk of the lattice. Although, under certain conditions,[8] the effect of the boundary can be neglected in the limit of large N. Alternatively we can apply *periodic boundary conditions*. This is done by imposing the condition that $n^{(\ell)}$ in (16.2.1) is an integer modulo $N^{(\ell)}$, for $\ell = 1, 2, \ldots, d$, and

$$N = \prod_{\ell=1}^{d} N^{(\ell)}. \tag{16.2.5}$$

The set of all vectors $\boldsymbol{r}_{\mathrm{R}}$ which satisfy the condition

$$\exp(\mathrm{i}\boldsymbol{r} \cdot \boldsymbol{r}_{\mathrm{R}}) = 1 \tag{16.2.6}$$

form the *reciprocal lattice* $\mathcal{N}_{\mathrm{R}}$ of $\mathcal{N}$ (Chaikin and Lubensky 1995, p. 45). The reciprocal lattice of a hypercubic lattice of spacing a is a hypercubic lattice of spacing $2\pi/a$. That of a triangular lattice is a triangular lattice and of a honeycomb lattice is a honeycomb lattice, but body-centred and face-centred cubic lattices form reciprocal pairs.

---

[8] Essentially those is which the van Hove thermodynamic limit (Sect. 8.1) can be taken.

16.2.2 Lattice Transformations

A fundamental part of the transformation methods considered in Chaps. 9 and 10 is a transformation of the lattice $\mathcal{N} \to \mathring{\mathcal{N}}$, which can be pictured as the addition (in a translationally symmetric way) of a site on each[9] of (i) the faces, or (ii) the edges, of $\mathcal{N}$, followed by either (a) removing the sites of $\mathcal{N}$ to leave the lattice $\mathring{\mathcal{N}}$ simply consisting of the added sites, or (b) leaving the sites of $\mathcal{N}$ in place so that $\mathring{\mathcal{N}}$ consists of all original and added sites. There are thus four pairs of choices:

(i) + (a) when $\mathring{\mathcal{N}} := \mathcal{N}^\star$ is the *dual lattice.*

(ii) + (a) when $\mathring{\mathcal{N}} := \mathcal{N}^\circledast$ is the *medial lattice.*

(i) + (b) when $\mathring{\mathcal{N}} := \mathcal{N} + \mathcal{N}^\star$ is a *face-decorated lattice.*

(ii) + (b) when $\mathring{\mathcal{N}} := \mathcal{N} + \mathcal{N}^\circledast$ is an *edge-decorated lattice.*

Each of these cases is now considered.

16.2.2.1 The Dual Lattice

An account of the construction of a dual graph $\mathfrak{g}^\star$ for a planar graph $\mathfrak{g}$ is given in Sect. 16.1.5. In general the points of the dual of such a graph can be placed, one to each face, at any locations in the faces. However, for the regular two-dimensional lattices listed above[10] the dual lattices $\mathcal{N}^\star$ are also regular lattices formed by placing the sites of $\mathcal{N}^\star$ at the points of coincidence of the orthogonal bisectors of the edges of $\mathcal{N}$.[11] It can be seen from Fig. 16.11a that the square lattice is *self-dual*, and, from Fig. 16.11b, that the dual of the honeycomb lattice is the triangular lattice and the dual of the triangular lattice is the honeycomb lattice. In each of these cases all first-neighbour pairs are connected by edges in the dual lattice. However, the dual of the kagomé lattice is the *dice lattice*. This is obtained from a triangular lattice by removing one-third of the edges to produce a mosaic of rhombi; meaning that its elementary polygons are not section graphs (Sect. 16.4.3). Although sites in adjacent faces of the kagomé lattice are first neighbours in the dice lattice, not all first neighbours in the dice lattice correspond to sites in adjacent faces of the kagomé lattice and are not connected by edges, so sites have coordination number three *or* six.

Let the sites of the dice lattice be separated into those originating from the centres of the hexagons of the kagomé lattice (with $z = 6$), which form a triangular lattice, and those originating from the triangles of the kagomé lattice (with $z = 3$), which form its honeycomb dual. In Sect. 16.3 we considered the graph $\mathfrak{g}_\diamond$ formed from the

[9] Variations of this where more than one site is added to a face or edge, or a site is added to only a translationally invariant subset of the edges or faces is obviously possible. A case where sites are added to only the downward-pointing triangles of a triangular lattice is described in Sect. 16.2.2.4 and used in the star-triangle transformation of Sect. 9.4.

[10] It is assumed that the lattices are finite with the cases of different boundary conditions considered below.

[11] The case of three-dimensional lattices is discussed below.

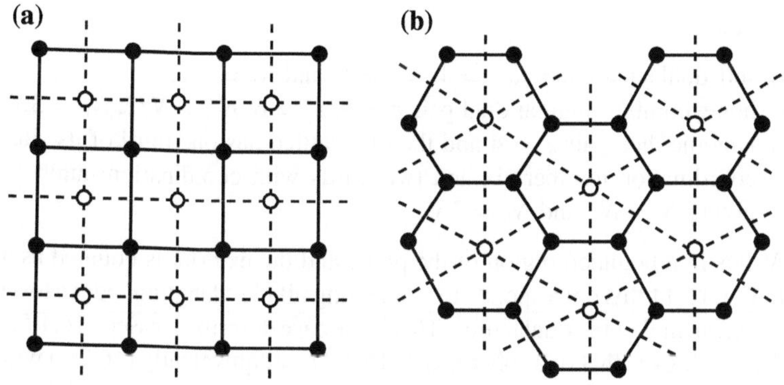

Fig. 16.11 Dual lattices: **a** square (self-dual); **b** honeycomb and triangular

vertices of a graph $\mathfrak{g}$ and its dual $\mathfrak{g}^\star$ and the edges connecting them. In this case $\mathfrak{g}_\Diamond$ for the honeycomb/triangle dual pair is just the dice lattice and, since its faces are rhombi, it follows from Theorem 16.3.2 and its corollary that *rapidity lines*[12] can be drawn. It is easy to see that these simply consist of the original kagomé lattice. A similar analysis of the square lattice shows that its rapidity lattice is also a square lattice. However, the lattice $\mathfrak{g}_\Diamond$ formed in the same way from the kagomé/dice pair does not have a rhombic embedding so, according to Theorem 16.3.2, there is no rapidity lattice.

The number of dual lattice sites $N^\star := |\mathcal{N}^\star|$ is the number of faces of $\mathcal{N}$ and this is related to the number of sites N and edges N_{E} of $\mathcal{N}$ by Euler's relationship (16.1.4) in the form

$$N + N^\star = N_{\mathrm{E}} + C. \tag{16.2.7}$$

Neglecting consideration of boundary conditions, it is clear that the transformation $\mathcal{N} \to \mathcal{N}^\star$ is self-inverse, in the sense that the dual of the dual of a lattice is the lattice itself; lattices form duality pairs, triangular/honeycomb, kagomé/dice, square/square. We now consider the implications of different kinds of boundary conditions. There are four possibilities:

(I) Toroidal boundary conditions are applied and each site of $\mathcal{N}$ has coordination number z. The formula (16.2.7) is satisfied with $C = 0$ and the self-inversion of $\mathcal{N} \to \mathcal{N}^\star$ is preserved. Now, apart from for the kagomé lattice, each site of $\mathcal{N}^\star$ has the same coordination number, denoted by $z^\star$, with[13]

$$N_{\mathrm{E}} = \tfrac{1}{2} zN = \tfrac{1}{2} z^\star N^\star = N_{\mathrm{E}}^\star, \tag{16.2.8}$$

$$(z + z^\star) = zz^\star. \tag{16.2.9}$$

[12] See Sect. 16.3 for an explication of rapidity lines and lattices.

[13] The relationship (16.2.9) between the coordination numbers of interior sites of a lattice and its dual has, of course, general validity.

For the[14]

(a) self-dual square lattice $z = 4$, $N^\star = N$ and $N_E = 2N$.
(b) honeycomb/triangular dual pair $z = 3$, $z^\star = 6$, $N^\star = N/2$, $N_E = 3N/2$.
(c) kagomé/dice pair $z = 4$ and the dice lattice has one-third of its sites with coordination number six and two-thirds with coordination number three, giving $N = N^\star$ and $N_E = 2N$.

(II) $\mathcal{N}$ lies in a bounded region of the plane and the exterior is counted as a face. In Fig. 16.11 this corresponds to connecting all the 'dangling' edges to a single 'exterior site' of the dual lattice. This is the case described in Sect. 16.1.5, where the topology is that of a sphere, and (16.1.3), or equivalently (16.2.7) with $C = 2$, applies. Here $N_E^\star = N_E$ but, of course, the coordination number of the exterior dual-lattice site is not that of the other dual-lattice sites. The transformation $\mathcal{N} \to \mathcal{N}^\star$ is still self-inverse in the sense that repeated application of the transformation produces the correct sequence of lattices (in the topological sense) and the numbers N and $N^\star$ are preserved.

(III) $\mathcal{N}$ lies in a bounded region of the plane with the exterior not counted as a face. In Fig. 16.11 this corresponds to excluding the 'dangling' edges leading to the exterior from the dual lattice. The formula (16.2.7) is satisfied with $C = 1$ and

$$\tfrac{1}{2} zN > N_E > \tfrac{1}{2} z^\star N^\star > N_E^\star . \qquad (16.2.10)$$

Repeated application of $\mathcal{N} \to \mathcal{N}^\star$ leads in the case of Fig. 16.11a to a succession of square lattices and in the cases of Fig. 16.11b to an alternation of honeycomb and triangular lattices, but there is a progressive reduction in the number of lattice sites.

(IV) $\mathcal{N}$ is finite and contained within a larger lattice of the same structure. A simple cycle of boundary sites surrounds $\mathcal{N}$, with N_L edges joining $\mathcal{N}$ to the boundary sites; $\mathcal{N}^{(+)}$ denotes the lattice formed by the union of $\mathcal{N}$ with the edges joining $\mathcal{N}$ to the boundary together with their terminating boundary sites. From the discussion in Sect. 16.1.5

$$N^{(+)} = N + N_L, \qquad (16.2.11)$$

$$N_E^{(+)} = N_E + N_L = \tfrac{1}{2}\{N z + N_L\}. \qquad (16.2.12)$$

The dual lattice $\mathcal{N}^\star$ is now formed from every site lying within the boundary and, from (16.1.13),

[14] Toroidal boundary conditions can be applied to the honeycomb and kagomé lattices only when N is divisible by two and three respectively and this will be assumed, where necessary, in all future discussion.

$$N + N^{\star} - 1 = N_{\mathrm{E}}^{(+)} = N_{\mathrm{E}}^{\star}. \tag{16.2.13}$$

Before leaving the topic of duality it is of interest to consider its extension to higher dimensions; in particular to the simple cubic lattice (Fig. 16.7), with $z = 6$. Two methods of designating the sites of the dual lattice suggest themselves. The first, advocated by Syozi (1972), is to place the sites of $\mathcal{N}^{\star}$ on the faces of $\mathcal{N}$ at the locations of the sites added to $\mathcal{N}$ to transform it to a face-centred cubic lattice (Fig. 16.9). $\mathcal{N}^{\star}$ is an octahedral lattice with $z^{\star} = 8$ and $N^{\star} = 3N$. Alternatively, as proposed by Savit (1980) and Chao (1982), the sites of $\mathcal{N}^{\star}$ could be placed in the cubes of $\mathcal{N}$ at the locations of the sites added to $\mathcal{N}$ to transform it to a body-centred cubic lattice (Fig. 16.8). In this case the simple cubic lattice is self-dual. In a similar way Chao (1982) proposed a duality relationship between the face-centred cubic lattice and the diamond lattice, using these transformations to study bond percolation on the lattices, and Savit (1980) has developed a general scheme of duality for d–dimensional hypercubic lattices.

16.2.2.2 The Medial Lattice

Suppose for simplicity that toroidal boundary conditions are applied to $\mathcal{N}$.[15] In Figs. 16.12 and 16.13 the sites of $\mathcal{N}$ and the medial lattice $\mathcal{N}^{\circledast}$, for which $|\mathcal{N}^{\circledast}| = N_{\mathrm{E}} = Nz/2$, are marked as black and grey respectively. The medial lattice of a square lattice is a square lattice (Fig. 16.12) which is the rapidity lattice for the lattice and its dual, for which it is, of course, also the medial lattice (Sect. 16.2.2.1). The medial lattice for the honeycomb lattice is a kagomé lattice (Fig. 16.13), which is the rapidity lattice for the triangular/honeycomb dual pair and also the medial lattice for the triangular lattice. Forming the medial lattice of a kagomé lattice yields, if only first-neighbour lattice edges are drawn, a set of disconnected triangles; while drawing both first- and second-neighbour edges yields a lattice with triangular, hexagonal and rectangular faces. This is consonant with the fact that a rapidity lattice does not exist for the kagomé/dice dual pair (Sect. 16.2.2.1).

16.2.2.3 Edge Decoration

In the context of both edge and face decoration the sites of $\mathcal{N}$ and the decorating sites ($\mathcal{N}^{\circledast}$ and $\mathcal{N}^{\star}$ respectively) are referred to as *primary and secondary sites*, respectively. The decoration of the edges of a lattice can be unambiguously implemented whatever the dimension of the lattice, but $\mathcal{N} + \mathcal{N}^{\circledast}$ may or may not be one of the regular lattice types. Both the square lattice (black sites) and its dual $\mathcal{N}^{\star}$ are edge-decorated by the same medial (square) lattice (Fig. 16.12), but the resulting lattice in each case is not one of the regular lattice types of Sect. 16.2.1. For the primary

[15] Which, as indicated above, carries the assumption for the honeycomb and kagomé lattices that N is divisible by two and three respectively.

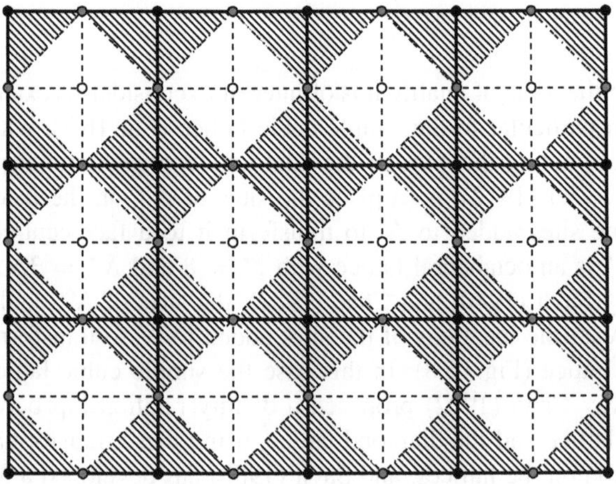

Fig. 16.12 Decoration of a square lattice, shown with *black sites* and *lines*. The dual square lattice is shown with *broken lines* and *white sites* and the square medial lattice with *chain lines* and *grey sites*

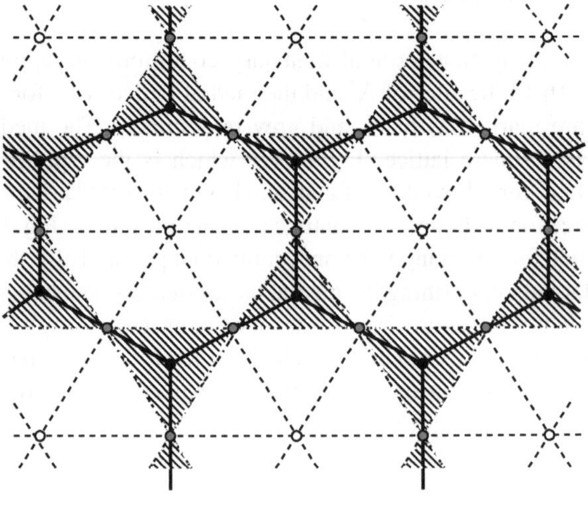

Fig. 16.13 Decoration of a honeycomb lattice, shown with *black sites* and *lines*. The dual triangular lattice is shown with *broken lines* and *white sites* and the kagomé medial lattice with *chain lines* and *grey sites*

honeycomb lattice (black sites) (Fig. 16.13) edge-decoration with the kagomé lattice (grey sites) does not yield a standard regular lattice. However, for the triangular lattice (white sites), edge decoration with the kagomé lattice yields a triangular lattice with four times as many sites.

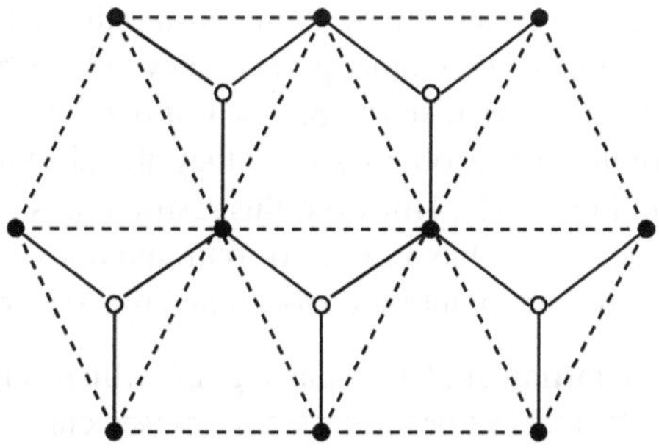

Fig. 16.14 Decoration of half the faces of the triangular lattice

16.2.2.4 Face Decoration

Unlike edge-decoration, face-decoration is confined to two-dimensional lattices.[16] Adding decoration sites to every face of a square lattice yields a square lattice with $2N$ sites, to every face of a triangular lattice yields a triangular lattice of $3N$ sites and to every face of a honeycomb lattice yields a triangular lattice of $3N/2$ sites.

If, as shown in Fig. 16.14, decoration sites are added to every downward-pointing triangle of a triangular lattice, and if the original edges (of broken lines) are removed and replace by new first-neighbour edges between black and white sites, then a honeycomb lattice is produced. Also (reversing the role of primary and secondary sites in Fig. 16.13) it can be seen that face decoration of the hexagons of a kagomé lattice produces a triangle lattice.

Dedecoration involves the removal of sites and is unproblematic if it is simply the reverse of a *known* decoration procedure. However, if this is not the case then there will often be more than one way of applying dedecoration. As we have already seen dedecoration using the white sites of the triangular lattice in Fig. 16.13 will yield another triangular lattice, whereas dedecoration using the black sites will yield a kagomé lattice. It can also be seen that the decorated square lattice of Fig. 16.12 can be obtained by taking square groups of four squares of a square lattice and removing the middle site. The removal of sites from a lattice is also an important technique in real-space renormalization methods (see Chap. 15) where it is called *decimation*.

16.3 Rapidity Graphs and Lattices

Consider a plane graph $\mathfrak{g}^{\circledast}$ consisting of a finite number of lines which do not self-intersect and which intersect with each other only in pairs and at most once. Recall that, according to the definition in Sect. 16.1.1, a plane graph simply consists of

[16] Although, of course, forms of decoration which are not edge-decoration can be devised and used for higher-dimensional lattices. Indeed, one might regard both the body-centered cubic lattice of Fig. 16.8 and the face-centred cubic lattice of Fig. 16.9 as decorations of the simple cubic lattice of Fig. 16.7.

vertices connected by non-crossing edges. It is clear from the construction process for $\mathfrak{g}^{\circledast}$ that it has another property beyond this basic definition, namely that a succession of edges are taken to be *parts of a single line*. For Baxter (1978) the identification of the lines is obvious since they are taken to be straight (as we have portrayed them in Fig. 16.15). However, this restriction is not necessary (Costa-Santos 2006); since each vertex has valency of four, the lines can be identified by taking each vertex as a crossing point (not a touching point) of two lines.

Definition 16.3.1 A plane graph which can be decomposed into lines which do not self-intersect or form a simple closed curve and which intersect with each other only in pairs, and at most once, is said to have the *rapidity property*, consist of *rapidity lines*, or be a *rapidity graph (or lattice)*.

Of course, it is not the case that all plane graphs with vertices of valency four are rapidity graphs. However, it must be the case that all two-dimensional regular lattices with vertices of valency (coordination number) four are rapidity lattices. As can be seen from Sect. 16.2.1 these are the square and kagomé lattices. Our interest in rapidity graphs is as hosts for eight-vertex models with arrow configurations defined as in Sect. 3.8.1. So it is clear that there is no loss of generality in representing the lines of $\mathfrak{g}^{\circledast}$ as straight; these are labelled L_i, $i = 1, 2, \ldots$. No assumption has been made about the lengths of these lines and it can be assumed that the graph $\mathfrak{g}^{\circledast}$ is contained within the interior of a simple closed curve $\mathfrak{C}$, which is not part of $\mathfrak{g}^{\circledast}$ and is represented in Fig. 16.15 as a circle. However we shall need to consider the graph $\mathcal{E}(\mathfrak{g}^{\circledast})$ obtained by extending, where necessary, the lines of $\mathfrak{g}^{\circledast}$ to include all possible vertices of non-parallel pairs. In the case of L_i and L_j the intersection is denoted by V_{ij}. Let *angle variables* α_{ij} and $\bar{\alpha}_{ij}$ be associated with the supplementary angles at V_{ij}, where

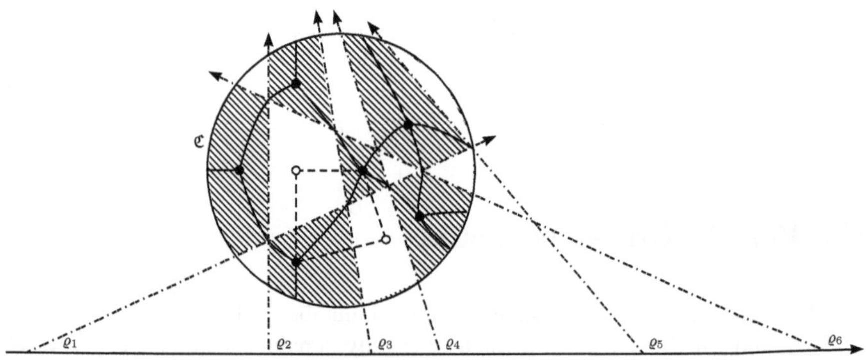

Fig. 16.15 The lines of $\mathcal{E}(\mathfrak{g}^{\circledast})$ are marked as *chain lines* and of $\mathfrak{g}$ as continuous lines with *black vertices*. The vertices of $\mathfrak{g}^{\star}$ (only two vertices marked, *white*) are in the unshaded regions, with edges (not marked) also passing through the intersection points of the edges of $\mathfrak{g}^{\circledast}$ and $\mathfrak{g}$. Four edges of $\mathfrak{g}_{\diamond}$, forming a rhombus are shown (by *broken lines*)

$$\alpha_{ij} + \bar{\alpha}_{ij} = \lambda, \tag{16.3.1}$$

for some parameter λ. For any polygon $\mathfrak{c}$ of m edges and vertices of $\mathcal{E}(\mathfrak{g}^{\circledast})$ the sum of the interior angle variables (referred to simply as the *interior sum*) is denoted by $\mathsf{IS}(\mathfrak{c})$. Denoting in a similar way the sum of the exterior angle variables (the *exterior sum*) by $\mathsf{ES}(\mathfrak{c})$, it follows, from (16.3.1), that

$$\mathsf{IS}(\mathfrak{c}) + \mathsf{ES}(\mathfrak{c}) = m\lambda. \tag{16.3.2}$$

A line in now needed which intersects all the lines of $\mathcal{E}(\mathfrak{g}^{\circledast})$. Such a line may already be part of $\mathcal{E}(\mathfrak{g}^{\circledast})$, but if it is not[17] it must be added to produce a graph $\mathcal{EB}(\mathfrak{g}^{\circledast})$. This line will be referred to as the *base line* and will be denoted by L_0, with the corresponding extension of the designation of angle variables to α_{i0} and $\bar{\alpha}_{i0}$, $i = 1, 2, \ldots$. The triangles formed by L_0 and all other pairs of non-parallel lines are called *base triangles* and, of course, every vertex of $\mathcal{E}(\mathfrak{g}^{\circledast})$ is a vertex of a base triangle. In Fig. 16.15 a simple $\mathfrak{g}^{\circledast}$ is shown with the base line drawn horizontally at the bottom of the figure. A direction is given to the base line (to the right in Fig. 16.15), then to the lines of $\mathcal{E}(\mathfrak{g}^{\circledast})$ so that their crossings of the base line with respect to its own direction are from right to left (upwards in Fig. 16.15). The two supplementary angles at a vertex V_{ij} can now be distinguished by line direction. Refining our notation slightly we designate α_{ij} as the angle variable for the angle occupied by the heavy line in Fig. 16.16a and $\bar{\alpha}_{ij}$ as that for the angle occupied by the heavy line in Fig. 16.16b. Let $\varrho_i := \alpha_{i0}$, $i = 1, 2, \ldots$. These quantities, associated each with one line of the lattice and completed by setting $\varrho_0 = 0$, are *rapidities*. Given that the interior sum of each base triangle is λ, α_{ij} for all i and j can now be expressed in terms of rapidities. If the indices of the lines (as in Fig. 16.15) increase along the base line from left to right and if the vertex V_{ij} for $i < j$ is above the base line[18]

$$\alpha_{ij} = \varrho_j - \varrho_i, \qquad \bar{\alpha}_{ij} = \lambda - \varrho_j + \varrho_i. \tag{16.3.3}$$

If V_{ij} is below the base line i and j and the relative slopes of the two lines in Fig. 16.16a are interchanged.

The following is a pivotal result for the discussion in Chap. 12:

Theorem 16.3.1 *If the interior sum of every base triangle is λ, the the interior sum of every polygon of m edges is $(m-2)\lambda$.*

Proof For such a polygon there can be only two crossings of rapidity lines of type (a). With the graph oriented as in Fig. 16.15 with the base line horizontal, these will

---

[17] As would be the case where every line of $\mathfrak{g}^{\circledast}$ is parallel to at least one other line.

[18] It should be noted that our definitions of the variables are formulated to achieve consistency with Baxter (1978). They differ from those used in Baxter (1986) and Costa-Santos (2006). Specifically the difference involves the interchange $\alpha_{ij} \leftrightarrow \lambda - \alpha_{ij}$, corresponding to an interchange between the edge variables of the dual lattice edges passing through every site of $\mathfrak{g}^{\circledast}$. This leads to the replacement of $(m-2)\lambda$ by 2λ in the statement of Theorem 16.3.1 (see Baxter 1986, Eq. (8.2)).

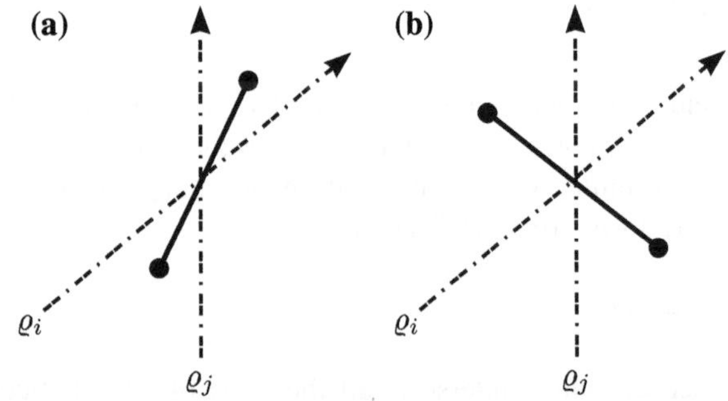

Fig. 16.16 Assignment of angle variables in terms of rapidities

be the lowest and highest vertices, representing respectively a source and a sink for the direction arrows. In the interior angle variables at the ends of a particular edge the rapidity appears with opposite signs, thus cancelling in $\mathrm{IS}(\mathfrak{c})$. So when the angle variables are summed the $(m-2)$ type (b) vertices contribute $(m-2)\lambda$. □

Rapidities are relative, in the sense that only their differences appear in the analysis. They have effectively been fixed by setting the rapidity of the base line to zero. A different choice of base line would lead to a set of rapidities differing by an additive factor.

The faces of $\mathfrak{g}^{\circledast}$ can be marked by shading or not shading so that no two shaded/unshaded faces share an edge. A point can then be placed in every shaded face, with the points joined by edges passing through the surrounding crossing points of the lines of $\mathfrak{g}^{\circledast}$. This yields a graph $\mathfrak{g}$, and repetition of the procedure for the unshaded faces yields a second graph $\mathfrak{g}^{\star}$, which is the dual of $\mathfrak{g}$. The graph $\mathfrak{g}^{\circledast}$ now comprises the set of *rapidity lines* for the dual pair $\mathfrak{g}$, $\mathfrak{g}^{\star}$.

The reverse of this procedure poses an interesting question, namely: given a graph $\mathfrak{g}$, is it possible to draw a set of lines which have the rapidity property and for which the intersections lie on the edges of $\mathfrak{g}$, with exactly one per edge? To answer this question the first step to to form a graph $\mathfrak{g}_{\diamondsuit}$ whose vertices are all the vertices of $\mathfrak{g}$ and its dual $\mathfrak{g}^{\star}$ and whose edges are, for every vertex of $\mathfrak{g}^{\star}$, the lines connecting it to the vertices of $\mathfrak{g}$ on the boundary of the face in which it is situated. Clearly every face of $\mathfrak{g}_{\diamondsuit}$ is (topologically speaking) a quadrilateral with the edges of $\mathfrak{g}$ and $\mathfrak{g}^{\star}$ as diagonals. A set of *train tracks* are now defined as sequences of faces of $\mathfrak{g}_{\diamondsuit}$ connected along opposite edges and the following theorem can be proved (Kenyon and Schlenker 2005):

Theorem 16.3.2 *The planar graph $\mathfrak{g}_{\diamondsuit}$ has a rhombic embedding*[19] *in the plane if and only if*

(i) *No train track crosses itself or forms a simple closed curve.*
(ii) *Two distinct train tracks cross each other at most once.*

---

[19] A *rhombic embedding* (also called a *isoradial embedding*) of quadratic graph is a continuous displacement of its vertices and edges which, while leaving all the edge/vertex connections unchanged, transforms every edge into a straight line of unit length; meaning that every face becomes a rhombus.

It is clear that to each train track there corresponds a unique curve drawn through the intersections of the diagonals of its constituent quadrilaterals and that the set of train tracks satisfy properties (i) and (ii) of Theorem 16.3.2 if and only if the set of curves are rapidity lines. Thus:

Corollary 16.3.1 *Rapidity lines, and hence a rapidity graph $\mathfrak{g}^{\circledast}$, exist for a graph $\mathfrak{g}$ if and only if the derived quadratic graph $\mathfrak{g}_{\diamond}$ has a rhombic embedding.*

Since an edge of $\mathfrak{g}$ and an edge of $\mathfrak{g}^{\star}$ passes through every vertex of $\mathfrak{g}^{\circledast}$ (as shown by the heavy lines in Fig. 16.16) they can be associated with the angle variables at the site. Given that the polygon $\mathfrak{c}$ is a face cycle, with the face containing a site of $\mathfrak{g}$ (say), the sum of the angle variables for the edges meeting at this vertex is equal to the interior sum $(m-2)\lambda$. And, if a polygon is formed with the edges of $\mathfrak{g}^{\star}$ passing through the vertices of $\mathfrak{c}$, the sum of the angle variables for the edges of this polygon is equal to the exterior sum 2λ.

The two regular two-dimensional four-coordinated lattices with the rapidity property are the square lattice and the kagomé lattice. Each is the medial lattice (and hence denoted by $\mathcal{N}^{\circledast}$) of a dual pair of lattices $\mathcal{N}$ and $\mathcal{N}^{\star}$ with sites situated on the faces of $\mathcal{N}^{\circledast}$ (Sect. 16.2.2.2). For the square lattice, $\mathcal{N}$ and $\mathcal{N}^{\star}$ are also square lattices (tilted through 45°), with $N^{\circledast} = 2N$, $N^{\star} = N$. For the kagomé lattice, $\mathcal{N}$ and $\mathcal{N}^{\star}$ are the honeycomb and triangular members of the dual pair, with $N^{\circledast} = \frac{3}{2}N$, $N^{\star} = \frac{1}{2}N$.

16.4 Lattice Graphs

The vertices of a *lattice graph* $\mathfrak{g}$ form a subset of the sites of the lattice $\mathcal{N}$, restricted so that edges can be drawn only between first-neighbour sites.[20] This is said to be a *weak embedding* of $\mathfrak{g}$ in $\mathcal{N}$. It not necessarily the case that there is an edge drawn between every pair of first-neighbour sites, but if this is so then the graph is called a *section graph*.[21] All low-temperature graphs are section graphs (Sect. 14.2.1) and if $\mathfrak{g}$ is a section graph in $\mathcal{N}$ then this is called a *strong embedding*.[22] A graph is said to be a *spanning graph of* $\mathcal{N}$ if it contains every vertex of $\mathcal{N}$. Of course, a spanning section graph (strongly embedded) in $\mathcal{N}$ is just $\mathcal{N}$ itself. Examples of weakly embedded spanning graphs are the augmented graphs defined in Sect. 16.4.1 and used in Sect. 3.8.2.1 and the dimer covering of Sect. 3.9 and Chap. 13.

---

[20] Although it is convenient to allow the topological equivalence between the triangular lattice and a square lattice with edges between all second-neighbour pairs with one of the two orientations.

[21] In some cases, like for example in the discussion of graphs on the body-centred cubic lattice in Example 16.4.7, the account given here is generalized to the case where there are two different kinds of edges, corresponding to first-neighbour and second-neighbour pairs of sites. A section graph is then one which has edges connecting all of its first-neighbour and second-neighbour pairs.

[22] In this terminology, rather unfortunately, strong embeddings are subclasses of weak embeddings.

16.4.1 Augmented Graphs and the Whitney Polynomial

Suppose the lattice graph $\mathfrak{g} \subseteq \mathcal{N}$ is augmented by adding all the sites of $\mathcal{N}\backslash\mathfrak{g}$ as single site components. No additional edges are added in this process and so $\mathfrak{e}(\mathfrak{g})$ and $\mathfrak{C}(\mathfrak{g})$ are unchanged with

$$\mathfrak{n}(\mathfrak{g}) \to N, \qquad \mathfrak{c}(\mathfrak{g}) \to \mathfrak{c}(\mathfrak{g}) + N - \mathfrak{n}(\mathfrak{g}). \tag{16.4.1}$$

An important quantity underlying many problems in statistical mechanics is the *Whitney polynomial* (Whitney 1932)

$$W(x, y) := \sum_{\{\mathfrak{g}\}}^{(\text{AG})} x^{\mathfrak{c}(\mathfrak{g})} y^{\mathfrak{e}(\mathfrak{g})} = \sum_{\{\mathfrak{g}\}} x^{N+\mathfrak{c}(\mathfrak{g})-\mathfrak{n}(\mathfrak{g})} y^{\mathfrak{e}(\mathfrak{g})}, \tag{16.4.2}$$

where '(AG)' indicates that the summation is over all graphs on the lattice augmented in the way described here.

16.4.2 Hopping Matrices and the Canonical Flux Distribution

Let $\mathcal{N}$ be a two-dimensional lattice of N sites labelled $j = 1, 2, \ldots, N$ and $\boldsymbol{T}$ be a Hermitian hopping matrix with elements satisfying (13.1.2) and (13.1.3). For convenience the length $\mathfrak{e}(\mathfrak{c}) = \mathfrak{n}(\mathfrak{c})$ of the elementary cycle $\mathfrak{c}$ on $\mathcal{N}$ (Sect. 16.1.1) is denoted by ℓ with an ordered labelling of vertices $(o_1, o_2, \ldots, o_\ell)$. The *flux* $\Phi(\boldsymbol{T}; \mathfrak{c})$ of $\boldsymbol{T}$ through $\mathfrak{c}$, with the orientation specified by the order of the vertices, is

$$\Phi(\boldsymbol{T}; \mathfrak{c}) := \operatorname{Arg}\{t(o_1, o_2) \cdots t(o_{\ell-1}, o_\ell) t(o_\ell, o_1)\}. \tag{16.4.3}$$

If the orientation of $\mathfrak{c}$ is reversed by specifying its vertices in the reverse order $(o_\ell, o_{\ell-1}, \ldots, o_1)$ then the flux changes sign. The cycle $\mathfrak{c}$ will circumscribe $\mathfrak{f}(\mathfrak{c})$ faces circumscribed by face cycles $\mathfrak{c}_k$, $k = 1, 2, \ldots, \mathfrak{f}(\mathfrak{c})$. With the orientations around the face cycles the same as that around $\mathfrak{c}$, it is obvious that

$$\Phi(\boldsymbol{T}; \mathfrak{c}) = \sum_{k=1}^{\mathfrak{f}(\mathfrak{c})} \Phi(\boldsymbol{T}; \mathfrak{c}_k). \tag{16.4.4}$$

Consider now a gauge transformation on the hopping matrix expressible in the equivalent forms

$$\begin{aligned} t'(i, j) &:= \exp(-\mathrm{i}\varphi(i)) t(i, j) \exp(\mathrm{i}\varphi(j)), \\ \boldsymbol{T}' &:= \boldsymbol{U}^\dagger \boldsymbol{T} \boldsymbol{U}, \qquad \langle i|\boldsymbol{U}|j\rangle := \delta^{\mathrm{Kr}}(i - j) \exp(\mathrm{i}\varphi(j)). \end{aligned} \tag{16.4.5}$$

It can then be shown that:

Theorem 16.4.1 *Two hopping matrices* $\boldsymbol{T}'$ *and* $\boldsymbol{T}$ *with the same hopping amplitudes have the same fluxes through all elementary cycles on* $\mathcal{N}$ *if and only if they are related by the gauge transformation* (16.4.5).

That the fluxes are invariant under the gauge transformation is obvious. That the identity of fluxes implies the existence of the gauge transformation was proved by Lieb and Loss (1993, p. 342) who also showed (p. 343) that:

Theorem 16.4.2 *If numbers* $\Phi^{(k)}$ *are assigned to the face cycles* $\mathfrak{c}_k$ *of* $\mathcal{N}$ *then there is an assignment of real numbers* $\theta(i, j)$ *to the first-neighbour pairs of the lattice with* $\theta(j, i) = -\theta(i, j)$ *and* $0 \leq |\theta(i, j)| < 2\pi$ *such that the hopping matrix* $\boldsymbol{T}$ *with elements*

$$t(i, j) = |t(i, j)| \exp[\mathrm{i}\theta(i, j)], \tag{16.4.6}$$

has the property

$$\Phi(\boldsymbol{T}; \mathfrak{c}_k) =, \qquad \forall\, \mathfrak{c}_k. \tag{16.4.7}$$

It follows from Theorem 16.4.1 that the numbers $\theta(i, j)$ are unique to within the gauge transformation (16.4.5).

The work of Lieb and Loss (1993) was motivated by a number of physical realizations, in each of which the ground-state energy of the system is dependent on the choice of fluxes $\{\Phi^{(k)}\}$. In particular, for what is called the *half-filled band* case, the ground-state energy is twice the sum of the negative eigenvalues of $-\boldsymbol{T}$ and part of the aim of their work was to determine a flux distribution which minimised this quantity. They were able to show in some special cases (with their work completed by Lieb (1994)) that this was achieved by taking the triangulation as described in Sect. 16.1.3 and (with the same orientation) assigning a flux $\frac{1}{2}\pi$ to each triangular face. Then the flux through $\mathfrak{c}_k$ is $\frac{1}{2}\pi\, \Upsilon(\mathfrak{c}_k)$, where, as defined in Sect. 16.1.3, $\Upsilon(\mathfrak{c}_k)$ is the triangulation number of $\mathfrak{c}_k$ and is independent of the mode of triangulation. It follows from Theorem 16.4.2 that there are hopping matrices $\boldsymbol{T}$ with

$$\Phi(\boldsymbol{T}; \mathfrak{c}) = \tfrac{1}{2}\pi \sum_{k=1}^{f(\mathfrak{c})} \Upsilon(\mathfrak{c}_k). \tag{16.4.8}$$

Such matrices are said to satisfy the *canonical flux distribution*. Given a set of hopping amplitudes, the $\boldsymbol{T}$ satisfying (16.4.8) is unique to within the gauge transformation.

16.4.3 Embeddings and Topologies

The *topology* of a graph $\mathfrak{g}$ comprises the number of its vertices and the way in which they are linked by edges. Two graphs with the same topology may have components with different *shapes*. Any graph without isolated vertices can be specified uniquely by giving the location of all its edges. The denotation used for such a graph is $(\mathfrak{e}, \mathfrak{k})$, where $\mathfrak{e}$ is the number of edges and the index $\mathfrak{k}$ is used to distinguish between shapes. Thus the one-component graph with three sites connected by two edges on the square lattice can have two shapes. These are given in Table 16.1 as (2, 1), with two collinear edges, and (2, 2), with two edges at right angles.[23] Graphs with the same topology and with components having the same shapes are called *isomorphic* meaning that one can be brought into coincidence with the other by some combination of translations, rotations and reflections acting independently on each component.[24] Graphs which are not isomorphic are said to be *distinct*. A selection of graphs on the square lattice which are useful for illustrating the discussion are given in Tables 16.1 and 16.2. The number of components and sites of $(\mathfrak{e}, \mathfrak{k})$ are denoted respectively by $\mathfrak{c}(\mathfrak{e}, \mathfrak{k})$ and $\mathfrak{n}(\mathfrak{e}, \mathfrak{k})$.

For section graphs the topology and shape is fixed once the relative location of the vertices is given. We thus are able to use an alternative notation $[\mathfrak{n}, \mathfrak{k}]$ exclusively for section graphs, where again $\mathfrak{k}$ distinguishes distinct graphs. The number of components and edges of the section graph $[\mathfrak{n}, \mathfrak{k}]$ are denoted respectively by $\mathfrak{c}[\mathfrak{n}, \mathfrak{k}]$ and $\mathfrak{e}[\mathfrak{n}, \mathfrak{k}]$. For section graphs and low-temperature expansions a useful parameter for $[\mathfrak{n}, \mathfrak{k}]$ is

$$\mathfrak{l}[\mathfrak{n}, \mathfrak{k}] := \mathfrak{n}z - 2\mathfrak{e}[\mathfrak{n}, \mathfrak{k}], \tag{16.4.9}$$

the number of links between a graph and exterior sites on the lattice. This is included in Tables 16.1 and 16.2. From (16.1.1) and (16.4.9)

$$\mathfrak{l}[\mathfrak{n}, \mathfrak{k}] = 2\mathfrak{c}[\mathfrak{n}, \mathfrak{k}] + \mathfrak{n}(z - 2) - 2\mathfrak{C}[\mathfrak{n}, \mathfrak{k}], \tag{16.4.10}$$

where, as we have noted above, the cyclomatic number $\mathfrak{C}[\mathfrak{n}, \mathfrak{k}]$ of the plane lattice graph $[\mathfrak{n}, \mathfrak{k}]$ is the number of its faces.

The use of round and square brackets to specify graphs is standard notation (see, for example, Sykes et al. 1966). Of course, all section graphs which do not have isolated vertices can be given in both notations (see Tables 16.1 and 16.2).

16.4.4 Lattice Constants

The number of ways that a graph $\mathfrak{g}$ can be placed on $\mathfrak{g}'$ is denoted by $\Omega(\mathfrak{g} \mid \mathfrak{g}')$, called *the lattice constant of the graph* $\mathfrak{g}$ *on the graph* $\mathfrak{g}'$. In general $\Omega(\mathfrak{g} \mid \mathfrak{g}')$ is the number of weak embeddings of $\mathfrak{g}$ on $\mathfrak{g}'$. If $\mathfrak{g}$ and $\mathfrak{g}'$ are section graphs then the embedding is strong and we indicate this by the notation $\Omega[\mathfrak{g} \mid \mathfrak{g}']$.

---

[23] The significance of this way of labeling graphs is described below.

[24] The 'shape of a graph' is, in the case of a multi-component graph, taken to mean the shapes of all the components.

Table 16.1 Some distinct graphs on the square lattice giving both specifications (e, k) and $[\mathrm{n}, \mathrm{k}]$, where possible

Graph	c	e	n	(e, k)	$\overline{\omega}(\mathrm{e}, \mathrm{k})$	[n, k]	$\overline{\omega}[\mathrm{n}, \mathrm{k}]$	$\mathfrak{l}[\mathrm{n}, \mathrm{k}]$
	1	0	1	–	–	[1, 1]	1	4
	2	0	2	–	–	[2, 1]	$\frac{1}{2}(N-5)$	8
	1	1	2	(1, 1)	2	[2, 2]	2	6
	3	0	3	–	–	[3, 1]	$\frac{1}{6}(N^2-15N+62)$	12
	2	1	3	–	–	[3, 2]	$2(N-8)$	10
	1	2	3	(2, 1)	2	[3, 3]	2	8
	1	2	3	(2, 2)	4	[3, 4]	4	8
	1	2	4	–	–	[4, 1]	$2(N-11)$	12
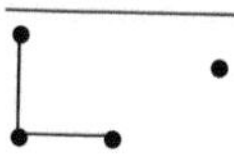	1	2	4	–	–	[4, 2]	$4(N-10)$	12
	2	2	4	(2, 3)	$2N-7$	[4, 3]	$2N-23$	12
	1	3	4	(3, 1)	2	[4, 4]	2	10
	1	3	4	(3, 2)	8	[4, 5]	8	10
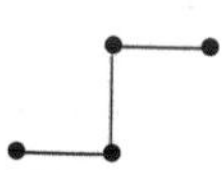	1	3	4	(3, 3)	4	[4, 6]	4	10
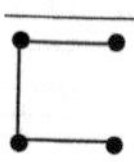	1	3	4	(3, 4)	4	—	—	
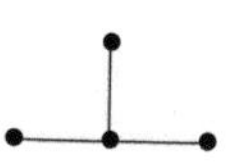	1	3	4	(3, 5)	4	[4, 7]	4	10
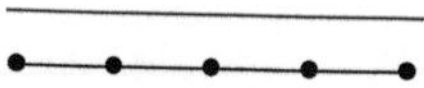	1	4	5	(4, 1)	2	[5, 1]	2	12

The graphs are listed in a partially ordered dictionary sequence using the pair $\{\mathrm{e}, \mathrm{n}\}$. The designation of k is arbitrary and simply reflects the choices of displayed graphs. $\mathfrak{l}[\mathrm{n}, \mathrm{k}]$, which is need in Chap. 14, is the number of edges which link the section graph to external sites

When we consider the placing of a graph $\mathfrak{g}$ on the lattice $\mathcal{N}$ itself it is clear that the lattice constant will depend not just on the number of lattice sites $N = |\mathcal{N}|$, but also on the 'shape' of the lattice. Different boundary conditions will lead to different lattice constants. The only useful way to resolve this problem is to assume periodic

Table 16.2 Continuation of Table 16.1

Graph	c	e	n	(e, k)	$\varpi(e, k)$	[n, k]	$\varpi[n, k]$	$\mathfrak{l}[n, k]$
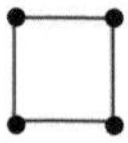	1	4	4	(4, 2)	1	[4, 8]	1	8
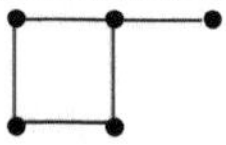	1	5	5	(5, 2)	8	[5, 2]	8	10
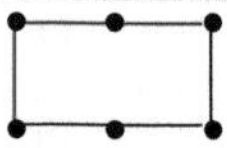	1	6	6	(6, 1)	2	–	–	
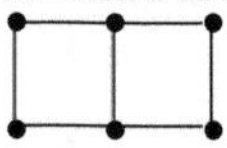	1	7	6	(7, 1)	2	[6, 1]	2	10
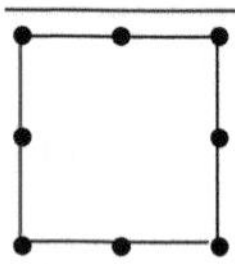	1	8	8	(8, 1)	1	–	–	
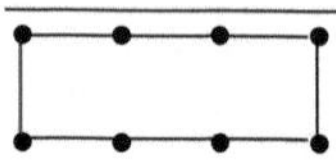	1	8	8	(8, 2)	2	–	–	
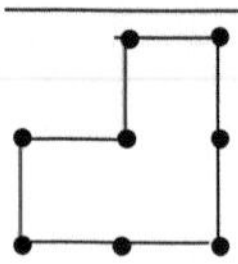	1	8	8	(8, 3)	4	–	–	
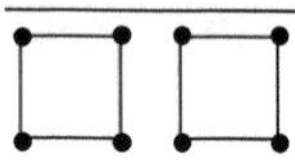	2	8	8	(8, 4)	$\frac{1}{2}(N-9)$	[8, 1]	$\frac{1}{2}(N-21)$	16
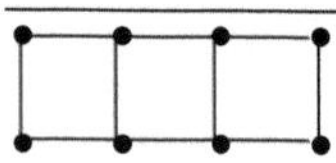	1	10	8	(10, 1)	2	[8, 2]	2	12
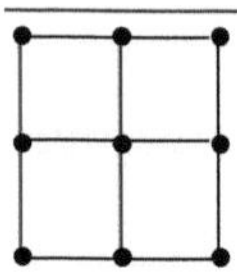	1	12	9	(12, 1)	1	[9, 1]	1	12

boundary conditions so that each site is of valency z and the number of edges and hence the lattice constant of [2, 2], a single edge connecting a first-neighbour pair of sites, is $\Omega[2, 2] = \frac{1}{2}zN$.[25]

[25] We simplify the case of graphs placed on the whole lattice to $\Omega(\mathfrak{g}) := \Omega(\mathfrak{g} \mid \mathcal{N})$. In the case where $\mathfrak{g}$ is (e, k) or $[n, k]$ we denote the lattice constants (on $\mathcal{N}$) by $\Omega(e, k)$ and $\Omega[n, k]$, respectively.

We define $\varpi(\mathrm{e}, \mathrm{k}) := \Omega(\mathrm{e}, \mathrm{k})/N$ with the *lattice constant per site* $\omega(\mathrm{e}, \mathrm{k})$ being the N-independent term in $\varpi(\mathrm{e}, \mathrm{k})$.[26] Of course, for a connected graph, $\varpi(\mathrm{e}, \mathrm{k}) = \omega(\mathrm{e}, \mathrm{k})$. A similar notation is used for section graphs.

If a section graph without isolated vertices has alternative designations (e, k) and $[\mathrm{n}, \mathrm{k}']$ then $\varpi(\mathrm{e}, \mathrm{k}) = \varpi[\mathrm{n}, \mathrm{k}]$ if the graph has one component (Tables 16.1 and 16.2). However, if the graph has more than one component the strong embedding of $[\mathrm{n}, \mathrm{k}]$ on $\mathcal{N}$ must be such that sites of the two components cannot be joined by edges; thus $\varpi(\mathrm{e}, \mathrm{k}) > \varpi[\mathrm{n}, \mathrm{k}]$, (see graphs (2, 3) of Table 16.1 and (8, 4) of Table 16.2).

The calculation of lattice constants for small graphs can be done quite easily. As a simple example, consider all the graphs with e edges. Since the lattice has $\frac{1}{2}Nz$ edges it must be the case that

$$\sum_{\{\mathrm{k}\}} \Omega(\mathrm{e}, \mathrm{k}) = \frac{(\frac{1}{2}Nz)!}{(\frac{1}{2}Nz - \mathrm{e})!\mathrm{e}!} \tag{16.4.11}$$

and in a similar way, with the counting based on sites rather than edges,

$$\sum_{\{\mathrm{k}\}} \Omega[\mathrm{n}, \mathrm{k}] = \frac{N!}{(N - \mathrm{n})!\mathrm{n}!} . \tag{16.4.12}$$

Example 16.4.1 The lattice constants for the three graphs (2, 1), (2, 2) and (2, 3), shown in Table 16.1 are easy to calculate and they provide a simple verification of (16.4.11), where the number of ways of arranging two edges on the lattice is $N(2N - 1)$. In the case of section graphs with $\mathrm{n} = 3$ the most difficult lattice constant is for [3, 1], three disconnected vertices. The easiest way to find this is from (16.4.12) which gives

$$\varpi[3, 1] = \tfrac{1}{6}(N - 1)(N - 2) - \varpi[3, 2] - \varpi[3, 3] - \varpi[3, 4] . \tag{16.4.13}$$

16.4.4.1 Chains

Graphs which are chains play an important part in the discussion and it is convenient to have a special notation to denote a chain of particular length irrespective of its topology on the lattice. Thus

- $(\mathrm{e}, \mathbb{C})$ denotes a chain of e edges.
- $(\mathrm{e}, \mathbb{E}\mathbb{C})$ denotes an elementary chain of e edges.
- $[\mathrm{n}, \mathbb{C}]$ denotes a chain of n sites which is a section graph.
- $[\mathrm{n}, \mathbb{E}\mathbb{C}]$ denotes an elementary chain of n sites which is a section graph.

---

[26] The term 'lattice constant' will whenever the meaning is clear be used to refer to 'per site' quantities.

Considering only the regular two-dimensional lattices shown in Sect. 16.2.1 and excluding the kagomé lattice (which has two types of lattice face), it is clear that all chains with $\mathrm{e} < \mathrm{z}^\star$ are elementary. This restriction is to prevent the chain closing in on itself, which could happen whenever e is at least as great as the number of edges around a lattice face, which is the coordination number $\mathrm{z}^\star$ of the dual lattice. In the case of chains of length $\mathrm{z}^\star$ there is one chain that needs to be excluded from the count and this is that which passes around a lattice face (a cycle with one site designated as the two ends of the chain), which can be placed on the lattice in $N^\star \mathrm{z}^\star = N\mathrm{z}$ ways. Thus to extend slightly the formula of Lavis and Bell (1999b, Sect. A.7.4)

$$\Omega(\mathrm{e}, \mathbb{E}\,\mathbb{C}) = \begin{cases} \frac{1}{2} N\mathrm{z}(\mathrm{z}-1)^{\mathrm{e}-1}, & \mathrm{e} < \mathrm{z}^\star, \\ \frac{1}{2} N\mathrm{z}(\mathrm{z}-1)^{\mathrm{e}-1} - N\mathrm{z}, & \mathrm{e} = \mathrm{z}^\star. \end{cases} \tag{16.4.14}$$

Clearly $\mathfrak{n}(\mathrm{e}, \mathbb{E}\,\mathbb{C}) = \mathrm{e} + 1$ and the formula (16.4.14) can be modified in such a way as to apply to section graphs. The additional consideration here is to prevent a chain containing all the sites around a lattice face, which can be placed on the lattice in $N^\star \mathrm{z}^\star = N\mathrm{z}$ ways. Such a chain is allowed with $\mathrm{e} = \mathrm{z}^\star - 1$ in formula (16.4.14), but for section graphs must be excluded since the strong embedding condition would force the inclusion of the final edge. Thus we have

$$\Omega[\mathfrak{n}, \mathbb{E}\,\mathbb{C}] = \begin{cases} \frac{1}{2} N\mathrm{z}(\mathrm{z}-1)^{\mathfrak{n}-2}, & \mathfrak{n} < \mathrm{z}^\star, \\ \frac{1}{2} N\mathrm{z}(\mathrm{z}-1)^{\mathfrak{n}-2} - N\mathrm{z}, & \mathfrak{n} = \mathrm{z}^\star. \end{cases} \tag{16.4.15}$$

Example 16.4.2 The two-link chains in Table 16.1 satisfy both (16.4.14) and (16.4.15). For the three-link chains $\mathfrak{n} = \mathrm{z}^\star = 4$, so the non-section graph $(3, 4)$ is excluded from the counting in (16.4.15).

Powerful computer programs are now available for calculating lattice constants (Oitmaa et al. 2006, Chap. 2).

16.4.5 Partially-Ordered Sequences of Graphs and the T Matrix

If $\mathfrak{g}'$ is a subgraph of $\mathfrak{g}$ we write $\mathfrak{g}' \subseteq \mathfrak{g}$.[27] The relation '$\subseteq$' defines a *partial ordering* of the connected graphs of $\mathcal{N}$. Thus, for the square lattice with $\mathfrak{n} \leq 4$, a partial ordering of the connected graphs is given by the order in which they appear in

[27] This includes the case where $\mathfrak{g}'$ and $\mathfrak{g}$ are isomorphic; when this is not the case we write $\mathfrak{g}' \subset \mathfrak{g}$ and say that $\mathfrak{g}'$ is a *proper subgraph* of $\mathfrak{g}$.

Table 16.1.[28] Let $\mathcal{S}$ be the partially-ordered sequence $\mathfrak{g}_1, \mathfrak{g}_2, \mathfrak{g}_3, \ldots \mathfrak{g}_n$, of n graphs such that if $\mathfrak{g}_i \subset \mathfrak{g}_j$ then $i < j$.[29]

For any graph $\mathfrak{g}$ and the sequence $\mathcal{S}$, there is a set $\Pi_n(\mathfrak{g}) := \{\Omega(\mathfrak{g}_i \mid \mathfrak{g}) : i = 1, 2, \ldots, n\}$ of lattice constants. It is clear that term by term the sets $\Pi_n(\mathfrak{g})$ and $\Pi_n(\mathfrak{g}')$, for two *non-intersecting* graphs $\mathfrak{g}$ and $\mathfrak{g}'$, satisfy the extensive property

$$\Omega(\mathfrak{g}_i \mid \mathfrak{g} \cup \mathfrak{g}') = \Omega(\mathfrak{g}_i \mid \mathfrak{g}) + \Omega(\mathfrak{g}_i \mid \mathfrak{g}') \tag{16.4.16}$$

and this is represented by

$$\Pi_n(\mathfrak{g} \cup \mathfrak{g}') = \Pi_n(\mathfrak{g}) + \Pi_n(\mathfrak{g}'). \tag{16.4.17}$$

Now suppose that $\mathcal{E}$ is a function defined on the graphs of $\mathcal{N}$ with the extensive property

$$\mathcal{E}(\mathfrak{g} \cup \mathfrak{g}') = \mathcal{E}(\mathfrak{g}) + \mathcal{E}(\mathfrak{g}'). \tag{16.4.18}$$

If it is further supposed that the dependence of $\mathcal{E}$ on $\mathfrak{g}$ is determined explicitly by the set $\Pi_n(\mathfrak{g})$, that is $\mathcal{E}(\mathfrak{g}) := E(\Pi_n(\mathfrak{g}))$, then, from (16.4.17) and (16.4.18),

$$E(\Pi_n(\mathfrak{g}) + \Pi_n(\mathfrak{g}')) = E(\Pi_n(\mathfrak{g})) + E(\Pi_n(\mathfrak{g}')). \tag{16.4.19}$$

We now use a result proved by Sykes et al. (1966, Theorem 3):

Theorem 16.4.3 *If* $\mathcal{E}(\mathfrak{g}) := E(\Pi_n(\mathfrak{g}))$ *satisfies* (16.4.19) *then*

$$\mathcal{E}(\mathfrak{g}) = \sum_{i=1}^{n} \theta_i \; \Omega\,(\mathfrak{g}_i \mid \mathfrak{g}), \tag{16.4.20}$$

for some set of parameters $\{\theta_i\}$.[30]

Now applying (16.4.20) to each of the graphs in the sequence $\mathcal{S}$

$$\mathcal{E}(\mathfrak{g}_j) = \sum_{i=1}^{n} \theta_i \Omega_{ij}, \tag{16.4.21}$$

where $\Omega_{ij} := \Omega(\mathfrak{g}_i \mid \mathfrak{g}_j)$. In matrix form this is

$$[\boldsymbol{\mathcal{E}}^{(n)}]^{\mathrm{T}} = [\boldsymbol{\theta}^{(n)}]^{\mathrm{T}} \boldsymbol{\Omega}^{(n)}, \tag{16.4.22}$$

[28] Permutations of the last five lines of the table give other partial orderings.

[29] We discuss the details of methods of generating such a sequence below.

[30] In the case where $\mathcal{E}(\mathfrak{g})$ is a function of other variables, for example those of a thermodynamic system, then that dependence is carried over into the set $\{\theta_i\}$.

where

$$\boldsymbol{\mathcal{E}}^{(n)} := \begin{pmatrix} \mathcal{E}(\mathfrak{g}_1) \\ \mathcal{E}(\mathfrak{g}_2) \\ \vdots \\ \mathcal{E}(\mathfrak{g}_n) \end{pmatrix}, \qquad \boldsymbol{\theta}^{(n)} := \begin{pmatrix} \theta_1 \\ \theta_2 \\ \vdots \\ \theta_n \end{pmatrix} \tag{16.4.23}$$

and the triangular array

$$\boldsymbol{\Omega}^{(n)} := \begin{pmatrix} \Omega_{11} & \Omega_{12} & \cdots & \Omega_{1\,n-1} & \Omega_{1\,n} \\ 0 & \Omega_{22} & \cdots & \Omega_{2\,n-1} & \Omega_{2\,n} \\ 0 & 0 & \cdots & \Omega_{3\,n-1} & \Omega_{3\,n} \\ 0 & 0 & \vdots & \vdots & \vdots \\ 0 & 0 & \vdots & 0 & \Omega_{n\,n} \end{pmatrix} \tag{16.4.24}$$

is the *T matrix* of Domb (1974b), where, of course, the diagonal elements $\Omega_{ii} = 1$.

For the lattice $\mathcal{N}$, $\Pi_n(\mathcal{N}) := \{N\omega_i : i = 1, 2, \ldots, n\}$, where $\omega_i := \omega(\mathfrak{g}_i)$, and, from (16.4.21) and (16.4.22),

$$\mathcal{E}(\mathcal{N}) = N[\boldsymbol{\theta}^{(n)}]^{\mathrm{T}}\boldsymbol{\omega}^{(n)} = N[\boldsymbol{\mathcal{E}}^{(n)}]^{\mathrm{T}}[\boldsymbol{\Omega}^{(n)}]^{-1}\boldsymbol{\omega}^{(n)} = N[\boldsymbol{\mathcal{E}}^{(n)}]^{\mathrm{T}}\boldsymbol{\chi}^{(n)}, \tag{16.4.25}$$

where

$$\boldsymbol{\omega}^{(n)} = \boldsymbol{\Omega}^{(n)}\boldsymbol{\chi}^{(n)} \tag{16.4.26}$$

and

$$\boldsymbol{\omega}^{(n)} := \begin{pmatrix} \omega_1 \\ \omega_2 \\ \vdots \\ \omega_n \end{pmatrix}, \qquad \boldsymbol{\chi}^{(n)} := \begin{pmatrix} \chi_1^{(n)} \\ \chi_2^{(n)} \\ \vdots \\ \chi_n^{(n)} \end{pmatrix}. \tag{16.4.27}$$

Finally from (16.4.25)

$$\mathcal{E}(\mathcal{N}) = N\sum_{i=1}^{n} \mathcal{E}(\mathfrak{g}_i)\,\chi_i^{(n)}\,. \tag{16.4.28}$$

Determination of the *weights* $\chi_j^{(n)}$, $j = 1, 2, \ldots, n$ of the sequence $\mathcal{S}$, defined by (16.4.26) and (16.4.27), play an important part in the use of the KHDeB method of Sect. 7.2, the finite-cluster method of Sect. 14.4 and the finite-lattice method of Sect. 14.5.

Two things are obvious about the set of weights: (i) $\chi_n^{(n)} = \omega_n$, and (ii) if members of $\mathcal{S}$ for which $\chi_j^{(n)} = 0$ are removed and the remaining weights recalculated then they remain unchanged. These properties can be confirmed for Example 16.4.4 with the weights recalculated for the sequence $\mathfrak{g}_3, \mathfrak{g}_5, \mathfrak{g}_6$. However, neither of these properties is of significant practical use in calculating weights. It is of greater importance to know whether a sequence $\mathcal{S}$ is *complete.* This is defined in the following way:

Definition 16.4.1 The partially-ordered sequence $\mathcal{S}$ of graphs $\mathfrak{g}_1, \mathfrak{g}_2, \ldots, \mathfrak{g}_n$ is complete if there does not exist a graph $\mathfrak{g}'$, which is not a member of $\mathcal{S}$, is a proper subgraph of at least one member of $\mathcal{S}$, and which when added to the sequence has a non-zero weight.

Of course, it is clear that not all the weights of a complete sequence are necessarily non-zero.

16.4.6 Generating the Partially-Ordered Sequence

Partially-ordered sequences of graphs together with the formalism of Sect. 16.4.5 are used in the discussion of the finite-cluster method of Sect. 14.4, the KHDeB cluster-variation method described in Sect. 7.2 and the finite-lattice method of Sect. 14.5. In the first of these the sequence of graphs consists simply of all the subgraphs of a particular graph and there is no restriction to section graphs. In the case of the KHDeB and finite-lattice methods the graphs are section graphs and the modes of generation of the sequences are both special cases of the *overlap procedure* defined in the following way:

(i) Choose an arbitrary subset $\mathfrak{S}$ of m connected section subgraphs of $\mathcal{N}$, none of which is a proper subgraph of any other.

(ii) Find all distinct subgraphs obtained by overlapping the members of $\mathfrak{S}$, then the overlap graphs of these overlaps and so on until the process is exhausted. All the graphs of this sequence will be section graphs. If any are not connected then included in the sequence are all of their components.

(iii) Let the number of members of $\mathcal{S}$ (including the members of $\mathfrak{S}$) be n and label them as specified in Sect. 16.4.5 for a partially-ordered sequence.

Then the following theorem is of importance:

Theorem 16.4.4 *A partially-ordered sequence generated by the overlap procedure is complete.*

Proof Let $\mathfrak{g}'$ be a graph which is a proper subgraph of at least one member of $\mathcal{S}$. Then there exists a $\mathfrak{g}_k \in \mathcal{S}$ such that $\mathfrak{g}' \subset \mathfrak{g}_k$ but $\mathfrak{g}' \not\subseteq \mathfrak{g}_j$, for all $j < k$. This means that $\mathfrak{g}'$ is not contained in the overlap of copies of $\mathfrak{g}_k$ and[31]

[31] The use of both round and square brackets in this equation indicates that, whereas the members of $\mathcal{S}$ are section graphs, this has not been assumed for $\mathfrak{g}'$.

$$\begin{aligned} \Omega(\mathfrak{g}' \mid \mathfrak{g}_j) &= \Omega(\mathfrak{g}' \mid \mathfrak{g}_k)\, \Omega\,[\mathfrak{g}_k \mid \mathfrak{g}_j], \qquad \forall\, \mathfrak{g}_j \in \mathcal{S}, \\ \omega(\mathfrak{g}') &= \Omega(\mathfrak{g}' \mid \mathfrak{g}_k)\, \omega\,[\mathfrak{g}_k]. \end{aligned} \tag{16.4.29}$$

The line of the weight-generating Eq. (16.4.26) for $\mathfrak{g}_k$ is

$$\chi_k^{(n)} + \Omega[\mathfrak{g}_k \mid \mathfrak{g}_{k+1}]\, \chi_{k+1}^{(n)} + \cdots + \Omega[\mathfrak{g}_k \mid \mathfrak{g}_n]\chi_n^{(n)} = \omega[\mathfrak{g}_k]. \tag{16.4.30}$$

If $\mathfrak{g}'$ were added to $\mathcal{S}$ and had weight $\chi(\mathfrak{g}')$, the relevant line in (16.4.26), from (16.4.26), would be

$$\begin{aligned} \chi(\mathfrak{g}') + \Omega(\mathfrak{g}' \mid \mathfrak{g}_k)\{\chi_k^{(n)} + \Omega[\mathfrak{g}_k \mid \mathfrak{g}_{k+1}]\, \chi_{k+1}^{(n)} + \cdots + \Omega[\mathfrak{g}_k \mid \mathfrak{g}_n]\chi_n^{(n)}\} \\ = \Omega(\mathfrak{g}' \mid \mathfrak{g}_k)\, \omega\,[\mathfrak{g}_k] \end{aligned} \tag{16.4.31}$$

and it follows, from (16.4.30), $\chi(\mathfrak{g}') = 0$; $\mathcal{S}$ is complete. □

A general procedure for obtaining weights, when the sequence consists of rectangular blocks generated by some overlapping procedure on the square lattice, was given by Enting (1978a). The first step is to count the two graphs arising from the vertical and horizontal orientations of an oblong block separately. Now $\omega_j = 1$ for all $j = 1, 2, \ldots, 9$. Suppose a rectangular block of x horizontal sites by y vertical sites on the square lattice is denoted by $[x \times y]$. Then it is not difficult to see that the elements of the T matrix $\boldsymbol{\Omega}^{(n)}$ are given by

$$\langle a \times b|\boldsymbol{\Omega}^{(n)}|x \times y\rangle = \begin{cases} (x - a + 1)(y - b + 1)\,, & \text{if } a \le x \text{ and } b \le y, \\ 0\,, & \text{otherwise.} \end{cases} \tag{16.4.32}$$

It can then be shown (Enting 1978a) that

$$\langle x \times y|[\boldsymbol{\Omega}^{(n)}]^{-1}|a \times b\rangle = \eta(x - a)\eta(y - b)\,, \tag{16.4.33}$$

where

$$\eta(p) = \delta^{\mathrm{Kr}}(p) - 2\delta^{\mathrm{Kr}}(p + 1) + \delta^{\mathrm{Kr}}(p + 2). \tag{16.4.34}$$

From (16.4.26),

$$\langle x \times y|\boldsymbol{\chi}^{(n)} = \sum_{\{[a \times b]\}} \eta(x - a)\eta(y - b)\,, \tag{16.4.35}$$

where the summation is over all the blocks $[a \times b]$ in the sequence. As we shall now see the method of evaluation of the right-hand side of (16.4.35) depends on the choice of $\mathfrak{S}$.

16.4.6.1 KHDeB Sequences

The simplest application of the overlap procedure for generating a partially-ordered sequence $\mathcal{S}$ is that used in the KHDeB method and also in the finite-lattice series method of Enting and Baxter (1977). Here $m = 1$ and $\mathfrak{G}$ consists of one member.

Example 16.4.3 For the triangular lattice with $\mathfrak{G}$ taken as a single triangle the sequence $\mathcal{S}$ is

$\mathfrak{g}_1$: a single site, $\mathfrak{g}_2$: a single edge, $\mathfrak{g}_3$: a single triangle.

Then

$$\boldsymbol{\Omega}^{(3)} = \begin{pmatrix} 1 & 2 & 3 \\ 0 & 1 & 3 \\ 0 & 0 & 1 \end{pmatrix}, \qquad \boldsymbol{\omega}^{(3)} = \begin{pmatrix} 1 \\ 3 \\ 2 \end{pmatrix}. \tag{16.4.36}$$

Solving (16.4.26) for (16.4.36) gives

$$\chi_1^{(3)} = 1, \ \ \chi_2^{(2)} = -3, \ \ \chi_3^{(3)} = 2. \tag{16.4.37}$$

Example 16.4.4 For the square lattice, with $\mathfrak{G}$ the single square [4, 8],[32] $\mathcal{S}$ is

$\mathfrak{g}_1 := [1, 1]$: a single site,

$\mathfrak{g}_2 := [2, 2]$: a first-neighbour pair,

$\mathfrak{g}_3 := [4, 8]$: a first-neighbour square

and

$$\boldsymbol{\Omega}^{(3)} = \begin{pmatrix} 1 & 2 & 4 \\ 0 & 1 & 4 \\ 0 & 0 & 1 \end{pmatrix}, \qquad \boldsymbol{\omega}^{(3)} = \begin{pmatrix} 1 \\ 2 \\ 1 \end{pmatrix}. \tag{16.4.38}$$

Solving (16.4.26) for (16.4.38) gives

$$\chi_1^{(3)} = 1, \ \ \chi_2^{(3)} = -2, \ \ \chi_3^{(6)} = 1. \tag{16.4.39}$$

If $\mathfrak{G}$ is the square block of four first-neighbour squares [9, 1], $\mathcal{S}$ is extended by adding

$\mathfrak{g}_4 := [3, 3]$: three co-linear first-neighbour sites,

$\mathfrak{g}_5 := [6, 1]$: a block of two first-neighbour squares,

$\mathfrak{g}_6 := [9, 1]$: the generating block of four first-neighbour squares.

giving

---

[32] In all the following discussion graphs on the square lattice are referred to by their designation $[\mathfrak{n}, \mathfrak{k}]$ as given in Tables 16.1 and 16.2.

$$\boldsymbol{\Omega}^{(6)} = \begin{pmatrix} 1 & 2 & 4 & 3 & 6 & 9 \\ 0 & 1 & 4 & 2 & 7 & 12 \\ 0 & 0 & 1 & 0 & 2 & 4 \\ 0 & 0 & 0 & 1 & 2 & 6 \\ 0 & 0 & 0 & 0 & 1 & 4 \\ 0 & 0 & 0 & 0 & 0 & 1 \end{pmatrix}, \qquad \boldsymbol{\omega}^{(6)} = \begin{pmatrix} 1 \\ 2 \\ 1 \\ 2 \\ 2 \\ 1 \end{pmatrix}. \tag{16.4.40}$$

Solving (16.4.26) for (16.4.40) gives

$$\chi_1^{(6)} = 0, \ \chi_2^{(6)} = 0, \ \chi_3^{(6)} = 1, \ \chi_4^{(6)} = 0, \ \chi_5^{(6)} = -2, \ \chi_6^{(6)} = 1. \tag{16.4.41}$$

It should be noted that the sequence generated by some $\mathfrak{g}_j \in \mathcal{S}$ for $j < n$ does not necessarily contain all $\mathfrak{g}_i \in \mathcal{S}$ for $i \leq j$. Here for example, $\mathfrak{g}_1$ and $\mathfrak{g}_2$ will be generated by $\mathfrak{g}_3$ or $\mathfrak{g}_4$, but $\mathfrak{g}_3$ will not be generated by $\mathfrak{g}_4$.

In the special case of (16.4.35) where the sequence is derived by choosing $\mathfrak{S}$ with one member $[k_{\mathrm{h}} \times k_{\mathrm{v}}]$ the summations can be taken independently over the ranges $a = 1, 2, \ldots, k_{\mathrm{h}}$ and $b = 1, 2, \ldots, k_{\mathrm{v}}$. From (16.4.34) and (16.4.35),

$$\sum_{m=1}^{k} \eta(q - m) = \delta^{\mathrm{Kr}}(q - k) - \delta^{\mathrm{Kr}}(q - k + 1), \tag{16.4.42}$$

$$\begin{aligned} \langle x \times y | \boldsymbol{\chi}^{(n)} \rangle &= [\delta^{\mathrm{Kr}}(x - k_{\mathrm{h}}) - \delta^{\mathrm{Kr}}(x - k_{\mathrm{h}} + 1)] \\ &\quad \times [\delta^{\mathrm{Kr}}(y - k_{\mathrm{v}}) - \delta^{\mathrm{Kr}}(y - k_{\mathrm{v}} + 1)]. \end{aligned} \tag{16.4.43}$$

It follows that there are four non-zero weights. In the list given by (16.4.41) two weights are combined in $\chi_5^{(6)}$ and, from (16.4.43),

$$\chi_5^{(6)} = \langle 2 \times 3 | \boldsymbol{\chi}^{(n)} \rangle + \langle 3 \times 2 | \boldsymbol{\chi}^{(n)} \rangle = -2\,. \tag{16.4.44}$$

16.4.6.2 The De Neef–Enting Sequence

A more complicated initial set of graphs is used with the overlap method to generate the partially-ordered sequence of graphs in the finite-lattice series method described in Sect. 14.5. This is a procedure specifically intended to be used for the square lattice, although it can be adapted for the kagomé, honeycomb and triangular lattices (Sect. 14.5). The set $\mathfrak{S}$ consists of all rectangular blocks of sites with $x + y = \ell$, for some specified integer ℓ, and then $\mathcal{S}$ includes all rectangular graphs with $x + y \leq \ell$. Equations (16.4.33)–(16.4.35) still apply but the summations in (16.4.35) are no longer independent. From (16.4.34) and (16.4.35),

$$\sum_{[a,b]} \eta(x-a)\eta(y-b) = \sum_{a=1}^{\ell-1}\sum_{b=1}^{\ell-a} \eta(x-a)\eta(y-b)$$
$$= \sum_{a=1}^{\ell-1} \eta(x-a)[\delta^{\mathrm{Kr}}(y-\ell+a) - \delta^{\mathrm{Kr}}(y-\ell+a+1)]$$
$$= \eta(x+y-\ell) - \eta(x+y-\ell+1), \qquad (16.4.45)$$
$$\langle x \times y|\mathbf{X}^{(n)} = \delta^{\mathrm{Kr}}(x+y-\ell) - 3\delta^{\mathrm{Kr}}(x+y-\ell+1) + 3\delta^{\mathrm{Kr}}(x+y-\ell+2) - \delta^{\mathrm{Kr}}(x+y-\ell+3). \qquad (16.4.46)$$

All rectangles with the same value of $x + y$ have the same weight which is zero unless

$$\begin{aligned}
x+y &= \ell, && \text{when} \quad \chi_i^{(n)} = +1,\\
x+y &= \ell-1, && \text{when} \quad \chi_i^{(n)} = -3,\\
x+y &= \ell-2, && \text{when} \quad \chi_i^{(n)} = +3,\\
x+y &= \ell-3, && \text{when} \quad \chi_i^{(n)} = -1.
\end{aligned} \qquad (16.4.47)$$

Example 16.4.5 With $\ell = 6$ the set $\mathfrak{S}$ consists of the block [3 × 3], which is the graph [9, 1] used to generate the sequence in Example 16.4.4, together with blocks [4 × 2], [2 × 4], [5 × 1] and [1 × 5]. Using this set the complete sequence with designations $\mathfrak{g}_i$, $[x \times y]$ and $[\mathfrak{n}, \mathfrak{k}]$ and with the values of $\chi_i^{(n)}$ given by (16.4.45) is listed in Table 16.3.

16.4.7 Incorporating Sublattices

As indicated in Sect. 16.2.1 the loose-packed (bipartite) lattices can be divided into two equal sublattices $\mathfrak{a}$, $\mathfrak{b}$ so that each first-neighbour pair consists of one $\mathfrak{a}$ site and one $\mathfrak{b}$ site. For models based on sublattice order, like the antiferromagnetic models described in Sect. 3.5.1.1, it is necessary to modify the above analysis in order to distinguish isomorphic graphs which differ in relation to the two sublattices. Thus, for the graphs of Example 16.4.4, when the square lattice (Fig. 16.3) is divided into two equivalent sublattices $\mathfrak{a}$ and $\mathfrak{b}$, the single-site graph classification $\mathfrak{g}_1$ needs to be split into two according to whether the site is an $\mathfrak{a}$ or a $\mathfrak{b}$, $\mathfrak{g}_4$ needs to be split into two according to whether the centre site is an $\mathfrak{a}$ or a $\mathfrak{b}$ and $\mathfrak{g}_6$ need to be split into two according to whether the centre and corner sites are $\mathfrak{a}$s or $\mathfrak{b}$s. Now with $m = 2$ and $\mathfrak{S}$ consisting of two blocks of four squares $\mathcal{S}$ has nine members and a calculation similar to that in Example 16.4.4 gives

$$\chi_1^{(9)} = -1, \quad \chi_2^{(9)} = -1, \quad \chi_3^{(9)} = 4, \quad \chi_4^{(9)} = 1, \quad \chi_5^{(9)} = -1, \quad \chi_6^{(9)} = -1,$$
$$\chi_7^{(9)} = -1, \quad \chi_8^{(9)} = \tfrac{1}{2}, \quad \chi_9^{(9)} = \tfrac{1}{2}. \qquad (16.4.48)$$

However, it is not necessary to restrict attention to interpenetrating sublattices.

Table 16.3 The de Neef–Enting sequence of graphs for $\ell = 6$

$\mathfrak{g}_i$	$[x \times y]$	$[\mathfrak{n}, \mathfrak{k}]$	$\chi_i^{(15)}$
$\mathfrak{g}_1$	$[1 \times 1]$	$[1, 1]$	0
$\mathfrak{g}_2$	$[2 \times 1]$	$[2, 2]$	-1
$\mathfrak{g}_3$	$[1 \times 2]$	$[2, 2]$	-1
$\mathfrak{g}_4$	$[3 \times 1]$	$[3, 3]$	$+3$
$\mathfrak{g}_5$	$[1 \times 3]$	$[3, 3]$	$+3$
$\mathfrak{g}_6$	$[2 \times 2]$	$[4, 8]$	$+3$
$\mathfrak{g}_7$	$[4 \times 1]$	$[4, 4]$	-3
$\mathfrak{g}_8$	$[1 \times 4]$	$[4, 4]$	-3
$\mathfrak{g}_9$	$[3 \times 2]$	$[6, 1]$	-3
$\mathfrak{g}_{10}$	$[2 \times 3]$	$[6, 1]$	-3
$\mathfrak{g}_{11}$	$[5 \times 1]$	$[5, 1]$	$+1$
$\mathfrak{g}_{12}$	$[1 \times 5]$	$[5, 1]$	$+1$
$\mathfrak{g}_{13}$	$[4 \times 2]$	$[8, 2]$	$+1$
$\mathfrak{g}_{14}$	$[2 \times 4]$	$[8, 2]$	$+1$
$\mathfrak{g}_{15}$	$[3 \times 3]$	$[9, 1]$	$+1$

Example 16.4.6 Consider a simple cubic lattice with a layered sublattice structure as in the model for a metamagnet described in Sect. 6.2. The equivalent sublattices are labelled $\mathfrak{a}$ and $\mathfrak{b}$. Let $\mathfrak{G}$ be the one graph consisting of a cube of eight first-neighbour sites, four $\mathfrak{a}$ sites on one face and four $\mathfrak{b}$ sites on the opposite face. This generates the sequence:

$\mathfrak{g}_1$: a single $\mathfrak{a}$ site, $\mathfrak{g}_2$: a single $\mathfrak{b}$ site,

$\mathfrak{g}_3$: a single $\mathfrak{a}$–$\mathfrak{a}$ pair, $\mathfrak{g}_4$: a single $\mathfrak{a}$–$\mathfrak{b}$ pair,

$\mathfrak{g}_5$: a single $\mathfrak{b}$–$\mathfrak{b}$ pair, $\mathfrak{g}_6$: a square of four $\mathfrak{a}$ sites,

$\mathfrak{g}_7$: a square of two $\mathfrak{a}$ and two $\mathfrak{b}$ sites, $\mathfrak{g}_8$: a square of four $\mathfrak{b}$ sites,

$\mathfrak{g}_9$: the cube $\mathfrak{G}$.

Then

$$\Omega^{(9)} = \begin{pmatrix} 1\,0\,2\,1\,0\,4\,2\,0\,4 \\ 0\,1\,0\,1\,2\,0\,2\,4\,4 \\ 0\,0\,1\,0\,0\,4\,1\,0\,4 \\ 0\,0\,0\,1\,0\,0\,1\,0\,4 \\ 0\,0\,0\,0\,1\,0\,1\,4\,4 \\ 0\,0\,0\,0\,0\,1\,0\,0\,1 \\ 0\,0\,0\,0\,0\,0\,1\,0\,4 \\ 0\,0\,0\,0\,0\,0\,0\,1\,1 \\ 0\,0\,0\,0\,0\,0\,0\,0\,1 \end{pmatrix}, \qquad \boldsymbol{\omega}^{(9)} = \begin{pmatrix} 1/2 \\ 1/2 \\ 1 \\ 1 \\ 1 \\ 1/2 \\ 1 \\ 1/2 \\ 1 \end{pmatrix}. \tag{16.4.49}$$

Solving (16.4.26) for (16.4.49) gives

$$\begin{aligned} &\chi_1^{(9)} = \tfrac{1}{2}, \ \chi_2^{(9)} = \tfrac{1}{2}, \ \chi_3^{(9)} = 2, \ \chi_4^{(9)} = 0, \ \chi_5^{(9)} = 2, \ \chi_6^{(9)} = -\tfrac{1}{2}, \\ &\chi_7^{(9)} = -3, \ \chi_8^{(9)} = -\tfrac{1}{2}, \ \chi_9^{(9)} = -1. \end{aligned} \tag{16.4.50}$$

We can also consider cases where there are four equivalent sublattices.

Example 16.4.7 Consider a body-centred cubic lattice (Fig. 16.8) divided, as shown in Fig. 7.2b into four equivalent face-centred cubic sublattices (Fig. 16.9) labelled $\mathfrak{a}$, $\mathfrak{b}$, $\mathfrak{c}$ and $\mathfrak{d}$. The pairs $\mathfrak{a}$–$\mathfrak{c}$, $\mathfrak{a}$–$\mathfrak{d}$, $\mathfrak{b}$–$\mathfrak{c}$, $\mathfrak{b}$–$\mathfrak{d}$ combine to form diamond lattices (Fig. 16.10). Let $\mathfrak{S}$ be the one (section) graph consisting of a tetrahedron of four sites, one in each sublattice connected by first-neighbour and second-neighbour edges.[33] This generates the sequence:

$\mathfrak{g}_1$: a single $\mathfrak{a}$ site,

$\mathfrak{g}_2$: a single $\mathfrak{b}$ site,

$\mathfrak{g}_3$: a single $\mathfrak{c}$ site,

$\mathfrak{g}_4$: a single $\mathfrak{d}$ site,

$\mathfrak{g}_5$: a single $\mathfrak{a}$–$\mathfrak{c}$ first-neighbour pair,

$\mathfrak{g}_6$: a single $\mathfrak{a}$–$\mathfrak{d}$ first-neighbour pair,

$\mathfrak{g}_7$: a single $\mathfrak{b}$–$\mathfrak{c}$ first-neighbour pair,

$\mathfrak{g}_8$: a single $\mathfrak{b}$–$\mathfrak{d}$ first-neighbour pair,

$\mathfrak{g}_9$: a single $\mathfrak{a}$–$\mathfrak{b}$ second-neighbour pair,

$\mathfrak{g}_{10}$: a single $\mathfrak{c}$–$\mathfrak{d}$ second-neighbour pair,

$\mathfrak{g}_{11}$: a $\mathfrak{a}$–$\mathfrak{b}$–$\mathfrak{c}$ triangular face,

$\mathfrak{g}_{12}$: a $\mathfrak{a}$–$\mathfrak{b}$–$\mathfrak{d}$ triangular face,

$\mathfrak{g}_{13}$: a $\mathfrak{a}$–$\mathfrak{c}$–$\mathfrak{d}$ triangular face,

$\mathfrak{g}_{14}$: a $\mathfrak{b}$–$\mathfrak{c}$–$\mathfrak{d}$ triangular face,

$\mathfrak{g}_{15}$: the tetrahedron $\mathfrak{S}$.

---

[33] This involves the generalization of the notion of a graph to the case of two types of edges mentioned in a footnote on p. 637.

Then

$$
\boldsymbol{\Omega}^{(15)} = \begin{pmatrix}
1&0&0&0&1&1&0&0&1&0&1&1&1&0&1\\
0&1&0&0&0&0&1&1&1&0&1&1&0&1&1\\
0&0&1&0&1&0&1&0&0&1&1&0&1&1&1\\
0&0&0&1&0&1&0&1&0&1&0&1&1&1&1\\
0&0&0&0&1&0&0&0&0&0&1&0&1&0&1\\
0&0&0&0&0&1&0&0&0&0&0&1&1&0&1\\
0&0&0&0&0&0&1&0&0&0&1&0&0&1&1\\
0&0&0&0&0&0&0&1&0&0&0&1&0&1&1\\
0&0&0&0&0&0&0&0&1&0&1&1&0&0&1\\
0&0&0&0&0&0&0&0&0&1&0&0&1&1&1\\
0&0&0&0&0&0&0&0&0&0&1&0&0&0&1\\
0&0&0&0&0&0&0&0&0&0&0&1&0&0&1\\
0&0&0&0&0&0&0&0&0&0&0&0&1&0&1\\
0&0&0&0&0&0&0&0&0&0&0&0&0&1&1\\
0&0&0&0&0&0&0&0&0&0&0&0&0&0&1
\end{pmatrix}, \qquad
\boldsymbol{\omega}^{(15)} = \begin{pmatrix}
1/4\\1/4\\1/4\\1/4\\1\\1\\1\\1\\3/2\\3/2\\3\\3\\3\\3\\6
\end{pmatrix}.
\tag{16.4.51}
$$

Solving (16.4.26) for (16.4.51) gives

$$
\begin{aligned}
&\chi_1^{(15)} = -\tfrac{1}{4}, \quad \chi_2^{(15)} = -\tfrac{1}{4}, \quad \chi_3^{(15)} = -\tfrac{1}{4}, \quad \chi_4^{(9)} = -\tfrac{1}{4}, \quad \chi_5^{(15)} = 1,\\
&\chi_6^{(15)} = 1, \quad \chi_7^{(15)} = 1, \quad \chi_8^{(15)} = 1, \quad \chi_9^{(15)} = \tfrac{3}{2}, \quad \chi_{10}^{(15)} = \tfrac{3}{2},\\
&\chi_{11}^{(15)} = -3, \quad \chi_{12}^{(15)} = -3, \quad \chi_{13}^{(15)} = -3, \quad \chi_{14}^{(15)} = -3, \quad \chi_{15}^{(15)} = 6.
\end{aligned}
\tag{16.4.52}
$$

16.4.8 The Guggenheim–McGlashan Approach

A generalization of the quasi-chemical method of Guggenheim (1935) was developed by Guggenheim and McGlashan (1951). This method can be formulated in terms of the partially-ordered sequence $\mathcal{S}$ used in the KHDeB approximation, but, while it retains the capability of including irreducible energy contributions from more complicated graphs than first-neighbour pairs, it is much simpler to use, and in consequence rather less accurate.

Consider a partially-ordered sequence of graphs $\mathcal{S} := \{\mathfrak{g}_1, \mathfrak{g}_2, \ldots, \mathfrak{g}_n\}$, where $\mathfrak{g}_1$, is the only type of single site graph, $\mathfrak{g}_2$ is the only type of first-neighbour pair graph and $n > 2$.[34] The crucial step in this approach is that the lattice $\mathcal{N}$ of N sites is replaced by a *pseudo-lattice* $\widetilde{\mathcal{N}}$ of N sites made up of graphs $\mathfrak{g}_n$ overlapping only by single sites $\mathfrak{g}_1$. The number of $\mathfrak{g}_n$ is taken so that the number of $\mathfrak{g}_2$ in $\mathcal{N}$ and $\widetilde{\mathcal{N}}$ is the same. Let $\widetilde{\omega}_j$, $j = 1, 2, \ldots, n$ be the lattice constants of $\widetilde{\mathcal{N}}$. Then

[34] We are for simplicity, at this stage, excluding sublattice ordering of the type described in Sect. 16.4.7.

$$\widetilde{\omega}_1 = \omega_1, \qquad \widetilde{\omega}_2 = \omega_2, \qquad \widetilde{\omega}_j = \Omega_{jn}\widetilde{\omega}_n, \qquad j = 2, \ldots, n, \tag{16.4.53}$$

from which it follows that

$$\widetilde{\omega}_j = \omega_2\, \Omega_{jn}\,/\Omega_{2n}, \qquad j = 2, 3, \ldots, n. \tag{16.4.54}$$

Replacing $\boldsymbol{\omega}^{(n)}$ in (16.4.26) with the vector with elements given by (16.4.54) we have

$$\frac{\omega_2}{\Omega_{2n}}\boldsymbol{\Omega}_n^{(n)} + \left[\omega_1 - \omega_2\frac{\Omega_{1n}}{\Omega_{2n}}\right]\boldsymbol{i}_1 = \boldsymbol{\Omega}^{(n)}\widetilde{\boldsymbol{\chi}}^{(n)}, \tag{16.4.55}$$

where $\boldsymbol{\Omega}_n^{(n)}$ is the nth column of $\boldsymbol{\Omega}^{(n)}$ and $\boldsymbol{i}_1$ is the unit vector with elements $\delta^{\mathrm{Kr}}(k, 1)$, $k = 1, 2, \ldots, n$. Since $[\boldsymbol{\Omega}^{(n)}]^{-1}$ is upper triangular with diagonal elements unity,

$$\begin{gathered}\widetilde{\chi}_1^{(n)} = \omega_1 - \omega_2\frac{\Omega_{1n}}{\Omega_{2n}}, \\ \widetilde{\chi}_j^{(n)} = 0, \qquad j = 2, \ldots, n-1, \qquad \widetilde{\chi}_n^{(n)} = \frac{\omega_2}{\Omega_{2n}}.\end{gathered} \tag{16.4.56}$$

It is easy to verify these results for Example 16.4.4. In particular, when $\mathfrak{S}$ is the single square,

$$\widetilde{\boldsymbol{\omega}}^{(3)} = \begin{pmatrix} 1 \\ 2 \\ 1/2 \end{pmatrix} \qquad \widetilde{\boldsymbol{\chi}}^{(3)} = \begin{pmatrix} -1 \\ 0 \\ 1/2 \end{pmatrix}. \tag{16.4.57}$$

The extension of this method to the case where the lattice is divided into a number of sublattices is straightforward. Now the pseudo-lattice $\widetilde{\mathcal{N}}$ is constructed so that the numbers of sites and first-neighbour pairs of all types are maintained constant. We illustrate this with the case described in Example 16.4.7, of a body-centred cubic lattice divided into four sublattices as shown in Fig. 7.2b. Then, from (16.4.51),

$$\begin{aligned}\widetilde{\omega}_j &= \omega_j = \tfrac{1}{4}, && j = 1, \ldots, 4, \\ \widetilde{\omega}_j &= \omega_j = 1, && j = 5, \ldots, 8, \\ \widetilde{\omega}_j &= \Omega_{j,15}\widetilde{\omega}_{15} = 1, && j = 5, \ldots, 15,\end{aligned} \tag{16.4.58}$$

giving

$$\begin{aligned}
\widetilde{\chi}_j^{(15)} &= -\tfrac{3}{4}, & j &= 1, \dots, 4, \\
\widetilde{\chi}_j^{(15)} &= 0, & j &= 5, \dots, 14, \\
\widetilde{\chi}_{15}^{(15)} &= 1.
\end{aligned} \tag{16.4.59}$$

Finally we may for this method need to adjust some of the energy parameters. This will not be the case if the only contributions to the energy are from single sites and first-neighbour pairs since the Guggenheim method counts correctly the number of sites and first-neighbour pairs. However, suppose ε_j is an energy contribution to some configurations of $\mathfrak{g}_j$, which does not occur for any subgraphs of $\mathfrak{g}_j$. Since there are $N\omega_j$ $\mathfrak{g}_j$ of these on $\mathcal{N}$ and $N\widetilde{\omega}_j$ $\mathfrak{g}_j$ on the Guggenheim pseudo-lattice $\widetilde{\mathcal{N}}$ the replacement

$$\varepsilon_j \to \frac{\omega_j}{\widetilde{\omega}_j} \varepsilon_j \tag{16.4.60}$$

is needed.

16.4.9 Further Results

The analysis in this section is needed for the calculation of the cluster-variation entropy in (7.2.4). Here we consider the case, of relevance to the KHDeB hierarchy, where the partially-ordered sequence of connected section graphs $\mathcal{S}$ is generated by one graph ($m = 1$, $\mathfrak{S}$ having one member). Let $\mathcal{S}$ have n members with $\mathfrak{g}_n$ the generating graph and let $\boldsymbol{\Omega}^{(k)}$ be the matrix formed from the leading $k \times k$ submatrix of $\boldsymbol{\Omega}^{(n)}$, with $\boldsymbol{\omega}^{(k)}$ the column vector formed by taking the first k elements of $\boldsymbol{\omega}^{(n)}$. Then, as in (16.4.26), $\boldsymbol{\chi}^{(k)}$ is defined to be the solution of

$$\boldsymbol{\omega}^{(k)} = \boldsymbol{\Omega}^{(k)} \boldsymbol{\chi}^{(k)}. \tag{16.4.61}$$

Because $\boldsymbol{\Omega}^{(n)}$ is an upper triangular matrix with diagonal elements equal to one, it is not difficult to see that the inverse of $\boldsymbol{\Omega}^{(k)}$ is the leading $k \times k$ submatrix of $[\boldsymbol{\Omega}^{(n)}]^{-1}$. That is

$$[\boldsymbol{\Omega}^{(k)}]^{-1} = \{[\boldsymbol{\Omega}^{(n)}]^{-1}\}^{(k)}. \tag{16.4.62}$$

So

$$\boldsymbol{\chi}^{(k)} = \{[\boldsymbol{\Omega}^{(n)}]^{-1}\}^{(k)} \boldsymbol{\omega}^{(k)}. \tag{16.4.63}$$

However, it is easy to see, using Example 16.4.4, that $\boldsymbol{\chi}^{(k)}$ is *not*, in general, the same as the vector formed from the first k elements of $\boldsymbol{\chi}^{(n)}$. The latter we shall denote by $[\boldsymbol{\chi}^{(n)}]^{(k)}$.

With $\boldsymbol{\Omega}_i^{(k)}$ representing the vector formed from the ith row of $\boldsymbol{\Omega}^{(k)}$, it follows from (16.4.61), that

$$\omega_i = \boldsymbol{\Omega}_i^{(k-1)} \cdot \boldsymbol{\chi}^{(k-1)}, \qquad \omega_i = \boldsymbol{\Omega}_i^{(k)} \cdot \boldsymbol{\chi}^{(k)}. \tag{16.4.64}$$

By considering the operation in (16.4.61) of the last row of $\boldsymbol{\Omega}^{(k)}$ on $\boldsymbol{\chi}^{(k)}$ it is clear that the last element of this vector is ω_k. So,

$$\boldsymbol{\Omega}_i^{(k-1)} \cdot \boldsymbol{\chi}^{(k-1)} = \boldsymbol{\Omega}_i^{(k-1)} \cdot [\boldsymbol{\chi}^{(k)}]^{(k-1)} + \Omega_{i\,k}\omega_k \tag{16.4.65}$$

and, from (16.4.64),

$$\boldsymbol{\Omega}^{(k-1)} \left(\boldsymbol{\chi}^{(k-1)} - [\boldsymbol{\chi}^{(k)}]^{(k-1)}\right) = \boldsymbol{\upsilon}^{(k)}, \tag{16.4.66}$$

where

$$\boldsymbol{\upsilon}^{(k)} := \begin{pmatrix} \Omega_{1\,k}\omega_k \\ \Omega_{2\,k}\omega_k \\ \vdots \\ \Omega_{k-1\,k}\omega_k \end{pmatrix} \tag{16.4.67}$$

is a $(k-1)$-dimensional vector.

Chapter 17
Algebra

17.1 Catastrophe Theory

According to Stewart (1997, p. 7), "catastrophe theory is not a single thread of ideas; it resembles more closely a web, with innumerable interconnected strands; these include physical intuition and experiment, geometry, algebra, calculus, topology, singularity theory and many others". The name catastrophe theory was coined by Thom (1972) with particular reference to the application of this web of ideas to biological morphogenesis and the development of languages (Thom 1974) and it was soon evident that it could be applied to a much wider range of problems. In particular thermodynamics and the theory of phase transitions (Fowler 1972; Schulman and Revzen 1972; Schulman 1973; Lavis and Bell 1977; Gaite 1992; Gaite et al. 1998). Of particular relevance is the relationship between Landau theory, as described in Chap. 5, and catastrophe theory. In the former, physical arguments are used to justify the replacement of the full free-energy density with a 'sufficient' number of terms of its Taylor expansion about a critical (or tricritical) point. The purpose of this appendix is to use catastrophe theory to bring a degree of mathematical precision to this important question of sufficiency.[1]

17.1.1 Equivalence and Determinancy

Our main interest will be in scalar-valued functions $\mathbb{R}^n \to \mathbb{R}$, however, for some purposes it will be necessary to give definitions for vector-valued functions $\mathbb{R}^n \to \mathbb{R}^m$.

---

[1] It is not our aim to give a comprehensive account of catastrophe theory (there are many of these, see e.g. Saunders 1980; Gilmore 1981; Poston and Stewart 1997). We simply wish to trace a thread through catastrophe theory leading to the results we require. In doing so we quote some definitions and theorems; for further details of these together with the proofs of the theorems (or references where they can be found) the reader is referred to Poston and Stewart.

D.A. Lavis, *Equilibrium Statistical Mechanics of Lattice Models*,
Theoretical and Mathematical Physics, DOI 10.1007/978-94-017-9430-5_17

Definition 17.1.1 A function $\boldsymbol{f} : \mathbb{R}^n \to \mathbb{R}^m$ is k-fold differentiable, or of class $\mathcal{C}^k$, if all its m components have continuous partial derivatives of all orders up to k. A class $\mathcal{C}^\infty$ function is said to be smooth.

Definition 17.1.2 Given a polynomial function $p : \mathbb{R}^n \to \mathbb{R}$, we denote by $\mathrm{J}^k p$ the truncated polynomial obtained by omitting all terms of degree greater than k. Given a $\mathcal{C}^k$ function $f : \mathbb{R}^n \to \mathbb{R}$, the k-jet $\mathrm{J}^k f$ of f is the polynomial obtained by truncating its Taylor series about $\boldsymbol{x} = \boldsymbol{0}$ at degree k.[2,3]

J^k is, of course, a linear operator with respect to sums of functions and for products of functions we use it in the sense that $\mathrm{J}^k fg$ means $\mathrm{J}^k(fg)$. It is also obvious that $\mathrm{J}^k fg = \mathrm{J}^k f(\mathrm{J}^k g) = \mathrm{J}^k(\mathrm{J}^k f)g = \mathrm{J}^k(\mathrm{J}^k f)(\mathrm{J}^k g)$.

Definition 17.1.3 Let $\mathcal{U}$ and $\mathcal{V}$ be open sets in $\mathbb{R}^n$ and $\mathbb{R}^m$ respectively with $\boldsymbol{f} : \mathcal{U} \to \mathbb{R}^m$ and $\boldsymbol{f}(\mathcal{U}) := \mathcal{V}$. Then $\boldsymbol{f}$ is a diffeomorphism on $\mathcal{U}$ if it is: (i) smooth, (ii) has an inverse $\boldsymbol{g} : \mathcal{V} \to \mathbb{R}^n$ which is also smooth. $\boldsymbol{f}$ is a local diffeomorphism at $\boldsymbol{x} = \boldsymbol{0}$ if it is a diffeomorphism on some $\mathcal{U}$ containing $\boldsymbol{x} = \boldsymbol{0}$.

Example 17.1.1 $\sin(x)$ is a local diffeomorphism at $x = 0$ since it is smooth with a smooth inverse in the interval $\left(-\frac{\pi}{2}, \frac{\pi}{2}\right)$.

Definition 17.1.4 Two smooth functions $f, g : \mathbb{R}^n \to \mathbb{R}$ are equivalent at $\boldsymbol{x} = \boldsymbol{0}$ if there exists a local diffeomorphism $\boldsymbol{\eta} : \mathbb{R}^n \to \mathbb{R}^n$ at $\boldsymbol{x} = \boldsymbol{0}$ and a constant γ so that, in some open set $\mathcal{U}$ containing $\boldsymbol{x} = \boldsymbol{0}$, $g(\boldsymbol{x}) = f(\boldsymbol{\eta}(\boldsymbol{x})) + \gamma$. If in addition

$$\left(\frac{\partial \eta_i}{\partial x_j}\right)_0 = \delta^{\mathrm{Kr}}(i - j), \qquad i, j = 1, 2, \ldots, n, \tag{17.1.1}$$

the functions are strongly equivalent.

For the application to Landau theory we shall need strong equivalence and we shall concentrate on that case.

Example 17.1.2 It can be shown that $f(x) := x^2$ and $g(x) := x^2 - x^4$ are strongly equivalent at $x = 0$ by taking

$$\eta(x) := x\sqrt{1 - x^2}, \qquad \text{which has an inverse in } \left(-\tfrac{1}{\sqrt{2}}, \tfrac{1}{\sqrt{2}}\right).$$

Definition 17.1.5 A smooth function $f : \mathbb{R}^n \to \mathbb{R}$ is k-determinate at $\boldsymbol{x} = \boldsymbol{0}$ if, for all $g : \mathbb{R}^n \to \mathbb{R}$ of order $k + 1$, $\mathrm{J}^k f + g$ is equivalent at $\boldsymbol{x} = \boldsymbol{0}$ to $\mathrm{J}^k f$. If the equivalence is strong then f is strongly k-determinate.

---

[2] For easy of discussion we shall, in this and following definitions, choose our reference point as the origin in the space $\mathbb{R}^n$ of $\boldsymbol{x}$. Translation to a point $\boldsymbol{x}_0 \neq \boldsymbol{0}$ is trivial.

[3] The terms of the polynomial p are monomials. The degree of a monomial is the sum of the degrees of each of its n variables. The degree of p is the largest of the degrees of its constituent monomials and the order of p is the smallest of the degrees of its constituent monomials. A homogeneous polynomials has terms all of the same degree, which is then both the degree and the order of the polynomial.

It will be useful to have a mechanism for determining whether any f is strongly k-determinate, and to explore this question the following notation will be useful:

Definition 17.1.6

(a) $\mathcal{E}_n^k$ is the vector space of polynomials $p : \mathbb{R}^n \to \mathbb{R}$ of degree less than or equal to k.
(b) $\mathcal{J}_n^k$ is the subspace of $\mathcal{E}_n^k$ consisting of polynomials with zero constant term, that is of order greater than or equal to one.
(c) $\mathcal{I}_n^k$ is the subspace of $\mathcal{J}_n^k$ consisting of polynomials $p \in \mathcal{J}_n^k$ with $(\nabla p)_0 = \mathbf{0}$, that is of order greater than or equal to two.
(d) $\mathcal{M}_n^k$ is the subspace of $\mathcal{I}_n^k$ consisting all homogeneous polynomials of degree (and order) k.

We use the notation $\mathcal{E}_1^k \stackrel{\text{B}}{=} \{1, x, x,^2 \dots, x^k\}$ meaning that $\mathcal{E}_1^k$ is the vector space of polynomials with basis given by the monomials x^i, $i = 0, 1, 2, \dots, k$, with a similar notation applied to other spaces. The dimension $\dim[\mathcal{M}_n^k]$ of $\mathcal{M}_n^k$ is the number of monomials of order k. Thus

$$\dim[\mathcal{M}_n^k] = \frac{(n-1+k)!}{(n-1)!k!}, \qquad \dim[\mathcal{E}_n^k] = \sum_{i=0}^{k} \frac{(n-1+i)!}{(n-1)!i!} = \frac{(n+k)!}{n!k!}, \tag{17.1.2}$$

$$\dim[\mathcal{J}_n^k] = \dim[\mathcal{E}_n^k] - 1, \qquad \dim[\mathcal{I}_n^k] = \dim[\mathcal{E}_n^k] - n - 1.$$

Definition 17.1.7 Let $\mathcal{Z}$ be any one of the vector spaces of Definition 17.1.6 and let $\mathcal{X}$ and $\mathcal{Y}$ be two subspaces of $\mathcal{Z}$. Then $\mathcal{X}$ and $\mathcal{Y}$ are transverse if the dimension of their intersection is as small as possible. That is

$$\dim[\mathcal{X} \cap \mathcal{Y}] = \max\{0, \dim[\mathcal{X}] + \dim[\mathcal{Y}] - \dim[\mathcal{Z}]\}. \tag{17.1.3}$$

Let $\mathcal{Z} \stackrel{\text{B}}{=} \{v_1, v_2, \dots, v_\zeta\}$ and $\mathcal{X} \stackrel{\text{B}}{=} \{v_1, v_2, \dots, v_\xi\}$, where $\zeta := \dim[\mathcal{Z}]$, $\xi := \dim[\mathcal{X}]$ and v_i, $i = 1, 2, \dots, \zeta$ are monomials.

Definition 17.1.8 When $\xi < \zeta$, $\{v_{\xi+1}, \dots, v_\zeta\}$ is a cobasis of $\mathcal{X}$ with respect to $\mathcal{Z}$. With $\overline{\mathcal{X}} \stackrel{\text{B}}{=} \{v_{\xi+1}, \dots, v_\zeta\}$, the codimension of $\mathcal{X}$ with respect to $\mathcal{Z}$ is given by

$$\operatorname{cod}[\mathcal{X}|\mathcal{Z}] := \dim[\overline{\mathcal{X}}] = \dim[\mathcal{Z}] - \dim[\mathcal{X}]. \tag{17.1.4}$$

It follows from (17.1.3) that $\mathcal{X}$ and $\mathcal{Y}$ are transverse if and only if either $\mathcal{Y} \subseteq \overline{\mathcal{X}}$ or $\overline{\mathcal{X}} \subseteq \mathcal{Y}$. In particular, of course $\mathcal{X}$ and $\overline{\mathcal{X}}$ are transverse.

Definition 17.1.9 Let $\mathcal{X}$ and $\mathcal{Y}$ be any subspaces of the vector spaces given in Definition 17.1.6. The product space $\mathcal{X} \star \mathcal{Y}$ consists of all linear combinations of products of one member of each of $\mathcal{X}$ and $\mathcal{Y}$. That is

$$\mathcal{X} \star \mathcal{Y} := \{x_1 y_1 + \cdots x_r y_r | \forall r, x_i \in \mathcal{X}, y_i \in \mathcal{Y}\}.$$

We denote by $\mathrm{J}^k(\mathcal{X} \star \mathcal{Y})$ the space of truncated members of $\mathcal{X} \star \mathcal{Y}$ and by $\Delta_k(f)$ the subspace of $\mathcal{E}_n^k$ spanned by

$$\left\{ \mathrm{J}^k \frac{\partial f}{\partial x_1}, \ldots, \mathrm{J}^k \frac{\partial f}{\partial x_n} \right\}.$$

It is clear that $\Delta_k(f) = \Delta_k(\mathrm{J}^{k+i} f)$, for $i \geq 1$. The following theorem allows us to ascertain whether the function f is strongly k-determinate:

Theorem 17.1.1 *A function $f : \mathbb{R}^n \to \mathbb{R}$ is strongly k-determinate at $\boldsymbol{x} = \boldsymbol{0}$ if and only if*

$$\mathcal{M}_n^{k+1} \subseteq \mathrm{J}^{k+1}(\mathcal{J}_n^{k+1} \star \Delta_{k-1}(f)) \tag{17.1.5}$$

and, since the order of the elements of $\mathcal{J}_n^{k+1}$ is at least two, $\Delta_{k-1}(f)$ in (17.1.5) *can be replaced by $\Delta_{k+i}(f)$, for $i = 0, 1, 2, \ldots$.*

From this result it is straightforward to show that:

Corollary 17.1.1 *If a function is strongly k-determinate at $\boldsymbol{x} = \boldsymbol{0}$ then it is strongly k'-determinate at $\boldsymbol{x} = \boldsymbol{0}$, for all $k' > k$.*

Definition 17.1.10 The smallest k for which f is k-determinate at $\boldsymbol{x} = \boldsymbol{0}$ is called the determinancy of f at $\boldsymbol{x} = \boldsymbol{0}$. A function whose determinancy is finite is called finitely-determined. Strong determinancy is defined in a similar way.

17.1.2 Critical Points, Codimension and Unfoldings

Definition 17.1.11 The smooth function $f : \mathbb{R} \to \mathbb{R}^n$ has an equilibrium point at $\boldsymbol{x} = \boldsymbol{0}$ if $(\nabla f)_0 = \boldsymbol{0}$. If in addition the Hessian matrix of second-order partial derivatives of f is singular then $\boldsymbol{x} = \boldsymbol{0}$ is a critical point.[4]

We are now concerned with the problem of classifying critical points.

Definition 17.1.12 $\mathcal{T}_k(f) := \mathrm{J}^k(\mathcal{E}_n^k \star \Delta_k(f))$ is the tangent space[5] to $\mathrm{J}^k f$ at $\boldsymbol{x} = \boldsymbol{0}$. This is a subspace of $\mathcal{J}_n^k$.

---

[4] In catastrophe theory what we have called an equilibrium point is designated a *critical point* with our critical point being a *degenerate* critical point. Our nomenclature is an attempt at uniformity with the rest of the book. However, even this is a compromise since a spinodal point, representing the meeting of a line of stable and unstable equilibrium points (Figs. 5.1a and 5.2), would according to the current definition be a critical point. In some ways it would be better to use the language of bifurcation theory (Iooss and Joseph 1989) in which thermodynamic critical points are singular bifurcations and a spinodal point is a regular turning-point bifurcation.

[5] The reasoning underlying this definition is given by Poston and Stewart (1997, p. 138).

Definition 17.1.13 The codimension cod$[f]$ of f is given by

$$\text{cod}[f] := \max_{k\geq 2}\{\text{cod}[\mathcal{T}_k(f)|\mathfrak{J}_n^k]\}. \tag{17.1.6}$$

Theorem 17.1.2 *If f has strong determinancy s, then,*

$$\text{cod}[f] = \text{cod}[\mathcal{T}_s(f)|\mathfrak{J}_n^s]. \tag{17.1.7}$$

From Theorem 17.1.1, $\mathcal{M}_n^{s+i} \subseteq \mathcal{T}_{s+i}(f)$ for $i = 1, 2, \ldots$ and it follows that cod$[\mathcal{T}_k(f)|\mathfrak{J}_n^k]$ increases with k until $k = s$, after which it remains constant yielding (17.1.7).

Theorem 17.1.3 *Given a function $f : \mathbb{R}^n \to \mathbb{R}$, with an equilibrium point at $\boldsymbol{x} = \boldsymbol{0}$, any function $F(\boldsymbol{x}) := \varepsilon g(\boldsymbol{x}) + f(\boldsymbol{x})$, for ε sufficiently small has at most* cod$[f]+1$ *equilibrium points near to $\boldsymbol{x} = \boldsymbol{0}$.*

Definition 17.1.14 Consider an m-parameter family of functions $F : \mathbb{R}^n \times \mathbb{R}^m \to \mathbb{R}$ with values $F(\boldsymbol{x}; \boldsymbol{c})$, where the vector $\boldsymbol{c} = (c_1, c_2, \ldots, c_m) \in \mathbb{R}^m$ of parameters define the members of the family. F is an m-unfolding of $f : \mathbb{R}^n \to \mathbb{R}$ if $F(\boldsymbol{x}; \boldsymbol{0}) = f(\boldsymbol{x})$.

Definition 17.1.15 An r-unfolding G of f is induced from the m-unfolding F of Definition 17.1.14 by:

(a) a smooth mapping $\boldsymbol{w} : \mathbb{R}^r \to \mathbb{R}^m$ so that
$\boldsymbol{t} := (t_1, t_2, \ldots, t_r) \to \boldsymbol{w}(\boldsymbol{t}) := (w_1(\boldsymbol{t}), w_2(\boldsymbol{t}), \ldots, w_m(\boldsymbol{t}))$.
(b) a smooth $\boldsymbol{t}$-dependent map $\boldsymbol{\eta} : \mathbb{R}^n \times \mathbb{R}^r \to \mathbb{R}$ such that, for each (fixed) $\boldsymbol{t}$ $\boldsymbol{x} \to \boldsymbol{\eta}(\boldsymbol{x}; \boldsymbol{t})$ is local diffeomorphism at $\boldsymbol{x} = \boldsymbol{0}$.
(c) a smooth map $\gamma : \mathbb{R}^r \to \mathbb{R}$,

if, in some open set $\mathcal{U}$ containing $\boldsymbol{x} = \boldsymbol{0}$ and in a neighbourhood of $\boldsymbol{t} = \boldsymbol{0}$,

$$G(\boldsymbol{x}; \boldsymbol{t}) = F(\boldsymbol{\eta}(\boldsymbol{x}; \boldsymbol{t})); \boldsymbol{w}(\boldsymbol{t})) + \gamma(\boldsymbol{t}).$$

If G is induced from F with $r = m$ then the unfoldings are equivalent. If in addition

$$\left(\frac{\partial \eta_i}{\partial x_j}\right)_0 = \delta^{\text{Kr}}(i - j), \qquad i, j = 1, 2, \ldots, n,$$

for all $\boldsymbol{t}$ in a neighbourhood of $\boldsymbol{t} = \boldsymbol{0}$ the unfoldings are strongly equivalent.

Definition 17.1.16 An m-unfolding F of f is versal if all other unfoldings of f can be induced from it. It is universal if no versal unfolding of f exists of dimension less than m.

Theorem 17.1.4 *Universal unfoldings of equivalent functions are equivalent.*

Definition 17.1.17 The subspace $\mathcal{V}^k(F)$ of $\mathfrak{J}_n^k$ is that spanned by the polynomials

$$v_j^k(\boldsymbol{x}) := \frac{\partial \mathrm{J}^k F(\boldsymbol{x}; 0, \ldots, 0, c_j, 0, \ldots, 0)}{\partial c_j}, \qquad j = 1, 2, \ldots, m. \tag{17.1.8}$$

Theorem 17.1.5 *An m-unfolding F of f, where f is k-determinate, is versal if and only if $\mathcal{V}^k(F)$ and $\mathcal{T}_k(f)$ are transverse. It is universal if and only if $m = \mathrm{cod}[f]$.*

Let f have an equilibrium point at $\boldsymbol{x} = \boldsymbol{0}$ and strong determinancy s then it follows from Theorems 17.1.2 and 17.1.5 that

$$F(\boldsymbol{x}; \boldsymbol{c}) := \sum_{j=1}^{\mathrm{cod}[f]} c_j v_j^s(\boldsymbol{x}) + f(\boldsymbol{x}) \tag{17.1.9}$$

is a universal unfolding of f if $\{v_1^s, v_2^s, \ldots, v_{\mathrm{cod}[f]}^s\}$ is a cobasis of $\mathcal{T}^s(f)$ in $\mathfrak{J}_n^s$; we assume that the parameters $\boldsymbol{c}$ are chosen so this is the case. Now f is equivalent to $\mathrm{J}^s f$ and to $\mathrm{J}^{s+1} f$. However, $\mathcal{T}_s(f) = \mathcal{T}_s(\mathrm{J}^{s+1} f)$ but not necessarily $\mathcal{T}_s(f) = \mathcal{T}_s(\mathrm{J}^s f)$. So

$$\tilde{F}(\boldsymbol{x}; \boldsymbol{c}) := \sum_{j=1}^{\mathrm{cod}[f]} c_j v_j^s(\boldsymbol{x}) + \mathrm{J}^{s+1} f(\boldsymbol{x}) \tag{17.1.10}$$

is a universal unfolding of $\mathrm{J}^{s+1} f$, which is equivalent to F. However

$$\tilde{F}'(\boldsymbol{x}; \boldsymbol{c}) := \sum_{j=1}^{\mathrm{cod}[f]} c_j v_j^s(\boldsymbol{x}) + \mathrm{J}^s f(\boldsymbol{x}) \tag{17.1.11}$$

is a universal unfolding of $\mathrm{J}^s f$, and thus equivalent to F, only if $\mathcal{T}_s(\mathrm{J}^s f)$ is transverse to $\mathcal{V}^s(F)$. A universal unfolding $F(\boldsymbol{x}; \boldsymbol{c})$ of $f(\boldsymbol{x})$, with a critical point at $\boldsymbol{x} = \boldsymbol{0}$ and small values of the unfolding parameters, is the most general perturbation of $f(\boldsymbol{x})$ and it will thus, from Theorem 17.1.3, have at most $\mathrm{cod}[f] + 1$ equilibrium points in a neighbourhood of $\boldsymbol{x} = \boldsymbol{0}$.

Example 17.1.3 Let f be a smooth function with a power series of the form

$$f(x) := a_0 + \sum_{i=s}^{\infty} a_i x^i, \qquad a_s \neq 0, \quad s \geq 2. \tag{17.1.12}$$

It is clear that the presence of the constant a_0 does not affect the calculation and

$$\mathrm{J}^{k-1}\frac{\mathrm{d}f}{\mathrm{d}x} = \begin{cases} \sum_{i=s}^{k} i a_i x^{i-1}, & k \geq s, \\ 0, & k < s. \end{cases} \tag{17.1.13}$$

Then

$$x^{s+1} = \mathrm{J}^{s+1}\frac{x^2}{s a_s}\mathrm{J}^{s-1}\frac{\mathrm{d}f}{\mathrm{d}x}. \tag{17.1.14}$$

So the function f has strong determinancy s and there exists a diffeomorphism $y(x)$ with $y'(0) = 1$ such that

$$f(y(x)) = a_0 + x^s. \tag{17.1.15}$$

The terms in the power series of $y(x)$ can be determined iteratively and in some particular cases an exact form is derivable. Thus, for example if $s = 2$ and $a_i = 1$, $i \geq 2$ then $y(x) = -\frac{1}{2}\left(x^2 - \sqrt{x^4 + 4x^2}\right)$.

From (17.1.13),

$$\mathcal{T}_k(f) \stackrel{\mathrm{B}}{=} \begin{cases} \{0\}, & k < (s-1), \\ \{x^{s-1}, x^s, \ldots, x^k\}, & k \geq (s-1) \end{cases} \tag{17.1.16}$$

and $\operatorname{cod}[f] = s - 2$. We note that the formulae for $\mathcal{T}_k(f)$ do not depend on $a_i \neq 0$ for $i > s$. The element x^j in the basis, for $s \leq j \leq k$, can be generated by the product of $x^{s-1} \in \Delta_s(f)$ and $x^{j-s+1} \in \mathcal{E}_1^k$. From this it follows that $\mathcal{T}_k(f) = \mathcal{T}_k(\mathrm{J}^{s+1}f) = \mathcal{T}_k(\mathrm{J}^s f)$. In particular consider the case $s = 4$ for which f has a critical point at $x = 0$ of codimension two. Suppose that $a_4 = 1$ and that the coefficients a_k, $k \geq 5$, are such that f has no further equilibrium points in a neighbourhood of $x = 0$. Now let $g(x) = x^j$, for some positive integer j, so that the equilibrium points for $F(x) = \varepsilon x^j + f(x)$ are given by

$$0 = j\varepsilon x^{j-1} + 4x^3 + \sum_{i=5}^{\infty} i a_i x^{i-1}. \tag{17.1.17}$$

It is clear that, for $j \geq 4$, F has a critical point at $x = 0$ and no other equilibrium points in a neighbour of the origin. The remaining cases are:

(i) $j = 3$: where there is a critical point at $x = 0$ and an equilibrium point at $x = -\frac{3}{4}\varepsilon$.
(ii) $j = 2$: where there is an equilibrium point at $x = 0$ and at $x = \pm\sqrt{-\varepsilon/2}$ for $\varepsilon < 0$.
(iii) $j = 1$: where there is an equilibrium point at the real cube root of $-\varepsilon/4$.

This is in accord with Theorem 17.1.3 which asserts that the number of equilibrium points must be less than or equal to three. Consider now the $(s-2)$-unfolding

$$F(x;\boldsymbol{c}) := a_0 + \sum_{j=1}^{s-2} c_j x^j + \sum_{i=s}^{\infty} a_i x^i \tag{17.1.18}$$

of (17.1.12). Then $\mathcal{V}^s(F) \stackrel{\mathrm{B}}{=} \{x, x^2, \ldots, x^{s-2}\}$ and since this is a cobasis of $\mathcal{T}^s(f) = \mathcal{T}^s(\mathrm{J}^{s+1} f) \stackrel{\mathrm{B}}{=} \{x^{s-1}, x^s\}$ in $\mathcal{J}_1^s$, (17.1.18) is a universal unfolding, equivalent to the universal unfolding

$$\tilde{F}(x;\boldsymbol{c}) := a_0 + \sum_{j=1}^{s-2} c_j x^j + a_s x^s + a_{s+1} x^{s+1} \tag{17.1.19}$$

of $\mathrm{J}^{s+1} f$ and since, as already indicated, $\mathcal{T}_s(\mathrm{J}^s f) = \mathcal{T}_s(\mathrm{J}^{s+1} f)$

$$\tilde{F}'(x;\boldsymbol{c}) := a_0 + \sum_{j=1}^{s-2} c_j x^j + a_s x^s \tag{17.1.20}$$

is also equivalent to F.

Example 17.1.4 For $n = 2$ the general cubic case is

$$f(x_1, x_2) := a_0 x_1^3 + a_1 x_1^2 x_2 + a_2 x_1 x_2^2 + a_3 x_2^3 + \mathrm{O}(x_1^i x_2^{4-i}). \tag{17.1.21}$$

To show that f has strong determinancy of 3 we must, according to Theorem 17.1.1, show that the the five monomials $x_1^4, x_1^3 x_2, x_1^2 x_2^2, x_1 x_2^3, x_2^4$ can be expressed as linear combinations of some of the six quantities $x_1^j x_2^{2-j} f_1(x_1, x_2)$, $x_1^j x_2^{2-j} f_2(x_1, x_2)$, $j = 0, 1, 2$, where

$$\begin{aligned} f_1(x_1, x_2) &:= \mathrm{J}^2 \frac{\partial f}{\partial x_1} = 3a_0 x_1^2 + 2a_1 x_1 x_2 + a_2 x_2^2, \\ f_2(x_1, x_2) &:= \mathrm{J}^2 \frac{\partial f}{\partial x_2} = a_1 x_1^2 + 2a_2 x_1 x_2 + 3a_3 x_2^2, \end{aligned} \tag{17.1.22}$$

This involves finding a solution to six equations in five variables. We have one surplus equation[6] and one needs to be eliminated. Which this will be will depend on the values of the parameters a_0, a_1, a_2, a_3. If we were to drop the equation for $x_2^2 f_2(x_1, x_2)$ then the problem is to solve $\boldsymbol{A}\boldsymbol{\chi} = \boldsymbol{\phi}$, where

---

[6] Unlike in the corresponding quartic case when we have exactly the right number; that is six equations and six monomials.

$$\boldsymbol{\chi} := \begin{pmatrix} x_1^4 \\ x_1^3 x_2 \\ x_1^2 x_2^2 \\ x_1 x_2^3 \\ x_2^4 \end{pmatrix}, \qquad \boldsymbol{\phi} := \begin{pmatrix} x_1^2 f_1(x_1, x_2) \\ x_1 x_2 f_1^{(2)}(x_1, x_2) \\ x_2^2 f_1(x_1, x_2) \\ x_1^2 f_2^{(} x_1, x_2) \\ x_1 x_2 f_2(x_1, x_2) \end{pmatrix}, \tag{17.1.23}$$

$$\boldsymbol{A} := \begin{pmatrix} 3a_0 & 2a_1 & a_2 & 0 & 0 \\ 0 & 3a_0 & 2a_1 & a_2 & 0 \\ 0 & 0 & 3a_0 & 2a_1 & a_2 \\ a_1 & 2a_2 & 3a_3 & 0 & 0 \\ 0 & a_1 & 2a_2 & 3a_3 & 0 \end{pmatrix}. \tag{17.1.24}$$

Thus a sufficient condition for f to be strongly 3-determinate is that

$$\mathrm{Det}[\boldsymbol{A}] = 3a_2(27a_0^2a_3^2 - 18a_0a_1a_2a_3 + 4a_0a_2^3 + 4a_1^3a_3 - a_1^2a_2^2) \neq 0. \tag{17.1.25}$$

Of course, this is not a necessary condition since, even if this particular matrix $\boldsymbol{A}$ is singular (if, for example, $a_2 = 0$), a non-singular matrix may be achievable by a different choice of five equations. However, suppose that the coefficients are such that (17.1.25) is true. Then f has strong determinancy 3 and we need to derive a universal unfolding. To do so we must compute a basis for $\mathcal{T}^3(f)$, which will consist of some or all of the monomials in x_1 and x_2 of degrees one, two and three. The inclusion of the cubic monomials could involve the quartic terms in f, but if they appear by virtue only of the cubic terms then, as in Example 17.1.3, we have the added bonus that $\mathcal{T}^3(f) = \mathcal{T}^3(\mathrm{J}^4 f) = \mathcal{T}^3(\mathrm{J}^3 f)$. For this to be the case the four monomials x_1^3, $x_1^2x_2$, $x_1x_2^2$ and x_2^3 need to be expressible as linear combinations of the four quantities $x_1 f_1(x_1, x_2)$, $x_2 f_1(x_1, x_2)$, $x_1 f_2(x_1, x_2)$ and $x_2 f_2(x_1, x_2)$. This, as in the calculation for determinancy, is equivalent to the determinant of a matrix being non-zero. In this case the condition is

$$27a_0^2a_3^2 - 18a_0a_1a_2a_3 + 4a_0a_2^3 + 4a_1^3a_3 - a_1^2a_2^2 \neq 0. \tag{17.1.26}$$

If this is satisfied then the only question concerns the quadratic contributions, since $\mathcal{T}^3(f)$ cannot contain linear terms. The two equations (17.1.22) are not sufficient to express all three monomials x_1^2, x_1x_2 and x_2^2 as linear combinations of $f_1(x_1, x_2)$ and $f_2(x_1, x_2)$. However, it is usually possible to express two of the monomials in terms of linear combinations of $f_1(x_1, x_2)$, $f_2(x_1, x_2)$ and the third monomial showing that $\mathrm{cod}[f] = 3$. In particular it is easy to verify that, if

$$3(a_0a_2 + a_1a_3) - a_1^2 - a_2^2 \neq 0, \tag{17.1.27}$$

x_1x_2 and $x_1^2 - x_2^2$ can be expressed in terms of $x_1^2 + x_2^2$. Thus

$$\mathcal{T}^3(f) \stackrel{\mathrm{B}}{=} \{x_1x_2, x_1^2 - x_2^2, x_1^3, x_1^2x_2, x_1x_2^2, x_2^3\} \tag{17.1.28}$$

with cobasis $\{x_1, x_2, x_1^2 + x_2^2\}$ and

$$F(x_1x_2; \boldsymbol{c}) := c_1x_1 + c_2x_2 + c_2(x_1^2 + x_2^2) + a_0x_1^3 + a_1x_1^2x_2 + a_2x_1x_2^2 + a_3x_2^3 + \mathrm{O}(x_1^i x_2^{4-i}) \qquad (17.1.29)$$

is a universal unfolding of (17.1.21). This is equivalent to $\mathrm{J}^4 F$ and will be equivalent to $\mathrm{J}^3 F$ if (17.1.26) is satisfied.

17.1.3 Symmetry Considerations

In the application of catastrophe theory to statistical mechanics, the function of interest, which is the free-energy density, may be restricted by certain symmetry properties. The simplest example of this is the mean-field model for ferromagnetism, where the one parameter is m the magnetization density and the free-energy density with zero field is invariant under the symmetry group $\mathcal{S}_2 : m \to -m$. The Landau expansion, apart from the linear term in the magnetic field, is thus an expansion in the invariants of $\mathcal{S}_2$, that is the even powers of m. In a similar way the free-energy density for the mean-field model for metamagnetism (Sects. 4.6, 5.5 and 6.2) is, in the absence of a staggered field, an even function of the order parameter q. A more complicated case is that of the 3-state Potts model described in Sect. 5.4.2, where there are now two parameters m and p which satisfy the symmetry group $\mathcal{S}_3$, with generators given by (5.4.4) and (5.4.7). The Landau expansion of the free-energy density now contains only powers of the invariants $\mathfrak{v}_{\mathrm{E}}(p, m)$ and $\mathfrak{v}_{\mathrm{O}}(p, m)$ given by (5.4.9) and (5.4.10) respectively. In general

$$f(\boldsymbol{x}) := \sum_{i=s}^{\infty} a_i v_i(\boldsymbol{x}), \qquad (17.1.30)$$

where $a_s \neq 0$ and the functions $v_i(\boldsymbol{x})$ are homogeneous in the components of $\boldsymbol{x}$, of increasing order with respect to i and invariant under the operations of the symmetry group. We shall also suppose that $v_i(\boldsymbol{0}) = 0$, with the order of $v_s(\boldsymbol{x})$ being greater that or equal to two, so that f has a critical point at $\boldsymbol{x} = \boldsymbol{0}$. Of the three parts of the analysis to: (i) obtain the strong determinancy, (ii) calculate the codimension, (iii) deriving a universal unfolding, only parts (ii) and (iii) are affected by the symmetry restriction. If k_i is the order of $v_i(\boldsymbol{x})$ and f has strong determinancy k_s then f is strongly equivalent to $a_s v_s(\boldsymbol{x})$, or indeed to $a_s v_s(\boldsymbol{x}) + a_{s+1} v_{s+1}(\boldsymbol{x})$ if we have need to retain a second term. The questions of codimension and unfolding will be treated by means of examples.

Example 17.1.5 Consider the modification

$$f(x) := a_0 + \sum_{i=s}^{\infty} a_{2i} x^{2i}, \qquad (17.1.31)$$

of (17.1.12) when f is an even function, which is thus invariant under the symmetry group $\mathcal{S}_2 : x \to -x$. As already seen in Example 17.1.3, f has strong determinancy $2s$. The definitions (17.1.6) and (17.1.7) for codimension must now be modified to take into account the restriction to even functions.

$$\overset{2}{\overline{\mathrm{cod}}}\,[f] = \dim[\overset{2}{\overline{\mathcal{J}_1^{2s}}}] - \dim[\overset{2}{\overline{\mathcal{T}_{2s}(f)}}], \tag{17.1.32}$$

where $\overset{2}{\overline{\mathcal{J}_1^{2s}}}$ and $\overset{2}{\overline{\mathcal{T}_{2s}(f)}}$ are the subspaces of $\mathcal{J}_1^{2s}$ and $\mathcal{T}_{2s}(f)$ consisting of polynomials invariant under $\mathcal{S}_2$; that is even polynomials. Thus, from (17.1.2) and (17.1.16),

$$\dim[\overset{2}{\overline{\mathcal{J}_1^{2s}}}] = s, \qquad \dim[\overset{2}{\overline{\mathcal{T}_{2s}(f)}}] = 1, \tag{17.1.33}$$

giving $\overset{2}{\overline{\mathrm{cod}}}\,[f] = s - 1$. The $(s-1)$-unfolding

$$F(x; \mathbf{c}) := a_0 + \sum_{j=1}^{s-1} a_{2j} x^{2j} + \sum_{i=s}^{\infty} a_{2i} x^{2i} \tag{17.1.34}$$

of (17.1.31) is universal in the sense that the space spanned by $\{x^2, x^4, \ldots, x^{2(s-1)}\}$ is transverse to $\overset{2}{\overline{\mathcal{T}_{2s}(f)}}$ in $\overset{2}{\overline{\mathcal{J}_1^{2s}}}$ and

$$\tilde{F}(x; \mathbf{c}) := a_0 + \sum_{j=1}^{s-2} a_{2j} x^{2j} + a_{2s} x^{2s} + a_{2s+2} x^{2s+2}, \tag{17.1.35}$$

is a universal unfolding of $\mathrm{J}^{2s+2} f$. Again the unfolding of $\mathrm{J}^{2s} f$ with the term of degree $2s + 2$ omitted from (17.1.35) is equivalent to (17.1.34).

Example 17.1.6 At the critical point in the 3-state Potts model the leading term in the Landau expansion of the free-energy density is the cubic invariant $v_{\mathrm{o}}(x_1, x_2) = x_1(x_1^2 - 3x_2^2)$. Thus we are interested in

$$f(x_1, x_2) := v_{\mathrm{o}}(x, x_2) + a[v_{\mathrm{E}}(x_1, x_2)]^2 + \text{ terms in combinations of } v_{\mathrm{o}}(x, x_2) \text{ and } v_{\mathrm{E}}(x, x_2) \text{ of increasing degree}, \tag{17.1.36}$$

where $v_{\mathrm{E}} = x_1^2 + x_2^2$. This is the special case of Example 17.1.4 with $a_0 = 1, a_1 = 0$, $a_2 = -3$ and $a_3 = 0$. Condition (17.1.25) is satisfied and f has strong determinancy of 3. The inequalities (17.1.26) and (17.1.27) are also satisfied so $\mathrm{cod}[f] = 3$. However, we are interest in

$$\overset{3}{\overline{\mathrm{cod}}}\,[f] = \dim[\overset{3}{\overline{\mathcal{J}_2^3}}] - \dim[\overset{3}{\overline{\mathcal{T}_3(f)}}], \tag{17.1.37}$$

where

$$\overset{3}{\overline{\mathfrak{J}_2^3}} \overset{\mathrm{B}}{=} \{v_{\mathrm{E}}(x_1, x_2), v_{\mathrm{O}}(x_1, x_2)\}, \qquad \overset{3}{\overline{\mathcal{T}_3(f)}} \overset{\mathrm{B}}{=} \{v_{\mathrm{O}}(x_1, x_2)\}, \tag{17.1.38}$$

are the subspaces of $\mathfrak{J}_2^3$ and $\mathcal{T}_3(f)$ consisting of polynomials invariant under $\mathcal{S}_3$. Thus $\overset{3}{\overline{\mathrm{cod}}}\,[f] = 1$ and

$$F(x_1, x_2; c) := cv_{\mathrm{E}}(x_1, x_2) + v_{\mathrm{O}}(x_, x_2) + a[v_{\mathrm{E}}(x_1, x_2)]^2 + \text{terms in combinations of } v_{\mathrm{O}}(x_, x_2) \text{ and } v_{\mathrm{E}}(x_, x_2) \text{ of increasing degree,} \tag{17.1.39}$$

is a universal of unfolding of f and this is equivalent to

$$\tilde{F}(x_1, x_2; c) := cv_{\mathrm{E}}(x_1, x_2) + v_{\mathrm{O}}(x_, x_2) + a[v_{\mathrm{E}}(x_1, x_2)]^2. \tag{17.1.40}$$

17.2 Matrix Algebra

All matrices in this section are taken to be square and of dimension M with possibly complex elements. $\boldsymbol{A}^{\mathrm{T}}$ denotes the transpose of $\boldsymbol{A}$ and $\overline{\boldsymbol{A}}$ the matrix obtained from $\boldsymbol{A}$ by taking the conjugate complex of every element. $\boldsymbol{A}^{\dagger} := \overline{\boldsymbol{A}}^{\mathrm{T}}$ is the *Hermitian transpose* of $\boldsymbol{A}$. If $\boldsymbol{A}^{\dagger} = \boldsymbol{A}$, then $\boldsymbol{A}$ is said to be *Hermitian* and, if $\boldsymbol{A}^{\dagger} = -\boldsymbol{A}$, to be *skew-Hermitian*. If $\mathrm{Det}\{\boldsymbol{A}\} \neq 0$ then $\boldsymbol{A}$ is *non-singular*, the rows and columns of $\boldsymbol{A}$ form sets of independent vectors, none of its eigenvalues is zero and the inverse $\boldsymbol{A}^{-1}$ of $\boldsymbol{A}$ exists. Henceforth, in this section we suppose that all the matrices are non-singular. A matrix $\boldsymbol{A}$ is *orthogonal* if it is real with $\boldsymbol{A}^{-1} = \boldsymbol{A}^{\mathrm{T}}$ and it is *unitary* if $\boldsymbol{A}^{-1} = \boldsymbol{A}^{\dagger}$. Thus a real unitary matrix is orthogonal. For an orthogonal or unitary matrix $\boldsymbol{A}$, $|\mathrm{Det}\{\boldsymbol{A}\}| = 1$.

The set of distinct eigenvalues of a matrix is its *spectrum* and the *algebraic multiplicity* of an eigenvalue is its multiplicity as a root of the characteristic equation. Clearly the sum of the algebraic multiplicities of all the eigenvalues is M. Eigenvectors associated with different eigenvalues are independent (Lancaster and Tismenetsky 1985, p. 150) and the *geometric multiplicity* of an eigenvalue is the number of independent eigenvectors with which it is associated. The geometric multiplicity of an eigenvalue does not exceed its algebraic multiplicity (Lancaster and Tismenetsky 1985, p. 160) and if they are equal it is called the *degeneracy* of the eigenvalue, with the eigenvalue said to be degenerate or non-degenerate according as its degeneracy is greater than or equal to one. A matrix is called *simple* if, for every eigenvalue, the geometric multiplicity is equal to the algebraic multiplicity; meaning that it has M independent eigenvectors. Non-simple matrices are called *defective*. If an eigenvalue of a simple matrix is non-degenerate then the eigenvector associated with it is (to within a multiplicative factor) uniquely given; if it has degeneracy ω then the eigenvectors associated with it span a subspace of dimension ω. Any set of vectors spanning this subspace can be chosen as the eigenvectors for this eigenvalue.

Matrices $\boldsymbol{A}$ and $\boldsymbol{A}'$ are *similar* if there exists a matrix $\boldsymbol{U}$ such that

$$\boldsymbol{A}' = \boldsymbol{U}^{-1}\boldsymbol{A}\boldsymbol{U}. \tag{17.2.1}$$

Such matrices have the same spectrum and if $\boldsymbol{x}$ is a right eigenvector of $\boldsymbol{A}$ associated with the eigenvalue Λ then $\boldsymbol{x}' := \boldsymbol{U}^{-1}\boldsymbol{x}$ is a right eigenvector of $\boldsymbol{A}'$ also associated with Λ (Lancaster and Tismenetsky 1985, p. 152). More generally if

$$\boldsymbol{A}' = \mu\boldsymbol{U}^{-1}\boldsymbol{A}\boldsymbol{U}, \tag{17.2.2}$$

for some scalar μ, then the eigenvalues of $\boldsymbol{A}'$ are μ times the eigenvalues of $\boldsymbol{A}$.

17.2.1 Diagonalizability

A matrix is said to be *diagonalizable* if it is similar to a diagonal matrix. The matrix $\boldsymbol{A}$ is diagonalizable if and only if it is simple (Lancaster and Tismenetsky 1985, p. 146), with

$$\boldsymbol{X}^{-1}\boldsymbol{A}\boldsymbol{X} = \boldsymbol{\Lambda}, \tag{17.2.3}$$

$$\boldsymbol{A} = \boldsymbol{X}\boldsymbol{\Lambda}\boldsymbol{X}^{-1}, \tag{17.2.4}$$

where $\boldsymbol{\Lambda}$ is the diagonal matrix with eigenvalues along the diagonal and the columns of $\boldsymbol{X}$ are the right eigenvectors ordered to make the correct associations with the eigenvalues. The rows of $\boldsymbol{Y} := \boldsymbol{X}^{-1}$ are the corresponding left eigenvectors.

A matrix $\boldsymbol{A}$ for which $[\boldsymbol{A}, \boldsymbol{A}^{\dagger}] = 0$ is said to be *normal*; unitary, Hermitian, skew-Hermitian and diagonal matrices are all normal. A matrix is normal if and only if it is diagonalizable using a unitary matrix (Lancaster and Tismenetsky 1985, p. 175); so that (17.2.3) becomes $\boldsymbol{X}^{\dagger}\boldsymbol{A}\boldsymbol{X} = \boldsymbol{\Lambda}$. Not all matrices are unitarily diagonalizable, but they are all unitarily similar to an upper triangular matrix (Lancaster and Tismenetsky 1985, p. 176).

It should be noted that diagonalizability is stronger than simply the condition that the matrix can be made diagonal by some product with other matrices; for a non-singular matrix this can be done in many ways.

Example 17.2.1 Let $\boldsymbol{A}$, $\boldsymbol{B}$ and the products $\boldsymbol{AB}$ and $\boldsymbol{BA}$ all be simple and $\boldsymbol{X}_{AB}$ and $\boldsymbol{X}_{BA}$ be, respectively, the matrices containing the right eigenvectors of $\boldsymbol{AB}$ and $\boldsymbol{BA}$ as columns with diagonal matrices $\boldsymbol{\Lambda}_{AB}$ and $\boldsymbol{\Lambda}_{BA}$ containing the respective sets of eigenvalues. Then it is simple to show that

$$\boldsymbol{\Lambda}_{AB}\boldsymbol{X}_{AB}^{-1}\boldsymbol{A}\boldsymbol{X}_{BA} = \boldsymbol{X}_{AB}^{-1}\boldsymbol{A}\boldsymbol{X}_{BA}\boldsymbol{\Lambda}_{BA}, \quad \boldsymbol{\Lambda}_{BA}\boldsymbol{X}_{BA}^{-1}\boldsymbol{B}\boldsymbol{X}_{AB} = \boldsymbol{X}_{BA}^{-1}\boldsymbol{B}\boldsymbol{X}_{AB}\boldsymbol{\Lambda}_{AB}, \tag{17.2.5}$$

from which it follows that $\boldsymbol{\Lambda}_{AB} = \boldsymbol{\Lambda}_{BA}$ and (unless these matrices are just some multiple of the unit matrix) that $\boldsymbol{X}_{BA}^{-1}\boldsymbol{B}\boldsymbol{X}_{AB}$ and $\boldsymbol{X}_{AB}^{-1}\boldsymbol{A}\boldsymbol{X}_{BA}$ are diagonal.

The complete eigen-problem can be summarized in the equations[7]

$$\begin{aligned} \boldsymbol{x}_m &:= \boldsymbol{X}|m\rangle, & \boldsymbol{y}_m &:= [\langle m|\boldsymbol{Y}]^{\mathrm{T}}, \\ \delta^{\mathrm{Kr}}(m-m')\Lambda_m &:= \langle m|\boldsymbol{\Lambda}|m'\rangle, \\ \boldsymbol{A}\boldsymbol{x}_m &= \boldsymbol{x}_m\Lambda_m, & [\boldsymbol{y}_m]^{\mathrm{T}}\boldsymbol{A} &= \Lambda_m[\boldsymbol{y}_m]^{\mathrm{T}}, \\ \boldsymbol{x}_m \cdot \boldsymbol{y}_{m'} &= \delta^{\mathrm{Kr}}(m-m'). \end{aligned} \tag{17.2.6}$$

For a normal matrix the eigenvectors are orthonormal, meaning that $\boldsymbol{y}_m = \overline{\boldsymbol{x}}_m$.

A matrix containing as columns a complete set of right eigenvectors of $\boldsymbol{A}$ is called an *eigenmatrix* of $\boldsymbol{A}$. A simple matrix has at least one eigenmatrix. Let $\mathbb{S}_{\mathrm{E}}(\boldsymbol{A})$ be the set of eigenmatrices of $\boldsymbol{A}$. The eigenvalues of $\boldsymbol{A}$ are all non-degenerate if and only if the eigenmatrix is unique, that is $\mathbb{S}_{\mathrm{E}}(\boldsymbol{A})$ has exactly one element. Otherwise there will be a set of eigenmatrices formed by taking different combinations of columns in the degenerate subspaces.

17.2.2 Commutativity

For the simple (non-identity, non-singular) matrices $\boldsymbol{A}$, $\boldsymbol{B}$ and $\boldsymbol{C}$:

(i) $[\boldsymbol{B}, \boldsymbol{C}] = \boldsymbol{0}$ if and only if $\boldsymbol{B}$ and $\boldsymbol{C}$ have a complete set of common eigenvectors; that is $\mathbb{S}_{\mathrm{E}}(\boldsymbol{B}, \boldsymbol{C})$, denoting the set of common eigenmatrices of $\boldsymbol{B}$ and $\boldsymbol{C}$, has at least one member (Lancaster and Tismenetsky 1985, p. 420). If $\boldsymbol{B}$ (say) has no degenerate eigenvalues then $\mathbb{S}_{\mathrm{E}}(\boldsymbol{B}, \boldsymbol{C})$ has exactly one member; though this does not mean that $\boldsymbol{C}$ has no degenerate eigenvalues.

(ii) In general non-commuting matrices $\boldsymbol{B}$ and $\boldsymbol{C}$ can have some common eigenvectors, but if $\boldsymbol{B}\boldsymbol{C} = \alpha\,\boldsymbol{C}\boldsymbol{B}$ for some scalar $\alpha \neq 1$ (*called* α*–related*) they will have none. A particular case is anticommuting matrices for which $\alpha = -1$.

(iii) All, none or exactly one of $[\boldsymbol{A}, \boldsymbol{B}] = \boldsymbol{0}$, $[\boldsymbol{A}, \boldsymbol{C}] = \boldsymbol{0}$ and $[\boldsymbol{A}, \boldsymbol{B}\boldsymbol{C}] = \boldsymbol{0}$ holds.

(iv) If $[\boldsymbol{A}, \boldsymbol{B}] = \boldsymbol{0}$ and $[\boldsymbol{A}, \boldsymbol{C}] = \boldsymbol{0}$ then:

  (a) Every non-degenerate eigenvector of $\boldsymbol{A}$ is an eigenvector of $\boldsymbol{B}$ and $\boldsymbol{C}$.

  (b) If every eigenvalue of $\boldsymbol{A}$ is non-degenerate then $[\boldsymbol{B}, \boldsymbol{C}] = \boldsymbol{0}$. But not conversely; that is, $[\boldsymbol{B}, \boldsymbol{C}] = \boldsymbol{0}$ does not imply that $\boldsymbol{A}$, or indeed $\boldsymbol{B}$ or $\boldsymbol{C}$, has no degenerate eigenvalues.

  (c) If $\boldsymbol{B}$ and $\boldsymbol{C}$ are α–related all eigenvalues of $\boldsymbol{A}$ are degenerate.

---

[7] Here and elsewhere in the book it is convenient to use the notation $\langle i|\boldsymbol{F}|j\rangle$ to denote the i, jth element of any matrix $\boldsymbol{F}$ with its ith row and jth column denoted respectively by $\langle i|\boldsymbol{F}$ and $\boldsymbol{F}|j\rangle$.

17.2.3 Reducibility

Let e_m be the unit vector with $\langle j|e_m = \delta^{\text{Kr}}(m-j)$. A *permutation matrix* $\boldsymbol{P}$ has columns consisting of the M unit vectors $\{e_m\}$ in some order. $\boldsymbol{P}$ is clearly orthogonal (and unitary). Premultiplication and postmutiplication of a matrix $\boldsymbol{A}$ by $\boldsymbol{P}$ permutes the rows and columns respectively. A matrix $\boldsymbol{A}$ is *reducible* if there exists a permutation matrix $\boldsymbol{P}$ such that

$$\boldsymbol{P}^{\text{T}}\boldsymbol{A}\boldsymbol{P} = \begin{pmatrix} \boldsymbol{A}_{11} & \boldsymbol{A}_{12} \\ \boldsymbol{0} & \boldsymbol{A}_{22} \end{pmatrix} \tag{17.2.7}$$

where the diagonal blocks $\boldsymbol{A}_{11}$ and $\boldsymbol{A}_{22}$ are square and of dimension less than M. If such a permutation matrix does not exist then $\boldsymbol{A}$ is said to be *irreducible*. The eigenvalues of $\boldsymbol{A}$, included according to their algebraic multiplicities, are the eigenvalues of $\boldsymbol{A}_{11}$ and $\boldsymbol{A}_{22}$. If $\boldsymbol{A}_{11}$ or $\boldsymbol{A}_{22}$ are themselves reducible they can be reduced by permutation to generalize (17.2.7) to a form with diagonal blocks which are irreducible.

It can be shown (Gantmacher 1959, p. 63) that a matrix is irreducible if and only if, for any pair of indices m and m', there exists a positive integer n such that $\langle m|\boldsymbol{V}^n|m'\rangle > 0$.

17.2.4 Theorems of Perron and Frobenius

Here we restrict attention to a real simple matrix $\boldsymbol{V}$, which is either positive or non-negative. Clearly a positive matrix is irreducible and for such, according to a theorem of Perron (1907) (Gantmacher 1959, p. 64):

Theorem 17.2.1 *The eigenvalue of a positive matrix* $\boldsymbol{V}$ *of largest magnitude is positive and non-degenerate with associated positive eigenvector.*

This is an important result for statistical mechanics since the transfer matrices of many systems, for example the Ising model, are positive matrices.

However, other models, in particular vertex models and models with some kind of first-neighbour exclusion, have transfer matrices which are merely non-negative. Then according to a theorem of Frobenius (1912) (Gantmacher 1959, p. 65):

Theorem 17.2.2 *A non-negative matrix* $\boldsymbol{V}$*, which is also irreducible, has a (real and positive) eigenvalue equal to the* spectral radius

$$\mathcal{SR}(\boldsymbol{V}) := \max_{\{m\}} |\Lambda_m|, \tag{17.2.8}$$

with associated positive eigenvector. The set of points in a complex plane which represent the spectrum is invariant under rotation about the origin with the smallest angle for which this is the case being $2\pi/p$*, where* $p > 0$ *is an integer called the* period *of the matrix.*

The matrix is said to be *periodic* or *aperiodic* according to whether $p > 1$ or $p = 1$. The non-negative matrix $\boldsymbol{V}$ is said to be *primitive* if $\boldsymbol{V}^n$ is positive for some positive integer n. Clearly a primitive matrix must be irreducible and it can be shown (Gantmacher 1959, p. 97) that $\boldsymbol{V}$ is primitive if and only if it is aperiodic; the period of a matrix is also call its *index of imprimitivity*. Theorem 17.2.1 applies with the restriction to a 'positive matrix' replaced by that to a 'primitive matrix'.

For a periodic matrix, p must be a divisor of M and the eigenvalues fall into sets

$$\Lambda_{sp+1}\exp(2\pi\mathrm{i}\,k/p), \qquad k = 0, 1, \dots, p-1, \qquad s = 0, 1, \dots, r-1, \tag{17.2.9}$$

where $M = rp$, with the set with $s = 0$ lying on the spectral circle. It follows that

$$\mathrm{Trace}\{\boldsymbol{V}^n\} = \Big[\sum_{s=0}^{r-1}\Lambda_{sp+1}\Big]\Big[\sum_{k=0}^{p-1}\exp(2\pi\mathrm{i}\,nk/p) = \widetilde{\delta^{\mathrm{Kr}}}(p;n)\sum_{s=0}^{r-1}\Lambda_{sp+1}, \tag{17.2.10}$$

where $\widetilde{\delta^{\mathrm{Kr}}}(p;n)$ is given by (18.1.10). The matrix $\boldsymbol{V}^n$ is traceless unless n is a multiple of p.[8]

In the characteristic equation

$$\mathrm{Det}[\Lambda\boldsymbol{I} - \boldsymbol{V}] := \sum_{m=0}^{M}\nu_m\Lambda^m = 0 \tag{17.2.11}$$

the coefficient ν_m is a product of $M - m$ elements of $\boldsymbol{V}$. Since, from Theorem 17.2.2, the spectrum is invariant under rotation through a smallest angle $2\pi/p$, it must be the case that $\exp(2\pi\mathrm{i}\,[M-m]/p) = 1$, for all m with $\nu_m \neq 0$. So p is the greatest common divisor of the set $\{M - m\}$, for all m with $\nu_m \neq 0$.

There remains the case where the non-negative matrix $\boldsymbol{V}$ is reducible. We first note that if the eigenvalue $\Lambda_1 = \mathcal{SR}(\boldsymbol{V})$ has degeneracy $\omega > 1$ then $\boldsymbol{V}$ must be reducible. Suppose that it is reduced in the form which is a generalisation of (17.2.7) with irreducible diagonal blocks $\boldsymbol{V}_{jj}$, $j = 1, 2, \dots, q$. Since the eigenvalues of $\boldsymbol{V}$, counting according to their degeneracies coincide with the eigenvalues of the diagonal blocks, $q \geq \omega$ with ω blocks (let them be $\boldsymbol{V}_{jj}$, $j = 1, 2, \dots, \omega$) having spectral radius $\mathcal{SR}(\boldsymbol{V})$ and a non-degenerate eigenvalue Λ_1. The eigenvalues on the spectral circle of $\boldsymbol{V}$ are just the total of all the eigenvalues on the spectral circles of $\boldsymbol{V}_{jj}$, $j = 1, 2, \dots, \omega$.

[8] For $n = 1$ this conclusion is part of Frobenius' theorem, where it shown that the elements of the matrix can be permuted to give a set of square zero matrices along the diagonal (Gantmacher 1959, p. 65).

17.2.5 Direct Products and Traces

In Chap. 12 we study matrices like $\boldsymbol{Q}_k$, whose rows and columns are indexed by vectors $\boldsymbol{\varsigma} := (\varsigma_1, \varsigma_2, \ldots, \varsigma_N)$, where $\varsigma_\ell = \pm 1$, and for which

$$\langle \boldsymbol{\varsigma} | \boldsymbol{Q}_k | \boldsymbol{\varsigma}' \rangle = \text{Trace}\left\{ \prod_{\ell=1}^{N} \boldsymbol{T}_k \begin{pmatrix} \varsigma'_\ell \\ \varsigma_\ell \end{pmatrix} \right\}. \tag{17.2.12}$$

Then it is straightforward to show (Lloyd 1953) that

$$\langle \boldsymbol{\varsigma} | \boldsymbol{Q}_1 \boldsymbol{Q}_2 | \boldsymbol{\varsigma}' \rangle = \text{Trace}\left\{ \prod_{\ell=1}^{N} \sum_{q=\pm 1} \boldsymbol{T}_1 \begin{pmatrix} q \\ \varsigma_\ell \end{pmatrix} \otimes \boldsymbol{T}_2 \begin{pmatrix} \varsigma'_\ell \\ q \end{pmatrix} \right\}. \tag{17.2.13}$$

It should be noted that there is no necessary relationship between the dimension of the matrices $\boldsymbol{Q}_k$ and $\boldsymbol{T}_k$, with, in the latter case, a possible variation with k. Suppose $\boldsymbol{T}_k$ is of dimension L_k with rows and columns indexed by i_k and j_k respectively. Then the most convenient way to index the rows an columns of a direct product is using double indices.

$$\left\langle {i_2 \atop i_1} \right| \boldsymbol{T}_1 \begin{pmatrix} q \\ \varsigma_\ell \end{pmatrix} \otimes \boldsymbol{T}_2 \begin{pmatrix} \varsigma'_\ell \\ q \end{pmatrix} \left| {j_2 \atop j_1} \right\rangle = \langle i_1 | \boldsymbol{T}_1 \begin{pmatrix} q \\ \varsigma_\ell \end{pmatrix} | j_1 \rangle \langle i_2 | \boldsymbol{T}_2 \begin{pmatrix} \varsigma'_\ell \\ q \end{pmatrix} | j_2 \rangle. \tag{17.2.14}$$

17.2.6 Defective Matrices

So far in the discussion we have considered only simple matrices where the geometric multiplicity of each eigenvalue is equal to its algebraic multiplicity. Consider now the case where a particular eigenvalue Λ has algebraic multiplicity ω_{A} and geometric multiplicity $\omega_{\text{G}} < \omega_{\text{A}}$. For this particular eigenvalue the reduction to diagonal form is replaced by ω_{G} blocks. These may be of dimension one, and simply contain the eigenvalue, or be of dimension greater than one, in which case they have Λ along the diagonal and one along the upper diagonal. The sum of the dimensions of the blocks is ω_{A}. Such a matrix is said to be of *Jordan normal form* and it can be shown (Lancaster and Tismenetsky 1985, Sect. 6.5) that:

Theorem 17.2.3 *Any complex square matrix is similar to a matrix of Jordan normal form, with the latter matrix being unique apart from the ordering of the blocks.*

It is also straightforward to show that:

Theorem 17.2.4 *If* $\boldsymbol{A}$ *and* $\boldsymbol{B}$ *are commuting complex square matrices, then every non-degenerate eigenvector of* $\boldsymbol{A}$ *is an eigenvector of* $\boldsymbol{B}$.

17.2.7 Groups of Matrices

The *orthogonal group* of n-dimensional orthogonal matrices is denoted by $\mathsf{O}(n)$. The *special orthogonal group* of n-dimensional orthogonal matrices with determinant $+1$ is denoted by $\mathsf{SO}(n)$. The effect of $\boldsymbol{A} \in \mathsf{O}(n)$ is a distance-preserving mapping on n-dimensional Euclidian space which leave the origin invariant. If $\mathrm{Det}\{\boldsymbol{A}\} = 1$, that is $\boldsymbol{A} \in \mathsf{SO}(n)$, then $\boldsymbol{A}$ is a rotation. If $\mathrm{Det}\{\boldsymbol{A}\} = -1$ then it is a combination of a rotation and a reflection. The *unitary group* of n-dimensional unitary matrices is denoted by $\mathsf{U}(n)$. $\mathsf{U}(1)$ is just multiplication by complex numbers of unit modulus, which can be re-expressed in two-dimensional matrix form showing that $\mathsf{U}(1)$ is isomorphic to $\mathsf{SO}(2)$.

17.3 Groups and Representations

A generic designation of a group[9] is used in the form $\mathcal{Q} := \{\mathrm{Q}^{(0)}, \mathrm{Q}^{(1)}, \mathrm{Q}^{(2)}, \ldots\}$ with the group operation represented multiplicatively and $\mathrm{Q}^{(0)} := \mathrm{I}$, the identity element. A *generating set* of a group is a subset such that every element of the group can be expressed as the combination (under the group operation) of finitely many elements of the subset and their inverses and we use the notation $\mathcal{Q} := \langle \mathrm{Q}_1, \mathrm{Q}_2, \ldots, \mathrm{Q}_k \rangle$ to denote the group in terms of its set of generators. Apart from the case of the n-vector model in Sect. 3.3, we are concerned simply with finite groups and the order of $\mathcal{Q}$ is denoted by $\mathfrak{n}(\mathcal{Q})$.

The use of group representation theory in statistical mechanics arises because of the presence of certain symmetries in the Hamiltonian. Specifically:

(A) *Translations, rotations or reflections of the lattice.* The rotations and reflections of a lattice form a *point group* which leaves one point P fixed.[10] A subgroup of all point groups is:

- A *cyclic group* $\mathcal{C}_n := \langle \mathrm{C}_n \rangle = \{\mathrm{I}, \mathrm{C}_n, \mathrm{C}_n^2, \ldots, \mathrm{C}_n^{n-1}\}$, where C_n is a rotation through $2\pi/n$ radians about an axis A through the specified point.[11]

For infinite three-dimensional lattices and two-dimensional lattices with A orthogonal to the plane of the lattice the only n-fold rotational symmetries consistent with translational symmetry are for $n = 1, 2, 3, 4, 6$ (Falicov 1966, pp. 79–81). However, we shall be interested in transfer matrix theory for lattices,

---

9 For the discussion in this section a working knowledge of group theory is assumed. For more extensive treatments of symmetry groups and their representations and for proofs of theorems see, for example, McWeeny (1963), Falicov (1966), Leech and Newman (1969), Serre (1977) or Lavis and Bell (1999b, Sect. A.3).

10 The choice of point is arbitrary since they are related by translation.

11 Any group with a single generator is a cyclic group. In this case the cyclic group is also a rotation group, meaning that it is physically realized as a set of rotation in three-dimensional space.

with periodic boundary conditions, forming a cylinder, where A is the axis of the cylinder (Sect. 17.3.1). In this case n can be arbitrarily large, so the discussion is formulated in this context. The other uniaxial point groups are:

- $\mathcal{C}_{n\text{h}} := \langle \mathrm{C}_n, \Sigma_\text{h} \rangle$, where Σ_h is a reflection in a plane Π_h through P and orthogonal to A.
- $\mathcal{J}_n := \langle \mathrm{J}_n \rangle$, where $\mathrm{J}_n = \mathrm{C}_n \Sigma_\text{h}$. This is the group of improper rotations.[12] Clearly $\mathrm{J}_n^m = \mathrm{C}_n^m$ when m is even and $\mathrm{J}_n^m = \mathrm{C}_n^m \Sigma_\text{h}$, when m is odd. Thus, if n is odd $\mathcal{J}_n = \mathcal{C}_{n\text{h}}$, but if n is even $\mathcal{J}_n$ is a proper subgroup of $\mathcal{C}_{n\text{h}}$.
- $\mathcal{C}_{n\text{v}} := \langle \mathrm{C}_n, \Sigma_0 \rangle$, where Σ_0 is a reflection in a plane Π_0 containing the axis P. This group contains reflections $\Sigma_m := \mathrm{C}_n^m \, \Sigma_0 \, \mathrm{C}_n^{-m}$ in the planes $\Pi_m := \mathrm{C}_n^m \Pi_0$, for $m = 0, 1, \ldots, \frac{n}{2} - 1$, when n is even and $m = 0, 1, \ldots, n-1$, when n is odd.

This completes the uniaxial point groups.[13] Now return to the cyclic group $\mathcal{C}_n$ and add an axis A_0 through P and in the plane Π_h about which there is a two-fold rotation U_0. This gives:

- The *dihedral group* $\mathcal{D}_n := \langle \mathrm{C}_n, \mathrm{U}_0 \rangle$. In a way similar to that for $\mathcal{C}_{n\text{v}}$, $\mathcal{D}_n$ contains two-fold rotations $\mathrm{U}_m := \mathrm{C}_n^m \mathrm{U}_0 \mathrm{C}_n^{-m}$ about the axes $\mathsf{A}_m := \mathrm{C}_n^m \mathsf{A}_0$, for $m = 0, 1, \ldots, \frac{n}{2} - 1$, when n is even and $m = 0, 1, \ldots, n-1$, when n is odd.

Reflections can now be added to the dihedral group:

- $\mathcal{D}_{n\text{h}} := \langle \mathrm{C}_n, \mathrm{U}_0, \Sigma_\text{h} \rangle$.
- $\mathcal{D}_{n\text{d}} := \langle \mathrm{C}_n, \mathrm{U}_0, \Sigma_0' \rangle$, where Σ_0' is a reflection through a plane Π_0' which bisects the angle between planes Π_0 and Π_1 defined for $\mathcal{C}_{n\text{v}}$. Again it is clear that the group contains planes of reflection bisecting the angles between the planes Π_m and Π_{m+1}, for $m = 0, 1, \ldots, \frac{n}{2} - 1$, when n is even and $m = 0, 1, \ldots, n-1$, when n is odd.

This completes the set of point groups which have at most one axis of n-fold rotation with $n > 2$. It also encompasses, with $n \le 6$, twenty-seven of the thirty-two crystallographic point groups.[14] The remaining point groups for three-dimensional lattices are the cubic groups with more than one n-fold axis with $n > 2$. There are five of these with $n \le 6$, thus completing the set of thirty-two crystallographic point groups (Falicov 1966, pp. 88–92).

All the group operations defined here are *isometries*, that is, they preserve distances, and an isometry is said to be *direct* if it preserves orientation or ordering.

---

[12] $\mathcal{S}_n$ is usually used in place of $\mathcal{J}_n$, but this clashes with the notation for the symmetric group.

[13] With $n = 1, 2, 3, 4, 6$, and given that $\mathcal{C}_{1\text{v}} = \mathcal{C}_{1\text{h}}$, $\mathcal{J}_1 = \mathcal{C}_{1\text{h}}$ and $\mathcal{J}_3 = \mathcal{C}_{3\text{h}}$ this gives $4 \times 5 - 3 = 17$ of the crystallographic point groups.

[14] With $n = 1, 2, 3, 4, 6$, and given that $\mathcal{D}_1 = \mathcal{C}_2$, $\mathcal{D}_\text{h} = \mathcal{C}_{2\text{v}}$, $\mathcal{D}_{1\text{d}} = \mathcal{C}_{2\text{h}}$ and $\mathcal{D}_{4\text{d}}$ and $\mathcal{D}_{6\text{d}}$ are excluded because they have respectively improper rotations with $n = 8$ and $n = 12$, this gives $3 \times 5 - 5 = 10$ additional crystallographic point groups.

In this context we can think of this in terms of the ordering (clockwise or anticlockwise) of sites round a lattice face with respect to a particular (unaltered direction). An isometry not preserving this property is called *indirect*. In the case of two-dimensional lattices with the only group operations in the plane of the lattice the symmetry group consists of translations and the point group $\mathcal{C}_{n\mathrm{v}}$, for some n. Translations and rotations are direct isometries and reflections are indirect isometries.

(B) *Operations on the microstates.* The largest group that can operate on a system of ν discrete microstates is the symmetric group $\mathcal{S}_\nu$ of all permutations. This applies to the zero-field standard ν-state Potts model (Sect. 3.6.4). In other cases the symmetry may be reduced, as for the zero-field ν-state planar clock model (Sect. 3.6) which is invariant under the cyclic group $\mathcal{C}_\nu$. The only case of continuous symmetry which will concern us is the n-vector model of Sect. 3.3 (and the special case $n = 2$ for the XY model in Sect. 3.6.1) for which the symmetry group in $\mathcal{O}(n)$ the group of orthogonal n-dimensional matrices (Sect. 17.2.7).

(C) *Operations on the independent couplings.* The primary instance of this is for operations on the direction of the magnetic field. In particular its reversal of direction, which for the Ising model simply means a change of sign.

The total symmetry group $\mathcal{Q}$ will, in general, be the direct product of the groups representing the different symmetries. However, there are exceptions to this, so that if $\mathcal{F} \times \mathcal{G}$ is the the direct production of two of the three symmetry types listed above, the Hamiltonian is invariant under only a proper subgroup of $\mathcal{F} \times \mathcal{G}$. The simplest example of this occurs in the spin-$\frac{1}{2}$ Ising model (Sect. 3.5.1) where the group of operations on the states and on the field consist of inversion (change of sign), but where the Hamiltonian is invariant only when the spins and the field are inverted *simultaneously*.

17.3.1 Representations

For a discussion of representations the idea of conjugacy is important. Two elements $\mathrm{Q}^{(\mathrm{s})}, \mathrm{Q}^{(\mathrm{s}')} \in \mathcal{Q}$ are said to be *conjugate* if there exists an element $\mathrm{Q}^{(\mathrm{s}'')} \in \mathcal{Q}$ such that $\mathrm{Q}^{(\mathrm{s})} = [\mathrm{Q}^{(\mathrm{s}'')}]^{-1}\mathrm{Q}^{(\mathrm{s}')}\mathrm{Q}^{(\mathrm{s}'')}$. The set of all elements conjugate to a particular element is called a *conjugacy class* of $\mathcal{Q}$ and any two elements are in the same conjugacy class if and only if they are conjugate; $\mathcal{Q}$ decomposes into a set of $\kappa(\mathcal{Q})$ mutually exclusive conjugacy classes $\mathcal{Q}_k, k = 1, 2, \ldots, \kappa$. Any element which commutes with all the other elements of the group is conjugate only to itself; if $\mathcal{Q}$ is commutative each element is a conjugacy class.

If, for all s, a square $M \times M$ dimensional matrix $\boldsymbol{R}(\mathrm{Q}^{(\mathrm{s})}) := \boldsymbol{Q}^{(\mathrm{s})}$ can be assigned to $\mathrm{Q}^{(\mathrm{s})}$ such that $\boldsymbol{Q}^{(\mathrm{s})}\boldsymbol{Q}^{(\mathrm{s}')} = \boldsymbol{Q}^{(\mathrm{s}'')}$ whenever $\mathrm{Q}^{(\mathrm{s})}\mathrm{Q}^{(\mathrm{s}')} = \mathrm{Q}^{(\mathrm{s}'')}$, then the set of matrices $\mathcal{R}(\mathcal{Q}) := \{\boldsymbol{I}, \boldsymbol{Q}^{(1)}, \ldots, \boldsymbol{Q}^{(\mathrm{n}-1)}\}$ is an M-dimensional *representation* of $\mathcal{Q}$. It must, of course, be the case that $\boldsymbol{R}(\mathrm{I}) = \boldsymbol{I}$, the M-dimensional unit matrix, and $\boldsymbol{R}([\mathrm{Q}^{(\mathrm{s})}]^{-1}) = [\boldsymbol{Q}^{(\mathrm{s})}]^{-1}$. For ease of notation $\boldsymbol{Q}^{(0)}$ is taken to denote $\boldsymbol{I}$, where necessary. It is clear that, for any group, one possible assignment is $\boldsymbol{R}(\mathsf{Q}^{(\mathsf{s})}) = (1)$, for ever s. This representation (usually denoted by A) is called the *one-dimensional symmetric representation*. If two representations, of the same dimension, $\mathcal{R}(\mathcal{Q})$ and $\mathcal{R}'(\mathcal{Q})$ are related by the similarity transformation $\boldsymbol{R}'(\mathsf{Q}^{(\mathsf{s})}) = \boldsymbol{U}^{-1}\boldsymbol{R}(\mathsf{Q}^{(\mathsf{s})})\boldsymbol{U}$, for all s, then the representations are said to be *equivalent*. A representation $\mathcal{R}(\mathcal{Q})$ is said to be *reducible*[15] if it is equivalent to another representation $\mathcal{R}'(\mathcal{Q})$ whose matrices have the block-diagonal form

$$\boldsymbol{R}'(\mathsf{Q}^{(\mathsf{s})}) = \boldsymbol{R}_1(\mathsf{Q}^{(\mathsf{s})}) \oplus \boldsymbol{R}_2(\mathsf{Q}^{(\mathsf{s})}) \oplus \boldsymbol{R}_3(\mathsf{Q}^{(\mathsf{s})}) \oplus \cdots, \tag{17.3.1}$$

where the blocks are of the same, non-zero, dimension for all $\mathsf{Q}^{(\mathsf{s})}$. It is clear that, for any i, the set of matrices $\boldsymbol{R}_i(\mathsf{Q}^{(\mathsf{s})})$, $\mathsf{s} = 0, 1, \ldots, \mathfrak{n}-1$ is itself a representation. Any representation which is not reducible is call *irreducible* and it follows that any reducible representation is equivalent to a representation which has matrices consisting of blocks which are irreducible representations.

Unitary matrices, yielding *unitary representations*, play a particularly important role in the theory of group representations. In particular, a representation in terms of permutation matrices is unitary. The following theorems are of importance:

Theorem 17.3.1 *Any representation of a finite group in terms of non-singular matrices is equivalent to a unitary representation.*

Theorem 17.3.2 *Any two equivalent unitary representations can be related by a unitary transformation.*

We shall henceforth assume that the matrices of all representations are unitary. In particular it follows from Theorem 17.3.1 that all the irreducible representations can be expressed in terms of unitary matrices. For one-dimensional irreducible representations this form is, of course, the unique expression of the representation. A one-dimensional matrix (η) belonging to a one-dimensional representation of a group must satisfy the unitary condition $\eta^{-1} = \overline{\eta}$, from which it follows that η must be a complex root of unity. For the representation $\mathcal{R}(\mathcal{Q})$,

$$\chi(\mathsf{Q}^{(\mathsf{s})}) := \text{Trace}\{\boldsymbol{Q}^{(\mathsf{s})}\} \tag{17.3.2}$$

is the *character* of $\mathsf{Q}^{(\mathsf{s})}$ and the set of characters $\chi(\mathsf{Q}^{(\mathsf{s})})$, $\mathsf{s} = 0, 1, \ldots, \mathfrak{n}-1$ is called the *character system* of $\mathcal{Q}$ in $\mathcal{R}(\mathcal{Q})$. Since $\text{Trace}\{\boldsymbol{P}^{-1}\boldsymbol{Q}^{(\mathsf{s})}\boldsymbol{P}\} = \text{Trace}\{\boldsymbol{Q}^{(\mathsf{s})}\}$, it follows that all equivalent representations have the same character set. The same type of argument also shows that all operators of the group in the same conjugacy class have the same character. We can conveniently regard all equivalent representations as essentially the *same* representation *realized* in different forms. This representation has a character set, with a character given for each conjugacy class. An important property of a group is the collection of character sets for the irreducible representations with the useful result that:

---

[15] 'Reducibility' for a representation has a slightly different meaning from that given in (17.2.7) for a single matrix.

Theorem 17.3.3 *The number of (inequivalent) irreducible representations of a group is equal to the number of conjugacy classes in the group.*

We denote the irreducible representations of $\mathcal{Q}$ by $\Theta(\mathrm{k})$, $\mathrm{k} = 1, 2, \ldots, \kappa$, where $\Theta(1) = A$ is the one-dimensional symmetric representation, and $\mathrm{e}(\mathrm{k})$ is the dimension of $\Theta(\mathrm{k})$. If we denote the (unitary) matrix for $\mathrm{Q}^{(\mathrm{s})}$ in the irreducible representation $\Theta(\mathrm{k})$ by $\boldsymbol{R}^{(\mathrm{k})}(\mathrm{Q}^{(\mathrm{s})})$ then the character of $\mathrm{Q}^{(\mathrm{s})}$ in the representation $\Theta(\mathrm{k})$ is $\chi^{(\mathrm{k})}(\mathrm{Q}^{(\mathrm{s})}) = \mathrm{Trace}\{\boldsymbol{R}^{(\mathrm{k})}(\mathrm{Q}^{(\mathrm{s})})\}$. It is clear that

$$\chi^{(\mathrm{k})}(\mathrm{Q}^{(0)}) = \mathrm{e}(\mathrm{k}), \qquad \mathrm{k} = 1, 2, \ldots, \kappa \tag{17.3.3}$$

and, since the modulus of the trace of a unitary matrix cannot exceed its dimension,

$$|\chi^{(\mathrm{k})}(\mathrm{Q}^{(\mathrm{s})})| \le \mathrm{e}(\mathrm{k}), \qquad \mathrm{s} = 1, 2, \ldots, \mathrm{n} - 1, \qquad \mathrm{k} = 1, 2, \ldots, \kappa, \tag{17.3.4}$$

with

$$|\chi^{(\mathrm{k})}(\mathrm{Q}^{(\mathrm{s})})| = 1, \qquad \mathrm{s} = 0, 1, \ldots, \mathrm{n} - 1, \quad \text{if } \mathrm{e}(\mathrm{k}) = 1. \tag{17.3.5}$$

It can be shown that:

Theorem 17.3.4 *If the elements of* $\boldsymbol{R}^{(\mathrm{k})}(\mathrm{Q}^{(\mathrm{s})})$ *are denoted by* $R_{ij}^{(\mathrm{k})}(\mathrm{Q}^{(\mathrm{s})})$ *for* $i, j = 1, 2, \ldots, \mathrm{e}(\mathrm{k})$ *and* $\omega(\mathrm{s})$ *is the number of elements in the conjugacy class containing* $\mathrm{Q}^{(\mathrm{s})}$ *then*

$$\sum_{\mathrm{s}=0}^{n-1} \overline{R}_{ij}^{(\mathrm{k})}(\mathrm{Q}^{(\mathrm{s})}) R_{i'j'}^{(\mathrm{k}')}(\mathrm{Q}^{(\mathrm{s})}) = \frac{\mathrm{n}}{\mathrm{e}(\mathrm{k})} \delta^{\mathrm{Kr}}(\mathrm{k} - \mathrm{k}') \delta^{\mathrm{Kr}}(i - i') \delta^{\mathrm{Kr}}(j - j'), \tag{17.3.6}$$

$$\sum_{\mathrm{k}=1}^{\kappa} \overline{\chi}^{(\mathrm{k})}(\mathrm{Q}^{(\mathrm{s})}) \chi^{(\mathrm{k})}(\mathrm{Q}^{(\mathrm{s}')}) = \begin{cases} \mathrm{n}/\omega(\mathrm{s}), & \mathrm{Q}^{(\mathrm{s})} \textit{ and } \mathrm{Q}^{(\mathrm{s}')} \textit{ conjugate}, \\ 0, & \mathrm{Q}^{(\mathrm{s})} \textit{ and } \mathrm{Q}^{(\mathrm{s}')} \textit{ not conjugate}. \end{cases} \tag{17.3.7}$$

Equations (17.3.6) and (17.3.7) are the *Schur orthogonality relations*. From (17.3.6),

$$\sum_{\mathrm{s}=0}^{\mathrm{n}-1} \overline{\chi}^{(\mathrm{k})}(\mathrm{Q}^{(\mathrm{s})}\mathrm{Q}^{(\mathrm{s}')}) \chi^{(k')}(\mathrm{Q}^{(\mathrm{s})}\mathrm{Q}^{(\mathrm{s}'')}) = \frac{\mathrm{n}}{\mathrm{e}(\mathrm{k})} \delta^{\mathrm{Kr}}(\mathrm{k} - \mathrm{k}') \chi^{(\mathrm{k})}([\mathrm{Q}^{(\mathrm{s}')}]^{-1} \mathrm{Q}^{(\mathrm{s}'')}). \tag{17.3.8}$$

Since $\mathrm{Q}^{(0)} = \mathrm{I}$ is in a conjugacy class on its own, (17.3.7), with $\mathrm{s} = \mathrm{s}' = 0$, yields

$$\sum_{\mathrm{k}=1}^{\kappa} \mathrm{e}^2(\mathrm{k}) = \mathrm{n}. \tag{17.3.9}$$

Given that $\mathcal{R}(\mathcal{Q})$ is reducible, it follows from the above discussion that there exists a unitary matrix which will reduce all the matrices $\boldsymbol{Q}^{(\mathrm{s})}$ to block-diagonal form with blocks corresponding to the irreducible representations. The number of blocks $\mathfrak{m}(k)$ corresponding to $\Theta(k)$ is called the *multiplicity* of $\Theta(k)$. Since the blocks for any $\Theta(k)$ can be collected together by the operation of a permutation matrix, which is itself unitary, there exists a unitary matrix $\boldsymbol{U}$ such that, for each $\boldsymbol{Q}^{(s)}$,

$$\boldsymbol{U}^{\dagger}\boldsymbol{Q}^{(s)}\boldsymbol{U} = \mathring{\boldsymbol{Q}}^{(s)} := \mathring{\boldsymbol{Q}}^{(s)}(1) \oplus \mathring{\boldsymbol{Q}}^{(s)}(2) \oplus \cdots \oplus \mathring{\boldsymbol{Q}}^{(s)}(\kappa) \tag{17.3.10}$$

where $\mathring{\boldsymbol{Q}}^{(s)}(k)$ is the block corresponding to the representation $\Theta(k)$. It is of dimension $w(k) = \mathfrak{m}(k)e(k)$ and itself consists of $\mathfrak{m}(k)$ identical blocks of dimension $e(k)$. That is

$$\mathring{\boldsymbol{Q}}^{(s)}(k) = \bigoplus_{\mathfrak{m}(k)} \boldsymbol{R}^{(k)}(Q^{(s)}) := \overbrace{\boldsymbol{R}^{(k)}(Q^{(s)}) \oplus \cdots \oplus \boldsymbol{R}^{(k)}(Q^{(s)})}^{\mathfrak{m}(k) \text{ times}}. \tag{17.3.11}$$

The realization of $\boldsymbol{Q}^{(s)}$ given by (17.3.10) and (17.3.11) is called its *canonical form*, with $\mathring{\boldsymbol{Q}}^{(s)}$ the *Jordan normal form* (Sect. 17.2.6) for $\boldsymbol{Q}^{(s)}$. For a one-dimensional representation $\Theta(k)$, $\mathring{\boldsymbol{Q}}^{(s)}(k)$ is diagonal with diagonal elements $\chi^{(k)}(Q^{(s)})$, which are the eigenvalues of $\mathring{\boldsymbol{Q}}^{(s)}(k)$ (and hence of $\mathring{\boldsymbol{Q}}^{(s)}$ and $\boldsymbol{Q}^{(s)}$). For a representation $\Theta(k)$ of dimension $e(k) > 1$, the matrices $\mathring{\boldsymbol{Q}}^{(s)}(k)$ are of Jordan form, with the sum of the $e(k)$ equal eigenvalues along the diagonal of $\boldsymbol{R}^{(k)}(Q^{(s)})$ giving $\chi^{(k)}(Q^{(s)})$. By taking the trace of (17.3.10) we have

$$\chi(Q^{(s)}) = \sum_{k=1}^{\kappa} \mathfrak{m}(k)\chi^{(k)}(Q^{(s)}). \tag{17.3.12}$$

It then follows from (17.3.3) and (17.3.8) that

$$\mathfrak{m}(k) = \frac{1}{\mathfrak{n}} \sum_{s=0}^{\mathfrak{n}-1} \overline{\chi}(Q^{(s)})\chi^{(k)}(Q^{(s)}). \tag{17.3.13}$$

The characters of the irreducible representations of a group are presented in the form of a character table.[16] If, therefore, $\mathcal{Q}$ is a standard group, or the direct product of standard groups, (17.3.13) can be used to determine the multiplicities of the irreducible representations in the arbitrary representation $\mathcal{R}(\mathcal{Q})$. Even if the character table of $\mathcal{Q}$ is not known we can still determine whether a particular representation is irreducible since, from (17.3.12) and (17.3.13),

$$\frac{1}{\mathfrak{n}} \sum_{s=0}^{\mathfrak{n}-1} \overline{\chi}(Q^{(s)})\chi(Q^{(s)}) = \sum_{k=1}^{\kappa} \mathfrak{m}^2(k). \tag{17.3.14}$$

[16] See the references given above or Conway et al. (1985).

The right-hand side of this equation is unity if and only if only one irreducible representation appears with multiplicity one in $\mathcal{R}(\mathcal{Q})$.

Apart from the ordering of the blocks, the canonical form for $\boldsymbol{Q}^{(s)}$ given by (17.3.10) and (17.3.11) is unique. Suppose that $\boldsymbol{W}$ is a unitary matrix such that

$$\boldsymbol{W}^{\dagger}\boldsymbol{Q}^{(s)}\boldsymbol{W} = \widetilde{\boldsymbol{Q}}^{(s)} := \widetilde{\boldsymbol{Q}}^{(s)}(1) \oplus \widetilde{\boldsymbol{Q}}^{(s)}(2) \oplus \cdots \oplus \widetilde{\boldsymbol{Q}}^{(s)}(\kappa), \qquad s = 0, 1, \ldots, n-1, \tag{17.3.15}$$

where

$$\frac{1}{n}\sum_{s=0}^{n-1} \text{Trace}\{\widetilde{\boldsymbol{Q}}^{(s)}(k)\}\overline{\chi}^{(k')}(Q^{(s)}) = \delta^{Kr}(k-k')m(k). \tag{17.3.16}$$

This represents a non-unique reduction of the matrices of $\mathcal{Q}$ to a set of κ reducible representations each of which involves only one irreducible representation. The final reduction to canonical form can be made if a set of unitary matrices $\boldsymbol{L}(k)$, $k = 1, 2, \ldots, \kappa$ can be found such that

$$\boldsymbol{L}^{\dagger}(k)\widetilde{\boldsymbol{Q}}^{(s)}(k)\boldsymbol{L}(k) = \mathring{\boldsymbol{Q}}^{(s)}(k), \qquad s = 0, 1, \ldots, n-1. \tag{17.3.17}$$

Then

$$\boldsymbol{U} = \boldsymbol{W}\boldsymbol{L}, \tag{17.3.18}$$

where

$$\boldsymbol{L} := \boldsymbol{L}(1) \oplus \boldsymbol{L}(2) \oplus \cdots \oplus \boldsymbol{L}(\kappa). \tag{17.3.19}$$

The matrix $\boldsymbol{W}$ can often be obtained in a straightforward way; computing the blocks of $\boldsymbol{L}$ is more difficult. However, the existence of the matrix $\boldsymbol{L}$, even without it being known is useful, as in the proof of the following result:

Theorem 17.3.5 *For the elements of the matrices* $\widetilde{\boldsymbol{Q}}^{(s)}(k)$ *of* (17.3.15)

$$\sum_{s=0}^{n-1} \overline{\widetilde{Q}_{ij}^{(s)}(k)\widetilde{Q}_{i'j'}^{(s)}(k')} = 0 \tag{17.3.20}$$

for all i, j, i' *and* j', *whenever* $k \neq k'$.

17.3.2 Permutation Representations and Equivalence Classes

As in Sect. 11.3 we now consider the group $\mathcal{Q} := \{I, Q^{(1)}, Q^{(2)}, \ldots, Q^{(n-1)}\}$ in relation to a set of 'states' $\boldsymbol{\tau}_m$, $m = 1, 2, \ldots, M$, such that each $Q^{(s)}$ maps a state $\boldsymbol{\tau}_m$ into another state $Q^{(s)}\boldsymbol{\tau}_m$, with $Q^{(s)}\boldsymbol{\tau}_m \neq Q^{(s)}\boldsymbol{\tau}_{m'}$ if $\boldsymbol{\tau}_m \neq \boldsymbol{\tau}_{m'}$.

The effect of the group $\mathcal{Q}$ on the states $\{\boldsymbol{\tau}_m\}$ is given, using the representation $\mathcal{R}(\mathcal{Q}) := \{\boldsymbol{I}, \boldsymbol{Q}^{(1)}, \ldots, \boldsymbol{Q}^{(\mathrm{n}-1)}\}$ of M-dimensional permutation matrices, by (11.3.1). This representation of the group $\mathcal{Q}$ is called the *permutation representation* associated with $\{\boldsymbol{\tau}_m\}$. If $M = \mathrm{n}$ and if the states can be labelled by the index $\mathrm{s} = 0, 1, \ldots, \mathrm{n}-1$ so that $\mathrm{Q}^{(\mathrm{s})}\boldsymbol{\tau}_{\mathrm{s}'} = \boldsymbol{\tau}_{\mathrm{s}''}$ whenever $\mathrm{Q}^{(\mathrm{s})}\mathrm{Q}^{(\mathrm{s}')} = \mathrm{Q}^{(\mathrm{s}'')}$, then the representation is called *regular*. It is not difficult to show that:

Theorem 17.3.6 *If the number of states $M = \mathrm{n}$ and if, for some arbitrarily chosen state denoted (without loss of generality) by $\boldsymbol{\tau}_1$, all the states $\mathrm{Q}^{(\mathrm{s})}\boldsymbol{\tau}_1$, $\mathrm{s} = 0, 1, \ldots, \mathrm{n}-1$, are distinct then the permutation representation associated with*

$$\boldsymbol{\tau}_{\mathrm{s}+1} = \mathrm{Q}^{(\mathrm{s})}\boldsymbol{\tau}_1, \qquad \mathrm{s} = 0, 1, \ldots, \mathrm{n}-1, \tag{17.3.21}$$

is regular.

It is clear that, for the regular representation, $\mathrm{Q}^{(\mathrm{s}')}\boldsymbol{\tau}_{\mathrm{s}} \neq \boldsymbol{\tau}_{\mathrm{s}}$ unless $\mathrm{s}' = 0$ and thus

$$\chi(\mathrm{Q}^{(\mathrm{s})}) = \mathrm{Trace}\{\boldsymbol{Q}^{(\mathrm{s})}\} = \mathrm{n}\,\delta^{\mathrm{Kr}}(\mathrm{s}), \qquad \mathrm{s} = 0, 1, \ldots, \mathrm{n}-1. \tag{17.3.22}$$

Then, from (17.3.3) and (17.3.13), $\mathrm{m}(\mathrm{k}) = \mathrm{e}(\mathrm{k})$; each irreducible representation occurs in the regular representation with a multiplicity equal to its dimension.

An *equivalence class* is a subset of the states $\boldsymbol{\tau}_m, m = 1, 2, \ldots, M$, which is closed under the operations of the group $\mathcal{Q}$ and is not further divisible into closed subsets. Suppose that the states are ordered in equivalence classes $\mathfrak{E}(\alpha)$, $\alpha = 1, 2, \ldots, \gamma$, where $\mathfrak{E}(\alpha) := \{\boldsymbol{\tau}_1(\alpha), \boldsymbol{\tau}_2(\alpha), \ldots, \boldsymbol{\tau}_{\mathrm{p}(\alpha)}(\alpha)\}$. The following result is useful in determining these equivalence classes:

Theorem 17.3.7 *The equivalence class $\mathfrak{E}(\alpha)$ can be generated by applying all the elements of $\mathcal{Q}$ to any one state of $\mathfrak{E}(\alpha)$. The number of states $\mathrm{p}(\alpha)$ in $\mathfrak{E}(\alpha)$ is a divisor of n.*

It follows from (11.3.1) that each $\boldsymbol{Q}^{(\mathrm{s})}$ has the block-diagonal form

$$\boldsymbol{Q}^{(\mathrm{s})} = \boldsymbol{Q}^{(\mathrm{s})}(1) \oplus \boldsymbol{Q}^{(\mathrm{s})}(2) \oplus \cdots \oplus \boldsymbol{Q}^{(\mathrm{s})}(\gamma), \tag{17.3.23}$$

with $\mathcal{R}(\alpha; \mathcal{Q}) := \{\boldsymbol{I}(\alpha), \boldsymbol{Q}^{(1)}(\alpha), \ldots, \boldsymbol{Q}^{(\mathrm{n}-1)}(\alpha)\}$, a $\mathrm{p}(\alpha)$-dimensional representation of $\mathcal{Q}$ operating on the equivalence class $\mathfrak{E}(\alpha)$, for $\alpha = 1, 2, \ldots, \gamma$.

A formula of the same form as (17.3.15) now applies to the matrices of each of these representations, namely[17]

$$\boldsymbol{W}^{\dagger}(\alpha)\boldsymbol{Q}^{(\mathrm{s})}(\alpha)\boldsymbol{W}(\alpha) = \widetilde{\boldsymbol{Q}}^{(\mathrm{s})}(\alpha) := \bigoplus_{\mathrm{k}=1}^{\mathrm{K}} \widetilde{\boldsymbol{Q}}^{(\mathrm{s})}(\alpha; \mathrm{k}). \tag{17.3.24}$$

---

[17] There is, of course, a similar specialization of (17.3.17), but we shall be concerned only with block-diagonalization with $\boldsymbol{W}(\alpha)$ and not the final reduction to canonical form.

The matrix $\boldsymbol{W}(\alpha)$ has the effect of reducing $\boldsymbol{Q}^{(\mathrm{s})}(\alpha)$ to block-diagonal form with the blocks corresponding to the irreducible representations in order. But to achieve the reduction given in (17.3.15), we need a further ordering between equivalence classes. Thus

$$\boldsymbol{W} = [\boldsymbol{W}(1) \oplus \boldsymbol{W}(2) \oplus \cdots \oplus \boldsymbol{W}(\gamma)]\boldsymbol{\Pi}, \tag{17.3.25}$$

where $\boldsymbol{\Pi}$ is a permutation matrix, chosen to achieve the final ordering described above. The following theorem is useful in determining the multiplicities of the irreducible representations within equivalence classes:

Theorem 17.3.8 *For any equivalence class $\mathfrak{E}(\alpha)$ the multiplicity $\mathfrak{m}(\alpha; \mathrm{k})$ of the irreducible representation $\Theta(\mathrm{k})$ of dimension $\mathrm{e}(\mathrm{k})$ satisfies the condition*

$$\mathfrak{m}(\alpha; \mathrm{k}) \leq \mathrm{e}(\mathrm{k}), \qquad \textit{with} \quad \mathfrak{m}(\alpha; 1) = \mathrm{e}(1) = 1. \tag{17.3.26}$$

17.3.3 Block Diagonalization Within an Equivalence Class

We now consider the problem of determining unitary matrices $\boldsymbol{W}(\alpha)$ which produce the block-diagonalization given by (17.3.24). For this the following result is useful:

Theorem 17.3.9 *The set of operators*

$$\mathsf{P}_{j\ell}^{(\mathrm{k})} := \frac{\mathrm{e}(\mathrm{k})}{\mathfrak{n}} \sum_{\mathrm{s}=0}^{\mathfrak{n}-1} \overline{R}_{j\ell}^{(\mathrm{k})}(\mathsf{Q}^{(\mathrm{s})})\mathsf{Q}^{(\mathrm{s})}, \tag{17.3.27}$$

for $j, \ell = 1, 2, \ldots, \mathrm{e}(\mathrm{k})$, $\mathrm{k} = 1, 2, \ldots, \kappa$, transform according to the formulae

$$\mathsf{Q}^{(\mathrm{s})}\mathsf{P}_{j\ell}^{(\mathrm{k})} = \sum_{i=1}^{\mathrm{e}(\mathrm{k})} R_{ij}^{(\mathrm{k})}(\mathsf{Q}^{(\mathrm{s})})\mathsf{P}_{i\ell}^{(\mathrm{k})}, \quad \mathsf{P}_{j\ell}^{(\mathrm{k})}\mathsf{Q}^{(\mathrm{s})} = \sum_{i=1}^{\mathrm{e}(\mathrm{k})} R_{\ell i}^{(\mathrm{k})}(\mathsf{Q}^{(\mathrm{s})})\mathsf{P}_{ji}^{(\mathrm{k})} \tag{17.3.28}$$

and satisfies the product rule

$$\mathsf{P}_{j'\ell'}^{(\mathrm{k}')}\mathsf{P}_{j\ell}^{(\mathrm{k})} = \delta^{\mathrm{Kr}}(\mathrm{k} - \mathrm{k}')\delta^{\mathrm{Kr}}(\ell' - j)\mathsf{P}_{j'\ell}^{(\mathrm{k})}. \tag{17.3.29}$$

Using (17.3.27), let

$$\mathsf{T}^{(\mathrm{k})} := \sum_{j=1}^{\mathrm{e}(\mathrm{k})} \mathsf{P}_{jj}^{(\mathrm{k})} = \frac{\mathrm{e}(\mathrm{k})}{\mathfrak{n}} \sum_{\mathrm{s}=0}^{\mathfrak{n}-1} \overline{\chi}^{(\mathrm{k})}(\mathsf{Q}^{(\mathrm{s})})\mathsf{Q}^{(\mathrm{s})}, \quad \mathrm{k} = 1, 2, \ldots, \kappa. \tag{17.3.30}$$

It follows from (17.3.28) and (17.3.29) that

$$\mathrm{Q}^{(s)}\mathrm{T}^{(k)} = \mathrm{T}^{(k)}\mathrm{Q}^{(s)}, \qquad s = 0, 1, \ldots, n-1, \qquad k = 1, 2, \ldots, \kappa, \quad (17.3.31)$$

$$\mathrm{T}^{(k)}\mathrm{T}^{(k')} = \delta^{\mathrm{Kr}}(k-k')\mathrm{T}^{(k')}, \qquad k = 1, 2, \ldots, \kappa. \quad (17.3.32)$$

With $\boldsymbol{T}^{(k)} := \boldsymbol{R}(\mathrm{T}^{(k)})$, it is obvious that formulae like (17.3.30)–(17.3.32) apply with operators replaced by their representation matrices. From (17.3.23) and (17.3.30)

$$\boldsymbol{T}^{(k)} = \boldsymbol{T}^{(k)}(1) \oplus \boldsymbol{T}^{(k)}(2) \oplus \cdots \oplus \boldsymbol{T}^{(k)}(\gamma), \quad (17.3.33)$$

where

$$\boldsymbol{T}^{(k)}(\alpha) := \frac{e(k)}{n}\sum_{s=0}^{n-1}\overline{\chi}^{(k)}(\mathrm{Q}^{(s)})\boldsymbol{Q}^{(s)}(\alpha), \qquad k = 1, 2, \ldots, \kappa. \quad (17.3.34)$$

Then

$$\boldsymbol{W}^{\dagger}(\alpha)\boldsymbol{T}^{(k)}(\alpha)\boldsymbol{W}(\alpha) = \bigoplus_{k'=1}^{\kappa}\widetilde{\boldsymbol{T}}^{(k)}(\alpha; k'), \quad (17.3.35)$$

where, from (17.3.24), (17.3.34) and the orthogonality formula (17.3.6),

$$\widetilde{\boldsymbol{T}}^{(k)}(\alpha; k') = \frac{e(k)}{n}\sum_{s=0}^{n-1}\overline{\chi}^{(k)}(\mathrm{Q}^{(s)})\widetilde{\boldsymbol{Q}}^{(s)}(\alpha; k') = \delta^{\mathrm{Kr}}(k-k')\boldsymbol{I}^{(k)}(\alpha), \quad (17.3.36)$$

$\boldsymbol{I}^{(k)}(\alpha)$ being a unit matrix of dimension $m(\alpha; k)$. From (17.3.35) and (17.3.36),

$$\boldsymbol{T}^{(k)}(\alpha)\boldsymbol{W}(\alpha) = \boldsymbol{W}(\alpha)\boldsymbol{J}^{(k)}(\alpha), \quad (17.3.37)$$

where $\boldsymbol{J}^{(k)}(\alpha)$ is a $p(\alpha)$-dimensional diagonal matrix with zeros along the diagonal except for $m(\alpha; k)$ ones in the rows corresponding to the irreducible representation $\Theta(k)$. From (11.3.1), (17.3.23) and (17.3.24),

$$\widetilde{\boldsymbol{Q}}^{(s)}(\alpha)[\boldsymbol{W}(\alpha)]^{\dagger}\begin{pmatrix} \boldsymbol{\tau}_1(\alpha) \\ \boldsymbol{\tau}_2(\alpha) \\ \vdots \\ \boldsymbol{\tau}_{p(\alpha)}(\alpha) \end{pmatrix} = [\boldsymbol{W}(\alpha)]^{\dagger}\begin{pmatrix} \mathrm{Q}^{(s)}\boldsymbol{\tau}_1(\alpha) \\ \mathrm{Q}^{(s)}\boldsymbol{\tau}_2(\alpha) \\ \vdots \\ \mathrm{Q}^{(s)}\boldsymbol{\tau}_{p(\alpha)}(\alpha) \end{pmatrix}, \quad s = 0, 1, \ldots, n-1, \quad (17.3.38)$$

and, from the block structure of $\widetilde{\boldsymbol{Q}}^{(s)}$ given by (17.3.24), the effect of $[\boldsymbol{W}(\alpha)]^{\dagger}$ is to form linear combinations of the states of $\mathfrak{E}(\alpha)$, which transform according to representations each of which reduces to a multiple of a single irreducible representation. Thus a group of $\mathfrak{w}(\alpha;\mathrm{k}) = \mathfrak{m}(\alpha;\mathrm{k})\mathfrak{e}(\mathrm{k})$ successive rows of $[\boldsymbol{W}(\alpha)]^{\dagger}$ (columns of $\boldsymbol{W}(\alpha)$) can be identified with the irreducible representation $\Theta(\mathrm{k})$. It follows from (17.3.37) that these columns are eigenvectors of $\boldsymbol{T}^{(\mathrm{k})}(\alpha)$ with eigenvalue one and all the other columns of $\boldsymbol{W}(\alpha)$ are eigenvectors of $\boldsymbol{T}^{(\mathrm{k})}(\alpha)$ with eigenvalue zero. However, from (17.3.32) and (17.3.33), each of the columns of $\boldsymbol{T}^{(\mathrm{k}')}(\alpha)$ is an eigenvector of $\boldsymbol{T}^{(\mathrm{k})}(\alpha)$ with eigenvalue one if $\mathrm{k}' = \mathrm{k}$ and zero if $\mathrm{k}' \neq \mathrm{k}$. The columns of the matrix $\boldsymbol{W}(\alpha)$ corresponding to the representation $\Theta(\mathrm{k})$ are, therefore, obtained by finding a subset of $\mathfrak{w}(\alpha;\mathrm{k})$ of the $\mathfrak{p}(\alpha)$ columns of $\boldsymbol{T}^{(\mathrm{k})}(\alpha)$ which satisfied the required orthonormality. This is possible since the dimension of the space spanned by the columns is the number of non-zero eigenvalues which, from (17.3.35), is $\mathfrak{w}(\alpha;\mathrm{k})$. The simplest way to do this is to find all the functions

$$\boldsymbol{\psi}^{(\mathrm{k})}(\alpha;m) := \mathrm{T}^{(\mathrm{k})}\boldsymbol{\tau}_m(\alpha) = \frac{\mathfrak{e}(\mathrm{k})}{\mathfrak{n}}\sum_{\mathrm{s}=0}^{\mathfrak{n}-1}\overline{\chi}^{(\mathrm{k})}(\mathrm{Q}^{(\mathrm{s})})\mathrm{Q}^{(\mathrm{s})}\boldsymbol{\tau}_m(\alpha). \tag{17.3.39}$$

A selection is now made from the sets of coefficients of these linear sums of states to satisfy the orthonormality condition. Let the members of this set of columns be $(W_{1j}^{(\mathrm{k})}(\alpha), W_{2j}^{(\mathrm{k})}(\alpha), \ldots, W_{\mathfrak{p}(\alpha)j}^{(\mathrm{k})}(\alpha))$, $j = 1, 2, \ldots, \mathfrak{w}(\alpha;\mathrm{k})$.

For the case of an irreducible representation for which $\mathfrak{e}(\mathrm{k}) = 1$ it follows from Theorem 17.3.8 that $\mathfrak{w}(\alpha;\mathrm{k}) = 0$ or 1. In the latter case it follows, from (17.3.30) and (17.3.31), that

$$\begin{aligned}\mathrm{Q}^{(\mathrm{s})}\boldsymbol{\psi}^{(\mathrm{k})}(\alpha;m) &= \frac{\mathfrak{e}(\mathrm{k})}{\mathfrak{n}}\sum_{\mathrm{s}'=0}^{\mathfrak{n}-1}\overline{\chi}^{(\mathrm{k})}(\mathrm{Q}^{(\mathrm{s}')}[\mathrm{Q}^{(\mathrm{s})}]^{-1})\mathrm{Q}^{(\mathrm{s}')}\boldsymbol{\tau}_m(\alpha)\\ &= \chi^{(\mathrm{k})}(\mathrm{Q}^{(\mathrm{s})})\boldsymbol{\psi}^{(\mathrm{k})}(\alpha;m).\end{aligned} \tag{17.3.40}$$

From Theorem 17.3.7 it follows that, given any $\boldsymbol{\tau}_{m'}(\alpha)$, there exists an operator $\mathrm{Q}^{(\mathrm{s}')}$ such that $\boldsymbol{\tau}_{m'}(\alpha) = \mathrm{Q}^{(\mathrm{s}')}\boldsymbol{\tau}_m(\alpha)$. Then, from (17.3.31), (17.3.39) and (17.3.40),

$$\boldsymbol{\psi}^{(\mathrm{k})}(\alpha;m') = \chi^{(\mathrm{k})}(\mathrm{Q}^{(\mathrm{s}')})\boldsymbol{\psi}^{(\mathrm{k})}(\alpha;m). \tag{17.3.41}$$

So, to within a multiplicative factor, all the functions $\boldsymbol{\psi}^{(\mathrm{k})}(\alpha;m)$ are the same. In two cases it is straightforward to compute the column elements explicitly:

17.3.3.1 For the One-Dimensional Representation

$\chi(\mathrm{Q}^{(\mathrm{s})}) = 1$, $\mathrm{s} = 0, 1, \ldots, \mathfrak{n}-1$ and

$$\boldsymbol{\psi}^{(1)}(\alpha;m) = \frac{1}{\mathfrak{n}}\sum_{\mathrm{s}=0}^{\mathfrak{n}-1}\mathrm{Q}^{(\mathrm{s})}\boldsymbol{\tau}_m(\alpha) = \frac{1}{\mathfrak{p}(\alpha)}\sum_{j=1}^{\mathfrak{p}(\alpha)}\boldsymbol{\tau}_j(\alpha). \tag{17.3.42}$$

Thus, on normalization,

$$W_{i1}^{(1)}(\alpha) = \frac{1}{\sqrt{\mathfrak{p}(\alpha)}}, \qquad i = 1, 2, \ldots, \mathfrak{p}(\alpha). \tag{17.3.43}$$

17.3.3.2 For a Regular Representation

$\mathfrak{p}(\alpha) = \mathfrak{n}$ and $\boldsymbol{\tau}_{s+1}(\alpha) = \mathrm{Q}^{(s)}\boldsymbol{\tau}_1(\alpha)$, giving

$$\boldsymbol{\psi}^{(k)}(\alpha; m) = \frac{\mathrm{e}(k)}{\mathfrak{n}} \sum_{s=0}^{\mathfrak{n}-1} \overline{\chi}^{(k)}(\mathrm{Q}^{(s)}[\mathrm{Q}^{(m-1)}]^{-1})\boldsymbol{\tau}_1(\alpha). \tag{17.3.44}$$

Regarding the states as a set of orthonormal vectors, it follows from (17.3.8) that

$$\boldsymbol{\psi}^{(k)}(\alpha; m) \cdot \overline{\boldsymbol{\psi}^{(k')}(\alpha; m')} = \delta^{\mathrm{Kr}}(k - k')\frac{\mathrm{e}(k)}{\mathfrak{n}}\chi^{(k)}(\mathrm{Q}^{(m-1)}[\mathrm{Q}^{(m'-1)}]^{-1}). \tag{17.3.45}$$

In this case $w(\alpha; k) = \mathrm{e}(k)^2$ for all irreducible representations. For a one-dimensional representation we take

$$W_{i1}^{(k)}(\alpha) := \frac{\overline{\chi}^{(k)}(\mathrm{Q}^{(i-1)})}{\sqrt{\mathfrak{n}}}. \tag{17.3.46}$$

When $\mathrm{e}(k) > 1$, an orthonormal set of columns can be obtained by choosing a subset of the group operators such that $\chi^{(k)}(\mathrm{Q}^{(s)}[\mathrm{Q}^{(s')}]^{-1}) = 0$ for all $s \neq s'$.

17.3.4 Symmetry Groups

At the beginning of Sect. 17.3 we discussed the symmetry group of the Hamiltonian formed as a direct product of a group applied to the lattice, to the states and possibly to an external parameter. Suppose that $\mathcal{F}$ and $\mathcal{G}$ are such groups with $\mathcal{Q} = \mathcal{F} \times \mathcal{G}$ being the symmetry group of the Hamiltonian. When constructing the representation $\mathcal{R}(\mathcal{Q})$ of $\mathcal{Q}$ from the representations $\mathcal{R}(\mathcal{F})$ and $\mathcal{R}(\mathcal{G})$ of $\mathcal{F}$ and $\mathcal{G}$ respectively it must be noted that this can be achieved either by setting

$$\boldsymbol{R}(\mathrm{F}^{(i)}\mathrm{G}^{(j)}) = \boldsymbol{R}(\mathrm{G}^{(j)}\mathrm{F}^{(i)}) = \boldsymbol{R}(\mathrm{F}^{(i)}) \otimes \boldsymbol{R}(\mathrm{G}^{(j)}) \tag{17.3.47}$$

or

$$\boldsymbol{R}(\mathrm{F}^{(i)}\mathrm{G}^{(j)}) = \boldsymbol{R}(\mathrm{G}^{(j)}\mathrm{F}^{(i)}) = \boldsymbol{R}(\mathrm{G}^{(j)}) \otimes \boldsymbol{R}(\mathrm{F}^{(i)}). \tag{17.3.48}$$

Since, in general,

$$\boldsymbol{R}(\mathsf{F}^{(i)}) \otimes \boldsymbol{R}(\mathsf{G}^{(j)}) \neq \boldsymbol{R}(\mathsf{G}^{(j)}) \otimes \boldsymbol{R}(\mathsf{F}^{(i)}), \tag{17.3.49}$$

(17.3.47) and (17.3.48) correspond to different representations of $\mathcal{R}(\mathcal{Q})$. It is important to choose *in advance* a particular group ordering for the representation and we shall make the choice given by (17.3.47). The irreducible representations $\Theta(\mathsf{k}', \mathsf{k}'')$ of $\mathcal{Q}$ are given by taking Kronecker products $\Theta_{\mathcal{F}}(\mathsf{k}') \otimes \Theta_{\mathcal{G}}(\mathsf{k}'')$ of all the irreducible representations $\Theta_{\mathcal{F}}(\mathsf{k}')$ and $\Theta_{\mathcal{G}}(\mathsf{k}'')$ of $\mathcal{F}$ and $\mathcal{G}$. By taking the trace of (17.3.47) for irreducible representations it follows from the property of Kronecker matrix products that

$$\chi^{(\mathsf{k}',\mathsf{k}'')}(\mathsf{F}^{(i)}\mathsf{G}^{(j)}) = \chi_{\mathcal{F}}^{(\mathsf{k}')}(\mathsf{F}^{(i)})\chi_{\mathcal{G}}^{(\mathsf{k}'')}(\mathsf{G}^{(j)}). \tag{17.3.50}$$

The character table of $\mathcal{Q}$ can, therefore, be constructed from those of $\mathcal{F}$ and $\mathcal{G}$ as shown in the particular case in Example 17.3.4.

17.3.4.1 Periodic Boundary Conditions

Consider a two-dimensional lattice where periodic boundary conditions are applied in one direction by wrapping the lattice on a cylinder. The translational symmetry of the lattice in the direction in which periodic boundary conditions have been applied becomes rotational symmetry, with the cyclic group $\mathcal{C}_\mathfrak{n}$ of rotations about the axis of the cylinder, where $\mathfrak{n}$ is either the number of rows of sites along the direction of the axis of the cylinder or some divisor of that number. Since the group is commutative each element is a separate conjugacy class and it follows from Theorem 17.3.3 that the group has $\mathfrak{n}$ irreducible representations and, from (17.3.9), that they are one-dimensional. If $\chi(\mathsf{C}_\mathfrak{n}) = \omega_\mathfrak{n}$ for an irreducible representation, $\chi(\mathsf{C}_\mathfrak{n}^\mathfrak{n}) = \omega_\mathfrak{n}^\mathfrak{n} = 1$. The characters of $\mathsf{C}_\mathfrak{n}$ in the $\mathfrak{n}$ irreducible representations are the complex $\mathfrak{n}$-th roots of unity and from them we can construct by multiplication all the characters of the other elements of $\mathcal{C}_\mathfrak{n}$.

Example 17.3.1 The character tables for $\mathcal{C}_2$ and $\mathcal{C}_6$ are

$\mathcal{C}_2$	I	C_2
A	1	1
B	1	-1

$\mathcal{C}_6$	I	C_6	C_6^2	C_6^3	C_6^4	C_6^5
A	1	1	1	1	1	1
B	1	-1	1	-1	1	-1
E_1	1	ω_6	ω_6^2	ω_6^3	ω_6^4	ω_6^5
$\overline{\mathsf{E}}_1$	1	ω_6^5	ω_6^4	ω_6^3	ω_6^2	ω_6
E_2	1	ω_6^2	ω_6^4	1	ω_6^2	ω_6^4
$\overline{\mathsf{E}}_2$	1	ω_6^4	ω_6^2	1	ω_6^4	ω_6^2

where $\omega_6 := \exp(2\pi\mathrm{i}/6)$ and the representations E_i and $\overline{\mathsf{E}}_i$, $i = 1, 2$ have conjugate complex characters. For the spin-$\frac{1}{2}$ Ising model the spin-inversion group $\mathcal{S}_2^{(m)} := \{\mathsf{I}, \mathsf{Q}^{(m)}\}$, discussed in Sect. 3.5.1, is, of course, isomorphic to $\mathcal{C}_2$.

In the special case of a square lattice with one lattice direction consisting of rings of $\mathfrak{n}$ sites and the other lattice direction along the cylinder there will also be $\mathfrak{n}$ reflectional symmetries[18] $\Sigma_1, \Sigma_2, \ldots, \Sigma_n$, about planes containing the axis of the cylinder. When n is odd these reflections are about planes which cut the surface of the cylinder along a line of sites and along a line equidistant between two lines of sites. When n is even the reflections fall into two sets: those about planes which cut the surface of the cylinder along two lines of sites, which we denote by Σ_i, with i odd and those about planes which cut the surface of the cylinder along two lines equidistant between two lines of sites, which we denote by Σ_i, with i even. The group is now $\mathcal{C}_{nv}$, which is not commutative and, for $n > 2$, the rotations no longer fall into separate conjugacy classes. However, for n odd all the reflections are in one conjugacy class and for n even they fall into two conjugacy classes according to whether the index of the reflection is even or odd.

Example 17.3.2 The character table for $\mathcal{C}_{2v}$ is

$\mathcal{C}_{2v}$	I	C_2	Σ_1	Σ_2
A_1	1	1	1	1
A_2	1	1	−1	−1
B_1	1	−1	1	−1
B_2	1	−1	−1	1

In general, of course, the group $\mathcal{C}_{2v}$ is not identical to $\mathcal{C}_2$. However, for the situation considered here where we have two sites in a ring, Σ_1 leave the two sites unchanged and Σ_2 interchanges them. Effectively $\Sigma_1 \equiv I$ and $\Sigma_2 \equiv C_2$.[19] The character tables for $\mathcal{C}_{3v}$ and $\mathcal{C}_{4v}$ are

$\mathcal{C}_{3v}$	$\mathcal{Q}_1$	$\mathcal{Q}_2$	$\mathcal{Q}_3$
A_1	1	1	1
A_2	1	1	−1
E	2	−1	0

$\mathcal{C}_{4v}$	$\mathcal{Q}_1$	$\mathcal{Q}_2$	$\mathcal{Q}_3$	$\mathcal{Q}_4$	$\mathcal{Q}_5$
A_1	1	1	1	1	1
A_2	1	1	1	−1	−1
B_1	1	1	−1	1	−1
B_2	1	1	−1	−1	1
E	2	−2	0	0	0

where, for $\mathcal{C}_{3v}$ the conjugacy classes are $\mathcal{Q}_1 := I$, $\mathcal{Q}_2 := \{C_3, C_3^2\}$ and $\mathcal{Q}_3 := \{\Sigma_1, \Sigma_2, \Sigma_3\}$ and for $\mathcal{C}_{4v}$ the conjugacy classes are $\mathcal{Q}_1 := I$, $\mathcal{Q}_2 := C_4^2$, $\mathcal{Q}_3 := \{C_4, C_4^3\}$, $\mathcal{Q}_4 := \{\Sigma_1, \Sigma_3\}$ and $\mathcal{Q}_5 := \{\Sigma_2, \Sigma_4\}$.

Example 17.3.3 The spin-$\frac{1}{2}$ Ising model on a square lattice with two rows wrapped on a cylinder has symmetry group $\mathcal{S}_2^{(m)} \times \mathcal{C}_2$, where, as indicated above, $\mathcal{S}_2^{(m)}$ is isomorphic to $\mathcal{C}_2$. The conjugacy classes, irreducible representations and character table for $\mathcal{C}_2$ are given in Example 17.3.1. The irreducible representations, in terms

[18] We are considering the square lattice for simplicity. Reflectional symmetries can arise for other lattices.

[19] Another way to think about it is that, if we use the permutation representation, it can contain only the irreducible representations A_1 and B_1. There is a caveat to these remarks. If the microstates on the lattice sites have some directional properties (like arrows) then the reflections need not reduce to the identity and a rotation.

of the products of the representations of $\mathcal{S}_2$ and $\mathcal{C}_2$, are $\mathsf{A} := \mathsf{A} \otimes \mathsf{A}$, $\mathsf{B}_1 := \mathsf{B} \otimes \mathsf{A}$, $\mathsf{B}_2 := \mathsf{A} \otimes \mathsf{B}$, and $\mathsf{B}_3 := \mathsf{B} \otimes \mathsf{B}$. The conjugacy classes each contain one element.

$\mathcal{S}_2^{(m)} \times \mathcal{C}_2$	I	$Q^{(m)}$	C_2	$Q^{(m)}C_2$
A	1	1	1	1
B_1	1	−1	1	−1
B_2	1	1	−1	−1
B_3	1	−1	−1	1

Example 17.3.4 The 3-state Potts model on a square lattice with two rows wrapped on a cylinder has symmetry group $\mathcal{S}_3 \times \mathcal{C}_2$. The character table is constructed using Table 5.1 for the character table of $\mathcal{S}_3$ and the character table for $\mathcal{C}_2$ given in Example 17.3.1. The irreducible representations, in terms of the products of the representations of $\mathcal{S}_3$ and $\mathcal{C}_2$, are $\mathsf{A} := \mathsf{A} \otimes \mathsf{A}$, $\mathsf{B}_1 := \mathsf{B} \otimes \mathsf{A}$, $\mathsf{B}_2 := \mathsf{A} \otimes \mathsf{B}$, $\mathsf{B}_3 := \mathsf{B} \otimes \mathsf{B}$, $\mathsf{E}_1 := \mathsf{E} \otimes \mathsf{A}$ and $\mathsf{E}_2 := \mathsf{E} \otimes \mathsf{B}$. The conjugacy classes Q_1, Q_2, Q_3 are those of $\mathcal{S}_3$ followed by the identity operator for $\mathcal{C}_2$ and Q_4, Q_5, Q_6 are similarly constructed with the identity of $\mathcal{C}_2$ replaced by C_2.

$\mathcal{S}_3 \times \mathcal{C}_2$	Q_1	Q_2	Q_3	Q_4	Q_5	Q_6
A	1	1	1	1	1	1
B_1	1	1	−1	1	1	−1
B_2	1	1	1	−1	−1	−1
B_3	1	1	−1	−1	−1	1
E_1	2	−1	0	2	−1	0
E_2	2	−1	0	−2	1	0

Example 17.3.5 The eight-vertex model on a square lattice with N rows wrapped on a cylinder has symmetry group $\mathcal{S}_2^{(e)} \times \mathcal{S}_N$, where $\mathcal{S}_2^{(e)}$ corresponds to the change of sign of edge variables in a row (reversal of arrow directions) and $\mathcal{S}_N$ corresponds to all permutations of a row of edge variables (which leaves invariant the number of up and down arrow directions). As we have seen $\mathcal{S}_2$ is isomorphic to $\mathcal{C}_2$ and $\mathcal{S}_3$ is isomorphic to $\mathcal{D}_3$ and $\mathcal{C}_{3v}$. However, the symmetric group $\mathcal{S}_4$, of order 24, is larger that $\mathcal{C}_{3v}$. The usual way to distinguish its conjugacy classes is in terms of cycles, thus, for example $\langle 2, 1, 1\rangle$ denotes one two-cycle combined with two one-cycles. The conjugacy classes are then: $Q_1 := \langle 1, 1, 1, 1\rangle = I$, with one element; $Q_2 := \langle 2, 1, 1\rangle$, with six elements, including Σ_1 and Σ_3 and with each of Σ_2 and Σ_4 broken into two separate two-cycles; $Q_3 := \langle 2, 2\rangle$, with three elements Σ_2, Σ_4 and Σ_1 combined with Σ_3; $Q_4 := \langle 3, 1\rangle$, with eight elements consisting of all three-cycles; $Q_5 := \langle 4\rangle$, with six elements including C_4, C_4^2 and C_4^3. The character table is

$\mathcal{S}_4$	Q_1	Q_2	Q_3	Q_4	Q_5
A	1	1	1	1	1
B	1	−1	1	1	−1
E	2	0	2	−1	0
T_1	3	−1	−1	0	1
T_2	3	1	−1	0	−1

17.4 The Conformal Group

We confine our attention to the d-dimensional Euclidean space of the vectors

$$\boldsymbol{r} := \begin{pmatrix} r^{(1)} \\ \vdots \\ r^{(d)} \end{pmatrix} \tag{17.4.1}$$

measured in units of the lattice spacing. The line element $\mathrm{d}r$ is given by

$$[\mathrm{d}r]^2 = \mathbf{dr} \cdot \mathbf{dr} := [\mathrm{d}r^{(1)}]^2 + [\mathrm{d}r^{(2)}]^2 + \cdots + [\mathrm{d}r^{(d)}]^2. \tag{17.4.2}$$

The *conformal group* is the set of transformations $\boldsymbol{r} \to \tilde{\boldsymbol{r}}$ which satisfy conditions of the form

$$[\mathrm{d}\tilde{r}]^2 = \Omega(\boldsymbol{r})[\mathrm{d}r]^2, \tag{17.4.3}$$

for some scalar fields $\Omega(\boldsymbol{r})$. Suppose that a transformation is given by

$$\tilde{r}^{(i)} = R^{(i)}(\boldsymbol{r}) \qquad i = 1, 2, \ldots, d. \tag{17.4.4}$$

Then, for the condition (17.4.3) to be satisfied, it must be the case that

$$\Omega(\boldsymbol{r})\,\delta^{\mathrm{Kr}}(j-k) = \sum_{i=1}^{d} \frac{\partial R^{(i)}}{\partial r^{(j)}} \frac{\partial R^{(i)}}{\partial r^{(k)}}, \tag{17.4.5}$$

and $[\Omega(\boldsymbol{r})]^{\frac{d}{2}}$ is the Jacobian of the transformation (17.4.4). It is clear that the following transformations satisfy (17.4.5):

(i) *Translations* where

$$R^{(i)}(\boldsymbol{r}) = r^{(i)} + c^{(i)}, \qquad i = 1, 2, \ldots, d, \tag{17.4.6}$$

for some constants $c^{(i)}$ and $\Omega(\boldsymbol{r}) = 1$.

(ii) *Rotations* where

$$R^{(i)}(\boldsymbol{r}) = \omega_i^{(1)} r^{(1)} + \cdots + \omega_i^{(d)} r^{(d)}, \qquad i = 1, \ldots, d, \tag{17.4.7}$$

the vectors

$$\boldsymbol{\omega}_i := \begin{pmatrix} \omega_i^{(1)} \\ \vdots \\ \omega_i^{(d)} \end{pmatrix} \tag{17.4.8}$$

are orthonormal and $\Omega(\boldsymbol{r}) = 1$.

(iii) *Dilatations* where

$$R^{(i)}(\boldsymbol{r}) = \lambda^{-1} r^{(i)}, \qquad i = 1, 2, \ldots, d, \tag{17.4.9}$$

and $\Omega(\boldsymbol{r}) = \lambda^{-2}$.

(iv) *Inversions* where

$$R^{(i)}(\boldsymbol{r}) = r^{(i)}/r^2, \qquad i = 1, 2, \ldots, d, \tag{17.4.10}$$

$r = |\boldsymbol{r}|$ and $\Omega(\boldsymbol{r}) = 1/r^4$.

It can be shown (Ginsparg 1990) that, for $d > 2$, the conformal group is finite-dimensional and generated by the transformations (i)–(iv). Alternatively, inversion can be replaced as a generator of the group by the *special conformal transformation*

$$R^{(i)}(\boldsymbol{r}) = \frac{r^{(i)} + c^{(i)} r^2}{1 + 2\boldsymbol{r} \cdot \boldsymbol{c} + c^2 r^2}, \tag{17.4.11}$$

for any constant vector $\boldsymbol{c} = (c^{(i)}, \ldots, c^{(d)})$, with $\Omega(\boldsymbol{r}) = (1 + 2\boldsymbol{r}.\boldsymbol{c} + c^2 r^2)^{-2}$. This can be seen to be composed of a translation, an inversion and then a further translation.

For $d = 2$ the notation $\boldsymbol{r} := (x, y)$ is used, with the transformation (17.4.4) given by

$$\tilde{x} = \mathrm{X}(x, y), \qquad \tilde{y} = \mathrm{Y}(x, y). \tag{17.4.12}$$

Equation (17.4.5) is satisfied if

$$\frac{\partial \mathrm{X}}{\partial x} = \frac{\partial \mathrm{Y}}{\partial y}, \qquad \frac{\partial \mathrm{X}}{\partial y} = -\frac{\partial \mathrm{Y}}{\partial x}. \tag{17.4.13}$$

These are the Cauchy–Riemann conditions for

$$\mathrm{Z}(z) := \mathrm{X}(x, y) + \mathrm{i}\mathrm{Y}(x, y) \tag{17.4.14}$$

to be an analytic function of the complex variable $z := x + \mathrm{i}y$. In this case, therefore, the conformal group is infinite-dimensional. From (17.4.5) the Jacobian of the transformation $z \to \mathsf{Z}(z)$ is

$$\Omega(z) = |\mathsf{Z}'(z)|^2. \tag{17.4.15}$$

17.5 Some Transformations in the Complex Plane

In the discussion of Ruelle's theorem in Sect. 8.4.1 a fundamental role is played by the polynomial $\xi(T, \{\mathfrak{z}_\kappa(\boldsymbol{r})\}; \mathcal{Y}_\kappa)$ given by Eq. (8.4.9), where $\mathcal{Y}_\kappa$ is a member of the covering $\mathcal{Y}_\mathcal{N}$ of the lattice $\mathcal{N}$. When $\mathcal{Y}_\kappa$ consists of two lattice sites $\boldsymbol{r}$ and $\boldsymbol{r}'$

$$\xi(T, \{\mathfrak{z}_\kappa(\boldsymbol{r})\}; \mathcal{Y}_\kappa) = f(\mathfrak{z}, \mathfrak{z}') := c_0 + c_1(\mathfrak{z} + \mathfrak{z}') + c_2\mathfrak{z}\mathfrak{z}', \tag{17.5.1}$$

where the real positive numbers c_0, c_1 and c_2 are the Boltzmann weights when the pair of sites is occupied by 0, 1 and 2 particles respectively and $\mathfrak{z} := \mathfrak{z}_\kappa(\boldsymbol{r})$, $\mathfrak{z}' := \mathfrak{z}_\kappa(\boldsymbol{r}')$. We need to obtain two closed sets $\mathbb{X}$ and $\mathbb{X}'$ in $\mathbb{C}_\mathfrak{z}$, not containing 0, such that $f(\mathfrak{z}, \mathfrak{z}') \neq 0$ when $\mathfrak{z} \notin \mathbb{X}$ and $\mathfrak{z}' \notin \mathbb{X}'$. In the Ising model spin formulation $c_0 = c_2 \neq c_1$ and in the lattice fluid formulation $c_0 = c_1 = 1$, $c_2 \neq 1$. In the former case we may, without loss of generality, set $c_0 = c_2 = 1$ and the two cases are related by a multiplicative transformation of the variables $\mathfrak{z}$ and $\mathfrak{z}'$.

We, therefore, begin by considering

$$f(\mathfrak{z}, \mathfrak{z}') := 1 + c(\mathfrak{z} + \mathfrak{z}') + \mathfrak{z}\mathfrak{z}', \tag{17.5.2}$$

where $f(\mathfrak{z}, \mathfrak{z}') = 0$ corresponds to the mapping

$$\mathfrak{z}' = -(1 + c\mathfrak{z})/(c + \mathfrak{z}). \tag{17.5.3}$$

With the change of variables $\zeta := \mathfrak{z} + c$, $\zeta' := \mathfrak{z}' + c$, this becomes

$$\zeta' = (c^2 - 1)/\zeta. \tag{17.5.4}$$

Excluding the case $c = 1$, which corresponds to no interaction for any site occupation, the fixed points

$$\zeta^{(\pm)} = \pm\sqrt{c^2 - 1} \tag{17.5.5}$$

of (17.5.4) lie on the real or imaginary axis of the complex plane $\mathbb{C}_\zeta$ of ζ according as $c > 1$ or $c < 1$, respectively. It is now not difficult to establish the following result:

Theorem 17.5.1 *Let $\mathcal{V}$ be the family of circles in $\mathbb{C}_\zeta$ passing through $\zeta = \zeta^{(\pm)}$.*[20] *Any circle $\mathbb{V} \in \mathcal{V}$ is divided into two open arcs $\mathbb{V}^{(\pm)}$ by the fixed points. Every $\zeta \neq \zeta^{(\pm)}$ in $\mathbb{C}_\zeta$ belongs to exactly one circle $\mathbb{V} \in \mathcal{V}$ and, if $\zeta \in \mathbb{V}^{(\pm)}$, then $\zeta' \in \mathbb{V}^{(\mp)}$.*

It follows that every circle $\mathbb{V} \in \mathcal{V}$ is invariant under the mapping (17.5.4). Since the arcs $\mathbb{V}_1^{(\pm)}$ of any other $\mathbb{V}_1 \in \mathcal{V}$ lie one in the interior and one in the exterior of $\mathbb{V}$, it follows that (17.5.4) is a mapping between the interior and exterior of $\mathbb{V}$. In $\mathbb{C}_\mathfrak{z}$ any circle through the points

$$\mathfrak{z}^{(\pm)} = -c \pm \sqrt{c^2 - 1} \tag{17.5.6}$$

is invariant under (17.5.3), with (17.5.6) as fixed points. The mapping is between interior and exterior points of the circle. We consider the two cases $c < 1$ and $c > 1$ separately:

(a) $c < 1$: In this case the fixed points are

$$\mathfrak{z}^{(\pm)} = -c \pm \mathrm{i}\sqrt{1 - c^2}. \tag{17.5.7}$$

The circle

$$\mathbb{V}(x) := \{\mathfrak{z} : |2c\mathfrak{z} + x| = \sqrt{4c^2(1-x) + x^2}\} \tag{17.5.8}$$

passes through the points (17.5.7) for any real x. Apart from $\mathbb{V}(1)$, which passes through the origin, any circle $\mathbb{V}(x)$ can be used to define the regions $\mathbb{X}$ and $\mathbb{X}'$. Thus we see that $f(\mathfrak{z}, \mathfrak{z}') \neq 0$, when $\mathfrak{z} \notin \mathbb{X}$ and $\mathfrak{z}' \notin \mathbb{X}$, if

$$\mathbb{X} := \begin{cases} \{\mathfrak{z} : |2c\mathfrak{z} + x| \leq \sqrt{4c^2(1-x) + x^2}\}, & \forall \quad x > 1, \\ \{\mathfrak{z} : |2c\mathfrak{z} + x| \geq \sqrt{4c^2(1-x) + x^2}\}, & \forall \quad x < 1. \end{cases} \tag{17.5.9}$$

In particular the unit circle $|\mathfrak{z}| = 1$ corresponds to $x = 0$ in (17.5.8) and gives a region in the second category of (17.5.9). The regions in the first category correspond to discs, with centres on the negative real axis $\mathbb{R}^{(-)}$ of $\mathbb{C}_\mathfrak{z}$, and lying entirely in $\Re(\mathfrak{z}) < 0$. The angles made with $\mathbb{R}^{(-)}$ by tangents from the origin to the boundary of the disc are $\pm \arccos(2c\sqrt{x-1}/x)$. These angles take a minimum absolute value when $x = 2$, when the points of contact of the tangents and the boundary are the fixed points (17.5.7) lying on $|\mathfrak{z}| = 1$.

(b) $c > 1$: In this case the fixed points (17.5.6) lie on $\mathbb{R}^{(-)}$. The circle

$$\mathbb{V}(y) := \{\mathfrak{z} : |\mathfrak{z} + c + \mathrm{i}y| = \sqrt{y^2 + c^2 - 1}\} \tag{17.5.10}$$

[20] Including the limiting circle consisting of the axis through the fixed points.

passes through the points (17.5.6) for any real y. The origin is exterior to all the circles of this family and thus the definition (17.5.9) can be replaced by

$$\mathbb{Y} := \{\mathfrak{z} : |\mathfrak{z} + c + \mathrm{i}y| \leq \sqrt{y^2 + c^2 - 1}\}, \qquad \forall \ y. \tag{17.5.11}$$

The angles made with $\mathbb{R}^{(-)}$ by tangents from the origin to the boundary of the disc are $\arccos\left(c/\sqrt{y^2+c^2}\right) \pm \arccos\left(1/\sqrt{y^2+c^2}\right)$. The larger of these angles take its minimum value when $y = 0$.

The other case of interest for (17.5.1) is

$$f(\mathfrak{z}, \mathfrak{z}') := 1 + \mathfrak{z} + \mathfrak{z}' + c\mathfrak{z}\mathfrak{z}'. \tag{17.5.12}$$

This can be expressed in the form

$$f(\mathfrak{z}, \mathfrak{z}') = 1 + (1/\sqrt{c})(\sqrt{c}\mathfrak{z} + \sqrt{c}\mathfrak{z}') + (\sqrt{c}\mathfrak{z})(\sqrt{c}\mathfrak{z}'), \tag{17.5.13}$$

and the previous results apply with c replaced by $1/\sqrt{c}$ and $\mathfrak{z}$ replaced by $\sqrt{c}\,\mathfrak{z}$. The two cases are:

(a) $c < 1$: From (17.5.11) and (17.5.13), $f(\mathfrak{z}, \mathfrak{z}') \neq 0$ when $\mathfrak{z} \notin \mathbb{Y}'$ and $\mathfrak{z}' \notin \mathbb{Y}'$ if

$$\mathbb{Y}' := \{\mathfrak{z} : |c\mathfrak{z} + 1 + \mathrm{i}y| \leq \sqrt{y^2 + 1 - c}\}, \qquad \forall \ y. \tag{17.5.14}$$

The angles made with $\mathbb{R}^{(-)}$ by tangents from the origin to the boundary of the disc are $\arccos\left(1/\sqrt{y^2+1}\right) \pm \arccos\left(\sqrt{c}/\sqrt{y^2+1}\right)$. The larger of these angles take its minimum value when $y = 0$.

(b) $c > 1$: From (17.5.9), $f(\mathfrak{z}, \mathfrak{z}') \neq 0$ when $\mathfrak{z} \notin \mathbb{X}'$ and $\mathfrak{z}' \notin \mathbb{X}'$ if

$$\mathbb{X}' := \begin{cases} \{\mathfrak{z} : |2\mathfrak{z} + x| \leq \sqrt{4c^{-1}(1-x) + x^2}\}, & \forall \quad x > 1, \\ \{\mathfrak{z} : |2\mathfrak{z} + x| \geq \sqrt{4c^{-1}(1-x) + x^2}\}, & \forall \quad x < 1. \end{cases} \tag{17.5.15}$$

The angles made with $\mathbb{R}^{(-)}$ by tangents from the origin to the boundary of the discs in the first category are $\pm \arccos(2\sqrt{x-1}/x\sqrt{c})$. These angles take a minimum absolute value when $x = 2$.

17.6 Algebraic Functions

Definition 17.6.1 An algebraic function $\lambda(w)$ is the solution of an equation

$$\mathcal{F}(\lambda, w) = 0, \qquad \text{where} \tag{17.6.1}$$

$$\mathcal{F}(\lambda, w) := \mathsf{f}_n(w)\lambda^n + \mathsf{f}_{n-1}(w)\lambda^{n-1} + \cdots + \mathsf{f}_0(w). \quad (17.6.2)$$

The coefficients $\mathsf{f}_j(w)$, $j = 0, 1, \ldots, n$, are polynomials in w with complex coefficients, having no common factor apart from a constant and $\mathcal{F}(\lambda, w)$ is not reducible[21] to a product of polynomials in λ, which themselves have coefficients which are polynomials in w.

$\mathcal{F}(\lambda, w)$ is called the *defining polynomial* of $\lambda(w)$ and the solutions $\lambda_1(w), \ldots, \lambda_n(w)$ are called the *branches* of $\lambda(w)$. A point at which the values of two or more of the branches of $\lambda(w)$ are equal is a value of w for which (17.6.1) has repeated roots. The set of such points is given by the simultaneous solution of the equations

$$\mathcal{F}(\lambda, w) = 0, \qquad \frac{\partial \mathcal{F}(\lambda, w)}{\partial \lambda} \equiv \mathcal{F}'(\lambda, w) = 0, \quad (17.6.3)$$

which is obtain by solving the equation

$$\mathrm{Res}\{\mathcal{F}, \mathcal{F}'; w\} = 0 \quad (17.6.4)$$

for the *resultant* $\mathrm{Res}\{\mathcal{F}; \mathcal{F}'; w\}$ of $\mathcal{F}(\lambda, w)$ and $\mathcal{F}'(\lambda, w)$.[22] Since in this case $\mathsf{f}_n(w)$ is a factor in the determinant equation, (17.6.4) can be replaced by

$$\mathrm{Dis}\{\mathcal{F}; w\} = 0, \quad (17.6.5)$$

where

$$\mathrm{Dis}\{\mathcal{F}; w\} := (-1)^{\frac{1}{2}n(n-1)}[\mathsf{f}_n(w)]^{-1}\mathrm{Res}\{\mathcal{F}, \mathcal{F}'; w\} \quad (17.6.6)$$

is called the *discriminant* (Hille 1962). Given that the coefficients $\mathsf{f}_j(w)$ are polynomials of degree m at most, it is clear that $\Delta(w) := \mathrm{Dis}\{\mathcal{F}; w\}$ is a polynomial in w of degree not exceeding $2m(n-1)$.

Algebraic functions feature in the methods described in this book in two way. Firstly is in the analysis of partition function zeros in Sect. 11.4 and secondly in the algebraic approximant method for the analysis of series expansions in Sect. 14.6.4 (see also Sect. 18.5.4). Most of this appendix is concerned with the former. However the importance for the latter resides in the relationship between differential and algebraic approximants. That the algebraic function $\lambda(w)$ is the solution of a linear differential equation with polynomial coefficients was observed and analysed in the nineteenth century (see Bostan et al. 2007, for references). More recently it has become a topic of some interest to computer science. The following result can be established (Cormier et al. 2002; Bostan et al. 2007):

---

21 The irreducibility of $\mathcal{F}(\lambda, w)$ is not always taken as part of the definition of an algebraic function (Bliss 1966). We shall however, following Hille (1962), assume this to be included.

22 The *resultant* of two polynomials is given by their Sylvester determinant (Bliss 1966).

Theorem 17.6.1 *When $\Delta(w) \neq 0$ the algebraic function $\lambda(w)$, defined by Eqs.* (17.6.1) *and* (17.6.2), *is the solution of a linear differential equation*

$$\mathrm{c}_s(w)[\Delta(w)]^s \frac{\mathrm{d}^s\lambda(w)}{\mathrm{d}w^s} + \cdots \mathrm{c}_1(w)\Delta(w)\frac{\mathrm{d}\lambda(w)}{\mathrm{d}w} + \mathrm{c}_0(w)\lambda(w) = 0, \tag{17.6.7}$$

where $s \leq n$.

In fact $\lambda(w)$ can be shown to also be the solution of an inhomogeneous differential equation.

Example 17.6.1 For the quadratic case, $n = 2$, the example given by Cormier et al. (2002, p. 367) is

$$\begin{aligned} &\mathrm{f}_2(w) = 1, \qquad \mathrm{f}_1(w) = -2w(1+2w), \\ &\mathrm{f}_0(w) = -w^2(w^5 - w^3 - 4w^2 - 4w - 1), \end{aligned} \tag{17.6.8}$$

for which

$$\Delta(w) = 4w^5(w^2 - 1) \tag{17.6.9}$$

and (17.6.7) is satisfied with $s = 2$ and

$$\begin{aligned} &\mathrm{c}_2(w) = 6w^3 + 5w^2 - 2w - 3, \\ &\mathrm{c}_1(w) = -2w^4(54w^5 + 35w^4 - 76w^3 - 54w^2 + 14w + 15), \\ &\mathrm{c}_0(w) = 8w^8(w^2 - 1)(84w^5 + 35w^4 - 120w^3 - 54w^2 + 20w + 15). \end{aligned} \tag{17.6.10}$$

The quadratic case has also been studied by Hunter and Baker (1979) who showed that $\lambda(w)$ satisfies the inhomogeneous equation

$$\mathrm{a}_1(w)\Delta(w)\frac{\mathrm{d}\lambda(w)}{\mathrm{d}w} + \mathrm{a}_0(w)\lambda(w) = \mathrm{b}(w), \tag{17.6.11}$$

where

$$\begin{aligned} &\mathrm{a}_0(w) = 2w^4(5 - 7w^2), \qquad \mathrm{a}_1(w) = 1, \\ &\mathrm{b}(w) = -2w^5(6w^3 + 5w^2 - 2w - 3). \end{aligned} \tag{17.6.12}$$

Eliminating the inhomogeneous terms between (17.6.11) and its derivative will, of course, yield the homogeneous second-order equation given by (17.6.7) and (17.6.10).

We now turn to material relevant to Sect. 11.4 and a theorem proved by Hille (1962):

Theorem 17.6.2 *The branches $\lambda_1(w), \lambda_2(w), \dots, \lambda_n(w)$ of the algebraic function $\lambda(w)$ defined by equations* (17.6.1) *and* (17.6.2) *are analytic functions in any simply-connected region of the complex plane which excludes (i) the zeros of $\mathfrak{f}_n(w)$, (ii) solutions of* (17.6.5), *(iii) the point at infinity.*

In general the zeros of $\mathfrak{f}_n(w)$ will be poles of one or more branches of $\lambda(w)$. However, in the application of the theory of this section in Sect. 11.4, the polynomials of interest have constant leading coefficients so the algebraic functions have no infinities. The solutions of (17.6.5) are points where two or more of the branches of $\lambda(w)$ have the same value. Suppose that w_0 is such a point and that the branches of $\lambda(w)$ are labelled in such a way that $\lambda_1(w), \lambda_2(w), \dots, \lambda_p(w)$, $(1 < p \leq n)$ is the complete set of branches for which

$$\lambda_1(w_0) = \lambda_2(w_0) = \cdots = \lambda_p(w_0). \tag{17.6.13}$$

There are now three possibilities:

(i) The branches which satisfy (17.6.13) can be ordered in such a way that, for small positive real r,

$$\lambda_1(w_0 + r\exp[2\ell i\pi]) = \lambda_{\ell+1}(w_0 + r), \tag{17.6.14}$$

where ℓ is an integer modulo p. It is clear that $\lambda_1(w)$ is an analytic function of the variable $(w - w_0)^{1/p}$ in a neighbourhood of w_0 and the power-series expansion

$$\sum_{q=0}^{\infty} c_q (w - w_0)^{q/p} \tag{17.6.15}$$

is applicable to all the branches $\lambda_1(w), \lambda_2(w), \dots, \lambda_p(w)$ in a neighbourhood of w_0, when $(w - w_0)^{1/p}$ takes its p different values. The point w_0 is a *branch-point* of $\lambda(w)$ with *cycle number p*. To distinguish this case from case (iii) below, such a branch-point is called *monocyclic*.

(ii) The branches which satisfy (17.6.13) also satisfy the condition

$$\lambda_\ell(w_0 + r) = \lambda_\ell(w_0 + r\exp[2i\ell\pi]), \qquad \ell = 1, 2, \dots, p, \tag{17.6.16}$$

for any integer r. This is a case where each of the branches is analytic at w_0, but they happen to have the same value there. Joyce (1988b) has referred to points of this type as *apparent singular points*.

(iii) The branches which satisfy (17.6.13) divide into distinct subsets containing $p_1, p_2, \dots, p_k$ members where

$$\sum_{s=1}^{k} p_s = p, \tag{17.6.17}$$

and, within the sth subset, the members have a cycle structure like that given by (17.6.14) and a power-series expansion like (17.6.15), with cycle number p_s. This can be thought of as a situation where k branch-points happen to coincide. We can also include within this case a combination of coincident branch-points and analytic components, where the latter correspond to instances with $p_s = 1$. A branch-point of this type is call *k-fold multicyclic.*

The reason why Theorem 17.6.2 contains the restriction to simply-connected regions is now clear. In a multi-connected region around a branch-point some of the branches of $\lambda(w)$ will not be single-valued. The usual way to define a region within which the branches of $\lambda(w)$ are analytic is to define *cuts* in the complex plane, which prevent the construction of any curve along which a branch of $\lambda(w)$ passes continuously from one of its values to another. Within such a cut plane each branch of $\lambda(w)$ will be analytic and if we take n copies of the complex plane and join them in a suitable way along the edges of the cuts a *Riemann surface* of *n sheets* can be constructed on which $\lambda(w)$ is itself analytic.[23]

The set of cuts needed to construct the Riemann surface for any algebraic function is not unique and we shall describe one of the methods for doing this based on the idea of *connection curves* introduced by Wood (1987). These are defined by all values of w which are simultaneous solutions of the equations

$$\mathcal{F}(\lambda, w) = 0, \qquad \mathcal{F}(\varpi\lambda, w) \equiv \mathcal{F}^{(\varpi)}(\lambda, w) = 0, \qquad \forall\ \varpi = \exp(\mathrm{i}\vartheta), \tag{17.6.18}$$

where $\vartheta \in [0, 2\pi)$. The connection curves are then given by

$$\mathrm{Res}\{\mathcal{F}, \mathcal{F}^{(\varpi)}; w\} = 0. \tag{17.6.19}$$

It is clear that the resultant of $\mathcal{F}(\lambda, w)$ and $\mathcal{F}^{(\varpi)}(\lambda, w)$ is identically zero when $\varpi = 1$ with

$$\begin{aligned}\mathrm{Res}\{\mathcal{F}, \mathcal{F}^{(\varpi)}; w\} &= \mathrm{Res}\{\mathcal{F}, \mathcal{F}^{(\varpi)} - \mathcal{F}; w\}\\ &= (\varpi - 1)^n \mathrm{Res}\left\{\mathcal{F}, \frac{\mathcal{F}^{(\varpi)} - \mathcal{F}}{\varpi - 1}; w\right\}\end{aligned} \tag{17.6.20}$$

and

$$\frac{\mathcal{F}_n^{(\varpi)}(\lambda, w) - \mathcal{F}(\lambda, w)}{\varpi - 1} = \lambda \mathcal{G}^{(\varpi)}(\lambda, w), \tag{17.6.21}$$

where

$$\begin{aligned}\mathcal{G}^{(\varpi)}(\lambda, w) = {} & \mathsf{f}_n(w)\, \phi^{(n)}\, (\varpi)\lambda^{n-1} + \mathsf{f}_{n-1}(w)\, \phi^{(n-1)}\, (\varpi)\lambda^{n-2}\\ & + \cdots + \mathsf{f}_1(w),\end{aligned} \tag{17.6.22}$$

---

[23] For an account of the theory of Riemann surfaces see Beardon (1984).

$$\phi^{(s)}(\varpi) = \frac{\varpi^s - 1}{\varpi - 1}. \tag{17.6.23}$$

The determinant $\mathrm{Res}\{\mathcal{F}, \frac{\mathcal{T}^{(\varpi)} - \mathcal{T}}{\varpi - 1}; w\}$ has only one non-zero element in the last column, which is $\mathfrak{f}_0(w)$ in the nth row. So

$$\mathrm{Res}\{\mathcal{F}, \mathcal{F}^{(\varpi)}; w\} = -\mathfrak{f}_0(w)(1 - \varpi)^n \mathrm{Res}\{\mathcal{F}, \mathcal{G}^{(\varpi)}; w\}. \tag{17.6.24}$$

The connection curves are thus given by

$$\mathrm{Res}\{\mathcal{F}, \mathcal{G}^{(\varpi)}; w\} = 0, \tag{17.6.25}$$

and, since

$$\lim_{\varpi \to 1} \phi^s(\varpi) = s, \tag{17.6.26}$$

it follows, from (17.6.22), that

$$\lim_{\varpi \to 1} \mathrm{Res}\{\mathcal{F}, \mathcal{G}^{(\varpi)}; w\} = \mathrm{Res}\{\mathcal{F}, \mathcal{F}'; w\}. \tag{17.6.27}$$

The solutions of (17.6.5), which are the branch-points and the apparent singular points, correspond to points on the connection curves when $\varpi = 1$.

17.7 Determinants of Cyclic Matrices

Let $\boldsymbol{Z}$ be an N-dimensional square matrix with elements $z(\boldsymbol{j}, \boldsymbol{\ell})$, indexed by two-dimensional vectors of the form

$$\boldsymbol{j} := (j_1, j_2), \qquad j_1 = 1, 2, \ldots, N_1, \qquad j_2 = 1, 2, \ldots, N_2, \tag{17.7.1}$$

with $N = N_1 N_2$. $\boldsymbol{Z}$ is taken to be *cyclic*, meaning that

$$\begin{aligned} &z(\boldsymbol{j}, \boldsymbol{\ell}) = z(\boldsymbol{\ell} - \boldsymbol{j}), \\ &z(\boldsymbol{j} + \boldsymbol{n}) = z(\boldsymbol{j}), \quad \boldsymbol{n} = (n_1 N_1, n_2 N_2), \quad \forall \text{ integers } n_1 \text{ and } n_2. \end{aligned} \tag{17.7.2}$$

For $\boldsymbol{\ell} := (\ell_1, \ell_2)$, the notation $z(\boldsymbol{\ell}) := z(\ell_1, \ell_2)$ is used. Let $\boldsymbol{\Omega}$ be the N-dimensional square matrix with

$$\langle \boldsymbol{j} | \boldsymbol{\Omega} | \boldsymbol{\ell} \rangle = \omega(\boldsymbol{j}, \boldsymbol{\ell}) := \frac{1}{\sqrt{N}} \exp\left[2\pi \mathrm{i} \left(\frac{j_1 \ell_1}{N_1} + \frac{j_2 \ell_2}{N_2} \right) \right], \tag{17.7.3}$$

where vectors $\boldsymbol{j}$ and $\boldsymbol{\ell}$ are of the form (17.7.1). It is not difficult to show that $\boldsymbol{\Omega}$ is unitary and

Theorem 17.7.1 *$\mathbf{Z}$ is diagonalized by the transformation $\boldsymbol{\Omega}^{-1}\mathbf{Z}\boldsymbol{\Omega}$ and*

$$\mathrm{Det}\{\mathbf{Z}\} = \prod_{\{\boldsymbol{k}\}} \zeta(\boldsymbol{k}), \tag{17.7.4}$$

where

$$\zeta(\boldsymbol{k}) := \sum_{\ell_1=1}^{N_1} \sum_{\ell_2=1}^{N_2} z(\ell_1, \ell_2) \exp\left[2\pi \mathrm{i}\left(\frac{\ell_1 k_1}{N_1} + \frac{\ell_2 k_2}{N_2}\right)\right]. \tag{17.7.5}$$

Proof The matrix $\mathbf{Z}' = \boldsymbol{\Omega}^{-1}\mathbf{Z}\boldsymbol{\Omega}$ has the elements

$$\begin{aligned} z'(\boldsymbol{j}, \boldsymbol{k}) &= \sum_{\{\boldsymbol{\ell}\}} \sum_{\{\boldsymbol{\ell}'\}} \overline{\omega(\boldsymbol{\ell}, \boldsymbol{j})} z(\boldsymbol{\ell}' - \boldsymbol{\ell}) \omega(\boldsymbol{\ell}', \boldsymbol{k}) \\ &= \sqrt{N} \sum_{\{\boldsymbol{\ell}\}} \overline{\omega(\boldsymbol{\ell}, \boldsymbol{j})} \omega(\boldsymbol{\ell}, \boldsymbol{k}) \sum_{\{\boldsymbol{\ell}'\}} z(\boldsymbol{\ell}' - \boldsymbol{\ell}) \omega(\boldsymbol{\ell}' - \boldsymbol{\ell}, \boldsymbol{k}). \end{aligned} \tag{17.7.6}$$

Since $z(\boldsymbol{j})\omega(\boldsymbol{j}, \boldsymbol{k}) = z(\boldsymbol{j} + \boldsymbol{n})\omega(\boldsymbol{j} + \boldsymbol{n}, \boldsymbol{j})$, for any of the vectors $\boldsymbol{n}$ defined in (17.7.2), $\boldsymbol{\ell}' - \boldsymbol{\ell}$ can be replaced by $\boldsymbol{\ell}'$ in the second summation, giving, from (17.7.4) and (17.7.5),

$$z'(\boldsymbol{j}, \boldsymbol{k}) = \sqrt{N} \sum_{\{\boldsymbol{\ell}\}} \overline{\omega(\boldsymbol{\ell}, \boldsymbol{j})} \omega(\boldsymbol{\ell}, \boldsymbol{k}) \sum_{\{\boldsymbol{\ell}'\}} z(\boldsymbol{\ell}') \omega(\boldsymbol{\ell}', \boldsymbol{k}) = \delta^{\mathrm{Kr}}(\boldsymbol{j} - \boldsymbol{k}) \zeta(\boldsymbol{k}). \tag{17.7.7}$$

(17.7.4) then follows from $\mathrm{Det}\{\mathbf{Z}\} = \mathrm{Det}\{\mathbf{Z}'\}$. □

These results can be generalized by replacing each element $z(\boldsymbol{j}, \boldsymbol{\ell})$ of $\mathbf{Z}$ by a $\eta \times \eta$ dimensional submatrix, or block, $\boldsymbol{z}(\boldsymbol{j}, \boldsymbol{\ell})$ with

$$\boldsymbol{z}(\boldsymbol{j}, \boldsymbol{\ell}) = \boldsymbol{z}(\boldsymbol{\ell} - \boldsymbol{j}), \qquad \boldsymbol{z}(\boldsymbol{j} + \boldsymbol{n}) = \boldsymbol{z}(\boldsymbol{j}). \tag{17.7.8}$$

Such a matrix $\mathbf{Z}$ is called a *cyclic block matrix*. With $\omega(\boldsymbol{j}, \boldsymbol{\ell})$ replaced by $\boldsymbol{\omega}(\boldsymbol{j}, \boldsymbol{\ell}) = \omega(\boldsymbol{j}, \boldsymbol{\ell})\boldsymbol{I}_\eta$, where $\boldsymbol{I}_\eta$ is the unit matrix of dimension η, the blocks $\boldsymbol{z}'(\boldsymbol{j}, \boldsymbol{\ell})$ of $\mathbf{Z}' = \boldsymbol{\Omega}^{-1}\mathbf{Z}\boldsymbol{\Omega}$ are given by

$$\boldsymbol{z}'(\boldsymbol{j}, \boldsymbol{k}) = \delta^{\mathrm{Kr}}(\boldsymbol{j} - \boldsymbol{k}) \boldsymbol{\zeta}(\boldsymbol{k}), \tag{17.7.9}$$

where

$$\boldsymbol{\zeta}(\boldsymbol{k}) := \sum_{\{\boldsymbol{\ell}\}} \boldsymbol{z}(\boldsymbol{\ell}) \exp\left[2\pi \mathrm{i}\left(\frac{\ell_1 k_1}{N_1} + \frac{\ell_2 k_2}{N_2}\right)\right]. \tag{17.7.10}$$

Equation (17.7.4) is then replaced by

$$\mathrm{Det}\{\mathbf{Z}\} = \prod_{\{\boldsymbol{k}\}} \mathrm{Det}\{\boldsymbol{\zeta}(\boldsymbol{k})\}. \tag{17.7.11}$$

Now consider the limit $N_s \to \infty$, for $s = 1, 2$. In (17.7.10) the vector $\boldsymbol{k}$ is replaced by the vector $\boldsymbol{\theta} := (\theta_1, \theta_2)$, where $\theta_s := 2\pi i k_s/N_s$, $s = 1, 2$. In the limit $N_s \to \infty$, θ_s becomes a continuous variable in the range $[0, 2\pi]$. From (17.7.11)

$$\lim_{N_1\to\infty} \lim_{N_2\to\infty} \frac{\ln[\mathrm{Det}\{\mathbf{Z}\}]}{N_1 N_2} = \frac{1}{(2\pi)^2} \int_0^{2\pi} \mathrm{d}\theta_1 \int_0^{2\pi} \mathrm{d}\theta_2 \ln[\mathrm{Det}\{\boldsymbol{\zeta}(\theta_1, \theta_2)\}]. \tag{17.7.12}$$

Chapter 18
Analysis

18.1 Fourier Transforms in d Dimensions

It is convenient to use the generic term *transform* to refer to both the discrete (series) and continuous (integral) cases. Here the former is related to lattices with attention confined to a hypercubic lattice. The corresponding restriction in the continuous case is to a hypercubic volume.

18.1.1 Discrete Finite Lattices

Consider a d-dimensional hypercubic lattice $\mathcal{N}$ with lattice spacing a. From (16.2.1) the sites of the lattice are

$$\boldsymbol{r} = a[n^{(1)}\hat{\boldsymbol{r}}^{(1)} + n^{(2)}\hat{\boldsymbol{r}}^{(2)} + \cdots + n^{(d)}\hat{\boldsymbol{r}}^{(d)}], \tag{18.1.1}$$

where $\hat{\boldsymbol{r}}^{(\ell)}$, $\ell = 1, 2, \ldots, d$ are the orthogonal unit axis vectors. Periodic boundary condition are applied as defined in Sect. 16.2.1, so that $n^{(\ell)}$ is an integer modulo $N^{(\ell)}$ with the number of lattice sites N given by (16.2.5). Let $f(\boldsymbol{r})$ be a function defined at each point of $\mathcal{N}$. Because of the periodic boundary conditions

$$f(\boldsymbol{r}) = f(\boldsymbol{r} + \boldsymbol{r}_{\mathrm{L}}), \tag{18.1.2}$$

where

$$\boldsymbol{r}_{\mathrm{L}} := a[m^{(1)}N^{(1)}\hat{\boldsymbol{r}}^{(1)} + m^{(2)}N^{(2)}\hat{\boldsymbol{r}}^{(2)} + \cdots + m^{(d)}N^{(d)}\hat{\boldsymbol{r}}^{(d)}], \tag{18.1.3}$$

for any integers $m^{(1)}, \ldots, m^{(d)}$. The Fourier transform $f^*(\boldsymbol{k})$ of $f(\boldsymbol{r})$ is given by[1]

[1] It is sometimes convenient to use the notation $\mathfrak{F}(f(\boldsymbol{r}))$ to denote the Fourier transform.

D.A. Lavis, *Equilibrium Statistical Mechanics of Lattice Models*,
Theoretical and Mathematical Physics, DOI 10.1007/978-94-017-9430-5_18

$$f^*(\boldsymbol{k}) := \sum_{\{\boldsymbol{r}\}} f(\boldsymbol{r}) \exp(-\mathrm{i}\boldsymbol{k} \cdot \boldsymbol{r}), \tag{18.1.4}$$

$$f(\boldsymbol{r}) = \frac{1}{N} \sum_{\{\boldsymbol{k}\}} f^*(\boldsymbol{k}) \exp(\mathrm{i}\boldsymbol{k} \cdot \boldsymbol{r}), \tag{18.1.5}$$

with the wave-vectors $\boldsymbol{k}$ given by

$$\boldsymbol{k} := \frac{2\pi}{a} \left[\frac{\kappa^{(1)}}{N^{(1)}} \hat{\boldsymbol{r}}^{(1)} + \frac{\kappa^{(2)}}{N^{(2)}} \hat{\boldsymbol{r}}^{(2)} + \cdots + \frac{\kappa^{(d)}}{N^{(d)}} \hat{\boldsymbol{r}}^{(d)} \right], \tag{18.1.6}$$

where in (18.1.5) $\kappa^{(\ell)}$ ranges over all distinct integer values modulo $N^{(\ell)}$, for $\ell = 1, 2, \ldots, d$. From (16.2.6)

$$f^*(\boldsymbol{k}) = f^*(\boldsymbol{k} + \boldsymbol{r}_{\mathrm{R}}), \tag{18.1.7}$$

where $\boldsymbol{r}_{\mathrm{R}}$ is any reciprocal lattice vector. When $d = 1$, it follows from (16.2.1), (18.1.3)–(18.1.6) that a function $f(n)$ with periodicity N has transform $f^*(\kappa)$ given by

$$f^*(\kappa) = \sum_{n=0}^{N-1} f(n) \exp\left(-\frac{2\pi \mathrm{i} \kappa n}{N}\right), \tag{18.1.8}$$

$$f(n) = \frac{1}{N} \sum_{\kappa=0}^{N-1} f^*(\kappa) \exp\left(\frac{2\pi \mathrm{i} \kappa n}{N}\right). \tag{18.1.9}$$

The periodic distribution

$$\widetilde{\delta^{\mathrm{Kr}}}(N; n) := \sum_{j=-\infty}^{\infty} \delta^{\mathrm{Kr}}(n - Nj) = \begin{cases} 1, & \text{if } n \text{ is a multiple of } N, \\ 0, & \text{otherwise,} \end{cases} \tag{18.1.10}$$

can be transformed, using (18.1.8) and (18.1.9), into

$$\widetilde{\delta^{\mathrm{Kr}}}(N; n) = \frac{1}{N} \sum_{k=0}^{N-1} \exp\left(\frac{2\pi \mathrm{i} \kappa n}{N}\right), \tag{18.1.11}$$

with $\widetilde{\delta^{\mathrm{Kr}}}(N; n) = \delta^{\mathrm{Kr}}(n)$, if $0 \le n < N$.

18.1.2 A Continuous Finite Volume

The case where $\boldsymbol{r} \in \mathcal{V}$, a hyper-cubic volume $\mathcal{V}$ of edge-lengths $L^{(1)}, \ldots, L^{(d)}$ with periodic boundary conditions and

$$|\mathcal{V}| = \widetilde{V} = \prod_{i=1}^{d} L^{(i)}, \tag{18.1.12}$$

can be obtained as the limit of the lattice case when $a \to 0$, $N_i \to \infty$ with $N^{(i)}a = L^{(i)}$, for $i = 1, 2, \ldots, d$. With $f^*(\boldsymbol{k})$ replaced by $Nf^*(\boldsymbol{k})$, Eq. (18.1.4) is replaced by an integral according to the prescription

$$\frac{1}{N}\sum_{\{\boldsymbol{r}\}} = \frac{a^d}{\widetilde{V}}\sum_{\{\boldsymbol{r}\}} \to \frac{1}{\widetilde{V}}\int_{\mathcal{V}} \mathrm{d}\mathcal{V}, \tag{18.1.13}$$

giving, in place of (18.1.4) and (18.1.5),

$$f^*(\boldsymbol{k}) := \frac{1}{\widetilde{V}}\int_{\mathcal{V}} f(\boldsymbol{r})\exp(-\mathrm{i}\boldsymbol{k}\cdot\boldsymbol{r})\,\mathrm{d}\mathcal{V}, \tag{18.1.14}$$

$$f(\boldsymbol{r}) = \sum_{\{\boldsymbol{k}\}} f^*(\boldsymbol{k})\exp(\mathrm{i}\boldsymbol{k}\cdot\boldsymbol{r}), \tag{18.1.15}$$

where the definition of the values of $\boldsymbol{k}$ given by (18.1.6) is replaced by

$$\boldsymbol{k} := 2\pi\left[\frac{\kappa^{(1)}}{L^{(1)}}\hat{\boldsymbol{r}}^{(1)} + \frac{\kappa^{(2)}}{L^{(2)}}\hat{\boldsymbol{r}}^{(2)} + \cdots + \frac{\kappa^{(d)}}{L^{(d)}}\hat{\boldsymbol{r}}^{(d)}\right], \quad k^{(i)} = 0, \pm 1, \pm 2, \ldots. \tag{18.1.16}$$

Taking the gradient of (18.1.15) gives

$$\nabla f(\boldsymbol{r}) = \mathrm{i}\sum_{\{\boldsymbol{k}\}} \boldsymbol{k} f^*(\boldsymbol{k})\exp(\mathrm{i}\boldsymbol{k}\cdot\boldsymbol{r}) \tag{18.1.17}$$

and then taking the divergence of (18.1.17) gives

$$\nabla^2 f(\boldsymbol{r}) = -\sum_{\{\boldsymbol{k}\}} k^2 f^*(\boldsymbol{k})\exp(\mathrm{i}\boldsymbol{k}\cdot\boldsymbol{r}). \tag{18.1.18}$$

Thus

$$k^2 f^*(\boldsymbol{k}) = -\frac{1}{\widetilde{V}}\int_{\mathcal{V}} \nabla^2 f(\boldsymbol{r})\exp(-\mathrm{i}\boldsymbol{k}\cdot\boldsymbol{r})\,\mathrm{d}\mathcal{V}. \tag{18.1.19}$$

For the case $d = 1$,

$$f^*(n) = \frac{1}{L}\int_{-L/2}^{L/2} f(r)\exp\left(-\frac{2\pi\mathrm{i}\kappa r}{L}\right)\mathrm{d}r, \tag{18.1.20}$$

$$f(r) = \sum_{\kappa=-\infty}^{\infty} f^*(\kappa)\exp\left(\frac{2\pi\mathrm{i}\kappa r}{L}\right). \tag{18.1.21}$$

In this instance we shall also need the convolution formula

$$\mathfrak{F}\left(\int_{-L/2}^{L/2} f(s)g(r-s)\mathrm{d}s\right) = L\, f^*(n)\, g^*(n). \tag{18.1.22}$$

The periodic distribution

$$\widetilde{\delta^{\mathrm{D}}}(L; r) := \sum_{j=-\infty}^{\infty} \delta^{\mathrm{D}}(r - jL) \tag{18.1.23}$$

of Dirac delta functions can be transformed using (18.1.20) and (18.1.21) into

$$\widetilde{\delta^{\mathrm{D}}}(L; r) = \frac{1}{L}\sum_{\kappa=-\infty}^{\infty} \exp\left(\frac{2\pi\mathrm{i}\kappa r}{L}\right), \tag{18.1.24}$$

and $\widetilde{\delta^{\mathrm{D}}}(L; r) = \delta^{\mathrm{D}}(r)$ if $0 \le r < L$. With the change of variable $r := xL/(2\pi)$, (18.1.20) and (18.1.21) give the usual Fourier series formulae for a function periodic over the interval $[-\pi, \pi]$. The condition for the convergence of the series (18.1.15) and (18.1.21) is that $f(\boldsymbol{r})$ is piecewise smooth on every finite interval in all its component variables (Courant and Hilbert 1962).

Suppose that $\varphi(x)$ is a continuous and continuously differentiable function of x so that the series

$$\sum_{n=-\infty}^{\infty} \varphi(2\pi n + y), \qquad \sum_{n=-\infty}^{\infty} \frac{\partial\varphi(2\pi n + y)}{\partial y} \tag{18.1.25}$$

converge absolutely and uniformly for $-\pi \le y < \pi$. Then the second series in (18.1.25) is the derivative of the first which can be transformed using (18.1.20) and (18.1.21) to give

$$\sum_{n=-\infty}^{\infty} \varphi(2\pi n + y) = \frac{1}{2\pi}\sum_{m=-\infty}^{\infty}\sum_{n=-\infty}^{\infty}\int_{-\pi}^{\pi} \varphi(2\pi n + x)\exp[\mathrm{i}m(y - x)]\mathrm{d}x. \tag{18.1.26}$$

Setting $y = 0$ and using the transform

$$\sum_{n=-\infty}^{\infty}\int_{-\pi}^{\pi}\varphi(2\pi n+x)\exp(-imx)\mathrm{d}x=\int_{-\infty}^{\infty}\varphi(x)\exp(-imx)\mathrm{d}x$$

gives

$$\sum_{n=-\infty}^{\infty}\varphi(2\pi n)=\frac{1}{2\pi}\sum_{m=-\infty}^{\infty}\int_{-\infty}^{\infty}\varphi(x)\exp(-imx)\mathrm{d}x, \tag{18.1.27}$$

which is the *Poisson summation formula* (Courant and Hilbert 1962).

18.1.3 A Continuous Infinite Volume

Suppose that $f^*(\boldsymbol{k})$ in Sect. 18.1.2 is replaced by $f^*(\boldsymbol{k})/\widetilde{V}$ and consider the limiting situation $L^{(i)}\to\infty$, for $i=1,2,\ldots,d$. The sum in (18.1.15) becomes an integral according to the prescription

$$\frac{1}{\widetilde{V}}\sum_{\{\boldsymbol{k}\}}=\frac{1}{(2\pi)^d}\prod_{i=1}^{d}\left(\frac{2\pi}{L_i}\right)\sum_{\{\boldsymbol{k}\}}\to\frac{1}{(2\pi)^d}\int_{\mathcal{K}}\mathrm{d}\mathcal{K}, \tag{18.1.28}$$

giving

$$f^*(\boldsymbol{k}):=\int_{\mathcal{R}}f(\boldsymbol{r})\exp(-\mathrm{i}\boldsymbol{k}\cdot\boldsymbol{r})\,\mathrm{d}\mathcal{R}, \tag{18.1.29}$$

$$f(\boldsymbol{r})=\frac{1}{(2\pi)^d}\int_{\mathcal{K}}f^*(\boldsymbol{k})\exp(\mathrm{i}\boldsymbol{k}\cdot\boldsymbol{r})\,\mathrm{d}\mathcal{K}, \tag{18.1.30}$$

where $\mathcal{R}$ and $\mathcal{K}$ are the infinite d-dimensional spaces of the vectors $\boldsymbol{r}$ and $\boldsymbol{k}$ respectively. For $d=1$ this becomes

$$f^*(\boldsymbol{k}):=\int_{-\infty}^{\infty}f(\boldsymbol{r})\exp(-\mathrm{i}k\,r)\,\mathrm{d}r, \tag{18.1.31}$$

$$f(\boldsymbol{r})=\frac{1}{(2\pi)}\int_{-\infty}^{\infty}f^*(\boldsymbol{k})\exp(\mathrm{i}k\,r)\,\mathrm{d}k. \tag{18.1.32}$$

The function f is dependent only on the magnitude r of the vector $\boldsymbol{r}$ if and only if f^* is a function only of the magnitude k of $\boldsymbol{k}$. In this case, for $d>1$, the integrals in (18.1.29) and (18.1.30) can be expressed in polar variables in the form

$$f^*(k) = \int_0^\infty \mathrm{d}r\, f(r) r \int_0^\pi \mathrm{d}\theta\, \mathcal{S}_{d-1}(r\sin\theta)\exp(-\mathrm{i}kr\cos\theta), \tag{18.1.33}$$

$$f(r) = \frac{1}{(2\pi)^d}\int_0^\infty \mathrm{d}k\, f^*(k) k \int_0^\pi \mathrm{d}\theta\, \mathcal{S}_{d-1}(k\sin\theta)\exp(\mathrm{i}rk\cos\theta), \tag{18.1.34}$$

where

$$\mathcal{S}_d(q) := 2\pi^{\frac{d}{2}} q^{d-1}/\Gamma\left(\tfrac{d}{2}\right), \tag{18.1.35}$$

is the surface area of a d-dimensional hypersphere of radius q.[2] To evaluate the integrals (18.1.33) and (18.1.34) we need the *Poisson integral formula*

$$\int_0^\pi \sin^{2\nu}\theta \exp(\pm \mathrm{i}z\cos\theta)\mathrm{d}\theta = \pi^{\frac{1}{2}}\Gamma\left(\nu + \tfrac{1}{2}\right)(2/z)^\nu \mathrm{J}_\nu(z), \qquad \Re(\nu) > -\tfrac{1}{2}, \tag{18.1.36}$$

for the *Bessel function of the first kind* $\mathrm{J}_\nu(z)$. Substituting from (18.1.35) into (18.1.33) and (18.1.34) and using (18.1.36) we have, for all $d > 1$,

$$f^*(k) = k\int_0^\infty f(r)\left(\frac{2\pi r}{k}\right)^{\frac{d}{2}} \mathrm{J}_{\frac{d}{2}-1}(kr)\mathrm{d}r, \tag{18.1.37}$$

$$f(r) = r\int_0^\infty f^*(k)\left(\frac{k}{2\pi r}\right)^{\frac{d}{2}} \mathrm{J}_{\frac{d}{2}-1}(kr)\mathrm{d}k\,. \tag{18.1.38}$$

Since $\mathrm{J}_{-\frac{1}{2}}(z) = (2/z\pi)^{\frac{1}{2}}\cos(z)$ these equations become (18.1.31) and (18.1.32) when $d = 1$.

18.1.4 Integrals Involving Bessel Functions

For Fourier transforms like (18.1.37) and (18.1.38) we need to evaluate integrals containing Bessel functions of the first kind. Cases of this occur in Sects. 3.6.1 and 5.6.3 with the integrals to be solved contained in the definitions of $\mathfrak{F}(\ell_0, k_0; x)$ and $\mathfrak{G}(k_0; x, y; z)$, given respectively by (3.6.24) and (5.6.34).
For real x and y it can be shown that

$$\lim_{k_0\to\infty} \mathfrak{G}(k_0; x, y; 0) = \Omega(x, y), \qquad \text{for } 1 - y < x < \tfrac{5}{2}, \tag{18.1.39}$$

[2] In one dimension the conditions that $f(\boldsymbol{r})$ and $f^*(\boldsymbol{k})$ are dependent only on the magnitudes of the vectors reduce to them being even functions. $\mathcal{S}_0(q)$ has just two points and can be defined as $\delta^{\mathrm{D}}(q)$, when (18.1.33) and (18.1.34) reduce to (18.1.31) and (18.1.32).

where

$$\Omega(x,y) := \frac{2^{x-2}\,\Gamma\left(\frac{1}{2}y + \frac{1}{2}x - \frac{1}{2}\right)}{\Gamma\left(\frac{1}{2}y - \frac{1}{2}x + \frac{3}{2}\right)}, \tag{18.1.40}$$

and that the corresponding limit exists for (real) $z \neq 0$ when $-1 - y < x < \frac{5}{2}$. In particular

$$\mathfrak{G}(\infty; x, x-1; z) = z^{x-1}\mathrm{K}_{x-1}(z), \quad \text{for } x < \tfrac{5}{2}, \tag{18.1.41}$$

where $\mathrm{K}_x(z)$ is the *Bessel function of complex argument,* for which we shall need the information that

$$\mathrm{K}_x(z) \simeq (\pi/2z)^{\frac{1}{2}} \exp(-z) \quad \forall\, x, \quad \text{as } |z| \to \infty, \tag{18.1.42}$$

$$\mathrm{K}_0(z) \simeq -\ln(z/2), \qquad K_1(z) \simeq 1/z, \quad \text{as } |z| \to 0, \tag{18.1.43}$$

with (18.1.42) exact for $x = \pm\frac{1}{2}$. To complete the requirements of Sect. 5.6.3 we need an asymptotic expression for $\mathfrak{G}(k_0; x, y; z)$ for $x \geq \frac{5}{2}$. A series with ascending powers of z can be generated by iterating

$$\mathfrak{G}(k_0; x, y; z) = \mathfrak{G}(k_0; x, y; 0) + z^2\mathfrak{G}(k_0; x-2, y; z) \tag{18.1.44}$$

until, whatever the value of $x > \frac{5}{2}$, the last term $z^{2n}\mathfrak{G}(k_0; x-2n, y; z)$ is such that $0 < x - 2n < \frac{5}{2}$. The most singular term in this series is the first, giving

$$\mathfrak{G}(k_0; x, y; z) \simeq \lim_{k_0\to\infty} \mathfrak{G}(k_0; x, y; 0). \tag{18.1.45}$$

The range of (18.1.39) can be extended with

$$\begin{aligned}\mathfrak{G}(k_0; x, y; 0) = {} & \Omega(x,y) + k_0(x+y-3)\mathrm{J}_y(k_0) + \mathrm{S}_{x-3,y-1}(k_0)\\ & - k_0\mathrm{J}_{y-1}(k_0)\mathrm{S}_{x-2,y}(k_0), \qquad \text{for } 1 - y < x,\end{aligned} \tag{18.1.46}$$

where $\mathrm{S}_{x,y}(z)$ is a *Lommel function.* Using the asymptotic forms

$$\mathrm{J}_y(z) \simeq \sqrt{\frac{2}{\pi z}} \cos\left(z - \tfrac{1}{2}\pi y - \tfrac{1}{4}\pi\right), \qquad \mathrm{S}_{x,y}(z) \simeq z^{x-1}, \tag{18.1.47}$$

as $z \to \infty$, gives

$$\begin{aligned}\mathfrak{G}(k_0; x, y; 0) = {} & \Omega(x,y) - (x+y-3)\sqrt{\frac{2k_0}{\pi}} \cos\left(k_0 - \tfrac{1}{2}\pi[y-1]\right) + k_0^{x-4}\\ & + k_0^{x-\frac{5}{2}}\sqrt{\frac{2}{\pi}} \cos\left(x - \tfrac{1}{4}\pi[2y-1]\right), \qquad \text{for } 1 - y < x.\end{aligned} \tag{18.1.48}$$

From (3.6.23), interest in $\mathfrak{F}(\ell_0, k_0; x)$ can be confined to the values $x = -2, -1, 0, 1, \ldots$ for which, from (18.1.46) and (18.1.47),

$$\begin{aligned}\mathfrak{F}(0, k_0; x) &= \frac{k_0^{x+1}}{x+1} - \mathfrak{G}(k_0; x+2, 0; 0)\\ &\simeq \frac{k_0^{x+1}}{x+1} - \Omega(x+2, 0) - k_0^{x-\frac{7}{2}} \cos\left(k_0 - \tfrac{1}{4}\right), \quad x = 0, 1, \ldots \end{aligned} \tag{18.1.49}$$

For the remaining cases the range of integration is divided into parts to give

$$\begin{aligned}\mathfrak{F}(\ell_0, k_0; -1) &= \int_{\ell_0}^{k_0} \frac{1 - \mathrm{J}_0(k)}{k}\, dk\\ &= \int_{\ell_0}^{1} \frac{1 - \mathrm{J}_0(k)}{k}\, dk + \ln(k_0) - \int_1^{k_0} \frac{\mathrm{J}_0(k)}{k}\, dk, \end{aligned} \tag{18.1.50}$$

$$\begin{aligned}\mathfrak{F}(\ell_0, k_0; -2) &= \int_{\ell_0}^{k_0} \frac{1 - \mathrm{J}_0(k)}{k^2}\, dk\\ &= \int_{\ell_0}^{1} \frac{1 - \mathrm{J}_0(k)}{k^2}\, dk + \frac{k_0 - 1}{k_0} - \int_1^{k_0} \frac{\mathrm{J}_0(k)}{k^2}\, dk, \end{aligned} \tag{18.1.51}$$

where

$$\lim_{\ell_0 \to 0} \int_{\ell_0}^{1} \frac{1 - \mathrm{J}_0(k)}{k}\, dk = \tfrac{1}{8}\, {}_2\mathrm{F}_3\left(1, 1; 2, 2, 2; -\tfrac{1}{4}\right) \approx 0.1212, \tag{18.1.52}$$

$$\lim_{k_0 \to \infty} \int_1^{k_0} \frac{\mathrm{J}_0(k)}{k}\, dk = \ln(2) - \gamma + \tfrac{1}{8}\, {}_2\mathrm{F}_3\left(1, 1; 2, 2, 2; -\tfrac{1}{4}\right) \approx 0.2371, \tag{18.1.53}$$

$$\begin{aligned}\lim_{\ell_0 \to 0} \int_{\ell_0}^{1} \frac{1 - \mathrm{J}_0(k)}{k^2}\, dk &= \lim_{k_0 \to \infty} \int_1^{k_0} \frac{\mathrm{J}_0(k)}{k^2}\, dk\\ &= \mathrm{J}_1(1)\, {}_1\mathrm{F}_2(1; \tfrac{3}{2}, \tfrac{3}{2}; -\tfrac{1}{4}) - \tfrac{1}{3}\, \mathrm{J}_0(1)\, {}_1\mathrm{F}_2(1; \tfrac{3}{2}, \tfrac{5}{2}, -\tfrac{1}{4})\\ &\quad + 2\mathrm{J}_0(1) - \mathrm{J}_1(1) - 1 \approx 0.2449, \end{aligned} \tag{18.1.54}$$

${}_p\mathrm{F}_q(\alpha_1, \ldots, \alpha_p; \beta_1, \ldots, \beta_q; z)$ is the *generalized hypergeometric function* and $\gamma \approx 0.5772$ is the *Euler-Mascheroni constant*.

18.1.5 Lattice Green's Functions

The inverse Fourier transform

$$G(d; \boldsymbol{r}) := \frac{1}{L^d} \sum_{\{\boldsymbol{k}\}} \frac{[\exp(i\boldsymbol{k} \cdot \boldsymbol{r}) - 1]}{k^2} \tag{18.1.55}$$

is a *Green's function* for the d-dimensional hypercubic lattice $\mathcal{N}$. For convergence in the continuum limit the wave-vector magnitude is confined to the range $\ell_1 < k < k_1$. Then, from (18.1.34) and (18.1.38),

$$\begin{aligned} G(d;r) &= \frac{1}{(2\pi)^d}\int_{\ell_1}^{k_1} \mathrm{d}k\, k^{-1}\int_0^{\pi} \mathrm{d}\theta\, \mathcal{S}_{d-1}(k\sin\theta)[\exp(irk\cos\theta)-1] \\ &= \frac{1}{2\pi}\int_{\ell_1}^{k_1} k^{-1}\left(\frac{k}{2\pi r}\right)^{\frac{d}{2}-1} \mathrm{J}_{\frac{d}{2}-1}(kr)\mathrm{d}k \\ &= \frac{1}{(2\pi)^{\frac{d}{2}}}\int_{\ell_1}^{k_1} k^{d-3}\left\{\frac{\mathrm{J}_{\frac{d}{2}-1}(kr)}{(kr)^{\frac{d}{2}-1}} - \frac{1}{2^{\frac{d}{2}-1}\Gamma\left(\frac{d}{2}\right)}\right\}\mathrm{d}k. \end{aligned} \tag{18.1.56}$$

Using (3.6.24),

$$G(2;r) = -\tfrac{1}{2\pi}\mathfrak{F}\big(\frac{2\pi r}{R}, \frac{2\pi r}{a}; -1\big), \tag{18.1.57}$$

where for use in the two-dimensional XY model this formula has been confined to the range $a \le r \le R$, so that $\ell_1 := 2\pi/R$, $k_1 := 2\pi/a$. At $r = R$ it is convenient to impose continuity by setting $G(2;r) := G(2;R)$ for $r \ge R$. A similar strategy for $r < a$ however leads, as we see, to difficulties in application to the XY model. We therefore let $G(2;r) := G(2;a) - \mu/4\pi^2$, for $r < a$ and some, as yet to be determined μ. Thus, from (18.1.50), (18.1.52) and (18.1.53), for large R and small a,

$$G(2;r) = \begin{cases} -\frac{1}{4\pi^2}\{\mu + 2\pi\ln(\pi) + 2\pi\gamma\}, & r \le a, \\ -\frac{1}{2\pi}\left\{\ln\left(\frac{r}{a}\right) + \ln(\pi) + \gamma\right\}, & a \le r \le R, \\ -\frac{1}{2\pi}\left\{\ln\left(\frac{R}{a}\right) + \ln(\pi) + \gamma\right\}, & r \ge R. \end{cases} \tag{18.1.58}$$

18.2 Doubly-Periodic and Quasi-Periodic Functions

Definition 18.2.1 A function $f(u)$, defined for $u \in \mathbb{C}_u$, is periodic, period ω, if

$$f(u+\omega) = f(u). \tag{18.2.1}$$

If $f(u)$ has periods $\omega_1, \omega_2, \ldots, \omega_n$, then it is clearly periodic with period $m_1\omega_1 + m_2\omega_2 + \cdots + m_n\omega_n$ for any integers $m_1, m_2, \ldots, m_n$. The period ω is *primitive* if $f(u)$ is not periodic, period ω/m, for some integer $m > 1$. Clearly every period is a sum of multiples of primitive periods. The following theorem is due to Jacobi:

Theorem 18.2.1 *There does not exist a nonconstant function with more than two primitive periods. There is a nonconstant function with two given primitive periods if and only if the ratio of these periods is not real.*

Since $u \in \mathbb{C}_u$ is representable as a vector this theorem is equivalent to the result that there are only two independent vectors in a plane and any point in the plane can be given in terms of two independent vectors.

Definition 18.2.2 The class of functions with primitive periods ω and ω' is *doubly-periodic* and is denoted by $\mathrm{DP}(\omega, \omega')$. The *period parallelogram* $\mathbb{W}(\omega, \omega')$ for members of this class has vertices[3] $0, \omega, \omega + \omega', \omega' \in \mathbb{C}_u$.

The values of $f(u) \in \mathrm{DP}(\omega, \omega')$ are repeated in parallelograms obtained by periodic translations of $\mathbb{W}(\omega, \omega')$. Before proceeding further we extend the definition of double-periodicity to:

Definition 18.2.3 The class $\mathrm{DP}(\omega, \omega'; \tau, \tau')$ of functions satisfying

$$f(u + \omega) = \tau f(u), \qquad f(u + \omega') = \tau' f(u), \tag{18.2.2}$$

where τ and τ' take possible values ± 1 will now be called doubly-periodic.

For this extended definition we retain the same period parallelogram but the values of $f(u) \in \mathrm{DP}(\omega, \omega'; \tau, \tau')$ are repeated with alternating signs, if τ and τ' are equal to -1, in parallelograms obtained by periodic translations of $\mathbb{W}(\omega, \omega')$. Clearly if, for example, $f(u) \in \mathrm{DP}(\omega, \omega'; -1, 1)$ then $f(u) \in \mathrm{DP}(2\omega, \omega'; 1, 1) = \mathrm{DP}(2\omega, \omega')$ and $f(u)$ is said to be *antiperiodic* in ω and *periodic* in ω'. It is further the case that if $f(u) \in \mathrm{DP}(\omega, \omega'; \tau, \tau')$ then $f'(u)/f(u) \in \mathrm{DP}(\omega, \omega')$.

Members of the class $\mathrm{DP}(\omega, \omega'; \tau, \tau')$ have a number of important properties.

Theorem 18.2.2 *If $f(u) \in \mathrm{DP}(\omega, \omega'; \tau, \tau')$ is analytic*[4] *in and on the boundary of $\mathbb{W}(\omega, \omega')$ it is a constant.*

This is a straightforward consequence of Liouville's theorem.[5]

Theorem 18.2.3 *If $f(u) \in \mathrm{DP}(\omega, \omega'; \tau, \tau')$ is meromorphic with exactly n poles β_j, $j = 1, \ldots, n$ inside $\mathbb{W}(\omega, \omega')$ and none on its boundary and m zeros α_j, $j = 1, 2, \ldots, m$ inside $\mathbb{W}(\omega, \omega')$ with none on the boundary,*[6] *then $m = n$ and*

$$\sum_{j=1}^{n} \alpha_j = \sum_{j=1}^{n} \beta_j + \tfrac{1}{4}(4\ell + 1 - \tau')\omega + \tfrac{1}{4}(4\ell' + 1 - \tau)\omega', \tag{18.2.3}$$

for some integers ℓ and ℓ'.[7]

---

[3] We assume without loss of generality that this ordering of vertices is anticlockwise round the parallelogram.

[4] See the footnote on p. 161 for the usage of the term 'analytic' for complex-valued functions. Clearly a doubly-periodic function which is analytic in and on the boundary of $\mathbb{W}(\omega, \omega')$ is an entire function.

[5] Liouville's theorem in complex variable theory (not to be confused with his proof of the conservation of measure under a Hamiltonian flow which plays an important role in statistical mechanics) states that, if a function is entire and bounded, it is constant.

[6] Here and elsewhere in this section it is assumed that multiple poles and zeros are counted according to their multiplicity and that, if necessary, the period parallelogram is displaced slightly so that no poles or zeros lie on its boundary.

[7] A version of the second part of this theorem is proved by Akhiezer (1990, pp. 8–9).

Proof Let $\mathcal{W}$ be the contour around the boundary of the parallelogram $\mathbb{W}(\omega_1, \omega_2)$ in the anticlockwise direction. It is divided into four sequential parts $\mathcal{W}_i, i = 1, \ldots, 4$ corresponding to the sides of $\mathbb{W}(\omega_1, \omega_2)$, where $\mathcal{W}_1$ is the edge from 0 to ω. Then, from Cauchy's theorem,

$$2\pi \mathrm{i}(m-n) = \int_{\mathcal{W}} \frac{f'(u)}{f(u)} du = \sum_{i=1}^{4} \int_{\mathcal{W}_i} \frac{f'(u)}{f(u)} du = 0, \tag{18.2.4}$$

since, because of the periodicity, the contribution to the contour integral from $\mathcal{W}_1$ cancels with $\mathcal{W}_3$ and $\mathcal{W}_2$ with $\mathcal{W}_4$. So $m = n$. In a similar way, using

$$\tau = \exp\left(\pm \frac{\mathrm{i}\pi[4\ell + 1 - \tau]}{2}\right), \tag{18.2.5}$$

for any integer ℓ,

$$\begin{aligned} 2\pi \mathrm{i}\Big\{\sum_{j=1}^{n}(\alpha_j - \beta_j)\Big\} &= \sum_{i=1}^{4} \int_{\mathcal{W}_i} \frac{u f'(u)}{f(u)} du = -\omega' \ln\left\{\frac{f(\omega)}{f(0)}\right\} + \omega \ln\left\{\frac{f(\omega')}{f(0)}\right\} \\ &= \tfrac{1}{2}\omega\pi \mathrm{i}(4\ell + 1 - \tau') + \tfrac{1}{2}\mathrm{i}\,\omega'\pi \mathrm{i}(4\ell' + 1 - \tau). \end{aligned} \tag{18.2.6}$$

□

Theorem 18.2.4 *If $f(u)$ and $g(u) \in \mathrm{DP}(\omega_1, \omega_2; \tau_1, \tau_2)$ are meromorphic with exactly the same poles and zeros then $f(u) = cg(u)$ for some constant c.*

Proof Consider

$$G(u) := \frac{f'(u)}{f(u)} - \frac{g'(u)}{g(u)} \in \mathrm{DP}(\omega_1, \omega_2; 1, 1). \tag{18.2.7}$$

It follows from Theorem 18.2.3, and Cauchy's theorem that both $G(u)$ and $uG(u)$ are analytic and from Theorem 18.2.2 that they are constants. So $G(u) = 0$. □

As we have seen all entire functions in $\mathrm{DP}(\omega, \omega'; \tau, \tau')$ are constant. However, this is not the case for all members (ν and ν' not both zero) of the larger class of functions defined by:

Definition 18.2.4 The class $\mathrm{QP}(\omega, \omega'; \tau, \tau'; \nu, \mu)$ of entire functions for which

$$f(u+\omega) = \tau f(u), \qquad f(u+\omega') = \tau' \mathfrak{A}(\omega, \omega'; \nu, \mu; u) f(u), \tag{18.2.8}$$

with *automorphy factor*

$$\mathfrak{A}(\omega, \omega'; \nu, \mu; u) := \exp\left(-\frac{\pi \mathrm{i}(\nu\omega' + 2\mu\, u)}{\omega}\right), \tag{18.2.9}$$

where ν and μ are integers is called *quasi-periodic*.

Analogues of the theorems for doubly-periodic functions can be obtained with proofs following those for Theorems 18.2.3 and 18.2.4 modified by

$$\frac{f'(u+\omega')}{f(u+\omega')} = \frac{f'(u)}{f(u)} - \frac{2\mu\pi \mathrm{i}}{\omega}. \tag{18.2.10}$$

Theorem 18.2.5 *If $f(u) \in \mathrm{QP}(\omega, \omega'; \tau, \tau'; \nu, \mu)$ has exactly m zeros α_j, $j = 1, 2, \ldots, m$ inside $\mathbb{W}(\omega_1, \omega_2)$ with none on the boundary, then*

$$m = \mu, \tag{18.2.11}$$

and

$$\sum_{j=1}^{m} \alpha_j = \tfrac{1}{4}\omega(4\ell + 1 - \tau' - 2\mu) + \tfrac{1}{4}\omega'(4\ell' + 1 - \tau - 2\mu), \tag{18.2.12}$$

for some integers ℓ and ℓ'.

Theorem 18.2.6 *If $f(u)$ and $g(u) \in \mathrm{QP}(\omega, \omega'; \tau, \tau'; \nu, \mu)$ have exactly the same zeros then $f(u) = cg(u)$ for some constant c.*

It is clear that, if $f(u) \in \mathrm{QP}(\omega, \omega'; \tau, \tau'; \nu, \mu)$ and $g(u) \in \mathrm{QP}(; \omega, \omega'; \bar{\tau}, \bar{\tau}'; \bar{\nu}, \bar{\mu})$, then $f(u)g(u) \in \mathrm{QP}(; \omega, \omega'; \tau\bar{\tau}, \tau'\bar{\tau}'; \nu+\bar{\nu}, \mu+\bar{\mu})$. Equations (18.2.11) and (18.2.12) for the two functions sum to give that for their product.[8]

18.3 Elliptic Integrals and Functions

For comprehensive presentations of the material in this section the reader is referred to Hancock (1958), Lawden (1989) and Akhiezer (1990).

18.3.1 Elliptic Integrals

Elliptic integrals of the first and second kind, are

[8] $\frac{1}{2}(1 - \tau\bar{\tau})$ and $\frac{1}{2}(2 - \tau - \bar{\tau})$ have the same parity.

$$F(\theta|\mathrm{k}) := \int_0^\theta \frac{\mathrm{d}\omega}{\sqrt{1-\mathrm{k}^2\sin^2(\omega)}}, \qquad E(\theta|\mathrm{k}) := \int_0^\theta \sqrt{1-\mathrm{k}^2\sin^2(\omega)}\,\mathrm{d}\omega, \tag{18.3.1}$$

respectively, where k is the *elliptic modulus* which for the moment we suppose to be real and in the range (0, 1). With θ replaced by $\mathrm{i}\theta$, formulae (18.3.1) become

$$F(\mathrm{i}\theta|\mathrm{k}) := \mathrm{i}\int_0^\theta \frac{\mathrm{d}\omega}{\sqrt{1+\mathrm{k}^2\sinh^2(\omega)}}, \qquad E(\mathrm{i}\theta|\mathrm{k}) := \mathrm{i}\int_0^\theta \sqrt{1+\mathrm{k}^2\sinh^2(\omega)}\,\mathrm{d}\omega. \tag{18.3.2}$$

We note for use in Chap. 12 that, if

$$G(\theta|\mathrm{k}) := \int_0^\theta \ln\left[1+\sqrt{1-\mathrm{k}^2\sin^2(\omega)}\right]\mathrm{d}\omega, \tag{18.3.3}$$

then

$$\mathrm{k}\frac{\mathrm{d}G(\theta|\mathrm{k})}{\mathrm{dk}} = \theta - F(\theta|\mathrm{k}). \tag{18.3.4}$$

The *complete elliptic integrals of the first and second kind*, are

$$\begin{aligned}\mathcal{K}(\mathrm{k}) &:= F\left(\tfrac{1}{2}\pi|\mathrm{k}\right) = \int_0^{\frac{\pi}{2}} \frac{\mathrm{d}\omega}{\sqrt{1-\mathrm{k}^2\sin^2(\omega)}},\\ \mathcal{E}(\mathrm{k}) &:= E\left(\tfrac{1}{2}\pi|\mathrm{k}\right) = \int_0^{\frac{\pi}{2}} \sqrt{1-\mathrm{k}^2\sin^2(\omega)}\,\mathrm{d}\omega,\end{aligned} \tag{18.3.5}$$

with $\mathcal{K}(0) = \mathcal{E}(0) = \frac{1}{2}\pi$, $\mathcal{E}(1) = 1$. $F(\theta|\mathrm{k})$ diverges as $\theta \to \infty$ along the real axis. However, from (18.3.2),

$$\mathcal{K}'(\mathrm{k}) := -\mathrm{i}\,F(\mathrm{i}\infty|\mathrm{k}) = \int_0^\infty \frac{\mathrm{d}\omega}{\sqrt{1+\mathrm{k}^2\sinh^2(\omega)}} \tag{18.3.6}$$

exists and, using the substitution $\sinh(\omega) = \tan(\gamma)$, it is not difficult to show that

$$\mathcal{K}'(\mathrm{k}) = \mathcal{K}(\mathrm{k}'), \qquad \text{where} \qquad \mathrm{k}' := \sqrt{1-\mathrm{k}^2} \tag{18.3.7}$$

is the *complementary modulus*, which like k is either non-negative or purely imaginary with a non-negative imaginary part. From (18.3.1) and (18.3.5), by differentiating under the integral sign,

$$\frac{\mathrm{d}\mathcal{K}(\mathrm{k})}{\mathrm{dk}} = \frac{\mathcal{E}(\mathrm{k}) - \mathrm{k}'^2\mathcal{K}(\mathrm{k})}{\mathrm{kk}'^2}, \qquad \frac{\mathrm{d}\mathcal{E}(\mathrm{k})}{\mathrm{dk}} = \frac{\mathcal{E}(\mathrm{k}) - \mathcal{K}(\mathrm{k})}{\mathrm{k}}. \tag{18.3.8}$$

From (18.3.7) and (18.3.8) it can be verified that $\mathcal{K}(\mathrm{k})$ and $\mathcal{K}'(\mathrm{k})$ are independent solutions of the differential equation

$$\frac{\mathrm{d}}{\mathrm{dk}}\left(\mathrm{kk}'^2\frac{\mathrm{d}f}{\mathrm{dk}}\right) = \mathrm{k}f. \tag{18.3.9}$$

Although the functions $\mathcal{K}(\mathrm{k})$ and $\mathcal{K}'(\mathrm{k})$ have been defined only for $\mathrm{k} \in (0, 1)$ the solutions of (18.3.9) provide analytic continuations of the functions into the region $|\mathrm{k}| < 1$. The further extension into the rest of the complex k plane is achieved by using the integral formula (18.3.5). It can be shown that $\mathcal{K}(\mathrm{k})$ is analytic apart from logarithmic singularities at $\mathrm{k} = \pm 1$ and consequently, from (18.3.7), that $\mathcal{K}'(\mathrm{k})$ is analytic apart from a logarithmic singularity at $\mathrm{k} = 0$. These singularities are encapsulated by the formula

$$\mathcal{K}'(\mathrm{k}) = \frac{2}{\pi}\mathcal{K}(\mathrm{k})\ln\left(\frac{4}{\mathrm{k}}\right) + w(\mathrm{k}^2), \tag{18.3.10}$$

where $w(z)$ has a power series expansion in z with radius of convergence one. $\mathcal{K}(\mathrm{k})$ is a single-valued function in the complex k plane with branch-cuts on the real axis in the intervals $(-\infty, -1]$ and $[1, \infty)$. It is real-valued on the interval $(-1, 1)$ of the real axis and on the whole of the imaginary axis. The *nome* of k is given by

$$\mathcal{Q}(\mathrm{k}) := \exp[-\pi\mathcal{K}(\mathrm{k}')/\mathcal{K}(\mathrm{k})] \tag{18.3.11}$$

and, from (18.3.7) and (18.3.8),

$$\frac{\mathrm{d}\mathcal{Q}(\mathrm{k})}{\mathrm{dk}} = \frac{\mathcal{Q}(\mathrm{k})\,\pi^2}{2\mathcal{K}^2(\mathrm{k})\mathrm{k}\,\mathrm{k}'^2}. \tag{18.3.12}$$

Henceforth, whenever it is convenient, the more compact notation $\mathcal{K} := \mathcal{K}(\mathrm{k})$, $\mathcal{K}' := \mathcal{K}(\mathrm{k}')$ and $\mathcal{Q} := \mathcal{Q}(\mathrm{k})$ is used.

The complete elliptic integral $\mathcal{K}$ and the moduli k and k′ can be expressed in the forms

$$\begin{gathered}\mathcal{K} = \frac{\pi}{2}\prod_{n=1}^{\infty}(1-\mathcal{Q}^{2n})^2(1+\mathcal{Q}^{2n-1})^4,\\ \mathrm{k} = 4\mathcal{Q}^{\frac{1}{2}}\prod_{n=1}^{\infty}\left(\frac{1+\mathcal{Q}^{2n}}{1+\mathcal{Q}^{2n-1}}\right)^4, \qquad \mathrm{k}' = \prod_{n=1}^{\infty}\left(\frac{1-\mathcal{Q}^{2n-1}}{1+\mathcal{Q}^{2n-1}}\right)^4.\end{gathered} \tag{18.3.13}$$

18.3.2 Jacobi Theta Functions

One way to define the *Jacobi theta functions* $\mathsf{H}(u|\mathrm{k})$ and $\Theta(u|\mathrm{k})$ is[9]

$$\mathsf{H}(u|\mathrm{k}) := 2\mathcal{Q}^{\frac{1}{4}} \sin\left(\frac{\pi\, u}{2\mathcal{K}}\right) \prod_{n=1}^{\infty}\left(1-\mathcal{Q}^{2n}\right)\left\{1-2\mathcal{Q}^{2n}\cos\left(\frac{\pi\, u}{\mathcal{K}}\right)+\mathcal{Q}^{4n}\right\}, \tag{18.3.14}$$

$$\Theta(u|\mathrm{k}) := \prod_{n=1}^{\infty}\left(1-\mathcal{Q}^{2n}\right)\left\{1-2\mathcal{Q}^{2n-1}\cos\left(\frac{\pi\, u}{\mathcal{K}}\right)+\mathcal{Q}^{4n-2}\right\}, \tag{18.3.15}$$

from which it is clear that they are entire with simple zeros, that $\mathsf{H}(u|\mathrm{k})$ and $\Theta(u|\mathrm{k})$ are, respectively, odd and even functions of u and, using (18.3.13), that

$$\mathsf{H}(0|\mathrm{k}) = 0, \quad \Theta(\mathcal{K}|\mathrm{k}) = \sqrt{\frac{2\mathcal{K}}{\pi}}, \quad \frac{\mathsf{H}(\mathcal{K}|\mathrm{k})}{\Theta(\mathcal{K}|\mathrm{k})} = \sqrt{\mathrm{k}}, \quad \frac{\Theta(0|\mathrm{k})}{\Theta(\mathcal{K}|\mathrm{k})} = \sqrt{\mathrm{k}'}. \tag{18.3.16}$$

It is also useful to know that

$$\mathsf{H}'(0|\mathrm{k}) = \frac{\pi}{2\mathcal{K}}\mathsf{H}(\mathcal{K}|\mathrm{k})\,\Theta\,(\mathcal{K}|\mathrm{k})\,\Theta\,(0|\mathrm{k}), \tag{18.3.17}$$

$$\mathsf{H}(2u|\mathrm{k}) = \frac{2\mathsf{H}(u|\mathrm{k})\,\Theta\,(u|\mathrm{k})\mathsf{H}(u+\mathcal{K}|\mathrm{k})\,\Theta\,(u+\mathcal{K}|\mathrm{k})}{\mathsf{H}(\mathcal{K}|\mathrm{k})\,\Theta\,(\mathcal{K}|\mathrm{k})\,\Theta\,(0|\mathrm{k})}. \tag{18.3.18}$$

It can be shown that

$$\mathsf{H}(u+\mho(2\ell,\ell'|\mathrm{k})|\mathrm{k}) = (-1)^{\ell}(-\mathrm{i})^{\ell'}\mathfrak{F}_{\ell'}(u+2\ell\mathcal{K}|\mathrm{k}) \times \begin{cases} \mathsf{H}(u|\mathrm{k}), & \ell' \text{ even},\\ \Theta(u|\mathrm{k}), & \ell' \text{ odd}, \end{cases} \tag{18.3.19}$$

$$\Theta(u+\mho(2\ell,\ell'|\mathrm{k})|\mathrm{k}) = (-\mathrm{i})^{\ell'}\mathfrak{F}_{\ell'}(u+2\ell\mathcal{K}|\mathrm{k}) \times \begin{cases} \Theta(u|\mathrm{k}), & \ell' \text{even},\\ \mathsf{H}(u|\mathrm{k}), & \ell' \text{odd}, \end{cases} \tag{18.3.20}$$

[9] The other two theta functions, which are avoided in this text, are $\mathsf{H}_1(u|\mathrm{k}) := \mathsf{H}(u+\mathcal{K}|\mathrm{k})$ and $\Theta_1(u|\mathrm{k}) := \Theta(u+\mathcal{K}|\mathrm{k})$. In the interests of clarity it should be observed that the notation $\vartheta_i(u)$, $i = 1,\ldots,4$, used in many texts for the theta functions, is related to this notation by $\mathsf{H}(u|\mathrm{k}) := \vartheta_1(\frac{\pi u}{2\mathcal{K}}|\mathcal{Q})$, $\mathsf{H}_1(u|\mathrm{k}) := \vartheta_2(\frac{\pi u}{2\mathcal{K}}|\mathcal{Q})$, $\Theta_1(u|\mathrm{k}) := \vartheta_3(\frac{\pi u}{2\mathcal{K}}|\mathcal{Q})$, $\Theta(u|\mathrm{k}) := \vartheta_4(\frac{\pi u}{2\mathcal{K}}|\mathcal{Q})$.

where

$$\mho(\ell, \ell'|\mathrm{k}) := \ell\,\mathcal{K} + \ell'\,\mathrm{i}\,\mathcal{K}', \tag{18.3.21}$$

$$\mathfrak{F}_\ell(u|\mathrm{k}) := \exp\left(\frac{-\mathrm{i}\pi\,\ell(\ell\,\mathrm{i}\,\mathcal{K}' + 2\,u)}{4\mathcal{K}}\right) = \mathcal{Q}^{-\frac{1}{4}\ell^2} \exp\left(\frac{-\pi\,\ell\,\mathrm{i}\,u}{2\mathcal{K}}\right), \tag{18.3.22}$$

with the special cases

$$\mathsf{H}(u \pm \mathrm{i}\mathcal{K}'|\mathrm{k}) = \mp\mathrm{i}\,\mathfrak{F}_1(\pm u|\mathrm{k})\,\Theta\,(u|\mathrm{k}), \quad \Theta(u \pm \mathrm{i}\mathcal{K}'|\mathrm{k}) = \mp\mathrm{i}\,\mathfrak{F}_1(\pm u|\mathrm{k})\mathsf{H}(u|\mathrm{k}), \tag{18.3.23}$$

$$\mathsf{H}(u \pm 2\mathcal{K}|\mathrm{k}) = -\mathsf{H}(u|\mathrm{k}), \quad \Theta(u \pm 2\mathcal{K}|\mathrm{k}) = \Theta(u|\mathrm{k}), \tag{18.3.24}$$

$$\mathsf{H}(u \pm 2\mathrm{i}\mathcal{K}'|\mathrm{k}) = -\mathfrak{F}_2(\pm u|\mathrm{k})\mathsf{H}(u|\mathrm{k}), \quad \Theta(u \pm 2\mathrm{i}\mathcal{K}'|\mathrm{k}) = -\mathfrak{F}_2(\pm u|\mathrm{k})\,\Theta\,(u|\mathrm{k}). \tag{18.3.25}$$

We also have

$$\begin{aligned}
&\mathsf{H}(\mathrm{i}\,u|\mathrm{k}) = \mathrm{i}\,\mathfrak{X}(u|\mathrm{k})\mathsf{H}(u|\mathrm{k}'), \quad \mathsf{H}(\mathrm{i}\,u + \mathcal{K}|\mathrm{k}) = \mathfrak{X}(u|\mathrm{k})\,\Theta\,(u|\mathrm{k}'),\\
&\Theta(\mathrm{i}\,u|\mathrm{k}) = \mathfrak{X}(u|\mathrm{k})\mathsf{H}(u + \mathcal{K}'|\mathrm{k}'), \quad \Theta(\mathrm{i}\,u + \mathcal{K}|\mathrm{k}) = \mathfrak{X}(u|\mathrm{k})\,\Theta\,(u + \mathcal{K}'|\mathrm{k}'),
\end{aligned} \tag{18.3.26}$$

where

$$\mathfrak{X}(u|\mathrm{k}) := \sqrt{\frac{\mathcal{K}}{\mathcal{K}'}}\,\exp\left(\frac{\pi\,u^2}{4\,\mathcal{K}\,\mathcal{K}'}\right), \tag{18.3.27}$$

in which the constant is evaluated using (18.3.16).

According to the notation of Sect. 18.2,

$$\begin{aligned}
&\mathsf{H}(u|\mathrm{k}) \in \mathrm{QP}(2\mathcal{K}, 2\mathrm{i}\,\mathcal{K}'; -1, -1; 1, 1), \quad \Theta(u|\mathrm{k}) \in \mathrm{QP}(2\mathcal{K}, 2\mathrm{i}\,\mathcal{K}'; 1, -1; 1, 1),\\
&\mathsf{H}(u + \mathcal{K}|\mathrm{k}) \in \mathrm{QP}(2\mathcal{K}, 2\mathrm{i}\,\mathcal{K}'; -1, 1; 1, 1), \quad \Theta(u + \mathcal{K}|\mathrm{k}) \in \mathrm{QP}(2\mathcal{K}, 2\mathrm{i}\,\mathcal{K}'; 1, 1; 1, 1),
\end{aligned} \tag{18.3.28}$$

in which, from (18.3.14) and (18.3.15), for $\mathrm{k} \in [0, 1]$:

(i) On the real-u axis and on $\Re\{u\} = \mathcal{K}$ both $\mathsf{H}(u|\mathrm{k})$ and $\Theta(u|\mathrm{k})$ are real-valued.
(ii) On the imaginary-u axis $\mathsf{H}(u|\mathrm{k})$ is purely imaginary and $\Theta(u|\mathrm{k})$ is real-valued.

Theorem 18.2.5 predicts a single zero for each elliptic function in the period parallelogram $\mathbb{W}(2\mathcal{K}, 2\mathrm{i}\mathcal{K}')$ (which now becomes a rectangle)[10] and

$$\begin{array}{lll} \mathrm{H}(u|\mathrm{k}) & \text{has zeros at} & u = \mho(2\ell, 2\ell'|\mathrm{k}), \\ \Theta(u|\mathrm{k}) & \text{has zeros at} & u = \mho(2\ell, 2\ell'+1|\mathrm{k}), \end{array} \quad \text{for all integers } \ell \text{ and } \ell'. \tag{18.3.29}$$

This then leads to the important theorem:

Theorem 18.3.1 *For some complex quantities α_j, β_j, $j = 1, 2, \ldots, r$, lying in the period rectangle $\mathbb{W}(2\mathcal{K}, 2\mathrm{i}\,\mathcal{K}')$,*

$$W(u|\mathrm{k}) := \exp\left(-\frac{\pi \mathrm{i} \aleph u}{2\mathcal{K}}\right) \prod_{j=1}^{r} \frac{\mathrm{H}(u-\alpha_j|\mathrm{k})}{\mathrm{H}(u-\beta_j|\mathrm{k})} \tag{18.3.30}$$

is, to within a multiplicative quantity possibly depending on k, *the unique member of* $\mathrm{DP}(2\mathcal{K}, 2\mathrm{i}\,\mathcal{K}'; \tau, \tau')$ *with zeros α_j, $j = 1, 2, \ldots, r$ and poles β_j, $j = 1, 2, \ldots, r$ if*

$$\aleph = \tfrac{1}{2}(4\ell' + 1 - \tau), \quad \sum_{j=1}^{r}(\alpha_j - \beta_j) = \mathrm{i}\aleph\mathcal{K}' + \tfrac{1}{2}\mathcal{K}(4\ell + 1 - \tau'), \tag{18.3.31}$$

for some integers ℓ and ℓ'.

Proof From (18.3.24) and (18.3.25),

$$\begin{aligned} W(u+2\mathcal{K}|\mathrm{k}) &= \exp(-\pi \mathrm{i} \aleph) W(u|\mathrm{k}), \\ W(u+2\mathrm{i}\mathcal{K}'|\mathrm{k}) &= \exp\left(\frac{\pi[\aleph\mathcal{K}' + \mathrm{i}\sum_{j=1}^{m}(\alpha_j - \beta_j)]}{\mathcal{K}}\right) W(u|\mathrm{k}). \end{aligned} \tag{18.3.32}$$

Using (18.2.5), the conditions that $W(u|\mathrm{k}) \in \mathrm{DP}(2\mathcal{K}, 2\mathrm{i}\,\mathcal{K}'; \tau, \tau')$ are (18.3.31). That it is, to within a multiplicative quantity possibly dependent on k, the unique member of $\mathrm{DP}(2\mathcal{K}, 2\mathrm{i}\,\mathcal{K}'; \tau, \tau')$ with these zeros and poles follows from Theorem 18.2.4. □

The uniqueness Theorems 18.2.4 and 18.2.6 can be used to establish identities between theta functions. In particular

$$\begin{aligned} \mathrm{H}(v+u|\mathrm{k})\mathrm{H}(t+u|\mathrm{k})\,\Theta\,(t+v|\mathrm{k})\,\Theta\,(0|\mathrm{k}) \\ = \mathrm{H}(v|\mathrm{k})\mathrm{H}(t|\mathrm{k})\,\Theta\,(u|\mathrm{k})\,\Theta\,(v+t+u|\mathrm{k}) \\ + \Theta(v|\mathrm{k})\,\Theta\,(t|\mathrm{k})\mathrm{H}(u|\mathrm{k})\mathrm{H}(v+t+u|\mathrm{k}), \end{aligned} \tag{18.3.33}$$

---

[10] In the case of $\Theta(u|\mathrm{k})$ it must be displaced slightly so that the zero at $u = \mathrm{i}\mathcal{K}'$ does not lie on the boundary.

$$\begin{aligned}\Theta(u+v|\mathrm{k})\,\Theta(t+u|\mathrm{k})\,\Theta(t+v|\mathrm{k})\,\Theta(0|\mathrm{k}) \\ &= \Theta(v|\mathrm{k})\,\Theta(t|\mathrm{k})\,\Theta(u|\mathrm{k})\,\Theta(v+t+u|\mathrm{k}) \\ &+ \mathrm{H}(v|\mathrm{k})\mathrm{H}(t|\mathrm{k})\mathrm{H}(u|\mathrm{k})\mathrm{H}(v+t+u|\mathrm{k}),\end{aligned} \tag{18.3.34}$$

$$\begin{aligned}\mathrm{H}(\mathcal{K}+u|\mathrm{k})\mathrm{H}(\mathcal{K}-v|\mathrm{k})+\Theta(\mathcal{K}+u|\mathrm{k})\,\Theta(\mathcal{K}-v|\mathrm{k}) \\ &= 2\mathcal{Q}^{\frac{1}{4}}[\mathrm{H}(\mathcal{K}|\mathrm{k})\,\Theta(\mathcal{K}|\mathrm{k})]^{-1} \\ &\times \mathrm{H}(\tfrac{1}{2}[\mathrm{i}\,\mathcal{K}'+u+v]+\mathcal{K}|\mathrm{k})\mathrm{H}(\tfrac{1}{2}[\mathrm{i}\,\mathcal{K}'-u-v]+\mathcal{K}|\mathrm{k}) \\ &\times \mathrm{H}(\tfrac{1}{2}[\mathrm{i}\,\mathcal{K}'+u-v]+\mathcal{K}|\mathrm{k})\mathrm{H}(\tfrac{1}{2}[\mathrm{i}\,\mathcal{K}'-u+v]+\mathcal{K}|\mathrm{k}).\end{aligned} \tag{18.3.35}$$

18.3.3 Jacobi Elliptic Functions

The *Jacobi amplitude* $\mathrm{am}(u|\mathrm{k})$, defined by

$$F(\mathrm{am}(u|\mathrm{k})|\mathrm{k}) = u, \tag{18.3.36}$$

is the inverse of the elliptic integral of the first kind. It is clear that it is an odd function of u and, from (18.3.1)–(18.3.7), it can be shown that

$$\begin{aligned}&\mathrm{am}(u|0) = u, \qquad \mathrm{am}(u|1) = \arcsin(\tanh(u)), \\ &\mathrm{am}(\mathrm{i}\,u|\mathrm{k}) = \arcsin(\mathrm{i}\,\tan(\mathrm{am}(u|\mathrm{k}'))), \\ &\lim_{\epsilon\to\pm 0}\mathrm{am}(\epsilon+2\mathcal{K}|\mathrm{k}) = \mp\pi, \qquad \lim_{\epsilon\to\pm 0}\mathrm{am}(\epsilon+2\,\mathrm{i}\,\mathcal{K}'|\mathrm{k}) = \pm\pi.\end{aligned} \tag{18.3.37}$$

One way to define the *Jacobi elliptic functions* $\mathrm{sn}(u|\mathrm{k})$, $\mathrm{cn}(u|\mathrm{k})$ and $\mathrm{dn}(u|\mathrm{k})$ is

$$\begin{aligned}\mathrm{sn}(u|\mathrm{k}) &:= \sin(\mathrm{am}(u|\mathrm{k})), \\ \mathrm{cn}(u|\mathrm{k}) &:= \cos(\mathrm{am}(u|\mathrm{k})) = \sqrt{1-\mathrm{sn}^2(u|\mathrm{k})}, \\ \mathrm{dn}(u|\mathrm{k}) &:= \frac{\mathrm{d}\,\mathrm{am}(u|\mathrm{k})}{\mathrm{d}u} = \sqrt{1-\mathrm{k}^2\mathrm{sn}^2(u|\mathrm{k})},\end{aligned} \tag{18.3.38}$$

giving the useful relations

$$\mathrm{am}(u|\mathrm{k}) = -\mathrm{i}\,\ln[\mathrm{cn}(u|\mathrm{k})+\mathrm{i}\,\mathrm{sn}(u|\mathrm{k})], \quad \mathrm{am}(u|\mathrm{k}) = \int_0^u \mathrm{dn}(v|\mathrm{k})\mathrm{d}\,v \tag{18.3.39}$$

and the derivatives

$$\begin{aligned}&\mathrm{sn}'(u|\mathrm{k}) = \mathrm{cn}(u|\mathrm{k})\mathrm{dn}(u|\mathrm{k}), \qquad \mathrm{cn}'(u|\mathrm{k}) = -\mathrm{sn}(u|\mathrm{k})\mathrm{dn}(u|\mathrm{k}), \\ &\mathrm{dn}'(u|\mathrm{k}) = -\mathrm{k}^2\mathrm{sn}(u|\mathrm{k})\mathrm{cn}(u|\mathrm{k}).\end{aligned} \tag{18.3.40}$$

From (18.3.37),

$$\mathrm{sn}(u|0) = \sin(u), \qquad \mathrm{cn}(u|0) = \cos(u), \qquad \mathrm{dn}(u|0) = 1. \tag{18.3.41}$$

We shall henceforth, and in Chap. 12, use, whenever convenient, a useful notation which was invented by Glaisher to express reciprocals and quotients of elliptic functions (Whittaker and Watson 1927, Sect. 22.122). For reciprocals the letters of the function are reversed and for quotients the first letters of the numerator and denominator are used to denote the quantity. Thus, for example,

$$\mathrm{ns}(u|\mathrm{k}) := \frac{1}{\mathrm{sn}(u|\mathrm{k})}, \qquad \mathrm{sc}(u|\mathrm{k}) := \frac{\mathrm{sn}(u|\mathrm{k})}{\mathrm{cn}(u|\mathrm{k})}. \tag{18.3.42}$$

In the spirit of Glaisher we also find it useful to define

$$\mathrm{cds}(u|\mathrm{k}) := \frac{\mathrm{cn}(u|\mathrm{k})\mathrm{dn}(u|\mathrm{k})}{\mathrm{sn}(u|\mathrm{k})}. \tag{18.3.43}$$

An alternative way to defined the elliptic functions is[11]

$$\mathrm{sn}(u|\mathrm{k}) := \frac{\mathrm{H}(u|\mathrm{k})}{\sqrt{\mathrm{k}}\,\Theta\,(u|\mathrm{k})}, \quad \mathrm{cn}(u|\mathrm{k}) := \sqrt{\frac{\mathrm{k}'}{\mathrm{k}}}\frac{\mathrm{H}(u+\mathcal{K}|\mathrm{k})}{\Theta(u|\mathrm{k})},$$
$$\mathrm{dn}(u|\mathrm{k}) := \frac{\sqrt{\mathrm{k}'}\,\Theta\,(u+\mathcal{K}|\mathrm{k})}{\Theta(u|\mathrm{k})}, \tag{18.3.44}$$

thus making it clear:

(i) from (18.3.28), that $\mathrm{sn}(u|\mathrm{k}) \in \mathrm{DP}(2\mathcal{K}, 2\mathrm{i}\,\mathcal{K}'; -1, 1)$, $\mathrm{cn}(u|\mathrm{k}) \in \mathrm{DP}(2\mathcal{K}, 2\mathrm{i}\,\mathcal{K}'; -1, -1)$ and $\mathrm{dn}(u|\mathrm{k}) \in \mathrm{DP}(2\mathcal{K}, 2\mathrm{i}\,\mathcal{K}'; 1, -1)$;
(ii) that, for real $u \in (0, \mathcal{K})$, $\mathrm{sn}(u|\mathrm{k})$, $\mathrm{cn}(u|\mathrm{k})$ and $\mathrm{dn}(u|\mathrm{k})$ are real-valued and positive, with $\mathrm{sn}(u|\mathrm{k})$ monotonically increasing and $\mathrm{cn}(u|\mathrm{k})$ and $\mathrm{dn}(u|\mathrm{k})$ monotonically decreasing;
(iii) that, for real $u \in (0, \mathcal{K}')$, $-\mathrm{i}\,\mathrm{sn}(\mathrm{i}\,u|\mathrm{k})$, $\mathrm{cn}(\mathrm{i}\,u|\mathrm{k})$ and $\mathrm{dn}(\mathrm{i}\,u|\mathrm{k})$ are real-valued, positive and monotonically increasing.

From (18.3.19)–(18.3.22),

$$\mathrm{sn}(u \pm \mho(2\ell, \ell'|\mathrm{k})|\mathrm{k}) = (-1)^{\ell} \times \begin{cases} \mathrm{sn}(u|\mathrm{k}), & \ell' \text{ even}, \\ \mathrm{k}^{-1}\mathrm{ns}(u|\mathrm{k}), & \ell' \text{ odd}, \end{cases} \tag{18.3.45}$$

$$\mathrm{cn}(u \pm \mho(2\ell, \ell'|\mathrm{k})|\mathrm{k}) = (-1)^{\ell}(\mathrm{i})^{-\ell'} \times \begin{cases} \mathrm{cn}(u|\mathrm{k}), & \ell' \text{ even}, \\ \mathrm{k}^{-1}\mathrm{ds}(u|\mathrm{k}), & \ell' \text{ odd}, \end{cases} \tag{18.3.46}$$

[11] It should be noted, for use in Chap. 12, that the expression for $\mathrm{sn}(u|\mathrm{k})$, but not $\mathrm{cn}(u|\mathrm{k})$ and $\mathrm{dn}(u|\mathrm{k})$, takes the same form in terms of Baxter's modified theta functions.

$$\mathrm{dn}(u \pm \mho(2\ell, \ell'|\mathrm{k})|\mathrm{k}) = (\mathrm{i})^{-\ell'} \times \begin{cases} \mathrm{dn}(u|\mathrm{k}), & \ell' \text{ even}, \\ \mathrm{cs}(u|\mathrm{k}), & \ell' \text{ odd}, \end{cases} \tag{18.3.47}$$

with special cases

$$\mathrm{sn}(u - \mho(\ell, \ell'|\mathrm{k})|\mathrm{k}) = (-1)^{\ell}\mathrm{sn}(u + \mho(\ell, \ell'|\mathrm{k})|\mathrm{k}), \tag{18.3.48}$$
$$\mathrm{sn}(u \pm 2\,\mho_0|\mathrm{k}) = \mathrm{sn}(u|\mathrm{k}), \tag{18.3.49}$$
$$\mathrm{sn}\left(u + \tfrac{1}{2}\mathrm{i}\,\mathcal{K}'|\mathrm{k}\right)\,\mathrm{sn}\left(u - \tfrac{1}{2}\mathrm{i}\,\mathcal{K}'|\mathrm{k}\right) = 1/\mathrm{k}, \tag{18.3.50}$$
$$\mathrm{cds}\left(u + \tfrac{1}{2}\mathrm{i}\,\mathcal{K}'|\mathrm{k}\right) = -\mathrm{cds}\left(u - \tfrac{1}{2}\mathrm{i}\,\mathcal{K}'|\mathrm{k}\right). \tag{18.3.51}$$

Integrating the last of Eqs. (18.3.47) and setting the constant using (18.3.37) gives

$$\mathrm{am}(u + 2\mathcal{K}|\mathrm{k}) = -\mathrm{am}(u + 2\,\mathrm{i}\,\mathcal{K}'|\mathrm{k}) = \mathrm{am}(u|\mathrm{k}) \pm \pi, \tag{18.3.52}$$

where the upper a lower signs correspond to the left and right of the discontinuity at $u = 0$. From (18.3.26) and (18.3.44),

$$\mathrm{sn}(\mathrm{i}\,u|\mathrm{k}) = \mathrm{i}\,\mathrm{sc}(u|\mathrm{k}'), \quad \mathrm{cn}(\mathrm{i}\,u|\mathrm{k}) = \mathrm{nc}(u|\mathrm{k}'), \quad \mathrm{dn}(\mathrm{i}\,u|\mathrm{k}) = \mathrm{dc}(u|\mathrm{k}'), \tag{18.3.53}$$

and, from (18.3.16), (18.3.25) and (18.3.44),

$$\begin{aligned} &\mathrm{sn}(0|\mathrm{k}) = 0, && \mathrm{cn}(0|\mathrm{k}) = 1, && \mathrm{dn}(0|\mathrm{k}) = 1, \\ &\mathrm{sn}\left(\tfrac{1}{2}\mathrm{i}\,\mathcal{K}'|\mathrm{k}\right) = \frac{\mathrm{i}}{\sqrt{\mathrm{k}}}, && \mathrm{cn}\left(\tfrac{1}{2}\mathrm{i}\,\mathcal{K}'|\mathrm{k}\right) = \sqrt{1+\mathrm{k}^{-1}}, && \mathrm{dn}\left(\tfrac{1}{2}\mathrm{i}\,\mathcal{K}'|\mathrm{k}\right) = \sqrt{1+\mathrm{k}}, \end{aligned} \tag{18.3.54}$$

and hence, from (18.3.45)–(18.3.47),

$$\mathrm{sn}(\mathrm{i}\,\mathcal{K}'|\mathrm{k}) = \mathrm{cn}(\mathrm{i}\,\mathcal{K}'|\mathrm{k}) = \mathrm{dn}(\mathrm{i}\,\mathcal{K}'|\mathrm{k}) = \infty, \tag{18.3.55}$$

but

$$\mathrm{cs}(\mathrm{i}\,\mathcal{K}'|\mathrm{k}) = -\mathrm{i}, \qquad \mathrm{ds}(\mathrm{i}\,\mathcal{K}'|\mathrm{k}) = -\mathrm{i}\,\mathrm{k}. \tag{18.3.56}$$

All three elliptic functions are meromorphic with simple poles at the roots of $\Theta(u|\mathrm{k})$, that is when $u = \mho(2\ell, 2\ell' + 1|\mathrm{k})$, for all integer values of ℓ and ℓ'. From (18.3.38), $\mathrm{cn}(\mho(2\ell, 2\ell')|\mathrm{k}) = 1$ and it follows from (18.3.39) that the zeros of $\mathrm{am}(u|\mathrm{k})$ coincide with those of $\mathrm{sn}(u|\mathrm{k})$ which are at $u = \mho(2\ell, 2\ell'|\mathrm{k})$.

From (18.3.33), (18.3.34) and (18.3.44),

$$\mathrm{sn}(u \pm v|\mathrm{k}) = \frac{\mathrm{sn}(u|\mathrm{k})\mathrm{cn}(v|\mathrm{k})\mathrm{dn}(v|\mathrm{k}) \pm \mathrm{sn}(v|\mathrm{k})\mathrm{cn}(u|\mathrm{k})\mathrm{dn}(u|\mathrm{k})}{1 - \mathrm{k}^2\mathrm{sn}^2(u|\mathrm{k})\mathrm{sn}^2(v|\mathrm{k})}, \tag{18.3.57}$$

$$\mathrm{cn}(u \pm v|\mathrm{k}) = \frac{\mathrm{cn}(u|\mathrm{k})\mathrm{cn}(v|\mathrm{k}) \mp \mathrm{sn}(u|\mathrm{k})\mathrm{sn}(v|\mathrm{k})\mathrm{dn}(u|\mathrm{k})\mathrm{dn}(v|\mathrm{k})}{1 - \mathrm{k}^2\mathrm{sn}^2(u|\mathrm{k})\mathrm{sn}^2(v|\mathrm{k})}, \tag{18.3.58}$$

$$\mathrm{dn}(u \pm v|\mathrm{k}) = \frac{\mathrm{dn}(u|\mathrm{k})\mathrm{dn}(v|\mathrm{k}) \mp \mathrm{k}^2\mathrm{sn}(u|\mathrm{k})\mathrm{sn}(v|\mathrm{k})\mathrm{cn}(u|\mathrm{k})\mathrm{cn}(v|\mathrm{k})}{1 - \mathrm{k}^2\mathrm{sn}^2(u|\mathrm{k})\mathrm{sn}^2(v|\mathrm{k})}, \tag{18.3.59}$$

from which it follows that

$$\mathrm{sn}(u + v|\mathrm{k})\mathrm{sn}(u - v) = \frac{\mathrm{sn}^2(u|\mathrm{k}) - \mathrm{sn}^2(v|\mathrm{k})}{1 - \mathrm{k}^2\mathrm{sn}^2(u|\mathrm{k})\mathrm{sn}^2(v|\mathrm{k})}, \tag{18.3.60}$$

$$\mathrm{dn}(u \pm v|\mathrm{k}) - \mathrm{dn}(u|\mathrm{k})\mathrm{dn}(v|\mathrm{k}) = \mp\mathrm{k}^2\mathrm{cn}(u \pm v|\mathrm{k})\mathrm{sn}(u|\mathrm{k})\mathrm{sn}(v|\mathrm{k}), \tag{18.3.61}$$

$$\mathrm{dn}(u + v|\mathrm{k})\mathrm{sn}(v|\mathrm{k}) = \mathrm{sn}(u + v|\mathrm{k})\mathrm{cn}(u|\mathrm{k})\mathrm{dn}(v|\mathrm{k}) - \mathrm{cn}(u + v|\mathrm{k})\mathrm{sn}(u|\mathrm{k}), \tag{18.3.62}$$

$$\mathrm{cn}(u|\mathrm{k})\mathrm{dn}(u|\mathrm{k})\mathrm{sn}(2u|\mathrm{k}) = \mathrm{sn}(u|\mathrm{k})[\mathrm{dn}(2u|\mathrm{k}) + \mathrm{cn}(2u|\mathrm{k})], \tag{18.3.63}$$

$$\begin{aligned} &\frac{\mathrm{cn}(u + v|\mathrm{k}) \mp \mathrm{i}\,\mathrm{sn}(u + v|\mathrm{k})}{\mathrm{dn}(u + v|\mathrm{k}) \pm \mathrm{i}\,\mathrm{k}\,\mathrm{sn}(u + v|\mathrm{k})} \\ &= \frac{[\mathrm{cn}(u|\mathrm{k}) \mp \mathrm{i}\,\mathrm{sn}(u|\mathrm{k})\mathrm{dn}(v|\mathrm{k})][\mathrm{cn}(v|\mathrm{k}) \mp \mathrm{i}\,\mathrm{sn}(v|\mathrm{k})\mathrm{dn}(u|\mathrm{k})]}{[\mathrm{dn}(u|\mathrm{k}) \pm \mathrm{i}\,\mathrm{k}\mathrm{sn}(u|\mathrm{k})\mathrm{cn}(v|\mathrm{k})][\mathrm{dn}(v|\mathrm{k}) \pm \mathrm{i}\,\mathrm{k}\,\mathrm{sn}(v|\mathrm{k})\mathrm{cn}(u|\mathrm{k})]}, \end{aligned} \tag{18.3.64}$$

$$\mathrm{cds}^2(u|\mathrm{k}) - \mathrm{cds}^2(v|\mathrm{k}) = \frac{[\mathrm{sn}^2(v|\mathrm{k}) - \mathrm{sn}^2(u|\mathrm{k})][1 - \mathrm{k}^2\mathrm{sn}^2(u|\mathrm{k})\mathrm{sn}^2(v|\mathrm{k})]}{\mathrm{sn}^2(u|\mathrm{k})\mathrm{sn}^2(v|\mathrm{k})}, \tag{18.3.65}$$

$$\begin{aligned} &\mathrm{sn}^2(u|\mathrm{k}) = \frac{1 - \mathrm{cn}(2u|\mathrm{k})}{1 + \mathrm{dn}(2u|\mathrm{k})}, \qquad \mathrm{cn}^2(u|\mathrm{k}) = \frac{\mathrm{cn}(2u|\mathrm{k}) + \mathrm{dn}(2u|\mathrm{k})}{1 + \mathrm{dn}(2u|\mathrm{k})}, \\ &\mathrm{dn}^2(u|\mathrm{k}) = \frac{\mathrm{cn}(2u|\mathrm{k}) + \mathrm{dn}(2u|\mathrm{k})}{1 + \mathrm{cn}(2u|\mathrm{k})}, \end{aligned} \tag{18.3.66}$$

$$\begin{aligned} \mathrm{k}\,\mathrm{sn}^2\left(u \pm \tfrac{1}{2}\,\mathrm{i}\,\mathcal{K}'|\mathrm{k}\right) &= \frac{-\mathrm{cn}(2u|\mathrm{k}) \pm \mathrm{i}\,\mathrm{sn}(2u|\mathrm{k})}{\mathrm{dn}(2u|\mathrm{k}) \pm \mathrm{i}\,\mathrm{k}\,\mathrm{sn}(2u|\mathrm{k})}, \\ \mathrm{k}\,\mathrm{cn}^2\left(u \pm \tfrac{1}{2}\,\mathrm{i}\,\mathcal{K}'|\mathrm{k}\right) &= \frac{\mathrm{dn}(2u|\mathrm{k}) + \mathrm{k}\,\mathrm{cn}(2u|\mathrm{k})}{\mathrm{cn}(2u|\mathrm{k}) \pm \mathrm{i}\,\mathrm{sn}(2u|\mathrm{k})}, \\ \mathrm{dn}^2\left(u \pm \tfrac{1}{2}\,\mathrm{i}\,\mathcal{K}'|\mathrm{k}\right) &= \frac{\mathrm{dn}(2u|\mathrm{k}) + \mathrm{k}\,\mathrm{cn}(2u|\mathrm{k})}{\mathrm{dn}(2u|\mathrm{k}) \pm \mathrm{i}\,\mathrm{k}\,\mathrm{sn}(2u|\mathrm{k})}, \end{aligned} \tag{18.3.67}$$

$$\mathrm{cds}\left(u \pm \tfrac{1}{2}\,\mathrm{i}\,\mathcal{K}'|\mathrm{k}\right) = \mp\mathrm{i}\,[\mathrm{dn}(2u|\mathrm{k}) + \mathrm{k}\mathrm{cn}(2u|\mathrm{k})]. \tag{18.3.68}$$

From (18.3.1), (18.3.37) and (18.3.38)

$$\int_0^{2\mathcal{K}} du\, \mathrm{dn}^2(u|\mathrm{k}) = 2\mathcal{E}. \tag{18.3.69}$$

Also

$$\int_0^{2\mathcal{K}} du \int_0^{2\mathcal{K}} dv\, \mathrm{dn}(u|\mathrm{k})\mathrm{dn}(u-v|\mathrm{k})\mathrm{dn}(v|\mathrm{k}) = 2\mathrm{k}'^2\mathcal{K}^2 + \tfrac{1}{2}\pi^2, \tag{18.3.70}$$

$$\int_0^{2\mathcal{K}} du\, \mathrm{dn}(u|\mathrm{k})\mathrm{dn}\left(u + \tfrac{1}{2}\mathrm{i}\mathcal{K}'|\mathrm{k}\right) = \tfrac{1}{2}\sqrt{1+\mathrm{k}}[\pi + 2(1-\mathrm{k})\mathcal{K}], \tag{18.3.71}$$

which are a little more difficult to prove and need the use of (18.3.59).

Suppose that θ, α and k are related by

$$\sinh(\theta) = \frac{\mathrm{i}}{\mathrm{k}\,\mathrm{sn}(\mathrm{i}\alpha|\mathrm{k})}. \tag{18.3.72}$$

Since, from (18.3.38), $\theta = -\mathrm{i}\,\mathrm{am}(\mathrm{i}[\mathcal{K}' - \alpha]|\mathrm{k})$, it follows, from (18.3.2), (18.3.6), (18.3.7) and (18.3.36), that, if $0 < \mathrm{k} < 1$ and θ is real,

$$\alpha = \int_\theta^\infty \frac{d\omega}{\sqrt{1+\mathrm{k}^2\sinh^2(\omega)}} \tag{18.3.73}$$

is a real, but not unique, solution for α.

18.3.4 Transformations in the Elliptic Modulus

It is often useful to transform the modulus of elliptic functions using combinations of the *complementary and Landen transformations*, $\mathcal{C}$ and $\mathcal{L}$ respectively, given by

$$\mathrm{k} \to \mathcal{C}(\mathrm{k}) := \sqrt{1-\mathrm{k}^2}, \quad \mathrm{k} \to \mathcal{L}(\mathrm{k}) := \frac{1-\mathrm{k}}{1+\mathrm{k}}, \quad \text{with} \quad \mathcal{C}\mathcal{L}(\mathrm{k}) = \frac{2\sqrt{\mathrm{k}}}{1+\mathrm{k}}. \tag{18.3.74}$$

It is clear that these transformations are self-inverse, but that they do not commute.

The complementary modulus $\mathrm{k}' := \mathcal{C}(\mathrm{k})$ has already been encountered and in Chap. 12 we shall make use of four elliptic moduli $\mathrm{k}_0, \ldots, \mathrm{k}_3$ related by Landen transformations. With the compact notation $\mathcal{K}_j := \mathcal{K}(\mathrm{k}_j)$, $\mathcal{K}'_j := \mathcal{K}(\mathrm{k}'_j)$, $\mathcal{Q}_j := \mathcal{Q}(\mathrm{k}_j)$, $\mathcal{E}_j := \mathcal{E}(\mathrm{k}_j)$, $j = 0, \ldots, 3$:

(i) $$\mathrm{k}_1 := \frac{1 - \mathrm{k}'_0}{1 + \mathrm{k}'_0} = \mathcal{L}\,\mathcal{C}(\mathrm{k}_0), \tag{18.3.75}$$

for which

$$\begin{aligned} &(1 + \mathrm{k}'_0)(1 + \mathrm{k}_1) = 2, \\ &\mathcal{K}_1 = \tfrac{1}{2}(1 + \mathrm{k}'_0)\mathcal{K}_0, \quad \mathcal{K}'_1 = (1 + \mathrm{k}'_0)\mathcal{K}'_0, \\ &\mathcal{K}'_1/\mathcal{K}'_0 = 2\mathcal{K}_1/\mathcal{K}_0, \quad \mathcal{Q}_1 = \mathcal{Q}_0^2. \end{aligned} \tag{18.3.76}$$

For k_0 and k_1 we shall also need the relationship

$$\mathcal{E}_1 = \frac{\mathcal{E}_0 + \mathrm{k}'_0\mathcal{K}_0}{1 + \mathrm{k}'_0} \tag{18.3.77}$$

between complete elliptic integrals of the second kind.

(ii) $$\mathrm{k}_2 := \mathcal{C}\,\mathcal{L}(\mathrm{k}_0) = \frac{2\sqrt{\mathrm{k}_0}}{1 + \mathrm{k}_0}, \tag{18.3.78}$$

for which

$$\begin{aligned} &(1 + \mathrm{k}_0)(1 + \mathrm{k}'_2) = 2, \\ &\mathcal{K}_2 = (1 + \mathrm{k}_0)\mathcal{K}_0, \quad \mathcal{K}'_2 = \tfrac{1}{2}(1 + \mathrm{k}_0)\mathcal{K}'_0, \\ &\mathcal{K}'_2/\mathcal{K}'_0 = \mathcal{K}_2/2\mathcal{K}_0, \quad \mathcal{Q}_2 = \mathcal{Q}_0^{\frac{1}{2}}. \end{aligned} \tag{18.3.79}$$

(iii) $$\mathrm{k}_3 := \mathcal{L}\,\mathcal{C}\,\mathcal{L}(\mathrm{k}_1) = \frac{1 + \mathrm{k}_1 - 2\sqrt{\mathrm{k}_1}}{1 + \mathrm{k}_1 + 2\sqrt{\mathrm{k}_1}}. \tag{18.3.80}$$

This is the *Bazhanov–Mangazeev* transformation (Bazhanov and Mangazeev 2007) for which

$$(1 - \mathrm{k}_1)(1 - \mathrm{k}_3) = 4\sqrt{\mathrm{k}_1\,\mathrm{k}_3}, \qquad \mathcal{K}'_3/\mathcal{K}_3 = 4\mathcal{K}_1/\mathcal{K}'_1. \tag{18.3.81}$$

For any u_0 let

$$u_1 := \frac{u_0}{1 + \mathrm{k}_1} = \tfrac{1}{2}u_0(1 + \mathrm{k}'_0) \quad \text{so that} \quad \frac{u_0}{\mathcal{K}_0} = \frac{u_1}{\mathcal{K}_1}. \tag{18.3.82}$$

Then elliptic functions are related by

$$\operatorname{sn}(u_0|\mathrm{k}_0) = \frac{(1+\mathrm{k}_1)\operatorname{sn}(u_1|\mathrm{k}_1)}{1+\mathrm{k}_1\operatorname{sn}^2(u_1|\mathrm{k}_1)}, \quad \operatorname{cn}(u_0|\mathrm{k}_0) = \frac{\operatorname{cn}(u_1|\mathrm{k}_1)\operatorname{dn}(u_1|\mathrm{k}_1)}{1+\mathrm{k}_1\operatorname{sn}^2(u_1|\mathrm{k}_1)},$$
$$\operatorname{dn}(u_0|\mathrm{k}_0) = \frac{1-\mathrm{k}_1\operatorname{sn}^2(u_1|\mathrm{k}_1)}{1+\mathrm{k}_1\operatorname{sn}^2(u_1|\mathrm{k}_1)}. \tag{18.3.83}$$

18.3.5 The Modified Amplitude Function

$$\operatorname{Am}(u|\mathrm{k}) := -\mathrm{i}\ln\left[\mathrm{i}\sqrt{\mathrm{k}}\operatorname{sn}\left(u - \tfrac{1}{2}\mathrm{i}\,\mathcal{K}'|\mathrm{k}\right)\right], \tag{18.3.84}$$

is the *modified amplitude function* (Baxter 1982b, p. 462). It is clear that it is not meromorphic since it has logarithmic branch-points at the zeros and poles of $\operatorname{sn}(u - \frac{1}{2}\mathrm{i}\,\mathcal{K}')$, that is, when $u = 2(2\ell+1)\mathcal{K} + (2\ell' \pm \frac{1}{2})\mathrm{i}\mathcal{K}'$ for all integers ℓ and ℓ'. From (18.3.45),

$$\begin{aligned}&\operatorname{Am}(u+2\mathcal{K}|\mathrm{k}) = \operatorname{Am}(u|\mathrm{k}) \pm \pi,\\ &\operatorname{Am}(u+\mathrm{i}\,\mathcal{K}'|\mathrm{k}) = -\operatorname{Am}(u|\mathrm{k}) \pm \pi,\ \operatorname{Am}(u+2\,\mathrm{i}\,\mathcal{K}_0'|\mathrm{k}) = \operatorname{Am}(u|\mathrm{k}),\end{aligned} \tag{18.3.85}$$

where the choice of signs, reflects the branch point structure.

For real values of u, $\operatorname{Am}(u|\mathrm{k})$ is real and the choice of signs can be made so that its values lie in the range $[-\pi, \pi]$. Then it monotonically increases from $-\pi$ to π as u increases from $2(2\ell-1)\mathcal{K}$ to $2(2\ell+1)\mathcal{K}$, with zeros at $4\ell\mathcal{K}$.

For the imaginary axis consider the function $-\mathrm{i}\operatorname{Am}(\mathrm{i}\,u|\mathrm{k}_0)$. For real values of u this is real and monotonically increasing from $-\infty$ to ∞ as u increases from $(2\ell' - \frac{1}{2})\mathcal{K}'$ to $(2\ell' + \frac{1}{2})\mathcal{K}'$, with zeros at $2\ell'\mathcal{K}'$. Otherwise on the real axis it is complex. From the elliptic moduli k_0 and k_1 related by the Landen transformation given in Sect. 18.3.4

$$\operatorname{cn}(u|\mathrm{k}_0) + \mathrm{i}\operatorname{sn}(u|\mathrm{k}_0) = \mathrm{i}\sqrt{\mathrm{k}_1}\operatorname{sn}\left(u_1 - \tfrac{1}{2}\mathrm{i}\,\mathcal{K}_1'|\mathrm{k}_1\right) \tag{18.3.86}$$

and, from (18.3.39),

$$\operatorname{am}(u|\mathrm{k}_0) = \operatorname{Am}(u_1|\mathrm{k}_1), \tag{18.3.87}$$

$$\mathrm{i}\operatorname{am}(\mathrm{i}[\mathcal{K}_0' - u]|\mathrm{k}_0) = -\mathrm{i}\operatorname{Am}(\mathrm{i}[u_1 - \tfrac{1}{2}\mathcal{K}_1']|\mathrm{k}_1) = \ln[-\mathrm{i}\sqrt{\mathrm{k}_1}\operatorname{sn}(\mathrm{i}\,u_1|\mathrm{k}_1)]. \tag{18.3.88}$$

The latter function is real and monotonically increasing from $-\infty$ to ∞ as u increases from zero to $2\mathcal{K}_0'$ with a zero at $u = \mathcal{K}_0'$. It also follows from (18.3.53), (18.3.83) and (18.3.88) that

$$\exp(\mathrm{i}\,\mathrm{am}(\mathrm{i}[\mathcal{K}_0' - u]|\mathrm{k}_0)) = \begin{cases} -\mathrm{i}[1 - \mathrm{dn}(\mathrm{i}\,u|\mathrm{k}_0)]/[\mathrm{k}_0\,\mathrm{sn}(\mathrm{i}\,u|\mathrm{k}_0)], \\ [\mathrm{dn}(u|\mathrm{k}_0') - \mathrm{cn}(u|\mathrm{k}_0')]/[\mathrm{k}_0\,\mathrm{sn}(u|\mathrm{k}_0')], \\ -\mathrm{i}\sqrt{\mathrm{k}_1}\,\mathrm{sn}(\mathrm{i}\,u_1|\mathrm{k}_1), \\ \sqrt{\mathrm{k}_1}\,\mathrm{sn}(u_1|\mathrm{k}_1'). \end{cases} \tag{18.3.89}$$

18.3.6 *Nome Series*

The product formulae for the theta functions can be used to obtain *nome series* for the elliptic functions. In particular

$$\mathrm{dn}(u|\mathrm{k}) = \frac{\pi}{2\mathcal{K}} + \frac{2\pi}{\mathcal{K}} \sum_{n=1}^{\infty} \frac{\mathcal{Q}^n}{1 + \mathcal{Q}^{2n}} \cos\left[\frac{n\pi u}{\mathcal{K}}\right]. \tag{18.3.90}$$

Substituting $u = 0$ into (18.3.90) and using (18.3.54) gives

$$\sum_{n=1}^{\infty} \frac{\mathcal{Q}^n}{1 + \mathcal{Q}^{2n}} = \frac{\mathcal{K}}{2\pi} - \frac{1}{4}. \tag{18.3.91}$$

By substituting from (18.3.90) into (18.3.69) and noting that the cosines in (18.3.90) are orthogonal over the range of this integral, it follows that

$$\sum_{n=1}^{\infty} \frac{\mathcal{Q}^{2n}}{(1 + \mathcal{Q}^{2n})^2} = \frac{\mathcal{E}\,\mathcal{K}}{2\pi^2} - \frac{1}{8} \tag{18.3.92}$$

and, by replacing k by k_1 in (18.3.71), substituting from (18.3.90) and performing the integration, it can be shown that

$$\begin{aligned} \sum_{n=1}^{\infty} \frac{\mathcal{Q}_0^{3n}(1 + \mathcal{Q}_0^{2n})}{(1 + \mathcal{Q}_0^{4n})^2} &= \frac{1}{4\pi^2}\mathcal{K}_1\sqrt{1 + \mathrm{k}_1}\left[\pi + 2(1 - \mathrm{k}_1)\mathcal{K}_1\right] - \frac{1}{4} \\ &= \frac{1}{4\pi^2}\mathcal{K}_0(\pi + 2\mathrm{k}_0'\,\mathcal{K}_0)\sqrt{\frac{1 + \mathrm{k}_0'}{2}} - \frac{1}{4}. \end{aligned} \tag{18.3.93}$$

18.3.7 Special Results and Functions for Chap. 12

• The functions

$$\begin{aligned} \mathsf{P}_{\mathsf{H}}(v,u|\mathrm{k}) &:= \mathsf{H}(\mathrm{i}[u+v]|\mathrm{k})\mathsf{H}(\mathrm{i}[u-v]|\mathrm{k}), \\ \mathsf{P}_{\Theta}(v,u|\mathrm{k}) &:= \Theta(\mathrm{i}[u+v]|\mathrm{k})\,\Theta\,(\mathrm{i}[u-v]|\mathrm{k}), \\ \mathsf{Q}(v,u|\mathrm{k}) &:= \mathsf{H}(\mathrm{i}[u+v]|\mathrm{k})\,\Theta(\mathrm{i}[u-v]|\mathrm{k}), \end{aligned} \tag{18.3.94}$$

have the properties

$$\begin{aligned} \mathsf{P}_{\mathsf{H}}(v+\mathcal{K}',u|\mathrm{k}) &= \mathfrak{F}_1(2\mathrm{i}v|\mathrm{k})\mathsf{P}_{\Theta}(v,u|\mathrm{k}), \\ \mathsf{P}_{\Theta}(v+\mathcal{K}',u|\mathrm{k}) &= \mathfrak{F}_1(2\mathrm{i}v|\mathrm{k})\mathsf{P}_{\mathsf{H}}(v,u|\mathrm{k}), \\ \mathsf{Q}(v+\mathcal{K}',u|\mathrm{k}) &= \mathfrak{F}_1(2\mathrm{i}v|\mathrm{k})\mathsf{Q}(v,u|\mathrm{k}), \end{aligned} \tag{18.3.95}$$

$$\mathsf{P}_{\mathsf{H}}(v-\tfrac{1}{2}\mathcal{K}',u+\tfrac{1}{2}\mathcal{K}'|\mathrm{k}) = \mathrm{i}\,\mathfrak{F}_1(\mathrm{i}[u-v]|\mathrm{k})\mathsf{Q}(v,u|\mathrm{k}), \tag{18.3.96}$$

and, from (18.2.8), (18.2.9) and (18.3.22), $\mathsf{P}_{\mathsf{H}}(v,u|\mathrm{k}) \in \mathrm{QP}(2\mathrm{i}\,\mathcal{K},2\mathcal{K}';1,1;-2,-2)$, $\mathsf{P}_{\Theta}(v,u|\mathrm{k}) \in \mathrm{QP}(2\mathrm{i}\,\mathcal{K},2\mathcal{K}';1,1;-2,-2)$ and $\mathsf{Q}(v,u|\mathrm{k}) \in \mathrm{QP}(2\mathrm{i}\,\mathcal{K}, 2\mathcal{K}';-1,1;-2,-2)$, with respect to u.

• The formula

$$\ln\left[\frac{\Theta(u+v|\mathrm{k})}{\Theta(u-v|\mathrm{k})}\right] = \sum_{n=1}^{\infty}\frac{2\sin(n\pi\,u/\mathcal{K})\sin(n\pi\,v/\mathcal{K})}{n\sinh(n\pi\mathcal{K}'/\mathcal{K})} \tag{18.3.97}$$

is obtained by taking the logarithm of (18.3.15) expanding the individual logarithmic factors and summing over m. It then follows in a straightforward way that

$$\ln\left[\frac{\Theta(u+v|\mathrm{k})\,\Theta\,(u+\mathcal{K}+v|\mathrm{k})}{\Theta(u-v|\mathrm{k})\,\Theta\,(u+\mathcal{K}-v|\mathrm{k})}\right] = \sum_{n=1}^{\infty}\frac{2\sin(2n\pi\,u/\mathcal{K})\sin(2n\pi\,v/\mathcal{K})}{n\sinh(2n\pi\mathcal{K}'/\mathcal{K})} \tag{18.3.98}$$

and

$$\ln\left[\frac{\mathsf{H}(u+v|\mathrm{k})}{\mathsf{H}(u-v|\mathrm{k})}\right] = \sum_{n=1}^{\infty}\frac{2\sin(n\pi\,u/\mathcal{K})\sin(n\pi\,v/\mathcal{K})\cosh(n\pi\,\mathcal{K}'/\mathcal{K})}{n\sinh(n\pi\mathcal{K}'/\mathcal{K})}, \quad \text{if } u>v>0. \tag{18.3.99}$$

• The function $\mathsf{Y}(u', u|\mathrm{k})$, defined by (12.2.57), can, with a slight change of notation, be expressed in terms of v and u so that

$$\mathsf{Y}(v, u|\mathrm{k}) := \mathsf{Y}(u+v, u-v|\mathrm{k}) = \{-\mathrm{sn}(\mathrm{i}[\mathcal{K}' - u - v]|\mathrm{k})\,\mathrm{sn}(\mathrm{i}[u-v]|\mathrm{k})\}^{-\frac{1}{2}}$$
$$= \{\mathrm{sn}(\mathrm{i}[\mathcal{K}' + u + v]|\mathrm{k})\,\mathrm{sn}(\mathrm{i}[u-v]|\mathrm{k})\}^{-\frac{1}{2}} = \sqrt{\frac{\mathrm{k}\mathsf{Q}(v, u|\mathrm{k})}{\mathsf{Q}(-v, u|\mathrm{k})}}, \qquad (18.3.100)$$

$$\mathsf{Y}(-v, u|\mathrm{k}) = \mathsf{Y}(v, u + \mathcal{K}'|\mathrm{k}), \qquad (18.3.101)$$

$$\mathsf{Y}(v, u|\mathrm{k}) = \mathsf{Y}(v, u + 2\mathcal{K}'|\mathrm{k}) = \mathsf{Y}(v, u + 2\mathrm{i}\,\mathcal{K}|\mathrm{k}), \qquad (18.3.102)$$

$$\frac{\mathsf{Y}(v, u|\mathrm{k})}{\mathsf{Q}(v, u|\mathrm{k})} = \frac{\mathsf{Y}(-v, u|\mathrm{k})}{\mathsf{Q}(-v, u|\mathrm{k})} = \sqrt{\frac{\mathrm{k}}{\mathsf{Q}(v, u|\mathrm{k})\mathsf{Q}(-v, u|\mathrm{k})}}. \qquad (18.3.103)$$

• The function

$$\mathsf{A}(v, \zeta|\mathrm{k}) := \mathrm{Am}(\zeta + \mathrm{i}\,v|\mathrm{k}) + \mathrm{Am}(\zeta - \mathrm{i}\,v|\mathrm{k}),$$
$$= -\mathrm{i}\,\ln[-\mathrm{k}\,\mathrm{sn}(\zeta - \mathrm{i}\,v - \tfrac{1}{2}\mathrm{i}\,\mathcal{K}'|\mathrm{k})\,\mathrm{sn}(\zeta + \mathrm{i}\,v - \tfrac{1}{2}\mathrm{i}\,\mathcal{K}'|\mathrm{k})], \quad (18.3.104)$$

is real-valued even in v and odd in ζ, with values in the range $[-2\pi, 2\pi]$ for $-2\mathcal{K} \leq \zeta \leq 2\mathcal{K}$. Its periodicity properties are given by

$$\mathsf{A}(v, \zeta + 2\mathcal{K}|\mathrm{k}) = \mathsf{A}(v, \zeta|\mathrm{k}) \pm 2\pi,$$
$$\mathsf{A}(v, \zeta + \mathrm{i}\,\mathcal{K}'|\mathrm{k}) = -\mathsf{A}(v, \zeta|\mathrm{k}) \pm 2\pi, \qquad \mathsf{A}(v, \zeta + 2\mathrm{i}\,\mathcal{K}'|\mathrm{k}) = \mathsf{A}(v, \zeta|\mathrm{k}). \qquad (18.3.105)$$

• Using the notation for rapidities given by (12.2.22), let

$$\mathsf{F}^{(s)}(v, u|\mathrm{k}) := \left[-\frac{\sinh(2K^{(\mathrm{f})}(u+v))}{\sinh(2K^{(\mathrm{b})}(u-v))}\right]^{n_2} + s\left[-\frac{\sinh(2K^{(\mathrm{b})}(u-v))}{\sinh(2K^{(\mathrm{f})}(u+v))}\right]^{n_2}, \qquad (18.3.106)$$

for $s = \pm 1$,[12] and, from (12.2.4), in terms of elliptic functions,

$$\mathsf{F}^{(s)}(v, u|\mathrm{k}) = \left[-\frac{\mathrm{sn}(\mathrm{i}[\mathcal{K}' - u - v]|\mathrm{k})}{\mathrm{sn}(\mathrm{i}[u-v]|\mathrm{k})}\right]^{n_2} + s\left[-\frac{\mathrm{sn}(\mathrm{i}[u-v]|\mathrm{k})}{\mathrm{sn}(\mathrm{i}[\mathcal{K}' - u - v]|\mathrm{k})}\right]^{n_2}, \qquad (18.3.107)$$

[12] This function just gives the 2^{2n_2-1}-dimensionally degenerate eigenvalues of the matrix given in (12.2.60).

from which we note that $\mathsf{F}^{(s)}(v, u|\mathrm{k}) \in \mathrm{DP}(2\mathrm{i}\,\mathcal{K}, 2\mathcal{K}'; 1, 1)$. From (18.3.50)

$$\mathsf{F}^{(s)}(v, u|\mathrm{k}) = \frac{s\,\mathsf{G}^{(s)}(v, u|\mathrm{k})}{[\mathsf{P}(v, u|\mathrm{k})]^{n_2}} \tag{18.3.108}$$

where $\mathsf{P}(v, u|\mathrm{k})$ is given by (12.2.81) and

$$\mathsf{G}^{(s)}(v, u|\mathrm{k}) := s + [\mathsf{P}(v, u|\mathrm{k})]^{2n_2}. \tag{18.3.109}$$

We are interested in the poles and zeros of $\mathsf{G}^{(s)}(v, u|\mathrm{k})$. Since $s = \exp(\mathrm{i}\,\pi[1-s]/2)$, we have the factorization

$$\mathsf{G}^{(s)}(v, u|\mathrm{k}) = \prod_{j=1}^{2n_2} \mathsf{g}(v, u; \psi_j^{(s)}|\mathrm{k}), \tag{18.3.110}$$

where[13]

$$\psi_j^{(s)} := \frac{\pi(4j - s - 1)}{8n_2}, \qquad j = 1, 2, \ldots, 2n_2 \tag{18.3.111}$$

and

$$\mathsf{g}(v, u; \psi|\mathrm{k}) := \exp(2\,\mathrm{i}\,\psi) - \mathsf{P}(v, u|\mathrm{k}) \tag{18.3.112}$$

$$= \exp\left(-\frac{\pi(\mathcal{K}' + 2v)}{2\mathcal{K}}\right) \frac{\mathsf{P}_\Theta(v, u|\mathrm{k}) \exp(2\,\mathrm{i}\,\psi) - \mathsf{P}_\mathrm{H}(v, u|\mathrm{k})}{\mathsf{P}_\mathrm{H}(v + \mathcal{K}', u|\mathrm{k})}. \tag{18.3.113}$$

This is the ratio of two $\mathrm{QP}(2\mathrm{i}\,\mathcal{K}, 2\mathcal{K}', 1, 1; -2, -2)$ functions making it $\mathrm{DP}(2\mathrm{i}\,\mathcal{K}, 2\mathcal{K}'; 1, 1)$. It has two first-order poles in $\mathbb{W}(2\mathrm{i}\,\mathcal{K}, 2\mathcal{K}')$ which coincide with the zeros of $\mathsf{P}_\Theta(v, u|\mathrm{k})$. The zeros of $\mathsf{g}(v, u; \psi|\mathrm{k})$ are determined by using (18.3.104) to re-express it in the form

$$\mathsf{g}(v, u; \psi|\mathrm{k}) = \exp(2\,\mathrm{i}\,\psi) + \exp[\mathrm{i}\,\mathsf{A}(v, \mathrm{i}\,u + \tfrac{1}{2}\mathrm{i}\,\mathcal{K}'|\mathrm{k})]. \tag{18.3.114}$$

Let $\zeta(v; \psi)$ be a solution of the equation

$$\mathsf{A}(v, \zeta|\mathrm{k}) = 2\psi - \pi, \qquad 0 \le \psi \le \pi. \tag{18.3.115}$$

It is straightforward to see that $\zeta(v; \psi)$ is an even function of v, and an increasing function of $0 < \psi < \pi$, with

[13] Note that $\prod_{j=1}^{2n_2} \exp\left(2\,\mathrm{i}\,\psi_j^{(s)}\right) = s$.

$$\zeta(v;\pi-\psi)=-\zeta(v;\psi),\quad \zeta(v;0)=-\mathcal{K},\quad \zeta\left(v;\tfrac{\pi}{2}\right)=0,\quad \zeta(v;\pi)=\mathcal{K}. \tag{18.3.116}$$

From (18.3.105),

$$u=-\mathrm{i}\,\zeta(v;\psi)-\tfrac{1}{2}\mathcal{K}' \quad\text{and}\quad u=\mathrm{i}\,\zeta(v;\psi)+\tfrac{1}{2}\mathcal{K}', \tag{18.3.117}$$

are roots of $\mathsf{g}(v;u;\psi)=0$ in $\mathbb{W}'(2\mathrm{i}\mathcal{K},2\mathcal{K}')$ and, from Theorem 18.3.1 and (18.3.95),

$$\mathsf{g}(v,u;\psi|\mathrm{k})=c(v;\psi|\mathrm{k})\frac{\mathrm{H}(\mathrm{i}\,u-\zeta(v;\psi)+\frac{1}{2}\,\mathrm{i}\,\mathcal{K}'|\mathrm{k})\mathrm{H}(\mathrm{i}\,u+\zeta(v;\psi)-\frac{1}{2}\,\mathrm{i}\,\mathcal{K}'|\mathrm{k})}{\mathsf{P}_{\Theta}(v,u|\mathrm{k})}, \tag{18.3.118}$$

where, from (18.3.112) and (18.3.118),

$$c(v;\psi|\mathrm{k}):=\frac{\exp(2\mathrm{i}\,\psi)\,\Theta\,(2\,\mathrm{i}\,v|\mathrm{k})\,\Theta\,(0|\mathrm{k})}{\mathrm{H}(\mathrm{i}\,v-\zeta(v;\psi)+\frac{1}{2}\,\mathrm{i}\,\mathcal{K}'|\mathrm{k})\mathrm{H}(\mathrm{i}\,v+\zeta(v;\psi)-\frac{1}{2}\,\mathrm{i}\,\mathcal{K}'|\mathrm{k})}. \tag{18.3.119}$$

18.3.8 Baxter's Modified Theta Functions

The modified theta functions

$$\Phi_{\mathfrak{B}}(u|\mathrm{k}):=\mathfrak{B}(u|\mathrm{k})\,\Phi\,(u|\mathrm{k}),\qquad \Phi=\mathrm{H},\Theta, \tag{18.3.120}$$

where

$$\mathfrak{B}(u|\mathrm{k}):=\mathfrak{E}\left(\frac{m'[u-\mathcal{K}]^2}{4\,\mathcal{K}}\middle|\mathrm{k}\right), \tag{18.3.121}$$

$$\mathfrak{E}(u|\mathrm{k}):=\exp\left(\frac{\mathrm{i}\pi u}{\mho_{\mathfrak{B}}}\right),\qquad \mho_{\mathfrak{B}}:=\mho(2m,m'|\mathrm{k}) \tag{18.3.122}$$

and $\mho(\ell,\ell'|\mathrm{k})$ is defined in (18.3.21), were introduced by Baxter (1973a). Of course, there is no loss of generality in assuming that $2m$ and m' are coprime in the definition of Baxter's modified theta functions. However, the case where m' is even gives a non-trivially different situation in the roots of unity formula (12.3.128). $\mathrm{H}(u|\mathrm{k})$ and $\Theta(u|\mathrm{k})$ are respectively odd and even functions, but

$$\mathrm{H}_{\mathfrak{B}}(-u|\mathrm{k})=-\mathfrak{E}(m'u|\mathrm{k})\mathrm{H}_{\mathfrak{B}}(u|\mathrm{k}),\quad \Theta_{\mathfrak{B}}(-u|\mathrm{k})=\mathfrak{E}(m'u|\mathrm{k})\,\Theta_{\mathfrak{B}}\,(u|\mathrm{k}). \tag{18.3.123}$$

From (18.3.19) and (18.3.20),

$$\mathrm{H}_{\mathfrak{B}}(u+\mho(2\ell,\ell'|\mathrm{k})|\mathrm{k}) = (-1)^{\ell}(-\mathrm{i})^{\ell'}\mathfrak{G}(\ell,\ell';u|\mathrm{k}) \times \begin{cases} \mathrm{H}_{\mathfrak{B}}(u|\mathrm{k}), & \text{if } \ell' \text{ even},\\ \Theta_{\mathfrak{B}}(u|\mathrm{k}), & \text{if } \ell' \text{ odd}, \end{cases} \tag{18.3.124}$$

$$\Theta_{\mathfrak{B}}(u+\mho(2\ell,\ell'|\mathrm{k})|\mathrm{k}) = (-\mathrm{i})^{\ell'}\mathfrak{G}(\ell,\ell';u|\mathrm{k}) \times \begin{cases} \Theta_{\mathfrak{B}}(u|\mathrm{k}), & \text{if } \ell' \text{ even},\\ \mathrm{H}_{\mathfrak{B}}(u|\mathrm{k}), & \text{if } \ell' \text{ odd}, \end{cases} \tag{18.3.125}$$

where

$$\mathfrak{G}(\ell,\ell';u|\mathrm{k}) := \frac{\mathfrak{B}(u+\mho(2\ell,\ell'|\mathrm{k}))\mathfrak{F}_{\ell'}(u+2\ell\mathcal{K}'|\mathrm{k})}{\mathfrak{B}(u|\mathrm{k})}. \tag{18.3.126}$$

The following special cases are needed in Chap. 12:

- $$\begin{aligned} \mathrm{H}_{\mathfrak{B}}(u+2\mathcal{K}|\mathrm{k}) &= -\mathfrak{E}(m'u)\mathrm{H}_{\mathfrak{B}}(u|\mathrm{k}),\\ \Theta_{\mathfrak{B}}(u+2\mathcal{K}|\mathrm{k}) &= \mathfrak{E}(m'u)\,\Theta_{\mathfrak{B}}(u|\mathrm{k}), \end{aligned} \tag{18.3.127}$$

  $$\begin{aligned} \mathrm{H}_{\mathfrak{B}}(u+\mathrm{i}\,\mathcal{K}'|\mathrm{k}) &= \mathfrak{E}(m[\mathcal{K}-\tfrac{1}{2}\mathrm{i}\mathcal{K}'-u])\,\Theta_{\mathfrak{B}}(u|\mathrm{k}),\\ \Theta_{\mathfrak{B}}(u+\mathrm{i}\,\mathcal{K}'|\mathrm{k}) &= \mathfrak{E}(m[\mathcal{K}-\tfrac{1}{2}\mathrm{i}\mathcal{K}'-u])\mathrm{H}_{\mathfrak{B}}(u|\mathrm{k}). \end{aligned} \tag{18.3.128}$$

- If m_0 is the greatest common divisor of $2m$ and m', with

  $$\overline{m} = 2m/m_0, \qquad \overline{m}' = m'/m_0, \tag{18.3.129}$$

  then $2(m'\ell - m\ell') = m_0(2\overline{m}'\ell - \overline{m}\ell')$. Thus, with $\ell = \overline{m}$, $\ell' = 2\overline{m}'$ and

  $$\mho_{\mathfrak{B}} := \mho_{\mathfrak{B}}/m_0 = \overline{m}\mathcal{K} + \mathrm{i}\,\overline{m}'\mathcal{K}', \tag{18.3.130}$$

  $$\begin{aligned} \mathrm{H}_{\mathfrak{B}}(u\pm 2\mho_{\mathfrak{B}}|\mathrm{k}) &= (-1)^{\overline{m}+\overline{m}\,\overline{m}'}\mathrm{H}_{\mathfrak{B}}(u|\mathrm{k}),\\ \Theta_{\mathfrak{B}}(u\pm 2\mho_{\mathfrak{B}}|\mathrm{k}) &= (-1)^{\overline{m}\,\overline{m}'}\,\Theta_{\mathfrak{B}}(u|\mathrm{k}). \end{aligned} \tag{18.3.131}$$

- If $\overline{m}$ is even and $\overline{m}'$ is odd, let $\ell = \overline{m}/2$, $\ell' = \overline{m}'$, giving

  $$\begin{aligned} \mathrm{H}_{\mathfrak{B}}(u\pm \mho_{\mathfrak{B}}|\mathrm{k}) &= (\mathrm{i})^{\overline{m}}\exp\left(\tfrac{1}{4}\mathrm{i}\pi\overline{m}\,\overline{m}'\right)\Theta_{\mathfrak{B}}(u|\mathrm{k}),\\ \Theta_{\mathfrak{B}}(u\pm \mho_{\mathfrak{B}}|\mathrm{k}) &= \exp\left(\tfrac{1}{4}\mathrm{i}\pi\overline{m}\,\overline{m}'\right)\mathrm{H}_{\mathfrak{B}}(u|\mathrm{k}). \end{aligned} \tag{18.3.132}$$

- Setting $\ell = n$, $\ell' = 2n'$ with

$$2mn' - m'n = m_0, \qquad \overline{m}n' - \overline{m}'n = 1, \tag{18.3.133}$$

gives

$$\mho(2(n-1), 2n'|\mathrm{k}) = 2(1+\overline{m}')\mathcal{O}'_{\mathfrak{B}} - 2n'\mathcal{O}_{\mathfrak{B}}, \tag{18.3.134}$$

where

$$\mathcal{O}'_{\mathfrak{B}} := n\mathcal{K} + \mathrm{i}\,n'\mathcal{K}' \tag{18.3.135}$$

and

$$\begin{aligned} \mathrm{H}_{\mathfrak{B}}(u \pm 2\mathcal{O}'_{\mathfrak{B}}|\mathrm{k}) &= (-1)^{n+n'+nn'} \exp\left(\frac{-\mathrm{i}\pi\{[1\pm\overline{m}']\mathcal{O}'_{\mathfrak{B}} \pm u\}}{\mathcal{O}_{\mathfrak{B}}}\right) \mathrm{H}_{\mathfrak{B}}(u|\mathrm{k}), \\ \Theta_{\mathfrak{B}}(u \pm 2\mathcal{O}'_{\mathfrak{B}}|\mathrm{k}) &= (-1)^{n'+nn'} \exp\left(\frac{-\mathrm{i}\pi\{[1\pm\overline{m}']\mathcal{O}'_{\mathfrak{B}} \pm u\}}{\mathcal{O}_{\mathfrak{B}}}\right) \Theta_{\mathfrak{B}}(u|\mathrm{k}). \end{aligned} \tag{18.3.136}$$

From (18.2.8) and (18.3.22) it can be seen that

$$\begin{aligned} \mathrm{H}_{\mathfrak{B}}(u|\mathrm{k}) &\in \mathrm{QP}(2\mathcal{O}_{\mathfrak{B}}, 2\mathcal{O}'_{\mathfrak{B}}; (-1)^{\overline{m}+\overline{m}\,\overline{m}'}, (-1)^{n+n'+nn'}; 1+\overline{m}', 1), \\ \Theta_{\mathfrak{B}}(u|\mathrm{k}) &\in \mathrm{QP}(2\mathcal{O}_{\mathfrak{B}}, 2\mathcal{O}'_{\mathfrak{B}}; (-1)^{\overline{m}\,\overline{m}'}, (-1)^{n'+nn'}; 1+\overline{m}', 1), \end{aligned} \tag{18.3.137}$$

which can be compared to the corresponding results (18.3.28) for the standard theta functions. It may be seen directly or from Theorem 18.2.5 that $\mathrm{H}_{\mathfrak{B}}(u|\mathrm{k})$ and $\Theta_{\mathfrak{B}}(u|\mathrm{k})$ each has one zero in $\mathbb{W}(2\mathcal{O}_{\mathfrak{B}}, 2\mathcal{O}'_{\mathfrak{B}})$ located respectively at

$$u = \tfrac{1}{2}\mathcal{O}_{\mathfrak{B}}(4\ell - 1 - (-1)^{n+n'+nn'}) + \tfrac{1}{2}\mathcal{O}'_{\mathfrak{B}}(4\ell' - 1 - (-1)^{\overline{m}+\overline{m}\,\overline{m}'}), \tag{18.3.138}$$
$$u = \tfrac{1}{2}\mathcal{O}_{\mathfrak{B}}(4\ell - 1 - (-1)^{n'+nn'}) + \tfrac{1}{2}\mathcal{O}'_{\mathfrak{B}}(4\ell' - 1 - (-1)^{\overline{m}\,\overline{m}'}), \tag{18.3.139}$$

for some choices of ℓ and ℓ'. Then corresponding to the Theorem 18.3.1 for doubly-periodic functions we have:

Theorem 18.3.2 *For some complex quantities α_j, $j = 1, 2, \ldots, r$, lying in the period rectangle $\mathbb{W}(2\mathcal{O}_{\mathfrak{B}}, 2\mathrm{i}\,\mathcal{O}'_{\mathfrak{B}})$,*

$$W(u|\mathrm{k}) := \exp\left(-\frac{\pi\mathrm{i}\aleph u}{2\mathcal{O}_{\mathfrak{B}}}\right)\prod_{j=1}^{r} \mathrm{H}_{\mathfrak{B}}(u - \alpha_j|\mathrm{k}) \tag{18.3.140}$$

is, to within a multiplicative quantity possibly depending on k*, the unique entire member of* $\mathrm{QP}(2\mathcal{O}_{\mathfrak{B}}, 2\mathrm{i}\,\mathcal{O}'_{\mathfrak{B}}, \tau, \tau'; r, r)$ *with zeros α_j, $j = 1, 2, \ldots, r$, if*

$$\aleph = \tfrac{1}{2}(4\ell' + 1 - \tau) + r\overline{m}(1 + \overline{m}'),$$

$$\sum_{j=1}^{r} \alpha_j = \aleph\,\mathcal{O}'_{\mathfrak{B}} + \tfrac{1}{2}\mathcal{O}_{\mathfrak{B}}(4\ell + 1 - \tau') + r\{\mathcal{O}_{\mathfrak{B}}(n + n' + nn') + \mathcal{O}'_{\mathfrak{B}}\overline{m}'\},$$

(18.3.141)

for some integers ℓ and ℓ'.

Proof From (18.3.133) and (18.3.138)

$$W(u + 2\mathcal{O}_{\mathfrak{B}}|\mathrm{k}) = (-1)^{r[\overline{m}+\overline{m}\,\overline{m}']}\exp(-\pi\mathrm{i}\,\aleph)W(u|\mathrm{k}),$$

$$W(u + 2\mathcal{O}'_{\mathfrak{B}}|\mathrm{k}) = (-1)^{r[n+n'+nn']}\exp\left(-\frac{\pi\mathrm{i}\,\aleph\,\mathcal{O}'_{\mathfrak{B}}}{\mathcal{O}_{\mathfrak{B}}}\right)$$

$$\times\exp\left(\frac{-\mathrm{i}\pi\{r[1+\overline{m}']\mathcal{O}'_{\mathfrak{B}} + ru - \sum_{j=1}^{r}\alpha_j\}}{\mathcal{O}_{\mathfrak{B}}}\right)W(u|\mathrm{k}).$$

(18.3.142)

Using (18.2.5), the conditions that $W(u|\mathrm{k}) \in \mathrm{QP}(2\mathcal{O}_{\mathfrak{B}}, 2\mathrm{i}\,\mathcal{O}'_{\mathfrak{B}}, \tau, \tau'; r, r)$ are (18.3.141). That it is, to within a multiplicative quantity possibly dependent on k, the unique member of $\mathrm{QP}(2\mathcal{O}_{\mathfrak{B}}, 2\mathrm{i}\,\mathcal{O}'_{\mathfrak{B}}, \tau, \tau'; r, r)$ with these zeros follows from Theorem 18.2.6. □

Formulae (18.3.33) and (18.3.34) hold without change for the modified theta functions. This is not the case for (18.3.35). Now it is convenient to define

$$\mathfrak{L}(v, u|\mathrm{k}) := \mathrm{H}_{\mathfrak{B}}(\mathcal{K}+u|\mathrm{k})\mathrm{H}_{\mathfrak{B}}(\mathcal{K}-v|\mathrm{k})+\Theta_{\mathfrak{B}}(\mathcal{K}+u|\mathrm{k})\Theta_{\mathfrak{B}}(\mathcal{K}-v|\mathrm{k}), \quad (18.3.143)$$

when the modified version of (18.3.35) is expressible in the form

$$\mathfrak{L}(v, u|\mathrm{k}) = \mathfrak{R}(\mathrm{k})\,\mathfrak{H}(u + v|\mathrm{k})\,\mathfrak{H}(u - v|\mathrm{k}), \quad (18.3.144)$$

where

$$\mathfrak{R}(\mathrm{k}) := 2\mathfrak{Q}_1^{\frac{1}{4}}[\mathrm{H}_{\mathfrak{B}}(\mathcal{K}|\mathrm{k})\,\Theta_{\mathfrak{B}}\,(\mathcal{K}|\mathrm{k})]^{-1}\mathfrak{E}\left(\frac{m'\mathcal{K}'^2}{4\mathcal{K}}\middle|\mathrm{k}\right), \quad (18.3.145)$$

$$\mathfrak{H}(u|\mathrm{k}) := \mathrm{H}_{\mathfrak{B}}(\tfrac{1}{2}[\mathrm{i}\,\mathcal{K}' + u] + \mathcal{K}|\mathrm{k})\mathrm{H}_{\mathfrak{B}}(\tfrac{1}{2}[\mathrm{i}\,\mathcal{K}' - u] + \mathcal{K}|\mathrm{k}). \quad (18.3.146)$$

It is clear that

$$\mathfrak{L}(v, u|\mathrm{k}) = \mathfrak{L}(v, u|\mathrm{k}) = \mathfrak{L}(-u, v|\mathrm{k}) = \mathfrak{L}(u, -v|\mathrm{k}), \tag{18.3.147}$$

$$\mathfrak{L}(u - \mho_{\mathfrak{B}}, v|\mathrm{k}) = (-1)^{mm'}\mathfrak{L}(u, v - \mho_{\mathfrak{B}}|\mathrm{k}). \tag{18.3.148}$$

For some α and integers j and k

$$\frac{\mathfrak{L}(v+\alpha j, u+\alpha k|\mathrm{k})}{\mathfrak{L}(u+\alpha j, v+\alpha k|\mathrm{k})} = \frac{\mathfrak{H}(u-v-\alpha[j-k]|\mathrm{k})}{\mathfrak{H}(u-v+\alpha[j-k]|\mathrm{k})}. \tag{18.3.149}$$

Let

$$\begin{aligned}
\Psi(+1,+1; v, u|\mathrm{k}) &:= \mathrm{H}(-2u|\mathrm{k})\,\Theta(u+v|\mathrm{k})\,\Theta(u-v|\mathrm{k}),\\
\Psi(+1,-1; v, u|\mathrm{k}) &:= \mathrm{H}(-2u|\mathrm{k})\,\mathrm{H}(u+v|\mathrm{k})\,\mathrm{H}(u-v|\mathrm{k}),\\
\Psi(-1,+1; v, u|\mathrm{k}) &:= \Theta(-2u|\mathrm{k})\mathrm{H}(u+v|\mathrm{k})\,\Theta(u-v|\mathrm{k}),\\
\Psi(-1,-1; v, u|\mathrm{k}) &:= \Theta(-2u|\mathrm{k})\,\Theta(u+v|\mathrm{k})\mathrm{H}(u-v|\mathrm{k}),
\end{aligned} \tag{18.3.150}$$

with the corresponding expressions when the theta functions are replaced by their modified forms given, for $\alpha, \beta = \pm 1$, by

$$\Psi_{\mathfrak{B}}(\alpha, \beta; v, u|\mathrm{k}) := \mathfrak{E}\left(\frac{m'[3\mathcal{K}^2 + 6u^2 + 2v^2]}{4\mathcal{K}}\Big|\mathrm{k}\right)\Psi(\alpha, \beta; v, u|\mathrm{k}). \tag{18.3.151}$$

These functions have the properties

$$\begin{aligned}
\Psi_{\mathfrak{B}}(\alpha, \beta; -v, u|\mathrm{k}) &= \Psi_{\mathfrak{B}}(\alpha, \alpha\beta; v, u; \mathrm{k}),\\
\Psi_{\mathfrak{B}}(\alpha, \beta; v, -u|\mathrm{k}) &= -\Psi_{\mathfrak{B}}(\alpha, \alpha\beta; v, u; \mathrm{k}),
\end{aligned} \tag{18.3.152}$$

$$\begin{aligned}
\Psi_{\mathfrak{B}}(\alpha, \beta; v+2\mathcal{K}, u|\mathrm{k}) &= \alpha\, \mathfrak{E}(2m'[\mathcal{K}+v]|\mathrm{k})\Psi_{\mathfrak{B}}(\alpha, \beta; v, u|\mathrm{k}),\\
\Psi_{\mathfrak{B}}(\alpha, \beta; v+\mathrm{i}\,\mathcal{K}', u|\mathrm{k}) &= \mathfrak{E}(-m[\mathrm{i}\,\mathcal{K}'+2v]|\mathrm{k})\Psi_{\mathfrak{B}}(\alpha, -\beta; v, u|\mathrm{k}),\\
\Psi_{\mathfrak{B}}(\alpha, \beta; v+2\mathrm{i}\,\mathcal{K}', u|\mathrm{k}) &= \mathfrak{E}(-4m[\mathrm{i}\,\mathcal{K}'+v]|\mathrm{k})\Psi_{\mathfrak{B}}(\alpha, \beta; v, u|\mathrm{k}),
\end{aligned} \tag{18.3.153}$$

$$\Psi_{\mathfrak{B}}(\alpha, \beta; v+\tfrac{1}{2}\mathrm{i}\mathcal{K}', u+\tfrac{1}{2}\mathrm{i}\mathcal{K}'|\mathrm{k}) = -\mathfrak{E}(-m[\mathrm{i}\mathcal{K}'+v+3u]|\mathrm{k})\Psi_{\mathfrak{B}}(-\alpha, \beta; v, u|\mathrm{k}). \tag{18.3.154}$$

In particular, the function

$$\mathfrak{h}_{\mathfrak{B}}(v|\mathrm{k}) := \Psi_{\mathfrak{B}}(-1,+1;v,0|\mathrm{k}) = \Theta_{\mathfrak{B}}(0|\mathrm{k})\mathrm{H}_{\mathfrak{B}}(v|\mathrm{k})\,\Theta_{\mathfrak{B}}(v|\mathrm{k}), \qquad (18.3.155)$$

is odd and has periodicity properties given by (18.3.153). In the special case $m' = 0$ the modified theta functions reduce to their standard form and $\mathfrak{h}_{\mathfrak{B}}(v|\mathrm{k})$ becomes

$$\mathfrak{h}(u|\mathrm{k}) := \Theta(0|\mathrm{k})\mathrm{H}(u|\mathrm{k})\,\Theta(-u|\mathrm{k}), \qquad (18.3.156)$$

with

$$\mathfrak{h}(u+2\mathcal{K}|\mathrm{k}) = -\mathfrak{h}(u) = \mathfrak{h}(-u|\mathrm{k}), \qquad (18.3.157)$$

$$\mathfrak{h}(u+\mathrm{i}\,\mathcal{K}'|\mathrm{k}) = -\frac{1}{\sqrt{\mathfrak{Q}}}\exp\left(-\frac{\pi\,\mathrm{i}\,u}{\mathcal{K}}\right)\mathfrak{h}(u|\mathrm{k}). \qquad (18.3.158)$$

The complementary transformation is particularly useful for manipulating the function

$$\frac{\mathfrak{h}(u+\mathcal{K}-\mathrm{i}\,v|\mathrm{k})}{\mathfrak{h}(u+\mathcal{K}+\mathrm{i}\,v|\mathrm{k})} = \frac{\mathrm{H}(u+\mathcal{K}-\mathrm{i}\,v|\mathrm{k})\,\Theta(u+\mathcal{K}-\mathrm{i}\,v|\mathrm{k})}{\mathrm{H}(u+\mathcal{K}+\mathrm{i}\,v|\mathrm{k})\,\Theta(u+\mathcal{K}+\mathrm{i}\,v|\mathrm{k})}$$
$$= \exp\left(\frac{2\pi\,\mathrm{i}\,u\,v}{\mathcal{K}\mathcal{K}'}\right)\frac{\Theta(v+\mathrm{i}\,u|\mathrm{k}')\,\Theta(v+\mathcal{K}'+\mathrm{i}\,u|\mathrm{k}')}{\Theta(v-\mathrm{i}\,u|\mathrm{k}')\,\Theta(v+\mathcal{K}'-\mathrm{i}\,u|\mathrm{k}')}. \qquad (18.3.159)$$

From (18.3.98),

$$\mathfrak{f}(v,u|\mathrm{k}) := \ln\left[\frac{\mathfrak{h}(u+\mathcal{K}-\mathrm{i}\,v|\mathrm{k})}{\mathfrak{h}(u+\mathcal{K}+\mathrm{i}\,v|\mathrm{k})}\right] = \frac{2\pi\,\mathrm{i}\,u\,v}{\mathcal{K}\mathcal{K}'}$$
$$+\sum_{n=1}^{\infty}\frac{2\mathrm{i}\,\sin(2n\pi\,v/\mathcal{K}')\sinh(2n\pi\,u/\mathcal{K}')}{n\sinh(2n\pi\mathcal{K}/\mathcal{K}')}. \qquad (18.3.160)$$

Using a dot to denote differentiation with respect to v[14]

$$\dot{\mathfrak{f}}(v,u|\mathrm{k}) = \frac{2\pi\mathrm{i}}{\mathcal{K}'}\sum_{n=-\infty}^{\infty}\frac{\exp(2n\pi\mathrm{i}\,v/\mathcal{K}')\sinh(2n\pi\,u/\mathcal{K}')}{\sinh(2n\pi\mathcal{K}/\mathcal{K}')}, \qquad (18.3.161)$$

which is periodic in v of period $\mathcal{K}'$. From (18.1.21), with $L = \mathcal{K}'$, we see that its Fourier transform is

$$\dot{\mathfrak{f}}^{\star}(u,n|\mathrm{k}) = \frac{2\pi\mathrm{i}\sinh(2n\pi\,u/\mathcal{K}')}{\mathcal{K}'\sinh(2n\pi\mathcal{K}/\mathcal{K}')}. \qquad (18.3.162)$$

[14] The $n = 0$ term is understood as the result of taking the limit $n \to 0$.

Again, for $\alpha, \beta = \pm 1$, let

$$W_{\mathfrak{B}}(\alpha, \beta; v, u|\mathrm{k}) := \tfrac{1}{2}[\Psi_{\mathfrak{B}}(\alpha, +1; v, u|\mathrm{k}) + \beta\,\Psi_{\mathfrak{B}}(\alpha, -1; v, u|\mathrm{k})]. \quad (18.3.163)$$

Then it can be shown that the effect of the Bazhanov–Mangazeev transformation with

$$u_3 := \frac{2\,\mathrm{i}\,\mathcal{K}_3[\mathcal{K}_1 - u_1]}{\mathcal{K}'_1}, \qquad v_3 := \frac{2\,\mathrm{i}\,\mathcal{K}_3[\mathcal{K}_1 - v_1]}{\mathcal{K}'_1}, \qquad (18.3.164)$$

is

$$\begin{aligned}
W_{\mathfrak{B}}(+1, +1; v_3, u_3|\mathrm{k}_3) &= \Pi_{\mathfrak{B}}(v_1, u_1|\mathrm{k}_1) W_{\mathfrak{B}}(-1, +1; v_1, u_1|\mathrm{k}_1),\\
W_{\mathfrak{B}}(+1, -1; v_3, u_3|\mathrm{k}_3) &= \Pi_{\mathfrak{B}}(v_1, u_1|\mathrm{k}_1) W_{\mathfrak{B}}(+1, +1; v_1, u_1|\mathrm{k}_1),\\
W_{\mathfrak{B}}(-1, +1; v_3, u_3|\mathrm{k}_3) &= \Pi_{\mathfrak{B}}(v_1, u_1|\mathrm{k}_1) W_{\mathfrak{B}}(-1, -1; v_1, u_1|\mathrm{k}_1),\\
W_{\mathfrak{B}}(-1, -1; v_3, u_3|\mathrm{k}_3) &= \Pi_{\mathfrak{B}}(v_1, u_1|\mathrm{k}_1) W_{\mathfrak{B}}(+1, -1; v_1, u_1|\mathrm{k}_1),
\end{aligned} \qquad (18.3.165)$$

where

$$\Pi_{\mathfrak{B}}(v_1, u_1|\mathrm{k}_1) := \frac{\mathfrak{C}\left(\dfrac{m'[6u_3^2 + 2v_3^2 + 3\mathcal{K}_3^2]}{4\mathcal{K}_3}\Big|\mathrm{k}_3\right)}{\mathfrak{C}\left(\dfrac{m'[6u_1^2 + 2v_1^2 + 3\mathcal{K}_1^2]}{4\mathcal{K}_1}\Big|\mathrm{k}_1\right)}\,\Pi(v_1, u_1|\mathrm{k}_1), \qquad (18.3.166)$$

$$\Pi(v_1, u_1|\mathrm{k}_1) := \mathrm{i}\sqrt{\frac{\mathcal{K}'_1}{2\mathcal{K}_1}\frac{c(\mathrm{k}_3)}{c(\mathrm{k}_1)}}\exp\left(\frac{\pi[v_1(v_1 - 2\mathcal{K}_1) + 3u_1(u_1 - 2\mathcal{K}_1) + 4\mathcal{K}_1^2]}{2\mathcal{K}_1\mathcal{K}'_1}\right), \qquad (18.3.167)$$

and

$$c(\mathrm{k}) := \mathrm{H}(\mathcal{K}(\mathrm{k})|\mathrm{k})\,\Theta\,(\mathcal{K}(\mathrm{k})|\mathrm{k})\,[\Theta(0|\mathrm{k})]^2. \qquad (18.3.168)$$

18.4 The Potts Delta Function

The properties of the *Potts delta function*

$$\Delta(\nu; \sigma) := \nu\,\delta^{\mathrm{Kr}}(\sigma) - 1, \qquad (18.4.1)$$

are important for the derivation of high-temperature series (Sect. 14.2.2).[15] In particular for the zero-μ series it is necessary to compute expressions of the form

$$\Sigma(\mathrm{e},\mathrm{k}) := \sum_{\{\sigma\}} \Pi(\mathrm{e},\mathrm{k}), \tag{18.4.2}$$

where[16]

$$\begin{aligned}\Pi(\mathrm{e},\mathrm{k}) := \nu^{-\mathrm{n}}&\Delta(\nu;\sigma(\boldsymbol{r}_1)-\sigma(\boldsymbol{r}'_1))\Delta(\nu;\sigma(\boldsymbol{r}_2)-\sigma(\boldsymbol{r}'_2))\\ &\times\Delta(\nu;\sigma(\boldsymbol{r}_3)-\sigma(\boldsymbol{r}'_3))\cdots\Delta(\nu;\sigma(\boldsymbol{r}_\mathrm{e})-\sigma(\boldsymbol{r}'_\mathrm{e}))\end{aligned} \tag{18.4.3}$$

is a product of Δ-functions on the e edges $(\boldsymbol{r}_1,\boldsymbol{r}'_1), (\boldsymbol{r}_2,\boldsymbol{r}'_2),\ldots,(\boldsymbol{r}_\mathrm{e},\boldsymbol{r}'_\mathrm{e})$ of a graph $(\mathrm{e},\mathrm{k}) \in \mathcal{N}$ and the summation is over all the state variables. In the case of a $\mu \neq 0$ series (18.4.2) is modified to

$$\Sigma(\mathrm{e},\mathrm{k};\mathrm{n}',\mathrm{k}') := \sum_{\{\sigma\}} \Pi(\mathrm{e},\mathrm{k}) \times \Lambda\{\mathrm{n}',\mathrm{k}'\}, \tag{18.4.4}$$

where[17]

$$\Lambda\{\mathrm{n}',\mathrm{k}'\} := \Delta(\nu;\sigma(\boldsymbol{r}_1))\Delta(\nu;\sigma(\boldsymbol{r}_2)\Delta(\nu;\sigma(\boldsymbol{r}_3))\cdots\Delta(\nu;\sigma(\boldsymbol{r}_{\mathrm{n}'})) \tag{18.4.5}$$

is a product of Δ-functions at n' vertices $\boldsymbol{r}_1,\boldsymbol{r}_2,\ldots,\boldsymbol{r}_{\mathrm{n}'}$ of $\mathcal{N}$ in a configuration denoted by $\{\mathrm{n}',\mathrm{k}'\}$ and the summation is over all the state variables for (e,k) and $\{\mathrm{n}',\mathrm{k}'\}$. These contributions to $\Sigma(\mathrm{e},\mathrm{k};\mathrm{n}',\mathrm{k}')$ will be referred to as *single-vertex terms*.

Before considering the general case we note that when $\nu = 2$ (the 'Ising case')

$$\sum_{\sigma(\boldsymbol{r}_0)=0}^{1}\prod_{k=1}^{n}\Delta(2;\sigma(\boldsymbol{r}_k)-\sigma(\boldsymbol{r}_0)) = (-1)^m[1+(-1)^n], \tag{18.4.6}$$

where m is the number of $\sigma(\boldsymbol{r}_k) = 0$. Thus $\Sigma(\mathrm{e},\mathrm{k}) = 0$ if (e,k) has odd vertices and $\Sigma(\mathrm{e},\mathrm{k};\mathrm{n}',\mathrm{k}') = 0$ if (e,k) has a single-vertex term on an even vertex or no single-vertex term on a odd vertex. Hence the name *magnetic* used to denote a graph with odd vertices (Sect. 16.4.3).[18]

Unfortunately the situation for general ν is more complicated and the aim of this section is to determine some conditions under which $\Sigma(\mathrm{e},\mathrm{k})$ and $\Sigma(\mathrm{e},\mathrm{k};\mathrm{n}',\mathrm{k}')$ are

---

[15] In all that follows σ (and all its primed and subscripted versions) is a Potts variable with integer values in the range $[0, \nu-1]$.

[16] This is a repetition of (14.2.37).

[17] This is a repetition of (14.2.42).

[18] These observations are useful because they predict a factor of $\nu-2$ in all the general computations of cases which would yield zero for $\nu = 2$.

zero; this being equivalent to a determination of graphs which can be excluded from the computation of high-temperature series for $\mu = 0$ and $\mu \neq 0$ respectively. For this analysis we need the following results

$$\sum_{\sigma=0}^{\nu-1} \Delta(\nu; \sigma_1 - \sigma) = 0, \tag{18.4.7}$$

$$\sum_{\sigma=0}^{\nu-1} \Delta(\nu; \sigma_1 - \sigma)\Delta(\nu; \sigma_2 - \sigma) = \nu\Delta(\nu; \sigma_1 - \sigma_2), \tag{18.4.8}$$

$$\sum_{\sigma=0}^{\nu-1} \Delta(\nu; \sigma_1 - \sigma)\Delta(\nu; \sigma_2 - \sigma)\Delta(\nu; \sigma_3 - \sigma)$$
$$= \nu[\Delta(\nu; \sigma_1 - \sigma_2)\Delta(\nu; \sigma_2 - \sigma_3) - \Delta(\nu; \sigma_1 - \sigma_3)] \tag{18.4.9}$$

and, more generally,[19]

$$\sum_{\sigma=0}^{\nu-1} \prod_{k=1}^{n} \Delta(\nu; \sigma_k - \sigma) = \nu B_n, \tag{18.4.10}$$

where

$$B_n := \sum_{k=1}^{n-1} (-1)^{k+1} A_{k,n} \qquad n > 1, \qquad B_1 = 0, \tag{18.4.11}$$

$$A_{k,n} := \prod_{i=k+1}^{n} \Delta(\nu; \sigma_i - \sigma_k), \qquad k < n, \qquad A_{n,n} = 1. \tag{18.4.12}$$

In the special cases when

(a) $\sigma_1 = \sigma_2 = \cdots = \sigma_n = \sigma'$,

$$\sum_{\sigma=0}^{\nu-1} [\Delta(\nu; \sigma' - \sigma)]^n = \nu(\nu - 1)Q(\nu; n - 1), \tag{18.4.13}$$

(b) $\sigma_1 = \sigma_2 = \cdots = \sigma_{n-1} = \sigma'$, $\sigma_n = \sigma''$,

---

[19] This result is derived by proving by induction that

$$\prod_{k=1}^{n} \Delta(\nu; \sigma_k - \sigma) = B_n + \sum_{k=1}^{n} (-1)^{k+1} A_{k,n} \Delta(\nu; \sigma_k - \sigma)$$

and then summing over the values of σ.

$$\sum_{\sigma=0}^{\nu-1}[\Delta(\nu;\sigma'-\sigma)]^{n-1}\Delta(\nu;\sigma''-\sigma) = \nu\Delta(\nu;\sigma'-\sigma'')Q(\nu;n-1), \quad (18.4.14)$$

where

$$Q(\nu;n) := \frac{(\nu-1)^n-(-1)^n}{\nu}. \quad (18.4.15)$$

18.4.1 The $\mu = 0$ Case

We now derive a number of properties of the sum (18.4.2) which allow us to restrict the number of graphs which need to be considered in the high-temperature zero-μ expansion in Sect. 14.2.2.

Property A.1: $\Sigma(\mathrm{e},\mathrm{k}) = 0$ if (e,k) has any vertices of valency one. The summation over a vertex of valency one involves only one Δ-function and the property follows from (18.4.7).

Property A.2: Summations in (18.4.2) over vertices of valency two in (e,k) can be performed independently of the other summations, yielding from (18.4.8) a factor ν. This means that in most cases vertices of valency two can be removed and ignored. An exception to this is the case where (e,k) is a simple cycle (Example 18.4.1, below).

Property A.3: Let

$$\Sigma_1(\mathrm{e},\mathrm{k}|\sigma(\boldsymbol{r}_1)=j) := \nu\sum_{\{\sigma\}_1}\Pi(\mathrm{e},\mathrm{k}), \quad (18.4.16)$$

where the summation is over all the vertex variables of (e,k) except one, which, without loss of generality, we can take to be $\sigma(\boldsymbol{r}_1)$, which is fixed at a particular value $j \in [0,\nu-1]$. Since the terms on the right-hand side of (18.4.2) transform into each other under the symmetry group $\mathbb{S}_\nu$, they form a set of ν equal groups labelled by the value of $\sigma(\boldsymbol{r}_1)$. Thus

$$\Sigma_1(\mathrm{e},\mathrm{k}) := \Sigma_1(\mathrm{e},\mathrm{k}|\sigma(\boldsymbol{r}_1)=j) = \Sigma(\mathrm{e},\mathrm{k}), \quad (18.4.17)$$

for any vertex of (e,k) and any state j. This means that by the time the last summation is performed in (18.4.2) all the vertex variables have disappeared and the summation just yields a factor of ν.

Property A.4: If (e,k) contains a type A articulation point (Sect. 16.1.1 and Fig. 16.1) then the two parts of the graph joined at the articulation point can be treated as separate components for the purposes of the summation (18.4.2). This follows from (18.4.17) by choosing the vertex for which the summation is omitted as the articulation point. Example 18.4.3 illustrates this property since the same result would be achieved for a two-component graph consisting of two simple cycles.

Property A.5: The summation $\Sigma(\mathrm{e}, \mathrm{k}) = 0$ for any graph containing a type B or C articulation point. In this case the graph has one or more bridges. If summations are performed over all variables of a part of the graph connected to the rest by a bridge, but omitting the vertex at the end of the bridge, the bridge will be left as a 'dangling' edge with a vertex of valency one. The result then follows from Property A.1.

As observed in Sect. 16.1.1 any general graph consists of a collection of stars linked by articulation points. We are interested in determining the graphs for which summations $\Sigma(\mathrm{e}, \mathrm{k})$ make a non-zero contribution to the high-temperature zero-μ series described in Sect. 14.2.2. According to the properties listed here graphs with vertices of valency one and those with articulation points of type B and C can be omitted since they make a zero contribution. The parts of a graph linked by articulation points of type A can be treated as separate components as long this is taken into account when calculating the lattice constants. The final conclusion of this discussion is a result due to Domb (1974a, b) that the only terms which need to be considered in the high-temperature zero-μ series are those due to star graphs.

Example 18.4.1 Let the graph be a simple cycle for which all the vertices are of valency two. If all except two are removed, using (18.4.8), the cycle is reduced to a double edge between these vertices and, applying (18.4.13), $\Sigma(\mathrm{e}, \mathrm{k}) = (\nu - 1)$.

Example 18.4.2 Let the graph consist of two simple cycles with one common edge. If all the edges are removed, using (18.4.8), except the shared edge the graph is reduced to a triple edge between these vertices and, applying (18.4.13), $\Sigma(\mathrm{e}, \mathrm{k}) = (\nu - 1)(\nu - 2)$. The graph has two odd vertices so, as indicated above, we expected the factor $\nu - 2$.

Example 18.4.3 Let the graph consist of two simple cycles linked at one vertex (an articulation point of type A). Then, as in Example 18.4.1, each cycle can be reduced to a double edge and the graph becomes two double edges linked at the articulation point. Using (18.4.13) $\Sigma(\mathrm{e}, \mathrm{k}) = (\nu - 1)^2$.

Example 18.4.4 Let the graph be (12, 1), the block of four squares shown in Table 16.1. Then, as in the previous examples, the four corner vertices of valency two can be removed leaving a diamond of four vertices. This, using (18.4.10) with $n = 4$, is followed by a summation over the centre vertex. After summation over the resulting cycle $\Sigma(12, 1) = (\nu - 1)(\nu - 2)(\nu^2 - 5\nu + 7)$. As an example of the discussion in Sect. 14.2.3 we note that [4, 8] in Table 16.1, with its vertices in the centres of the faces, is the dual of (12, 1) and that the number of ways of assigning state variables to the vertices of [4, 8] so that each edge contains an unlike pair with variable values in the range $\sigma = 1, 2, \ldots, \nu - 1$ is also $(\nu - 1)(\nu - 2)(\nu^2 - 5\nu + 7)$.[20]

[20] The state $\sigma = 0$ is excluded by the links to the boundary vertices.

18.4.2 The $\mu \neq 0$ Case

The properties from which it can deduced that $\Sigma(e, k) = 0$ no longer apply to $\Sigma(e, k; n', k')$. However, it is easy to see that

Property B.1: $\Sigma(e, k; n', k') = 0$ unless $\{n', k'\}$ is identical to, or a subset of the vertex configuration of (e, k) with $n' \leq n(e, k)$. This follows from (18.4.7).

Property B.2: $\Sigma(e, k; 1, k') = 0$. This follows from (18.4.17), where the initial summation excludes the site containing the single-vertex term. The final summation is then over the single-vertex Δ-function.

These properties are now used in some examples:

Example 18.4.5 Let the graph be (1, 1), the single pair of vertices connected by an edge shown in Table 16.1. It follows from Properties 16.1 and 17.1 that $\Sigma(1, 1; n', k') = 0$ if one or both of the vertices does not contain a single-vertex term. However, when both do it follows from (18.4.8) and (18.4.14) that $\Sigma(1, 1; 2, 1) = \nu - 1$.

Example 18.4.6 For the simple chain[21] $(e, \mathbb{SC})$, $\Sigma(e, \mathbb{SC}; n', k') = 0$ unless the two ends of the chain both contain single-vertex terms and a vertex not containing a single-vertex term can be removed. So we need consider only the case where each vertex contains a single-vertex term when it follows from (18.4.13) and (18.4.14) that

$$\Sigma(e, \mathbb{SC}; e + 1, k') = (\nu - 1)(\nu - 2)^{e-1}. \tag{18.4.18}$$

Example 18.4.7 Let the graph (e, k) be a simple cycle of e edges with two vertices containing single-vertex terms. The cycle can be reduced to a double edge with a single-vertex term at each end giving, from (18.4.14), $\Sigma(e, k; 2, k') = (\nu - 1)(\nu - 2)$.

Example 18.4.8 Consider now the simple cycle (e, k) with every vertex occupied by a single vertex term. From (18.4.9), (18.4.13) and (18.4.14)

$$\Sigma(e, k; e, 1) = (\nu - 1)(\nu - 2)^{e-1} - \Sigma(e - 1, k; e - 1, 1), \tag{18.4.19}$$

which can be iterated to give

$$\Sigma(e, k; e, 1) = (\nu - 2)\left[(\nu - 2)^{e-1} - (-1)^{e-1}\right]. \tag{18.4.20}$$

Example 18.4.9 Let the graph consist of two simple cycles linked by one bridge (and two type B articulation points). Each cycle can be reduced to a double edge and summing over the vertices at the free ends using (18.4.13) gives a factor of $(\nu - 1)^2$.

---

[21] See the notation defined in Sect. 16.4.4.

We are left with the bridge. This has reduced the problem to Example 18.4.5. It follows that the result will be zero unless both ends of the bridge hold a single-vertex term in which case $\Sigma(\mathrm{e}, \mathrm{k}; 2, 1) = (\nu - 1)^3$. A similar result arises when the graph consists of a simple cycle attached at one vertex to an edge terminating at a vertex of valency one (see the graph (5, 1) of Table 16.2). The only difference here is that one cycle and one factor of $\nu - 1$ is missing. Again the result is zero unless both ends of the dangling edge contain single-vertex terms, in which case $\Sigma(\mathrm{e}, \mathrm{k}; 2, 1) = (\nu-1)^2$.

18.5 Padé, Differential and Algebraic Approximants

Here we work with the function $F(w)$, with power series (14.1.1), and $H(w)$, with power series

$$H(w) = \sum_{j=0}^{\infty} h_j w^j, \tag{18.5.1}$$

for which, as for $F(w)$, the series truncated at w^ℓ is denoted by $H_\ell(w)$.

18.5.1 Padé Approximants

A Padé approximant is a rational function of a variable w constructed as an approximation to a function $F(w)$ so that the power series of $F(w)$ and the power series of the approximant agree up to a specified degree in w. First introduced in the eighteenth century, they were presented in a systematic way using *Padé tables* (see below) by the French mathematician Henri Padé (1863–1953) in his thesis of 1892.[22]

Definition 18.5.1 The function[23]

$$H(w) := [n/m] := \frac{P_n^{(m)}(w)}{Q_m^{(n)}(w)}, \tag{18.5.2}$$

where

$$P_n^{(m)}(w) := \sum_{j=0}^{n} p_j w^j, \qquad Q_m^{(n)}(w) := \sum_{j=0}^{m} q_j w^j, \quad \text{with } q_0 = 1, \tag{18.5.3}$$

[22] For an historical account of the development of Padé approximants see Brezinski (1996, pp. 311–313) and for a comprehensive account of the modern theory see Baker and Graves-Morris (1996).

[23] This is the Baker definition (Baker and Graves-Morris 1996, p. 21). The condition $q_0 \neq 0$ originally replaced $q_0 = 1$.

are coprime, is the n–m Padé approximant to the function $F(w)$ if[24]

$$F(w) - [n/m] = \mathrm{O}(w^{n+m+1}). \tag{18.5.4}$$

This definition does not, of course, imply the existence of $[n/m]$, for all n and m for a given $F(w)$. To explore this question we note that (18.5.4) and (18.5.2) are equivalent to

$$F(w)Q_m^{(n)}(w) - P_n^{(m)}(w) = \mathrm{O}(w^{n+m+1}), \tag{18.5.5}$$

$$H(w)Q_m^{(n)}(w) - P_n^{(m)}(w) = 0, \tag{18.5.6}$$

which means that

$$F_\ell(w) = H_\ell(w), \tag{18.5.7}$$

$$\ell := n + m \tag{18.5.8}$$

and the coefficients in $P_n^{(m)}(w)$ and $Q_m^{(n)}(w)$ are obtained, if they exist, by solving the $n + m + 1$ linear equations

$$f_0 = p_0, \tag{18.5.9}$$

$$\sum_{i=0}^{m} f_{k-i}q_i = p_k, \qquad k = 1, \ldots, n, \tag{18.5.10}$$

$$\sum_{i=0}^{m} f_{k-i}q_i = 0, \qquad k = n+1, \ldots, n+m, \tag{18.5.11}$$

using the known coefficients in $F_\ell(w)$, with $f_k := 0$ if $k < 0$.

Initially we shall suppose that $f_0 \neq 0$:
Let

$$X(n/m; w) := \begin{pmatrix} 1 & w & w^2 & \cdots & w^m \\ f_{n+1} & f_n & f_{n-1} & \cdots & f_{n-m+1} \\ f_{n+2} & f_{n+1} & \cdots & \cdots & \cdots \\ \vdots & \vdots & \vdots & \vdots & \vdots \\ f_{n+m} & f_{n+m-1} & \cdots & \cdots & f_{n+m-m} \end{pmatrix}, \tag{18.5.12}$$

[24] Notation: $g(w) = \mathrm{O}(w^k)$ if $g(w)/w^k$ tends to a finite limit as $w \to 0$. If $g(w) = \mathrm{O}(w^k)$ then $g(w) = \mathrm{O}(w^{k'})$ for $k' \leq k$.

and denote by $\boldsymbol{X}_k(n/m)$ the matrix obtained by deleting the first row and $(k+1)$th column of $\boldsymbol{X}(n/m; w)$. Thus

$$X_0(n/m) := \begin{pmatrix} f_n & f_{n-1} \cdots & f_{n-m+1} \\ f_{n+1} & \cdots \;\cdots & \cdots \\ \vdots & \vdots \;\; \vdots & \vdots \\ f_{n+m-1} & \cdots \;\cdots & f_n \end{pmatrix}. \tag{18.5.13}$$

From (18.5.12) and (18.5.13),[25]

$$X_0(0/m) = (f_0)^m \neq 0, \tag{18.5.14}$$

$$X_m(n/m) = X_0(n+1/m), \tag{18.5.15}$$

$$X(n/m; w) = \sum_{j=0}^{m} (-w)^j X_j(n/m), \qquad X_0(n/0) := 1. \tag{18.5.16}$$

Equations (18.5.9)–(18.5.11) can be expressed in the matrix form

$$\boldsymbol{p} = \boldsymbol{f}_n^{(0)} + \boldsymbol{X}_0(-1/n+1, m)\,\boldsymbol{q}, \tag{18.5.17}$$

$$\boldsymbol{X}_0(n/m)\boldsymbol{q} = -\boldsymbol{f}_{n+m}^{(n+1)}, \tag{18.5.18}$$

where

$$\boldsymbol{p} := \begin{pmatrix} p_0 \\ p_1 \\ \vdots \\ p_n \end{pmatrix}, \qquad \boldsymbol{q} := \begin{pmatrix} q_1 \\ q_2 \\ \vdots \\ q_m \end{pmatrix}, \qquad \boldsymbol{f}_j^{(k)} := \begin{pmatrix} f_k \\ f_{k+1} \\ \vdots \\ f_j \end{pmatrix}. \tag{18.5.19}$$

Thus it can be concluded that:

Theorem 18.5.1 *There exists a unique n–m Padé approximant to $F(w)$ if $X_0(n/m) \neq 0$.*

In that case, from (18.5.12), (18.5.16)–(18.5.18),

---

[25] Adopting the notation that for any square matrix $\boldsymbol{W}$, $W := \mathrm{Det}\{\boldsymbol{W}\}$.

$$q_k = (-1)^k \frac{X_k(n/m)}{X_0(n/m)}, \qquad k = 0, 1, \ldots, m, \tag{18.5.20}$$

$$Q_m^{(n)}(w) = \frac{X(n/m; w)}{X_0(n/m)}, \tag{18.5.21}$$

$$p_k = \frac{1}{X_0(n/m)} \sum_{j=0}^{m} (-1)^j f_{k-j} X_j(n/m),$$

$$= \frac{1}{X_0(n/m)} \begin{vmatrix} f_k & f_{k-1} & f_{k-2} & \cdots & f_{k-m} \\ f_{n+1} & f_n & f_{n-1} & \cdots & f_{n-m+1} \\ f_{n+2} & f_{n+1} & \cdots & \cdots & \cdots \\ \vdots & \vdots & \vdots & \vdots & \vdots \\ f_{n+m} & f_{n+m-1} & \cdots & \cdots & f_n \end{vmatrix}, \quad k = 0, 1, \ldots, n, \tag{18.5.22}$$

$$P_n^{(m)}(w) = \frac{1}{X_0(n/m)} \begin{vmatrix} F_n(w) & w\, F_{n-1}(w) & \cdots & w^m\, F_{n-m}(w) \\ f_{n+1} & f_n & \cdots & f_{n-m+1} \\ \vdots & \vdots & \cdots & \vdots \\ f_{n+m} & f_{n+m-1} & \cdots & f_n \end{vmatrix}. \tag{18.5.23}$$

Formulae (18.5.20)–(18.5.23) provide an ansatz for calculating Padé approximants when $X_0(n/m) \neq 0$. However, it is necessary to know, and straightforward to verify, that:

Theorem 18.5.2 *The polynomials* $Q_m^{(n}(w)$ *and* $P_n^{(m)}(w)$*, calculated using* (18.5.21) *and* (18.5.23) *respectively are coprime.*

The array

$$\begin{array}{ccccc} [0/0] & [0/1] & [0/2] & [0/3] & \cdots \\ [1/0] & [1/1] & [1/2] & [1/3] & \cdots \\ [2/0] & [2/1] & [2/2] & [2/3] & \cdots \\ [3/0] & [3/1] & [3/2] & [3/3] & \cdots \\ \vdots & \vdots & \vdots & \vdots & \cdots \end{array} \tag{18.5.24}$$

is called the *Padé table* of $F(w)$ and the X_0-table is just the corresponding array of values of $X_0(n/m)$. This latter table can tell us a lot about the Padé table.

In particular, of course, a zero entry in the X_0-table corresponds to a place in the Padé table where a *unique* approximant does not exist. In this respect the following result is important (Baker and Graves-Morris 1996, pp. 24–27):

Theorem 18.5.3 *Given that sufficient elements of the X_0-table have been calculated and that $f_0 \neq 0$, then every zero entry, if such exists, lies either in a square block of zero entries bounded by non-zero entries or in the infinite segment bounded above and to the left by non-zero entries and extending to infinity below and to the right.*

Using this result for the X_0-table we now consider the Padé table. First note, from (18.5.15), (18.5.20) and (18.5.22), that

$$p_0 = f_0, \qquad p_n = \frac{X_0(n/m+1)}{X_0(n/m)},$$
$$q_0 = 1, \qquad q_m = (-1)^m \frac{X_0(n+1/m)}{X_0(n/m)}. \tag{18.5.25}$$

When $p_n \neq 0$ and $q_m \neq 0$, $[n/m]$ is said to be of *full nominal degree* and thus a unique $[n/m]$ of full nominal degree exists if $X_0(n/m) \neq 0$, $X_0(n/m+1) \neq 0$ and $X_0(n+1/m) \neq 0$. With reference to the X_0-table this with be true for all points (n, m) except those lying inside a square of zero values or on the left-hand and top edges of such a square. This leads to *Padé's principal theorem* (Baker and Graves-Morris 1996, pp. 27–28).

Theorem 18.5.4 *Given a $k \times k$ square of zero entries in the X_0-table, surrounded by non-zero entries and with top left-hand corner $(n+1, m+1)$:*

(i) *$[n, m]$ exists and is of full nominal degree.*
(ii) *$[n+r/m+s] = [n, m]$ for $r \geq 0$, $s \geq 0$ and $r+s \leq k$.*
(iii) *$[n+r/m+s]$ does not exist for $1 \leq r \leq k$, $1 \leq s \leq k$, $r+s > k$.*

(ii) and (iii) correspond, respectively, to the situations where there is an infinite number of solutions to the linear equations (18.5.10) and (18.5.11), and where (18.5.10) and (18.5.11) are inconsistent.

Corollary 18.5.1 *The X_0-table contains an infinite segment as defined in Theorem 18.5.3 if and only if $F(w)$ is a rational function.*

Proof (*of corollary*) Consider the limit $k \to \infty$ for the square in Theorem 18.5.4. This gives an infinite segment of zero entries in the X_0-table with top left-hand corner $(n+1, m+1)$, $[n/m]$ of full nominal degree and $[n+r/m+s] = [n/m]$, for all $r \geq 0$ and $s \geq 0$. From (18.5.4),

$$F(w) - [n/m] = \mathrm{O}(w^{n+r+m+s+1}) \tag{18.5.26}$$

and taking the limit $(r+s) \to \infty$ it follows that $F(w)$ is the rational function $[n/m]$. Conversely if $F(w)$ is a rational function of coprime polynomials with numerator and denominator degrees n and m respectively, then it is given exactly by its Padé approximant $[n/m]$ and in the Padé table $[n+r/m+s] = [n/m]$, for all $r \geq 0$ and $s \geq 0$. □

A function $F(w)$ is said to be *normal* if its X_0-table contains no zero entries.

Consider now the case $f_0 = \cdots f_{s-1} = 0$:

It is straightforward to show that:

Theorem 18.5.5 *If*

$$F(w) = Cw^s G(w), \quad \textit{where} \quad G(w) := \sum_{j=0}^{\infty} g_j w^j \quad \textit{and} \quad a_0 \neq 0, \tag{18.5.27}$$

then

$$[n/m]^{(F)} = Cw^s [n-s/m]^{(G)}, \qquad n \geq s, \tag{18.5.28}$$

where Padé approximants for F and G are distinguished by superscripts.

18.5.2 Dlog Padé Approximants

Using the transformation (14.6.12), let

$$G(w) := \mathcal{DL}\{F(w)\}, \quad \text{where} \quad G(w) := \sum_{j=0}^{\infty} g_j w^j. \tag{18.5.29}$$

This transformation is effected by solving the linear equations

$$\sum_{j=0}^{k} f_{k-j}\, g_j = (k+1) f_{k+1}, \qquad k = 0, 1, \ldots. \tag{18.5.30}$$

Suppose that all the coefficients of $F_\ell(w)$ are known. Then the first ℓ of these linear equations can be expressed in the form

$$X_0(0/\ell)\boldsymbol{g} = \boldsymbol{f}', \tag{18.5.31}$$

where

$$\boldsymbol{g} := \begin{pmatrix} g_0 \\ g_1 \\ \vdots \\ g_{\ell-1} \end{pmatrix}, \qquad \boldsymbol{f}' := \begin{pmatrix} f_1 \\ 2f_2 \\ \vdots \\ nf_\ell \end{pmatrix}. \tag{18.5.32}$$

Since $X_0(0/\ell) = f_0^\ell$ the coefficients of $G_{\ell-1}(w)$ are derivable from the coefficients of $F_\ell(w)$ by inverting the lower-triangular matrix $\boldsymbol{X}_0(0/\ell)$.

Definition 18.5.2 The n–m Dlog Padé approximant for $F(w)$, denoted by $[n/m]_{\text{DL}}$, is the 'ordinary' n–m Padé approximant for $G(w)$.

The uniqueness of the Dlog Padé follows from Theorem 18.5.1. The method of calculation follows that given in Sect. 18.5.1 with $F_\ell(w)$ replaced by $G_{\ell-1}(w)$. Let

$$[n/m]_{\text{DL}} := \frac{\mathring{P}_n^{(m)}(w)}{\mathring{Q}_m^{(n)}(w)}, \tag{18.5.33}$$

where

$$\mathring{P}_n^{(m)}(w) := \sum_{j=0}^{n} \mathring{p}_j w^j, \qquad \mathring{Q}_m^{(n)}(w) := \sum_{j=0}^{m} \mathring{q}_j w^j, \quad \text{with } \mathring{q}_0 = 1. \tag{18.5.34}$$

Then

$$F'(w)\mathring{Q}_m^{(n)}(w) - F(w)\mathring{P}_n^{(m)}(w) = \mathrm{O}(w^{n+m+1}). \tag{18.5.35}$$

Given that $H(w)$ satisfies

$$H'(w)\mathring{Q}_m^{(n)}(w) - H(w)\mathring{P}_n^{(m)}(w) = 0, \tag{18.5.36}$$

(18.5.35) and (18.5.36) replace (18.5.5) and (18.5.6) for ordinary approximants, but now the recurrence relations derived from (18.5.35) for terms of degree w^ℓ, for $\ell = 0, 1, \ldots, n+m$, contain f_ℓ, for $\ell = 0, 1, \ldots, n+m+1$. So the solution $H(w)$ of the first-order differential equation (18.5.36), with $H(0) = F(0)$, is such that

$$F_\ell(w) = H_\ell(w), \tag{18.5.37}$$

$$\ell := n+m+1. \tag{18.5.38}$$

18.5.3 Differential Approximants

The coefficients in the expansion of (18.5.36) satisfy the recurrence relationships

$$\sum_{i=0}^{m} \mathring{q}_i\, h_{k-1} - \sum_{i=0}^{n} (k+1-i)\mathring{p}_i\, h_{k+1-i} = 0, \qquad k = 0, 1, \ldots . \tag{18.5.39}$$

Given that, apart from $\mathring{q}_0 = 1$, the $n+m+1$ coefficients of $\mathring{Q}_m^{(n)}(w)$ and $\mathring{P}_n^{(m)}(w)$ are unknown they can be found using the identity (18.5.37). After which (18.5.39) can be used to successively determine further coefficients in the power series of $H(w)$. Here we have seen a change of interest from the Dlog Padé approximant $[n/m]_{\mathrm{DL}}$ to $H(w)$, which analytically continues the known $F_\ell(w)$ and

(a) has coefficients which satisfy the recurrence relations (18.5.39) for all k,
(b) satisfies the differential equation (18.5.36).

These represent the two strands in the development of the differential approximant method[26] by Joyce (1972) (see also Joyce and Guttmann 1973; Guttmann 1989).

Definition 18.5.3 The function $H(w)$ which satisfies

$$\sum_{i=0}^{k} A_i(w)\mathcal{D}^i\{H(w)\} = B(w), \quad \text{where} \quad \mathcal{D} := w\frac{\mathrm{d}}{\mathrm{d}w}, \tag{18.5.40}$$

and

$$\begin{aligned} A_i(w) &:= \sum_{j=0}^{m_i} a_{ij} w^j, \qquad i = 0, 1, \ldots, k, \quad \text{with} \quad a_{k0} := 1, \\ B(w) &:= \sum_{j=0}^{n} b_j w^j, \qquad \text{with} \quad B(w) \neq 0, \end{aligned} \tag{18.5.41}$$

is the kth order inhomogeneous differential approximant to $F(w)$, denoted by $[n/m_0, \ldots, m_k]_{\mathrm{DA}}$, if

$$\sum_{i=0}^{k} A_i(w)\mathcal{D}^i\{F(w)\} - B(w) = \mathrm{O}(w^{\eta+1}), \tag{18.5.42}$$

$$\eta := k + n + \sum_{i=0}^{k} m_i. \tag{18.5.43}$$

[26] Initially called the 'recurrence relation method'.

In the corresponding homogeneous case, $B(w) = 0$, (18.5.43) is replaced by

$$\eta := k - 1 + \sum_{i=0}^{k} m_i. \tag{18.5.44}$$

It can, therefore, be conveniently denoted by $[-1/m_0, \ldots, m_k]_{\mathrm{DA}}$.[27]

The recurrence relations corresponding to (18.5.40) are, for the inhomogeneous case,

$$\sum_{i=0}^{k}\sum_{j=0}^{m_i}(s-j)^i a_{ij}\, h_{s-j} = \begin{cases} b_s, & 0 \le s \le n, \\ 0, & s > n, \end{cases} \tag{18.5.45}$$

and, for the homogeneous case,

$$\sum_{i=0}^{k}\sum_{j=0}^{m_i}(s-j)^i a_{ij}\, h_{s-j} = 0. \tag{18.5.46}$$

Those for $s = 0, 1, \ldots, \eta$, needed to determine the $\eta + 1$ unknown parameters in (18.5.42), contain h_j, $j = 0, 1, \ldots, \eta$. For the homogeneous case a_{00} is zero if $f_0 \neq 0$ and can be chosen arbitrarily if $f_0 = 0$.

The solution $H(w)$ of the k-order differential equation (18.5.40), with

$$\left(\frac{\mathrm{d}^s H(w)}{\mathrm{d}w^s}\right)_{w=0} := s!\, f_s, \qquad s = 0, 1, \ldots, k-1, \tag{18.5.47}$$

is the differential approximant $[n/m_0, \ldots, m_k]_{\mathrm{DA}}$ for $F(w)$. It can be obtained directly using one of the algebraic computing packages or by Frobenius' method. For the latter we need to determine the singularities of (18.5.40). First note that

$$\mathcal{D}^i = \sum_{r=1}^{i} \theta(i, r) w^r \frac{\mathrm{d}^r}{\mathrm{d}w^r}, \tag{18.5.48}$$

where the coefficients $\theta(i, r)$ are given by the recursion equations

$$\begin{aligned} &\theta(i, 1) = 1, \qquad \theta(i, i) = 1, \\ &\theta(i, r) = \theta(i-1, r-1) + r\theta(i-1, r), \quad r = 2, \ldots, i-1. \end{aligned} \tag{18.5.49}$$

Substituting into (18.5.40) gives

---

[27] $[0/m_0, \ldots, m_k]_{\mathrm{DA}}$ is the case $B(w) = b_0$.

$$w^k A_k(w)\frac{\mathrm{d}^k H(w)}{\mathrm{d}w^k} + \sum_{r=0}^{k-1} w^r \mathring{A}_r(w)\frac{\mathrm{d}^r H(w)}{\mathrm{d}w^r} = B(w), \tag{18.5.50}$$

where

$$\mathring{A}_r(w) := \sum_{i=r}^{k} A_i(w)\theta(i,r), \quad r = 1,\ldots,k-1, \qquad \mathring{A}_0(w) := A_0(w). \tag{18.5.51}$$

The roots of $A_k(w)$ (all non-zero), 0 and ∞ are the *singularities* of (18.5.40). If w' is a root of $w^k A_k(w)$ of degree τ then it is a *regular singularity* if $(w-w')^{k-r-\tau} w^r \mathring{A}_r(w)$ is regular at $w = w'$, for $r = 0,\ldots,k-1$. This is certainly the case for $w = 0$ and ∞, and it is not difficult to show that:

Theorem 18.5.6 *$(w-w')^{k-r-\tau}\mathring{A}_r(w)$ is regular, for $r = 0,\ldots,k-1$, at $w = w'$ if and only if $(w-w')^{k-r-\tau} A_r(w)$ is regular, for $r = 0,\ldots,k-1$, at $w = w'$.*

Corollary 18.5.2 *A zero $w = w'$ of $A_k(w)$ of degree τ is a regular singularity of (18.5.40) if $A_r(w)$, for all $k > r > k-\tau$, has $w = w'$ as a zero of degree at least $r+\tau-k$.*

The equation is said to be of *Fuchsian type* if all its singularities are regular and this can be determined from $A_r(w)$, $r = 0,\ldots,k$, without the need to compute $\mathring{A}_r(w)$, $r = 0,\ldots,k-1$.

Suppose that $w = w'$ is a root of $A_k(w)$ of degree one and not a root of $A_r(w)$, $r = 0,\ldots,k-1$; then, from Corollary 18.5.2, it is a regular singularity. Let the solution in a neighbourhood of $w = w'$ be

$$H(w) \simeq (w-w')^{-\gamma}U(w) + V(w), \tag{18.5.52}$$

where $U(w)$ and $V(w)$ are regular in a neighbourhood of $w = w'$ with expansions

$$U(w) := \sum_{j=0}^{\infty} u_j(w-w')^j, \qquad V(w) := \sum_{j=0}^{\infty} v_j(w-w')^j. \tag{18.5.53}$$

Then[28]

$$\frac{\mathrm{d}^r H(w)}{\mathrm{d}w^r} = \sum_{j=0}^{\infty} \frac{\Gamma(j-1-\gamma)}{\Gamma(j-r+1-\gamma)} u_j(w-w')^{j-r+1-\gamma} + \sum_{j=0}^{\infty} \frac{\Gamma(j-1)}{\Gamma(j-r+1)} v_j(w-w')^{j-r+1}. \tag{18.5.54}$$

[28] The Γ-functions take care of the zero terms since, for any integer n, if $1 < n \le r-1$, $\frac{\Gamma(n-1)}{\Gamma(n-r+1)} = 0$. Also if $n \le 1$, $\frac{\Gamma(n-1)}{\Gamma(n-r+1)} = \frac{\Gamma(1-n)}{\Gamma(r-n-1)}$.

Substituting into (18.5.50), using (18.5.49) and (18.5.51), and equating the coefficient of the term of order $(w - w')^{2-k-\gamma}$ to zero gives

$$\gamma = 1 - k + \frac{A_{k-1}(w')}{w' A_k'(w')}. \tag{18.5.55}$$

This formula gives a singularity except when γ is a negative integer. Then the first term in (18.5.52) can be absorbed into the second, resetting γ to zero and indicating that $H(w)$ is regular at $w = w'$.

A test function, given by Hunter and Baker (1979) and quoted by Guttmann (1989) serves as an illustration.

Example 18.5.1 Suppose that

$$\begin{aligned} F_9(w) := {} & 2 + \tfrac{1}{2}w + \tfrac{19}{8}w^2 + \tfrac{97}{48}w^3 + \tfrac{961}{384}w^4 + \tfrac{10363}{3840}w^5 + \tfrac{135199}{46080}w^6 \\ & + \tfrac{2026897}{645120}w^7 + \tfrac{34459681}{10321920}w^8 + \tfrac{654728563}{185794560}w^9. \end{aligned} \tag{18.5.56}$$

Then the maximum second-order homogeneous approximant has $\eta = 1 + m_0 + m_1 + m_2 = 9$. Let $m_0 := 3$, $m_1 := 3$ and $m_2 := 2$. With $a_{20} := 1$ it is now straightforward to solve the system of ten linear equations yielding the differential approximant equation in the form

$$\begin{aligned} & \left[\tfrac{2}{5}w^2 - \tfrac{7}{5}w + 1\right]\mathfrak{D}^2\{H(w)\} + \left[\tfrac{2}{5}w^3 - \tfrac{6}{5}w^2 + \tfrac{3}{10}w - 1\right]\mathfrak{D}\{H(w)\} \\ & \qquad + \left[\tfrac{3}{5}w^3 - \tfrac{21}{10}w^2\right]H(w) = 0. \end{aligned} \tag{18.5.57}$$

The roots of $A_2(w)$ are $w = w_1 := 1$, $w = w_2 := \frac{5}{2}$ and, from (18.5.55), $\gamma_1 = \frac{3}{2}$, $\gamma_2 = -2$; $w = 1$ is a regular singularity and $w = \frac{5}{2}$ is a regular point. The solution to (18.5.57) with $H(0) := 2$ and $H'(0) := \frac{1}{2}$ is easily shown to be

$$H(w) = (1 - w)^{-\frac{3}{2}} + \exp(-w). \tag{18.5.58}$$

It should also be noted that (18.5.57) is equivalent to the quadratic recurrence relationship

$$\begin{aligned} & (s+3)(s+2)h_{s+3} - \tfrac{1}{10}(s+2)(14s+25)h_{s+2} \\ & \qquad + \tfrac{1}{10}(4s^2 - 4s - 29)h_{s+1} + \tfrac{1}{5}(2s+3)h_s = 0 \end{aligned} \tag{18.5.59}$$

between the coefficients of the power series for $H(w)$. For $s = 1, \ldots, 6$ this recurrence relationship is also satisfied by the coefficients of $F_9(w)$.

For a discussion of the case of confluent singularities, of biased approximants and of the derivation of the leading amplitude the reader is referred to Guttmann (1989, pp. 89–94).

18.5.4 Algebraic Approximants

Differential approximants represent one way to extend Padé approximants. Another was proposed by Shafer (1974) who generalized (18.5.5) and (18.5.6) to

$$C_2(w)\{F(w)\}^2 + C_1(w)F(w) + C_0(w) = \mathrm{O}(w^{n_0+n_1+n_2+2}), \tag{18.5.60}$$

$$C_2(w)\{H(w)\}^2 + C_1(w)H(w) + C_0(w) = 0, \tag{18.5.61}$$

where

$$C_i(w) := \sum_{j=0}^{n_i} c_{ij} w^j, \quad i = 0, 1, 2. \tag{18.5.62}$$

He also suggested the generalization to an s-degree algebraic equation, so that:

Definition 18.5.4 The function $H(w)$ which satisfies

$$\sum_{i=0}^{s} C_i(w)\{H(w)\}^i = 0, \tag{18.5.63}$$

where $C_i(w)$, given by extending the range of (18.5.62) to $i = 0, \ldots, s$, are the *Hermite-Padé polynomials*, is the sth order algebraic approximant to $F(w)$, denoted by $[n_0, \ldots, n_s]_{\mathrm{AA}}$, if

$$\sum_{i=0}^{s} C_i(w)\{F(w)\}^i = \mathrm{O}(w^{\eta+1}), \tag{18.5.64}$$

$$\eta := s - 1 + \sum_{i=0}^{s} n_i. \tag{18.5.65}$$

Substituting from (18.5.1) and (18.5.62) into (18.5.63) gives[29]

$$c_{0j} + \omega_j = 0 \tag{18.5.66}$$

where

$$\omega_j := \sum_{i=1}^{s} \sum_{r=0}^{n_i} \sum_{k_1=0}^{\infty} \cdots \sum_{k_i=0}^{\infty} \delta^{\mathrm{Kr}}(r + k_1 + \cdots + k_i - j)\, c_{ir} \prod_{m=1}^{i} h_{k_m}. \tag{18.5.67}$$

The $\eta + 1$ equations with $0 \le j \le \eta$ are satisfied by $h_k := f_k$, which can be solved to determine the $\eta + 1$ constants $\{c_{ij}\}$. The problem now is to solve the sth degree

[29] With $c_{0j} := 0$ for $j > n_0$.

equation for $H(w)$. This is straightforward for the quadratic case and possible, but complicated, for the cubic and quartic cases. The alternative is to derive recurrence relations giving the successive coefficients h_j for $j = \eta, \eta + 1, \ldots$. Since $n_0 < \eta$, $\omega_j = 0$ for $j > \eta$. The largest h_{k_m} in ω_j is h_j and this will occur only in terms where all the other $h_{k_m} = f_0$. These terms are given by $h_j \tau$, where

$$\tau := \sum_{i=1}^{s} i c_{i0} f_0^{i-1}. \tag{18.5.68}$$

If all h_m are known for $m = 0, \ldots, j-1$,

$$\theta_j := \omega_j - h_j \tau \tag{18.5.69}$$

is known, and thus

$$h_j = -\theta_j / \tau. \tag{18.5.70}$$

This formula was derived by Homeier (2011) who gave examples of error-checking, by determining the succession of coefficients h_j using fewer than the known number of f_j. An example is provided by the work of Joyce (1988b) on the hard-hexagon model (Sect. 14.6.4.2).

Example 18.5.2 Suppose that

$$\begin{aligned} F_{13}(w) := {} & -\tfrac{1}{2} + \tfrac{15}{4} w - \tfrac{15}{8} w^2 - \tfrac{85}{16} w^3 - \tfrac{15}{32} w^4 + \tfrac{3315}{64} w^5 + \tfrac{52185}{128} w^6 \\ & + \tfrac{709515}{256} w^7 + \tfrac{9543585}{512} w^8 + \tfrac{129548515}{1024} w^9 + \tfrac{1777942185}{2048} w^{10} \\ & + \tfrac{24646968315}{4096} w^{11} + \tfrac{344714623985}{8192} w^{12} + \tfrac{4859013288915}{6384} w^{13}. \end{aligned} \tag{18.5.71}$$

Then the maximum second-degree approximant has $\eta = 1 + m_0 + m_1 + m_2 = 13$. Let $m_0 := 4$, $m_1 := 4$ and $m_2 := 4$. With $c_{20} := 1$ it is now straightforward to solve the system of fourteen linear equations, yielding the algebraic approximant equation

$$\begin{aligned} (1 - \tfrac{3}{2} w)(1 - w)^3 [H(w)]^2 + \tfrac{1}{2}(1 - 12w + 45w^2 - 66w^3 + 33w^4) H(w) & \\ + \tfrac{1}{2} w^3 (1 - 3w) = 0. & \end{aligned} \tag{18.5.72}$$

The singular points are given by the roots $w = 1$ and $w = \frac{2}{3}$ of $C_2(w)$ and the roots $w = \frac{1}{10}(5 \pm \sqrt{5})$ and $w = \frac{1}{6}(3 \pm \sqrt{5})$ of the discriminant (Sect. 17.6). The radius of the circle of convergence is $w = w_c := \frac{1}{10}(5 - \sqrt{5})$. Singular points of quadratic algebraic functions have cycle number two. By making the change of variable $x := \sqrt{w - w_c}$ and substituting the Taylor expansion for $H(x^2 + w_c)$ into (18.5.72), it can be shown that

$$H(w) \simeq H_0 + H_3(w - w_c)^{\frac{3}{2}}, \tag{18.5.73}$$

in a neighbourhood of w_c, where

$$H_0 := H(w_c) = -\tfrac{1}{2}(11 - 5\sqrt{5}), \qquad H_3 := \sqrt{\frac{76875\sqrt{5}-171875}{2}}. \tag{18.5.74}$$

Of course, in this quadratic case, direct solutions

$$H(w) = -\frac{(1 - 12w + 45w^2 - 66w^3 + 33w^4) \pm (1 - 5w + 5w^2)^{\frac{3}{2}}(1 - 9w + 9w^2)^{\frac{1}{2}}}{2(2 - 3w)(1 - w)^3}. \tag{18.5.75}$$

of (18.5.72) are easily obtained, together with the singularities and their exponents.

References and Author Index

A seminal resource for statistical mechanics is the series of twenty volumes *Phase Transitions and Critical Phenomena* published by Academic Press, London. The first six volumes[1] which appeared in the years 1972–1975 were edited by C. Domb and M.S. Green and, after the death of Professor Green, Volumes 7–20, which appeared in the years 1983–2001, were edited by C. Domb and J. Lebowitz. Many articles from these volumes are referenced below, where for brevity Volume n of this series is denoted by PTCP(n).

At certain points in the text there is the need to use the definitions of particular mathematical quantities or the results of extensive manipulations. The canonical sources for these are Abramowitz and Segun (1965) and Gradshteyn and Ryzhik (1980). However, most are now easily available on the internet or by using computer algebra packages. So specific references are omitted.

Page numbers where the references are cited in the text are given in square brackets.

Abraham, D. B., & Heilmann, O. J. (1980). Interacting dimers on the simple cubic lattice as a model for liquid crystals. *Journal of Physics A: Mathematical and General*, *13*, 1051–1062. [253].

Abramowitz, M., & Segun, I. A. (1965). *Handbook of mathematical functions*. New York, USA: Dover. [757].

Adler, J., Brandt, A., Janke, W., & Shmulyian, S. (1995). Three-state Potts model on the triangular lattice. *Journal of Physics A: Mathematical and General*, *28*, 5117–5129. [188].

Adler, J., & Enting, I. G. (1984). The two-dimensional spin-1 Ising system and related models. *Journal of Physics A: Mathematical and General*, *17*, 2233–2345. [555].

Adler, J., Enting, I. G., & Privman, V. (1983). Two three-state Potts models: The universality of confluent corrections to scaling. *Journal of Physics A: Mathematical and General*, *16*, 1967–1973. [551, 553].

Affleck, I. (1986). Universal term in the free energy at a critical point and the conformal anomaly. *Physical Review Letters*, *56*, 746–748. [162].

[1] Volume 4 was never published, but Volume 5 appeared in two parts 5a and 5b.

D.A. Lavis, *Equilibrium Statistical Mechanics of Lattice Models*,
Theoretical and Mathematical Physics, DOI 10.1007/978-94-017-9430-5

Aharony, A. (1974). Two-scale-factor universality and the ε expansion. *Physical Review B*, *9*, 2107–2109. [151].

Akhiezer, N. I. (1990). *Elements of elliptic functions* (Vol. 79). Translations of Mathematical Monographs. Providence, Rhode Island: American Mathematical Society. [712, 714].

Alexander, S. (1975). Lattice gas transitions of He on grafoil. A continuous transition with cubic terms. *Physics Letters A*, *54*, 353–354. [490, 605, 615].

Amit, D. J. (1978). *Field theory, the renormalization group and critical phenomena*. New York, USA: McGraw-Hill. [167, 520].

Anderson, G. R., & Wheeler, J. C. (1978a). Directionality dependence of lattice models for solutions with closed-loop coexistence curves. *Journal of Chemical Physics*, *69*, 2082–2088. [331].

Anderson, G. R., & Wheeler, J. C. (1978b). Theory of lower critical solution points in aqueous mixtures. *Journal of Chemical Physics*, *69*, 3403–3413. [332].

Andrews, T. (1869). The Bakerian Lecture: On the continuity of the gaseous and liquid states of matter. *Philosophical Transactions of the Royal Society*, *159*, 575–590. [171].

Arisue, H., & Tabata, K. (1997). Low-temperature series for the square lattice Potts model by the improved finite-lattice method. *Journal of Physics A: Mathematical and General*, *30*, 3313–3327. [518, 553].

Arrott, A., & Noakes, J. E. (1967). Approximate equation of state for nickel near its critical temperature. *Physical Review Letters*, *19*, 786–789. [149].

Asano, T. (1968). Generalization of the Lee-Yang theorem. *Progress of Theoretical Physics*, *40*, 1328–1336. [254].

Asano, T. (1970). Theorems on the partition functions of the Heisenberg ferromagnets. *Journal of the Physical Society of Japan*, *29*, 350–359. [254, 274].

Ashcroft, N. W., & Mermin, N. D. (1976). *Solid state physics*. New York, USA: Holt-Saunders. [530].

Ashkin, J., & Teller, E. (1943). Statistics of two-dimensional lattices with four components. *Physical Review*, *64*, 178–184. [53].

Au-Yang, H., & Fisher, M. E. (1975). Bounded and inhomogeneous Ising models. II Specific heat scaling function for a strip. *Physical Review B*, *11*, 3469–3487. [156].

Au-Yang, H., & Perk, J. H. H. (1995). The chiral Potts models revisited. *Journal of Statistical Physics*, *78*, 17–78. [59].

Babujian, H. M. (1982). Exact solution of the one-dimensional isotropic Heisenberg chain with arbitrary spin *s*. *Physics Letters*, *90A*, 479–482. [32].

Baker, G. A., Jr. (1961). Application of the Padé approximant method to the investigation of some magnetic properties of the Ising model. *Physical Review*, *124*, 768–774. [519, 559].

Baker, G. A., Jr. (2000). Defects and the convergence of Padé approximants. *Acta Applicandae Mathematica*, *61*, 37–52. [561].

Baker, G. A., Jr. (2005). Counter-examples to the Baker-Gammel-Wills conjecture and patchwork convergence. *Journal of Computational and Applied Mathematics*, *179*, 1–14. [561].

Baker, G. A., Jr., & Gaunt, D. S. (1967). Ising-model critical indices below the critical temperature. *Physical Review*, *155*, 545–552. [338].

Baker, G. A., Jr., & Graves-Morris, P. R. (1996). *Padé approximants* (2nd ed.). Cambridge, UK: C.U.P. [743, 747].

Baker, G. A., Jr., & Hunter, D. L. (1973). Methods of series analysis. II Generalized and extended methods with application to the Ising model. *Physical Review B*, *7*, 3377–3392. [556, 560].

Barber, M. N. (1983). Finite-size scaling. In *PTCP(8)* (pp. 145–266). [153, 157, 158, 609].

Barber, M. N., & Fisher, M. E. (1973a). Critical phenomena in systems of finite thickness. I The spherical model. *Annals of Physics*, *77*, 1–78. [156].

Barber, M. N., & Fisher, M. E. (1973b). Critical phenomena in systems of finite thickness. III Specific heat of an ideal boson film. *Physical Review A*, *8*, 1124–1135. [156].

Barbosa, M. A., & Henriques, V. B. (2008). Frustration and anomalous behaviour in the Bell-Lavis model of liquid water. *Physical Review E: Statistical, Nonlinear, and Soft Matter*, *77*, 051204-1–051204-15. [59].

Barker, J. A., & Fock, W. (1953). Theory of upper and lower critical solution temperatures. *Discussions of the Faraday Society*, *15*, 188–195. [332].

Bartis, J. T., & Hall, C. K. (1974). A lattice model of gas-gas equilibria in binary mixtures. *Physica*, *78*, 1–21. [332].

Batchelor, M. T. (2007). The Bethe ansatz after 75 years. *Physics Today*, *60*(1), 36–40. [32].

Baxter, R. J. (1972a). One-dimensional anisotropic Heisenberg chain. *Annals of Physics*, *70*, 323–337. [33, 439].

Baxter, R. J. (1972b). Partition function of the eight-vertex lattice model. *Annals of Physics*, *70*, 193–228. [61, 257, 391, 434, 438, 439, 442, 450, 452, 456, 459, 476, 477, 478].

Baxter, R. J. (1973a). Eight-vertex model in lattice statistics and one-dimensional anisotropic Heisenberg chain. I. Some fundamental eigenvectors. *Annals of Physics*, *76*, 1–24. [257, 434, 442, 452, 731].

Baxter, R. J. (1973b). Eight-vertex model in lattice statistics and one-dimensional anisotropic Heisenberg chain. II. Equivalence to a generalized ice-type lattice model. *Annals of Physics*, *76*, 25–47. [257, 434, 442].

Baxter, R. J. (1973c). Eight-vertex model in lattice statistics and one-dimensional anisotropic Heisenberg chain. III. Eigenvectors of the transfer matrix and Hamiltonian. *Annals of Physics*, *76*, 48–71. [257, 434, 442].

Baxter, R. J. (1973d). Potts model at the critical temperature. *Journal of Physics C: Solid State Physics*, *6*, L445–L448. [79, 187, 489].

Baxter, R. J. (1973e). Spontaneous staggered polarization of the F model. *Journal of Physics C: Solid State Physics*, *6*, L94–L96. [490].

Baxter, R. J. (1973f). Spontaneous staggered polarization of the F model. *Journal of Statistical Physics*, *9*, 145–182. [490].

Baxter, R. J. (1974). Ising model on a triangular lattice with three-spin interactions. II Free energy and correlation length. *Australian Journal of Physics*, *27*, 369–381. [379].

Baxter, R. J. (1978). Solvable eight-vertex model on an arbitrary lattice. *Philosophical Transactions of the Royal Society A*, *289*, 315–346. [23, 60, 62, 63, 64, 257, 384, 385, 391, 393, 394, 395, 431, 634, 635].

Baxter, R. J. (1980). Hard hexagons: Exact solution. *Journal of Physics A: Mathematical and General*, *13*, L61–L70. [282].

Baxter, R. J. (1981). Rogers-Ramanujan identities in the hard hexagon model. *Journal of Statistical Physics*, *26*, 427–452. [282].

Baxter, R. J. (1982a). Critical antiferromagnetic square lattice Potts model. *Proceedings of the Royal Society*, *383*, 43–54. [80].

Baxter, R. J. (1982b). *Exactly solved models in statistical mechanics*. London, UK: Academic Press. [33, 35, 61, 62, 149, 230, 282, 300, 385, 394, 395, 402, 407, 421, 434, 435, 450, 452, 477, 481, 490, 491, 514, 602, 726].

Baxter, R. J. (1986). Free-fermion, checkerboard and Z-invariant lattice models in statistical mechanics. *Proceedings of the Royal Society of London. Series A*, *404*, 1–33. [71, 72, 257, 395, 399, 402, 421, 635].

Baxter, R. J. (1988). Free energy of the solvable chiral Potts model. *Journal of Statistical Physics*, *52*, 639–667. [59].

Baxter, R. J. (1990). Chiral Potts model: Eigenvalues of the transfer matrix. *Physics Letters A*, *146*, 110–114. [59].

Baxter, R. J. (2005a). Derivation of the order parameter of the chiral Potts model. *Physical Review Letters*, *94*, 130602. [59].

Baxter, R. J. (2005b). The order parameter of the chiral Potts model. *Journal of Statistical Physics*, *120*, 1–36. [59].

Baxter, R. J. (2006). The challenge of the chiral Potts model. *Journal of Physics: Conference Series*, *42*, 11–24. [59].

Baxter, R. J., Bazhanov, V. V., & Perk, J. H. H. (1990). Functional relations for transfer matrices of the chiral Potts model. *International Journal of Modern Physics B*, *4*, 803–870. [59].

Baxter, R. J., & Enting, I. G. (1978). 339th solution of the Ising model. *Journal of Physics A: Mathematical and General*, *11*, 2463–2473. [256].

Baxter, R. J., Enting, I. G., & Tsang, S. K. (1980). Hard-square lattice gas. *Journal of Statistical Physics*, *22*, 465–489. [282].

Baxter, R. J., Kelland, S. B., & Wu, F. Y. (1976). Equivalence of the Potts model or Whitney polynomial with an ice-type model. *Journal of Physics A: Mathematical and General*, *9*, 397–406. [75, 78].

Baxter, R. J., Perk, J. H. H., & A-Yang, H. (1988). New solutions of the star-triangle relations for the chiral Potts model. *Physics Letters A*, *128*, 138–142. [59].

Baxter, R. J., Temperley, H. N. V., & Ashley, S. E. (1978). Triangular Potts model at its transition temperature, and related problems. *Proceedings of the Royal Society A*, *358*, 535–559. [187].

Baxter, R. J., & Wu, F. Y. (1973). Exact solution of an Ising model with three-spin interactions on a triangular lattice. *Physical Review Letters*, *31*, 1294–1303. [379, 561].

Baxter, R. J., & Wu, F. Y. (1974). Ising model on a triangular lattice with three-spin interactions. I The eigenvalue problem. *Australian Journal of Physics*, *27*, 357–367. [379].

Bazhanov, V. V., & Mangazeev, V. V. (2007). Analytic theory of the eight-vertex model. *Nuclear Physics B*, *775*, 225–282. [442, 725].

Beardon, A. F. (1984). *A primer on Riemann surfaces*. LMS Lecture Note Series (Vol. 78). Cambridge, UK: C.U.P. [699].

Bell, G. M. (1972). Statistical mechanics of water: Lattice model with directed bonding. *Journal of Physics C: Solid State Physics*, *5*, 889–905. [80, 246, 247, 248, 251, 252, 254].

Bell, G. M. (1974a). Ising ferrimagnetic models: I. *Journal of Physics C: Solid State Physics*, *7*, 1174–1188. [90].

Bell, G. M. (1974b). Ising ferrimagnetic models II. *Journal of Physics C: Solid State Physics*, *7*, 1189–1205. [319, 322].

Bell, G. M., & Lavis, D. A. (1970). Two-dimensional bonded lattice fluids. II orientable molecule model. *Journal of Physics A: Mathematical and General*, *3*, 568–581. [59, 80, 246, 252, 254].

Bell, G. M., Mingins, J., & Taylor, R. A. G. (1978). Second-order phase changes in phospholipid monolayer at the oil water interface and related phenomena. *Journal of the Chemical Society, Faraday Transactions II*, *74*, 223–234. [515].

Bell, G. M., & Sallouta, H. (1975). An interstitial model for fluid water: Accurate and approximate results. *Molecular Physics*, *29*, 1621–1637. [332, 337, 338, 339].

Bell, G. M., & Salt, D. W. (1976). Three-dimensional lattice model for the water/ice system. *Journal of the Chemical Society, Faraday Transactions II*, *72*, 76–86. [251].

Berezinskii, V. L. (1971). Destruction of long-range order in one-dimensional and two-dimensional systems having a continuous symmetry group I. Classical systems. *Soviet Physics JETP*, *32*, 493–500. [50].

Berezinskii, V. L. (1972). Destruction of long-range order in one-dimensional and two-dimensional systems with a continuous symmetry group II. Quantum systems. *Soviet Physics JETP*, *34*, 610–616. [50].

Berge, C. (1962). *The theory of graphs*. London: Methuen. [619, 621].

Berlin, T. H., & Kac, M. (1952). The spherical model of a ferromagnet. *Physical Review*, *86*, 821–835. [34, 35].

Bethe, H. A. (1931). Theorie der Metalle. Erster Teil. Eigenwerte und Eigenfunktionen der lineären atomischen Kette [English translation in Mattis (1993), pp. 689–716]. *Zeitschrift für Physik*, *71*, 205–226. [32, 169, 229].

Bethe, H. A. (1935). Statistical theory of superlattices. *Proceedings of the Royal Society A*, *150*, 552–575. [169, 229].

Betts, D. D. (1974). The XY model. In *PTCP(3)* (pp. 569–652). [560].

Bhanot, G., Creutz, M., Glässer, U., Horvath, I., Lacki, J., Schilling, K., et al. (1993). Low-temperature expansions for Potts models. *Physical Review B*, *48*, 6183–6191. [563].

Bidaux, R., Carrara, P., & Vivet, B. (1967). Antiferromagnetisme dans un champ magnetique I. traitement de champ moleculaire. *Journal of Physics and Chemistry of Solids*, *28*, 2453–2469. [226].

Binder, K. E. (1983). Critical behaviour at surfaces. In *PTCP(8)* (pp. 1–144). [153].

Binder, K. E. (1986). *Monte Carlo methods in statistical physics*. Topics in Current Physics (Vol. 7). Berlin: Springer. [616].

Binder, K. E. (1987). Theory of first-order transitions. *Reports on Progress in Physics*, *50*, 783–859. [175].

Binder, K. E., & Landau, D. P. (1984). Finite-size scaling at first-order phase transitions. *Physical Review B*, *30*, 1477–1485. [153].

Binney, J. J., Dowrick, N. J., Fisher, A. J., & Newman, M. E. J. (1993). *The theory of critical phenomena*. Oxford, UK: O.U.P. [25, 520].

Bishop, D. J., & Reppy, J. D. (1978). Study of the superfluid transition in two-dimensional ^{4}He films. *Physical Review Letters*, *40*, 1727–1730. [49].

Björnberg, J. E., & Grimmett, G. R. (2009). The phase transition of the quantum Ising model is sharp. *Journal of Statistical Physics*, *136*, 231–273. [31].

Black, J. L., & Emery, V. J. (1981). Critical properties of two-dimensional models. *Physical Review B*, *23*, 429–432. [490].

Bliss, G. A. (1966). *Algebraic functions*. New York, USA: Dover. [696].

Blöte, H. W. J., Cardy, J. L., & Nightingale, M. P. (1986). Conformal invariance, the central charge and universal finite-size amplitudes at criticality. *Physical Review Letters*, *56*, 742–745. [162].

Blöte, H. W. J., & Nightingale, M. P. (1982). Critical behaviour of the two-dimensional Potts model with a continuous number of states; a finite-size scaling analysis. *Physica A*, *112*, 405–465. [490, 613].

Blöte, H. W. J., Nightingale, M. P., & Derrida, B. (1981). Critical exponents of two-dimensional Potts and bond percolation models. *Journal of Physics A: Mathematical and General*, *14*, L45–L49. [613, 615].

Blume, M., Emery, V. J., & Griffiths, R. B. (1971). Ising model for the λ transition and phase separation on He^3-He^4 mixtures. *Physical Review A*, *4*, 1071–1077. [44, 92].

Bostan, A., Chyzak, F., Lecerf, G., Salvy, B., & Schost, E. (2007). Differential equations for algebraic functions. In C. W. Brown (Ed.), *Proceedings of the 2007 international symposium on symbolic and algebraic computing* (pp. 25–32). ACM Press. [696].

Boukraa, S., Guttmann, A. J., Hassani, S., Jensen, I., Maillard, J.-M., Nickel, B., & Zenine, N. (2008). Experimental mathematics on the magnetic susceptibility of the square lattice Ising model. *Journal of Physics A: Mathematical and Theoretical*, *41*, 455202 (51 pp). [563].

Bragg, W. L., & Williams, E. (1934). The effect of thermal agitation on atomic arrangement in alloys. *Proceedings of the Royal Society A*, *145*, 699–730. [44, 169].

Brak, R., & Guttmann, A. J. (1990). Algebraic approximants: A new method of series analysis. *Journal of Physics A: Mathematical and General*, *23*, L1331–L1337. [519, 564].

Brascamp, H. J., & Kunz, H. (1974). Zeros of the partition function for the Ising model in the complex temperature plane. *Journal of Mathematical Physics*, *15*, 65–66. [254].

Bretz, M. (1977). Ordered helium films on highly uniform graphite–finite-size effects, critical parameters and the three-state Potts model. *Physical Review Letters*, *38*, 501–505. [606].

Brézin, E. (1982). An investigation of finite-size scaling. *Journal de Physique*, *43*, 15–22. [609].

Brézin, E. (1983). Finite-size scaling. *Annals of the New York Academy of Sciences*, *410*, 339–349. [157].

Brezinski, C. (1996). Extrapolation algorithms and Padé approximations: A historical survey. *Applied Numerical Mathematics*, *20*, 299–318. [743].

Brezinski, C. (2000). Convergence acceleration during the 20th century. *Journal of Computational and Applied Mathematics*, *122*, 1–21. [558].

Briggs, K. M., Enting, I. G., & Guttmann, A. J. (1994). Series studies of the Potts model. II: Bulk series for the square lattice. *Journal of Physics A: Mathematical and General*, *27*, 1503–1523. [523, 553, 565, 566].

Brush, S. G. (1967). History of the Lenz-Ising model. *Reviews of Modern Physics*, *39*, 883–893. [37].

Buckingham, M. J. (1972). Thermodynamics. In *PTCP(2)* (pp. 1–38). [3, 124, 133].

Burkhardt, T. W. (1976a). Application of Kadanoff's lower-bound renormalization transformation to the Blume-Capel model. *Physical Review B*, *14*, 1196–1201. [609].

Burkhardt, T. W. (1976b). Kadanoff's lower-bound renormalization transformation. *Physical Review B*, *13*, 3187–3191. [609].

Burkhardt, T. W., & Eisenriegler, E. (1978). Renormalization group approach to surface critical behaviour in the Ising model. *Physical Review B*, *17*, 318–323. [609].

Burkhardt, T. W., & Knops, H. J. F. (1977). Renormalization group results for the Blume-Capel model in two and three dimensions. *Physical Review B*, *15*, 1602–1605. [609].

Burkhardt, T. W., Knops, H. J. F., & den Nijs, M. (1976). Renormalization group results for the three-state Potts model. *Journal of Physics A: Mathematical and General*, *9*, L179–L181. [609].

Burkhardt, T. W., & Swendsen, R. H. (1976). Critical temperatures of the spin-*s* Ising model. *Physical Review B*, *13*, 3071–3073. [613].

Burkhardt, T. W., & van Leeuwen, J. M. J. E. (1982). *Real-space renormalization*. Topics in Current Physics (Vol. 30). Berlin: Springer. [520, 615].

Butara, P., & Comi, M. (1994). High-temperature study of the Kosterlitz-Thouless phase transition in the XY model on the triangular lattice. *Physical Review B*, *50*, 3052–3057. [560].

Butera, P., Comi, M., & Guttmann, A. J. (1993). Critical exponents of the three-dimensional classical plane rotator model on the SC lattice from a high temperature series analysis. *Physical Review B*, *48*, 13987–13990. [49, 560].

Butera, P., Comi, M., & Marchesini, G. (1989). Critical behaviour of the two-dimensional XY model: An analysis of extended high-temperature series. *Physical Review B*, *40*, 534–545. [560].

Callender, C. (1999). Reducing thermodynamics to statistical mechanics: The case of entropy. *Journal of Philosophy*, *96*, 348–373. [1].

Camp, W. J., & Van Dyke, J. P. (1975). Isotropic magnets in two dimensions. *Journal of Physics C: Solid State Physics*, *8*, 336–352. [560].

Cardy, J. (2005). SLE for theoretical physicists. *Annals of Physics*, *318*, 81–118. [163].

Cardy, J. L. (1984). Conformal invariance and universality in finite-size scaling. *Journal of Physics A: Mathematical and General*, *17*, L385–L387. [159, 162, 593].

Cardy, J. L. (1987). Conformal invariance. In *PTCP(11)* (pp. 55–126). [3, 160, 162].

Cardy, J. L. (1988). *Finite-size scaling*. Current Physics-Sources and Comments (Vol. 2). Amsterdam, New York: North-Holland. [153].

Cardy, J. L. (1996). *Scaling and renormalization in statistical physics*. Cambridge, UK: C.U.P. [30, 34, 53, 490, 520].

Cardy, J. L., Nauenberg, M., & Scalapino, D. J. (1980). Scaling theory of the Potts-model multicritical point. *Physical Review B*, *22*, 2560–2568. [491].

Chaikin, P. M., & Lubensky, T. C. (1995). *Principles of condensed matter physics*. Cambridge, UK: C.U.P. [49, 53, 593, 627].

Chao, N.-C. (1982). Duality in three-dimensions. *Journal of Physics C: Solid State Physics*, *15*, L1263–L1267. [631].

Christe, P., & Henkel, M. (1993). *Introduction to conformal invariance and its applications to critical phenomena* (Vol. 16). Lecture Notes in Physics. Berlin: Springer. [3, 159].

Coleman, S. (1973). There are no Goldstone bosons in two dimensions. *Communications in Mathematical Physics*, *31*, 259–264. [31].

Conway, J. H., Curtis, R. T., Norton, S. P., Parker, R. A., & Wilson, R. A. (1985). *Atlas of finite groups*. Oxford, UK: O.U.P. [681].

Corless, R., Gonnet, G. H., Hare, D. E. G., Jeffrey, D. J., & Knuth, D. E. (1996). On the Lambert W function. *Advances in Computational Mathematics*, *5*, 329–359. [318].

Cormier, O., Singer, M. F., Trager, B. M., & Ulmer, F. (2002). Linear differential operators for polynomial equations. *Journal of Symbolic Computation*, *34*, 355–398. [696, 697].

Costa-Santos, R. (2006). Geometrical aspects of the Z-invariant Ising model. *European Physical Journal B: Condensed Matter and Complex Systems, 53*, 85–90. [634, 635].

Courant, R., & Hilbert, D. (1962). *Methods of Mathematical Physics*. New York, USA: Interscience. [706, 707].

Darboux, G. (1878). Mémoire sur l'approximation des fonctions de trés-grand nombres, et sur une classe étendue de développements en série. *Journal de Mathématiques Pures et Appliquées, 3*, 5–56. [556].

de Gennes, P. G. (1972). Exponents for the excluded volume problem as derived by the Wilson method. *Physics Letters A, 38*, 339–340. [34].

de Gennes, P. G., & Prost, J. (1993). *The physics of liquid crystals*. Oxford, UK: O.U.P. [2].

de Neef, T. (1975). Some applications of series expansions in magnetism. Ph.D. Thesis, Technische Wetenschappen aan de Technische Hogeschool, Eindhoven. [518, 524, 543].

de Neef, T., & Enting, I. G. (1977). Series expansions from the finite-lattice method. *Journal of Physics A: Mathematical and General, 10*, 801–805. [230, 518, 553].

de Podesta, M. (2013). Redefining temperature. *Physics World, 26*(8), 28–32. [5].

den Nijs, M. P. M. (1979). A relationship between the temperature exponents of the eight-vertex and q-state Potts models. *Journal of Physics A: Mathematical and General, 12*, 1857–1868. [490].

des Cloizeaux, J., & Pearson, J. J. (1962). Spin-wave spectrum of the antiferromagnetic linear chain. *Physical Review, 128*, 2131–2135. [32].

Devaney, R. L. (1989). *Chaotic dynamic systems*. California, USA: Addison-Wesley. [570].

Dingle, R. B. (1973). *Asymptotic expansions: Their derivation and interpretation*. London: Academic Press. [556].

Dobrushin, R. L. (1965). Existence of a phase transition in two-dimensional and three-dimensional Ising models. *Theory of Probability and Its Applications, 10*, 193–213. [253].

Dobrushin, R. L. (1968a). Gibbsian random fields for lattice systems with pairwise interactions. *Functional Analysis and Its Applications, 2*, 292–301. [253, 261, 265].

Dobrushin, R. L. (1968b). The problem of uniqueness of a Gibbsian random field and the problem of phase transitions. *Functional Analysis and Its Applications, 2*, 302–312. [253, 265].

Domb, C. (1960). On the theory of cooperative phenomena in crystals. *Philosophical Magazine Supplement, 9*, 149–361. [301, 429, 579, 583, 584, 605].

Domb, C. (1974a). Configurational studies of the Potts models. *Journal of Physics A: Mathematical and General, 7*, 1335–1348. [54, 55, 532, 741].

Domb, C. (1974b). Graph theory and embeddings. In *PTCP(3)* (pp. 1–95). [532, 540, 646, 741].

Domb, C. (1974c). Ising model. In *PTCP(3)* (pp. 357–484). [222, 223].

Domb, C. (1976). Finite-cluster patition functions for the d-vector model. *Journal of Physics A: Mathematical and General, 9*, 983–998. [543].

Domb, C., & Guttmann, A. J. (1970). Low-temperature series for the Ising model. *Journal of Physics C: Solid State Physics, 3*, 1652–1660. [558].

Domb, C., & Hunter, D. L. (1965). On the critical behaviour of ferromagnets. *Proceedings of the Physical Society, 86*, 1147–1151. [3, 315].

Domb, C., & Sykes, M. F. (1956). On metastable approximations in cooperative assemblies. *Proceedings of the Physical Society A, 235*, 247–259. [518].

Domb, C., & Sykes, M. F. (1957a). The calculation of lattice constants in crystal statistics. *Philosophical Magazine, 2*, 733–749. [518].

Domb, C., & Sykes, M. F. (1957b). On the susceptibility of a ferromagnetic above the Curie point. *Proceedings of the Royal Society A, 240*, 214–228. [518].

Duxbury, P. M., Yeomans, J., & Beale, P. D. (1984). Wavevector scaling and the phase diagram of the chiral clock model. *Journal of Physics A: Mathematical and General, 17*, L179–L184. [59].

Ehrenfest, P. (1933). Phase changes in the ordinary and extended sense classified according to the corresponding singularity of the thermodynamic potential. *Proceedings of the National Academy of Sciences Amsterdam, 36*, 153–157. [2].

Eisenberg, D., & Kauzmann, W. (1969). *The structure and properties of water*. Oxford, UK: Clarendon Press. [2, 61, 247].

Ellis, R. S., Otto, P. T., & Touchette, H. (2005). Analysis of phase transitions in the mean-field Blume-Emery-Griffiths model. *Annals of Applied Probability*, *15*, 2203–2254. [44].

English, P. S., Hunter, D. L., & Domb, C. (1979). Extension of the high-temperature free energy series for the classical vector model of ferromagnetism in general spin dimensionality. *Journal of Physics A: Mathematical and General*, *12*, 2111–2130. [543].

Enting, I. G. (1978a). Generalized Möbius functions for rectangles on the square lattice. *Journal of Physics A: Mathematical and General*, *11*, 563–568. [648].

Enting, I. G. (1978b). Some algebraic techniques for obtaining low-temperature series expansions. *Australian Journal of Physics*, *31*, 515–522. [551].

Enting, I. G. (1980a). Series analysis for the three-state Potts model. *Journal of Physics A: Mathematical and General*, *13*, L133–L136. [553].

Enting, I. G. (1980b). Universality in a generalised Potts model. *Journal of Physics A: Mathematical and General*, *13*, L409–L412. [555].

Enting, I. G. (1996). Series expansions from the finite lattice method. *Nuclear Physics B: Proceedings Supplements*, *47*, 180–187. [518, 533].

Enting, I. G., & Baxter, R. J. (1977). A special series expansion technique for the square lattice. *Journal of Physics A: Mathematical and General*, *10*, L117–L119. [649].

Enting, I. G., & Baxter, R. J. (1980). An investigation of the high-field series expansions for the square lattice Ising model. *Journal of Physics A: Mathematical and General*, *13*, 3723–3734. [524, 565].

Enting, I. G., Guttmann, A. J., & Jensen, I. (1994). Low-temperature series expansions for the spin-1 Ising model. *Journal of Physics A: Mathematical and General*, *27*, 6987–7006. [555].

Enting, I. G., & Wu, F. Y. (1982). Triangular lattice Potts models. *Journal of Statistical Physics*, *28*, 351–373. [184, 555].

Erez, A., & Meir, Y. (2010). Thermal phase transition in two-dimensional disordered superconductors. *Europhysics Letters*, *91*(47003), 1–5. [49].

Essam, J. W. (1972). Percolation and cluster size. In *PTCP(2)* (pp. 197–270). [318].

Essam, J. W., & Fisher, M. E. (1963). Padé approximant studies of the lattice gas and Ising ferromagnet below the critical point. *Journal of Chemical Physics*, *38*, 802–812. [133].

Essam, J. W., & Garelick, H. (1967). Critical behaviour of a soluble model of dilute ferromagnetism. *Proceedings of the Physical Society*, *92*, 136–149. [318].

Essam, J. W., & Sykes, M. F. (1963). The crystal statistics for the diamond lattice. *Physica*, *29*, 378–388. [338].

Fabricius, K. (2007). A new q-matrix in the eight-vertex model. *Journal of Physics A: Mathematical and Theoretical*, *40*, 4075–4086. [442, 450, 456].

Fabricius, K., & McCoy, B. M. (2003). New developments in the eight-vertex model. *Journal of Statistical Physics*, *111*, 323–327. [434, 442, 450, 469].

Fabricius, K., & McCoy, B. M. (2004). Functional equations and fusion matrices for the eight-vertex model. *Publication of RIMS Kyoto University*, *40*, 905–932. [469].

Fabricius, K., & McCoy, B. M. (2005). New developments in the eight-vertex model II: Chains of odd length. *Journal of Statistical Physics*, *120*, 37–70. [434, 442, 472].

Fabricius, K., & McCoy, B. M. (2007). The tq equation of the eight-vertex model for complex elliptic roots of unity. *Journal of Physics A: Mathematical and Theoretical*, *40*, 14893–14926. [442, 450, 452, 456, 465].

Fabricius, K., & McCoy, B. M. (2009). New q matrices and their functional equations for the eight-vertex model at elliptic roots of unity. *Journal of Statistical Physics*, *134*, 643–668. [442, 452].

Falicov, L. M. (1966). *Group theory and its applications*. Chicago, USA: University of Chicago Press. [676, 677].

Fan, C. (1972). On the critical properties of the Ashkin-Teller model. *Physics Letters A*, *39*, 136–136. [53].

Fan, C., & Wu, F. Y. (1970). General lattice statistical model of phase transitions. *Physical Review B*, *2*, 723–733. [71, 492].

Felderhof, B. U. (1973a). Direct diagonalization of the transfer matrix of the zero-field free-fermion model. *Physica*, *65*, 421–451. [71].

Felderhof, B. U. (1973b). Diagonalization of the transfer matrix of the free-fermion model II. *Physica*, *66*, 279–297. [71].

Feldmann, H., Guttmann, A. J., Jensen, I., Schrock, R., & Tsai, S.-H. (1998). Study of the Potts model on the honeycomb and triangular lattices: Low-temperature series and partition function zeros. *Journal of Physics A: Mathematical and General*, *31*, 2287–2310. [554].

Ferdinand, A. E., & Fisher, M. E. (1969). Bounded and inhomogeneous Ising models. I Specific heat anomaly on a finite lattice. *Physical Review*, *185*, 832–846. [156].

Ferer, M., Moore, M. A., & Wortis, M. (1973). Critical indices and amplitudes of classical planar models in finite field for temperatures greater that t_c. *Physical Review*, *8*, 5205–5212. [151].

Ferer, M., & Velgakis, M. J. (1983). High-temperature critical behaviour of two-dimensional planar models: A series investigation. *Physical Review B*, *27*, 314–325. [560].

Fields, J. L. (1967). A uniform treatment of Darboux's method. *Archive for Rational Mechanics and Analysis*, *27*, 289–305. [557].

Fiore, C. E., Szortyka, M. M., Barbosa, M. C., & Henriques, V. B. (2009). Liquid polymorphism, order-disorder transitions and anomalous behaviour: A Monte Carlo study of the Bell-Lavis model for water. *Journal of Chemical Physics*, *131*, 9–164506. [60].

Fischer, P., Lebech, B., Meier, G., Rainford, B. D., & Vogt, O. (1978). Magnetic phase transitions of CeSb: I Zero applied magnetic field. *Journal of Physics C: Solid State Physics*, *11*, 345–364. [2].

Fisher, M. E. (1959). Transformations of Ising model. *Physical Review*, *113*, 969–981. [312].

Fisher, M. E. (1961). Statstical mechanics of dimers on a plane lattice. *Physical Review*, *124*, 1664–1672. [257, 503, 504].

Fisher, M. E. (1963). Perpendicular susceptibility of the Ising model. *Journal of Mathematical Physics*, *4*, 124–135. [223].

Fisher, M. E. (1965). The nature of critical points. In W. Brittin (Ed.), *Lectures in theoretical physics* (Vol. 7c, pp. 1–157). Boulder, USA: University of Colorado Press. [254, 374].

Fisher, M. E. (1966). On the dimer solution of planar Ising models. *Journal of Mathematical Physics*, *7*, 1776–1781. [86].

Fisher, M. E. (1967). Theory of equilibrium critical phenomena. *Reports on Progress in Physics*, *30*, 615–730. [2, 156].

Fisher, M. E. (1968). Renormalization of critical exponents by hidden variables. *Physical Review*, *176*, 257–272. [96, 318].

Fisher, M. E. (1969). Rigorous inequalities for critical-point correlation exponents. *Physical Review*, *180*, 594–600. [135].

Fisher, M. E. (1971). The theory of critical point singularities. In M. Green (Ed.), *Critical Phenomena, Proceedings of the 51st Enrico Fermi Summer School, Varenna, Italy* (pp. 1–99). London, UK: Academic Press. [153, 156].

Fisher, M. E. (1983). Scaling, universality and renormalization group theory. In F. Hahne (Ed.), *Critical phenomena* (pp. 1–139). Berlin: Springer. [29, 139, 147, 149].

Fisher, M. E., & Au-Yang, H. (1979). Inhomogeneous differential approximants for power series. *Journal of Physics A*, *12*, 1677–1692. [563].

Fisher, M. E., & Barber, M. N. (1972). Scaling theory for finite-size effects in the critical region. *Physical Review Letters*, *28*, 1516–1519. [153, 380].

Fisher, M. E., & Berker, A. N. (1982). Scaling for first-order transitions in thermodynamic and finite systems. *Physical Review B*, *26*, 2507–2513. [105, 134, 153].

Fisher, M. E., & Ferdinand, A. E. (1967). Interfacial, boundary and size effects at critical points. *Physical Review Letters*, *19*, 169–172. [153, 156, 380].

Fisher, M. E., & Privman, V. (1985). First-order transitions breaking O(n) symmetry: Finite-size scaling. *Physical Review B*, *32*, 447–464. [153].

Fisher, M. E., & Scesney, P. E. (1970). Visibility of critical-exponent renormalization. *Physical Review A*, *2*, 825–835. [317].

Fisher, M. E., & Sykes, M. F. (1962). Antiferromagnetic susceptibilities on the simple cubic and body-centred cubic Ising lattices. *Physica*, *28*, 939–956. [222].

Fleming, P. D., & Gibbs, J. H. (1974). An adaption of the lattice gas to the water problem. *Journal of Statistical Physics*, *10*, 157–173 & 351–378. [252].

Forder, H. G. (1949). The Euler-Maclauren formula. *The Mathematical Gazette*, *33*, 172–176. [426].

Fowler, D. (1972). The Riemann-Hugoniot catastrophe and van der Waals equation. In C. H. Waddington (Ed.), *Towards a theoretical biology* (Vol. 4, pp. 1–7). Edinburgh: Edinburgh University Press. [659].

Fox, P. F., & Guttmann, A. J. (1973). Low-temperature critical behaviour of the Ising model with $s > \frac{1}{2}$. *Journal of Physics C: Solid State Physics*, *6*, 913–930. [518].

Friedan, D., Qiu, Z., & Shenker, S. (1984). Conformal invariance, unitarity and critical exponents in two dimensions. *Physical Review Letters*, *52*, 1575–1578. [490].

Frobenius, G. (1912). *Über Matrizen aus nicht negativen Elementen*. Berlin: Sitzungsberichte Preussische Akademie der Wissenschaft (pp. 456–477). [673].

Gaite, J. (1992). Symmetry in catastophe theory, phase diagrams and two-dimensional conformal field theories. *Journal of Physics A: Mathematical and General*, *25*, 3051–3062. [185, 659].

Gaite, J., Margalef-Roig, J., & Miret-Artés, S. (1998). Analysis of a three-component model phase diagram by catastrophe theory. *Physical Review B*, *57*, 13527–13534. [659].

Gammel, J. L. (1973). Review of two recent generalizations of the Padé approximant. In P. R. Graves-Morris (Ed.), *Padé approximants and their applications* (pp. 3–9). London: Academic Press. [563].

Gantmacher, F. R. (1959). *Applications of the theory of matrices*. New York, USA: Interscience. [673, 674].

Gaunt, D. S., & Fisher, M. E. (1965). Hard-sphere lattice gases. I Plane-square lattice. *Journal of Chemical Physics*, *43*, 2840–2863. [282].

Gaunt, D. S., & Guttmann, A. J. (1965). Asymptotic analysis of coefficients. In *PTCP(3)* (pp. 181–243). [555].

Georges, A., Hansel, D., Le Doussal, P., & Maillard, J. M. (1986). The vicinity of disorder varieties: A systematic expansion. *Journal of Physics A: Mathematical and General*, *19*, 1001–1005. [326].

Georgii, H.-O. (1988). *Gibbs measures and phase transitions*. Berlin, New York: Walter de Gruyter. [263].

Gibbs, J. W. (1902). *Elementary principles in statistical mechanics*. Yale University Press. Reprinted by New York: Dover (1960). [14].

Gilmore, R. (1981). *Catastrophe theory for scientists and engineers*. New York: Wiley. [659].

Ginsparg, P. (1990). Applied conformal field theory. In E. Brezin & J. Zinn-Justin (Eds.), *Fields, strings and critical phenomena. Proceedings of 49th Summer School in Theoretical Physics*, Les Houches, 1988 (pp. 6–168). Amsterdam, New York: North-Holland. [692].

Ginzburg, V. L. (1960). Some remarks on phase transitions of the second kind and the microscopic theory of ferroelectric materials. *Soviet Physics JETP*, *2*, 1824–1834. [204].

Ginzburg, V. L., & Landau, L. D. (1950). On the theory of superconductivity. In D. ter Haar (Ed.), *Collected papers of L.D. Landau* (pp. 546–568). Reprinted in (1965), Oxford, UK: Pergamon. [167].

Glasser, M. L. (1969). Evaluation of the partition functions for some two-dimensional ferroelectric models. *Physical Review*, *184*, 539–542. [488].

Goldenfeld, N. (1992). *Lectures on phase transitions and the renormalization group*. Reading, Mass: Addison-Wesley. [520, 593].

Goodwin, E. T. (1949). The evaluation of integrals of the form $\int_{-\infty}^{\infty} f(x)\mathrm{e}^{-x^2}\mathrm{d}x$. *Proceedings of the Cambridge Philosophical Society*, *45*, 241–245. [426].

Gradshteyn, I. S., & Ryzhik, I. M. (1980). *Table of integrals, series, and products* (4th ed.). London, UK: Academic Press. [757].

Green, H. S., & Hurst, C. A. (1964). *Order-disorder phenomena*. London: Interscience. [497].

Green, M. S., Vicentini-Missoni, M., & Levelt-Sengers, J. M. H. (1967). Scaling-law equation of state for gases in the critical region. *Physical Review Letters*, *18*, 1113–1117. [149].

Grest, G. S. (1981). Monte Carlo study of the antiferromagnetic Potts model on frustrated lattices. *Journal of Physics A: Mathematical and General, 14*, L217–L221. [158].

Griffiths, R. B. (1964). Peierls proof of spontaneous magnetization in a two-dimensional Ising model. *Physical Review A, 136*, 437–439. [253].

Griffiths, R. B. (1965). Ferromagnets and simple fluids near the critical point: Some thermodynamic inequalities. *Journal of Chemical Physics, 43*, 1958–1968. [125].

Griffiths, R. B. (1970). Thermodynamics near the two-fluid critical mixing point in He^3-He^4. *Physical Review Letters, 24*, 715–717. [94].

Griffiths, R. B. (1972). Rigorous results and theorems. In *PTCP(1)* (pp. 7–109), UK. [260].

Griffiths, R. B. (1981). Mathematical properties of renormalization group transformations. *Physica A, 106*, 59–69. [573].

Griffiths, R. B., & Pearce, P. A. (1978). Position-space renormalization group transformations: Some proofs and some problems. *Physical Review Letters, 41*, 917–920. [573].

Griffiths, R. B., & Pearce, P. A. (1979). Mathematical properties of position-space renormalization group transformations. *Journal of Statistical Physics, 20*, 499–545. [573].

Griffiths, R. B., & Wheeler, J. C. (1970). Critical points in multicomponent systems. *Physical Review A, 2*, 1047–1064. [136].

Guggenheim, E. A. (1935). The statistical mechanics of regular solutions. *Proceedings of the Royal Society A, 148*, 304–312. [229, 654].

Guggenheim, E. A., & McGlashan, M. L. (1951). Statistical mechanics of regular mixtures. *Proceedings of the Royal Society A, 206*, 335–353. [229, 230, 248, 654].

Gunton, J. D., San Miguel, M., & Sahai, P. S. (1983). Dynamics of first-order transitions. In *PTCP(8)* (pp. 269–466). [175].

Guttmann, A. J. (1969). Determination of critical behaviour in lattice statistics from series expansions III. *Journal of Physics C: Solid State, 2*, 1900–1907. [558].

Guttmann, A. J. (1975). On the recurrence relation method of series analysis. *Journal of Physics A: Mathematical and General, 8*, 1081–1088. [556].

Guttmann, A. J. (1978). The critical behaviour of two-dimensional isotropic spin systems. *Journal of Physics A: Mathematical and General, 11*, 545–553. [560].

Guttmann, A. J. (1989). Asymptotic analysis of power-series expansions. In PTCP(13) (pp. 1–234). [518, 519, 522, 555, 558, 562, 750, 753].

Guttmann, A. J., & Enting, I. G. (1988). The number of convex polygons on the square and honeycomb lattices. *Journal of Physics A, 21*, L467–L474. [562].

Guttmann, A. J., & Enting, I. G. (1993). Series studies of the Potts model: I The simple cubic Ising model. *Journal of Physics A: Mathematical and General, 26*, 807–821. [555, 563].

Guttmann, A. J., & Enting, I. G. (1994). Series studies of the Potts model: III The 3-state model on the simple cubic lattice. *Journal of Physics A: Mathematical and General, 27*, 5801–5812. [555, 566].

Guttmann, A. J., & Joyce, G. S. (1972). On a new method of series analysis in lattice statistics. *Journal of Physics A: Mathematical and General, 5*, L81–L84. [519, 559].

Hancock, H. (1958). *Lectures on the theory of elliptic functions*. New York: Dover. [714].

Hankey, A., & Stanley, H. E. (1972). Systematic application of generalized homogeneous functions to static scaling, dynamic scaling and universality. *Physical Review B, 6*, 3515–3542. [3, 96, 99, 100].

Hankey, A., Stanley, H. E., & Chang, T. S. (1972). Geometric predictions of scaling at tricritical points. *Physical Review Letters, 29*, 278–281. [3, 142].

Heilmann, O. J. (1972). Existence of phase transitions in certain lattice gases with repulsive potential. *Lettere Al Nuovo Cimento, 3*, 95–98. [253].

Heilmann, O. J. (1980). Existence of an ordered phase for the repulsive lattice gas on the fcc lattice. *Journal of Physics A: Mathematical and General, 13*, 1803–1810. [253].

Heilmann, O. J., & Huckaby, D. A. (1979). Phase transitions in lattice gas models for water. *Journal of Statistical Physics, 20*, 371–383. [254, 264, 265].

Heilmann, O. J., & Lieb, E. H. (1972). Theory of monomer-dimer systems. *Communications in Mathematical Physics*, *25*, 190–232. [254].

Hemmer, P. C., & Stell, G. (1970). Fluids with several phase transitions. *Physical Review Letters*, *24*, 1284–1287. [43].

Henkel, M. (1999). *Conformal invariance and critical phenomena*. Berlin: Springer. [159, 162].

Hijmans, J., & de Boer, J. (1955a). An approximation method for order-disorder problems I. *Physica*, *21*, 471–484. [230, 234, 235, 236].

Hijmans, J., & de Boer, J. (1955b). An approximation method for order-disorder problems II. *Physica*, *21*, 485–498. [230].

Hijmans, J., & de Boer, J. (1955c). An approximation method for order-disorder problems III. *Physica*, *21*, 499–516. [230, 241].

Hilhorst, H. J., Schick, M., & van Leeuwen, J. M. J. (1978). Differential form of real-space renormalization: Exact results for two-dimensional Ising models. *Physical Review Letters*, *40*, 1605–1608. [588].

Hilhorst, H. J., Schick, M., & van Leeuwen, J. M. J. (1979). Exact renormalization group equations for the two-dimensional Ising model. *Physical Review B*, *19*, 2749–2763. [588].

Hille, E. (1959). *Analytic function theory* (Vol. 1). New York, USA: Chelsea. [556].

Hille, E. (1962). *Analytic function theory* (Vol. 2). New York, USA: Chelsea. [696, 697].

Hindermann, A., Kunz, H., & Wu, F. Y. (1978). Exact results for the Potts model in two dimensions. *Journal of Statistical Physics*, *19*, 623–632. [306].

Ho, J. T., & Litster, J. D. (1969). Magnetic equation of state of $CrBr_3$ near the critical point. *Physical Review Letters*, *22*, 603–606. [149].

Hohenberg, P. C. (1967). Existence of long-range order in one and two dimensions. *Physical Review*, *158*, 383–386. [31].

Homeier, H. H. H. (2011). Series prediction based on algebraic approximants. *ISRN Applied Mathematics, 2011*, 7 (Article ID 958968). [755].

Houlrik, J. M., & Knak Jensen, S. J. (1986). Phase diagram of the three-state chiral clock model studied by Monte Carlo renormalization-group calculations. *Physical Review B*, *34*, 325–329. [59].

Houtappel, R. M. F. (1950). Order-disorder in hexagonal lattices. *Physica*, *16*, 425–455. [257, 318].

Hsu, S.-C., & Gunton, J. D. (1977). Study of the cumulant expansion for renormalization group transformations on two- and three-dimensional Ising models. *Physical Review B*, *15*, 2688–2693. [599].

Hu, Z.-N., & Hou, B.-Y. (1996). Three-dimensional vertex model in statistical mechanics from the Baxter-Bazhanov model. *Journal of Statistical Physics*, *82*, 633–655. [60].

Huang, H. Y., Popkov, V., & Wu, F. Y. (1997). Exact solution of a three-dimensional dimer system. *Physical Review Letters*, *78*, 409–412. [256].

Hulthén, L. (1938). Über das austauschproblem eines kristalles. *Arkiv för Matematik, Astronomi och Fysik*, *26A*, 1–106. [32].

Hunter, D. L., & Baker, G. A., Jr. (1973). Method of series analysis. I. Comparison of current methods used in the theory of critical phenomena. *Physical Review B*, *7*, 3346–3376. [556].

Hunter, D. L., & Baker, G. A., Jr. (1979). Methods of series analysis, III Integral approximant methods. *Physical Review B*, *19*, 3808–3821. [563, 597, 753].

Huse, D. A. (1981). Simple three-state model with infintely many phases. *Physical Review B*, *24*, 5180–5194. [59].

Husimi, K., & Syôzi, I. (1950). The statistics of honeycomb and triangular lattice I. *Progress of Theoretical Physics*, *5*, 177–186. [257].

Iooss, G., & Joseph, D. D. (1989). *Elementary stability and bifurcation theory*. Berlin: Springer. [662].

Isakov, S. N. (1984). Nonanalytic features of the first order phase transition in the Ising model. *Communications in Mathematical Physics*, *95*, 427–443. [565].

Ising, E. (1925). Beitrag zur Theorie des Ferromagnetismus. *Zeitschrift für Physik*, *31*, 253–258. [37, 256].

Itzykson, C. G. (1988). *Conformal invariance and applications to statistical mechanics*. Singapore: World Scientific. [159].

Itzykson, C. G., Pearson, R. B., & Zuber, J. B. (1983). Distribution of zeros in Ising and gauge models. *Nuclear Physics B*, *220*, 415–433. [379, 380].

Izyumov, Y. A., & Skryabin, Y. N. (1988). *Statstical mehanics of magnetically ordered systems*. New York: Consultants Bureau. [32].

Jensen, I., & Guttmann, A. J. (1996). Extrapolation procedure for the low-temperature series for the square lattice spin-1 Ising model. *Journal of Physics A: Mathematical and General*, *29*, 3817–3836. [518].

Jensen, I., Guttmann, A. J., & Enting, I. G. (1996). Low-temperature series expansions for the square lattice Ising model with $s > 1$. *Journal of Physics A: Mathematical and General*, *29*, 3805–3815. [555].

Jensen, I., Guttmann, A. J., & Enting, I. G. (1997). The Potts model on kagomé and honeycomb lattices. *Journal of Physics A: Mathematical and General*, *30*, 8067–8083. [554].

Jimbo, M., & Miwa, T. (1995). *Algebraic analysis of solvable lattice models*. Providence, Rhode Island: AMS for CBMS. [32].

Jones, G. L. (1966). Complex temperatures and phase transitions. *Journal of Mathematical Physics*, *7*, 2000–2005. [254].

José, J. V., Kadanoff, L. P., Kirkpatrick, S., & Nelson, D. R. (1977). Renormalization, vortices, and symmetry-breaking perturbations in the two-dimensional planar model. *Physical Review B*, *16*, 1217–1241. [53].

Josephson, B. D. (1967). Inequality for the specific heat. *Proceedings of the Physical Society*, *92*, 269–275 & 276–284. [135].

Joyce, G. S. (1966). Spherical model with long-range ferromagnetic interactions. *Physical Review*, *146*, 349–358. [35].

Joyce, G. S. (1967). Classical Heisenberg model. *Physical Review*, *155*, 478–495. [35, 543].

Joyce, G. S. (1969). Absence of ferromagnetism or antiferromagnetism in the isotropic Heisenberg model with long-range interactions. *Journal of Physics C: Solid State Physics*, *2*, 1531–1533. [36].

Joyce, G. S. (1972). Critical properties of the spherical model. In *PTCP(2)* (pp. 375–442). [750].

Joyce, G. S. (1975a). Analytic properties of the Ising model with triplet interactions on the triangular lattice. *Proceedings of the Royal Society A*, *343*, 45–62. [379, 561].

Joyce, G. S. (1975b). On the magnetization of the triangular lattice Ising model with triplet interactions. *Proceedings of the Royal Society A*, *345*, 277–293. [379, 561].

Joyce, G. S. (1988a). Exact results for the activity and isothermal compressibility of the hard-hexagon model. *Journal of Physics A: Mathematical and General*, *21*, L983–L988. [282].

Joyce, G. S. (1988b). On the hard-hexagon model and the theory of modular functions. *Philosophical Transactions of the Royal Society A*, *325*, 643–702. [282, 564, 698, 775].

Joyce, G. S., & Bowers, R. G. (1966). Cluster series for the infinite spin Heisenberg model. *Proceedings of the Physical Society*, *88*, 1053–1055. [43].

Joyce, G. S., & Guttmann, A. J. (1973). A new method of series analysis. In P. R. Graves-Morris (Ed.), *Padé approximants and their applications* (pp. 163–167). London, UK: Academic Press. [562, 750].

Jüngling, K. (1975). Exact solution of a nonplanar two-dimensional Ising model with short range two-spin interaction. *Journal of Physics C: Solid State Physics*, *8*, L169–L171. [70, 71].

Jüngling, K., & Obermair, G. (1974). Note on universality and the eight-vertex model. *Journal of Physics C: Solid State Physics*, *7*, L363–L365. [70].

Kac, M. (1964). The work of T. H. Berlin in statistical mechanics.a personal reminiscence. *Physics Today*, *17*(10), 40–42. [45].

Kac, M., & Thompson, C. S. (1971). Spherical model and the infinite spin dimensionality limit. *Physica Norvegica*, *5*, 163–168. [35].

Kadanoff, L. P. (1966). Scaling laws for Ising models near T_c. *Physics*, *2*, 263–272. [96, 100, 182, 520].

Kadanoff, L. P. (1969a). Correlations along a line in the two-dimensional Ising model. *Physical Review*, *188*, 859–863. [120].
Kadanoff, L. P. (1969b). Operator algebra and the determination of critical indices. *Physical Review Letters*, *23*, 1430–1433. [119, 120].
Kadanoff, L. P. (1975). Variational principles and approximate renormalization group calculations. *Physical Review Letters*, *34*, 1005–1008. [3, 600, 607, 609].
Kadanoff, L. P. (1976a). Notes on Migdal's recursion formulas. *Annals of Physics*, *100*, 359–394. [599, 600, 601].
Kadanoff, L. P. (1976b). Scaling, universality and operator algebras. In *PTCP(5a)* (pp. 1–34). [95].
Kadanoff, L. P. (2006). Theories of matter: infinities and renormalization. See under Papers about Science. http://jfi.uchicago.edu/leop/. [520].
Kadanoff, L. P. (2009). More is the same; phase transitions and mean field theories. *Journal of Statistical Physics*, *137*, 777–797. [167].
Kadanoff, L. P. (2013). Relating theories via renormalization. *Studies in History and Philosophy of Modern Physics*, *44*, 22–39. [520].
Kadanoff, L. P., Götze, W., Hamblen, D., Hecht, R., Lewis, E. A. S., Palciauskas, V. V., et al. (1967). Static phenomena near critical points: Theory and experiments. *Reviews of Modern Physics*, *19*, 395–431. [96].
Kadanoff, L. P., & Houghton, A. (1975). Numerical evaluations of the critical properties of the two-dimensional Ising model. *Physical Review B*, *11*, 377–386. [572, 573, 607].
Kadanoff, L. P., Houghton, A., & Yalabik, M. C. (1976). Variational approximations for renormalization group transformations. *Journal of Statistical Physics*, *14*, 171–203. [600, 607, 609].
Kadanoff, L. P., & Wegner, F. J. (1971). Some critical properties of the eight-vertex model. *Physical Review B*, *4*, 3989–3993. [54, 67, 120, 482, 491].
Kager, W., Nienhuis, B., & Kadanoff, L. P. (2004). Exact solutions for Loewner evolutions. *Journal of Statistical Physics*, *115*, 805–821. [165].
Kasteleyn, P. W. (1961). The statistics of dimers on a lattice I. The number of dimer arrangements on a quadratic lattice. *Physica*, *27*, 1209–1225. [257, 496, 497, 504].
Kasteleyn, P. W. (1963). Dimer statistics and phase transitions. *Journal of Mathematical Physics*, *4*, 287–293. [508, 510].
Kasteleyn, P. W. (1967). Graph theory and crystal statistics. In F. Harary (Ed.), *Graph theory and theoretical physics* (pp. 43–110). London, UK: Academic Press. [497].
Kasteleyn, P. W., & Fortuin, C. M. (1969). Phase transitions in lattice systems with random local properties. *Journal of the Physical Society of Japan*, *26*, 11–14. [56].
Kasteleyn, P. W., & van Kranendonk, J. (1956a). Constant-coupling approximation for Heisenberg ferromagnetism. *Physica*, *22*, 317–337. [229].
Kasteleyn, P. W., & van Kranendonk, J. (1956b). Constant-coupling approximation for antiferromagnetism. *Physica*, *22*, 367–385. [229].
Kasteleyn, P. W., & van Kranendonk, J. (1956c). Constant-coupling approximation for Ising spin systems. *Physica*, *22*, 387–396. [229].
Katsura, S. (1962). Statistical mechanics of the anisotropic linear Heisenberg model. *Physical Review*, *127*, 1508–1518. [33].
Kaufman, B. (1949). Crystal statistics. II Partition function evaluated by spinor analysis. *Physical Review*, *76*, 1232–1243. [257].
Kaufman, B., & Onsager, L. (1949). Crystal statistics. III Short-range order in a binary Ising lattice. *Physical Review*, *76*, 1244–1252. [257].
Kaufman, M., & Griffiths, R. B. (1981). Exactly soluable Ising models on hierarchical lattices. *Physical Review B*, *24*, 496–498. [588].
Kenna, R. (1998). Fisher zeros and singular behaviour of the two-dimensional Potts model in the thermodynamic limit. *Journal of Physics A: Mathematical and General*, *31*, 9419–9427. [255].
Kennedy, T. (1985). Long range order in the anisotropic quantum ferromagnetic Heisenberg model. *Communications in Mathematical Physics*, *100*, 447–462. [31].

Kenyon, R., & Schlenker, J.-M. (2005). Rhombic embeddings of planar graphs with faces of degree 4. *Transactions of the American Mathematical Society*, *357*, 3443–3458. [636].

Khinchin, A. I. (1949). *Mathematical foundations of statistical mechanics*. New York: Dover. [14].

Kikuchi, R. (1951). A theory of cooperative phenomena. *Physical Review*, *81*, 988–1003. [229, 236, 241].

Kim, D., & Joseph, R. I. (1975). High-temperature series study of the q-component Potts model in two and three dimensions. *Journal of Physics A: Mathematical and General*, *8*, 891–904. [518, 543, 553].

Kincaid, J. M., & Cohen, E. G. D. (1975). Phase diagrams of liquid helium mixtures and metamagnets: Experiment and mean field theory. *Physics Reports*, *22*, 57–143. [90, 93, 225].

Kinzel, W., & Schick, M. (1981). Phenomenological scaling approach to the triangular Ising antiferromagnet. *Physical Review B*, *23*, 3435–3441. [611, 614, 615].

Knops, H. J. F. (1977). Exact relation between the solid-on-solid model and the xy model. *Physical Review Letters*, *39*, 766–769. [53, 609].

Kosterlitz, J. M. (1974). The critical properties of the two-dimensional xy model. *Journal of Physics C: Solid State Physics*, *7*, 1046–1060. [588, 591, 594].

Kosterlitz, J. M., & Thouless, D. J. (1972). Long range order and metastability in two dimensional solids and superfluids. *Journal of Physics C: Solid State Physics*, *5*, L124–L126. [50].

Kosterlitz, J. M., & Thouless, D. J. (1973). Ordering, metastability and phase transitions in two-dimensional systems. *Journal of Physics C: Solid State Physics*, *6*, 1181–1203. [50, 52, 560].

Kouvel, J. S., & Comly, J. B. (1968). Magnetic equation of state for nickel near its Curie point. *Physical Review Letters*, *20*, 1237–1239. [149].

Kouvel, J. S., & Rodbell, D. S. (1967). Magnetic equation of state for CrO_2 and nickel near their Curie points. *Physical Review Letters*, *18*, 215–218. [149].

Kramers, H. A., & Wannier, G. H. (1941a). Statistics of the two-dimensional ferromagnet Part I. *Physical Review*, *60*, 252–262. [256, 283].

Kramers, H. A., & Wannier, G. H. (1941b). Statistics of the two dimensional ferromagnet Part II. *Physical Review*, *60*, 263–276. [256].

Kurmann, J., Thomas, H., & Müller, G. (1982). Antiferromagnetic long-range order in the anisotropic quantum spin chain. *Physica A*, *112*, 235–255. [33].

Lancaster, P., & Tismenetsky, M. (1985). *The theory of matrices* (2nd ed.). London: Academic Press. [670, 671, 672, 675].

Landau, L. D. (1937). On the theory of phase transitions. In D. ter Haar (Ed.), *Collected papers of L.D. Landau* (pp. 193–216). Reprinted in (1965). Oxford, UK: Pergamon. [94, 176].

Lanford, O. E., & Ruelle, D. (1969). Observables at infinity and states with short range correlations in statistical mechanics. *Communications in Mathematical Physics*, *13*, 194–215. [261].

Lavis, D. A. (1973). The steam-water-ice system: A two-dimensional bonded lattice model. The first-order approximation. *Journal of Physics C: Solid State Physics*, *6*, 45–1530. [59].

Lavis, D. A. (1975). Symmetry properties and the two-phase coexistence curve of a two-dimensional bonded lattice fluid. *Journal of Physics A: Mathematical and General*, *8*, 51–1933. [59].

Lavis, D. A. (1976). An exact matrix calculation for a two-dimensional model of the steam-water-ice system. Bulk and boundary properties. *Journal of Physics A: Mathematical and General*, *9*, 2077–2095. [59, 153, 256, 427].

Lavis, D. A. (1996). The derivation of the free energy of the Ising model from that of the eight-vertex model. *Reports on Mathematical Physics*, *37*, 147–155. [284, 380, 484].

Lavis, D. A. (2005). Boltzmann and Gibbs: An attempted reconciliation. *Studies in History and Philosophy of Modern Physics*, *36*, 245–273. [1].

Lavis, D. A. (2008). Boltzmann, Gibbs and the concept of equilibrium. *Philosophy of Science*, *75*, 682–696. [1].

Lavis, D. A. (2011). An objectivist account of probabilities in statistical mechanics. In C. Beisbart & S. Hartmann (Eds.), *Probabilities in physics*. Oxford: Oxford University Press (Chap. 2). [1].

Lavis, D. A., & Bell, G. M. (1977). Thermodynamic phase changes and catastrophe theory. *Bulletin I. M. A.*, *13*(2), 34–42. [176, 659].

Lavis, D. A., & Bell, G. M. (1999a). *Statistical mechanics of lattice systems (Vol. 1): Closed-Form and Exact Solutions*. Berlin: Springer. [256, 429, 511].

Lavis, D. A., & Bell, G. M. (1999b). *Statistical mechanics of lattice systems (Vol. 2): Exact, Series and Renormalization Group Methods*. Berlin: Springer. [294, 371, 434, 524, 537, 620, 644, 676].

Lavis, D. A., & Christou, N. I. (1977). The dielectric constant of a bonded lattice model for water. *Journal of Physics A: Mathematical and General, 10*, 2153–2169. [59, 251].

Lavis, D. A., & Christou, N. I. (1979). A two-dimensional bonded lattice model for water. *Journal of Physics A: Mathematical and General, 12*, 1869–1890. [252].

Lavis, D. A., & Quinn, A. G. (1983). An Ising ferrimagnetic model on a triangular lattice. *Journal of Physics C: Solid State Physics, 16*, 347–3562. [90, 606, 613].

Lavis, D. A., & Quinn, A. G. (1987). Transfer matrix and phenomenological renormalization methods applied to a triangular Ising ferrimagnetic model. *Journal of Physics C: Solid State Physics, 20*, 2129–2138. [90, 613, 615].

Lavis, D. A., & Southern, B. W. (1984). Renormalization group study of a three-dimensional lattice model with directional bonding. *Journal of Statistical Physics, 35*, 489–506. [252, 254, 607].

Lavis, D. A., Southern, B. W., & Bell, G. M. (1982). Phase transitions in monolayers at air/water and oil/water interfaces. *Journal of Physics C: Solid State Physics, 15*, 1077–1088. [606].

Lawden, D. F. (1989). *Elliptic functions and applications*. Berlin: Springer. [714].

Lawler, G. F. (2011). Scaling limits and the Schramm-Loewner evolution. *Probability Surveys, 8*, 442–495. [163].

Lawler, G. F., Schramm, O., & Werner, W. (2004). Conformal invariance of planar loop-erased random walks and uniform spanning trees. *Annals of Probability, 32*, 939–995. [166].

Lawrie, I. D., & Sarbach, S. (1984). Theory of tricritical points. In *PTCP(9)* (pp. 2–161). [94, 139].

Le Guennec, P. (1998). On the concept of chirality. *Journal of Mathematical Chemistry, 23*, 429–439. [57].

Lee, T. D., & Yang, C. N. (1952). Statistical theory of equations of state and phase transitions. II Lattice gas and Ising model. *Physical Review, 87*, 410–419. [43, 254, 273].

Leech, J. W., & Newman, D. J. (1969). *How to use groups*. London, UK: Methuen. [185, 676].

Levelt Sengers, J. M. H. (1974). From van der Waals' equation to the scaling laws. *Physica, 73*, 73–106. [94, 152].

Lieb, E. H. (1967a). Exact solution of the problem of the entropy of two-dimensional ice. *Physical Review Letters, 18*, 692–694. [2, 32, 67, 73, 257, 488].

Lieb, E. H. (1967b). Residual entropy of square ice. *Physical Review, 162*, 162–171. [2, 32, 61, 73, 257, 511].

Lieb, E. H. (1967c). Exact solution of the F model of an antiferroelectric. *Physical Review Letters, 18*, 1046–1048. [2, 32, 61, 73, 257].

Lieb, E. H. (1967d). Exact solution of the two-dimensional Slater KDP model of a ferroelectric. *Physical Review Letters, 19*, 108–110. [2, 32, 61, 73, 257].

Lieb, E. H. (1969). Two dimensional ice and ferroelectric models. In K. Mantappa & W. Brittin (Eds.), *Lectures in theoretical physics* (Vol. 11, pp. 329–354). New York, USA: Gordon and Breach. [74].

Lieb, E. H. (1994). Flux phase of the half-filled band. *Physical Review Letters, 73*, 2158–2161. [639].

Lieb, E. H., & Loss, M. (1993). Fluxes, Laplacians and Kasteleyn's theorem. *Duke Mathematical Journal, 71*, 337–363. [495, 639].

Lieb, E. H., & Ruelle, D. (1972). A property of zeros of the partition function for Ising spin systems. *Journal of Mathematical Physics, 13*, 781–784. [254].

Lieb, E. H., & Sokal, A. D. (1981). A general Lee-Yang theorem for one-component and multi-component ferromagnets. *Communications in Mathematical Physics, 80*, 153–179. [254].

Lieb, E. H., & Wu, F. Y. (1972). Two-dimensional ferroelectric models. In *PTCP(1)* (pp. 331–487). [93, 96, 257, 488, 515, 516].

Lieb, E. H., & Yngvason, J. (1998). A guide to entropy and the second law of thermodynamics. *Notices of the American Mathematical Society, 45*, 571–581. [1].

Lieb, E. H., & Yngvason, J. (1999). The physics and mathematics of the second law of thermodynamics. *Physics Reports, 310*, 1–96 (For Fig. 8 see ibid. (1999), 314, 699). [1].

Lieb, E. H., & Yngvason, J. (2000). A fresh look at entropy and the second law of thermodynamics. *Physics Today*, *53*, 32–37. [1].

Lieb, E. H., & Yngvason, J. (2013). The entropy concept for non-equilibrium states. *Proceedings of the Royal Society A*, *469*, 20130408. [1].

Lieb, E., Schultz, T., & Mattis, D. (1961). Two soluble models of an antiferromagnetic chain. *Annals of Physics*, *16*, 407–466. [32, 33].

Lloyd, E. H. (1953). The direct product of matrices. *The Mathematical Gazette*, *37*, 29–33. [675].

Löwner, K. (1923). Untersuchungen über schlichte konforme Abbildungen des Einheitskreises I. *Mathematische Annalen*, *89*, 103–121. [163, 164, 165].

Luban, M. (1976). Generalized Landau theories. In *PTCP(5a)* (pp. 35–86). [167].

Ma, S. K. (1976a). *Modern theory of critical phenomena*. Mass, USA: Benjamin. [106, 202, 520, 616].

Ma, S. K. (1976b). Renormalization group by Monte Carlo methods. *Physical Review Letters*, *37*, 461–464. [616].

Martin, D. H. (1967). *Magnetism in solids*. London, UK: Iliffe. [30, 40, 90, 212, 223].

Martin, P. P. (1991). *Potts models and related problems in statistical mechanics*. Philadelphia, USA: World Scientific. [55].

Martin, P. P., & Maillard, J. M. (1986). Zeros of the partition function for the triangular lattice three-state Potts model. *Journal of Physics A: Mathematical and General*, *19*, L547–L551. [254].

Mattis, D. C. (1981). *The theory of magnetism I*. Berlin: Springer. [30].

Mattis, D. C. (1985). *The theory of magnetism II*. Berlin: Springer. [223].

Mattis, D. C. (1993). *The many-body problem: An encyclopedia of exactly solved models in one dimension*. Singapore: World Scientific. [760].

Maxwell, J. C. (1875). On the dynamical evidence of the molecular constitution of bodies. *Nature, 11*, 357–374. (Reprinted in *The Scientific Papers of James Clerk Maxwell*, Vol. 2, pp. 418–438, W. D. Niven (Ed.), 1890, C.U.P.). [171].

McCoy, B. M. (2010). *Advanced statistical mechanics*. Oxford: Oxford University Press. [434, 442, 456, 459, 465, 467, 469, 472].

McCoy, B. M., & Wu, T. T. (1973). *The two-dimensional Ising model*. Cambridge, Mass, USA: Harvard U.P. [483].

McWeeny, R. (1963). *Symmetry*. Oxford, UK: Pergamon. [676].

Mermin, N. D. (1971a). Lattice gas with short-range pair interactions and a singular coexistence-curve diameter. *Physical Review Letters*, *26*, 957–959. [43, 332, 337].

Mermin, N. D. (1971b). Solvable model of a vapour-liquid transition with a singular coexistence-curve diameter. *Physical Review Letters*, *26*, 169–172. [43, 337].

Mermin, N. D., & Rehr, J. J. (1971). Generality of the singular diameter of the liquid-vapour coexistence curve. *Physical Review Letters*, *26*, 1155–1156. [43].

Mermin, N. D., & Wagner, H. (1966). Absence of ferromagnetism or antiferromagnetism in one- and two-dimensional isotropic Heisenberg models. *Physical Review Letters*, *17*, 1133–1136. [31].

Metcalf, B. D. (1973). Phase diagram of a nearest-neighbour triangular antiferromagnet in an external field. *Physics Letters A*, *45*, 1–3. [605, 606].

Migdal, A. A. (1975). Phase transitions in gauge and spin-lattice theory. *Soviet Physics JETP*, *42*, 743–746. [600, 601].

Minnhagen, P. (1987). The two-dimensional Coulomb gas, vortex unbinding and superfluid-superconducting films. *Reviews of Modern Physics*, *59*, 1001–1066. [49, 53].

Mittag, L., & Stephen, M. J. (1974). Mean-field theory of the many component Potts model. *Journal of Physics A: Mathematical and General*, *7*, L109–L112. [187].

Montroll, E. W. (1964). Lattice statistics. In E. Beckenbach (Ed.), *Applied combinatorial mathematics* (pp. 96–143). New York, USA: Wiley. [500, 501].

Moore, M. A. (1969). Additional evidence for a phase transition in the plane-rotator and classical Heisenberg models for two-dimensional lattices. *Physical Review Letters*, *23*, 861–863. [49].

Mukamel, D., & Blume, M. (1974). Ising model for tricritical points in ternary mixtures. *Physical Review A, 10*, 610–617. [45].

Mulholland, G. W., & Rehr, J. J. (1974). Coexistence curve properties of Mermin's decorated lattice gas. *Journal of Chemical Physics, 60*, 1297–1306. [332, 339, 342, 343].

Munkres, J. R. (1984). *Elements of algebraic topology*. London, UK: Addison-Wesley. [265].

Nagle, J. F. (1966a). Lattice statistics of hydrogen bonded crystals. I the residual entropy of ice. *Journal of Mathematical Physics, 7*, 1484–1491. [257, 488].

Nagle, J. F. (1966b). Lattice statistics of hydrogen bonded crystals. II. The Slater KDP model and the Rys F-model. *Journal of Mathematical Physics, 7*, 1492–1496. [73].

Nagle, J. F. (1969). Study of the F model using low-temperature series. *Journal of Chemical Physics, 50*, 2813–2818. [562].

Nagle, J. F. (1975). Chain model theory of lipid monolayer transitions. *Journal of Chemical Physics, 63*, 1255–1261. [515].

Nagle, J. F. (1986). Theory of lipid monolayer and bilayer chain-melting phase transitions. *Faraday Discussions of the Chemical Society, 81*, 151–162. [515].

Nagle, J. F., & Temperley, H. N. V. (1968). Combinatorial theorem for graphs on a lattice. *Journal of Mathematical Physics, 9*, 1020–1026. [284].

Nagle, J. F., Yokoi, C. S. O., & Bhattacharjee, S. M. (1989). Dimer modules on anisotropic lattices. In *PTCP(13)* (pp. 235–297). [86, 96, 510, 515].

Nakono, H. (1968). Ordering in certain statistical systems of Ising spins. *Progress of Theoretical Physics, 39*, 1121–1132. [326].

Nauenberg, M. (1975). Renormalization group solution of the one-dimensional Ising model. *Journal of Mathematical Physics, 16*, 703–705. [583].

Nauenberg, M., & Nienhuis, B. (1974a). Critical surface for square Ising spin lattice. *Physical Review Letters, 33*, 944–946. [603].

Nauenberg, M., & Nienhuis, B. (1974b). Renormalization group approach to the solution of general Ising models. *Physical Review Letters, 33*, 1598–1601. [571, 603].

Nauenberg, M., & Scalapino, D. J. (1980). Singularities and scaling functions at the Potts-model multicritical point. *Physical Review Letters, 44*, 837–840. [491].

Nelson, D. R. (1983). Defect-mediated phase transitions. In *PTCP(7)* (pp. 1–99). [53].

Nelson, D. R., & Fisher, M. E. (1975). Soluble renormalization groups and scaling fields for low-dimensional Ising systems. *Annals of Physics, 91*, 226–274. [572, 579, 585].

Newell, G. F. (1950). Crystal statistics of a two-dimensional triangular Ising lattice. *Physical Review, 79*, 876–882. [257].

Newman, C. M. (1975). Inequalities for Ising models and field theories which obey the Lee-Yang theorem. *Communications in Mathematical Physics, 41*, 1–9. [254].

Niemeijer, T., & van Leeuwen, J. M. J. (1973). Wilson theory for spin systems on a triangular lattice. *Physical Review Letters, 31*, 1411–1414. [571].

Niemeijer, T., & van Leeuwen, J. M. J. (1974). Wilson theory for two-dimensional Ising spin systems. *Physica, 71*, 17–40. [571, 595].

Niemeijer, T., & van Leeuwen, J. M. J. (1976). Renormalization theory for Ising-like spin systems. In *PTCP(6)* (pp. 425–505). [115, 572, 595, 603].

Nienhuis, B. (1987). Coulomb gas formulation of two-dimensional phase transitions. In *PTCP(11)* (pp. 1–53). [53].

Nienhuis, B., & Nauenberg, M. (1975). First-order phase transitions in renormalization group theory. *Physical Review Letters, 35*, 477–479. [105].

Nienhuis, B., Riedel, E. K., & Schick, M. (1980). Magnetic exponents of the two-dimensional q-state Potts model. *Journal of Physics A: Mathematical and General, 13*, L189–L192. [490].

Nightingale, M. P. (1976). Scaling theory and finite systems. *Physica A, 83*, 561–572. [609, 611, 612].

Nightingale, M. P. (1977). Non-universality for Ising-like spin systems. *Physics Letters A, 59*, 486–488. [611].

Nightingale, M. P., & Blöte, H. W. J. (1980). Finite-size scaling and critical point exponents of the Potts model. *Physica A, 104*, 352–357. [613].

Nightingale, M. P., &'T Hooft, A. H. (1974). Scaling theory and logarithmic singularities. *Physica, 77*, 390–402. [3, 100, 109].

Nishimori, H., & Griffiths, R. B. (1983). Structure and motion of the Lee-Yang zeros. *Journal of Mathematical Physics, 24*, 2637–2647. [254].

Niss, M. (2005). History of the Lenz-Ising model 1920–1950: From ferromagnetic to cooperative phenomena. *Archive for History of Exact Sciences, 59*, 267–318. [37].

Oitmaa, J., Hamer, C., & Zheng, W. (2006). *Series expansion methods for strongly interacting lattice models*. Cambridge UK: CUP. [530, 644].

Onsager, L. (1944). Crystal statistics. I Two-dimensional model with order-disorder transition. *Physical Review, 65*, 117–149. [95, 256, 306, 372, 421, 423].

Opechowski, W. (1937). On the exchange interaction in magnetic crystals. *Physica, 4*, 181–199. [517].

O'Reilly, D. E. (1973). Scaled particle-quasilattice model of liquid water. *Physical Review A, 7*, 1659–1661. [252].

O'Rourke, M. J., Baxter, R. J., & Bazhanov, V. V. (1995). Numerical results for the three-state critical Potts model on finite rectangular lattices. *Journal of Statistical Physics, 78*, 665–680. [254].

Ostlund, S. (1981). Incommensurate and commensurate phases in asymmetric clock models. *Physical Review B, 24*, 398–405. [59].

Park, H. (1994). Three-state Potts model on a triangular lattice. *Physical Review B, 49*, 881–887. [188].

Patashinskii, A. Z., & Pokrovskii, V. L. (1966). Behaviour of ordered systems near the transition point. *Soviet Physics JETP, 23*, 292–297. [3].

Pauling, L. (1935). The structure and entropy of ice and other crystals with some randomness of atomic arrangement. *Journal of the American Chemical Society, 57*, 2680–2684. [488].

Pawley, G. S., Swendsen, R. H., Wallace, D. J., & Wilson, K. G. (1984). Monte Carlo renormalization group calculations of critical behaviour in the simple cubic Ising model. *Physical Review B, 29*, 4030–4040. [616].

Pearce, P. A., & Thompson, C. J. (1977). The spherical limit for n-vector correlations. *Journal of Statistical Physics, 17*, 189–196. [35].

Pearson, R. (1980). Conjecture for the extended Potts model magnetic eigenvalue. *Physical Review B, 22*, 2579–2580. [490].

Peierls, R. (1936). On Ising's model of ferromagnetism. *Proceedings of the Cambridge Philosophical Society, 32*, 477–481. [253, 264].

Penrose, O. (1979). Foundations of statistical mechanics. *Reports on Progress in Physics, 42*, 200–1937. [261].

Perram, J. W. (1971). Interstitial models of water in the random mixing approximation. *Molecular Physics, 20*, 1077–1085. [337].

Perron, O. (1907). Grundlagen für eine Theoriedes Jacobischen Kettenbruchalgorithmus. *Mathematische Annalen, 64*, 11–76. [673].

Pfeuty, P., & Toulouse, G. (1977). *Introduction to the renormalization group and to critical phenomena*. London, UK: Wiley. [104].

Pierson, S. W. (1997). A review of the real space renormalization group analysis of the two-dimensional Coulomb gas, the Kosterlitz-Thouless-Berezinski transition and extensions to a layered vortex gas. *Philosophical Magazine B, 76*, 715–730. [588].

Pippard, A. B. (1957). *Elements of classical thermodynamics*. Cambridge, UK: C.U.P. [1, 2].

Polyakov, A. M. (1970). Conformal symmetry of critical fluctuations, Soviet Phys. *Soviet Physics JETP Letters, 12*, 381–383. [3, 159–161].

Polyakov, A. M. (1987). *Gauge fields and strings*. Chur: Harwood Academic Publishers. [2].

Popkov, V., & Nienhuis, B. (1997). The 3D multi-layered solvable six-vertex model: The phase diagram. *Journal of Physics A: Mathematical and General, 30*, 99–116. [256].

Poston, T., & Stewart, I. N. (1997). *Catastrophe theory and its applications*. New York, USA: Dover. [659, 662].

Potts, R. B. (1952). Some generalized order-disorder transformations. *Proceedings of the Cambridge Philosophical Society, 48*, 106–169. [54, 55, 490].

Prigogine, I. (1967). *Introduction to the thermodynamics of irreversible processes*. London: Interscience. [1].

Privman, V., & Fisher, M. E. (1983). Finite-size effects at first-order transitions. *Journal of Statistical Physics, 33*, 385–417. [153].

Privman, V., & Fisher, M. E. (1984). Universal critical amplitudes in finite-size scaling. *Physical Review B, 30*, 322–327. [151, 154, 158].

Pynn, R., & Skjeltorp, A. E. (1984). *Multicritical phenomena*. New York, USA: Plenum. [94].

Rapaport, D. C. (1972). The Ising ferromagnet with impurities: A series expansion approach, I. *Journal of Physics C: Solid State Physics, 5*, 1830–1858. [313].

Ree, F. H., & Chesnut, D. A. (1966). Phase transition of a hard-core lattice gas. The square lattice with nearest-neighbour exclusion. *Journal of Chemical Physics, 45*, 3983–4003. [153, 354, 361].

Rehr, J. J., Joyce, G. S., & Guttmann, A. J. (1980). A recurrence technique for confluent singularity analysis of power series. *Journal of Physics A, 13*, 1587–1602. [563].

Resnick, D. J., Garland, J. C., Boyd, J. T., Shoemaker, S., & Newrock, R. S. (1981). Kosterlitz-Thouless transition in proximity-coupled superconducting arrays. *Physical Review Letters, 47*, 1542–1545. [49].

Riedel, E. K. (1972). Scaling approach to tricritical phase transitions. *Physical Review Letters, 28*, 675–678. [142].

Riedel, E. K., & Wegner, F. J. (1974). Effective critical and tricritical exponents. *Physical Review B, 9*, 294–315. [105, 108].

Roan, S.-S. (2007). The q operator and the functional relations of the eight-vertex model at root-of-unity $\eta = 2mk/n$ for odd n. *Journal of Physics A: Mathematical and Theoretical, 40*, 11019–11044. [442].

Rohrer, H. (1975). Properties of $GdAlO_3$ near the spin flop bicritical point. *Physical Review Letters, 34*, 1638–1641. [93].

Roomany, H. H., & Wyld, H. W. (1981). Finite-lattice Hamiltonian results for the phase structure of the $z(q)$ models and the q-state Potts models. *Physical Review B, 23*, 1357–1361. [615].

Rossat-Mignod, J., Burlet, P., Bartholin, H., Vogt, O., & Lagnier, R. (1980). Specific heat analysis of the magnetic phase diagram of CeSb. *Journal of Physics C: Solid State Physics, 13*, 6381–6389. [2].

Rowlinson, J. S. (1969). Thomas Andrews and the critical point. *Nature, 224*, 541–543. [171].

Rowlinson, J. S. (1973). Legacy of van der Waals. *Nature, 244*, 414–417. [175].

Rowlinson, J. S. (1979). English translation of J.D. van der Waals' "the thermodynamic theory of capilliarity under the hypothesis of a continuous variation of density". *Journal of Statistical Physics, 20*, 197–244. [778].

Rowlinson, J. S. (1988). English translation (with introductory essay) of J. D. van der Waals: On the continuity of the gaseous and liquid states of matter. *Studies in Statistical Mechanics* (Vol. 14). Amsterdam, New York: North-Holland. [175].

Rowlinson, J., & Swinton, F. L. (1982). *Liquids and liquid mixtures*. London, UK: Butterworths. [337].

Ruelle, D. (1969). *Statistical mechanics*. New York, USA: Benjamin. [260–261].

Ruelle, D. (1971). Extension of the Lee-Yang circle theorem. *Physical Review Letters, 26*, 303–304. [254, 273, 275].

Runnels, L. K., & Combs, L. L. (1971). Exact finite method of lattice statistics. I square and triangular lattice gases of hard molecules. *Journal of Chemical Physics, 45*, 2482–2492. [153, 282].

Runnels, L. K., & Hubbard, J. B. (1972). Applications of the Yang-Lee-Ruelle theory to hard-core lattice gases. *Journal of Statistical Physics, 6*, 1–20. [281–282].

Rushbrooke, G. S. (1963). On the thermodynamics of the critical region for the Ising model. *Journal of Chemical Physics*, *39*, 842–843. [125].

Rys, F. (1963). A two-dimensional classic configuration model. *Helvetica Physica Acta*, *36*, 537–559. [73].

Sacco, J. E., & Wu, F. Y. (1975). 32-vertex model on the triangular lattice. *Journal of Physics A: Mathematical and General*, *8*, 1780–1787. [60].

Saito, Y. (1981). Monte Carlo study of the three-state Potts model with two- and three-body interactions. *Journal of Physics A: Mathematical and General*, *15*, 1885–1892. [188].

Salas, J., & Sokal, A. D. (1997). Logarithmic corrections and finite-size scaling in the two-dimensional 4-state Potts model. *Journal of Statistical Physics*, *88*, 567–615. [491].

Sarbach, S., & Fisher, M. E. (1978). Tricriticality and the failure of scaling in the many-component limit. *Physical Review B*, *18*, 2350–2363. [204].

Saunders, P. T. (1980). *An introduction to catastrophe theory*. Cambridge, UK: C.U.P. [659].

Savit, R. (1980). Duality in field theories and statistical systems. *Reviews of Modern Physics*, *52*, 453–487. [631].

Schäfer, L. (1976). Conformal covariance in the framework of Wilson's renormalization group approach. *Journal of Physics A: Mathematical and General*, *9*, 377–395. [160, 606].

Schick, M. A., Walker, J. S., & Wortis, M. (1976). Antiferromagnetic triangular Ising model. *Physics Letters A*, *58*, 479–480. [603, 605].

Schick, M., & Griffths, R. B. (1977). Antiferromagnetic ordering in the three-state Potts model. *Journal of Physics A: Mathematical and General*, *10*, 2123–2131. [60, 187, 605, 606].

Schick, M., Walker, J. S., & Wortis, M. (1977). Phase diagram of the triangular Ising model: Renormalization group calculation with application to adsorbed monolayers. *Physical Review B*, *16*, 2205–2219. [603, 604, 606].

Schramm, O. (2000). Scaling limits of loop-erased random walks and uniform spanning trees. *Israel Journal of Mathematics*, *118*, 221–288. [163, 165].

Schulman, L. S. (1973). Tricritical points and type-three phase transitions. *Physical Review B*, *7*, 1960–1966. [659].

Schulman, L. S., & Revzen, M. (1972). Phase transitions and catastrophes. *Collective Phenomena*, *1*, 43–49. [659].

Serre, J.-P. (1977). *Linear representations of finite groups*. Berlin: Springer. [676].

Sewell, G. L. (1989). *Quantum theory of collective phenomena* (2nd ed.). Oxford, UK: O.U.P. [263].

Shafer, R. E. (1974). On quadratic approximation. *SIAM Journal for Numerical and Analytical*, *11*, 447–460. [754].

Shinmi, M., & Huckaby, D. (1986). A lattice gas model for carbon tetrachloride. *Journal of Chemical Physics*, *84*, 951–955. [252].

Shockley, W. (1938). Theory of order for the copper gold alloy system. *Journal of Chemical Physics*, *6*, 130–144. [44].

Simon, B. (1993). *The statistical mechanics of lattice gases* (Vol. I). Princeton: Princeton University Press. [263].

Sinha, K., & Kumar, N. (1980). *Interactions in magnetically ordered solids*. Oxford, UK: O.U.P. [30].

Sklar, L. (1993). *Physics and chance*. Cambridge, UK: C.U.P. [1].

Slater, J. C. (1941). Theory of the transition KH_2P0_4. *Journal of Chemical Physics*, *9*, 16–33. [73].

Smirnov, S. (2001). Critical percolation in the plane: Conformal invariance, Cardy's formula, scaling limits. *Comptes Rendus de l'Académie des Sciences-Series I Mathematics*, *333*(2001), 239–244. [166].

Sneddon, L. (1978). Critical properties of two-dimensional spin *s* Ising systems with lattice anisotropy. *Journal of Physics C: Solid State Physics*, *11*, 2823–2828. [611, 613, 614].

Sneddon, L., & Barber, M. N. (1977). Decimation transformations for two-dimensional Ising spin systems. *Journal of Physics C: Solid State Physics*, *10*, 2653–2664. [574, 599].

Southern, B. W. (1978). Kadanoff's variational renormalization group method: The Ising model on the square and triangular lattices. *Journal of Physics A: Mathematical and General, 11*, L1–L4. [609].

Southern, B. W., & Lavis, D. A. (1979). A model for adsorbed monolayers of orientable molecules. *Journal of Physics C: Solid State Physics, 12*, 5333–5343. [606].

Southern, B. W., & Lavis, D. A. (1980). Renormalization group study of a two-dimensional lattice model with directional bonding. *Journal of Physics A: Mathematical and General, 13*, 251–262. [578, 606].

Stahl, H. (1997). The convergence of diagonal Padé approximants and the Padé conjecture. *Journal of Computational and Applied Mathematics, 86*, 287–296. [561].

Stanley, H. E. (1968a). Dependence of critical properties on dimensionality of spins. *Physical Review Letters, 20*, 589–592. [34].

Stanley, H. E. (1968b). Spherical model as the limit of infinite spin dimensionality. *Physical Review, 176*, 718–722. [35, 36].

Stanley, H. E. (1974). D-vector model. In *PTCP(3)* (pp. 485–567). [34].

Stanley, H. E. (1987). *Introduction to phase transitions and critical phenomena*. Oxford, UK: O.U.P. [133].

Stauffer, D., Ferer, M., & Wortis, M. (1972). Universality of second-order phase transitions: The scale factor for the correlation length. *Physical Review Letters, 29*, 345–349. [151].

Stephenson, J. (1970). Ising model with antiferromagnetic next-nearest-neighbour coupling: Spin correlations and disorder points. *Physical Review B, 1*, 4405–4409. [326].

Stillinger, E. H. (1975). Theory and molecular models for water. *Advances in Chemical Physics, 31*, 1–101. [252].

Sutherland, B. (1967). Exact solution of a two-dimensional model for hydrogen-bonded crystals. *Physical Review Letters, 19*, 103–104. [74].

Sutherland, B. (1970). Two-dimensional hydrogen bonded crystals without the ice rule. *Journal of Mathematical Physics, 11*, 3183–3186. [2, 33, 299, 441].

Suzuki, M. (1974). New universality of critical exponents. *Progress of Theoretical Physics, 51*, 1992–1993. [492].

Suzuki, M. (1976). Relationship between d-dimensional quantal spin ssystems and $(d + 1)$-dimensional Ising systems. *Progress of Theoretical Physics, 56*, 1454–1469. [2, 31].

Suzuki, M. (1977). Static and dynamic finite-size scaling theory based on the renormalization group approach. *Progress of Theoretical Physics, 58*, 1142–1150. [609].

Suzuki, M., & Fisher, M. E. (1971). Zeros of the partition function for the Heisenberg, ferroelectric and general Ising models. *Journal of Mathematical Physics, 12*, 235–246. [60, 254].

Swendsen, R. H. (1979a). Monte Carlo renormalization group. *Physical Review Letters, 42*, 859–861. [616].

Swendsen, R. H. (1979b). Monte Carlo renormalization group studies of the $d = 2$ Ising model. *Physical Review B, 20*, 2080–2087. [517].

Sykes, M. F., Essam, J. W., & Gaunt, D. S. (1965). Derivation of low-temperature expansions for the Ising model of a ferromagnet and an antiferromagnet. *Journal of Mathematical Physics, 6*, 283–298. [517].

Sykes, M. F., Essam, J. W., Heap, B. R., & Hiley, B. J. (1966). Lattice constant systems and graph theory. *Journal of Mathematical Physics, 7*, 1557–1572. [517, 640, 645].

Sykes, M. F., Gaunt, D. S., Essam, J. W., & Elliott, C. J. (1973). Derivation of low-temperature expansions for the Ising model. VI Three-dimensional lattices-temperature grouping. *Journal of Physics A: Mathematical and General, 6*, 1507–1516. [517].

Sykes, M. F., Gaunt, D. S., Essam, J. W., Heap, B. R., Elliott, C. J., & Mattingly, S. R. (1973). Derivation of low-temperature expansions for the Ising model. V Three-dimensional lattices-field grouping. *Journal of Physics A: Mathematical and General, 6*, 1498–1506. [517].

Sykes, M. F., Gaunt, D. S., Essam, J. W., & Hunter, D. L. (1973). Derivation of low-temperature expansions for the Ising model. II General theory. *Journal of Mathematical Physics, 14*, 1060–1065. [517].

Sykes, M. F., Gaunt, D. S., Martin, J. L., Mattingly, S. R., & Essam, J. W. (1973). Derivation of low-temperature expansions for the Ising model. IV Two-dimensional lattices-temperature grouping. *Journal of Mathematical Physics*, *14*, 1071–1073. [517].

Sykes, M. F., Gaunt, D. S., Mattingly, S. R., Essam, J. W., & Elliott, C. J. (1973). Derivation of low-temperature expansions for the Ising model. III Two-dimensional lattices-field grouping. *Journal of Mathematical Physics*, *14*, 1066–1070. [517].

Sykes, M. F., Gaunt, D. S., Roberts, P. D., & Wyles, J. A. (1972a). High temperature series for the susceptibility of the Ising model. I Two dimensional lattices. *Journal of Physics A: General Physics*, *5*, 624–639. [518].

Sykes, M. F., Gaunt, D. S., Roberts, P. D., & Wyles, J. A. (1972b). High temperature series for the susceptibility of the Ising model. II Three dimensional lattices. *Journal of Physics A: General Physics*, *5*, 640–652. [518].

Sykes, M. F., McKenzie, S., Watts, M. G., & Gaunt, D. S. (1975). Derivation of low-temperature expansions for the Ising model. IX High-field polynomials for the honeycomb-triangular system. *Journal of Physics A: Mathematical and General*, *8*, 1461–1468. [517].

Sykes, M. F., & Watts, M. G. (1975). Derivation of low-temperature series expansions for the Ising model with triplet interactions on the plane triangular lattice. *Journal of Physics A: Mathematical and General*, *8*, 1469–1479. [518].

Sykes, M. F., Watts, M. G., & Gaunt, D. S. (1975a). Derivation of low-temperature expansions for the Ising model. VII The honeycomb-triangular code system. *Journal of Physics A: Mathematical and General*, *8*, 1441–1447. [517].

Sykes, M. F., Watts, M. G., & Gaunt, D. S. (1975b). Derivation of low-temperature expansions for the Ising model. VIII Ferromagnetic and antiferromagnetic polynomials for the honeycomb-triangular system. *Journal of Physics A: Mathematical and General*, *8*, 1448–1460. [517].

Syôzi, I. (1950). The statistics of honeycomb and triangular lattice. *Progress of Theoretical Physics*, *5*, 341–351. [517].

Syozi, I. (1972). Transformation of Ising models. In *PTCP(1)* (pp. 270–329). [318, 319, 326, 631].

Syozi, I., & Nakono, H. (1955). Statistical models of ferrimagnetism. *Progress of Theoretical Physics*, *13*, 69–78. [319].

Takhtadzhyan, L. A., & Faddeev, L. D. (1979). The quantum method of the inverse problem and the Heisenberg XYZ model. *Uspekhi Matematicheskikh Nauk*, *34*, 13–63. [33].

Temperley, H. N. V. (1950). Statistical mechanics of the two-dimensional assembly. *Proceedings of the Royal Society A*, *202*, 202–207. [257].

Temperley, H. N. V., & Fisher, M. E. (1961). Dimer problem in statistical mechanics-an exact result. *Philosophical Magazine*, *6*, 1061–1063. [257, 497].

Temperley, H. N. V., & Lieb, E. H. (1971). Relations between the percolation and colouring problems on the plane square lattice. Some exact results for the percolation problem. *Proceedings of the Royal Society*, *322*, 251–280. [55, 75].

Thibaudier, C., & Villain, J. (1972). On the three dimensional eight-vertex model. *Journal of Physics C: Solid State Physics*, *5*, 3429–3437. [60].

Thom, R. (1972). *Stabilité Structuralle et Morphogenèse*. Reading, Mass: W. A. Benjamin. [659].

Thom, R. (1974). *Modèles Mathématiques de la Morphogenèse* (10–18 ed.). Paris: UGE. [659].

Thompson, B. V., & Lavis, D. A. (1967). The Gaussian average method applied to phase transitions in dilute magnetic alloys. *Proceedings of the Physical Society*, *91*, 645–657. [45].

Thompson, C. S. (1988). *Classical equilibrium statistical mechanics*. Oxford, UK: O.U.P. [264].

Thomson, J. (1871). Considerations on the abrupt change at boiling or condensing in reference to the continuity of the fluid state of matter. *Proceedings of the Royal Society*, *20*, 1–8. [171, 174].

Thomson, W. T. (1904). *Baltimore lectures on molecular dynamics and the wave theory of light*. Baltimore: C.J. Clay. [57].

Titchmarch, E. C. (1939). *The theory of functions* (2nd ed.). Oxford, UK: O.U.P. [271].

Tjon, J. A. (1974). Numerical study of the renormalization group equations in the four-cell approximation. *Physics Letters A*, *49*, 289–290. [603].

van der Waals, J. D. (1873). Over de Continuiteit van der Gas- en Vloeistoftoestand. Ph.D. Thesis, Leiden. [For an English translation see Rowlinson (1988)]. [94, 171, 173, 175].

van der Waals, J. D. (1893). The thermodynamic theory of capilliarity under the hypothesis of a continuous variation of density. Verhandel Konink Akad Weten Amsterdam, 1(8). [For an English translation see Rowlinson (1979)]. [775].

van Enter, A., & Zegarlinski, B. (1997). Non-periodic long-range order for one-dimensional pair interactions. *Journal of Physics A: Mathematical and General*, *30*, 501–505. [353].

van Leeuwen, J. M. J. (1975). Renormalization theory for spin systems. In E. Cohen (Ed.), *Fundamental problems in statistical mechanics* (Vol. 3, pp. 81–101). Amsterdam, New York: North-Holland. [573].

Villian, J. (1975). Theory of one- and two-dimensional magnets with an easy magnetization plane. II. The planar, classical, two-dimensional magnet. *Journal de Physique*, *36*, 581–590. [53].

Wang, J.-S., Swendsen, R. H., & Kotecky, R. (1989). Antiferromagnetic Potts models. *Physical Review Letters*, *63*, 109–112. [187].

Wang, Z. F., Southern, B. W., & Lavis, D. A. (2003). Monte Carlo study of an extended 3-state Potts model on the triangular lattice. *Physical Review B*, *67*, 054415-1–054415-6. [60].

Wannier, G. H. (1950). Antiferromagnetism: The triangular Ising net. *Physical Review*, *79*, 357–364. [257].

Wegner, F. J. (1972). Duality relation between the Ashkin-Teller and eight-vertex models. *Journal of Physics C: Solid State Physics*, *5*, L131–L132. [54].

Wegner, F. J. (1973). A transformation including the weak-graph theorem and the duality transformation. *Physica*, *68*, 570–578. [254, 304].

Weres, O., & Rice, S. A. (1972). A new model for liquid water. *Journal of the American Chemical Society*, *94*, 8983–9002. [252].

Wheeler, J. C. (1975). "Exactly solvable" two component lattice solution with upper and lower critical solution temperatures. *Journal of Chemical Physics*, *62*, 433–439. [331].

Wheeler, J. C. (1977). Decorated lattice-gas models of critical phenomena in fluids and fluid mixtures. *Annual Review of Physical Chemistry*, *28*, 411–443. [43, 331, 332].

Whitehouse, J. S., Christou, N. I., Nicholson, D., & Parsonage, N. G. (1984). A grand ensemble Monte Carlo investigation of the Bell lattice model for water. *Journal of Physics A: Mathematical and General*, *17*, 1671–1682. [252].

Whitney, H. (1932). The coloring of graphs. *Annals of Mathematics*, *33*, 688–718. [638].

Whittaker, E. T., & Watson, G. N. (1927). *A course of modern analysis* (4th ed.). Cambridge, England: Cambridge University Press. [721].

Widom, B. (1964). Degree of the critical isotherm. *Journal of Chemical Physics*, *41*, 1633–1634. [133].

Widom, B. (1965). Equation of state in the neighbourhood of the critical point. *Journal of Chemical Physics*, *43*, 3898–3905. [3, 616].

Widom, B. (1967). Plait points in two- and three-component liquid mixtures. *Journal of Chemical Physics*, *49*, 3324–3333. [318].

Widom, B., & Rowlinson, J. S. (1970). New model for the study of liquid-vapour phase transitions. *Journal of Chemical Physics*, *52*, 1670–1684. [43].

Williams, N. O., & Lavis, D. A. (1996). Finite-size scaling of branch-points in lattice models. *Physics Letters A*, *217*, 275–279. [380].

Wilson, G. L., & Bell, G. M. (1978). Lattice model for aqueous solutions of non-electrolytes. *Journal of the Chemical Society, Faraday Transactions II*, *74*, 1702–1722. [251].

Wilson, K. G. (1971). Renormalization group and critical phenomena. I. Renormalization group and the Kadanoff picture. *Physical Review B*, *4*, 3174–3183. [519].

Wilson, K. G. (1975). The renormalization group: Critical phenomena and the Kondo problem. *Reviews of Modern Physics*, *47*, 773–840. [520, 573, 586, 588].

Wilson, R. J., & Watkins, J. J. (1990). *Graphs: An introductory approach*. London, UK: Wiley. [619, 621].

Witten, L. (1954). A generalization of Yang and Lee's theory of condensation. *Physical Review*, *93*, 1131–1135. [260].

Wong, R., & Zhao, Y.-Q. (2005). On a uniform treatment of Darboux's method. *Constructive Approximation*, *21*, 225–255. [557].

Wood, D. W. (1985). The exact location of partition function zeros, a new method for statistical mechanics. *Journal of Physics A: Mathematical and General*, *18*, L917–L921. [255, 374].

Wood, D. W. (1987). The algebraic construction of partition function zeros: Universality and algebraic cycles. *Journal of Physics A: Mathematical and General*, *20*, 3471–3493. [256, 371, 376, 379, 699].

Wood, D. W., & Ball, J. K. (1990). On a point of order. *Journal of Statistical Physics*, *58*, 599–615. [256, 376].

Wood, D. W., & Goldfinch, M. (1980). Vertex models for the hard-square and hard-hexagon gases, and critical parameters for the scaling transformation. *Journal of Physics A: Mathematical and General*, *13*, 2781–2794. [282].

Wood, D. W., & Osbaldstin, A. H. (1982). Phase equilibria and the scaling transformation. *Journal of Physics A: Mathematical and General*, *15*, 3579–3591. [615].

Wood, D. W., & Osbaldstin, A. H. (1983). Multiple phase coexistence and the scaling transformation. *Journal of Physics A: Mathematical and General*, *16*, 1019–1033. [615].

Wood, D. W., & Turnbull, R. W. (1988). $z^2 - 11z - 1$ as an algebraic invariant for the hard-hexagon model. *Journal of Physics A: Mathematical and General*, *21*, L989–L994. [256].

Wood, D. W., Turnbull, R. W., & Ball, J. K. (1987). Algebraic approximations to the locus of partition function zeros. *Journal of Physics A: Mathematical and General*, *20*, 3495–3521. [376, 379].

Wood, D. W., Turnbull, R. W., & Ball, J. K. (1989). An observation on the partition function zeros of the hard hexagon model. *Journal of Physics A: Mathematical and General*, *22*, L105–L109. [256].

Wortis, M. (1963). Bound states of two spin waves in the Heisenberg ferromagnet. *Physical Review*, *132*, 85–97. [32].

Wu, F. Y. (1968). Remarks on the modified potassium dihydrogen phosphate model of a ferroelectric. *Physical Review*, *168*, 539–543. [506].

Wu, F. Y. (1971). Ising model with four-spin interactions. *Physical Review B*, *4*, 2312–2314. [54, 67].

Wu, F. Y. (1974). Eight-vertex model on the honeycomb lattice. *Journal of Mathematical Physics*, *15*, 687–691. [60, 379].

Wu, F. Y. (1982). The Potts model. *Reviews of Modern Physics*, *54*, 235–268. [55, 184, 188, 306, 555].

Yang, C. N., & Lee, T. D. (1952). Statistical theory of equations of state and phase transitions I Theory of condensation. *Physical Review*, *87*, 404–409. [43, 254, 260, 271, 273].

Yang, C. N., & Yang, C. P. (1966a). Ground-state energy of a Heisenberg-Ising chain. *Physical Review*, *147*, 303–306. [32].

Yang, C. N., & Yang, C. P. (1966b). One-dimensional chain of anisotropic spin-spin interactions. I. Proof of Bethe's hypothesis for ground state in a finite system. *Physical Review*, *150*, 321–327. [32, 474].

Yang, C. N., & Yang, C. P. (1966c). One-dimensional chain of anisotropic spin-spin interactions. II. Properties of the ground-state energy per lattice site for an infinite system. *Physical Review*, *150*, 327–339. [32].

Yang, C. N., & Yang, C. P. (1966d). One-dimensional chain of anisotropic spin-spin interactions III. *Applications. Physical Review*, *151*, 258–264. [32].

Yeomans, J. M. (1992). *Statistical mechanics of phase transitions*. Oxford, UK: O.U.P. [133].

Young, A. P., & Lavis, D. A. (1979). Critical behaviour of a two-dimensional bonded lattice fluid. *Journal of Physics A: Mathematical and General*, *12*, 229–243. [60, 606].

Young, S. (1892). On the generalizations of van der Waals regarding "corresponding" temperatures, pressures and volumes. *Philosophical Magazine*, *33*, 153–185. [95].

Zener, C. (1951). Interaction between the *d* shells in the transition metals. *Physical Review, 81*, 440–444. [30].

Ziman, J. M. (1979). *Models of disorder*. Cambridge, UK: C.U.P. [2, 97].

Zollweg, J. A., & Mulholland, G. W. (1972). On the law of the rectilinear diameter. *Journal of Chemical Physics, 57*, 1021–1025. [332, 337].

Index

D.A. Lavis, *Equilibrium Statistical Mechanics of Lattice Models*,
Theoretical and Mathematical Physics, DOI 10.1007/978-94-017-9430-5

C

H

M

N

O

P

MIX
Papier aus verantwortungsvollen Quellen
Paper from responsible sources
FSC® C105338

If you have any concerns about our products,
you can contact us on
ProductSafety@springernature.com

In case Publisher is established outside the EU,
the EU authorized representative is:
Springer Nature Customer Service Center GmbH
Europaplatz 3, 69115 Heidelberg, Germany

Printed by Libri Plureos GmbH
in Hamburg, Germany